Ivan Singer

# Bases in Banach Spaces II

Editura Academiei
Republicii Socialiste România, București
1981
Springer-Verlag
Berlin Heidelberg New York

Prof. Dr. Ivan Singer
Department of Mathematics
National Institute for Scientific and Technical Creation
Bd. Păcii 220, R — 79622 Bucureşti
Institute of Mathematics
Str. Academiei 14, R — 70109 Bucureşti

Sole distribution rights for all non-socialist countries
granted to Springer-Verlag Berlin Heidelberg New York

ISBN-13: 978-3-642-67846-2 e-ISBN-13: 978-3-642-67844-8
DOI: 10.1007/978-3-642-67844-8


Softcover reprint of the hardcover 1st edition 1981

# Preface

Since the appearance, in 1970, of Vol. I of the present monograph [370], the theory of bases in Banach spaces has developed substantially. Therefore, the present volume contains only Ch. III of the monograph, instead of Ch. III, IV and V, as was planned initially (cp. the table of contents of Vol. I). Since this volume is a continuation of Vol. I of the same monograph, we shall refer to the results of Vol. I directly as results of Ch. I or Ch. II (without specifying Vol. I). On the other hand, sometimes we shall also mention that certain results will be considered in Vol. III (Ch. IV, V).

In spite of the many new advances made in this field, the statement in the Preface to Vol. I, that "the existing books on functional analysis contain only a few results on bases", remains still valid, with the exception of the recent book [248 a] of J. Lindenstrauss and L. Tzafriri. Since we have learned about [248 a] only in 1978, in this volume there are only references to previous works, instead of [248 a]; however, this will cause no inconvenience, since the intersection of the present volume with [248 a] is very small. Let us also mention the appearance, since 1970, of some survey papers on bases in Banach spaces (V. D. Milman [287], [288], C. W. McArthur [275], M. I. Kadec [204], § 3 and others).

The fact that the basis problem (and even the approximation problem) was solved in the negative, by Per Enflo [99] (due to its special importance, we shall present this result in § 0), has increased the usefulness of a monograph on bases and their generalizations in Banach spaces.

This volume attempts to present the results known today on generalizations of bases in Banach spaces and some unsolved problems concerning them. The style is, deliberately, the same as that of Vol. I, except that the section of Notes and remarks and the Bibliography are larger. The works which have appeared after the main part of this volume had been completed, are usually encompassed by the Notes and remarks; also, this section is more detailed, since for most of the results we indicate the paper and the place in it, where the result occurs.

We hope that, similarly to Vol. I, the present volume will be useful to specialists, stimulating further research, and to a large circle of those who want to use it for applications to other problems (for example, orthogonal series, summability, functional equations, etc.). Also, in order to make the book suitable for study, the necessary tools from functional analysis have been carefully explained (either by giving them, as lemmas, with their proofs, or by giving references to treatises containing their proofs). Some of our unpublished results and remarks have been also included, without any special mention.

In order to give some applications of bases and their generalizations to the study of Banach spaces, we shall publish, hopefully soon after the appearance of this volume, a part of Ch. IV of the present monograph, in the Lecture Notes in Mathematics series of the Springer Verlag.

We wish to thank here our friend, Professor Czeslaw Bessaga, for reading a large part of the manuscript and making valuable suggestions and observations. We extend our thanks, for valuable remarks made in discussions and letters, to our colleagues and friends Professors W. J. Davis, D. van Dulst, T. Figiel, D. J. H. Garling, V. I. Gurariĭ, W. B. Johnson, M. I. Kadec, S. Kwapien, P. Masani, R. J. Nessel, K. I. Oskolkov and W. H. Ruckle. Furthermore, we thank our colleagues in the University of Amsterdam and Purdue University, who attended, in 1973/74, over a period of 5 months and 3 months respectively, our seminars on § 0 and on parts of §§ 8—11, for their stimulating interest and remarks. Also, we received useful comments from the participants to our lectures on parts of §§ 10, 11, at the Semester on approximation theory of the Stefan Banach International Mathematical Center in Warsaw, in 1975.

During the writing of this volume we have benefited by excellent working conditions at the Institute of Mathematics and at the National Institute for Scientific and Technical Creation in Bucharest, as well as at various Universities which we visited for periods of several months and we wish to express our gratitude to all who contributed to ensure these conditions. Our special thanks are due to Dr. Ing. Constantin Teodorescu, General Director of the National Institute for Scientific and Technical Creation and to Dr. Zoia Ceauşescu, Head of the Department of Mathematics at this Institute, without whose support this volume could not have been completed.

February, 1978 IVAN SINGER

After this book has been typeset, we added an Appendix, containing further notes and remarks and further bibliography. We extend our thanks to Professors B. V. Godun, M. I. Kadec, A. N. Pličko and P. Terenzi, for correspondence concerning some parts of the Appendix. Finally, our thanks are due to Editura Academiei and to Springer Verlag, for undertaking the task of publishing this volume and for their help during its preparation.

May, 1980 IVAN SINGER

# Contents

Chapter III

# Generalizations of the Notion of a Basis

## § 0. Banach spaces which do not have the approximation property

We recall that a Banach space $E$ is said to have the *approximation property* if the identity operator $I_E: E \to E$ can be approximated, uniformly on every compact subset of $E$, by continuous linear operators of finite rank (i.e., of finite-dimensional range), that is, if for every compact subset $Q \subset E$ and every $\varepsilon > 0$ there exists an endomorphism*) $v = v_{Q,\varepsilon} \in L(E, E)$ of finite rank, such that

$$\|v(x) - x\| < \varepsilon \qquad (x \in Q). \qquad (0.1)$$

We have seen in Ch. I, § 17, theorem 17.3, that every Banach space $E$ with a basis has the approximation property. The same argument also shows that a similar result holds**) for "basis" replaced by various generalizations of bases which will be introduced in the sequel (involving a sequence of finite rank operators on $E$ converging pointwise to the identity operator $I_E$). Thus, the examples of separable Banach spaces failing to have the approximation property, which will be constructed in the present section, will give a (negative) solution to the basis problem (Ch. I, § 1, problem 1.1) and to the similar existence problems for these generalizations of bases.

The existence of separable Banach spaces without the approximation property has also other applications. For example, this fact implies, as we shall see below, that there exist Banach spaces with bases, whose conjugate space is separable and fails to have the approximation property. This latter result will be used in § 9 to show that there exists a separable Banach space $E$ with the approximation property, having no basis (and which even fails to have any of the generalizations of bases mentioned above) and having $E^*$ separable.

---

*) We recall that $L(E, E)$ denotes the Banach space of all endomorphisms (continuous linear mappings of $E$ into $E$), with the norm $\|u\| = \sup_{\substack{x \in E \\ \|x\| \leq 1}} \|u(x)\|$.

**) However, we shall see in § 9, example 9.2, that the converse results are not true.

We shall give two different proofs of the existence of subspaces of $c_0$ and $l^p$ $(2 < p < \infty)$ which fail to have the approximation property (theorems 0.1 and 0.2 below). Although these proofs have, essentially, the same underlying idea, each of them has its own interest. The first one will be of a constructive character, using a combinatorial lemma, while the second one will be a proof of existence, using a probabilistic lemma.

In both proofs we shall make use of the following sufficient condition for a Banach space $E$ to fail the approximation property:

**Lemma 0.1.** a) *Let $E$ be a Banach space. Assume that there exists a sequence $\{\varphi_n\}_{n=0}^{\infty}$ of linear functionals on $L(E, E)$, such that*

$$\varphi_n(I_E) = 1 \qquad (n = 3, 4, 5, \ldots), \tag{0.2}$$

$$\lim_{n\to\infty} \varphi_n(v) = 0 \qquad (v \in L(E, E),\ \dim v(E) < \infty), \tag{0.3}$$

*a sequence $\{\mathscr{A}_n\}_{n=0}^{\infty}$ of finite subsets of $E$ and a sequence $\alpha_n > 0$ $(n = 0, 1, 2, 3, \ldots)$ with $\sum_{n=0}^{\infty} \alpha_n < \infty$, such that*

$$|\varphi_n(u) - \varphi_{n-1}(u)| \leqslant \alpha_n \max_{x\in\mathscr{A}_n} \|u(x)\| \qquad (u \in L(E, E),\ n = 0,1,2, \ldots), \tag{0.4}$$

$$\max_{x\in\mathscr{A}_n} \|x\| \to 0 \ \textit{as}\ n \to \infty, \tag{0.5}$$

*where $\varphi_{-1} = 0$. Then $E$ does not have the approximation property.*

b) *In particular, the last assumption* (0.4), (0.5) *is satisfied if $\varphi_0 = \varphi_1 = \varphi_2 = 0$ and if there exist decompositions*

$$\varphi_n = \frac{1}{2^n} \sum_{k=1}^{2^n} \varphi_{n,k}, \quad \varphi_{n-1} = \frac{1}{2^n} \sum_{k=1}^{2^n} \psi_{n,k} \qquad (n = 3,4,5, \ldots) \tag{0.6}$$

*where $\varphi_{n,k}$ and $\psi_{n,k}$ $(k = 1, \ldots, 2^n;\ n = 3,4,5, \ldots)$ are linear functionals on $L(E, E)$, and a family $\{\mathscr{A}_{n,k}\}_{\substack{1\leqslant k\leqslant 2^n\\ 3\leqslant n<\infty}}$ of finite subsets of $E$ such that*

$$|\varphi_{n,k}(u) - \psi_{n,k}(u)| \leqslant \max_{x\in\mathscr{A}_{n,k}} \|u(x)\|$$

$$(u \in L(E, E),\ k = 1, \ldots, 2^n;\ n = 3,4, \ldots), \tag{0.7}$$

*and such that the numbers $C_n = \max_{x\in \bigcup_{k=1}^{2^n} \mathscr{A}_{n,k}} \|x\|$ $(n = 3,4,5, \ldots)$ satisfy*

$$\sum_{n=3}^{\infty} C_n < \infty. \tag{0.8}$$

*Proof.* a) Put

$$Q_0 = \{0\} \cup \bigcup_{n=0}^{\infty} \mathscr{A}_n. \tag{0.9}$$

Then, by (0.5), $Q_0$ is compact. Furthermore, by (0.4),

$$|\varphi_n(u) - \varphi_{n-1}(u)| \leqslant \alpha_n \sup_{x \in Q_0} \|u(x)\| \quad (u \in L(E, E), n = 0, 1, \ldots), \tag{0.10}$$

whence, since $\alpha_n > 0$, $\sum_{n=0}^{\infty} \alpha_n < \infty$, the limits

$$\varphi(u) = \lim_{n\to\infty} \varphi_n(u) \qquad (u \in L(E, E)) \tag{0.11}$$

exist and satisfy the inequalities

$$|\varphi(u)| \leqslant \sum_{n=0}^{\infty} \alpha_n \sup_{x \in Q_0} \|u(x)\| \qquad (u \in L(E, E)). \tag{0.12}$$

Now let $v$ be an arbitrary continuous linear operator of finite rank on $E$ (i.e., $v \in L(E, E)$, $\dim v(E) < \infty$) and let $u = I_E - v$. Then, by (0.11) and (0.2), (0.3),

$$\varphi(u) = \lim_{n\to\infty} \varphi_n(u) = \lim_{n\to\infty} \varphi_n(I_E) - \lim_{n\to\infty} \varphi_n(v) = 1, \tag{0.13}$$

whence, by (0.12),

$$1 = |\varphi(u)| \leqslant \sum_{n=0}^{\infty} \alpha_n \sup_{x \in Q_0} \|x - v(x)\|,$$

so there is an $x_0 \in Q_0$ such that $\|x_0 - v(x_0)\| \geqslant \frac{1}{\sum_{n=0}^{\infty} \alpha_n}$. Thus, since $v \in L(E, E)$ was an arbitrary operator of finite rank, $E$ does not have the approximation property.

b) By (0.8), we can choose*) a sequence $\{\alpha_n\}_{n=3}^{\infty}$ of positive numbers such that $\sum_{n=3}^{\infty} \alpha_n < \infty$, $\lim_{n\to\infty} \frac{C_n}{\alpha_n} = 0$. Put

$$\mathscr{A}_n = \bigcup_{k=1}^{2^n} \frac{1}{\alpha_n} \mathscr{A}_{n,k} \qquad (n = 3, 4, \ldots). \tag{0.14}$$

---

*) Indeed, since $\sum_{j=3}^{\infty} C_j < \infty$, there exist integers $m_n \geqslant 2$ such that $\sum_{j=m_n+1}^{\infty} C_j < \frac{1}{n \cdot 2^{n+1}}$ $(n = 1, 2, \ldots)$. Hence, putting $\alpha_j = nC_j$ $(j = m_n + 1, \ldots$

Then $\mathscr{A}_n$ is finite and by (0.6), (0.7) we have

$$|\varphi_n(u) - \varphi_{n-1}(u)| = \left|\frac{1}{2^n}\sum_{k=1}^{2^n}(\varphi_{n,k}(u) - \psi_{n,k}(u))\right| \leqslant \frac{1}{2^n}\sum_{k=1}^{2^n}\max_{x\in\mathscr{A}_{n,k}}\|u(x)\| \leqslant$$

$$\leqslant \max_{x\in\bigcup_{k=1}^{2^n}\mathscr{A}_{n,k}}\|u(x)\| = \alpha_n \max_{x\in\mathscr{A}_n}\|u(x)\| \quad (u\in L(E, E),\ n = 3, 4, \ldots),$$

which, together with $\varphi_0 = \varphi_1 = \varphi_2 = 0$, gives (0.4) (with any finite sets $\mathscr{A}_0, \mathscr{A}_1, \mathscr{A}_2$). Also,

$$\max_{x\in\mathscr{A}_n}\|x\| = \max_{x\in\bigcup_{k=1}^{2^n}\frac{1}{\alpha_n}\mathscr{A}_{n,k}}\|x\| \leqslant \frac{C_n}{\alpha_n} \to 0 \text{ as } n\to\infty,$$

and thus we have (0.5), which completes the proof.

*Remark 0.1.* a) One can also give the following geometric interpretation of the above proof of part a): Since each $\varphi_n$ is linear on $L(E, E)$, so is $\varphi$ defined by (0.11) (we have used this above, in (0.13)). Moreover, by (0.12) $\varphi$ is also continuous for the topology of uniform convergence on compact subsets of $E$ (since a neighborhood base of 0 for that topology is*) the family $V_{Q,r}(0) = \{u\in L(E, E) | \sup_{u\in Q}\|u(x)\| < r\}$, where $Q\subset E$ is compact and $r > 0$, and since for each $\varepsilon > 0$ there exists $V_{Q_\varepsilon, r_\varepsilon}(0)$, namely, $Q_\varepsilon = Q_0$ and $r_\varepsilon = \frac{\varepsilon}{\sum_{n=0}^{\infty}\alpha_n}$, such that $|\varphi(u)| < \varepsilon$ for all $u\in V_{Q_\varepsilon, r_\varepsilon}(0)$). Since by (0.2) and (0.3), $\varphi(I_E) = \lim_{n\to\infty}\varphi_n(I_E) = 1$ and $\varphi(v) = \lim_{n\to\infty}\varphi_n(v) = 0$ for each $v\in L(E, E)$ with $\dim v(E) < \infty$, $\varphi$ separates $I_E$ from the closure of the linear subspace $\{v\in L(E, E) \mid \dim\ v(E) < \infty\}$ of $L(E, E)$ in the topology of compact convergence, which is equivalent to the fact that $E$ does not have the approximation property.

---

$\ldots, m_{n+1};\ n = 1, 2, \ldots)$, we have $\frac{C_j}{\alpha_j} = \frac{1}{n}$ $(j = m_n + 1, \ldots, m_{n+1})$, which $\to 0$ as $j\to\infty$, and $\sum_{j=m_1+1}^{\infty}\alpha_j = \sum_{n=1}^{\infty} n \sum_{j=m_n+1}^{m_{n+1}} C_j \leqslant \sum_{n=1}^{\infty}\frac{1}{2^{n+1}} = \frac{1}{2}$, so it remains to take arbitrary $\alpha_3, \ldots, \alpha_{m_1} > 0$.

*) See e.g. [355], p. 79; this topology is also called the "topology of compact convergence". Condition (0.4) means that, for each $n$, $\varphi_n$ is so "near" to $\varphi_{n-1}$, as to guarantee the existence and the continuity of $\varphi$ for this topology.

b) Let us observe that any decompositions $\varphi_n = \frac{1}{2^n} \sum_{k=1}^{2^n} \varphi_{n,k}$ ($n = 3, 4, 5, \ldots$) imply the obvious decompositions $\varphi_{n-1} = \frac{1}{2^n} \sum_{k=1}^{2^n} \psi'_{n,k}$ ($n = 3, 4, 5, \ldots; \varphi_2 = 0$), where

$$\psi'_{n,k} = \begin{cases} 2\varphi_{n-1,k} & \text{for } k = 1, \ldots, 2^{n-1} \\ 0 & \text{for } k = 2^n; \end{cases}$$

however, we shall work with other decompositions (0.6).

Let us explain now how we shall define the subspaces $E$ of $c_0$ and $l^p$ ($2 < p < \infty$), the linear functionals $\varphi_n$ on $L(E, E)$ and the finite subsets $\mathscr{A}_n$ of $E$ satisfying the conditions of lemma 0.1. We recall that if $E$ is a Banach space and if $(x_i, f_i)_{i \in \mathscr{M}}$ ($\{x_i\}_{i \in \mathscr{M}} \subset E$, $\{f_i\}_{i \in \mathscr{M}} \subset E^*$) is a countable biorthogonal system, then for any finite non-empty subset $M$ of the index set $\mathscr{M}$, the *average trace* over $M$ is the linear functional on $L(E, E)$ defined by

$$\mathscr{T}(M; u) = \frac{1}{|M|} \sum_{i \in M} f_i(u(x_i)) \qquad (u \in L(E, E)), \qquad (0.15)$$

where $|M|$ denotes*) the cardinality of the set $M$.

The desired subspaces $E$ of $c_0$ and $l^p$ ($2 \leqslant p < \infty$) will be of the form $E = [x_i]_{i \in \mathscr{M}}$, where $(x_i, f_i)_{i \in \mathscr{M}}$ ($\{x_i\}_{i \in \mathscr{M}} \subset E$, $\{f_i\}_{i \in \mathscr{M}} \subset E^*$) is a suitable countable $E$-complete biorthogonal system (the index set $\mathscr{M}$ will be chosen later). Each functional $\varphi_n(u)$ (except $\varphi_0 = \varphi_1 = \varphi_2 = 0$ in the first proof) will be of the form $\mathscr{T}(M_n; u)$, for a suitable sequence $\{M_n\}$ of pairwise disjoint finite subsets**) of $\mathscr{M}$ such that $\cup M_n = \mathscr{M}$ (thus, the nearness condition (0.4) will mean that the "jump" of $\mathscr{T}$ from $M_{n-1}$ to $M_n$ is***) small enough) and each $\mathscr{A}_n$ will be a finite set of finite linear combinations of the $x_i$'s.

Let us first observe that by (0.15) and biorthogonality we have, for any finite non-empty subset $M \subset \mathscr{M}$,

$$\mathscr{T}(M; I_E) = \frac{1}{|M|} \sum_{i \in M} f_i(x_i) = 1,$$

and hence the linear functionals $\varphi_n(u) = \mathscr{T}(M_n; u)$ will certainly satisfy (0.2). Furthermore, if $v \in L(E, E)$ is an operator of rank 1 of the form

$$v(x) = f_0(x) x_{i_0} \qquad (x \in E), \qquad (0.16)$$

---

*) In the other sections we shall use the notation card $M$.

**) In theorems 0.1 and 0.2 we shall have $|M_n| = 2^{2n-1}$ and $|M_n| = 2^n$ ($n = 1, 2, \ldots$), respectively.

***) For a consequence of this fact see remark 0.5.

where $f_0 \in E^*$, $i_0 \in I$, then, by biorthogonality,

$$\varphi_n(v) = \mathscr{T}(M_n; v) = \frac{1}{|M_n|} \sum_{i \in M_n} f_i(f_0(x_i)x_{i_0}) =$$

$$= \frac{1}{|M_n|} \sum_{i \in M_n} f_0(x_i)\delta_{i,i_0} \qquad (n = 0,1,2, \ldots).$$

However, when $n$ is large enough, we have $i_0 \notin M_n$, so $\delta_{i,i_0} = 0$ $(i \in M_n)$, and hence $\varphi_n(v) = 0$. Consequently, since the set of all operators of rank 1 of the form (0.16) is complete, for the norm topology of $L(E, E)$, in the set of all finite rank operators on $E$, in order to prove (0.3) it is enough to show that $\sup_{0 \leqslant n < \infty} \|\varphi_n\| < \infty$. But, by (0.10) and $\alpha_j > 0$, $\sum_{j=0}^{\infty} \alpha_j = 1$, we have

$$|\varphi_n(u)| = \left| \sum_{j=0}^{n} (\varphi_j(u) - \varphi_{j-1}(u)) \right| \leqslant \sum_{j=0}^{n} \alpha_j \sup_{x \in Q_0} \|u(x)\| \leqslant$$

$$\leqslant \left( \sum_{j=0}^{\infty} \alpha_j \sup_{x \in Q_0} \|x\| \right) \|u\| \qquad (u \in L(E, E),\ n = 0, 1, 2, \ldots),$$

whence $\sup_{0 \leqslant n < \infty} \|\varphi_n\| < \infty$,*) which will prove (0.3), provided that we shall have (0.4) and (0.5) assured. Thus, it will be sufficient to concentrate our efforts to achieve (0.4) and (0.5).

In both proofs below it will be convenient to replace $c_0$ and $l^p$ $(2 < p < \infty)$, by the (isometric) spaces $c_0(\Gamma)$ and $l^p(\Gamma)$ $(2 < p < \infty)$ for a suitable countable set $\Gamma$, so the elements of the desired space $E$ will be**) scalar-valued functions on $\Gamma$. We shall take the $x_i$'s to be certain functions on $\Gamma$ with finite supports supp $x_i$ and such that

$$|x_i(g)| = 1 \qquad (g \in \operatorname{supp} x_i;\ i \in \mathscr{M}), \tag{0.17}$$

whence $x_i \in c_0(\Gamma) \subset l^p(\Gamma)$, $\|x_i\|_{c_0(\Gamma)} = 1$, $\|x_i\|_{l^p(\Gamma)} = |\operatorname{supp} x_i|^{\frac{1}{p}}$ $(1 \leqslant p < \infty)$, and then we shall put $E = [x_i]_{i \in \mathscr{M}}$ in $c_0(\Gamma)$, respectively

*) Actually, we shall see in a moment that the biorthogonal system $(x_i, f_i)_{i \in \mathscr{M}}$ will be chosen so as to satisfy $\|x_i\| \|f_i\| = 1$ $(i \in \mathscr{M})$, whence $|\varphi_n(u)| = |\mathscr{T}(M_n, u)| \leqslant \|u\|$ and thus, by (0.2), $\|\varphi_n\| = 1$ for all $n$.

**) For any countable set $\Gamma = \{\gamma_n\}$, $c_0(\Gamma)$ is the Banach space of all functions $x(.)$ on $\Gamma$ such that $\{x(\gamma_n)\}_{n=1}^{\infty} \in c_0$, with $\|x(.)\|_{c_0(\Gamma)} = \|\{x(\gamma_n)\}\|_{c_0} = \sup_{1 \leqslant n < \infty} |x(\gamma_n)|$; the case of $l^p(\Gamma)$ is similar.

in $l^p(\Gamma)$. Naturally, the supports of the $x_i$'s cannot be pairwise disjoint (since if they were, then $\{x_i\}_{i\in\mathcal{M}}$ would be a bloc basic sequence with respect to the unit vector basis of $c_0(\Gamma)$ and $l^p(\Gamma)$, and hence a basis of $E=[x_i]_{i\in\mathcal{M}}$), but we shall choose the $x_i$'s in such a way that the sequence $\{x_i\}_{i\in\mathcal{M}}$ will be orthogonal in $l^2(\Gamma)$. Therefore, defining

$$f_i(x)=\frac{1}{\|x_i\|^2_{l^2(\Gamma)}}(x,x_i)=\frac{1}{|\operatorname{supp} x_i|}\sum_{g\in\operatorname{supp} x_i} x(g)\overline{x_i(g)}\quad (x\in E,\ i\in\mathcal{M}),$$

we shall have $f_i\in E^*$ and $(x_i,f_i)_{i\in\mathcal{M}}$ will be an $E$-complete biorthogonal system; also, clearly,

$$\|f_i\|=\frac{1}{|\operatorname{supp} x_i|}\|x_i\|_{l^q(\Gamma)}=|\operatorname{supp} x_i|^{\frac{1}{q}-1},$$

where $\frac{1}{p}+\frac{1}{q}=1$ $(1\leqslant p\leqslant\infty;\ \frac{1}{\infty}=0)$, whence $\|x_i\|\,\|f_i\|=1$ $(i\in\mathcal{M})$. Then, by the above choice of the functionals $\varphi_n$,

$$\varphi_n(u)-\varphi_{n-1}(u)=\frac{1}{|M_n|}\sum_{i\in M_n}D_i\sum_{g\in\Gamma}u(x_i)(g)\overline{x_i(g)}-$$

$$-\frac{1}{|M_{n-1}|}\sum_{i\in M_{n-1}}D_i\sum_{g\in\Gamma}u(x_i)(g)\overline{x_i(g)}=$$

$$=\sum_{g\in\Gamma_n}u\left(\frac{1}{|M_n|}\sum_{i\in M_n}D_i\overline{x_i(g)}x_i-\frac{1}{|M_{n-1}|}\sum_{i\in M_{n-1}}D_i\overline{x_i(g)}x_i\right)(g)$$

$$(u\in L(E,E),\ n=0,1,\ldots),$$

where $\Gamma_n=\bigcup_{i\in M_n\cup M_{n-1}}\operatorname{supp} x_i$ and $D_i=\frac{1}{|\operatorname{supp} x_i|}$. This formula can be further simplified by choosing the $x_i$'s so that $D_i=\frac{1}{|\operatorname{supp} x_i|}$ is constant on each $M_n$, say $D_i=\delta_n$ $(i\in M_n)$. Then, putting

$$y_g^n=\frac{\delta_n}{|M_n|}\sum_{i\in M_n}\overline{x_i(g)}x_i-\frac{\delta_{n-1}}{|M_{n-1}|}\sum_{i\in M_{n-1}}\overline{x_i(g)}x_i\qquad (g\in\Gamma_n),\quad (0.18)$$

we obtain

$$|\varphi_n(u) - \varphi_{n-1}(u)| = |\sum_{g\in\Gamma_n} u(y_g^n)(g)| \leqslant \sum_{g\in\Gamma_n} |u(y_g^n)(g)| \leqslant$$

$$\leqslant \sum_{g\in\Gamma_n} \|u(y_g^n)\|_{l^p(\Gamma)} \leqslant |\Gamma_n| \max_{g\in\Gamma_n} \|u(y_g^n)\|_{l^p(\Gamma)} \quad (u\in L(E,E),\ n=0,1,\ldots),$$

where $1\leqslant p\leqslant\infty$. Thus, it is intuitively clear that if we could take the elements of the finite sets $\mathscr{A}_n$ to be of a suitable similar form*) to the $\frac{|\Gamma_n|}{\alpha_n} y_g^n$'s, then (0.4) would be satisfied and we would have to concentrate only to obtain such elements with norms small enough to assure (0.5) as well. Actually, in order to define such elements, in the first proof we shall also exploit that each $\varphi_n$ admits two different decompositions, as in (0.6) (the functionals $\varphi_{n,k}(u)$, $\psi_{n,k}(u)$ will be average traces $\mathscr{T}(M_{n,k}; u)$, $\mathscr{T}(L_{n,k}; u)$ over some suitable subsets $M_{n,k}$, $L_{n,k}$ of $M_n$ and $M_{n-1}$, respectively), while in the second proof we shall be able to write each $f_i$ in two different forms and to exploit both of them. This will lead to (0.4), (0.5) for $\|.\|_{l^p(\Gamma)}$, where $2<p<\infty$.

To this end, we shall take $\Gamma$ to be a countable union of certain pairwise disjoint finite groups $G_n$ and we shall take each $x_i$ to be a certain linear combination, with coefficients $\pm 1$, of a constant finite number**) of characters (or, equivalently, Walsh functions) on different groups $G_n$, extended by 0 on $\Gamma\setminus G_n$. Then, in order to prove the existence of elements of similar form to (0.18), with norms growing not too fast, we shall need (in each proof) a lemma on the existence of a "good" partition of the set of all characters of the groups $G_n$ entering in the definition of $\Gamma$. Also, in the first proof we shall use a combinatorial lemma which will enable us to "stretch out" $\Gamma$ so much as to obtain disjoint supports for the five summands in the definition of the functions $x_i$ and supports with small intersection for the summands in the elements of the sets $\mathscr{A}_n$ (thus, this lemma will help to make our choice of $\Gamma$ and of the $x_i$'s and $\mathscr{A}_n$'s), while in the second proof we shall use, instead, a probabilistic lemma, to show that "good" partitions of sets of characters and "good" choices of the above mentioned coefficients $\pm 1$ in the definition of the $x_i$'s exist.

Let us pass now to the first proof. We start with the lemma on the partition of the set of characters, mentioned above. For convenience,

---

*) In fact, in both proofs we shall take the elements of $\mathscr{A}_n$ to be linear combinations of the $x_i$'s with $i\in M_n\cup M_{n-1}$.

**) In the first proof this number will be five, while in the second proof it will be two.

in the sequel we shall use the notation $\|.\|_p$ instead of $\|.\|_{l^p(G)}$ (where $G$ is any set).

**Lemma 0.2.** *Let $n$ be a positive integer, let $G$ be the (abelian) group $\{-1,1\}^n$ (of all functions from $\{1, 2, \ldots, n\}$ into $\{-1, 1\}$, with pointwise multiplication) and let $H \subset C(G)$ be the set of all characters*) of $G$. Then there exist two disjoint subsets $W^+$, $W^-$ of $H$ with cardinalities satisfying*

$$|W^+| = |W^-| = \frac{1}{2}|G| = 2^{n-1} \tag{0.19}$$

*(hence $W^+ \cup W^- = H$), such that*

$$\|\sum_{w\in W^+} w - \sum_{w\in W^-} w\|_\infty \leqslant 2^{1+\left[\frac{n}{2}\right]}. \tag{0.20}$$

*Proof.* Let $k = \left[\frac{n-1}{2}\right]$ and let $s = n - 2k$, hence $s \geqslant 1$ and $n = 2k + s$. Define $r_j \in H$ by

$$r_j(g) = g(j) \qquad (g \in G, j = 1, \ldots, n) \tag{0.21}$$

and $y_0 \in C(G)$ by

$$y_0 = \prod_{j=1}^{k} (1 - r_{2j-1} - r_{2j} - r_{2j-1}r_{2j}) \prod_{t=1}^{s} (1 - r_{2k+t}). \tag{0.22}$$

Now, by (0.21), each product $\prod_{j\in S} r_j$, where $S \subset \{1, 2, \ldots, n\}$, is a character $w = w_S \in H$ and the set of all such products coincides with the set $H$ of all characters (since it contains $2^n$ distinct elements, so it has the same cardinality as $H$). Since at "term by term multiplication" in (0.22) each such product occurs exactly once, we obtain $y_0 = \sum_{w\in H} c_w w$, where each coefficient $c_w$ is either 1 or $-1$. Put

$$W^+ = \{w \in H | c_w = 1\}, \; W^- = \{w \in H | c_w = -1\}, \tag{0.23}$$

*) I.e., of all homomorphisms of $G$ into $\{\zeta \,|\, |\zeta| = 1\}$. Since $G = \{-1, 1\}^n$, all characters have real values, i.e., $\pm 1$.

and let $e$ denote the unit of $G$. Since $r_{2k+s}(e) = r_n(e) = e(n) = 1$, the last factor in (0.22) is 0 at $e$, so $y_0(e) = 0$, whence, since $w(e) = 1$ $(w \in H)$, we obtain

$$|W^+| - |W^-| = \sum_{w \in H} c_w = \sum_{w \in H} c_w w(e) = y_0(e) = 0,$$

which proves (0.19). Finally, for any $g \in G$, by (0.22) $y_0(g)$ is a product of $k + s = 1 + \left[\frac{n}{2}\right]$ numbers, each of them being 2, $-2$ or 0, whence, since $y_0 = \sum_{w \in H} c_w w = \sum_{w \in W^+} w - \sum_{w \in W^-} w$, we obtain (0.20), which completes the proof of lemma 0.2.

*Remark 0.2.* Let $2^{[1,n]}$ denote the (abelian) group of all subsets of the set $[1, n] = \{1,2, \ldots, n\}$, with the group operation $S_1 S_2 = S_1 \div \div S_2 = (S_1 \setminus S_2) \cup (S_2 \setminus S_1)$ (the symmetric difference of the subsets $S_1$, $S_2$ of $[1, n]$). Then $G = \{-1, 1\}^n$ is isomorphic to $2^{[1,n]}$, by the mapping $g \to S_g = \{j \in [1, n] \mid g(j) = -1\}$, and $C(G)$ is isomorphic to $l^\infty_{2^n}$ (the Banach space of all scalar functions $\xi = \{\xi(S)\}_{S \in 2^{[1,n]}}$ with the norm $\|\xi\| = \max_{S \in 2^{[1,n]}} |\xi(s)|)$, by the mapping $x \to \xi_x$, where $\xi_x(S_g) = x(g)$ $(g \in G)$. This latter isomorphism carries $H$ onto the set of all Walsh functions $\beta_S(S_g) = (-1)^{|S \cap S_g|}$ and $\{r_j\}_{j=1}^n$ onto the set of all Rademacher functions $\rho_j(S_g) = 1$ for $S_g \not\ni j$ and $-1$ for $S_g \ni j$ $(j = 1, \ldots, n)$. This motivates the notations $w$ and $r_j$ used above.

Let us give now the combinatorial lemma mentioned above:

**Lemma 0.3.** *Let $n \geqslant 4$ be a positive integer and let $(A_i)_{i \in I}$, $(B_j)_{j \in J}$ be two families of mutually disjoint sets with cardinalities satisfying*

$$|A_i| = \frac{1}{2}|I| = 2^{n-2} \qquad (i \in I), \tag{0.24}$$

$$|B_j| = \frac{1}{2}|J| = 2^{n-1} \qquad (j \in J) \tag{0.25}$$

*(i.e., there are $2^{n-1}$ sets $A_i$, each of cardinality $2^{n-2}$ and there are $2^n$ sets $B_j$, each of cardinality $2^{n-1}$). Then there exists a function $p \colon \bigcup_{j \in J} B_j \to \bigcup_{i \in I} A_i$ such that*

$$|p(B_j) \cap A_i| = 1 \qquad (i \in I,\ j \in J), \tag{0.26}$$

$$|p^{-1}(a)| = 4 \qquad \Big(a \in \bigcup_{i \in I} A_i\Big), \tag{0.27}$$

$$|p(B_j) \cap p(B_{j'})| \leqslant 2 \qquad (j, j' \in J,\ j \neq j') \tag{0.28}$$

(*i.e.*, *the elements of any fixed* $B_j$ *are carried into elements of different**) *sets* $A_i$, *each element of any fixed* $A_i$ *is "covered" exactly***) 4 *times, and the images of any two distinct sets* $B_j$, $B_{j'}$, *can have at most* 2 *common elements*).

*Proof.* Clearly, it is enough to prove that there exist $(A_i)_{i\in I}$, $(B_j)_{j\in J}$ and $p$ as above (since then, using one-to-one mappings onto any other given $(A_i^0)_{i\in I}$, $(B_j^0)_{j\in J}$ one obtains also a mapping $p_0\colon \bigcup_{j\in J} B_j^0 \to \bigcup_{i\in I} A_i^0$ with the required properties).

Let $F$ be a finite field with $2^{n-2}$ elements***) and let $D$ be a 4-element subset of $F$. It will be convenient to regard the indices $i$ of the $A_i$'s as the pairs $(x, \delta)$ with $x\in F$, $\delta\in\{0, 1\}$ and the elements of an $A_i$, where $i=(x, \delta)$, as the triples $(x, \delta, y)$ with $y\in F$, that is, to put

$$I = F\times\{0, 1\} = \{(x, \delta)\mid x\in F, \delta\in\{0, 1\}\}, \tag{0.29}$$

$$A_i = \{i\}\times F = \{(x, \delta, y)\mid y\in F\} \qquad (i=(x, \delta)\in I). \tag{0.30}$$

Similarly, we put

$$J = F\times D = \{(z, d)\mid z\in F, d\in D\}, \tag{0.31}$$

$$B_j = \{j\}\times I = \{(z, d, x, \delta)\mid x\in F, \delta\in\{0, 1\}\} \quad (j=(z, d)\in J). \tag{0.32}$$

Let us define

$$p((z, d, x, \delta)) = (x, \delta, z+dx) \qquad ((z, d, x, \delta)\in \bigcup_{j\in J} B_j), \tag{0.33}$$

where (until the end of proof of the lemma) all algebraic operations are operations in the field. We shall show that $p$ has the required properties.

We have

$$p(B_j) = \{(x, \delta, z+dx)\mid (x, \delta)\in I\} \qquad (j=(z, d)\in J),$$

and hence, by (0.30), for any fixed $i=(x, \delta)\in I$ and $j=(z, d)\in J$, $p(B_j)\cap A_i$ is the singleton $\{(x, \delta, z+dx)\}$, which proves (0.26).

---

*) Note that there are $2^{n-1}$ elements in $B_j$ and $2^{n-1}$ sets $A_i$, hence at this process for any fixed $j$ each set $A_i$ is used. Consequently, we have $p(b)\neq p(b')$ for $b, b'\in B_j$, $b\neq b'$, that is, $p$ is one-to-one on each $B_j$.

**) This is a natural requirement, since by our assumption $\bigcup_{j\in J} B_j = 2^n\cdot 2^{n-1} = 4\cdot 2^{n-1}\cdot 2^{n-2} = 4\left|\bigcup_{i\in I} A_i\right|$.

***) See e.g. [404].

Next, we claim that for each fixed $d \in D$ we have

$$p(B_{(z,d)}) \cap p(B_{(z_1,d)}) = \varnothing \qquad (z, z_1 \in F,\ z \neq z_1), \tag{0.34}$$

$$\bigcup_{z \in F} p(B_{(z,d)}) = \bigcup_{i \in I} A_i \tag{0.35}$$

(i.e., the sets $p(B_{(z,d)})$, with $z \in F$, form a disjoint partition of $\bigcup_{i \in I} A_i$). Indeed, if $(x, \delta, y) \in p(B_{(z,d)}) \cap p(B_{(z_1,d)})$, then $z + dx = y = z_1 + dx$, whence $z = z_1$, which proves (0.34). On the other hand, for each $(x, \delta, y) \in \bigcup_{i \in I} A_i$ we have $(x, \delta, y) \in p(B_{(y-dx,d)})$ (since $y = (y - dx) + dx$), which proves (0.35). Now, from (0.34) and (0.35) it follows that for fixed $d \in D$, each $(x, \delta, y) \in \bigcup_{i \in I} A_i$ belongs to exactly one of the sets $p(B_{(z,d)})$. Consequently, each $(x, \delta, y) \in \bigcup_{i \in I} A_i$ belongs to exactly four of the sets $p(B_{(z,d)})$ (since $|D| = 4$), which proves (0.27).

Finally, assume that $(x, \delta, y) \in p(B_j) \cap p(B_{j'})$, where $j = (z, d)$, $j' = (z', d') \in J$, $j \neq j'$. Then, by (0.34), $d \neq d'$ and, by (0.33), $y = z + dx = z' + d'x$, whence $(d - d')x = z' - z$. Hence, since $F$ is a field, we obtain

$$x = (d - d')^{-1}(z' - z).$$

Thus, the assumption $(x, \delta, y) \in p(B_j) \cap p(B_{j'})$ with $j \neq j'$ implies that $x$ and $y = z + dx$ are uniquely determined and only $\delta$ can vary for the elements $(x, \delta, y)$ of such an intersection. Since $\delta \in \{0,1\}$, this proves (0.28), completing the proof of lemma 0.3.

*Remark 0.3.* a) In the above proof the fact that $F$ is a field has been used only to prove (0.28). One could also use arithmetic progressions instead of fields, which would yield a slightly weaker estimate in (0.28) (namely, 2 would be replaced by 6) and would imply some corresponding minor changes in the subsequent computations. b) Let us also mention the following consequence of (0.26) and (0.28), which will be used in the sequel:

$$|p^{-1}(p(B_j)) \cap B_{j'}| \leqslant 2 \qquad (j, j' \in J, j \neq j'). \tag{0.36}$$

Indeed, if $b, b', b'' \in p^{-1}(p(B_j)) \cap B_{j'}$, then $p(b), p(b'), p(b'') \in p(B_j) \cap p(B_{j'})$, whence, by (0.28), at least two of these latters must coincide, say $p(b) = p(b')$. Then, by (0.26), we must have also $b = b'$ (indeed, we have already observed that (0.26) implies $p(b) \neq p(b')$ for $b, b' \in B_{j'}$, $b \neq b'$), which proves (0.36).

Now we are ready for the construction. We shall take as the set $\Gamma$ the disjoint union $\Gamma = \bigcup_{n=3}^{\infty} \bigcup_{k=1}^{2^n} G_{n,k}$, where $G_{n,k}$ is the set of all elements of the group $\{-1,1\}^n$ ($k = 1,2, \ldots, 2^n$; $n = 3,4, \ldots$). Let $X$ denote the set of all real functions on $\Gamma$ with finite supports and for each $n \geqslant 3$, $k \leqslant 2^n$ and $\varepsilon \in \{+, -\}$ let

$$W_{n,k}^{\varepsilon} = \{x \in X \mid x|_{G_{n,k}} \in W^{\varepsilon}, x|_{\Gamma \setminus G_{n,k}} = 0\}, \tag{0.37}$$

where $W^{\varepsilon}$ is that of lemma 0.2 above for $G = G_{n,k}$. Put

$$W = \bigcup_{n=3}^{\infty} \bigcup_{k=1}^{2^n} W_{n,k}^{+}. \tag{0.38}$$

Then, since $|W_{n,k}^{+}| = |W_{n,k}^{-}| = 2^{n-1}$, by lemma 0.3 there exists a function

$$h: \bigcup_{n=4}^{\infty} \bigcup_{k=1}^{2^n} W_{n,k}^{-} \to W \tag{0.39}$$

with $h(W_{n,k}^{-}) \subset \bigcup_{l=1}^{2^{n-1}} W_{n-1,l}^{+}$ ($n \geqslant 4$; $k \leqslant 2^n$), such that

$$|h(W_{n,k}^{-}) \cap W_{n-1,l}^{+}| = 1 \qquad (n \geqslant 4;\ k \leqslant 2^n;\ l \leqslant 2^{n-1}) \tag{0.40}$$

(so $h|_{W_{n,k}^{-}}$ is one-to-one, as was observed above),

$$|h^{-1}(w)| = 4 \qquad (w \in W), \tag{0.41}$$

$$|h(W_{n,k}^{-}) \cap h(W_{n,k'}^{-})| \leqslant 2 \qquad (n \geqslant 4;\ k,k' \leqslant 2^n;\ k \neq k'), \tag{0.42}$$

whence, by remark 0.3 b), also

$$|h^{-1}(h(W_{n,k}^{-})) \cap W_{n,k'}^{-}| \leqslant 2 \qquad (n \geqslant 4;\ k, k' \leqslant 2^n;\ k \neq k'); \tag{0.43}$$

indeed, one can define $h_n: \bigcup_{k=1}^{2^n} W_{n,k}^{-} \to \bigcup_{l=1}^{2^{n-1}} W_{n-1,l}^{+}$ as in lemma 0.3 and then put $h(w) = h_n(w)$ for $w \in \bigcup_{k=1}^{2^n} W_{n,k}^{-}$, $n = 4,5, \ldots$ We fix such a mapping $h$ and define

$$x_w = w + \sum_{y \in h^{-1}(w)} y \qquad (w \in W); \tag{0.44}$$

thus, the index set $\mathscr{M}$ (mentioned after remark 0.1) will be the set $W$ defined by (0.38).

By (0.41), each $x_w$ is a sum of five functions on $\Gamma$ with finite supports, so $x_w \in X$ ($w \in W$). Let us observe that the summands in $\sum_{y \in h^{-1}(w)} y$ belong to different sets $W^-_{n+1,l_i}$, whence their supports are different sets $G_{n+1,l_i}$ ($i = 1, \ldots, 4$), so they are pairwise disjoint. Indeed, if two distinct elements $y_1, y_2 \in h^{-1}(W^+_{n,k}) \subset \bigcup_{l=1}^{2^{n+1}} W^-_{n+1,l}$ belonged to the same set $W^-_{n+1,l_i}$, then we would have $h(y_1), h(y_2) \in h(W^-_{n+1,l_i}) \cap W^+_{n,k}$, whence, by (0.40), $h(y_1)=h(y_2)$, in contradiction with the fact that $h|_{W^-_{n+1,l_i}}$ is one-to-one. Thus, the supports of the five summands in (0.44) are pairwise disjoint and hence we have indeed (0.17). Also, it is clear that

$$\operatorname{supp} x_w = G_{n,k} \cup \bigcup_{i=1}^{4} G_{n+1,l_i} \quad (w \in W^+_{n,k};\ k = 1, \ldots, 2^n;\ n = 3,4, \ldots),$$

where $1 \leqslant l_i \leqslant 2^{n+1}$ and where $h^{-1}(w) \in \bigcup_{i=1}^{4} W^-_{n+1,l_i}$, so the $x_w$'s are not disjointly supported*⁾. Nevertheless, the (5-element) sets of summands $\{w\} \cup \{h^{-1}(w)\}$ occurring in the definition (0.44) of the $x_w$'s are pairwise disjoint, that is,

$$(w_1 \cup h^{-1}(w_1)) \cap (w_2 \cup h^{-1}(w_2)) = \emptyset \quad (w_1, w_2 \in W,\ w_1 \neq w_2). \tag{0.45}$$

Indeed, $w_1 \notin h^{-1}(w_2)$ and $w_2 \notin h^{-1}(w_1)$, since $w_1, w_2 \in W = \bigcup_{n=3}^{\infty} \bigcup_{k=1}^{2^n} W^+_{n,k}$ and $h^{-1}(w_1), h^{-1}(w_2) \subset \bigcup_{n=4}^{\infty} \bigcup_{k=1}^{2^n} W^-_{n,k}$. Also, if $y \in h^{-1}(w_1) \cap h^{-1}(w_2)$, then $h(y) = w_1$, $h(y) = w_2$, so $w_1 = w_2$, which proves (0.45).

Hence, since the characters of each group $G_{n,k}$ are mutually orthogonal (for the natural scalar product $(x, z) = \sum_{g \in G_{n,k}} x(g)\overline{z(g)}$) and since all $G_{n,k}$ are mutually disjoint, if follows that $\{x_w\}_{w \in W}$ is indeed an orthogonal system in $l^2(\Gamma)$ (with the natural scalar product $(x, z) = \sum_{g \in \Gamma} x(g)\overline{z(g)}$), as was announced (after formula (0.17)).

We are now ready to prove

---

*⁾ In fact, if $w, w' \in W^+_{n,k}$, then $\operatorname{supp} x_w \cap \operatorname{supp} x_{w'} = G_{n,k}$; for, the intersection contains $G_{n,k}$ and we have observed above that there is no pair $y \in h^{-1}(w)$, $y' \in h^{-1}(w')$, with $y \neq y'$, such that $y, y'$ belong to the same $W^-_{n+1,l_i}$. If $w \in W^+_{n,k}$, $w' \in W^+_{n,k'}$, where $k \neq k'$, then $\operatorname{supp} x_w \cap \operatorname{supp} x_{w'}$ is the union of the supports of all such pairs $y, y'$ (there may exist 0, 1, 2, 3, or 4 such pairs). Finally, if $w \in W^+_{n,k}$, $w' \in W^+_{n',k'}$, where $n \neq n'$, then $\operatorname{supp} x_w \cap \operatorname{supp} x_{w'} = \emptyset$.

**Theorem 0.1.** *Let* $E = [x_w]_{w \in W}$ *in* $c_0(\Gamma)$ *or in* $l^p(\Gamma)$ $(2 < p < \infty)$. *Then* $E$ *does not have the approximation property.*

*Proof.* Let us observe that $\overline{x_w(g)} = x_w(g)$ $(g \in \Gamma)$, since all five summands in the definition (0.44) of $x_w$ are real functions on $\Gamma$. Thus, the biorthogonal functionals $f_w \in E^*$ are now

$$f_w(x) = \frac{1}{\|x_w\|_2^2} \sum_{g \in \Gamma} x(g) x_w(g) \qquad (x \in E,\ w \in W). \tag{0.46}$$

Define linear functionals $\varphi_n$ on $L(E, E)$ by

$$\varphi_0 = \varphi_1 = \varphi_2 = 0,\ \varphi_n(u) = \mathscr{T}\Big(\bigcup_{k=1}^{2^n} W_{n,k}^+;\ u\Big)\ (u \in L(E, E);\ n = 3,4, \ldots), \tag{0.47}$$

where $\mathscr{T}$ is the average trace; thus, the sets $M_n$ of our heuristic introduction are now $\bigcup_{k=1}^{2^n} W_{n,k}^+$ $(n = 3,4,5, \ldots)$. Define linear functionals $\varphi_{n,k}$, $\psi_{n,k}$ on $L(E, E)$ by*)

$$\varphi_{n,k}(u) = \mathscr{T}(W_{n,k}^+;\ u)\ (u \in L(E, E),\ k = 1, \ldots, 2^n;\ n = 3,4,5, \ldots), \tag{0.48}$$

$$\psi_{3,k} = 0\ \ (k = 1, \ldots, 2^3),\ \ \psi_{n,k}(u) = \mathscr{T}(h(W_{n,k}^-);\ u)$$
$$(u \in L(E, E),\ k = 1, \ldots, 2^n;\ n = 4,5, \ldots). \tag{0.49}$$

Then, since the sets $W_{n,k}^+$ $(k = 1, \ldots, 2^n)$ are pairwise disjoint, we have

$$\varphi_n(u) = \frac{1}{2^n . 2^{n-1}} \sum_{w \in \bigcup_{k=1}^{2^n} W_{n,k}^+} f_w(u(x_w)) = \sum_{k=1}^{2^n} \frac{1}{2^n . 2^{n-1}} \sum_{w \in W_{n,k}^+} f_w(u(x_w)) =$$
$$= \frac{1}{2^n} \sum_{k=1}^{2^n} \varphi_{n,k}(u) \qquad (u \in L(E, E),\ n = 3,4,5, \ldots),$$

i.e. the first part of (0.6). Furthermore, since $\varphi_2 = 0$ and $\psi_{3,k} = 0$ $(k = 1, \ldots, 2^3)$, the second part of (0.6) is satisfied for $n = 3$. On the other hand, for $n \geqslant 4$ and for any $u \in L(E, E)$ we have

$$\frac{1}{2^n} \sum_{k=1}^{2^n} \psi_{n,k}(u) = \frac{1}{2^n} \sum_{k=1}^{2^n} \frac{1}{2^{n-1}} \sum_{w \in h(W_{n,k}^-)} f_w(u(x_w)) =$$
$$= \frac{1}{2^n . 2^{n-1}} \sum_{k=1}^{2^n} \sum_{w \in h(W_{n,k}^-)} f_w(u(x_w)),$$

*) We recall that $h|_{W_{n,k}^-}$ is one-to-one (see (0.40)).

whence, since $\bigcup_{k=1}^{2^n} h(W_{n,k}^-) = \bigcup_{l=1}^{2^{n-1}} W_{n-1,l}^+$ in such a way that each element of $\bigcup_{l=1}^{2^{n-1}} W_{n-1,l}^+$ is the image by $h$ of exactly 4 distinct elements of $\bigcup_{k=1}^{2^n} W_{n,k}^-$ (by (0.41)), and since $\frac{4}{2^n . 2^{n-1}} = \frac{1}{2^{n-1} . 2^{n-2}}$, we obtain

$$\frac{1}{2^n} \sum_{k=1}^{2^n} \psi_{n,k}(u) = \frac{1}{2^{n-1} . 2^{n-2}} \sum_{w \in \bigcup_{l=1}^{2^{n-1}} W_{n-1,l}^+} f_w(u(x_w)) = \varphi_{n-1}(u),$$

which proves (0.6).

Now we shall define the finite subsets $\mathscr{A}_{n,k}$ of $E$. For $n = 3$ it is enough to take

$$\mathscr{A}_{3,k} = \left\{ \frac{1}{4} \left( \sum_{w' \in W_{3,k}^+} \|f_{w'}\| \right) x_w \mid w \in W_{3,k}^+ \right\} \quad (k = 1, \ldots, 2^3), \tag{0.50}$$

since then for any $u \in L(E, E)$ and $k = 1, \ldots, 2^3$ we have

$$|\varphi_{3,k}(u) - \psi_{3,k}(u)| = |\varphi_{3,k}(u)| = \frac{1}{2^2} \left| \sum_{w \in W_{3,k}^+} f_w(u(x_w)) \right| \leqslant$$

$$\leqslant \frac{1}{2^2} \left( \sum_{w' \in W_{3,k}^+} \|f_{w'}\| \right) \max_{w \in W_{3,k}^+} \|u(x_w)\| = \max_{x \in \mathscr{A}_{3,k}} \|u(x)\|,$$

i.e., (0.7) for $n = 3$. Now, put*)

$$\mathscr{A}_{n,k} = \left\{ \frac{1}{2^{n-1}} \left( \sum_{w \in W_{n,k}^+} w(g) x_w - \sum_{w \in W_{n,k}^-} w(g) x_{h(w)} \right) \middle| \, g \in G_{n,k} \right\}$$

$$(k = 1, \ldots, 2^n;\ n = 4,5, \ldots). \tag{0.51}$$

Let us prove (0.7). For the sake of simplicity, we shall use from now on the notations of lemma 0.2 with $G = G_{n,k}$ (so $W^+ = W_{n,k}^+$, $W^- = W_{n,k}^-$). In order to unify the notation, we introduce the function

$$q(w) = \begin{cases} w \text{ for } w \in W^+ (= W_{n,k}^+), \\ h(w) \text{ for } w \in W^- (= W_{n,k}^-). \end{cases} \tag{0.52}$$

*) We recall that $w(g) = \pm 1$ for each $g \in G_{n,k}$ and each $w \in W_{n,k}^+ \cup W_{n,k}^-$.

Furthermore, we shall denote by $\tau$ the linear mapping of $C(G)$ into $E$ which carries each $w \in H$ into $x_{q(w)}$, i.e.

$$\tau(\sum_{w \in H} b_w w) = \sum_{w \in H} b_w x_{q(w)}. \tag{0.53}$$

Also, let us define, for each $g \in G$, an operator $t_g \in L(C(G), C(G))$ by

$$t_g(x)(g') = x(g^{-1}g') \quad (x \in C(G), g' \in G). \tag{0.54}$$

Then for each $w \in H$ and $g, g' \in G$ we have

$$t_g(w)(g') = w(g^{-1}g') = w(g^{-1})w(g') = \frac{1}{w(g)} w(g') = w(g)w(g'),$$

whence

$$t_g(w) = w(g)w \qquad (w \in H, \ g \in G). \tag{0.55}$$

By (0.52), (0.23), (0.53) and (0.55) we have

$$\frac{1}{2^{n-1}}\Big(\sum_{w \in W_{n,k}^+} w(g)x_w - \sum_{w \in W_{n,k}^-} w(g)x_{h(w)}\Big) = \frac{1}{2^{n-1}} \sum_{w \in H} c_w w(g) x_{q(w)} =$$

$$= \frac{1}{2^{n-1}} \sum_{w \in H} c_w w(g)\tau(w) = \frac{1}{2^{n-1}} \sum_{w \in H} c_w \tau t_g(w) = \frac{1}{2^{n-1}} \tau t_g \Big(\sum_{w \in H} c_w w\Big) =$$

$$= \frac{1}{2^{n-1}} \tau t_g(y_0) \qquad (g \in G), \tag{0.56}$$

where $y_0 = \sum_{w \in H} c_w w = \sum_{w \in W^+} w - \sum_{w \in W^-} w$, and hence $\mathscr{A}_{n,k}$ can be written in the form

$$\mathscr{A}_{n,k} = \left\{ \frac{1}{2^{n-1}} \tau t_g(y_0) \,\middle|\, g \in G \right\}. \tag{0.57}$$

Let us also introduce the notation

$$\rho(x)(g) = \begin{cases} x(g) & \text{for } g \in G \\ 0 & \text{for } g \in \Gamma \setminus G \end{cases} \qquad (x \in E). \tag{0.58}$$

We claim that

$$f_{q(w)}(x) = \frac{1}{|G|}(\rho(x), w) \qquad (x \in E,\ w \in H). \tag{0.59}$$

Indeed, assume first that $x = x_{w'}$, where $w' \in W$. Then, by biorthogonality, the left-hand side of (0.59) is $f_{q(w)}(x_{w'}) = \delta_{q(w),w'}$. On the other hand, the right-hand side of (0.59) for $x = x_{w'}$ is

$$\frac{1}{|G|}(\rho(x_{w'}),\ w) = \frac{1}{|G|}\sum_{g\in G} x_{w'}(g)w(g) =$$

$$= \frac{1}{|G|}\sum_{g\in G}\left(w'(g) + \sum_{y\in h^{-1}(w')} y(g)\right) w(g).$$

Now, if $w \in W^+$, then $\sum_{g\in G}(\sum_{y\in h^{-1}(w')} y(g))w(g) = 0$ $\left(\text{since } h^{-1}(w') \subset \subset \bigcup_{n=4}^{\infty}\bigcup_{k=1}^{2^n} W_{n,k}^-\right)$ and $w = q(w)$, whence in this case

$$\frac{1}{|G|}(\rho(x_{w'}),w) = \frac{1}{|G|}\sum_{g\in G} w'(g)w(g) = \delta_{w,w'} = \delta_{q(w),w'} = f_{q(w)}(x_{w'}),$$

so (0.59) holds for $x = x_{w'}$ and $w \in W^+$. Assume now that $w \in W^-$, whence $\sum_{g\in G} w'(g)w(g) = 0$ $\left(\text{since } w' \in W = \bigcup_{n=3}^{\infty}\bigcup_{k=1}^{2^n} W_{n,k}^+\right)$ and $h(w) = = q(w)$. If $w' \neq h(w)$, then $w \notin h^{-1}(w')$, so again $\sum_{g\in G}(\sum_{y\in h^{-1}(w')} y(g))w(g) = = 0$, whence

$$\frac{1}{|G|}(\rho(x_{w'}),\ w) = 0 = \delta_{h(w),w'} = \delta_{q(w),w'} = f_{q(w)}(x_{w'});$$

finally, if $w' = h(w)$, then $w \in h^{-1}(w') \subset W^- = W_{n,k}^-$, whence

$$\frac{1}{|G|}(\rho(x_{w'}), w) = \frac{1}{|G|}\sum_{g\in G}(\sum_{y\in h^{-1}(w')} y(g))\, w(g) =$$

$$= 1 = \delta_{h(w),w'} = f_{q(w)}(x_{w'}),$$

which proves the claim (0.59) for $x = x_{w'}$, whence also for all $x \in E = = [x_{w'}]_{w'\in W}$.

By (0.59), (0.53) and (0.55) we have

$$f_{q(w)}(u(x_{q(w)})) = \frac{1}{|G|}(\rho u(x_{q(w)}),\ w) = \frac{1}{|G|}(\rho u\tau(w), w) =$$

$$= \frac{1}{|G|}\sum_{g\in G} u\tau(w)(g)w(g) = \frac{1}{|G|}\sum_{g\in G} u\tau t_g(w)(g) \quad (u \in L(E, E),\ \ w \in W),$$

whence, by (0.48), (0.49), (0.19) and (0.23),

$$\varphi_{n,k}(u) - \psi_{n,k}(u) = \frac{1}{2^{n-1}}\sum_{w\in H} c_w f_{q(w)}(u(x_{q(w)})) =$$

$$= \frac{1}{2^{n-1}|G|}\sum_{g\in G} u\tau t_g(\sum_{w\in H} c_w w)(g) =$$

$$= \frac{1}{2^{n-1}|G|}\sum_{g\in G}(u\tau t_g y_0)(g) \qquad (u \in L(E, E);\ k = 1, \ldots, 2^n;$$

$$n = 4,5, \ldots).$$

Consequently, by formula (0.57) for $\mathscr{A}_{n,k}$, we obtain

$$|\varphi_{n,k}(u) - \psi_{n,k}(u)| \leqslant \frac{1}{2^{n-1}|G|}\sum_{g\in G}|(u\tau t_g y_0)(g)| \leqslant \frac{1}{2^{n-1}|G|}\sum_{g\in G}\|u\tau t_g y_0\|_p \leqslant$$

$$\leqslant \frac{1}{2^{n-1}}\max_{g\in G}\|u\tau t_g y_0\|_p \leqslant \max_{x\in\mathscr{A}_{n,k}}\|u(x)\|_p \qquad (u \in L(E, E);\ k = 1, \ldots, 2^n;$$

$$n = 4,5, \ldots),$$

(where $1 \leqslant p \leqslant \infty$), which proves (0.7).

Finally, let us prove (0.8). By the definition (0.44) of the $x_w$'s, for any $g \in G$ we can write (it will not be necessary to specify all of the coefficients)

$$\sum_{w\in W^+} w(g)x_w - \sum_{w\in W^-} w(g)x_{h(w)} =$$

$$= \sum_{w\in W^+} w(g)(w + \sum_{y\in h^{-1}(w)} y) - \sum_{w\in W^-} w(g)\,(h(w) + \sum_{y\in h^{-1}(h(w))} y) =$$

$$= \sum_{w\in W^+} w(g)w + \sum_{y\in h^{-1}(W^+)} \pm y - \sum_{w\in W^-} \pm h(w) -$$

$$- \sum_{w\in W^-} w(g)w - \sum_{y\in h^{-1}(h(W^-))\setminus W^-} \pm y = z_g^1 + z_g^2 - z_g^3 - z_g^4 - z_g^5. \tag{0.60}$$

Let us observe that for each fixed $g \in G = G_{n,k}$ the supports of the functions $z_g^1 - z_g^4$, $z_g^2$, $z_g^3$ and $z_g^5$ are mutually disjoint. Indeed, since the summands in $z_g^1 - z_g^4$, $z_g^2$, $z_g^3$ and $z_g^5$ (modulo the signs $\pm$) belong to the sets

$$W_{n,k}^+ \cup W_{n,k}^-, \; h^{-1}(W_{n,k}^+) \subset \bigcup_{l=1}^{2^{n+1}} W_{n+1,l}^-, \; h(W_{n,k}^-) \subset \bigcup_{l=1}^{2^{n-1}} W_{n-1,l}^+$$

and

$$h^{-1}(h(W_{n,k}^-)) \diagdown W_{n,k}^- \subset \left(\bigcup_{l=1}^{2^n} W_{n,l}^-\right) \diagdown W_{n,k}^- = \bigcup_{\substack{l=1 \\ l \neq k}}^{2^n} W_{n,l}^-$$

respectively, we have, by (0.37),

$$\begin{cases} \operatorname{supp}\,(z_g^1 - z_g^4) \subset G_{n,k}, \; \operatorname{supp}\, z_g^2 \subset \bigcup_{l=1}^{2^{n+1}} G_{n+1,l}, \\ \operatorname{supp}\, z_g^3 \subset \bigcup_{l=1}^{2^{n-1}} G_{n-1,l}, \; \operatorname{supp}\, z_g^5 \subset \bigcup_{\substack{l=1 \\ l \neq k}}^{2^n} G_{n,l}, \end{cases} \tag{0.61}$$

so they are mutually disjoint. Consequently,

$$\Big\| \sum_{w \in W^+} w(g) x_w - \sum_{w \in W^-} w(g) x_{h(w)} \Big\|_\infty = \max(\|z_g^1 - z_g^4\|_\infty, \|z_g^2\|_\infty, \|z_g^3\|_\infty, \|z_g^5\|_\infty).$$

Now, by (0.55), $\|t_g\| \leqslant 1$ (see (0.54)) and (0.20), we have

$$\|z_g^1 - z_g^4\|_\infty = \Big\| \sum_{w \in W^+} w(g) w - \sum_{w \in W^-} w(g) w \Big\|_\infty = \Big\| t_g\Big( \sum_{w \in W^+} w - \sum_{w \in W^-} w\Big) \Big\|_\infty \leqslant$$

$$\leqslant \Big\| \sum_{w \in W^+} w - \sum_{w \in W^-} w \Big\|_\infty \leqslant 2^{1+\left[\frac{n}{2}\right]} \leqslant 2.2^{\frac{n}{2}}. \tag{0.62}$$

Furthermore, the argument of the above proof that the supports of the five summands in (0.44) are pairwise disjoint, also shows that the summands in $z_g^2 = \sum_{y \in h^{-1}(W^+)} \pm y$ have disjoint supports. Similarly, using again (0.40), it follows that the summands in $z_g^3 = \sum_{w \in W^-} \pm h(w)$ have disjoint supports. Therefore,

$$\|z_g^2\|_\infty = \|z_g^3\|_\infty = 1. \tag{0.63}$$

Finally, by (0.43), the set $h^{-1}(h(W_{n,k}^{-}))\setminus W_{n,k}^{-} = h^{-1}(h(W_{n,k}^{-})) \cap \cap \bigcup_{\substack{l=1 \\ l\neq k}}^{2^n} W_{n,l}^{-}$ contains at most two elements $y$ from each $W_{n,l}^{-}$, where $l \neq k$. Hence, for each $G_{n,l}$, where $l = 1, \ldots, 2^n$, the supports of at most two summands in $z_g^5 = \sum_{y\in h^{-1}(h(W^-))\setminus W^-} \pm y$ are contained in $G_{n,l}$ and therefore

$$\|z_g^5\|_\infty \leqslant 2. \tag{0.64}$$

From the above relations we obtain

$$\|\sum_{w\in W^+} w(g)x_w - \sum_{w\in W^-} w(g)x_{h(w)}\|_\infty \leqslant 2.2^{\frac{n}{2}} \qquad (g\in G). \tag{0.65}$$

Consequently, if $E = [x_w]_{w\in W}$ in $c_0(\Gamma)$, then by the definition (0.51) of $\mathscr{A}_{n,k}$ we have

$$C_n = \max_{x\in \bigcup_{k=1}^{2^n} \mathscr{A}_{n,k}} \|x\|_\infty \leqslant \frac{1}{2^{n-1}} \cdot 2.2^{\frac{n}{2}} = \frac{4}{(\sqrt{2})^n} \qquad (n = 4,5, \ldots), \tag{0.66}$$

whence $\sum_{n=3}^{\infty} C_n \leqslant C_3 + \sum_{n=4}^{\infty} \frac{4}{(\sqrt{2})^n} < \infty$, which proves (0.8). This completes the proof of theorem 0.1 for $E = [x_w]_{w\in W}$ in $c_0(\Gamma)$.

In order to prove (0.8) for $E = [x_w]_{w\in W}$ in $l^p(\Gamma)$, where $2 < p < \infty$, we shall use the observation that if an $x \in l^p(\Gamma)$ has at most $m$ non-zero coordinates, then $\|x\|_p \leqslant (m\|x\|_\infty^p)^{\frac{1}{p}} = m^{\frac{1}{p}}\|x\|_\infty$. Applying this observation to $x = z_g^1 - z_g^4$ and (0.62) and taking into account (0.60), (0.61), (0.63), (0.64), we obtain

$$\|\sum_{w\in W^+} w(g)x_w - \sum_{w\in W^-} w(g)x_{h(w)}\|_p \leqslant \|z_g^1 - z_g^4\|_p + \|z_g^2 - z_g^3 - z_g^5\|_p \leqslant$$

$$\leqslant (2^n)^{\frac{1}{p}}\, 2.2^{\frac{n}{2}} + \Big(\sum_{g'\in \bigcup_{l=1}^{2^{n+1}} G_{n+1,l}} 1^p + \sum_{g'\in \bigcup_{l=1}^{2^{n-1}} G_{n-1,l}} 1^p + \sum_{g'\in \bigcup'_{\substack{l=1\\ l\neq k}}^{2^n} G_{n,l}} 2^p\Big)^{\frac{1}{p}} =$$

$$= 2.2^{n\left(\frac{1}{p}+\frac{1}{2}\right)} + (2^{2n+2} + 2^{2n-2} + 2^n(2^n - 1)2^p)^{\frac{1}{p}}. \tag{0.67}$$

Consequently, if $E = [x_w]_{w \in W}$ in $l^p(\Gamma)$ $(2 < p < \infty)$, then by (0.51) and by $2 < p < \infty$ we obtain

$$C_n = \max_{x \in \bigcup_{k=1}^{2^n} \mathscr{A}_{n,k}} \|x\|_p \leqslant \frac{1}{2^{n-1}} \left( 2.2^{n\left(\frac{1}{p}+\frac{1}{2}\right)} + 2^{\frac{2n}{p}} \left(4 + \frac{1}{4} + 2^p\right)^{\frac{1}{p}} \right) =$$

$$= 4.2^{n\left(\frac{1}{p}-\frac{1}{2}\right)} + 2^{2n\left(\frac{1}{p}-\frac{1}{2}\right)} 2 \left(4 + \frac{1}{4} + 2^p\right)^{\frac{1}{p}} \leqslant B_p\, 2^{n\left(\frac{1}{p}-\frac{1}{2}\right)}$$

$$(n = 4,5, \ldots,), \tag{0.68}$$

whence, again since $2 < p < \infty$, it follows that $\sum_{n=3}^{\infty} C_n < \infty$. Thus, we have (0.8), which completes the proof of theorem 0.1.

Let us pass now to the second proof of the existence of subspaces of $c_0$ and $l^p$ $(2 < p < \infty)$ which do not have the approximation property. We start with the probabilistic lemma mentioned in the above.

**Lemma 0.4.** a) *Let $\alpha_1, \ldots, \alpha_N$ be fixed complex numbers and let $\rho_1, \ldots, \rho_N$ be chosen independently at random, taking the values 2 and $-1$ with probabilities $\frac{1}{3}$ and $\frac{2}{3}$, respectively. Then there exists an absolute constant $A_0$ such that*

$$P\left\{\left|\sum_{j=1}^{N} \alpha_j \rho_j\right| > A_0 \left(\sum_{j=1}^{N} |\alpha_j|^2 \log N\right)^{\frac{1}{2}}\right\} < \frac{A_0}{N^3}. \tag{0.69}$$

b) *Let $\alpha_1, \ldots, \alpha_N$ be fixed complex numbers and let $\varepsilon_1, \ldots, \varepsilon_N$ be chosen independently at random, taking the values 1 and $-1$ with equal probability $\frac{1}{2}$. Then there exists an absolute constant $A_0'$ such that*

$$P\left\{\left|\sum_{j=1}^{N} \alpha_j \varepsilon_j\right| > A_0' \left(\sum_{j=1}^{N} |\alpha_j|^2 \log N\right)^{\frac{1}{2}}\right\} < \frac{A_0'}{N^3}. \tag{0.70}$$

*Proof.* a) We shall use the following inequality:

$$\frac{1}{3}\,e^{2t} + \frac{2}{3}\,e^{-t} \leqslant e^{2t^2} \qquad (-\infty < t < \infty). \tag{0.71}$$

In order to prove this inequality, let us first observe that

$$e^{2t} + 2e^{-t} = \sum_{n=0}^{\infty} \frac{(2t)^n + 2(-t)^n}{n!} =$$

$$= \sum_{n=0}^{\infty} \left( \frac{(2t)^{2n} + 2(-t)^{2n}}{(2n)!} + \frac{(2t)^{2n+1} + 2(-t)^{2n+1}}{(2n+1)!} \right) =$$

$$= 3 + 3t^2 + t^3 +$$

$$+ \sum_{n=2}^{\infty} \left( \frac{2^{2n} + 2}{(2n)!} + \frac{2^{2n+1} - 2}{(2n+1)!}\, t \right) t^{2n} \quad (-\infty < t < \infty). \tag{0.72}$$

Now, (0.71) is obviously true for $t \geqslant 1$ (since then $2t \leqslant 2t^2$ and $-t < 2t^2$). On the other hand, if $t < 1$, then $3t^2 + t^3 \leqslant 6t^2$ and

$$\frac{2^{2n} + 2}{(2n)!} + \frac{2^{2n+1} - 2}{(2n+1)!}\, t < \frac{2^{2n+1}}{(2n)!} + \frac{2^{2n+1}}{(2n+1)!} = \frac{2^{2n+1}(2n+2)}{(2n+1)!} <$$

$$< \frac{3.2^{2n} \,.\, 2(n+1)}{n!\,(n+1)\,(n+2)\ldots(2n+1)} < \frac{3.2^n}{n!}\,\frac{4^n}{(n+2)\ldots(2n+1)} <$$

$$< \frac{3.2^n}{n!} \qquad (n \geqslant 2),$$

whence, by (0.72), we obtain

$$e^{2t} + 2e^{-t} \leqslant 3 + 6t^2 + \sum_{n=2}^{\infty} \frac{3.2^n}{n!}\, t^{2n} = 3\, e^{2t^2},$$

which proves the inequality (0.71).

Now, in order to prove (0.69), assume first that the $\alpha_j$'s are real and $\sum_{j=1}^{N} \alpha_j^2 = 1$ and let us denote by $\mathscr{E}$ the expectation*). Then for any

---

*) We recall that if $\gamma$ is a random variable taking the values $\gamma_j$ with probabilities $P(\gamma = \gamma_j)$, then $\mathscr{E}(\gamma) = \sum_j \gamma_j P(\gamma = \gamma_j)$.

$\lambda>0$ we have, since $e^{|t|}\leqslant e^{t}+e^{-t}$ and since $\rho_1,\ldots,\rho_N$ are independent,

$$\mathscr{E}\left(e^{\lambda\left|\sum_{j=1}^{N}\alpha_j\rho_j\right|}\right)\leqslant\mathscr{E}\left(e^{\lambda\sum_{j=1}^{N}\alpha_j\rho_j}\right)+\mathscr{E}\left(e^{-\lambda\sum_{j=1}^{N}\alpha_j\rho_j}\right)=$$

$$=\mathscr{E}\left(\prod_{j=1}^{N}e^{\lambda\alpha_j\rho_j}\right)+\mathscr{E}\left(\prod_{j=1}^{N}e^{-\lambda\alpha_j\rho_j}\right)=$$

$$=\prod_{j=1}^{N}\mathscr{E}(e^{\lambda\alpha_j\rho_j})+\prod_{j=1}^{N}\mathscr{E}(e^{-\lambda\alpha_j\rho_j})=$$

$$=\prod_{j=1}^{N}\left(\frac{1}{3}e^{2\lambda\alpha_j}+\frac{2}{3}e^{-\lambda\alpha_j}\right)+\prod_{j=1}^{N}\left(\frac{1}{3}e^{-2\lambda\alpha_j}+\frac{2}{3}e^{\lambda\alpha_j}\right),$$

whence, by (0.71) and $\sum_{j=1}^{N}\alpha_j^2=1$,

$$\mathscr{E}\left(e^{\lambda\left|\sum_{j=1}^{N}\alpha_j\rho_j\right|}\right)\leqslant\prod_{j=1}^{N}e^{2(\lambda\alpha_j)^2}+\prod_{j=1}^{N}e^{2(\lambda\alpha_j)^2}=2e^{2\lambda^2\sum_{j=1}^{N}\alpha_j^2}=2e^{2\lambda^2}.$$

Consequently,

$$\mathscr{E}\left(e^{\lambda\left|\sum_{j=1}^{N}\alpha_j\rho_j\right|-2\lambda^2-3\log N}\right)=e^{-2\lambda^2-3\log N}\,\mathscr{E}\left(e^{\lambda\left|\sum_{j=1}^{N}\alpha_j\rho_j\right|}\right)\leqslant$$

$$\leqslant 2e^{-3\log N}=\frac{2}{N^3},$$

whence, putting $\lambda=\sqrt{3\log N}$, we obtain

$$\mathscr{E}\left(e^{\sqrt{3\log N}\left|\sum_{j=1}^{n}\alpha_j\rho_j\right|-9\log N}\right)\leqslant\frac{2}{N^3}. \tag{0.73}$$

We shall use now the observation that for any random variable $\gamma$ and any $\lambda>0$ we have*) $P(\gamma>0)\leqslant\mathscr{E}(e^{\lambda\gamma})$. Applying this to $\gamma=\left|\sum_{j=1}^{N}\alpha_j\rho_j\right|-$

*) Indeed, we have $P(\gamma>0)=\mathscr{E}(\chi_{(0,\infty)}(\gamma))$, where $\chi_{(0,\infty)}(\gamma)=1$ for $\gamma\in(0,\infty)$ and 0 for $\gamma\in(-\infty,0]$, whence, since $\chi_{(0,\infty)}(\gamma)\leqslant e^{\lambda\gamma}$ for all $\gamma$ and $\lambda>0$, we obtain $P(\gamma>0)\leqslant\mathscr{E}(e^{\lambda\gamma})$.

$-3\sqrt{3}\,(\log N)^{\frac{1}{2}}$ and $\lambda=\sqrt{3\log N}$ and taking into account (0.73), we obtain

$$P\left\{\left|\sum_{j=1}^{N}\alpha_j\rho_j\right|-3\sqrt{3}(\log N)^{\frac{1}{2}}>0\right\}\leqslant\frac{2}{N^3}\leqslant\frac{3\sqrt{3}}{N^3},\tag{0.74}$$

which proves (0.69) (with $A_0=3\sqrt{3}$) for $\alpha_1,\ldots,\alpha_N$ real and $\sum_{j=1}^{N}\alpha_j^2=1$. Since for any real $\alpha_1,\ldots,\alpha_N$ which are not all 0 we have

$$P\left\{\left|\sum_{j=1}^{N}\alpha_j\rho_j\right|>A_0\left(\sum_{j=1}^{N}|\alpha_j|^2\log N\right)^{\frac{1}{2}}\right\}=$$

$$=P\left\{\left|\sum_{j=1}^{N}\frac{\alpha_j}{\sqrt{\sum_{k=1}^{N}|\alpha_k|^2}}\rho_j\right|>A_0\,(\log N)^{\frac{1}{2}}\right\},$$

it follows that (0.69) holds for all real $\alpha_1,\ldots,\alpha_N$.

Assume now that the $\alpha_j$'s are complex, say $\alpha_j=s_j+it_j$ $(j=1,\ldots,N)$, hence $|\alpha_j|^2=|s_j|^2+|t_j|^2$ $(j=1,\ldots,N)$ and $\left|\sum_{j=1}^{N}\alpha_j\rho_j\right|^2=\left|\sum_{j=1}^{N}s_j\rho_j\right|^2+$ $+\left|\sum_{j=1}^{N}t_j\rho_j\right|^2$. Then, by the first inequality in (0.74), we have

$$P\left\{\left|\sum_{j=1}^{N}\alpha_j\rho_j\right|^2>27\sum_{j=1}^{N}|\alpha_j|^2\log N\right\}\leqslant$$

$$\leqslant P\left\{\text{either}\left|\sum_{j=1}^{N}s_j\rho_j\right|^2>27\sum_{j=1}^{N}|s_j|^2\log N\text{ or}\left|\sum_{j=1}^{N}t_j\rho_j\right|^2>\right.$$

$$\left.>27\sum_{j=1}^{N}|t_j|^2\log N\right\}\leqslant P\left\{\left|\sum_{j=1}^{N}s_j\rho_j\right|^2>27\sum_{j=1}^{N}|s_j|^2\log N\right\}+$$

$$+P\left\{\left|\sum_{j=1}^{N}t_j\rho_j\right|^2>27\sum_{j=1}^{N}|t_j|^2\log N\right\}\leqslant\frac{2}{N^3}+\frac{2}{N^3}=\frac{4}{N^3}\leqslant\frac{3\sqrt{3}}{N^3},$$

which proves (0.69) (with $A_0=3\sqrt{3}$) for all complex $\alpha_1,\ldots,\alpha_N$.

b) The proof of part b) is similar, using instead of (0.71) the inequality

$$\frac{e^t + e^{-t}}{2} = \cosh t = \sum_{n=0}^{\infty} \frac{t^{2n}}{(2n)!} \leqslant \sum_{n=0}^{\infty} \frac{t^{2n}}{n!} = e^{t^2} \qquad (-\infty < t < \infty). \tag{0.75}$$

This completes the proof of lemma 0.4.

Let us give now the lemma on the partition of the set of characters, corresponding to lemma 0.2:

**Lemma 0.5.** *For each non-negative integer n let $G_n$ be an abelian group with $3.2^n$ elements and let $H_n$ be the set of all characters of $G_n$. Then for each n there exist two disjoint subsets $W_n^+$, $W_n^-$ of $H_n$ with cardinalities satisfying*

$$|W_n^+| = 2^n, \ |W_n^-| = 2^{n+1} = 2.2^n, \tag{0.76}$$

*(hence $W_n^+ \cup W_n^- = H_n$), such that*

$$\left\| 2 \sum_{w \in W_n^+} w - \sum_{w \in W_n^-} w \right\|_\infty \leqslant A_1 (n+1)^{\frac{1}{2}} 2^{\frac{n}{2}}, \tag{0.77}$$

*where $A_1$ is some absolute constant.*

*Proof.* Let $H_n = \{w_1, \ldots, w_{3.2^n}\}$. We claim that there exist numbers $\rho_1, \ldots, \rho_{3.2^n}$ with $\rho_j = 2$ or $-1$ $(j = 1, \ldots, 3.2^n)$ such that

$$\left| \sum_{j=1}^{3.2^n} \rho_j w_j(g) \right| \leqslant A_1 (n+1)^{\frac{1}{2}} 2^{\frac{n}{2}} \qquad (g \in G_n). \tag{0.78}$$

Indeed, for any fixed $g \in G_n$ we have, by lemma 0.4 a) with $N = 3.2^n$ and $\alpha_j = w_j(g)$ (hence $|\alpha_j| = |w_j(g)| = 1$ for $j = 1, \ldots, 3.2^n$),

$$P\left\{ \left| \sum_{j=1}^{3.2^n} \rho_j w_j(g) \right| > A_0 \left( \sum_{j=1}^{3.2^n} \log 3.2^n \right)^{\frac{1}{2}} \right\} < \frac{A_0}{27.2^{3n}},$$

whence, since $|G_n| = 3.2^n$, we obtain, for $n$ large enough*),

$$P\left\{ \left| \sum_{j=1}^{3.2^n} \rho_j w_j(g) \right| > A_0 \left( \sum_{j=1}^{3.2^n} \log 3.2^n \right)^{\frac{1}{2}} \text{ for some } g \in G_n \right\} \leqslant$$

$$\leqslant \sum_{g \in G_n} P\left\{ \left| \sum_{j=1}^{3.2^n} \rho_j w_j(g) \right| > A_0 \left( \sum_{j=1}^{3.2^n} \log 3.2^n \right)^{\frac{1}{2}} \right\} < \frac{3.2^n}{27.2^{3n}} A_0 < 1,$$

*) Actually, we can take $A_0 = 3\sqrt{3}$ (see the above proof of lemma 0.4 (a)) and then the inequality holds for all $n \geqslant 0$. However, we shall not need this remark.

i.e., for $n$ large enough,

$$P\left\{\left|\sum_{j=1}^{3.2^n} \rho_j w_j(g)\right| \leqslant A_0 \left(\sum_{j=1}^{3.2^n} \log 3.2^n\right)^{\frac{1}{2}} \text{ for all } g \in G_n\right\} > 0.$$

Consequently, for $n$ large enough, say $n > n_0$, there exist numbers $\rho_1, \ldots, \rho_{3.2^n}$ with $\rho_j = 2$ or $-1$ ($j = 1, \ldots, 3.2^n$), such that

$$\left|\sum_{j=1}^{3.2^n} \rho_j w_j(g)\right| \leqslant A_0 \left(\sum_{j=1}^{3.2^n} \log 3.2^n\right)^{\frac{1}{2}} =$$

$$= A_0 (3.2^n (\log 3 + n \log 2))^{\frac{1}{2}} \leqslant A_1 2^{\frac{n}{2}} (n+1)^{\frac{1}{2}} \qquad (g \in G_n).$$

But, since each $G_n$ is a finite set, the same inequalities also hold for $n = 0,1,2, \ldots, n_0$ by increasing $A_1$ if necessary. This proves the claim (0.78).

Now we shall show that by changing some of the $\rho_j$'s (so as to be still either 2 or $-1$) and increasing $A_1$, we can obtain both (0.78) and

$$\sum_{j=1}^{3.2^n} \rho_j = 0, \tag{0.79}$$

which will complete the proof (since then among $\rho_1, \ldots, \rho_{3.2^n}$ there must be exactly $2^n$ which are $= 2$ and exactly $2.2^n = 2^{n+1}$ which are $= -1$).

Applying (0.78) in particular to the unit $g = e$ of $G$, we obtain, since $w_j(e) = 1$ ($j = 1, \ldots, 3.2^n$), that

$$\left|\sum_{j=1}^{3.2^n} \rho_j\right| = \left|\sum_{j=1}^{3.2^n} \rho_j w_j(e)\right| \leqslant A_1 (n+1)^{\frac{1}{2}} 2^{\frac{n}{2}}. \tag{0.80}$$

Let us denote by $m$ and $m'$ respectively the number of those $\rho_j$ which are $= 2$, respectively $= -1$. Then $\sum_{j=1}^{3.2^n} \rho_j = 2m - m'$ and $m' = 3.2^n - m$, whence

$$\sum_{j=1}^{3.2^n} \rho_j = 2m - m' = 3m - 3.2^n = 3(m - 2^n). \tag{0.81}$$

If $m = 2^n$, we are done. Assume now that $m > 2^n$. Select any set $S$ of indices such that $|S| = m - 2^n$ and $\rho_j = 2$ for all $j \in S$ and put

$$\tilde{\rho}_j = \begin{cases} 1 - \rho_j = -1 & \text{for } j \in S \\ \rho_j & \text{for } j \in \{1, \ldots, 3.2^n\} \setminus S. \end{cases} \tag{0.82}$$

Then, by (0.82), $|w_j(g)| = 1$, (0.81) and (0.80), we obtain

$$\left|\sum_{j=1}^{3.2^n} \rho_j w_j(g) - \sum_{j=1}^{3.2^n} \tilde{\rho}_j w_j(g)\right| = \left|\sum_{j\in S} (2\rho_j - 1) w_j(g)\right| =$$

$$= \left|\sum_{j\in S} 3\, w_j(g)\right| \leqslant 3|S| = 3(m - 2^n) = \left|\sum_{j=1}^{3.2^n} \rho_j\right| \leqslant A_1(n+1)^{\frac{1}{2}} 2^{\frac{n}{2}} \quad (g \in G_n),$$

whence, by (0.78),

$$\left|\sum_{j=1}^{3.2^n} \tilde{\rho}_j w_j(g)\right| \leqslant 2A_1(n+1)^{\frac{1}{2}} 2^{\frac{n}{2}} \qquad (g \in G_n),$$

and thus we have (0.78) for $\rho_j$ and $A_1$ replaced by $\tilde{\rho}_j$ and $2A_1$, respectively. Also, by (0.82) and (0.81),

$$\sum_{j=1}^{3.2^n} \rho_j - \sum_{j=1}^{3.2^n} \tilde{\rho}_j = \sum_{j\in S} (2\rho_j - 1) = \sum_{j\in S} 3 = 3|S| = 3(m - 2^n) = \sum_{j=1}^{3.2^n} \rho_j,$$

whence $\sum_{j=1}^{3.2^n} \tilde{\rho}_j = 0$, i.e., we have (0.79) for $\rho_j$ replaced by $\tilde{\rho}_j$.

Finally, in the case when $m - 2^n < 0$, the argument is similar (one has to replace any $2^n - m$ numbers $\rho_j$ which are $-1$ by $\tilde{\rho}_j = 2$), which completes the proof of lemma 0.5.

In the following proof of the existence of subspaces of $c_0(G)$ and $l^p(G)$ $(2 < p < \infty)$ which fail the approximation property, we shall take as the set $G$ the disjoint union $G = \bigcup_{n=0}^{\infty} G_n$, where $G_n$ is an abelian group with $3.2^n$ elements $(n = 0,1,2,\ldots)$. For each $n$ let us denote by $\sigma_1^n, \ldots, \sigma_{2^n}^n$ and $\tau_1^n, \ldots, \tau_{2^{n+1}}^n$ the elements of $W_n^+$ and $W_n^-$, respectively, where $W_n^+$, $W_n^-$ are as in lemma 0.5, so we can write (0.77) in the form

$$\left|2\sum_{j=1}^{2^n} \sigma_j^n(g) - \sum_{j=1}^{2^{n+1}} \tau_j^n(g)\right| \leqslant A_1 (n+1)^{\frac{1}{2}} 2^{\frac{n}{2}} \qquad (g \in G_n). \qquad (0.77')$$

Let $X$ denote the set of all functions on $G$ with finite support and for each $j = 1, \ldots, 2^n$ and each $n = 0,1,2,\ldots$ define $x_j^n \in X$ by

$$x_j^n(g) = \begin{cases} \tau_j^{n-1}(g) & \text{if } g \in G_{n-1} \qquad (n \geqslant 1) \\ \varepsilon_j^n \sigma_j^n(g) & \text{if } g \in G_n \\ 0 & \text{if } g \in G \setminus (G_{n-1} \cup G_n), \end{cases} \qquad (0.83)$$

where $\varepsilon_j^n = \pm 1$ will be chosen later. Note that if we extend $\sigma_j^n$ and $\tau_j^n$ beyond $G_n$ by 0 (similarly to (0.37)), and denote them still by $\sigma_j^n$ and $\tau_j^n$ respectively, then we can write $x_j^n$ also in the following form, corresponding to (0.44):

$$x_1^0 = \varepsilon_1^0 \sigma_1^0, \; x_j^n = \tau_j^{n-1} + \varepsilon_j^n \sigma_j^n \quad (j = 1, \ldots, 2^n;\; n = 1,2, \ldots); \quad (0.83')$$

let us also note that by (0.76) there are exactly $2^n$ distinct $\tau_j^{n-1}$'s and $2^n$ distinct $\sigma_j^n$'s and all of them are used in (0.83′). Thus, now the index set $\mathcal{M}$ (mentioned after remark 0.1) will be the set of pairs $\mathcal{M} = \bigcup_{n=0}^{\infty} \bigcup_{j=1}^{2^n} (n,j)$. The supports of the two summands in (0.83′) are disjoint and hence we have indeed (0.17). Also, it is clear that

$$\operatorname{supp} x_j^n = G_{n-1} \cup G_n \qquad (j = 1, \ldots, 2^n;\; n = 0,\, 1,\, 2, \ldots),$$

so the $x_j^n$'s are not disjointly supported*). Nevertheless, since the characters of each group $G_n$ are mutually orthogonal and since the groups $G_n$ are pairwise disjoint, from (0.83) it follows that $\{x_j^n\}_{(n,j)\in\mathcal{M}}$ is an orthogonal system in $l^2(G)$ (with the natural scalar product $(x, z) = \sum_{g\in G} x(g)\overline{z(g)}$), as was announced (after formula (0.17)).

**Theorem 0.2.** *There exist signs $\varepsilon_j^n = \pm 1$ such that if $E = [x_j^n]_{(n,j)\in\mathcal{M}}$ in the complex space $E = c_0(G)$ or $l^p(G)$ $(2 < p < \infty)$, then $E$ does not have the approximation property.*

*Proof.* It will be convenient to express the biorthogonal functionals $f_j^n \in E^*$ in the following two different ways**):

$$f_j^n(x) = \frac{1}{3.2^n} \sum_{g\in G_n} x(g)\varepsilon_j^n \overline{\sigma_j^n(g)} \quad (x \in E,\; j = 1, \ldots, 2^n;\; n = 0,1,2, \ldots), \tag{0.84}$$

$$f_j^n(x) = \frac{1}{3.2^{n-1}} \sum_{g\in G_{n-1}} x(g)\, \overline{\tau_j^{n-1}(g)} \quad (x \in E,\; j = 1, \ldots, 2^n;\; n = 1,2, \ldots). \tag{0.85}$$

---

*) In fact, by (0.83),

$$\operatorname{supp} x_j^n \cap \operatorname{supp} x_k^{n+l} = \begin{cases} G_n \cup G_{n-1} & \text{if } l = 0 \\ G_n & \text{if } l = 1 \\ \emptyset & \text{if } l \geq 2 \end{cases}$$

$$(j = 1, \ldots, 2^n;\; k = 1, \ldots, 2^{n+l};\; n = 0, 1, 2, \ldots).$$

**) Note that, since $G_n$ is an arbitrary abelian group with $|G_n| = 3.2^n$, the characters of $G_n$ need not be real. Note also that, for $n = 0$, $f_1^0$ occurs only in (0.84).

Let us first show that for both (0.84) and (0.85) we have $f_j^n(x_i^m) = \delta_{jn}^{im}$; since $[x_i^m] = E$, this will also prove that (0.84) and (0.85) define the same continuous linear functionals on $E$.

By (0.84) for $x = x_j^n$ and (0.83) we have

$$f_j^n(x_j^n) = \frac{1}{3.2^n} \sum_{\in G_n} x_j^n(g)\, \varepsilon_j^n\, \overline{\sigma_j^n(g)} = \frac{1}{3.2^n} \sum_{g \in G_n} (\varepsilon_j^n)^2\, |\sigma_j^n(g)|^2 = 1,$$

and similarly, by (0.85) for $x = x_j^n$ and (0.83) we have

$$f_j^n(x_j^n) = \frac{1}{3.2^{n-1}} \sum_{g \in G_{n-1}} x_j^n(g)\, \overline{\tau_j^{n-1}(g)} = \frac{1}{3.2^{n-1}} \sum_{g \in G_{n-1}} |\tau_j^{n-1}(g)|^2 = 1.$$

Consider now (0.84) for $x = x_i^m$, that is,

$$f_j^n(x_i^m) = \frac{1}{3.2^n}\, \varepsilon_j^n \sum_{g \in G_n} x_i^m(g)\, \overline{\sigma_j^n(g)}. \tag{0.86}$$

If $n \notin \{m-1, m\}$, then $G_n \cap (G_{m-1} \cup G_m) = \emptyset$, whence, since by (0.83) $x_i^m(g) = 0$ for $g \notin G_{m-1} \cup G_m$, we infer $f_j^n(x_i^m) = 0$. If $n = m-1$, then, since by (0.83) $x_i^m(g) = \tau_i^{m-1}(g) = \tau_i^n(g)$ for $g \in G_{m-1} = G_n$, from (0.86) and the orthogonality of the characters of $G_n$ we infer

$$f_j^n(x_i^m) = f_j^n(x_i^{n+1}) = \frac{1}{3.2^n}\, \varepsilon_j^n \sum_{g \in G_n} \tau_i^n(g)\, \overline{\sigma_j^n(g)} = 0.$$

Finally, if $n = m$, then, since by (0.83) $x_i^m(g) = \varepsilon_i^m \sigma_i^m(g) = \varepsilon_i^n \sigma_i^n(g)$ for $g \in G_m = G_n$, from (0.86) and the orthogonality of the characters of $G_n$ we infer

$$f_j^n(x_i^m) = f_j^n(x_i^n) = \frac{1}{3.2^n}\, \varepsilon_j^n \varepsilon_i^n \sum_{g \in G_n} \sigma_i^n(g)\, \overline{\sigma_j^n(g)} = 0 \text{ for } i \neq j.$$

If we consider now (0.85) for $x = x_i^m$, that is,

$$f_j^n(x_i^m) = \frac{1}{3.2^{n-1}} \sum_{g \in G_{n-1}} x_i^m(g)\, \overline{\tau_j^{n-1}(g)}, \tag{0.87}$$

then, by similar computations we obtain again that $f_j^n(x_i^m) = 0$ for all $n \neq m$ and all $i, j$ as well as for $n = m$ and $i \neq j$. This proves our assertion about (0.84), (0.85).

Define linear functionals $\varphi_n$ on $L(E, E)$ by*)

$$\varphi_n(u) = \mathscr{T}\left(\bigcup_{j=1}^{2^n}(n, j); u\right) = \frac{1}{2^n}\sum_{j=1}^{2^n} f_j^n(u(x_j^n)) \quad (u \in L(E, E),\ n=0,1,2,\ldots); \tag{0.88}$$

thus, the sets $M_n$ of our heuristic introduction are now $\bigcup_{j=1}^{2^n}(n, j)$ $(n = 0,1,2, \ldots)$. Using (0.85), we obtain

$$\varphi_{n+1}(u) = \frac{1}{2^{n+1}}\sum_{j=1}^{2^{n+1}} f_j^{n+1}(u(x_j^{n+1})) = \frac{1}{2^{n+1}}\sum_{j=1}^{2^{n+1}} \frac{1}{3.2^n}\sum_{g\in G_n} u(x_j^{n+1})(g)\,\overline{\tau_j^n(g)} = \tag{0.89}$$

$$= \frac{1}{3.2^{2n+1}}\sum_{g\in G_n} u\left(\sum_{j=1}^{2^{n+1}} \overline{\tau_j^n(g)}\, x_j^{n+1}\right)(g) \quad (u \in L(E, E),\ n = 0,1,2, \ldots).$$

Similarly, using (0.84) we get

$$\varphi_n(u) = \frac{1}{2^n}\sum_{j=1}^{2^n} f_j^n(u(x_j^n)) = \frac{1}{2^n}\sum_{j=1}^{2^n}\frac{1}{3.2^n}\sum_{g\in G_n} u(x_j^n)(g)\,\overline{\varepsilon_j^n\sigma_j^n(g)} =$$

$$= \frac{1}{3.2^{2n}}\sum_{g\in G_n} u\left(\sum_{j=1}^{2^n} \overline{\varepsilon_j^n\sigma_j^n(g)}\, x_j^n\right)(g) \qquad (u \in L(E, E),\ n = 0,1,2, \ldots). \tag{0.90}$$

Hence, by (0.89) and (0.90), for each $n = 0,1,2, \ldots$ we have

$$\varphi_{n+1}(u) - \varphi_n(u) = \frac{1}{3.2^n}\sum_{g\in G_n} u(z_g^n)(g) \qquad (u \in L(E, E)), \tag{0.91}$$

---

*) Note that by (0.88) (or (0.89) and (0.85)) we can write $\varphi_n = \frac{1}{2^n}\sum_{j=1}^{2^n}\varphi_{n,j}$, where

$$\varphi_{n,j}(u) = \mathscr{T}(\{(n,j)\}; u) = f_j^n(u(x_j^n)) \quad (u \in L(E, E);\ j = 1, \ldots, 2^n;\ n = 0, 1, 2, \ldots);$$

however, (0.90) and (0.84) give only the obvious decomposition functionals $\psi'_{n,j}$ corresponding to $\varphi_{n,j}$, mentioned in remark 0.1 b).

where[*]

$$z_g^n = \frac{1}{2^n}\left(\frac{1}{2}\sum_{j=1}^{2^{n+1}} \overline{\tau_j^n(g)}\, x_j^{n+1} - \sum_{j=1}^{2^n} \varepsilon_j^n \overline{\sigma_j^n(g)}\, x_j^n\right) \qquad (g \in G_n). \qquad (0.92)$$

Now put

$$\mathscr{A}_0 = \{x_1^0\}, \quad \mathscr{A}_{n+1} = \{(n+1)^2 z_g^n |\ g \in G_n\} \qquad (n = 0,1,2,\ldots). \qquad (0.93)$$

Then, by (0.91), for any $u \in L(E, E)$ and $n = 0,1,2,\ldots$ we have

$$|\varphi_{n+1}(u) - \varphi_n(u)| \leqslant \frac{1}{3 \cdot 2^n} \sum_{g \in G_n} |u(z_g^n)(g)| \leqslant \frac{1}{3 \cdot 2^n} \sum_{g \in G_n} \|u(z_g^n)\|_p \leqslant$$

$$\leqslant \max_{g \in G_n} \|u(z_g^n)\|_p = \frac{1}{(n+1)^2} \max_{x \in \mathscr{A}_{n+1}} \|u(x)\|_p \qquad (1 \leqslant p \leqslant \infty),$$

and, by (0.90), $|\varphi_0(u)| = \frac{1}{3}\left|\sum_{g \in G_0} u(\varepsilon_1^0 \overline{\sigma_1^0(g)}\, x_1^0)(g)\right| \leqslant \|u(x_1^0)\|_p$ ($u \in L(E,E)$,
$1 \leqslant p \leqslant \infty$), i.e., we have (0.4) with $\alpha_0 = 1$, $\alpha_{n+1} = \frac{1}{(n+1)^2}$ ($n =$
$= 0,1,2,\ldots$) (hence $\sum_{n=0}^{\infty} \alpha_n < \infty$). Thus, it remains to prove (0.5).

Using the definition (0.83) of $x_j^{n+1}$, $x_j^n$ and taking into account that

---

[*] Note that, by (0.92) and (0.83), we have

$$z_g^n = \frac{1}{2^n}\left(\frac{1}{2}\sum_{j=1}^{2^{n+1}} \overline{x_j^{n+1}(g)}\, x_j^{n+1} - \sum_{j=1}^{2^n} \overline{x_j^n(g)}\, x_j^n\right) \qquad (g \in G_n),$$

while (0.18) for $n+1$ would be

$$y_g^{n+1} = \frac{1}{2^n(3 \cdot 2^{n-1} + 3 \cdot 2^n)}\left(\frac{1}{4}\sum_{j=1}^{2^{n+1}} \overline{x_j^{n+1}(g)}\, x_j^{n+1} - \right.$$

$$\left. - \sum_{j=1}^{2^n} \overline{x_j^n(g)}\, x_j^n\right) \qquad (g \in G_{n+1} \cup G_n \cup G_{n-1}),$$

since now $\delta_n = \frac{1}{|\operatorname{supp} x_j^n|} = \frac{1}{3 \cdot 2^{n-1} + 3 \cdot 2^n}$, $|M_n| = 2^n$, $\Gamma_{n+1} = G_{n+1} \cup G_n \cup G_{n-1}$.

$(\varepsilon_j^n)^2 = 1$ and that $\tau_j^n$, $\sigma_j^n$ are characters of $G_n$, we obtain, for any $g \in G_n$,

$$z_g^n(g') = \begin{cases} \dfrac{1}{2^{n+1}} \displaystyle\sum_{j=1}^{2^{n+1}} \overline{\tau_j^n(g)\varepsilon_j^{n+1}}\sigma_j^{n+1}(g') & \text{if } g' \in G_{n+1} \\ \dfrac{1}{2^n}\left(\dfrac{1}{2}\displaystyle\sum_{j=1}^{2^{n+1}} \tau_j^n(g^{-1}g') - \sum_{j=1}^{2^n} \sigma_j^n(g^{-1}g')\right) & \text{if } g' \in G_n \\ -\dfrac{1}{2^n}\displaystyle\sum_{j=1}^{2^n} \varepsilon_j^n\overline{\sigma_j^n(g)}\tau_j^{n-1}(g') & \text{if } g' \in G_{n-1},\ n > 0 \\ 0 & \text{if } g' \in G \setminus (G_{n-1} \cup G_n \cup G_{n+1}). \end{cases} \tag{0.94}$$

By (0.94) and (0.77′) we have, for any $g \in G_n$,

$$|z_g^n(g')| \leqslant \frac{1}{2^n} A_1(n+1)^{\frac{1}{2}} 2^{\frac{n}{2}} = A_1(n+1)^{\frac{1}{2}} 2^{-\frac{n}{2}} \qquad (g' \in G_n). \tag{0.95}$$

We claim that we can choose the signs $\varepsilon_j^n = \pm 1$ in such a way that for any $g \in G_n$ (0.95) remains valid for all $g' \in G$ if we increase $A_1$, i.e. that

$$\|z_g^n\|_\infty \leqslant A_2(n+1)^{\frac{1}{2}} 2^{-\frac{n}{2}} \qquad (g \in G_n;\ n = 0, 1, 2, \ldots), \tag{0.96}$$

where $A_2$ is some absolute constant.

Indeed, for any fixed $g \in G_{n-1}$, $g' \in G_n$ we have, by lemma 0.4 b) with $N = 2^n$ and $\alpha_j = \overline{\tau_j^{n-1}(g)}\sigma_j^n(g')$ (thus, $|\alpha_j| = 1$ for $j = 1, \ldots, 2^n$),

$$P_1^{(g,g')} = P\left\{\left|\sum_{j=1}^{2^n} \varepsilon_j^n \overline{\tau_j^{n-1}(g)}\sigma_j^n(g')\right| > A_0'\left(\sum_{j=1}^{2^n} \log 2^n\right)^{\frac{1}{2}}\right\} < \frac{A_0'}{2^{3n}}$$

and hence, considering $\bar{\alpha}_j$, for any fixed $g \in G_n$, $g' \in G_{n-1}$ we get

$$P_2^{(g,g')} = P\left\{\left|\sum_{j=1}^{2^n} \varepsilon_j^n \overline{\sigma_j^n(g)}\tau_j^{n-1}(g')\right| > A_0'\left(\sum_{j=1}^{2^n} \log 2^n\right)^{\frac{1}{2}}\right\} < \frac{A_0'}{2^{3n}},$$

whence, since $|G_{n-1}| = 3.2^{n-1}, |G_n| = 3.2^n$, we obtain, for $n$ large enough,

$$P\left\{\text{either}\left|\sum_{j=1}^{2^n} \varepsilon_j^n \overline{\tau_j^{n-1}(g)}\sigma_j^n(g')\right| > A_0'\left(\sum_{j=1}^{2^n} \log 2^n\right)^{\frac{1}{2}} \text{ for}\right.$$

some $g \in G_{n-1}$, $g' \in G_n$ or $\left|\sum_{j=1}^{2^n} \varepsilon_j^n \overline{\sigma_j^n(g)} \tau_j^{n-1}(g')\right| > A_0' \left(\sum_{j=1}^{2^n} \log 2^n\right)^{\frac{1}{2}}$ for some $g \in G_n$, $g' \in G_{n-1}\Big\} = \sum_{g \in G_{n-1}} \sum_{g' \in G_n} P_1^{(g,g')} + \sum_{g \in G_n} \sum_{g' \in G_{n-1}} P_2^{(g,g')} <$

$$< \left(\frac{9.2^{2n-1}}{2^{3n}} + \frac{9.2^{2n-1}}{2^{3n}}\right) A_0' < 1,$$

i.e., for $n$ large enough,

$$P\left\{\left|\sum_{j=1}^{2^n} \varepsilon_j^n \overline{\tau_j^{n-1}(g)} \sigma_j^n(g')\right| \leqslant A_0' \left(\sum_{j=1}^{2^n} \log 2^n\right)^{\frac{1}{2}} \text{ for all } g \in G_{n-1}, g' \in G_n \text{ and}\right.$$

$$\left.\left|\sum_{j=1}^{2^n} \varepsilon_j^n \overline{\sigma_j^n(g)} \tau_j^{n-1}(g')\right| \leqslant A_0' \left(\sum_{j=1}^{2^n} \log 2^n\right)^{\frac{1}{2}} \text{ for all } g \in G_n,\ g' \in G_{n-1}\right\} > 0.$$

Consequently, for $n$ large enough, say $n > n_0$, there exist signs $\varepsilon_j^n = \pm 1$ $(j = 1, \ldots, 2^n)$ such that

$$\left|\sum_{j=1}^{2^n} \varepsilon_j^n \overline{\tau_j^{n-1}(g)} \sigma_j^n(g')\right| \leqslant A_0' \left(\sum_{j=1}^{2^n} \log 2^n\right)^{\frac{1}{2}} =$$

$$= A_0'(2^n n \log 2)^{\frac{1}{2}} = A_0'' 2^{\frac{n-1}{2}} n^{\frac{1}{2}} \qquad (g \in G_{n-1},\ g' \in G_n),$$

$$\left|\sum_{j=1}^{2^n} \varepsilon_j^n \overline{\sigma_j^n(g)} \tau_j^{n-1}(g')\right| \leqslant A_0' \left(\sum_{j=1}^{2^n} \log 2^n\right)^{\frac{1}{2}} =$$

$$= A_0'(2^n n \log 2)^{\frac{1}{2}} \leqslant A_0'' 2^{\frac{n}{2}} (n+1)^{\frac{1}{2}} \qquad (g \in G_n,\ g' \in G_{n-1}),$$

where $A_0'' = (2 \log 2)^{\frac{1}{2}} A_0'$.*) But, since each $G_n$ is a finite set, the same inequalities also hold for $n = 0,1,2, \ldots, n_0$ by increasing $A_0''$ if necessary, which, by (0.94) and (0.95), proves the claim (0.96).

Now, from (0.93) and (0.96) it follows that

$$\max_{x \in \mathscr{A}_{n+1}} \|x\|_\infty \leqslant A_2 (n+1)^{\frac{5}{2}} 2^{-\frac{n}{2}} \to 0 \text{ as } n \to \infty, \tag{0.97}$$

---

*) Alternatively, one can also obtain this by using only $P_1^{(g,g')}$, since the last inequalities (for $g \in G_n$, $g' \in G_{n-1}$) follow obviously from the preceding ones.

i.e. (0.5). This completes the proof of theorem 0.2 for $E = [x_j^n]_{(n,j)\in\mathcal{M}}$ in $c_0(G)$.

In order to prove (0.5) for $E = [x_j^n]_{(n,j)\in\mathcal{M}}$ in $l^p(G)$, where $2<p<\infty$, we shall use again (0.96) and the observation that if an $x \in l^p(G)$ has at most $m$ non-zero coordinates, then $\|x\|_p \leqslant (m\|x\|_\infty^p)^{\frac{1}{p}} = m^{\frac{1}{p}}\|x\|_\infty$. Applying this observation to $x = z_g^n$ and (0.96) and taking into account that by (0.94) supp $z_g^n \subset G_{n+1} \cup G_n \cup G_{n-1}$, we obtain

$$\|z_g^n\|_p \leqslant (3.2^{n+1} + 3.2^n + 3.2^{n-1})^{\frac{1}{p}} A_2(n+1)^{\frac{1}{2}} 2^{-\frac{n}{2}} \leqslant$$

$$\leqslant A_3^{(p)}(n+1)^{\frac{1}{2}} 2^{n\left(\frac{1}{p}-\frac{1}{2}\right)}. \tag{0.98}$$

From (0.93) and (0.98) it follows that

$$\max_{x\in\mathcal{A}_{n+1}} \|x\|_p \leqslant A_3^{(p)}(n+1)^{\frac{5}{2}} 2^{n\left(\frac{1}{p}-\frac{1}{2}\right)} \qquad (n = 0,1,2,\ldots), \tag{0.99}$$

whence, since $2 < p < \infty$, we infer (0.5), which completes the proof of theorem 0.2.

*Remark 0.4*. While theorem 0.1 is valid both for complex and real scalars, theorem 0.2 above shows only that the complex space $E$ does not have the approximation property. However, this also implies that the real Banach space $E_{(r)}$ obtained from $E$ by the restriction of the field of scalars*) does not have the approximation property. Indeed, let $Q$ be an arbitrary compact set in $E$ and let $\varepsilon>0$. If $E_{(r)}$ has the approximation property, then, since $Q \cup iQ$ is compact in $E_{(r)}$, there exists a finite rank operator $u$ on $E_{(r)}$ such that $\|u(x) - x\| < \varepsilon$ for all $x \in Q \cup iQ$. But then $v(x) = u(x) - iu(ix)$ $(x \in E)$ is a continuous linear operator of finite rank on $E$ and

$$\left\|\frac{v(x)}{2} - x\right\| = \left\|\frac{u(x)-x}{2} - \frac{iu(ix) - i(ix)}{2}\right\| \leqslant$$

$$\leqslant \frac{1}{2}(\|u(x) - x\| + \|u(ix) - ix\|) \qquad (x \in Q \cup iQ),$$

---

*) I.e., $E$ regarded as a Banach space over the subfield $R$ of real numbers (see Ch. I, § 1).

whence, in particular, for all $x \in Q$, so $E$ has the approximation property, in contradiction with our assumption. This proves that $E_{(r)}$ does not have the approximation property.

Now, if $E \subset l^p$, where $2 < p < \infty$, then $E_{(r)} \subset l^p_{(r)}$, whence, since $l^p_{(r)}$ is isomorphic to $l^p_R$=the real space $l^p$ (indeed,*) by Ch. I, § 1, proposition 1.1, if $\{e_n\}$ is the unit vector basis of $l^p$, then $\{e_n\} \cup \{ie_n\}$ is a basis of $l^p_{(r)}$, whence, e.g. by Ch. II, § 18, lemma 18.5 b), $l^p_{(r)} \cong (l^p_R)^2 \cong l^p_R$), it follows that the image of $E_{(r)}$ under this isomorphism is a subspace of $l^p_R$ which does not have the approximation property. A similar remark is valid also for the real space $c_0$.

*Remark 0.5*. If $X$ is a finite-dimensional Banach space and $(x_i, h_i)_{i \in I}$ ($\{x_i\} \subset X$, $\{h_i\} \subset X^*$) is an $X$-complete biorthogonal system, and if $A \subset I$ is such that the average trace $\mathcal{T}$ with respect to $(x_i, h_i)_{i \in I}$ satisfies

$$|\mathcal{T}(A; u) - \mathcal{T}(I \setminus A; u)| \leqslant \alpha \|u\| \qquad (u \in L(X, X)), \qquad (0.100)$$

then for any projection $v$ of $X$ onto $[x_i]_{i \in A}$ we have

$$\|v\| \geqslant \frac{1}{\alpha} \qquad (0.101)$$

$\Big($whence $X = [x_i]_{i \in A} \oplus [x_i]_{i \in I \setminus A}$, where each projection of $X$ onto $[x_i]_{i \in A}$ and each projection of $X$ onto $[x_i]_{i \in I \setminus A}$ is of norm $\geqslant \frac{1}{\alpha}\Big)$. Indeed, we have $\mathcal{T}(A; v) = \frac{1}{|A|} \sum_{i \in A} h_i(v(x_i)) = \frac{1}{|A|} \sum_{i \in A} h_i(x_i) = 1$ and, since $v(X) = [x_j]_{j \in A}$, by biorthogonality we have $\mathcal{T}(I \setminus A; v) = \frac{1}{|I \setminus A|} \sum_{i \in I \setminus A} h_i(v(x_i)) = 0$, whence, by (0.100),

$$1 = |\mathcal{T}(A; v) - \mathcal{T}(I \setminus A; v)| \leqslant \alpha \|v\|,$$

which proves (0.101). In particular, if $X$ is a subspace of a Banach space $E$, then the same conclusion holds for each projection of $E$ onto $[x_i]_{i \in A}$ or onto $[x_i]_{i \in I \setminus A}$. Now, by the above proof of theorem 0.1, we have

$$|\mathcal{T}(W^+_{n,k}; u) - \mathcal{T}(h(W^-_{n,k}); u)| = |\varphi_{n,k}(u) - \psi_{n,k}(u)| \leqslant (\max_{x \in \mathcal{A}_{n,k}} \|x\|_p) \|u\| \leqslant$$

$$\leqslant B_p 2^{n\left(\frac{1}{p} - \frac{1}{2}\right)} \|u\| \quad (u \in L(E, E);\ k = 1, \ldots, 2^n;\ n = 4, 5, \ldots),$$

*) We recall (see Ch. I, § 13) that we write $E_1 \cong E_2$ if $E_1$ and $E_2$ are isomorphic.

so the above conditions are satisfied for each $X=X_{n,k}=[x_w]_{w\in W^+_{n,k}}\oplus\oplus[x_w]_{w\in h(W^-_{n,k})}$, and hence the norm of every projection $v$ of $E$ onto $[x_w]_{w\in W^+_{n,k}}$ satisfies $\|v\|\geqslant\frac{1}{B_p}\,2^{n\left(\frac{1}{2}-\frac{1}{p}\right)}$. On the other hand, it is known*) that for any finite-dimensional subspace $F$ of any Banach space $E$ there exists a projection $v$ of $E$ onto $F$, of norm $\|v\|\leqslant\sqrt{\dim F}$, so there exists a projection $v$ of the space $E$ of theorem 0.1 onto $[x_w]_{w\in W^+_{n,k}}$, of norm $\|v\|\leqslant\sqrt{2^{n-1}}=\frac{1}{\sqrt{2}}\,2^{\frac{n}{2}}$. Similar remarks can be also made for $|\varphi_n(u)-\varphi_{n-1}(u)|$ in theorems 0.1 and 0.2. More generally, if $|\varphi_n(u)-\varphi_{n-1}(u)|$ satisfies (0.4) and if $\varphi_n(u)=\mathscr{T}(M_n;u)$ as in the heuristic introduction made before lemma 0.2, then

$$|\mathscr{T}(M_n;u)-\mathscr{T}(M_{n-1};u)|=|\varphi_n(u)-\varphi_{n-1}(u)|\leqslant\alpha_n\max_{x\in\mathscr{A}_n}\|u(x)\|$$

$$(u\in L(E,E),\ n=0,1,2,\ldots),$$

so the above conditions are satisfied for each $X=X_n=[x_i]_{i\in M_n}\oplus\oplus[x_i]_{i\in M_{n-1}}$ and hence the norm of every projection $v$ of $E$ onto $[x_i]_{i\in M_n}$ (or $[x_i]_{i\in M_{n-1}}$) satisfies $\|v\|\geqslant\frac{1}{\alpha_n\max_{x\in\mathscr{A}_n}\|x\|}$. On the other hand, there exists a projection $v$ of $E=[x_i]_{i\in\mathscr{M}}$ onto $[x_i]_{i\in M_n}$ (or $[x_i]_{i\in M_{n-1}}$) of norm $\|v\|\leqslant\sqrt{\dim[x_i]_{i\in M_n}}=\sqrt{|M_n|}$ (respectively, $\|v\|\leqslant\sqrt{|M_{n-1}|}$). Consequently, we obtain the estimate

$$\alpha_n\max_{x\in\mathscr{A}_n}\|x\|\geqslant\max\left(\frac{1}{\sqrt{|M_n|}},\ \frac{1}{\sqrt{|M_{n-1}|}}\right)\quad(n=0,1,2,\ldots),$$

whence, by (0.5) and $\sum_{n=0}^{\infty}\alpha_n<\infty$,

$$\sum_{n=0}^{\infty}\frac{1}{\sqrt{|M_n|}}<\infty. \tag{0.102}$$

It is natural to raise the following more general version of Ch. I, § 1, problem 1.3:

*) See [207], formula (5).

*Problem 0.1.**) Does every Banach space $X$ which is not isomorphic to a Hilbert space contain a subspace $E$ which does not have the approximation property?

From theorems 0.1 and 0.2 it follows that the answer is affirmative for every Banach space $X$ containing a subspace isomorphic to $c_0$ or $l^p$ for some $p$ satisfying $2 < p < \infty$. However, the answer is not known**) already for $X = l^p$, where $1 \leqslant p < 2$. The above proofs of theorems 0.1 and 0.2 fail in this case at their final parts, since for $1 \leqslant p < 2$ from (0.68) it does not follow (0.8) and from (0.99) there does not follow (0.5).

Now we shall use theorem 0.1 to construct a Banach space $E$ with a basis, such that the conjugate space $E^*$ is separable and does not have the approximation property; then, in particular, $E^*$ has no basis, so this example will show that the answer to Ch. II, § 5, problem***) 5.1 b) is negative. To this end, we recall the following lemma (which will have also other applications, in Vol. III; actually, here we shall use only the main part of it):

**Lemma 0.6.** *Let $E$ be a separable Banach space. Then there exists a separable Banach space $F$ such that*

i) *$F$ has a monotone shrinking basis.*

ii) *There exists a continuous linear mapping $v$ of $F^*$ onto $E$, such that $v^*$ is a linear isometry (of $E^*$ into $F^{**}$).*

iii) *$F^{**} = \pi(F) \oplus v^*(E^*)$, where $\pi$ is the canonical embedding of $F$ into $F^{**}$, and the projection of $F^{**}$ onto $v^*(E^*)$ along $\pi(F)$ is of norm 1.*

*Proof.* Let $\{x_n\}$ be a dense sequence on the boundary $\operatorname{Fr} S_E = \{x \in E \mid \|x\| = 1\}$ of the unit ball $S_E = \{x \in E \mid \|x\| \leqslant 1\}$. Let $A_0 = A_0(\{x_n\})$ consist of all sequences of scalars $\{\alpha_n\} \subset K$ such that

$$\|\{\alpha_n\}\| = \sup \left( \sum_{j=1}^{k} \left\| \sum_{i=m_{j-1}+1}^{m_j} \alpha_i x_i \right\|^2 \right)^{\frac{1}{2}} < \infty, \tag{0.103}$$

where the sup is taken over all $k = 1, 2, \ldots$ and all finite sequences of integers $0 = m_0 < m_1 < \ldots < m_k$. Then $A_0$ is a Banach space.

Let $e_n$ be the $n$-th unit vector in $A_0$ $(n = 1, 2, \ldots)$. Then, by $\left\| \sum_{i=1}^{n} \alpha_i e_i \right\| = \|\{\alpha_1, \ldots, \alpha_n, 0, 0, \ldots\}\|$ and (0.103), $\{e_n\}$ is a monotone boundedly complete basis of $A_0$. Hence, by Ch. II, § 6, theorem 6.2, implication $1^\circ \Rightarrow 3^\circ$, the canonical mapping $u$ of $A_0$ into $[h_n]^*$, where $\{h_n\} \subset A_0^*$ is the a.s.c.f. to $\{e_n\}$, is a linear isometry of $A_0$ *onto* $[h_n]^*$. Also, by (0.103), we have $\|e_n\| = 1$ $(n = 1, 2, \ldots)$ and hence $\|u(e_n)\| = 1$ $(n = 1, 2, \ldots)$. Now let

$$F = [h_n]. \tag{0.104}$$

---

*) Recently, this problem has been solved in the negative (see the Notes and remarks).

**) Recently, this problem has been solved in the affirmative (see the Notes and remarks).

***) This problem was also raised in Ch. I, § 14, in the comments to table 14.1.

Then, by Ch. II, § 1, proposition 1.4 and Ch. II, § 6, corollary 6.1, $\{h_n\}$ is a monotone shrinking basis of $F$, so we have i). Furthermore, for each $\{\alpha_n\} \in A_0$ the series $\sum_{i=1}^{\infty} \alpha_i x_i$ converges (since it is a Cauchy series, by (0.103)) and

$$v: u(\{\alpha_n\}) \to \sum_{i=1}^{\infty} \alpha_i x_i \tag{0.105}$$

is a linear mapping of norm $\leqslant 1$ of $F^*$ into $E$ (again by (0.103)). Since $\|u(e_n)\| = 1$ $(n = 1,2, \ldots)$ and

$$v(u(e_n)) = x_n \qquad (n = 1,2, \ldots) \tag{0.106}$$

and since $\{x_n\}$ is dense in Fr $S_E$, it follows*) that $v$ maps $F^*$ *onto* $E$ and $S_{F^*}$ onto $S_E$, and hence $v^*$ is a linear isometry of $E^*$ into $F^{**}$. Thus, we have ii).

Now let $h = \sum_{i=1}^{\infty} \beta_i h_i \in F$ and $f \in E^*$ be arbitrary. Then, from $\|e_n\| = 1$ $(n = 1,2, \ldots)$ and Ch. I, § 3, corollary 3.1 we infer

$$|\beta_n| \leqslant \frac{1}{\inf\limits_{1 \leqslant j < \infty} \|h_j\|} \|\beta_n h_n\| \to 0 \quad \text{as } n \to \infty,$$

whence, since by biorthogonality and (0.106) we have

$$\pi(h)(u(e_n)) = u(e_n)(h) = h(e_n) = \beta_n \qquad (n = 1,2, \ldots),$$
$$v^*(f)(u(e_n)) = f(v(u(e_n))) = f(x_n) \qquad (n = 1,2, \ldots),$$

and since $\{x_n\}$ is dense in Fr $S_E$, it follows that

$$\|\pi(h) + v^*(f)\| \geqslant \sup_{1 \leqslant n < \infty} |(\pi(h) + v^*(f))(u(e_n))| = \sup_{1 \leqslant n < \infty} |\beta_n + f(x_n)| \geqslant$$

$$\geqslant \overline{\lim_{n\to\infty}} |f(x_n)| = \|f\| = \|v^*(f)\| \quad \left(h = \sum_{i=1}^{\infty} \beta_i h_i \in F, \ f \in E^*\right)$$

Consequently, $\pi(F) \cap v^*(E^*) = \{0\}$ and the mapping

$$\pi(h) + v^*(f) \to v^*(f)$$

---

*) See e.g. [140], Ch. I, § 14 ,corollary 1.

is a projection of norm 1 of $\pi(F)\oplus v^*(E^*)$ onto $v^*(E^*)$. Thus, in order to complete the proof of iii) it will be sufficient to show that the closed linear subspace $\pi(F)\oplus v^*(E^*)$ exhausts all of $F^{**}$. For this purpose it will be sufficient*) to show that

$$\operatorname{dist}(\Phi, \pi(F)\oplus v^*(E^*)) \leqslant \frac{7}{8} \qquad (\Phi \in F^{**}, \|\Phi\| = 1). \quad (0.107)$$

Let us define a set $M \subset \pi(F)$ as follows: We shall say that $\Psi \in M$ if there are integers $k$ and $0 = m_0 < m_1 < \ldots < m_k$ and functionals $g_1, \ldots, g_k \in E^*$ such that

$$\Psi = \sum_{j=1}^{k} \sum_{i=m_{j-1}+1}^{m_j} g_j(x_i)\pi(h_i), \quad \sum_{j=1}^{k} \|g_j\|^2 \leqslant 1; \quad (0.108)$$

the integers $m_j$ appearing in (0.108) will be called the *division points* of $\Psi$. The representation (0.108) is not unique in general, but this will lead to no confusion. Let us also note that the set $M$ is circled (i.e., $\beta M \subset M$ for each scalar $\beta$ with $|\beta| = 1$).

We claim that

$$S_{\pi(F)} = \overline{\text{co}}\, M, \quad (0.109)$$

the norm-closed convex hull of $M$ in $\pi(F)$. Indeed,

$$\pi(h_i)(u(\{\alpha_n\})) = u(\{\alpha_n\})(h_i) = h_i(\{\alpha_n\}) = \alpha_i$$
$$(u(\{\alpha_n\}) \in F^*;\ i = 1, 2, \ldots), \quad (0.110)$$

whence, by the Hölder inequality,

$$|\Psi(u(\{\alpha_n\}))| = \left|\sum_{j=1}^{k} \sum_{i=m_{j-1}+1}^{m_j} g_j(x_i)\,\alpha_i\right| = \left|\sum_{j=1}^{k} g_j\left(\sum_{i=m_{j-1}+1}^{m_j} \alpha_i x_i\right)\right| \leqslant$$
$$\leqslant \sum_{j=1}^{k} \|g_j\| \left\|\sum_{i=m_{j-1}+1}^{m_j} \alpha_i x_i\right\| \leqslant \left(\sum_{j=1}^{k} \|g_j\|^2\right)^{\frac{1}{2}} \left(\sum_{j=1}^{k} \left\|\sum_{i=m_{j-1}+1}^{m_j} \alpha_i x_i\right\|^2\right)^{\frac{1}{2}} \leqslant$$
$$\leqslant \|\{\alpha_n\}\| = \|u(\{\alpha_n\})\| \quad (\Psi \in M;\ u(\{\alpha_n\}) \in F^*); \quad (0.111)$$

*) Indeed, if $\pi(F)\oplus v^*(E^*)$ were a proper subspace of $F^{**}$, then for every $\varepsilon > 0$ there would exist a $\Phi \in F^{**}$ with $\|\Phi\| = 1$, such that $\operatorname{dist}(\Phi, \pi(F)\oplus v^*(E^*)) \geqslant 1 - \varepsilon$.

thus, $M \subset S_{\pi(F)}$, whence also $\overline{\text{co}}\ M \subset S_{\pi(F)}$. Assume now that $\Psi_0 \in S_{\pi(F)} \setminus \overline{\text{co}}\ M$. Then, since $\overline{\text{co}}\ M$ is a circled closed convex set, there exists, by the separation theorem*), a functional $u(\{\alpha_n^0\}) \in F^*$ with $\|u(\{\alpha_n^0\})\| = 1$, such that

$$1 \geqslant \|\Psi_0\| \geqslant |\Psi_0(u(\{\alpha_n^0\})| > \sup_{\Psi \in M} |\Psi(u\{\alpha_n^0\})|. \tag{0.112}$$

But, by (0.103), for any $\varepsilon > 0$ there exist integers $0 = m_0 < m_1 < \ldots < m_k$ such that

$$\sum_{j=1}^{k} \left\| \sum_{i=m_{j-1}+1}^{m_j} \alpha_i^0 x_i \right\|^2 \geqslant \|\{\alpha_n^0\}\|^2 - \varepsilon = \|u(\{\alpha_n^0\})\|^2 - \varepsilon = 1 - \varepsilon.$$

By a corollary of the Hahn-Banach theorem, there exist functionals $g_1, \ldots, g_k \in E^*$ such that

$$g_j\left(\sum_{i=m_{j-1}+1}^{m_j} \alpha_i^0 x_i\right) = \|g_j\| \left\| \sum_{i=m_{j-1}+1}^{m_j} \alpha_i^0 x_i \right\| \qquad (j = 1, \ldots, k),$$

$$\|g_j\| = \left\| \sum_{i=m_{j-1}+1}^{m_j} \alpha_i^0 x_i \right\| \qquad (j = 1, \ldots, k),$$

whence

$$\sum_{j=1}^{k} \|g_j\|^2 = \sum_{j=1}^{k} \left\| \sum_{i=m_{j-1}+1}^{m_j} \alpha_i^0 x_i \right\|^2 \leqslant \|\{\alpha_n^0\}\|^2 = \|u(\{\alpha_n^0\})\|^2 = 1,$$

and thus we can define $\Psi \in M$ by (0.108). Then, by (0.110), we obtain

$$|\Psi(u(\{\alpha_n^0\}))| = \left| \sum_{j=1}^{k} g_j \left( \sum_{i=m_{j-1}+1}^{m_j} \alpha_i^0 x_i \right) \right| =$$

$$= \sum_{j=1}^{k} \|g_j\| \left\| \sum_{i=m_{j-1}+1}^{m_j} \alpha_i^0 x_i \right\| = \sum_{j=1}^{k} \left\| \sum_{i=m_{j-1}+1}^{m_j} \alpha_i^0 x_i \right\|^2 \geqslant 1 - \varepsilon,$$

in contradiction with (0.112). This proves the claim (0.109). Consequently, since $F^*$ is separable, $\overline{\text{co}}\ M = \pi(S_F)$ *is***) *sequentially* $w^*$*-dense in* $S_{F^{**}}$.

---

*) See e.g. [70], Ch. I, § 6, theorem 5 or [409], p. 109, theorem 3 (which remains valid also for complex scalars).

**) See e.g. [87], p. 424, theorem 5 and p. 426, theorem 1.

Now let $\Phi \in F^{**}$, $\|\Phi\| = 1$, be arbitrary; we shall prove that $\Phi$ satisfies (0.107). By above, there exists a sequence

$$\Omega_n = \sum_{l=1}^{\gamma_n} \lambda_{l,n} \Psi_{l,n}, \quad \lambda_{l,n} \geqslant 0, \quad \sum_{l=1}^{\gamma_n} \lambda_{l,n} = 1, \quad \Psi_{l,n} \in M, \tag{0.113}$$

such that $\Phi = w^*\text{-}\lim\limits_{n\to\infty} \Omega_n$. We shall consider two cases:

*Case 1°.* There exists an integer $i_0$ such that for every $i > i_0$, $\overline{\lim\limits_{n\to\infty}} \sum_{l=1}^{\gamma_n}{}' \lambda_{l,n} > \frac{1}{8}$, where in $\sum'$ we sum only over those indices $l$ for which $\Psi_{l,n}$ does not have a division point between $i_0$ and $i$ (the set of indices which enter into $\sum'$ depends on $i_0$ and $i$). Then there exists a subsequence of $\{\Omega_n\}$, which, for simplicity, we shall denote again by $\{\Omega_n\}$, such that $\Omega_n = \Omega_n' + \Omega_n''$ with $\Omega_n' = \sum_{l=1}^{\gamma_n}{}' \lambda_{l,n} \Psi_{l,n}$, where each $\Psi_{l,n}$ has no point of division between $i_0$ and some $i_n$, $\sum_{l=1}^{\gamma_n}{}' \lambda_{l,n} \geqslant \frac{1}{8}$ and $\lim\limits_{n\to\infty} i_n = \infty$. Hence

$$\|\Omega_n''\| = \|\Omega_n - \Omega_n'\| = \|\sum_{l=1}^{\gamma_n}{}'' \lambda_{l,n} \Psi_{l,n}\| \leqslant \sum_{l=1}^{\gamma_n}{}'' \lambda_{l,n} \leqslant \frac{7}{8}, \tag{0.114}$$

where in $\sum''$ we sum over all those indices $l \leqslant \gamma_n$ which do not appear in $\sum'$. We may assume, without loss of generality (by passing to a subsequence, if necessary), that $w^*\text{-}\lim\limits_{n\to\infty} \Omega_n''$ exists, say $w^*\text{-}\lim\limits_{n\to\infty} \Omega_n'' = \Omega'' \in F^{**}$. Then, by (0.114),

$$\|\Omega''\| \leqslant \underline{\lim_{n\to\infty}} \|\Omega_n''\| \leqslant \frac{7}{8}. \tag{0.115}$$

Now, for $\Psi_{l,n}$ entering into $\Omega_n'$, let $0 = m_0(l,n) < m_1(l,n) < \ldots$ $\ldots < m_{k(l,n)}(l,n)$ denote the division points of $\Psi_{l,n}$. Then, since by our assumption $\Psi_{l,n}$ has no division points between $i_0$ and $i_n$, there exists an index $j_0(l,n)$ such that $m_{j_0(l,n)-1}(l,n) + 1 \leqslant i_0 < i_n \leqslant m_{j_0(l,n)}(l,n)$. Hence, for any $p$ such that $i_0 \leqslant p \leqslant i_n$ we have, taking into account (0.110),

$$\Omega_n'(u(e_p)) = \left( \sum_{l=0}^{\gamma_n}{}' \lambda_{l,n} \sum_{j=1}^{k(l,n)} \sum_{l=m_{j-1}(l,n)+1}^{m_j(l,n)} g_j^{(l,n)}(x_i)\pi(h_i) \right)(u(e_p)) =$$

$$= \sum_{l=1}^{\gamma_n}{}' \lambda_{l,n} \sum_{j=1}^{k(l,n)} \sum_{i=m_{j-1}(l,n)+1}^{m_j(l,n)} g_j^{(l,n)}(x_i)\delta_{ip} = \sum_{l=1}^{\gamma_n}{}' \lambda_{l,n} g_{j_0(l,n)}^{(l,n)}(x_p).$$

Let $f_n = \sum_{l=1}^{\gamma_n}{}' \lambda_{l,n} g_{j_0(l,n)}^{(l,n)}$, so $\Omega'_n(u(e_p)) = f_n(x_p)$ $(n = 1,2, \ldots)$. Since $\|f_n\| \leqslant \sum_{l=1}^{\gamma_n}{}' \lambda_{l,n} \leqslant 1$, there is no loss of generality to assume that $f = w^*\text{-}\lim_{n\to\infty} f_n$ exists. Then, taking into account (0.106), we get

$$(\Phi - \Omega'')(u(e_p)) = \lim_{n\to\infty} (\Omega_n - \Omega''_n)(u(e_p)) = \lim_{n\to\infty} \Omega'_n(u(e_p)) =$$

$$= \lim_{n\to\infty} f_n(x_p) = f(x_p) = v^*(f)(u(e_p)) \qquad (p \geqslant i_0),$$

whence, denoting by $\tilde{\Gamma}$ the $\sigma(F^{**}, F^*)$-closure of any $\Gamma \subset F^{**}$, we obtain

$$\Phi - \Omega'' - v^*(f) \in ([u(e_p)]_{p \geqslant i_0})^{\perp} = (([\pi(h_i)]_{i=1}^{i_0-1})_{\perp})^{\perp} = \widetilde{[\pi(h_i)]_{i=1}^{i_0-1}} =$$

$$= [\pi(h_i)]_{i=1}^{i_0-1} \subset \pi(F),$$

and thus

$$\Phi - \Omega'' \in \pi(F) \oplus v^*(E^*).$$

Consequently, by (0.115),

$$\text{dist}\ (\Phi, \pi(F) \oplus v^*(E^*)) \leqslant \|\Phi - (\Phi - \Omega'')\| \leqslant \frac{7}{8},$$

which proves that in case 1° we have (0.107).

*Case 2°*. There exists no integer $i_0$ as in case 1°. Since $\{h_n\}$ is a monotone shrinking basis of $F$ with the a.s.c.f. $\{u(e_n)\}$, by Ch. II, § 4, theorem 4.2 we have $1 = \|\Phi\| = \lim_{n\to\infty} \left\| \sum_{i=1}^{n} \Phi(u(e_i))h_i \right\| = \lim_{n\to\infty} \left\| \sum_{i=1}^{n} \Phi(u(e_i))\pi(h_i) \right\|$ and hence we can choose $i_0$ such that

$$\left\| \sum_{i=1}^{i_0} \Phi(u(e_i))\, \pi(h_i) \right\| \geqslant \frac{7}{8}. \tag{0.116}$$

Then, since we are in case 2°, for this $i_0$ there exists an $i_1 > i_0$ such that for all sufficiently large $n$, $\sum_{l=1}^{\gamma_n}{}' \lambda_{l,n} \leqslant \frac{1}{8}$, where $\sum'$ has the same meaning as in case 1°, with $i = i_1$. For every $l \leqslant \gamma_n$ which does not enter into $\sum'$ define

$$\Psi_{l,n} = \Pi_{l,n} + \Xi_{l,n} \in M, \tag{0.117}$$

where $\Pi_{l,n} \in \pi(F)$ is defined by

$$\Pi_{l,n} = \sum_{i=1}^{i(l,n)}{}' \Psi_{l,n}(u(e_i))\pi(h_i), \tag{0.118}$$

with $i(l, n)$ being the first division point of $\Psi_{l,n}$ after $i_0$ (by definition, $i(l, n) \leqslant i_1$). Thus, if the division points of $\Psi_{l,n}$ are $0 = m_0(l, n) < < m_1(l, n) < \ldots < m_{k(l,n)}(l, n)$, then there exists a $j_0(l, n) \leqslant k(l, n)$ such that $m_{j_0(l,n)}(l, n) = i(l, n)$ and we can write

$$\Psi_{l,n} = \Pi_{l,n} + \Xi_{l,n} = \sum_{j=1}^{j_0(l,n)} \sum_{i=m_{j-1}(l,n)+1}^{m_j(l,n)} g_j^{(l,n)}(x_i)\pi(h_i) + $$

$$+ \sum_{j=j_0(l,n)+1}^{k(l,n)} \sum_{i=m_{j-1}(l,n)+1}^{m_j(l,n)} g_j^{(l,n)}(x_i)\pi(h_i).$$

Consequently, for any $u(\{\alpha_p\}), u(\{\beta_p\}) \in F^*$ of norm $\leqslant 1$ we obtain, similarly to (0.111),

$$|\Pi_{l,n}(u(\{\alpha_p\}))|^2 + |\Xi_{l,n}(u(\{\beta_p\}))|^2 \leqslant$$

$$\leqslant \sum_{j=1}^{j_0(l,n)} \|g_j^{(l,n)}\|^2 \sum_{j=1}^{j_0(l,n)} \left\| \sum_{i=m_{j-1}(l,n)+1}^{m_j(l,n)} \alpha_i x_i \right\|^2 +$$

$$+ \sum_{j=j_0(l,n)+1}^{k(l,n)} \|g_j^{(l,n)}\|^2 \sum_{j=j_0(l,n)+1}^{k(l,n)} \left\| \sum_{i=m_{j-1}(l,n)+1}^{m_j(l,n)} \beta_i x_i \right\|^2 \leqslant$$

$$\leqslant \|\{\alpha_p\}\|^2 \sum_{j=1}^{j_0(l,n)} \|g_j^{(l,n)}\|^2 + \|\{\beta_p\}\|^2 \sum_{j=j_0(l,n)+1}^{k(l,n)} \|g_j^{(l,n)}\|^2 \leqslant$$

$$\leqslant \sum_{j=1}^{k(l,n)} \|g_j^{(l,n)}\|^2 \leqslant 1,$$

whence

$$\|\Pi_{l,n}\|^2 + \|\Xi_{l,n}\|^2 \leqslant 1, \tag{0.119}$$

and this holds for all $l \leqslant \gamma_n$ which do not enter into $\sum'$ (i.e., for which $\Psi_{l,n}$ has a division point between $i_0$ and $i_1$).

From (0.119) and $\sum_{l=1}^{\gamma_n}{}''\lambda_{l,n}\leqslant 1$ (see (0.113)), where the sum $\sum''$ is taken over all $l \leqslant \gamma_n$ which do not enter into $\sum'$, we infer

$$\left(\sum_{l=1}^{\gamma_n}{}''\lambda_{l,n}\|\Pi_{l,n}\|\right)^2+\left(\sum_{l=1}^{\gamma_n}{}''\lambda_{l,n}\|\Xi_{l,n}\|\right)^2=$$

$$=\sum_{l=1}^{\gamma_n}{}''\lambda_{l,n}^2(\|\Pi_{l,n}\|^2+\|\Xi_{l,n}\|^2)+\sum_{l\neq r}{}''\lambda_{l,n}\lambda_{r,n}(\|\Pi_{l,n}\|\ \|\Pi_{r,n}\|+$$

$$+\|\Xi_{l,n}\|\ \|\Xi_{r,n}\|)\leqslant\sum_{l=1}^{\gamma_n}{}''\lambda_{l,n}^2(\|\Pi_{l,n}\|^2+\|\Xi_{l,n}\|^2)+$$

$$+\sum_{l<r}{}''\lambda_{l,n}\lambda_{r,n}(\|\Pi_{l,n}\|^2+\|\Pi_{r,n}\|^2+\|\Xi_{l,n}\|^2+\|\Xi_{r,n}\|^2)=$$

$$=\left(\sum_{l=1}^{\gamma_n}{}''\lambda_{l,n}\right)\left(\sum_{l=1}^{\gamma_n}{}''\ \lambda_{l,n}(\|\Pi_{l,n}\|^2+\|\Xi_{l,n}\|^2)\right)\leqslant 1,$$

whence

$$\sum_{l=1}^{\gamma_n}{}''\lambda_{l,n}\|\Xi_{l,n}\|\leqslant\left(1-\sum_{l=1}^{\gamma_n}{}''\lambda_{l,n}\|\Pi_{l,n}\|^2\right)^{\frac{1}{2}}. \tag{0.120}$$

Now put

$$\Pi_n=\sum_{l=1}^{\gamma_n}{}''\lambda_{l,n}\Pi_{l,n},\ \Xi_n=\sum_{l=1}^{\gamma_n}{}''\lambda_{l,n}\,\Xi_{l,n}\qquad(n=1,2,\ldots),$$

$$\Omega_n'=\Omega_n-(\Pi_n+\Xi_n)\qquad(n=1,2,\ldots).$$

Then, by (0.113), (0.117), $\|\Psi_{l,n}\|\leqslant 1$ and $\sum_{l=1}^{\gamma_n}{}'\lambda_{l,n}\leqslant\frac{1}{8}$, we have

$$\|\Omega_n'\|=\left\|\sum_{l=1}^{\gamma_n}\lambda_{l,n}\Psi_{l,n}-\sum_{l=1}^{\gamma_n}{}''\lambda_{l,n}\Psi_{l,n}\right\|=\left\|\sum_{l=1}^{\gamma_n}{}'\lambda_{l,n}\Psi_{l,n}\right\|\leqslant\frac{1}{8}$$

and by (0.117), (0.118), the biorthogonality of $(u(e_i),\ \pi(h_i))$, and $i_0\leqslant i(l,n)$, we have

$$\Xi_n(u(e_i))=\sum_{l=1}^{\gamma_n}{}''\lambda_{l,n}(\Psi_{l,n}-\Pi_{l,n})(u(e_i))=0\qquad(i=1,\ldots,i_0).$$

Since $F^*$ is separable, without loss of generality we may assume that $\Omega' = w^*\text{-}\lim_{n\to\infty} \Omega'_n$, $\Pi = w^*\text{-}\lim_{n\to\infty} \Pi_n$ and $\Xi = w^*\text{-}\lim_{n\to\infty} \Xi_n$ exist. Then, by the above,

$$\Omega' = \Phi - (\Pi + \Xi),$$

$$\|\Omega'\| \leqslant \varliminf_{n\to\infty} \|\Omega'_n\| \leqslant \frac{1}{8},$$

$$\Xi(u(e_i)) = 0 \qquad (i = 1, \ldots, i_0),$$

whence, by Ch. II, § 4, theorem 4.2 and by (0.116),

$$\varliminf_{n\to\infty} \sum_{l=1}^{\gamma_n}{}'' \lambda_{l,n}\|\Pi_{l,n}\| = \varliminf_{n\to\infty} \|\Pi_n\| \geqslant \|\Pi\| \geqslant$$

$$\geqslant \left\|\sum_{i=1}^{i_0} \Pi(u(e_i))\pi(h_i)\right\| = \left\|\sum_{i=1}^{i_0} (\Omega' - \Phi)(u(e_i))\pi(h_i)\right\| \geqslant$$

$$\geqslant \left\|\sum_{i=1}^{i_0} \Phi(u(e_i))\pi(h_i)\right\| - \left\|\sum_{i=1}^{i_0} \Omega'(u(e_i))\pi(h_i)\right\| \geqslant \frac{7}{8} - \|\Omega'\| \geqslant \frac{3}{4},$$

and therefore, by (0.120),

$$\|\Xi\| \leqslant \varliminf_{n\to\infty} \|\Xi_n\| \leqslant \varliminf_{n\to\infty} \sum_{l=1}^{\gamma_n}{}'' \lambda_{l,n}\|\Xi_{l,n}\| \leqslant$$

$$\leqslant \varliminf_{n\to\infty} \left(1 - \sum_{l=1}^{\gamma_n}{}'' \lambda_{l,n}\|\Pi_{l,n}\|^2\right)^{\frac{1}{2}} \leqslant \left(1 - \left(\frac{3}{4}\right)^2\right)^{\frac{1}{2}} = \frac{\sqrt{7}}{4}.$$

Consequently, since $\Pi \in [\pi(h_i)]_{i=1}^{i_1} \subset \pi(F)$ (by (0.118) and $i(l, n) \leqslant i_1$ for all $l$, $n$), we obtain

$$\operatorname{dist}(\Phi, \pi(F) \oplus v^*(E^*)) \leqslant \operatorname{dist}(\Phi, \pi(F)) \leqslant \|\Phi - \Pi\| =$$

$$= \|\Omega' + \Xi\| \leqslant \frac{1}{8} + \frac{\sqrt{7}}{4} < \frac{7}{8},$$

which proves that in case 2° we have (0.107). This completes the proof of lemma 0.6. Note that in this proof the space $F$ was effectively constructed.

Now we are ready to give the example announced above.

*Example 0.1.* By theorem 0.1, let $B$ be a subspace of $c_0$ (hence $B^*$ is separable), which does not have the approximation property (hence $B^*$

does not have*) it either). Let $F$ be a separable Banach space corresponding to $B$ by lemma 0.6 above (so $F$ has a shrinking basis and $F^{**}$ is isomorphic to $F \times B^*$) and let $E = F^*$. Then $E$ has a basis, $E^* \cong F \times B^*$ is separable and $E^* \cong F \times B^*$ does not have the approximation property. (Indeed, it is easy to see**) that every complemented subspace of a Banach space with the approximation property has the same property, while the image of $B^*$ in $E^*$, under the isomorphism $F \times B^* \cong E^*$, does not have it.)

# I. Countable Generalizations of Bases

## § 1. Basic sequences. Bibasic systems

We recall (see Ch. I, § 4, definition 4.5) that a sequence $\{x_n\}$ in a Banach space $E$ is called a *basic sequence* if $\{x_n\}$ is a basis of the closed linear subspace $[x_n]$ of $E$ spanned by the sequence $\{x_n\}$. We shall assume that $\dim E = \infty$ and that $\{x_n\}$ is infinite.

It is obvious that a sequence $\{x_n\} \subset E$ is a basis of $E$ if and only if $\{x_n\}$ is a basic sequence such that $[x_n] = E$. Hence, as was already remarked in Ch. I, § 7 (before corollary 7.4), from the characterizations of bases given in Ch.I one can obtain, by omitting the condition $[x_n] = E$, various characterizations of basic sequences. We leave to the reader the explicit formulations of these characterizations.

Let us recall the following result on selection of basic sequences, given in Ch. II, §15, proposition 15.1:

**Proposition 1.1.** *Let $X$ be a Banach space with a basis $\{x_n\}$ and let $\{f_n\} \subset X^*$ be the a.s.c.f.***) to $\{x_n\}$. If a sequence $\{y_n\} \subset X$ satisfies the conditions*

$$\inf_{1 \leqq n < \infty} \|y_n\| > 0, \tag{1.1}$$

$$\lim_{n \to \infty} f_i(y_n) = 0 \qquad (i = 1,2, \ldots), \tag{1.2}$$

*then $\{y_n\}$ has a subsequence $\{y_{p_n}\}$ which is a basic sequence, equivalent to a block basic sequence with respect to $\{x_n\}$.*

---

*) See e.g. § 9, theorem 9.3, implication 11°⇒4° and proposition 9.8, or [139], prooposition 36.

**) See e.g. § 9, proof of proposition 9.5 and the footnote to it.

***) We recall (see Ch. I, § 3) that a.s.c.f. stands for: associated sequence of coefficient functionals. Also, we recall (see Ch. I, § 8) that two sequences $\{x_n\} \subset E_1$, $\{z_n\} \subset E_2$ are said to be equivalent, and we write $\{x_n\} \sim \{z_n\}$, if $\left\{\{\alpha_n\} \subset K \,\middle|\, \sum_{i=1}^{\infty} \alpha_i x_i \text{ converges}\right\} = \left\{\{\alpha_n\} \subset K \,\middle|\, \sum_{i=1}^{\infty} \alpha_i z_i \text{ converges}\right\}$.

**Corollary 1.1.** *Let* $\{y_n\}$ *be a sequence in a Banach space E, satisfying* (1.1) *and such that* $y_n \overset{w}{\to} 0$. *Then* $\{y_n\}$ *has a subsequence* $\{y_{p_n}\}$ *which is a basic sequence.*

*Proof.* Since $[y_n]$ is separable, it can be embedded into a Banach space $X$ with a basis $\{x_n\}$ (e.g., into $C([0, 1])$) and the image of $\{y_n\}$ obviously satisfies (1.1) and (1.2). Therefore the image of $\{y_n\}$, whence also $\{y_n\}$, contains a subsequence which is a basic sequence.

**Corollary 1.2.** *If E is a subspace of a Banach space X with a basis* $\{x_n\}$, *then E contains a basic sequence which is equivalent to a block basic sequence with respect to* $\{x_n\}$.

*Proof.* Let $\{f_n\} \subset X^*$ be the a.s.c.f. to the basis $\{x_n\}$. Then, since*)

$$\dim E = \infty, \ \operatorname{codim}_X [f_1, \ldots f_n]_\perp = n, \tag{1.3}$$

by Ch. II, § 4, lemma 4.1 there exist elements

$$y_n \in E \cap [f_1, \ldots, f_n]_\perp \qquad (n = 1,2, \ldots) \tag{1.4}$$

such that $\|y_n\| = 1$ $(n = 1, 2, \ldots)$. Hence, by proposition 1.1, $\{y_n\}$ has a subsequence $\{y_{p_n}\}$ which is a basic sequence, equivalent to a block basic sequence with respect to $\{x_n\}$.

From corollary 1.2 it results the following theorem on the existence of basic sequences, which was also observed in Ch. I, § 14 (after table 14.1):

**Theorem 1.1.** *In every Banach space E there exists a basic sequence* $\{x_n\}$.

*Proof.* It is sufficient to embed an arbitrary separable subspace of $E$ into $X = C([0, 1])$ and to apply corollary 1.2.

We shall give a slightly stronger result in theorem 1.2 below, with a more direct proof. For this purpose, let us first recall the following lemma, which was used, implicitly, in Ch. I, § 20, proof of theorem 20.2**):

**Lemma 1.1.** *Let E be a Banach space, G a finite-dimensional subspace of E and* $0 < \varepsilon < 1$. *Then there exists a subspace Y of finite codimension in E such that* $\widehat{(G; Y)} > 1 - \varepsilon$ *(or, equivalently, such that* $G \cap Y = \{0\}$ *and that the natural projection of* $G \oplus Y$ *onto G has norm* $< 1 + \dfrac{\varepsilon}{1 - \varepsilon}$*).*

*) We recall that by our convention $E$ is assumed to be infinite dimensional and that $[f_1, \ldots, f_n]_\perp = \{x \in X \mid f_1(x) = \ldots = f_n(x) = 0\}$.

**) For a related result, see also Ch. II, § 16, lemma 16.4. We recall (see Ch. I, § 7) that $\widehat{(G; Y)} = \operatorname{dist}(\{x \in G \mid \|x\| = 1\}, Y)$.

*Proof.* Since $\dim G < \infty$, the set $\sigma_G = \{x \in G \mid \|x\| = 1\}$ is compact and hence it has a finite $\varepsilon'$-net $z_1, \ldots, z_m$, where $0 < \varepsilon' < \varepsilon$. Let $f_1, \ldots, f_m \in E^*$ be functionals with $\|f_i\| = 1$, $f_i(z_i) = 1$ $(i = 1, \ldots, m)$ and let $Y = \{y \in E \mid f_i(y) = 0 \ (i = 1, \ldots, m)\}$. Then $\operatorname{codim} Y < \infty$ and for every $x \in \sigma_G$ and $y \in Y$ we have, with a suitable $z_{i_0}$,

$$\|x - y\| \geqslant \|z_{i_0} - y\| - \|x - z_{i_0}\| > |f_{i_0}(z_{i_0} - y)| - \varepsilon' =$$

$$= |f_{i_0}(z_{i_0})| - \varepsilon' = 1 - \varepsilon',$$

whence $\widehat{(G; Y)} = \inf\limits_{\substack{x \in \sigma_G \\ y \in Y}} \|x - y\| \geqslant 1 - \varepsilon' > 1 - \varepsilon$, which completes the proof.

**Theorem 1.2.** *Let $E$ be a Banach space. Then for every $\varepsilon$ with $0 < \varepsilon < 1$ there exists in $E$ a basic sequence $\{x_n\}$ of index**) $\gamma_{\{x_n\}} \geqslant 1 - \varepsilon$ $\left(\right.$*or, equivalently, of norm* $\nu_{\{x_n\}} \leqslant 1 + \dfrac{\varepsilon}{1 - \varepsilon}\left.\right)$.

*Proof.* Let $x_1 \in E \setminus \{0\}$ be arbitrary. Then by lemma 1.1 for $G = P_{(1)} = [x_1]$, there exists a subspace $Y_1 \subset E$ with $\operatorname{codim} Y_1 < \infty$, such that $\widehat{(P_{(1)}; Y_1)} > 1 - \varepsilon$. Let $x_2 \in Y_1 \setminus \{0\}$ be arbitrary. Then by lemma 1.1 for $G = P_{(2)} = [x_1, x_2]$, there exists a subspace $Y$ of $E$ with $\operatorname{codim} Y < \infty$, such that $\widehat{(P_{(2)}; Y)} > 1 - \varepsilon$. Let $Y_2 = Y_1 \cap Y$. Then $\operatorname{codim} Y_2 < \infty$, $Y_2 \subset Y_1$ and $\widehat{(P_{(2)}; Y_2)} > 1 - \varepsilon$. Continuing in this way, we obtain an infinite sequence of elements $\{x_n\} \subset E$ and an infinite sequence of subspaces $\{Y_n\}$ of $E$ with $\operatorname{codim} Y_n < \infty$ $(n = 1, 2, \ldots)$, such that (a) $Y_1 \supset Y_2 \supset \ldots$; (b) $0 \neq x_{n+1} \in Y_n$ $(n = 1, 2, \ldots)$; (c) $\widehat{(P_{(n)}; Y_n)} > 1 - \varepsilon$, where $P_{(n)} = [x_1, \ldots, x_n]$ $(n = 1, 2, \ldots)$. Then for any scalars $\alpha_{n+1}, \ldots, \alpha_{n+m}$ we have $\sum\limits_{i=n+1}^{n+m} \alpha_i x_i \in Y_n + Y_{n+1} + \ldots + Y_{n+m-1} \subset Y_n$, whence $P^{(n)} = [x_{n+1}, x_{n+2}, \ldots] \subset Y_n$ and hence, by (c), $\operatorname{dist}(\sigma_{P_{(n)}}, P^{(n)}) > 1 - \varepsilon$ $(n = 1, 2, \ldots)$. Consequently, by Ch. I, § 7, theorem 7.1, $\{x_n\}$ is a basic sequence with $\gamma_{\{x_n\}} \geqslant 1 - \varepsilon$, which completes the proof.

*Remark 1.1.* If we replace in (c) above $\varepsilon$ by $\varepsilon_n$, where $0 < \varepsilon_n < 1$ $(n = 1, 2, \ldots)$, $\lim\limits_{n \to \infty} \varepsilon_n = 0$, we obtain that *in every Banach space $E$ there*

---

*) For the definitions of the index $\gamma_{\{x_n\}}$ and of the norm $\nu_{\{x_n\}}$ see Ch. I, § 7

*exists a basic sequence* $\{x_n\}$ *with* $\|s_n\| \leqslant 1 + \frac{\varepsilon_n}{1-\varepsilon_n}$ *for all* $n = 1,2, \ldots$ (*hence* $\lim_{n\to\infty} \|s_n\| = 1$), *where* $\{s_n\}$ *is the sequence of partial sum operators associated to the basis* $\{x_n\}$ *of* $[x_n]$. Let us also note that one can give the following alternative proof of theorem 1.2, based again on lemma 1.1: Let $\varepsilon_n$ with $0 < \varepsilon_n < 1$ $(n = 1,2, \ldots)$ be such that $\prod_{n=1}^{\infty} (1 - \varepsilon_n) \geqslant 1 - \varepsilon$. Choose $P_{(1)} = [x_1]$ and $Y_1$ as in the above proof, with codim $Y_1 < \infty$, $\widehat{(P_{(1)}; Y_1)} > 1 - \varepsilon_1$. Let $x_2 \in Y_1 \setminus \{0\}$ be arbitrary and, by lemma 1.1 for $G = P_{(2)} = [x_1, x_2]$, let $Y_2 \subset E$ with codim $Y_2 < \infty$ be such that $\widehat{(P_{(2)}; Y_2)} > 1 - \varepsilon_2$. Continuing in this way, we obtain an infinite sequence of elements $\{x_n\} \subset E$ and infinite sequence of subspaces $\{Y_n\}$ of $E$ with codim $Y_n < \infty$ $(n = 1,2, \ldots)$ such that (a) $0 \neq x_{n+1} \in Y_n$ $(n=1,2, \ldots)$; (b) $\widehat{(P_{(n)}; Y_n)} > 1 - \varepsilon_n$ $(n=1,2,\ldots)$. Then $\prod_{n=1}^{\infty} \widehat{(P_{(n)}; [x_{n+1}])} \geqslant \prod_{n=1}^{\infty} (1 - \varepsilon_n) \geqslant 1 - \varepsilon$, where $[x_{n+1}]$ is the one-dimensional subspace of $E$ spanned by $x_{n+1}$. Hence, by Ch. I, § 7, corollary 7.1, $\{x_n\}$ is a basic sequence in $E$, of index $\gamma_{\{x_n\}} \geqslant 1 - \varepsilon$, which completes the proof.

The proof of proposition 1.1, i.e., of Ch. II, § 15, proposition 15.1, can be refined so as to yield the following sharper result:

**Proposition 1.2.** *Under the hypotheses of proposition* 1.1, *for every* $\eta$ *with* $0 < \eta < 1$ *there exists a subsequence* $\{y_{p_{n+1}}\}$ *of* $\{y_n\}$ *which is a basic sequence, "*$\eta$*-equivalent"**) *to a block basic sequence* $\{z_n\}$ *with respect to* $\{x_n\}$ *and of norm* $v_{\{y_{p_{n+1}}\}} \leqslant \frac{1+\eta}{1-\eta} v_{\{x_n\}}$ (*hence, in particular, if* $\{x_n\}$ *is a monotone basis, then* $v_{\{y_{p_{n+1}}\}} \leqslant \frac{1+\eta}{1-\eta}$).

*Proof.* By omitting, if necessary, a finite number of the elements $y_{p_{i+1}}$, one can replace the last relation of the proof of proposition 1.1 (i.e., of Ch. II, § 15, proposition 15.1) by $\sum_{i=1}^{\infty} \|h_i\| \, \|z_i - y_{p_{i+1}}\| < \eta$, where $0 < \eta < 1$ is arbitrary. Then, by Ch. I, §10, proof of theorem 10.1 and Ch. I, § 8, proof of theorem 8.1, implication d) 2° ⇒ d) 1°, there exists an isomorphism $u$: $[z_n] \to [y_{p_{n+1}}]$ with $\|u\| \leqslant 1 + \eta$, $\|u^{-1}\| \leqslant$

*) I.e., such that the mapping $u: z_n \to y_{p_{n+1}}$ yields an isomorphism of $[z_n]$ onto $[y_{p_{n+1}}]$, with $\max(\|u\|, \|u^{-1}\|) \leqslant \frac{1}{1-\eta}$.

$\leqslant \dfrac{1}{1-\eta}$, such that $u(z_n) = y_{p_{n+1}}$ $(n = 1,2, \ldots)$. Then for any finite sequence of scalars $\alpha_1, \ldots, \alpha_{n+m}$ we have

$$\left\|\sum_{i=1}^{n} \alpha_i y_{p_{i+1}}\right\| \leqslant (1+\eta)\left\|\sum_{i=1}^{n} \alpha_i z_i\right\| \leqslant (1+\eta)\, v_{\{z_n\}}\left\|\sum_{i=1}^{n+m} \alpha_i z_i\right\| \leqslant$$

$$\leqslant \frac{1+\eta}{1-\eta}\, v_{\{z_n\}}\left\|\sum_{i=1}^{n+m} \alpha_i y_{p_{i+1}}\right\|,$$

whence $v_{\{y_{p_{n+1}}\}} \leqslant \dfrac{1+\eta}{1-\eta}\, v_{\{z_n\}} \leqslant \dfrac{1+\eta}{1-\eta}\, v_{\{x_n\}}$ (by Ch. I, § 7, corollary 7.4), which completes the proof.

*Remark 1.2.* a) Proposition 1.2 implies again theorem 1.2, by taking a separable subspace $E_0$ of $E$ and embedding $E_0$ into a Banach space $X$ with a monotone basis (e.g., into $C([0,1])$, since in Ch. II, § 1, it has been observed that the Schauder basis of $C([0,1])$ is monotone).

b) By proposition 1.2, given any $\eta$ with $0 < \eta < 1$, we can replace in corollary 1.2 "equivalent" by: $\eta$-equivalent.

Now we shall give necessary and sufficient conditions for a sequence in a Banach space $E$ to contain a subsequence which is a basic sequence*⁾. For this purpose, let us first prove

**Proposition 1.3.** *Let $M$ be a bounded set in a Banach space $E$ and let $\Phi_0 \in E^{**}$, with* dist $(\Phi_0, \pi(M)) = d > 0$ *(where $\pi$ denotes the canonical mapping of $E$ into $E^{**}$), be a limit point of $\pi(M)$ for the weak* topology $\sigma(E^{**}, E^*)$. Then there exist a sequence $\{x_n\} \subset M$ and a functional $f_0 \in E^*$ such that*

(a) $\lim\limits_{n\to\infty} f_0(x_n) = \Phi_0(f_0) \geqslant \dfrac{1}{2}\,\|\Phi_0\|,$

(b) $\{\pi(x_n) - \Phi_0\}$ *is a basic sequence in* $E^{**}$,

(c) *if* $\Phi_0 \neq 0$, *then* $\Phi_0 \notin [\pi(x_n) - \Phi_0]$.

*Proof.* Let $\varepsilon_n$ with $0 < \varepsilon_n < 1$ $(n = 0,1,2, \ldots)$ be such that

$$\prod_{i=p}^{q} (1-\varepsilon_i) \geqslant 1 - \varepsilon_0 \quad (1 \leqslant p < q < \infty) \tag{1.5}$$

and let $f_0 \in E^*$ be an arbitrary functional such that

$$\Phi_0(f_0) \geqslant \frac{1}{2}\,\|\Phi_0\|. \tag{1.6}$$

---

*⁾ Some sufficient conditions have been given in Ch. II, § 1, proposition 1.2, for $E = C([0, 1])$ and in Ch. II, § 15, proposition 15.1 (recalled in this section as proposition 1.1) for spaces $E$ with bases. We recall our convention that by "basic sequence" we mean here "infinite basic sequence".

Since $\Phi_0$ is a $\sigma(E^{**}, E^*)$-limit point of $\pi(M)$, there exists an element $x_1 \in M$ such that

$$|f_0(x_1) - \Phi_0(f_0)| < 1. \tag{1.7}$$

Assume now that for some $n \geqslant 1$ we have constructed $x_1, \ldots, x_n \in M$ with the properties

$$|f_0(x_k) - \Phi_0(f_0)| < \frac{1}{k} \qquad (k = 1, \ldots, n), \tag{1.8}$$

$$\left\| \sum_{i=1}^{q} \alpha_i(\pi(x_i) - \Phi_0) \right\| \geqslant \prod_{i=p}^{q-1} (1 - \varepsilon_i) \left\| \sum_{i=1}^{p} \alpha_i(\pi(x_i) - \Phi_0) \right\| \tag{1.9}$$

for all scalars $\alpha_1, \ldots, \alpha_q$ and for $1 \leqslant p < q \leqslant n$ (in the case when $n = 1$, we merely assume that $x_1 \in M$ satisfies (1.7)). Let $E_n = [\pi(x_1) - \Phi_0, \ldots, \pi(x_n) - \Phi_0] \subset E^{**}$. Since $\dim E_n < \infty$, the set $\sigma_{E_n} = \{\Psi \in E_n | \, \|\Psi\| = 1\}$ is compact and hence it has a finite $\frac{\varepsilon_n}{3}$-net $\Psi_1, \ldots, \Psi_{m(n)}$. Choose $h_1, \ldots, h_{m(n)} \in E^*$ with $\|h_i\| = 1$ such that $|\Psi_i(h_i)| > 1 - \frac{\varepsilon_n}{3}$ $(i = 1, \ldots, m(n))$. Since $\Phi_0$ is a $\sigma(E^{**}, E^*)$-limit point of $\pi(M)$, there exists an element $x_{n+1} \in M$ such that

$$|f_0(x_{n+1}) - \Phi_0(f_0)| < \frac{1}{n+1}, \tag{1.10}$$

$$|h_i(x_{n+1}) - \Phi_0(h_i)| < \frac{d\varepsilon_n}{6} \qquad (i = 1, \ldots, m(n)). \tag{1.11}$$

We claim that

$$\|\Psi + \alpha(\pi(x_{n+1}) - \Phi_0)\| \geqslant (1 - \varepsilon_n)\|\Psi\| \quad (\Psi \in E_n, \ \alpha \in K). \tag{1.12}$$

Indeed, assume first that $\|\Psi\| = 1$. If $|\alpha| > \frac{2}{d}$, we have

$$\|\alpha(\pi(x_{n+1}) - \Phi_0) + \Psi\| > \frac{2}{d} \|\pi(x_{n+1}) - \Phi_0\| - \|\Psi\| \geqslant 1 \geqslant (1 - \varepsilon_n)\|\Psi\|,$$

while if $|\alpha| \leqslant \frac{2}{d}$, then, choosing an index $i$ with $1 \leqslant i \leqslant m(n)$ such that $\|\Psi - \Psi_i\| < \frac{\varepsilon_n}{3}$, we obtain

$$\|\alpha(\pi(x_{n+1}) - \Phi_0) + \Psi\| \geqslant |[\alpha(\pi(x_{n+1}) - \Phi_0) + \Psi](h_i)| \geqslant$$
$$\geqslant |\Psi_i(h_i)| - |\alpha(\pi(x_{n+1}) - \Phi_0)(h_i)| - \|h_i\| \, \|\Psi - \Psi_i\| \geqslant$$
$$\geqslant 1 - \frac{\varepsilon_n}{3} - \frac{2}{d}\frac{d\varepsilon_n}{6} - \frac{\varepsilon_n}{3} = 1 - \varepsilon_n = (1 - \varepsilon_n)\|\Psi\|.$$

Thus, we have (1.12) for every $\Psi \in E_n$ with $\|\Psi\| = 1$, whence also for every $\Psi \in E_n$, which proves our claim.

Now, putting in (1.12) $\alpha = \alpha_{n+1}$ and $\Psi = \sum_{i=1}^{n} \alpha_i(\pi(x_i) - \Phi_0)$ and combining this with (1.9) it follows that (1.9) also holds for all $1 \leqslant p < < q \leqslant n + 1$. Thus we obtain, by induction, an infinite sequence $\{x_n\} \subset M$ satisfying (1.8) and (1.9) for all $n = 1,2, \ldots$ Obviously, the conditions (1.8) for $n = 1,2, \ldots$ and (1.6) imply (a). Furthermore, (1.9) for $n = = 1,2, \ldots$ and (1.5) imply, by Ch. I, § 7, theorem 7.1, that $\{\pi(x_n) - \Phi_0\}$ is a basic sequence in $E^{**}$. Finally, if $\Phi_0 \neq 0$, then there exists an index $k_0$ such that $\Phi_0 \notin [\pi(x_n) - \Phi_0]_{n=k_0}^{\infty}$ (because $\{\pi(x_n) - \Phi_0\}$ is a basic sequence, whence $\bigcap_{k=1}^{\infty} [\pi(x_n) - \Phi_0]_{n=k}^{\infty} = \{0\}$). Consequently, the sequence $\{x_n\}_{n=k_0}^{\infty} = \{x_{k_0+n-1}\}_{n=1}^{\infty}$ satisfies (a)—(c), which completes the proof of proposition 1.3.

**Theorem 1.3.** *Let $E$ be a Banach space and $\{x_n\}$ a sequence in $E$ with $0 < \inf_{1 \leqslant n < \infty} \|x_n\| \leqslant \sup_{1 \leqslant n < \infty} \|x_n\| < \infty$. In order that $\{x_n\}$ contain a subsequence which is a basic sequence, it is necessary and sufficient that one of the following two conditions be satisfied:*

(i) *$\{x_n\}$ is not conditionally weakly compact**);

(ii) *0 is a weak limit point of $\{x_n\}$.*

*Moreover, we have* (i) *if and only if one can select from $\{x_n\}$ a basic subsequence of type $l_+$.*

*Proof. Necessity.* Let $\{x_{n_k}\}$ be a basic subsequence of $\{x_n\}$. If (i) is not satisfied, i.e., if $\{x_n\}$, whence also $\{x_{n_k}\}$, is conditionally weakly compact, let $x_0$ be a weak limit point**) of $\{x_{n_k}\}$. Then, by Ch. II, § 7, proposition 7.2, $x_0 = 0$ and thus we have (ii).

---

*) I.e., the weak closure of $\{x_n\}$ is not weakly compact.

**) Such a point $x_0$ exists by the Eberlein-Šmulian theorem (see e.g. [87], p. 430, theorem 1).

*Sufficiency*. Assume that we have (i). Since $\{\pi(x_n)\} \subset E^{**}$ is bounded, whence conditionally $\sigma(E^{**}, E^*)$-compact, it has a $\sigma(E^{**}, E^*)$-limit point and hence, by (i), also a $\sigma(E^{**}, E^*)$-limit point $\Phi_0 \in E^{**} \setminus \pi(E)$. Since dist $(\Phi_0, \{\pi(x_n)\}) \geqslant$ dist $(\Phi_0, \pi(E)) > 0$, there exist a functional $f_0 \in E^*$ and a subsequence $\{x_{n_k}\}$ of $\{x_n\}$ satisfying (a) – (c) of proposition 1.3 with $n_k$ instead of $n$. Let $Z = [\Phi_0, \pi(x_{n_k})]$. Since $\Phi_0 \notin [\pi(x_{n_k})]$ and $\Phi_0 \notin [\pi(x_{n_k}) - \Phi_0]$, we have

$$\operatorname{codim}_Z [\pi(x_{n_k})] = \operatorname{codim}_Z [\pi(x_{n_k}) - \Phi_0] = 1, \tag{1.13}$$

and hence there exist projections $u, v$ of $Z$ onto $[\pi(x_{n_k})]$ and $[\pi(x_{n_k}) - \Phi_0]$ respectively, such that $u(\Phi_0) = v(\Phi_0) = 0$. Obviously $\Phi - u(\Phi) = \lambda \Phi_0$ for some scalar $\lambda$ depending on $\Phi$ and for $\Phi \in Z$. Thus, if $\Phi \in [\pi(x_{n_k}) - \Phi_0]$, then $\Phi = v(\Phi) = v(u(\Phi))$. By symmetry, if $\pi(x) \in [\pi(x_{n_k})]$, then $u(v(\pi(x))) = \pi(x)$. Consequently, $u$ maps isomorphically $[\pi(x_{n_k}) - \Phi_0]$ onto $[\pi(x_{n_k})]$. Since

$$u(\pi(x_{n_k}) - \Phi_0) = \pi(x_{n_k}) \qquad (k = 1,2, \ldots) \tag{1.14}$$

and since $\{\pi(x_{n_k}) - \Phi_0\}$ is a basic sequence, it follows that $\{\pi(x_{n_k})\}$, whence also $\{x_{n_k}\}$, is a basic sequence. By (a) we have (omitting, if necessary, a finite number of terms of the sequence $\{x_{n_k}\}$) $\inf_{1 \leqslant k < \infty} f_0(x_{n_k}) > 0$, so $\{x_{n_k}\}$ is of type $l_+$, by Ch. II, § 10, theorem 10.1 (implication $5° \Rightarrow 1°$). Conversely, if $\{x_n\}$ has a basic subsequence $\{x_{n_k}\}$ of type $l_+$, hence non-$wc_0$ (by Ch. II, § 12, theorem 12.1), then by Ch. II, § 7, theorem 7.2 $\{x_{n_k}\}$, whence also $\{x_n\}$, is not conditionally weakly compact, so (i) holds.

Let us consider now the case when we have (ii). We may assume, without loss of generality, that $E$ is separable and hence that there exists on $E$ a countable total set of functionals $V = \{f_n\} \subset E^*$. Then by (ii) there exists a subsequence $\{x_{n_k}\}$ of $\{x_n\}$ such that

$$|f_j(x_{n_k})| < \frac{1}{k} \qquad (j = 1, \ldots, k; k = 1,2, \ldots), \tag{1.15}$$

whence $x_{n_k} \xrightarrow{V} 0$.*) Thus, in order to complete the proof it will be sufficient to prove

**Proposition 1.4.** *Let $E$ be a Banach space, $V$ a total subset of $E^*$ and $\{x_n\}$ a sequence in $E$ with $0 < \inf_{1 \leqslant n < \infty} \|x_n\| \leqslant \sup_{1 \leqslant n < \infty} \|x_n\| < \infty$, such that $x_n \xrightarrow{V} 0$. Then $\{x_n\}$ contains a basic subsequence.*

*Proof.* If $\{x_n\}$ is not conditionally weakly compact, then, by case (i) of theorem 1.3, proved above, $\{x_n\}$ has a basic subsequence. If $\{x_n\}$

*) I. e. (see Ch. I, § 13), $f(x_{n_k}) \to 0$ $(f \in V)$.

is conditionally weakly compact, then, by the theorem of Eberlein*), $\{x_n\}$ contains a subsequence $\{x_{n_k}\}$ converging weakly to an element $x \in E$. However, by $x_n \xrightarrow{V} 0$ and since $V$ is total on $E$, we must have $x = 0$. Consequently, $x_n \xrightarrow{w} 0$, but $\inf_{1 \leq n < \infty} \|x_n\| > 0$, whence, by corollary 1.1, $\{x_n\}$ contains a basic subsequence, which completes the proof of proposition 1.4 and of theorem 1.3.

The above arguments also show that if $0 < \inf_{1 \leq n < \infty} \|x_n\| \leq \sup_{1 \leq n < \infty} \|x_n\| < \infty$ and if $V$ is a total subset of $E^*$, then in order that $\{x_n\}$ contain a subsequence which is a basic sequence, it is *sufficient* that one of the following two conditions be satisfied:

($\mathrm{i}_V$) *$\{x_n\}$ is not conditionally $\sigma(E, V)$-compact;*

($\mathrm{ii}_V$) *0 is a $\sigma(E, V)$-limit point of $\{x_n\}$.*

However, this is no longer *necessary* in order that $\{x_n\}$ contain a subsequence which is a basic sequence, even if we assume that $V$ is a subspace of $E^*$ with**) $r(V) > 0$. Indeed, in Ch. II, § 8, example 8.2 we have seen that if $B = c_0$, then there exists a basic sequence $\{f_{2n-1}\} \subset B^*$ with $f_{2n-1} \xrightarrow{w^*} f_1$; thus, one can take ***) $E = B^* \equiv l^1$ and $V =$ the canonical image of $B$ in $E^* = B^{**} \equiv m$.

**Corollary 1.3.** *Let $E$ be a Banach space and $(x_n, f_n)$ ($\{x_n\} \subset E$, $\{f_n\} \subset E^*$) a biorthogonal system such that $\{f_n\}$ is total on $E$. Then there exists a (finite or infinite) partition of $\{x_n\}$ into subsequences which are basic sequences.*

*Proof.* We may assume, without loss of generality (considering, if necessary, the biorthogonal system $\left(\frac{x_n}{\|x_n\|}, \|x_n\| f_n\right)$ instead of $(x_n, f_n)$) that $\|x_n\| = 1$ $(n = 1,2, \ldots)$. Then for $V = \{f_n\}$ the sequence $\{x_n\}$ satisfies the conditions of proposition 1.4 and hence it contains a basic subsequence, say $\{x_{n_k^{(1)}}\}$. We may assume that $x_1 \in \{x_{n_k^{(1)}}\}$ (since $\{x_1, x_{n_k^{(1)}}\}$ is still a basic sequence, by $f_i(x_j) = \delta_{ij}$). Let $\{l_k^{(1)}\} = \mathcal{N} \setminus \{n_k^{(1)}\}$, where $\mathcal{N} = \{1,2,3, \ldots\}$. If $\{l_k^{(1)}\}$ is finite, then, obviously, $\{x_n\}$ itself is a basic sequence. If $\{l_k^{(1)}\}$ is infinite, then the space $E_1 = [x_{l_k^{(1)}}]$ and the biorthogonal system $(x_{l_k^{(1)}}, f_{l_k^{(1)}} | E_1)$ satisfy the hypotheses of corollary 1.3. Indeed, if $x \in E_1$ and $f_{l_k^{(1)}}(x) = 0$ $(k = 1,2, \ldots)$, then, by $x \in E_1$, we also have $f_{n_k^{(1)}}(x) = 0$ $(k = 1,2, \ldots)$ and thus $f_n(x) = 0$ $(n = 1,2, \ldots)$, whence $x = 0$. Consequently, as

*) See e.g. [87], p. 430, theorem 1.

**) For the definition of the characteristic $r(V)$ see Ch. I, § 12.

***) We recall (see Ch. I, § 12) that we write $E_1 \equiv E_2$ if $E_1$ and $E_2$ are linearly isometric.

above, $\{x_{l_k^{(1)}}\}$ contains a basic subsequence, say $\{x_{n_k^{(2)}}\}$. We may assume that $x_2 \in \{x_{n_k^{(1)}}\} \cup \{x_{n_k^{(2)}}\}$ (since $\{x_2, x_{n_k^{(2)}}\}$ is still a basic sequence). Letting $\{l_k^{(2)}\} = \mathcal{N} \setminus (\{n_k^{(1)}\} \cup \{n_k^{(2)}\})$ and continuing in this way, we obtain a partition of $\{x_n\}$ into basic subsequences, which completes the proof.

**Corollary 1.4.** *Let $E$ be a Banach space and let $\{x_n\} \subset E$, $x_n \neq 0$ $(n = 1,2, \ldots)$. The following statements are equivalent:*

1°. *$\{x_n\}$ has a subsequence which is a basic sequence.*

2°. *There exist a total set $V \subset E^*$ and a sequence of indices $\{n_k\}$ such that $\dfrac{x_{n_k}}{\|x_{n_k}\|} \xrightarrow{V} 0$.*

3°. *There exist a linear subspace $V$ of $E^*$ with $r(V) > 0$ and a sequence of indices $\{n_k\}$ such that $\alpha_k x_{n_k} \xrightarrow{V} 0$ for every sequence of scalars $\{\alpha_k\}$.*

*Proof.* Assume that we have 1°, say $\{y_n\}$ is a subsequence of $\{x_n\}$ such that $\{y_n\}$ is a basis of $Y = [y_n] \subset E$. Let $V_0 \subset Y^*$ be the linear subspace of $Y^*$ spanned by the a.s.c.f. $\{\varphi_n\}$ to $\{y_n\}$ and let

$$V = \{f \in E^* \mid f|_Y \in V_0\}. \tag{1.16}$$

Then $V$ is a linear subspace of $E^*$. We shall prove that $r(V) > 0$. By virtue of Ch. I, § 12, theorem 12.2, we have $r(V_0) > 0$. Let $x \in E$ be arbitrary. We shall consider two cases:

a) There exists an element $y \in Y$ such that $\|x - y\| < \dfrac{r(V_0)}{4} \|x\|$.

Then let $\varphi_0 \in V_0$ with $\|\varphi_0\| \leqslant 1$ be such that $|\varphi_0(y)| > \dfrac{3}{4} r(V_0) \|y\|$;

such a $\varphi_0 \in V_0$ exists, by $\|y\| \leqslant \dfrac{1}{r(V_0)} \sup\limits_{\substack{\varphi \in V_0 \\ \|\varphi\| \leqslant 1}} |\varphi(y)|$ (see Ch. I, § 12, formula (12.13)). By the Hahn-Banach theorem, let $f \in V$ be an extension of $\varphi_0$ such that $\|f\| = \|\varphi_0\| \leqslant 1$. Then

$$|f(x)| \geqslant |f(y)| - |f(x - y)| > \frac{3}{4} r(V_0) \|y\| - \frac{1}{4} r(V_0) \|x\|,$$

which, together with $\|y\| \geqslant \|x\| - \|x - y\| > \|x\| - \dfrac{r(V_0)}{4} \|x\| \geqslant \dfrac{3}{4} \|x\|$,

implies

$$|f(x)| > \left(\frac{3}{4}\right)^2 r(V_0) \|x\| - \frac{1}{4} r(V_0) \|x\| > \frac{1}{4} r(V_0) \|x\|. \tag{1.17}$$

b) $\inf_{y \in Y} \|x - y\| \geqslant \frac{1}{4} r(V_0) \|x\|$. Then, by a well-known corollary of the Hahn-Banach theorem, there exists an $f \in Y^\perp \subset V$ such that $f(x) = 1$, $\|f\| \leqslant \frac{1}{\operatorname{dist}(x, Y)} \leqslant \frac{4}{r(V_0)} \frac{1}{\|x\|}$, whence, for $g = \frac{f}{\|f\|} \in V$ we have $\|g\| = 1$ and

$$|g(x)| = \frac{|f(x)|}{\|f\|} \geqslant \frac{r(V_0)}{4} \|x\|. \tag{1.18}$$

From (1.17) and (1.18) we infer that in both cases

$$r(V) \geqslant \frac{r(V_0)}{4} > 0. \tag{1.19}$$

Now let $f \in V$ be arbitrary. Then by (1.16) we have $f|_Y = \sum_{i=1}^{m} \beta_i \varphi_i$ with suitable scalars $\beta_1, \ldots, \beta_m$, whence, by biorthogonality, for any sequence of scalars $\{\alpha_n\}$ we obtain $f(\alpha_n y_n) = 0 \ (n \geqslant m + 1)$, so $\alpha_n y_n \xrightarrow{V} 0$. Thus $1° \Rightarrow 3°$.

The implication $3° \Rightarrow 2°$ is obvious.

Finally, if we have 2°, then, by proposition 1.4, $\left\{\frac{x_n}{\|x_n\|}\right\}$ has a basic subsequence $\left\{\frac{x_{n_k}}{\|x_{n_k}\|}\right\}$, whence $\{x_{n_k}\}$ is a basic subsequence of $\{x_n\}$. Thus $2° \Rightarrow 1°$, which completes the proof of corollary 1.4.

**Corollary 1.5.** *Let $A$ be a bounded set in a Banach space $E$, such that $A$ is not conditionally compact in the norm topology. Then there exists an infinite subset $B$ of $A$ such that the space*) $[B]$ has a basis.*

*Proof.* Since $A$ is not conditionally compact, there exists a sequence $\{x_n\} \subset A$ having no convergent subsequence; hence $\inf_{1 \leqslant n < \infty} \|x_n\| > 0$. If $\{x_n\}$ is not conditionally weakly compact, then, by theorem 1.3 (i), $\{x_n\}$ contains a basic subsequence $B = \{x_{n_k}\}$. If $\{x_n\}$ is conditionally weakly compact, then, by Eberlein's theorem, $\{x_n\}$ contains a subsequence $\{x_{n_k}\}$ converging weakly to an element $x_0 \in E$. If $x_0 = 0$, then by theorem 1.3 (ii), $\{x_{n_k}\}$ contains a basic subsequence $B = \{x_{n_{k_m}}\}$. If $x_0 \neq 0$, then, again by theorem 1.3 (ii), $\{x_0 - x_{n_k}\}$ contains a basic subsequence $\{x_0 - x_{n_{k_m}}\} = \{x_0 - y_m\}$. Then, as in the above proof of proposition 1.3 (c), we may assume that $x_0 \notin [x_0 - y_m]$ and hence

*) We recall that $[B]$ denotes the closed linear subspace of $E$ spanned by the set $B$.

$\{x_0, x_0 - y_m\}$ is a basic sequence. Let $B = \{y_m\}$. Then $B$ is an infinite subset of $A$. Furthermore, since $y_m \xrightarrow{w} x_0$, we have $x_0 \in [y_m]$, whence $[y_m] = [x_0, x_0 - y_m]$. Thus $[B] = [y_m]$ has a basis (namely, $\{x_0, x_0 - - y_m\}$), which completes the proof of corollary 1.5.

We shall show (in example 1.1 below) that in general for a bounded set $A \subset E$ some additional assumption is indeed necessary in order that the conclusion of corollary 1.5 be valid. For this purpose, we need

*Definition 1.1.* Let $E$ be a Banach space. A sequence $\{x_n\} \subset E$ is said to be *hypercomplete* in $E$ if every infinite subsequence $\{x_{n_k}\}$ of $\{x_n\}$ is complete in $E$.

For instance, if $\alpha_n > 0$ $(n = 1,2, \ldots)$ and $\lim\limits_{n\to\infty} \alpha_n = \alpha$, $0 < \alpha < \infty$, then by the theorem of Müntz*) the sequence $\{\tilde{x}_n\}$, where

$$x_n(t) = t^{\alpha_n} \qquad (t \in [0, 1],\ n = 1,2, \ldots) \tag{1.20}$$

is hypercomplete in the space $E = L^1([0, 1])$.

If $E = l^1$, then the sequence $x_n = \{\xi_j^{(n)}\}_{j=1}^{\infty}$ $(n = 1,2, \ldots)$, where

$$\xi_j^{(n)} = \mathrm{e}^{-\left(1+\frac{1}{n}\right)j} \qquad (j, n = 1,2, \ldots), \tag{1.21}$$

is hypercomplete in $E$. Indeed, assume that some subsequence $\{x_{n_k}\}$ of $\{x_n\}$ is not complete in $E = l^1$. Then there exists a sequence $\{\beta_j\} \in \in m \setminus \{0\}$ such that

$$\sum_{j=1}^{\infty} \beta_j \, \mathrm{e}^{-\left(1+\frac{1}{n_k}\right)j} = 0 \qquad (k = 1,2, \ldots). \tag{1.22}$$

However, the function $\Phi(\zeta) = \sum\limits_{j=1}^{\infty} \beta_j \mathrm{e}^{-j\zeta}$ is analytic in the right half-plane and thus it cannot have a sequence of zeros $\zeta_k = 1 + \dfrac{1}{n_k}$ $(k = 1,2, \ldots)$ converging to $z = 1$, and this contradiction to (1.22) completes the proof.

Clearly, if a Banach space $E$ has a hypercomplete sequence, then it is separable. The converse also holds, namely, we have

**Proposition 1.5.** *In every separable Banach space E there exist hypercomplete sequences.*

*Proof.* Let $\{x_n\}$ be a hypercomplete sequence in $L^1([0, 1])$ or $l^1$ (e.g. (1.20), respectively (1.21)) and let $u$ be a continuous linear mapping of

*) See e.g. [194], Ch. III, § 6; in [194] instead of "complete" and "hypercomplete" the terms "closed" and "densely-closed" are used.

$L^1([0, 1])$ or $l^1$ onto*) $E$. Then the sequence $\{u(x_n)\}$ is clearly hypercomplete in $E$, which completes the proof.

*Remark 1.3*. One can also give the following more direct proof of proposition 1.5: Let $\{x_j\}$ be a complete sequence in $E$, with $\sup_{1\leqslant j<\infty} \|x_j\| < \infty$. Then the sequence $\{y_n\} \subset E$ defined by

$$y_n = \sum_{j=1}^{n} \frac{1}{j^n} x_j \qquad (n = 1,2, \ldots) \tag{1.23}$$

is hypercomplete in $E$. Indeed, let $\{y_{n_k}\}$ be an arbitrary subsequence of $\{y_n\}$ and let $f \in E^*$ be such that

$$f(y_{n_k}) = \sum_{j=1}^{n_k} \frac{1}{j^{n_k}} f(x_j) = 0 \qquad (k = 1,2, \ldots). \tag{1.24}$$

Then, putting $C = \|f\| \sup_{1\leqslant j<\infty} \|x_j\|$, we obtain from (1.24)

$$|f(x_1)| \leqslant C \sum_{j=2}^{n_k} \frac{1}{j^{n_k}} \leqslant C \frac{n_k - 1}{2^{n_k}} \qquad (k = 1,2, \ldots),$$

whence $f(x_1) = 0$. Therefore, again by (1.24),

$$|f(x_2)| \leqslant C \sum_{j=3}^{n_k} \frac{2^{n_k}}{j^{n_k}} \leqslant C \frac{n_k - 2}{\left(\frac{3}{2}\right)^{n_k}} \qquad (k = 1,2, \ldots),$$

whence $f(x_2) = 0$. Continuing in this way, we obtain $f(x_n) = 0$ ($n = 1,2, \ldots$) whence, since $[x_n] = E$, it follows that $f = 0$. Consequently, $[y_{n_k}] = E$, which completes the proof.

Now we can give

*Example 1.1*. Let $E$ be a separable Banach space which has no basis (such a space $E$ exists, by § 0, theorem 0.1 or 0.2). Then $E$ contains an infinite bounded set $A$ such that for each infinite subset $B \subset A$ the space $[B]$ has no basis. Indeed, by proposition 1.5 there exists a normalized hypercomplete sequence $\{x_n\}$ in $E$, so one can take $A = \{x_n\}$. Observe that by corollary 1.5 this $A = \{x_n\}$ is necessarily conditionally compact; one could also take $A_1 = \left\{\frac{1}{n} x_n\right\}$, which converges to 0.

*) See e.g. [12], theorem e) or [223], p. 283, theorem 1.

Note also that the converse of the statement of example 1.1 is not valid, since every separable Banach space $E$ can be embedded isomorphically into $E_1 = C([0,1])$, which has a basis.

The special classes of bases, introduced in Ch. II, imply the introduction of as many special classes of basic sequences. Namely, it is natural to give

*Definition 1.2.* A basic sequence $\{x_n\}$ in a Banach space $E$ is said to be *monotone* (*normal*, etc.) if $\{x_n\}$ is a monotone (respectively normal, etc.) basis of the subspace $[x_n]$.

In Ch. IV we shall see that the existence of basic sequences belonging to certain special classes (similarly to the existence, in Banach spaces with bases, of bases belonging to certain special classes) characterizes certain structure properties of Banach spaces (e.g., reflexivity). However, now we shall show that for certain special classes of basic sequences one may consider the problem of their existence in every (infinite dimensional) Banach space $E$; hence we do not assume that $E$ is separable, nor that $E$ has a basis.

We have seen in Ch. II, § 1, theorem 1.4 and corollary 1.1, that there exist (infinite dimensional) Banach spaces $E$ with bases, having no monotone basis. However, the following problem is unsolved:

*Problem 1.1.* Does there exist in every Banach space $E$ a monotone basic sequence? What about a strictly monotone basic sequence?

Theorem 1.2 may be considered as stating the existence, in every Banach space $E$, of a basic sequence which is "almost monotone", that is, within any $\varepsilon>0$. Remark 1.1 establishes the existence of "asymptotically monotone" basic sequences, that is, "within $\{\varepsilon_n\}$ with $\varepsilon_n > 0$, $\lim_{n\to\infty} \varepsilon_n = 0$", and hence such that $\lim_{n\to\infty} \|s_n\| = 1$.

A possible candidate to yield a negative answer to problem 1.1 could be the subspace $E_0 = [x_{n_k}] = [t^{n_k}]$ of $C([0,1])$, constructed in Ch. II, § 1, corollary 1.1 (in spite of the fact that by the proofs of Ch. II, § 1, corollary 1.1 and Ch. II, § 1, proposition 1.2 it follows that for any $\varepsilon > 0$ the basis $\{x_{n_k}\}$ of $E_0$ contains a basic subsequence of index $\geqslant 1 - \varepsilon$). In fact, from Ch. II, § 1, theorem 1.4 it follows that if a subspace $G$ of $E_0$, with $\dim G = \infty$, contains an element of the positive cone associated to the basis $\{x_{n_k}\}$ of $E_0$, then $G$ has no monotone basis, and that for every subspace $G$ of $E_0$ with $\dim G = \infty$ the subspace $G_1 = [G \cup \cup \{x_{n_1}\}]$ of $E_0$ has no monotone basis.

**Proposition 1.6.** *Let $\{x_n\}$ be a basic sequence in a Banach space $E$. Then there exists an equivalent norm on $E$, in which the basic sequence $\{x_n\}$ is monotone.*

*Proof.* As was observed in Ch. II, § 1, after proposition 1.3, there exists an equivalent norm $|||x|||$ on the subspace $[x_n]$, in which $\{x_n\}$ is a monotone basis of $[x_n]$. Hence, by Ch. II, § 13, lemma 13.6 (applied to $F = [x_n]$ and $F_1 = [x_n]$ endowed with the norm $|||x|||$), there exists an equivalent norm on $E$, extending the norm $|||x|||$, which completes the proof.

Let us observe that similar extensions are valid, by the same argument, for the other results of Vol. I on the existence of certain equivalent norms on Banach spaces with bases (e.g. $T$-norms, etc.).

An affirmative answer to the problem of existence of non-monotone and non-strictly monotone basic sequences (as well as other classes of basic sequences) follows easily by using the results of Ch. II and

*Remark 1.4.* If every Banach space with a basis has also a basis belonging to a certain special class, then every Banach space has a basic sequence belonging to that special class.

Indeed, by theorem 1.1 every Banach space contains a subspace with a basis and by our assumption this subspace has also a basis belonging to that special class.

In order to prove the existence of normal basic sequences in every Banach space we shall need the following result of M. G. Kreĭn, M. A. Krasnoselskiĭ and D. P. Milman [225]:

**Lemma 1.2.** *Let $E$ be a normed linear space and $G_1$, $G_2$ two linear subspaces of $E$ such that*

$$\dim G_1 < \infty, \dim G_1 < \dim G_2. \tag{1.25}$$

*Then there exists an element $y \in G_2$ such that*

$$\|y\| = 1, \operatorname{dist}(y, G_1) = 1. \tag{1.26}$$

For a proof we refer the reader e.g. to the monograph [371], Ch. II, § 6, proof of lemma 6.1, or to the papers [225], [130].

**Theorem 1.4.** *Let $E$ be a Banach space. Then for every $\varepsilon$ with $0 < \varepsilon < 1$ there exists in $E$ a normal basic sequence $\{x_n\}$ of index $\gamma_{\{x_n\}} \geqslant 1 - \varepsilon$.*

*Proof.* It will be sufficient to construct sequences $\{x_n\} \subset E$, $\{f_n\} \subset E^*$ and a sequence of subspaces $\{Y_n\}$ of $E$ with codim $Y_n < \infty$ $(n = 1,2, \dots)$ such that*) (a) $Y_1 \supset Y_2 \supset \dots$; (b) $\|x_n\| = \|f_n\| = f_n(x_n) = 1$, $x_{n+1} \in \in ([f_i]_{i=1}^n)_\perp \cap Y_n$, $f_{n+1} \in P_{(n)}^\perp$ $(n = 1,2, \dots)$; (c) $(\widehat{P_{(n)}; Y_n}) > 1 - \varepsilon$, where $P_{(n)} = [x_i]_{i=1}^n$ $(n = 1,2, \dots)$. Indeed, then we obtain, as in the proof of theorem 1.2, that $\{x_n\}$ is a basic sequence with $\gamma_{\{x_n\}} \geqslant 1 - \varepsilon$, and by (b) this basic sequence is normal.

Take $x_1 \in E$ and $f_1 \in E^*$ such that $\|x_1\| = \|f_1\| = 1$. Then, by lemma 1.1 for $G = P_{(1)} = [x_1]$, there exists a subspace $Y_1 \subset E$ with codim $Y_1 < < \infty$, such that $(\widehat{P_1; Y_1}) > 1 - \varepsilon$. Assume now that for some $n \geqslant 1$ we have constructed $\{x_i\}_{i=1}^n \subset E$, $\{f_i\}_{i=1}^n \subset E^*$ and $\{Y_i\}_{i=1}^n$. Then, by

*) We recall that $P_{(n)}^\perp = \{f \in E^* \mid f(x) = 0 \ (x \in P_{(n)})\}$.

lemma 1.2 for $G_1 = P_{(n)} = [x_i]_{i=1}^n$, $G_2 = ([f_i]_{i=1}^n)_\perp \cap Y_n$, there exists an element $x_{n+1} \in ([f_i]_{i=1}^n)_\perp \cap Y_n$ such that $\|x_{n+1}\| = 1$, dist $(x_{n+1}, P_{(n)}) = 1$. Hence, by a corollary of the Hahn-Banach theorem, there exists $f_{n+1} \in E^*$ such that $\|f_{n+1}\| = 1$, $f_{n+1} \in P_{(n)}^\perp$ and $f_{n+1}(x_{n+1}) = 1$. Finally, by lemma 1.1 for $G = P_{(n+1)}$, there exists a subspace $Y_{n+1} \subset E$ with codim $Y_{n+1} < \infty$, such that $\widehat{(P_{(n+1)}; Y_{n+1})} > 1 - \varepsilon$, which completes the proof.

*Remark 1.5.* Similarly to remark 1.1, one can also prove that *in every Banach space E there exists a normal basic sequence* $\{x_n\}$ *with* $\|s_n\| \leqslant \leqslant 1 + \frac{1}{n}$ $(n = 1,2, \ldots)$, *where* $\{s_n\}$ *is the sequence of partial sum operators associated to the basis* $\{x_n\}$ *of* $[x_n]$.

On the other hand, the existence of non-normal basic sequences follows from remark 1.4.

For every integer $k \geqslant 0$, the answer to the problems of existence of $k$-shrinking and $k$-boundedly complete basic sequences is negative, as shown by the spaces $E = l^1$ and $E = c_0$, respectively (see Ch. II, § 4, example 4.2. and § 6, example 6.2).

For $k = 0$, the answer to the problems of existence of non $k$-shrinking and non $k$-boundedly complete basic sequences is also negative, as shown by the example of an arbitrary reflexive Banach space (see Ch. II, § 4, example 4.3 and § 6, example 6.3). On the other hand, for $k > 0$ the answer to these problems is affirmative, i.e., in every Banach space there exist non $k$-shrinking and non $k$-boundedly complete basic sequences. Indeed, this follows from remark 1.4 and from the fact that in every Banach space with a basis there exist, for any integer $k > 0$, a non $k$-shrinking basis and a non $k$-boundedly complete basis (see Vol. III, Ch. IV).

Leaving to the reader to consider the problems of existence of the special classes of basic sequences corresponding to the other classes of bases introduced in Ch. II, Part I, let us pass now to those introduced in Ch. II, Part II.

*Problem 1.2.* Does there exist in every Banach space $E$ an unconditional basic sequence?

An affirmative answer to problem 1.2 would simplify very much the proofs of some well-known results on Banach spaces. For example, in this case the Dvoretzky-Rogers theorem*), according to which in every (infinite dimensional) Banach space $E$ there exists an unconditionally convergent series $\sum_{i=1}^{\infty} y_i$ such that $\sum_{i=1}^{\infty} \|y_i\| = \infty$, could be proved as follows: Let $\{x_n\}$ be an unconditional basic sequence in $E$, with $\|x_n\| = 1$ $(n = 1,2, \ldots)$. Then, if $\{x_n\}$ is not equivalent to

*) See e.g. [70], Ch. IV, § 1, theorem 2.

the unit vector basis of $l^1$, there exists a convergent (hence unconditionally convergent) expansion $\sum_{i=1}^{\infty}\alpha_i x_i$ such that $\sum_{i=1}^{\infty}\|\alpha_i x_i\| = \sum_{i=1}^{\infty} |\alpha_i| = \infty$; on the other hand, if $\{x_n\}$ is equivalent to the unit vector basis of $l^1$, then $E$ is isomorphic to $l^1$ and hence there exists in $E$ an unconditionally convergent series $\sum_{i=1}^{\infty} y_i$ with $\sum_{i=1}^{\infty} \|y_i\| = \infty$ (e.g., by Ch. II, § 8, example 8.1, there exists in $E \cong l^1$ a normalized unconditional basic sequence $\{z_n\}$ which is not equivalent to the unit vector basis of $l^1$ and hence also an unconditionally convergent expansion $\sum_{i=1}^{\infty} \alpha_i z_i$ with $\sum_{i=1}^{\infty} \|\alpha_i z_i\| = \sum_{i=1}^{\infty} |\alpha_i| = \infty$). For another example, let us mention that C. Bessaga and A. Pelczynski [25] have proved the following result: Every separable Banach space which contains an unconditional basic sequence is homeomorphic to the space $l^2$. Consequently, an affirmative answer to problem 1.2, combined with this result of Bessaga and Pelczynski, would imply the theorem of M. I. Kadec [201], [202]*) according to which all separable infinite dimensional Banach spaces are homeomorphic to each other. Also, we shall see in Vol. III, Ch. IV, that an affirmative answer to problem 1.2 would imply an affirmative answer to some unsolved problems.

Since by Ch. II, § 17, corollary 17.2, every block basic sequence with respect to an unconditional basis is an unconditional basic sequence, from proposition 1.1 and corollaries 1.1 and 1.2 we obtain the following results related to problem 1.2**):

**Proposition 1.7.** *Let $E$ be a subspace of a Banach space $X$ with an unconditional basis $\{x_n\}$ and let $\{y_n\}$ be a sequence in $E$, satisfying* (1.1) *and* (1.2) (*or, in particular,* (1.1) *and* $y_n \xrightarrow{w} 0$). *Then $\{y_n\}$ has a subsequence $\{y_{p_n}\}$ which is an unconditional basic sequence, equivalent to a block basic sequence with respect to $\{x_n\}$.*

**Corollary 1.6.** *Every subspace of a Banach space $X$ with an unconditional basis $\{x_n\}$ contains an unconditional basic sequence, which is equivalent to a block basic sequence with respect to $\{x_n\}$.*

The answer to the problem of existence of conditional basic sequences is affirmative. Indeed, by remark 1.4 and Ch. II, § 23, theorem 23.2, *in every Banach space $E$ there exists a conditional basic sequence.*

---

*) See also [26] and the references therein.

**) For some recent results related to problem 1.2, see also the Notes and remarks.

*Problem 1.3.* Does there exist in every Banach space $E$ an orthogonal basic sequence? What about a strictly orthogonal, hyperorthogonal or strictly hyperorthogonal basic sequence?

The answer to the similar problem for subsymmetric (and hence also for symmetric and for perfectly homogeneous) basic sequences is negative*).

Similarly to bases (see Ch. I, §§ 13, 14), the notion of a basic sequence has a natural extension to a general topological linear space $U$. In particular, when $U$ is a conjugate Banach space $E^*$ endowed with the weak* topology $\sigma(E^*, E)$ and when the coefficient functionals are continuous, we arrive at the following notion:

*Definition 1.3.* A sequence $\{f_n\}$ in a conjugate Banach space $E^*$ is called a *w*-Schauder basic sequence* if it is a Schauder basis of $\widetilde{[f_n]}$ for the weak* topology $\sigma(E^*, E)$, where $\widetilde{[f_n]}$ denotes the $\sigma(E^*, E)$-closed linear subspace of $E^*$ spanned by $\{f_n\}$ (this coincides with the $\sigma(E^*, E)$-closure of $[f_n]$; in general, for $A \subset E^*$ we shall denote by $\widetilde{A}$ the $\sigma(E^*, E)$-closure of $A$). In other words, $\{f_n\}$ is a $w^*$-Schauder basic sequence in $E^*$ if there exists a sequence $\{x_n\} \subset E$ such that $(x_n, f_n)$ is a biorthogonal system and that

$$\mathscr{S}_n(f) = \sum_{i=1}^{n} f(x_i) f_i \xrightarrow{w^*} f \qquad (f \in \widetilde{[f_k]}). \tag{1.27}$$

*Remark 1.6.* Although the sequence $\{x_n\}$ in definition 1.3 is not uniquely determined by $\{f_n\}$, if $\{z_n\} \subset E$ is such that $(z_n, f_n)$ is a biorthogonal system, then

$$f(x_n) = f(z_n) \qquad (f \in [f_k],\ n = 1,2, \ldots), \tag{1.28}$$

and hence $\mathscr{S}_n(f) = \sum_{i=1}^{n} f(x_i) f_i = \sum_{i=1}^{n} f(z_i) f_i$ $(f \in [f_k],\ n = 1,2, \ldots)$; indeed, by biorthogonality and (1.27) we have, for each $j = 1,2, \ldots$

$$f(x_j) = \sum_{i=1}^{n} f(x_i) f_i(z_j) = (\mathscr{S}_n(f))(z_j) \to f(z_j) \qquad (n \geqslant j,\ n \to \infty),$$

whence (1.28).

From Ch. I, § 14, theorem 14.1 it follows that *every subsequence $\{f_{k_n}\}$ of a $w^*$-Schauder basis $\{f_n\}$ of a conjugate Banach space $E^*$ is a $w^*$-Schauder basic sequence.* Indeed, if $\{x_n\}$ is a basis of $E$, with the a.s.c.f. $\{f_n\} \subset E^*$, then $(x_{k_n}, f_{k_n})$ is a biorthogonal system and for each $f \in \widetilde{[f_{k_n}]} = ([f_{k_n}]_\perp)^\perp = ([x_j]_{j \in \mathscr{N} \setminus \{k_n\}})^\perp$ we have $\sum_{i=1}^{n} f(x_{k_i}) f_{k_i} = \sum_{i=1}^{k_n} f(x_i) f_i \xrightarrow{w^*} f.$

*) See the Notes and remarks.

Some elementary characterizations of $w^*$-Schauder basic sequences and of some special classes of such sequences are collected in

**Proposition 1.8.** *Let $E$ be a Banach space and $(x_n, f_n)$ ($\{x_n\} \subset E$, $\{f_n\} \subset E^*$) a biorthogonal system and let $\omega$ denote the canonical mapping of $E$ onto $E/[f_n]_\perp$. Then*

a) *The following statements are equivalent:*

1°. *$\{f_n\}$ is a $w^*$-Schauder basic sequence.*

2°. *$\{(\omega^*)^{-1}(f_n)\}$ is a $w^*$-Schauder basis**) *of $(E/[f_n]_\perp)^*$.*

3°. *$\{\omega(x_n)\}$ is a basis of $E/[f_n]_\perp$.*

b) *In the situation of* a) *the a.s.c.f. $\{h_n\} \subset (E/[f_n]_\perp)^*$ to $\{\omega(x_n)\}$ satisfies*

$$\omega^*(h_n) = f_n \qquad (n = 1,2, \ldots). \tag{1.29}$$

*Consequently, every $w^*$-Schauder basic sequence $\{f_n\}$ is a basic sequence.*

c) *The following statements are equivalent:*

1°. *$\{f_n\}$ is a boundedly complete***) *$w^*$-Schauder basic sequence.*

2° *$\{f_n\}$ is a basic sequence satisfying*

$$\widetilde{[f_n]} = [f_n]. \tag{1.30}$$

3°. *$\{\omega(x_n)\}$ is a shrinking basis of $E/[f_n]_\perp$.*

d) *The following statements are equivalent:*

1°. *$\{f_n\}$ is a shrinking $w^*$-Schauder basic sequence.*

2°. *$\{\omega(x_n)\}$ is a boundedly complete basis of $E/[f_n]_\perp$.*

*Proof.* a) and b). The equivalence a) 1°$\Leftrightarrow$ 2° follows from the fact that $\omega^*$ is***) a linear isometry and a $w^*$-isomorphism of $(E/[f_n]_\perp)^*$ onto $([f_n]_\perp)^\perp = \widetilde{[f_n]} \subset E^*$ (Note that the implication a) 2° $\Rightarrow$ 1° holds for any $\{f_n\} \subset E^*$, so a) 2° implies the existence of $\{x_n\} \subset E$ with $(x_n, f_n)$ biorthogonal).

The equivalence a) 2° $\Leftrightarrow$ 3° and the first statement in b) follow from Ch. I, § 14, theorem 14.1 and from the relations

$$(\omega^*)^{-1}(f_i)\,(\omega(x_j)) = f_i(x_j) = \delta_{ij} \qquad (i, j = 1,2, \ldots); \tag{1.31}$$

furthermore, by (1.29) and since $\omega^*$ is a linear isometry, the second part of b) is a consequence of Ch. I, § 12, theorem 12.1.

---

*) See Ch. I, § 14, definition 14.2.

**) Throughout this monograph we shall use "boundedly complete" and "shrinking" only in the sense of the norm-topology.

***) See e.g. [70], Ch. II, § 1, lemma 1.

c) and d). The equivalences c) 1°$\Leftrightarrow$3° and d) 1° $\Leftrightarrow$ 2° follow from a), b) and the duality between shrinking and boundedly complete bases (Ch. II, § 6, corollary 6.1). Thus, it remains to prove the equivalence c) 1° $\Leftrightarrow$ 2°.

Assume that we have c) 1° and let $f \in \widetilde{[f_n]}$. Then by remark 1.6 and (1.27) we have $\sup_{1 \leqslant n < \infty} \left\| \sum_{i=1}^{n} f(x_i) f_i \right\| < \infty$, whence, since $\{f_n\}$ is boundedly complete, $\sum_{i=1}^{\infty} f(x_i) f_i$ is norm-convergent, obviously to $f$ (by (1.27)). Consequently, $f \in [f_n]$, which proves (1.30). Thus, c) 1° $\Rightarrow$ 2°.

Conversely, assume now that we have c) 2° and let $\{\alpha_n\} \subset K$ be such that $\sup_{1 \leqslant n < \infty} \left\| \sum_{i=1}^{n} \alpha_i f_i \right\| < \infty$. Then the sequence $\left\{ \sum_{i=1}^{n} \alpha_i f_i \right\}$ has a $w^*$-cluster point, say $f$, and, clearly, $f \in \widetilde{[f_n]}$. Now, since $(x_n, f_n)$ is a biorthogonal system,

$$\left( \sum_{i=1}^{n} \alpha_i f_i \right)(x_j) = \alpha_j \qquad (j = 1,2, \ldots; n \geqslant j),$$

whence

$$f(x_j) = \alpha_j \qquad (j = 1, 2, \ldots). \tag{1.32}$$

On the other hand, by $f \in \widetilde{[f_n]}$ and (1.30) we have $f \in [f_n]$. Consequently, since $\{f_n\}$ is a basic sequence, $\left\{ \sum_{i=1}^{n} \alpha_i f_i \right\} = \left\{ \sum_{i=1}^{n} f(x_i) f_i \right\}$ is norm-convergent to $f$, which proves that $\{f_n\}$ is boundedly complete. Furthermore, by (1.30) and since $\{f_n\}$ is basic, we have (1.27), so $\{f_n\}$ is $w^*$-Schauder basic. Thus, c) 2° $\Rightarrow$ 1°, completing the proof of proposition 1.8.

*Remark 1.7.* Note that if $(x_n, f_n)$ is a biorthogonal system and $\omega: E \to E/[f_n]_\perp$ the canonical mapping*⁾, *$\{\omega(x_n)\}$ is complete in $E/[f_n]_\perp$ if and only if $[x_n] + [f_n]_\perp$ is dense in $E$.* Indeed, if $[\omega(x_n)] = E/[f_n]_\perp$ and $x \in E$, $\varepsilon > 0$, then there exist scalars $\alpha_1, \ldots, \alpha_m$ such that $\left\| \omega\left( x - \sum_{i=1}^{m} \alpha_i x_i \right) \right\| = \left\| \omega(x) - \sum_{i=1}^{m} \alpha_i \omega(x_i) \right\| < \frac{\varepsilon}{2}$, whence also an element $y \in [f_n]_\perp$ with $\left\| x - \sum_{i=1}^{m} \alpha_i x_i - y \right\| \leqslant \left\| \omega\left( x - \sum_{i=1}^{m} \alpha_i x_i \right) \right\| + \frac{\varepsilon}{2} < \varepsilon$, so $[x_n] + [f_n]_\perp$ is dense in $E$; the converse implication is obvious. Hence, by the implication a) 1° $\Rightarrow$ 3° of proposition 1.8 it follows that *if $\{f_n\} \subset E^*$ is a*

*⁾ Hence, by (1.31), $\{\omega(x_n)\}$ is a minimal sequence in $E/[f_n]_\perp$.

*$w^*$-Schauder basic sequence in $E^*$ and if $\{x_n\} \subset E$ is such that $(x_n, f_n)$ is a biorthogonal system, then $[x_n] + [f_n]_\perp$ is dense in $E$.*

Some equivalence properties of $w^*$-Schauder basic sequences are given in

**Proposition 1.9.** a) *Every basis of a Banach space (and hence every basic sequence) is equivalent to a suitable $w^*$-Schauder basic sequence (or, what is equivalent*), to a suitable $w^*$-Schauder basis).*

b) *Let $\{f_n\}$ be a $w^*$-Schauder basic sequence in a conjugate Banach space $E^*$ and let $\{y_n\}$ be a basis of a Banach space $F$, with the a.s.c.f. $\{g_n\} \subset F^*$ such that $\{f_n\} \sim \{g_n\}$. Then $E/[f_n]_\perp$ is isomorphic to $F$ and $\{f_n\} \overset{w^*}{\sim} \{g_n\}$, that is, for any sequence $\{\alpha_n\}$ of scalars, $\sum\limits_{i=1}^{\infty} \alpha_i f_i$ is $w^*$-convergent if and only if $\sum\limits_{i=1}^{\infty} \alpha_i g_i$ is $w^*$-convergent.*

*Proof.* a) Let $\{y_n\}$ be a basis of a Banach space $F$, with the a.s.c.f. $\{g_n\} \subset F^*$ and let $u$ denote the canonical mapping of $F$ into $[g_n]^*$. Then, by Ch. I, § 12, theorem 12.2, $u$ is an isomorphism and hence, by Ch.I, § 8, theorem 8.1, $\{u(y_n)\} \subset [g_n]^*$ is a basic sequence equivalent to $\{y_n\}$. Moreover, by Ch. I, § 12, theorem 12.1, $\{g_n\}$ is a basis of $[g_n]$, with the a.s.c.f. $\{u(y_n)\}$ and hence, by Ch. I, § 14, theorem 14.1, $\{u(y_n)\}$ is a $w^*$-Schauder basis of $[g_n]^*$.

b) By proposition 1.8 a), b), $E/[f_n]_\perp$ has a basis $\{X_n\}$ with the a.s.c.f. $\{h_n\} \subset (E/[f_n]_\perp)^*$ satisfying (1.29), whence, since $\omega^*$ is a linear isometry, $\{h_n\} \sim \{f_n\} \sim \{g_n\}$. Consequently, by Ch. I, § 12, proposition 12.1, $\{X_n\} \sim \{y_n\}$ and hence, by Ch. I, § 8, theorem 8.1, there exists an isomorphism $v$ of $E/[f_n]_\perp$ onto $F$ such that $v(X_n) = y_n$ $(n = 1,2, \ldots)$. Then $v^*$ is a $w^*$-isomorphism of $F^*$ onto $(E/[f_n]_\perp)^*$, satisfying**) $v^*(g_n) = h_n$ $(n = 1,2, \ldots)$, whence $\{h_n\} \overset{w^*}{\sim} \{g_n\}$. Furthermore, by (1.29) and since $\omega^*$ is a $w^*$-isomorphism, we have $\{f_n\} \overset{w^*}{\sim} \{h_n\}$. Consequently, $\{f_n\} \overset{w^*}{\sim} \{g_n\}$, which completes the proof of proposition 1.9.

Now we shall prove the following $w^*$-Schauder basic analogue to corollary 1.1 on selection of basic sequences:

**Theorem 1.5.** *Let $E$ be a separable Banach space and let $\{f_n\}$ be a sequence in the conjugate space $E^*$, such that*

$$\inf_{1 \leqslant n < \infty} \|f_n\| > 0, \tag{1.33}$$

$$f_n \overset{w^*}{\to} 0. \tag{1.34}$$

---

*) By proposition 1.8 a).

**) See Ch. I, § 12, proof of proposition 12.1.

*Then* $\{f_n\}$ *has a subsequence* $\{f_{k_n}\}$ *which is a w*-Schauder basic sequence such that for any* $\{x_n\} \subset E$ *with* $(x_n, f_{k_n})$ *biorthogonal and for the operators* $\mathscr{S}_m: [f_{k_n}] \to [f_{k_n}]$ *defined by*

$$\mathscr{S}_m(f) = \sum_{i=1}^{m} f(x_i) f_{k_i} \qquad (f \in [f_{k_n}],\ m = 1,2, \ldots) \tag{1.35}$$

*we have*

$$\lim_{m \to \infty} \|\mathscr{S}_m\| = 1. \tag{1.36}$$

*Proof.* The proof will be a variation of the "product" technique for constructing basic sequences, used in remark 1.1 and in the proof of proposition 1.3.

Since for any $x \in E$ we have, by (1.33) and (1.34), $\left|\frac{f_n}{\|f_n\|}(x)\right| \leqslant \frac{1}{\inf\limits_{1 \leqslant k < \infty} \|f_k\|} |f_n(x)| \to 0$ as $n \to \infty$, we may assume $\left(\text{replacing } f_n \text{ by } \frac{f_n}{\|f_n\|}\right)$, that

$$\|f_n\| = 1 \qquad (n = 1,2, \ldots). \tag{1.37}$$

Let $\varepsilon_n$ with $0 < \varepsilon_n < 1$ $(n = 1,2, \ldots)$ be such that $\sum_{n=1}^{\infty} \varepsilon_n < \infty$ $\left(\text{hence } \prod_{n=1}^{\infty} \frac{1}{1 - \varepsilon_n} < \infty\right)$. Then we can construct, inductively, an increasing sequence of positive integers $\{k_n\}$ and an increasing sequence $\{A_n\}$ of finite subsets of Fr $S_E = \{x \in E \mid \|x\| = 1\}$, with $\left[\bigcup_{i=1}^{\infty} A_i\right] = E$, such that for each $n = 1,2, \ldots$ the following two conditions are satisfied:

(a) for every $\varphi \in [f_{k_1}, \ldots, f_{k_n}]^*$ with $\|\varphi\| = 1$, there exists an $x \in A_n$ satisfying

$$|f(x) - \varphi(f)| \leqslant \frac{\varepsilon_n}{3} \|f\| \qquad (f \in [f_{k_1}, \ldots, f_{k_n}]); \tag{1.38}$$

(b) we have

$$|f_{k_{n+1}}(x)| < \frac{\varepsilon_n}{3} \qquad (x \in A_n). \tag{1.39}$$

Indeed, let $k_1 = 1$ and assume that we have already constructed $k_1 < \ldots < k_n$ and $A_1 \subset \ldots \subset A_{n-1}$. Let $g_1, \ldots, g_l \in [f_{k_1}, \ldots, f_{k_n}]$

be an $\frac{\varepsilon_n}{18}$-net for $\operatorname{Fr} S_{[f_{k_1},\ldots,f_{k_n}]}$, let $\varphi \in [f_{k_1}, \ldots, f_{k_n}]^*$, $\|\varphi\| = 1$ and let $\Phi \in E^{**}$ be an extension of $\varphi$, with $\|\Phi\| = \|\varphi\| = 1$. Then, since the canonical image of $\operatorname{Fr} S_E$ is $w^*$-dense*) in $S_{E^{**}}$, there exists an $x \in E$ with $\|x\| = 1$ such that $|\Phi(g_i) - g_i(x)| < \frac{\varepsilon_n}{18}$ $(i = 1, \ldots, l)$. For this $x$ and any $f \in [f_{k_1}, \ldots, f_{k_n}]$ we have, with a suitable $i = i(f) \leqslant 1$,

$$\left|\varphi\left(\frac{f}{\|f\|}\right) - \frac{f}{\|f\|}(x)\right| = \left|\Phi\left(\frac{f}{\|f\|}\right) - \frac{f}{\|f\|}(x)\right| \leqslant \left|\Phi\left(\frac{f}{\|f\|}\right) - \Phi(g_i)\right| +$$

$$+ |\Phi(g_i) - g_i(x)| + \left|g_i(x) - \frac{f}{\|f\|}(x)\right| \leqslant \|\Phi\| \left\|\frac{f}{\|f\|} - g_i\right\| +$$

$$+ |\Phi(g_i) - g_i(x)| + \|x\| \left\|g_i - \frac{f}{\|f\|}\right\| < 3\frac{\varepsilon_n}{18} = \frac{\varepsilon_n}{6},$$

whence

$$|\varphi(f) - f(x)| \leqslant \frac{\varepsilon_n}{6}\|f\| \qquad (f \in [f_{k_1}, \ldots, f_{k_n}]). \tag{1.40}$$

In this way, for each $\varphi \in [f_{k_1}, \ldots, f_{k_n}]^*$ with $\|\varphi\| = 1$ we can find an $x = x_\varphi \in \operatorname{Fr} S_E$ satisfying (1.40). Now, since $E$ is separable, let $\{y_n\}$ be a sequence in $\operatorname{Fr} S_E$ such that $[y_n] = E$ and let

$$A_n = \{x_{\varphi_1}, \ldots, x_{\varphi_m}, y_1, \ldots, y_n\} \cup A_{n-1}, \tag{1.41}$$

where $\varphi_1, \ldots, \varphi_m$ is an $\frac{\varepsilon_n}{6}$-net for $\operatorname{Fr} S_{[f_{k_1},\ldots,f_{k_n}]^*}$. Then for any $\varphi \in [f_{k_1}, \ldots, f_{k_n}]^*$ with $\|\varphi\| = 1$ there exists an $i = i(\varphi) \leqslant m$ such that for all $f \in [f_{k_1}, \ldots, f_{k_n}]$ we have

$$|\varphi(f) - f(x_{\varphi_i})| \leqslant |\varphi(f) - \varphi_i(f)| + |\varphi_i(f) - f(x_{\varphi_i})| \leqslant$$

$$\leqslant \|\varphi - \varphi_i\| \|f\| + |\varphi_i(f) - f(x_{\varphi_i})| \leqslant 2\frac{\varepsilon_n}{6}\|f\| = \frac{\varepsilon_n}{3}\|f\|,$$

which proves (a) (with $x = x_{\varphi_i}$) for $A_n$. Finally, since $A_n$ is finite, by (1.34) there is an index $k_{n+1} > k_n$ such that we have (1.39), i.e., (b) for $k_{n+1}$.

We shall complete the proof by showing that $\{f_{k_n}\}$ is a $w^*$-Schauder basic sequence having the required property (1.36).

---

*) See e.g. [87], p. 424, theorem 5, combined with the fact that $\operatorname{Fr} S_E$ is $w$-dense in $S_E$.

Let us first prove that $\{f_{k_n}\}$ is a basic sequence. Let $\alpha_1, \ldots, \alpha_n$ be scalars such that $\left\|\sum_{i=1}^{n} \alpha_i f_{k_i}\right\|=1$. Choose $\varphi \in [f_{k_1}, \ldots, f_{k_n}]^*$ with $\|\varphi\|=1$, such that $\varphi\left(\sum_{i=1}^{n} \alpha_i f_{k_i}\right)=1$ and choose $x \in A_n$ satisfying (a) for this $\varphi$. Then

$$\left|\left(\sum_{i=1}^{n} \alpha_i f_{k_i}\right)(x)\right| \geqslant \left|\left|\sum_{i=1}^{n} \alpha_i(f_{k_i}(x) - \varphi(f_{k_i}))\right| - \left|\sum_{i=1}^{n} \alpha_i \varphi(f_{k_i})\right|\right| =$$

$$= \left|\left|\left(\sum_{i=1}^{n} \alpha_i f_{k_i}\right)(x) - \varphi\left(\sum_{i=1}^{n} \alpha_i f_{k_i}\right)\right| - 1\right| \geqslant$$

$$\geqslant 1 - \frac{\varepsilon_n}{3}\left\|\sum_{i=1}^{n} \alpha_i f_{k_i}\right\| = 1 - \frac{\varepsilon_n}{3},$$

and hence, by (b), for any scalar $\alpha$ with $|\alpha| \leqslant 2$ we obtain

$$\left\|\sum_{i=1}^{n} \alpha_i f_{k_i} + \alpha f_{k_{n+1}}\right\| \geqslant \left|\left(\sum_{i=1}^{n} \alpha_i f_{k_i}\right)(x) + \alpha f_{k_{n+1}}(x)\right| \geqslant$$

$$\geqslant \left|\left|\left(\sum_{i=1}^{n} \alpha_i f_{k_i}\right)(x)\right| - |\alpha|\,|f_{k_{n+1}}(x)|\right| \geqslant$$

$$\geqslant 1 - \frac{\varepsilon_n}{3} - 2\,\frac{\varepsilon_n}{3} = 1 - \varepsilon_n .$$

On the other hand, for any scalar $\alpha$ with $|\alpha| \geqslant 2$ we have, using also (1.37),

$$\left\|\sum_{i=1}^{n} \alpha_i f_{k_i} + \alpha f_{k_{n+1}}\right\| \geqslant \left|\left\|\sum_{i=1}^{n} \alpha_i f_{k_i}\right\| - |\alpha|\,\|f_{k_{n+1}}\|\right| \geqslant 1.$$

Thus, by homogeneity, for any scalars $\alpha_1, \ldots, \alpha_{n+1}$ we have

$$\left\|\sum_{i=1}^{n} \alpha_i f_{k_i}\right\| \leqslant \frac{1}{1-\varepsilon_n}\left\|\sum_{i=1}^{n+1} \alpha_i f_{k_i}\right\|,$$

and hence for any scalars $\alpha_1, \ldots, \alpha_{n+m}$ we obtain

$$\left\|\sum_{i=1}^{n} \alpha_i f_{k_i}\right\| \leqslant \left(\sum_{j=n}^{n+m-1} \frac{1}{1-\varepsilon_j}\right)\left\|\sum_{i=1}^{n+m} \alpha_i f_{k_i}\right\| \leqslant \left(\sum_{j=1}^{\infty} \frac{1}{1-\varepsilon_j}\right)\left\|\sum_{i=1}^{n+m} \alpha_i f_{k_i}\right\|. \quad (1.42)$$

Consequently, by Ch. I, § 7, theorem 7.1, $\{f_{k_n}\}$ is a basic sequence. Moreover, if $\{\Psi_n\} \subset [f_{k_n}]^*$ is the a.s.c.f. to $\{f_{k_n}\}$ and if $\{S_m\}$ is the sequence of partial sum operators associated to $\{f_{k_n}\}$, that is, $S_m(f) = \sum_{i=1}^{m} \Psi_i(f) f_{k_i}$ for all $f \in [f_{k_n}]$, then, by Ch. I, § 12, formula (12.2),

$$S_m^*(\Psi) = \sum_{i=1}^{m} \Psi(f_{k_i})\Psi_i \qquad (\Psi \in [\Psi_n]), \tag{1.43}$$

whence, by (1.42),

$$1 \leqslant \|S_m^*\| \leqslant \prod_{n=m}^{\infty} \frac{1}{1-\varepsilon_n}, \tag{1.44}$$

and hence

$$\lim_{m\to\infty} \|S_m^*\| = 1. \tag{1.45}$$

Let $u$ denote the canonical mapping of $E$ into $[f_{k_n}]^*$, i.e.

$$(u(x))(f) = f(x) \qquad (x \in E,\ f \in [f_{k_n}])$$

(this is not one-to-one when $\{f_{k_n}\}$ is not total on $E$). We shall now prove that

$$u(E) = [\Psi_n] \tag{1.46}$$

and that $u$ is a "metric homomorphism" of $E$ onto $[\Psi_n]$, that is, the mapping $u_0$: $E/\mathrm{Ker}\, u \to [\Psi_n]$ induced by $u$ is an isometry of $E/\mathrm{Ker}\, u$ onto $[\Psi_n]$.

If $x \in A_n$ for some $n$, then $x \in A_n \subset A_{n+1} \subset \ldots$, whence, by (b) and $\sum_{i=1}^{\infty} \varepsilon_n < \infty$, we obtain

$$\sum_{i=1}^{\infty} |f_{k_i}(x)| < \infty$$

and therefore, since $\sup_{1\leqslant i<\infty} \|\Psi_i\| < \infty$ (by $\|f_{k_i}\| = 1$ for $i = 1,2, \ldots$ and by Ch. I, § 3, corollary 3.1), it follows that the series $\sum_{i=1}^{\infty} f_{k_i}(x)\Psi_i$ converges. Since $\{f_{k_n}\}$ is a basis of $[f_{k_n}]$ with the a.s.c.f. $\{\Psi_n\}$, for any $f \in [f_{k_n}]$ we have $f = \sum_{i=1}^{\infty} \Psi_i(f) f_{k_i}$, whence, since $\sum_{i=1}^{\infty} f_{k_i}(x)\Psi_i$ converges,

$$(u(x))(f) = f(x) = \left(\sum_{i=1}^{\infty} \Psi_i(f) f_{k_i}\right)(x) = \left(\sum_{i=1}^{\infty} f_{k_i}(x)\Psi_i\right)(f) \quad (f \in [f_{k_n}]),$$

and thus $u(x)=\sum_{i=1}^{\infty} f_{k_i}(x)\Psi_i \in [\Psi_n]$. Consequently, since $\left[\bigcup_{n=1}^{\infty} A_n\right]=E$ and since $u$ is continuous, it follows that

$$u(E) \subset [\Psi_n]. \tag{1.47}$$

Thus, in order to prove (1.46) and that $u$ is a metric homomorphism of $E$ onto $[\Psi_n]$, it will be sufficient*) to prove that

(c) For every $\Psi \in \operatorname{lin}\{\Psi_n\}$ (= the linear span of $\{\Psi_n\}$), with $\|\Psi\|=1$, and every $\varepsilon>0$, there exists an $x \in E$ with $\|x\|=1$ such that

$$\|u(x)-\Psi\|<4\varepsilon. \tag{1.48}$$

Now, in order to prove (c), let $0<\varepsilon<1$ and choose $N$ such that

$$\sum_{i=n}^{\infty} \varepsilon_i < \varepsilon \qquad (n>N), \tag{1.49}$$

$$\|S_n\| \leqslant 1+\varepsilon \qquad (n>N); \tag{1.50}$$

this is possible by $\sum_{i=1}^{\infty} \varepsilon_i < \infty$ and (1.45). Fix an $n>N$ and put

$$\|\Psi\|_1 = \|\Psi|_{[f_{k_1},\dots,f_{k_n}]}\| \qquad (\Psi \in [\Psi_1,\dots,\Psi_n]). \tag{1.51}$$

Then, by (1.50), for any $\Psi \in [\Psi_1,\dots,\Psi_n]$ we have

$$\|\Psi\|_1 \leqslant \|\Psi\| = \left\|\sum_{i=1}^{n} \Psi(f_{k_i})\Psi_i\right\| = \sup_{\substack{f\in[f_{k_j}]\\ \|f\|\leqslant 1}} \left|\sum_{i=1}^{n} \Psi(f_{k_i})\Psi_i(f)\right| =$$

$$= \sup_{\substack{f\in[f_{k_j}]\\ \|f\|\leqslant 1}} \left|\Psi\left(\sum_{i=1}^{n} \Psi_i(f) f_{k_i}\right)\right| \leqslant \|\Psi\|_1 \sup_{\substack{f\in[f_{k_j}]\\ \|f\|\leqslant 1}} \|S_n(f)\| \leqslant (1+\varepsilon)\|\Psi\|_1. \tag{1.52}$$

Fix $\Psi \in [\Psi_1,\dots,\Psi_n]$ with $\|\Psi\|=1$ and put

$$\bar{\Psi} = \frac{\Psi}{\|\Psi\|_1}. \tag{1.53}$$

---

*) See e.g. [140], Ch. I, § 14, corollary 1.

Then, choosing $x \in A_n$ satisfying (a) for $\varphi = \overline{\Psi}|_{[f_{k_1}, \ldots, f_{k_n}]}$, we have

$$\left\| \sum_{i=1}^{n} f_{k_i}(x)\Psi_i - \overline{\Psi} \right\|_1 = \sup_{\substack{f \in [f_{k_1}, \ldots, f_{k_n}] \\ \|f\|=1}} \left| \left( \sum_{i=1}^{n} \Psi_i(f) f_{k_i} \right)(x) - \overline{\Psi}(f) \right| =$$

$$= \sup_{\substack{f \in [f_{k_1}, \ldots, f_{k_n}] \\ \|f\|=1}} |f(x) - \varphi(f)| \leqslant \frac{\varepsilon_n}{3},$$

whence, by (1.52) and (1.49),

$$\left\| \sum_{i=1}^{n} f_{k_i}(x)\Psi_i - \overline{\Psi} \right\| \leqslant 2 \left\| \sum_{i=1}^{n} f_{k_i}(x)\Psi_i - \overline{\Psi} \right\|_1 \leqslant 2\,\frac{\varepsilon_n}{3} < \frac{2}{3}\,\varepsilon.$$

Furthermore, by (1.50),

$$\|\Psi_i\| = \sup_{\substack{f \in [f_{k_i}] \\ \|f\| \leqslant 1}} \|\Psi_i(f) f_{k_i}\| = \|S_i - S_{i-1}\| \leqslant 4 \quad (i > N),$$

whence, by (b) and (1.49),

$$\left\| \sum_{i=n+1}^{\infty} f_{k_i}(x)\Psi_i \right\| < 4 \sum_{i=n}^{\infty} \frac{\varepsilon_i}{3} < \frac{4}{3}\,\varepsilon.$$

Consequently,

$$\|u(x) - \overline{\Psi}\| = \left\| \sum_{i=1}^{\infty} f_{k_i}(x)\Psi_i - \overline{\Psi} \right\| < \frac{2}{3}\,\varepsilon + \frac{4}{3}\,\varepsilon = 2\varepsilon.$$

On the other hand, again by (1.52), $1 = \|\Psi\| \leqslant (1+\varepsilon)\,\|\Psi\|_1$, whence

$$\|\overline{\Psi} - \Psi\| = \left\| \frac{\Psi}{\|\Psi\|_1} - \Psi \right\| = \left| \frac{1}{\|\Psi\|_1} - 1 \right| \|\Psi\| \leqslant$$

$$\leqslant \varepsilon(1+\varepsilon)\,\|\Psi\|_1 \leqslant \varepsilon(1+\varepsilon) < 2\varepsilon,$$

and therefore we have (1.48). This proves (c) and hence also (1.46) and that $u$ is a metric homomorphism of $E$ onto $[\Psi_n]$.

Now, from this it follows easily that $\{f_{k_n}\}$ is a $w^*$-Schauder basic sequence. Indeed, by (1.46) let $x_n \in u^{-1}(\Psi_n)$ $(n = 1,2, \ldots)$ be arbitrary. Then

$$f_{k_i}(x_j) = (u(x_j))(f_{k_i}) = \Psi_j(f_{k_i}) = \delta_{ij} \qquad (i, j = 1,2, \ldots),$$

and thus $(x_n, f_{k_n})$ is a biorthogonal system. Furthermore, since the mapping $u_0$: $E/\text{Ker } u \to [\Psi_n]$ induced by $u$ is an isometry, $\{u_0^{-1}(\Psi_n)\}$ is a basis of $E/\text{Ker } u$. However, we have

$$\text{Ker } u = \{x \in E \mid (u(x))(f) = 0 \quad (f \in [f_{k_n}])\} =$$

$$= \{x \in E \mid f(x) = 0 \ (f \in [f_{k_n}])\} = [f_{k_n}]_\perp, \tag{1.54}$$

and by $u_0(\omega(x_n)) = u(x_n) = \Psi_n$ (where $\omega$ denotes the canonical mapping $E \to E/[f_{k_n}]_\perp$) we have

$$\omega(x_n) = u_0^{-1}(\Psi_n) \qquad (n = 1,2, \ldots), \tag{1.55}$$

and thus $\{\omega(x_n)\}$ is a basis of $E/[f_{k_n}]_\perp$. Consequently, by proposition 1.8 a), implication 3° ⇒ 1°, $\{f_{k_n}\}$ is a $w^*$-Schauder basic sequence.

Finally, in order to prove (1.36), it will be sufficient, by virtue of (1.45), to prove that

$$\|\mathcal{S}_m\| = \|S_m^*\| \qquad (m = 1,2, \ldots), \tag{1.56}$$

where $\mathcal{S}_m: [\widetilde{f_{k_n}}] \to [\widetilde{f_{k_n}}]$ and $S_m^*$: $[\Psi_n] \to [\Psi_n]$ are defined by (1.35) and (1.43) respectively. Let $\{h_n\}$ be the a.s.c.f. and $\{s_m\}$ the sequence of partial sum operators associated to the basis $\{\omega(x_n)\}$ of $E/[f_{k_n}]_\perp$. Then by Ch. I, § 12, formula (12.2),

$$s_m^*(h) = \sum_{i=1}^{m} h(\omega(x_i))h_i = \sum_{i=1}^{m} (\omega^*(h))(x_i)h_i \qquad (h \in (E/[f_{k_n}]_\perp)^*)$$

whence, by proposition 1.8 b),

$$\omega^* s_m^*(h) = \sum_{i=1}^{n} (\omega^*(h))\,(x_i)\,\omega^*(h_i) = \sum_{i=1}^{m} (\omega^*(h))\,(x_i)\,f_{k_i} =$$

$$= \mathcal{S}_m(\omega^*(h)) \quad (h \in (E/[f_{k_n}]_\perp)^*)$$

and thus, since $\omega^*$ is an isometry of $(E/[f_{k_n}]_\perp)^*$ onto $[\widetilde{f_{k_n}}]$,

$$\|\mathcal{S}_m\| = \|\omega^* s_m^* (\omega^*)^{-1}\| = \|s_m^*\| \qquad (m = 1,2, \ldots). \tag{1.57}$$

On the other hand, by the definition of $u_0$ and by (1.55),

$$S_m^* u_0 \omega(x) = S_m^* u(x) = \sum_{i=1}^{m} (u(x))(f_{k_i})\Psi_i = \sum_{i=1}^{m} f_{k_i}(x) u_0 \omega(x_i) =$$

$$= u_0\left(\sum_{i=1}^{m} h_i(\omega(x))\omega(x_i)\right) = u_0(s_m(\omega(x))) \quad (\omega(x) \in E/[f_{k_n}]_\perp)$$

whence, since $u_0$ is an isometry of $E/[f_{k_n}]_\perp$ onto $[\Psi_n]$, we obtain

$$\|S_m^*\| = \|u_0 s_m u_0^{-1}\| = \|s_m\| = \|s_m^*\| \qquad (m = 1,2, \ldots). \quad (1.58)$$

From (1.57) and (1.58) it follows (1.56), whence (1.36), which completes the proof of theorem 1.5.

*Remark 1.8*. For an arbitrary (not necessarily separable) Banach space $E$ the above argument yields the following result: *If* $\{f_n\}$ *is a sequence in* $E^*$ *satisfying* (1.33) *and having* 0 *as the only weak* cluster point, then* $\{f_n\}$ *has a basic subsequence* $\{f_{k_n}\}$ *such that*

$$u(E) \supset [\Psi_n], \qquad (1.59)$$

*where* $u$ *is the canonical mapping of* $E$ *into* $[f_{k_n}]^*$ *and* $\{\Psi_n\} \subset [f_{k_n}]^*$ *is the a.s.c.f. to* $\{f_{k_n}\}$. Indeed, the separability of $E$ was used only to find $\{A_n\}$ such that $\left[\bigcup_{i=1}^{\infty} A_n\right] = E$, from which it followed that $u(E) \subset [\Psi_n]$. When $E$ is non-separable and $\{f_n\}$ is as above, one can still choose $\{k_n\}$ and $\{A_n\}$ to satisfy (a) and (b), whence, exactly as in the above proof, it follows that (c) holds,which implies (1.59). Let us also observe that this result (applied in $E^{**}$ instead of $E^*$) implies again corollary 1.1.

*Remark 1.9*. For separable $E$, theorem 1.5 gives, in a certain sense, the best possible result, since *if* $E$ *is any Banach space and* $\{f_n\} \subset E^*$ *is a* $w^*$*-Schauder basic sequence with* $\sup_{1 \leqslant n < \infty} \|f_n\| < \infty$, *then* $f_n \xrightarrow{w^*} 0$. Indeed, this follows from proposition 1.8 a), b) and Ch. II, § 7, proposition 7.4, taking into account that $\omega^*$ is a linear isometry and a $w^*$-isomorphism of $(E/[f_n]_\perp)^*$ onto $\widetilde{[f_n]}$.

In the above we have given conditions for a sequence in a Banach space $E$ (respectively, in particular, in a conjugate Banach space $E^*$) to contain a basic subsequence. Let us consider now, more generally, the problem of existence of block basic sequences. The notion of block basic sequence with respect to a basis can be extended in a natural way to that of a block basic sequence with respect to an arbitrary sequence, as follows:

*Definition 1.4*. A sequence $\{z_n\}$ in a Banach space $E$ is called a *block basic sequence with respect to a sequence* $\{y_n\} \subset E$ if it is a basic sequence of the form

$$z_n = \sum_{i=m_{n-1}+1}^{m_n} \beta_i y_i \neq 0 \qquad (n = 1,2, \ldots), \tag{1.60}$$

where $\{m_n\}$ is an increasing sequence of positive integers and $m_0 = 0$.

In particular, if $\{y_n\}$ is a basic sequence, then this reduces to Ch. I, § 7, definition 7.3; as observed there, in this case any sequence $\{z_n\}$ of the form (1.60) is necessarily a basic sequence.

**Theorem 1.6.** *Let $X$ be a Banach space with a basis $\{x_n\}$. Then every sequence $\{y_n\} \subset X$ with* $\dim [y_n] = \infty$ *has a block basic sequence which is equivalent to a block basic sequence with respect to* $\{x_n\}$.

*Proof.* We may assume (omitting, if necessary, a suitable subsequence of $\{y_n\}$), that

$$\dim [y_{\frac{n(n+1)}{2}}, \ldots, y_{\frac{(n+1)(n+2)}{2}-1}] = n + 1 \qquad (n = 1,2, \ldots). \tag{1.61}$$

Let $\{f_n\} \subset X^*$ be the a.s.c.f. to the basis $\{x_n\}$. Then, since $\operatorname{codim} [f_1, \ldots, f_n]_\perp = n$, by Ch. II, § 4, lemma 4.1 there exist elements

$$z_n \in [y_{\frac{n(n+1)}{2}}, \ldots, y_{\frac{(n+1)(n+2)}{2}-1}] \cap [f_1, \ldots, f_n]_\perp \quad (n=1,2, \ldots) \tag{1.62}$$

such that $\|z_n\| = 1$ $(n = 1,2, \ldots)$. By virtue of proposition 1.1, $\{z_n\}$ has a subsequence $\{z_{p_n}\}$ which is a basic sequence, equivalent to a block basic sequence with respect to $\{x_n\}$. Since $\{z_{p_n}\}$ is obviously a block basic sequence with respect to $\{y_n\}$, this completes the proof of theorem 1.6.

**Corollary 1.7.** *Let $E$ be a Banach space. Then every sequence $\{y_n\} \subset E$ with* $\dim [y_n] = \infty$ *has a block basic sequence.*

*Proof.* It is sufficient to embed $[y_n]$ into $X = C([0, 1])$ and to apply theorem 1.6.

**Corollary 1.8.** *Let $X$ be a Banach space with a shrinking (respectively boundedly complete, respectively unconditional) basis $\{x_n\}$. Then every sequence $\{y_n\} \subset X$ with* $\dim [y_n] = \infty$ *has a shrinking (respectively boundedly complete, respectively unconditional) block basic sequence*

*Proof.* This follows from theorem 1.6 and the following observations: By Ch. II, § 4, theorem 4.2 (equivalence $1° \Leftrightarrow 2°$) and Ch. II, § 17, corollary 17.2, every block basic sequence with respect to a shrinking (respectively, unconditional) basis is a shrinking (respectively, unconditional) basic sequence. Finally, *every block basic sequence*

$\{z_n\}$ *with respect to a boundedly complete basis* $\{x_n\}$ *is a boundedly complete basic sequence*. Indeed, let $z_n = \sum_{i=m_{n-1}+1}^{m_n} \beta_i x_i \neq 0$ $(n=1,2, \ldots)$ and let

$$\sup_{1 \leqslant n < \infty} \left\| \sum_{j=1}^{n} \alpha_j z_j \right\| = \sup_{1 \leqslant n < \infty} \left\| \sum_{j=1}^{n} \sum_{l=m_{j-1}+1}^{m_j} \alpha_j \beta_l x_l \right\| = \sup_{1 \leqslant n < \infty} \left\| \sum_{l=1}^{m_n} \gamma_l x_l \right\| < \infty,$$

where $\gamma_l = \alpha_j \beta_l$ $(l = m_{j-1} + 1, \ldots, m_j;\ j = 1,2, \ldots)$. Then

$$\sup_{1 \leqslant k < \infty} \left\| \sum_{l=1}^{k} \gamma_l x_l \right\| \leqslant v_{\{x_n\}} \sup_{1 \leqslant n < \infty} \left\| \sum_{l=1}^{m_n} \gamma_l x_l \right\| < \infty$$

(where $v_{\{x_n\}}$ is the norm of the basis $\{x_n\}$), whence, since $\{x_n\}$ is boundedly complete, $\sum_{j=1}^{\infty} \alpha_j z_j = \sum_{l=1}^{\infty} \gamma_l x_l$ converges. Thus $\{z_n\}$ is boundedly complete, which concludes the proof.

**Corollary 1.9.** *Let X be a Banach space with a basis. If X contains a shrinking (respectively boundedly complete, respectively unconditional) basic sequence, then every basis of X has a shrinking (respectively, boundedly complete, respectively unconditional) block basic sequence.*

*Proof.* This again follows from theorem 1.6 and the observations made in the above proof of corollary 1.10.

Let us also observe that by the preceding results problem 1.2 is equivalent to the following:

*Problem 1.4.* Let $X$ be a Banach space with a basis $\{x_n\}$. Does $\{x_n\}$ have an unconditional block basic sequence?

Indeed, if the answer to poblem 1.4 is affirmative, then, since every Banach space contains a subspace with a basis (by theorem 1.1), the answer to problem 1.2 is also affirmative. Conversely, if the answer to problem 1.2 is affirmative, then, by corollary 1.9, so is the answer to problem 1.4.

As observed in this argument, theorem 1.1 can be also interpreted as stating that every Banach space $E$ contains a subspace with a basis. It is natural to ask the following dual question:

*Problem 1.5.* Does every Banach space $E$ have a quotient space with a basis? Or, in other words: Does there exist a continuous linear mapping of any Banach space $E$ onto a Banach space with a basis? Or, what is equivalent (by proposition 1.8 a)): Does every conjugate Banach space $E^*$ contain a $w^*$-Schauder basic sequence?

For separable Banach spaces $E$ the answer is affirmative:

**Theorem 1.7.** *Every separable Banach space E has a quotient space with a basis.*

*Proof.* For any separable Banach space $E$, the conjugate space $E^*$ contains a sequence $\{f_n\}$ satisfying (1.33) and (1.34), and hence also a

$w^*$-Schauder basic sequence (by theorem 1.5). Consequently, by proposition 1.8 a), $E$ has a quotient space with a basis, which completes the proof.

Since for every Banach space $E$ the conjugate space $E^*$ contains*) a sequence $\{f_n\}$ satisfying (1.33) and (1.34), the above argument shows that if theorem 1.5 remained valid for non-separable Banach spaces, then the answer to problem 1.5 would be affirmative.

Let us also mention that problem 1.5 is equivalent to the following:

*Problem 1.5′*. Does every Banach space $E$ have a separable quotient space?

Indeed, if there exists a continuous linear map $u_1$ of $E$ onto a separable space $E_1$ then, since by theorem 1.7 there is a continuous linear map $u_2$ of $E_1$ onto a space $E_2$ with a basis, we infer that $u_2u_1$ is a continuous linear map of $E$ onto $E_2$. We shall give some results related to this problem in Vol. III, Ch. IV. Here we only observe that *for a Banach space E the following statements are equivalent:*

1°. *There exists a sequence $\{G_n\}$ of subspaces of E such that*

$$\text{(i) } G_1 \subset G_2 \subset G_3 \subset \ldots; \quad \text{(ii) } G_1 \neq G_2 \neq G_3 \ldots; \quad \text{(iii) } \overline{\bigcup_{n=1}^{\infty} G_n} = E.$$

2°. *E has a separable quotient space $E/G$.*

Indeed, assume that we have 1°. Choose $f_n \in E^*$ with $\|f_n\|=1$ such that $f_n \in G_n^\perp$ $(n = 1,2,\ldots)$. Then a slight modification of the above proof of theorem 1.5 yields that $\{f_n\}$ has a subsequence $\{f_{k_n}\}$ which is a $w^*$-Schauder basic sequence, whence, by proposition 1.8, $E/[f_{k_n}]_\perp$ has a basis and thus 1°⇒2°. Namely, the sets $A_n$ will no longer be finite, but they will satisfy $\left[\bigcup_{i=1}^{\infty} A_i\right]=E$ and (a), (b) of that proof.**) Indeed, let $k_1 = 1$ and assume that we have already constructed $k_1 < \ldots < k_n$ and $A_1 \subset \ldots \subset A_{n-1}$. Let $x_{\varphi_1}, \ldots, x_{\varphi_m}$ be as in the proof of theorem 1.5 and let

$$A_n = \{x_{\varphi_1}, \ldots, x_{\varphi_m}\} \cup G_n \cup A_{n-1}.$$

Then, by (i) and (ii), $A_{n-1} \subset A_n$ and $A_{n-1} \neq A_n$. Furthermore, as in the proof of theorem 1.5, we have (a). Finally, since $\{x_{\varphi_1}, \ldots, x_{\varphi_m}\} \cup A_{n-1}$ is finite and since $f_n, f_{n+1}, \ldots \in G_n^\perp$ $(n = 1,2,\ldots)$ and***) $f_n \xrightarrow{w^*} 0$,

*) See [294], [192].

**) This is enough to yield the desired conclusion, since in the proof of theorem 1.5 the finiteness of $A_n$ was used only to get (b).

***) Indeed, $\lim_{n\to\infty} f_n(x)=0$ for all $x \in \bigcup_{n=1}^{\infty} G_n$, whence, since $\|f_n\|=1$ and $\overline{\bigcup_{n=1}^{\infty} G_n} = E$, we obtain $f_n \xrightarrow{w^*} 0$.

there is an index $k_{n+1} > k_n$ such that we have (1.39). Also, by (iii), we obtain $\left[\bigcup_{i=1}^{\infty} A_i\right] = E$. Thus, $1^\circ \Rightarrow 2^\circ$.

Conversely, assume now that we have $2^\circ$. Let $\{X_n\}$ be a finitely linearly independent complete sequence in $E/G$. Then the sequence of subspaces of $E$ defined by

$$G_1 = G, \; G_{n+1} = \{x \in E \mid x + G \in [X_1, \ldots, X_n]\} \qquad (n = 1,2, \ldots)$$

satisfies (i), (ii) and (iii). Thus, $2^\circ \Rightarrow 1^\circ$, which completes the proof of our assertion.

*Problem 1.5″*. Does every (or, at least, every separable) Banach space have a quotient space with an unconditional basis? Or, equivalently, does every conjugate Banach space $E^*$ (respectively, of a separable Banach space $E$) contain an unconditional $w^*$-Schauder basic sequence?

Let us return now to basic sequences in the initial space $E$. In Ch. I, § 4, problem 4.1, it has been asked whether in a Banach space $E$ with a basis $\{x_n\}$ every basic sequence $\{y_n\}$ can be extended to a basis*) of $E$ and in Ch. I, § 7, theorem 7.2 it was established that the answer is affirmative when $\{y_n\}$ is a block basic sequence with respect to some basis of $E$. (Hence, the problem becomes equivalent to the following: is every basic sequence a block basic sequence with respect to a suitable basis of $E$?). Another positive result in this direction is given by

**Theorem 1.8.** *Let $E$ be a Banach space with a basis $\{x_n\}$ and let $\{f_n\} \subset E^*$ be the a.s.c.f. to $\{x_n\}$. If a sequence $\{y_n\} \subset E$ satifies the conditions* (1.1), (1.2) *of proposition* 1.1, *then $\{y_n\}$ has a subsequence $\{y_{p_n}\}$ which can be extended to a basis of $E$.*

*Proof.* Let $0 < \varepsilon_n < \delta \leqslant \|y_n\|$ $(n = 1,2, \ldots)$ be such that

$$\frac{72\, \nu^2(\nu+1)}{\delta} \sum_{n=1}^{\infty} \varepsilon_n < 1, \tag{1.63}$$

where $\nu = \nu_{\{x_n\}}$ is the norm of the basis $\{x_n\}$. By proposition 1.1 and its proof**), there exist a subsequence $\{y_{p_n}\}$ of $\{y_n\}$ and a block basic sequence $\{z_n\}$ with respect to $\{x_n\}$ such that

$$\|y_{p_n} - z_n\| \leqslant \frac{\varepsilon_n}{2} \qquad (n = 1,2, \ldots). \tag{1.64}$$

---

*) Recently, this problem has been solved in the negative (see the Notes and remarks).

**) See Ch. II, § 15, proof of proposition 15.1.

We shall show that $\{y_{p_n}\}$ can be extended to a basis of $E$, which will complete the proof. By Ch. I, § 7, theorem 7.2 and its proof, the block basic sequence $\{z_n\}$ can be extended to a basis $\{z'_n\}$ of the whole space $E$, with*) $\nu_{\{z'_n\}} \leqslant 36\, \nu^2_{\{x_n\}}(\nu_{\{x_n\}} + 1)$, say

$$z'_{m_n} = z_n \qquad (n = 1,2, \ldots). \tag{1.65}$$

Since $\varepsilon_n < \delta \leqslant \|y_n\|$ $(n = 1,2, \ldots)$, we have

$$\|z_n\| \geqslant | \, \|y_{p_n}\| - \|y_{p_n} - z_n\| \, | \geqslant \delta - \frac{\varepsilon_n}{2} > \delta - \frac{\delta}{2} = \frac{\delta}{2} \qquad (n = 1,2, \ldots),$$

whence, for the a.s.c.f. $\{h_n\} \subset E^*$ to $\{z'_n\}$ we obtain

$$|h_{m_n}(x)| = \frac{\|h_{m_n}(x) z_n\|}{\|z_n\|} \leqslant \frac{2.36\, \nu^2(\nu + 1)}{\frac{\delta}{2}} \|x\| \qquad (x \in E, \ n = 1,2, \ldots),$$

and thus

$$\|h_{m_n}\| \leqslant \frac{144\ \nu^2(\nu + 1)}{\delta} \qquad (n = 1,2, \ldots). \tag{1.66}$$

Now put

$$y'_{m_n} = y_{p_n} \qquad (n = 1,2, \ldots), \tag{1.67}$$

$$y'_j = z'_j \qquad (j \notin \{m_n\}). \tag{1.68}$$

Then, taking into account (1.63)—(1.68), we obtain

$$\sum_{j=1}^{\infty} \|h_j\| \, \|z'_j - y'_j\| = \sum_{n=1}^{\infty} \|h_{m_n}\| \, \|z_n - y_{p_n}\| \leqslant \frac{72\, \nu^2(\nu + 1)}{\delta} \sum_{n=1}^{\infty} \varepsilon_n < 1,$$

whence, by Ch. I, § 10, theorem 10.1, $\{y'_j\}$ is a basis of $E$, which completes the proof of theorem 1.8.

**Corollary 1.10.** *Let $\{y_n\}$ be a sequence in a Banach space $E$ with a basis, satisfying* (1.1) *and such that $y_n \xrightarrow{w} 0$ (e.g., in particular, a shrinking basic sequence). Then $\{y_n\}$ has a subsequence which can be extended to a basis of $E$.*

*Proof.* Obviously, $\{y_n\}$ satisfies the conditions of theorem 1.8.

---

*) Actually, with more care in the proof of Ch. I, § 7, corollary 7.3, one can also obtain better estimates.

**Corollary 1.11.** *Let $E$ be a Banach space with a basis $\{x_n\}$. Then every sequence $\{y_n\} \subset E$ with* $\dim [y_n] = \infty$ *has a block basic sequence which can be extended to a basis of $E$.*

*Proof.* As in the proof of theorem 1.6, we may assume (1.61) and then there exists a sequence $\{z_n\} \subset E$ with $\|z_n\| = 1$ $(n = 1,2, \ldots)$ satisfying (1.62), where $\{f_n\} \subset E^*$, $f_i(x_j) = \delta_{ij}$ $(i, j = 1, 2, \ldots)$. Hence, by theorem 1.8, $\{z_n\}$ has a basic subsequence (which is obviously a block basic sequence with respect to $\{y_n\}$) which can be extended to a basis of $E$. This completes the proof.

If $\{x_n\}$ is a basic sequence in a Banach space $E$, with the a.s.c.f. $\{\varphi_n\} \subset [x_n]^*$, then for any sequence $\{f_n\} \subset E^*$ of extensions of $\{\varphi_n\}$ (i.e., $f_n|_{[x_j]} = \varphi_n$ for $n = 1,2, \ldots$) we have that $(x_n, f_n)$ is a biorthogonal system. Conversely, if $\{x_n\} \subset E$ is a basic sequence and $\{f_n\} \subset E^*$ is any sequence such that $(x_n, f_n)$ is a biorthogonal system, then, clearly, $\{f_n|_{[x_j]}\} \subset [x_j]^*$ is the a.s.c.f. to $\{x_n\}$. We shall consider now the problem of obtaining more information about these sequences $\{f_n\} \subset E^*$.

Let us first observe that for any basic sequence $\{x_n\}$ in a Banach space $E$ one can define, in a natural way, a subspace $V'$ of $E^*$ containing all these sequences $\{f_n\} \subset E^*$ (thus, $V'$ depends only on $\{x_n\}$, but not on $\{f_n\}$), namely

$$V' = \{f \in E^* \mid \|f\|_n = \|f|_{[x_{n+1}, x_{n+2}, \ldots]}\| \to 0 \text{ as } n \to \infty\}; \tag{1.69}$$

moreover, note that $V'$ has sense for any (not necessarily basic) sequence $\{x_n\}$.

**Proposition 1.10.** *Let $E$ be a Banach space, let $(x_n, \varphi_n)$ ($\{x_n\} \subset$ $\subset E$, $\{\varphi_n\} \subset [x_n]^*$) be a biorthogonal system and let $V' \subset E^*$ be defined by* (1.69). *Then*

a) *We have*

$$V' = \{f \in E^* \mid f|_{[x_j]} \in [\varphi_n]\}, \tag{1.70}$$

*and hence*

$$[\varphi_n] = \{f|_{[x_j]} \mid f \in V'\} = \iota^*(V'), \tag{1.71}$$

*where $\iota$ is the natural embedding of $[x_n]$ into $E$.*

b) *Consequently, if $\{x_n\}$ is a basic sequence, we have*

$$r(V') \geqslant \frac{1}{4} r([\varphi_n]) > 0. \tag{1.72}$$

*Proof.* a) By Ch. II, § 4, proposition 4.2 (applied in $[x_n]$) we have

$$[\varphi_n] = \{\varphi \in [x_j]^* \mid \|\varphi\|_n = \|\varphi|_{[x_{n+1}, x_{n+2}, \ldots]}\| \to 0 \text{ as } n \to \infty\}, \tag{1.73}$$

which, together with (1.69), implies (1.70). Furthermore,

$$(\imath^*(f))(x) = f(\imath(x)) = f|_{[x_j]}(x) \qquad (x \in [x_j],\ f \in E^*),$$

whence

$$\imath^*(f) = f|_{[x_j]} \qquad (f \in E^*), \qquad (1.74)$$

which, together with (1.70), implies $\imath^*(V') \subset [\varphi_n]$. Conversely, for any $\varphi \in [\varphi_n]$ there exists an $f \in E^*$ such that $\imath^*(f) = \varphi \in [\varphi_n]$, whence $f \in V'$ (by (1.70)), and thus $[\varphi_n] \subset \imath^*(V')$, which proves (1.71).

b) (1.72) is a consequence of (1.70) and the proof of corollary 1.4, formula (1.19). This completes the proof of proposition 1.10.

**Proposition 1.11.** *Let $E$ be a Banach space, let $(x_n, f_n)$ ($\{x_n\} \subset E$, $\{f_n\} \subset E^*$) be a biorthogonal system and let $V' \subset E^*$ be defined by* (1.69). *Then*

$$\overline{[x_n]^\perp + [f_n]} = V', \qquad (1.75)$$

*and hence, if $\{x_n\}$ is a basic sequence, we have*

$$r([x_n]^\perp + [f_n]) > 0. \qquad (1.76)$$

*Proof.* Obviously, $V'$ is closed, so

$$\overline{[x_n]^\perp + [f_n]} \subset V'.$$

Conversely, let $f \in V'$. Then, by formula (1.70) of proposition 1.10, we have $f|_{[x_j]} \in [\varphi_n]$, where $\varphi_n = f_n|_{[x_j]}$ ($n = 1, 2, \ldots$), and hence, for suitable scalars $\beta_i^{(n)}$,

$$\lim_{n\to\infty} \|f|_{[x_j]} - \sum_{i=1}^{m_n} \beta_i^{(n)} f_i|_{[x_j]}\| = 0.$$

Let $h_n \in E^*$ be an extension of $\left(f - \sum_{i=1}^{m_n} \beta_i^{(n)} f_i\right)\Big|_{[x_j]}$, with the same norm, so $\lim_{n\to\infty} \|h_n\| = 0$. Then, clearly, $f - \sum_{i=1}^{m_n} \beta_i^{(n)} f_i - h_n \in [x_j]^\perp$ and

$$\left(f - \sum_{i=1}^{m_n} \beta_i^{(n)} f_i - h_n\right) + \sum_{i=1}^{m_n} \beta_i^{(n)} f_i = f - h_n \to f \text{ as } n \to \infty,$$

whence $f \in \overline{[x_n]^\perp + [f_n]}$, which proves (1.75). This, together with (1.72), implies (1.76), completing the proof of proposition 1.11.

In particular, from the above it follows that *if* $\{x_n\}$ *is a shrinking basic sequence* (i.e., $[\varphi_n] = [x_n]^*$ and hence, by (1.70), $V' = E^*$), *then* $\overline{[x_n]^\perp + [f_n]} = E^*$; clearly, the converse is also true.

We shall consider now the following problem: If $\{x_n\} \subset E$ is a basic sequence, with the a.s.c.f. $\{\varphi_n\} \subset [x_n]^*$ (which is a basic sequence, by Ch. I, § 12, theorem 12.1) and $f_n \in E^*$ is any extension of $\varphi_n$ $(n = 1,2, \ldots)$ (or, equivalently, if $\{f_n\} \subset E^*$ is any sequence such that $(x_n, f_n)$ is a biorthogonal system), is $\{f_n\}$ a basic sequence, and if yes, what can we say about relations between $\{f_n\}$ and $\{x_n\}$ or $\{f_n\}$ and $\{\varphi_n\}$? Let us begin with

**Proposition 1.12.** *For every basic sequence* $\{x_n\}$ *in a Banach space* $E$, *which is not a basis of* $E$, *there exists a non-basic sequence* $\{f_n\} \subset E^*$ *such that* $(x_n, f_n)$ *is a biorthogonal system.*

*Proof.* Clearly, we may assume that $\inf_{1 \leqslant n < \infty} \|x_n\| > 0$ $\left(\text{by considering } \left\{\frac{x_n}{\|x_n\|}\right\}\right)$. Then, by Ch. I, § 3, corollary 3.1, for the a.s.c.f. $\{\varphi_n\} \subset [x_n]^*$ to $\{x_n\}$ we have $\sup_{1 \leqslant n < \infty} \|\varphi_n\| < \infty$. Let $g_n \in E^*$ be any extension of $\varphi_n$ such that $\|g_n\| = \|\varphi_n\|$ $(n = 1,2, \ldots)$, hence $\sup_{1 \leqslant n < \infty} \|g_n\| < \infty$. Since by our assumption $[x_n] \neq E$, there exists $h \in [x_n]^\perp$, $h \neq 0$. Put

$$f_n = g_n + nh \qquad (n = 1,2, \ldots). \tag{1.77}$$

Then we have $f_i(x_j) = \delta_{ij}$ $(i, j = 1,2, \ldots)$. Furthermore,

$$\left\|h - \frac{1}{n} f_n\right\| = \frac{1}{n} \|g_n\| \to 0 \text{ as } n \to \infty,$$

whence $h \in [f_n]$. Now let $u$ be the canonical mapping of $E$ into $[f_n]^*$ (not necessarily one-to-one). Then $(f_n, u(x_n))$ is a biorthogonal system, but

$$(u(x_n))(h) = h(x_n) = 0 \qquad (n = 1,2, \ldots),$$

which (since $h \in [f_n]$, $h \neq 0$) shows that $\{u(x_n)\}$ is not total on $[f_n]$. Consequently, $\{f_n\}$ is not a basis of $[f_n]$, which completes the proof. By a similar argument, $\{f_n\}$ has no basic subsequence.

*Remark 1.10. When* $\inf_{1 \leqslant n < \infty} \|x_n\| > 0$, the sequence $\{f_n\}$ constructed in (1.77) above is unbounded, but *for reflexive separable Banach spaces* $E$ *one can also find a bounded sequences* $\{f_n\}$ *with the required properties*, as follows: By Ch. II, § 7, proposition 7.4 (which used only $\inf_{1 \leqslant n < \infty} \|x_n\| > 0$), we have $\varphi_n \xrightarrow{w^*} 0$, whence, by § 7, lemma 7.8, there exists a total sequence

$\{f_n\} \subset E^*$ such that $f_n|_G = \varphi_n$ $(n = 1,2, \ldots)$, $\sup_{1 \leqslant n < \infty} \|f_n\| < \infty$. Since $E$ is reflexive and separable, it results that $[f_n] = E^*$, whence, since $(x_n, f_n)$ is biorthogonal and $[x_n] \neq E$, it follows (by Ch. I, § 12, corollary 12.1) that $\{f_n\}$ is not a basis of $E^* = [f_n]$, which completes the proof. However, in general such a result no longer holds for non-reflexive Banach spaces, as shown by

*Example 1.2.* Let $E = c_0$ and let

$$x_n = e_{n+1} \qquad (n = 1,2, \ldots), \tag{1.78}$$

where $\{e_n\}$ is the unit vector basis of $c_0$. Then $\{x_n\}$ is a normalized basic sequence, which is not a basis, but every sequence $\{f_n\} \subset E^*$ with $\sup_{1 \leqslant n < \infty} \|f_n\| < \infty$ and $(x_n, f_n)$ biorthogonal, is a basic sequence. Indeed, let $\{h_n\} \subset E^*$ be the a.s.c.f. to $\{e_n\}$, that is, the sequence of coordinate functionals on $E = c_0$. Then, since $[x_n]^\perp = [e_{n+1}]^\perp = [h_1]$ is one-dimensional, and since $h_{i+1}(x_j) = h_{i+1}(e_{j+1}) = \delta_{ij}$, every sequence $\{f_n\} \subset E^*$ for which $(x_n, f_n)$ is biorthogonal can be written in the form

$$f_n = \beta_n h_1 + h_{n+1} \qquad (n = 1,2, \ldots), \tag{1.79}$$

where $\beta_n$ are suitable scalars. Since $\|f_n\| = |\beta_n| + 1$ $(n = 1,2, \ldots)$, we have $\sup_{1 \leqslant n < \infty} \|f_n\| < \infty$ if and only if $\sup_{1 \leqslant n < \infty} |\beta_n| < \infty$. However, if $\sup_{1 \leqslant n < \infty} |\beta_n| < \infty$, then for any scalars $\alpha_1, \ldots, \alpha_{n+m}$ we have

$$\left\| \sum_{i=1}^{n} \alpha_i f_i \right\| = \left\| \sum_{i=1}^{n} \alpha_i (\beta_i h_1 + h_{i+1}) \right\| = \left| \sum_{i=1}^{n} \alpha_i \beta_i \right| + \sum_{i=1}^{n} |\alpha_i| \leqslant$$

$$\leqslant ( \sup_{1 \leqslant j < \infty} |\beta_j| + 1) \sum_{i=1}^{n} |\alpha_i| \leqslant ( \sup_{1 \leqslant j < \infty} |\beta_j| + 1) \left( \sum_{i=1}^{n+m} |\alpha_i| + \left| \sum_{i=1}^{n+m} \alpha_i \beta_i \right| \right) =$$

$$= ( \sup_{1 \leqslant j < \infty} |\beta_j| + 1) \left\| \sum_{i=1}^{n+m} \alpha_i f_i \right\|,$$

whence, by Ch. I, § 7, theorem 7.1, $\{f_n\}$ is a basic sequence.

The following problem arises naturally:

*Problem 1.6.**) Let $\{x_n\}$ be a basic sequence in a Banach space $E$. a) Does there exist a basic sequence $\{f_n\} \subset E^*$ such that $(x_n, f_n)$ is a biorthogonal system? b) What about a $w^*$-Schauder basic sequence $\{f_n\}$ with this property?

Clearly, if $\{x_n\}$ can be extended to a basis $\{x_n\} \cup \{z_n\}$ of $E$, then the answer to both questions is affirmative.

---

*) Concerning problems 1.6 a), 1.6′ and 1.6″, see the Appendix.

Let us also mention the following slight generalization of problem 1.6 a):

*Problem 1.6'*. If $\{X_n\}$ is a basic sequence in (or basis of) a quotient space $E/G$ of a Banach space $E$, do there exist elements $x_n \in X_n$ $(n = 1,2, \ldots)$ such that $\{x_n\}$ is a basic sequence (respectively, a basis of $E$, when $E$ has a basis)?

The particular case of problem 1.6 a) is obtained by taking in the place of $\{X_n\}$ the image in $E^*/[x_n]^\perp$ of the a.s.c.f. $\{\varphi_n\} \subset [x_n]^*$ to a basic sequence $\{x_n\} \subset E$ by the canonical linear isometry $[x_n]^* \equiv E^*/[x_n]^\perp$, since the extension of a $\varphi_n \in [x_j]^*$ to an $f_n \in E^*$ amounts to the selection of an element $f_n$ from the equivalence class $\{g_n + [x_j]^\perp\} \in E^*/[x_j]^\perp$, where $g_n$ is an arbitrary extension of $\varphi_n$ $(n = 1,2, \ldots)$. If we regard the problem of extension of basic sequences to bases (Ch. I, § 4, problem 4.1) as that of extending a basis of a subspace to a basis of the whole space*), then problem 1.6' may be also considered dual to this problem, namely, as "lifting" a basic sequence in (or a basis of) a quotient space to a basic sequence in (respectively, a basis of) the whole space.

It is worth while to formulate separately the following particular case of problem 1.6' (which is still a slight generalization of problem 1.6 a)):

*Problem 1.6''*. If $\{\varphi_n\}$ is a basic sequence in the conjugate space $G^*$ of a subspace $G$ of a Banach space $E$, does there exists a basic sequence $\{f_n\} \subset E^*$ such that $f_n|_G = \varphi_n$ $(n = 1,2, \ldots)$?

Some other extension problems for the a.s.c.f. $\{\varphi_n\} \subset [x_n]^*$ to a basic sequence $\{x_n\} \subset E$ (or, equivalently, problems of finding, for $\{x_n\} \subset E$ basic, a sequence $\{f_n\} \subset E^*$ with some other properties, such that $(x_n, f_n)$ is a biorthogonal system) will be also considered in § 6, problem 6.2' and § 7, theorems 7.5 and 7.8; these properties are concerned with the "size" of $[f_n]_\perp$ and the boundedness of the sequence $\{f_n\}$.

In view of the above (in particular, of proposition 1.12), it is natural to give

*Definition 1.5.* Let $E$ be a Banach space. A biorthogonal system $(x_n, f_n)$ $(\{x_n\} \subset E, \{f_n\} \subset E^*)$ is called a *bibasic system* if both $\{x_n\}$ and $\{f_n\}$ are basic sequences (in $E$ and $E^*$, respectively).

We have the following sharpening of theorem 1.1:

**Theorem 1.9.** *For every Banach space $E$ there exists a bibasic system $(x_n, f_n)$ $(\{x_n\} \subset E, \{f_n\} \subset E^*)$ such that*

$$\sup_{1 \leqslant n < \infty} \|x_n\| < \infty, \quad \sup_{1 \leqslant n < \infty} \|f_n\| < \infty. \tag{1.80}$$

---

*) As we already mentioned, this problem has been solved recently in the negative.

*Proof.* Let $x_1 \in E$, $f_1 \in E^*$ be such that $\|x_1\| = \|f_1\| = f_1(x_1) = 1$. Choose $z_1, \ldots, z_{q_1} \in E$ such that the subspace $Z_1 = [x_1, z_1, \ldots, z_{q_1}]$ of $E$ satisfies*)

$$\sup_{\substack{z \in Z_1 \\ \|z\| \leqslant 1}} |f(z)| > \frac{1}{2} \|f\| \qquad (f \in [f_1]); \tag{1.81}$$

since $\dim Z_1 < \infty$, we can**) also choose $h_1, \ldots, h_{p_1} \in E^*$ such that the subspace $V_1 = [f_1, h_1, \ldots, h_{p_1}]$ of $E^*$ satisfies

$$\sup_{\substack{h \in V_1 \\ \|h\| \leqslant 1}} |h(z)| > \frac{1}{2} \|z\| \qquad (z \in Z_1). \tag{1.82}$$

Then the natural projections $u_1$, $v_1$ of $Z_1 \oplus (V_1)_\perp$ onto $Z_1$ and $[f_1] \oplus Z_1^\perp$ onto $[f_1]$, respectively, have norm $\leqslant 2$, since

$$\|z - x\| \geqslant \sup_{\substack{h \in V_1 \\ \|h\| \leqslant 1}} |h(z - x)| = \sup_{\substack{h \in V_1 \\ \|h\| \leqslant 1}} |h(z)| > \frac{1}{2} \|z\| \qquad (z \in Z_1,\ x \in (V_1)_\perp), \tag{1.83}$$

$$\|\alpha_1 f_1 - g\| \geqslant \sup_{\substack{z \in Z_1 \\ \|z\| \leqslant 1}} |\alpha_1 f_1(z) + g(z)| = \sup_{\substack{z \in Z_1 \\ \|z\| \leqslant 1}} |\alpha_1 f_1(z)| > \frac{1}{2} \|\alpha_1 f_1\|$$

$$(\alpha_1 \in K,\ g \in Z_1^\perp).$$

Now let $x_2 \in (V_1)_\perp$, $\|x_2\| = 1$. If $z \in Z_1$, $\|z\| \geqslant \frac{2}{3}$, then by (1.83) for $x = x_2$, we have $\|z - x_2\| > \frac{1}{3}$. On the other hand, if $z \in Z_1$, $\|z\| < \frac{2}{3}$, then $\|z - x_2\| \geqslant | \|z\| - \|x_2\| | = | \|z\| - 1| > \frac{1}{3}$. Consequently,

$$\operatorname{dist}(x_2, Z_1) \geqslant \frac{1}{3},$$

---

*) Of course, at this step one could simply take $z_1 = \ldots = z_{q_1} = x_1$ and then $h_1 = \ldots = h_{p_1} = f_1$.

**) Indeed, it is enough to pick a finite $\frac{1}{4}$-net $\{y_1, \ldots, y_{p_1}\}$ of $\sigma_{Z_1} = \{z \in Z_1 \mid \|z\| = 1\}$ and then $h_1, \ldots, h_{p_1} \in E^*$ with $\|h_i\| = 1$, $|h_i(y_i)| \geqslant \frac{3}{4}$ $(i = 1, \ldots, p_1)$.

and hence, by a corollary of the Hahn-Banach theorem, there exists an $f_2 \in Z_1^\perp$ such that $f_2(x_2) = 1$, $\|f_2\| \leqslant \dfrac{1}{\operatorname{dist}(x_2, Z_1)} \leqslant 3$.

Assume now that we have obtained $\{x_1, \ldots, x_n\}$, $\{f_1, \ldots, f_n\}$, $\{z_1, \ldots, z_{q_1}, \ldots, z_n, \ldots, z_{q_n}\}$ and $\{h_1, \ldots, h_{p_1}, \ldots, h_n, \ldots, h_{p_n}\}$ such that $Z_n = [x_1, \ldots, x_n, z_1, \ldots, z_{q_n}]$ satisfies

$$\sup_{\substack{z \in Z_n \\ \|z\| \leqslant 1}} |f(z)| > \frac{1}{2}\|f\| \qquad (f \in [f_1, \ldots, f_n]), \tag{1.84}$$

and that $V_n = [f_1, \ldots, f_n, h_1, \ldots, h_{p_n}]$ satisfies

$$\sup_{\substack{h \in V_n \\ \|h\| \leqslant 1}} |h(z)| > \frac{1}{2}\|z\| \qquad (z \in Z_n). \tag{1.85}$$

Then, as before, (1.84) and (1.85) imply that the natural projections $u_n$, $v_n$ of $Z_n \oplus (V_n)_\perp$ onto $Z_n$ and $[f_1, \ldots, f_n] \oplus Z_n^\perp$ onto $[f_1, \ldots, f_n]$, respectively, have norm $\leqslant 2$. Thus, if $x_{n+1} \in (V_n)_\perp$, $\|x_{n+1}\| = 1$, then, as above,

$$\operatorname{dist}(x_{n+1}, Z_n) \geqslant \frac{1}{3},$$

and hence, by a corollary of the Hahn-Banach theorem, there exists an $f_{n+1} \in Z_n^\perp$ such that $f_{n+1}(x_{n+1}) = 1$, $\|f_{n+1}\| \leqslant \dfrac{1}{\operatorname{dist}(x_{n+1}, Z_n)} \leqslant 3$. Now choose $z_{q_n+1}, \ldots, z_{q_{n+1}} \in E$, so that $Z_{n+1} = [x_1, \ldots, x_{n+1}, z_1, \ldots, z_{q_{n+1}}]$ satisfies (1.84) for $[f_1, \ldots, f_{n+1}]$ and then choose $h_{p_n+1}, \ldots, h_{p_{n+1}} \in E^*$ so that $V_{n+1} = [f_1, \ldots, f_{n+1}, h_1, \ldots, h_{p_{n+1}}]$ satisfies (1.85) for $Z_{n+1}$. In this way we obtain a biorthogonal system $(x_n, f_n)$ with $\|x_n\| = 1$, $\|f_n\| \leqslant 3$ $(n = 1,2, \ldots)$. Finally, for any $n \geqslant 1$ we have $x_{n+m} \in (V_{n+m-1})_\perp \subset (V_n)_\perp$, $f_{n+m} \in Z_{n+m-1}^\perp \subset Z_n$ $(m = 1,2, \ldots)$, whence for any scalars $\alpha_1, \ldots, \alpha_{n+m}$

$$\left\|\sum_{i=1}^{n} \alpha_i x_i\right\| \leqslant \|u_n\| \left\|\sum_{i=1}^{n+m} \alpha_i x_i\right\| \leqslant 2\left\|\sum_{i=1}^{n+m} \alpha_i x_i\right\|,$$

$$\left\|\sum_{i=1}^{n} \alpha_i f_i\right\| \leqslant \|v_n\| \left\|\sum_{i=1}^{n+m} \alpha_i f_i\right\| \leqslant 2\left\|\sum_{i=1}^{n+m} \alpha_i f_i\right\|,$$

and hence, by Ch. I, § 7, theorem 7.1, $(x_n, f_n)$ is a bibasic system, which completes the proof. Note that this proof may be also regarded as a variant of the proof of theorem 1.2; indeed, if we put $P_{(n)} = [x_1, \dots, x_n]$, $Y_n = (V_n)_\perp$ $(n = 1,2, \dots)$, then the conditions (a), (b) and (c) of that proof will be satisfied $\left(\text{with } \frac{1}{2} \text{ instead of } 1-\varepsilon, \text{ because now } (\widehat{P_{(n)}; Y_n}) \geqslant \geqslant (\widehat{Z_n; Y_n}) \geqslant \frac{1}{2}\right)$, and a similar property holds for $P'_n = [f_1, \dots, f_n]$, $Y'_n = Z_n^\perp$ $(n = 1,2, \dots)$.

*Remark 1.11. If E is separable*, one can prove, with a different method, a somewhat stronger result, namely, in this case *there exists a bibasic system* $(x_k, f_k)$ *such that* $\|x_k\| = 1$, $\|f_k\| \leqslant 2$ *for* $k = 1,2, \dots$ *and that* $\{f_k\}$ *is* $w^*$*-Schauder basic* (and hence, by remark 1.7, $[x_k] + [f_k]_\perp$ *is dense in* $E$). Indeed, by theorem 1.4, there exists in $E$ a normal basic sequence $\{y_n\}$. Let $\{\psi_n\} \subset [y_n]^*$ be the a.s.c.f. to $\{y_n\}$ and let $g_n \in E^*$ be an extension of $\psi_n$, with $\|g_n\| = \|\psi_n\| = 1$ $(n = 1,2, \dots)$. Then, since $E$ is separable, $\{g_n\}$ has a subsequence $\{g_{n_m}\}$ such that $g_{n_m} \xrightarrow{w^*} g \in E^*$. Clearly, $\overline{\lim\limits_{m\to\infty}} \|g_{n_m} - g\| > 0$ (since otherwise we would have $g_{n_m} \to g$ in the norm topology of $E^*$, contradicting the relations $\|g_n - g_m\| \geqslant |(g_n - g_m)(y_m)| = 1$ for $n \neq m$) and hence, using again that $E$ is separable, we can select, by theorem 1.5, a subsequence $\{g_{n_{m_k}} - g\}$ of $\{g_{n_m} - g\}$, which is a $w^*$-Schauder basic sequence. Consequently, putting $x_k = y_{n_{m_k}}$, $f_k = g_{n_{m_k}} - g$ $(k = 1,2, \dots)$, we obtain a bibasic system $(x_k, f_k)$ with the required properties.

If $(x_n, f_n)$ is a bibasic system, then $\{f_n\}$ need not be equivalent to the a.s.c.f. $\{\varphi_n\} = \{f_n|_{[x_j]}\} \subset [x_n]^*$ to $\{x_n\}$. Moreover, in contrast with the particular case when $\{x_n\}$ is a basis of the whole space (see Ch. II, § 6, corollary 6.1), for a general bibasic system there need not be any duality relation between the shrinkingness or boundedly completeness of $\{x_n\}$ and $\{f_n\}$. For the construction of examples illustrating these statements the following observation will be useful:

**Proposition 1.13.** *Let $G$, $F$ be two Banach spaces with bases $\{y_n\}$ and $\{z_n\}$, respectively. Then the sequence $\{x_n\} \subset G \times F$ defined by*

$$x_n = \{y_n, z_n\} \qquad (n = 1,2, \dots) \tag{1.86}$$

*is a basic sequence in $G \times F$.*

*First proof.* By Ch.I, § 4, proposition 4.2, the sequence*) $\{\{y_n, 0\} \cup \cup \{0, z_n\}\}$ is a basis of $G \times F$. Hence, by Ch. I, § 4, proposition 4.4, the block-perturbation $\{\{y_n, 0\} \cup \{y_n, z_n\}\}$ is a basis of $G \times F$. Con-

*) We recall (see Ch. I, § 4, definition 4.6) that this notation stands for $\{y_1, 0\}$, $\{0, z_1\}$, $\{y_2, 0\}$, $\{0, z_2\}, \dots$

sequently, by Ch. I, § 4, proposition 4.1, (1.86) is a basic sequence in $G \times F$.

*Second proof.* For any scalars $\alpha_1, \ldots, \alpha_{n+m}$ we have

$$\left\| \sum_{i=1}^{n} \alpha_i x_i \right\| = \sqrt{\left\| \sum_{i=1}^{n} \alpha_i y_i \right\|^2 + \left\| \sum_{i=1}^{n} \alpha_i z_i \right\|^2} \leqslant$$

$$\leqslant \max (\nu_{\{y_n\}}, \nu_{\{z_n\}}) \sqrt{\left\| \sum_{i=1}^{n+m} \alpha_i y_i \right\|^2 + \left\| \sum_{i=1}^{n+m} \alpha_i z_i \right\|^2} = \left\| \sum_{i=1}^{n+m} \alpha_i x_i \right\|,$$

and hence, by Ch. I, § 7, theorem 7.1, $\{x_n\}$ is a basic sequence.

Now we can give the examples mentioned above.

*Example 1.3.* Let $E = c_0 \times c_0$ and let

$$x_n = \left\{ e_n, \sum_{i=1}^{n} e_i \right\}, f_n = \{h_n, 0\} \qquad (n = 1,2, \ldots), \quad (1.87)$$

where $\{e_n\}$ is the unit vector basis of $c_0$ and $\{h_n\}$ the sequence of coordinate functionals on $c_0$. Then, by proposition 1.13 and Ch. II, § 14, example 14.1, $(x_n, f_n)$ is a bibasic system. Clearly $\{f_n\}$ is boundedly complete. However, $\{x_n\}$ is not shrinking, since for $g = \{0, h_1\} \in E^*$ we have $g(x_n) = 1$ $(n = 1,2, \ldots)$, whence $\|g|_{[x_{n+1}, x_{n+2}, \ldots]}\| \geqslant 1$ $(n = 1,2, \ldots)$. This shows that for $(x_n, f_n)$ bibasic and $\{f_n\}$ boundedly complete, $\{x_n\}$ need not be shrinking. By interchanging the roles of $\{x_n\}$ and $\{f_n\}$, we also see that for $(x_n, f_n)$ bibasic with $\{x_n\}$ boundedly complete, $\{f_n\}$ need not be shrinking.

*Example 1.4.* Let $E = c_0 \times l^2$ and let

$$x_n = \left\{ e_n, \frac{1}{2^n} e'_n \right\}, f_n = \{0, 2^n h_n\} \qquad (n = 1,2, \ldots), (1.88)$$

where $\{e_n\}$, $\{e'_n\}$ are the unit vector bases of $c_0$ and $l^2$, respectively, and $h'_n$ the sequence of coordinate functionals on $l^2$. Then, by proposition 1.13, $(x_n, f_n)$ is a bibasic system and, clearly, $\{f_n\}$ is shrinking. However, $\{x_n\}$ is not boundedly complete, since

$$\left\| \sum_{i=1}^{n} x_i \right\| = \sqrt{\left\| \sum_{i=1}^{n} e_i \right\|_{c_0}^2 + \left\| \sum_{i=1}^{n} \frac{1}{2^i} e'_i \right\|_{l^2}^2} = \sqrt{1 + \sum_{i=0}^{n} \frac{1}{2^{2i}}} < 2 \quad (n=1,2,\ldots),$$

but $\sum_{i=1}^{\infty} x_i$ does not converge (because $\|x_n\| > 1$ for $n = 1,2, \ldots$). Thus for $(x_n, f_n)$ bibasic and $\{f_n\}$ shrinking, $\{x_n\}$ need not be boundedly complete. By interchanging the roles of $\{x_n\}$ and $\{f_n\}$, we also see that for $(x_n, f_n)$ bibasic with $\{x_n\}$ shrinking, $\{f_n\}$ need not be boundedly complete. Furthermore, by Ch. II, § 6, corollary 6.1, these examples also show that for $(x_n, f_n)$ bibasic, $\{f_n\}$ need not be equivalent to the a.s.c.f. $\{\varphi_n\} = \{f_n|_{[x_j]}\} \subset [x_n]^*$ to $\{x_n\}$. Another example of this latter situation, even with $\{f_n\}$ unconditional and $\{\varphi_n\}$ conditional, appears in Ch. II, § 17, example 17.2.

In theorem 1.10 below we shall give several necessary and sufficient conditions in order that for a bibasic system $(x_n, f_n)$, the sequence $\{f_n\}$ be equivalent to the a.s.c.f. $\{\varphi_n\} = \{f_n|_{[x_j]}\} \subset [x_n]^*$ to $\{x_n\}$. For this we need some preparation. We recall that if $G$, $V$ are linear subspaces of $E$ and $E^*$, respectively, then the *characteristic of* $V$ *with respect to* $G$ is defined by

$$r_G(V) = \inf_{\substack{x \in G \\ x \neq 0}} \sup_{\substack{f \in V \\ \|f\| \leqslant 1}} \left| f\left(\frac{x}{\|x\|}\right) \right| \tag{1.89}$$

and, dually, the *characteristic of* $G$ *with respect to* $V$ is defined by

$$r_V(G) = \inf_{\substack{f \in V \\ f \neq 0}} \sup_{\substack{x \in G \\ \|x\| \leqslant 1}} \left| \frac{f}{\|f\|}(x) \right|. \tag{1.90}$$

In particular, for $G = E$ we have $r_E(V) = r(V)$, the usual characteristic of $E$ (by Ch. I, § 12, formula (12.15)). It is also clear that $r_G(V) > 0$, respectively, $r_V(G) > 0$, if and only if there exists a constant $C > 0$ such that

$$\sup_{\substack{f \in V \\ \|f\| \leqslant 1}} |f(x)| \geqslant C\|x\| \qquad (x \in G), \tag{1.91}$$

respectively, such that

$$\sup_{\substack{x \in G \\ \|x\| \leqslant 1}} |f(x)| \geqslant C\|f\| \qquad (f \in V). \tag{1.92}$$

Similarly, extending the notion of totality, $V$ is said to be *total over* $G$, if $\{x \in G \mid f(x) = 0 \ (f \in V)\} = \{0\}$, i.e., if $G \cap V_\perp = \{0\}$ and, dually, $G$ is called *total over* $V$, if $V \cap G^\perp = \{0\}$.

**Lemma 1.3.** *Let $E$ be a Banach space and $G$, $V$ (closed linear) subspaces of $E$ and $E^*$ respectively, such that $G$ is total over $V$. The following statements are equivalent:*

1°. $r_V(G) > 0$.

2°. *$G^\perp + V$ is closed in $E^*$.*

3°. *The restriction mapping $\rho: f \to f|_G$ from $V$ to $G^*$ is an isomorphism of $V$ into $G^*$.*

*Proof.* We have

$$\|f\| \geqslant \|\rho(f)\| = \|f|_G\| = \sup_{\substack{x \in G \\ \|x\| \leqslant 1}} |f(x)| \qquad (f \in V),$$

and thus the equivalence 1° $\Leftrightarrow$ 3° is obvious (by 1.92)).

Furthermore, $G^\perp + V$ is closed in $E^*$ if and (by the closed graph theorem) only if*) the natural projection of $G^\perp + V$ onto $V$ is continuous, which happens if and only if there exists a constant $C > 0$ such that

$$\inf_{h \in G^\perp} \|f + h\| \geqslant C\|f\| \qquad (f \in V).$$

Hence, since by the Hahn-Banach theorem $\|\rho(f)\| = \|f|_G\| = \inf_{h \in G^\perp} \|f + h\|$, we see that 2° $\Leftrightarrow$ 3°, which completes the proof of lemma 1.3.

It will be convenient to consider, for an arbitrary biorthogonal system $(x_n, f_n)$ ($\{x_n\} \subset E$, $\{f_n\} \subset E^*$), the linear subspaces

$$\mathscr{E}_1 = \left\{ x \in E \,\middle|\, \sum_{i=1}^{\infty} f_i(x) x_i \text{ converges} \right\}, \tag{1.93}$$

$$\mathscr{E}_2 = \left\{ x \in E \,\middle|\, \sup_{1 \leqslant n < \infty} \left\| \sum_{i=1}^{n} f_i(x) x_i \right\| < \infty \right\} \tag{1.94}$$

of $E$; clearly, $\mathscr{E}_1 \subset \mathscr{E}_2$. In the particular case when $[x_n] = E$, these linear subspaces have been introduced in Ch. I, § 5, formula (5.3) and (5.4) and it has been proved (Ch. I, § 5, theorem 5.1) that in this case $\{x_n\}$ is a basis of $E$ if and only if $\mathscr{E}_1 = \mathscr{E}_2$ ($= E$). When $[x_n] \neq E$ and $\{x_n\}$ is a boundedly complete basic sequence, we still have that $\mathscr{E}_1 = \mathscr{E}_2$, but in general, for arbitrary basic sequences $\{x_n\}$, we have $\mathscr{E}_1 \neq \mathscr{E}_2$; for example, if $\{x_n\}$ is the sequence of unit vectors in $E = m$, then $\mathscr{E}_1 = c_0$ and $\mathscr{E}_2 = m$.

**Proposition 1.14.** *Let $E$ be a Banach space and $(x_n, f_n)$ ($\{x_n\} \subset E$, $\{f_n\} \subset E^*$) a biorthogonal system.*

---

*) See e.g. [87], p. 480.

a) *If* $\{x_n\}$ *is a basic sequence, then*

$$\mathscr{E}_1 = [x_n] + [f_n]_\perp, \tag{1.95}$$

$$\overline{\mathscr{E}}_1 \subset \{x \in E \mid \|x\|_{[f_{n+1}, f_{n+2}, \ldots]} \to 0 \text{ as } n \to \infty\}, \tag{1.96}$$

$$\left\{\Phi \in E^{**} \,\middle|\, \sum_{i=1}^{\infty} \Phi(f_i)x_i \text{ converges}\right\} = \pi([x_n]) + [f_n]^\perp, \tag{1.97}$$

*where* $\mathscr{E}_1$ *is defined by* (1.93), $\|x\|_V = \sup_{\substack{f \in V \\ \|f\| \leqslant 1}} |f(x)|$ *and* $\pi$ *is the canonical embedding of* $E$ *into* $E^{**}$.

b) *If* $\{f_n\}$ *is a basic sequence, then*

$$\left\{f \in E^* \,\middle|\, \sum_{i=1}^{\infty} f(x_i) f_i \text{ converges}\right\} = [x_n]^\perp + [f_n]. \tag{1.98}$$

*Proof.* Since part b) is a consequence of part a) applied to the biorthogonal system $(f_n, \pi(x_n))$, we have only to prove part a).

Assume that $\{x_n\}$ is a basic sequence. Then, clearly, $[x_n]+[f_n]_\perp \subset \mathscr{E}_1$. Conversely, given $x \in \mathscr{E}_1$, let $y = \sum_{i=1}^{\infty} f_i(x)x_i$. Then $y \in [x_n]$ and $f_n(x) = f_n(y)$ $(n = 1,2, \ldots)$, so $x - y \in [f_n]_\perp$ and hence

$$x = y + (x - y) \in [x_n] + [f_n]_\perp,$$

which proves (1.95).

Furthermore, by proposition 1.11 applied to the biorthogonal system $(f_n, \pi(x_n))$, we have

$$\overline{[f_n]^\perp + [\pi(x_n)]} = \{\Phi \in E^{**} \mid \|\Phi \mid_{[f_{n+1}, f_{n+2}, \ldots]}\| \to 0 \text{ as } n \to \infty\}, \tag{1.99}$$

whence, by (1.95) and since $\pi$ is a linearly isometric embedding of $E$ into $E^{**}$, we obtain (1.96) (taking on both sides of (1.99) the intersection with $\pi(E)$ and then applying $\pi^{-1}$).

Finally, by (1.95) applied to the biorthogonal system $(\pi(x_n), \pi_1(f_n))$, where $\pi_1$ denotes the canonical mapping of $E^*$ into $E^{***}$, we have

$$\pi([x_n]) + [f_n]^\perp = [\pi(x_n)] + [\pi_1(f_n)]_\perp =$$

$$= \left\{\Phi \in E^{**} \,\middle|\, \sum_{i=1}^{\infty} (\pi_1(f_i))\,(\Phi)\pi(x_i) \text{ converges}\right\} =$$

$$= \left\{\Phi \in E^{**} \,\middle|\, \sum_{i=1}^{\infty} \Phi(f_i)\pi(x_i) \text{ converges}\right\} =$$

$$= \left\{\Phi \in E^{**} \,\middle|\, \sum_{i=1}^{\infty} \Phi(f_i)x_i \text{ converges}\right\},$$

i.e. (1.97), which completes the proof of proposition 1.14. Note that in the particular case when $[x_n] = E$, formula (1.97) has been proved in Ch. I, § 12, theorem 12.5 f), with a different method.

**Corollary 1.12.** *Let $E$ be a Banach space and let $(x_n, f_n)$ ($\{x_n\} \subset E$, $\{f_n\} \subset E^*$) be a biorthogonal system.*

*a) If $\{x_n\}$ is a basic sequence, then*

*(i) $\mathscr{E}_1 = E$ if and only if $E = [x_n] \oplus [f_n]_\perp$.*

*(ii) $\left\{\Phi \in E^{**} \,\middle|\, \sum_{i=1}^{\infty} \Phi(f_i)x_i \text{ converges}\right\} = E^{**}$ if and only if $E^{**} = \pi([x_n]) \oplus [f_n]^\perp$.*

*Consequently, in this case we have also $E = [x_n] \oplus [f_n]_\perp$.*

*b) If $\{f_n\}$ is a basic sequence, then*

*(iii) $\left\{f \in E^* \,\middle|\, \sum_{i=1}^{\infty} f(x_i) f_i \text{ converges}\right\} = E^*$ if and only if $E^* = [x_n]^\perp \oplus [f_n]$.*

*Proof.* Since part b) is a consequence of part a) applied to the biorthogonal system $(f_n, \pi(x_n))$, we have only to prove part a).

Now, if $\{x_n\}$ is a basic sequence, the sufficiency parts in (i), (ii) are obvious. Furthermore, the necessity parts are consequences of proposition 1.14 a). Finally, in the situation of (ii) we have, clearly, $\mathscr{E}_1 = E$, and hence, by (i), $E = [x_n] \oplus [f_n]_\perp$, which completes the proof.

Let us prove now

**Theorem 1.10.** *Let $E$ be a Banach space, $\{x_n\}$ a basic sequence in $E$ with the a.s.c.f. $\{\varphi_n\} \subset [x_n]^*$, and $\{f_n\} \subset E^*$ a basic sequence such that $f_n \mid_{[x_j]} = \varphi_n$ for $n = 1,2, \ldots$ (hence $(x_n, f_n)$ is a bibasic system). The following statements are equivalent:*

1°. $\{f_n\} \sim \{\varphi_n\}$.

2°. $r_{[f_n]}([x_n]) > 0$.

3°. *$[x_n]^\perp + [f_n]$ is closed in $E^*$ (hence $V' = [x_n]^\perp \oplus [f_n]$, where $V' \subset E^*$ is defined by (1.69)).*

4°. *The restriction mapping $\rho: f \to f\,|_{[x_n]}$ from $[f_n]$ to $[x_n]^*$ is an isomorphism of $[f_n]$ into $[x_n]^*$.*

5°. $r_{[x_n]}([f_n]) > 0$.

6°. *$\pi([x_n]) + [f_n]^\perp$ is closed in $E^{**}$.*

7°. *The canonical mapping $u$ of $[x_n]$ into $[f_n]^*$ is an isomorphism.*

8°. *$\sup_{1 \leqslant n < \infty} \|s_n\| < \infty$, where*

$$s_n(x) = \sum_{i=1}^{n} f_i(x)x_i \qquad (x \in E,\ n = 1,2, \ldots). \tag{1.100}$$

9°. *$\sup_{1 \leqslant n < \infty} \left\| \sum_{i=1}^{n} f_i(x)x_i \right\| < \infty$ for all $x \in E$ (i.e., $\mathscr{E}_2 = E$).*

9'. $\sup_{1 \leqslant n < \infty} \left\| \sum_{i=1}^{n} f(x_i) f_i \right\| < \infty$ *for all* $f \in E^*$.

9''. $\sup_{1 \leqslant n < \infty} \left\| \sum_{i=1}^{n} \Phi(f_i) x_i \right\| < \infty$ *for all* $\Phi \in E^{**}$.

10°. $E^* = [x_n]^\perp \oplus \Gamma$, *where* $\Gamma$ *is a closed linear subspace of* $E^*$, *containing* $[f_n]$.

11°. $E^{**} = [x_n]^{\perp\perp} \oplus \Gamma^\perp$, *where* $\Gamma$ *is as in* 10°.

12°. $E^{**} = [f_n]^\perp \oplus \Gamma_1$, *where* $\Gamma_1$ *is a closed linear subspace of* $E^{**}$, *containing* $\pi([x_n])$.

*These statements imply*

13°. $[x_n] + [f_n]_\perp$ *is closed in* $E$.

*Statements* 1°—12° *are implied by — and, if and only if* $\{x_n\}$ *is a shrinking basic sequence, statements* 1°—12° *are equivalent to — the following statements:*

14°. *For every* $f \in E^*$ *the series* $\sum_{i=1}^{\infty} f(x_i) f_i$ *converges.*

15°. $E^* = [x_n]^\perp \oplus [f_n]$.

*Proof.* If we have 1° then, by Ch. I, § 8, theorem 8.1. d) there exists an isomorphism $v$ of $[f_n]$ onto $[\varphi_n]$ such that $v(f_n) = \varphi_n$ $(n = 1,2, \ldots)$. Hence, since $\varphi_n = f_n|_{[x_j]}$ $(n = 1,2, \ldots)$, it follows that $v(f) = f|_{[x_j]}$ for all $f \in [f_n]$, so $v = \rho$. Thus, 1° $\Rightarrow$ 4°.

Conversely, if we have 4°, then, since $\rho(f_n) = f_n|_{[x_j]} = \varphi_n$ $(n=1,2,\ldots)$ it follows that $\{f_n\} \sim \{\varphi_n\}$. Thus, 4° $\Rightarrow$ 1°.

Since $\{f_n\}$ is a basic sequence, we have $[f_n] \cap [x_n]^\perp = \{0\}$ and hence the equivalences 2° $\Leftrightarrow$ 3° $\Leftrightarrow$ 4° follow from lemma 1.3 (for $G = [x_n]$, $V = [f_n]$). The equality $V' = [x_n]^\perp \oplus [f_n]$ in 3° is a consequence of (1.75).

Assume now that we have 4°. Since $\{x_n\}$ is a basic sequence, we have $r_{[x_n]}([\varphi_n]) > 0$ (by Ch. I, § 12, theorem 12.2). Hence, by 4°, $r_{[x_n]}([f_n]) > 0$. Thus, 4° $\Rightarrow$ 5°.

Assume now that we have 5°. Then $r_{[\pi(x_n)]}([f_n]) = r_{[x_n]}([f_n]) > 0$, whence, applying the implication 2° $\Rightarrow$ 5° proved above, to the bibasic system $(f_n, \pi(x_n))$, we obtain $r_{[f_n]}([\pi(x_n)]) > 0$, and therefore $r_{[f_n]}([x_n]) > 0$. Thus, 5° $\Rightarrow$ 2°.

Since $\{x_n\}$, and thus $\{\pi(x_n)\}$, is a basic sequence, we have $\pi([x_n]) \cap \cap [f_n]^\perp = \{0\}$ and hence the equivalences 5° $\Leftrightarrow$ 6° $\Leftrightarrow$ 7° follow from lemma 1.3 applied for $[f_n] \subset E^*$ and $[\pi(x_n)] \subset E^{**}$, taking into account that $r_{[\pi(x_n)]}([f_n]) = r_{[x_n]}([f_n])$ and that $u = \rho_1 \pi$, where $\rho_1$ is the restriction map from $\pi(E) \subset E^{**}$ to $[f_n]^*$.

Assume now again that we have 5° and let $C > 0$ be a constant such that we have (1.91) for $G = [x_n]$, $V = [f_n]$, that is,

$$\sup_{\substack{f \in [f_n] \\ \|f\| \leqslant 1}} |f(x)| \geqslant C \|x\| \qquad (x \in [x_n]).$$

Then, since $s_n(x) \in [x_j]$ ($x \in E$, $n = 1,2, \ldots$), we obtain

$$\|s_n(x)\| \leqslant \frac{1}{C} \sup_{\substack{f \in [f_n] \\ \|f\| \leqslant 1}} |f(s_n(x))| = \frac{1}{C} \sup_{\substack{f \in [f_n] \\ \|f\| \leqslant 1}} |(s_n^*(f))(x)| \leqslant$$

$$\leqslant \frac{1}{C} \|s_n^*|_{[f_j]}\| \, \|x\| = \frac{v_{\{f_j\}}}{C} \|x\| \qquad (x \in E,\ n = 1,2, \ldots),$$

whence $\sup_{1 \leqslant n < \infty} \|s_n\| \leqslant \frac{v_{\{f_j\}}}{C} < \infty$, where $v_{\{f_j\}}$ is the norm of $\{f_j\}$. Thus, 5° ⇒ 8°.

Conversely, assume that we have 8° and let $x \in [x_n]$, $\|x\| = 1$, $0 < \varepsilon < \frac{1}{2}$ and $\|x - s_n(x)\| < \varepsilon$. Then, if $f \in E^*$, $\|f\| = 1$, $|f(x)| > 1 - \varepsilon$, we have

$$|f(x)| - |f(s_n(x))| \leqslant |f(x) - f(s_n(x))| \leqslant \|x - s_n(x)\| < \varepsilon,$$

whence

$$|s_n^*(f)(x)| = |f(s_n(x))| \geqslant |f(x)| - \varepsilon > 1 - 2\varepsilon.$$

Since by 8° we have $\|s_n^*(f)\| \leqslant \sup_{1 \leqslant j < \infty} \|s_j^*\| = M < \infty$, it follows that

$$\sup_{\substack{g \in [f_n] \\ \|g\| \leqslant M}} |g(x)| \geqslant 1,$$

whence

$$\sup_{\substack{g \in [f_n] \\ \|g\| \leqslant 1}} |g(x)| \geqslant \frac{1}{M} > 0,$$

and therefore $r_{[x_n]}([f_n]) > 0$. Thus, 8° ⇒ 5°.

The equivalences 8° ⇔ 9° ⇔ 9′ ⇔ 9″ are a consequence of the principle of uniform boundedness and of Ch. I, § 12, formula (12.2) and (12.48) for $s_n^*$ and $s_n^{**}$.

Assume now that we have 8°. Then $\sup_{1 \leqslant n < \infty} \|s_n^*\| < \infty$ and hence there exists a subnet $\{s_{n_\alpha}^*\}$ of $\{s_n^*\}$ converging in the $w^*$-operator topology to an operator $\tau \in L(E^*, E^*)$. We claim that $\tau$ is a projection on $E^*$. Indeed, let $f \in E^*$ be arbitrary and put $g = \tau(f)$. Then

$$g(x_j) = \tau(f)(x_j) = \lim_{n_\alpha} s_{n_\alpha}^*(f)(x_j) =$$
$$= \lim_{n_\alpha} \sum_{i=1}^{n_\alpha} f(x_i) f_i(x_j) = f(x_j) \qquad (j = 1,2, \ldots), \tag{1.101}$$

whence

$$\tau^2(f) = \tau(\tau(f)) = \tau(g) = w^*\text{-}\lim_{n_\alpha} \sum_{i=1}^{n_\alpha} g(x_i)f_i = w^*\text{-}\lim_{n_\alpha} \sum_{i=1}^{n_\alpha} f(x_i)f_i = \tau(f),$$

which proves the claim. Moreover, by (1.101) we also have

$$(I-\tau)(f) = f - \tau(f) = f - g \in [x_j]^\perp \qquad (f \in E^*),$$

that is, $(I-\tau)(E^*) \subset [x_j]^\perp$. Conversely, for any $f \in [x_j]^\perp$ we have

$$\tau(f) = w^*\text{-}\lim_{n_\alpha} \sum_{i=1}^{n_\alpha} f(x_i) f_i = 0,$$

whence $f = (I-\tau)(f) \in (I-\tau)(E^*)$, and therefore

$$(I-\tau)(E^*) = [x_j]^\perp. \tag{1.102}$$

Consequently, $E_n^* = [x_j]^\perp \oplus \Gamma$, where $\Gamma = \tau(E^*)$. Since

$$\tau(f_j) = w^*\text{-}\lim_{n_\alpha} \sum_{i=1}^{n_\alpha} f_j(x_i) f_i = f_j \qquad (j = 1,2, \ldots),$$

we have also $f_j \in \Gamma$ $(j = 1,2, \ldots)$, whence $[f_j] \subset \Gamma$. Thus, $8° \Rightarrow 10°$.
The implication $10° \Rightarrow 11°$ is obvious and the implication $11° \Rightarrow 10°$ is well known (see § 8, lemma 8.2). The implication $10° \Rightarrow 3°$ is also obvious (since by 10° the restriction to $[x_n]^\perp + [f_n]$ of the natural projection $E^* = [x_n]^\perp \oplus \Gamma \to [x_n]^\perp$ is continuous).

Assume now that we have 5°. Then $r_{[\pi(x_n)]}([f_n]) = r_{[x_n]}([f_n]) > 0$, whence, applying the implication $2° \Rightarrow 10°$ proved above, to the biorthogonal... basic system $(f_n, \pi(x_n))$, we obtain 12°. Conversely, similarly to the implication $10° \Rightarrow 3°$ above, the implication $12° \Rightarrow 6°$ is obvious. Thus $1° \Leftrightarrow \ldots \Leftrightarrow 12°$.
Assume now that we have 6°. Then, since

$$\pi([x_n] + [f_n]_\perp) = \pi([x_n]) + \pi([f_n]_\perp) =$$
$$= \pi([x_n]) + ([f_n]^\perp \cap \pi(E)) = (\pi([x_n]) + [f_n])^\perp \cap \pi(E),$$

and since $\pi$ is a linear isometry, it follows that $[x_n] + [f_n]_\perp$ is closed in $E$. Thus, $6° \Rightarrow 13°$.
Furthermore, the implication $15° \Rightarrow 3°$ is obvious and the equivalence $14° \Leftrightarrow 15°$ is nothing else than corollary 1.12 b). Now, if $\{x_n\}$ is a shrinking basic sequence, then $\sum_{i=1}^{\infty} \varphi(x_i)\varphi_i$ converges for all $\varphi \in [x_n]^*$

and hence, if we have also 4°, then $\sum_{i=1}^{\infty}\varphi(x_i)f_i$ converges for all $\varphi\in[x_n]^*$, so $\sum_{i=1}^{\infty} f(x_i)f_i$ converges for all $f\in E^*$. Thus, for $\{x_n\}$ shrinking, 4° $\Rightarrow$ 14° (note also that for $\{x_n\}$ shrinking the implication 3° $\Rightarrow$ 15° follows from the remark made after proposition 1.11). Conversely, if we have 14° (or, equivalently, 15°), then $\sum_{i=1}^{\infty}\varphi(x_i)f_i$ converges for all $\varphi\in[x_n]^*$, whence, since the restriction mapping $\rho: f\to f|_{[x_n]}$ from $[f_n]$ to $[x_n]^*$ is always continuous (because it is linear and of norm $\leqslant 1$), it follows that $\sum_{i=1}^{\infty}\varphi(x_i)\varphi_i$ converges for all $\varphi\in[x_n]^*$, so $\{x_n\}$ is shrinking. Thus, if we have 14° (hence, in particular, if 4° $\Rightarrow$ 14°), then $\{x_n\}$ is shrinking, which completes the proof of theorem 1.10.

*Remark 1.12.* For any biorthogonal system $(x_n, f_n)$ ($\{x_n\}\subset E$, $\{f_n\}\subset E^*$) such that $\{x_n\}$ is a basic sequence, each of the conditions 1°, 2°, 4°, 8°—11°, 14° and 15° imply that $\{f_n\}$ is a basic sequence, so $(x_n, f_n)$ is a bibasic system, but the conditions 3°, 5°—7°, 12°, 13° and 15° with + instead of $\oplus$, do not imply this property. Indeed, if $\{x_n\}$ is a basic sequence, then $\sum_{i=1}^{\infty}\varphi(x_i)\varphi_i$ converges for every $\varphi\in[\varphi_n]$ (by Ch. I, § 12, theorem 12.1), and hence, if we have 1°, then $\sum_{i=1}^{\infty}\varphi(x_i)f_i$ converges for all $\varphi\in[\varphi_n]$. Consequently, $\sum_{i=1}^{\infty}f(x_i)f_i$ converges for all $f\in[f_n]$ (because for every $f\in[f_n]$ we have $f|_{[x_j]}\in[\varphi_n]$), whence $\{f_n\}$ is a basic sequence (e.g. by Ch. I, § 4, theorem 4.1, part 3°$\Rightarrow$1°, applied to the biorthogonal system $(f_n, u(x_n))$, where $u$ is the canonical mapping of $[x_n]$ into $[f_n]^*$). Furthermore, obviously 4°, whence also 2°, implies that $\{f_n\}$ is a basic sequence. Now if we have 8° or 9° or 14°, then $\sup_{1\leqslant n<\infty}\|s_n^*\|<\infty$, whence, since $s_n^*(f)=\sum_{i=1}^{n}f(x_i)f_i$ for all $f\in[f_n]$ (even for all $f\in E^*$), it follows that $\{f_n\}$ is a basic sequence (e.g. again by Ch. I, § 4, theorem 4.1). Finally, if we have 11° or 10° (or in particular, 15°), then we have the canonical isomorphisms $\Gamma\cong E^*/[x_n]^\perp\equiv[x_n]^*$ given by $f\to f+[x_n]^\perp\to f|_{[x_n]}$, so their composition is the restriction map $\rho$ and therefore (by $[f_n]\subset\Gamma$) we have 4°, whence $\{f_n\}$ is a basic sequence. On the other hand, let $E$ be a reflexive separable Banach space, $\{x_n\}$ a basic sequence in $E$ which is not a basis of $E$, $\{\varphi_n\}\subset[x_n]^*$ the a.s.c.f. to $\{x_n\}$ and $\{f_n\}\subset E^*$ total on $E$, with $f_n|_{[x_j]}=\varphi_n$ $(n=1,2,\dots)$; such a sequence $\{f_n\}$ exists by § 7, lemma 7.5. Then, since $E$ is reflexive and separable, we have $[f_n]=E^*$ and hence we have 3°, 5°—7°, 12°, 13° and $E^*=[x_n]^\perp+[f_n]$ (but

$[x_n]^\perp \cap [f_n] = [x_n]^\perp \neq \{0\}$), but, as observed in remark 1.10, $\{f_n\}$ is not a basic sequence. We mention that examples with $\{f_n\}$ satisfying 3°, 5°—7° and 13°, but not basic, follow also from proposition 1.15 below.

*Remark 1.13.* Every subspace $\Gamma$ of $E^*$ satisfying condition 10° of theorem 1.10 contains also the closed linear subspace*)

$$B = \left\{ f \in E^* \,\middle|\, f = \sum_{i=1}^{\infty}{}^* f(x_i) f_i \right\} \quad (\supset [f_n]). \tag{1.103}$$

Indeed, if $f \in B$, then for the projection $\tau\colon E^* \to \Gamma$ defined in the above proof we have

$$\tau(f) = w^*\text{-}\lim_{n_\alpha} \sum_{i=1}^{n_\alpha} f(x_i) f_i = \sum_{i=1}^{\infty}{}^* f(x_i) f_i = f,$$

whence $f \in \tau(E^*) = \Gamma$. Now, to see that $B$ is closed, let $\{g_n\} \subset B, f \in E^*$, $\|g_n - f\| \to 0$. Then for each $x \in E$ we have, by 8° of theorem 1.10,

$$\begin{aligned} |(s_n^*(f))(x) - f(x)| &\leqslant |(s_n^*(f))(x) - (s_n^*(g_i))(x)| + \\ &+ |(s_n^*(g_i))(x) - g_i(x)| + |g_i(x) - f(x)| \leqslant \\ &\leqslant \sup_{1 \leqslant j < \infty} \|s_j^*\| \, \|x\| \, \|f - g_i\| + |(s_n^*(g_i))(x) - g_i(x)| + \|x\| \, \|f - g_i\| \end{aligned}$$

and hence, for any $\varepsilon > 0$, taking $i$ such that $\|f - g_i\| < \dfrac{\varepsilon}{2(1 + \sup_{1 \leqslant j < \infty} \|s_j\|) \|x\|}$ and then taking $n_0 = n_0(i)$ such that $|(s_n^*(g_i))(x) - g_i(x)| < \dfrac{\varepsilon}{2}$ $(n \geqslant n_0)$, we obtain

$$|(s_n^*(f))(x) - f(x)| < \varepsilon \qquad (n \geqslant n_0).$$

Thus, $f = w^*\text{-}\lim_{n \to \infty} s_n^*(f) = \sum_{i=1}^{\infty}{}^* f(x_i) f_i$, whence $f \in B$, which proves that $B$ is closed.

Note also that we have, obviously, $B \subset \widetilde{[f_n]}$ and, by definition 1.3, we have $B = \widetilde{[f_n]}$ if and only if $\{f_n\}$ is a $w^*$-Schauder basic sequence in $E^*$.

*Remark 1.14.* If we have 8° of theorem 1.10, then $E$ contains finite-dimensional subspaces $G_n$ of arbitrarily large dimension $n$ (namely, $G_n = [x_i]_{i=1}^n$), which admit uniformly bounded projections. We recall that it is not known whether every (or every separable) Banach space has this property. Note also that if $\dim [x_n] = \operatorname{codim} [x_n] = \infty$, then $\dim [x_n]^\perp = \dim (E/[x_n])^* = \infty$, $\dim \Gamma \geqslant \dim [f_n] = \infty$, so 10° of theorem 1.10 gives a decomposition of $E^*$ into the direct sum of two infinite di-

*) We denote by $\Sigma^*$ the sum in the topology $\sigma(E^*, E)$.

mensional subspaces (and 11°, 12° give such decompositions for $E^{**}$, the latter only when codim $[f_n] = \infty$). We recall that it is not known whether every (or every separable) Banach space admits such a decomposition. Furthermore, if 1°—15° of theorem 1.10 are satisfied, the sum in 13° need not be non-trivial and we need not have $[x_n] + [f_n]_\perp = E$, as shown by the sequence $\{x_n\}$ of unit vectors in $E = m$, with $\{f_n\}$ = the sequence of coordinate functionals on $m$ (here $[f_n]_\perp = \{0\}$ and $[x_n] + [f_n]_\perp = [x_n] \neq E$); in connection with 13° of theorem 1.10 let us also mention that it is not known whether every Banach space contains a (closed linear) subspace which is the direct sum of two infinite dimensional subspaces*). In corollary 1.13 below we shall see examples in which 1°—15° of theorem 1.10 are not satisfied, i.e. examples of basic sequences $\{x_n\} \subset E$ such that the a.s.c.f. $\{\varphi_n\} \subset [x_n]^*$ admits no extension $\{f_n\} \subset E^*$ with $\{f_n\} \sim \{\varphi_n\}$.

If $\{f_n\}$ is not assumed to be a basic sequence, the converse implication 13° ⇒ 1° in theorem 1.10 is not valid, as shown by a basic sequence $\{x_n\}$ in a reflexive separable Banach space $E$, which is not a basis of $E$, and $\{f_n\}$ = a total extension of the a.s.c.f. $\{\varphi_n\} \subset [x_n]^*$ to $\{x_n\}$ (the example given in remarks 1.10 and 1.12). However, since in this example $\{f_n\}$ is not a basic sequence, it is natural to ask whether the converse implication 13° ⇒ 1° in theorem 1.10 is valid under the additional assumption that $\{f_n\}$ is a basic sequence. By remark 1.7 and theorem 1.11 below, implication 3° ⇒ 1°, the answer is affirmative if $\{f_n\}$ is a $w^*$-Schauder basic sequence and thus, in particular, if $[f_n]$ is $w^*$-closed; note that in this latter case $\{x_n\}$ is necessarily shrinking (indeed, by $\widetilde{[f_n]} = [f_n]$ and proposition 1.8 c), $\{f_n\}$ is boundedly complete, whence, by $\{f_n\} \sim \{\varphi_n\}$, so is $\{\varphi_n\}$ and therefore $\{x_n\}$ is shrinking; or, alternatively, by $\widetilde{[f_n]} = [f_n]$ and 4° of theorem 1.11 below we have 15° of theorem 1.10, whence $\{x_n\}$ is shrinking). Another case when the answer to the above question is affirmative is when codim $[x_n] < \infty$ (then, clearly, $[x_n] + [f_n]_\perp$ is closed, since $[x_n]$ is a closed subspace of finite codimension in $[x_n] + [f_n]_\perp$). Indeed, we have

**Proposition 1.15.** *Let $E$ be a Banach space and $(x_n, f_n)$ ($\{x_n\} \subset E$, $\{f_n\} \subset E^*$) a biorthogonal system such that $\{x_n\}$ is a basic sequence and* codim $[x_n] < \infty$. *Then*

a) $r_{[x_n]}([f_n]) > 0$.

b) *The following statements are equivalent:*

1°. $\{f_n\} \sim \{\varphi_n\}$ $(= \{f_n|_{[x_j]}\})$.

2°. $[x_n]^\perp \cap [f_n] = \{0\}$.

3°. $\{f_n\}$ *is a basic sequence.*

*Proof.* Let us first prove part b). The implication 1° ⇒ 3° was observed in remark 1.12, and the implication 3° ⇒ 2° is obvious. Now, if

*) Of course, an affirmative answer to problem 1.2 would solve this question in the affirmative.

we have 2°, then, since $[x_n]^\perp + [f_n]$ is closed in $E^*$ (because $\dim [x_n]^\perp < \infty$), from lemma 1.3 it follows that the restriction mapping $\rho: f \to f|_{[x_j]}$ from $[f_n]$ to $[x_n]^*$ is an isomorphism, so $\{f_n\} \sim \{\varphi_n\}$.

Let us prove now part a). If $[x_n]^\perp \cap [f_n] = \{0\}$, then, by part b) proved above and by theorem 1.10, we have $r_{[x_n]}([f_n]) > 0$. On the other hand, if $[x_n]^\perp \cap [f_n] \neq \{0\}$, then, since $\dim ([x_n]^\perp \cap [f_n]) \leqslant \leqslant \dim [x_n]^\perp < \infty$, there exists a subspace $\Gamma$ of $[f_n]$ such that

$$[f_n] = \Gamma \oplus ([x_n]^\perp \cap [f_n]), \tag{1.104}$$

so we can write $f_n = g_n + (f_n - g_n)$, where $g_n \in \Gamma$ and $f_n - g_n \in [x_j]^\perp \cap \cap [f_j] \subset [x_j]^\perp$ $(n = 1,2,\dots)$. Then $g_i(x_j) = f_i(x_j) = \delta_{ij}$ $(i, j = 1,2,\dots)$ and, by (1.104), $[g_n] \cap [x_n]^\perp \subset \Gamma \cap [x_n]^\perp = \Gamma \cap ([x_n]^\perp \cap [f_n]) = \{0\}$. Consequently, by the above, $r_{[x_n]}([g_n]) > 0$ and hence $r_{[x_n]}([f_n]) > 0$, which completes the proof of proposition 1.15.

In connection with 13° of theorem 1.10, let us also give

**Proposition 1.16.** *Let $E$ be a Banach space and $(x_n, f_n)$ ($\{x_n\} \subset E, \{f_n\} \subset \subset E^*$) a biorthogonal system such that $\{x_n\}$ is a basic sequence, with the a.s.c.f. $\{\varphi_n\} = \{f_n|_{[x_j]}\}$. The following statements are equivalent:*

1°. *$[x_n] + [f_n]_\perp$ is closed in $E$.*

2°. *$\{\omega(x_n)\}$ is a basic sequence in $E/[f_n]_\perp$, equivalent to $\{x_n\}$, where $\omega$ denotes the canonical mapping of $E$ onto $E/[f_n]_\perp$.*

3°. *$r_{[x_n]}(\widetilde{[f_n]}) > 0$.*

4°. *$\pi([x_n]) + (\widetilde{[f_n]})^\perp$ is closed in $E^{**}$.*

5°. *The canonical mapping $v$ of $[x_n]$ into $(\widetilde{[f_n]})^*$ is an isomorphism.*

*Proof.* Observe first that we have $[x_n] \cap [f_n]_\perp = \{0\}$, since $\{x_n\}$ is a basic sequence.

Assume now that we have 1°. Then, by the closed graph theorem, the natural projection $[x_n] \oplus [f_n]_\perp \to [x_n]$ is continuous*), so there exists a constant $C > 0$ such that

$$\|x\| \leqslant C\|x + y\| \qquad (x \in [x_n], y \in [f_n]_\perp),$$

whence

$$\|x\| \leqslant C \inf_{y \in [f_n]_\perp} \|x + y\| = C\|\omega(x)\| \leqslant C\|x\| \qquad (x \in [x_n]). \tag{1.105}$$

Consequently, $\omega|_{[x_n]}$ is an isomorphism, and hence we have 2°. Thus, 1° ⇒ 2°.

Conversely, if we have 2°, then $\omega|_{[x_n]}$ is an isomorphism, so there exists a constant $C > 0$ such that we have (1.105). Consequently, the natural projection $[x_n] + [f_n]_\perp \to [x_n]$ is continuous and hence $[x_n] + + [f_n]_\perp$ is closed. Thus, 2° ⇒ 1°.

---

*) See e.g. [87], p. 480.

Finally, the equivalence 2° ⇔ 3° is a consequence (via (1.105)) of the equality*)

$$\inf_{y \in [f_n]_\perp} \|x + y\| = \sup_{\substack{f \in \widetilde{[f_n]} \\ \|f\| \leq 1}} |f(x)| \qquad (x \in E \setminus [f_n]_\perp),$$

and the proof of the equivalences 3° ⇔ 4° ⇔ 5° is similar to that of the equivalences 5° ⇔ 6° ⇔ 7° of theorem 1.10. This completes the proof of proposition 1.16.

**Theorem 1.11.** *Let $E$ be a Banach space and $(x_n, f_n)$ ($\{x_n\} \subset E, \{f_n\} \subset E^*$) a biorthogonal system such that $\{x_n\}$ is a basic sequence, with the a.s.c.f. $\{\varphi_n\} = \{f_n|_{[x_j]}\}$. The following statements are equivalent:*

1°. *$\{f_n\}$ is a w*-Schauder basic sequence such that $\{f_n\} \sim \{\varphi_n\}$.*

2°. *For every $x \in E$ the series $\sum_{i=1}^{\infty} f_i(x) x_i$ converges (i.e. $\mathscr{E}_1 = E$, where $\mathscr{E}_1$ is defined by* (1.93)).

3°. *$E = [x_n] \oplus [f_n]_\perp$.*

4°. *$E^* = [x_n]^\perp \oplus \widetilde{[f_n]}$.*

5°. *$E^{**} = [x_n]^{\perp\perp} \oplus (\widetilde{[f_n]})^\perp = \widetilde{\pi([x_n])} \oplus (\widetilde{[f_n]})^\perp$.*

6°. *$\{\omega(x_n)\}$ is a basis of $E/[f_n]_\perp$, equivalent to $\{x_n\}$, where $\omega$ denotes the canonical mapping of $E$ onto $E/[f_n]_\perp$.*

*These statements are implied by — and, if $\{x_n\}$ is a boundedly complete basic sequence, they are equivalent to — the following statements:*

7°. *For every $\Phi \in E^{**}$ the series $\sum_{i=1}^{\infty} \Phi(f_i) x_i$ converges.*

8°. *$E^{**} = \pi([x_n]) \oplus [f_n]^\perp$.*

*Statements* 1°—8° *imply — and, if $\{x_n\}$ is a boundedly complete basic sequence, they are equivalent to — the following statement:*

9°. *$\{f_n\} \sim \{\varphi_n\}$.*

*Proof.* Since $\{x_n\}$ is a basic sequence, we have $[x_n] \cap [f_n]_\perp = \{0\}$. Now, if we have 1°, then, by remark 1.7, $[x_n] + [f_n]_\perp$ is dense in $E$ and, by the implication 1° ⇒ 13° of theorem 1.10, $[x_n] + [f_n]_\perp$ is also closed in $E$, so $E = [x_n] \oplus [f_n]_\perp$. Thus, 1° ⇒ 3°.

The equivalences 2° ⇔ 3° and 7° ⇔ 8° are nothing else than corollary 1.12 a). The implications 3° ⇒ 4° ⇒ 5° are obvious**) and the implications 5° ⇒ 4° ⇒ 3° are well known (see § 8, lemma 8.2).

Assume now that we have 3°. Then $E/[f_n]_\perp \cong [x_n]$ canonically, and hence $\{\omega(x_n)\}$ is a basis of $E/[f_n]_\perp$, where $\omega$ is the canonical mapping of $E$ onto $E/[f_n]_\perp$. Consequently, by proposition 1.8 a), $\{f_n\}$ is a $w^*$-Schauder basic sequence. Furthermore, by the implication 3° ⇒ 2° observed above and since 2° obviously implies 9° of theorem 1.10, we

*) See e.g. [371], Ch. I, § 1.

**) See e.g. [87], p. 481, lemma 3.

have $\{f_n\} \sim \{\varphi_n\}$. Thus, $3° \Rightarrow 1°$. Also, the beginning of this argument shows that $3° \Rightarrow 6°$. Conversely, if we have 6°, then $\omega|_{[x_n]}$ is an isomorphism of $[x_n]$ onto $E/[f_n]_\perp$ (by Ch. I, § 8, theorem 8.1), and hence $(\omega|_{[x_n]})^{-1}\omega$ is a projection of $E$ onto $[x_n]$, along $[f_n]_\perp$. Thus, $6° \Rightarrow 3°$.

The implications $7° \Rightarrow 2°$ and $1° \Rightarrow 9°$ are obvious.

Finally, assume that $\{x_n\}$ is a boundedly complete basic sequence and that we have 9°. Then, by theorem 1.10, implication $1° \Rightarrow 9''$, and since $\{x_n\}$ is boundedly complete, it follows that for every $\Phi \in E^{**}$ the series $\sum_{i=1}^{\infty} \Phi(f_i)x_i$ converges. Thus, $9° \Rightarrow 7°$ (whence also $1° \Rightarrow 9° \Rightarrow 7°$ and $9° \Rightarrow 7° \Rightarrow 2°$) when $\{x_n\}$ is boundedly complete, which concludes the proof of theorem 1.11.

*Remark 1.15.* Dually to theorem 1.10, implication 14° (or 15°) $\Rightarrow$ $\Rightarrow \{x_n\}$ is shrinking , we have for theorem 1.11 that 7° (or 8°) $\Rightarrow \{x_n\}$ is boundedly complete. Indeed, if we have 7° (or 8°) of theorem 1.11, then $(f_n, \pi(x_n))$ is a bibasic system satisfying 14° of theorem 1.10, so $\{f_n\}$ is shrinking, whence, by $\{f_n\} \sim \{\varphi_n\}$ and Ch. II, § 6, corollary 6.1, $\{x_n\}$ is boundedly complete. However, 1°–6° and 9° of theorem 1.11 *do not* imply that $\{x_n\}$ is boundedly complete, since they are obviously satisfied by any basis $\{x_n\}$ of a Banach space $E$. Actually, in the implications $9° \Rightarrow 1°, \ldots, 6°$ of theorem 1.11 (and in corollary 1.13 below) the assumption that $\{x_n\}$ be boundedly complete is essential, as shown by the sequence of unit vectors $\{x_n\}$ in $E = m$, with $\{f_n\}$ = the sequence of coordinate functionals on $m$.

*Remark 1.16.* Theorem 1.11 shows that in remark 1.15 we have also an example of a bibasic system $(x_n, f_n)$ with $\{f_n\} \sim \{\varphi_n\} = \{f_n|_{[x_j]}\}$, for which there exists no extension $\{h_n\} \subset E^*$ of $\{\varphi_n\}$ such that $\{h_n\}$ is a $w^*$-Schauder basic sequence and $\{h_n\} \sim \{\varphi_n\}$. Moreover, let us observe that *for a basic sequence* $\{x_n\} \subset E$ *there exists such a sequence* $\{h_n\} \subset E^*$ *if and only if* $[x_n]$ *is complemented in E.* Indeed, the condition is necessary by theorem 1.11; conversely, if $[x_n] \oplus F = E$ and $\{\varphi_n\} \subset$ $\subset [x_n]^*$ is the a.s.c.f. to $\{x_n\}$, then, putting $h_n(x+y) = \varphi_n(x)$ $(x \in$ $\in [x_n],\ y \in F)$, we obtain that $(x_n, h_n)$ is a biorthogonal system satisfying $[h_n]_\perp = F$ (because we have $x \in [x_n]$, $y \in F$, $h_n(x+y) = \varphi_n(x) = 0$ for $n = 1,2, \ldots$ if and only if $x = 0$, $x + y = y \in F$), hence $[x_n] \oplus$ $\oplus\ [h_n]_\perp = E$ and theorem 1.11 (implication $3° \Rightarrow 1°$) applies.

The implication $9° \Rightarrow 3°$ (for $\{x_n\}$ boundedly complete) of theorem 1.11 permits us to give a large class of basic sequences $\{x_n\} \subset E$ such that the a.s.c.f. $\{\varphi_n\} \subset [x_n]^*$ admits no extension $\{f_n\} \subset E^*$ with $\{f_n\} \sim \{\varphi_n\}$. Indeed, obviously we have

**Corollary 1.13.** *Let* $\{x_n\}$ *be a boundedly complete basic sequence in a Banach space E, such that* $[x_n]$ *is not complemented in E and let* $\{\varphi_n\} \subset$ $\subset [x_n]^*$ *be the a.s.c.f. to* $\{x_n\}$*. Then there exists no sequence* $\{f_n\} \subset E^*$ *with* $f_n|_{[x_j]} = \varphi_n$ $(n = 1,2, \ldots)$, *such that* $\{f_n\} \sim \{\varphi_n\}$.

For example, it is well known*) that $E = C([0,1])$ has no complemented subspace isomorphic to a conjugate Banach space, and $E = m$ has no complemented separable subspace, whence every boundedly complete basic sequence in $E = C([0,1])$ or $E = m$ satisfies the assumption of corollary 1.13. Naturally, one can also find basic sequences $\{x_n\}$ in some reflexive spaces $E$, satisfying the assumption of corollary 1.13.

In view of the above, it is natural to ask

*Problem 1.7.* Let $E$ be a Banach space. Does there exist an infinite bibasic system $(x_n, f_n)$ ($\{x_n\} \subset E, \{f_n\} \subset E^*$) such that $\{f_n\}$ is equivalent to the a.s.c.f. $\{\varphi_n\} = \{f_n|_{[x_j]}\}$ to $\{x_n\}$?

## § 2. Deficient basic sequences. Images and inverse images of bases by continuous linear mappings

We have seen in Ch. I, § 10, theorem 10.2 and Ch. II, § 11, theorem 11.3, that if a sequence $\{y_n\}$ in a Banach space $E$ is $KL$-near (respectively, weakly quadratically near) to a basis (respectively, a Besselian basis) of $E$, then $\{y_n\}$ can be transformed to become a basis of $E$, by changing suitable $k$ elements of it, where $k = \operatorname{codim}_E[y_n] < \infty$, and hence, in particular,**) $\{y_n\}$ can be transformed into a basic sequence by omitting a finite number of its elements. This suggests to study sequences with one or both of these properties.

*Definition 2.1.* A sequence $\{x_n\}$ in a Banach space $E$ is called a *deficient basic sequence* if it can be transformed into a basic sequence by omitting a finite number of its elements. We shall use the notation

$$\alpha_{\{x_n\}} = \min \{k \mid \{x_n\} \text{ becomes a basic sequence after omitting suitable } k \text{ elements}\}. \tag{2.1}$$

In particular, every basic sequence is a deficient basic sequence with $\alpha_{\{x_n\}} = 0$, so by § 1, theorem 1.1, deficient basic sequences exist in every Banach space. Conversely, a deficient basic sequence $\{x_n\}$ is a basic sequence (if and) only if $\alpha_{\{x_n\}} = 0$.

**Proposition 2.1.** *Let $\{x_n\}$ be a deficient basic sequence in a Banach space $E$ and let $\{x_n\}_{n \neq i_1, \ldots, i_k}$ be a basic sequence obtained by omitting $k$ elements $x_{i_1}, \ldots, x_{i_k}$ (hence $\alpha_{\{x_n\}} \leqslant k$). We have $[x_n]_{n \neq i_1, \ldots, i_k} = [x_n]$ if and only if $\alpha_{\{x_n\}} = k$.*

---

*) See e.g. [299], p. 221, corollary 3 and p. 222, corollary 5.

**) We recall that every subsequence of a basis is a basic sequence (see Ch. I., § 4, proposition 4.1.).

*Proof.* If $[x_n]_{n\neq i_1,\ldots,i_k}\neq[x_n]$, then among the elements $x_{i_1},\ldots,x_{i_k}$ there exists at least one which does not belong to $[x_n]_{n\neq i_1,\ldots,i_k}$, say $x_{i_1}$. Then, since $\{x_n\}_{n\neq i_1,\ldots,i_k}$ is basic, the sequence $x_{i_1}\cup\{x_n\}_{n\neq i_1,\ldots,i_k}=$ $=\{x_n\}_{n\neq i_2,\ldots,i_k}$ is a basic sequence, too, and hence $\alpha_{\{x_n\}}\leqslant k-1$. Thus, the assumption $\alpha_{\{x_n\}}=k$ implies $[x_n]_{n\neq i_1,\ldots,i_k}=[x_n]$.

Conversely, assume now that $[x_n]_{n\neq i_1,\ldots,i_k}=[x_n]$. Let $\alpha=\alpha_{\{x_n\}}$ and let $\{x_n\}_{n\neq j_1,\ldots,j_\alpha}$ be a basic sequence. Then the sequence $\{x_n\}_{n\neq i_1,\ldots,i_k,j_1,\ldots,j_\alpha}$ is obtained from $\{x_n\}_{n\neq i_1,\ldots,i_k}$ by omitting $\alpha-l$ elements, where

$$l=\operatorname{card}\{i_1,\ldots,i_k\}\cap\{j_1,\ldots,j_\alpha\}. \tag{2.2}$$

Hence, since $\{x_n\}_{n\neq i_1,\ldots,i_k}$ is a basic sequence,

$$\dim\left([x_n]_{n\neq i_1,\ldots,i_k}/[x_n]_{n\neq i_1,\ldots,i_k,j_1,\ldots,j_\alpha}\right)=\alpha-l \tag{2.3}$$

and, similarly, since $\{x_n\}_{n\neq j_1,\ldots,j_\alpha}$ is a basic sequence,

$$\dim\left([x_n]_{n\neq j_1,\ldots,j_\alpha}/[x_n]_{n\neq i_1,\ldots,i_k,j_1,\ldots,j_\alpha}\right)=k-l. \tag{2.4}$$

Consequently, taking into account that by the first part of this proof and by our assumption we have

$$[x_n]_{n\neq j_1,\ldots,j_\alpha}=[x_n]=[x_n]_{n\neq i_1,\ldots,i_k},$$

it follows that $\alpha-l=k-l$, whence $\alpha=k$, which completes the proof.

**Corollary 2.1.** *A deficient basic sequence $\{x_n\}$ is $\omega$-linearly independent**) *if and only if $\alpha_{\{x_n\}}=0$ (i.e., if and only if $\{x_n\}$ is a basic sequence).*

*Proof.* Obviously, every deficient basic sequence $\{x_n\}$ with $\alpha_{\{x_n\}}=0$ is a basic sequence, whence $\omega$-linearly independent.

Conversely, assume that $\{x_n\}$ is a deficient basic sequence with $\alpha_{\{x_n\}}>0$. Then, by proposition 2.1, some proper subsequence $\{x_n\}_{n\neq j_1,\ldots,j_\alpha}$ of $\{x_n\}$ is a basis of $[x_n]$. Hence $x_{j_1}$ has an expansion $x_{j_1}=\sum_{j\neq j_1,\ldots,j_\alpha}\gamma_j x_j$ and thus $\{x_n\}$ is not $\omega$-linearly independent, which completes the proof.

**Proposition 2.2.** *Let $\{x_n\}$ be a deficient basic sequence in a Banach space E and let*

$$A_0=\left\{\{\gamma_n\}\in s\;\middle|\;\sum_{i=1}^{\infty}\gamma_i x_i=0\right\}, \tag{2.5}$$

*) I.e. (see Ch. I, § 6, definition 6.1), the relations $\{\alpha_n\}\subset K$, $\sum_{i=1}^{\infty}\alpha_i x_i=0$ imply $\alpha_i=0$ $(i=1,2,\ldots)$.

*where s denotes the linear space of all sequences of scalars. Then*

$$\dim A_0 = \alpha_{\{x_n\}}. \tag{2.6}$$

*Proof.* By proposition 2.1 we may assume, without restriction of the generality, that $\{x_n\}_{n=\alpha+1}^{\infty}$ is a basis of $[x_n]$, where $\alpha = \alpha_{\{x_n\}}$. Hence there exist expansions

$$x_j = \sum_{i=\alpha+1}^{\infty} \beta_i^{(j)} x_i \qquad (j = 1, \ldots, \alpha). \tag{2.7}$$

Define elements $z_1, \ldots, z_\alpha \in s$ by

$$z_j = \{\underbrace{0, \ldots, 0}_{j-1}, 1, \underbrace{0, \ldots, 0}_{\alpha-j}, -\beta_{\alpha+1}^{(j)}, -\beta_{\alpha+2}^{(j)}, \ldots\} \quad (j=1, \ldots, \alpha). \tag{2.8}$$

Then, by (2.7), $z_j \in A_0$ $(j = 1, \ldots, \alpha)$. We shall show that $\{z_j\}_{j=1}^{\alpha}$ is a basis of $A_0$. Indeed, if $\sum_{j=1}^{\alpha} \lambda_j z_j = 0$, then, writing this relation for the first $\alpha$ coordinates of $\sum_{j=1}^{\alpha} \lambda_j z_j$, we obtain $\lambda_1 = \ldots = \lambda_\alpha = 0$. Now let $\{\gamma_n\} \in A_0$ be arbitrary. Then

$$-\sum_{j=1}^{\alpha} \gamma_j x_j = \sum_{i=\alpha+1}^{\infty} \gamma_i x_i,$$

whence, by (2.7),

$$-\sum_{i=\alpha+1}^{\infty} \sum_{j=1}^{\alpha} \gamma_j \beta_i^{(j)} x_i = \sum_{i=\alpha+1}^{\infty} \gamma_i x_i$$

which, since $\{x_n\}_{n=\alpha+1}^{\infty}$ is a basic sequence, implies

$$-\sum_{j=1}^{\alpha} \gamma_j \beta_i^{(j)} = \gamma_i \qquad (i = \alpha + 1, \alpha + 2, \ldots).$$

Hence, by (2.8), we obtain $\sum_{j=1}^{\alpha} \gamma_j z_j = \left\{\gamma_1, \ldots, \gamma_\alpha, -\sum_{j=1}^{\alpha} \gamma_j \beta_{\alpha+1}^{(j)}, \ldots\right\} = \{\gamma_n\}$ and thus $\{z_j\}_{j=1}^{\alpha}$ is a basis of $A_0$. Consequently, we have (2.6), which completes the proof of proposition 2.2.

*Remark 2.1.* The above argument also gives another proof of Ch. I, § 10, theorem 10.2 c) (using b) of the same theorem).

*Remark 2.2.* The sequence space (2.5) can be defined for any sequence $\{x_n\} \subset E$. The condition $\dim A_0 < \infty$ does not imply that

$\{x_n\}$ is a deficient basic sequence, since already for an arbitrary $\omega$-linearly independent sequences $\{x_n\}$ we have $\dim A_0 = 0$.

Let us introduce now the following two numbers, which give some insight on the behaviour of a deficient basic sequence $\{x_n\}$ and of its span $[x_n]$, with respect to the whole space $E$:

$$\beta_{\{x_n\}} = \operatorname{codim}_E [x_n] = \dim E/[x_n], \tag{2.9}$$

$$\varkappa_{\{x_n\}} = \beta_{\{x_n\}} - \alpha_{\{x_n\}}. \tag{2.10}$$

Obviously, $0 \leqslant \beta_{\{x_n\}} \leqslant \infty$. For a deficient basic sequence $\{x_n\}$ we have $0 \leqslant \alpha_{\{x_n\}} < \infty$, whence $\varkappa_{\{x_n\}} > -\infty$; furthermore, it also follows that $\varkappa_{\{x_n\}} < \infty$ if and only if $\beta_{\{x_n\}} < \infty$. It may also occur that $\varkappa_{\{x_n\}} < 0$ (for example, if $[x_n] = E$ and $\alpha_{\{x_n\}} > 0$). The number $\varkappa_{\{x_n\}}$ has the following interpretation:

**Proposition 2.3.** *For a deficient basic sequence $\{x_n\}$ in a Banach space E we have $\varkappa_{\{x_n\}} = m - k$, with $m \geqslant \beta_{\{x_n\}}$ and $k \geqslant \alpha_{\{x_n\}}$ (not uniquely determined), if and only if by omitting suitable k elements of $\{x_n\}$ and adding suitable m elements to the remaining sequence, we obtain a basis of E.*

*Proof.* If $\beta_{\{x_n\}} - \alpha_{\{x_n\}} = \varkappa_{\{x_n\}} = m - k$, with $\alpha_{\{x_n\}} \leqslant k$, then we can write $k = \alpha_{\{x_n\}} + l$, where $l \geqslant 0$ and hence $m = \beta_{\{x_n\}} - \alpha_{\{x_n\}} + k = \beta_{\{x_n\}} + l$; since by omitting suitable $\alpha_{\{x_n\}}$ elements of $\{x_n\}$ and adding suitable $\beta_{\{x_n\}}$ elements to the remaining sequence we obtain a basis of $E$, it follows that the same is true also for $k = \alpha_{\{x_n\}} + l$ and $m = \beta_{\{x_n\}} + l$ elements respectively (omitting and adding the same $l$ extra elements).

Conversely, assume now that omitting suitable $k$ elements of $\{x_n\}$ and adding suitable $m$ elements to the remaining sequence, we obtain*) a basis of $E$, say $\{y_i\}_{i=1}^m \cup \{x_n\}_{n \neq i_1, \dots, i_k}$. By definition, omitting suitable $\alpha = \alpha_{\{x_n\}}$ elements of $\{x_n\}$ and adding suitable $\beta = \beta_{\{x_n\}}$ elements to the remaining sequence, we obtain a basis of $E$, say $\{z_i\}_{i=1}^{\beta} \cup \{x_n\}_{n \neq j, \dots, j_\alpha}$. Consequently, defining $l$ by (2.2), and taking into account (2.3) and (2.4), we obtain

$$\alpha - l + m = \operatorname{codim}_E [x_n]_{n \neq i_1, \dots, i_k, j_1, \dots, j_\alpha} = k - l + \beta,$$

whence $m - k = \beta - \alpha = \varkappa_{\{x_n\}}$, which completes the proof.

**Lemma 2.1.****) *Let $X$ be a Banach space, $G$ a subspace of $X$ of codimension $m$, where $0 < m < \infty$, and let $y_1, \dots, y_k$ be elements in $X$*

*) By (2.1), this already implies that $\alpha_{\{x_n\}} \leqslant k$.

**) For a somewhat related result see Ch. II, § 4, lemma 4.1.

*(some of them possibly in G) such that* $[G \cup \{y_i\}_{i=1}^k] = X$ *(hence* $m \leqslant k$*). Then among* $y_1, \ldots, y_k$ *there exist m linearly independent elements, say* $y_{i_1}, \ldots, y_{i_m}$ *such that*

$$[y_{i_l}]_{l=1}^m \cap G = \{0\}. \tag{2.11}$$

*Proof.* There exists $y_{i_1} \notin G$, since otherwise we would have $X = [G \cup \{y_i\}_{i=1}^k] = G$, in contradiction with the assumption $m = \operatorname{codim}_X G > 0$. Then, obviously, $y_{i_1} \neq 0$ and $[y_{i_1}] \cap G = \{0\}$. Assume now that we have found $p$ linearly independent elements $y_{i_1}, \ldots, y_{i_p} \subset \{y_i\}_{i=1}^k$, where $1 \leqslant p \leqslant m-1$, such that $[y_{i_l}]_{l=1}^p \cap G = \{0\}$. Then there exists $y_{i_{p+1}} \notin G \oplus [y_{i_l}]_{l=1}^p$, since otherwise we would have $X = [G \cup \{y_i\}_{i=1}^k] = G \oplus [y_{i_l}]_{l=1}^p$ whence $m = \operatorname{codim}_X G = p \leqslant m-1$, a contradiction. Then $y_{i_1}, \ldots, y_{i_{p+1}}$ are linearly independent and $[y_{i_l}]_{l=1}^{p+1} \cap G = \{0\}$. Indeed, otherwise there would exist a linear combination $\sum_{l=1}^{p+1} \alpha_l y_{i_l} = g \in G$, whence, since $\alpha_{p+1} \neq 0$ (by $[y_{i_l}]_{l=1}^p \cap G = \{0\}$), we would obtain $y_{i_{p+1}} = \frac{1}{\alpha_{l+1}} g - \frac{1}{\alpha_{l+1}} \sum_{l=1}^p \alpha_l y_{i_l} \in G \oplus [y_{i_l}]_{l=1}^p$, in contradiction with the choice of $y_{i_{p+1}}$. This completes the proof.

**Proposition 2.4.** *Let* $\{x_n\}$ *be a deficient basic sequence in a Banach space E and let* $y_1, \ldots, y_k \in E$. *Then* $\{y_i\}_{i=1}^k \cup \{x_n\}$ *is a deficient basic sequence and*

$$\alpha_{\{y_i\}_{i=1}^k \cup \{x_n\}} \leqslant \alpha_{\{x_n\}} + k, \tag{2.12}$$

$$\beta_{\{y_i\}_{i=1}^k \cup \{x_n\}} \geqslant \beta_{\{x_n\}} - k, \tag{2.13}$$

$$\varkappa_{\{y_i\}_{i=1}^k \cup \{x_n\}} = \varkappa_{\{x_n\}} - k. \tag{2.14}$$

*Proof.* Since $\{y_i\}_{i=1}^k \cup \{x_n\}$ is obviously a deficient basic sequence, it will be sufficient to prove (2.12)–(2.14).

Let

$$m = \dim [\{y_i\}_{i=1}^k \cup \{x_n\}]/[x_n]. \tag{2.15}$$

Then, obviously, $0 \leqslant m \leqslant k$ and

$$\beta_{\{y_i\}_{i=1}^k \cup \{x_n\}} = \beta_{\{x_n\}} - m \geqslant \beta_{\{x_n\}} - k. \tag{2.16}$$

Furthermore, by lemma 2.1, among $y_1, \ldots, y_k$ there exist $m$ linearly independent elements, say $y_{i_1}, \ldots, y_{i_m}$ such that $[y_{i_l}]_{l=1}^m \cap [x_n] = \{0\}$, whence, by (2.15),

$$[y_{i_l}]_{l=1}^m \oplus [x_n] = [\{y_i\}_{i=1}^k \cup \{x_n\}]. \tag{2.17}$$

Since by proposition 2.1 $\{x_n\}$ becomes a basis of $[x_n]$ after omitting $\alpha_{\{x_n\}}$ elements, it follows that $\{y_i\}_{i=1}^k \cup \{x_n\}$ becomes a basis of $[\{y_i\}_{i=1}^k \cup \{x_n\}]$ after omitting those $\alpha_{\{x_n\}}$ elements and the elements $\{y_i\}_{i \neq i_1, \ldots, i_m}$, that is, $\alpha_{\{x_n\}} + k - m$ elements. Hence again by proposition 2.1,

$$\alpha_{\{y_i\}_{i=1}^k \cup \{x_n\}} = \alpha_{\{x_n\}} + k - m \leqslant \alpha_{\{x_n\}} + k, \tag{2.18}$$

and, by (2.16) and (2.18),

$$\varkappa_{\{y_i\}_{i=1}^k \cup \{x_n\}} = \beta_{\{x_n\}} - m - (\alpha_{\{x_n\}} + k - m) = \varkappa_{\{x_n\}} - k,$$

which completes the proof of proposition 2.4.

**Lemma 2.2.** *Let $\{x_n\}$ be a basis of a Banach space $E$ and let*

$$y_j = \sum_{i=1}^{\infty} \gamma_{ij} x_i \qquad (j = 1, \ldots, m) \tag{2.19}$$

*be $m$ linearly independent elements in $E$. Then*

$$\operatorname{rank} (\gamma_{ij})_{\substack{i=1,2,\ldots \\ j=1,\ldots,m}} = m. \tag{2.20}$$

*Proof.* Let $\{f_n\} \subset E^*$ be the a.s.c.f. to the basis $\{x_n\}$. Then

$$\gamma_{ij} = f_i(y_j) \qquad (i = 1,2, \ldots; j = 1, \ldots, m). \tag{2.21}$$

There exists an index $i_1$ such that $f_{i_1}(y_1) \neq 0$, since otherwise we would have $f_i(y_1) = 0$ $(i = 1,2, \ldots)$, whence $y_1 = 0$, in contradiction with the assumption that $y_1, \ldots, y_m$ are linearly independent.

Assume now that for $p - 1 < m$ there exist $i_1, \ldots, i_{p-1}$ such that $\det (f_{i_l}(y_j))_{l,j=1,\ldots,p-1} \neq 0$. Then we cannot have

$$\begin{vmatrix} f_{i_1}(y_1) & \ldots & f_{i_1}(y_{p-1}) & f_{i_1}(y_p) \\ \cdot & \cdot & \cdot & \cdot \\ f_{i_{p-1}}(y_1) & \ldots & f_{i_{p-1}}(y_{p-1}) & f_{i_{p-1}}(y_p) \\ f_i(y_1) & \ldots & f_i(y_{p-1}) & f_i(y_p) \end{vmatrix} = 0$$

for all $i = 1,2, \ldots$ since otherwise, denoting by $\Delta_j$ the cofactor of $f_i(y_j)$ $(j = 1, \ldots, p)$, we would have $f_i\left(\sum_{j=1}^{p} \Delta_j y_j\right) = 0$ $(i=1,2,\ldots)$, whence $\sum_{j=1}^{p} \Delta_j y_j = 0$, which, by $\Delta_p \neq 0$, would contradict the assumption of linear independence of $y_1, \ldots, y_m$. Thus there exists an index $i_p$ such that $\det (f_{i_l}(y_j))_{l,j=1,\ldots,p} \neq 0$, which, together with (2.21), completes the proof.

We recall that a continuous linear mapping $u$ of a Banach space $E_1$ into a Banach space $E_2$ is called*) a *$\Phi_+$-operator* if $u(E_1)$ is closed in $E_2$ and

$$\alpha_u = \dim \operatorname{Ker} u < \infty, \tag{2.22}$$

where $\operatorname{Ker} u = \{x \in E_1 \mid u(x) = 0\}$. For a $\Phi_+$-operator $u: E_1 \to E_2$ put

$$\beta_u = \operatorname{codim}_{E_2} u(E_1) = \dim E_2/u(E_1), \tag{2.23}$$

$$\varkappa_u = \beta_u - \alpha_u. \tag{2.24}$$

We shall give now some relations between $\Phi_+$-operators and deficient basic sequences.

**Proposition 2.5.** *Let $\{x_n\}$ be a basis of a Banach space $E_1$ and let $u$ be a $\Phi_+$-operator from $E_1$ into a Banach space $E_2$. Then $\{u(x_n)\}$ is a deficient basic sequence in $E_2$, satisfying*

$$\alpha_{\{u(x_n)\}} = \alpha_u, \quad \beta_{\{u(x_n)\}} = \beta_u. \tag{2.25}$$

*Proof.* Since $\{x_n\}$ is a basis of $E_1$, we have $[u(x_n)] = u(E_1)$, whence

$$\beta_{\{u(x_n)\}} = \dim E_2/[u(x_n)] = \dim E_2/u(E_1) = \beta_u.$$

Now let $\alpha = \alpha_u$ and let $y_1, \ldots, y_\alpha$ be a basis of Ker $u$, say

$$y_j = \sum_{i=1}^{\infty} \gamma_{ij} x_i \qquad (j = 1, \ldots, \alpha). \tag{2.26}$$

Then, by lemma 2.2, there exist indices $i_1, \ldots, i_\alpha$ such that

$$\det (\gamma_{i_l, j})_{l,j=1,\ldots,\alpha} \neq 0. \tag{2.27}$$

*) See the Notes and remarks.

We claim that

$$(\text{Ker } u) \cap [x_n]_{n \neq i_1, \ldots, i_\alpha} = \{0\}. \tag{2.28}$$

Indeed, assume that $\sum_{j=1}^{\alpha} \beta_j y_j = \sum_{i \neq i_1, \ldots, i_\alpha} \lambda_i x_i$ for some scalars $\beta_j$ and $\lambda_i$. Then

$$\sum_{j=1}^{\alpha} \beta_j \sum_{i=1}^{\infty} \gamma_{ij} x_i = \sum_{i=1}^{\infty} \left( \sum_{j=1}^{\alpha} \beta_j \gamma_{ij} \right) x_i = \sum_{i \neq i_1, \ldots, i_\alpha} \lambda_i x_i,$$

whence $\sum_{j=1}^{\alpha} \beta_j \gamma_{i_l, j} = 0$ $(l = 1, \ldots, \alpha)$ and hence, by (2.27), $\beta_j = 0$ $(j = 1, \ldots, \alpha)$, which proves (2.28). Since dim Ker $u = \alpha_u = \alpha$, it follows that

$$E_1 = (\text{Ker } u) \oplus [x_n]_{n \neq i_1, \ldots, i_\alpha} \tag{2.29}$$

and hence $u$ maps isomorphically $[x_n]_{n \neq i_1, \ldots, i_\alpha}$ onto $u(E_1)$. Consequently, $\{u(x_n)\}_{n \neq i_1, \ldots, i_\alpha}$ is a basis of $u(E_1) = [u(x_n)]$, whence $\{u(x_n)\}$ is a deficient basic sequence and, by proposition 2.1,

$$\alpha_{\{u(x_n)\}} = \alpha = \alpha_u,$$

which completes the proof of proposition 2.5.

If $E_1 = E_2 = l^2$, then we have the following converse to proposition 2.5:

**Proposition 2.6.** *If $\{y_n\}$ is a deficient basic sequence in $l^2$, then there exist a basis $\{x_n\}$ of $l^2$ and a $\Phi_+$-operator $u: l^2 \to l^2$ such that $u(x_n) = y_n$ $(n = 1,2, \ldots)$.*

*Proof.* By proposition 2.1 we may assume, without loss of generality, that $\{y_n\}_{n=\alpha+1}^{\infty}$ is a basis of $[y_n]$, where $\alpha = \alpha_{\{y_n\}}$. Let $G$ be an arbitrary subspace of $l^2$ with $\text{codim}_{l^2} G = \alpha$. Since $l^2$ is a Hilbert space, there exists a linear isometry $v$ of $[y_n]$ onto $G$ and hence $\{v(y_n)\}_{n=\alpha+1}^{\infty}$ is a basis of $G$. Let $x_1, \ldots, x_\alpha$ be arbitrary elements in $l^2$ such that $G \oplus [x_i]_{i=1}^{\alpha} = l^2$ and let

$$x_n = v(y_n) \qquad (n = \alpha + 1, \alpha + 2, \ldots).$$

Then, by the above, $\{x_n\}$ is a basis of $l^2$. For any finite sequence of scalars $\beta_1, \ldots, \beta_m$ put

$$u\left( \sum_{i=1}^{m} \beta_i x_i \right) = \sum_{i=1}^{m} \beta_i y_i. \tag{2.30}$$

Then the restriction of $u$ to $[x_1, \ldots, x_\alpha]$ is continuous and the restriction of $u$ to the linear span of $\{x_n\}_{n=\alpha+1}^{\infty}$ is $v^{-1}$, whence continous. Consequently, since $\{x_n\}$ is a basis of $l^2$, $u$ can be extended, by continuity, to all of $l^2$. Furthermore, $u(l^2) = [y_n]$ is closed and, since $u|_{[x_n]_{n=\alpha+1}^{\infty}} = v^{-1}$ is one-to-one, we have $[x_n]_{n=\alpha+1}^{\infty} \cap \operatorname{Ker} u = \{0\}$, whence $\alpha_u = \dim \operatorname{Ker} u \leqslant \alpha = \alpha_{\{y_n\}} < \infty$, so $u$ is a $\Phi_+$-operator. Since obviously $u(x_n) = y_n$ $(n = 1,2, \ldots)$, the proof is thus complete.

*Remark 2.3*. The same conclusion holds, with a similar argument, when $E_1 = E_2$ is a Banach space and $\{y_n\}$ is a deficient basic sequence with $\varkappa_{\{y_n\}} = 0$ (hence $\operatorname{codim}_{E_2} [y_n] = \alpha_{\{y_n\}}$) or when $E_1 = E_2$ is a Banach space which is isomorphic to its hyperplanes*) and $\{y_n\}$ is a deficient basic sequence with $\varkappa_{\{y_n\}} < \infty$. Indeed, in the first case one can take $G = [y_n]$ and $v =$ the identical mapping, while in the second case one can take $G$ to be a subspace of codimension $\alpha = \alpha_{\{y_n\}}$ and $v$ to be an isomorphism of $[y_n]$ onto $G$.

**Lemma 2.3.** *Let $u$ be a $\Phi_+$-operator from a Banach space $E_1$ into a Banach space $E_2$ and let $G$ be a subspace of $E_1$. Then the restriction $u|_G$: $G \to E_2$ is a $\Phi_+$-operator.*

*Proof.* Since $\dim \operatorname{Ker} u < \infty$, it is obvious that $\dim \operatorname{Ker} u|_G < \infty$ and thus it remains to prove that $u(G)$ is closed in $E_2$. Since by our assumption $u(E_1)$ is closed in $E_2$, it will be sufficient to prove that $u(G)$ is closed in $u(E_1)$.

Let $y_n = u(g_n)$ $(n = 1,2, \ldots)$ be an arbitrary sequence in $u(G)$, such that $\lim_{n\to\infty} y_n = y$. Then, since $u(E_1)$ is closed in $E_2$, we have $y = u(x) \in u(E_1)$, where $x \in E_1$. Since $\dim \operatorname{Ker} u < \infty$, let $Q$ be a complementary subspace to $\operatorname{Ker} u$ in $E_1$ and let $P$ be a complementary subspace to $\operatorname{Ker} u \cap G$ in $\operatorname{Ker} u$, that is, $E_1 = \operatorname{Ker} u \oplus Q$ and $\operatorname{Ker} u = (\operatorname{Ker} u \cap G) \oplus P$, whence

$$E_1 = (\operatorname{Ker}\ u \cap G) \oplus P \oplus Q. \tag{2.31}$$

Then $u|_Q$ is an isomorphism of $Q$ onto $u(Q) = u(E_1)$ and we have (unique) decompositions $g_n = z_n + p_n + q_n$, $x = z + p + q$, where $z_n, z \in \operatorname{Ker} u \cap G$, $p_n, p \in P$ and $q_n, q \in Q$ $(n = 1,2, \ldots)$. Since $u(q_n) = u(g_n) \to u(x) = u(q)$ and since $u|_Q$ is an isomorphism, it follows that $\lim_{n\to\infty} q_n = q$. On the other hand, we claim that

$$\sup_{1 \leqslant n < \infty} \|p_n + q_n\| < \infty. \tag{2.32}$$

---

*) Hence all subspaces of $E_1$ of finite codimension are isomorphic to each other. It is an unsolved problem, raised by S. Banach [11], whether every Banach space has this property, but it was proved in [362] that the usual concrete Banach spaces do have this property.

Indeed, if not, we may assume that $\lim_{n\to\infty} \|p_n + q_n\| = \infty$. Then, since $\dim P \leqslant \dim \operatorname{Ker} u < \infty$ and since $\sup_{1\leqslant n<\infty} \frac{\|p_n\|}{\|p_n + q_n\|} < \infty$ (because $p_n$ is the projection of $p_n + q_n \in P \oplus Q$ onto $P$), there exists a subsequence $\left\{\frac{p_{n_k}}{\|p_{n_k} + q_{n_k}\|}\right\}$ converging to an element $p_0 \in P$, whence, since $\lim_{k\to\infty} \frac{q_{n_k}}{\|p_{n_k} + q_{n_k}\|} = 0$ (by $\lim_{n\to\infty} q_n = q$) and since $g_n, z_n \in G$, we obtain

$$p_0 = \lim_{k\to\infty} \frac{p_{n_k} + q_{n_k}}{\|p_{n_k} + q_{n_k}\|} = \lim_{k\to\infty} \frac{g_{n_k} - z_{n_k}}{\|g_{n_k} - z_{n_k}\|} \in P \cap G \subset \operatorname{Ker} u \cap G, \text{ which,}$$

since $p_0 \in P$ and $\|p_0\| = 1$, contradicts $P \cap (\operatorname{Ker} u \cap G) = \{0\}$. This proves the claim (2.32).

Consequently, we have $\sup_{1\leqslant n<\infty} \|p_n\| < \infty$ (because $p_n$ is the projection of $p_n + q_n \in P \oplus Q$ onto $P$), whence, since $\dim P \leqslant \dim \operatorname{Ker} u < \infty$, there exists a subsequence $\{p_{n_k}\}$ of $\{p_n\}$ converging to an element $p' \in P \subset \operatorname{Ker} u$. Then we obtain $p' + q = \lim_{k\to\infty} (p_{n_k} + q_{n_k}) = \lim_{k\to\infty} (g_{n_k} - z_{n_k}) \in G$ and

$$y = u(x) = u(z + p + q) = u(q) = u(p' + q),$$

whence $y \in u(G)$, which completes the proof of lemma 2.3.

**Theorem 2.1.** *Let $\{x_n\}$ be a deficient basic sequence in a Banach space $E_1$ and let $u$ be a $\Phi_+$-operator from $E_1$ into a Banach space $E_2$. Then $\{u(x_n)\}$ is a deficient basic sequence in $E_2$, satisfying*

$$\varkappa_{\{u(x_n)\}} = \varkappa_u + \varkappa_{\{x_n\}}. \tag{2.33}$$

*Proof.* Let $n_1 = \dim \operatorname{Ker} u \cap [x_n]$ and, by proposition 2.1, let $\{x_n\}_{n\neq i_1,\dots,i_\alpha}$ be a basis of $[x_n]$, where $\alpha = \alpha_{\{x_n\}}$. Then by lemma 2.3 $u|_{[x_n]}$ is a $\Phi_+$-operator from $[x_n]$ into $E_2$ and thus, by proposition 2.5, $\{u(x_n)\}_{n\neq i_1,\dots,i_\alpha}$, whence also $\{u(x_n)\}$, is a deficient basic sequence in $E_2$ and from $\{u(x_n)\}_{n\neq i_1,\dots,i_\alpha}$ one can omit $\alpha_{\{u(x_n)\}_{n\neq i_1,\dots,i_\alpha}} = \alpha_{u|[x_n]} = \dim \operatorname{Ker} u|_{[x_n]} = n_1$ elements so that the remaining elements constitute a basis of $[u(x_n)]_{n\neq i_1,\dots,i_\alpha}$. Consequently, from $\{u(x_n)\}$ one can omit $\alpha + n_1$ elements in such a way that the remaining elements form a basis of $[u(x_n)] = [u(x_n)]_{n\neq i_1,\dots,i_\alpha}$ and thus, by proposition 2.1,

$$\alpha_{\{u(x_n)\}} = \alpha + n_1 = \alpha_{\{x_n\}} + n_1. \tag{2.34}$$

Let $P$ be an arbitrary complementary subspace to $\operatorname{Ker} u \cap [x_n]$ in $\operatorname{Ker} u$, that is, $\operatorname{Ker} u = (\operatorname{Ker} u \cap [x_n]) \oplus P$, whence

$$\dim P = \alpha_u - n_1. \tag{2.35}$$

Observe now that if $\beta_u = \dim E_2/u(E_1) = \infty$, then $\beta_{\{u(x_n)\}} = \dim E_2/[u(x_n)] = \infty$ and hence (2.33) is true. On the other hand, if $\beta_{\{x_n\}} = \dim E_1/[x_n] = \infty$, then again $\beta_{\{u(x_n)\}} = \infty$ and thus (2.33) is true. Indeed, assume that $\beta_{\{u(x_n)\}} < \infty$. Then $\operatorname{codim}_{u(E_1)} [u(x_n)] \leqslant \beta_{\{u(x_n)\}} < \infty$ and hence $u(E_1) = [u(x_n)] \oplus [u(z_i)]_{i=1}^m$, with suitable $z_1, \ldots, z_m \in E_1$. Since $u(x_n) \in u([x_n])$ and since by lemma 2.3 $u([x_n])$ is closed in $E_2$, we have $[u(x_n)] \subset u([x_n])$ and hence $u(E_1) = u([x_n]) + [u(z_i)]_{i=1}^m$. Thus for every $x \in E_1$ we have $u(x) = u(y) + \sum_{i=1}^{m} \alpha_i u(z_i)$, where $y \in [x_n]$, whence $x - y - \sum_{i=1}^{m} \alpha_i z_i \in \operatorname{Ker} u$. Consequently, if $\{y_i\}_{i=1}^{\alpha_u}$ is a basis of $\operatorname{Ker} u$, we can write $x = y + \sum_{i=1}^{m} \alpha_i z_i + \sum_{i=1}^{\alpha_u} \gamma_i y_i$, where $y \in [x_n]$ and $\alpha_i$, $\gamma_i$ are suitable scalars, whence $\beta_{\{x_n\}} = \operatorname{codim}_{E_1} [x_n] < \infty$, which proves our assertion.

Thus we may assume that $\beta_u < \infty$ and $\beta_{\{x_n\}} < \infty$. Let $Q$ be a complementary subspace to $[x_n] \oplus P$ in $E_1$, that is, $E_1 = [x_n] \oplus P \oplus Q$, and let $n_2 = \dim Q$. Obviously, $\dim P \oplus Q = \beta_{\{x_n\}}$, whence, by (2.35), we obtain

$$n_1 - n_2 = \alpha_u - \beta_{\{x_n\}}. \tag{2.36}$$

Since $Q \cap \operatorname{Ker} u = \{0\}$, the mapping $u|_Q$ is an isomorphism, and hence $\dim u(Q) = \dim Q = n_2$. Now let $Q_1$ be a complementary subspace to $\operatorname{Ker} u \cap [x_n]$ in $[x_n]$, that is, $[x_n] = (\operatorname{Ker} u \cap [x_n]) \oplus Q_1$, hence $u([x_n]) = u(Q_1)$ and $E_1 = (\operatorname{Ker} u \cap [x_n]) \oplus Q_1 \oplus P \oplus Q = \operatorname{Ker} u \oplus \oplus Q_1 \oplus Q$. Then $u(Q_1) \cap u(Q) = \{0\}$, since the relations $u(q_1) = u(q)$, where $q_1 \in Q_1$, $q \in Q$, imply $q_1 - q \in (Q_1 \oplus Q) \cap \operatorname{Ker} u = \{0\}$, whence $q_1 = q \in Q_1 \cap Q = \{0\}$ and hence $u(q_1) = u(q) = u(0) = 0$. Consequently, we have

$$u(E_1) = u(Q_1 \oplus Q) = u(Q_1) \oplus u(Q) = u([x_n]) \oplus u(Q),$$

whence, writing $E_2 = u(E_1) \oplus M$, where $\dim M = \beta_u < \infty$, we obtain

$$\beta_{\{u(x_n)\}} = \dim E_2/[u(x_n)] = \dim u(Q) \oplus M = n_2 + \beta_u. \tag{2.37}$$

Subtracting (2.34) from (2.37) and taking into account (2.36), we obtain

$$\varkappa_{\{u(x_n)\}} = n_2 + \beta_u - \alpha_{\{x_n\}} - n_1 = \beta_u - \alpha_{\{x_n\}} + \beta_{\{x_n\}} - \alpha_u = \varkappa_u + \varkappa_{\{x_n\}},$$

which completes the proof of theorem 2.1.

For the proof of the next corollary we shall need

**Lemma 2.4.** *Let $s$ be a compact linear operator from a Banach space $E$ into itself and let $I_E$ be the identity operator on $E$. Then $I_E - s$ is a $\Phi_+$-operator, satisfying*

$$\varkappa_{I_E - s} = 0. \tag{2.38}$$

*Proof.* It is well known*) that $(I_E - s)(E)$ is closed and that $\dim \operatorname{Ker} (I_E - s) < \infty$ and thus $I_E - s$ is a $\Phi_+$-operator.

Furthermore, it is well known**) that $(I_E - s)^*(E^*)$ is closed, whence $(I_E - s)(E) = [\operatorname{Ker} (I_E - s)^*]_\perp$ and that $\dim \operatorname{Ker} (I_E - s)^* = \dim \operatorname{Ker} (I_E - s)$. Consequently, since $(E/(I_E - s)(E))^* \equiv (I_E - s)(E)^\perp$, we obtain $\dim (E/(I_E - s)(E))^* = \dim (I_E - s)(E)^\perp = \dim \operatorname{Ker} (I_E - s)^* = \dim \operatorname{Ker} (I_E - s) = \alpha_{I_E - s} < \infty$, whence

$$\beta_{I_E - s} = \dim \ E/(I_E - s)(E) = \dim \ (E/(I_E - s)(E))^* = \alpha_{I_E - s},$$

i.e., (2.38), which completes the proof.

Now we can give the following corollary of theorem 2.1:

**Corollary 2.2.** *Let $\{x_n\}$ be a deficient basic sequence in a Banach space $E$ and let $s$ be a compact linear operator from $E$ into itself. Then $\{x_n - s(x_n)\}$ is a deficient basic sequence in $E$, satisfying*

$$\varkappa_{\{x_n - s(x_n)\}} = \varkappa_{\{x_n\}}. \tag{2.39}$$

*In particular, if $\{x_n\}$ is a basis of $E$ and $\{x_n - s(x_n)\}$ is $\omega$-linearly independent, then $\{x_n - s(x_n)\}$ is a basis of $E$.*

*Proof.* The first statement follows by applying theorem 2.1 to $u = I_{E-s}$ and taking into account lemma 2.4.

Assume now that $\{x_n\}$ is a basis of $E$ and $\{x_n - s(x_n)\}$ is $\omega$-linearly independent. Then $\varkappa_{\{x_n - s(x_n)\}} = \varkappa_{\{x_n\}} = 0$ and, by corollary 2.1, $\alpha_{\{x_n - s(x_n)\}} = 0$. Consequently, $\beta_{\{x_n - s(x_n)\}} = 0$, whence $[x_n - s(x_n)] = E$ and $\{x_n - s(x_n)\}$ is a basis of $E$, which completes the proof.

*Remark 2.4.* The last statement of corollary 2.2 is also contained, implicitly, in Ch. I, § 10, proof of theorem 10.2 (implication 2° ⇒ 5°); for a related result see also Ch. II, § 11, proof of theorem 11.3 (implication 2° ⇒ 6°), where $\{x_n\}$ is a Besselian basis of $E$ but $\{x_n - s(x_n)\}$ is only $l^2$-linearly independent.

In the sequel we shall need the following sharpening of the first statement of lemma 2.4:

---

*) See [11], p. 151, theorem 11 and p. 152, theorem 12.

**) See [11], p. 151, theorem 11, p. 150, theorem 9 and p. 154 theorem 15.

**Lemma 2.5.** *Let $G$ be a subspace of a Banach space $E$, let $s$ be a compact linear operator from $G$ into $E$ and let $u$ be the injection operator from $G$ into $E$ (that is, $u = I_E|_G$). Then $u-s$ is a $\Phi_+$-operator.*

*Proof.* The proof of the fact that $\dim \operatorname{Ker}(u-s) < \infty$ is the same as that of the relation $\dim \operatorname{Ker}(I_G - v) < \infty$, where $v$ is any compact linear operator from $G$ into itself*).

Now let $Q$ be an arbitrary complementary subspace to $\operatorname{Ker}(u-s)$ in $G$, that is, $G = \operatorname{Ker}(u-s) \oplus Q$. In order to prove that $(u-s)(G) = (u-s)(Q)$ is closed in $E$, it will be sufficient to prove that $(u-s)|_Q$ is an isomorphism. Obviously, $(u-s)|_Q$ is one to one and continuous. Assume that there exists a sequence $\{x_n\} \subset Q$ with $\|x_n\| = 1$ $(n = 1,2, \ldots)$, such that $\lim_{n\to\infty} (u-s)(x_n) = \lim_{n\to\infty} (x_n - s(x_n)) = 0$. Since $s\colon G \to E$ is compact, the sequence $\{s(x_n)\}$ has a convergent subsequence, say $\{s(x_{n_k})\}$. Then $\lim_{k\to\infty} x_{n_k} = \lim_{k\to\infty} u(x_{n_k}) = \lim_{k\to\infty} s(x_{n_k}) = x_0$ exists and, obviously, $x_0 \in Q$, $\|x_0\| = 1$, $(u-s)(x_0) = 0$. Hence $x_0 \in (Q \cap \operatorname{Ker}(u-s)) \setminus \{0\}$, in contradiction with the definition of $Q$, which completes the proof.

The introduction of deficient basic sequences has been motivated (at the beginning of this section) by some stability theorems for bases. Let us give now some stability theorems for deficient basic sequences. To this end, we recall the following stability theorem for bases, mentioned at the end of Ch. II, § 11 ("extended" theorem 11.3):

*Let $F$ be a Banach space, $(e_n, h_n)$ $(\{e_n\} \subset F, \{h_n\} \subset F^*)$ a biorthogonal system, $\{x_n\}$ an $(F, \{e_n\})$-Besselian basis**) of a Banach space $E$, and $\{y_n\}$ a sequence in $E$, weakly $([h_n], \{h_n\})$-near***) to $\{x_n\}$. Then the sequence $\{y_n\}$ can be modified to become an $(F, \{e_n\})$-Besselian basis of $E$, by changing suitable $k$ elements of it, where $k = \operatorname{codim}_E [y_n] < \infty$. In particular, if $\{e_n\}$ is an unconditional basis of $F$ and $\{y_n\}$ is $([h_n], \{h_n\})$-near****) to $\{x_n\}$, the same conclusion holds.*

It is natural to ask what can we say if $\{x_n\}$ is only a deficient basic sequence (in particular, a basic sequence) and $\{y_n\}$ is near to $\{x_n\}$ in one of the above two senses, about the behaviour of the sequences $\{x_n\}$,

---

*) See [11], p. 152, theorem 12.

**) I.e. (see Ch. II, § 11, definition 11.5) such that $\{x_n\} \succ \{e_n\}$ $\left(\text{that is, the convergence of } \sum_{i=1}^{\infty} \alpha_i x_i \text{ in } E \text{ implies the convergence of } \sum_{i=1}^{\infty} \alpha_i e_i \text{ in } F\right)$.

***) I.e. (see Ch. II, § 11, definition 11.6) $\sum_{i=1}^{\infty} f(x_i - y_i) h_i$ converges uniformly with respect to $f \in E^*$, $\|f\| \leqslant 1$.

****) I.e. (see Ch. II, § 11, definition 11.6) $\sum_{i=1}^{\infty} \|x_i - y_i\| h_i$ converges.

$\{y_n\}$ or of the subspaces $[x_n]$, $[y_n]$ with respect to the whole space $E$, in particular, about relations between the numbers $\alpha$, $\beta$ and $\varkappa$ associated to the sequences $\{x_n\}$ and $\{y_n\}$. In this direction, we shall prove now

**Theorem 2.2.** *Let $F$ be a Banach space, $(e_n, h_n)$ ($\{e_n\} \subset F$, $\{h_n\} \subset F^*$) a biorthogonal system, $\{x_n\}$ an $(F, \{e_n\})$-Besselian deficient basic sequence*) in a Banach space $E$ and $\{y_n\}$ a sequence in $E$, weakly $([h_n], \{h_n\})$-near to $\{x_n\}$. Then $\{y_n\}$ is a deficient basic sequence in $E$, satisfying*

$$\varkappa_{\{y_n\}} = \varkappa_{\{x_n\}}. \tag{2.40}$$

*In particular, if $\{e_n\}$ is an unconditional basis of $F$ and $\{y_n\}$ is $([h_n], \{h_n\})$-near to $\{x_n\}$, the same conclusion holds.*

*Proof.* By proposition 2.1, let $\{x_n\}_{n \neq i_1, \ldots, i_\alpha}$ be a basis of $[x_n]$, where $\alpha = \alpha_{\{x_n\}}$. Then, as in the proof of the above stability theorem (see Ch. II, proof of theorem 11.3), we obtain that

$$s(x) = \sum_{n \neq i_1, \ldots, i_\alpha} f_n(x)\,(x_n - y_n) \tag{2.41}$$

is a well defined compact linear operator from $[x_n]$ into $E$. Hence, by lemma 2.5, $u - s$ is a $\Phi_+$-operator from $[x_n]$ into $E$, where $u = I_E|_{[x_n]}$. Therefore, by theorem 2.1, $\{y_n\} = \{x_n - s(x_n)\} = \{(u - s)(x_n)\}$ is a deficient basic sequence in $E$, such that if $\varkappa_{\{x_n\}} = \infty$ then also $\varkappa_{\{y_n\}} = \infty$.

Assume now that $\varkappa_{\{x_n\}} < \infty$, whence $\beta = \beta_{\{x_n\}} < \infty$ and let $z_1, \ldots, z_\beta$ be such that $\{z_j\}_{j=1}^{\beta} \cup \{x_n\}_{n \neq i_1, \ldots, i_\alpha}$ is a basis of $E$. Then, by the above stability theorem, $\{z_j\}_{j=1}^{\beta} \cup \{y_n\}_{n \neq i_1, \ldots, i_\alpha}$ can be modified to become an $(F, \{e_n\})$-Besselian basis of $E$, by changing suitable $k$ elements of it, where $k = \operatorname{codim}_E [\{z_j\}_{j=1}^{\beta} \cup \{y_n\}_{n \neq i_1, \ldots, i_\alpha}]$. Thus $\{y_n\}$ is a deficient basic sequence such that omitting $\alpha$ elements, adding $\beta$ elements and then omitting $k$ elements and adding $k$ elements, we obtain a basis of $E$. Hence, by proposition 2.3,

$$\varkappa_{\{y_n\}} = \beta - \alpha + k - k = \varkappa_{\{x_n\}},$$

which completes the proof of theorem 2.2.

*Remark 2.5.* By taking in theorem 2.2 various particular $F$ and $(e_n, h_n)$, we obtain extensions of Ch. I, § 10, theorem 10.2 b) (hence also an extension of**) the Krein-Milman-Rutman theorem), of Ch. II, § 11, theorem 11.3 and of other stability theorems, to deficient basic sequences.

For the proof of the next stability theorem we shall need

**Lemma 2.6.** *Let $G$ be a subspace of a Banach space $E$, let $s$ be a continuous linear operator from $G$ into $E$, such that $0 < \|s\| < 1$, and let $u$ be*

---

*) In the obvious sense (see definition 2.4 below).

**) I.e., of Ch. I, § 10, theorem 10.3.

*the injection operator from G into E. Then $u - s$ is an isomorphism (and hence a $\Phi_+$-operator).*

*Proof.* We have

$$\|(u-s)(x)\| \geqslant |\|u(x)\| - \|s(x)\|| \geqslant \|x\| - \|s\|\,\|x\| =$$

$$= (1 - \|s\|)\,\|x\| \quad (x \in G),$$

whence $u - s$ is an isomorphism from $G$ into $E$, which completes the proof.

Let us recall now the following stability theorem for bases, mentioned at the end of Ch. II, § 11 ("extended" theorem 11.2):

*Let $F$ be a Banach space, $(e_n, h_n)$ ($\{e_n\} \subset F$, $\{h_n\} \subset F^*$) a biorthogonal system, $\{x_n\}$ an $(F, \{e_n\})$-Besselian basis of a Banach space $E$, and $\{y_n\}$ a sequence in $E$, such that*

$$\sup_{1 \leqslant n < \infty} \sup_{\substack{f \in E^* \\ \|f\| \leqslant 1}} \left\| \sum_{i=1}^{n} f(x_i - y_i)\, h_i \right\| = M < c, \tag{2.42}$$

*where $c > 0$ is any constant with the property that*

$$c \left\| \sum_{i=1}^{n} \gamma_i e_i \right\| \leqslant \left\| \sum_{i=1}^{n} \gamma_i x_i \right\| \tag{2.43}$$

*for all finite sequences of scalars $\gamma_1, \ldots, \gamma_n$. Then $\{y_n\}$ is a basis of $E$, equivalent to $\{x_n\}$ (whence $(F, \{e_n\})$-Besselian). In particular, if $\{e_n\}$ is an unconditional basis of $F$ and $\{y_n\} \subset E$ is such that*

$$\sup_{1 \leqslant n < \infty} \left\| \sum_{i=1}^{n} \|x_i - y_i\|\, h_i \right\| = M' < \frac{c}{\lambda^{(u)}_{\{e_n\}}}, \tag{2.44}$$

*where $\{h_n\}$ and $c$ are as above and where**) $\lambda^{(u)}_{\{e_n\}} = \sup_{1 \leqslant n < \infty} \sup_{\substack{z \in F \\ \|z\| \leqslant 1}} \left\| \sum_{i=1}^{n} \beta_i h_i(z) e_i \right\|$,

*then the same conclusion holds.*

We have the following extension of this stability theorem to deficient basic sequences, which also gives some information about the relation between $\alpha_{\{x_n\}}$, $\beta_{\{x_n\}}$ and $\alpha_{\{y_n\}}$, $\beta_{\{y_n\}}$ (observe that theorem 2.2 contains no such information):

---

*) $\lambda^{(u)}_{\{e_n\}}$ is nothing else than the constant inf $M_5$ occurring in Ch. II, § 17, theorem 17.1.

**Theorem 2.3.** *Let $F$ be a Banach space, $(e_n, h_n)$ ($\{e_n\} \subset F$, $\{h_n\} \subset F^*$) a biorthogonal system, $\{x_n\}$ an $(F, \{e_n\})$-Besselian deficient basic sequence in a Banach space $E$ and $\{y_n\}$ a sequence in $E$, satisfying* (2.42), *where $c > 0$ is any constant with the property* (2.43). *Then $\{y_n\}$ is a deficient basic sequence in $E$, satisfying*

$$\alpha_{\{y_n\}} \leqslant \alpha_{\{x_n\}}, \quad \beta_{\{y_n\}} \leqslant \beta_{\{x_n\}}, \quad \varkappa_{\{y_n\}} = \varkappa_{\{x_n\}}. \tag{2.45}$$

*In particular, if $\{e_n\}$ is an unconditional basis of $F$ and $\{y_n\} \subset E$ satisfies* (2.44), *then the same conclusion holds.*

*Proof.* Define a linear mapping $s_0$ of the linear span of $\{x_n\}$ into $E$ by

$$s_0\left(\sum_{i=1}^{n} \gamma_i x_i\right) = \sum_{i=1}^{n} \gamma_i (x_i - y_i). \tag{2.46}$$

Let $\gamma_1, \ldots, \gamma_n$ be arbitrary scalars and take a $g_n \in E^*$ with $\|g_n\| = 1$ such that $g_n\left(\sum_{i=1}^{n} \gamma_i (x_i - y_i)\right) = \left\|\sum_{i=1}^{n} \gamma_i (x_i - y_i)\right\|$. Then, by (2.42) and (2.43),

$$\left\|s_0\left(\sum_{i=1}^{n} \gamma_i x_i\right)\right\| = \left\|\sum_{i=1}^{n} \gamma_i (x_i - y_i)\right\| = \sum_{i=1}^{n} \gamma_i g_n (x_i - y_i) =$$

$$= \left(\sum_{i=1}^{n} g_n (x_i - y_i) h_i\right)\left(\sum_{j=1}^{n} \gamma_j e_j\right) \leqslant \left\|\sum_{i=1}^{n} g_n (x_i - y_i) h_i\right\| \left\|\sum_{j=1}^{n} \gamma_j e_j\right\| \leqslant$$

$$\leqslant \frac{M}{c}\left\|\sum_{i=1}^{n} \gamma_i x_i\right\|,$$

whence $s_0$ can be extended, by continuity, to a continuous linear mapping $s$ of $[x_n]$ into $E$, of norm $\|s\| \leqslant \frac{M}{c} < 1$. Consequently, by lemma 2.6, $u - s$ is an isomorphism (and hence a $\Phi_+$-operator) from $[x_n]$ into $E$, where $u = I_E|_{[x_n]}$. Therefore, by theorem 2.1, $\{y_n\} = \{x_n - s(x_n)\} = \{(u - s)(x_n)\}$ is a deficient basic sequence in $E$, such that if $\varkappa_{\{x_n\}} = \infty$ then also $\varkappa_{\{y_n\}} = \infty$.

Assume now that $\varkappa_{\{x_n\}} < \infty$, whence $\beta = \beta_{\{x_n\}} < \infty$ and let $z_1, \ldots, z_\beta$ be such that $\{z_j\}_{j=1}^{\beta} \cup \{x_n\}_{n \neq i_1, \ldots, i_\alpha}$ is a basis of $E$, where $\alpha = \alpha_{\{x_n\}}$ and where $\{x_n\}_{n \neq i_1, \ldots, i_\alpha}$ is a basis of $[x_n]$ (by proposition 2.1). Then, by the stability theorem recalled above, $\{z_j\}_{j=1}^{\beta} \cup \{y_n\}_{n \neq i, \ldots, i_\alpha}$ is a basis of $E$, equivalent to $\{z_j\}_{j=1}^{\beta} \cup \{x_n\}_{n \neq i_1, \ldots, i_\alpha}$. Hence $\{y_n\}_{n \neq i_1, \ldots, i_\alpha}$

is a basic sequence and thus $\alpha_{\{y_n\}} \leqslant \alpha = \alpha_{\{x_n\}}$. Furthermore, $\{y_n\}$ is a deficient basic sequence such that omitting $\alpha$ elements and adding $\beta$ elements we obtain a basis of $E$, whence, by proposition 2.3,

$$\varkappa_{\{y_n\}} = \beta - \alpha = \varkappa_{\{x_n\}}.$$

Now, from the above relations it follows that

$$\beta_{\{y_n\}} = \varkappa_{\{y_n\}} + \alpha_{\{y_n\}} \leqslant \varkappa_{\{x_n\}} + \alpha_{\{x_n\}} = \beta_{\{x_n\}},$$

which proves (2.45).

Finally, the last statement is a particular case of the above*), since by $\lambda^{(u)}_{\{h_n\}} \leqslant \lambda^{(u)}_{\{e_n\}}$ and (2.44) we have

$$\left\| \sum_{i=1}^{n} f(x_i - y_i) h_i \right\| \leqslant \lambda^{(u)}_{\{h_n\}} \left\| \sum_{i=1}^{n} \|x_i - y_i\| h_i \right\| \leqslant \lambda^{(u)}_{\{e_n\}} M' < c$$

$$(f \in E^*,\ \|f\| \leqslant 1,\ n = 1,2, \ldots),$$

and thus the proof of theorem 2.3 is complete.

*Remark 2.6.* In particular, if $\alpha_{\{x_n\}} = 0$, that is, if $\{x_n\}$ is a basic sequence, then from (2.45) we obtain that $\alpha_{\{y_n\}} = 0$ (that is, $\{y_n\}$ is basic) and $\beta_{\{y_n\}} = \beta_{\{x_n\}}$ (that is, $\operatorname{codim}_E [y_n] = \operatorname{codim}_E [x_n]$). On the other hand, if $\beta_{\{x_n\}} = 0$, that is, if $[x_n] = E$, then from (2.45) we obtain that $\beta_{\{y_n\}} = 0$ (that is, $[y_n] = E$) and $\alpha_{\{y_n\}} = \alpha_{\{x_n\}}$. Furthermore, by taking in theorem 2.3 various particular $F$ and $(e_n, h_n)$, we obtain extensions of stability theorems of Ch. I and Ch. II, to deficient basic sequences.

We shall give now stability theorems involving nearness conditions of Krein-Milman-Rutman type, which complement and extend some other results of Ch. I, § 10 (e.g., Ch. I, § 10, theorems 10.3 and 10.4). We recall that two Banach spaces $E_1$ and $E_2$ are said to be *ε-isometric*,

---

*) It can be proved also directly, with a similar argument, by using the inequalities

$$\left\| \sum_{i=1}^{n} \gamma_i (x_i - y_i) \right\| \leqslant \sum_{i=1}^{n} |\gamma_i| \, \|x_i - y_i\| = \left( \sum_{i=1}^{n} \|x_i - y_i\| h_i \right) \left( \sum_{j=1}^{n} |\gamma_j| e_j \right) \leqslant$$

$$\leqslant \left\| \sum_{i=1}^{n} \|x_i - y_i\| h_i \right\| \left\| \sum_{j=1}^{n} |\gamma_j| e_j \right\| \leqslant \frac{M' \lambda^{(u)}_{\{e_n\}}}{c} \left\| \sum_{i=1}^{n} \gamma_i x_i \right\|.$$

where $0 < \varepsilon < 1$, if there exists an isomorphism $u$ of $E_1$ onto $E_2$ such that

$$(1 - \varepsilon)\|x\| \leqslant \|u(x)\| \leqslant (1 + \varepsilon)\|x\| \qquad (x \in E_1); \quad (2.47)$$

any such isomorphism $u$ is called an *$\varepsilon$-isometry.*

**Theorem 2.4.** *Let $\{x_n\}$ be a minimal sequence in a Banach space $E$ and let $0 < \varepsilon < 1$. Then there exists a sequence of constants $\gamma_n > 0$ $(n = 1,2, \ldots)$ such that for every sequence $\{y_n\} \subset E$ satisfying*

$$\|x_n - y_n\| \leqslant \gamma_n \qquad (n = 1,2, \ldots) \quad (2.48)$$

*the linear mapping $u$ which carries $x_n$ into $y_n$ $(n = 1,2, \ldots)$ is an $\varepsilon$-isometry of $[x_n]$ onto $[y_n]$.*

*Proof.* Since $\{x_n\}$ is minimal, there exists a sequence $\{f_n\} \subset E^*$ such that $(x_n, f_n)$ is a biorthogonal system. Put*)

$$\gamma_n = \frac{\varepsilon}{2^{n+1}\|f_n\|} \qquad (n = 1,2, \ldots) \quad (2.49)$$

and let $x = \sum_{i=1}^{n} \alpha_i x_i$. Then

$$\left\| \sum_{i=1}^{n} \alpha_i(x_i - y_i) \right\| = \left\| \sum_{i=1}^{n} f_i(x)(x_i - y_i) \right\| \leqslant \sum_{i=1}^{n} |f_i(x)| \gamma_i \leqslant$$

$$\leqslant \left( \sum_{i=1}^{n} \gamma_i \|f_i\| \right) \|x\| \leqslant \sum_{i=1}^{n} \frac{\varepsilon}{2^{i+1}} \|x\| < \varepsilon \|x\| = \varepsilon \left\| \sum_{i=1}^{n} \alpha_i x_i \right\|,$$

whence

$$(1 - \varepsilon)\|x\| = \left\| \sum_{i=1}^{n} \alpha_i x_i \right\| - \varepsilon \left\| \sum_{i=1}^{n} \alpha_i x_i \right\| \leqslant \left\| \sum_{i=1}^{n} \alpha_i x_i \right\| -$$

$$- \left\| \sum_{i=1}^{n} \alpha_i(x_i - y_i) \right\| \leqslant \left\| \sum_{i=1}^{n} \alpha_i y_i \right\| = \|u(x)\| \leqslant \left\| \sum_{i=1}^{n} \alpha_i x_i \right\| +$$

$$+ \left\| \sum_{i=1}^{n} \alpha_i(x_i - y_i) \right\| \leqslant (1 + \varepsilon)\|x\|.$$

---

*) This argument is similar to that of Ch. I, § 10, remark 10.5.

Since the set of all finite linear combinations $\sum_{i=1}^{n} \alpha_i x_i$ is dense in $[x_n]$, it follows that these inequalities also hold for every $x \in [x_n]$, which completes the proof of theorem 2.4.

*Remark 2.7.* Formula (2.49) and the above argument also show that for each $n$ the subspaces $[x_k]_n^\infty$ and $[y_k]_n^\infty$ are $\sum_{i=n}^{\infty} \frac{\varepsilon}{2^{i+1}} = \frac{\varepsilon}{2^n}$ -isometric. Let us also mention that *if $\{x_n\}$ is not minimal, then for every sequence $\gamma_n > 0$ $(n = 1,2, \ldots)$ there exists a sequence $\{y_n\} \subset E$ satisfying* (2.48) *and such that the linear mapping $u$ which carries $x_n$ into $y_n$ is unbounded.* Indeed, if $x_k \in [x_n]_{n \neq k}$, it is enough to take

$$y_n = x_n \qquad (n = 1, \ldots, k-1, k+1, \ldots),$$

$$y_k \neq x_k, \qquad \|x_k - y_k\| = \gamma_k.$$

For, then clearly $\{y_n\}$ satisfies (2.48). Furthermore, by $x_k \in [x_n]_{n \neq k}$, for any given $N$ there exists a finite linear combination of the form $z^{(N)} = x_k - \sum_{i \neq k} \alpha_i^{(N)} x_i$, such that

$$\|z^{(N)}\| \leqslant \frac{\gamma_k}{N}.$$

But then

$$\|u(z^{(N)})\| = \left\| y_k - \sum_{i \neq k} \alpha_i^{(N)} x_i \right\| \geqslant$$

$$\geqslant \|y_k - x_k\| - \left\| x_k - \sum_{i \neq k} \alpha_i^{(N)} x_i \right\| \geqslant \gamma_k - \frac{\gamma_k}{N} = \left(1 - \frac{1}{N}\right) \gamma_k,$$

whence

$$\|u\| \geqslant \frac{\|u(z^{(N)})\|}{\|z^{(N)}\|} \geqslant \frac{\left(1 - \frac{1}{N}\right)\gamma_k}{\frac{\gamma_k}{N}} = N - 1,$$

which, since $N$ was arbitrary, proves that $\|u\| = \infty$.

*Definition 2.2.* A sequence $\{x_n\}$ in a Banach space $E$ is called a *deficient minimal* sequence if it can be transformed into a minimal sequence by omitting a finite number of its elements.

**Theorem 2.5.** *Let $E$ be a Banach space and $(x_n, f_n)$ $(\{x_n\} \subset E, \{f_n\} \subset \subset E^*)$ a biorthogonal system (respectively, let $\{x_n\}$ be a basis of $E$, with the a.s.c.f. $\{f_n\}$). In order that a sequence $\gamma_n > 0$ $(n = 1,2, \ldots)$ have the property that every sequence $\{y_n\} \subset E$ satisfying*

$$\|x_n - y_n\| \leqslant \gamma_n \qquad (n = 1,2, \ldots) \tag{2.50}$$

*is a deficient minimal (respectively, a deficient basic) sequence, it is sufficient that there exist a positive integer $n_0$ such that*

$$b_{n_0} = \sup_{|\varepsilon_i|=1} \sup_{n_0 \leqslant n < \infty} \left\| \sum_{i=n_0}^{n} \varepsilon_i \gamma_i f_i \Big|_{[x_j]} \right\| < 1 \tag{2.51}$$

*and it is necessary that there exist a positive integer $n_0$ such that*

$$b_{n_0} \leqslant 1. \tag{2.52}$$

*Proof. Sufficiency.* If we have (2.51), then, by Ch. I, § 10, theorem 10.4, for every sequence $\{y_n\}$ satisfying (2.50) we have $\{x_n\}_{n_0}^{\infty} \approx \{y_n\}_{n_0}^{\infty}$ and hence $\{y_n\}$ is a deficient minimal (respectively, a deficient basic sequence.

*Necessity.* Assume that (2.52) is not satisfied for any $n_0$, that is, $b_n > 1$ $(n = 1,2, \ldots)$. Then, since $b_1 > 1$, by Ch. I, § 10, proof of theorem 10.4, there exists a finite sequence of the form

$$y_i = x_i - \varepsilon_i^{(0)} \gamma_i z_0 \qquad (i = 1, \ldots, n_1),$$

where $z_0 \in E$, $\|z_0\| \leqslant 1$, $|\varepsilon_i^{(0)}| \leqslant 1$ $(i = 1, \ldots, n_1)$ (hence $\|x_i - y_i\| \leqslant \gamma_i$ for $i=1, \ldots, n_1$), such that $y_1, \ldots, y_{n_1}$ is not linearly independent. Considering now the sequence $\{x_n\}_{n_1+1}^{\infty}$ and continuing in this way, we obtain an infinite sequence $\{y_n\}$ satisfying (2.50) and consisting of an infinite number of linearly dependent finite sequences, whence $\{y_n\}$ is not deficient minimal, which completes the proof of theorem 2.5.

**Corollary 2.3.** *Let $E$ be a Banach space and let $(x_n, f_n)$ $(\{x_n\} \subset E, \{f_n\} \subset E^*)$ be an $E$-complete biorthogonal system (respectively, let $\{x_n\}$ be a basis of $E$, with the a.s.c.f. $\{f_n\}$) such that $[f_n]$ contains no subspace isomorphic to $c_0$. In order that a sequence $\gamma_n > 0$ $(n = 1,2, \ldots)$ have the property that every sequence $\{y_n\} \subset E$ satisfying (2.50) is a deficient minimal (respectively, a deficient basic) sequence, it is necessary and sufficient that $\sum_{i=1}^{\infty} \gamma_i f_i$ be unconditionally convergent.*

*Proof. Sufficiency.* If $\sum_{i=1}^{\infty} \gamma_i f_i$ is unconditionally convergent, then, by Ch. I, § 10, corollary 10.3 (implication $4° \Rightarrow 2°$) there exists a positive integer $n_0$ such that for every sequence $\{y_n\} \subset E$ satisfying (2.50) we have $\{x_n\}_{n_0}^{\infty} \approx \{y_n\}_{n_0}^{\infty}$, whence $\{y_n\}$ is a deficient minimal (respectively, a deficient basic) sequence.

*Necessity.* If a sequence $\gamma_n > 0$ $(n = 1,2, \ldots)$ has the property stated in corollary 2.3, then, by theorem 2.5 above, there exists a positive integer $n_0$ such that $b_{n_0} \leqslant 1$, and hence, by Ch. I, § 10, proof of corollary 10.3 (implication $3° \Rightarrow 4°$), $\sum_{i=1}^{\infty} \gamma_i f_i$ is unconditionally convergent, which completes the proof.

For the sake of completeness, let us prove now the following lemma, which was mentioned without proof in Ch. I, p.205, and which will be used below:

**Lemma 2.7.** *Let $\{x_n\}$, $\{y_n\}$ be sequences in Banach spaces E and F respectively, such that $x_n \neq 0$, $y_n \neq 0$ $(n = 1,2, \ldots)$. We have $\{x_n\} \sim \{y_n\}$ if and only if*

$$\left\{\{\alpha_n\} \subset K \,\middle|\, \sup_{1 \leqslant n < \infty} \left\| \sum_{i=1}^{n} \alpha_i x_i \right\| < \infty \right\} = \left\{\{\alpha_n\} \subset K \,\middle|\, \sup_{1 \leqslant n < \infty} \left\| \sum_{i=1}^{n} \alpha_i y_i \right\| < \infty \right\}. \tag{2.53}$$

*Proof.* If $\{x_n\} \sim \{y_n\}$, then, by Ch. I, § 8, theorem 8.1 c) (implication $1° \Rightarrow 2°$), there exist a positive integer $n_0$ and a constant $C > 0$ such that we have

$$\left\| \sum_{i=n_0}^{n_0+m} \alpha_i y_i \right\| \leqslant C \sup_{n_0 \leqslant k \leqslant n_0+m} \left\| \sum_{i=n_0}^{k} \alpha_i x_i \right\|, \tag{2.54}$$

$$\left\| \sum_{i=n_0}^{n_0+m} \alpha_i x_i \right\| \leqslant C \sup_{n_0 \leqslant k \leqslant n_0+m} \left\| \sum_{i=n_0}^{k} \alpha_i y_i \right\| \tag{2.55}$$

for all finite sequence of scalars $\alpha_{n_0}, \ldots, \alpha_{n_0+m}$. Assume now that $\sup_{1 \leqslant n < \infty} \left\| \sum_{i=1}^{n} \alpha_i x_i \right\| = M < \infty$. Then for every $k \geqslant n_0$ we have

$$\left\| \sum_{i=n_0}^{k} \alpha_i x_i \right\| = \left\| \sum_{i=1}^{k} \alpha_i x_i - \sum_{i=1}^{n_0-1} \alpha_i x_i \right\| \leqslant (n_0 - 1) \max_{1 \leqslant i \leqslant n_0-1} \|\alpha_i x_i\| + M = M_0,$$

whence, by (2.54),

$$\sup_{1 \leqslant n < \infty} \left\| \sum_{i=1}^{n} \alpha_i y_i \right\| \leqslant (n_0 - 1) \max_{1 \leqslant i \leqslant n_0-1} \|\alpha_i y_i\| + CM_0 < \infty.$$

Similarly, $\sup_{1 \leqslant n < \infty} \left\| \sum_{i=1}^{n} \alpha_i y_i \right\| < \infty$ implies $\sup_{1 \leqslant n < \infty} \left\| \sum_{i=1}^{n} \alpha_i x_i \right\| < \infty$ and thus we have (2.53).

Conversely, assume now that we have (2.53) and let $A_2(\{x_n\})$, $A_2(\{y_n\})$ be the Banach spaces of sequences of scalars introduced in Ch. I, § 5, remark 5.4. Then by (2.53) we have $A_2(\{x_n\}) = A_2(\{y_n\}) = A_2$ and hence, by the argument given in Ch. I, § 12, proof of formula (12.66), the norms on $A_2$ induced by $A_2(\{x_n\})$ and $A_2(\{y_n\})$ are equivalent, that is, there exist two constants $C_1, C_2 > 0$ such that for any sequence $\{\alpha_n\} \in A_2$ we have

$$C_1 \sup_{1 \leqslant n < \infty} \left\| \sum_{i=1}^{n} \alpha_i y_i \right\| \leqslant \sup_{1 \leqslant n < \infty} \left\| \sum_{i=1}^{n} \alpha_i x_i \right\| \leqslant C_2 \sup_{1 \leqslant n < \infty} \left\| \sum_{i=1}^{n} \alpha_i y_i \right\| .$$

Consequently, by Ch. I, § 8, theorem 8.1 c) (implication $2^\circ \Rightarrow 1^\circ$) we have $\{x_n\} \sim \{y_n\}$, which completes the proof of lemma 2.7.

It is convenient to introduce the following terminology, somewhat similar to that of Ch. I, § 10, definition 10.3:

*Definition 2.3.* Let $\gamma_n > 0$ $(n = 1,2, \ldots)$. A minimal (respectively, a basic) sequence $\{x_n\}$ in a Banach space $E$ is called

a) *$\{\gamma_n\}$-stable*, if every sequence $\{y_n\} \subset E$ satisfying (2.50) is a minimal (respectively, a basic) sequence;

b) *deficient $\{\gamma_n\}$-stable*, if every sequence $\{y_n\} \subset E$ satisfying (2.50) is a deficient minimal (respectively, a deficient basic) sequence.

**Theorem 2.6.** *Let $\{x_n\}$, $\{y_n\}$ be unconditional basic sequences in Banach spaces $E$ and $F$, respectively. If for every sequence $\gamma_n > 0$ $(n = 1,2, \ldots)$ the basic sequences $\{x_n\}$, $\{y_n\}$ are simultaneously deficient $\{\gamma_n\}$-stable or not deficient $\{\gamma_n\}$-stable, then $\{x_n\} \approx \{y_n\}$.*

*Proof.* Let $\{\varphi_n\} \subset [x_n]^*$, $\{\psi_n\} \subset [y_n]^*$ be the a.s.c.f. to $\{x_n\}$ and $\{y_n\}$, respectively. Then, by Ch. II, § 17, theorem 17.7, $\{\varphi_n\}$ and $\{\psi_n\}$ are unconditional basic sequences. Let $\{\alpha_n\} \subset K$ be a sequence of scalars such that $\sup_{1 \leqslant n < \infty} \left\| \sum_{i=1}^{n} \alpha_i \varphi_i \right\| < \infty$. Then, by Ch. II, § 16, theorem 16.1, we have $\sup_{|\varepsilon_i|=1} \sup_{1 \leqslant n < \infty} \left\| \sum_{i=1}^{n} \varepsilon_i \alpha_i \varphi_i \right\| = M < \infty$, whence for $\gamma_n = \frac{|\alpha_n|}{2M}$ $(n = 1,2, \ldots)$ we infer $\sup_{|\varepsilon_i|=1} \sup_{1 \leqslant n < \infty} \left\| \sum_{i=1}^{n} \varepsilon_i \gamma_i \varphi_i \right\| \leqslant \frac{1}{2} < 1$. Consequently, by theorem 2.5, $\{x_n\}$ is deficient $\{\gamma_n\}$-stable and hence, by our hypothesis, $\{y_n\}$ is deficient $\{\gamma_n\}$-stable, too. Therefore, again by theorem 2.5, we have $\sup_{|\varepsilon_i|=1} \sup_{n_0 \leqslant n < \infty} \left\| \sum_{i=n_0}^{n} \varepsilon_i \gamma_i \psi_i \right\| \leqslant 1$, whence $\sup_{1 \leqslant n < \infty} \left\| \sum_{i=1}^{n} \alpha_i \psi_i \right\| < \infty$.

Similarly, $\sup_{1\leqslant n<\infty}\left\|\sum_{i=1}^{n}\alpha_i\psi_i\right\|<\infty$ implies $\sup_{1\leqslant n<\infty}\left\|\sum_{i=1}^{n}\alpha_i\varphi_i\right\|<\infty$ and hence, by lemma 2.7, $\{\varphi_n\}\sim\{\psi_n\}$. Consequently, by Ch. I, § 12, proposition 12.1, $\{x_n\}\sim\{y_n\}$, whence, by Ch. I, § 8, theorem 8.1 d), we obtain $\{x_n\}\approx\{y_n\}$, which completes the proof of theorem 2.6.

**Corollary 2.4.** *Let $1\leqslant p<\infty$ and let $\{x_n\}$ be an unconditional basic sequence in a Banach space $E$, such that $\{x_n\}$ is deficient $\{\gamma_n\}$-stable if and only if $\sum_{i=1}^{\infty}\gamma_i^p<\infty$. Then $\{x_n\}\approx\{e_n\}$, the unit vector basis of $l^q$, where $\frac{1}{p}+\frac{1}{q}=1$ (respectively, of $c_0$, in the case when $q=\infty$).*

*Proof.* Let $\{h_n\}\subset(l^q)^*$ be the a.s.c.f. to the unit vector basis $\{e_n\}$ of $l^q$ (respectively, of $c_0$, if $q=\infty$). Then $[h_n]$ contains no subspace isomorphic to $c_0$, whence, by corollary 2.3, $\{e_n\}$ is deficient $\{\gamma_n\}$-stable if and only if $\sum_{i=1}^{\infty}\gamma_i h_i$ is unconditionally convergent, that is, if and only if $\sum_{i=1}^{\infty}\gamma_i^p<\infty$. Consequently, by theorem 2.6, we have $\{x_n\}\approx\{e_n\}$, which completes the proof.

*Remark 2.8.* If $q=1$ (hence $p=\infty$) and $\gamma_n>0$ $(n=1,2,\ldots)$, $\sup_{n_0\leqslant n<\infty}\gamma_n<1$ for some $n_0$, then, by Ch. II, § 11, teorem 11.5, $\{e_n\}_{n_0}^{\infty}$ is $\{\gamma_n\}_{n_0}^{\infty}$-stable, whence $\{e_n\}$ is deficient $\{\gamma_n\}$-stable. Conversely, if the unit vector basis $\{e_n\}$ of $l^1$ is deficient $\{\gamma_n\}$-stable, then, by theorem 2.5, there exists a positive integer $n_0$ such that $b_{n_0}=\sup_{|\varepsilon_i|=1}\sup_{n_0\leqslant n<\infty}\left\|\sum_{i=n_0}^{n}\varepsilon_i\gamma_i h_i\right\|=$ $=\sup_{|\varepsilon_i|=1}\sup_{n_0\leqslant n<\infty}\sup_{n_0\leqslant i\leqslant n}|\varepsilon_i\gamma_i|=\sup_{n_0\leqslant n<\infty}\gamma_n\leqslant 1$, but not necessarily $\sup_{n_0\leqslant n<\infty}\gamma_n<1$.

Corollary 2.4 shows that in certain cases the set of all "permissible translations", that is, the set of all sequences $\{\gamma_n\}$ (where $\gamma_n>0$ for $n=1,2,\ldots$) such that $\{x_n\}$ is deficient $\{\gamma_n\}$-stable, determines completely the behaviour of $\{x_n\}$ (as concerns properties invariant under isomorphisms).

One can introduce various special classes of deficient basic sequences corresponding to the special classes of basic sequences introduced in § 1. Namely, it is natural to give

*Definition 2.4.* A deficient basic sequence $\{x_n\}$ in a Banach space $E$ is said to be *monotone* (*normal*, etc.) if it can be transformed into a monotone (normal, etc.) basic sequence by omitting a finite number of its elements.

We shall give below a number of results on unconditional deficient basic sequences. For this purpose, we need some preparation.

In § 1 we have considered hypercomplete sequences, i.e. sequences $\{x_n\}$ such that $[x_{n_k}] = E$ for every infinite sequence of indices $\{n_k\}$. An opposite extremal case, in a certain sense, is considered in

**Proposition 2.7.** *Let $\{x_n\}$ be a sequence in a Banach space $E$, such that for every infinite sequence of indices $\{i_n\}$ we have*

$$[x_n]_{n \neq i_1, i_2, \ldots} \neq [x_n]. \tag{2.56}$$

*Then $\{x_n\}$ is a deficient minimal sequence.*

*Proof.* Assume that $\{x_n\}$ is not deficient minimal, that is, every subsequence obtained from $\{x_n\}$ by omitting a finite number of its elements is non-minimal. We shall show that there exists an infinite sequence of indices $\{i_n\}$ such that

$$[x_n]_{n \neq i_1, i_2, \ldots} = [x_n], \tag{2.57}$$

which will complete the proof.

Since $\{x_n\}$ is non-minimal, there exists an index $i_1$ such that $x_{i_1} \in [x_n]_{n \neq i_1} = [x_n]$ and hence there is an index $m_1 \geqslant i_1$ such that dist $(x_{i_1}, [x_n]_{\substack{n=1\\ n \neq i_1}}^{m_1}) < \frac{1}{2}$ $\Big($because there is a polynomial $p = \sum_{\substack{i=1\\ i \neq i_1}}^{m_1} \gamma_i x_i$ such that $\|x_{i_1} - p\| < \frac{1}{2}\Big)$. Since by our assumption $\{x_n\}_{n=m_1+1}^{\infty}$ is non-minimal, there exists an index $i_2 \geqslant m_1 + 1$ such that $x_{i_2} \in [x_n]_{\substack{n=m_1+1\\ n \neq i_2}}^{\infty} = [x_n]_{n=m_1+1}^{\infty}$ and hence there is an index $m_2 \geqslant i_2$ such that dist $(x_{i_2}, [x_n])_{\substack{n=1\\ n \neq i_1, i_2}}^{m_2} < \frac{1}{4}$.

Furthermore, since $[x_n]_{n \neq i_1, i_2} = [x_n]$ (because $[x_n] = [x_n]_{n \neq i_1} = [x_n]_{\substack{n=1\\ n \neq i_1}}^{m_1} + [x_n]_{n=m_1+1}^{\infty} = [x_n]_{\substack{n=1\\ n \neq i_1}}^{m_1} + [x_n]_{\substack{n=m_1+1\\ n \neq i_2}}^{\infty} = [x_n]_{n \neq i_1, i_2}$), we can choose the above index $m_2 \geqslant i_2$ in such a way that we also have dist $(x_{i_1}, [x_n]_{\substack{n=1\\ n \neq i_1, i_2}}^{m_2}) < \frac{1}{4}$. Continuing in this way, we obtain two infinite sequences of indices $\{i_n\}$, $\{m_n\}$ with $m_{n-1} + 1 \leqslant i_n \leqslant m_n$ $(n = 1, 2, \ldots;\ m_0 = 0)$, such that

$$\text{dist}\,(x_{i_j}, [x_n]_{\substack{n=1\\ n \neq i_1, \ldots, i_k}}^{m_k}) < \frac{1}{2^k} \quad (j = 1, \ldots, k;\ k = 1, 2, \ldots). \tag{2.58}$$

We claim that for the sequence $\{i_n\}$ constructed above we have (2.57). Indeed, it is sufficient to show that $x_{i_j} \in [x_n]_{n \neq i_1, i_2, \ldots}$ for all $j = 1, 2, \ldots$ However, this is an obvious consequence of (2.58), which completes the proof of proposition 2.7.

**Proposition 2.8.** *Let $\{x_n\}$ be a sequence in a Banach space $E$, with $x_n \neq 0$ $(n = 1,2, \ldots)$, such that every convergent series of the form $\sum_{i=1}^{\infty} \alpha_i x_i$ is unconditionally convergent. Then either* a) *$\{x_n\}$ is an unconditional deficient basic sequence or* b) *$\{x_n\}$ admits a block basic sequence**) *$\{z_n\}$, equivalent to the unit vector basis of $c_0$.*

*Proof.* We shall consider two cases:

*Case* 1°. There exists an infinite sequence of indices $\{i_n\}$ such that

$$\widehat{([x_{i_n}]_{n=j}^{\infty}; [x_{l_n}]_{n=j}^{\infty})} = 0 \qquad (j = 1,2, \ldots), \tag{2.59}$$

where $\{l_n\}$ denotes the set of indices complementary to $\{i_n\}$.

Let $\{\varepsilon_n\}$ be an arbitrary sequence of positive numbers, such that $\sum_{n=1}^{\infty} \varepsilon_n < \infty$. Then, by (2.59), there exists a sequence $\{y_n\} \subset E$ with the following properties:

$$y_{2n-1} = \sum_{\substack{k=m_{n-1}+1 \\ k \neq l_1, l_2, \ldots}}^{m_n} \gamma_k x_k, \quad \|y_{2n-1}\| = 1 \qquad (n = 1,2, \ldots), \tag{2.60}$$

$$y_{2n} = \sum_{\substack{k=m_{n-1}+1 \\ k \neq i_1, i_2, \ldots}}^{m_n} \gamma_k x_k \qquad (n = 1,2, \ldots), \tag{2.61}$$

$$\|y_{2n-1} - y_{2n}\| < \varepsilon_n \qquad (n = 1,2, \ldots), \tag{2.62}$$

where $\{m_n\}_{n=1}^{\infty}$ is an increasing sequence of positive integers and $m_0 = 0$ and where $\gamma_k$ are scalars. Put

$$c_{2n-1} = \max_{m_{n-1}+1 \leqslant j \leqslant m_n} \left\| \sum_{\substack{k=j \\ k \neq l_1, l_2, \ldots}}^{m_n} \gamma_k x_k \right\| = \left\| \sum_{\substack{k=j_{2n-1}^0 \\ k \neq l_1, l_2, \ldots}}^{m_n} \gamma_k x_k \right\|, \tag{2.63}$$

$$c_{2n} = \max_{m_{n-1}+1 \leqslant j \leqslant m_n} \left\| \sum_{\substack{k=j \\ k \neq i_1, i_2, \ldots}}^{m_n} \gamma_k x_k \right\| = \left\| \sum_{\substack{k=j_{2n}^0 \\ k = i_1, i_2, \ldots}}^{m_n} \gamma_k x_k \right\|, \tag{2.64}$$

$$d_n = \max (c_{2n-1}, c_{2n}) \qquad (n = 1,2, \ldots). \tag{2.65}$$

---

*) See § 1, definition 1.3.

Then by $\|y_{2n-1}\| = 1$ we have $d_n \geqslant 1$ $(n = 1,2, \ldots)$ and hence, by (2.62) and $\sum_{n=1}^{\infty} \varepsilon_n < \infty$, the series $\sum_{n=1}^{\infty} \xi_n \frac{y_{2n-1} - y_{2n}}{d_n} = \sum_{i=1}^{\infty} \alpha_i x_i$ converges for any $\{\xi_n\} \in c_0$. Consequently, by our assumption, this series $\sum_{i=1}^{\infty} \alpha_i x_i$ is unconditionally convergent. Now choose an infinite sequence of indices $\{n_p\}$ such that either $d_{n_p} = c_{2n_p-1}$ $(p = 1,2, \ldots)$ or $d_{n_p} = c_{2n_p}$ $(p = 1,2, \ldots)$; assume, without loss of generality, the first case (in the second case the argument is similar). The unconditional convergence of the above series implies that the series $\sum_{p=1}^{\infty} \xi_{n_p} \frac{y_{2n_p-1}}{c_{2n_p-1}}$ is unconditionally convergent. Put

$$y'_{2n-1} = \frac{1}{c_{2n-1}} \sum_{\substack{k=j_{2n-1}^0 \\ k \neq l_1, l_2, \ldots}}^{m_n} \gamma_k x_k \quad (n = 1,2, \ldots). \tag{2.66}$$

Then from the above it follows, taking again into account Ch. II, § 16, lemma 16.1, that for every $\{\xi_{n_p}\} \in c_0$ the series $\sum_{p=1}^{\infty} \xi_{n_p} y'_{2n_p-1}$ is unconditionally convergent, too. Consequently, by Ch. I, § 17, corollary 17.5, the series $\sum_{p=1}^{\infty} y'_{2n_p-1}$ is weakly unconditionally Cauchy. Since by (2.63) we have $\|y'_{2n-1}\| = 1$ $(n = 1,2, \ldots)$, from Ch. II, § 15, proof of lemma 15.8, $\{y'_{2n_p-1}\}$, whence also $\{x_n\}$, admits a block basic sequence $\{z_n\}$ equivalent to the unit vector basis of $c_0$.

*Case* 2°. There exists no infinite sequence of indices $\{i_n\}$ such that we have (2.59), that is, for every infinite sequence $\{i_n\}$ there exists an index $j_0$ such that

$$\widehat{([x_{i_n}]_{n=j_0}^{\infty};\ [x_{l_n}]_{n=j_0}^{\infty})} > 0, \tag{2.67}$$

where $\{l_n\}$ is the set of indices complementary to $\{i_n\}$. Then*) $\operatorname{codim}_{[x_n]} [x_{l_n}]_{n=j_0}^{\infty} = \infty$, whence

$$[x_n]_{n \neq i_1, i_2, \ldots} = [x_{l_n}]_{n=1}^{\infty} \neq [x_n],$$

---

*) Indeed, since $\{i_n\}$ is infinite and since every convergent series of the form $\sum_{n=j_0}^{\infty} \alpha_{i_n} x_{i_n}$ is unconditionally convergent, we have $\dim [x_{i_n}]_{n=j_0}^{\infty} = \infty$.

and therefore, by proposition 2.7, $\{x_n\}$ is deficient minimal. Let $\{x'_n\}$ be a minimal sequence obtained from $\{x_n\}$ by omitting a finite number of its elements. Then for every pair of complementary sets of indices $\{i_n\}$, $\{l_n\}$ such that either $\{i_n\}$ or $\{l_n\}$ is finite, we have, by Ch. I, § 6, theorem 6.1, implication $1° \Rightarrow 9°$,

$$\widehat{([x'_{i_n}]; [x'_{l_n}])} > 0. \tag{2.68}$$

We claim that the same inequality also holds every pair of infinite complementary sets of indices $\{i_n\}$, $\{l_n\}$. Indeed, let us show first that $[x'_{i_n}] \cap [x'_{l_n}] = \{0\}$. Assume that $x \in [x'_{i_n}] \cap [x'_{l_n}]$. Then, by Ch. I, § 6, theorem 6.1, implication $1° \Rightarrow 6°$, we can write $x = y_1 + y_2 = z_1 + z_2$, where $y_1 \in [x'_{i_n}]_{n=1}^{j_0-1}$, $y_2 \in [x'_{i_n}]_{n=j_0}^{\infty}$, $z_1 \in [x'_{l_n}]_{n=1}^{j_0-1}$, $z_2 \in [x'_{l_n}]_{n=j_0}^{\infty}$, and where $j_0$ is as in (2.67). Hence we obtain

$$y_1 - z_1 = z_2 - y_2 \in ([x'_{i_n}]_{n=1}^{j_0-1} \oplus [x'_{l_n}]_{n=1}^{j_0-1}) \cap ([x'_{i_n}]_{n=j_0}^{\infty} \oplus [x'_{l_n}]_{n=j_0}^{\infty}),$$

and therefore, by the minimality of $\{x'_n\}$, $y_1 = z_1$, $y_2 = z_2$. Consequently, by $[x'_{i_n}]_{n=1}^{j_0-1} \cap [x'_{l_n}]_{n=1}^{j_0-1} = \{0\}$ and (2.67) we obtain $y_1 = z_1 = 0$, $y_2 = z_2 = 0$, and thus $[x'_{i_n}] \cap [x'_{l_n}] = \{0\}$. On the other hand, by (2.67) the sum $[x'_{i_n}]_{n=j_0}^{\infty} + [x'_{l_n}]_{n=j_0}^{\infty}$, whence also $[x'_{i_n}] + [x'_{l_n}]$, is closed, which proves the claim (2.68).

Now, by Ch. II, § 1, the footnote on p. 239, (2.68) is equivalent to the condition that there exists a continuous linear projection $u$ of $[x'_n]$ onto $[x'_{i_n}]$ such that $u([x'_{l_n}]) = 0$, or, in other words, to the condition $[x'_n] = [x'_{i_n}] \oplus [x'_{l_n}]$. Since this holds for every pair of complementary sets of indices $\{i_n\}$, $\{l_n\}$, from § 7, theorem 7.3 it follows that $\{x'_n\}$ is an unconditional basic sequence, which completes the proof of proposition 2.8.

An immediate consequence of proposition 2.8 is the following characterization of unconditional deficient basic sequences:

**Corollary 2.5.** *Let $E$ be a Banach space containing no subspace isomorphic to $c_0$. A sequence $\{x_n\} \subset E$ such that $x_n \neq 0$ $(n = 1,2, \ldots)$ is an unconditional deficient basic sequence if and only if every convergent series of the form $\sum_{i=1}^{\infty} \alpha_i x_i$ is unconditionally convergent.*

Case b) of proposition 2.8 may occur without case a), as shown by

*Example 2.1.* Let $E = c_0$ and let

$$x_{2n-1} = x_{2n} = e_n \qquad (n = 1,2, \ldots), \tag{2.69}$$

where $\{e_n\}$ denotes the unit vector basis of $c_0$. Then every convergent series of the form $\sum_{i=1}^{\infty} \alpha_i x_i$ is unconditionally convergent, but $\{x_n\}$ is not a deficient minimal sequence (and hence not an unconditional deficient basic sequence).

Indeed, if $\sum_{i=1}^{\infty} \alpha_i x_i$ converges, then, since $\|x_n\| = 1$ $(n = 1,2, \ldots)$, we have $\lim_{n\to\infty} \alpha_n = 0$ and therefore the series $\sum_{i=1}^{\infty} \alpha_{2i-1} x_{2i-1}$ and $\sum_{i=1}^{\infty} \alpha_{2i} x_{2i}$, whence also the series $\sum_{i=1}^{\infty} \alpha_i x_i$, are unconditionally convergent.

We have seen in Ch. I, § 8, proposition 8.1, that for every sequence $\{x_n\}$ in a Banach space $E$, with $x_n \neq 0$ $(n = 1,2, \ldots)$, there exists a basis $\{e_n\}$ of a suitable space $A_1$, such that $\{x_n\} \sim \{e_n\}$. Now we shall show that here we cannot replace "basis" by "boundedly complete unconditional basis". Actually, we have the following property of sequences equivalent to a boundedly complete unconditional basis:

**Theorem 2.7.** *Let $\{y_n\}$ be a boundedly complete unconditional basis of a Banach space $F$ and let $\{x_n\}$ be a sequence in a Banach space $E$, such that $\{x_n\} \sim \{y_n\}$. Then $\{x_n\}$ is a boundedly complete unconditional deficient basic sequence. Consequently, one can omit from the sequences $\{x_n\}$ and $\{y_n\}$ a finite number of elements (with the same indices) such that the remaining sequences be strictly equivalent; furthermore, $[x_n]$ is isomorphic to a subspace of finite codimension of $F$.*

*Proof.* If $\sum_{i=1}^{\infty} \alpha_i x_i$ converges, then $\sum_{i=1}^{\infty} \alpha_i y_i$ converges, whence every series $\sum_{i=1}^{\infty} \varepsilon_i \alpha_i y_i$ converges, where $\varepsilon_i = \pm 1$ $(i = 1,2, \ldots)$, and hence every series $\sum_{i=1}^{\infty} \varepsilon_i \alpha_i x_i$ converges, where $\varepsilon_i = \pm 1$ $(i = 1,2, \ldots)$. Thus, every convergent series of the form $\sum_{i=1}^{\infty} \alpha_i x_i$ is unconditionally convergent. Furthermore, by the assumption $\{x_n\} \sim \{y_n\}$, and since $\{y_n\}$ is a basis (hence $y_n \neq 0$ for $n = 1,2, \ldots$), only a finite number of elements of the sequence $\{x_n\}$ can be $= 0$ (because otherwise there would exist an infinite subsequence $\{y_{j_n}\}$ of $\{y_n\}$ such that $\sum_{n=1}^{\infty} \alpha_{j_n} y_{j_n}$ converges for all $\alpha_{j_n}$), say $x_{i_1}, x_{i_2}, \ldots, x_{i_k}$. Therefore we can apply proposition 2.8 to the sequence $\{x_n\}_{n \neq i_1, \ldots, i_k}$. Let us show that case b) of proposition 2.8 cannot occur. Indeed, assume that $\{x_n\}$ admits a block basic sequence

$$z_n = \sum_{i=m_{n-1}+1}^{m_n} \alpha_i x_i \qquad (n = 1,2, \ldots), \tag{2.70}$$

equivalent to the unit vector basis $\{e_n\}$ of $c_0$. Then the series $\sum_{n=1}^{\infty} \xi_n z_n =$

$$= \sum_{n=1}^{\infty} \sum_{i=m_{n-1}+1}^{m_n} \xi_n \alpha_i x_i \text{ converges if and only if the series} \sum_{n=1}^{\infty} \sum_{i=m_{n-1}+1}^{m_n} \xi_n \alpha_i y_i =$$

$$= \sum_{n=1}^{\infty} \xi_n \sum_{i=m_{n-1}+1}^{m_n} \alpha_i y_i \text{ converges and thus } \left\{ \sum_{i=m_{n-1}+1}^{m_n} \alpha_i y_i \right\} \sim \{z_n\} \sim \{e_n\},$$

which, by Ch. II, § 12, theorem 12.2, contradicts the assumption that $\{y_n\}$ is boundedly complete. This proves that case b) of proposition 2.8 cannot occur and hence, by case a), $\{x_n\}$ is an unconditional deficient basic sequence.

Let $\{x_n\}_{n \neq i_1, \ldots, i_k, \ldots, i_{k+\alpha}}$ be an unconditional basis of $[x_n]$, obtained from $\{x_n\}_{n \neq i_1, \ldots, i_k}$ by omitting $\alpha = \alpha_{\{x_n\}_{n \neq i_1, \ldots, i_k}}$ elements (by proposition 2.1). Then $\{x_n\}_{n \neq i_1, \ldots, i_{k+\alpha}} \sim \{y_n\}_{n \neq i_1, \ldots, i_{k+\alpha}}$, whence, by Ch. I, § 8, theorem 8.1, $\{x_n\}_{n \neq i_1, \ldots, i_{k+\alpha}} \approx \{y_n\}_{n \neq i_1, \ldots, i_{k+\alpha}}$ and thus $\{x_n\}$ is boundedly complete and $[x_n] = [x_n]_{n \neq i_1, \ldots, i_{k+\alpha}}$ is isomorphic to the subspace $[y_n]_{n \neq i_1, \ldots, i_{k+\alpha}}$ of finite codimension $k + \alpha$ of $F$, which completes the proof of theorem 2.7.

Let us mention separately

**Corollary 2.6.** *Let $\{x_n\}$ be a sequence in a Banach space $E$, such that $\{x_n\}$ is equivalent to the unit vector basis $\{e_n\}$ of $l^p$, where $1 \leqslant p < \infty$. Then $\{x_n\}$ is a boundedly complete unconditional deficient basic sequence and one can omit from $\{x_n\}$ a finite number of elements in such a way that the remaining sequence be strictly equivalent to $\{e_n\}$; furthermore, $[x_n]$ is isomorphic to $l^p$.*

*Proof.* This follows by applying theorem 2.7 to $F = l^p$, $\{y_n\} = \{e_n\}$ and observing that $\{e_n\}$ is strictly equivalent to any subsequence of the form $\{e_n\}_{n \neq i_1, \ldots, i_k}$ (e.g. by Ch. II, § 18, proposition 18.1).

The equivalence of two bases $\{x_n\}$, $\{y_n\}$ implies, by Ch. I, § 8, theorem 8.1 d), their strict equivalence and hence the behaviour of $\{y_n\}$ determines completely that of $\{x_n\}$ (as concerns properties invariant under isomorphisms). Although Ch. I, § 8, proposition 8.1 shows that the properties of a sequence $\{x_n\}$ with $x_n \neq 0$ $(n = 1,2, \ldots)$, which is not a basis, are not influenced at all, in general, by the fact that $\{x_n\}$ is equivalent to a basis $\{y_n\}$, corollary 2.6 above shows that for certain bases $\{y_n\}$ the condition $\{x_n\} \sim \{y_n\}$ for a sequence $\{x_n\}$ which is not a basis still continues to determine completely the isomorphic behaviour of $\{x_n\}$.

In the preceding (propositions 2.5, 2.6 and theorem 2.1 above) we have considered the images and inverse images of bases (and, more generally, of deficient basic sequences), by a $\Phi_+$-operator $u$. Replacing now the $\Phi_+$-operator by a general continuous linear operator, i.e., considering the images and inverse images of bases by a continuous linear operator $u$, we obtain clearly, generalizations of the notion of a basis. However, they are not new classes of sequences, as shown by

**Proposition 2.9.** *Let $\{x_n\}$ be a sequence in a Banach space E. Then*
a) *There exists a basis $\{e_n\}$ of a Banach space F such that**⁾ $\{e_n\} \gg \{x_n\}$.
b) *The following statements are equivalent:*
1°. *There exists a basis $\{e_n\}$ of a Banach space F such that $\{x_n\} \gg \{e_n\}$.*
2°. *There exists a minimal sequence $\{e_n\}$ in a Banach space F such that $\{x_n\} \gg \{e_n\}$.*
3°. *$\{x_n\}$ is a minimal sequence in E.*

*Proof.* a) Let $F$ be a Banach space with a basis $\{e_n\}$ such that for the a.s.c.f. $\{h_n\} \subset F^*$ we have $\|h_i\| \leqslant \frac{1}{\|x_i\|}$ for all $i = 1,2, \ldots$ with $x_i \neq 0$. Then

$$u(y) = \sum_{i=1}^{\infty} \frac{1}{2^i} h_i(y)x_i \qquad (y \in F) \qquad (2.71)$$

is a well defined continuous linear mapping of $F$ into $E$, satisfying

$$u(2^n e_n) = x_n \qquad (n = 1,2, \ldots), \qquad (2.72)$$

which, since $\{2^n e_n\}$ is also a basis of $F$, proves a).
b) The implication 1° ⇒ 2° is obvious and the implication 2° ⇒ 3° is nothing else than Ch. I, § 8, proposition 8.2 d).
Assume now that we have 3°. Let $\{f_n\} \subset E^*$, $f_i(x_j) = \delta_{ij}$ $(i, j = 1,2, \ldots)$ and let $F$ be a Banach space with a basis $\{e_n\}$ such that $\|e_i\| \leqslant \frac{1}{\|f_i\|}$ $(i = 1,2, \ldots)$. Then

$$u(x) = \sum_{i=1}^{\infty} \frac{1}{2^i} f_i(x)e_i \qquad (x \in E) \qquad (2.73)$$

is a well defined continuous linear mapping of $E$ into $F$, satisfying

$$u(x_n) = \frac{1}{2^n} e_n \qquad (n = 1,2, \ldots), \qquad (2.74)$$

which, since $\left\{\frac{1}{2^n} e_n\right\}$ is also a basis of $F$, proves that 3°⇒1°. This completes the proof of proposition 2.9.

---

*⁾ I.e. (see Ch. I, § 8, definition 8.1) such that there exists a continuous linear mapping $u$ of $[e_n]$ into $[x_n]$ satisfying $u(e_n) = x_n$ $(n = 1, 2, \ldots)$.

*Remark 2.9.* For sequences $\{x_n\}$ with $x_n \neq 0$ $(n = 1,2, \ldots)$ part a) is already contained in Ch. I, § 8, proposition 8.1, which also shows that we can find the $\{e_n\} \gg \{x_n\}$ so as to satisfy $\|e_n\| = \|x_n\|$ $(n = 1,2, \ldots)$.

*Remark 2.10.* Proposition 2.9 also shows that the relations $\{e_n^{(1)}\} \gg \{x_n\} \gg \{e_n^{(2)}\}$, where $\{e_n^{(1)}\}$, $\{e_n^{(2)}\}$ are bases, do not imply the existence of a basis $\{e_n\}$ such that $\{x_n\} \approx \{e_n\}$.

*Remark 2.11.* If we replace in proposition 2.9 "basis $\{e_n\}$" by "unconditional basis $\{e_n\}$", we obtain the same classes of sequences, since the proof remains unchanged, observing that in this case $\{2^n e_n\}$ and $\left\{\frac{1}{2^n} e_n\right\}$ are also unconditional bases of $F$.

In the above proof of proposition 2.9 the basis $\{e_n\}$ has been tailored to the given sequence $\{x_n\}$, by the conditions $\|h_i\| \leqslant \frac{1}{\|x_i\|}$ and $\|e_i\| \leqslant \frac{1}{\|f_i\|}$, respectively. It is natural to ask whether one can also find bases $\{e_n\}$, common for all sequences $\{x_n\}$ belonging to certain classes of sequences. The following is an affirmative result of this type:

**Proposition 2.10.** a) *For any sequence $\{x_n\}$ in a Banach space $E$, with $\sup_{1 \leqslant n < \infty} \|x_n\| < \infty$, we have $\{e_n\} \gg \{x_n\}$, where $\{e_n\}$ is the unit vector basis of $l^1$.*

b) *For any minimal sequence $\{x_n\} \subset E$, with biorthogonal sequence $\{f_n\} \subset E^*$ satisfying $\sup_{1 \leqslant n < \infty} \|f_n\| < \infty$, we have $\{x_n\} \gg \{e_n\}$, where $\{e_n\}$ is the unit vector basis of $c_0$.*

*Proof.* a) Let $\{x_n\}$ be a sequence in a Banach space $E$, with $\sup_{1 \leqslant n < \infty} \|x_n\| < \infty$, and let $\{e_n\}$ be the unit vector basis of $l^1$. Then

$$u\left(\sum_{i=1}^{\infty} \alpha_i e_i\right) = \sum_{i=1}^{\infty} \alpha_i x_i \qquad \left(\{\alpha_n\} = \sum_{i=1}^{\infty} \alpha_i e_i \in l^1\right) \tag{2.75}$$

is a well defined continuous linear mapping of $l^1$ into $E$, satisfying $u(e_n) = x_n$ $(n = 1,2, \ldots)$, and thus $\{e_n\} \gg \{x_n\}$.

b) Let $\{x_n\}$ be a complete minimal sequence in a Banach space $E$, such that for the biorthogonal sequence $\{f_n\} \subset E^*$ we have $\sup_{1 \leqslant n < \infty} \|f_n\| < \infty$, and let $\{e_n\}$ be the unit vector basis of $c_0$. Then the mapping $u$ of the linear span of $\{x_n\}$ into $c_0$, defined by

$$u\left(\sum_{i=1}^{n} \alpha_i x_i\right) = \sum_{i=1}^{n} \alpha_i e_i \tag{2.76}$$

satisfies

$$\left\| u\left(\sum_{i=1}^{n} \alpha_i x_i\right) \right\| = \left\| \sum_{i=1}^{n} \alpha_i e_i \right\| = \max_{1 \leqslant i \leqslant n} |\alpha_i| =$$

$$= \max_{1 \leqslant i \leqslant n} \left| f_i\left(\sum_{j=1}^{n} \alpha_j x_j\right) \right| \leqslant \sup_{1 \leqslant i < \infty} \|f_i\| \left\| \sum_{j=1}^{n} \alpha_j x_j \right\|,$$

whence it can be extended to a continuous linear mapping $\tilde{u}: E \to c_0$ satisfying $\tilde{u}(x_n) = e_n$ $(n = 1,2, \ldots)$. Thus, $\{x_n\} \gg \{e_n\}$, which completes the proof.

*Remark 2.12.* In the above we have proved more, namely that there exists a mapping $u: l^1 \to E$ with $u(e_n) = x_n$ $(n = 1,2, \ldots)$ and $\|u\| \leqslant \sup_{1 \leqslant n < \infty} \|x_n\|$ (respectively, a mapping $\tilde{u}: E \to c_0$ with $\tilde{u}(x_n) = e_n$ $(n = 1,2, \ldots)$ and $\|\tilde{u}\| \leqslant \sup_{1 \leqslant n < \infty} \|f_n\|$, where $f_i(x_j) = \delta_{ij}$). Actually, we have here the equality $\|u\| = \sup_{1 \leqslant n < \infty} \|x_n\|$ (respectively, $\|\tilde{u}\| = \sup_{1 \leqslant n < \infty} \|f_n\|$), since for every $u$: $l^1 \to E$ with $u(e_n) = x_n$ (respectively, $\tilde{u}: E \to c_0$ with $\tilde{u}(x_n) = e_n$) we have

$$\|x_n\| = \|u(e_n)\| \leqslant \|u\| \, \|e_n\| = \|u\| \qquad (n = 1,2, \ldots)$$

(respectively, $\|f_n\| = \|\tilde{u}^*(h_n)\| \leqslant \|\tilde{u}^*\| \, \|h_n\| = \|\tilde{u}\|$ $(n = 1,2, \ldots)$, where $\{h_n\} \subset c_0^*$, $h_i(e_j) = \delta_{ij}$).

## § 3. Complete sequences

In this section we shall give some results on arbitrary complete sequences.

If $\{x_n\}$ is a normalized complete sequence in a Banach space $E$, then for every $x \in E$ there exists a sequence of finite linear combinations $\{p_n(x)\} = \sum_{i=1}^{m_n} \alpha_i^{(n)}(x) x_i$ such that $x = \lim_{n \to \infty} p_n(x)$. It is natural to ask whether one can find these $p_n(x)$ such that, in addition, $\sup_{1 \leqslant n < \infty} |\alpha_i^{(n)}(x)| < \infty$ for all $i = 1,2, \ldots$ In theorem 3.1 below we shall show that the answer is affirmative. Let us first prove the following weaker result:

**Lemma 3.1.** *Let $\{x_n\}$ be a normalized complete sequence in a Banach space $E$. There exists a sequence of positive numbers $\{C_i\}$ (depending* *on* $\{x_n\}$) *such that for each* $x \in E$ *with* $\|x\| = 1$, *each* $\varepsilon > 0$ *and each* $k \in \mathcal{N}$ *there exists a finite linear combination* $p_{\varepsilon,k}(x) = \sum_{i=1}^{m} \alpha_i^{(\varepsilon,k)}(x)x_i$ (*where* $m = m_{\varepsilon,k}(x)$) *satisfying*

$$\|x - p_{\varepsilon,k}(x)\| < \varepsilon, \tag{3.1}$$

$$|\alpha_i^{(\varepsilon,k)}(x)| \leqslant C_i \qquad (i = 1, \ldots, k). \tag{3.2}$$

*Proof.* Let us denote

$$d_i = \operatorname{dist}(x_i, \operatorname{lin}\{x_j\}_{j=i+1}^{\infty}) \qquad (i = 1, 2, \ldots), \tag{3.3}$$

where $\operatorname{lin}\{x_j\}_{j=i+1}^{\infty}$ is the linear subspace of $E$ spanned by $\{x_j\}_{j=i+1}^{\infty}$, and let $\{d_{i_n}\}$ be the sequence of all non-zero $d_i$'s. Put

$$C_i = 0 \qquad (i \in N \setminus \{i_n\}), \tag{3.4}$$

$$C_{i_1} = \frac{1}{d_{i_1}} \cdot \tag{3.5}$$

Since $[x_n] = E$ and $d_1 = \ldots = d_{i_1-1} = 0$, for each $x \in E$ with $\|x\| = 1$ and each $\varepsilon > 0$ there exists a polynomial $p_{\varepsilon,i_1}(x) = \sum_{i=1}^{m_1} \alpha_i^{(\varepsilon,i_1)}(x)x_i$ with $\|p_{\varepsilon,i_1}(x)\| = 1$, such that

$$\|x - p_{\varepsilon,i_1}(x)\| < \varepsilon, \tag{3.6}$$

$$\alpha_i^{(\varepsilon,i_1)}(x) = 0 \qquad (i = 1, \ldots, i_1 - 1). \tag{3.7}$$

Then we have, whenever $\alpha_{i_1}^{(\varepsilon,i_1)}(x) \neq 0$,

$$1 = \|p_{\varepsilon,i_1}(x)\| = \left\| \alpha_{i_1}^{(\varepsilon,i_1)}(x)x_{i_1} + \sum_{i=i_1+1}^{m_1} \alpha_i^{(\varepsilon,i_1)}(x)x_i \right\| =$$

$$= |\alpha_{i_1}^{(\varepsilon,i_1)}(x)| \left\| x_{i_1} + \sum_{i=i_1+1}^{m_1} \frac{\alpha_i^{(\varepsilon,i_1)}(x)}{\alpha_{i_1}^{(\varepsilon,i_1)}(x)} x_i \right\| \geqslant |\alpha_{i_1}^{(\varepsilon,i_1)}(x)| d_{i_1}$$

whence

$$|\alpha_{i_1}^{(\varepsilon,i_1)}(x)| \leqslant \frac{1}{d_{i_1}} = C_{i_1}, \tag{3.8}$$

and clearly this remains also valid if $\alpha_{i_1}^{(\varepsilon, i_1)}(x) = 0$. Thus, by (3.7) and (3.8),

$$|\alpha_i^{(\varepsilon, i_1)}(x)| \leqslant C_i \qquad (i = 1, \ldots, i_1).$$

Assume now that we have already constructed $C_{i_1}, \ldots, C_{i_{n-1}}$ having the property required in lemma 3.1. Put

$$C_{i_n} = \frac{1 + \sum_{s=1}^{n-1} C_{i_s}}{d_{i_n}}. \tag{3.9}$$

Then, by our assumption and by $d_{i_{n-1}+1} = \ldots = d_{i_n - 1} = 0$, for each $x \in E$ with $\|x\| = 1$ and each $\varepsilon > 0$ there exists a polynomial $p_{\varepsilon, i_n}(x) = \sum_{i=1}^{m_n} \alpha_i^{(\varepsilon, i_n)}(x) x_i$ with $\|p_{\varepsilon, i_n}(x)\| = 1$, such that

$$\|x - p_{\varepsilon, i_n}(x)\| < \varepsilon, \tag{3.10}$$

$$|\alpha_i^{(\varepsilon, i_n)}(x)| \leqslant C_i \qquad (i = 1, \ldots, i_{n-1}), \tag{3.11}$$

$$\alpha_i^{(\varepsilon, i_n)}(x) = 0 \qquad (i = i_{n-1} + 1, \ldots, i_n - 1). \tag{3.12}$$

Then we have, whenever $\alpha_{i_n}^{(\varepsilon, i_n)}(x) \neq 0$,

$$1 = \|p_{\varepsilon, i_n}(x)\| = \Big\| \alpha_{i_n}^{(\varepsilon, i_n)}(x) x_{i_n} + \sum_{i = i_n + 1}^{m_n} \alpha_i^{(\varepsilon, i_n)}(x) x_i +$$

$$+ \sum_{i=1}^{i_n - 1} \alpha_i^{(\varepsilon, i_n)}(x) x_i \Big\| \geqslant |\alpha_{i_n}^{(\varepsilon, i_n)}(x)| d_{i_n} -$$

$$- \sum_{i=1}^{i_{n-1}} |\alpha_i^{(\varepsilon, i_n)}(x)| \geqslant |\alpha_{i_n}^{(\varepsilon, i_n)}(x)| d_{i_n} - \sum_{s=1}^{n-1} C_{i_s},$$

whence

$$|\alpha_{i_n}^{(\varepsilon, i_n)}(x)| \leqslant \frac{1 + \sum_{s=1}^{n-1} C_{i_s}}{d_{i_n}} = C_{i_n}, \tag{3.13}$$

and clearly this remains also valid if $\alpha_{i_n}^{(\varepsilon, i_n)}(x) = 0$. Thus, by (3.11)—(3.13), we have

$$|\alpha_i^{(\varepsilon, i_n)}(x)| \leqslant C_i \qquad (i = 1, \ldots, i_n),$$

which, since for $k \in \mathcal{N}$ with $i_{n-1}+1 \leqslant k \leqslant i_n - 1$ (where $i_0 = 0$) one can take $p_{\varepsilon,k}(x) = p_{\varepsilon,i_n}(x)$, completes the proof of lemma 3.1.

Now we can prove

**Theorem** 3.1. *Let $\{x_n\}$ be a normalized complete sequence in a Banach space E. There exists a sequence of positive numbers $\{\lambda_i\}$ (depending on $\{x_n\}$) such that for each $x \in E$ with $\|x\| = 1$ and each $\varepsilon > 0$ there exists a finite linear combination $p_\varepsilon(x) = \sum_{i=1}^{m} \alpha_i^{(\varepsilon)}(x) x_i$ (where $m = m_\varepsilon(x)$) satisfying*

$$\|x - p_\varepsilon(x)\| < \varepsilon, \tag{3.14}$$

$$|\alpha_i^{(\varepsilon)}(x)| \leqslant \lambda_i \qquad (i = 1, \ldots, m). \tag{3.15}$$

*Proof.* Let $\{y_n\} \subset E$ be a normalized sequence which is dense in $\sigma_E = \{y \in E |\ \|y\| = 1\}$. By lemma 3.1, there exists a polynomial $z_1 = \sum_{i=1}^{m_1} \alpha_i^{(1)} x_i$ such that $\|y_1 - z_1\| < 1$, $|\alpha_1^{(1)}| \leqslant C_1$. Next, again by lemma 3.1, there exists a polynomial $z_2 = \sum_{i=1}^{m_2} \alpha_i^{(2)} x_i$, with $m_2 > m_1$, such that $\|y_2 - z_2\| < \frac{1}{2}$, $|\alpha_i^{(2)}| \leqslant C_i$ $(i = 1, \ldots, m_1)$. Continuing in this way indefinitely, we obtain an increasing sequence $\{m_n\} \subset \mathcal{N}$ and a sequence of polynomials $\{z_n\} = \left\{\sum_{i=1}^{m_n} \alpha_i^{(n)} x_i\right\}$ such that

$$\|y_n - z_n\| < \frac{1}{n} \qquad (n = 1,2, \ldots), \tag{3.16}$$

$$|\alpha_1^{(1)}| \leqslant C_1,\ |\alpha_i^{(n)}| \leqslant C_i \quad (i = 1, \ldots, m_{n-1};\ n = 2,3, \ldots). \tag{3.17}$$

Define now $\{\lambda_i\}$ by

$$\lambda_i = \max\ (C_i, |\alpha_i^{(n)}|) \qquad (m_{n-1} + 1 \leqslant i \leqslant m_n;\ n = 1,2, \ldots;\ \ m_0 = 0). \tag{3.18}$$

We shall show that the sequence $\{\lambda_i\}$ has the required property. Indeed, by (3.17) and (3.18), for each $z_n = \sum_{i=1}^{m_n} \alpha_i^{(n)} x_i$ we have

$$|\alpha_i^{(n)}| \leqslant \lambda_i \qquad (i = 1, \ldots, m_n). \tag{3.19}$$

Now let $x \in E$, $\|x\| = 1$ and $\varepsilon > 0$. Then, since $\{y_n\}$ is dense in $\sigma_E$, there exists $y_N$ such that $\|x - y_N\| < \frac{\varepsilon}{2}$ and that $N > \frac{2}{\varepsilon}$. Put

$$p_\varepsilon(x) = z_N. \tag{3.20}$$

Then, by (3.16),

$$\|y_N - p_\varepsilon(x)\| = \|y_N - z_N\| < \frac{1}{N} < \frac{\varepsilon}{2},$$

whence

$$\|x - p_\varepsilon(x)\| \leqslant \|x - y_N\| + \|y_N - p_\varepsilon(x)\| < \frac{\varepsilon}{2} + \frac{\varepsilon}{2} = \varepsilon,$$

which, together with (3.19), completes the proof of theorem 3.1.

Another question which arises naturally is whether for complete sequences we have a stability theorem of Krein-Milman-Rutman type*), i.e., whether for every complete sequence in a Banach space $E$ there exists a sequence of constants $\gamma_n > 0$ $(n = 1,2, \ldots)$ such that each sequence $\{y_n\} \subset E$ satisfying $\|x_n - y_n\| \leqslant \gamma_n$ $(n = 1,2, \ldots)$ is complete in $E$ (by § 2, remark 2.7, one cannot expect to get the stronger conclusion that there exists an isomorphism $u\colon [x_n] \to [y_n]$ satisfying $u(x_n) = y_n$ for $n = 1,2, \ldots$ and $\|I_E - u\| < 1$). We shall show now, using theorem 3.1, that the answer is affirmative.

**Theorem 3.2.** *Let $\{x_n\}$ be a complete sequence in a Banach space $E$. There exists a sequence of positive numbers $\{\gamma_n\}$ (depending on $\{x_n\}$) such that every sequence $\{y_n\} \subset E$ satisfying*

$$\|x_n - y_n\| \leqslant \gamma_n \qquad (n = 1,2, \ldots) \tag{3.21}$$

*is complete in $E$.*

*Proof.* Clearly, we may assume that $x_n \neq 0$ $(n = 1,2, \ldots)$, by taking $\gamma_n$ to be arbitrary for those $n$ for which $x_n = 0$. Moreover, we may assume that $\|x_n\| = 1$ $(n = 1,2, \ldots)$, since if $\{\gamma_n\}$ has the required stability property for $\left\{\frac{x_n}{\|x_n\|}\right\}$, then $\{\gamma_n \|x_n\|\}$ has that property for $\{x_n\}$.

Thus, let $\|x_n\| = 1$ $(n = 1,2, \ldots)$ and let $\{\gamma_n\}$ be any sequence of positive numbers such that

$$\sum_{i=1}^{\infty} \lambda_i \gamma_i = \lambda < 1, \tag{3.22}$$

*) See Ch. I, § 10, theorem 10.3.

where $\{\lambda_i\}$ is as in theorem 3.1. We shall show that $\{\gamma_n\}$ has the required stability property.

Assume, a contrario, that there exists a sequence $\{y_n\} \subset E$ with $[y_n] \neq E$, satisfying (3.21). Let $\delta > 0$ be such that $\lambda < \lambda + \delta < 1$. Then there exists*) an element $x \in E$ with $\|x\| = 1$ such that

$$\text{dist}\,(x, [y_n]) > \lambda + \delta. \tag{3.23}$$

By theorem 3.1, there exists a polynomial $z = \sum_{i=1}^{m} \alpha_i x_i$ satisfying

$$\|x - z\| < \delta, \tag{3.24}$$

$$|\alpha_i| \leqslant \lambda_i \qquad (i = 1, \ldots, m). \tag{3.25}$$

Then for $y = \sum_{i=1}^{m} \alpha_i y_i \in [y_n]$ we have, by (3.25), (3.21) and (3.22),

$$\|z - y\| = \left\| \sum_{i=1}^{m} \alpha_i (x_i - y_i) \right\| \leqslant \sum_{i=1}^{m} |\alpha_i| \, \|x_i - y_i\| \leqslant \sum_{i=1}^{m} \lambda_i \gamma_i \leqslant \lambda,$$

whence, by (3.24), we obtain

$$\|x - y\| \leqslant \|x - z\| + \|z - y\| \leqslant \delta + \lambda,$$

in contradiction with (3.23). This completes the proof.

In contrast with completeness, the property of being an incomplete sequence is not stable in the above sense. For example, let us mention here the following result on "instability of incompleteness":

**Proposition 3.1.** *Let $\{x_n\}$ be a (complete or incomplete) sequence in a separable Banach space E, which is not deficient minimal**). Then for every sequence of positive numbers $\{\gamma_n\}$ there exists a complete sequence $\{y_n\}$ in E satisfying* (3.21).

*Proof.* Since $\{x_n\}$ is not deficient minimal, by § 2, proposition 2.7 there exists an infinite sequence of indices $\{i_n\}$ such that

$$[x_{i_n}] = [x_n]_{n \neq i_1, i_2, \ldots} = [x_n], \tag{3.26}$$

*) See e.g. [409], p. 84, theorem (of F. Riesz).

**) See § 2, definition 2.2. Note that incomplete sequences which are not deficient minimal exist in every separable Banach space $E$.

where $\{l_n\}=\mathcal{N}\setminus\{i_n\}$. Let $\{z_n\}$ be a normalized complete sequence in $E$. Put

$$y_{i_n}=x_{i_n}+\gamma_{i_n}z_n \qquad (n=1,2,\ldots), \tag{3.27}$$

$$y_{l_n}=x_{l_n} \qquad (n=1,2,\ldots). \tag{3.28}$$

Then, clearly, $\{y_n\}$ satisfies (3.21). Also, if $f\in E^*$, $f(y_n)=0$ ($n=1,2,\ldots$), then by (3.28) and (3.26) $f(x_n)=0$ ($n=1,2,\ldots$), whence, by (3.27), $\gamma_{i_n}f(z_n)=-f(x_{i_n})=0$ ($n=1,2,\ldots$). Since $\gamma_{i_n}\neq 0$ ($n=1,2,\ldots$), it follows that $f(z_n)=0$ ($n=1,2,\ldots$), whence, since $[z_n]=E$, we get $f=0$. Thus, $[y_n]=E$, which completes the proof of proposition 3.1.

*Remark 3.1.* For minimal sequences the incompleteness is already stable, that is, for every incomplete minimal sequence $\{x_n\}\subset E$ (such sequences exist in every separable Banach space $E$) there exists a sequence of positive numbers $\{\gamma_n\}$ such that every sequence $\{y_n\}\subset E$ satisfying (3.21) is incomplete in $E$. Indeed, this follows from Ch. I, § 10, theorem 10.3 and Ch. I, § 9, theorem 9.2.

We conclude this section with the following decomposition property of the elements of a Banach space with respect to a complete sequence:

**Theorem 3.3.** *Let $\{x_n\}$ be a complete sequence in a Banach space E and let $\varepsilon_1\geqslant\varepsilon_2\geqslant\ldots>0$, $\lim_{n\to\infty}\varepsilon_n=0$. Then for every element $x\in E$ there exist two elements $y_1, y_2\in E$ such that*

$$x=y_1+y_2, \tag{3.29}$$

$$\varliminf_{n\to\infty}\frac{\mu_n(y_1)}{\varepsilon_n}=0,\ \varliminf_{n\to\infty}\frac{\mu_n(y_2)}{\varepsilon_n}=0, \tag{3.30}$$

*where $\mu_n(y)=\inf_{z\in P_{(n)}}\|y-z\|$ (with $P_{(n)}=[x_1,\ldots,x_n]$).*

*Proof.* Given $x\in E$, let us construct a sequence $\{p_n\}$ of finite linear combinations of the elements $x_i$ as follows: Let $p_1=x_1$. If $p_n\in P_{(m_n)}$, choose $p_{n+1}\in P_{(m_{n+1})}$ with $m_{n+1}>m_n$, such that $\|x-p_{n+1}\|<\frac{\varepsilon_{m_n}}{2^{n+1}}$. The sequence $\{p_n\}$ being thus constructed, put $p_0=0$ and

$$y_1=\sum_{i=1}^{\infty}(p_{2i}-p_{2i-1}),\ y_2=\sum_{i=1}^{\infty}(p_{2i-1}-p_{2i-2}). \tag{3.31}$$

Then, clearly, $x = \lim_{n\to\infty} p_n = y_1 + y_2$. Furthermore, for $p = \sum_{i=1}^{n} (p_{2i} - p_{2i-1})$ we have $p \in P_{(m_{2n})}$, whence, since $\varepsilon_1 \geqslant \varepsilon_2 \geqslant \ldots$, we obtain

$$\mu_{m_{2n}}(y_1) \leqslant \|y_1 - p\| \leqslant \sum_{i=n+1}^{\infty} \|p_{2i} - p_{2i-1}\| \leqslant \sum_{i=n+1}^{\infty} (\|x - p_{2i}\| +$$

$$+ \|x - p_{2i-1}\|) < \sum_{i=n+1}^{\infty} \left( \frac{\varepsilon_{m_{2i-1}}}{2^{2i}} + \frac{\varepsilon_{m_{2i-2}}}{2^{2i-1}} \right) < \frac{\varepsilon_{m_{2n}}}{2^{2n}} \qquad (n = 1,2, \ldots),$$

and therefore $\underline{\lim}_{n\to\infty} \frac{\mu_{m_{2n}}(y_1)}{\varepsilon_{m_{2n}}} = 0$. Finally, the second equality in (3.30) follows similarly, which completes the proof of theorem 3.3.

## § 4. Bases with respect to a class of sequences of indices

*Definition 4.1.* Let $\mathfrak{M}$ be a class*) of subsequences of the set $\mathscr{N} = \{1,2,3, \ldots\}$. A sequence $\{x_n\}$ in a Banach space $E$ is said to be a *basis with respect to* $\mathfrak{M}$ of the space $E$, if $\{x_n\}$ is complete in $E$ and for every $\{i_n\} \in \mathfrak{M}$ the subsequence $\{x_{i_n}\}$ of $\{x_n\}$ is a basic sequence.

Let us give some examples of bases with respect to the class $\Lambda$ of all lacunary sequences of positive integers. We recall that a sequence $\{i_n\} \subset \mathscr{N}$ is called *lacunary* if

$$\inf_{1 \leqslant n < \infty} \frac{i_{n+1}}{i_n} = \lambda > 1 . \tag{4.1}$$

By § 11, corollary 11.5, *every Cesàro basis* $\{x_n\}$ of a Banach space $E$ is a basis with respect to the class $\Lambda$ of all lacunary sequences of positive integers (some Cesàro bases in concrete spaces, which are not bases, are given in § 11, examples 11.1—11.3). Another example of a basis with respect to the class $\Lambda$, which is only $\omega$-linearly independent, but not a minimal sequence, is given in

**Theorem 4.1.** *Let* $E = C([0,1])$, *let*

$$x_n(t) = t^n \qquad (t \in [0,1],\ n = 1,2, \ldots) \tag{4.2}$$

*) We use here the term "class" in the sense: family, collection.

*and let $\{i_n\}$ be an increasing sequence of positive integers. The following statements are equivalent:*

1°. *The sequence $\{i_n\}$ is lacunary.*

2°. *We have*

$$\inf_{j \neq k} \|x_{i_j} - x_{i_k}\| > 0. \tag{4.3}$$

3°. *The sequence $\{x_{i_n}\}$ is "uniformly minimal", i.e.**)

$$\inf_{1 \leqslant n < \infty} \operatorname{dist}(x_{i_n}, E^{(i_n)}) > 0, \tag{4.4}$$

*where* $E^{(i_n)} = [x_{i_1}, \ldots, x_{i_{n-1}}, x_{i_{n+1}}, \ldots]$ $(n = 1,2, \ldots)$.

4°. *$\{x_{i_n}\}$ is a basic sequence.*

5°. *$\{x_{i_n}\}$ is a basic sequence, equivalent to "the standard conditional basis" of $c_0$, i.e.***) *to*

$$y_n = \sum_{i=1}^{n} e_i = \{\underbrace{1, \ldots, 1}_{n},0,0, \ldots\} \qquad (n = 1,2, \ldots), \tag{4.5}$$

*where* $e_n = \{\delta_{nj}\}_{j=1}^{\infty}$ $(n = 1,2, \ldots)$.

*Consequently, in $E = C([0,1])$ the sequence $\{x_n\}$ defined by* (4.2) *is a basis with respect to the class $\Lambda$ of all lacunary sequences of positive integers.*

*Proof.* 1° ⇒ 3°. Let $\{i_n\}$ be lacunary. It will be sufficient to prove that there exists a constant $D = D(\lambda)$, where $\lambda = \inf_{1 \leqslant n < \infty} \frac{i_{n+1}}{i_n}$, such that for any finite sequence of scalars $\alpha_1, \ldots, \alpha_m$ we have

$$\left\| \sum_{j=1}^{m} \alpha_j x_{i_j} \right\| \geqslant D \sup_{1 \leqslant j \leqslant m} |\alpha_j|; \tag{4.6}$$

indeed, this will imply (4.4), since $\sup_{1 \leqslant j \leqslant m} |\alpha_j| \geqslant 1$ whenever $\alpha_{i_n} = 1$ for some $i_n \leqslant m$.

For any polynomial $p = p(t) = \sum_{l=1}^{s} c_l t^l \neq 0$ we have (using the substitutions $t^l = \tau$ for $l = 1, \ldots, s$)

$$\left\| \sum_{j=1}^{m} \alpha_j p(t^{i_j}) \right\| = \left\| \sum_{j=1}^{m} \alpha_j \sum_{l=1}^{s} c_l t^{l i_j} \right\| = \left\| \sum_{l=1}^{s} c_l \sum_{j=1}^{m} \alpha_j t^{l i_j} \right\| \leqslant$$

$$\leqslant \sum_{l=1}^{s} |c_l| \left\| \sum_{j=1}^{m} \alpha_j t^{l i_j} \right\| = \sum_{l=1}^{s} |c_l| \left\| \sum_{j=1}^{m} \alpha_j \tau^{i_j} \right\| \leqslant C(p) \left\| \sum_{j=1}^{m} \alpha_j x_{i_j} \right\|,$$

*) In Ch. I, § 7, remark 7.2 we have used the term "uniformly minimal" in a slightly different (stronger) sense.

**) See Ch. II, § 14, example 14.1.

where $C(p) = s \max_{1 \leqslant l \leqslant s} |c_l| \neq 0$. Consequently,

$$\left\| \sum_{j=1}^{m} \alpha_j x_{i_j} \right\| \geqslant \frac{1}{C(p)} \left\| \sum_{j=1}^{m} \alpha_j p(t^{i_j}) \right\| = \frac{1}{C(p)} \max_{0 \leqslant t \leqslant 1} \left| \sum_{j=1}^{m} \alpha_j p(t^{i_j}) \right| \geqslant$$

$$\geqslant \frac{1}{C(p)} \sup_{1 \leqslant k < \infty} \left| \sum_{j=1}^{m} \alpha_j p\left(e^{-\frac{i_j}{i_k}}\right) \right| . \tag{4.7}$$

Now let us choose the polynomial $p_\lambda(t)$ in the following way:

a) $p_\lambda(t)$ is non-decreasing for $0 \leqslant t \leqslant e^{-\lambda}$ and non-increasing for $e^{-\frac{1}{\lambda}} \leqslant t \leqslant 1$; $p_\lambda(e^{-1}) = 1$;

b) we have

$$|p_\lambda(t)| \leqslant \frac{1}{4 \sum_{j=1}^{\infty} e^{-\lambda^j}} t \qquad (0 \leqslant t \leqslant e^{-\lambda}), \tag{4.8}$$

$$|p_\lambda(t)| \leqslant \frac{1}{4 \sum_{j=1}^{\infty} \frac{1}{\lambda^j}} (1-t) \qquad (e^{-\frac{1}{\lambda}} \leqslant t \leqslant 1); \tag{4.9}$$

the existence of such a polynomial is obvious. By a) and b) we have $p_\lambda(0) = p_\lambda(1) = 0$ and $p_\lambda(t) \geqslant 0$ for $0 \leqslant t \leqslant e^{-\lambda}$ and $e^{-\frac{1}{\lambda}} \leqslant t \leqslant 1$. Put

$$m_{jk} = \begin{cases} p_\lambda\left(e^{-\frac{i_j}{i_k}}\right) & \text{for } j \neq k \\ 0 & \text{for } j = k \end{cases} \qquad (j, k = 1, 2, \ldots). \tag{4.10}$$

Then, since for $j > k$ we have, by (4.1), $\frac{i_j}{i_k} \geqslant \lambda^{j-k} \geqslant \lambda$, whence

$e^{-\frac{i_j}{i_k}} \leqslant e^{-\lambda^{j-k}} \leqslant e^{-\lambda}$, we obtain $0 \leqslant m_{jk} = p_\lambda\left(e^{-\frac{i_j}{i_k}}\right) \leqslant p_\lambda(e^{-\lambda^{j-k}})$ and thus

$$\sum_{j=k+1}^{\infty} m_{jk} \leqslant \sum_{j=k+1}^{\infty} p_\lambda(e^{-\lambda^{j-k}}) \leqslant \frac{1}{4 \sum_{j=1}^{\infty} e^{-\lambda^j}} \sum_{j=k+1}^{\infty} e^{-\lambda^{j-k}} = \frac{1}{4} .$$

On the other hand, since for $j<k$ we have $\frac{i_j}{i_k} \leqslant \lambda^{j-k} \leqslant \frac{1}{\lambda}$, whence $1 \geqslant e^{-\frac{i_j}{i_k}} \geqslant e^{-\lambda^{j-k}} \geqslant e^{-\frac{1}{\lambda}}$, we obtain again $0 \leqslant m_{jk} = p_\lambda\left(e^{-\frac{i_j}{i_k}}\right) \leqslant$ $\leqslant p_\lambda(e^{-\lambda^{j-k}})$ and thus

$$\sum_{j=1}^{k-1} m_{jk} \leqslant \sum_{j=1}^{k-1} p_\lambda(e^{-\lambda^{j-k}}) \leqslant \frac{1}{4\sum_{j=1}^{\infty}\frac{1}{\lambda^j}} \sum_{j=1}^{k-1}(1-e^{-\lambda^{j-k}}) \leqslant$$

$$\leqslant \frac{1}{4\sum_{j=1}^{\infty}\frac{1}{\lambda^j}} \sum_{j=1}^{k-1}\lambda^{j-k} < \frac{1}{4}.$$

Consequently,

$$\sum_{j=1}^{\infty}|m_{jk}| = \sum_{j=1}^{\infty} m_{jk} < \frac{1}{2} \qquad (k=1,2,\ldots). \tag{4.11}$$

Consider now the linear mapping $u: x=\{\xi_n\} \to y=\{\eta_n\}$ of $l^\infty$ into itself defined by

$$\eta_k = \sum_{j=1}^{\infty} m_{jk}\xi_j \qquad (k=1,2,\ldots). \tag{4.12}$$

By (4.11) we have

$$\|u(x)\| = \sup_{1\leqslant k<\infty}|\eta_k| = \sup_{1\leqslant k<\infty}\left|\sum_{j=1}^{\infty} m_{jk}\xi_j\right| \leqslant \sup_{1\leqslant j<\infty}|\xi_j| \sup_{1\leqslant k<\infty}\sum_{j=1}^{\infty}|m_{jk}| <$$

$$< \frac{1}{2}\sup_{1\leqslant j<\infty}|\xi_j| = \frac{1}{2}\|x\| \qquad (x=\{\xi_n\}\in l^\infty),$$

and thus $\|u\| \leqslant \frac{1}{2}$. Consequently, the mapping $I+u$ (where $I$ is the identical mapping of $l^\infty$ onto itself) has a bounded linear inverse and from the inequalities

$$\|(I+u)(x)\| \geqslant |\,\|x\| - \|u(x)\|\,| \geqslant \frac{1}{2}\|x\| \qquad (x\in l^\infty)$$

it follows that $\|(I+u)^{-1}\| \leqslant 2$. Hence, by (4.7), (4.10), $p_\lambda(e^{-1}) = 1$ and (4.12), we obtain

$$\left\|\sum_{j=1}^{m} \alpha_j x_{i_j}\right\| \geqslant \frac{1}{C(p_\lambda)} \sup_{1 \leqslant k < \infty} \left|\sum_{j=1}^{m} \alpha_j p_\lambda\left(e^{-\frac{i_j}{i_k}}\right)\right| =$$

$$= \frac{1}{C(p_\lambda)} \sup_{1 \leqslant k < \infty} \left|\sum_{j=1}^{m} \alpha_j (m_{jk} + \delta_{jk})\right| =$$

$$= \frac{1}{C(p_\lambda)} \|(I+u)(\{\alpha_1, \ldots, \alpha_m, 0, 0, \ldots\})\|_{l^\infty} \geqslant \frac{1}{2C(p_\lambda)} \sup_{1 \leqslant j \leqslant m} |\alpha_j|,$$

which proves (4.6). Thus, $1° \Rightarrow 3°$.

$1° \cap 3° \Rightarrow 4°$. Assume that $\{i_n\}$ is lacunary. Let $n$, $m$ be positive integers and let

$$x(t) = \sum_{k=1}^{n} \alpha_k t^{i_k}, \; y(t) = \sum_{k=n+1}^{n+m} \alpha_k t^{i_k}, \; \|x\| = 1. \tag{4.13}$$

In order to evaluate $\|x+y\|$ we shall consider two cases:

a) $|\alpha_k| < a \leqslant \dfrac{\lambda - 1}{2\lambda}$ $(k = 1, \ldots, n)$, where $a$ will be chosen later.

Then, since by (4.1) $i_n \geqslant \lambda^{n-k} i_k$ $(k = 1, \ldots, n)$, we have

$$\max_{0 \leqslant t \leqslant 1} |x'(t)| = \max_{0 \leqslant t \leqslant 1} \left|\sum_{k=1}^{n} i_k \alpha_k t^{i_k - 1}\right| \leqslant a \sum_{k=1}^{n} i_k \leqslant a \sum_{j=0}^{n-1} \frac{i_n}{\lambda^j} \leqslant$$

$$\leqslant a\, i_n \sum_{j=0}^{\infty} \frac{1}{\lambda^j} = \frac{a\lambda}{\lambda - 1} i_n \leqslant \frac{1}{2} i_n.$$

Let $\tau = \max\limits_{\substack{0 \leqslant t \leqslant 1 \\ |x(t)| = 1}} t$ and let

$$\tau_0 = \begin{cases} \tau - \dfrac{1}{i_n} & \text{if } \tau > 1 - \dfrac{1}{i_n} \\[2ex] \tau & \text{if } \tau \leqslant 1 - \dfrac{1}{i_n}. \end{cases} \tag{4.14}$$

Then

$$|x(\tau_0)| \geqslant |x(\tau)| - \max_{0\leqslant t\leqslant 1} |x'(t)|(\tau-\tau_0) \geqslant 1 - \frac{1}{2} i_n \cdot \frac{1}{i_n} = \frac{1}{2}.$$

On the other hand, since $\tau_0 \leqslant 1 - \frac{1}{i_n}$ and $\left(1-\frac{1}{i_n}\right)^{i_n} \leqslant \frac{1}{e}$, we have

$$|y(\tau_0)| \leqslant a \sum_{k=n+1}^{n+m} \left(1-\frac{1}{i_n}\right)^{i_k} = a \sum_{k=n+1}^{n+m} \left[\left(1-\frac{1}{i_n}\right)^{i_n}\right]^{\frac{i_k}{i_n}} \leqslant$$

$$\leqslant a \sum_{k=n+1}^{n+m} \left(\frac{1}{e}\right)^{\frac{i_k}{i_n}} \leqslant a \sum_{k=n+1}^{n+m} \left(\frac{1}{e}\right)^{\lambda^{k-n}} \leqslant a \sum_{j=1}^{\infty} \left(\frac{1}{e}\right)^{\lambda^j} = aN(\lambda).$$

Consequently,

$$\|x+y\| \geqslant |x(\tau_0)+y(\tau_0)| \geqslant \frac{1}{2} - aN(\lambda), \tag{4.15}$$

where $N(\lambda)$ depends only on $\lambda$.

b) There exists an index $k_0 \leqslant n$ such that $|\alpha_{k_0}| \geqslant a$. Then, by 3° (see also the above proof of the implication 1° ⇒ 3°), we have

$$\|x+y\| = \left\| \sum_{k=1}^{n+m} \alpha_k x_{i_k} \right\| = |\alpha_{k_0}| \left\| x_{i_{k_0}} + \sum_{\substack{k=1\\k\neq k_0}}^{n+m} \frac{\alpha_k}{\alpha_{k_0}} x_{i_k} \right\| \geqslant a\rho(\lambda), \tag{4.16}$$

where $\rho(\lambda)$ depends only on $\lambda$.

Since the solution of the equation

$$\frac{1}{2} - \xi N(\lambda) = \xi\rho(\lambda)$$

is $\xi = \frac{1}{2(N(\lambda)+\rho(\lambda))}$, let us put

$$a = \min\left\{\frac{1}{2(N(\lambda)+\rho(\lambda))}, \frac{\lambda-1}{2\lambda}\right\}. \tag{4.17}$$

Then, by (4.13), (4.15) and (4.16),

$$\left\| \sum_{k=1}^{n+m} \alpha_k x_{i_k} \right\| = \|x+y\| \geqslant \rho(\lambda) a,$$

whence, by $\left\| \sum_{k=1}^{n} \alpha_k x_{i_k} \right\| = \|x\| = 1$ and Ch. I, § 7, theorem 7.1 (implication $5° \Rightarrow 1°$), $\{x_{i_n}\}$ is a basic sequence.

$4° \Rightarrow 5°$. Assume that $\{x_{i_n}\}$ is a basic sequence. If $\sum_{k=1}^{\infty} \alpha_k x_{i_k}$ converges, then

$$\left| \sum_{k=n+1}^{n+m} \alpha_k \right| = \left| \sum_{k=n+1}^{n+m} \alpha_k x_{i_k}(1) \right| \leqslant \left\| \sum_{k=n+1}^{n+m} \alpha_k x_{i_k} \right\| < \varepsilon \qquad (n > N(\varepsilon)),$$

whence $\sum_{k=1}^{\infty} \alpha_k$ converges. Conversely, if $\sum_{k=1}^{\infty} \alpha_k$ converges, then, since for every $t \in (0,1)$ the sequence $\{x_{i_k}(t)\} = \{t^{i_k}\}$ is monotonely decreasing, from a criterion of Abel*) it follows that the series $\sum_{k=1}^{\infty} \alpha_k x_{i_k}$ is uniformly convergent.

On the other hand, for the standard conditional basis $\{y_n\}$ of $c_0$ (defined by (4.5)) we have seen in Ch. II, § 14, example 14.1 that $\sum_{i=1}^{\infty} \alpha_i y_i$ converges if and only if $\sum_{i=1}^{\infty} \alpha_i$ converges. Consequently, $\{x_{i_n}\} \sim \{y_n\}$.

The implications $5° \Rightarrow 4° \Rightarrow 3° \Rightarrow 2°$ are obvious.

$2° \Rightarrow 1°$. Assume that we have $\inf_{j \neq k} \|x_{i_j} - x_{i_k}\| = a > 0$. Then, using the substitution $t^{i_j} = \tau$, we obtain

$$0 < a \leqslant \|t^{i_j} - t^{i_k}\|_{C([0,1])} = \|\tau - \tau^{\frac{i_k}{i_j}}\|_{C([0,1])}. \tag{4.18}$$

Now, if the sequence $\{i_n\}$ were not lacunary, then the latter expression could have arbitrarily small values (for suitable $j \neq k$), in contradiction with (4.18). This completes the proof of theorem 4.1.

*Remark 4.1.* Combining the implication $4° \Rightarrow 5°$ of theorem 4.1 with Ch. II, § 1, corollary 1.1, it follows that *there exists an equivalent norm on the space $c_0$ such that $c_0$ endowed with this new norm has no*

---

*) See e.g. [112], p. 429.

*monotone basis.* A stronger property of this isomorph of $c_0$ will be given in § 12, example 12.2.

Similarly to theorem 4.1, one can prove

**Theorem 4.2.** *Let $E = L^p([0,1])$, where $1 \leqslant p < \infty$, let*

$$x_n(t) = \sqrt[p]{n + \frac{1}{p}}\, t^n \qquad (t \in [0,1],\ n = 1,2, \ldots) \tag{4.19}$$

*and let $\{i_n\}$ be an increasing sequence of positive integers. The following statements are equivalent:*

1°. *The sequence $\left\{i_n + \frac{1}{p}\right\}_{n=1}^{\infty}$ is lacunary.*

2°. *We have*

$$\inf_{j \neq k} \|x_{i_j} - x_{i_k}\| > 0. \tag{4.20}$$

3°. *The sequence $\{x_{i_n}\}$ is uniformly minimal.*

4°. *$\{x_{i_n}\}$ is a basic sequence.*

5°. *$\{x_{i_n}\}$ is a basic sequence, equivalent to the unit vector basis of $l^p$.*

*Consequently, in $E = L^p([0,1])$ the sequence*) $\{\tilde{x}_n\}$ defined by* (4.19) *is a basis with respect to the class of all sequences $\{i_n\}$ for which $\left\{i_n + \frac{1}{p}\right\}_{n=1}^{\infty}$ is lacunary.*

For an increasing sequence of positive integers $\{i_n\}$ we shall denote by $\mathfrak{M}_{\{i_n\}}$ the class of all increasing sequences $\{j_n\} \subset \mathscr{N}$ such that $\{j_k\}$ fails to contain at least one number of the pair $(i_n, i_n + 1)$ $(n = 1,2, \ldots)$.

**Theorem 4.3.** *For every increasing sequence $\{i_n\} \subset \mathscr{N}$ there exists, in every separable Banach space $E$, a basis with respect to the class $\mathfrak{M}_{\{i_n\}}$.*

*Proof.* Let $\{y_n\}$ be a normalized complete sequence in $E$ and, by § 1, theorem 1.1, let $\{x_n\}$ be a basic sequence in $E$. We shall construct the desired basis with respect to $\mathfrak{M}_{\{i_n\}}$ inductively, as follows. Put

$$z_k = x_k \qquad (k = 1, \ldots, i_1),$$

$$z_{i_1+1} = x_{i_1} + \gamma_1 y_1,$$

where $\gamma_1 > 0$. Assume now that we have defined $z_1, \ldots, z_{i_n}$ of the form $x_k$ or $x_k + \gamma_j y_j$ and that in this process we have used $x_1, \ldots, x_p$ and

*) We recall that $\tilde{x}_n$ denotes the equivalence class of the function $x_n$ (with respect to the Lebesgue measure).

$\gamma_1 y_1, \ldots, \gamma_{l-1} y_{l-1}$, where $\gamma_1, \ldots, \gamma_{l-1} > 0$ (no $\gamma_j y_j$, when $n=1$). We shall construct $z_{i_n+1}, \ldots, z_{i_{n+1}}$, which will complete the induction.

*Case 1°.* If $z_{i_n} = x_p$ and $i_{n+1} = i_n + 1$, put

$$z_{i_{n+1}} = z_{i_n+1} = x_p + \gamma_l y_l,$$

where $\gamma_l > 0$.

*Case 2°.* If $z_{i_n} = x_p$ and $i_{n+1} \geqslant i_n + 2$, put

$$z_{i_n+1} = x_p + \gamma_l y_l, \quad z_{i_n+k+1} = x_{p+k} \qquad (k = 1, \ldots, i_{n+1} - i_n - 1)$$

where $\gamma_l > 0$.

*Case 3°.* If $z_{i_n} = x_p + \gamma_{l-1} y_{l-1}$, put

$$z_{i_n+k} = x_{p+k} \qquad (k = 1, \ldots, i_{n+1} - i_n).$$

We shall show that the sequence $\{z_n\}$ constructed in this way is a basis with respect to $\mathfrak{M}_{\{i_n\}}$, provided that $\gamma_n > 0$ $(n = 1,2, \ldots)$ are sufficiently small, which will complete the proof. Indeed, $\gamma_1 y_1 = z_{i_1+1} - z_{i_1}$ and, in cases 1° and 2°, $\gamma_l y_l = z_{i_n+1} - z_{i_n}$. Also, observe that in case 3° we must have (by 1°—3° for $n-1$ instead of $n$) $i_n = i_{n-1} + 1$ and $z_{i_n-1} = z_{i_{n-1}} = x_p$, whence $\gamma_{l-1} y_{l-1} = z_{i_n} - z_{i_n-1}$. Thus, $\gamma_l y_l \in [z_n]$ for $l = 1,2, \ldots$ whence, since $\gamma_l > 0$ $(l = 1,2, \ldots)$ and since by our assumption $[y_n] = E$, it follows that $[z_n] = E$. Furthermore, each $z_n$ is of the form $x_k$ or $x_k + \gamma_j y_j$, but by the above construction*) and the definition of $\mathfrak{M}_{\{i_n\}}$, for every $\{j_n\} \in \mathfrak{M}_{\{i_n\}}$ and every $p$, $l = 1,2, \ldots$ the sequence $\{z_{j_n}\}$ fails to contain at least one element of the pair $(x_p, x_p + \gamma_l y_l)$. Consequently, by Ch. I, § 4, proposition 4.1 and Ch. I, § 10, theorem 10.3, $\{z_{j_n}\}$ is a basic sequence if $\gamma_n > 0$ $(n = 1,2, \ldots)$ are sufficiently small, which completes the proof of theorem 4.3.

In particular, for $i_n = n$ $(n = 1,2, \ldots)$ we obtain

**Corollary 4.1.** *In every separable Banach space E there exists a complete sequence $\{z_n\}$ such that every subsequence $\{z_{j_n}\}$ of $\{z_n\}$ with $j_{n+1} - j_n > 1$ $(n = 1,2, \ldots)$ is a basic sequence (hence, in particular, $\{z_n\}$ is a basis with respect to the class $\Lambda$ of all lacunary sequences of positive integers).*

Note that in this case, the construction of the above proof of theorem 4.3 gives $\{z_n\}$ as

$$z_1 = x_1, \; z_2 = x_1 + \gamma_1 y_1, \ldots, z_{2n-1} = x_n, \; z_{2n} = x_n + \gamma_n y_n, \ldots$$

where $\{y_n\}$ is a normalized complete sequence in $E$, $\{x_n\}$ is a basic sequence in $E$ and $\gamma_n > 0$ $(n = 1,2, \ldots)$ are sufficiently small.

---

*) We use again the observation that in case 3° we must have $i_n = i_{n-1} + 1$ and $z_{i_n-1} = z_{i_{n-1}} = x_p$.

## § 5. Pseudo-bases. Semi-bases

The definition of the notion of a basis (Ch. I, § 1, definition 1.1) can be split into two parts: 1) the existence, for every $x \in E$, of a sequence of scalars*) $\{\alpha_n\} \subset K$ such that $x = \sum_{i=1}^{\infty} \alpha_i x_i$ and 2) the uniqueness of such a sequence of scalars. Considering sequences which have only one of these properties, one obtains generalizations of the notion of a basis.

*Definition 5.1.* a) A sequence $\{x_n\}$ in a Banach space $E$, with $x_n \neq 0$ $(n = 1,2, \ldots)$, is called a *pseudo-basis* of $E$ if for every $x \in E$ there exists a sequence of scalars $\{\alpha_n\} \subset K$ such that

$$x = \sum_{i=1}^{\infty} \alpha_i x_i. \tag{5.1}$$

b) A pseudo-basis $\{x_n\}$ of $E$ is said to be *unconditional* if every $x \in E$ has an expansion (5.1) with $\sum_{i=1}^{\infty} \alpha_i x_i$ unconditionally convergent.

Note that definition 5.1 does not require that the elements $x_n$ be distinct.

Obviously, every pseudo-basis of $E$ is complete in $E$ and hence every Banach space $E$ with a pseudo-basis is separable.

The existence of an unconditional pseudo-basis in every separable Banach space $E$ and examples of pseudo-bases which are not bases follow from

**Theorem 5.1.** *Let $E$ be a separable Banach space. Then*

a) *The space $E$ has a dense finitely linearly independent***) *unconditional pseudo-basis.*

b) *Every sequence $\{x_n\}$ with $x_n \neq 0$ $(n = 1,2, \ldots)$, which is dense in the unit cell $S_E = \{x \in E \mid \|x\| \leqslant 1\}$, is an unconditional pseudo-basis of the space $E$. For every such sequence $\{x_n\}$ there exists a subset $\mathscr{L}$ of the space $l^1$ with the following property: for every $x \in E$ there exists a unique sequence of scalars $\{\alpha_n\} \in \mathscr{L}$ such that we have* (5.1) *and that the mapping $x \to \{\alpha_n\}$ be a homeomorphism of $E$ onto $\mathscr{L}$.*

*Proof.* a) Let $\{z_n\}$ be a countable dense set in $E$, such that $z_n \neq 0$ $(n = 1,2, \ldots)$. Put $x_1 = z_1$. Assume that $x_1, \ldots, x_n$ have been chosen.

---

*) We recall (see Ch. I, § 1) that we denote by $K$ the field of scalars, which can be either the field of complex numbers or the field of real numbers.

**) I.e. (see Ch. I, § 6, definition 6.1) such that every finite subsequence is linearly independent.

If $z_{n+1} \notin [x_1, \ldots, x_n]$, let $x_{n+1} = z_{n+1}$. If $z_{n+1} \in [x_1, \ldots, x_n]$, choose $x_{n+1} \notin [x_1, \ldots, x_n]$ such that $\|x_{n+1} - z_{n+1}\| < \frac{1}{n+1}$. Continuing in this way, we obtain a dense finitely linearly independent sequence $\{x_n\}$. Moreover, $\{x_n\}$ is also an unconditional pseudo-basis of $E$, by part b) below (and since every supersequence of an unconditional pseudo-basis is also an unconditional pseudo-basis).

b) Let $\{x_n\}$ be a sequence with $x_n \neq 0$ $(n = 1,2, \ldots)$, which is dense in $S_E$. To show that $\{x_n\}$ is an unconditional pseudo-basis of $E$, we shall use the argument of the proof of the classical Banach-Mazur theorem*) on quotient spaces of $l^1$. Namely, define a mapping $u: l^1 \to E$ by

$$u(\{\xi_n\}) = \sum_{i=1}^{\infty} \xi_i x_i \qquad (\{\xi_n\} \in l^1). \tag{5.2}$$

Since $\sum_{i=1}^{\infty} |\xi_i| < \infty$ for $\{\xi_n\} \in l^1$, the series $\sum_{i=1}^{\infty} \xi_i x_i$ is unconditionally convergent and we have $\|u(x)\| \leqslant \sum_{i=1}^{\infty} |\xi_i| = \|\{\xi_i\}\|$. Thus, $u$ is a continuous linear mapping of $l^1$ into $E$. Moreover, $u$ maps $l^1$ *onto* $E$ and hence $\{x_n\}$ is an unconditional pseudo-basis of $E$. Indeed, let $x \in E$, $x \neq 0$ be arbitrary. Then, since $\{x_n\}$ is dense in $S_E$, we may choose, successively, $x_{n_1}, x_{n_2}, \ldots$ such that $\left\| \frac{x}{\|x\|} - x_{n_1} \right\| < \frac{1}{2}$, $\left\| \frac{x}{\|x\|} - x_{n_1} - \frac{1}{2} x_{n_2} \right\| < \frac{1}{4}$, $\left\| \frac{x}{\|x\|} - x_{n_1} - \frac{1}{2} x_{n_2} - \frac{1}{4} x_{n_3} \right\| < \frac{1}{8}, \ldots$ Then we have $\frac{x}{\|x\|} = \sum_{i=1}^{\infty} \frac{1}{2^{i-1}} x_{n_i}$, whence $x = \sum_{i=1}^{\infty} \frac{\|x\|}{2^{i-1}} x_{n_i} = u(\{\xi_n\})$, where $\xi_{n_i} = \frac{\|x\|}{2^{i-1}}$ $(i = 1,2, \ldots)$, $\xi_n = 0$ $(n \neq n_1, n_2, \ldots)$, which proves our assertion.

Now, since $u$ is a continuous linear mapping of $l^1$ onto $E$, the set-valued function $u^{-1}: E \to 2^{l^1}$ admits, by a theorem of R. G. Bartle and L. M. Graves**), a continuous selection $\varphi$, i.e., there exists a continuous function $\varphi: E \to l^1$ such that $\varphi(x) \in u^{-1}(x)$ $(x \in E)$. Then $\varphi$ is one to one on $E$ (since the relations $x, y \in E$, $\varphi(x) = \varphi(y)$ imply $x = u(\varphi(x)) = u(\varphi(y)) = y$). Let $\mathscr{L} = \varphi(E) \subset l^1$. Then the inverse mapping $\varphi^{-1} = u|_{\mathscr{L}}: \mathscr{L} \to E$ is also continuous and thus $\varphi$ is a homeomorphism of $E$ onto $\mathscr{L}$, which completes the proof of theorem 5.1.

---

*) See [12], theorem e) or [223], p. 283, theorem (1).

**) See [14], theorem 4 or [281], Introduction, corollary of theorem 3.2″.

*Remark 5.1.* Theorem 5.1 b) shows that a subsequence $\{x_{i_n}\}$ of an unconditional pseudo-basis $\{x_n\}$ of $E$ need not be a "pseudo-basic sequence" (i.e., a pseudo-basis of $[x_{i_n}]$). Indeed, let $\{x_{i_n}\} \subset S_E$ be a sequence, with $x_{i_n} \neq 0$ $(n = 1,2, \ldots)$, which is not a pseudo-basis*) of its closed linear span $[x_{i_n}]$ and let $\{x_n\}$ be a dense sequence in $S_E$, with $x_n \neq 0$ $(n = 1,2, \ldots)$, containing $\{x_{i_n}\}$ as a subsequence; then, by theorem 5.1 b), $\{x_n\}$ is a unconditional pseudo-basis of $E$.

In the situation of theorem 5.1 b) it follows that the coefficients $\alpha_i$ in the expansion (5.1), with $\{\alpha_n\} \in \mathscr{L}$, are continuous functionals of $x$; in general, these functionals are not linear**). Now we shall show that this holds not only for the pseudo-bases $\{x_n\}$ which are dense sequences in $S_E$, but for every pseudo-basis.

**Theorem 5.2.** *Let $\{x_n\}$ be a pseudo-basis of a Banach space $E$. Then*

a) *For the Banach space of sequence of scalars*

$$A_1 = \left\{ \{\alpha_n\} \subset K \,\middle|\, \sum_{i=1}^{\infty} \alpha_i x_i \text{ converges} \right\} \tag{5.3}$$

$\left(\textit{with the norm } \|\{\alpha_n\}\| = \sup_{1 \leqslant n < \infty} \left\| \sum_{i=1}^{n} \alpha_i x_i \right\| \right)$, *introduced in Ch.* I, § 3, *proposition* 3.1, *the mapping*

$$w \colon \{\alpha_n\} \to \sum_{i=1}^{\infty} \alpha_i x_i \qquad (\{\alpha_n\} \in A_1) \tag{5.4}$$

*is a continuous linear mapping of $A_1$ onto $E$, of norm $\|w\| = 1$. The converse is also true: if $w$ maps $A_1$ onto $E$, then $\{x_n\}$ is a pseudo-basis of $E$, whenever $x_n \neq 0$ $(n = 1,2, \ldots)$.*

b) *There exists a subset $\mathscr{L}$ of $A_1$ with the same property as the subset $\mathscr{L}$ of $l^1$ in theorem* 5.1 b).

c) *There exists on $E$ a sequence of continuous (not necessarily linear) functionals $\{\alpha_n(.)\}$ such that*

$$x = \sum_{i=1}^{\infty} \alpha_i(x) x_i \qquad (x \in E). \tag{5.5}$$

*Proof.* a) By the definition of $A_1$ the mapping (5.4) is well defined and

$$w(\{\alpha_n\}) = \left\| \sum_{i=1}^{\infty} \alpha_i x_i \right\| \leqslant \sup_{1 \leqslant n < \infty} \left\| \sum_{i=1}^{n} \alpha_i x_i \right\| = \|\{\alpha_n\}\| \qquad (\{\alpha_n\} \in A_1),$$

---

*) For example, one can take $\{x_{i_n}\}$ to be a minimal sequence which is not a basic sequence.

**) For the case when they can be taken continuous linear functionals of $x$, see § 9.

whence $\|w\| \leq 1$. Since

$$\|w(e_n)\| = \|x_n\| = \sup_{1 \leq k < \infty} \left\| \sum_{i=1}^{k} \delta_{in} x_i \right\| = \|\{\delta_{in}\}_{i=1}^{\infty}\| = \|e_n\| \quad (n = 1,2,\ldots),$$

we have $\|w\| = 1$. Furthermore, since $\{x_n\}$ is a pseudo-basis, $w$ maps $A_1$ *onto* $E$. The converse statement in part a) is obvious from the definition of a pseudo-basis.

The proof of part b) is analogous to the second part of the above proof of theorem 5.1 b), with $l^1$ replaced by $A_1$ and using part a) proved above. Finally, part c) is an immediate consequence of b), which completes the proof of theorem 5.2.

*Remark 5.2.* Theorem 5.2 a) should be compared with Ch. I, § 3, proposition 3.2 a), which concerns the particular case when $\{x_n\}$ is a basis of $E$ and with § 9, theorem 9.1, equivalence 1° ⇔ 3°, which concerns the particular case when $\{x_n\}$ is a quasi-basis of $E$.

**Theorem 5.3.** *Let $\{x_n\}$ be an unconditional pseudo-basis of a Banach space E. Then we have analogous statements to* a), b), c) *of theorem* 5.2, *with $A_1$ replaced by the Banach space of sequences of scalars*

$$A_1^{(u)} = \left\{ \{\alpha_n\} \subset K \;\middle|\; \sum_{i=1}^{\infty} \alpha_i x_i \text{ is unconditionally convergent} \right\} \tag{5.6}$$

$\left(\textit{with the norm } \|\{\alpha_n\}\| = \sup_{|\beta_1|, |\beta_2|, \ldots \leq 1} \sup_{1 \leq l < \infty} \left\| \sum_{i=1}^{l} \beta_i \alpha_i x_i \right\| \right)$ *introduced in Ch.* II, § 18, *proof of proposition* 18.5 a) *and with unconditional convergence in* (5.1) *and* (5.5).

The proof is analogous to that of theorem 5.2.

*Definition 5.2.* A sequence $\{x_n\}$ in a Banach space $E$ is called a *semi-basis* of $E$ if $\{x_n\}$ is complete in $E$ and the relations $\{\alpha_n\}, \{\beta_n\} \subset K$, $\sum_{i=1}^{\infty} \alpha_i x_i = \sum_{i=1}^{\infty} \beta_i x_i$ imply $\alpha_n = \beta_n$ $(n = 1,2, \ldots)$.

Obviously, *$\{x_n\}$ is a semi-basis of E if and only if $\{x_n\}$ is a complete $\omega$-linearly independent sequence in E.* Therefore the existence of a semi-basis in every separable Banach space $E$ and examples of semi-bases which are not bases follow from Ch. I, § 4, remark 4.1 (see also Ch. I, § 6, example 6.1), taking into account that every minimal sequence is $\omega$-linearly independent.

## § 6. Minimal sequences. $\{\lambda_i\}$-linearly independent sequences. Complete minimal sequences. Maximal biorthogonal systems

Since every basic sequence is minimal, it is clear from § 1, theorem 1.1, that minimal sequences exist in every Banach space $E$. Also, by Ch. I, § 4, remark 4.1, in every separable Banach space $E$, if $\{y_n\} \subset E$ is a finitely linearly independent sequence such that $[y_n] = E$, then there exists a complete minimal sequence $\{x_n\} \subset E$ such that

$$[x_1, \ldots, x_n] = [y_1, \ldots, y_n] \quad (n = 1,2, \ldots). \tag{6.1}$$

This can be seen also directly, as follows: For each $n$ there exists a functional $f_n \in E^*$ such that $f_n(y_i) = 0$ $(i = 1, \ldots, n-1)$, $f_n(y_n) = 1$ (since $y_n \notin [y_1, \ldots, y_{n-1}]$). Then for the sequence

$$x_1 = y_1, \; x_n = y_n - \sum_{i=1}^{n-1} f_i(y_n)x_i \qquad (n = 2,3, \ldots) \tag{6.2}$$

we have $f_i(x_j) = \delta_{ij}$ $(i, j = 1,2, \ldots)$, and hence $\{x_n\}$ is a complete minimal sequence in $E$, satisfying (6.1). In § 8 we shall give stronger results.

Similarly to § 1, it is natural to ask the question of the existence of (infinite) minimal subsequences of a given sequence, more precisely, the following problem: Does every $\omega$-linearly independent sequence in a Banach space contain a minimal subsequence? The answer to this problem is negative; moreover, in every separable (respectively, in every) Banach space there exists an $\omega$-linearly independent complete (respectively, an $\omega$-linearly independent) sequence containing no minimal subsequence, as shown by

*Example 6.1.* Let $E$ be a separable Banach space and, by § 1, proposition 1.5, let $\{x_n\}$ be a hypercomplete sequence in $E$. Furthermore, let $\{y_n\}$ be a finitely linearly independent infinite subsequence of $\{x_n\}$ (it is easy to choose such a subsequence: let $y_1 = x_1$, let $y_2 = x_{n_2}$, where $n_2$ is the smallest index such that $x_{n_2} \notin [x_1, \ldots, x_{n_2-1}]$, and continue in this way indefinitely). Finally, let $\{z_n\}$ be an $\omega$-linearly independent infinite subsequence of $\{y_n\}$ (by corollary 6.1 below). Then $\{z_n\}$ is an $\omega$-linearly independent complete sequence in $E$, having no minimal subsequence. Indeed, from § 1, definition 1.1 it follows that every (infinite) subsequence of $\{z_n\}$ (in particular, $\{z_n\}$ itself, too) is still hypercomplete in $E$ and hence non-minimal.

In § 1, theorem 1.3 and corollary 1.4, we have given necessary and sufficient conditions for a given sequence $\{x_n\} \subset E$ to contain a basic subsequence. It would be also useful to find necessary and sufficient conditions for the existence of minimal subsequences.

It is natural to raise the question, whether every finitely linearly independent sequence $\{x_n\}$ in a Banach space $E$ contains an infinite subsequence $\{x_{n_i}\}$ which is linearly independent in a stronger sense, intermediate between finitely linear independence and minimality. We shall show that, if $\sup\limits_{1\leqslant n<\infty} \|x_n\| < \infty$, this is indeed the case for the following notion of linear independence:

*Definition 6.1.* Let $\lambda_i > 0$ $(i = 1,2, \ldots)$. A sequence $\{x_n\}$ in a Banach space $E$ is said to be *$\{\lambda_i\}$-linearly independent* if the relations

$$|\alpha_i^{(n)}| \leqslant \lambda_i \ (i, n = 1,2, \ldots), \lim_{n\to\infty} \sum_{i=1}^{\infty} \alpha_i^{(n)} x_i = 0 \tag{6.3}$$

imply

$$\lim_{n\to\infty} \alpha_i^{(n)} = 0 \qquad (i = 1,2, \ldots). \tag{6.4}$$

**Proposition 6.1.** *Let $\lambda_i > 0$ $(i = 1,2, \ldots)$ and let $E$ be a Banach space. Then*

a) *Every minimal sequence in $E$ is $\{\lambda_i\}$-linearly independent.*

b) *Every $\{\lambda_i\}$-linearly independent sequence in $E$ is finitely linearly independent.*

c) *If $\inf\limits_{1\leqslant k<\infty} \lambda_k > 0$, then every normalized $\{\lambda_i\}$-linearly independent sequence in $E$ is $\omega$-linearly independent.*

*Proof.* a) Let $\{x_n\}$ be a minimal sequence in $E$ and let $\{f_n\} \subset E^*$, $f_i(x_j) = \delta_{ij}$ $(i, j = 1,2, \ldots)$. If we have (6.3), then

$$\lim_{n\to\infty} \alpha_k^{(n)} = \lim_{n\to\infty} f_k\left(\sum_{i=1}^{\infty} \alpha_i^{(n)} x_i\right) = f_k\left(\lim_{n\to\infty} \sum_{i=1}^{\infty} \alpha_i^{(n)} x_i\right) = f_k(0) = 0$$
$$(k = 1,2, \ldots).$$

c) Let $\inf\limits_{1\leqslant k<\infty} \lambda_k > 0$ and let $\{x_n\}$ be a normalized $\{\lambda_i\}$-linearly independent sequence in $E$, which is not $\omega$-linearly independent. Then there exists a sequence of scalars $\{\alpha_n\} \subset K$ with $\sup\limits_{1\leqslant k<\infty} |\alpha_k| \neq 0$ such that $\sum\limits_{i=1}^{\infty} \alpha_i x_i = 0$, whence $\sup\limits_{1\leqslant k<\infty} |\alpha_k| = \sup\limits_{1\leqslant k<\infty} \|\alpha_k x_k\| < \infty$. Put

$$\gamma = \frac{\inf\limits_{1\leqslant k<\infty} \lambda_k}{\sup\limits_{1\leqslant k<\infty} |\alpha_k|}, \tag{6.5}$$

$$\alpha_i^{(n)} = \gamma \alpha_i \qquad (i, n = 1,2, \ldots). \tag{6.6}$$

Then we have

$$|\alpha_i^{(n)}| = |\gamma\alpha_i| \leqslant \inf_{1\leqslant k<\infty} \lambda_k \leqslant \lambda_i \qquad (i,\ n = 1,2, \ldots),$$

$$\lim_{n\to\infty} \sum_{i=1}^{\infty} \alpha_i^{(n)} x_i = \sum_{i=1}^{\infty} \gamma\alpha_i x_i = \gamma \sum_{i=1}^{\infty} \alpha_i x_i = 0,$$

whence, since $\{x_n\}$ is $\{\lambda_i\}$-linearly independent, we obtain

$$\gamma\alpha_i = \lim_{n\to\infty} \alpha_i^{(n)} = 0 \qquad (i = 1,2, \ldots).$$

Since $\gamma \neq 0$, it follows that $\alpha_i = 0$ $(i = 1,2, \ldots)$, in contradiction with the assumption $\sup_{1\leqslant k<\infty} |\alpha_k| \neq 0$.

b) Let $\{x_n\}$ be a $\{\lambda_i\}$-linearly independent sequence in $E$, which is not finitely linearly independent. Then there exists a finite sequence of scalars $\alpha_1, \ldots, \alpha_m$ with $\sup_{1\leqslant k\leqslant m} |\alpha_k| \neq 0$, such that $\sum_{i=1}^{m} \alpha_i x_i = 0$. Put

$$\gamma = \frac{\inf_{1\leqslant k\leqslant m} \lambda_k}{\sup_{1\leqslant k\leqslant m} |\alpha_k|}, \tag{6.7}$$

$$\alpha_i^{(n)} = \begin{cases} \gamma\alpha_i & \text{for } i = 1, \ldots, m \\ 0 & \text{for } i = m+1,\ m+2, \ldots \end{cases} \qquad (n = 1,2, \ldots). \tag{6.8}$$

Then $\gamma \neq 0$ and hence, as in the above proof of part c), we obtain that $\alpha_i = 0$ $(i = 1, \ldots, m)$ in contradiction with the assumption $\sup_{1\leqslant k\leqslant m} |\alpha_k| \neq 0$. This completes the proof of proposition 6.1.

*Remark 6.1.* We have seen in the above proof of part a) that if $\{x_n\} \subset E$ is minimal, then already the second condition in (6.3) implies (6.4). The converse statement is also true (that is, if $\{x_n\} \subset E$ is such that $\lim_{n\to\infty} \sum_{i=1}^{\infty} \alpha_i^{(n)} x_i = 0$ implies $\lim_{n\to\infty} \alpha_i^{(n)} = 0$ for $i = 1,2, \ldots$ then $\{x_n\}$ is minimal); indeed, this follows from Ch. I, § 6, theorem 6.1, implication 4° ⇒ 1° (by taking $\alpha_i^{(n)} = 0$ for $i = m_n + 1, m_n + 2, \ldots$; $n = 1,2, \ldots$). Thus we have another characterization of minimal sequences.

The converse statements in proposition 6.1 are not valid, as shown by

*Example 6.2.* a) For any $\lambda_i > 0$ $(i = 1,2, \ldots)$, in every Banach space $E$ there exists a $\{\lambda_i\}$-linearly independent sequence which is not minimal. Indeed, by example 6.1, let $\{x_n\}$ be a normalized $\omega$-linearly independent sequence in $E$, containing no minimal subsequence. By theorem 6.1 below, let $\{z_n\}$ be a $\{\lambda_i\}$-linearly independent subsequence of $\{x_n\}$. Then $\{z_n\}$ is $\{\lambda_i\}$-linearly independent, but not minimal.

b) Let $\{x_n\}$ be a normalized finitely linearly independent sequence in a Banach space $E$, which is not $\omega$-linearly independent (see e.g. Ch.I, § 6, example 6.2). Then, by proposition 6.1 c), $\{x_n\}$ is not $\{\lambda_i\}$-linearly independent, whenever $\inf_{1 \leqslant k < \infty} \lambda_k > 0$.

c) Let $\{x_n\}$ be a normalized complete $\omega$-linearly independent sequence in a separable Banach space $E$, which is not a basis of $E$ and let $x$ be an element of $E$ which admits no expansion of the form $x = \sum_{i=1}^{\infty} \alpha_i x_i$; by § 3, theorem 3.1, we can write $x = \lim_{n\to\infty} \sum_{i=1}^{m_n} \alpha_i^{(n)} x_i$, where $\sup_{1 \leqslant n < \infty} |\alpha_i^{(n)}| < \infty$ $(i = 1,2, \ldots)$. Then the sequence $\{y_n\} \subset E$ defined by

$$y_1 = x, \; y_n = x_{n-1} \qquad (n = 2,3, \ldots)$$

is $\omega$-linearly independent, but not $\{\lambda_i\}$-linearly independent for any $\lambda_1 \geqslant 1$, $\lambda_i \geqslant \sup_{1 \leqslant n < \infty} |\alpha_{i-1}^{(n)}|$ $(i = 2,3,\ldots)$. Indeed, $\{y_n\}$ is obviously $\omega$-linearly independent; furthermore, putting

$$\beta_1^{(n)} = 1, \; \beta_i^{(n)} = \begin{cases} -\alpha_{i-1}^{(n)} & \text{for } i = 2,3, \ldots, m_n + 1 \\ 0 & \text{for } i = m_n + 2, m_n + 3, \ldots \end{cases} \qquad (n = 1,2, \ldots),$$

we have $|\beta_i^{(n)}| \leqslant \lambda_i$ $(i,n = 1,2,\ldots)$, $\lim_{n\to\infty} \sum_{i=1}^{\infty} \beta_i^{(n)} y_i = \lim_{n\to\infty} \left( x - \sum_{i=1}^{m_n} \alpha_i^{(n)} x_i \right) = 0$, but $\lim_{n\to\infty} \beta_1^{(n)} = 1$, and thus $\{y_n\}$ is not $\{\lambda_i\}$-linearly independent.

**Theorem 6.1** *Let $\lambda_i > 0$ $(i = 1,2, \ldots)$ and let $\{x_n\}$ be a finitely linearly independent sequence in a Banach space $E$, with $\sup_{1 \leqslant n < \infty} \|x_n\| < \infty$. Then $\{x_n\}$ has a $\{\lambda_i\}$-linearly independent subsequence.*

*Proof.* We may assume that $E$ is separable, since we may restrict our attention to the subspace $[x_n]$. Then $E$ can be embedded isometrically*) into $C([0,1])$. Furthermore, $C([0,1]) \subset L^2([0,1])$ and $\|x\|_{L^2} \leqslant \|x\|_C$ for all $x \in C([0,1])$, and therefore $\{\lambda_i\}$-linear independence in $L^2([0,1])$ implies $\{\lambda_i\}$-linear independence in $C([0,1])$. Consequently, it is sufficient to prove theorem 6.1 for $E = L^2([0,1])$.

---

*) See e.g. [11], p. 185, theorem 9.

We may also assume that $\|x_n\| = 1$ $(n = 1,2, \ldots)$. Indeed, if theorem 6.1 is true for this case, and if $\{x_n\}$ is a finitely linearly independent sequence in $E$, with $\sup_{1 \leqslant n < \infty} \|x_n\| = C < \infty$, then $x_n \neq 0$ $(n = 1,2, \ldots)$ and $\left\{\frac{x_n}{\|x_n\|}\right\}$ is finitely linearly independent, whence, by our assumption, it contains a $\{C\lambda_i\}$-linearly independent subsequence $\left\{\frac{x_{n_i}}{\|x_{n_i}\|}\right\}$. Now, if $|\alpha_{n_i}^{(k)}| \leqslant \lambda_i$, $\lim_{k\to\infty} \sum_{i=1}^{\infty} \alpha_{n_i}^{(k)} x_{n_i} = 0$, then $|\|x_{n_i}\|\alpha_{n_i}^{(k)}| \leqslant C\lambda_i$, $\lim_{k\to\infty} \sum_{i=1}^{\infty} \|x_{n_i}\|\alpha_{n_i}^{(k)} \frac{x_{n_i}}{\|x_{n_i}\|} = 0$, whence $\|x_{n_i}\| \lim_{k\to\infty} \alpha_{n_i}^{(k)} = \lim_{k\to\infty} \|x_{n_i}\|\alpha_{n_i}^{(k)} = 0$, and hence, since $x_{n_i} \neq 0$ $(i = 1,2, \ldots)$, it follows that $\lim_{k\to\infty} \alpha_{n_i}^{(k)} = 0$, which proves that $\{x_{n_i}\}$ is a $\{\lambda_i\}$-linearly independent subsequence of $\{x_n\}$.

Finally, observe that if we prove the theorem for $\mu_i \geqslant \lambda_i$, then it is also proved for $\lambda_i$. Hence we may set

$$\mu_i = \max\{1, \lambda_1, \ldots, \lambda_i\} \qquad (i = 1,2, \ldots), \tag{6.9}$$

and then $\mu_i \geqslant 1$ $(n = 1,2, \ldots)$ and $\{\mu_i\}$ is non-decreasing.

Assume now that theorem 6.1 is false, that is, for every (infinite) subsequence $\{x_{l_k}\}$ of $\{x_n\}$ there exists a sequence of sequences $\{c_k^{(m)}\}$ with

$$|c_k^{(m)}| \leqslant \mu_k \ (k, m = 1,2, \ldots), \lim_{m\to\infty} \sum_{k=1}^{\infty} c_k^{(m)} x_{l_k} = 0, \tag{6.10}$$

such that $\overline{\lim_{m\to\infty}} |c_{k_0}^{(m)}| \neq 0$ for some fixed $k_0$. Then we can select a subsequence of sequences $\{c_k^{(m_i)}\}$ such that the limits

$$\lim_{i\to\infty} c_k^{(m_i)} = c_k \qquad (k = 1,2, \ldots)$$

exist and $c_{k_0} \neq 0$. For convenience of notation we assume

$$\lim_{m\to\infty} c_k^{(m)} = c_k \qquad (k = 1,2, \ldots). \tag{6.11}$$

Since $c_{k_0} \neq 0$ and $|c_k| \leqslant \mu_k$ $(k = 1,2, \ldots)$, there exists *a least* $k_1 \geqslant k_0$ such that

$$|c_k| < 2^{k-k_0} \mu_k |c_{k_0}| \qquad (k > k_1). \tag{6.12}$$

If $k_1 = k_0$, then (6.12) becomes

$$|c_k| < 2^{k-k_1} \mu_k |c_{k_1}| \quad (k = k_1 + 1, k_1 + 2, \ldots). \tag{6.13}$$

On the other hand, if $k_1 > k_0$, then, since $k_1$ is the least integer $\geqslant k_0$ with property (6.12), this property does not hold for $k_1 - 1 \geqslant k_0$, whence

$$|c_{k_1}| \geqslant 2^{k_1-k_0}\mu_{k_1}|c_{k_0}|,$$

and hence by (6.12) and $\mu_k \geqslant 1$,

$$|c_k| < 2^{k-k_0}\mu_k \frac{1}{2^{k_1-k_0}\mu_{k_1}}|c_{k_1}| \leqslant 2^{k-k_1}\mu_k|c_{k_1}| \quad (k = k_1+1,\ k_1+2, \ldots),$$

i.e.again we have (6.13).

Now, since $|c_k^{(m)}| \leqslant \mu_k \neq 0$ $(k, m = 1,2, \ldots)$, there exists a $k_2 \geqslant k_1$ such that

$$|c_k^{(m)}| \leqslant \mu_k < 2^{k-k_1}\mu_k|c_{k_1}| \qquad (k > k_2;\ m = 1,2, \ldots)$$

and hence also an $m_0'$ independent of $k$ such that

$$|c_k^{(m)}| \leqslant \mu_k < 2^{k-k_1}\mu_k|c_{k_1}^{(m)}| \qquad (k > k_2;\ m > m_0').$$

Furthermore, by (6.13) there exists an $m_0''$ independent of $k$ such that

$$|c_k^{(m)}| < 2^{k-k_1}\mu_k|c_{k_1}| \qquad (k = k_1+1, \ldots, k_2;\ m > m_0''),$$

$$|c_k^{(m)}| < 2^{k-k_1}\mu_k|c_{k_1}^{(m)}| \qquad (k = k_1+1, \ldots, k_2;\ m > m_0'');$$

obviously, these relations remain also valid for $k = k_1$, with $\leqslant$ instead of $<$.

Consequently, for $m_0 = \max(m_0', m_0'')$, we have

$$|c_k^{(m)}| \leqslant 2^{k-k_1}\mu_k|c_{k_1}| \qquad (k \geqslant k_1;\ m > m_0), \tag{6.14}$$

$$|c_k^{(m)}| \leqslant 2^{k-k_1}\mu_k|c_{k_1}^{(m)}| \qquad (k \geqslant k_1;\ m > m_0). \tag{6.15}$$

Let $\{z_n\}$ be the orthonormal sequence obtained from $\{x_n\}$ by the Gram-Schmidt process; then

$$x_n = \sum_{m=1}^{n} a_{nm} z_m, \tag{6.16}$$

with $a_{nn} \neq 0$ $(n = 1,2, \ldots)$ and $|a_{nm}| \leqslant 1$ $(m = 1, \ldots, n;\ n = 1,2, \ldots)$, because $\|x_n\| = 1$ $(n = 1,2, \ldots)$.

Since $\{a_{nm}\}$ is bounded for fixed $m$, we can select a subsequence $\{x_{n_i}\}$ such that the limits

$$\lim_{i\to\infty} a_{n_i m} = b_m \qquad (m = 1,2, \ldots) \tag{6.17}$$

exist; obviously, $|b_m| \leqslant 1$ $(m = 1,2, \ldots)$.

We shall consider two cases:

*Case* 1°. There exists an infinite subsequence $\{n_{i_j}\}$ of $\{n_i\}$ such that $b_{n_{i_j}} = 0$ $(j = 1,2,\ldots)$. In order to simplify notation, we assume that

$$\lim_{p\to\infty} a_{n_p, n_i} = b_{n_i} = 0 \qquad (i = 1,2, \ldots), \tag{6.18}$$

by omitting all terms with $n_i \neq n_{i_j}$ from our subsequence. Let us select a subsequence $\{x_{l_k}\}$ of $\{x_{n_i}\}$ as follows: take $l_1 = n_1$, $l_k = n_{i_k}$ $(k = 2,3, \ldots)$, hence

$$x_{l_1} = x_{n_1},\ x_{l_k} = x_{n_{i_k}} \qquad (k = 2,3, \ldots), \tag{6.19}$$

where

$$|a_{n_{i_k}, n_j}| < \frac{|a_{n_j, n_j}|}{4^k \mu_k} \qquad (j = 1, \ldots, i_{k-1};\ k = 2,3, \ldots). \tag{6.20}$$

Then, by the second relation in (6.10), we have

$$\left\| \sum_{k=1}^{\infty} c_k^{(m)} x_{l_k} \right\| = \varepsilon_m \to 0 \text{ as } m \to \infty, \tag{6.21}$$

for $\{x_{l_k}\}$ defined by (6.19), (6.20) and for $\{c_k^{(m)}\}$ as above. Now, by (6.16) we can write

$$\varepsilon_m = \left\| \sum_{k=1}^{\infty} c_k^{(m)} \sum_{j=1}^{l_k} a_{l_k, j} z_j \right\| = \left\| \sum_{p=1}^{\infty} \sum_{j=l_{p-1}+1}^{l_p} \left( \sum_{k=p}^{\infty} c_k^{(m)} a_{l_k, j} \right) z_j \right\| \tag{6.22}$$

(where $l_0 = 0$), and hence, in particular (since $\{z_n\}$ is orthonormal), for $p = k_1$, $j = l_{k_1}$ we get

$$\left| \sum_{k=k_1}^{\infty} c_k^{(m)} a_{l_k, l_{k_1}} \right| \leqslant \varepsilon_m \qquad (m = 1,2, \ldots). \tag{6.23}$$

However, by (6.14) and (6.20) we have

$$\left| \sum_{k=k_1+1}^{\infty} c_k^{(m)} a_{l_k, l_{k_1}} \right| \leqslant \sum_{k=k_1+1}^{\infty} \frac{2^{k-k_1} \mu_k |c_{k_1}|\, |a_{l_{k_1}, l_{k_1}}|}{4^k \mu_k} =$$

$$= 2^{-2k_1} |c_{k_1}|\ |a_{l_{k_1}, l_{k_1}}| \qquad (m > m_0). \tag{6.24}$$

By (6.11) and (6.21), choose $m > m_0$ so large that

$$|c_{k_1}^{(m)} - c_{k_1}| < 2^{-4k_1} |c_{k_1}|, \tag{6.25}$$

$$\varepsilon_m < 2^{-4k_1} |c_{k_1}| \, |a_{l_{k_1}, l_{k_1}}|. \tag{6.26}$$

Then from (6.23)—(6.26) we obtain

$$\varepsilon_m \geqslant \left| \sum_{k=k_1}^{\infty} c_k^{(m)} a_{l_k, l_{k_1}} \right| \geqslant |c_{k_1}^{(m)}| \, |a_{l_{k_1}, l_{k_1}}| - \left| \sum_{k=k_1+1}^{\infty} c_k^{(m)} a_{l_k, l_{k_1}} \right| \geqslant$$

$$\geqslant |c_{k_1}| \, |a_{l_{k_1}, l_{k_1}}| - 2^{-4k_1} |c_{k_1}| \, |a_{l_{k_1}, l_{k_1}}| - 2^{-2k_1} |c_{k_1}| \, |a_{l_{k_1}, l_{k_1}}| =$$

$$= (1 - 2^{-4k_1} - 2^{-2k_1}) |c_{k_1}| \, |a_{l_{k_1}, l_{k_1}}| > 2^{-4k_1} |c_{k_1}| \, |a_{l_{k_1}, l_{k_1}}| > \varepsilon_m,$$

which is impossible.

*Case* 2°. We have $b_{n_i} \neq 0$ except for a finite number of $i$. We may assume, without loss of generality (by omitting, if necessary, a finite number of elements from $\{x_n\}$), that

$$\lim_{p \to \infty} a_{n_p, n_i} = b_{n_i} \neq 0 \qquad (i = 1, 2, \ldots). \tag{6.27}$$

Let us select a subsequence $\{x_{l_k}\}$ of $\{x_{n_i}\}$ by formula (6.19), where

$$|a_{n_{i_k}, n_j} - b_{n_j}| < \frac{|b_{n_j}|}{4^k \mu_k} \qquad (j = 1, \ldots, i_{k-1}; \; k = 2, 3, \ldots). \tag{6.28}$$

Then, by (6.27), (6.28) and (6.15) we have*)

$$M = \left| \frac{1}{b_{l_{k_1}}} \sum_{k=k_1}^{\infty} c_k^{(m)} a_{l_k, l_{k_1}} - \frac{1}{b_{l_{k_1}+1}} \sum_{k=k_1+1}^{\infty} c_k^{(m)} a_{l_k, l_{k_1}+1} \right| =$$

$$= \left| \frac{1}{b_{l_{k_1}}} \sum_{k=k_1}^{\infty} c_k^{(m)} (a_{l_k, l_{k_1}} - b_{l_{k_1}}) + c_{k_1}^{(m)} - \right.$$

$$\left. - \frac{1}{b_{l_{k_1}+1}} \sum_{k=k_1+1}^{\infty} c_k^{(m)} (a_{l_k, l_{k_1}+1} - b_{l_{k_1}+1}) \right| \geqslant |c_{k_1}^{(m)}| -$$

$$- \frac{1}{|b_{l_{k_1}}|} \sum_{k=k_1}^{\infty} |c_k^{(m)}| \frac{|b_{l_{k_1}}|}{4^k \mu_k} - \frac{1}{|b_{l_{k_1}+1}|} \sum_{k=k_1+1}^{\infty} |c_k^{(m)}| \frac{|b_{l_{k_1}+1}|}{4^k \mu_k} =$$

$$= |c_{k_1}^{(m)}| - \frac{|c_{k_1}^{(m)}|}{4^{k_1}} - 2 \sum_{k=k_1+1}^{\infty} \frac{|c_k^{(m)}|}{4^k \mu_k} \geqslant$$

$$\geqslant |c_{k_1}^{(m)}| \left( 1 - \frac{1}{4^{k_1}} - \sum_{k=k_1+1}^{\infty} \frac{2.2^{k-k_1} \mu_k}{4^k \mu_k} \right) > \frac{1}{2} |c_{k_1}^{(m)}| > \frac{1}{4} |c_{k_1}| > 0$$

*) See the Appendix.

for all $m > m_0$. On the other hand, by (6.22) (for $p = k_1$, $j = l_{k_1}$ and $p = k_1 + 1$, $j = l_{k_1+1}$) and (6.21) we have

$$M \leqslant \left( \frac{1}{|b_{l_{k_1}}|} + \frac{1}{|b_{l_{k_1+1}}|} \right) \varepsilon_m < \frac{1}{4} |c_{k_1}|$$

for all sufficiently large $m$, a contradiction, which completes the proof of theorem 6.1.

From theorem 6.1 and proposition 6.1 c) it follows

**Corollary 6.1.** *Every finitely linearly independent sequence $\{x_n\}$ in a Banach space $E$ contains an $\omega$-linearly independent subsequence.*

Similarly to the problem of extension of basic sequences (Ch. I, § 4, problem 4.1), it is natural to consider the problem of extension of minimal sequences to "maximal" minimal sequences $\{x_n\}$, i.e. to minimal sequences $\{x_n\}$ such that there exists no minimal sequence containing $\{x_n\}$ as a proper subsequence. Obviously, a minimal sequence in a Banach space $E$ is maximal in this sense if and only if it is complete in $E$ and hence, if such a sequence exists, then the space $E$ is separable; conversely, as we have seen at the beginning of this section, such a sequence exists in every separable Banach space. Now we shall show that the extension problem mentioned above has an affirmative answer.

**Theorem 6.2.** *Every minimal sequence $\{y_n\}$ in a separable Banach space $E$ can be extended to a minimal sequence $\{y_n\} \cup \{z_n\}$ which is complete in $E$.*

*Proof.* We may assume that $\dim [y_j] = \infty$ (if $\dim [y_j] < \infty$, it is enough to take any complete minimal sequence $\{z_n\}$ in some complementary subspace to $[y_j]$). By the beginning of this section, let $\{Z_n\}$ be a complete minimal sequence in the quotient space $E/[y_j]$. Then, since $\dim [y_j] = \infty$, we have $\dim Z_n = \infty$ $(n = 1,2,\ldots)$ and hence, by Ch. II, § 4, lemma 4.1, there exist elements

$$z_n \in Z_n \cap \{x \in E \mid g_1(x) = \ldots = g_n(x) = 0\}, \quad z_n \neq 0 \quad (n = 1,2,\ldots),$$

where $\{g_n\} \subset E^*$, $g_n(y_j) = \delta_{nj}$ $(n, j=1,2,\ldots)$. One can also see directly the existence of such elements $z_n$; indeed, it is enough to take arbitrary $x_n \in Z_n$ and to put

$$z_n = x_n - \sum_{i=1}^{n} g_i(x_n) y_i \qquad (n = 1,2,\ldots), \tag{6.29}$$

since then $z_n \in x_n + [y_j] = Z_n$, $z_n \neq 0$ (because otherwise $x_n = \sum_{i=1}^{n} g_i(x_n) y_i \in [y_j]$, whence $Z_n = 0$, in contradiction with the assumption that $\{Z_n\}$ is minimal) and $g_j(z_n) = g_j(x_n) - \sum_{i=1}^{n} g_i(x_n)g_j(y_i) = = g_j(x_n) - g_j(x_n) = 0 \quad (j = 1, \ldots, n)$.

We claim that the sequence $\{y_n\} \cup \{z_n\}$ has the required properties. Indeed, if $x \in E$ and $\varepsilon > 0$, then there exist scalars $\alpha_1, \ldots, \alpha_m$ such that $\left\|\omega\left(x - \sum_{j=1}^{m} \alpha_j z_j\right)\right\| = \left\|\omega(x) - \sum_{j=1}^{m} \alpha_j Z_j\right\| < \frac{\varepsilon}{4}$, where $\omega$ is the canonical mapping of $E$ onto $E/[y_j]$, whence also an element $y \in [y_j]$ such that $\left\|x - \sum_{j=1}^{m} \alpha_j z_j - y\right\| < \frac{\varepsilon}{2}$ and scalars $\beta_1, \ldots, \beta_l$ such that $\left\|y - \sum_{j=1}^{l} \beta_j y_j\right\| < \frac{\varepsilon}{2}$; then $\left\|x - \sum_{j=1} \alpha_j z_j - \sum_{j=1}^{l} \beta_j y_j\right\| < \varepsilon$, which proves that $\{y_n\} \cup \{z_n\}$ is complete in $E$. Furthermore, let $\{\psi_n\} \subset (E/[y_j])^*$, $\psi_n(Z_j) = \delta_{nj}$ $(n, j = 1,2, \ldots)$, let $\{h_n\}$ be the image of $\{\psi_n\}$ under the canonical linear isometry $(E/[y_j])^* \equiv [y_j]^\perp \subset E^*$, that is,

$$h_n(x) = \psi_n(x + [y_j]) \qquad (x \in E,\ n = 1,2, \ldots), \tag{6.30}$$

and let

$$f_n = g_n - \sum_{i=1}^{n-1} g_n(z_i)h_i \qquad (n = 1,2, \ldots). \tag{6.31}$$

Then

$$h_n(y_j) = \psi_n(y_j + [y_k]) = \psi_n([y_k]) = 0 \qquad (n,\ j = 1,2, \ldots),$$

$$h_n(z_j) = \psi_n(z_j + [y_k]) = \psi_n(Z_j) = \delta_{nj} \qquad (n,\ j = 1,2, \ldots),$$

$$f_n(y_j) = g_n(y_j) - \sum_{i=1}^{n-1} g_n(z_i)h_i(y_j) = \delta_{nj} \qquad (n,\ j = 1,2, \ldots),$$

$$f_n(z_j) = g_n(z_j) - \sum_{i=1}^{n-1} g_n(z_i)h_i(z_j) =$$

$$= \begin{cases} g_n(z_j) - g_n(z_j) = 0 & (j = 1, \ldots, n-1;\ n = 1,2, \ldots) \\ g_n(z_j) = 0 & (j = n,\ n+1, \ldots;\ n = 1,2, \ldots). \end{cases}$$

Thus, $(\{y_n\} \cup \{z_n\}, \{f_n\} \cup \{h_n\})$ is a biorthogonal system and hence $\{y_n\} \cup \{z_n\}$ is minimal, which completes the proof of theorem 6.2.

In § 1, definition 1.4 we have introduced block basic sequences with respect to a sequence, which generalize the notion of block basic sequences with respect to a basis (Ch. I, § 7, definition 7.3). By generalizing further this concept, we arrive, in a natural way, at

*Definition 6.2.* A sequence $\{y_n\}$ in a Banach space $E$ is called a *block sequence* with respect to a sequence $\{x_n\} \subset E$ if it is of the form

$$y_n = \sum_{i=m_{n-1}+1}^{m_n} \alpha_i x_i \neq 0 \qquad (n = 1,2, \ldots), \tag{6.32}$$

where $\{m_n\}$ is an increasing sequence of positive integers and $m_0 = 0$.

**Proposition 6.2.** *Let $\{x_n\}$ be a minimal sequence in a Banach space $E$ and let $\{y_n\}$ be a block sequence* (6.32) *with respect to $\{x_n\}$. Then $\{y_n\}$ is a minimal sequence.*

*Proof.* By (6.32) and $x_i \neq 0$ $(i = 1,2, \ldots)$, for every $n$ there exists an index $i_n$ with $m_{n-1} + 1 \leqslant i_n \leqslant m_n$, such that $\alpha_{i_n} \neq 0$. If $\{f_n\} \subset E^*$, $f_i(x_j) = \delta_{ij}$ $(i, j = 1,2, \ldots)$, then for

$$g_n = \frac{1}{\alpha_{i_n}} f_{i_n} \qquad (n = 1,2, \ldots) \tag{6.33}$$

we have $\{g_n\} \subset E^*$, $g_i(y_j) = \delta_{ij}$ $(i, j = 1,2, \ldots)$, which completes the proof. Moreover, let us also observe that the functionals $g_n$ defined by (6.33) satisfy

$$g_n \in [f_i]_{i=m_{n-1}+1}^{m_n} \qquad (n = 1,2, \ldots). \tag{6.34}$$

We have the following result on extension of minimal block sequences, analogous to (but simpler to prove than) Ch. I, § 7, theorem 7.2 on extension of block basic sequences:

**Proposition 6.3.** *Let $\{x_n\}$ be a complete minimal sequence in a Banach space $E$ and let $\{y_n\}$ be a block sequence* (6.32) *with respect to $\{x_n\}$. Then there exists a complete minimal sequence $\{z_n\}$ in $E$ such that*

$$z_{m_n} = y_n \qquad (n = 1,2, \ldots), \tag{6.35}$$

$$[z_i]_{i=m_{n-1}+1}^{m_n} = [x_i]_{i=m_{n-1}+1}^{m_n} \qquad (n = 1,2, \ldots), \tag{6.36}$$

*Consequently, for the sequences of functionals*) $\{f_n\}$, $\{h_n\} \subset E^*$ biorthogonal to $\{x_n\}$ and $\{z_n\}$ respectively, we have*

$$[h_i]_{i=m_{n-1}+1}^{m_n} = [f_i]_{i=m_{n-1}+1}^{m_n} \qquad (n = 1,2, \ldots), \tag{6.37}$$

$$[h_i] = [f_i]. \tag{6.38}$$

---

*) These sequences are uniquely determined (see proposition 6.4).

*Proof.* By (6.32) and $x_i \neq 0$ $(i = 1,2, \ldots)$, for every $n$ there exists an index $i_n$ with $m_{n-1} + 1 \leqslant i_n \leqslant m_n$, such that $\alpha_{i_n} \neq 0$. Put

$$z_i = \begin{cases} x_{\sigma_n(i)} & \text{for } m_{n-1} + 1 \leqslant i \leqslant m_n - 1 \\ y_n & \text{for } i = m_n \end{cases} \qquad (n = 1,2, \ldots), \tag{6.39}$$

where $\sigma_n$ is any one-to-one mapping of the set $\{m_{n-1} + 1, \ldots, m_n - 1\}$ onto the set $\{m_{n-1} + 1, \ldots, i_n - 1, i_n + 1, \ldots m_n\}$. Then $\{z_n\}$ satisfies (6.35) and $[z_i]_{i=m_{n-1}+1}^{m_n} \subset [x_i]_{i=m_{n-1}+1}^{m_n}$ $(n = 1,2, \ldots)$. Also, $\{z_n\}$ is minimal, since for the sequence $\{h_n\} \subset E^*$ defined by

$$h_i = \begin{cases} f_{\sigma_n(i)} - \dfrac{\alpha_{\sigma_n(i)}}{\alpha_{i_n}} f_{i_n} & \text{for } m_{n-1} + 1 \leqslant i \leqslant m_n - 1 \\ \dfrac{1}{\alpha_{i_n}} f_{i_n} & \text{for } i = m_n \end{cases} \qquad (n = 1,2, \ldots) \tag{6.40}$$

we have $h_i(z_j) = \delta_{ij}$ $(i, j = 1,2, \ldots)$. Thus, in particular, $z_{m_{n-1}+1}, \ldots, z_{m_n}$ are linearly independent and hence we have (6.36) and $[z_n] = E$. Finally, by (6.40) we have (6.37) and (6.38), which completes the proof of proposition 6.3.

*Remark 6.2.* As an application of blocks with respect to minimal sequences let us mention that convergence of series of blocks can be used to characterize bases, as follows: *A complete minimal sequence* $\{x_n\}$ *in a Banach space E is a basis of E if and only if for every increasing sequence of positive integers* $\{m_n\}$ *the convergence of* $\sum_{n=1}^{\infty} \sum_{i=m_{n-1}+1}^{m_n} \alpha_i x_i$ *(where* $m_0 = 0$*) implies the convergence of* $\sum_{i=1}^{\infty} \alpha_i x_i$. Indeed, the necessity is obvious by biorthogonality and the proof of the sufficiency is essentially the same as the argument given in Ch. I, § 5, remark 5.2.

The problem of extension of minimal sequences to complete minimal sequences, solved in the affirmative by theorem 6.2, may be also regarded as the problem of extending a minimal sequence which is complete in a subspace (namely, in its closed linear span), to a minimal sequence which is complete in the whole space. It is natural to consider also the following dual problem: Given a complete minimal sequence $\{X_n\}$ in a quotient space $E/G$ of a separable Banach space $E$, is it possible to find elements $x_n \in X_n$ $(n = 1,2, \ldots)$ such that $\{x_n\}$ be a complete minimal sequence in $E$? The answer is affirmative, but we shall give it only later, in theorem 6.4 a) below, together with (actually, as a consequence of) a slightly stronger result (theorem 6.4 b)).

In the above we have used Ch.I, § 6, theorem 6.1, that a sequence $\{x_n\} \subset E$ is minimal if and only if there exists a sequence $\{f_n\} \subset E^*$ such that $(x_n, f_n)$ be a biorthogonal system. In this respect, let us observe

**Proposition 6.4.** *For a minimal sequence $\{x_n\} \subset E$, there exists a unique sequence $\{f_n\} \subset E^*$ such that $f_i(x_j) = \delta_{ij}$ $(i, j = 1,2, \ldots)$ if and only if $[x_n] = E$.*

*Proof.* If $[x_n] = E$ and $f_i(x_j) = g_i(x_j) = \delta_{ij}$ $(i, j = 1,2, \ldots)$, then $(f_i - g_i)(x_j) = 0$ $(i, j=1,2, \ldots)$, whence $f_i = g_i$ $(i = 1,2, \ldots)$. Conversely, if $f_i(x_j) = \delta_{ij}$ $(i, j = 1,2, \ldots)$, $[x_n] \neq E$ and $f_0 \in E^*$, $f_0 \neq 0$, $f_0(x_n) = 0$ $(n = 1,2, \ldots)$, then $\{g_n\}$, where $g_1 = f_1 - f_0$, $g_n = f_n$ $(n = 2,3, \ldots)$ satisfies $g_1 \neq f_1$ and $g_i(x_j) = \delta_{ij}$ $(i, j = 1,2, \ldots)$, which completes the proof.

*Definition 6.3.* Let $\{x_n\}$ be a minimal sequence in a Banach space $E$ and let*) $\{\varphi_n\} \subset [x_n]^*$, $\varphi_i(x_j) = \delta_{ij}$ $(i, j = 1,2, \ldots)$. Then $\{\varphi_n\}$ is called *the sequence of functionals associated to the minimal sequence* $\{x_n\}$, or shortly, *the associated sequence of functionals* (we shall write: a.s.f.).

In particular, if $\{x_n\}$ is a basis of $E$ (and hence a minimal sequence), then the a.s.f. to $\{x_n\}$ is nothing else than the a.s.c.f. to $\{x_n\}$.

One can introduce various special classes of minimal sequences, generalizing the special classes of basic sequences introduced in § 1. Thus, it may seem natural to say that a minimal sequence $\{x_n\}$, with the a.s.f. $\{\varphi_n\}$, is *monotone* if for every $x \in [x_n]$ the number $\|s_n(x)\|$ is a monotone non-decreasing function of $n$, where $\{s_n\}$ is the sequence of partial sum operators associated to the biorthogonal system $(x_n, \varphi_n)$; however, by taking elements $x$ of the form $\sum_{i=1}^{m} \alpha_i x_i$, it follows that every monotone minimal sequence in this sense is already a monotone basic sequence.

*Definition 6.4.* A minimal sequence $\{x_n\}$ in a Banach space $E$, with the a.s.f. $\{\varphi_n\} \subset [x_n]^*$, is said to be

a) *semi-bounded*, if $\sup_{1 \leq n < \infty} \|x_n\| < \infty$;

b) **-semi-bounded*, if $\sup_{1 \leq n < \infty} \|\varphi_n\| < \infty$;

c) *strictly bounded*, if it is both semi-bounded and *-semi-bounded;

d) *normalized*, if $\|x_n\| = 1$ $(n = 1,2, \ldots)$;

e) *normal*, if $\|x_n\| = \|\varphi_n\| = 1$ $(n = 1,2, \ldots)$.

By Ch. I, § 3, corollary 3.1, a basic sequence $\{x_n\}$ is a *-semi-bounded (respectively, a strictly bounded) minimal sequence if and only if

*) If $\{f_n\}$ is any sequence in $E^*$ such that $f_i(x_j) = \delta_{ij}$ $(i, j = 1, 2, \ldots)$, then, obviously,

$$\varphi_n = f_n|_{[x_j]} \qquad (n = 1, 2, \ldots).$$

$\inf_{1 \leqslant n < \infty} \|x_n\| > 0$ (respectively, $0 < \inf_{1 \leqslant n < \infty} \|x_n\| \leqslant \sup_{1 \leqslant n < \infty} \|x_n\| < \infty$). Obviously, for every *-semi-bounded minimal sequence we have $\inf_{1 \leqslant n < \infty} \|x_n\| > 0$ (because $\|x_n\| \|\varphi_n\| \geqslant |\varphi_n(x_n)| = 1$), but the converse is not true. Actually, we have

**Proposition 6.5.** *Let $\{x_n\}$ be a minimal sequence in a Banach space $E$, with the a.s.f. $\{\varphi_n\} \subset [x_n]^*$. The following statements are equivalent:*

1°. *$\{x_n\}$ is *-semi-bounded, i.e.* $\sup_{1 \leqslant n < \infty} \|\varphi_n\| < \infty$.

2°. $\lim_{n \to \infty} \varphi_n(x) = 0$ *for all* $x \in [x_n]$.

3°. *There exists a constant* $\alpha > 0$ *such that*

$$\operatorname{dist}(x_n, E^{(n)}) \geqslant \alpha \qquad (n = 1,2, \ldots), \tag{6.41}$$

*where* $E^{(n)} = [x_1, \ldots, x_{n-1}, x_{n+1}, \ldots]$ $(n = 1,2, \ldots)$.*)

*4° There exists a constant* $M > 0$ *such that for every finite sequence of scalars* $\alpha_1, \ldots, \alpha_n$ *we have*

$$\max_{1 \leqslant i \leqslant n} |\alpha_i| \leqslant M \left\| \sum_{i=1}^{n} \alpha_i x_i \right\|. \tag{6.42}$$

*Proof.* 1° ⇒ 2°. If we have 1°, then, since $\lim_{n \to \infty} \varphi_n(p) = 0$ for every $p = \sum_{i=1}^{m} \alpha_i x_i$, it follows that we also have 2°.

The implication 2° ⇒ 1° is a consequence of the principle of uniform boundedness.

The equivalence 1° ⇔ 3° follows from Ch. II, § 2, corollary 2.1.

1° ⇒ 4°. If we have 1°, then for any scalars $\alpha_1, \ldots, \alpha_n$

$$|\alpha_i| = \left| \varphi_i \left( \sum_{j=1}^{n} \alpha_j x_j \right) \right| \leqslant \sup_{1 \leqslant k < \infty} \|\varphi_k\| \left\| \sum_{j=1}^{n} \alpha_j x_j \right\| \qquad (i = 1, \ldots, n).$$

4° ⇒ 3°. If we have 4°, then for $\alpha_n = 1$ we obtain

$$1 \leqslant M \inf_{\substack{\alpha_1, \ldots, \alpha_{n+p} \\ 1 \leqslant p < \infty}} \left\| x_n - \sum_{i=1}^{n-1} \alpha_i x_i - \sum_{i=n+1}^{n+p} \alpha_i x_i \right\| =$$

$$= M \operatorname{dist}(x_n, E^{(n)}) \qquad (n = 1,2, \ldots),$$

*) In other words, the sequence $\{x_n\}$ is "uniformly minimal" in the sense of § 4, theorem 4.1.

and thus we have 3° with $\alpha = \frac{1}{M}$, which completes the proof of proposition 6.5. Let us also note that the implication 1° ⇒ 2° generalizes Ch. II, § 7, proposition 7.4.

It is natural to ask whether in every separable Banach space $E$ there exists a strictly bounded complete minimal sequence $\{x_n\}$. The answer is affirmative even for $M$-bases $\{x_n\}$, i.e., even if we require in addition that the a.s.f. $\{f_n\} \subset E^*$ be total on $E$, as we shall see in § 8, theorem 8.5. However, we shall show in this section that if we do not require the totality of $\{f_n\}$, then we can obtain better bounds*) for $\sup_{1 \leqslant n < \infty} \|x_n\|$ and $\sup_{1 \leqslant n < \infty} \|f_n\|$. For this purpose, we shall first use the (weakened version of the) theorem of Dvoretzky **) to prove the following auxiliary result:

**Proposition 6.6.** *For every separable Banach space $E$ there exists a biorthogonal system $(x_n, f_n)$ ($\{x_n\} \subset E$, $\{f_n\} \subset E^*$) such that*

(i) $\|x_n\| = \|f_n\| = f_n(x_n) = 1$ $(n = 1,2, \ldots)$;

(ii) $\{x_n\}$ *is a basic sequence;*

(iii) $\{x_{\frac{(n-1)n}{2}+j}\}_{j=1}^n$ *is* $\frac{1}{2}$*-equivalent****) *to the unit vector basis* $\{e_{\frac{(n-1)n}{2}+j}\}_{j=1}^n$ *of* $l_n^2$ $(n = 1,2, \ldots)$, *i.e., for each $n$ the linear mapping $v_n$ which carries $x_{\frac{(n-1)n}{2}+j}$ into $e_{\frac{(n-1)n}{2}+j}$ $(j = 1, \ldots, n)$ is an isomorphism of* $[x_{\frac{(n-1)n}{2}+j}]_{j=1}^n$ *onto $l_n^2$ satisfying*

$$\|v_n\| \leqslant 2, \quad \|v_n^{-1}\| \leqslant 2; \tag{6.43}$$

(iv) $[x_n] + [f_n]_\perp$ *is dense in $E$.*

*Proof.* In view of (i) and (ii), the proof will be a modification of §1, proof of theorem 1.4. Let $\{z_n\}_0^\infty$ be a dense sequence in $E$ with $z_0 = 0$, let $\varepsilon > 0$ and let

$$m_n = \frac{n(n+1)}{2} \qquad (n = 0,1,2, \ldots). \tag{6.44}$$

It will be sufficient to construct sequences $\{x_n\} \subset E$, $\{f_n\} \subset E^*$ and a sequence of subspaces of finite codimension $Y_0 = E \supset Y_1 \supset$

---

*) In the mean time it has been proved that the same better bounds can be also obtained for $M$-bases (see the Notes and remarks to § 8).

**) See Ch. II, § 8, formula (8.4). A proof of this theorem, which is valid both for real and for complex Banach spaces, will be given in Vol. III, Ch. IV.

***) See § 1, proposition 1.2. For a related notion see § 2, formula (2.47).

$\supset Y_2 \supset \ldots$ satisfying (i), (iii) and

$$x_{m_{n-1}+j} \in ([f_i]_{i=1}^{m_{n-1}+j-1})_\perp \cap Y_{n-1} \quad (j = 1, \ldots, n;\ n = 1,2, \ldots), \tag{6.45}$$

$$f_{m_{n-1}+j} \in ([x_i]_{i=1}^{m_{n-1}+j-1} \cup [z_i]_{i=0}^{n-1})^\perp \quad (j = 1, \ldots, n;\ n = 1,2, \ldots), \tag{6.46}$$

$$\widehat{([x_i]_{i=1}^{m_n};\ Y_n)} > 1 - \varepsilon \qquad (n = 1,2, \ldots). \tag{6.47}$$

Indeed, then by (i), (6.45) and (6.46), $(x_n, f_n)$ is biorthogonal. Furthermore, from (6.45) it follows that for any scalars $\alpha_1, \ldots, \alpha_{m_{n+p}}$ we have $\sum_{i=m_n+1}^{m_{n+p}} \alpha_i x_i \in Y_n + Y_{n+1} + \ldots + Y_{n+p-1} \subset Y_n$, whence, by (6.47), $\left\| \sum_{i=1}^{m_{n+p}} \alpha_i x_i \right\| \geqslant (1 - \varepsilon) \left\| \sum_{i=1}^{m_n} \alpha_i x_i \right\|$. But, by (iii), for $1 \leqslant j \leqslant n$ we have

$$\left\| \sum_{i=m_{n-1}+1}^{m_{n-1}+j} \alpha_i x_i \right\| = \left\| v_n^{-1} \left( \sum_{i=m_{n-1}+1}^{m_{n-1}+j} \alpha_i e_i \right) \right\| \leqslant 2 \left\| \sum_{i=m_{n-1}+1}^{m_{n-1}+j} \alpha_i e_i \right\| \leqslant$$

$$\leqslant 2 \left\| \sum_{i=m_{n-1}+1}^{m_n} \alpha_i e_i \right\| = 2 \left\| v_n \left( \sum_{i=m_{n-1}+1}^{m_n} \alpha_i x_i \right) \right\| \leqslant 4 \left\| \sum_{i=m_{n-1}+1}^{m_n} \alpha_i x_i \right\|,$$

and hence*) $\{x_n\}$ is a basic sequence. Finally, by (6.46),

$$z_{n-1} = \sum_{i=1}^{m_{n-1}} f_i(z_{n-1}) x_i + \left( z_{n-1} - \sum_{i=1}^{m_{n-1}} f_i(z_{n-1}) x_i \right) \in [x_j] + [f_j]_\perp$$

$$(n = 1,2, \ldots),$$

so (iv) holds.

Take $x_1 \in E$ and $f_1 \in E^*$ such that $\|x_1\| = \|f_1\| = f_1(x_1) = 1$. Then, by § 1, lemma 1.1 for $G = [x_1]$, there exists a subspace $Y_1 \subset E$ with codim $Y_1 < \infty$, such that $\widehat{([x_1];\ Y_1)} > 1 - \varepsilon$. Assume now that for some $n \geqslant 2$ we have constructed $(x_i, f_i)_{i=1}^{m_{n-1}}$ and $\{Y_i\}_{i=1}^{n-1}$. Put $k_n = m_n + 5(n-1)$. Then, by the theorem of Dvoretzky, there exists an isomophism $u_n$ of some $k_n$-dimensional subspace $Z_n$ of $([f_i]_{i=1}^{m_{n-1}})_\perp \cap Y_{n-1}$ onto $l^2_{k_n}$, with

*) See e.g. § 13, corollary 13.3, proposition 13.11 and theorem 13.4.

$\|u_n\| \leqslant 2$, $\|u_n^{-1}\| = 1$. We shall construct $(x_i, f_i)_{i=m_{n-1}+1}^{m_n}$ with $x_i \in Z_n$ $(i = m_{n-1}+1, \ldots, m_n)$, satisfying (i), (6.45), (6.46) and such that $\{u_n(x_i)\}_{i=m_{n-1}+1}^{m_n}$ *is orthogonal;* this latter property will imply (iii), since then $\left\{\frac{u_n(x_i)}{\|u_n(x_i)\|}\right\}_{i=m_{n-1}+1}^{m_n}$ is orthonormal and hence for any scalars $\{x_i\}_{i=m_{n-1}+1}^{m_n}$ we have (since $\|x_i\|=1$)

$$\left\|v_n\left(\sum_{i=m_{n-1}+1}^{m_n} \alpha_i x_i\right)\right\| = \left\|\sum_i \alpha_i e_i\right\| = \left(\sum_i |\alpha_i|^2\right)^{\frac{1}{2}} \leqslant \left(\sum_i |\alpha_i|^2 \|u_n(x_i)\|^2\right)^{\frac{1}{2}} =$$

$$= \left\|\sum_i \alpha_i \|u_n(x_i)\| \frac{u_n(x_i)}{\|u_n(x_i)\|}\right\| = \left\|u_n\left(\sum_i \alpha_i x_i\right)\right\| \leqslant 2 \left\|\sum_i \alpha_i x_i\right\|,$$

$$\left\|\sum_{i=m_{n-1}+1}^{m_n} \alpha_i x_i\right\| \leqslant \left\|u_n\left(\sum_i \alpha_i x_i\right)\right\| = \left\|\sum_i \alpha_i \|u_n(x_i)\| \frac{u_n(x_i)}{\|u_n(x_i)\|}\right\| =$$

$$= \left(\sum_i |\alpha_i|^2 \|u_n(x_i)\|^2\right)^{\frac{1}{2}} \leqslant 2\left(\sum_i |\alpha_i|^2\right)^{\frac{1}{2}} = 2\left\|v_n\left(\sum_i \alpha_i x_i\right)\right\|,$$

i.e. (6.43). Assume that for some $j$ with $1 \leqslant j \leqslant n$ we have constructed $(x_i, f_i)_{i=m_{n-1}+1}^{m_{n-1}+j-1}$ (for $j = 1$ this means that we have not yet constructed more than the previous $(x_i, f_i)_{i=1}^{m_{n-1}}$). Let $W = l_{k_n}^2 \ominus [u_n(x_i)]_{i=m_{n-1}+1}^{m_{n-1}+j-1}$, the orthogonal complement of $[u_n(x_i)]_{i=m_{n-1}+1}^{m_{n-1}+j-1}$ in $l_{k_n}^2$ (for $j = 1$ let $W = l_{k_n}^2$) and let $G_1 = [x_i]_{i=1}^{m_{n-1}+j-1} \cup [z_i]_{i=0}^{n-1}$, $G_2 = u_n^{-1}(W) \cap ([f_i]_{i=m_{n-1}+1}^{m_{n-1}+j-1})_\perp$. Then $\dim G_1 \leqslant m_{n-1} + j - 1 + (n-1) \leqslant m_n + 2(n-1)$ and*) $\dim G_2 \geqslant$ $\geqslant k_n - (j-1) - (j-1) \geqslant k_n - 2(n-1) = m_n + 3(n-1)$, and hence, by § 1, lemma 1.2, there exists $x_{m_{n-1}+j} \in G_2$ such that $\|x_{m_{n-1}+j}\| = 1$, $\operatorname{dist}(x_{m_{n-1}+j}, G_1) = 1$. Then $u_n(x_{m_{n-1}+j}) \in W$, so $u_n(x_{m_{n-1}+j})$ is orthogonal to $\{u_n(x_i)\}_{i=m_{n-1}+1}^{m_{n-1}+j-1}$. Also, by $u_n^{-1}(W) \subset u_n^{-1}(l_{k_n}^2) = Z_n \subset ([f_i]_{i=1}^{m_{n-1}})_\perp \cap$ $\cap Y_{n-1}$ and by $x_{m_{n-1}+j} \in G_2$, we have (6.45). Next, by a corollary of the Hahn-Banach theorem, there exists $f_{m_{n-1}+j} \in E^*$ such that $\|f_{m_{n-1}+j}\| = 1$, $f_{m_{n-1}+j} \in G_1^\perp$ (so we have (6.46)) and $f_{m_{n-1}+j}(x_{m_{n-1}+j}) = 1$ (so we have (i)). Finally, by § 1, lemma 1.1 for $G = [x_i]_{i=1}^{m_n}$, there exists a subspace $Y_n \subset E$ with $\operatorname{codim} Y_n < \infty$, satisfying (6.47), which completes the proof of proposition 6.6.

Now we can prove the following result on the existence of strictly bounded (even "almost normal") complete minimal sequences:

**Theorem 6.3.** *For every separable Banach space $E$ and every $\varepsilon > 0$ there exists**) a complete minimal sequence $\{x_n\} \subset E$ such that*

$$\|x_n\| \leqslant 1 + \varepsilon, \quad \|f_n\| = 1 \qquad (n = 1, 2, \ldots), \qquad (6.48)$$

*where $\{f_n\} \subset E^*$ is the a.s.f. to $\{x_n\}$.*

---

*) By $\dim u_n^{-1}(W) = k_n - (j-1)$, $\operatorname{codim}_E ([f_i]_{i=m_{n-1}+1}^{m_{n-1}+j-1})_\perp = j-1$.

**) As we already mentioned, there exists even an $M$-basis satisfying (6.48) (see the Notes and remarks to § 8).

*Proof.* By proposition 6.6, let $(z_n, f_n)$ ($\{z_n\} \subset E$, $\{f_n\} \subset E^*$) be a biorthogonal system satisfying (i)—(iv) of that proposition (with $\{x_n\}$ replaced by $\{z_n\}$). If $[f_n]_\perp = \{0\}$, then by (iv) we have $[z_n] = E$ and hence, by (i), the theorem is proved with $\{z_n\} = \{x_n\}$ and $\varepsilon = 0$. Thus, assume that $[f_n]_\perp \neq \{0\}$ and let $\{y_n\} \subset [f_n]_\perp$ with $\|y_n\| = 1$ $(n = 1,2, \ldots)$ be such that $[y_n] = [f_n]_\perp$; we may assume that $\{y_n\}$ is infinite (since otherwise the proof will be similar, with obvious simplifications). Let $\{n_i^{(k)}\}_{i=1}^{\infty}$ $(k = 1,2, \ldots)$ be an infinite partition of $\mathcal{N} = \{1,2,3, \ldots\}$ into infinite subsequences, such that for each $k$ and $m$ there exists $j$ so that $\{x_{n_i^{(k)}}\}_{i=j+1}^{j+m}$ is $\frac{1}{2}$-equivalent to the unit vector basis $\{e_i\}_{i=1}^m$ of $l_m^2$. Then for each $k$ the sequence $\{z_{n_i^{(k)}}\}_{i=1}^{\infty}$ is not equivalent to the unit vector basis of $l^1$, so there exists a sequence of scalars $\{\alpha_i^{(k)}\}_{i=1}^{\infty}$ such that $\sum_{i=1}^{\infty} \alpha_i^{(k)} z_{n_i^{(k)}}$ converges and $\sum_{i=1}^{\infty} |\alpha_i^{(k)}| = \infty$. Let $\varepsilon > 0$ and put

$$x_{n_i^{(k)}} = z_{n_i^{(k)}} - (\varepsilon \operatorname{sign} \alpha_i^{(k)}) y_k \qquad (i, k = 1,2, \ldots). \tag{6.49}$$

Then, since $\{y_k\} \subset [f_n]_\perp$, the system $(x_{n_i^{(k)}}, f_{n_i^{(k)}})$ is biorthogonal. Furthermore, we have

$$\|x_{n_i^{(k)}}\| \leqslant \|z_{n_i^{(k)}}\| + \varepsilon = 1 + \varepsilon \qquad (i, k = 1,2, \ldots), \tag{6.50}$$

and thus (6.48) holds. Finally, let $f \in E^*$ be such that $f(x_{n_i^{(k)}}) = 0$ $(i, k = 1,2, \ldots)$. Then for every $k$ and $m$,

$$f\left(\sum_{i=1}^{m} \alpha_i^{(k)} z_{n_i^{(k)}}\right) = \varepsilon \sum_{i=1}^{m} |\alpha_i^{(k)}| f(y_k), \tag{6.51}$$

whence, since $\sup_{1 \leqslant m < \infty} \left| f\left(\sum_{i=1}^{m} \alpha_i^{(k)} z_{n_i^{(k)}}\right)\right| \leqslant \|f\| \sup_{1 \leqslant m < \infty} \left\|\sum_{i=1}^{m} \alpha_i^{(k)} z_{n_i^{(k)}}\right\| < \infty$ $\left(\text{because } \sum_{i=1}^{\infty} \alpha_i^{(k)} z_{n_i^{(k)}} \text{ converges}\right)$ and $\lim_{m\to\infty} \varepsilon \sum_{i=1}^{m} |\alpha_i^{(k)}| = \infty$, it follows that we must have $f(y_k) = 0$ $(k = 1,2, \ldots)$, which, by (6.49), implies that $f(z_{n_i^{(k)}}) = 0$ $(i, k = 1,2, \ldots)$. Thus, $f$ vanishes on $[z_{n_i^{(k)}}]_{i,k=1}^{\infty} + [y_k]_{k=1}^{\infty} = [z_n] + [f_n]_\perp$, which is dense in $E$, so $f = 0$. Consequently, $[x_n] = E$, which completes the proof of theorem 6.3.

*Remark 6.3.* Note that part (ii) of proposition 6.6 has not been used in the above proof. However, it will be used in § 7, proof of theorem 7.7 (actually, even there, an $M$-basic sequence $\{x_n\}$ also works).

For every special class of minimal sequences for which it is known that complete sequences belonging to that class exist in every separable Banach space $E$, it is natural to raise the extension problem. Thus, we arrive at

*Problem 6.1.* Can every strictly bounded minimal sequence in a separable Banach space $E$ be extended to a complete strictly bounded minimal sequence in $E$?

Even the following weaker problem seems to be open:

*Problem 6.2.* Let $\{x_n\}$ be a strictly bounded minimal sequence in a separable Banach space $E$. Does there exist an extension of $\{x_n\}$ to a complete minimal sequence $\{x_n\} \cup \{y_n\}$ in $E$, with the a.s.f. $\{f_n\} \cup \cup \{g_n\} \subset E^*$ such that $\sup_{1 \leqslant n < \infty} \|f_n\| < \infty$?

Let us observe that this problem is equivalent to the following one, related to proposition 6.6:

*Problem 6.2'.* Let $\{x_n\}$ be a strictly bounded minimal sequence (or, in particular, a bounded basic sequence) in a separable Banach space $E$, with the a.s.f. $\{\varphi_n\} \subset [x_n]^*$. Does there exist an extension of each $\varphi_n$ to a functional $f_n \in E^*$ $(n = 1,2, \ldots)$ such that $\sup_{1 \leqslant n < \infty} \|f_n\| < \infty$ and that $[x_n] + [f_n]_\perp$ is dense in $E$?

Indeed, if the answer to problem 6.2 is affirmative, then $[f_n]_\perp \supset \supset [y_n]$, whence $[x_n] + [f_n]_\perp \supset [x_n] + [y_n]$, which is dense in $E$. Conversely, if the answer to problem 6.2' is affirmative, then $f_n \in ([f_j]_\perp)^\perp, f_n|_{[x_j]} = = \varphi_n$ $(n = 1,2, \ldots)$, whence, by the proof of § 8, theorem 8.3 (with $G = [x_n]$, $F = [f_n]_\perp$), there exists a sequence $\{y_n\} \subset [f_n]_\perp$ with $[y_n] = = [f_n]_\perp$ such that $\{x_n\} \cup \{y_n\}$ is a complete minimal sequence in $E$.

It is also natural to raise the following slight generalization of problem 6.2':

*Problem 6.2''.* Let $G$ be a subspace of a separable Banach space $E$, and let $\{\varphi_n\} \subset G^*$, $\sup_{1 \leqslant n < \infty} \|\varphi_n\| < \infty$. Does there exist a sequence $\{f_n\} \subset E^*$ such that $f_n|_G = \varphi_n$ $(n = 1,2, \ldots)$, $\sup_{1 \leqslant n < \infty} \|f_n\| < \infty$ and that $G + [f_n]_\perp$ is dense in $E$?

Let us consider now the dual problems mentioned after remark 6.2, namely, those of "lifting" a complete minimal sequence or a strictly bounded complete minimal sequence, from a quotient space $E/G$ to the whole space E. We have the following "selection theorem":

**Theorem 6.4.** *Let $G$ be a subspace of infinite codimension of a separable normed linear space $E$. Then*

a) *If $\{X_n\}$ is a complete sequence in $E/G$, there exist elements $x_n \in X_n$ $(n = 1,2, \ldots)$ such that $\{x_n\}$ is a complete sequence in E. If in addition $\{X_n\}$ is minimal, then $\{x_n\}$ is also minimal.*

b) *If* $\{X_n\}$ *is a complete minimal sequence in* $E/G$, *such that* $X_n \to 0$ *(in the norm topology of* $E/G$*), then there exist elements* $x_n \in X_n$ $(n = 1,2, \ldots)$ *such that* $\{x_n\}$ *is complete in E and that*

$$\sup_{1 \leqslant n < \infty} \|x_n\| < \infty. \tag{6.52}$$

*Proof.* a) The first statement is a consequence of part b). Indeed, if $\{X_n\}$ is a complete sequence in $E/G$, with $X_n \neq 0$ $(n = 1,2, \ldots)$, then $\left\{\frac{1}{n\|X_n\|} X_n\right\}$ is a complete sequence in $E/G$, converging to 0, whence, by part b), there exist elements $y_n \in \frac{1}{n\|X_n\|} X_n$ $(n = 1,2, \ldots)$ such that $[y_n] = E$ and then the elements $x_n = n\|X_n\| y_n$ satisfy $x_n \in X_n$ $(n = 1,2,\ldots)$ and $[x_n] = E$.

The second statement in part a) is a consequence of the following more general remark: *If* $\{X_n\} \in E/G$ *is minimal and* $x_n \in X_n$ $(n = 1,2, \ldots)$ *are arbitrary, then* $\{x_n\}$ *is minimal.* Indeed, if $\{\psi_n\} \subset (E/G)^*$, $\psi_i(X_j) = \delta_{ij}$ $(i, j = 1,2, \ldots)$ and if $\{f_n\} \subset G^\perp$ is the image of $\{\psi_n\}$ by the canonical linear isometry $(E/G)^* \equiv G^\perp$, i.e.

$$f_n(x) = \psi_n(x + G) \qquad (x \in E,\ n = 1,2, \ldots), \tag{6.53}$$

then for any $x_n \in X_n$ $(n = 1,2, \ldots)$ we have

$$f_i(x_j) = \psi_i(x_j + G) = \psi_i(X_j) = \delta_{ij} \qquad (i, j = 1,2, \ldots) \tag{6.54}$$

and thus $\{x_n\}$ is minimal.

b) Assume that $\{X_n\}$ is a complete sequence in $E$, such that $X_n \to 0$. Let $\{y_n\} \subset G$ be such that $\|y_n\| < \frac{1}{n}$ $(n = 1,2, \ldots), [y_n] = G$ (for simplicity of notation, we shall assume that $\dim G = \infty$, so $\{y_n\}$ is infinite), let $z_n \in X_n$ be such that $\|z_n\| \leqslant 2\|X_n\|$ $(n = 1,2, \ldots)$, and let $\{n_i^{(k)}\}_{i=1}^{\infty}$ $(k = 1,2, \ldots)$ be an infinite partition of $\mathcal{N} = \{1,2,3, \ldots\}$ into infinite subsequences. Put

$$x_{n_i^{(k)}} = z_{n_i^{(k)}} + \frac{1}{\|X_{n_i^{(k)}}\|} \|z_{n_i^{(k)}}\| y_k \qquad (i, k = 1,2, \ldots). \tag{6.55}$$

Then

$$\|x_{n_i^{(k)}}\| \leqslant 2\|X_{n_i^{(k)}}\| + \frac{2}{k} \qquad (i,\ k = 1,2, \ldots), \tag{6.56}$$

whence, since $X_n \to 0$, we obtain (6.52) (moreover, observe that by (6.56) we also have $\lim_{k\to\infty} x_{n_i^{(k)}} = 0$ for $i = 1,2, \ldots$).

We claim that $[x_n]^\perp \subset G^\perp$. Indeed, let $f \in [x_n]^\perp$, hence $f(x_n) = 0$ $(n = 1,2, \ldots)$. Then, by (6.55),

$$\|X_{n_1^{(k)}}\| \frac{z_{n_1^{(k)}}}{\|z_{n_1^{(k)}}\|} = \|X_{n_2^{(k)}}\| \frac{z_{n_2^{(k)}}}{\|z_{n_2^{(k)}}\|} = \ldots = -f(y_k) \quad (k = 1,2, \ldots), \tag{6.57}$$

whence, since $X_n \to 0$, it follows that $f(y_k) = 0$ $(k = 1,2, \ldots)$. Consequently, since $[y_k] = G$, we obtain $f|_G = 0$, i.e. $f \in G^\perp$, which proves that $[x_n]^\perp \subset G^\perp$.

Now we shall show that $[x_n]^\perp = \{0\}$, whence $[x_n] = E$, which will complete the proof. Let $f \in [x_n]^\perp \subset G^\perp$ be arbitrary and let $\psi \in (E/G)^*$ be the image of $f$ by the canonical linear isometry $G^\perp \equiv (E/G)^*$, i.e.

$$\psi(x + G) = f(x) \qquad (x + G \in E/G). \tag{6.58}$$

Then, since $X_n = x_n + G$ $(n = 1,2, \ldots)$, we have

$$\psi(X_n) = f(x_n) = 0 \qquad (n = 1,2, \ldots),$$

whence, since $[X_n] = E/G$, it follows that $\psi = 0$. Consequently, by (6.58), $f = 0$, which completes the proof of theorem 6.4.

It is natural to ask whether theorem 6.4 b) remains valid if we replace the assumption $X_n \to 0$ by $\sup_{1\leqslant n<\infty} \|X_n\| < \infty$. In other words, if $G$ is a subspace of infinite codimension of a separable normed linear space $E$ and $\{X_n\}$ is a complete sequence in $E/G$, such that $\sup_{1\leqslant n<\infty} \|X_n\| < \infty$ (or, in particular, a strictly bounded complete minimal sequence), do there exist elements $x_n \in X_n$ $(n = 1,2, \ldots)$ such that $\{x_n\}$ is complete in $E$ and $\sup_{1\leqslant n<\infty} \|x_n\| < \infty$ (respectively, such that $\{x_n\}$ is a strictly bounded complete minimal sequence)? The answer is negative, as shown by

*Example 6.3*. Let $E = l^1 \oplus G$ (hence $E/G$ is isomorphic to $l^1$) and let

$$X_n = e_n + G \in E/G \qquad (n = 1,2, \ldots), \tag{6.59}$$

where $\{e_n\}$ is the unit vector basis of $l^1$. Then $\{X_n\}$ is a complete bounded sequence (even a strictly bounded basis) in $E/G$, but there do not exist elements $x_n \in X_n$ $(n = 1,2, \ldots)$ such that $\{x_n\}$ is complete in $E$ and $\sup_{1\leqslant n<\infty} \|x_n\| < \infty$. Indeed, if $x_n = e_n + y_n$, where $y_n \in G$ $(n=1,2,\ldots)$, then

$$\sup_{1\leqslant n<\infty} \|y_n\| \leqslant \|u\| \sup_{1\leqslant n<\infty} \|e_n + y_n\| < \infty, \tag{6.60}$$

where $u$ is the projection of $E$ onto $G$ along $l^1$, whence

$$\sup_{1 \leq n < \infty} |\varphi(y_n)| < \infty \qquad (\varphi \in G^*). \tag{6.61}$$

Consequently, taking any $\varphi \in G^*$ with $\varphi \neq 0$ and then $\psi \in (l^1)^* = m$ with

$$\psi(e_n) = -\varphi(y_n) \qquad (n = 1,2, \ldots), \tag{6.62}$$

for the functional $f = \psi + \varphi \in (l^1)^* \oplus G^* \equiv E^*$ we shall have $f(x_n) = \psi(e_n) + \varphi(y_n) = 0$ $(n = 1,2, \ldots)$, but $f \neq 0$, whence $[x_n] = [e_n + y_n] \neq E$.

Let us pass now to other special classes of minimal sequences.

By § 1, theorem 1.4, in every Banach space $E$ there exist normal minimal sequences. However, if we require, in addition, completeness, we arrive at

*Problem 6.3*. Does there exist in every separable Banach space $E$ a normal complete minimal sequence? Or, in other words: Can we replace $1 + \varepsilon$ by 1 in (6.48)?

*Definition 6.5*. A minimal sequence $\{x_n\}$ in a Banach space $E$ is said to be *shrinking*, if for the a.s.f. $\{\varphi_n\} \subset [x_n]^*$ we have

$$[\varphi_n] = [x_n]^*. \tag{6.63}$$

The problem of existence*) of shrinking minimal sequences has a negative answer, since e.g. the Banach space $E = l^1$ has no shrinking minimal sequence.

It might seem natural to say that a minimal sequence $\{x_n\}$ in a Banach space $E$ is *strongly boundedly complete* if the relation $\sup_{1 \leq n < \infty} \left\| \sum_{i=1}^{n} \alpha_i x_i \right\| < \infty$ implies the convergence of $\sum_{i=1}^{\infty} \alpha_i x_i$; however, from Ch. I, § 5, theorem 5.1, implication 3° ⇒ 1°, it follows that every strongly boundedly complete minimal sequence in this sense is already a boundedly complete basic sequence. Another possible generalization of boundedly complete basic sequences is the following:

*Definition 6.6*. A minimal sequence $\{x_n\}$ in a Banach space $E$ is said to be *quasi-boundedly complete* if for every sequence of scalars $\{\alpha_n\} \subset K$ with the property $\sup_{1 \leq n < \infty} \left\| \sum_{i=1}^{n} \alpha_i x_i \right\| < \infty$ there exists an element $x \in [x_n]$ such that

$$\varphi_n(x) = \alpha_n \qquad (n = 1,2, \ldots). \tag{6.64}$$

Obviously, every strongly boundedly complete minimal sequence is also quasi-boundedly complete. The converse is not true, as shown by

*) We use the term "problem of existence" and related terms in the sense explained in the Introduction to Ch. II.

*Example 6.4*. Let $E$ be a reflexive Banach space and let $\{x_n\}$ be a minimal sequence in $E$, which is not a basic sequence. Then, by the above remark, $\{x_n\}$ is not strongly boundedly complete (since it is not basic). However, $\{x_n\}$ is quasi-boundedly complete. Indeed, if $\sup_{1 \leqslant n < \infty} \left\| \sum_{i=1}^{n} \alpha_i x_i \right\| < \infty$, then, since $[x_n]$ is reflexive, there exists a subsequence of $\left\{ \sum_{i=1}^{n} \alpha_i x_i \right\}$, say $\left\{ \sum_{i=1}^{n_k} \alpha_i x_i \right\}$, which is $\sigma([x_n], [x_n]^*)$-convergent to an element $x \in [x_n]$, whence $\varphi_j(x) = \lim_{k \to \infty} \varphi_j \left( \sum_{i=1}^{n_k} \alpha_i x_i \right) = \alpha_j$ $(j = 1, 2, \ldots)$.

*Remark 6.4*. Such an example is no longer possible if $\{x_n\}$ is a basic sequence, since *every quasi-boundedly complete basic sequence is boundedly complete* (because $x \in [x_n]$ and (6.64) imply that the series $\sum_{i=1}^{\infty} \alpha_i x_i = \sum_{i=1}^{\infty} \varphi_i(x) x_i$ converges).

For minimal sequences $\{x_n\}$ which admit a total sequence $\{f_n\} \subset E^*$ of biorthogonal functionals, the element $x \in [x_n]$ in (6.64) is uniquely determined. For such sequences $\{x_n\} \subset E$ we shall give another convenient generalization of boundedly completeness in § 7, definition 7.7.

The statement on the quasi-boundedly completeness of $\{x_n\}$ in example 6.4 admits the following generalization:

**Proposition 6.7.** *Let $\{x_n\}$ be a minimal sequence in a Banach space $E$, such that the subspace $[x_n]$ is canonically isomorphic $[\varphi_n]^*$, where $\{\varphi_n\} \subset [x_n]^*$ is the a.s.f. to $\{x_n\}$. Then $\{x_n\}$ is quasi-boundedly complete. Hence, in particular, if $(x_n, f_n)$ $(\{x_n\} \subset E, \{f_n\} \subset E^*)$ is a biorthogonal system such that $[x_n] = E$ and $[f_n] = E^*$, then $\{f_n\}$ is a quasi-boundedly complete minimal sequence in $E^*$.*

*Proof*. Let $\sup_{1 \leqslant n < \infty} \left\| \sum_{i=1}^{n} \alpha_i x_i \right\| < \infty$. Then $\sup_{1 \leqslant n < \infty} \left\| \sum_{i=1}^{n} \alpha_i u(x_i) \right\| < \infty$, where $u\colon [x_n] \to [\varphi_n]^*$ is the canonical mapping*). Therefore, since $[\varphi_n]$ is separable, there exists a subsequence $\left\{ \sum_{i=1}^{n_k} \alpha_i u(x_i) \right\}$ of $\left\{ \sum_{i=1}^{n} \alpha_i u(x_i) \right\}$, which is $\sigma([\varphi_n]^*, [\varphi_n])$-convergent to an element $h \in [\varphi_n]^*$. Since by our assumption $h = u(x)$ for some $x \in [x_n]$, it follows that $\varphi_j(x) = u(x)(\varphi_j) =$

*) I.e. (see Ch. I, § 12, formula (12.5)), $u(x)(\varphi) = \varphi(x)$ $(x \in [x_n], \varphi \in [\varphi_n])$.

$$=h(\varphi_j) = \lim_{k\to\infty} \sum_{i=1}^{n_k} \alpha_i u(x_i)(\varphi_j) = \lim_{k\to\infty} \sum_{i=1}^{n_k} \alpha_i \varphi_j(x_i) = \alpha_j \quad (j = 1,2, \ldots),$$

which completes the proof.

The converse of proposition 6.7 is not valid, as shown by

*Example 6.5.* Let $E = (E_1 \times E_2 \times \ldots)_{l^1} \equiv l^1$, where $E_j = l^1$ ($j = 1,2, \ldots$) and for each $j = 1,2, \ldots$ let

$$x_1^{(j)} = e_1^{(j)}, \; x_n^{(j)} = e_1^{(j)} + \frac{1}{j} e_n^{(j)} \qquad (n = 2,3, \ldots), \tag{6.65}$$

$$f_1^{(j)} = h_1^{(j)} - j \sum_{i=2}^{\infty}{}^* h_i^{(j)}, \; f_n^{(j)} = j h_n^{(j)} \qquad (n = 2,3, \ldots), \tag{6.66}$$

where $\{e_n^{(j)}\}$ is the unit vector basis of $E_j = l^1$, $\{h_n^{(j)}\}$ the sequence of coordinate functionals on $E_j = l^1$ and $\sum^*$ the sum in the weak* topology. Since the set $\bigcup_{j=1}^{\infty} \bigcup_{n=1}^{\infty} \{\underbrace{0, \ldots, 0}_{j-1}, x_n^{(j)}, 0, \ldots\}$ in $E$ is countable, let $\{x_n\}$ be any "quadratic" numbering of it, i.e., such that $\bigcup_{j=1}^{n} \bigcup_{k=1}^{n} \{\underbrace{0, \ldots, 0}_{j-1}, x_k^{(j)}, 0, 0, \ldots\} = \{x_i\}_{i=1}^{n^2}$ for $n = 1,2, \ldots$ (e.g., as in Ch. I, § 8, formula (8.19) or Ch. I, § 18, formula (18.1)) and let $\{f_n\}$ be the corresponding numbering of the functionals $\{\underbrace{0, \ldots 0}_{j-1}, f_n^{(j)}, 0, \ldots\} \in (E_1^* \times E_2^* \times \ldots)_m \equiv E^* \equiv m$. Then, as in Ch. I, § 12, example 12.4, $(x_n, f_n)$ is an $E$-complete total biorthogonal system such that the characteristic of the subspace $[f_n]$ is $r([f_n]) = 0$. Therefore, by Ch. I, § 12, formula (12.15), the canonical mapping of $E$ into $[f_n]^*$ is not an isomorphism (and hence it is not *onto*). However, if $\sup_{1 \leq n < \infty} \left\| \sum_{i=1}^{n} \alpha_i x_i \right\| = M_0 < \infty$ and $\alpha_i = \alpha_k^{(j)}$, where $k, j$ are defined by the numbering $x_i = \{\underbrace{0, \ldots, 0}_{j-1}, x_k^{(j)}, 0, \ldots\}$, then the element

$$x = \left\{ \left\{ \sum_{k=1}^{\infty} \alpha_k^{(1)}, \alpha_2^{(1)}, \alpha_3^{(1)} \ldots \right\}, \ldots, \left\{ \sum_{k=1}^{\infty} \alpha_k^{(j)}, \frac{\alpha_2^{(j)}}{j}, \frac{\alpha_3^{(j)}}{j}, \ldots \right\}, \ldots \right\}$$

belongs to $E = [x_n]$ and satisfies (6.64) and thus $\{x_n\}$ is a quasi-boundedly complete minimal sequence (actually, a quasi-boundedly complete "$M$-basis" in the sense of § 8, since we have seen that $\{f_n\}$ is total on $E$). Indeed, since our numbering is "quadratic", we have

$$\sum_{j=1}^{p}\left(\left|\sum_{k=1}^{n}\alpha_k^{(j)}\right|+\frac{1}{j}\sum_{k=2}^{n}|\alpha_k^{(j)}|\right)=\sum_{j=1}^{p}\left\|\alpha_1^{(j)}e_1^{(j)}+\sum_{k=2}^{n}\left(\alpha_k^{(j)}e_1^{(j)}+\frac{1}{j}\alpha_k^{(j)}e_k^{(j)}\right)\right\|=$$

$$=\sum_{j=1}^{p}\left\|\sum_{k=1}^{n}\alpha_k^{(j)}x_k^{(j)}\right\|\leqslant\sum_{j=1}^{n}\left\|\sum_{k=1}^{n}\alpha_k^{(j)}x_k^{(j)}\right\|=\left\|\sum_{=1}^{n^2}\alpha_i x_i\right\|\leqslant M_0\quad(n\geqslant p),$$

whence, for $n\to\infty$, we obtain that $\sum_{k=1}^{\infty}|\alpha_k^{(j)}|$ (and therefore also $\sum_{k=1}^{\infty}\alpha_k^{(j)}$) converges and

$$\sum_{j=1}^{p}\left(\left|\sum_{k=1}^{\infty}\alpha_k^{(j)}\right|+\frac{1}{j}\sum_{k=2}^{\infty}|\alpha_k^{(j)}|\right)\leqslant M_0\qquad(p=1,2,\ldots).$$

Hence, for $p\to\infty$, we get $\sum_{j=1}^{\infty}\left(\left|\sum_{k=1}^{\infty}\alpha_k^{(j)}\right|+\frac{1}{j}\sum_{k=2}^{\infty}|\alpha_k^{(j)}|\right)<\infty$, and thus $x\in E=[x_n]$. Finally, writing $f_n^{(j)}$ for $\{\underbrace{0,\ldots,0}_{j-1},f_n^{(j)},0,\ldots\}$,

$$f_1^{(j)}(x)=\sum_{k=1}^{\infty}\alpha_k^{(j)}-j\sum_{i=2}^{\infty}\frac{\alpha_i^{(j)}}{j}=\alpha_1^{(j)},\ f_n^{(j)}(x)=j\frac{\alpha_n^{(j)}}{j}=\alpha_n^{(j)}$$

$$(n=2,3,\ldots;\ j=1,2,\ldots),$$

so $x$ satisfies (6.64), which proves our assertion on $\{x_n\}$.

In § 1, proof of corollary 1.8, we have observed that every block basic sequence with respect to a shrinking (respectively, boundedly complete) basic sequence is shrinking (respectively, boundedly complete). Similarly, for shrinking minimal sequences we have

**Proposition 6.8.** *Every block sequence with respect to a shrinking minimal sequence is a shrinking minimal sequence.*

*Proof.* Let $\{x_n\}$ be a shrinking minimal sequence in a Banach space $E$. Replacing $E$ by $[x_n]$ if necessary, we may assume, without loss of generality, that $[x_n]=E$. Let $\{f_n\}\subset E^*$, $f_i(x_j)=\delta_{ij}$ $(i,j=1,2,\ldots)$ (hence, by our hypothesis, $[f_n]=E^*$) and let $\{y_n\}$ be a block sequence with respect to $\{x_n\}$, say

$$y_n=\sum_{i=m_{n-1}+1}^{m_n}\alpha_i x_i\neq 0\qquad(n=1,2,\ldots).\qquad(6.67)$$

Take an extension of $\{y_n\}$ to a complete minimal sequence in $E$, say $\{z_n\}$, having the properties described in proposition 6.3. If $\{h_n\} \subset E^*$, $h_i(z_j) = \delta_{ij}$ $(i, j = 1,2, \ldots)$, put

$$g_n = h_{m_n} \qquad (n = 1,2, \ldots). \tag{6.68}$$

Then, by (6.67) and (6.68), we have $g_i(y_j) = h_{m_i}(z_{m_j}) = \delta_{ij}$ $(i,j = 1,2, \ldots)$ and $h_i(y_j) = h_i(z_m) = 0$ $(i \neq m_1, m_2, \ldots;\ j = 1,2, \ldots)$, whence

$$[h_n|_{[y_j]}] = [h_{m_n}|_{[y_j]}] = [g_n|_{[y_j]}].$$

But, since $[h_n] = [f_n] = E^*$, we have $[h_n|_{[y_j]}] = [y_n]^*$. Consequently, $[g_n|_{[y_j]}] = [y_n]^*$, so $\{y_n\}$ is a shrinking minimal sequence, which completes the proof of proposition 6.8.

For block sequences with respect to quasi-boundedly complete minimal sequences a similar result is no longer valid, as shown by

*Example 6.6.* Let $E = c_0$ and let

$$x_{2n-1} = e_{2n-1} - 2^n e_{2n} + 2^{n+1} e_{2n+2} \qquad (n = 1,2, \ldots), \tag{6.69}$$

$$x_{2n} = 2^n e_{2n} \qquad (n = 1,2, \ldots), \tag{6.70}$$

$$f_{2n-1} = h_{2n-1} \qquad (n = 1,2, \ldots), \tag{6.71}$$

$$f_2 = h_1 + \frac{1}{2} h_2, f_{2n} = - h_{2n-3} + h_{2n-1} + \frac{1}{2^n} h_{2n} \quad (n = 2,3, \ldots), \tag{6.72}$$

where $\{e_n\}$ is the unit vector basis of $c_0$ and $\{h_n\}$ the sequence of coordinate functionals on $E = c_0$. Then, clearly, $(x_n, f_n)$ is a biorthogonal system. Furthermore, since $e_{2n} = \frac{1}{2^n} x_{2n}$, $e_{2n-1} = x_{2n-1} + x_{2n} - x_{2n+2}$ $(n = 1,2, \ldots)$, and since $[e_n] = E$, we have $[x_n] = E$. Similarly, since $h_{2n-1} = f_{2n-1}$, $h_2 = 2(f_2 - f_1)$, $h_{2n} = 2^n(f_{2n} - f_{2n-1} + f_{2n-3})$ $(n = 2,3, \ldots)$, and since $[h_n] = E^*$, we have $[f_n] = E^*$. Consequently, by proposition 6.7, $\{f_n\}$ is a quasi-boundedly complete minimal sequence in $E^*$. However, the subsequence $\{f_{2n}\}$ of $\{f_n\}$ is not quasi-boundedly complete. Indeed, $\{f_{2n}\}$ is a monotone basic sequence in $E^*$, since for any scalars $\alpha_1, \ldots, \alpha_n$ we have

$$\left\| \sum_{i=1}^{n} \alpha_i f_{2i} \right\| = \left\| \alpha_1 \left( h_1 + \frac{1}{2} h_2 \right) + \sum_{i=2}^{n} \alpha_i \left( - h_{2i-3} + h_{2i-1} + \frac{1}{2^i} h_{2i} \right) \right\| =$$

$$= \left\| \sum_{i=1}^{n-1} (\alpha_i - \alpha_{i+1}) h_{2i-1} + \alpha_n h_{2n-1} + \sum_{i=1}^{n} \frac{1}{2^i} \alpha_i h_{2i} \right\| =$$

$$= \sum_{i=1}^{n-1} |\alpha_i - \alpha_{i+1}| + |\alpha_n| + \sum_{i=1}^{n} \frac{1}{2^i} |\alpha_i|, \tag{6.73}$$

whence, for any scalars $\alpha_1, \ldots, \alpha_{n+1}$,

$$\left\|\sum_{i=1}^{n} \alpha_i f_{2i}\right\| \leqslant \sum_{i=1}^{n} |\alpha_i - \alpha_{i+1}| + |\alpha_{n+1}| + \sum_{i=1}^{n+1} \frac{1}{2^i} |\alpha_i| = \left\|\sum_{i=1}^{n+1} \alpha_i f_{2i}\right\|.$$

Finally, by (6.73) for $\alpha_1 = \ldots = \alpha_n = 1$, we have $\left\|\sum_{i=1}^{n} f_{2i}\right\| = 1 + \sum_{i=1}^{n} \frac{1}{2^i}$ $(n = 1,2, \ldots)$, whence $\sup_{1 \leqslant n < \infty} \left\|\sum_{i=1}^{n} f_{2i}\right\| = 2$, but $\sum_{i=1}^{\infty} f_{2i}$ is not convergent, since $\|f_{2n}\| > 1$ $(n = 1,2, \ldots)$. Thus $\{f_{2n}\}$ is non-boundedly complete and hence, by remark 6.4, non-quasi boundedly complete.

*Definition 6.7.* A minimal sequence $\{x_n\}$ in a Banach space $E$ is said to be *of type $l_+$* if it is strictly bounded and there exists a constant $\eta > 0$ such that we have, for all finite sequences $\alpha_1, \ldots, \alpha_n \geqslant 0$,

$$\left\|\sum_{i=1}^{n} \alpha_i x_i\right\| \geqslant \eta \sum_{i=1}^{n} \alpha_i. \tag{6.74}$$

**Theorem 6.5.** *In every separable Banach space $E$ there exists a complete minimal sequence $\{x_n\}$ of type $l_+$.*

*Proof.* By theorem 6.3, there exists a biorthogonal system $(y_n, g_n)$ ($\{y_n\} \subset E$, $\{g_n\} \subset E^*$) such that $[y_n] = E$, $\sup_{1 \leqslant n < \infty} \|y_n\| < \infty$, $\sup_{1 \leqslant n < \infty} \|g_n\| < \infty$. Put

$$x_n = y_1 + y_{n+1} \qquad (n = 1,2, \ldots), \tag{6.75}$$

$$f_n = g_{n+1} \qquad (n = 1,2, \ldots). \tag{6.76}$$

Then $(x_n, f_n)$ is a biorthogonal system with $\sup_{1 \leqslant n < \infty} \|x_n\| < \infty$, $\sup_{1 \leqslant n < \infty} \|f_n\| < \infty$ and for every finite sequence $\alpha_1, \ldots, \alpha_n \geqslant 0$ we have

$$\left\|\sum_{i=1}^{n} \alpha_i x_i\right\| \geqslant \left|\frac{g_1}{\|g_1\|}\left(\sum_{i=1}^{n} \alpha_i x_i\right)\right| = \frac{1}{\|g_1\|}\left|\sum_{i=1}^{n} \alpha_i g_1(x_i)\right| = \frac{1}{\|g_1\|}\sum_{i=1}^{n} \alpha_i, \tag{6.77}$$

i.e. (6.64) with $\eta = \frac{1}{\|g_1\|}$. Therefore, if $[x_n] = E$, then $\{x_n\}$ is a complete minimal sequence of type $l_+$. If $[x_n] \neq E$, then there exists a

$g \in E^*$ such that $g \neq 0$, $g(x_n) = 0$ $(n = 1,2, \ldots)$, whence $g(y_1) = -g(y_2) = -g(y_3) = \ldots$ and thus, since $[y_n] = E$ and $g \neq 0$, it follows that $g(y_1) \neq 0$. Put

$$x_0 = y_1, \quad f_0 = \frac{1}{g(y_1)} g. \tag{6.78}$$

Then $[x_n]_0^\infty = E$ and $(x_n, f_n)_0^\infty$ is a biorthogonal system with $\sup_{0 \leq n < \infty} \|x_n\| < \infty$, $\sup_{0 \leq n < \infty} \|f_n\| < \infty$, such that for every finite sequence $\alpha_0, \alpha_1, \ldots, \alpha_n \geqslant 0$ we have (6.77) with $\sum_{i=1}^{n}$ replaced by $\sum_{i=0}^{n}$. Therefore $\{x_n\}_0^\infty$ is a complete minimal sequence of type $l_+$, which completes the proof of theorem 6.5.

Dropping the assumption that $E$ is separable (and hence the requirement that $\{x_n\}$ be complete $E$), we have, clearly,

**Corollary 6.2.** *Every Banach space $E$ contains a minimal sequence $\{x_n\}$ of type $l_+$.*

*Proof.* Apply theorem 6.5 in any separable subspace of $E$.

*Remark 6.5.* One can also give a simpler proof of corollary 6.2, avoiding the use of theorems 6.5, 6.3. Indeed, by § 1, theorem 1.1, $E$ has a basic sequence $\{z_n\}$. Then, by Ch. I, § 3, corollary 3.1, $\{y_n\} = \left\{\frac{z_n}{\|z_n\|}\right\}$ is a strictly bounded complete minimal sequence in $[y_n]$, whence $\{x_n\} = \{y_1 + y_{n+1}\}$ is a minimal sequence of type $l_+$ (by (6.77)), which completes the proof. Note that in this proof of corollary 6.2. the case $[x_n] \neq E$ of the proof of theorem 6.5 has also been omitted.

*Remark 6.6.* Such a result is no longer true if we require, in addition, that the a.s.f. $\{\varphi_n\} \subset [x_n]^*$ to $\{x_n\}$ be total on $[x_n]$. Indeed, if $\{\varphi_n\}$ is total on $[x_n]$, then $[\varphi_n]$ is $\sigma([x_n]^*, [x_n])$-dense in $[x_n]^*$, and hence, if $E$ is reflexive, we obtain $[\varphi_n] = [x_n]^*$. Consequently, since $\sup_{1 \leq n < \infty} \|x_n\| < \infty$ and $\lim_{n \to \infty} \varphi(x_n) = 0$ for every $\varphi$ of the form $\varphi = \sum_{i=1}^{m} \beta_i \varphi_i$, we obtain $\lim_{n \to \infty} \varphi(x_n) = 0$ for every $\varphi \in [\varphi_n] = [x_n]^*$. However, if $\{x_n\}$ is a minimal sequence of type $l_+$, then, by the same argument as in Ch. II, § 10, proof of theorem 10.1, implication $1° \Rightarrow 5°$, there exists*) a $\varphi \in [x_n]^*$ such that $\operatorname{Re} \varphi(x_n) \geqslant 1$ $(n = 1,2, \ldots)$, a contradiction which proves our assertion. Thus, if a Banach space $E$ has a minimal sequence $\{x_n\}$ of

*) Obviously, if $\{x_n\}$ is a strictly bounded minimal sequence, the converse is also true and thus the existence of such a functional $\varphi \in [x_n]^*$ characterizes minimal sequences of type $l_+$ among strictly bounded minimal sequences.

type $l_+$ such that the a.s.f. $\{\varphi_n\} \subset [x_n]^*$ to $\{x_n\}$ is total on $[x_n]$, then $E$ is non-reflexive. The converse is also true, since we shall see in Vol. III, Ch. IV, that every non-reflexive Banach space has even a basic sequence $\{x_n\}$ of type $l_+$.

This remark suggests to give also the following generalization of basic sequences of type $l_+$:

*Definition 6.8.* A minimal sequence $\{x_n\}$ in a Banach space $E$ is said to be *of type* $l_+^0$ if it is strictly bounded and $\{\omega(x_n)\}$ is a minimal sequence of type $l_+$ in $\omega([x_n])$, where $\omega$ denotes the canonical mapping of $[x_n]$ onto the quotient space $[x_n]/[\varphi_n]_\perp$ and where $\{\varphi_n\} \subset [x_n]^*$ is the a.s.f. to $\{x_n\}$.

We recall that $[\varphi_n]_\perp = \{x \in [x_n] \mid \varphi_n(x) = 0 \quad (n = 1,2, \ldots)\}$. Let us observe that if $\{x_n\}$ is a minimal sequence in $E$, then $\{\omega(x_n)\}$ is a minimal sequence in $\omega([x_n]) = [x_n]/[\varphi_n]_\perp$, with the a.s.f. $\{\psi_n\}$ total on $\omega([x_n])$; indeed, $\{\psi_n\}$ is nothing else than the image of $\{\varphi_n\}$ in $([x_n]/[\varphi_n]_\perp)^*$ under the canonical isometry $([\varphi_n]_\perp)^\perp \equiv ([x_n]/[\varphi_n]_\perp)^*$, i.e., $\psi_n(x+[\varphi_j]_\perp) = \varphi_n(x)$ for $x \in [x_n]$.

Every minimal sequence $\{x_n\}$ of type $l_+^0$ is also of type $l_+$ (since $\left\|\sum_{i=1}^n \alpha_i x_i\right\| \geqslant \left\|\omega\left(\sum_{i=1}^n \alpha_i x_i\right)\right\| = \left\|\sum_{i=1}^n \alpha_i \omega(x_i)\right\| \geqslant \eta \sum_{i=1}^n \alpha_i$ for all $\alpha_1, \ldots, \alpha_n \geqslant 0$), but the converse is not true; indeed, a minimal sequence of type $l_+$ in $E = l^2$ (such a sequence exists by corollary 6.3) is not of type $l_+^0$, since by remark 6.6 a reflexive Banach space contains no minimal sequence of type $l_+^0$.

*Definition 6.9.* A minimal sequence $\{x_n\}$ in a Banach space $E$ is said to be

a) *of type P, if* $\{x_n\}$ is strictly bounded and

$$\sup_{1 \leqslant n < \infty} \left\|\sum_{i=1}^n x_i\right\| < \infty; \tag{6.79}$$

b) *of type* $P^*$, if $\{x_n\}$ is strictly bounded and for the a.s.f. $\{\varphi_n\} \subset [x_n]^*$ we have

$$\sup_{1 \leqslant n < \infty} \left\|\sum_{i=1}^n \varphi_i\right\| < \infty. \tag{6.80}$$

*Remark 6.7.* Some of the results of Ch. II, § 9, § 10 and § 12, on bases of types $l_+$, $P$, $P^*$ can be extended to minimal sequences of these types, with the same proofs. In particular, it follows that *a reflexive Banach space E contains no minimal sequence* $\{x_n\}$ *of type P (or* $P^*$*).*

Indeed, by $\sup_{1 \leqslant n < \infty} \|\varphi_n\| < \infty$ and proposition 6.5 we have $\lim_{n\to\infty} \varphi_n(x) = 0$ for all $x \in [x_n]$. But, if $E$ is reflexive and $\{x_n\}$ satisfies (6.79), then $\left\{\sum_{i=1}^{n} x_i\right\}$ has a subsequence $\left\{\sum_{i=1}^{n_k} x_i\right\}$ converging weakly to an element $x \in [x_n]$. Then $\varphi_j(x) = \lim_{k\to\infty} \varphi_j\left(\sum_{i=1}^{n_k} x_i\right) = 1$ $(j = 1,2, \ldots)$, in contradiction with $\lim_{n\to\infty} \varphi_n(x) = 0$, which proves our assertion.

Finally, rather than introducing other special classes of minimal sequences, we shall consider now a special class of biorthogonal systems.

*Definition 6.10.* Let $E$ be a Banach space and let $(x_n, f_n)$, $(y_n, g_n)$ $(\{x_n\}, \{y_n\} \subset E,\ \{f_n\}, \{g_n\} \subset E^*)$ be two biorthogonal systems. The system $(x_n, f_n)$ is said to be an *extension* of $(y_n, g_n)$, if $\{y_n\}$ is a subsequence of $\{x_n\}$ and $\{g_n\}$ is a subsequence of $\{f_n\}$. A biorthogonal system $(x_n, f_n)$ $(\{x_n\} \subset E,\ \{f_n\} \subset E^*)$ is said to be *maximal* if it has no extension except $(x_n, f_n)$ itself.

For example, every $E$-complete biorthogonal system $(x_n, f_n)$ (Ch.I, § 4, definition 4.1) is maximal and every biorthogonal system $(x_n, f_n)$ such that $\{f_n\}$ is total on $E$, is maximal. This latter remark shows that maximal biorthogonal systems may exist also in non-separable Banach spaces (e.g. the system $(x_n, f_n)$ in $E = m$, where $\{x_n\}$ is the sequence of unit vectors and $\{f_n\}$ the sequence of coordinate functionals).

Thus, if $\{x_n\}$ is a maximal minimal sequence (that is, a complete minimal sequence) with the a.s.f. $\{f_n\} \subset E^*$, then $(x_n, f_n)$ is a maximal biorthogonal system, but the converse is not true, namely, a biorthogonal system $(x_n, f_n)$ may be maximal without $\{x_n\}$ being complete (e.g., when $\{f_n\}$ is total on $E$). Note that in this latter case there exists another sequence $\{g_n\} \subset E^*$ such that $(x_n, g_n)$ is a non-maximal biorthogonal system; indeed, if we take any $x_0 \in E \setminus [x_n]$ and any $f_0 \in [x_n]^{\perp}$ with $f_0(x_0) = 1$ and if we put $g_n = f_n - f_n(x_0) f_0$ $(n = 1,2, \ldots)$, then $(x_0 \cup \{x_n\}, f_0 \cup \{g_n\})$ is a biorthogonal system and hence $(x_n, g_n)$ is a non-maximal biorthogonal system.

In the above examples of maximal biorthogonal systems $(x_n, f_n)$ we have either $\operatorname{codim}_E [x_n] = 0$ or $\operatorname{codim}_{E^*} W = 0$, where $W = \widetilde{[f_n]}$, the $\sigma(E^*, E)$-closed linear subspace of $E^*$ spanned by $\{f_n\}$. There also exist maximal biorthogonal systems $(x_n, f_n)$ with $\operatorname{codim}_E [x_n] = \operatorname{codim}_{E^*} W = \infty$, even in separable Hilbert spaces, as shown by

*Example 6.7.* a) Let $E = l^2$, let $\{e_n\}$ be the unit vector basis of $E$ and let

$$x_n = e_1 + e_{n+1} \qquad (n = 1,2, \ldots), \tag{6.81}$$

$$f_n(x) = \xi_{n+1} \qquad (x = \{\xi_n\} \in l^2,\ n = 1,2, \ldots). \tag{6.82}$$

Then $(x_n, f_n)$ is a maximal biorthogonal system, since $[x_n] = E$ (indeed, the relations $f \in E^*$, $f(x_n) = 0$ for $n = 1,2, \ldots$ imply $f(e_1) = -f(e_2) = -f(e_3) = \ldots$ whence, since by remark 6.6 $\lim_{n\to\infty} f(e_n) = 0$ for every $f \in E^*$, we infer $f = 0$). Furthermore, $\operatorname{codim}_{E^*}[f_n] = 1$. Let us observe that $\{f_n\}$ is not total on $E$.

b) Let $E = (E_1 \times E_2 \times \ldots)_{l^2} \equiv l^2$, where $E_j = l^2$ $(j = 1,2, \ldots)$ and for each $j$ let $\{x_n^{(j)}\}$ be the sequence (6.81) in $E_j = l^2$. Since the set $\bigcup_{j=1}^{\infty}\bigcup_{n=1}^{\infty}\{\underbrace{0, \ldots, 0}_{j-1}, x_n^{(j)}, 0, \ldots\}$ in $E$ is countable, let $\{x_n\}$ be an arbitrary numbering of it, and let $\{f_n\}$ be the corresponding numbering of the functionals $\{\underbrace{0, \ldots, 0}_{j-1}, f_n^{(j)}, 0, \ldots\} \in (E_1^* \times E_2^* \times \ldots)_{l^2} \equiv E^* \equiv l^2$, where $\{f_n^{(j)}\}$ is the sequence (6.82) in $E_j^* \equiv l^2$. Then $(x_n, f_n)$ is a maximal biorthogonal system, with $[x_n] = E$ and $\operatorname{codim}_{E^*}[f_n] = \infty$.

c) Let $E = (E_1 \times E_2)_{l_2^2} \equiv l^2$, where $E_j = l^2$ $(j = 1,2)$. By example 6.7 b) above, let $(x_n, f_n)$ be a maximal biorthogonal system in $E_1$ with $[x_n] = E_1$ and $\operatorname{codim}_{E_1^*}[f_n] = \infty$ and let $(y_n, g_n)$ be a maximal biorthogonal system in $E_2$ with $[g_n] = E_2^*$ and $\operatorname{codim}_{E_2}[y_n] = \infty$. Then $(\{x_n\} \times \{y_n\}, \{f_n\} \times \{g_n\}) = (\{x_n, 0\} \cup \{0, y_n\}, \{f_n, 0\} \cup \{0, g_n\})$ is a maximal biorthogonal system in $E$, with $\operatorname{codim}_E [\{x_n\} \times \{y_n\}] = \infty$, $\operatorname{codim}_{E^*} [\{f_n\} \times \{g_n\}] = \infty$. Finally, it remains to observe that $E = l^2$ is reflexive and thus the $\sigma(E^*, E)$-closed linear span $W$ of $\{f_n\} \times \{g_n\}$ in $E^*$ coincides with the $\sigma(E^*, E^{**})$-closed linear span of $\{f_n\} \times \{g_n\}$ and hence*) with the norm-closed linear span $[\{f_n\} \times \{g_n\}]$.

Some useful characterizations of maximal biorthogonal systems are given in

**Proposition 6.9.** *Let $E$ be a Banach space, $(x_n, f_n)$ $(\{x_n\} \subset E, \{f_n\} \subset E^*)$ a biorthogonal system and $W = \widetilde{[f_n]}$, the $\sigma(E^*, E)$-closed linear subspace of $E^*$ spaned by $\{f_n\}$. The following statements are equivalent:*

1°. *$(x_n, f_n)$ is a maximal biorthogonal system.*

2°. $W_\perp = [f_n]_\perp \subset [x_n]$.

3°. $[x_n]^\perp \subset W$.

*Proof.* Since $W = ([f_n]_\perp)^\perp$ and $W_\perp = [f_n]_\perp$, 2° is equivalent to 3°.

Assume now that $(x_n, f_n)$ is not maximal, that is, there exist $x_0 \in E$ and $f_0 \in E^*$ such that $(x_n, f_n)_{n=0}^{\infty}$ is a biorthogonal system. Then

$$f_0(x_0) = 1, \; f_0(x_n) = f_n(x_0) = 0 \qquad (n = 1,2, \ldots),$$

and hence $x_0 \notin [x_n]$ and $x_0 \in [f_n]_\perp$. Thus, 2° $\Rightarrow$ 1°.

---

*) See e.g. [355], Ch. II, § 9, corollary 2 of theorem 9.2.

Conversely, assume that 2° is not satisfied, that is, there exists an element $x_0 \notin [x_n]$ such that $x_0 \in [f_n]_\perp$. Then, by a corollary of the Hahn-Banach theorem, there exists an $f_0 \in E^*$ such that

$$f_0(x_0) = 1, f_0(x_n) = 0 \qquad (n = 1,2, \ldots),$$

whence $(x_n, f_n)_{n=0}^{\infty}$ is a biorthogonal system and therefore $(x_n, f_n)$ is not maximal. Thus, $1° \Rightarrow 2°$, which completes the proof of proposition 6.9.

The problem of extension of biorthogonal systems to maximal biorthogonal systems has an affirmative answer:

**Theorem 6.6.** *Let $E$ be a separable Banach space and let $(y_n, g_n)$ ($\{y_n\} \subset E$, $\{g_n\} \subset E^*$) be a biorthogonal system. Then $(y_n, g_n)$ can be extended to a maximal biorthogonal system.*

*Proof.* If $[g_n]_\perp \subset [y_n]$, then, by proposition 6.9, $(y_n, g_n)$ is maximal and the proof is complete. If this inclusion does not hold, then, since $E$ is separable, let $\{x_n\}$ be a countable dense set in $[g_n]_\perp \setminus [y_n]$. Let $z_1 = x_1$. Then, since $z_1 \notin [y_n]$, there exists a functional $h_1 \in E^*$ such that

$$h_1(z_1) = 1, \; h_1(y_n) = 0 \qquad (n = 1,2, \ldots).$$

If $[g_n]_\perp \subset [z_1 \cup \{y_n\}]$, then by proposition 6.9 $(z_1 \cup \{y_n\}, h_1 \cup \cup \{g_n\})$ is a maximal biorthogonal system (because $[h_1 \cup \{g_n\}]_\perp \subset \subset [g_n]_\perp \subset [z_1 \cup \{y_n\}]$) and the proof is complete. If this inclusion does not hold, then there exists a smallest index $n_2$ such that $x_{n_2} \notin [z_1 \cup \cup \{y_n\}]$ (since otherwise we would have $[g_n]_\perp \setminus [y_n] = [x_n] \subset [z_1 \cup \{y_n\}]$ and $[y_n] \subset [z_1 \cup \{y_n\}]$, whence $[g_n]_\perp \subset [z_1 \cup \{y_n\}]$). Since $[h_1 \cup \{g_n\}]_\perp$ is a hyperplane in $[g_n]_\perp$ and $[z_1, x_{n_2}] \subset [g_n]_\perp$, there exists, by Ch. II, § 4, lemma 4.1, an element*)

$$z_2 \in [z_1, x_{n_2}] \cap [h_1 \cup \{g_n\}]_\perp, \; z_2 \neq 0.$$

Observe now that $z_2 \neq \alpha_1 z_1$ (since $z_2 \neq 0$ and $h_1(z_1) = 1$, $h_1(z_2) = 0$) and hence $z_2 \notin [z_1 \cup \{y_n\}]$ (since otherwise we would have $z_2 = \alpha_1 z_1 + \alpha_2 x_{n_2} = \beta_1 z_1 + \beta_2 y$ with $\alpha_2 \neq 0$ and $y \in [y_n]$, whence $x_{n_2} = \frac{1}{\alpha_2}(\beta_1 - \alpha_1) z_1 + \frac{\beta_2}{\alpha_2} y \in [z_1 \cup \{y_n\}]$, in contradiction with the above choice of $x_{n_2}$). Consequently, there exists a functional $h_2 \in E^*$ such that

$$h_2(z_2) = 1, \; h_2(z_1) = h_2(y_n) = 0 \qquad (n = 1,2, \ldots).$$

*) One can also see directly that such an element is $z_2 = x_{n_2} - h_1(x_{n_2}) z_1$.

Continuing in this way, we obtain (finite or infinite) sequences $\{z_n\} \subset$ $\subset E$ and $\{h_n\} \subset E^*$ such that $(\{y_n\} \cup \{z_n\}, \{g_n\} \cup \{h_n\})$ is a biorthogonal system and $x_{n_k}, \ldots, x_{n_{k+1}-1} \in [\{z_1, \ldots, z_k\} \cup \{y_n\}]$ $(k=1,2,\ldots;$ $n_1 = 1)$, whence $[g_n]_\perp \setminus [y_n] = [x_n] \subset [\{y_n\} \cup \{z_n\}]$. Then

$$[\{g_n\} \cup \{h_n\}]_\perp \subset [g_n]_\perp \subset [\{y_n\} \cup \{z_n\}],$$

and hence, by proposition 6.9, $(\{y_n\} \cup \{z_n\}, \{g_n\} \cup \{h_n\})$ is a maximal biorthogonal system, which completes the proof.

Comparing theorems 6.2 and 6.6 we see that in theorem 6.6 we extend both $\{y_n\} \subset E$ and $\{g_n\} \subset E^*$, but the extended sequence $\{y_n\} \cup \{z_n\}$ is not necessarily complete in $E$, while in theorem 6.2 we have extended $\{y_n\}$ to a complete minimal sequence $\{y_n\} \cup \{z_n\}$, but we had to change $\{g_n\}$ in $E^*$ by a suitable sequence $\{f_n\} \cup \{h_n\}$.

It is natural to ask, which minimal sequences $\{x_n\}$ in a separable Banach space $E$ can be "left side of" a maximal biorthogonal system, i.e., for which minimal sequences $\{x_n\} \subset E$ does there exist a sequence of functionals $\{f_n\} \subset E^*$ such that $(x_n, f_n)$ be a maximal biorthogonal system. We shall show that every infinite minimal sequence $\{x_n\}$ has this property. To this end, let us first give

**Lemma 6.1.** *Let $E$ be a normed linear space and $G$ an infinite dimensional subspace of $E$ such that $(E/G)^*$ contains a total sequence $\{\psi_n\}$ and let $\{\varphi_n\} \subset G^*$, $\varphi_n \neq 0$ $(n = 1,2, \ldots)$. Then there exists a sequence $\{f_n\} \subset E^*$ such that*

$$f_n|_G = \varphi_n \qquad (n = 1,2, \ldots), \tag{6.83}$$

$$[f_n]_\perp \subset G. \tag{6.84}$$

*Proof.* Let $\{h_n\} \subset G^\perp$ be the image of $\{\psi_n\}$ by the canonical linear isometry $(E/G)^* \equiv G^\perp$, i.e.

$$h_n(x) = \psi_n(x + G) \qquad (x \in E,\ n = 1,2, \ldots), \tag{6.85}$$

let $\{g_n) \subset E^*$, $g_n|_G = \varphi_n$ $(n = 1,2, \ldots)$ and let $\{n_i^{(k)}\}_{i=1}^\infty$ $(k = 1,2, \ldots)$ be an infinite partition of $\mathcal{N} = \{1,2,3, \ldots\}$ into infinite subsequences. Put

$$f_{n_i^{(k)}} = g_{n_i^{(k)}} + n_i^{(k)} \|g_{n_i^{(k)}}\| h_k \qquad (i, k = 1,2, \ldots). \tag{6.86}$$

Then, since $\{h_n\} \subset G^\perp$, we have $f_n|_G = g_n|_G = \varphi_n$ $(n = 1,2, \ldots)$, that is, (6.83). Furthermore, assume that $x \in E$, $f_n(x) = 0$ $(n = 1,2, \ldots)$. Then

$$\frac{1}{n_1^{(k)} \|g_{n_1^{(k)}}\|} g_{n_1^{(k)}}(x) = \frac{1}{n_2^{(k)} \|g_{n_2^{(k)}}\|} g_{n_2^{(k)}}(x) = \ldots = -h_k(x) \ (k = 1,2, \ldots), \tag{6.87}$$

whence, since $\lim_{i\to\infty} \frac{1}{n_i^{(k)}\|g_{n_i^{(k)}}\|} g_{n_i^{(k)}}(x) = 0$ $(k = 1,2, \ldots)$, it follows that $h_k(x) = 0$ $(k = 1,2, \ldots)$. Consequently, by (6.85) and since $\{\psi_n\}$ is total on $E/G$, we infer $x \in G$, and thus we have (6.84), which completes the proof of lemma 6.1.

Now we can prove

**Theorem 6.7.** *For every infinite minimal sequence $\{x_n\}$ in a separable Banach space $E$ there exists a sequence $\{f_n\} \subset E^*$ such that $(x_n, f_n)$ be a maximal biorthogonal system.*

*Proof.* Let $\{\varphi_n\} \subset [x_n]^*$ be the a.s.f. to $\{x_n\}$ and let $G = [x_n]$. Then, since $E$ is separable, so is $E/G$, and therefore $(E/G)^*$ contains*) a total sequence $\{\psi_n\}$. Hence, by lemma 6.1, there exists a sequence $\{f_n\} \subset E^*$ satisfying (6.83) and (6.84). Then by (6.83) we have $f_i(x_j) = \varphi_i(x_j) = \delta_{ij}$ $(i, j = 1,2, \ldots)$ and by (6.84) we have $[f_n]_\perp \subset [x_n]$. Consequently, by proposition 6.9, implication $2° \Rightarrow 1°$, $(x_n, f_n)$ is a maximal biorthogonal system, which completes the proof.

The following property of minimal sequences will be used in § 10, proof of theorem 10.2:

**Proposition 6.10.** *Let $\{x_n\}$ be an infinite minimal sequence in a Banach space $E$. Then for every $n$ and $\varepsilon > 0$ there exists a positive integer $N_0 = N_0(n, \varepsilon) > n$ such that for all $y \in [x_i]_1^n$ we have*

$$\operatorname{dist}(y, [x_i]_{n+1}^{N_0}) \leqslant (1+\varepsilon) \operatorname{dist}(y, [x_i]_{n+1}^{\infty}). \tag{6.88}$$

*Proof.* The functions

$$e_n^{(N)}(y) = \operatorname{dist}(y, [x_i]_{n+1}^{N}),\ e_n(y) = \operatorname{dist}(y, [x_i]_{n+1}^{\infty}) \quad (y \in [x_i]_1^n) \tag{6.89}$$

are continuous and satisfy

$$e_n^{(n+1)}(y) \geqslant e_n^{(n+2)}(y) \geqslant \ldots \geqslant e_n(y) \qquad (y \in [x_i]_1^n), \tag{6.90}$$

$$\lim_{N\to\infty} e_n^{(N)}(y) = e_n(y) \qquad (y \in [x_i]_1^n). \tag{6.91}$$

Consequently, by the classical theorem of Dini**), the convergence in (6.91) is uniform on $\{y \in [x_i]_1^n \mid \|y\| \leqslant 1\}$, i.e.

$$\lim_{N\to\infty} \sup_{\substack{y \in [x_i]_1^n \\ \|y\| \leqslant 1}} \{\operatorname{dist}(y, [x_i]_{n+1}^{N}) - \operatorname{dist}(y, [x_i]_{n+1}^{\infty})\} = 0$$

*) See e.g. [11], p. 124, theorem 4.

**) See e.g. [215], p. 239, theorem E.

and hence there exists a positive integer $N_0 = N_0(n, \varepsilon)$ such that we have (6.88) for all $y \in [x_i]_1^n$ with $\|y\| \leqslant 1$. Since $\operatorname{dist}(y, G) = \|y\| \operatorname{dist}\left(\frac{y}{\|y\|}, G\right)$ for every $y \in E \setminus \{0\}$ and every subspace $G$ of $E$, it follows that we have (6.88) for all $y \in [x_i]_1^n$, which completes the proof.

We conclude this section with the following decomposition property of the elements of a Banach space with respect to a complete minimal sequence (which is obtained from § 3, theorem 3.3):

**Theorem 6.8.** *Let $\{x_n\}$ be a complete minimal sequence in a Banach space $E$, with the a.s.f. $\{f_n\} \subset E^*$. Then for every element $x \in E$ there exist two elements $y_1, y_2 \in E$ such that*

$$x = y_1 + y_2 \tag{6.92}$$

*and that each of the series $\sum_{i=1}^{\infty} f_i(y_1)x_i$, $\sum_{i=1}^{\infty} f_i(y_2)x_i$ has a sequence of partial sums converging to $y_1$ and $y_2$, respectively.*

*Proof.* If $\{x_n\}$ is a basis of $E$, the statement is obvious.

Assume now that $\{x_n\}$ is not a basis of $E$ and let $\{s_n\}$ be the associated sequence of partial sum operators. Then, by Ch. I, § 4, theorem 4.1, $\sup_{1 \leqslant n < \infty} \|s_n\| = \infty$. Furthermore, if $\pi_n(y_k) \in P_{(n)} = [x_1, \ldots, x_n]$ is a polynomial of best approximation of $y_k$ (that is, $\|y_k - \pi_n(y_k)\| = \inf_{z \in P_{(n)}} \|y_k - z\| = \mu_n(y_k)$), then

$$\|y_k - s_n(y_k)\| \leqslant \|y_k - \pi_n(y_k)\| + \|s_n(\pi_n(y_k) - y_k)\| \leqslant (1 + \|s_n\|)\mu_n(y_k) \leqslant$$

$$\leqslant \frac{1}{\varepsilon_n} \mu_n(y_k) \qquad (k = 1, 2;\ n = 1, 2, \ldots), \tag{6.93}$$

where we have put

$$\varepsilon_n = \frac{1}{1 + \sup_{1 \leqslant j \leqslant n} \|s_j\|} \qquad (n = 1, 2, \ldots). \tag{6.94}$$

But, since $\varepsilon_1 \geqslant \varepsilon_2 \geqslant \ldots > 0$, $\lim_{n\to\infty} \varepsilon_n = 0$, there exist, by § 3, theorem 3.3, two elements $y_1, y_2 \in E$ satisfying (6.92) and $\varliminf_{n\to\infty} \frac{\mu_n(y_k)}{\varepsilon_n} = 0$ $(k =$

$= 1,2$). Hence, by (6.93), $\lim\limits_{n\to\infty} \|y_k - s_n(y_k)\| = 0$ ($k = 1,2$), which completes the proof of theorem 6.8.

For a certain special class of complete minimal sequences a stronger result will be given in § 13, theorem 13.2 (see also § 13, problem 13.4).

## § 7. Generalized bases

*Definition 7.1.* A sequence $\{x_n\}$ in a Banach space $E$ is called a *generalized basis* of $E$, if there exists a total sequence of functionals $\{f_n\} \subset E^*$ such that $(x_n, f_n)$ is a biorthogonal system. Any such total sequence $\{f_n\} \subset E^*$ is called an *admissible sequence* (for the generalized basis $\{x_n\}$). The one-to-one mapping $u: x \to \{f_n(x)\}$ of $E$ into the space $s$ of all sequences of scalars is called the *admissible mapping* determined by $\{f_n\}$.

For example, if $S$ is any $BK$-space*) containing all unit vectors $e_n$ ($n = 1,2, \dots$), the sequence $\{e_n\}$ is a generalized basis of $S$, having the coordinate functionals

$$h_n(\{\alpha_j\}) = \alpha_n \qquad (\{\alpha_j\} \in S;\ n = 1,2, \dots) \tag{7.1}$$

as an admissible sequence.

*Remark 7.1.* The above example leads us to another interpretation of generalized bases in the equivalent language of sequence spaces. Indeed, every Banach space $E$ on which there exists a total sequence of functionals $\{f_n\} \subset E^*$ (in particular, every separable Banach space $E$) is linearly isometric to the $BK$-space**)

$$S = S(E, \{f_n\}) = \{\{f_n(x)\} \mid x \in E\} \tag{7.2}$$

(where the norm is defined by $\|\{f_n(x)\}\| = \|x\|$), by the mapping***) $v: x \to \{f_n(x)\}$. Thus, $E$ can be identified with the $BK$-space $S$, by this mapping, and then the mapping $(v^*)^{-1}: E^* \to S^*$ (where $v^*$ is the adjoint of $v$), will identify $\{f_n\}$ with the sequence of coordinate functionals $\{h_n\}$; obviously, the existence of a generalized basis $\{x_n\}$ of $E$ having the given total sequence $\{f_n\} \subset E^*$ as an admissible sequence is equivalent to the condition $e_n \in S$ ($n = 1,2, \dots$).

---

*) See Ch. I, § 12 for the definition of a $BK$-space.

**) In the particular case when $\{x_n\}$ is a basis of $E$, with the a.s.c.f. $\{f_n\} \subset E^*$, the set $S$ coincides with the set $A_1 = \left\{\{\alpha_n\} \subset K \,\middle|\, \sum_{i=1}^{\infty} \alpha_i x_i \text{ converges}\right\}$, but on $A_1$ we have used a different norm (see Ch. I, § 3).

***) $v$ is the astriction of the admissible embedding $u: E \to s$.

*Remark 7.2.* The identification device of remark 7.1 can be carried much further and it reveals many parallelisms between the theory of total biorthogonal systems*) (in particular, of bases) and that of sequence spaces, showing that often these two theories obtain essentially the same results, in two different (but equivalent) languages. Moreover, sometimes this remark can be used to obtain a new result in one of the theories, corresponding to a known result in the other theory.

Indeed, let us give an example. We recall that *the n-th section* of a sequence $\{\alpha_j\} \in s$ is, by definition, the sequence

$$P_n(\{\alpha_j\}) = \{\alpha_1, \ldots, \alpha_n, 0, 0, \ldots\}. \tag{7.3}$$

A $BK$-space $S$ containing all unit vectors $e_n$ $(n = 1,2, \ldots)$ is said to be

a) an *AD-space*, if $[e_n] = S$;

b) a *BS-space*, if $\sup\limits_{1 \leq n < \infty} \|P_n(\{\alpha_j\})\| < \infty$ for all $\{\alpha_j\} \in S$ (hence, in particular, in this case $\{e_n\}$ is a basic sequence in $S$);

c) an *AK-space*, if $\lim\limits_{n \to \infty} P_n(\{\alpha_j\}) = \{\alpha_j\}$ for all $\{\alpha_j\} \in S$ (or, equivalently, if $\{e_n\}$ is a basis of $S$); by Ch. I, § 4, theorem 4.1, this happens if and only if $S$ is both an $AD$-space and a $BS$-space.

We shall consider coordinatewise multiplication of sequences: $\{\alpha_n\} \cdot \{\beta_n\} = \{\alpha_n \beta_n\}$. We shall write $ST$ for the set $\{\{\alpha_n\}\{\beta_n\} \mid \{\alpha_n\} \in S, \{\beta_n\} \in T\}$. A sequence space $S$ is said to be *T-invariant* if $TS(=ST) = S$.

It is easy to see what are the corresponding notions for total biorthogonal systems $(x_n, f_n)$ $(\{x_n\} \subset E, \{f_n\} \subset E^*)$ by the identification device of remark 7.1. Namely, if $S$ is the $BK$-space (7.2) associated to $(E, (x_n, f_n))$ then $P_n(\{f_j(x)\})$ is the sequence corresponding to $s_n(x)$, where $s_n$ is the, $n$-th partial sum operator and

a) $S$ is an $AD$-space if and only if $[x_n] = E$;

b) $S$ is a $BS$-space if and only if $\sup\limits_{1 \leq n < \infty} \|s_n(x)\| < \infty$ for all $x \in E$ (hence, in particular, in this case $\{x_n\}$ is a basic sequence in $E$);

c) $S$ is an $AK$-space if and only if $\{x_n\}$ is a basis of $E$;

d) $TS(= ST) \subset S$ if and only if $T \subset M(E, (x_n, f_n))$ (see Ch. I, § 5, definition 5.1).

In the theory of sequence spaces it has been proved**) that a $BK$-space $S$ containing all unit vectors $e_n$ $(n = 1,2, \ldots)$ is an $AK$-space if and only if it is $bv_0$-invariant, that is, $(bv_0)S = S$. By the preceding remarks, this result is equivalent to the following: If $(x_n, f_n)$ $(\{x_n\} \subset E, \{f_n\} \subset E^*)$ is a biorthogonal system such that $\{f_n\}$ is total on $E$, we have $\lim\limits_{n \to \infty} s_n(x) = x$ for all $x \in E$ if and only if $M(E, (x_n, f_n)) \supset bv_0$ and

*) I.e. biorthogonal systems $(x_n, f_n)$ such that $\{f_n\}$ is total on $E$.

**) See [123], corollary 1 of theorem 4; actually, in [123] $bv_0$ is replaced by its unit cell, which obviously gives the same condition.

for every $x \in E$ there exist $\{\gamma_j\} \in bv_0$ and $z \in E$ satisfying $f_j(x) = \gamma_j f_j(z)$ $(j = 1, 2, \ldots)$ (cp. Ch. I, § 5, theorem 5.2).

Furthermore, the following related result has been also proved in the theory of sequence spaces*): a *BK*-space $S$ containing all unit vectors $e_n$ $(n = 1, 2, \ldots)$ is an *AK*-space if and only if $S \subset (bv_0)\{\{\alpha_j\} \in S \mid \sup_{1 \leqslant n < \infty} \|P_n(\{\alpha_j\})\| < \infty\}$. By the above identifications, this leads to the following new characterization of bases among total biorthogonal systems:

**Theorem 7.1.** *Let $E$ be a Banach space and $(x_n, f_n)$ $(\{x_n\} \subset E$, $\{f_n\} \subset E^*)$ a biorthogonal system such that $\{f_n\}$ is total on $E$ (in other words, let $\{x_n\}$ be a generalized basis of $E$, with an admissible sequence $\{f_n\} \subset E^*$). The sequence $\{x_n\}$ is a basis of $E$ if and only if for every $x \in E$ there exist a $\{\gamma_j\} \in bv_0$ and an element $z \in E$ with $\sup_{1 \leqslant n < \infty} \left\| \sum_{i=1}^{n} f_i(z) x_i \right\| < \infty$, such that***)

$$f_j(x) = \gamma_j f_j(z) \qquad (j = 1, 2, \ldots), \tag{7.4}$$

*Proof.* Assume that $\{x_n\}$ is a basis of $E$ and let $x \in E$ be arbitrary. Then $\lim_{k \to \infty} s_k(x) = \lim_{k \to \infty} \sum_{i=1}^{k} f_i(x) x_i = x$, and hence there exists a sequence of positive integers $\{m_n\}$ such that

$$\|x - s_k(x)\| \leqslant \frac{1}{4^n} \qquad (k \geqslant m_n;\ n = 1, 2, \ldots). \tag{7.5}$$

Let $y_n = \sum_{i=m_{n-1}+1}^{m_n} f_i(x) x_i$ $(n = 1, 2, \ldots;\ m_0 = 0)$. Then, by (7.5), $\|y_n\| = \|s_{m_{n-1}}(x) - s_{m_n}(x)\| \leqslant \frac{2}{4^n}$ $(n = 1, 2, \ldots)$ and hence the series $\sum_{n=1}^{\infty} 2^{n-1} y_n$ converges. Let

$$z = \sum_{n=1}^{\infty} 2^{n-1} y_n, \tag{7.6}$$

$$\gamma_j = \frac{1}{2^{n-1}} \qquad (m_{n-1} + 1 \leqslant j \leqslant m_n;\ n = 1, 2, \ldots). \tag{7.7}$$

---

*) This follows e.g. from [123], lemma 1 and proof of proposition 7.

**) I.e., with the terminology of Ch. I, § 5, such that $\{\gamma_j\} \in M(z, (x_n, f_n))$ and $x = z_{\{\gamma_n\}}$.

Then $\{\gamma_j\} \in bv_0$ and, since $\{x_n\}$ is a basis of $E$, we have $\sup_{1 \leqslant n < \infty} \left\| \sum_{i=1}^{n} f_i(z) x_i \right\| < \infty$. Furthermore, by biorthogonality, $f_j(z) = 2^{n-1} f_j(x)$ for $m_{n-1} + 1 \leqslant j \leqslant m_n$ $(n = 1, 2, \ldots)$ and thus we have (7.4).

Conversely, assume now that for every $x \in E$ there exist a $\{\gamma_j\} \in bv_0$ and an element $z \in E$ with $\sup_{1 \leqslant n < \infty} \left\| \sum_{i=1}^{n} f_i(z) x_i \right\| < \infty$ such that we have (7.4), and let $x \in E$ be arbitrary. Then for any positive integers $p \leqslant q$ we have

$$\left\| \sum_{i=p}^{q} f_i(x) x_i \right\| = \left\| \sum_{i=p}^{q} \gamma_i f_i(z) x_i \right\| = \left\| \sum_{i=p}^{q} \gamma_i \left( \sum_{j=1}^{i} f_j(z) x_j - \sum_{j=1}^{i-1} f_j(z) x_j \right) \right\| =$$

$$= \left\| -\gamma_p \sum_{j=1}^{p-1} f_j(z) x_j + \sum_{i=p}^{q-1} (\gamma_i - \gamma_{i+1}) \sum_{j=1}^{i} f_j(z) x_j + \gamma_q \sum_{j=1}^{q} f_j(z) x_j \right\| \leqslant$$

$$\leqslant \left( |\gamma_p| + \sum_{i=p}^{q-1} |\gamma_i - \gamma_{i+1}| + |\gamma_q| \right) \sup_{1 \leqslant n < \infty} \left\| \sum_{j=1}^{n} f_j(z) x_j \right\|, \tag{7.8}$$

whence $\{s_n(x)\}$ is a Cauchy sequence. Since $\{f_n\}$ is total on $E$, by Ch. I, § 5, proposition 5.1 it follows that $\lim_{n \to \infty} s_n(x) = x$ and thus $\{x_n\}$ is a basis of $E$, which completes the proof.

The condition $\sup_{1 \leqslant n < \infty} \left\| \sum_{i=1}^{n} f_i(z) x_i \right\| < \infty$ in theorem 7.1 cannot be omitted, as shown by

*Example 7.1.* Let $E = c_0$ and let

$$x_{2n-1} = \frac{1}{2^{n-1}} e_{2n-1} - e_{2n}, \quad x_{2n} = e_{2n} \qquad (n = 1, 2, \ldots), \tag{7.9}$$

$$f_{2n-1} = 2^{n-1} h_{2n-1}, \quad f_{2n} = 2^{n-1} h_{2n-1} + h_{2n} \qquad (n = 1, 2, \ldots), \tag{7.10}$$

where $\{e_n\}$ is the unit vector basis of $E$ and $\{h_n\}$ the sequence of coordinate functionals on $E$. Then $(x_n, f_n)$ is a biorthogonal system with $\{f_n\}$ total on $E$ (also, $[x_n] = E$), and $\{x_n\}$ is not a basis of $E$, since $\inf_{1 \leqslant n < \infty} \|x_n\| = 1$, $\sup_{1 \leqslant n < \infty} \|f_n\| = \infty$ (see Ch. I, § 3, corollary 3.1 a)). However, for every $x \in E$ there exist a $\{\gamma_j\} \in bv_0$ and an element $z \in E$ such that we have (7.4). Indeed, let $x = \{\xi_n\} \in E$ be arbitrary. Then there exists an increasing sequence of positive integers $\{n_k\}$ such that

$$|\xi_i| < \frac{1}{k^2} \qquad (2n_{k-1} + 1 \leqslant i \leqslant 2n_k; \quad k = 2, 3, \ldots). \tag{7.11}$$

Put

$$\gamma_i = \frac{1}{k} \qquad (2n_{k-1}+1 \leqslant i \leqslant 2n_k;\ k=1,2,\ldots;\ n_0=0), \quad (7.12)$$

$$\zeta_i = k\xi_i \qquad (2n_{k-1}+1 \leqslant i \leqslant 2n_k;\ k=1,2,\ldots). \quad (7.13)$$

Then $\{\gamma_n\} \in bv_0$ and, by (7.11), we have $z = \{\zeta_n\} \in c_0 = E$. Furthermore, for each $k = 1,2,\ldots$ we have

$$f_{2n-1}(z) = 2^{n-1}h_{2n-1}(z) = 2^{n-1}\zeta_{2n-1} = k2^{n-1}\xi_{2n-1} = kf_{2n-1}(x)$$

$$(2n_{k-1}+1 \leqslant 2n-1 \leqslant 2n_k),$$

$$f_{2n}(z) = 2^{n-1}\zeta_{2n-1} + \zeta_{2n} = k2^{n-1}\xi_{2n-1} + k\xi_{2n} = kf_{2n}(x)$$

$$(2n_{k-1}+1 \leqslant 2n \leqslant 2n_k),$$

which proves (7.4). Let us also observe that, since $\{x_n\}$ is not a basis of $E$, there exists an element $x \in E$ such that $\sup_{1 \leqslant n < \infty} \left\| \sum_{i=1}^{n} f_i(x)x_i \right\| = \infty$ (by Ch. I, § 4, theorem 4.1) and then, by (7.8) for $p = 1$, we have $\sup_{1 \leqslant n < \infty} \left\| \sum_{j=1}^{n} f_j(z)x_j \right\| = \infty$.

One can make similar remarks for unconditional bases, replacing $bv_0$ by $c_0$ (see Ch. II, § 16, theorem 16.5) and the condition $\sup_{1 \leqslant n < \infty} \left\| \sum_{i=1}^{n} f_i(z)x_i \right\| < \infty$ by $\sup_{|\beta_1|,|\beta_2|,\ldots \leqslant 1} \sup_{1 \leqslant n < \infty} \left\| \sum_{i=1}^{n} \beta_i f_i(z)x_i \right\| < \infty$, which is equivalent to the condition that $\sum_{i=1}^{\infty} f_i(z)x_i$ be weakly unconditionally Cauchy (Ch. II, § 15, corollary 15.1). Indeed, then the necessity part of the unconditional analogue to theorem 7.1 is a consequence of the necessity part of theorem 7.1 (since $bv_0 \subset c_0$), while the sufficiency part follows by observing that $\sum_{i=1}^{\infty} f_i(x)x_i = \sum_{i=1}^{\infty} \gamma_i f_i(z)x_i$ is unconditionally norm convergent (by $\{\gamma_n\} \in c_0$ and Ch.I, § 17, corollary 17.5). A related result is known also in the theory of sequence spaces*).

---

*) See [122], theorem 3.

We shall give a characterization of unconditional convergence of biorthogonal expansions $\sum_{i=1}^{\infty} f_i(x)x_i$ with respect to generalized bases (and, in particular, of unconditional bases among generalized bases), in terms of multipliers, solving in the affirmative Ch. II, §16, problem 16.1 a) (and hence b)). For this purpose we shall need some preparation.

**Lemma 7.1.** *Let $A_1, \ldots, A_n$ be n subsets of a set $\mathscr{A}$, such that if $i > j$, then either $A_i \cap A_j = \emptyset$ or $A_i \subset A_j$. Then there exist n pairwise disjoint sets $A'_1, \ldots, A'_n \subset \mathscr{A}$ such that each $A'_i$ is contained in some $A_j$ and that $A_j = \bigcup_{A'_i \subset A_j} A'_i$ for all $j = 1, \ldots, n$.*

*Proof.* For $n = 1$ the lemma is obviously true. Assume now that the lemma is valid for all $k < n$ and let $A_1, \ldots, A_n \subset \mathscr{A}$ be as in the hypothesis. Then, by our induction assumption, there exist pairwise disjoint sets $A'_1, \ldots, A'_{n-1} \subset \mathscr{A}$ such that each $A'_i$ is contained in some $A_j \setminus A_n$ and that $A_j \setminus A_n = \bigcup_{A'_i \subset A_j \setminus A_n} A'_i$ for all $j = 1, \ldots, n-1$. Put $A'_n = A_n$ and let $n > j$. If $A_j \cap A_n = \emptyset$, then $A_j = A_j \setminus A_n = \bigcup_{A'_i \subset A_j \setminus A_n} A'_i = \bigcup_{A'_i \subset A_j} A'_i$. If $A_j \supset A_n$, then $A_j = A_n \cup (A_j \setminus A_n) = A'_n \cup \bigcup_{A'_i \subset A_j \setminus A_n} A'_i = \bigcup_{A'_k \subset A_j} A'_k$. Finally, each $A'_i$ is obviously contained in some $A_j$, which completes the proof.

**Lemma 7.2.** *Let $\{\mu_n\}$ be a sequence of bounded finitely additive set functions defined on a discrete set $\mathscr{A}$. If $\sup_{1 \leqslant n < \infty} |\mu_n(A)| < \infty$ for all $A \subset \mathscr{A}$, then*

$$\sup_{1 \leqslant n < \infty} \|\mu_n\| < \infty, \tag{7.14}$$

*where $\|\mu_n\|$ is the variation of $\mu_n$ on $\mathscr{A}$ (i.e., $\|\mu_n\| = \sup \sum_{i \in I} |\mu_n(A_i)|$, where the* sup *is taken over all finite families $(A_i)_{i \in I}$ of disjoint subsets of $\mathscr{A}$).*

*Proof.* Suppose that $\rho(A) = \sup_{1 \leqslant n < \infty} |\mu_n(A)| < \infty$ for all $A \subset \mathscr{A}$, but $\sup_{1 \leqslant n < \infty} \|\mu_n\| = \infty$. Then, since*) $\|\mu_n\| \leqslant 4 \sup_{A \subset \mathscr{A}} |\mu_n(A)|$ $(n = 1,2, \ldots)$, we have $\sup_{A \subset \mathscr{A}} \rho(A) = \infty$.

We claim that there exists a sequence $\{A_k\}$, $A_k \subset \mathscr{A}$, such that

$$\rho(A_k) > k \qquad (k = 1,2, \ldots), \tag{7.15}$$

$$A_k \cap A_m = \emptyset \qquad (k \neq m;\ k, m = 1,2, \ldots). \tag{7.16}$$

*) See e.g. [87], p. 97, lemma 5.

We shall prove this claim in several steps. First we shall prove that there exists a sequence $\{A_k\}$, $A_k \subset \mathscr{A}$, satisfying (7.15) and such that if $k > m$, then either $A_k \cap A_m = \emptyset$ or $A_k \subset A_m$. We shall construct such a sequence inductively. Let $A_1 \subset \mathscr{A}$ be such that $\rho(A_1) > 1$. Assume that $A_1, \ldots, A_k$ have been chosen such that $\rho(A_i) > i$ $(i = 1, \ldots, k)$ and such that for $k \geqslant i > j \geqslant 1$ we have either $A_i \cap A_j = \emptyset$ or $A_i \subset A_j$. Let $A \subset \mathscr{A}$ be such that $\rho(A) > (k+1)^2$. If $A \cap A_j = \emptyset$ $(j = 1,\ldots,k)$, let $A_{k+1} = A$. If not, then, by lemma 7.1, there exist $k$ pairwise disjoint $A'_1, \ldots, A'_k \subset \mathscr{A}$ such that each $A'_i$ is contained in some $A \cap A_j$ and that $A \cap A_j = \bigcup_{A'_i \subset A \cap A_j} A'_i$ for all $j = 1, \ldots, k$. Put $A'_{k+1} = A \setminus \bigcup_{i=1}^{k} A_i$.

Then $A = \bigcup_{i=1}^{k+1} A'_i$ and the $A'_i$ are pairwise disjoint, whence

$$(k+1)^2 < \rho(A) = \sup_{1 \leqslant n < \infty} |\mu_n(A)| =$$

$$= \sup_{1 \leqslant n < \infty} \left| \sum_{i=1}^{k+1} \mu_n(A'_i) \right| \leqslant \sup_{1 \leqslant n < \infty} \sum_{i=1}^{k+1} |\mu_n(A'_i)| \leqslant \sum_{i=1}^{k+1} \rho(A'_i),$$

and thus there exists an $A'_{i_0}$ such that $\rho(A'_{i_0}) > k+1$. Let $A_{k+1}$ be any such $A'_{i_0}$ and let $k+1 > j$. If $A_j \cap A_{k+1} = A_j \cap A'_{i_0} \neq \emptyset$, then*) $i_0 \leqslant k$ and $A_{k+1} = A'_{i_0} \subset A_j$. Thus, whenever $k+1 \geqslant i > j \geqslant 1$, we have either $A_i \cap A_j = \emptyset$ or $A_i \subset A_j$, completing the induction.

Now, by Ch. II, § 18, lemma 18.9 (with [0,1] replaced by $\mathscr{A}$), there exists a subsequence of $\{A_k\}$, which we shall denote again by $\{A_k\}$, such that we have either (7.16) or

$$A_1 \supset A_2 \supset A_3 \ldots \tag{7.17}$$

If we have (7.16), then the claim is proved. Assume now that we have (7.17). Let $n_1 = 1$ and by (7.15) choose $n_2, n_3, \ldots$ such that

$$\rho(A_{n_{k+1}}) \geqslant k + 1 + \rho(A_{n_k}) \qquad (k = 1,2,\ldots).$$

Then, putting

$$B_k = A_{n_k} \setminus A_{\ _{k+1}} \qquad (k = 1,2,\ldots), \tag{7.18}$$

we have $B_k \cap B_m = \emptyset$ for $k \neq m$ and since for every $k$ there exists a $j_0 = j_0(k)$ such that $|\mu_{j_0}(A_{n_{k+1}})| \geqslant \rho(A_{n_{k+1}}) - 1$, we have

$$\rho(B_k) = \rho(A_{n_k} \setminus A_{n_{k+1}}) \geqslant |\mu_{j_0}(A_{n_k} \setminus A_{n_{k+1}})| =$$

$$= |\mu_{j_0}(A_{n_k}) - \mu_{j_0}(A_{n_{k+1}})| \geqslant |\mu_{j_0}(A_{n_{k+1}})| - |\mu_{j_0}(A_{n_k})| \geqslant$$

$$\geqslant \rho(A_{n_{k+1}}) - 1 - |\mu_{j_0}(A_{n_k})| \geqslant k + 1 + \rho(A_{n_k}) - 1 -$$

$$- |\mu_{j_0}(A_{n_k})| \geqslant k,$$

*) Indeed, $A'_{i_0} \neq A'_{k+1}$, since $A_j \cap A'_{k+1} = \emptyset$.

and thus we have (7.15) and (7.16) for $B_k$ instead of $A_k$, which proves our claim.

Now, by (7.15), let $\{\nu_k\}$ be a subsequence of $\{\mu_k\}$, such that $\lim_{k\to\infty} |\nu_k(A_k)| = \infty$. For $P \subset \mathcal{N} = \{1, 2, 3, \ldots\}$, put

$$\lambda_n(P) = \frac{\nu_n(\bigcup_{k\in P} A_k)}{\nu_n(A_n)} \qquad (n = 1, 2, \ldots). \tag{7.19}$$

Then every $\lambda_n$ is a bounded finitely additive set function defined on the set $\mathcal{N}$ and for every set $P \subset \mathcal{N}$ we have $|\lambda_n(P)| \leqslant \frac{1}{|\nu_n(A_n)|} \rho(\bigcup_{k\in P} A_k)$, whence $\lim_{n\to\infty} \lambda_n(P) = 0$. Consequently, by a well-known lemma of Phillips*), we have $\lim_{n\to\infty} \lambda_n(\{n\}) = 0$. However, $\lambda_n(\{n\}) = \frac{1}{\nu_n(A_n)} \nu_n(A_n) = 1$ $(n = 1, 2, \ldots)$, a contradiction which completes the proof of lemma 7.2.

For $\mathcal{A} = \mathcal{N}$, one can also express lemma 7.2 in other terms, namely, as the following sharpening of the principle of uniform boundedness in the case of the space $E = m$ $(= l^\infty)$:

**Lemma 7.3.** *Let $\{g_n\} \subset m^*$ be a sequence of continuous linear functionals on the space $m$ such that $\sup_{1\leqslant n<\infty} |g_n(\{\alpha_j\})| < \infty$ for all sequences $\{\alpha_j\} \in m$ with $\alpha_j = 1$ or $0$ $(j = 1, 2, \ldots)$. Then*

$$\sup_{1\leqslant n<\infty} \|g_n\| < \infty. \tag{7.20}$$

*Proof.* The space $m^*$ is linearly isometric to**) the Banach space $ba(\mathcal{N})$ of all bounded finitely additive set functions on $\mathcal{N}$ (where the norm is defined to be the variation of the set function on $\mathcal{N}$), by the mapping $g \to \mu$, where

$$g(\{\alpha_j\}) = \int_{\mathcal{N}} \{\alpha_j\} d\mu(j) \qquad (\{\alpha_j\} \in m).$$

Thus, in particular, if $\alpha_j = 1$ or $0$ $(j = 1, 2, \ldots)$, then $g(\{\alpha_j\}) = \mu(A)$, where $A = \{j \in \mathcal{N} | \alpha_j = 1\}$, and conversely, given a set $A \subset \mathcal{N}$,

---

*) Let us recall this lemma of Phillips (see e.g. [70], Ch. II, § 2, lemma 1 or [317], p. 525): If $\{\lambda_n\}$ is a sequence of bounded finitely additive set functions on $\mathcal{N} = \{1,2,3, \ldots\}$, such that $\lim_{n\to\infty} \lambda_n(P) = 0$ for all $P \subset \mathcal{N}$, then $\lim_{n\to\infty} \sum_{k=1}^{\infty} |\lambda_n(\{k\})| = 0$ (hence, in particular, $\lim_{n\to\infty} \lambda_n(\{n\}) = 0$).

**) See e.g. [87], p. 296, theorem 16.

we have $\mu(A) = g(\{\alpha_j\})$, where $\alpha_j = 1$ for $j \in A$ and $\alpha_j = 0$ for $j \in \mathcal{N} \setminus A$. Consequently, lemma 7.2 for $\mathscr{A} = \mathcal{N}$ and lemma 7.3 are equivalent.

*Remark 7.3*. Obviously, the hypothesis of lemma 7.3 is equivalent to the following: $\sup_{1 \leqslant n < \infty} |g_n(\{\varepsilon_j\})| < \infty$ for all sequences $\{\varepsilon_j\} \in m$ with $\varepsilon_j = \pm 1$ $(j = 1, 2, \ldots)$.

**Lemma 7.4.** *Let $F$ be a Banach space and let $u: F \to m$ be a continuous linear mapping such that $u(F)$ contains all sequences $\{\alpha_j\} \in m$ with $\alpha_j = 1$ or $0$ $(j = 1, 2, \ldots)$ (or, equivalently, all $\{\varepsilon_j\} \in m$ with $\varepsilon_j = \pm 1$ for $j = 1, 2, \ldots$). Then $u(F) = m$.*

*Proof.* By our assumption, $u(F)$ is dense in $m$ and thus it is sufficient to prove that $u(F)$ is closed in $m$. If not, then $u^*(m^*)$ is not closed*) in $F^*$ and hence there exists **) a sequence $\{g_n\} \subset m^*$ such that $\lim_{n \to \infty} \|g_n\| = \infty$, $\|u^*(g_n)\| = 1$ $(n = 1, 2, \ldots)$.

Now let $\{\alpha_j\}$ be an arbitrary element in $m$ with $\alpha_j = 1$ or $0$ $(j = 1, 2, \ldots)$. Then, by our assumption, there exists an element $y_\alpha \in F$ such that $u(y_\alpha) = \{\alpha_j\}$, whence

$$\sup_{1 \leqslant n < \infty} |g_n(\{\alpha_j\})| = \sup_{1 \leqslant n < \infty} |g_n(u(y_\alpha))| = \sup_{1 \leqslant n < \infty} |u^*(g_n)(y_\alpha)| \leqslant \|y_\alpha\| < \infty.$$

Consequently, by lemma 7.3 we have $\sup_{1 \leqslant n < \infty} \|g_n\| < \infty$, a contradiction, which completes the proof of lemma 7.4.

Now we can prove

**Theorem 7.2.** *Let $E$ be a separable Banach space, $(x_n, f_n)$ ($\{x_n\} \subset E$, $\{f_n\} \subset E^*$) a biorthogonal system such that $\{f_n\}$ is total on $E$ (in other words, let $\{x_n\}$ be a generalized basis of a separable Banach space $E$, with an admissible sequence $\{f_n\} \subset E^*$) and let $x \in E$.*

a) *If $M(x, (x_n, f_n))$ contains all sequences $\{\alpha_j\}$ with $\alpha_j = 1$ or $0$ $(j = 1, 2, \ldots)$ (or, equivalently, all $\{\varepsilon_j\}$ with $\varepsilon_j = \pm 1$ for $j = 1, 2, \ldots$), then the series $\sum_{i=1}^{\infty} f_i(x) x_i$ is unconditionally convergent to $x$.*

b) *If $M(E, (x_n, f_n))$ contains all sequences $\{\alpha_j\}$ with $\alpha_j = 1$ or $0$ $(j = 1, 2, \ldots)$ (or, equivalently, all $\{\varepsilon_j\}$ with $\varepsilon_j = \pm 1$ for $j = 1, 2, \ldots$), then $\{x_n\}$ is an unconditional basis of $E$.*

*Proof.* a) Let $F = m \cap M(x, (x_n, f_n))$, endowed with the norm

$$\|\{\gamma_n\}\| = \sup_{1 \leqslant j < \infty} |\gamma_j| + \|x_{\{\gamma_n\}}\|, \tag{7.21}$$

---

*) See e.g. [87], p. 488, theorem 4.

**) See e.g. [87], p. 513, exercise 15 (ii); or, alternatively, observe that $u^*(m^*)$ is closed if and only if $u^*$ induces an isomorphism of $m^*/\mathrm{Ker}\, u^*$ into $F^*$.

where $x_{\{\gamma_n\}}$ is the unique element of $E$ for which $f_k(x_{\{\gamma_n\}}) = \gamma_k f_k(x)$ $(k = 1,2, \ldots)$. Then $F$ is a Banach space. Indeed, if $\{\gamma_n^{(p)}\}$ $(p = 1,2, \ldots)$ is a Cauchy sequence in $F$, then it is also a Cauchy sequence in $m$, and hence there exists a $\{\gamma_n\} \in m$ such that $\sup_{1 \leqslant n < \infty} |\gamma_n^{(p)} - \gamma_n| \to 0$ as $p \to \infty$; furthermore, $\|x_{\{\gamma_n^{(p)}\}} - x_{\{\gamma_n^{(q)}\}}\| = \|x_{\{\gamma_n^{(p)} - \gamma_n^{(q)}\}}\| \to 0$ as $p, q \to \infty$ and hence there exists an element $y \in E$ such that $\|x_{\{\gamma_n^{(p)}\}} - y\| \to 0$ as $p \to \infty$. Then, since

$$f_k(y) = \lim_{p\to\infty} f_k(x_{\{\gamma_n^{(p)}\}}) = \lim_{p\to\infty} \gamma_k^{(p)} f_k(x) = \gamma_k f_k(x) \qquad (k = 1, 2, \ldots),$$

we have $y = x_{\{\gamma_n\}}$, whence $\{\gamma_n\} \in M(x, (x_n, f_n))$. Obviously, $\|\{\gamma_n^{(p)}\} - \{\gamma_n\}\| = \|\{\gamma_n^{(p)} - \gamma_n\}\| = \sup_{1 \leqslant j < \infty} |\gamma_j^{(p)} - \gamma_j| + \|x_{\{\gamma_n^{(p)} - \gamma_n\}}\| \to 0$ as $p \to \infty$, which proves that $F$ is a Banach space.

Now, by (7.21) the inclusion map $u\colon F \to m$ is continuous and by our assumption $u(F)$ contains all sequences $\{\alpha_j\}$ with $\alpha_j = 1$ or $0$ $(j = 1,2, \ldots)$. Hence, by lemma 7.4, $m \cap M(x, (x_n, f_n)) = u(F) = m$ and thus $m \subset M(x, (x_n, f_n))$. Consequently, by Ch.II, §16, theorem 16.5 a), $\sum_{i=1}^{\infty} f_i(x)x_i$ is unconditionally convergent (obviously to $x$, since $\{f_n\}$ is total on $E$).

b) is an immediate consequence of a), which completes the proof of theorem 7.2.

Some assumption on $E$ in theorem 7.2, e.g. that $E$ is separable, is necessary, as shown by Ch.II, §16, example 16.7. Even in the particular case when $[x_n] = E$, the assumption that $\{f_n\}$ is total on $E$, is also necessary, as shown by

*Example* 7.2. Let $\{e_n\}$ be the unit vector basis of $E = l^2$ and let $\{h_n\} \subset E^*$, $h_i(e_j) = \delta_{ij}$ $(i, j = 1,2, \ldots)$. Furthermore, let $\{y_n\} \subset E$ be a sequence with the following properties:

(i) for each $n$ there are infinitely many indices $m_1^{(n)}, m_2^{(n)}, \ldots$ such that $y_{m_1^{(n)}} = y_{m_2^{(n)}} = \ldots = y_n$;

(ii) $y_{2j} = e_1 \qquad (j = 1, 2, \ldots)$;

(iii) $\{y_n\}_{n=1}^{\infty} = \{e_{2n-1}\}_{n=1}^{\infty}$.

Put

$$x_n = e_{2n} + y_n \qquad (n = 1, 2, \ldots), \tag{7.22}$$

$$f_n = h_{2n} \qquad (n = 1, 2, \ldots), \tag{7.23}$$

$$x = \sum_{n=1}^{\infty} \frac{1}{n} e_{4n}. \tag{7.24}$$

Then, by (iii), $(x_n, f_n)$ is a biorthogonal system, with $\{f_n\}$ not total on $E$. We claim that $[x_n] = E$. Indeed, let $f \in E^*$, $f(x_n) = 0$ $(n = 1, 2, \ldots \ldots)$. Then, by (i),

$$0 = f(x_{m_j^{(n)}}) = f(e_{2m_j^{(n)}}) + f(y_{m_j^{(n)}}) = f(e_{2m_j^{(n)}}) + f(y_n) \qquad (n, j = 1, 2, \ldots),$$

whence, since $\lim_{j\to\infty} f(e_{2m_j^{(n)}}) = 0$ $(n = 1, 2, \ldots)$, we obtain

$$-f(y_n) = f(e_{2m_1^{(n)}}) = f(e_{2m_2^{(n)}}) = \ldots = 0 \quad (n = 1, 2, \ldots).$$

Consequently, $f(e_{2n}) = f(e_{2n}) + f(y_n) = f(x_n) = 0$ $(n = 1, 2, \ldots)$ and, by (iii), also $f(e_{2n-1}) = 0$ $(n = 1, 2, \ldots)$, whence, since $[e_n] = E$, we get $f = 0$, which proves that $[x_n] = E$.

Furthermore, by (7.23) and since $\{e_n\}$ is an unconditional basis of $E$, for $x$ defined by (7.24) and for every $\{\gamma_n\} \in m$ there exists*) an element $x_{\{\gamma_n\}} \in E$ (which is not uniquely determined by $\{\gamma_n\}$) such that

$$f_k(x_{\{\gamma_n\}}) = h_{2k}(x_{\{\gamma_n\}}) = \gamma_k h_{2k}(x) = \gamma_k f_k(x) \quad (k = 1, 2, \ldots); \quad (7.25)$$

for example, one can take $x_{\{\gamma_n\}} = \sum_{n=1}^{\infty} \frac{1}{n} \gamma_{2n} e_{4n}$. However, $\sum_{i=1}^{\infty} f_i(x) x_i$ is not convergent, since by (7.22)—(7.24) and (ii) we have

$$\left\| \sum_{i=1}^{n} f_i(x) x_i \right\| = \left\| \sum_{i=1}^{n} h_{2i}(x)(e_{2i} + y_i) \right\| = \left\| \sum_{j=1}^{\left[\frac{n}{2}\right]} \frac{1}{j} (e_{4j} + y_{2j}) \right\| =$$

$$= \left\| \sum_{j=1}^{\left[\frac{n}{2}\right]} \frac{1}{j} (e_{4j} + e_1) \right\| = \left\{ \left( \sum_{j=1}^{\left[\frac{n}{2}\right]} \frac{1}{j} \right)^2 + \sum_{j=1}^{\left[\frac{n}{2}\right]} \frac{1}{j^2} \right\}^{\frac{1}{2}} \to \infty \text{ as } n \to \infty.$$

Since the above proof of theorem 7.2 makes use of powerful tools (lemma 7.4 and Ch. II, §16, theorem 16.5 a)), it will be of some interest to give here also a more elementary proof of theorem 7.2 b). As in Ch. I, § 5, proposition 5.4, for any $\{\gamma_n\} \in M(E, (x_n, f_n))$ we shall denote by $v_{\{\gamma_n\}}$ the continuous linear mapping of $E$ into $E$ defined by

$$v_{\{\gamma_n\}}(x) = x_{\{\gamma_n\}} \qquad (x \in E); \qquad (7.26)$$

---

*) Actually, for any $x = \sum_{n=1}^{\infty} \xi_n e_n \in l^2$ and $\{\gamma_n\} \in m$ the element $x_{\{\gamma_n\}} = \sum_{n=1}^{\infty} \gamma_{\left[\frac{n}{2}\right]} \xi_n e_n \in l^2$ satisfies (7.25). Thus, we have even $m \subset M(E, (x_n, f_n))$.

in particular, for $\gamma_1 = \ldots = \gamma_n = 1, \gamma_{n+1} = \gamma_{n+2} = \ldots = 0$ we have

$$v_{\{\gamma_n\}}(x) = s_n(x) = \sum_{i=1}^{n} f_i(x) x_i \qquad (x \in E).$$

If $A \subset \mathcal{N} = \{1, 2, 3, \ldots\}$ and $\{\alpha_n\} \in M(E, (x_n, f_n))$, where $\alpha_n = 1$ for $n \in A$ and $\alpha_n = 0$ for $n \in \mathcal{N} \setminus A$, then we shall also use the notation

$$v_A = v_{\{\alpha_n\}}. \tag{7.27}$$

We shall first give another proof for the following weaker result:

**Proposition 7.1.** *Let $E$ be a Banach space (not necessarily separable), and $(x_n, f_n)$ ($\{x_n\} \subset E, \{f_n\} \subset E^*$) a biorthogonal system such that $\{f_n\}$ is total on $E$ and that $M(E, (x_n, f_n))$ contains all sequences $\{\alpha_j\}$ with $\alpha_j = 1$ or $0$ ($j = 1, 2, \ldots$) (or, equivalently, all $\{\varepsilon_j\}$ with $\varepsilon_j = \pm 1$ for $j = 1, 2, \ldots$). Then*

$$\sup_{1 \leqslant n < \infty} \|s_n\| < \infty. \tag{7.28}$$

*Consequently, $\{x_n\}$ is an unconditional basic sequence*) (and hence, if $[x_n] = E$, then $\{x_n\}$ is an unconditional basis of $E$).*

*Proof.* Assume that $\sup_{1 \leqslant n < \infty} \|s_n\| = \infty$. Let $\{n_k\}$ be an increasing sequence of positive integers such that $\|s_{n_k}\| \geqslant k + \|s_{n_{k-1}}\|$ $\quad (k = 1, 2, \ldots;$ $n_0 = 0, s_0 = 0)$, whence

$$\lim_{k \to \infty} \|s_{n_k} - s_{n_{k-1}}\| = \infty. \tag{7.29}$$

Furthermore, let $\{M_p\}$ be an infinite sequence of pairwise disjoint infinite subsets of $\mathcal{N} = \{1, 2, 3, \ldots\}$ and let

$$I_k = \{n_{k-1} + 1, \ldots, n_k\} \quad (k = 1,2, \ldots; \; n_0 = 0), \tag{7.30}$$

$$A_p = \bigcup_{k \in M_p} I_k \qquad (p = 1, 2, \ldots), \tag{7.31}$$

$$F_p = [\{f_j\}_{j \in \mathcal{N} \setminus A_p}]_\perp = \{x \in E \mid f_j(x) = 0 \; (j \in \mathcal{N} \setminus A_p)\} \; (p=1,2, \ldots). \tag{7.32}$$

Then for every $k \in M_p$ and $x \in E$ we have

$$\|(s_{n_k} - s_{n_{k-1}})(x)\| = \left\| \sum_{i=n_{k-1}+1}^{n_k} f_i(x) x_i \right\| = \left\| \sum_{i=n_{k-1}+1}^{n_k} f_i(v_{A_p}(x)) x_i \right\| =$$

$$= \|(s_{n_k} - s_{n_{k-1}})(v_{A_p}(x))\| \leqslant \|(s_{n_k} - s_{n_{k-1}})|_{F_p}\| \, \|v_{A_p}\| \, \|x\|,$$

*) Note also that by § 1, theorem 1.10, implication $8° \Rightarrow 1°$, the a.s.c.f. $\{\varphi_n\} \subset [x_n]^*$ to $\{x_n\}$ satisfies $\{f_n\} \sim \{\varphi_n\}$.

whence, by (7.29), we infer

$$\sup_{k \in M_p} \| (s_{n_k} - s_{n_{k-1}}) |_{F_p} \| = \infty \qquad (p = 1, 2, \ldots). \tag{7.33}$$

By (7.33), choose $z_p \in F_p$ and $k_p \in M_p$ such that

$$\|z_p\| \leqslant \frac{1}{2^p} \qquad (p = 1, 2, \ldots), \tag{7.34}$$

$$\|(s_{n_{k_p}} - s_{n_{k_p-1}})(z_p)\| \geqslant 1 \qquad (p = 1, 2, \ldots), \tag{7.35}$$

and let

$$A = \bigcup_{p=1} I_{k_p}. \tag{7.36}$$

Since $A \cap A_p = \left(\bigcup_{j=1}^{\infty} I_{k_j}\right) \cap \left(\bigcup_{k \in M_p} I_k\right) = I_{k_p}$ (because by $M_j \cap M_p = \varnothing$ for $j \neq p$ we have $k_j \notin M_p$ for $j \neq p$), for every $y_p \in F_p$ we have

$$f_i(v_A(y_p)) = f_i((s_{n_{k_p}} - s_{n_{k_p-1}})(y_p)) \qquad (i = 1, 2, \ldots),$$

whence, since $\{f_n\}$ is total on $E$,

$$v_A(y_p) = (s_{n_{k_p}} - s_{n_{k_p-1}})(y_p).$$

Thus, by (7.34) and (7.35), the sequence $\left\{\sum_{p=1}^{l} z_p\right\}$ converges, but the sequence $\left\{v_A\left(\sum_{p=1}^{l} z_p\right)\right\} = \left\{\sum_{p=1}^{l} v_A(z_p)\right\} = \left\{\sum_{p=1}^{l} (s_{n_{k_p}} - s_{n_{k_p-1}})(z_p)\right\}$ does not converge, contradicting Ch. I, § 5, proposition 5.4, that $v_A$ is continuous. This proves (7.28). Hence, by Ch. I, § 4 theorem 4.1, $\{x_n\}$ is a basic sequence. Since the same argument remains valid for every permutation $\{x_{\sigma(n)}\}$ of $\{x_n\}$, it follows that $\{x_n\}$ is an unconditional basic sequence, which completes the proof.

*Remark 7.4.* One can give a much simpler proof of the fact that under the hypotheses of proposition 7.1 we have

$$\sup_{1 \leqslant n < \infty} \|s_n|_{[x_j]}\| < \infty, \tag{7.37}$$

whence $\{x_n\}$ is an unconditional basic sequence (and, if $[x_n] = E$, then $\{x_n\}$ is an unconditional basis of $E$). Indeed, if (7.37) does not hold, then, as observed also in Ch. I, § 5, proof of theorem 5.1, there exist increasing sequences of positive integers $\{k_p\}, \{n_p\}$ with $k_{p-1}+1 \leqslant \leqslant n_p \leqslant k_p$ $(p=1,2,\ldots; k_0=0)$ and a sequence $\{z_p\} \subset E$ such that

$$z_p \in [x_{k_{p-1}+1}, \ldots, x_{k_p}] \qquad (p=1,2,\ldots; k_0=0), \tag{7.38}$$

$$\|z_p\| \leqslant \frac{1}{2^p} \qquad (p=1,2,\ldots), \tag{7.39}$$

$$\|s_{n_p}(z_p)\| \geqslant 1 \qquad (p=1,2,\ldots). \tag{7.40}$$

Then the sequence $\left\{\sum_{p=1}^{l} z_p\right\}$ is convergent, but for

$$A = \{1, \ldots, n_1, k_1+1, \ldots, n_2, k_2+1, \ldots, n_3, \ldots\} \tag{7.41}$$

the sequence $\left\{v_A\left(\sum_{p=1}^{l} z_p\right)\right\} = \left\{\sum_{p=1}^{l} s_{n_p}(z_p)\right\}$ is not convergent, contradicting Ch. I, § 5 proposition 5.4, that $v_A$ is continuous. This completes the proof of (7.37).

Now we can give the

*Second proof of theorem 7.2* b). Assume that $\{x_n\}$ is not a basis of $E$. Then there exists an element $x \in E$ such that $\{s_n(x)\}$ does not converge to $x$ and hence $\{s_n(x)\}$ does not converge at all (if it converges, its limit must be $x$, because $\{f_n\}$ is total on $E$). Consequently, there exist two sequences of positive integers $\{m_k\}, \{n_k\}$ with $m_k+1 \leqslant n_k \leqslant m_{k+1}$ $(k=1,2,\ldots)$ and an $\varepsilon > 0$ such that

$$\|(s_{n_k} - s_{m_k})(x)\| > \varepsilon \qquad (k=1,2,\ldots). \tag{7.42}$$

Put

$$z_k = (s_{n_k} - s_{m_k})(x) = \sum_{i=m_k+1}^{n_k} f_i(x)x_i \qquad (k=1,2,\ldots). \tag{7.43}$$

Since by our hypothesis every sequence $\alpha_j$ with $\alpha_j = 1$ or $0$ $(j = = 1,2,\ldots)$ is in $M(E, (x_n, f_n)) \subset M(x, (x_n, f_n))$, for every such sequence there exists a unique element $y \in E$ such that

$$f_j(y) = \begin{cases} \alpha_k f_j(x) & \text{for } m_k+1 \leqslant j \leqslant n_k \quad (k=1,2,\ldots) \\ 0 & \text{for } n_{k-1} \leqslant j \leqslant m_k \quad (k=1,2,\ldots; n_0=1), \end{cases} \tag{7.44}$$

i.e. such that $y \sim \sum_{k=1}^{\infty} \alpha_k \sum_{j=m_k+1}^{n_k} f_j(x)x_j = \sum_{k=1}^{\infty} \alpha_k z_k$ (Ch. I, § 4, definition 4.3). Observe now that the correspondence $\{\alpha_n\} \rightarrow y$ is one-to-one, since for each $k$ there exists a $j_0 = j_0(k)$ with $m_k + 1 \leqslant j_0 \leqslant n_k$ such that $f_{j_0}(x) \neq 0$ (by (7.43) and $\|z_k\| > \varepsilon$). Hence, since the set $\{\{\alpha_j\} | \alpha_j = 1 \text{ or } 0 \ (j = 1,2, \ldots \ldots)\}$ is uncountable and since by our hypothesis $E$ is separable, there exists*) a sequence $\{y_p\}_0^{\infty} \subset E$ with $y_p \sim \sum_{k=1}^{\infty} \alpha_k^{(p)} z_k$ $(p = 0,1,2, \ldots)$, $\alpha_k^{(p)} = 1$ or $0$, such that $y_p \neq y_q$ for all $p \neq q$ and that $y_p \rightarrow y_0 \sim \sim \sum_{k=1}^{\infty} \alpha_k^{(0)} z_k$ as $p \rightarrow \infty$. By proposition 7.1, let $M = \sup_{1 \leqslant k < \infty} \|s_{n_k} - s_{m_k}\|$. Then there exists a positive integer $p_0 = p_0(\varepsilon)$ such that

$$\|(s_{n_k} - s_{m_k})(y_p - y_0)\| \leqslant M \|y_p - y_0\| < \varepsilon$$

$$(p > p_0(\varepsilon); \ k = 1, 2, \ldots). \tag{7.45}$$

However, by (7.44) we have

$$(s_{n_k} - s_{m_k})(y_p - y_0) = (\alpha_k^{(p)} - \alpha_k^{(0)}) z_k \quad (p, k = 1, 2, \ldots), \tag{7.46}$$

whence, since $\alpha_k^{(p)} = 1$ or $0$ $(k = 1, 2, \ldots; p = 0, 1, 2, \ldots)$,

$$\|(s_{n_k} - s_{m_k})(y_p - y_0)\| = \begin{cases} 0 & \text{if } \alpha_k^{(p)} = \alpha_k^{(0)} \\ |\alpha_k^{(p)} - \alpha_k^{(0)}| \, \|z_k\| = \|z_k\| & \text{if } \alpha_k^{(p)} \neq \alpha_k^{(0)}. \end{cases} \tag{7.47}$$

Since $y_p \neq y_0$ for all $p \neq 0$, there exists for each $p$ an index $k_p$ such that $\alpha_{k_p}^{(p)} \neq \alpha_{k_p}^{(0)}$. Consequently, by (7.47) and (7.42),

$$\|(s_{n_{k_p}} - s_{m_{k_p}})(y_p - y_0)\| = \|z_{k_p}\| > \varepsilon \qquad (p = 1, 2, \ldots),$$

in contradiction with (7.45). This proves that $\{x_n\}$ is a basis of $E$. Since the same argument remains valid for every permutation $\{x_{\sigma(n)}\}$ of $\{x_n\}$, it follows that $\{x_n\}$ is an unconditional basis of $E$, which completes the proof.

*Remark 7.5.* In the particular case when the linear subspace $V_0$ of $E^*$ spanned by $\{f_n\}$ is of characteristic $r(V_0) > 0$, theorem 7.2 reduces

---

*) Indeed, it is well known (see e.g. [226], Ch. II, § 18, Section III) that if $E$ is a separable metric space, then every uncountable set $A \subset E$ has at least one condensation point (i.e. a point $y_0 \in A$ such that in every neighbourhood of $y_0$ there exists an uncountable number of points of $A$); actually, the set of all points of $A$ which are not condensation points of $A$ is at most countable.

to Ch. II, § 16, theorem 16.6. The above methods yield also the following simpler proof of part a) (hence also b)) of that theorem: By $r(V_0) > 0$ and Ch. II, § 16, lemma 16.3, there exists *) an equivalent norm $|||x|||$ on $E$, with the property

$$(\mathrm{K}_1)\ \textit{If}\ \lim_{n\to\infty} f_k(y_n) = f_k(y_0)\ (k = 1, 2, \ldots),\ \textit{then}\ \underline{\lim}_{n\to\infty} |||y_n||| \geqslant |||y_0|||.$$

Let $x \in E$ be such that $M(x, (x_n, f_n))$ contains all sequences $\{\alpha_j\}$ with $\alpha_j = 1$ or $0$ $(j = 1, 2, \ldots)$. As above, it is sufficient to show that $\{s_n(x)\}$ converges. If not, then there exist two sequences of positive integers $\{m_k\}$, $\{n_k\}$ with $m_k + 1 \leqslant n_k \leqslant m_{k+1}$ $(k = 1, 2, \ldots)$ and an $\varepsilon > 0$ such that we have (7.42). By $(\mathrm{K}_1)$ and Ch. II, § 16, lemma 16.4, there exist subsequences of $\{m_k\}$ and $\{n_k\}$, which we shall denote again by $\{m_k\}$ and $\{n_k\}$ respectively, such that for each $k$ the natural projection $u_k$ of $[x_1, \ldots, x_{n_k}] \oplus [f_1, \ldots, f_{m_{k+1}}]_\perp$ onto $[x_1, \ldots, x_{n_k}]$ is of norm **) $|||u_k||| \leqslant 1 + \dfrac{\varepsilon}{1-\varepsilon}$. Observe now that for every sequence $\{\alpha_j\}$ with $\alpha_j = 1$ or $0$ $(j = 1,2, \ldots)$ the unique element $y \in E$ defined by (7.44) is in each of the subspaces $[x_1, \ldots, x_{n_k}] \oplus [f_1, \ldots, f_{m_{k+1}}]_\perp$, since for each $k$ we can write $y = y_k^{(1)} + y_k^{(2)}$, where $y_k^{(1)} = u_k(y) = \sum_{i=1}^{k} \alpha_i \sum_{j=m_i+1}^{n_i} f_j(x)x_j \in$ $\in [x_1, \ldots, x_{n_k}]$ and $y_k^{(2)} = y - y_k^{(1)} \in [f_1, \ldots, f_{m_{k+1}}]_\perp$ (because $y_k^{(2)} \sim$ $\sim \sum_{i=k+1}^{\infty} \alpha_i \sum_{j=m_i+1}^{n_i} f_j(x)x_j$); furthermore, $(u_k - u_{k-1})(y) = \alpha_k \sum_{j=m_k+1}^{n_k} f_j(x)x_j$. Therefore we can apply the same argument as in the above proof of theorem 7.2 b), with $s_{n_k} - s_{m_k}$ replaced by $u_k - u_{k-1}$, and thus we arrive at a contradiction, which completes the proof.

In Ch. II, § 16, theorem 16.8 we have seen that a basis $\{x_n\}$ of a Banach space $E$ is an unconditional basis of $E$ if and only if for every increasing sequence of indices $\{i_n\}$ the subspaces $[x_{i_n}]$ and $[x_j]_{j \in \mathcal{N} \setminus \{i_n\}}$ are complementary to each other (i.e., $E = [x_{i_n}] \oplus [x_j]_{j \in \mathcal{N} \setminus \{i_n\}}$). Now we are able to sharpen this result, weakening considerably the assumption that $\{x_n\}$ is a basis of $E$. Namely, we have

---

*) This can be seen also in a much simpler way, by observing that if $r(V_0) > 0$, then the norm $|||x||| = \|x\|_{V_0} = \sup_{\substack{f \in V_0 \\ \|f\| \leqslant 1}} |f(x)|$ $(x \in E)$ has the required properties. We observe that the converse is also true, since $(\mathrm{K}_1)$ implies that $r_{|||\cdot|||}(V_0) = 1$ (indeed, since $V_0$ is separable, $S_{(E, |||\cdot|||)}$ endowed with $\sigma(E, V_0)$ is metrizable and hence, by $(\mathrm{K}_1)$, closed, so Ch. I, § 12, formula (12.15) applies), whence $r(V_0) > 0$ in the initial norm.

**) Actually, the same proof works if we assume, instead of $r(V_0) > 0$, only the following: There exists a constant $C \geqslant 1$ such that for any integer $n$ there is an integer $m \geqslant n$ with the property that the natural projection of $[x_1, \ldots, x_n] \oplus [f_1, \ldots, f_m]_\perp$ onto $[x_1, \ldots, x_n]$ has norm $\leqslant C$.

**Theorem 7.3.** *A complete sequence $\{x_n\}$ in a Banach space E, such that $x_n \neq 0$ $(n = 1, 2, \ldots)$, is an unconditional basis of E if and only if for every increasing sequence of indices $\{i_n\}$ the subspaces $[x_{i_n}]$ and $[x_j]_{j \in \mathcal{N} \setminus \{i_n\}}$ are complementary to each other (i.e., $E = [x_{i_n}] \oplus [x_j]_{j \in \mathcal{N} \setminus \{i_n\}}$).*

*Proof.* The necessity part is a consequence of Ch. II, § 16, theorem 16.8.

Conversely, assume now that the condition is satisfied. Then, taking $\{i_n\}$ to be the finite sets $\{1, \ldots, n\}$ $(n = 1, 2, \ldots)$, from Ch. I, § 6, theorem 6.1 (implication $6° \Rightarrow 1°$) it follows that $\{x_n\}$ is a minimal sequence. By our assumption, for every set $A \subset \mathcal{N} = \{1, 2, 3, \ldots\}$ there exists a projection $v_A$ of $E$ onto $[x_i]_{i \in A}$ along $[x_j]_{j \in \mathcal{N} \setminus A}$ (i.e., such that $v_A(x_j) = 0$ for all $j \in \mathcal{N} \setminus A$). Observe now that for any finite set $I$ of indices and any scalars $\alpha_i$ $(i \in I)$ we have $\sum_{i \in I \cap A} \alpha_i x_i \in [x_i]_{i \in A}$ and $\sum_{i \in I \cap (\mathcal{N} \setminus A)} \alpha_i x_i \in [x_j]_{j \in \mathcal{N} \setminus A}$, whence $v_A(\sum_{i \in I} \alpha_i x_i) = \sum_{i \in I \cap A} \alpha_i x_i$. Consequently, applying to these projections $v_A$ the argument used in remark 7.4, it follows that $\{x_n\}$ is an unconditional basis of $E$, which completes the proof of theorem 7.3.

Concerning the existence of generalized bases, we have

**Theorem 7.4.** *A Banach space E has a generalized basis if and only if there exists a total sequence of functionals on E (or, equivalently, $E^*$ is $w^*$-separable). Hence, in particular, every separable Banach space E has a generalized basis.*

*Proof.* If $\{x_n\}$ is a generalized basis of $E$, with an admissible sequence $\{f_n\} \subset E^*$, then, by definition 7.1, $\{f_n\}$ is a total sequence of functionals on $E$.

Conversely, let $\{g_n\} \subset E^*$ be a total sequence of functionals on $E$; we may assume, without loss of generality, that $\{g_n\}$ is finitely linearly independent. Then there exists, for each $n$, an element $x_n \in E$ such that $g_i(x_n) = 0$ $(i = 1, \ldots, n-1)$, $g_n(x_n) = 1$; indeed, if $g_i(x) = 0$ $(i = 1, \ldots, n-1)$ would imply $g_n(x) = 0$, then we would have*) $g_n \in [g_1, \ldots, g_{n-1}]$, contradicting our assumption. We claim that $\{x_n\}$ is a generalized basis of $E$. Indeed, the sequence $\{f_n\} \subset E^*$ defined by

$$f_1 = g_1, \quad f_n = g_n - \sum_{i=1}^{n-1} g_n(x_i) f_i \qquad (n = 2, 3, \ldots) \tag{7.48}$$

is total on $E$ (since $f_n(x) = 0$ for $n = 1, 2, \ldots$ implies $g_n(x) = 0$ for $n = 1, 2, \ldots$, whence $x = 0$) and $(x_n, f_n)$ is a biorthogonal system. Since $\{g_n\}$ is total on $E$ if and only if its finite linear combinations with rational coefficients are $w^*$-dense in $E^*$ and since the conjugate space of every

---

*) See e.g. [87], p. 421, lemma 10.

separable Banach space is $w^*$-separable*), the proof of theorem 7.4 is complete.

Let us give now some properties of generalized bases. To every biorthogonal system one can associate a generalized basis in a natural way, as shown by

**Proposition 7.2.** *Let $E$ be a Banach space, let $(x_n, f_n)$ ($\{x_n\} \subset E$, $\{f_n\} \subset E^*$) be a biorthogonal system and let $\omega$ be the quotient map from $E$ onto $E/[f_n]_\perp$. Then $(\omega(x_n), \psi_n)$ is a generalized basis of $E/[f_n]_\perp$, where $\{\psi_n\}$ is the image of $\{f_n\}$ under the canonical linear isometry $([f_n]_\perp)^\perp \equiv \equiv (E/[f_n]_\perp)^*$, i.e.*

$$\psi_n(\omega(x)) = f_n(x) \qquad (x \in E, n = 1, 2, \ldots). \tag{7.49}$$

*Proof.* By (7.49) we have $\psi_i(\omega(x_j)) = f_i(x_j) = \delta_{ij}$ $(i, j = 1,2, \ldots)$, i.e. $(\omega(x_n), \psi_n)$ is**) a biorthogonal system. If $\psi_n(\omega(x)) = 0$ $(n = 1,2,\ldots)$ for some $x \in E$, then, by (7.49), $f_n(x) = 0$ $(n = 1,2, \ldots)$, i.e. $x \in [f_n]_\perp$, whence $\omega(x) = 0$, which completes the proof.

From § 6, proposition 6.4 it follows that for a generalized basis $\{x_n\}$ of a Banach space $E$ the admissible sequence $\{f_n\} \subset E^*$ (or, equivalently, the admissible mapping $u: x \to \{f_n(x)\}$ of $E$ into $s$) is uniquely determined if and only if $[x_n] = E$.

If $\{x_n\}$ is a generalized basis of $E$ such that $[x_n] \neq E$, then for every admissible mapping $u: E \to s$ we have $u([x_n]) \neq u(E)$, since $u$ is one-to-one.

We have observed in § 6 (after definition 6.10) that if $\{x_n\}$ is a generalized basis of $E$ and $\{f_n\} \subset E^*$ an admissible sequence, then $(x_n, f_n)$ is a maximal biorthogonal system, but the converse is not true.

*Definition 7.2.* A sequence $\{y_n\}$ in a Banach space $E$ is called a *generalized basic sequence* if $\{y_n\}$ is a generalized basis of a subspace $G$ of $E$, containing $[y_n]$ (where $[y_n] \neq G$ is also possible).

As in § 6 for minimal sequences, one may consider the problem of extension of a generalized basic sequence $\{y_n\} \subset E$ to a generalized basis of the whole space $E$. It is easy to see that if $E$ is separable, the answer is affirmative; indeed, if $\{y_n\} \subset E$ is a generalized basic sequence, then $\{y_n\}$ is an $M$-basis***) of $[y_n]$ in the sense of § 8, definition 8.1, and hence, by § 8, theorem 8.2, it can be extended to an $M$-basis of $E$, which is obviously a generalized basis of $E$, extending $\{y_n\}$. Moreover, we shall prove now the stronger result that if $\{y_n\}$ is infinite, the sequence $\{y_n\}$ itself is a generalized basis of the whole space $E$, since the functionals $\varphi_n \in G^*$ admit extensions $f_n \in E^*$ such that $\{f_n\}$ is total on $E$. For this purpose, let us first give

---

*) See e.g. [11], p. 124, theorem 4.

**) For a similar observation, see also § 6, the remark made after definition 6.8.

***) For, if $\{y_n\}$ is a generalized basis of $G \subset E$ and $\{\varphi_n\} \subset G^*$ an admissible sequence for $\{y_n\}$, then $\{\varphi_n|_{[y_j]}\}$, the a.s.f. to $\{y_n\}$, is total on $[y_n]$.

**Lemma 7.5.** *Let $E$ be a normed linear space and $G$ an infinite dimensional subspace of $E$ such that $G^*$ contains a total sequence $\{\varphi_n\}$ and $(E/G)^*$ contains a total sequence $\{\psi_n\}$. Then there exists a total sequence $\{f_n\} \subset E^*$ such that*

$$f_n|_G = \varphi_n \qquad (n = 1, 2, \ldots). \tag{7.50}$$

*Proof.* We may assume, without loss of generality, that $\varphi_n \neq 0$ $(n = 1,2, \ldots)$. Then, by § 6, lemma 6.1, there exists a sequence $\{f_n\} \subset E^*$ satisfying (7.50) and $[f_n]_\perp \subset G$. Now let $x \in E, f_n(x) = 0$ $(n = 1, 2, \ldots)$. Then $x \in [f_n]_\perp \subset G$ and $\varphi_n(x) = f_n(x) = 0$ $(n = 1,2, \ldots)$, whence, since $\{\varphi_n\}$ is total on $G$, we infer $x = 0$. Thus, $\{f_n\}$ is total on $E$, which completes the proof.

**Corollary 7.1.** *Under the hypotheses of lemma 7.5, for every sequence $\{g_n\} \subset E^*$ satisfying $g_n|_G = \varphi_n$ $(n = 1, 2, \ldots)$ there exists a sequence $\{h_n\} \subset G^\perp$ such that the sequence $\{g_n\} \cup \{h_n\}$ is total on $E$.*

*Proof.* If $\{f_n\} \subset E^*$ is as in lemma 7.5, then the sequence

$$h_n = g_n - f_n \qquad (n = 1, 2, \ldots) \tag{7.51}$$

has the required properties. Note that the image $\{h_n\}$ of $\{\psi_n\}$ by the canonical linear isometry $(E/G)^* \equiv G^\perp$, i.e.

$$h_n(x) = \psi_n(x + G) \qquad (x \in E, n = 1, 2, \ldots) \tag{7.52}$$

also has these properties. Indeed, if $g_n(x) = h_n(x) = 0$ $(n = 1, 2, \ldots)$, then, by (7.52) and since $\{\psi_n\}$ is total on $E/G$, we have $x \in G$, whence, by $\varphi_n(x) = g_n(x) = 0$ $(n = 1, 2, \ldots)$ and since $\{\varphi_n\}$ is total on $G$, it follows that $x = 0$.

From lemma 7.5 we infer

**Theorem 7.5.** *Let $E$ be a separable Banach space. Then every infinite generalized basic sequence $\{y_n\} \subset E$ (hence, in particular, every infinite basic sequence) is a generalized basis of the space $E$.*

*Proof.* Since $E$ is separable, so is $E/G$ and hence $(E/G)^*$ contains*) a total sequence $\{\psi_n\}$. The conclusion follows now from lemma 7.5.

*Remark 7.6.* The above method cannot be applied if we assume only that $E^*$ contains a total sequence, since in this case there may exist subspaces $G$ of $E$ for which the condition of lemma 7.5 is not satisfied. For example, although $m^*$ contains a total sequence, we shall prove now that $(m/c_0)^*$ contains no total sequence. Let us first recall

---

*) See e.g. [11], p. 124, theorem 4.

**Lemma 7.6.** *Let $I$ be a countable set. Then there exists a family $\{A_d\}_{d\in D}$ of subsets of $I$ with the following properties:*
(i) *The index set $D$ is uncountable.*
(ii) *Each set $A_d$ is infinite.*
(iii) *$A_{d_1} \cap A_{d_2}$ is finite for $d_1 \neq d_2$.*

*Proof.* Take $I$ to be the set of all rational numbers in (0, 1), $D$ the set of all irrational numbers in (0, 1) and, for each $d \in D$, let $A_d$ be an arbitrary infinite sequence in $I$ converging to $d$.

Now we can prove

**Lemma 7.7.** *The conjugate space $(m/c_0)^*$ contains no total sequence.*

*Proof.* Let us consider $m$ as $m(I)$, the Banach space of all bounded functions on a countable set $I$, with the sup norm. Let $\{A_d\}_{d\in D}$ be a family of subsets of $I$ as in lemma 7.6 and let $Y_d$ be the coset in $m/c_0$ which contains the characteristic function $\chi_{A_d}$ of the set $A_d$.

We claim that for each $\psi \in (m/c_0)^*$ the set $\{Y_d | \psi(Y_d) \neq 0\}$ is at most countable. Indeed, it is sufficient to show that for each $\psi \in (m/c_0)^*$ and each $n$ the set $B(\psi, n) = \left\{ Y_d \,\middle|\, |\psi(Y_d)| \geqslant \frac{1}{n} \right\}$ is finite. Let $Y_1, \ldots, Y_k \in B(\psi, n)$ be arbitrary and let $\beta_i = \operatorname{sign} \psi(Y_i) = \frac{\overline{\psi(Y_i)}}{\psi(Y_i)}$ $(i = 1, \ldots, k)$. Then $\left\| \sum_{i=1}^{k} \beta_i Y_i \right\|_{m/c_0} \leqslant 1$; for, if $Y_i = Y_{d_i} \ni \chi_{A_{d_i}}$ $(i = 1, \ldots, k)$, then by (ii), (iii), the disjoint sets $B_1 = A_{d_1}$, $B_i = A_{d_i} \setminus \bigcup_{j=1}^{i-1} A_{d_j}$ $(i = 2, \ldots, k)$ are nonvoid and by (iii) each $A_{d_i} \setminus B_i = \bigcup_{j=1}^{i-1} (A_{d_i} \cap A_{d_j})$ is finite, so $\chi_{B_i} - \chi_{A_{d_i}} \in c_0$, whence $\left\| \sum_{i=1}^{k} \beta_i Y_i \right\|_{m/c_0} = \left\| \sum_{i=1}^{k} \beta_i \chi_{A_{d_i}} + c_0 \right\|_{m/c_0} = \left\| \sum_{i=1}^{k} \beta_i \chi_{B_i} + c_0 \right\|_{m/c_0} \leqslant \left\| \sum_{i=1}^{k} \beta_i \chi_{B_i} \right\|_{m} = 1$. Consequently,

$$\|\psi\| \geqslant \left| \psi\left( \sum_{i=1}^{k} \beta_i Y_i \right) \right| = \sum_{i=1}^{k} |\psi(Y_i)| \geqslant \frac{k}{n},$$

whence $B(\psi, n)$ is finite, which proves the claim.

Now let $\{\psi_n\}$ be an arbitrary sequence in $(m/c_0)^*$. Then, by the above, the set $\bigcup_{n=1}^{\infty} \{Y_d | \psi_n(Y_d) \neq 0\}$ is at most countable and hence, by (i), there exists a $d_0 \in D$ such that $Y_{d_0} \neq 0$, $\psi_n(Y_{d_0}) = 0$ $(n = 1, 2, \ldots)$.

Thus, $\{\psi_n\}$ is not total on $m/c_0$, which completes the proof of lemma 7.7 and of remark 7.6.

Since by definition two bases $\{x_n\} \subset E$, $\{y_n\} \subset F$ are said to be equivalent if

$$\{\{f_n(x)\} | x \in E\} = \left\{\{\alpha_n\} \subset K \,\middle|\, \sum_{i=1}^{\infty} \alpha_i x_i \text{ converges in } E\right\} =$$

$$= \left\{\{\alpha_n\} \subset K \,\middle|\, \sum_{i=1}^{\infty} \alpha_i y_i \text{ converges in } F\right\} = \{\{g_n(y)\} | y \in F\}$$

(where $\{f_n\} \subset E^*$, $\{g_n\} \subset F^*$, $f_i(x_j) = g_i(y_j) = \delta_{ij}$), it is natural to give the following extension of this concept to biorthogonal systems and generalized bases:

*Definition 7.3.* Let $E, F$ be two Banach spaces and let $(x_n, f_n)$ ($\{x_n\} \subset E$, $\{f_n\} \subset E^*$) and $(y_n, g_n)$ ($\{y_n\} \subset F$, $\{g_n\} \subset F^*$) be biorthogonal systems. The systems $(x_n, f_n)$ and $(y_n, g_n)$ are said to be *similar* if

$$\{\{f_n(x)\} | x \in E\} = \{\{g_n(y)\} | y \in F\}. \tag{7.53}$$

A generalized basis $\{x_n\}$ of a Banach space $E$, with an admissible sequence $\{f_n\} \subset E^*$, is said to be *similar* to a generalized basis $\{y_n\}$ of a Banach space $F$, with an admissible sequence $\{g_n\} \subset F^*$, if the biorthogonal systems $(x_n, f_n)$ and $(y_n, g_n)$ are similar.

The condition of being similar is quite restrictive, namely, we have*)

**Theorem 7.6.** *If two generalized bases $\{x_n\}, \{y_n\}$ of Banach spaces $E$ and $F$, respectively, with admissible sequences $\{f_n\} \subset E^*$ and $\{g_n\} \subset F^*$, respectively, are similar, then $\{x_n\} \approx \{y_n\}$, i.e.***) *there exists an isomorphism $w$ of $E$ onto $F$ such that*

$$w(x_n) = y_n \qquad (n = 1,2, \ldots). \tag{7.54}$$

*Proof.* Assume that $\{x_n\}$ and $\{y_n\}$ are similar, i.e. (7.53). Introducing on the two sequence spaces occurring in (7.53) norms as in remark 7.1, they become $BK$-spaces linearly isometric to $E$ and $F$ respectively, by the mappings $v_E: x \to \{f_n(x)\}$ and $v_F: y \to \{g_n(y)\}$, respectively. By the argument used in Ch. I, § 12, proof of formula (12.66), the identical mapping $I$: $\{\{f_n(x)\} | x \in E\} \to \{\{g_n(y)\} | y \in F\}$ is an isomorphism between these $BK$-spaces. Consequently, $w = v_F^{-1} I v_E$ is

---

*) In the particular case when $\{x_n\}, \{y_n\}$ are bases of $E$ and $F$, respectively, theorem 7.6 amounts to Ch. I, § 8, theorem 8.1 d), equivalence 5° ⇔ 1°, and thus it may be regarded as an extension of this latter result to generalized bases.

**) See Ch. I, § 8, definition 8.1.

an isomorphism of $E$ onto $F$, satisfying (7.54) (because $v_E(x_n) = e_n = v_F(y_n)$ for $n = 1, 2, \ldots$), which completes the proof.

From Ch. I, § 8, theorem 8.1 d) (equivalence 5° $\Leftrightarrow$ 1°) it follows that if $\{x_n\} \subset E$, $\{y_n\} \subset F$ are generalized bases of $E$ and $F$, with admissible sequences $\{f_n\} \subset E^*$, $\{g_n\} \subset F^*$, then $[x_n]$ as a generalized basis of $[x_n]$ and $\{y_n\}$ as a generalized basis of $[y_n]$, with admissible sequences $\{f_n|_{[x_j]}\}$ and $\{g_n|_{[y_j]}\}$ respectively, are similar if and only if $\{x_n\} \approx \{y_n\}$, i.e. if and only if there exists an isomorphism $w$ *of* $[x_n]$ *onto* $[y_n]$ satisfying (7.54). Hence, in particular, if the condition of theorem 7.6 is satisfied, we have $\{\{f_n(x)\}|x \in [x_n]\} = \{\{g_n(y)\}|y \in [y_n]\}$. However, we cannot make the stronger assertion that the converse of theorem 7.6 is true, as shown by

*Example 7.3.* Let $E = F = c$, $\{x_n\} = \{e_n\}$ = the sequence of unit vectors in $E$, $\{h_n\} \subset E^*$ the sequence of coordinate functionals on $E$ and

$$f_{2k-1}(x) = \xi_{2k-1} - \lim_{n\to\infty} \xi_n \qquad (x = \{\xi_n\} \in c;\ k = 1, 2, \ldots), \tag{7.55}$$

$$f_{2k} = h_{2k} \qquad (k = 1, 2, \ldots). \tag{7.56}$$

Then $\{x_n\}$ and $\{e_n\}$, with the admissible sequences $\{f_n\}$ and $\{h_n\}$, respectively, are generalized bases of $c$ and $\{x_n\} \approx \{e_n\}$, since the identity mapping $w = I_c$ of $c$ onto $c$ is an isomorphism satisfying (7.54), but $\{\{h_n(x)\}|\ x \in c\} = c$ and $\{\{f_n(x)\}|\ x \in c\} = \{x = \{\xi_n\} \in s|\ \lim_{k\to\infty} \xi_{2k-1} = 0$ and $\lim_{k\to\infty} \xi_{2k}$ exists$\} \neq c$.

In theorem 7.6 the assumption that $\{f_n\} \subset E^*$ and $\{g_n\} \subset F^*$ are total, is essential, as shown by

*Example 7.4.* Let $E = l^2$, $\{e_n\}$ the unit vector basis of $E$, $\{h_n\} \subset E^*$ the sequence of coordinate functionals on $E$ and $(x_n, f_n)$ the biorthogonal system considered in § 6, example 6.7 a), hence $[x_n] = E$ and $\{f_n\}$ is not total on $E$. Then $(e_n, h_n)$ and $(x_n, f_n)$ are similar biorthogonal systems, since $\{\{f_n(x)\} = \{\xi_{n+1}\}|x = \{\xi_n\} \in E\} = l^2 = \{\{h_n(x)\}|\ x \in E\}$, but we do not have $\{e_n\} \approx \{x_n\}$, since $\{e_n\}$ is a basis of $E$, while $\{x_n\}$ is not (because $\{f_n\}$ is not total on $E$).

For the general case of similar biorthogonal systems, from theorem 7.6 and proposition 7.2 we infer

**Proposition 7.3.** *Let $E$, $F$ be two Banach spaces, let $(x_n, f_n)$ ($\{x_n\} \subset E$, $\{f_n\} \subset E^*$) and $(y_n, g_n)$ ($\{y_n\} \subset F$, $\{g_n\} \subset F^*$) be similar biorthogonal systems and let $\omega_1\colon E \to E/[f_n]_\perp$ and $\omega_2\colon F \to F/[g_n]_\perp$ be the quotient maps. Then there exists an isomorphism $w$ of $E/[f_n]_\perp$ onto $F/[g_n]_\perp$ such that*

$$w(\omega_1(x_n)) = \omega_2(y_n) \qquad (n = 1,2, \ldots). \tag{7.57}$$

Dually, for two biorthogonal systems $(x_n, f_n)$ ($\{x_n\} \subset E$, $\{f_n\} \subset E^*$) and $(y_n, g_n)$ ($\{y_n\} \subset F$, $\{g_n\} \subset F^*$) or two minimal sequences $\{x_n\} \subset E$, $\{y_n\} \subset F$, one can introduce the notion of **-similarity* by the condition

$$\{\{f(x_n)\} | f \in E^*\} = \{\{g(y_n)\} | g \in F^*\}, \tag{7.58}$$

but this does not lead to a new concept, since by the Hahn-Banach theorem we have (7.58) if and only if $\{\{\varphi(x_n)\} | \varphi \in [x_n]^*\} = \{\{\psi(y_n)\} | \psi \in \in [y_n]^*\}$, which, by Ch. I, § 8, theorem 8.1 d) (equivalence $3° \Leftrightarrow 1°$), happens if and only if $\{x_n\} \approx \{y_n\}$.

The equivalences $1° \Leftrightarrow 3°$, $1° \Leftrightarrow 4°$ of Ch. I, § 12, theorem 12.7 (on infinite matrices which preserve bases) suggest to give

*Definition 7.4*. Let $E, F$ be two Banach spaces and let $(x_n, f_n)$ ($\{x_n\} \subset E$, $\{f_n\} \subset E^*$) and $(y_n, g_n)$ ($\{y_n\} \subset F$, $\{g_n\} \subset F^*$) be biorthogonal systems. The systems $(x_n, f_n)$ and $(y_n, g_n)$ are said to be *quasi-similar* if either there exists a matrix $a = (a_{ij})$ such that

$$v_a(\{\alpha_n\}) = \left\{\sum_{j=1}^{\infty} a_{ij}\alpha_j\right\}_{i=1}^{\infty} \tag{7.59}$$

defines a one-to-one mapping of $\{\{f_n(x)\} | x \in E\}$ onto $\{\{g_n(y)\} | y \in F\}$, or there exists a matrix $b = (b_{ij})$ such that $v_b$ defines a one-to-one mapping of $\{\{g_n(y)\} | y \in F\}$ onto $\{\{f_n(x)\} | x \in E\}$. A generalized basis $\{x_n\}$ of a Banach space $E$, with an admissible sequence $\{f_n\} \subset E^*$, is said to be *quasi-similar* to a generalized basis $\{y_n\}$ of a Banach space $F$, with an admissible sequence $\{g_n\} \subset F^*$, if the biorthogonal systems $(x_n, f_n)$ and $(y_n, g_n)$ are quasi-similar.

Replacing in the above proof of theorem 7.6 the identity operator $I$ by $v_a$ or $v_b$, it follows that *if two generalized bases $\{x_n\}$, $\{y_n\}$ of Banach spaces $E$ and $F$ respectively, with admissible sequences $\{f_n\} \subset E^*$ and $\{g_n\} \subset F^*$, respectively, are quasi-similar, then there exists an isomorphism $w$ of $E$ onto $F$.* In this case $w$ need not satisfy (7.54); precisely, if $v_a$ is one-to-one and onto, then for each $n = 1,2, \ldots$ we have $w(x_n) = v_F^{-1}(v_a(e_n)) = = v_F^{-1}(\{a_{1n}, a_{2n}, \ldots\})$ (= the unique element $y \in F$ such that $g_j(y) = a_{jn}$ for $j = 1,2, \ldots$). A partial converse is also true, namely, *if there exists an isomorphism $w$ of $E$ onto $F$ and if $\{x_n\}$ is a basis of $E$ with the a.s.c.f. $\{f_n\} \subset E^*$ and $\{y_n\}$ a generalized basis of $F$ with an admissible sequence $\{g_n\} \subset F^*$, then they are quasi-similar* (thus, quasi-similarity does not preserve completeness of sequences). Indeed, the mapping $v_F w v_E^{-1}$ is an isomorphism of $\{\{f_n(x)\} | x \in E\}$ onto $\{\{g_n(y)\} | y \in F\}$ and

$$v_F w v_E^{-1}(\{f_n(x)\}) = \{g_n(w v_E^{-1}(\{f_n(x)\}))\} =$$

$$= \left\{g_n\left(w\left(\sum_{k=1}^{\infty} f_k(x) x_k\right)\right)\right\} = \left\{\sum_{k=1}^{\infty} g_n(w(x_k)) f_k(x)\right\} = v_a(\{f_n(x)\}) \quad (x \in E),$$

where $a = (g_n(w(x_k)))_{n,k=1,2,\ldots}$.

One can introduce some special classes of generalized bases $\{x_n\}$, by using the admissible sequence $\{f_n\} \subset E^*$ instead of the a.s.f. $\{\varphi_n\} \subset \subset [x_n]^*$.

*Definition* 7.5. A generalized basis $\{x_n\}$ of a Banach space $E$ is said to be

a) *strictly bounded*, if $\sup_{1 \leqslant n < \infty} \|x_n\| < \infty$ and if $\{x_n\}$ has an admissible sequence $\{f_n\} \subset E^*$ such that $\sup_{1 \leqslant n < \infty} \|f_n\| < \infty$;

b) *normal*, if $\|x_n\| = 1$ $(n = 1,2, \dots)$ and if $\{x_n\}$ has an admissible sequence $\{f_n\} \subset E^*$ such that $\|f_n\| = 1$ $(n = 1,2, \dots)$.

Dually to § 6, theorem 6.3, we have the following result on the existence of strictly bounded (even "almost normal") generalized bases:

**Theorem 7.7.** *For every separable Banach space E and every* $\varepsilon > 0$ *there exists**) *a biorthogonal system* $(x_n, f_n)$ $(\{x_n\} \subset E, \{f_n\} \subset E^*)$, *with* $\{f_n\}$ *total on E, such that*

$$\|x_n\| = 1, \quad \|f_n\| \leqslant 1 + \varepsilon \quad (n = 1,2, \dots). \tag{7.60}$$

*Proof.* If $\dim E = \infty$, then by § 6, proposition 6.6, there exists an infinite biorthogonal system $(x_n, g_n)$ $(\{x_n\} \subset E, \{g_n\} \subset E^*)$ such that**)

(a) $\|x_n\| = \|g_n\| = g_n(x_n) = 1 \quad (n = 1,2, \dots)$;

(b) $\{x_n\}$ is a basic sequence;

(c) $[x_n] + [g_n]_\perp$ is dense in $E$.

By (c) and $\|g_n\| = 1$ $(n = 1,2, \dots)$ we have $g_n \xrightarrow{w^*} 0$ (since $g_n|_{[x_j]} \xrightarrow{w^*} 0$ by § 6, proposition 6.5). Since $E$ is separable, so is $E/[x_n]$, and hence there exists***) a sequence $\{\psi_n\} \subset (E/[x_n])^*$ with $\|\psi_n\| = 1$ $(n = 1, 2, \dots)$, which is total on $E/[x_n]$. Let $\{h_n\} \subset [x_n]^\perp$ be the image of $\{\psi_n\}$ by the canonical linear isometry $(E/[x_n])^* \equiv [x_n]^\perp$, i.e.

$$h_n(x) = \psi_n(x + [x_j]) \quad (x \in E, \ n = 1,2, \dots), \tag{7.61}$$

and let $\{n_i^{(k)}\}_{i=1}^\infty$ $(k = 1,2, \dots)$ be an infinite partition of $\mathcal{N} = \{1,2,3, \dots\}$ into infinite subsequences. Put

$$f_{n_i^{(k)}} = g_{n_i^{(k)}} - \varepsilon h_k \quad (i, k = 1,2, \dots), \tag{7.62}$$

*) As we already mentioned in § 6, there exists even an $M$-basis satisfying (7.60) (see the Notes and remarks to § 8).

**) Note that this proof does not make use of (iii) of § 6, proposition 6.6 (and hence of the theorem of Dvoretzky).

***) See e.g. [11], p. 124, theorem 4.

where $\varepsilon$ is any positive number, given in advance. Then $(x_{n_i^{(k)}}, f_{n_i^{(k)}})$ is a biorthogonal system satisfying (7.60), and thus it remains only to prove that $\{f_n\}$ is total on $E$. Let $x \in E$, $f_{n_i^{(k)}}(x) = 0$ $(i, k = 1,2, \dots)$. Then, by (7.62),

$$\varepsilon h_k(x) = g_{n_i^{(k)}}(x) \qquad (i, k = 1,2, \dots),$$

whence, by $g_n \xrightarrow{w^*} 0$ we obtain (taking $i \to \infty$) that

$$h_k(x) = 0, \ g_{n_i^{(k)}}(x) = 0 \qquad (i, k = 1,2, \dots).$$

Therefore $\psi_n(x + [x_j]) = 0$ $(n = 1,2, \dots)$, whence, since $\{\psi_n\}$ is total on $E/[x_j]$, it follows that $x \in [x_j]$. Consequently, by (b), we obtain $x = \sum_{i=1}^{\infty} g_i(x) x_i = 0$, which completes the proof of theorem 7.7 (If $\dim E < \infty$, the theorem is true by Ch. II, § 2, theorem 2.1).

Using a different method, one can also prove another result, which implies again the existence of strictly bounded generalized bases (with slightly weaker bounds on $\sup_{1 \leqslant n < \infty} \|x_n\|$, $\sup_{1 \leqslant n < \infty} \|f_n\|$ than in (7.60)). To this end, let us first give the following sharpening of lemma 7.5:

**Lemma 7.8.** *Let $E$ be a separable normed linear space, $G$ an infinite dimensional subspace of $E$ and $\{\varphi_n\} \subset G^*$ a total sequence of functionals on $G$, such that $\varphi_n \xrightarrow{w^*} 0$. Then there exists a total sequence $\{f_n\} \subset E^*$ such that*

$$f_n|_G = \varphi_n \quad (n = 1,2, \dots), \tag{7.63}$$

$$\sup_{1 \leqslant n < \infty} \|f_n\| < \infty. \tag{7.64}$$

*Proof.* Let $\{g_n\} \subset E^*$, $g_n|_G = \varphi_n$, $\|g_n\| = \|\varphi_n\|$ $(n = 1,2, \dots)$, and let $\{n_i^{(k)}\}_{i=1}^{\infty}$ $(k = 1,2, \dots)$ be an infinite partition of $\mathcal{N} = \{1,2,3, \dots\}$ into infinite subsequences. Then, since $\sup_{1 \leqslant n < \infty} \|g_n\| = \sup_{1 \leqslant n < \infty} \|\varphi_n\| < \infty$ and since $E$ is separable, for each $k = 1,2, \dots$ there exists a subsequence $\{g_{n_{i_m}^{(k)}}\}_{i=1}^{\infty}$ of $\{g_{n_i^{(k)}}\}_{i=1}^{\infty}$, $w^*$-converging to some $\bar{g}_k \in E^*$; obviously,

$$\|\bar{g}_k\| \leqslant \sup_{1 \leqslant i < \infty} \|g_{n_i^{(k)}}\| \leqslant \sup_{1 \leqslant n < \infty} \|g_n\| = \sup_{1 \leqslant n < \infty} \|\varphi_n\| \quad (k = 1,2, \dots). \tag{7.65}$$

Furthermore, since $E$ is separable, so is $E/G$, and therefore $(E/G)^*$ contains a total sequence*) $\{\psi_n\}$; we may assume (considering, if necessary,

---

*) If $\dim E/G < \infty$, then only a finite number of the $\psi_n$'s will be linearly independent.

$\left\{\frac{\varepsilon\psi_n}{\|\psi_n\|}\right\}$ instead of $\{\psi_n\}$) that

$$\sup_{1 \leqslant n < \infty} \|\psi_n\| \leqslant \varepsilon, \tag{7.66}$$

where $\varepsilon$ is any positive number, given in advance. Let $\{h_n\} \subset G^\perp$ be the image of $\{\psi_n\}$ by the canonical linear isometry $(E/G)^* \equiv G^\perp$ (see (7.52)). Put

$$f_{n_i^{(k)}} = g_{n_i^{(k)}} - \bar{g}_k + h_k \qquad (i, k = 1,2, \ldots). \tag{7.67}$$

Then, since $g_n|_G = \varphi_n$ $(n = 1,2, \ldots)$ and $\varphi_n \xrightarrow{w^*} 0$, we have

$$\bar{g}_k(x) = \lim_{m\to\infty} g_{n_{i_m(k)}^{(k)}}(x) = \lim_{m\to\infty} \varphi_{n_{i_m(k)}^{(k)}}(x) = 0 \qquad (x \in G,\ k = 1, 2, \ldots),$$

that is, $\bar{g}_k \in G^\perp$ $(k = 1,2, \ldots)$, whence, since $g_n|_G = \varphi_n$ and $h_n \in G^\perp$ $(n = 1,2, \ldots)$, we obtain

$$f_{n_i^{(k)}}|_G = (g_{n_i^{(k)}} - \bar{g}_k + h_k)|_G = \varphi_{n_i^{(k)}} \qquad (i, k = 1,2, \ldots),$$

i.e. (7.63). Furthermore, by (7.67), (7.65) and (7.66) we have

$$\sup_{1\leqslant n<\infty} \|f_n\| \leqslant 2 \sup_{1\leqslant n<\infty} \|g_n\| + \varepsilon = 2 \sup_{1\leqslant n<\infty} \|\varphi_n\| + \varepsilon < \infty, \tag{7.68}$$

whence (7.64).

Finally, to prove that $\{f_n\}$ is total on $E$, let $x \in E$, $f_n(x) = 0$ $(n = 1, 2, \ldots)$. Then, by (7.67),

$$g_{n_i^{(k)}}(x) - \bar{g}_k(x) = -h_k(x) \qquad (i, k = 1,2, \ldots),$$

whence

$$0 = \lim_{m\to\infty} (g_{n_{i_m(k)}^{(k)}}(x) - \bar{g}_k(x)) = -h_k(x) \qquad (k = 1,2, \ldots),$$

and therefore, by (7.52) and since $\{\psi_n\}$ is total on $E/G$, we infer $x \in G$. Consequently, by (7.63), $\varphi_n(x) = f_n(x) = 0$ $(n = 1,2, \ldots)$, whence, since $\{\varphi_n\}$ is total on $G$, it follows that $x = 0$, which completes the proof of lemma 7.8.

*Remark 7.6.* By slightly changing the above proof, one can also show that the condition $\varphi_n \xrightarrow{w^*} 0$ in lemma 7.8 can be replaced by the weaker condition $\sup_{1\leqslant n<\infty} \|\varphi_n\| < \infty$ and that the $f_n$ can be chosen so as to satisfy $\sup_{1\leqslant n<\infty} \|f_n\| \leqslant \sup_{1\leqslant n<\infty} \|\varphi_n\| + \varepsilon$. Indeed, to this end it is clearly enough to replace (7.66) by $\sup_{1\leqslant n<\infty} \|\psi_n\| \leqslant \frac{\varepsilon}{2}$ and then (7.67) by

$$f_n = \begin{cases} g_n \text{ for } n \in \mathcal{N} \setminus \{n_{i_m(k)}^{(k)}\}_{k,\, m=1}^{\infty} \\ g_n + h_k \text{ for } n \in \{n_{i_{2m}(k)}^{(k)}\}_{m=1}^{\infty}\,;\ k = 1,2, \ldots \\ g_n + 2h_k \text{ for } n \in \{n_{i_{2m-1}(k)}^{(k)}\}_{m=1}^{\infty};\ k=1,2, \ldots \end{cases} \tag{7.69}$$

However, lemma 7.8 itself is already sufficient to prove the following sharpening of theorem 7.5, which implies again the existence of strictly bounded generalized bases:

**Theorem 7.8.** *Every infinite strictly bounded generalized basic sequence $\{x_n\}$ in a separable Banach space $E$ is a strictly bounded generalized basis of $E$.*

*Proof.* Let $\{\varphi_n\} \subset [x_n]^*$ be the a.s.f. to $\{x_n\}$. Then, by our assumption*), $\sup_{1\leqslant n<\infty} \|\varphi_n\| < \infty$ and $\{\varphi_n\}$ is total on $[x_n]$ and, by § 6, proposition 6.5, we have $\varphi_n \xrightarrow{w^*} 0$. Consequently, by lemma 7.8, there exists a total sequence $\{f_n\} \subset E^*$ satisfying (7.63) and (7.64), whence also $f_i(x_j) = \delta_{ij}$ $(i, j = 1,2, \ldots)$, which completes the proof.

Dually to § 6, problem 6.3, we have

*Problem 7.1.* Does every separable Banach space $E$ have a normal generalized basis?

Let us pass now to other special classes of generalized bases.

*Definition 7.6.* A generalized basis $\{x_n\}$ of a Banach space $E$ is said to be *shrinking* if $\{x_n\}$ has an admissible sequence $\{f_n\} \subset E^*$ such that $[f_n] = E^*$.

By the Hahn-Banach theorem, *every shrinking generalized basis is a shrinking minimal sequence* in the sense of § 6, definition 6.5 (that is, $[f_n] = E^*$ implies $[\varphi_n] = [f_n|_{[x_j]}] = [x_n]^*$), but the converse is not true, as shown e.g. by the sequence of unit vectors $\{x_n\}$ in $E = m$. Obviously, if $[x_n] = E$, then the two notions coincide.

*Definition 7.7.* A generalized basis $\{x_n\}$ of a Banach space $E$ is said to be *quasi-boundedly complete* if $\{x_n\}$ has an admissible sequence $\{f_n\} \subset E^*$

---

*) See the remarks made before lemma 7.5.

such that for every sequence of scalars $\{\alpha_n\} \subset K$ with the property $\sup_{1 \leqslant n < \infty} \left\| \sum_{i=1}^{n} \alpha_i x_i \right\| < \infty$ there exists an element $x \in E$ such that

$$f_n(x) = \alpha_n \quad (n = 1,2, \ldots). \tag{7.70}$$

Since $\{f_n\}$ is total on $E$, the element $x$ is unique.

Obviously, *every generalized basis* $\{x_n\}$ *which is a quasi-boundedly complete minimal sequence in the sense of* § 6, *definition* 6.6, *is also a quasi-boundedly complete generalized basis*, but the converse is not true, as shown by the sequence of unit vectors $\{x_n\}$ in $E = m$, with $\{f_n\} =$ the sequence of coordinate functionals on $E$. Indeed, for every $\{\alpha_n\} \subset K$ with $\sup_{1 \leqslant n < \infty} \left\| \sum_{i=1}^{n} \alpha_i x_i \right\| < \infty$, the element $x = \{\alpha_n\} \in m$ satisfies (7.70), but, by § 6, remark 6.4, $\{x_n\}$ is not a quasi-boundedly complete minimal sequence. Clearly, if $[x_n] = E$, then the two notions coincide.

Let us also give another generalization of boundedly completeness:

*Definition 7.8.* A generalized basis $\{x_n\}$ of a Banach space $E$ is said to be *boundedly complete* if $\{x_n\}$ has an admissible sequence $\{f_n\} \subset E^*$ such that every bounded $\sigma(E, [f_n])$-Cauchy*) sequence $\{z_j\} \subset [x_n]$ is $\sigma(E, [f_n])$-convergent to **) an element $x$ *of* $[x_n]$.

**Theorem 7.9.** *Let* $\{x_n\}$ *be a generalized basis of a Banach space E. The following statements are equivalent:*

1°. $\{x_n\}$ *is boundedly complete.*

2°. $\{x_n\}$ *has an admissible sequence* $\{f_n\} \subset E^*$ *such that for every bounded sequence* $\{z_j\} \subset [x_n]$ *for which the limits*

$$\lim_{j \to \infty} f_n(z_j) = \alpha_n \qquad (n = 1,2, \ldots) \tag{7.71}$$

*exist, there is an element* $x \in [x_n]$ *satisfying*

$$f_n(x) = \alpha_n \quad (n = 1,2, \ldots). \tag{7.72}$$

3°. $[x_n]$ *is canonically isomorphic to* $[\varphi_n]^*$, *where* $\{\varphi_n\} \subset [x_n]^*$ *is the a.s.f. to* $\{x_n\}$.

*Proof.* Assume 1° and let $\{z_j\} \subset [x_n]$ be a bounded sequence such that the limits (7.71) exist. Then $\{z_j\}$ is also $\sigma(E, [f_n])$-Cauchy and hence, by 1°, $\{z_j\}$ is $\sigma(E, [f_n])$-convergent to an element $x \in [x_n]$, which obviously satisfies (7.72). Thus, 1° $\Rightarrow$ 2°.

---

*) I.e., such that $\lim_{j \to \infty} f(z)$ exists for each $f \in [f_n]$.

**) Since $\{f_n\}$ is total on $E$, the $\sigma(E, [f_n])$-limit is unique.

Conversely, assume 2° and let $\{z_j\} \subset [x_n]$ be a bounded $\sigma(E, [f_n])$-Cauchy sequence. Then $\{z_j\}$ satisfies (7.71) with suitable scalars $\alpha_n$ and hence, by 2°, there exists an element $x \in [x_n]$ satisfying $f_n(x) = \alpha_n = \lim_{j\to\infty} f_n(z_j)$ $(n = 1,2, \ldots)$. Hence, since $\sup_{1 \leq j < \infty} \|z_j\| < \infty$, it follows that $f(x) = \lim_{j\to\infty} f(z_j)$ for all $f \in [f_n]$. Thus, 2° ⇒ 1°.

Assume now that we have 3° and let $\{z_j\} \subset [x_n]$ be a bounded sequence satisfying (7.71), where $\{f_n\} \subset E^*$ is any admissible sequence for $\{x_n\}$. Then, with the same argument as in § 6, proof of proposition 6.7 $\left(\text{replacing } \sum_{i=1}^{n} \alpha_i x_i \text{ by } z_j\right)$, it follows that there exists an element $x \in [x_n]$ satisfying (7.72). Thus, 3° ⇒ 2°.

Finally, assume that we have 2° and let $\Psi \in [\varphi_n]^*$ be arbitrary. Then for any scalars $\beta_1, \ldots, \beta_k$ we have $\left\|\sum_{i=1}^{k} \beta_i \Psi(\varphi_i)\right\| \leqslant \|\Psi\| \left\|\sum_{i=1}^{k} \beta_i \varphi_i\right\|$ and hence, by a classical theorem of Helly*), for each $j$ there exists an element $z_j \in [x_n]$ such that

$$\Psi(\varphi_n) = \varphi_n(z_j) = f_n(z_j) \qquad (n = 1, \ldots, j), \tag{7.73}$$

$$\|z_j\| \leqslant \|\Psi\| + 1. \tag{7.74}$$

Thus, $\{z_j\} \subset [x_n]$ is a bounded sequence satisfying (7.71) with $\alpha_n = \Psi(\varphi_n)$ $(n = 1,2, \ldots)$ and therefore, by 2°, there exists an element $x \in [x_n]$ satisfying

$$\Psi(\varphi_n) = \alpha_n = f_n(x) = \varphi_n(x) \qquad (n = 1,2, \ldots),$$

whence also

$$\Psi(\varphi) = \varphi(x) = u(x)(\varphi) \qquad (\varphi \in [\varphi_n]),$$

where $u\colon [x_n] \to [\varphi_n]^*$ is the canonical mapping. This proves that $u$ maps $[x_n]$ *onto* $[\varphi_n]^*$ and hence, by the inversion theorem of Banach**), $u$ is an isomorphism. Thus, 2° ⇒ 3°, which completes the proof of theorem 7.9.

Some more characterizations of boundedly completeness will be given in § 8, theorem 8.6 (see also § 8, remark 8.7).

Note that by the above proof of theorem 7.9, in definition 7.8 and in 2° of theorem 7.9 the words "$\{x_n\}$ has an admissible sequence $\{f_n\} \subset E^*$ such that..." can be replaced by "if $\{f_n\} \subset E^*$ is any admissible sequence for $\{x_n\}$, then..."

---

*) See e.g. [70], Ch. II, § 4, theorem 3.

**) See e.g. [11], p. 41, theorem 5.

**Corollary 7.2.** a) *If a generalized basis* $\{x_n\}$ *of a Banach space E is boundedly complete, then* $\{x_n\}$ *is a quasi-boundedly complete minimal sequence (and hence also a quasi-boundedly complete generalized basis).*

b) *If* $\{x_n\}$ *is a basic sequence in E, and hence*) a generalized basis of E, then* $\{x_n\}$ *is boundedly complete in the sense of definition* 7.8 *if and only if it is a boundedly complete basic sequence.*

*Proof.* a) Let $\{\alpha_n\} \subset K$, $\sup\limits_{1 \leqslant n < \infty} \left\| \sum\limits_{i=1}^{n} \alpha_i x_i \right\| < \infty$. Then, since $\lim\limits_{j \to \infty} f_n\left(\sum\limits_{i=1}^{j} \alpha_i x_i\right) = \alpha_n$ $(n = 1,2, \ldots)$, by the implication $1° \Rightarrow 2°$ of theorem 7.9 there exists an element $x \in [x_n]$ satisfying (7.72).

b) The necessity part results from part a) and § 6, remark 6.4.

The sufficiency part is an immediate consequence of the implication $3° \Rightarrow 1°$ of theorem 7.9 and of Ch. II, § 6, theorem 6.2, implication $1° \Rightarrow 3°$ (applied to $[x_n]$), but can be proved also more directly, as follows: Let $\{z_j\} \subset [x_n]$ be a sequence with $\sup\limits_{1 \leqslant j < \infty} \|z_j\| = M < \infty$, satisfying (7.71), whence $\sum\limits_{k=1}^{n} \alpha_k x_k = \lim\limits_{j \to \infty} \sum\limits_{k=1}^{n} f_k(z_j) x_k$ $(n = 1,2, \ldots)$. Since $\{x_n\}$ is a basic sequence, we have $\left\| \sum\limits_{i=1}^{n} f_k(z_j) x_k \right\| \leqslant v_{\{x_k\}} \|z_j\| \leqslant v_{\{x_k\}} M$ $(j, n = 1,2, \ldots)$, and hence $\sup\limits_{1 \leqslant n < \infty} \left\| \sum\limits_{k=1}^{n} \alpha_k x_k \right\| < \infty$. Therefore, since $\{x_n\}$ is a boundedly complete basic sequence, the series $\sum\limits_{i=1}^{\infty} \alpha_i x_i$ converges to an element $x \in [x_n]$, satisfying, obviously, (7.72). Thus, we have 2° of theorem 7.9, whence $\{x_n\}$ is boundedly complete in the sense of definition 7.8, which completes the proof.

The converse of corollary 7.2 a) is not valid, as shown by theorem 7.9 and § 6, example 6.5. Let us also mention that the unit vectors $\{x_n\}$ in the space $E = m$, with the admissible sequence $\{f_n\}$ = the coordinate functionals on $E$, constitute a quasi-boundedly complete generalized basis of $E$ (but not a quasi-boundedly complete minimal sequence), which is not boundedly complete, since e.g. the sequence $z_j = \underbrace{\{1, \ldots, 1}_{j},0,0, \ldots\} = \sum\limits_{k=1}^{j} x_k \in [x_n]$ $(j = 1,2, \ldots)$ is bounded and $\sigma(E, [f_n])$-convergent to $\{1,1,1, \ldots\} \notin [x_n]$; the fact that $\{x_n\}$ is not boundedly complete follows also from corollary 7.2 b), since $\{x_n\}$ is a non-boundedly complete basic sequence.

---

*) By virtue of theorem 7.5.

## § 8. $M$-bases. Strong $M$-bases. Series summable $M$-bases

*Definition 8.1.* A sequence $\{x_n\}$ in a Banach space $E$ is called an *M-basis* (or *Markuševič basis*) of $E$, if it is complete in $E$ and a generalized basis of $E$, i.e., if $[x_n] = E$ and there exists a total sequence of functionals $\{f_n\} \subset E^*$ such that $(x_n, f_n)$ is a biorthogonal system.

*Remark 8.1.* One can also define, in the same way, the notion of $M$-basis of an arbitrary normed linear space $E$. In general, *the results of this section* (and of § 6, § 7) *remain valid for this case, since their proofs do not use the completeness of $E$.* Moreover, the existence of $M$-bases in separable normed linear spaces (and some other results) can be deduced from the corresponding Banach space result, by applying this latter to the completion $E^\wedge$ of $E$ and then using the Krein-Milman-Rutman stability theorem for $M$-bases in Banach spaces (see the remark made after example 8.3). However, since the framework for the present monograph is that of Banach spaces (this was partially motivated by some remarks made in Ch. I, § 16), we shall restrict ourselves to $M$-bases in Banach spaces.

In the sequence space terminology of § 7, remark 7.1, a sequence $\{x_n\} \subset E$, with biorthogonal functionals $\{f_n\} \subset E^*$, is an $M$-basis of $E$ if and only if $[e_n] = S$, where $S$ is the $BK$-space (7.2).

There exist complete minimal sequences which are not $M$-bases and there exist generalized bases which are not $M$-bases, as shown by § 6, example 6.7 a) and, respectively, the unit vectors in $E = m$.

Obviously, every Banach space with an $M$-basis $\{x_n\}$ is separable. Conversely, the argument of Ch. I, § 4, remark 4.1 yields the existence of an $M$-basis in every separable Banach space $E$; moreover, it shows that if $\{y_n\}$ is a finitely linearly independent sequence in $E$ and $[y_n] = E$, then there exists an $M$-basis $\{x_n\}$ of $E$ such that

$$[x_1, \ldots, x_n] = [y_1, \ldots, y_n] \quad (n = 1, 2, \ldots). \tag{8.1}$$

Now we shall prove the existence, in every separable Banach space $E$, of $M$-bases $\{x_n\}$ satisfying other additional conditions, involving $\{x_n\}$ or the a.s.f. $\{f_n\} \subset E^*$ (theorem 8.1 below). To this end, let us first give

**Proposition 8.1.** *Let $E$ be a Banach space and let $\{y_n\} \subset E$, $\{g_n\} \subset E^*$ be finitely linearly independent sequences such that**)

$$y = \sum_{j=1}^{p} \alpha_j y_j,\ g_n(y) = 0\ (n = 1, 2, \ldots)\ \textit{imply}\ y = 0, \tag{8.2}$$

$$g = \sum_{i=1}^{q} \beta_i g_i,\ g(y_n) = 0\ (n = 1, 2, \ldots)\ \textit{imply}\ g = 0. \tag{8.3}$$

*) In other words, $(\text{lin}\,\{y_n\}) \cap \{g_n\}_\perp = \{0\}$ and $\{y_n\}^\perp \cap \text{lin}\,\{g_n\} = \{0\}$, where lin $M$ denotes the linear subspace spanned by the set $M$.

*Then there exist permutations* $\sigma, \tau$ *of* $\mathcal{N} = \{1,2,3, \ldots\}$ *such that*

$$\det (g_{\sigma(i)}(y_j))_{i,j=1,\ldots,n} \neq 0 \quad (n = 1,2, \ldots), \tag{8.4}$$

$$\det (g_i(y_{\tau(j)}))_{i,j=1,\ldots,n} \neq 0 \quad (n = 1,2, \ldots). \tag{8.5}$$

*Proof.* Since $y_1 \neq 0$, there exists, by (8.2), a least index $n_1$ such that $g_{n_1}(y_1) \neq 0$. Furthermore, there exists a least index $n_2 \in \mathcal{N} \setminus \{n_1\}$ such that $\det (g_{n_i}(y_j))_{i,j=1,2} \neq 0$; indeed, if there existed no such $n_2$, then for the element

$$x_2 = \frac{1}{g_{n_1}(y_1)} \begin{vmatrix} g_{n_1}(y_1) & g_{n_1}(y_2) \\ y_1 & y_2 \end{vmatrix} = y_2 - \frac{g_{n_1}(y_2)}{g_{n_1}(y_1)} y_1 \tag{8.6}$$

we would have $g_n(x_2) = 0$ $(n = 1,2, \ldots)$, whence, by (8.2), $x_2 = 0$, which would contradict the linear independence of $\{y_1, y_2\}$. Similarly, there exists a least index $n_3 \in \mathcal{N} \setminus \{n_1, n_2\}$ such that $\det (g_{n_i}(y_j))_{i,j=1,2,3} \neq 0$; indeed, otherwise for the element

$$x_3 = \frac{1}{\det (g_{n_i}(y_j))_{i,j=1,2}} \begin{vmatrix} g_{n_1}(y_1) & g_{n_1}(y_2) & g_{n_1}(y_3) \\ g_{n_2}(y_1) & g_{n_2}(y_2) & g_{n_2}(y_3) \\ y_1 & y_2 & y_3 \end{vmatrix} \tag{8.7}$$

we would have $g_n(x_3) = 0$ $(n = 1,2, \ldots)$, whence, by (8.2), $x_3 = 0$, which would contradict the linear independence of $y_1, y_2, y_3$. Continuing in this way indefinitely, we obtain a sequence $\{g_{n_i}\} \subset \{g_n\}$ and a sequence

$$x_1 = y_1,\ x_k = \frac{1}{\det(g_{n_i}(y_j))_{i,j=1,\ldots,k-1}} \begin{vmatrix} g_{n_1}(y_1) & \cdots & g_{n_1}(y_k) \\ \cdot & \cdots & \cdot \\ g_{n_{k-1}}(y_1) & \cdots & g_{n_{k-1}}(y_k) \\ y_1 & \cdots & y_k \end{vmatrix} \quad (k = 2,3, \ldots) \tag{8.8}$$

such that

$$g_{n_1}(x_1) = g_{n_1}(y_1) \neq 0,\ g_{n_k}(x_k) = \frac{\det (g_{n_i}(y_j))_{i,j=1,\ldots,k}}{\det (g_{n_i}(y_j))_{i,j=1,\ldots,k-1}} \neq 0 \ (k = 2,3, \ldots), \tag{8.9}$$

$$g_i(x_k) = 0 \qquad (i = 1, \ldots, n_k - 1;\ k = 1,2, \ldots). \tag{8.10}$$

We claim that $\{n_i\} = \mathcal{N}$, i.e. $\{g_{n_i}\} = \{g_n\}$. Indeed, assume, a contrario, that there exists a least $l \in \mathcal{N} \setminus \{n_i\}_{i=1}^{\infty}$, that is, a least $l \neq n_1, n_2, \ldots$

Then there exists a positive integer $r$ such that $l$ is also the least element of $\mathcal{N}\setminus\{n_i\}_{i=1}^r$, whence $l \leqslant n_k-1$ for $k = r+1, r+2, \ldots$ and thus, by (8.10),

$$g_l(x_k) = 0 \qquad (k = r+1, r+2, \ldots); \tag{8.11}$$

also, by the construction (8.8) of $x_k$ we have, obviously,

$$g_{n_1}(x_k) = \ldots = g_{n_r}(x_k) = 0 \quad (k = r+1, r+2, \ldots). \tag{8.12}$$

Consider now the system of equations

$$\sum_{i=1}^{r} \beta_i g_{n_i}(x_k) = g_l(x_k) \qquad (k = 1, \ldots, r). \tag{8.13}$$

At least one $g_l(x_{k_0}) \neq 0$, with $1 \leqslant k_0 \leqslant r$, since otherwise by (8.11) we would have $g_l(x_k) = 0$ $(k = 1,2, \ldots)$, whence, by (8.8), $g_l(y_k) = 0$ $(k = 1,2, \ldots)$ and thus, by (8.3), $g_l = 0$, in contradiction with the linear independence of $\{g_n\}$. Since by (8.8) and (8.9) det $(g_{n_i}(x_k))_{i,k=1,\ldots,r} = \prod_{k=1}^{r} g_{n_k}(x_k) \neq 0$, it follows that the system (8.13) has a solution $\beta_1, \ldots, \beta_r$. Then, by (8.11) – (8.13), we have

$$\left(g_l - \sum_{i=1}^{r} \beta_i g_{n_i}\right)(x_k) = 0 \qquad (k = 1,2, \ldots),$$

whence, by (8.8), $\left(g_l - \sum_{i=1}^{r} \beta_i g_{n_i}\right)(y_k) = 0$ $(k = 1,2, \ldots)$ and thus, by (8.3), $g_l - \sum_{i=1}^{r} \beta_i g_{n_i} = 0$, in contradiction with the linear independence of $\{g_n\}$. Thus, we have $\{n_i\} = \mathcal{N}$ and hence the mapping $\sigma(i) = n_i$ $(i = 1,2, \ldots)$ is a permutation of $\mathcal{N}$, which, by (8.9), satisfies (8.4).

Finally, the existence of a permutation $\tau$ of $\mathcal{N}$ satisfying (8.5) follows from this result applied to the sequences $\{g_n\} \subset E^*$ and $\{\pi(y_n)\} \subset E^{**}$, where $\pi$ denotes the canonical embedding of $E$ into $E^{**}$. This completes the proof of proposition 8.1.

**Proposition 8.2.** *Let $E$ be a Banach space and let $\{y_n\} \subset E$, $\{g_n\} \subset E^*$ be sequences satisfying*

$$\Delta_n = \det\,(g_i(y_j))_{i,j=1,\ldots,n} \neq 0 \quad (n = 1,2, \ldots). \tag{8.14}$$

*Then there exists a unique biorthogonal system* $(x_n, f_n)$ $(\{x_n\} \subset E, \{f_n\} \subset E^*)$ *such that*

$$x_1 = y_1, x_n - y_n \in [y_1, \ldots, y_{n-1}] \qquad (n = 2,3, \ldots), \quad (8.15)$$

$$[f_1, \ldots, f_n] = [g_1, \ldots, g_n] \qquad (n = 1,2, \ldots). \quad (8.16)$$

*This biorthogonal system is given by the formulae*

$$x_1 = y_1, \; x_n = \frac{1}{\Delta_{n-1}} \begin{vmatrix} g_1(y_1) & \ldots & g_1(y_n) \\ \cdot & \cdot & \cdot \\ g_{n-1}(y_1) & \ldots & g_{n-1}(y_n) \\ y_1 & \ldots & y_n \end{vmatrix} \qquad (n = 2,3, \ldots), \quad (8.17)$$

$$f_1 = \frac{1}{\Delta_1} g_1, \; f_n = \frac{1}{\Delta_n} \begin{vmatrix} g_1(y_1) & \ldots & g_1(y_{n-1}) & g_1 \\ \cdot & \cdot & \cdot & \cdot \\ g_n(y_1) & \ldots & g_n(y_{n-1}) & g_n \end{vmatrix} \qquad (n = 2,3, \ldots), \quad (8.18)$$

*and it satisfies the recurrence relations*

$$x_1 = y_1, \; x_n = y_n - \sum_{i=1}^{n-1} f_i(y_n) x_i \qquad (n = 2,3, \ldots), \quad (8.19)$$

$$f_1 = \frac{1}{g_1(x_1)} g_1, f_n = \frac{1}{g_n(x_n)} \left( g_n - \sum_{i=1}^{n-1} g_n(x_i) f_i \right) \quad (n = 2,3, \ldots). \quad (8.20)$$

*Proof.* Let us first show that if there exists a biorthogonal system $(x_n, f_n)$ satisfying (8.15), (8.16) then it satisfies (8.19), and (8.20) and hence it is unique. If we have (8.15) and (8.16) for some biorthogonal system $(x_n, f_n)$, then there exist scalars $\alpha_{in}$ and $\beta_{ni}$ with $\beta_{nn} \neq 0$ $(n = 1,2, \ldots)$, such that

$$y_n = x_n + \sum_{i=1}^{n-1} \alpha_{in} x_i, \; g_n = \sum_{i=1}^{n} \beta_{ni} f_i \quad (n = 1,2, \ldots). \quad (8.21)$$

Hence, by biorthogonality, we obtain

$$f_k(y_n) = \alpha_{kn}, \; g_n(x_k) = \beta_{nk} \qquad (k = 1, \ldots, n-1; \; n = 1,2, \ldots),$$

$$g_n(x_n) = \beta_{nn} \neq 0 \qquad (n = 1,2, \ldots),$$

which, together with (8.21), gives (8.19), (8.20) and proves the uniqueness of $(x_n, f_n)$.

Let us observe now that if we have (8.14), then the system $(x_n, f_n)$ defined by (8.17) and (8.18) satisfies (8.15) and (8.16), so it remains to show that it is biorthogonal. In order to unify the notation, put $\Delta_0 = 1$. For each $n = 1,2, \ldots$ we have

$$g_i(x_n) = \frac{1}{\Delta_{n-1}} \begin{vmatrix} g_1(y_1) & \cdots & g_1(y_n) \\ \cdot & \cdots & \cdot \\ g_{n-1}(y_1) & \cdots & g_{n-1}(y_n) \\ g_i(y_1) & \cdots & g_i(y_n) \end{vmatrix} = \begin{cases} 0 & \text{for } i=1, \ldots, n-1 \\ \dfrac{\Delta_n}{\Delta_{n-1}} & \text{for } i = n, \end{cases} \tag{8.22}$$

$$f_n(y_i) = \frac{1}{\Delta_n} \begin{vmatrix} g_1(y_1) & \cdots & g_1(y_{n-1}) & g_1(y_i) \\ \cdot & \cdots & \cdot & \cdot \\ g_n(y_1) & \cdots & g_n(y_{n-1}) & g_n(y_i) \end{vmatrix} = \begin{cases} 0 & \text{for } i = 1, \ldots, n-1 \\ 1 & \text{for } i = n. \end{cases} \tag{8.23}$$

Consequently, for each $n = 1,2, \ldots$ we obtain

$$f_k(x_n) = \frac{1}{\Delta_k} \begin{vmatrix} g_1(y_1) \ldots g_1(y_{k-1}) & g_1(x_n) \\ \cdot & \cdot \\ g_k(y_1) \ldots g_k(y_{k-1}) & g_k(x_n) \end{vmatrix} = \begin{cases} 0 & \text{for } k = 1, \ldots, n-1 \\ \dfrac{1}{\Delta_n} \Delta_{n-1} \dfrac{\Delta_n}{\Delta_{n-1}} = 1 & \text{for } k = n \end{cases}$$

$$f_n(x_k) = \frac{1}{\Delta_{k-1}} \begin{vmatrix} g_1(y_1) & \cdots & g_1(y_k) \\ \cdot & \cdots & \cdot \\ g_{k-1}(y_1) & \cdots & g_{k-1}(y_k) \\ f_n(y_1) & \cdots & f_n(y_k) \end{vmatrix} = \begin{cases} 0 & \text{for } k = 1, \ldots, n-1 \\ \dfrac{1}{\Delta_{n-1}} \Delta_{n-1} = 1 & \text{for } k = n, \end{cases}$$

which completes the proof of proposition 8.2.

The fact that the system $(x_n, f_n)$ defined by (8.17), (8.18) satisfies the recurrence relations (8.19), (8.20) means that

$$y_n - \frac{g_1(y_n)}{g_1(y_1)} y_1 - \sum_{k=2}^{n-1} \frac{1}{\Delta_{k-1}\Delta_k} \begin{vmatrix} g_1(y_1) \ldots g_1(y_{k-1}) & g_1(y_n) \\ \cdot & \cdot \\ g_k(y_1) \ldots g_k(y_{k-1}) & g_k(y_n) \end{vmatrix} \cdot$$

$$\cdot \begin{vmatrix} g_1(y_1) & \cdots & g_1(y_k) \\ \cdot & \cdots & \cdot \\ g_{k-1}(y_1) & \cdots & g_{k-1}(y_k) \\ y_1 & \cdots & y_k \end{vmatrix} = \frac{1}{\Delta_{n-1}} \begin{vmatrix} g_1(y_1) & \cdots & g_1(y_n) \\ \cdot & \cdots & \cdot \\ g_{n-1}(y_1) & \cdots & g_{n-1}(y_n) \\ y_1 & \cdots & y_n \end{vmatrix} \quad (n = 2,3, \ldots), \tag{8.24}$$

$$\frac{\Delta_{n-1}}{\Delta_n}\left(g_n - \frac{g_n(y_1)}{g_1(y_1)}g_1 - \sum_{k=2}^{n-1}\frac{1}{\Delta_{k-1}\Delta_k}\begin{vmatrix} g_1(y_1) & \dots & g_1(y_k) \\ \dots & \dots & \dots \\ g_{k-1}(y_1) & \dots & g_{k-1}(y_k) \\ g_n(y_1) & \dots & g_n(y_k)\end{vmatrix}\cdot\right.$$

$$\left.\cdot\begin{vmatrix} g_1(y_1) & \dots & g_1(y_{k-1}) & g_1 \\ \dots & \dots & \dots & \dots \\ g_k(y_1) & \dots & g_k(y_{k-1}) & g_k\end{vmatrix}\right) = \frac{1}{\Delta_n}\begin{vmatrix} g_1(y_1) & \dots & g_1(y_{n-1}) & g_1 \\ \dots & \dots & \dots & \dots \\ g_n(y_1) & \dots & g_n(y_{n-1}) & g_n\end{vmatrix} \quad (n = 2,3, \dots); \tag{8.25}$$

alternatively, one can also prove these relations directly, by induction. Let us also point out that (8.19), (8.20), are essentially nothing else than § 6, formula (6.2) and § 7, formula (7.48), respectively.

Now we can prove

**Theorem 8.1.** *Let $E$ be a separable Banach space and let $\{g_n\} \subset E^*$ be a finitely linearly independent sequence of functionals, which is total on $E$. Then*

a) *If $\{y_n\} \subset E$ is a finitely linearly independent sequence such that $[y_n] = E$, there exists an $M$-basis $\{x_n\}$ of $E$ satisfying* (8.1) *and*

$$[f_n] = [g_n], \tag{8.26}$$

*where $\{f_n\} \subset E^*$ is the a.s.f. to $\{x_n\}$.*

b) *There exists an $M$-basis $\{x_n\}$ of $E$, such that*

$$[f_1, \dots, f_n] = [g_1, \dots, g_n] \quad (n = 1,2, \dots), \tag{8.27}$$

*where $\{f_n\} \subset E^*$ is the a.s.f. to $\{x_n\}$, or, equivalently, such that*

$$[x_{n+1}, x_{n+2}, \dots] = [g_1, \dots, g_n]_\perp \quad (n = 1,2, \dots). \tag{8.28}$$

*Proof.* a) By proposition 8.1, choose a permutation $\sigma$ of $\mathcal{N}$ such that we have (8.4). Next, by proposition 8.2, define a biorthogonal system $(x_n, f_n)$ satisfying (8.1) and $[f_1, \dots, f_n] = [g_{\sigma(1)}, \dots, g_{\sigma(n)}]$ $(n = 1,2,\dots)$. Then $\{x_n\}$ is an $M$-basis of $E$, which satisfies (8.1) and (8.26).

b) Since $E$ is separable, there exists a finitely linearly independent sequence $\{y_n\} \subset E$ such that $[y_n] = E$. By proposition 8.1, choose a permutation $\tau$ of $\mathcal{N}$ such that we have (8.5). Next, by proposition 8.2, define a biorthogonal system $(x_n, f_n)$ satisfying $[x_1, \ldots, x_n] = [y_{\tau(1)}, \ldots \ldots, y_{\tau(n)}]$ $(n = 1,2, \ldots)$ and (8.16). Then $\{x_n\}$ is an $M$-basis of $E$, which satisfies (8.27).

Finally, let us prove the equivalence of (8.27) and (8.28). If we have (8.27), then, by Ch. II, § 4, formula (4.7), we have

$$[x_{n+1}, x_{n+2}, \ldots] = [f_1, \ldots, f_n]_\perp = [g_1, \ldots, g_n]_\perp \quad (n = 1,2, \ldots),$$

i.e. (8.28). Conversely, if we have (8.28), then, by Ch. II, § 4, formula (4.7) we obtain

$$[f_1, \ldots, f_n]_\perp = [g_1, \ldots, g_n]_\perp \quad (n = 1,2, \ldots),$$

whence it follows*) that we also have (8.27), which completes the proof of theorem 8.1.

*Remark 8.2.* Since $\{g_n\} \subset E^*$ is total if and only if $\bigcap_{n=1}^{\infty} [g_1, \ldots, g_n]_\perp = \{0\}$, and since $\{g_n\}$ is finitely linearly independent if and only if $\operatorname{codim}_E [g_1, \ldots, g_n]_\perp = n$ $(n = 1,2, \ldots)$, theorem 8.1 b) admits also the following equivalent formulation: *Let $E$ be a separable Banach space and let $\{F_n\}$ be a sequence of subspaces of $E$ such that $F_1 \supset F_2 \supset \ldots, \bigcap_{n=1}^{\infty} F_n = \{0\}$ and $\operatorname{codim}_E F_n = n$ $(n = 1,2, \ldots)$. Then there exists an $M$-basis $\{x_n\}$ of $E$ such that*

$$[x_{n+1}, x_{n+2}, \ldots] = F_n \quad (n = 1,2, \ldots). \tag{8.29}$$

*Remark 8.3.* The argument of remark 8.2 and Ch. II, § 4, formula (4.7) also show that *an $M$-basis can be characterized as a complete minimal sequence $\{x_n\}$ such that $\bigcap_{n=1}^{\infty} [x_{n+1}, x_{n+2}, \ldots] = \{0\}$.* On the other hand, let us mention that by § 6, proposition 6.4, an $M$-basis can be also characterized as a generalized basis $\{x_n\}$ having a uniquely determined admissible sequence $\{f_n\} \subset E^*$.

*Definition 8.2.* An $M$-basis $\{x_n\}$ of a Banach space $E$ is said to be *norming* if $r([f_n]) > 0$, where $\{f_n\} \subset E^*$ is the a.s.f. to $\{x_n\}$ and $r([f_n])$ is the characteristic of the subspace $[f_n]$.

We shall show now the existence of norming $M$-bases in every separable Banach space. Let us first recall

**Lemma 8.1.** *Let $E$ be a separable Banach space. Then there exists a separable subspace $V$ of $E^*$ such that $r(V) = 1$.*

---

*) See e.g. [87], p. 421, lemma 10.

*Proof.* Let $\{y_n\}$ be a sequence which is dense in $E$. Then, by a corollary of the Hahn-Banach theorem, there exists a sequence $\{g_n\} \subset E^*$ such that $\|g_n\| = 1$, $g_n(y_n) = \|y_n\|$ $(n = 1,2, \ldots)$. We claim that the subspace $V = [g_n]$ of $E^*$ has the required property. Indeed, let $x \in E$ and $\varepsilon > 0$ be arbitrary. Then there exists a $y_n$ such that $\|x - y_n\| < \frac{\varepsilon}{2}$, whence

$$0 \leqslant \|x\| - |g_n(x)| \leqslant |\,\|x\| - \|y_n\|\,| + |\,\|y_n\| - g_n(x)| =$$

$$= |\,\|x\| - \|y_n\|\,| + |g_n(y_n - x)| \leqslant \|x - y_n\| + \|g_n\|\,\|y_n - x\| < \varepsilon.$$

Consequently, since $x \in E$ and $\varepsilon > 0$ were arbitrary, we obtain

$$\|x\| \leqslant \sup_{1 \leqslant n < \infty} |g_n(x)| \leqslant \sup_{\substack{f \in V \\ \|f\| \leqslant 1}} |f(x)| \leqslant \|x\| \quad (x \in E),$$

whence $r(V) = 1$, which completes the proof of lemma 8.1.

From theorem 8.1 ( a) or b)) and lemma 8.1 it follows

**Corollary 8.1.** *Every separable Banach space E has a norming M-basis* $\{x_n\}$.

By Ch. I, § 12, theorem 12.2 a), every basis is a norming $M$-basis. Another class of norming $M$-bases is given in

**Proposition 8.3.** *Let $\{x_n\}$ be an M-basis of a Banach space E. Then for the sequence of coordinate functionals $\{h_n\}$ on the space of multipliers*)* $M(E, (x_n, f_n))$ *we have* $r([h_n]) = 1$. *Consequently, the sequence of unit vectors $\{e_n\}$ is a norming M-basis of* $[e_n] \subset M(E, (x_n, f_n))$.

*Proof.* Let $P$ denote the set of all finite linear combinations $\sum_{i=1}^{n} \alpha_i x_i$. Then for every $p = \sum_{i=1}^{n} \alpha_i x_i \in P$ with $\|p\| \leqslant 1$, $f \in E^*$ with $\|f\| \leqslant 1$ and $\{\gamma_n\} \in M(E, (x_n, f_n))$ we have

$$\left|\sum_{i=1}^{n} \alpha_i f(x_i) h_i(\{\gamma_k\})\right| = \left|\sum_{i=1}^{n} \alpha_i f(x_i)\gamma_i\right| = \left|f\left(\sum_{i=1}^{n} \gamma_i \alpha_i x_i\right)\right| =$$

$$= |f(p_{\{\gamma_n\}})| \leqslant \|f\|\,\|p_{\{\gamma_n\}}\| \leqslant \|f\|\,\|p\|\,\|\{\gamma_n\}\| \leqslant \|\{\gamma_n\}\|,$$

whence $\left\|\sum_{i=1}^{n} \alpha_i f(x_i) h_i\right\| \leqslant 1$. Consequently, for every $\{\gamma_n\} \in M(E, (x_n, f_n))$

*) See Ch. I, § 5, definition 5.1. and proposition 5.4.

we obtain

$$\|\{\gamma_n\}\| = \sup_{\substack{x\in E\\ \|x\|\leqslant 1}} \|x_{\{\gamma_n\}}\| = \sup_{\substack{p=\sum_{i=1}^{n}\alpha_i x_i\in P\\ \|p\|\leqslant 1}} \|p_{\{\gamma_k\}}\| = \sup_{\substack{p=\sum_{i=1}^{n}\alpha_i x_i\in P\\ \|p\|\leqslant 1}} \left\|\sum_{i=1}^{n}\gamma_i\alpha_i x_i\right\| =$$

$$= \sup_{\substack{=\sum_{i=1}^{n}\alpha_i x_i\in P\\ \|p\|\leqslant 1}} \sup_{\substack{f\in E^*\\ \|f\|\leqslant 1}} \left|\sum_{i=1}^{n}\gamma_i\alpha_i f(x_i)\right| =$$

$$= \sup_{\substack{p=\sum_{i=1}^{n}\alpha_i x_i\in P\\ \|p\|\leqslant 1}} \sup_{\substack{f\in E^*\\ \|f\|\leqslant 1}} \left|\sum_{i=1}^{n}\alpha_i f(x_i)h_i(\{\gamma_k\})\right| \leqslant$$

$$\leqslant \sup_{\substack{h\in[h_n]\\ \|h\|\leqslant 1}} |h(\{\gamma_n\})| \leqslant \|\{\gamma_n\}\|,$$

whence $r([h_n]) = 1$, which implies also $r([h_n|_{[e_j]}]) = 1$. This completes the proof of proposition 8.3.

*Definition 8.3.* A sequence $\{x_n\}$ in a Banach space $E$ is called an *M-basic sequence* if $\{x_n\}$ is an $M$-basis of the closed linear subspace $[x_n]$ of $E$.

Obviously, $\{x_n\}$ is an $M$-basic sequence if and only if it is a generalized basis of $[x_n]$. In particular, every generalized basis of the space $E$ is an $M$-basic sequence, but the converse is not true, as shown by the example of a basic sequence in a space $E$ such that the conjugate space $E^*$ is not $w^*$-separable (such a space $E$ has no generalized basis, by § 7, theorem 7.4). However, if $E$ is separable, then, by § 7, theorem 7.5, the converse is also true, i.e., *in a separable Banach space $E$ a sequence $\{x_n\}$ is an M-basic sequence if and only if it is a generalized basis of $E$.*

**Proposition 8.4.** *Every subsequence $\{x_{n_k}\}$ of an M-basis $\{x_n\}$ of a Banach space $E$ is an M-basic sequence.*

*Proof.* If $\{f_n\}\subset E^*$ is the a.s.f. to $\{x_n\}$, we have to prove that $\{f_{n_k}|_{[x_{n_i}]}\}$ is total on $[x_{n_i}]$. Assume that $x\in[x_{n_i}]$, $f_{n_k}(x)=0$ $(k=1,2,\ldots)$. Then, since $x\in[x_{n_i}]$, by biorthogonality we also have $f_j(x)=0$ $(j\in\mathcal{N}\setminus\{n_k\})$. Hence, since $\{f_n\}$ is total on $E$, we infer $x=0$, which completes the proof.

From proposition 8.4 it follows, in particular, that if $\{x_n\}$ is an $M$-basis of a Banach space $E$ and $\{n_k\}\subset\mathcal{N}$, then both $\{x_{n_k}\}$ and $\{x_j\}_{j\in\mathcal{N}\setminus\{n_k\}}$ are $M$-basic sequences. Conversely, *if $\{x_n\}$ is a complete minimal sequence in E, which can be partitioned into two M-basic se-*

*quences* $\{x_{n_k}\}$ *and* $\{x_j\}_{j\in\mathcal{N}\setminus\{n_k\}}$, *where* $\{n_k\}$ *(or* $\mathcal{N}\setminus\{n_k\}$*) is finite, then* $\{x_n\}$ *is an M-basis of E.* Indeed, let $x\in E$, $f_i(x)=0$ $(i=1,2,\ldots)$, where $\{f_n\}\subset E^*$ is the a.s.f. to $\{x_n\}$. Then, by Ch. I, § 6, Theorem 6.1 (implication 1°⇒6°) we can write $x=y_1+y_2$, where $y_1\in[x_{n_k}]$, $y_2\in[x_j]_{j\in\mathcal{N}\setminus\{n_k\}}$. Since $f_i(y_1)=-f_i(y_2)=0$ for $i\in\{n_k\}$ and since $\{n_k\}$ is finite, we have $y_1=\sum_{i\in\{n_k\}} f_i(y_1)x_i=0$. Since $f_j(y_2)=-f_j(y_1)=0$ for $j\in\mathcal{N}\setminus\{n_k\}$ and since $\{x_j\}_{j\in\mathcal{N}\setminus\{n_k\}}$ is an $M$-basic sequence, we have also $y_2=0$. Hence $x=y_1+y_2=0$, which proves our assertion. However, such a result is no longer true if both $\{n_k\}$ and $\mathcal{N}\setminus\{n_k\}$ are infinite, even if we assume that $\{x_{n_k}\}$ and $\{x_j\}_{j\in\mathcal{N}\setminus\{n_k\}}$ are basic sequences, that is, there exist complete minimal sequences $\{x_n\}$ which are not $M$-bases but may be partitioned into two (infinite) basic subsequences, as shown by

*Example 8.1.* Let $E=l^1$ and let

$$x_{2n-1}=e_1+e_{2n}+e_{2n+1}\qquad (n=1,2,\ldots),\qquad (8.30)$$

$$x_0=e_1-e_2,\ x_{2n}=e_1-e_{2n+1}-e_{2n+2}\qquad (n=1,2,\ldots),\qquad (8.31)$$

$$f_{2n-1}=\sum_{k=1}^{\infty}{}^{*}\,(-1)^{k+1}h_{2n+k}\qquad (n=1,2,\ldots),\qquad (8.32)$$

$$f_{2n}=\sum_{k=1}^{\infty}{}^{*}\,(-1)^{k}\,h_{2n+1+k}\qquad (n=0,1,2,\ldots),\qquad (8.33)$$

where $\{e_n\}$ is the unit vector basis of $E$, $\{h_n\}$ the sequence of coordinate functionals on $E$ and $\sum^*$ the sum in the weak* topology. Then $(x_n,f_n)_0^\infty$ is a biorthogonal system. Furthermore, $x_{2n-1}+x_{2n}=2e_1+e_{2n}-e_{2n+2}$ $(n=1,2,\ldots)$, whence

$$\frac{1}{2n}\sum_{i=1}^{2n}x_i=\frac{1}{2n}\sum_{j=1}^{n}(x_{2j-1}+x_{2j})=\frac{1}{2n}\sum_{j=1}^{n}(2e_1+e_{2j}-e_{2j+2})=$$
$$=e_1+\frac{1}{2n}e_2-\frac{1}{2n}e_{2n+2}\to e_1\ \text{as}\ n\to\infty,$$

and therefore $e_1\in[x_j]_1^\infty$. Since $e_2=e_1-x_0$, it follows that $e_2\in[x_j]_0^\infty$ and hence, inductively, $e_n\in[x_j]_0^\infty$ $(n=1,2,\ldots)$, which proves that $[x_n]_0^\infty=E$. Thus $\{x_n\}_0^\infty$ is a complete minimal sequence in $E$, which is not an $M$-basis of $E$ since $f_n(e_1)=0$ $(n=0,1,2,\ldots)$. However, both $\{x_{2n-1}\}_1^\infty$ and $\{x_{2n}\}_0^\infty$ are basic sequences, equivalent to the unit vector basis of $l^1$. Indeed, for any scalars $\alpha_j$ we have

$$\left\|\sum_{j=1}^{n}\alpha_jx_{2j-1}\right\|=\left\|\sum_{j=1}^{n}\alpha_j(e_1+e_{2j}+e_{2j+1})\right\|=$$

$$=\left\|\left(\sum_{j=1}^{n}\alpha_j\right)e_1+\sum_{j=1}^{n}\alpha_j e_{2j}+\sum_{j=1}^{n}\alpha_j e_{2j+1}\right\|=\left|\sum_{j=1}^{n}\alpha_j\right|+2\sum_{j=1}^{n}|\alpha_j|,$$

$$\left\|\sum_{j=0}^{n}\alpha_j x_{2j}\right\|=\left\|\alpha_0(e_1-e_2)+\sum_{j=1}^{n}\alpha_j(e_1-e_{2j+1}-e_{2j+2})\right\|=$$

$$=\left\|\left(\sum_{j=0}^{n}\alpha_j\right)e_1-\alpha_0 e_2-\sum_{j=1}^{n}\alpha_j e_{2j+1}-\sum_{j=1}^{n}\alpha_j e_{2j+2}\right\|=$$

$$=\left|\sum_{j=0}^{n}\alpha_j\right|+|\alpha_0|+2\sum_{j=1}^{n}|\alpha_j|,$$

whence

$$2\sum_{j=1}^{n}|\alpha_j|\leqslant\left\|\sum_{j=1}^{n}\alpha_j x_{2j-1}\right\|\leqslant 3\sum_{j=1}^{n}|\alpha_j|,$$

$$\sum_{j=0}^{n}|\alpha_j|\leqslant\left\|\sum_{j=0}^{n}\alpha_j x_{2j}\right\|\leqslant 3\sum_{j=0}^{n}|\alpha_j|,$$

which completes the proof of our assertions.

However, if we require that $\{x_{n_k}\}$ and $\{x_j\}_{j\in\mathcal{N}\setminus\{n_k\}}$ be $M$-basic sequences for *all* infinite $\{n_k\}\subset\mathcal{N}$ such that $\mathcal{N}\setminus\{n_k\}$ is infinite, then the situation is different. Namely, we have

**Proposition 8.5.** *A complete minimal sequence* $\{x_n\}$ *in a Banach space E is an M-basis of E if and only if for every infinite sequence* $\{n_k\}$ *of positive integers such that* $\mathcal{N}\setminus\{n_k\}$ *is infinite, either* $\{x_{n_k}\}$ *or* $\{x_j\}_{j\in\mathcal{N}\setminus\{n_k\}}$ *is an M-basic sequence (hence, if and only if both* $\{x_{n_k}\}$ *and* $\{x_j\}_{j\in\mathcal{N}\setminus\{n_k\}}$ *are M-basic sequences).*

*Proof.* The necessity part is immediate by proposition 8.4. Conversely, assume now that $\{x_n\}$ is a complete minimal sequence in $E$, which is not an $M$-basis of $E$. Then, by remark 8.3, there exists an element $x\in\bigcap_{n=1}^{\infty}[x_n, x_{n+1}, \ldots]\setminus\{0\}$. Since $x\in[x_i]_1^{\infty}=E$, there is an index $m_1$ such that $\operatorname{dist}(x,[x_i]_1^{m_1})<1$. Since $x\in[x_i]_{m_1+1}^{\infty}$, there is an index $m_2>m_1$ such that $\operatorname{dist}(x,[x_i]_{m_1+1}^{m_2})<\frac{1}{2}$. Continuing in this way indefinitely, we obtain an increasing sequence of positive integers $\{m_n\}$ such that

$$\operatorname{dist}(x,[x_i]_{m_{n-1}+1}^{m_n})<\frac{1}{n}\qquad(n=1,2,\ldots;m_0=0).\quad(8.34)$$

Put $\{n_k\} = \bigcup_{n=1}^{\infty} I_{2n-1}$, where $I_n = \{m_{n-1}+1, \ldots, m_n\}$ $(n=1,2,\ldots)$. Then both $\{n_k\}$ and $\mathcal{N}\setminus\{n_k\} = \bigcup_{n=1}^{\infty} I_{2n}$ are infinite and by (8.34) we have $x \in \bigcap_{k=1}^{\infty} [x_{n_k}, x_{n_{k+1}}, \ldots]$ whence, by remark 8.3, $\{x_{n_k}\}$ is not an $M$-basic sequence and, similarly, $\{x_j\}_{j\in\mathcal{N}\setminus\{n_k\}}$ is not $M$-basic, which completes the proof.

We recall that two subspaces $G$, $F$ of a normed linear space $E$ are said to be *quasi-complementary* if $G \cap F = \{0\}$ and $G + F$ is dense in $E$; in this case, each of $G$ and $F$ are called a *quasi-complement* of the other of them.

**Proposition 8.6.** *If $\{x_n\}$ is an $M$-basis of a Banach space $E$, and $\{n_k\}$ a sequence of positive integers, then the subspaces $[x_{n_k}]$ and $[x_j]_{j\in\mathcal{N}\setminus\{n_k\}}$ are quasi-complementary.*

*Proof.* If $x \in [x_{n_k}] \cap [x_j]_{j\in\mathcal{N}\setminus\{n_k\}}$ and $\{f_n\} \subset E^*$ is the a.s.f. to $\{x_n\}$, then, since $x \in [x_{n_k}]$, we have $f_j(x) = 0$ $(j \in \mathcal{N}\setminus\{n_k\})$ and, since $x \in [x_j]_{j\in\mathcal{N}\setminus\{n_k\}}$, we have $f_{n_k}(x) = 0$ $(k = 1,2,\ldots)$. Hence, since $\{f_n\}$ is total on $E$, we infer $x = 0$, and thus $[x_{n_k}] \cap [x_j]_{j\in\mathcal{N}\setminus\{n_k\}} = \{0\}$. Finally, since $[\{x_{n_k}\} \cup \{x_j\}_{j\in\mathcal{N}\setminus\{n_k\}}] = [x_n] = E$, it follows that $[x_{n_k}] + [x_j]_{j\in\mathcal{N}\setminus\{n_k\}}$ is dense in $E$, which completes the proof.

The assumption in proposition 8.6 that $\{x_n\}$ is an $M$-basis of $E$ cannot be replaced by the weaker assumption that $\{x_n\}$ is a complete minimal sequence in $E$, as shown by

*Example 8.2.* Let $E = l^2$ and let $\{x_n\}$ be the complete minimal sequence in $E$ considered in § 6, example 6.7 a). Then for every infinite sequence of positive integers $\{n_k\}$ such that $\mathcal{N}\setminus\{n_k\}$ is infinite, the subspaces $[x_{n_k}]$ and $[x_j]_{j\in\mathcal{N}\setminus\{n_k\}}$ of $E$ are non-quasi-complementary, namely, $e_1 \in [x_{n_k}] \cap [x_j]_{j\in\mathcal{N}\setminus\{n_k\}}$. Indeed, assume that for some $f \in E^*$ we have $f(x_{n_k}) = 0$ $(k = 1,2,\ldots)$. Then there exists a sequence $\{\eta_n\} \in l^2$ such that $f(x) = \sum_{i=1}^{\infty} \eta_i \xi_i$ $(x = \{\xi_n\} \in l^2)$, whence

$$0 = f(x_{n_k}) = f(e_1 + e_{n_k+1}) = \eta_1 + \eta_{n_k+1} \qquad (k = 1,2,\ldots).$$

Thus, $\eta_{n_1+1} = \eta_{n_2+1} = \ldots = -\eta_1$, whence, since by $\{\eta_n\} \in l^2$ we have $\lim_{n\to\infty} \eta_n = 0$, we infer $\eta_1 = 0$. Therefore, $f(e_1) = 0$, which proves that $e_1 \in [x_{n_k}]$. Since $\{n_k\}$ was arbitrary, we also have $e_1 \in [x_j]_{j\in\mathcal{N}\setminus\{n_k\}}$ whenever $\mathcal{N}\setminus\{n_k\}$ is infinite, which proves our assertion.

It is natural to ask whether the converse of proposition 8.6 holds. If $\{x_n\}$ is any complete minimal sequence in $E$ and $\{n_k\}$ a finite sequence of positive integers (or such that $\mathcal{N}\setminus\{n_k\}$ is finite), then by Ch. I, § 6, theorem 6.1 (implication $1° \Rightarrow 6°$) $[x_{n_k}]$ and $[x_j]_{j\in\mathcal{N}\setminus\{n_k\}}$ are complementary (and hence quasi-complementary). Moreover, there exist complete minimal sequences $\{x_n\}$ which are not $M$-bases, such that $[x_{n_k}]$ and $[x_j]_{j\in\mathcal{N}\setminus\{n_k\}}$ are quasi-complementary for some infinite $\{n_k\}$ with $\mathcal{N}\setminus\{n_k\}$ infinite, as shown by example 8.1, in which $[x_{2n-1}]_1^\infty \cap [x_{2n}]_0^\infty = 0$ (indeed, $x = \sum_{j=1}^\infty \alpha_j x_{2j-1} = \sum_{j=0}^\infty \beta_j x_{2j}$ implies $\left(\sum_{j=1}^\infty \alpha_j\right) e_1 + \sum_{j=1}^\infty \alpha_j e_{2j} + \sum_{j=1}^\infty \alpha_j e_{2j+1} = \left(\sum_{j=0}^\infty \beta_j\right) e_1 - \beta_0 e_2 - \sum_{j=1}^\infty \beta_j e_{2j+1} - \sum_{j=1}^\infty \beta_j e_{2j+2}$, whence $-\beta_{j-1} = \alpha_j = -\beta_j$ for $j = 1,2,\ldots$ which, by $\lim_{j\to\infty} \beta_j = 0$, implies $\beta_j = 0$ for $j = 0,1,2,\ldots$ and hence $x = 0$). A somewhat simpler example is $E = c_0$, $x_{2n-1} = e_{2n-1}$, $x_{2n} = e_2 + e_{2n+2}$ $(n = 1,2,\ldots)$, with the a.s.f. $f_{2n-1} = h_{2n-1}$, $f_{2n} = h_{2n+2}$ $(n = 1,2,\ldots)$, which satisfies $[x_{2n-1}] \cap [x_{2n}] = [e_{2n-1}] \cap [e_{2n}] = \{0\}$. However, if we assume that for *all* sequences $\{n_k\}$ of positive integers the subspaces $[x_{n_k}]$ and $[x_j]_{j\in\mathcal{N}\setminus\{n_k\}}$ are quasi-complementary (or, equivalently, $[x_{n_k}] \cap [x_j]_{j\in\mathcal{N}\setminus\{n_k\}} = \{0\}$), then the situation is different, namely, we have

**Proposition 8.7.** *A complete minimal sequence $\{x_n\}$ in a Banach space $E$ is an $M$-basis of $E$ if and only if for every infinite sequence $\{n_k\}$ of positive integers such that $\mathcal{N}\setminus\{n_k\}$ is infinite, the subspaces $\{x_{n_k}\}$ and $\{x_j\}_{j\in\mathcal{N}\setminus\{n_k\}}$ are quasi-complementary* (*or, what amounts to the same thing,* $[x_{n_k}] \cap [x_j]_{j\in\mathcal{N}\setminus\{n_k\}} = \{0\}$).

*Proof.* The necessity part follows from proposition 8.6. Conversely, assume now that $\{x_n\}$ is a complete minimal sequence in $E$, which is not an $M$-basis of $E$ and construct $x \neq 0$ and $\{n_k\}$ with $\mathcal{N}\setminus\{n_k\}$ infinite, as in the proof of proposition 8.5. Then, by (8.34), $x \in [x_{n_k}] \cap [x_j]_{j\in\mathcal{N}\setminus\{n_k\}}$ and thus $[x_{n_k}]$ and $[x_j]_{j\in\mathcal{N}\setminus\{n_k\}}$ are not quasi-complementary, which completes the proof.

Let us also give the following version of this result:

**Proposition 8.7′.** *A complete sequence $\{x_n\}$ in a Banach space $E$, with $x_n \neq 0$ $(n = 1,2,\ldots)$, is an $M$-basis of $E$ if and only if for every* (*finite or infinite*) *sequence of positive integers $\{n_k\}$ such that $\mathcal{N}\setminus\{n_k\}$ is infinite, we have $[x_{n_k}] \cap [x_j]_{j\in\mathcal{N}\setminus\{n_k\}} = \{0\}$.*

*Proof.* This follows from proposition 8.7, since the minimality of $\{x_n\}$ is equivalent to $[x_1] \cap [x_j]_{j\neq 1} = [x_2] \cap [x_j]_{j\neq 2} = \ldots = \{0\}$.

The problem of extension of $M$-basic sequences to $M$-bases has an affirmative answer:

**Theorem 8.2.** *Let $\{y_n\}$ be an $M$-basic sequence in a separable Banach space $E$ and let $F$ be any quasi-complement of $[y_n]$ in $E$. Then there exists a sequence $\{z_n\} \subset F$ such that $\{y_n\} \cup \{z_n\}$ is an $M$-basis of $E$.*

*Proof.* Since $[y_n] + F$ is dense in $E$, $\omega(F)$ is dense in $E/[y_n]$, where $\omega$ is the canonical mapping of $E$ onto $E/[y_n]$ (indeed, if $x \in E$ and $\varepsilon > 0$, then there exist $y \in [y_n]$, $z \in F$ such that $\|x - (y + z)\| < \varepsilon$, whence $\|\omega(x) - \omega(z)\| \leqslant \|x - z - y\| < \varepsilon$). Hence, by theorem 8.1 and Ch. I,

§ 10, theorem 10.3, there exists an $M$-basis $\{Z_n\}$ of $E/[y_n]$ such that $\{Z_n\} \subset \omega(F)$. Let $\{\psi_n\} \subset (E/[y_n])^*$ be the a.s.f. to $\{Z_n\}$. Extend $\{y_n\}$, as in § 6, proof of theorem 6.2, to a complete minimal sequence $\{y_n\} \cup \cup \{z_n\}$, with the a.s.f. $\{f_n\} \cup \{h_n\}$. Then, since $\{f_n|_{[y_j]}\}$ is total on $[y_n]$ and $\{\psi_n\}$ is total on $E/[y_n]$, it follows, by § 7, proof of corollary 7.1, that $\{f_n\} \cup \{h_n\}$ is total on $E$ and thus $\{y_n\} \cup \{z_n\}$ is an $M$-basis of $E$, which completes the proof of theorem 8.2.*)

From § 6, proposition 6.3 it follows that *every block sequence* $\{y_n\}$ *with respect to an M-basis* $\{x_n\}$ *of a Banach space E can be extended to an M-basis* $\{z_n\}$ *of E, satisfying* (6.35)—(6.38).

From the preceding results it follows immediately

**Corollary 8.2.** *Every subspace G of a separable Banach***) *space E admits a quasi-complement in E.*

*Proof.* By theorem 8.1 $G$ has an $M$-basis $\{y_n\}$ and by theorem 8.2 $\{y_n\}$ can be extended to an $M$-basis $\{y_n\} \cup \{z_n\}$ of $E$. Finally, by proposition 8.6, $F = [z_n]$ is a quasi-complement of $G = [y_n]$ in $E$, which completes the proof of corollary 8.2.

According to theorem 8.2, every $M$-basic sequence $\{y_n\}$ in a separable Banach space $E$ can be extended, for any quasi-complement $F$ of $[y_n]$, to an $M$-basis $\{y_n\} \cup \{z_n\}$ of $E$ with $\{z_n\} \subset F$. We shall show (in corollary 8.3 below) that this is no longer true if we require, in addition, that $[z_n] = F$. To this end, let us first prove

**Theorem 8.3.** *Let* $\{y_n\}$ *be an M-basic sequence in a separable Banach space E and let* $F \subset E$ *be an arbitrary quasi-complement of* $G = [y_n]$. *The following statements are equivalent:*

1°. *There exists a sequence* $\{z_n\} \subset F$ *with* $[z_n] = F$ *such that* $\{y_n\} \cup \cup \{z_n\}$ *is an M-basis of E.*

2°. *There exists a sequence* $\{z_n\} \subset F$ *with* $[z_n] = F$ *such that* $\{y_n\} \cup \{z_n\}$ *is a minimal sequence in E (which is, obviously, complete in E).*

3°. *There exists a sequence* $\{g_n\} \subset E^*$ *such that*

$$g_n \in F^{\perp}, \ g_n|_G = \varphi_n \quad (n = 1,2, \ldots), \tag{8.35}$$

*where* $\{\varphi_n\} \subset G^*$, $\varphi_i(y_j) = \delta_{ij}$ $(i, j = 1,2, \ldots)$ *(obviously,* $\{g_n\}$ *is uniquely determined by these conditions).*

*Proof.* The implications $1° \Rightarrow 2° \Rightarrow 3°$ are obvious.

Assume now that we have 3°. Let $\{\psi_n\} \subset (E/G)^*$ be a sequence of functionals which is total on $E/G$, let $\{h_n\}$ be the image of $\{\psi_n\}$ under

*) See the Appendix.
**) This holds also for every normed linear space (see remark 8.1).

the canonical linear isometry $(E/G)^* \equiv G^\perp \subset E^*$, that is,

$$h_n(x) = \psi_n(x + G) \quad (x \in E, n = 1,2, \ldots), \tag{8.36}$$

and let

$$\chi_n = h_n|_F \quad (n = 1,2, \ldots). \tag{8.37}$$

Then, by theorem 8.1 b), there exists an $M$-basis $\{z_n\}$ of $F$ having the a.s.f. $\{\theta_n\} \subset F^*$ of the form

$$\theta_n = \chi_n + \sum_{i=1}^{n-1} \beta_{ni}\chi_i \quad (n = 1,2, \ldots), \tag{8.38}$$

with suitable scalars $\beta_{ni}$. We claim that $\{y_n\} \cup \{z_n\}$ is an $M$-basis of $E$. Indeed, obviously, $[\{y_n\} \cup \{z_n\}] = E$. Now, put

$$f_n = h_n + \sum_{=1}^{n-1} \beta_{ni}h_i \quad (n = 1,2, \ldots). \tag{8.39}$$

Then, by 3° and $\varphi_i(y_j) = \delta_{ij}$ $(i,j = 1,2, \ldots)$, we have $g_i(z_j) = 0$ and $g_i(y_j) = \delta_{ij}$ $(i,j = 1,2, \ldots)$. Furthermore, by $f_n \in G^\perp$ (since $h_n \in G^\perp$), $f_n|_F = \theta_n$ $(n = 1,2, \ldots)$ and $\theta_i(z_j) = \delta_{ij}$ $(i,j = 1,2, \ldots)$, we have $f_i(y_j) = 0$ and $f_i(z_j) = \delta_{ij}$ $(i, j = 1,2, \ldots)$ and thus $(\{y_n\} \cup \{z_n\}, \{g_n\} \cup \{f_n\})$ is a biorthogonal system. Finally, assume that $g_n(x) = f_n(x) = 0$ $(n = 1,2, \ldots)$. Then, by (8.39), $h_n(x) = 0$ $(n = 1,2, \ldots)$, whence, by (8.36) and since $\{\psi_n\}$ is total on $E/G$, we infer $x \in G$ and hence, by (8.35) and since $\{\varphi_n\}$ is total on $G$, it follows that $x = 0$. This proves that $\{g_n\} \cup \{f_n\}$ is total on $E$, so $\{y_n\} \cup \{z_n\}$ is an $M$-basis of $E$. Thus, $3° \Rightarrow 1°$, which completes the proof of theorem 8.3.

*Remark 8.4.* Condition 3° can be also written in the form

$$\{g_n^{(0)} + G^\perp\} \cap F^\perp \neq \emptyset \qquad (n = 1,2, \ldots), \tag{8.40}$$

where $\{g_n^{(0)}\} \subset E^*$ is any sequence such that $g_n^{(0)}|_G = \varphi_n$ $(n = 1,2, \ldots)$. As observed also in 3°, for each $n$ the set in (8.40) contains at most one element.

We recall

**Lemma 8.2.** *Let $G$, $F$ be two subspaces of a Banach space $E$, such that $E^* = G^\perp \oplus F^\perp$. Then $E = G \oplus F$.*

*Proof.* Define a continuous linear mapping $u$ of $E$ into $D = E/G \times E/F$ by

$$u(x) = \{\omega_G(x), \omega_F(x)\} \qquad (x \in E), \tag{8.41}$$

where $\omega_G$ and $\omega_F$ denote the canonical mappings $E \to E/G$ and $E \to E/F$, respectively. Then, denoting by $i_G$, $i_F$ the canonical embeddings $E/G \to D$ and $E/F \to D$, respectively, we have

$$(u^*(h))(x) = h(u(x)) = h(\{\omega_G(x), \omega_F(x)\}) =$$
$$= h(\{\omega_G(x), 0\}) + h(\{0, \omega_F(x)\}) = (h \circ i_G)(\omega_G(x)) + (h \circ i_F)(\omega_F(x)) =$$
$$= (\omega_G^*(h \circ i_G))(x) + (\omega_F^*(h \circ i_F))(x) \qquad (x \in E, h \in D^*),$$

whence

$$u^*(h) = \omega_G^*(h \circ i_G) + \omega_F^*(h \circ i_F) \qquad (h \in D^*). \tag{8.42}$$

But, the mapping $h \to \{h \circ i_G, h \circ i_F\}$ is an isomorphism*) of $D^* = (E/G \times E/F)^*$ onto $(E/G)^* \times (E/F)^*$. Furthermore, the mapping $\psi \to \omega_G^*(\psi) = \psi \circ \omega_G$ is nothing else than the canonical linear isometry of $(E/G)^*$ onto $G^\perp$ and a similar remark is true for $\theta \to \omega_F^*(\theta)$, whence $\{\psi, \theta\} \to \{\omega_G^*(\psi), \omega_F^*(\theta)\}$ is an isomorphism of $(E/G)^* \times (E/F)^*$ onto $G^\perp \times F^\perp$. Finally, since $E^* = G^\perp \oplus F^\perp$, the mapping $(f, g) \to f + g$ is an isomorphism**) of $G^\perp \times F^\perp$ onto $E^*$. Since by (8.42) $u^*$ is the composition of these three mappings, it follows that $u^*$ is an isomorphism of $D^*$ onto $E^*$ and hence $u$ is an isomorphism of $E$ *onto* $D = E/G \times E/F$ (indeed, by the Hahn-Banach theorem and since $u^*$ is one-to-one, $u(E)$ is dense in $D$; hence, since $E$ is complete and since $u$ is an isomorphism***) of $E$ into $D$, it follows that $u$ maps $E$ onto $D$).

Now, if $x \in G \cap F$, then $u(x) = \{\omega_G(x), \omega_F(x)\} = 0$, whence, since $u$ is one-to-one, $x = 0$. Thus, $G \cap F = \{0\}$. Furthermore, since $u$ maps $E$ onto $D = E/G \times E/F$, for every $x \in E$ there exists an element $y \in E$ such that $u(y) = \{0, \omega_F(x)\}$, i.e., such that $\omega_G(y) = 0$, $\omega_F(y) = \omega_F(x)$, whence $y \in G$ and $x - y \in F$. Since $x = y + (x - y)$, it follows that $E = G \oplus F$, which completes the proof of lemma 8.2.

We are now ready to give the following corollary of theorem 8.3:

**Corollary 8.3.** *For every pair of quasi-complementary subspaces $G, F$ of a separable Banach space $E$, with $G \oplus F \neq E$, there exists an $M$-basis $\{y_n\}$ of $G$ with the property that there is no sequence $\{z_n\} \subset F$ with $[z_n] = F$ such that $\{y_n\} \cup \{z_n\}$ is a minimal sequence (or, in particular, an $M$-basis of $E$).*

*Proof.* Since $G \oplus F \neq E = \overline{G + F}$, there exists, by lemma 8.2, a functional $g_0 \in E^*$ such that $g_0 \notin G^\perp + F^\perp$. Then for $\varphi_0 = g_0|_G \in G^*$ there exists no $g \in E^*$ satisfying $g \in F^\perp$, $g|_G = \varphi_0$ (since otherwise we would have $g_0 = (g_0 - g) + g \in G^\perp + F^\perp$, a contradiction). By theorem 8.1 b) there exists an $M$-basis $\{y_n\}$ of $G$ with the a.s.f. $\{\varphi_n\} \subset G^*$ satisfying

*) See e.g. [11], p. 192, theorem 14.
**) See Ch. I, § 4, lemma 4.1.
***) See e.g. [87], p. 513, exercise 15.

$\varphi_1 = \varphi_0$. By theorem 8.3 (implication $2° \Rightarrow 3°$), for this $M$-basis $\{y_n\}$ of $G$ there exists no sequence $\{z_n\} \subset F$ with $[z_n] = F$ such that $\{y_n\} \cup \{z_n\}$ is a minimal sequence, which completes the proof of corollary 8.3.

Nevertheless, we have the following positive result:

**Corollary 8.4.** *For every pair of quasi-complementary subspaces $G$, $F$ of a separable Banach space $E$ there exist an $M$-basis $\{y_n\}$ of $G$ and an $M$-basis $\{z_n\}$ of $F$ such that $\{y_n\} \cup \{z_n\}$ is an $M$-basis of $E$.*

*Proof.* By the proof of theorem 8.3, for every quasi-complementary pair $G$, $F$ there exist*) an $M$-basis $\{z_n\}$ of $F$ and a sequence of functionals $\{f_n\} \subset E^*$ such that

$$f_n \in G^\perp, \quad f_n|_F = \theta_n \qquad (n = 1,2, \ldots), \tag{8.43}$$

where $\{\theta_n\} \subset F^*$, $\theta_i(z_j) = \delta_{ij}$ $(i, j = 1,2, \ldots)$. Hence, by theorem 8.3 (implication $3° \Rightarrow 1°$), applied for $G$ interchanged with $F$, there exists a sequence $\{y_n\} \subset G$ with $[y_n] = G$ such that $\{y_n\} \cup \{z_n\}$ is an $M$-basis of $E$. By proposition 8.4, $\{y_n\}$ is an $M$-basis of $G$, which completes the proof of corollary 8.4.

In the proof of theorem 8.2 (respectively, of § 6, theorem 6.2), we constructed the extension $\{y_n\} \cup \{z_n\}$ of the $M$-basic (respectively, minimal) sequence $\{y_n\} \subset E$ with the aid of another minimal sequence $\{x_n\} \subset E$ ($\{x_n\}$ was minimal, since we had $h_n(x_j) = \psi_n(x_j + [y_k]) = \psi_n(Z_j) = \delta_{nj}$ for $n, j = 1,2, \ldots$), via the "perturbation" (6.29). Let us mention the following somewhat related result:

**Proposition 8.8.** *Let $\{y_n\} \cup \{x_n\}$ be a minimal sequence in a Banach space $E$. Then*

a) *For any scalars $\alpha_k$ and any positive integers $n_k$ the sequence*

$$z_k = \alpha_k y_k + x_{n_k} \qquad (k = 1,2, \ldots) \tag{8.44}$$

*is minimal.*

b) *If $\{y_n\} \cup \{x_n\}$ is an $M$-basic sequence, then $\{z_k\}$ is an $M$-basic sequence.*

c) *If $\{z_k\}$ is an $M$-basic sequence, then*

$$[y_n] \cap [z_k] = \{0\}. \tag{8.45}$$

*Proof.* a) Let $\{g_n\} \cup \{h_n\} \subset E^*$, $g_i(y_j) = h_i(x_j) = \delta_{ij}$, $g_i(x_j) = h_i(y_j) = 0$ $(i, j = 1,2, \ldots)$. Then $h_{n_i}(z_k) = h_{n_i}(\alpha_k y_k + x_{n_k}) = \delta_{ik}$ $(i, k = 1,2, \ldots)$.

*) Indeed, observe that this part of the proof of theorem 8.3, implication $3° \Rightarrow 1°$, did not involve $\{y_n\}$, $\{g_n\}$ and (8.35).

b) Assume that $\{g_n\} \cup \{h_n\}$ is as in a) and that $\{g_n\} \cup \{h_n\}|_{[\{y_n\} \cup \{x_n\}]}$ is total on $[\{y_n\} \cup \{x_n\}]$ and let $x \in [z_k]$, $h_{n_k}(x) = 0$ $(k = 1,2, \ldots)$. Then we can write

$$x = \lim_{p\to\infty} \sum_{k=1}^{m_p} \beta_k^{(p)} z_k = \lim_{p\to\infty} \left( \sum_{k=1}^{m_p} \beta_k^{(p)} \alpha_k y_k + \sum_{k=1}^{m_p} \beta_k^{(p)} x_{n_k} \right),$$

with suitable scalars $\beta_k^{(p)}$. Consequently, by biorthogonality,

$$h_j(x) = 0 \qquad (j \in \mathcal{N} \setminus \{n_k\}),$$

$$g_j(x) = \lim_{p\to\infty} \beta_j^{(p)} \alpha_j = h_{n_j}(x) \alpha_j = 0 \quad (j = 1,2, \ldots),$$

and hence, since $x \in [\{y_n\} \cup \{x_n\}]$, it follows, by our assumptions, that $x = 0$.

c) Assume that $\{h_{n_i}|_{[z_k]}\}$ is total on $[z_k]$, where $\{h_{n_i}\}$ is as in a), and let $x \in [y_n] \cap [z_k]$. Then, since $x \in [y_n]$, by biorthogonality we have $h_{n_i}(x) = 0$ $(i = 1,2, \ldots)$, whence, since $x \in [z_k]$, it follows, by our assumption, that $x = 0$. Thus, we have (8.45), which completes the proof of proposition 8.8.

We also mention the following method of construction of $M$-bases with the aid of a given basis, related to Ch. I, § 4, proposition 4.3:

**Proposition 8.9.** *Let $\{x_n\}$ be a basis of a Banach space E, with the a.s.c.f. $\{f_n\} \subset E^*$ and let*

$$y_n = \sum_{i=1}^{n} \alpha_i x_i \qquad (n = 1,2, \ldots), \tag{8.46}$$

*where $\alpha_n \neq 0$ $(n = 1,2, \ldots)$. Then $\{y_n\}$ is a complete minimal sequence in E, with the a.s.f.*

$$g_n = \frac{1}{\alpha_n} f_n - \frac{1}{\alpha_{n+1}} f_{n+1} \qquad (n = 1,2, \ldots). \tag{8.47}$$

*Furthermore, the following statements are equivalent:*

1°. *$\{y_n\}$ is an M-basis of E.*

2°. *$\{y_n\}$ has no weak sequential limit point in E.*

3°. *$\sum_{i=1}^{\infty} \alpha_i x_i$ does not converge (in the norm-topology).*

*Hence, in particular,* $\left\{\sum_{i=1}^{n} x_i\right\}$ *is an M-basis of E if* $\overline{\lim_{n\to\infty}}\|x_n\| > 0$.

*The statements* 1°—3° *are implied by — and, if E is reflexive, equivalent to — the following:*

4°. *We have*

$$\lim_{n\to\infty}\|y_n\| = \infty. \tag{8.48}$$

*Consequently, if E is reflexive and* $\sup_{1\leq n<\infty}\|x_n\| < \infty$, $\sup_{1\leq n<\infty}|\alpha_n| < \infty$, *then* $\{y_n\}$ *is not a basis of E.*

*Proof.* Since $x_1 = \frac{1}{\alpha_1} y_1$, $x_n = \frac{1}{\alpha_n}(y_n - y_{n-1})$ $(n = 2,3, \ldots)$, we have $[y_n] = E$. Obviously, $g_i(y_j) = \delta_{ij}$ $(i, j = 1,2, \ldots)$.

Assume now that $\{y_n\}$ is an $M$-basis of $E$ and has a weak sequential limit point $x \in E$. Then

$$g_j(x) = \lim_{n\to\infty} g_j(y_{k_n(j)}) = 0 \qquad (j = 1,2, \ldots),$$

where $\{y_{k_n(j)}\}_{n=1}^{\infty}$ is some subsequence of $\{y_n\}$ depending on $g_j$, whence, since $\{g_n\}$ is total on $E$, we infer $x = 0$. Consequently,

$$\alpha_j = f_j(y_{l_n(j)}) \to f_j(x) = 0 \text{ as } n \to \infty \qquad (j = 1,2, \ldots),$$

where $\{y_{l_n(j)}\}_{n=1}^{\infty}$ is some subsequence of $\{y_n\}$ depending on $f_j$, whence $\alpha_j = 0$ $(j = 1,2, \ldots)$, contradicting the assumption. Thus, 1° ⇒ 2°.

The implication 2° ⇒ 3° is obvious.

Assume now that $\{y_n\}$ is not an $M$-basis of $E$. Then there is an $x \in E$ with $x \neq 0$ such that $g_n(x) = 0$ $(n = 1,2, \ldots)$, whence, by (8.47), $\frac{1}{\alpha_n} f_n(x) = \frac{1}{\alpha_{n+1}} f_{n+1}(x)$ for $n = 1,2, \ldots$ and hence

$$f_n(x) = \frac{\alpha_n}{\alpha_1} f_1(x) \neq 0 \qquad (n = 1,2, \ldots).$$

Since $\{x_n\}$ is a basis of $E$ with the a.s.c.f. $\{f_n\} \subset E^*$, we obtain

$$x = \sum_{i=1}^{\infty} f_i(x)x_i = \sum_{i=1}^{\infty} \frac{\alpha_i}{\alpha_1} f_1(x)x_i = \frac{f_1(x)}{\alpha_1} \sum_{i=1}^{\infty} \alpha_i x_i,$$

and therefore $\sum_{i=1}^{\infty} \alpha_i x_i$ converges. Thus, 3° ⇒ 1° and hence 1° ⇔ 2° ⇔ 3°.

The implication 4° $\Rightarrow$ 3° is obvious and, if $E$ is reflexive, the implication 2° $\Rightarrow$ 4° is also obvious.

Finally, assume that $E$ is reflexive and $\sup_{1\leqslant n<\infty} \|x_n\|<\infty$, $\sup_{1\leqslant n<\infty} |\alpha_n|<\infty$. If $\{y_n\}$ is not an $M$-basis of $E$, then it is not a basis of $E$. If $\{y_n\}$ is an $M$-basis of $E$, then, by 1°$\Rightarrow$4° we have $\left\|\frac{y_n}{\alpha_{n+1}}\right\| \geqslant \frac{1}{\sup_{1\leqslant j<\infty} |\alpha_j|} \|y_n\| \to \infty$ as $n \to \infty$, whence, by the necessity part of Ch. I, § 4, proposition 4.3 (and its proof), $\{y_n\}$ is not a basis of $E$. This completes the proof of proposition 8.9.

In general 3° does not imply 4°, as shown by any basis $\{x_n\}$ of type $P$ of a Banach space $E$.

The following theorem gives some relations between $M$-bases and complemented subspaces:

**Theorem 8.4.** *Let $\{x_n\}$ be an M-basis of a Banach space E, with the a.s.f. $\{f_n\} \subset E^*$, and let $\{n_k\}$ be an infinite sequence of positive integers. If $r_{[x_{n_k}]}([f_{n_k}]) > 0$, then*

$$E = [x_{n_k}] \oplus [x_j]_{j\in\mathcal{N}\setminus\{n_k\}}. \tag{8.49}$$

*If $\{x_n\}$ is a norming M-basis, the converse is also valid.*

*Proof.* Let $\omega_0$ be the canonical mapping of $E$ onto $E/[x_j]_{j\in\mathcal{N}\setminus\{n_k\}}$. Then, by $r_{[x_{n_k}]}([f_{n_k}]) = C > 0$, we have

$$C\|x\| \leqslant \sup_{\substack{f\in[f_{n_k}]\\ \|f\|\leqslant 1}} |f(x)| \leqslant \sup_{\substack{f\in([x_j]_{j\in\mathcal{N}\setminus\{n_k\}})^{\perp}\\ \|f\|\leqslant 1}} |f(x)| = \|\omega_0(x)\| \leqslant \|x\| \quad (x\in[x_{n_k}]),$$

and hence $\omega_0|_{[x_{n_k}]}$ is an isomorphism. Furthermore, since $[x_n]=E$, we have $\omega_0([x_{n_k}]) = [\omega_0(x_{n_k})] = E/[x_j]_{j\in\mathcal{N}\setminus\{n_k\}}$, and hence $(\omega_0|_{[x_{n_k}]})^{-1}\omega_0$ is a projection of $E$ onto $[x_{n_k}]$ along $[x_j]_{j\in\mathcal{N}\setminus\{n_k\}}$. Thus, we have (8.49), which proves the first assertion.

For the second assertion, assume that $r([f_n]) > 0$ and that we have (8.49) and let $v$ denote the natural projection of $E$ onto $[x_{n_k}]$ along $[x_j]_{j\in\mathcal{N}\setminus\{n_k\}}$. Then, by biorthogonality,

$$(v^*(f_{n_k}))(x) = f_{n_k}(v(x)) = f_{n_k}(x) \qquad (x\in E,\ k = 1,2, \ldots),$$

$$(v^*(f_j))(x) = f_j(v(x)) = 0 \qquad (x\in E, j\in\mathcal{N}\setminus\{n_k\}),$$

so $v^*(f_{n_k}) = f_{n_k}$ $(k=1,2, \ldots)$ and $v^*(f_j) = 0$ $(j\in\mathcal{N}\setminus\{n_k\})$, whence

$$v^*([f_n]) \subset [f_{n_k}].$$

Consequently, by $r([f_n]) = C_1 > 0$, for every $x \in [x_{n_k}]$ we have

$$C_1 \| x \| \leqslant \sup_{\substack{f \in [f_n] \\ \|f\| \leqslant 1}} |f(x)| = \sup_{\substack{f \in [f_n] \\ \|f\| \leqslant 1}} |f(v(x))| = \sup_{\substack{f \in [f_n] \\ \|f\| \leqslant 1}} |(v^*(f))(x)| =$$

$$= \| v^* \| \sup_{\substack{f \in [f_n] \\ \|f\| \leqslant 1}} \left| \left( \frac{v^*}{\| v^* \|} (f) \right) (x) \right| \leqslant \| v^* \| \sup_{\substack{h \in [f_{n_k}] \\ \|h\| \leqslant 1}} |h(x)|,$$

so $r_{[x_{n_k}]}([f_{n_k}]) > 0$, which completes the proof of theorem 8.4.

In the particular case when $(x_{n_k}, f_{n_k})$ is a bibasic system, several equivalent conditions to $r_{[x_{n_k}]}([f_{n_k}]) > 0$ are given in § 1, theorem 1.10. If, in addition, $\{x_{n_k}\}$ is a part of a *basis* $\{x_n\}$ of $E$, with the a.s.c.f. $\{f_n\} \subset E^*$, then $[x_j]_{j \in \mathcal{N} \setminus \{n_k\}} = [f_{n_k}]_\perp$ and hence the conclusion (8.49) of the first assertion of theorem 8.4 amounts to condition 3° of § 1, theorem 1.11. Note that already the assumption (of theorem 8.4) that $\{x_n\}$ is an $M$-basis of $E$, with the a.s.f. $\{f_n\} \subset E^*$, implies that $[x_{n_k}] + [f_{n_k}]_\perp$ is dense in $E$ (since $[f_{n_k}]_\perp \supset [x_j]_{j \in \mathcal{N} \setminus \{n_k\}}$).

**Corollary 8.5.** *Let $\{x_n\}$ be an $M$-basis of a Banach space $E$, with the a.s.f. $\{f_n\} \subset E^*$, such that for every infinite sequence $\{n_k\}$ of positive integers, $r_{[x_{n_k}]}([f_{n_k}]) > 0$. Then $\{x_n\}$ is an unconditional basis of $E$.*

*Proof.* By theorem 8.4, for every infinite sequence $\{n_k\}$ we have (8.49). Hence, by § 7, theorem 7.3, $\{x_n\}$ is an unconditional basis of $E$, which completes the proof.

It is natural to ask whether for every $M$-basis $\{x_n\}$ of a Banach $E$, with the a.s.f. $\{f_n\} \subset E^*$, there exists an infinite sequence $\{n_k\}$, with $\mathcal{N} \setminus \{n_k\}$ infinite, such that $r_{[x_{n_k}]}([f_{n_k}]) > 0$. The answer is negative, even if $\{x_n\}$ is a basis of $E$, as shown by

*Example 8.3.* Let $\{x_n\}$ be the standard conditional basis of $E = c_0$, i.e. (see Ch. II, § 14, example 14.1)

$$x_n = \sum_{i=1}^{n} e_i \qquad (n = 1,2, \ldots), \tag{8.50}$$

$$f_n = h_n - h_{n+1} \qquad (n = 1,2, \ldots), \tag{8.51}$$

where $\{e_n\}$ is the unit vector basis of $E = c_0$ and $\{h_n\}$ is the sequence of coordinate functionals on $E = c_0$. Then, as we have seen in Ch. II, § 17, example 17.2, every infinite subsequence $\{x_{n_k}\}$ of $\{x_n\}$ is equivalent to the basis $\{x_n\}$, but $\{f_{2n}\}$ is not equivalent to $\{\varphi_{2n}\} = \{f_{2n}|_{[x_{2j}]}\}$. Similarly, we shall prove now that for any infinite $\{n_k\} \subset \mathcal{N}$ with $\mathcal{N} \setminus \{n_k\}$ infinite, $\{f_{n_k}\}$ is not equivalent to $\{\varphi_{n_k}\} = \{f_{n_k}|_{[x_{n_j}]}\}$ and hence, by § 1, theorem 1.10, implication $5° \Rightarrow 1°$, $r_{[x_{n_k}]}([f_{n_k}]) = 0$. Since $\{x_{n_k}\} \sim \sim \{x_n\}$, we have $\{\varphi_{n_k}\} \sim \{f_n\}$ (by Ch. I, § 12, proposition 12.1) and therefore it will be sufficient to prove that $\{f_{n_k}\}$ is not equivalent to $\{f_n\}$. By the computations of Ch. II, § 14, example 14.2, for any finite linear combination $\sum_{i=m}^{m+p} \alpha_i f_i$ we have

$$\left\| \sum_{i=m}^{m+p} \alpha_i f_i \right\| = \left\| \sum_{i=m}^{m+p} \alpha_i (h_i - h_{i+1}) \right\| = \left\| \alpha_m h_m + \right.$$

$$\left. + \sum_{i=m}^{m+p-1} (\alpha_{i+1} - \alpha_i) h_{i+1} - \alpha_{m+p} h_{m+p+1} \right\| = |\alpha_m| + \tag{8.52}$$

$$+ \sum_{i=m}^{m+p-1} |\alpha_{i+1} - \alpha_i| + |\alpha_{m+p}|.$$

Now, since $\{n_k\}$ and $\mathcal{N} \setminus \{n_k\}$ are infinite, there exists an infinite subsequence $\{n_{k_m}\}$ of $\{n_k\}$ such that

$$n_{k_m} + 1 \notin \{n_k\} \qquad (m = 1,2, \ldots). \tag{8.53}$$

Then $n_{k_m} + 1 \neq n_{k_m+1}$ $(m = 1,2, \ldots)$, and hence, by (8.52), for any $f = \sum_{k=1}^{\infty} \alpha_{n_k} f_{n_k} \in [f_{n_k}]$ we have

$$\|f\| = \left\| \sum_{k=1}^{\infty} \alpha_{n_k} f_{n_k} \right\| = \left\| \sum_{k=1}^{\infty} \alpha_{n_k} (h_{n_k} - h_{n_k+1}) \right\| =$$

$$= \left\| \sum_{m=1}^{\infty} \sum_{k=k_{m-1}+1}^{k_m} \alpha_{n_k} (h_{n_k} - h_{n_k+1}) \right\| = \sum_{m=1}^{\infty} \left\| \sum_{k=k_{m-1}+1}^{k_m} \alpha_{n_k} (h_{n_k} - h_{n_k+1}) \right\| =$$

$$= \sum_{m=1}^{\infty} \left\| \sum_{k=k_{m-1}+1}^{k_m} \alpha_{n_k} f_{n_k} \right\|,$$

where $k_0 = 0$. Consequently, putting

$$B_m = [f_{n_k}]_{k=k_{m-1}+1}^{k_m} \qquad (m = 1,2, \ldots), \tag{8.54}$$

we have*) $[f_{n_k}] \equiv (B_1 \times B_2 \times \ldots)_{l^1}$. On the other hand, $[f_n]$ itself is not isomorphic to $(B_1 \times B_2 \times \ldots)_{l^1}$, since by (8.52)

$$\left\| \sum_{k=1}^{\infty} \alpha_{n_k} f_k \right\| = |\alpha_{n_1}| + \sum_{k=1}^{\infty} |\alpha_{n_{k+1}} - \alpha_{n_k}|,$$

$$\sum_{m=1}^{\infty} \left\| \sum_{k=k_{m-1}+1}^{k_m} \alpha_{n_k} f_k \right\| = \sum_{m=1}^{\infty} \left( |\alpha_{n_{k_{m-1}+1}}| + \sum_{k=k_{m-1}+1}^{k_m-1} |\alpha_{n_{k+1}} - \alpha_{n_k}| + |\alpha_{n_{k_m}}| \right),$$

where $\{n_{k_m}\}$ is infinite. Consequently, $\{f_{n_k}\}$ is not equivalent to $\{f_n\}$.

Let us observe that for $M$-bases we have stability theorems analogous to those proved in Ch. I, § 9 and § 10 for complete minimal sequences (in particular, for bases), because if $u$ is an isomorphism of $E$ onto $F$, then $(u^{-1})^*$ is an isomorphism of $E^*$ onto $F^*$, which carries total sequences $\{f_n\} \subset E^*$ into total sequences $\{h_n\} \subset F^*$; indeed, the relations $y \in F$, $h_n(y) = 0$ $(n = 1,2, \ldots)$ imply $f_n(u^{-1}(y)) = ((u^{-1})^* f_n)(y) = h_n(y) = 0$ $(n = 1,2, \ldots)$, whence, since $\{f_n\}$ is total on $E$, $u^{-1}(y) = 0$, and thus $y = 0$.

One can also give some results of strong duality, corresponding to those of Ch. I, § 12. For example, we have

**Proposition 8.10.** *If $\{x_n\}$ is a complete minimal sequence (or, in particular, an M-basis) in a Banach space E, then the a.s.f. $\{f_n\} \subset E^*$ to $\{x_n\}$ is an M-basic sequence in $E^*$.*

*Proof.* Let $u$ be the canonical mapping of $E$ into $[f_n]^*$ (note that if $\{f_n\}$ is not total on $E$, then $u$ is not one-to-one). Then $(f_n, u(x_n))$ is an $[f_n]$-complete biorthogonal system. Furthermore, if $f \in [f_n]$, $(u(x_n))(f) = 0$ $(n = 1,2, \ldots)$, then $f(x_n) = 0$ $(n = 1,2, \ldots)$, whence, by $[x_n] = E$, we infer $f = 0$. Thus, $\{f_n\}$ is an $M$-basis of $[f_n]$, which completes the proof.

**Corollary 8.6.** *Let E be a Banach space and let $(x_n, f_n)$ $(\{x_n\} \subset E$, $\{f_n\} \subset E^*)$ be a biorthogonal system such that $[f_n] = E^*$. Then $\{x_n\}$ is an M-basis of E.*

*Proof.* Let $\pi$ be the canonical embedding of $E$ into $E^{**}$. Then $\{f_n\}$ is an $M$-basis of $E^*$, with the a.s.f. $\{\pi(x_n)\}$. Hence, by proposition 8.10, $\{\pi(x_n)\}$ is an $M$-basis of $[\pi(x_n)] = \pi(E)$. Since $\pi$ is a linear isometry, it follows that $\{x_n\}$ is an $M$-basis of $E$.

The converse of proposition 8.10 is not valid, as shown by

*Example 8.4.* Let $E = c_0$ and let

$$x_n = e_{2n-1} \qquad (n = 1,2, \ldots), \tag{8.55}$$

---

*) For the definition of $(B_1 \times B_2 \times \ldots)_{l^1}$ see e.g. Ch. II, § 18 (before lemma 18.5).

where $\{e_n\}$ is the unit vector basis of $E$. Then $[x_n] \neq E$, but $\{x_n\}$ admits a biorthogonal sequence $\{f_n\} = \{h_{2n-1}\} \subset E^*$ (where $\{h_n\}$ is the sequence of coordinate functionals) such that $\{f_n\}$ is a basis (hence an $M$-basis) of $[f_n]$.

In Ch. II, § 4, formula (4.7) we have seen that if $\{x_n\}$ is a complete minimal sequence in a Banach space $E$, with the a.s.f. $\{f_n\} \subset E^*$, then

$$[f_1, \ldots, f_n]_\perp = [x_{n+1}, x_{n+2}, \ldots]; \tag{8.56}$$

therefore it is natural to ask what can we say about the relations between $[f_{n_k}]_\perp$ and $[x_j]_{j \in \mathcal{N} \setminus \{n_k\}}$, where $\{n_k\}$ is any sequence of positive integers. Let us observe that if $\mathcal{N} \setminus \{n_k\}$ is finite and

$$[f_{n_k}]_\perp = [x_j]_{j \in \mathcal{N} \setminus \{n_k\}}, \tag{8.57}$$

then $\{f_n\}$ must be total on $E$ and thus $\{x_n\}$ is an $M$-basis of $E$. Indeed, let $x \in E$, $f_n(x) = 0$ $(n = 1,2, \ldots)$. Then $x \in [f_{n_k}]_\perp = [x_j]_{j \in \mathcal{N} \setminus \{n_k\}}$, whence since $\mathcal{N} \setminus \{n_k\}$ is finite and $f_j(x) = 0$ $(j \in \mathcal{N} \setminus \{n_k\})$, it follows that $x = \sum_{j \in \mathcal{N} \setminus \{n_k\}} f_j(x) x_j = 0$, which proves our assertion.

If we assume that $\{x_n\}$ is a complete minimal sequence in $E$ satisfying (8.57) for every sequence of positive integers $\{n_k\}$ such that $\mathcal{N} \setminus \{n_k\}$ is infinite, then $\{x_n\}$ need not be an $M$-basis of $E$, as shown by

*Example* 8.5. Let $E = l^2$ and let $\{x_n\}$ be the complete minimal sequence considered in example 8.2. Then $\{x_n\}$ is not an $M$-basis of $E$, but it satisfies (8.57) for every $\{n_k\}$ with $\mathcal{N} \setminus \{n_k\}$ infinite. Indeed, let $x_0 = \{\xi_n^{(0)}\} \in [f_{n_k}]_\perp$ and let $f = \{\eta_n\} \in E^*$, $f(x_j) = 0$ $(j \in \mathcal{N} \setminus \{n_k\})$. Then $\xi_{n_k+1}^{(0)} = f_{n_k}(x_0) = 0$ $(k = 1,2, \ldots)$, and, as we have seen in example 8.2, $\eta_{j+1} = -\eta_1 = 0$ $(j \in \mathcal{N} \setminus \{n_k\})$, because $\mathcal{N} \setminus \{n_k\}$ is infinite. Hence

$$f(x_0) = \eta_1 \xi_1^{(0)} + \sum_{k=1}^{\infty} \eta_{n_k+1} \xi_{n_k+1}^{(0)} + \sum_{j \in \mathcal{N} \setminus \{n_k\}} \eta_{j+1} \xi_{j+1}^{(0)} = 0,$$

which proves that $x_0 \in [x_j]_{j \in \mathcal{N} \setminus \{n_k\}}$. Thus,

$$[f_{n_k}]_\perp \subset [x_j]_{j \in \mathcal{N} \setminus \{n_k\}}, \tag{8.58}$$

whence since the opposite inclusion is obvious, we obtain (8.57).

It is now natural to introduce the following special class of $M$-bases:

*Definition 8.4.* An $M$-basis $\{x_n\}$ of a Banach space $E$ is said to be *strong*, if for every sequence $\{n_k\}$ of positive integers we have (8.57) (or, equivalently, (8.58)), where $\{f_n\} \subset E^*, f_i(x_j) = \delta_{ij}$ $(i, j = 1,2, \ldots)$.

For example, *every basis $\{x_n\}$ of a Banach space $E$ is a strong $M$-basis of $E$.* Indeed, if $\{f_n\} \subset E^*$ is the a.s.c.f. to $\{x_n\}$ and $x \in [f_{n_k}]_\perp$, then $x = \sum_{i=1}^{\infty} f_i(x)x_i = \sum_{j \in \mathcal{N} \setminus \{n_k\}} f_j(x)x_j \in [x_j]_{j \in \mathcal{N} \setminus \{n_k\}}$ and thus we have (8.58).

Some characterizations of strong $M$-bases are given in

**Proposition 8.11.** *Let $\{x_n\}$ be an $M$-basis of a Banach space $E$, with the a.s.f. $\{f_n\} \subset E^*$. The following statements are equivalent:*

1°. *$\{x_n\}$ is a strong $M$-basis.*

2°. *For every sequence of positive integers $\{n_k\}$ we have*

$$([f_{n_k}]_\perp)^\perp = ([x_j]_{j \in \mathcal{N} \setminus \{n_k\}})^\perp \tag{8.59}$$

*(or, equivalently, $([f_{n_k}]_\perp{}^\perp \supset ([x_j]_{j \in \mathcal{N} \setminus \{n_k\}})^\perp)$.*

3°. *For every sequence of positive integers $\{n_k\}$ the image of $\{f_{n_k}\}$ under the canonical linear isometry $([x_j]_{j \in \mathcal{N} \setminus \{n_k\}})^\perp \equiv (E/[x_j]_{j \in \mathcal{N} \setminus \{n_k\}})^*$, i.e. the sequence $\{\psi_{n_k}\} \subset (E/[x_j]_{j \in \mathcal{N} \setminus \{n_k\}})^*$ defined by*

$$\psi_{n_k}(x + [x_j]_{j \in \mathcal{N} \setminus \{n_i\}}) = f_{n_k}(x) \quad (x \in E, k = 1,2, \ldots) \tag{8.60}$$

*is total on $E/[x_j]_{j \in \mathcal{N} \setminus \{n_k\}}$ (or, in other words, $\{x_{n_k} + [x_j]_{j \in \mathcal{N} \setminus \{n_i\}}\}$ is an $M$-basis of $E/[x_j]_{j \in \mathcal{N} \setminus \{n_i\}}$).*

4°. *For every $x \in E$ and $f \in E^*$ such that the set $\{i \in \mathcal{N} | f(x_i)f_i(x) \neq 0\}$ is finite or empty, we have*

$$\sum_{i=1}^{\infty} f(x_i)f_i(x) = f(x). \tag{8.61}$$

*Proof.* The implication 1° ⇒ 2° is obvious. Conversely, assume that 1° is not satisfied, i.e., there exists an element $x_0 \in [f_{n_k}]_\perp$ such that $x_0 \notin [x_j]_{j \in \mathcal{N} \setminus \{n_k\}}$. Then, by a corollary of the Hahn-Banach theorem, there exists a functional $f_0 \in E^*$ such that $f_0(x_j) = 0$ $(j \in \mathcal{N} \setminus \{n_k\})$, $f_0(x_0) = 1$, whence $f_0 \in ([x_j]_{j \in \mathcal{N} \setminus \{n_k\}})^\perp$, $f_0 \notin ([f_{n_k}]_\perp)^\perp$, i.e., 2° is not satisfied. Thus, 2° ⇒ 1°. Since for every biorthogonal system $(x_n, f_n)$ we have

$$([x_j]_{j \in \mathcal{N} \setminus \{n_k\}})^\perp \supset ([f_{n_k}]_\perp)^\perp \supset [f_{n_k}], \tag{8.62}$$

condition (8.59) is equivalent to $([f_{n_k}]_\perp)^\perp \supset ([x_j]_{j \in \mathcal{N} \setminus \{n_k\}})^\perp$.

Assume now that we have 1° and let $x \in E$, $\psi_{n_k}(x + [x_j]_{j \in \mathcal{N} \setminus \{n_i\}}) = 0$ $(k = 1,2, \ldots)$. Then by (8.60) we have $f_{n_k}(x) = 0$ $(k = 1,2, \ldots)$, that is, $x \in [f_{n_k}]_\perp$, whence, by 1°, $x \in [x_j]_{j \in \mathcal{N} \setminus \{n_k\}}$, which proves that $\{\psi_{n_k}\}$ is total on $E/[x_j]_{j \in \mathcal{N} \setminus \{n_k\}}$. Thus, 1° ⇒ 3°.

Conversely, assume now that we have 3° and let $x \in [f_{n_k}]_{\perp}$. Then by (8.60) we have $\psi_{n_k}(x + [x_j]_{j \in \mathcal{N} \setminus \{n_i\}}) = 0$ $(k = 1,2, \ldots)$, whence, by 3°, $x \in [x_j]_{j \in \mathcal{N} \setminus \{n_k\}}$, which proves (8.58). Thus, 3° ⇒ 1°.

Assume now again that we have 1° and let $x \in E, f \in E^*$ be such that the set $\{i \in \mathcal{N} \mid f(x_i) f_i(x) \neq 0\}$ is finite or empty. Let $\{n_k\}$ be the set of all indices such that $f_{n_k}(x) = 0$; thus, $f_j(x) \neq 0$ for all $j \in \mathcal{N} \setminus \{n_k\}$. Then $x \in [f_{n_k}]_{\perp}$ and hence, by 1°, $x \in [x_j]_{j \in \mathcal{N} \setminus \{n_k\}}$. By our assumption, there exists a finite or empty set $J \subset \mathcal{N} \setminus \{n_k\}$ such that $f(x_j) = 0$ for all $j \in (\mathcal{N} \setminus \{n_k\}) \setminus J$. If $J$ is non-empty, then, by Ch. I, § 6, theorem 6.1, we have $x = \sum_{j \in J} f_j(x) x_j + y$, where $y \in [x_j]_{j \in (\mathcal{N} \setminus \{n_k\}) \setminus J}$, and hence

$$f(x) = \sum_{j \in J} f_j(x) f(x_j) + f(y) = \sum_{j=1}^{\infty} f_j(x) f(x_j);$$

if $J$ is empty, then, since $x \in [x_j]_{j \in \mathcal{N} \setminus \{n_k\}}$, we have $f(x) = 0 = \sum_{i=1}^{\infty} f(x_i) f_i(x)$. Thus, 1° ⇒ 4°.

Finally, assume that we have 4°. Let $\{n_k\}$ be an arbitrary sequence of positive integers and let $x \in [f_{n_k}]_{\perp}$. In order to prove that $x \in [x_j]_{j \in \mathcal{N} \setminus \{n_k\}}$, assume that $f \in E^*$, $f(x_j) = 0$ for all $j \in \mathcal{N} \setminus \{n_k\}$. Then, since $f_{n_k}(x) = 0$ $(k = 1,2, \ldots)$, the set $\{i \in \mathcal{N} \mid f(x_i) f_i(x) \neq 0\}$ is empty and hence, by 4°, we have

$$f(x) = \sum_{i=1}^{\infty} f(x_i) f_i(x) = 0,$$

which proves (by a corollary of the Hahn-Banach theorem) that $x \in [x_j]_{j \in \mathcal{N} \setminus \{n_k\}}$. Thus, 4° ⇒ 1°, which completes the proof of proposition 8.11.

From § 6, example 6.6 and proposition 8.16 below it follows that there exist $M$-bases which are not strong $M$-bases. Moreover, we shall prove now

**Proposition 8.12.** *Let $E$ be a separable Banach space and let $\{e_n\}$ be an $M$-basis of $E$, with the a.s.f. $\{h_n\} \subset E^*$. Then $E$ has an $M$-basis $\{x_n\}$ which is not a strong $M$-basis and which satisfies*

$$[x_1, \ldots, x_n] = [e_1, \ldots, e_n] \qquad (n = 1,2, \ldots), \tag{8.63}$$

$$[f_1, \ldots, f_n] \subset [h_1, \ldots, h_{n+3}] \qquad (n = 1,2, \ldots), \tag{8.64}$$

$$[h_1, \ldots, h_n] \subset [f_1, \ldots, f_{n+3}] \qquad (n = 1,2, \ldots), \tag{8.65}$$

*where $\{f_n\} \subset E^*$ is the a.s.f. to $\{x_n\}$.*

*Proof.* We may assume, without loss of generality (considering, if necessary, $\{e_{2n-1}\|h_{2n-1}\|\} \cup \left\{\frac{e_{2n}}{\|e_{2n}\|}\right\}$ instead of $\{e_n\}$), that

$$\sup_{1\leq n<\infty} \|e_{2n}\| < \infty, \quad \sup_{1\leq n<\infty} \|h_{2n-1}\| < \infty. \tag{8.66}$$

Now let

$$x_{2n-1} = -2^{n-1}e_{2n-1} \qquad (n = 1,2, \ldots), \tag{8.67}$$

$$x_2 = -e_1 + \frac{1}{2} e_2, \; x_{2n} = 2^{n-2} e_{2n-3} - 2^{n-1} e_{2n-1} + \frac{1}{2^n} e_{2n} \; (n=2,3,\ldots), \tag{8.68}$$

$$f_{2n-1} = -\frac{1}{2^{n-1}} h_{2n-1} - 2^n h_{2n} + 2^{n+1} h_{2n+2} \qquad (n = 1,2, \ldots), \tag{8.69}$$

$$f_{2n} = 2^n h_{2n} \qquad (n = 1,2, \ldots). \tag{8.70}$$

Then, clearly, $\{x_n\}$ is an $M$-basis of $E$, with the a.s.f. $\{f_n\}$ and we have (8.63), (8.64) and (8.65) (the last one because $h_{2n-1} = 2^{n-1}(-f_{2n-1} - - f_{2n} + f_{2n+2})$ for $n = 1,2, \ldots$). Let

$$x_0 = \sum_{i=1}^{\infty} \frac{1}{2^i} e_{2i}. \tag{8.71}$$

Then $x_0 \in E$ (by (8.66)) and $f_{2n-1}(x_0) = -\frac{2^n}{2^n} + \frac{2^{n+1}}{2^{n+1}} = 0$ $(n = = 1,2, \ldots)$, so $x_0 \in [f_{2n-1}]_{\perp}$. However, $x_0 \notin [x_{2n}]$, since for

$$f_0 = h_2 + \sum_{i=1}^{\infty} \frac{1}{2^i} h_{2i-1} \tag{8.72}$$

we have $f_0 \in E^*$ (by (8.66)), $f_0(x_0) = \frac{1}{2} \neq 0$, $f_0(x_2) = \frac{1}{2} - \frac{1}{2} = 0$ and

$$f_0(x_{2n}) = \frac{2^{n-2}}{2^{n-1}} - \frac{2^{n-1}}{2^n} = 0 \qquad (n = 2,3, \ldots).$$

Thus, the $M$-basis $\{x_n\}$ of $E$ is not strong, which completes the proof of proposition 8.12.

From proposition 8.11 and the inclusions (8.62) it follows that a sufficient condition for an $M$-basis $\{x_n\}$ to be strong is that for every sequence $\{n_k\}$ of positive integers we have

$$[f_{n_k}] = ([x_j]_{j \in \mathcal{N} \setminus \{n_k\}})^{\perp} \tag{8.73}$$

(or, equivalently $[f_{n_k}] \supset ([x_j]_{j \in \mathcal{N} \setminus \{n_k\}})^{\perp}$); moreover, in this case each subspace $[f_{n_k}]$ is obviously $\sigma(E^*, E)$-closed and $\{f_n\}$ is a strong $M$-basis of $[f_n]$. The $M$-basis $\{x_n\}$ of $E$ defined by (8.67), (8.68) does not satisfy (8.73) for $n_k = 2k - 1$ $(k = 1,2, \ldots)$, since we have seen that for $f_0 \in E^*$ defined by (8.72) we have $f_0 \in [x_{2k}]^\perp$, but even $f_0 \notin \widetilde{[f_{2k-1}]}$ (because for $x_0 \in E$ defined by (8.71), $f_0(x_0) = -\frac{1}{2} \neq 0$, $f_{2k-1}(x_0) = 0$ $(k = 1,2, \ldots)$).

*Problem 8.1.* Does there exist in every separable Banach space $E$ a strong $M$-basis $\{x_n\}$? What about an $M$-basis $\{x_n\}$, with the a.s.f. $\{f_n\} \subset E^*$, satisfying (8.73) for every sequence $\{n_k\}$ of positive integers?

In the case of $M$-bases, the special classes of minimal sequences and of generalized bases $\{x_n\}$ defined in § 6 and § 7 by using the a.s.f. $\{\varphi_n\} \subset [x_n]^*$, respectively an admissible sequence $\{f_n\} \subset E^*$ for $\{x_n\}$, coincide, because now $[x_n] = E$. Thus, we obtain, in a natural way, various special classes of $M$-bases.

We shall show now (in theorem 8.5 below) that strictly bounded $M$-bases exist in every separable Banach space $E$. To this end, let us first prove two propositions.

**Proposition 8.13.** *Let $E$ be a separable Banach space and let $V$ be a separable total subspace of the conjugate space $E^*$. Then there exists an $M$-basis $\{x_n\}$ of $E$ such that*

$$[f_n] \supset V, \tag{8.74}$$

$$\|x_{3k}\| = \|f_{3k}\| = 1 \qquad (k = 1,2, \ldots), \tag{8.75}$$

*where $\{f_n\} \subset E^*$ is the a.s.f. to $\{x_n\}$.*

*Proof.* If $\dim E < \infty$, then by Ch. II, § 2, theorem 2.2, $E$ has even a normal basis $\{x_n\}$.

Assume now that $\dim E = \infty$. Then, since $E$ and $V$ are separable, there exist sequences of subspaces $\{G_n\}$ of $E$ and $\{\Gamma_n\}$ of $V$ such that

$$\dim G_n = \dim \Gamma_n = n \qquad (n = 1,2, \ldots), \tag{8.76}$$

$$G_n \subset G_{n+1},\ \Gamma_n \subset \Gamma_{n+1} \qquad (n = 1,2, \ldots), \tag{8.77}$$

$$\overline{\bigcup_{n=1}^{\infty} G_n} = E, \quad \overline{\bigcup_{n=1}^{\infty} \Gamma_n} = V. \tag{8.78}$$

Thus, in order to complete the proof, it will be sufficient to construct a biorthogonal system $(x_n, f_n)$ $(\{x_n\} \subset E,\ \{f_n\} \subset E^*)$ satisfying (8.75) and

$$P_{(3k-2)} = [x_1, \ldots, x_{3k-2}] \supset G_k, \quad \Pi_{(3k-1)} = [f_1, \ldots, f_{3k-1}] \supset \Gamma_k \tag{8.79}$$

$$(k = 1,2, \ldots).$$

Take $x_1 \in G_1 \setminus \{0\}$ and $f_1 \in E^*$ with $f_1(x_1) = 1$. Assume that for some $n \geqslant 2$ we have defined $x_1, \ldots, x_{n-1} \in E$ and $f_1, \ldots, f_{n-1} \in E^*$ biorthogonal and so as to satisfy (8.79) and (8.75) for all indices $\leqslant n-1$. We shall consider separately three cases:

1°. $n = 3k-2$. If $P_{(n-1)} \supset G_k$, take an arbitrary $x_n \in [f_1, \ldots, f_{n-1}]_\perp \setminus \{0\}$. Then $x_n \notin P_{(n-1)}$, whence there is $f_n \in P_{(n-1)}^\perp$ with $f_n(x_n) = 1$. Thus, $f_i(x_j) = \delta_{ij}$ $(i, j = 1, \ldots, n)$ and

$$P_{(3k-2)} = P_{(n)} \supset P_{(n-1)} \supset G_k.$$

If $P_{(n-1)} \not\supset G_k$, say $z \in G_k \setminus P_{(n-1)}$, put

$$x_n = z - \sum_{i=1}^{n-1} f_i(z) x_i, \tag{8.80}$$

so $f_i(x_n) = 0$ $(i = 1, \ldots, n-1)$. Since $z \notin P_{(n-1)}$, we have $x_n \notin P_{(n-1)}$, whence there exists $f_n \in P_{(n-1)}^\perp$ with $f_n(x_n) = 1$. Thus, $f_i(x_j) = \delta_{ij}$ $(i, j = 1, \ldots, n)$. Furthermore, by the induction hypothesis we have

$$G_{k-1} \subset P_{(3k-5)} = P_{(n-3)} \subset P_{(n-1)},$$

whence, since $z \in G_k \setminus P_{(n-1)}$, we obtain $z \in G_k \setminus G_{k-1}$. Therefore, by $\dim G_k = \dim G_{k-1} + 1$, we have $G_k = G_{k-1} \oplus [z]$, whence, since $G_{k-1} \subset P_{(n-1)} \subset P_{(n)}$ and $z \in P_{(n)}$, it follows that

$$P_{(3k-2)} = P_{(n)} \supset G_{k-1} \oplus [z] = G_k.$$

2°. $n = 3k-1$. In this case the argument is dual to the above. (If $\Pi_{(n-1)} \not\supset \Gamma_k$, say $h \in \Gamma_k \setminus \Pi_{(n-1)}$, we put $f_n = h - \sum_{j=1}^{n-1} h(x_j) f_j$; then, since $h \notin \Pi_{(n-1)}$, we have $f_n \notin \Pi_{(n-1)}$, whence, since $\Pi_{(n-1)}$ is $w^*$-closed, there exists $x_n \in (\Pi_{(n-1)})_\perp$ with $f_n(x_n) = 1$. By the induction hypothesis now $\Gamma_{k-1} \subset \Pi_{(3k-4)} = \Pi_{(n-3)} \subset \Pi_{(n-1)}$, whence, similarly to the above, $\Pi_{(3k-1)} \supset \Gamma_k$.)

3°. $n = 3k$. By § 1, lemma 1.2, there exists $x_n \in [f_1, \ldots, f_{n-1}]_\perp$ such that $\|x_n\| = 1$, $\operatorname{dist}(x_n, P_{(n-1)}) = 1$. Hence, by a corollary of the Hahn-Banach theorem, there exists $f_n \in P_{(n-1)}^\perp$ with $f_n(x_n) = 1$, $\|f_n\| = 1$. Then $f_i(x_j) = \delta_{ij}$ $(i, j = 1, \ldots, n)$.

Thus, in each case $x_1, \ldots, x_n \in E$ and $f_1, \ldots, f_n \in E^*$ satisfy $f_i(x_j) = \delta_{ij}$, (8.79) and (8.75) for all indices $\leqslant n$. This completes the proof of proposition 8.13.

*Remark* 8.5. a) Condition (8.75) can be replaced by

$$\|x_{n_k}\| = \|f_{n_k}\| = 1 \qquad (k = 1, 2, \ldots), \tag{8.75'}$$

where $\{n_k\}$ is any given infinite subsequence of $\mathcal{N} = \{1, 2, 3, \ldots\}$ such that $\mathcal{N} \setminus \{n_k\}$ is infinite. Indeed, it is enough to take a permutation $\sigma$

of $\mathcal{N}$ such that $\sigma(n_k) = 3k$ $(k = 1,2, \ldots)$ and then to consider the biorthogonal system $(\tilde{x}_n, \tilde{f}_n)$, where $\tilde{x}_n = x_{\sigma(n)}, \tilde{f}_n = f_{\sigma(n)}$ $(n = 1,2, \ldots)$, with $(x_n, f_n)$ as in proposition 8.13.

b) In particular, if $r(V) > 0$, then by (8.74) $r([f_n]) > 0$, i.e. $\{x_n\}$ is a norming $M$-basis of $E$.

**Proposition 8.14.** *Let $E$ be a Banach space and $n$ a positive integer and let $x_1, \ldots, x_{2^n} \in E$ and $f_1, \ldots, f_{2^n} \in E^*$ be such that $f_i(x_j) = \delta_{ij}$ $(i,j = 1, \ldots, 2^n)$. Then there exists a unitary real $2^n \times 2^n$ matrix $(\beta_{ij}^{(n)})_{i,j=1}^{2^n}$ such that the elements $z_i \in E$ and the functionals $h_i \in E^*$ defined by*

$$z_i = \sum_{j=1}^{2^n} \beta_{ij}^{(n)} x_j \in E, \; h_i = \sum_{j=1}^{2^n} \beta_{ij}^{(n)} f_j \in E^* \; (i = 1, \ldots, 2^n) \tag{8.81}$$

*satisfy*

$$\max_{1 \leqslant i \leqslant 2^n} \|z_i\| < (1 + \sqrt{2}) \max_{2 \leqslant j \leqslant 2^n} \|x_j\| + \frac{1}{\sqrt{2^n}} \|x_1\|, \tag{8.82}$$

$$\max_{1 \leqslant i \leqslant 2^n} \|h_i\| < (1 + \sqrt{2}) \max_{2 \leqslant j \leqslant 2^n} \|f_j\| + \frac{1}{\sqrt{2^n}} \|f_1\|, \tag{8.83}$$

$$h_i(z_j) = \delta_{ij} \qquad (i,j = 1, \ldots, 2^n), \tag{8.84}$$

$$[z_i]_{i=1}^{2^n} = [x_i]_{i=1}^{2^n}, \; [h_i]_{i=1}^{2^n} = [f_i]_{i=1}^{2^n}. \tag{8.85}$$

*Proof.* We recall*) that the normalized Haar basis of $l_{2^n}^2$ is defined by

$$y_1 = \sum_{i=1}^{2^n} \beta_{i,1}^{(n)} e_i, \; y_{2^k+l} = \sum_{i=1}^{2^n} \beta_{i,2^k+l}^{(n)} e_i \; (l = 1, \ldots, 2^k; \; k=0, 1, \ldots, n-1), \tag{8.86}$$

where $\{e_i\}_{i=1}^{2^n}$ is the unit vector basis of $l_{2^n}^2$ (so $(l_{2^n}^2, \{e_i\})$ is a $2^n$-dimensional symmetric space) and where

$$\beta_{i,1}^{(n)} = \frac{1}{\sqrt{2^n}} \text{ for } i = 1, \ldots, 2^n$$

$$\beta_{i,2^k+l}^{(n)} = \begin{cases} \dfrac{1}{\sqrt{2^{n-k}}} \text{ for } (2l-2)\, 2^{n-k-1} + 1 \leqslant i \leqslant (2l-1)\, 2^{n-k-1}, \\ -\dfrac{1}{\sqrt{2^{n-k}}} \text{ for } (2l-1)\, 2^{n-k-1} + 1 \leqslant i \leqslant 2l.2^{n-k-1}, \\ 0 \quad \text{for } 1 \leqslant i \leqslant (2l-2)\, 2^{n-k-1} \text{ and } 2l.2^{n-k-1}+1 \leqslant i \leqslant 2^n. \end{cases} \tag{8.87}$$

*) See Ch. II, § 22, definition 22.5.

We shall show that the above matrix $(\beta_{i,j}^{(n)})_{i,j=1}^{2^n}$ has the required properties. Indeed, $(l_{2^n}^2, \{e_i\})$ is isometric to the subspace of $L^2([0, 1])$ spanned by the characteristic functions $\chi_i(.)$ of the intervals $\left(\frac{i-1}{2^n}, \frac{i}{2^n}\right)$ $(i = 1, \ldots, 2^n)$, by a mapping*) $u$ which carries $y_1, \ldots, y_{2^n}$ onto the first $2^n$ functions of the normalized Haar system in $L^2([0, 1])$. Therefore, by Ch. I, § 2, formula (2.19), $\{y_i\}_{i=1}^{2^n}$ is an orthogonal basis of $(l_{2^n}^2, \{e_i\})$ and hence, since $(\beta_{ij}^{(n)})_{i,j=1}^{2^n}$ transforms the orthogonal basis $\{e_i\}_{i=1}^{2^n}$ of $(l_{2^n}^2, \{e_i\})$ onto $\{y_i\}_{i=1}^{2^n}$, $(\beta_{ij}^{(n)})_{i,j=1}^{2^n}$ is a unitary matrix. Consequently, by (8.81) we have (8.85) and

$$h_p(z_q) = \left(\sum_{i=1}^{2^n} \beta_{pi}^{(n)} f_i\right)\left(\sum_{j=1}^{2^n} \beta_{qj}^{(n)} x_j\right) = \sum_{i=1}^{2^n}\sum_{j=1}^{2^n} \beta_{pi}^{(n)} \beta_{qj}^{(n)} \delta_{ij} =$$

$$= \sum_{i=1}^{2^n} \beta_{pi}^{(n)} \beta_{qi}^{(n)} = \delta_{pq} \qquad (p, q = 1, \ldots, 2^n),$$

that is, (8.84). Finally, for each $i = 1, \ldots, 2^n$ we have

$$\sum_{j=1}^{2^n} |\beta_{ij}^{(n)}| = \sum_{k=0}^{n-1} \frac{1}{\sqrt{2^{n-k}}} < \frac{\frac{1}{\sqrt{2}}}{1-\frac{1}{\sqrt{2}}} = \frac{1}{\sqrt{2}-1} = 1+\sqrt{2},$$

whence, by (8.81), we obtain

$$\|z_i\| = \left\|\sum_{j=1}^{2^n} \beta_{ij}^{(n)} x_i\right\| \leqslant \sum_{j=2}^{2^n} |\beta_{ij}^{(n)}| \max_{2\leqslant p\leqslant 2^n} \|x_p\| + |\beta_{i1}^{(n)}|\,\|x_1\| <$$

$$< (1+\sqrt{2}) \max_{2\leqslant p\leqslant 2^n} \|x_p\| + \frac{1}{\sqrt{2^n}} \qquad (i = 1, \ldots, 2^n),$$

that is, (8.82) and, similarly, (8.83). This completes the proof of proposition 8.14.

Now we are ready to prove

**Theorem 8.5.** *Let $E$ be a separable Banach space, $V$ a separable total subspace of the conjugate space $E^*$ and $\varepsilon > 0$. Then there exists an $M$-basis $\{z_n\}$ of $E$ such that*

$$[h_n] \supset V, \tag{8.88}$$

$$\sup_{1\leqslant n<\infty} \|z_n\| < 1+\sqrt{2}+\varepsilon, \quad \sup_{1\leqslant n<\infty} \|h_n\| < 1+\sqrt{2}+\varepsilon, \tag{8.89}$$

*where $\{h_n\} \subset E^*$ is the a.s.f. to $\{z_n\}$. Hence, in particular, $E$ has a norming $M$-basis satisfying* (8.89).

*) Namely, $u: e_i \to \sqrt{2^n}\chi_i(.)$ for $i = 1, \ldots, 2^n$ (see e.g. Ch. II, § 18, proposition 18.3 and Ch. II, § 2, formula (2.3) with $p = 2$).

*Proof.* Let $\{x_n\}$ be an $M$-basis of $E$, with the a.s.f. $\{f_n\} \subset E^*$, as in proposition 8.13 above. Let $\{l_k\}$ denote the sequence $\{1,2,4,5, \ldots \ldots, 3s-2, 3s-1, \ldots\}$. Choose, successively, an increasing sequence of positive integers $\{m_k\}$ such that

$$1+\sqrt{2}+\varepsilon > \max\left(1+\sqrt{2}+\frac{1}{\sqrt{2^{m_k}}}\|x_{l_k}\|,\ 1+\sqrt{2}+\frac{1}{\sqrt{2^{m_k}}}\|f_{l_k}\|\right) \tag{8.90}$$

$$(k=1,2,\ldots).$$

Next, choose a permutation $\sigma$ of $\mathcal{N} = \{1,2,3, \ldots\}$ such that

$$\sigma(q_{k-1}+1) = l_k \quad (k=1,2,\ldots), \tag{8.91}$$

where $q_0 = 0$, $q_k = 2^{m_1} + 2^{m_2} + \ldots + 2^{m_k}$ $(k=1,2,\ldots)$, and put

$$\tilde{x}_n = x_{\sigma(n)},\ \tilde{f}_n = f_{\sigma(n)} \quad (n=1,2,\ldots). \tag{8.92}$$

Finally, put

$$z_{q_{k-1}+i} = \sum_{j=1}^{2^{m_k}} \beta_{ij}^{(m_k)} \tilde{x}_{q_{k-1}+j} \qquad (i=1,\ldots,2^{m_k};\ k=1,2,\ldots), \tag{8.93}$$

$$h_{q_{k-1}+i} = \sum_{j=1}^{2^{m_k}} \beta_{ij}^{(m_k)} \tilde{f}_{q_{k-1}+j} \qquad (i=1,\ldots,2^{m_k};\ k=1,2,\ldots), \tag{8.94}$$

where $\{\beta_{ij}^{(m_k)}\}_{i,j=1}^{2^{m_k}}$ is the unitary real $2^{m_k} \times 2^{m_k}$ matrix of proposition 8.14 for $\{\tilde{x}_{q_{k-1}+j}\}_{j=1}^{2^{m_k}} \subset E$, $\{\tilde{f}_{q_{k-1}+j}\}_{j=1}^{2^{m_k}} \subset E^*$. Then, by (8.82) for $\tilde{x}_{q_{k-1}+j}$, $\tilde{f}_{q_{k-1}+j}$ and by (8.92), (8.91) and (8.75),

$$\max_{1 \leqslant i \leqslant 2^{m_k}} \|z_{q_{k-1}+i}\| < (1+\sqrt{2}) \max_{2 \leqslant j \leqslant 2^{m_k}} \|x_{q_{k-1}+j}\| + \frac{1}{\sqrt{2^{m_k}}}\|\tilde{x}_{q_{k-1}+1}\| =$$

$$= 1+\sqrt{2}+\frac{1}{\sqrt{2^{m_k}}}\|x_{l_k}\| \qquad (k=1,2,\ldots),$$

and, similarly, using (8.83),

$$\max_{1 \leqslant i \leqslant 2^{m_k}} \|h_{q_{k-1}+i}\| < 1+\sqrt{2}+\frac{1}{\sqrt{2^{m_k}}}\|f_{l_k}\| \qquad (k=1,2,\ldots).$$

Hence, by (8.90), we obtain (8.89). Furthermore, by (8.84) we have $h_i(z_j) = \delta_{ij}$ $(i,j=1,2,\ldots)$. Finally, by (8.85) and (8.74) we obtain $[z_n] = [\tilde{x}_n] = E$, $[h_n] = [\tilde{f}_n] \supset V$, which completes the proof of theorem 8.5.

As we already mentioned (in connection with § 6, theorem 6.3 and § 7, theorem 7.7), it has been shown recently that in every separable

Banach space $E$ there exists, for each $\varepsilon > 0$, an $M$-basis $\{z_n\}$ of $E$ such that $\sup_{1\leqslant n<\infty} \|z_n\| < 1+\varepsilon$ and $\sup_{1\leqslant n<\infty} \|h_n\| < 1+\varepsilon$, where $\{h_n\}$ is the a.s.f. to $\{z_n\}$.

*Problem 8.2.* a) Does there exist in every separable Banach space $E$ a normal*) $M$-basis? b) How about a normal $M$-basis satisfying, in addition, (8.88)?

The following proposition, together with theorem 8.5, shows that if we require these conditions to be satisfied only in a suitable equivalent norm on $E$, then the answer is affirmative:

**Proposition 8.15.** *Let $E$ be a separable Banach space. Then for every normalized strictly bounded $M$-basis $\{z_n\}$ of $E$ there exists an equivalent norm $|||\cdot|||$ on $E$ in which $\{z_n\}$ is a normal $M$-basis, that is,*

$$|||z_n||| = |||h_n||| = 1 \qquad (n = 1,2,\ldots), \tag{8.95}$$

*where $\{h_n\} \subset E^*$ is the a.s.f. to $\{z_n\}$.*

*Proof.* The norm defined by

$$|||x||| = \max\,(\|x\|, \sup_{1\leqslant n<\infty} |h_n(x)|) \qquad (x \in E) \tag{8.96}$$

has the required properties. Indeed, we have

$$\|x\| \leqslant |||x||| \leqslant (\sup_{1\leqslant n<\infty} \|h_n\|)\|x\| \qquad (x \in E),$$

so $|||\cdot|||$ is equivalent to the initial norm on $E$. Furthermore, since $\{z_n\}$ is normalized and since $(z_n, h_n)$ is biorthogonal, we have

$$|||z_n||| = \max\,(\|z_n\|, \sup_{1\leqslant i<\infty} |h_i(z_n)|) = 1 \qquad (n = 1,2,\ldots).$$

Finally, since

$$\sup_{1\leqslant n<\infty} |h_n(x)| \leqslant |||x||| \qquad (x \in E),$$

and since $|||h_n||| \geqslant \dfrac{1}{|||z_n|||}|h_n(z_n)| = 1$, we have $|||h_n||| = 1$ $(n=1,2,\ldots)$, which completes the proof of proposition 8.15.

*Remark 8.6.* The above argument also gives a simpler proof for Ch. II, § 2, theorem 2.3.

---

*) See § 6, definition 6.4. Obviously, an affirmative answer to problem 8.2 would imply the same for § 6, problem 6.3 and § 7, problem 7.1.

From proposition 8.4 and the remark made after theorem 8.2 it follows that every block sequence $\{y_n\}$ *with respect to an M-basis* $\{x_n\}$ *of a Banach space E is an M-basic sequence.* Also, by § 6, proposition 6.8, *every block sequence* $\{y_n\}$ *with respect to a shrinking M-basis* $\{x_n\}$ *is a shrinking M-basic sequence.* For subsequences of (and hence also for block sequences with respect to) boundedly complete $M$-bases*) $\{x_n\}$ a similar result is no longer valid, as shown by § 6, example 6.6, since the $M$-basis $\{f_n\}$ of $E^* \equiv l^1$ considered in that example is not only quasi-boundedly complete, but also boundedly complete (by § 7, theorem 7.9, implication 3° ⇒ 1°). In this direction we have the following positive result:

**Proposition 8.16.** *Let* $\{x_n\}$ *be a boundedly complete strong M-basis of a Banach space E. Then every subsequence* $\{x_{n_k}\}$ *of* $\{x_n\}$ *is a boundedly complete M-basic sequence.*

*Proof.* By proposition 8.4, $\{x_{n_k}\}$ is an $M$-basic sequence. Let $\{y_j\} \subset [x_{n_k}]$ be a bounded sequence such that $\lim_{j\to\infty} f_{n_k}(y_j) = \alpha_k$ $(k = 1,2, \ldots)$. Then, by biorthogonality, $f_i(y_j) = 0$ $(i \in \mathcal{N}\setminus\{n_k\},\ j = 1,2, \ldots)$, whence $\lim_{j\to\infty} f_i(y_j) = 0$ $(i \in \mathcal{N}\setminus\{n_k\})$. Therefore, since $\{x_n\}$ is boundedly complete, there exists, by § 7, theorem 7.9, implication 1° ⇒ 2°, an element $x \in E$ such that

$$f_{n_k}(x) = \alpha_k \qquad (k = 1,2, \ldots), \tag{8.97}$$

$$f_i(x) = 0 \qquad (i \in \mathcal{N}\setminus\{n_k\}). \tag{8.98}$$

Then, by (8.98) and since $\{x_n\}$ is a strong $M$-basis, we have $x \in [x_{n_k}]$, which, together with (8.97) and § 7, theorem 7.9, implication 2° ⇒ 1°, shows that $\{x_{n_k}\}$ is boundedly complete. This concludes the proof of proposition 8.16.

The boundedly complete $M$-bases are in the following relation of duality with shrinking $M$-bases:

**Proposition 8.17.** *Let* $\{x_n\}$ *be an M-basis of a Banach space E, with the a.s.f.* $\{f_n\} \subset E^*$.

a) *If* $\{x_n\}$ *is boundedly complete, then* $\{f_n\}$ *is a shrinking M-basis of* $[f_n]$.

b) *If* $\{x_n\}$ *is shrinking, then* $\{f_n\}$ *is a boundedly complete M-basis of* $[f_n]$.

*Proof.* a) By proposition 8.10, $\{f_n\}$ is an $M$-basis of $[f_n]$, with the a.s.f. $\{u(x_n)\} \subset [f_n]^*$, where $u$ is the canonical mapping of $E$ into $[f_n]^*$. Furthermore, by § 7, theorem 7.9, implication 1° ⇒ 3°, we have $[u(x_n)] = u(E) = [f_n]^*$, and thus $\{f_n\}$ is a shrinking $M$-basis of $[f_n]$.

---

*) See § 7, definition 7.8.

b) If $[f_n] = E^*$, then $\{f_n\}$ is an $M$-basis of $E^*$ with the a.s.f. $\{\pi(x_n)\}$, where $\pi$ is the canonical embedding of $E$ into $E^{**}$, and the canonical mapping $v$ of $[f_n]$ into $[\pi(x_n)]^* = \pi(E)^*(\equiv E^*)$ is an isometry onto. Hence, by § 7, theorem 7.9, implication $3° \Rightarrow 1°$, $\{f_n\}$ is a boundedly complete $M$-basis of $[f_n] = E^*$, which completes the proof.

In contrast with the situation for bases (see Ch. II, § 6, corollary 6.1), the converse statement in proposition 8.17 a) above is not valid, as shown by

*Example 8.6.* Let $E = (E_1 \times E_2 \times \ldots)_{l^1} \equiv l^1$, where $E_j = l^1$ $(j = 1,2, \ldots)$ and let $\{x_n\}$ be the $M$-basis of $E$ considered in Ch. I, § 12, example 12.4. Then, as we observed in that example, $r([f_n]) = 0$ and hence the canonical mapping $u: E \to [f_n]^*$ (where $\{f_n\}$ is the a.s.f. to $\{x_n\}$) is not an isomorphism; therefore, by § 7, theorem 7.9, $\{x_n\}$ is not boundedly complete. However, in each $E_j^* \equiv m$ the subspace $[f_n^{(j)}]$ admits an isomorphism $u_j$ onto $c_0$, satisfying $\frac{1}{4}\|f\| \leqslant \|u_j(f)\| \leqslant \leqslant 4\|f\|$ for all $f \in [f_n^{(j)}]$; indeed, this follows by the argument of Ch. I, § 7, proof of lemma 7.1 (because $[f_n^{(j)}] \cap [h_n^{(j)}] = [h_n^{(j)}]_{n=2}^{\infty}$ is of codimension 1 in both $[f_n^{(j)}]$ and $[h_n^{(j)}]_{n=1}^{\infty} \equiv c_0$; since these spaces are not reflexive, we take $\frac{1}{4}$ and 4 instead of $\frac{1}{3}$ and 3, respectively). Hence, by Ch. II, § 18, lemma 18.7 a), $[f_n]$ is isomorphic to $(c_0 \times c_0 \times \ldots)_{c_0} \equiv c_0$ (the proof of this latter isometry is similar to that of Ch. II, § 18, lemma 18.5 b)). Therefore, since $\{f_n\}$ is a bounded unconditional basis of $[f_n]$, it follows that $\{f_n\}$ is shrinking (e.g., by Ch. II, § 17, corollary 17.3 or by Ch. II, § 18, theorem 18.2).

We know no example disproving the converse of proposition 8.17 b), that is, of a non-shrinking $M$-basis $\{x_n\}$ of a Banach space $E$, such that $\{f_n\}$ is a boundedly complete $M$-basis of $[f_n]$. Note that in any such example the characteristic $r([f_n])$ of the subspace $[f_n]$ must be $= 0$. Indeed, since $\{f_n\}$ is a boundedly complete $M$-basis of $[f_n]$, $\{u(x_n)\}$ is a shrinking $M$-basis of $[u(x_n)]$, where $u$ is the canonical mapping of $E$ into $[f_n]^*$. Now, if $r([f_n]) > 0$, then, by Ch. I, § 12, formula (12.15), $u$ is an isomorphism of $E$ onto $u(E) \subset [f_n]^*$ and hence $\{x_n\}$ is a shrinking $M$-basis of $E$.

Proposition 8.17 a) and example 8.6 show that in order to obtain a characterization of boundedly complete $M$-bases $\{x_n\}$ in terms of strong duality properties, one has to add something more to the condition that the a.s.f. $\{f_n\}$ be a shrinking $M$-basis of $[f_n]$. Such a characterization, and some other characterizations of boundedly completeness, are given in

**Theorem 8.6.** *Let $\{x_n\}$ be an $M$-basis of a Banach space $E$, with the a.s.f. $\{f_n\} \subset E^*$. The following statements are equivalent:*

1°. *$\{x_n\}$ is boundedly complete (i.e., every bounded $\sigma(E, [f_n])$-Cauchy sequence $\{y_j\} \subset E$ is $\sigma(E, [f_n])$-convergent to some $x \in E$).*

2°. *For every bounded sequence $\{y_j\} \subset E$ such that the limits*

$$\lim_{j\to\infty} f_n(y_j) = \alpha_n \qquad (n = 1,2, \ldots) \tag{8.99}$$

*exist, there is an element $x \in E$ such that*

$$f_n(x) = \alpha_n \qquad (n = 1,2, \ldots). \tag{8.100}$$

3°. *$E$ is canonically isomorphic to $[f_n]^*$.*

4°. *$\{f_n\}$ is a shrinking M-basis of $[f_n]$ and $\{x_n\}$ is a norming M-basis of $E$ (i.e., $r([f_n]) > 0$).*

5°. *The relations*

$$\{\alpha_n\} \subset K, \ \sup_{1\leqslant n<\infty} \operatorname{dist}\left(\sum_{i=1}^{n} \alpha_i x_i, [x_i]_{n+1}^{\infty}\right) < \infty \tag{8.101}$$

*imply that there exists an element $x \in E$ satisfying* (8.100).

*Proof.* The equivalences $1° \Leftrightarrow 2° \Leftrightarrow 3°$ are nothing else than the particular case $[x_n] = E$ of § 7, theorem 7.9.

Assume now that we have 3°. Then we have 1° and therefore, by proposition 8.17 a), $\{f_n\}$ is a shrinking $M$-basis of $[f_n]$. Furthermore, by 3° the canonical mapping $u$ of $E$ into $[f_n]^*$ is an isomorphism and hence, by Ch. I, § 12, formula (12.15), $r([f_n]) > 0$. Thus, $3° \Rightarrow 4°$.

Conversely, assume that we have 4°. Then, by $r([f_n]) > 0$ and Ch. I, § 12, formula (12.15), $u$ is an isomorphism of $E$ into $[f_n]^*$. Furthermore, since $\{f_n\}$ is a shrinking $M$-basis of $[f_n]$, we have $[u(x_n)] = [f_n]^*$ and hence $u(E)$ is dense in $[f_n]^*$. Consequently, since $E$ is complete, it follows that $u(E) = [f_n]^*$. Thus, $4° \Rightarrow 3°$.

Assume again that we have 3° and let $\{\alpha_n\}$ satisfy (8.101). Then we can choose $y_n \in [x_i]_{n+1}^{\infty}$ $(n = 1,2, \ldots)$ such that $\sup_{1\leqslant n<\infty} \left\|\sum_{i=1}^{n} \alpha_i x_i + y_n\right\| < \infty$, whence

$$\sup_{1\leqslant n<\infty} \left\|u\left(\sum_{i=1}^{n} \alpha_i x_i + y_n\right)\right\| < \infty,$$

and therefore, since $[f_n]$ is separable, there exists a subsequence $\left\{u\left(\sum_{i=1}^{n_k} \alpha_i x_i + y_{n_k}\right)\right\}$ of $\left\{u\left(\sum_{i=1}^{n} \alpha_i x_i + y_n\right)\right\}$, converging to some $\Psi \in [f_n]^*$ in the weak* topology $\sigma([f_n]^*, [f_n])$. Then, since by 3° $\Psi = u(x)$

for some $x \in E$, we have

$$f(x) = u(x)(f) = \Psi(f) = \lim_{k\to\infty} u\left(\sum_{i=1}^{n_k} \alpha_i x_i + y_{n_k}\right)(f) =$$

$$= \lim_{k\to\infty} f\left(\sum_{i=1}^{n_k} \alpha_i x_i + y_{n_k}\right) \qquad (f \in [f_n]),$$

whence, in particular, for $f = f_n$ we obtain (8.100). Thus, 3° ⇒ 5°.

Finally, assume that we have 5° and let $\{y_j\} \subset E$ be a bounded sequence such that the limits (8.99) exist. Then $\lim_{j\to\infty} s_n(y_j) = \lim_{j\to\infty} \sum_{i=1}^{n} f_i(y_j)x_i = \sum_{i=1}^{n} \alpha_i x_i$, whence we can choose a subsequence $\{y_{j_n}\}$ of $\{y_j\}$ such that

$$\left\| \sum_{i=1}^{n} \alpha_i x_i - s_n(y_{j_n}) \right\| < \frac{1}{2^n} \qquad (n = 1,2, \ldots),$$

and therefore

$$\operatorname{dist}\left(\sum_{i=1}^{n} \alpha_i x_i, [x_i]_{n+1}^{\infty}\right) \leqslant \left\| \sum_{i=1}^{n} \alpha_i x_i + (I - s_n)(y_{j_n}) \right\| \leqslant \|y_{j_n}\| + \frac{1}{2^n}$$

$$(n = 1,2, \ldots).$$

Hence, since $\sup_{1 \leqslant j < \infty} \|y_j\| < \infty$, from 5° it follows that there exists an $x \in E$ satisfying (8.100). Thus 5° ⇒ 2°, which completes the proof of theorem 8.6.

*Remark 8.7.* Similarly to the equivalence 1° ⇔ 4°, one can also give the following reformulation of proposition 8.17 b) and the observation made after example 8.6: *Let $\{x_n\}$ be an M-basis of a Banach space E, with the a.s.f. $\{f_n\} \subset E^*$. The following statements are equivalent:*

1°. *$\{x_n\}$ is shrinking.*

2°. *$\{f_n\}$ is a boundedly complete M-basis of $[f_n]$ and $r([f_n]) > 0$.*

Let us also note that if $\{x_n\}$ is a generalized basis of $E$, then, applying theorem 8.6 to the subspace $E_0 = [x_n]$, we obtain characterizations of the boundedly completeness of $\{x_n\}$, which complement those given in § 7, theorem 7.9.

*Remark 8.8.* Condition 5° of theorem 8.6 is somewhat similar to that occurring in the definition of boundedly complete bases, $\left\| \sum_{i=1}^{n} \alpha_i x_i \right\|$

being replaced by dist $\left(\sum_{i=1}^{n} \alpha_i x_i, [x_i]_{n+1}^{\infty}\right)$. Note that this is always a *monotone non-decreasing* function of $n$; indeed, for every $y \in [x_i]_{n+2}^{\infty}$ we have

$$\left\| \sum_{i=1}^{n+1} \alpha_i x_i + y \right\| = \left\| \sum_{i=1}^{n} \alpha_i x_i + \alpha_{n+1} x_{n+1} + y \right\| \geqslant \operatorname{dist}\left(\sum_{i=1}^{n} \alpha_i x_i, \ [x_i]_{n+1}^{\infty}\right),$$

whence the assertion follows.

We also mention the following related proposition:

**Proposition 8.18.** *Let $\{x_n\}$ be an M-basis of a Banach space $E$, with the a.s.f. $\{f_n\} \subset E^*$. Then*

$$\|x\|_{[f_n]} = \sup_{\substack{f \in [f_n] \\ \|f\| \leqslant 1}} |f(x)| = \sup_{1 \leqslant n < \infty} \operatorname{dist}(s_n(x), [x_i]_{n+1}^{\infty}) \quad (x \in E). \tag{8.102}$$

*Hence, in particular, if $r([f_n]) = 1$, then*

$$\|x\| = \sup_{1 \leqslant n < \infty} \operatorname{dist}\,(s_n(x), \ [x_i]_{n+1}^{\infty}) \qquad (x \in E). \tag{8.103}$$

*Proof.* Let $x \in E$ and let $g_k = \sum_{i=1}^{m_k} \beta_i^{(k)} f_i$ $(k = 1,2, \ldots)$ be a sequence of finite linear combinations of the $f_n$ (depending on $x$), with $\|g_k\| \leqslant 1$ $(k = 1,2, \ldots)$, such that

$$\|x\|_{[f_n]} = \lim_{k \to \infty} |g_k(x)|. \tag{8.104}$$

Then for every $y \in [x_i]_{m_k+1}^{\infty}$ we have, by biorthogonality,

$$|g_k(x)| = \left| g_k\left(\sum_{i=1}^{m_k} f_i(x) x_i\right)\right| = \left| g_k\left(\sum_{i=1}^{m_k} f_i(x) x_i + y\right)\right| \leqslant \left\| \sum_{i=1}^{m_k} f_i(x) x_i + y \right\|,$$

whence

$$|g_k(x)| \leqslant \operatorname{dist}\left(\sum_{i=1}^{m_k} f_i(x) x_i, \quad [x_i]_{m_k+1}^{\infty}\right) \leqslant$$

$$\leqslant \sup_{1 \leqslant n < \infty} \operatorname{dist}\,(s_n(x), [x_i]_{n+1}^{\infty}) \quad (k = 1,2, \ldots),$$

and therefore, by (8.104),

$$\|x\|_{[f_n]} \leqslant \sup_{1 \leqslant n < \infty} \operatorname{dist}\,(s_n(x), [x_i]_{n+1}^{\infty}). \tag{8.105}$$

On the other hand, for each $n$ there exists, by a corollary of the Hahn-Banach theorem, a functional $h_n \in ([x_i]_{n+1})^{\perp} = [f_1, \ldots, f_n]$ of norm $\|h_n\| = 1$, such that $h_n(x) = h_n(s_n(x)) = \operatorname{dist}(s_n(x), [x_i]_{n+1}^{\infty})$. Hence

$$\sup_{1 \leqslant n < \infty} \operatorname{dist}(s_n(x), [x_i]_{n+1}^{\infty}) = \sup_{1 \leqslant n < \infty} |h_n(x)| \leqslant$$

$$\leqslant \sup_{1 \leqslant n < \infty} \sup_{\substack{f \in [f_1, \ldots, f_n] \\ \|f\| \leqslant 1}} |f(x)| = \sup_{\substack{f \in [f_n] \\ \|f\| \leqslant 1}} |f(x)| = \|x\|_{[f_n]},$$

which, together with (8.105), gives (8.102). In particular, if $r([f_n]) = 1$, then $\|x\| = \|x\|_{[f_n]}$ for all $x \in E$, whence we obtain (8.103), which completes the proof of proposition 8.18.

*Remark 8.9.* One can also extend to $M$-bases the notions of $k$-shrinking basis and $k$-boundedly complete basis given in Ch. II, § 4 and § 6. Indeed, for $k \geqslant 0$, an $M$-basis $\{x_n\}$ of $E$, with the a.s.f. $\{f_n\} \subset E^*$, may be called *$k$-shrinking* if $\operatorname{codim}_{E^*} [f_n] = k$ and *$k$-boundedly complete* if $\{f_n\}$ is a $k$-shrinking $M$-basis of $[f_n]$ (which amounts to $\operatorname{codim}_{[f_n]^*}[u(x_n)] = k$, where $u$ is the canonical mapping of $E$ into $[f_n]^*$) and $r([f_n]) > 0$ (see theorem 8.6, equivalence $1° \Leftrightarrow 4°$).

In Ch. I, § 5, proposition 5.4 we have seen that if $\{x_n\}$ is a generalized basis of a Banach space $E$, with an admissible sequence $\{f_n\} \subset E^*$, then the mapping $v: \{\gamma_n\} \to v_{\{\gamma_n\}}$, where $v_{\{\gamma_n\}}(x) = x_{\{\gamma_n\}}$ $(x \in E)$, is an isometrical algebraic isomorphism of the Banach algebra $M(E, (x_n, f_n))$ into $L(E, E)$. In the case when $\{x_n\}$ is an $M$-basis of $E$, one can say more, namely

**Proposition 8.19.** *Let $\{x_n\}$ be an $M$-basis of a Banach space $E$, with the a.s.f. $\{f_n\} \subset E^*$. Then*

a) *The image of $M(E, (x_n, f_n))$ in $L(E, E)$ by the embedding $\{\gamma_n\} \to v_{\{\gamma_n\}}$ is the set $D(E, E)$ of all $u \in L(E, E)$ satisfying*

$$f_i(u(x_j)) = 0 \quad (i \neq j;\ i, j = 1, 2, \ldots). \tag{8.106}$$

b) *The inverse mapping $v^{-1}: D(E, E) \to M(E, (x_n, f_n))$ is given by*

$$v^{-1}(u) = \{f_j(u(x_j))\}_{j=1}^{\infty} \quad (u \in D(E, E)). \tag{8.107}$$

*Proof.* If $\{\gamma_n\} \in M(E, (x_n, f_n))$, then by Ch. I, § 5, formula (5.32) we have

$$f_i(v_{\{\gamma_n\}}(x_j)) = f_i(\gamma_j x_j) = \gamma_j f_i(x_j) = 0 \quad (i \neq j;\ i, j = 1, 2, \ldots),$$

and thus $v_{\{\gamma_n\}} \in D(E, E)$. Therefore $v_{M(E, (x_n, f_n))} \subset D(E, E)$; observe that we have not yet used the assumption $[x_n] = E$.

Conversely, let $u \in D(E, E)$ be arbitrary. Put

$$\gamma_i = f_i(u(x_i)) \qquad (i = 1,2, \ldots), \tag{8.108}$$

whence

$$u^*(f_i)(x_i) = f_i(u(x_i)) = \gamma_i = \gamma_i f_i(x_i) \qquad (i = 1,2, \ldots).$$

Then, since by (8.106) and biorthogonality we have

$$u^*(f_i)(x_j) = f_i(u(x_j)) = 0 = \gamma_i f_i(x_j) \qquad (i \neq j;\ i,j = 1,2, \ldots),$$

and since $[x_n] = E$, it follows that

$$u^*(f_i) = \gamma_i f_i \qquad (i = 1,2, \ldots). \tag{8.109}$$

Consequently,

$$f_i(u(x)) = u^*(f_i)(x) = \gamma_i f_i(x) = f_i(v_{\{\gamma_n\}}(x)) \qquad (x \in E, i = 1,2, \ldots),$$

whence, since $\{f_n\}$ is total on $E$, we infer $u = v_{\{\gamma_n\}}$. Thus $v$ maps $M(E, (x_n, f_n))$ onto $D(E, E)$ and we have (8.107), which completes the proof of proposition 8.19.

Formula (8.107) suggests to associate with $(x_n, f_n)$ some new sequence spaces, by taking instead of $D(E, E)$ other subsets*) of $L(E, E)$ and placing a suitable norm on the space of all sequences (8.107) resulting in this way. As an example, we shall take the subset of $L(E, E)$ consisting of all nuclear mappings of $E$ into $E$. We recall that a mapping $u \in L(E, E)$ is said to be *nuclear* if it can be represented in the form

$$u(x) = \sum_{i=1}^{\infty} g_i(x) y_i \qquad (x \in E), \tag{8.110}$$

where $\{y_n\} \subset E$, $\{g_n\} \subset E^*$ and

$$\sum_{i=1}^{\infty} \|g_i\| \|y_i\| < \infty. \tag{8.111}$$

The Banach space of all nuclear mappings $u: E \to E$, with the norm (called the *nuclear norm*)

$$\|u\|_N = \inf \sum_{i=1}^{\infty} \|g_i\| \|y_i\|, \tag{8.112}$$

---

*) For a similar device see Ch. II, § 16, proof of theorem 16.7 (formula (16.78)).

where the inf is taken over all representations of $u$ in the form (8.110), is denoted by $N(E, E)$; obviously, $\|u\| = \sup_{\substack{x\in E\\ \|x\|\leqslant 1}} \|u(x)\| \leqslant \|u\|_N$ $(u \in N(E, E))$.

*Definition 8.5.* Let $\{x_n\}$ be an $M$-basis of a Banach space $E$, with the a.s.f. $\{f_n\} \subset E^*$. The space

$$\mathscr{S}(E, (x_n, f_n)) = \{\{f_j(u(x_j))\}_{j=1}^{\infty} \mid u \in N(E, E)\}, \qquad (8.113)$$

with the norm

$$\|\{\alpha_j\}\| = \inf_{\substack{w\in N(E,E)\\ \{\alpha_j\}=\{f_j(w(x_j))\}}} \|w\|_N \qquad (\{\alpha_j\} \in \mathscr{S}(E, (x_n, f_n))), \quad (8.114)$$

is called *the series space of* $(x_n, f_n)$.

Obviously, we have

$$\|\{f_j(u(x_j))\}_{j=1}^{\infty}\| = \inf \sum_{i=1}^{\infty} \|g_i\| \|y_i\|, \qquad (8.115)$$

where the inf is taken over all $\{g_n\} \subset E^*$ and $\{y_n\} \subset E$ satisfying (8.111) and such that

$$f_j(u(x_j)) = \sum_{i=1}^{\infty} g_i(x_j) f_j(y_i) \qquad (j = 1,2, \ldots). \qquad (8.116)$$

By (8.113) and (8.116), the elements of $\mathscr{S}(E, (x_n, f_n))$ are sequences of scalar series, which motivates the term "series space"; another motivation is the remark made before definition 8.6 below.

**Proposition 8.20.** *Let $\{x_n\}$ be an M-basis of a Banach space E, with the a.s.f. $\{f_n\} \subset E^*$. Then*

a) *$\mathscr{S}(E, (x_n, f_n))$ is a BK-space linearly isometric to $N(E, E)/N_0(E, E)$, where*

$$N_0(E, E) = \{u_0 \in N(E, E) \mid f_j(u_0(x_j)) = 0 \quad (j = 1,2, \ldots)\}. \qquad (8.117)$$

b) *$\mathscr{S}(E, (x_n, f_n))$ contains all sequences of the form*

$$\{f_0(x_j) f_j(x_0)\}, \qquad (8.118)$$

*where $x_0 \in E$, $f_0 \in E^*$. Furthermore, we have*

$$\|\{f_0(x_j) f_j(x_0)\}_{j=1}^{\infty}\| \leqslant \|f_0\| \|x_0\| \qquad (x_0 \in E,\ f_0 \in E^*). \quad (8.119)$$

c) *The sequence of unit vectors* $\{e_n\}$ *is an M-basis of* $\mathscr{S}(E, (x_n, f_n))$, *and we have*

$$\frac{1}{\|x_n\|\|f_n\|} \leqslant \|e_n\| \leqslant \|x_n\|\|f_n\| \qquad (n = 1,2, \ldots), \qquad (8.120)$$

$$\frac{1}{\|x_n\|\|f_n\|} \leqslant \|h_n\| \leqslant \|x_n\|\|f_n\| \qquad (n = 1,2, \ldots), \qquad (8.121)$$

*where* $\{h_n\}$ *is the a.s.f. to* $\{e_n\}$, *i.e., the sequence of coordinate functionals on* $\mathscr{S}(E, (x_n, f_n))$.

*Proof.* a) The functionals

$$\varphi_j(u) = f_j(u(x_j)) \qquad (u \in N(E, E),\ j = 1,2, \ldots) \qquad (8.122)$$

are continuous on $N(E, E)$, because

$$|\varphi_j(u)| = |f_j(u(x_j))| \leqslant \|f_j\| \|x_j\| \|u\| \leqslant \|f_j\| \|x_j\| \|u\|_N$$

$$(u \in N(E, E),\ j = 1,2, \ldots), \qquad (8.123)$$

and hence $N_0(E, E)$, the intersection of the null-spaces of these functionals, is closed. Since by (8.114) we have

$$\|\{f_j(u(x_j))\}\| = \inf_{u_0 \in N_0(E, E)} \|u + u_0\|_N \qquad (u \in N(E, E)), \qquad (8.124)$$

it follows that $\mathscr{S}(E, (x_n, f_n))$ is linearly isometric to the Banach space $N(E, E)/N_0(E, E)$, and hence $\mathscr{S}(E, (x_n, f_n))$ is a Banach space.

Now let $\{h_n\}$ be the sequence of coordinate functionals on $\mathscr{S}(E, (x_n, f_n))$. Then, by (8.123), for any $u \in N(E, E)$, $u_0 \in N_0(E, E)$ and $n = 1,2, \ldots$ we have

$$|h_n(\{f_j(ux_j))\})| = |f_n(u(x_n))| = |f_n(u + u_0)(x_n)| \leqslant \|f_n\| \|x_n\| \|u + u_0\|_N,$$

whence, by (8.124),

$$|h_n(\{f_j(u(x_j))\})| \leqslant \|f_n\| \|x_n\| \|\{f_j(u(x_j))\}\| \qquad (u \in N(E, E),\ n = 1,2, \ldots), \qquad (8.125)$$

and thus $\mathscr{S}(E, (x_n, f_n))$ is a *BK*-space.

b) For any $x_0 \in E$ and $f_0 \in E^*$ the mapping $u_0 \in L(E, E)$ defined by

$$u_0(x) = (x_0 \otimes f_0)(x) = f_0(x)x_0 \qquad (x \in E) \qquad (8.126)$$

is nuclear and hence

$$\{f_0(x_j)f_j(x_0)\}_{j=1}^{\infty} = \{f_j(u_0(x_j))\}_{j=1}^{\infty} \in \mathscr{S}(E, (x_n, f_n)),$$

$$\|\{f_0(x_j)f_j(x_0)\}_{j=1}^{\infty}\| \leqslant \|u_0\|_N = \|x_0\| \|f_0\|.$$

c) From b) for $x_0 = x_n$ and $f_0 = f_n$, we obtain

$$e_k = \{f_k(x_j)f_j(x_k)\}_{j=1}^{\infty} \in \mathscr{S}(E, (x_n, f_n)) \qquad (k = 1,2, \dots). \tag{8.127}$$

Now let $\{\alpha_j\} \in \mathscr{S}(E, (x_n, f_n))$ and $\varepsilon > 0$ be arbitrary. Then there exists $u \in N(E, E)$ such that

$$\alpha_j = f_j(u(x_j)) \qquad (j = 1,2, \dots).$$

Let (8.110), (8.111) be an arbitrary representation of $u$. Then

$$\left\| u - \sum_{i=1}^{n} y_i \otimes g_i \right\|_N \leqslant \sum_{i=n+1}^{\infty} \|y_i\| \|g_i\| < \frac{\varepsilon}{2} \qquad (n > n_0(\varepsilon)).$$

Furthermore, since $[x_n] = E$, there exist finite linear combinations $p_1, \dots, p_n$ of the elements $x_k$ such that

$$\left\| \sum_{i=1}^{n} y_i \otimes g_i - \sum_{i=1}^{n} p_i \otimes g_i \right\|_N \leqslant \sum_{i=1}^{n} \|y_i - p_i\| \|g_i\| < \frac{\varepsilon}{2},$$

whence $\left\| u - \sum_{i=1}^{n} p_i \otimes g_i \right\|_N < \varepsilon$ for $n > n_0(\varepsilon)$. Therefore, by (8.114),

$$\left\| \{\alpha_j\}_{j=1}^{\infty} - \left\{ \sum_{i=1}^{n} g_i(x_j) f_j(p_i) \right\}_{j=1}^{\infty} \right\| =$$

$$= \left\| \left\{ f_j \left( \left( u - \sum_{i=1}^{n} p_i \otimes g_i \right) (x_j) \right) \right\}_{j=1}^{\infty} \right\| \leqslant \left\| u - \sum_{i=1}^{n} p_i \otimes g_i \right\|_N < \varepsilon$$

$$(n > n_0(\varepsilon)),$$

which, by (8.127) and since $\left\{ \sum_{i=1}^{n} g_i(x_j) f_j(p_i) \right\}_{j=1}^{\infty}$ is a finite linear combination of the unit vectors $e_k$ (because $f_j(p_i) = 0$ for $i = 1, \dots, n$ and all sufficiently large $j$), proves that $[e_n] = \mathscr{S}(E, (x_n, f_n))$. Hence, by a), $\{e_n\}$ is an $M$-basis of $\mathscr{S}(E, (x_n, f_n))$.

Finally, the second inequality in (8.120) follows from (8.119) and (8.127). Also the second inequality in (8.121) results from (8.125). Therefore

$$1 = |h_n(e_n)| \leqslant \|e_n\| \, \|h_n\| \leqslant \min(\|e_n\| \, \|x_n\| \, \|f_n\|, \ \|h_n\| \, \|x_n\| \, \|f_n\|) \qquad (n=1,2,\ldots),$$

and hence we also have the first inequalities both in (8.120) and (8.121), which completes the proof of proposition 8.20.

**Proposition 8.21.** *Let $\{x_n\}$ be an M-basis of a Banach space $E$, with the a.s.f. $\{f_n\} \subset E^*$. Then*

$$M(E, (x_n, f_n)) \subset M(\mathscr{S}(E, (x_n, f_n)), \ (e_n, h_n)), \tag{8.128}$$

*where $\{e_n\}$ is the sequence of unit vectors in $\mathscr{S}(E, (x_n, f_n))$ and $\{h_n\}$ the sequence of coordinate functionals on $\mathscr{S}(E, (x_n, f_n))$.*

*Proof.* Let $\{\gamma_n\} \in M(E, (x_n, f_n))$ and $\{\alpha_j\} \in \mathscr{S}(E, (x_n, f_n))$ be arbitrary. Let $u \in N(E, E)$ be such that

$$\alpha_j = f_j(u(x_j)) \qquad (j = 1,2, \ldots),$$

let (8.110), (8.111) be an arbitrary representation of $u$ and for each $i$ let $(y_i)_{\{\gamma_n\}} \in E$ be such that

$$f_k((y_i)_{\{\gamma_n\}}) = \gamma_k f_k(y_i) \qquad (k = 1,2, \ldots).$$

Then we have

$$\sum_{i=1}^{\infty} \|g_i\| \, \|(y_i)_{\{\gamma_n\}}\| \leqslant \|\{\gamma_n\}\| \sum_{i=1}^{\infty} \|g_i\| \, \|y_i\| < \infty,$$

and therefore we can define a mapping $w \in N(E, E)$ by

$$w(x) = \sum_{i=1}^{\infty} g_i(x)(y_i)_{\{\gamma_n\}} \qquad (x \in E),$$

whence

$$\{\beta_j\} = \{f_j(w(x_j))\}_{j=1}^{\infty} \in \mathscr{S}(E, (x_n, f_n)).$$

Since

$$h_k(\{\beta_j\}) = \beta_k = f_k(w(x_k)) = \sum_{i=1}^{\infty} g_i(x_k) f_k((y_i)_{\{\gamma_n\}}) =$$

$$= \gamma_k \sum_{i=1}^{\infty} g_i(x_k) f_k(y_i) = \gamma_k f_k(u(x_k)) = \gamma_k \alpha_k = \gamma_k h_k(\{\alpha_j\}) \qquad (k = 1,2, \ldots),$$

we have $\{\beta_j\} = \{\alpha_j\}_{\{\gamma_n\}}$ and $\{\gamma_j\} \in M(\mathscr{S}(E, (x_n, f_n)), (e_n, h_n))$, which completes the proof.

From proposition 8.21 and the characterization of bases (respectively, of unconditional bases) in terms of multipliers, given in Ch. I, § 5, theorem 5.2 (respectively, in Ch. II, § 16, theorem*) 16.5 c)), it follows

**Corollary 8.7.** *If $\{x_n\}$ is a basis (respectively, an unconditional basis) of a Banach space $E$, with the a.s.c.f. $\{f_n\} \subset E^*$, then the sequence of unit vectors $\{e_n\}$ is a basis (respectively, an unconditional basis) of $\mathscr{S}(E,(x_n, f_n))$.*

Moreover, in corollary 8.9 b) below we shall see that in this case $\{e_n\}$ is a basis of type $P^*$ of $\mathscr{S}(E, (x_n, f_n))$.

In § 7, remark 7.1 we have observed that every Banach space $E$ on which there exists a total sequence of functionals $\{f_n\} \subset E^*$ is linearly isometric to the $BK$-space $S = S(E, \{f_n\}) = \{\{f_n(x)\} \mid x \in E\}$ (where the norm is defined by $\|\{f_n(x)\}\| = \|x\|$), by the mapping $x \to \{f_n(x)\}$. In particular, if $\{x_n\}$ is a complete sequence in a Banach space $E$ (i.e., $[x_n] = E$), then its canonical image $\{\pi(x_n)\} \subset E^{**}$ is total on $E^*$ hence $E^*$ is linearly isometric to the $BK$-space

$$(E, \{x_n\})^d = S(E^*, \{\pi(x_n)\}) = \{\{f(x_n)\} \mid f \in E^*\} \tag{8.129}$$

(where the norm is defined by $\|\{f(x_n)\}\| = \|f\|$), by the mapping $f \to \{f(x_n)\}$. The sets $(E, \{x_n\})^d$ defined by (8.129) have also occurred in Ch. I, § 12, theorems 12.5 a), 12.6 (formula (12.65)) and 12.7 (condition 9°) and, implicitly, in Ch. I, § 8, theorem 8.2.

**Proposition 8.22.** *Let $\{x_n\}$ be an $M$-basis of a Banach space $E$, with the a.s.c.f. $\{f_n\} \subset E^*$. Then*

a) *If $\{h_n\}$ denotes the sequence of coordinate functionals on $M(E, (x_n, f_n))$, and if $\varphi_n = h_n|_{[e_j]}$ $(n = 1,2, \ldots)$, we have*

$$([\varphi_n], \{\varphi_n\})^d \cup ([h_n], \{h_n\})^d \subset M(E, (x_n, f_n)). \tag{8.130}$$

b) *If $\{e_n\}$ is the sequence of unit vectors in $\mathscr{S}(E, (x_n, f_n))$, we have*

$$(\mathscr{S}(E, (x_n, f_n)), \{e_n\})^d \subset M(E, (x_n, f_n)). \tag{8.131}$$

*Proof.* a) Let $\{\gamma_n\} \in ([\varphi_n], \{\varphi_n\})^d$ be arbitrary and let $\Psi \in [\varphi_n]^*$ be such that

$$\gamma_n = \Psi(\varphi_n) \qquad (n = 1,2, \ldots).$$

*) Naturally, one can also use § 7, theorem 7.2.

Then for any scalars $\beta_1, \ldots, \beta_m$ we have $\left\| \sum_{i=1}^{m} \beta_i \Psi(\varphi_i) \right\| \leqslant \leqslant \|\Psi\| \left\| \sum_{i=1}^{m} \beta_i \varphi_i \right\|$ and hence, by Helly's theorem*), we can find a sequence $\{\gamma_n^{(j)}\} \in [e_n] \subset M(E, (x_n, f_n))$ $(j = 1,2, \ldots)$ such that

$$\gamma_k = \Psi(\varphi_k) = \lim_{j\to\infty} \varphi_k(\{\gamma_n^{(j)}\}) = \lim_{j\to\infty} \gamma_k^j \qquad (k = 1,2, \ldots),$$

$$\|\{\gamma_n^{(j)}\}\| \leqslant \|\Psi\| + 1 \qquad (j = 1,2, \ldots).$$

Since $\{\gamma_n^{(j)}\} \in M(E, (x_n, f_n))$, for each $x \in E$ and $j \in \mathcal{N}$ there exists an element $x_{\{\gamma_n^{(j)}\}} \in E$ such that

$$f_k(x_{\{\gamma_n^{(j)}\}}) = \gamma_k^{(j)} f_k(x) \qquad (k = 1,2, \ldots).$$

Let

$$v_j(x) = x_{\{\gamma_n^{(j)}\}} \qquad (x \in E, j = 1,2, \ldots).$$

Then, by Ch. I, § 5, proposition 5.4 (since $v_{\{\gamma_n^{(j)}\}} = v_j$), we have $v_j \in L(E, E)$ and

$$v_j(x_k) = \gamma_k^{(j)} x_k \to \gamma_k x_k \text{ as } j \to \infty \qquad (k = 1,2, \ldots),$$

$$\|v_j\| = \|\{\gamma_n^{(j)}\}\| \leqslant \|\Psi\| + 1 \qquad (j = 1,2, \ldots).$$

Consequently, there exists**) $v_0 \in L(E, E)$ such that $v_j(x) \to v_0(x)$ $(x \in E)$. Then, clearly,

$$f_k(v_0(x)) = \lim_{j\to\infty} f_k(v_j(x)) = \lim_{j\to\infty} f_k(x_{\{\gamma_n^{(j)}\}}) =$$

$$= \lim_{j\to\infty} \gamma_k^{(j)} f_k(x) = \gamma_k f_k(x) \qquad (x \in E, \ k = 1,2, \ldots),$$

whence $\{\gamma_n\} \in M(E, (x_n, f_n))$ (and $v_0(x) = x_{\{\gamma_n\}}$). The proof for $([h_n], \{h_n\})^d$ is entirely similar (with $\{\gamma_n^{(j)}\} \in M(E, (x_n, f_n))$ only).

b) Let $\{\gamma_n\} \in (\mathcal{S}(E, (x_n, f_n)), \{e_n\})^d$ be arbitrary and let $h \in \mathcal{S}(E, (x_n, f_n))^*$ be such that

$$\gamma_n = h(e_n) \qquad (n = 1,2, \ldots).$$

*) See e.g. [70], Ch. II, § 4, theorem 3.
**) See e.g. [87], p. 55, theorem 18.

Then for every finite linear combination $p = \sum_{i=1}^{n} \alpha_i x_i \in P$ we have $p_{\{\gamma_n\}} = \sum_{i=1}^{n} \gamma_i \alpha_i x_i$ and hence, by (8.119),

$$\|p_{\{\gamma_n\}}\| = \left\| \sum_{i=1}^{n} h(e_i) f_i(p) x_i \right\| = \sup_{\substack{f\in E^* \\ \|f\|\leqslant 1}} \left| h\left( \sum_{i=1}^{n} f_i(p) f(x_i) e_i \right) \right| \leqslant$$

$$\leqslant \|h\| \sup_{\substack{f\in E^* \\ \|f\|\leqslant 1}} \left\| \sum_{i=1}^{n} f_i(p) f(x_i) e_i \right\|_{\mathscr{S}} = \|h\| \sup_{\substack{f\in E^* \\ \|f\|\leqslant 1}} \|\{f(x_j) f_j(p)\}_{j=1}^{\infty}\|_{\mathscr{S}} \leqslant$$

$$\leqslant \|h\| \sup_{\substack{f\in E^* \\ \|f\|\leqslant 1}} \|f\| \; \|p\| = \|h\| \; \|p\|.$$

Thus, $p \to p_{\{\gamma_n\}}$ is a continuous linear mapping of $P$ into $P$ and hence it can be extended continuously to a mapping $v_{\{\gamma_n\}}: E \to E$. Now, let $x \in E$ be arbitrary. Then, since $[x_n] = E$, there exists a sequence of finite linear combinations $p_j \in P$ such that $\lim_{j\to\infty} p_j = x$. By the preceding, we have

$$f_k(v_{\{\gamma_n\}}(x)) = \lim_{j\to\infty} f_k(v_{\{\gamma_n\}}(p_j)) = \lim_{j\to\infty} f_k((p_j)_{\{\gamma_n\}}) =$$

$$= \gamma_k \lim_{j\to\infty} f_k(p_j) = \gamma_k f_k(x) \qquad (k = 1,2, \ldots),$$

whence $\{\gamma_n\} \in M(E, (x_n, f_n))$ (and $v_{\{\gamma_n\}}(x) = x_{\{\gamma_n\}}$), which completes the proof of proposition 8.22.

**Proposition 8.23.** *Let $\{x_n\}$ be an $M$-basis of a Banach space $E$, with the a.s.f. $\{f_n\} \subset E^*$. Then*

$$M(E, (x_n, f_n)) \subset M((E, \{x_n\})^d, (e_n, h_n)), \tag{8.132}$$

*where $\{e_n\}$ is the sequence of unit vectors in $(E, \{x_n\})^d$ and $\{h_n\}$ the sequence of coordinate functionals on $(E, \{x_n\})^d$.*

*Proof.* Let $\{\gamma_n\} \in M(E, (x_n, f_n))$ and $\{\delta_n\} \in (E, \{x_n\})^d$ be arbitrary and let $f \in E^*$ be such that

$$\delta_n = f(x_n) \qquad (n = 1,2, \ldots).$$

Put

$$f^{\{\gamma_n\}}(x) = f(x_{\{\gamma_n\}}) \qquad (x \in E). \tag{8.133}$$

Then $|f^{\{\gamma_n\}}(x)| \leqslant \|f\| \, \|x_{\{\gamma_n\}}\| \leqslant \|f\| \, \|\{\gamma_n\}\| \, \|x\|$ for all $x \in E$, whence $f^{\{\gamma_n\}} \in E^*$. Consequently, for

$$\{\beta_n\} = \{f^{\{\gamma_j\}}(x_n)\} \in (E, \{x_n\})^d$$

we have, by Ch. I, § 5, formula (5.32),

$$h_k(\{\beta_n\}) = \beta_k = f^{\{\gamma_n\}}(x_k) = f((x_k)_{\{\gamma_n\}}) = f(\gamma_k x_k) =$$

$$= \gamma_k f(x_k) = \gamma_k \delta_k = \gamma_k h_k(\{\delta_n\}) \qquad (k = 1,2, \ldots),$$

whence $\{\beta_n\} = \{\delta_n\}_{\{\gamma_n\}}$ and $\{\gamma_n\} \in M((E, \{x_n\})^d, (e_n, h_n))$, which completes the proof of proposition 8.23.

Proposition 8.20 shows that, roughly speaking, $\mathscr{S}(E, (x_n, f_n))$ is the smallest Banach sequence space which contains all sequences of the form (8.118). Since these sequences are the terms of the numerical expansions

$$f_0(x_0) \sim \sum_{j=1}^{\infty} f_0(x_j) f_j(x_0), \tag{8.134}$$

it is natural to give

*Definition 8.6.* An $M$-basis $\{x_n\}$ of a Banach space $E$, with the a.s.f. $\{f_n\} \subset E^*$, is said to be *series summable*, if there exists a functional $h_0 \in \mathscr{S}(E, (x_n, f_n))^*$ such that

$$h_0(\{f(x_j) f_j(x)\}_{j=1}^{\infty}) = f(x) \qquad (x \in E, f \in E^*), \tag{8.135}$$

i.e., a functional which gives to each series $\sum_{j=1}^{\infty} f(x_j) f_j(x)$ its "correct sum".

Some characterizations of series summable $M$-bases are given in

**Proposition 8.24.** *Let $\{x_n\}$ be an $M$-basis of a Banach space $E$, with the a.s.f. $\{f_n\} \subset E^*$. The following statements are equivalent:*

1°. *$\{x_n\}$ is a series summable $M$-basis.*

2°. *We have*

$$e = \{1,1,1, \ldots\} \in (\mathscr{S}(E, (x_n, f_n)), \{e_n\})^d, \tag{8.136}$$

*i.e., there exists a functional $h_0 \in \mathscr{S}(E, (x_n, f_n))^*$ such that*

$$h_0(e_n) = 1 \qquad (n = 1,2, \ldots). \tag{8.137}$$

3°. *We have*

$$(\mathscr{S}(E, (x_n, f_n)), \{e_n\})^d = M(E, (x_n, f_n)). \tag{8.138}$$

*These statements are implied by the following:*
4°. *$\{e_n\}$ is an $M$-basis of type $P^*$ of $\mathscr{S}(E, (x_n, f_n))$.*

*Proof.* If $h_0 \in \mathscr{S}(E, (x_n, f_n))^*$ satisfies (8.135), then

$$h_0(e_n) = h_0(\{f_n(x_j)f_j(x_n)\}_{j=1}^{\infty}) = f_n(x_n) = 1 \qquad (n = 1,2, \ldots),$$

which proves that 1° $\Rightarrow$ 2°.

Conversely, assume now that we have 2°. Given $\{\alpha_n\} \in (E, \{x_n\})^d$, say $\alpha_n = f(x_n)$ $(n = 1,2, \ldots)$, where $f \in E^*$, define a linear mapping $w_{\{\alpha_n\}}: E \to \mathscr{S}(E, (x_n, f_n))$ by

$$w_{\{\alpha_n\}}(x) = \{\alpha_j f_j(x)\}_{j=1}^{\infty} = \{f(x_j)f_j(x)\}_{j=1}^{\infty} \qquad (x \in E). \tag{8.139}$$

We claim that $w_{\{\alpha_n\}}$ is closed. Indeed, let $y_k, x \in E$, $\lim_{k\to\infty} y_k = x$ and $\lim_{k\to\infty} w_{\{\alpha_n\}}(y_k) = \lim_{k\to\infty} \{\alpha_j f_j(y_k)\}_{j=1}^{\infty} = \{\beta_j\} \in \mathscr{S}(E, (x_n, f_n))$. Then, since $\mathscr{S}(E, (x_n, f_n))$ is a $BK$-space, $\lim_{k\to\infty} \alpha_j f_j(y_k) = \beta_j$ $(j = 1,2, \ldots)$, whence, by $\lim_{k\to\infty} \alpha_j f_j(y_k) = \alpha_j f_j(\lim_{k\to\infty} y_k) = \alpha_j f_j(x)$ $(j = 1,2, \ldots)$, we infer $\{\beta_j\} = \{\alpha_j f_j(x)\} = w_{\{\alpha_n\}}(x)$, which proves our claim that $w_{\{\alpha_n\}}$ is closed and hence continuous.*) Consequently, the functional $f^{\{\alpha_n\}}$ on $E$ defined by

$$f^{\{\alpha_n\}}(x) = h_0(w_{\{\alpha_n\}}(x)) = h_0(\{\alpha_j f_j(x)\}_{j=1}^{\infty}) \qquad (x \in E), \tag{8.140}$$

where $h_0$ is defined by 2°, is in $E^*$. Hence, since for every finite linear combination $p = \sum_{i=1}^{l} \beta_i x_i \in P$ we have, by 2°,

$$f^{\{\alpha_n\}}(p) = f^{\{\alpha_n\}}\left(\sum_{i=1}^{l} \beta_i x_i\right) = h_0\left(\left\{\alpha_j f_j\left(\sum_{i=1}^{l} \beta_i x_i\right)\right\}_{j=1}^{\infty}\right) =$$

$$= h_0\left(\sum_{j=1}^{l} \alpha_j \beta_j e_j\right) = \sum_{j=1}^{l} \alpha_j \beta_j h_0(e_j) = \sum_{j=1}^{n} \alpha_j \beta_j = \sum_{j=1}^{n} f(x_j)\beta_j = f(p),$$

and since $P$ is dense in $E$, it follows that $f^{\{\alpha_n\}} = f$. Therefore

$$f(x) = f^{\{\alpha_n\}}(x) = h_0(\{\alpha_j f_j(x)\}) = h_0(\{f(x_j)f_j(x)\}) \qquad (x \in E),$$

which, since $f \in E^*$ was arbitrary, proves that 2° $\Rightarrow$ 1°.

---

*) Note that until now we have not used 2°, so this statement is true for any $M$-basis $\{x_n\}$ with the a.s.f. $\{f_n\}$. The continuity of $w_{\{\alpha_n\}}$ follows also from (8.119).

Assume now again that we have 2°. In order to prove that in this case we have 3°, it is sufficient, by proposition 8.22 b), to prove the inclusion

$$M(E, (x_n, f_n)) \subset (\mathcal{S}(E, (x_n, f_n)), \{e_n\})^d. \tag{8.141}$$

Since by propositions 8.21 and 8.23 we have

$$M(E,(x_n,f_n)) \subset M(\mathcal{S}(E,(x_n,f_n)),(e_n,h_n)) \subset M((\mathcal{S}(E,(x_n,f_n)), \{e_n\})^d,(e_n,h_n)),$$

in order to prove (8.141) it will be sufficient to prove the inclusion

$$M((\mathcal{S}(E, (x_n,f_n)), \{e_n\})^d, (e_n, h_n)) \subset \mathcal{S}(E, (x_n, f_n)), \{e_n\})^d. \tag{8.142}$$

Let $\{\gamma_n\} \in M((\mathcal{S}(E, (x_n,f_n)), \{e_n\})^d, (e_n, h_n))$ be arbitrary. Then, by 2°, there exists $e_{\{\gamma_n\}} \in (\mathcal{S}(E, (x_n, f_n)), \{e_n\})^d$ such that

$$h_k(e_{\{\gamma_n\}}) = \gamma_k h_k(e) = \gamma_k = h_k(\{\gamma_n\}) \qquad (k = 1,2, \ldots),$$

whence $\{\gamma_n\} = e_{\{\gamma_n\}} \in (\mathcal{S}(E, (x_n,f_n)), \{e_n\})^d$, which proves (8.142). Thus, $2° \Rightarrow 3°$.

The converse implication $3° \Rightarrow 2°$ is obvious, since $e \in M(E, (x_n, f_n))$.

Finally, assume that we have 4°. Then $\mathcal{S}(E, (x_n,f_n))$ is separable and $\sup\limits_{1 \leqslant n < \infty} \left\| \sum\limits_{i=1}^{n} h_i \right\| < \infty$, whence there exists a subsequence $\left\{ \sum\limits_{i=1}^{n_k} h_i \right\}$ of $\left\{ \sum\limits_{i=1}^{n} h_i \right\}$, which is $w^*$-convergent to an element $h_0 \in \mathcal{S}(E, (x_n, f_n))^*$, satisfying, obviously, (8.137). Thus*), $4° \Rightarrow 2°$, which completes the proof.

**Corollary 8.8.** *For every separable Banach space E and every series summable M-basis $\{x_n\}$ of E, with the a.s.f. $\{f_n\} \subset E^*$, such that*

$$\sup_{1 \leqslant n < \infty} \|x_n\| \, \|f_n\| < \infty, \tag{8.143}$$

*the space***) $\mathcal{S}(E, (x_n, f_n)) \equiv N(E, E)/N_0(E, E)$ *is non-reflexive.*

---

*) Actually, this argument, together with proposition 8.20 c), shows that the weaker assumption $\sup\limits_{1 \leqslant n < \infty} \left\| \sum\limits_{i=1}^{n} h_i \right\| < \infty$ also implies 2° (without assuming that $\sup\limits_{1 \leqslant n < \infty} \|e_n\| < \infty$).

**) See proposition 8.20 a).

*Proof*. By (8.143) and proposition 8.20 c) we have $\sup_{1 \leqslant n < \infty} \|e_n\| < \infty$, whence, by § 6, remark 6.6, $\lim_{n\to\infty} h(e_n) = 0$ for all $h \in \mathscr{S}(E, (x_n, f_n))^*$, whenever $\mathscr{S}(E, (x_n, f_n)) \equiv N(E, E)/N_0(E, E)$ is reflexive. Hence, in this case, by proposition 8.24, implication $1° \Rightarrow 2°$, $\{x_n\}$ is not series summable, which completes the proof.

**Corollary 8.9.** a) *Every basis $\{x_n\}$ of a Banach space $E$ is a series summable $M$-basis.**)

b) *For every basis $\{x_n\}$ of a Banach space $E$, with the a.s.c.f. $\{f_n\} \subset E^*$, the sequence $\{e_n\}$ is a basis of type $P^*$ of $\mathscr{S}(E, (x_n, f_n))$.*

c) *For every unconditional basis $\{x_n\}$ of $E$, with the a.s.c.f. $\{f_n\} \subset E^*$, the sequence $\{e_n\}$ is a basis of $\mathscr{S}(E, (x_n, f_n))$, equivalent to the unit vector basis of $l^1$ (and hence $\mathscr{S}(E, (x_n, f_n)) \equiv N(E, E)/N_0(E, E)$ is isomorphic to $l^1$).*

*Proof*. a) Let $\{x_n\}$ be a basis of $E$, with the a.s.c.f. $\{f_n\} \subset E^*$. Define a linear functional $h_0$ on the linear span $P_{\mathscr{S}}$ of $\{e_n\}$ in $\mathscr{S}(E, (x_n, f_n))$ by

$$h_0\left(\sum_{j=1}^{n} \beta_j e_j\right) = \sum_{j=1}^{n} \beta_j \qquad \left(\sum_{j=1}^{n} \beta_j e_j \in P_{\mathscr{S}}\right). \tag{8.144}$$

Let $\sum_{j=1}^{n} \beta_j e_j \in P_{\mathscr{S}}$ be arbitrary. Then, by the definition of $\mathscr{S}(E, (x_n, f_n))$, there exist sequences $\{y_n\} \subset E$, $\{g_n\} \subset E^*$ with $\sum_{i=1}^{\infty} \|g_i\| \, \|y_i\| < \infty$, such that

$$\sum_{i=1}^{\infty} g_i(x_j) f_j(y_i) = \begin{cases} \beta_j & \text{for } j = 1, \ldots, n \\ 0 & \text{for } j = n+1, n+2, \ldots \end{cases} \tag{8.145}$$

Hence

$$\left| h_0\left(\sum_{j=1}^{n} \beta_j e_j\right)\right| = \left|\sum_{j=1}^{n} \beta_j\right| = \left|\sum_{j=1}^{n} \sum_{i=1}^{\infty} g_i(x_j) f_j(y_i)\right| =$$

$$= \left|\sum_{i=1}^{\infty} g_i\left(\sum_{i=1}^{n} f_j(y_i) x_j\right)\right| = \left|\sum_{i=1}^{\infty} g_i(s_n(y_i))\right| \leqslant \sup_{1 \leqslant k < \infty} \|s_k\| \sum_{i=1}^{\infty} \|g_i\| \|y_i\|,$$

where $s_k(x) = \sum_{i=1}^{k} f_i(x) x_i$ and $\sup_{1 \leqslant k < \infty} \|s_k\| < \infty$ since $\{x_n\}$ is a basis of $E$.

*) A stronger result will be given in § 11, proposition 11.8 (combined with § 11, proposition 11.11).

Taking the infimum over all representations (8.145) of $\sum_{j=1}^{n} \beta_j e_j$, we obtain

$$\left| h_0\left(\sum_{j=1}^{n} \beta_j e_j\right)\right| \leqslant \sup_{1\leqslant k<\infty} \|s_k\| \left\| \sum_{j=1}^{n} \beta_j e_j \right\|_{\mathscr{S}(E,(x_n,f_n))}$$

Consequently, $h_0$ can be extended to a continuous linear functional $h_0 \in \mathscr{S}(E,(x_n,f_n))^*$, satisfying, obviously, (8.137), and thus, by proposition 8.24, implication $2° \Rightarrow 1°$, $\{x_n\}$ is a series summable $M$-basis of $E$.

b) Since $\{x_n\}$ is a basis of $E$, by Ch. I, § 3, theorem 3.1 we have $\sup_{1\leqslant n<\infty} \|x_n\|\|f_n\| < \infty$, whence, by proposition 8.20 c), $\sup_{1\leqslant n<\infty} \|e_n\| < \infty$ in $\mathscr{S}(E,(x_n,f_n))$. Furthermore, by corollary 8.7, $\{e_n\}$ is a basis of $\mathscr{S}(E,(x_n,f_n))$, and therefore, by part a), (8.137) and Ch. II, § 9, theorem 9.2 b), implication $4° \Rightarrow 1°$, $\{e_n\}$ is a basis of type $P^*$ of $\mathscr{S}(E,(x_n,f_n))$.

Finally, c) is an immediate consequence of part b) and Ch. II, § 17, corollary 17.1 b), which completes the proof.

**Proposition 8.25.** *Every series summable M-basis of a Banach space E is a strong M-basis of E.*

*Proof.* Let $\{x_n\}$ be a series summable $M$-basis of a Banach space $E$, with the a.s.f. $\{f_n\} \subset E^*$ and let $x \in E$, $f \in E^*$ be such that the set $J = \{j \in \mathscr{N} \mid f(x_j)f_j(x) \neq 0\}$ is finite or empty. Then, by proposition 8.24, implication $1° \Rightarrow 2°$ and by definition 8.6, formula (8.135), we have

$$\sum_{j=1}^{\infty} f(x_j)f_j(x) = \sum_{j\in J} f(x_j)f_j(x) = h_0\Big(\sum_{j\in J} f(x_j)f_j(x)e_j\Big) =$$
$$= h_0(\{f(x_j)f_j(x)\}_{j=1}^{\infty}) = f(x),$$

and hence, by proposition 8.11, implication $4° \Rightarrow 1°$, $\{x_n\}$ is a strong $M$-basis of $E$. This completes the proof of proposition 8.25.

From propositions 8.25 and 8.12 it follows that *every separable Banach space E has an M-basis which is not series summable*. It is not known whether the converse of proposition 8.25 holds, i.e.:

*Problem 8.3*. Is every strong $M$-basis of $E$ series summable?

We have the following sufficient condition for the series summability of an $M$-basis $\{x_n\}$, which will be used in § 11:

**Proposition 8.26.** *Let $\{x_n\}$ be an M-basis of a Banach space E, with the a.s.f. $\{f_n\} \subset E^*$. If $(E,\{x_n\})^d$ is a BK-algebra containing $e = \{1,1,\ldots\}$, then $\{x_n\}$ is series summable.*

*Proof.* We shall first show that *if $(E,\{x_n\})^d$ is a BK-algebra, then*

(i) *$(E,\{x_n\})^d \subset M(E,(x_n,f_n))$ and the inclusion mapping is continuous;*

(ii) *there is a one-to-one continuous mapping $w$ of $\mathscr{S}(E,(x_n,f_n))$ into E, such that*

$$f_k(w(\{\alpha_n\})) = \alpha_k \qquad (k = 1,2,\ldots). \tag{8.146}$$

To prove (i), let $\{\gamma_n\} \in (E, \{x_n\})^d$ and let $f \in E^*$ be such that $f(x_n) = \gamma_n$ ($n = 1,2, \ldots$). Define a linear mapping $v_{\{\gamma_n\}}$ from $P$ into $P$ (the linear span of $\{x_n\}$) by

$$v_{\{\gamma_n\}}\left(\sum_{i=1}^{m} \alpha_i x_i\right) = \sum_{i=1}^{m} \gamma_i \alpha_i x_i = \sum_{i=1}^{m} \alpha_i f(x_i) x_i \quad \left(\sum_{i=1}^{m} \alpha_i x_i \in P\right). \tag{8.147}$$

Since $(E, \{x_n\})^d$ is a $BK$-algebra, for any $f, g \in E^*$ we have*) $\{f\cdot g(x_n)\} = \{f(x_n)g(x_n)\} \in (E, \{x_n\})^d$ and there is a constant $C > 0$ such that

$$\|\{f\cdot g(x_n)\}\| = \|\{f(x_n)g(x_n)\}\| \leqslant C\|\{f(x_n)\}\|\,\|\{g(x_n)\}\| \qquad (f, g \in E^*).$$

Hence, for $f \in E^*$ as above and for any $g \in E^*$ we have

$$\left|g\left(v_{\{\gamma_n\}}\left(\sum_{i=1}^{m} \alpha_i x_i\right)\right)\right| = \left|\sum_{i=1}^{m} \alpha_i f(x_i) g(x_i)\right| = \left|f\cdot g\left(\sum_{i=1}^{m} \alpha_i x_i\right)\right| \leqslant$$

$$\leqslant \|\{f\cdot g(x_n)\}\| \left\|\sum_{i=1}^{m} \alpha_i x_i\right\| \leqslant C\|\{\gamma_n\}\|\,\|\{g(x_n)\}\| \left\|\sum_{i=1}^{m} \alpha_i x_i\right\|,$$

so $v_{\{\gamma_n\}}$ is continuous on $P$ and $\|v_{\{\gamma_n\}}\| \leqslant C\|\{\gamma_n\}\|$. Consequently, $v_{\{\gamma_n\}}$ can be extended to $v_{\{\gamma_n\}} \in L(E, E)$ satisfying the same inequality. Then, as in the proof of proposition 8.22 b), it is clear that $\{\gamma_n\} \in M(E, (x_n, f_n))$ and that $v_{\{\gamma_n\}}(x) = x_{\{\gamma_n\}}$ ($x \in E$). Finally, for any $\{\gamma_n\} \in (E, \{x_n\})^d$,

$$\|\{\gamma_n\}\|_{M(E, (x_n, f_n))} = \|v_{\{\gamma_n\}}\| \leqslant C\|\{\gamma_n\}\|_{(E, \{x_n\})^d},$$

which proves (i).

Now, to prove (ii), let $\{\alpha_n\} \in \mathscr{S}(E, (x_n, f_n))$, let $u \in N(E, E)$ be such that $\alpha_n = f_n(u(x_n))$ ($n = 1,2, \ldots$) and let (8.110), (8.111) be an arbitrary representation of $u$. Furthermore, define $\{\gamma_n^{(i)}\}_{n=1}^{\infty} \in (E, \{x_n\})^d$ ($i = 1,2, \ldots$) by

$$\gamma_n^{(i)} = g_i(x_n) \qquad (i, n = 1,2, \ldots), \tag{8.148}$$

and let

$$w(\{\alpha_n\}) = \sum_{i=1}^{\infty} v_{\{\gamma_n^{(i)}\}}(y_i), \tag{8.149}$$

---

*) I.e., there exists a functional $f\cdot g \in E^*$ such that $f\cdot g(x_n) = f(x_n)g(x_n)$ ($n = 1, 2, \ldots$).

where $v_{\{\gamma_n^{(i)}\}}$ is as in part (i) above. Then the series (8.149) converges absolutely, since

$$\sum_{i=1}^{\infty} \|v_{\{\gamma_n^{(i)}\}}(y_i)\| \leqslant C \sum_{i=1}^{\infty} \|\{\gamma_n^{(i)}\}\| \, \|y_i\| = C \sum_{i=1}^{\infty} \|g_i\| \, \|y_i\| < \infty,$$

and the element $w(\{\alpha_n\}) \in E$ does not depend on the representation of $\{\alpha_n\}$ as $\{f_n(u(x_n))\}$, since we have (8.146) by

$$f_k(w(\{\alpha_n\})) = \sum_{i=1}^{\infty} f_k(v_{\{\gamma_n^{(i)}\}}(y_i)) = \sum_{i=1}^{\infty} \gamma_k^{(i)} f_k(y_i) =$$

$$= \sum_{i=1}^{\infty} g_i(x_k) f_k(y_i) = f_k(u(x_k)) = \alpha_k \qquad (k = 1,2, \ldots)$$

and since $\{f_n\}$ is total on $E$. Moreover, we have $\|w(\{\alpha_n\})\| \leqslant C \sum_{i=1}^{\infty} \|g_i\| \, \|y_i\|$, whence $\|w(\{\alpha_n\})\| \leqslant C \|u\|_{N(E,\, E)}$ and hence

$$\|w(\{\alpha_n\})\| \leqslant C \|\{\alpha_n\}\|_{\mathscr{S}(E,\, (x_n, f_n))},$$

so $w$ is continuous. Finally, if $w(\{\alpha_n\}) = 0$, then by (8.146) $\{\alpha_n\} = 0$, so $w$ is one-to-one, which proves (ii).

Now we are ready to complete the proof of proposition 8.26. Indeed, the condition $(E, \{x_n\})^d \ni e$ means that there exists a functional $f_0 \in E^*$ such that

$$f_0(x_n) = 1 \qquad (n = 1,2, \ldots). \tag{8.150}$$

Define $h_0 \in \mathscr{S}(E, (x_n, f_n))^*$ by $h_0 = f_0 \circ w$, i.e.

$$h_0(\{\alpha_n\}) = f_0(w(\{\alpha_n\})) \qquad (\{\alpha_n\} \in \mathscr{S}(E, (x_n, f_n))), \tag{8.151}$$

where $w$ is as in part (ii) above. Then, by (8.146),

$$f_k(w(e_n)) = \delta_{kn} = f_k(x_n) \qquad (k, n = 1,2, \ldots),$$

whence, since $\{f_n\}$ is total on $E$,

$$w(e_n) = x_n \qquad (n = 1,2, \ldots), \tag{8.152}$$

and therefore, by (8.151) and (8.150),

$$h_0(e_n) = f_0(w(e_n)) = f_0(x_n) = 1 \qquad (n = 1,2, \ldots),$$

which completes the proof of proposition 8.26.

*Remark 8.10.* a) In the proof of proposition 8.24, implication $2° \Rightarrow 1°$, it was observed that for any $M$-basis $\{x_n\} \subset E$ with the a.s.f. $\{f_n\} \subset E^*$ and any $f \in E^*$ there is a continuous linear mapping $w_f$ of $E$ into $\mathscr{S}(E, (x_n, f_n))$ defined by

$$w_f(x) = \{f(x_j)f_j(x)\}_{j=1}^{\infty} \qquad (x \in E); \tag{8.153}$$

hence, *if* $(E, \{x_n\})^d$ *is a BK-algebra containing* $e$ and $f_0 \in E^*$ satisfies (8.150), we have $w_{f_0}: E \to \mathscr{S}(E, (x_n, f_n))$ defined by

$$w_{f_0}(x) = \{f_j(x)\}_{j=1}^{\infty} \qquad (x \in E), \tag{8.154}$$

and, by (ii), a mapping $w: \mathscr{S}(E, (x_n, f_n)) \to E$ defined by (8.146), which is clearly the inverse of $w_{f_0}$. Thus, in this case $\mathscr{S}(E, (x_n, f_n))$ *is isomorphic to E.* b) *If* $e \in (E, \{x_n\})^d$, *then* $M(E, (x_n, f_n)) \subset (E, \{x_n\})^d$ *and the inclusion map is continuous.* Indeed, if $f_0 \in E^*$ satisfies (8.150) and if $\{\gamma_n\} \in M(E, (x_n, f_n))$, then for $f_0^{\{\gamma_n\}} \in E^*$ defined by (8.133) (with $f = f_0$) we obtain, by Ch. I, § 5, formula (5.32),

$$f_0^{\{\gamma_j\}}(x_n) = f_0((x_n)_{\{\gamma_j\}}) = f_0(\gamma_n x_n) = \gamma_n f_0(x_n) = \gamma_n \qquad (n = 1, 2, \ldots),$$

so $\{\gamma_n\} \in (E, \{x_n\})^d$. Also,

$$\|\{f_0^{\{\gamma_j\}}(x_n)\}\| = \|f_0^{\{\gamma_n\}}\| = \sup_{\substack{x \in E \\ \|x\| \leqslant 1}} |f_0(x_{\{\gamma_n\}})| \leqslant \|f_0\| \sup_{\substack{x \in E \\ \|x\| \leqslant 1}} \|x_{\{\gamma_n\}}\| =$$

$$= \|f_0\| \, \|\{\gamma_n\}\|_{M(E, (x_n, f_n))},$$

so the inclusion map is continuous. Consequently, by (i) (or, alternatively, by a) above and proposition 8.24, implication $1° \Rightarrow 3°$) it follows that *if* $(E, \{x_n\})^d$ *is a BK-algebra containing* $e$, *then* $(E, \{x_n\})^d = M(E, (x_n, f_n))$ *and their norms are equivalent.*

The converse of proposition 8.26 is not valid, as shown by

*Example 8.7.* Let $E = l^p$, where $1 < p < \infty$ and let $\{x_n\}$ be the unit vector basis of $E$. Then $\{x_n\}$ is a series summable $M$-basis, but $(E, \{x_n\})^d$ is not a $BK$-algebra and $e \notin (E, \{x_n\})^d$. Note also that in this example $M(E, (x_n, f_n)) = l^\infty$ (by Ch. II, § 16, theorem 16.5 b)) and $(E, \{x_n\})^d \neq l^\infty$ $\left(\text{since } E^* \equiv l^q, \text{ where } \frac{1}{p} + \frac{1}{q} = 1\right)$, so $M(E, (x_n, f_n)) \not\subset (E, x_n)^d$. Also, $E$ is not isomorphic to $\mathscr{S}(E, (x_n, f_n))$ (in fact, $\mathscr{S}(E, (x_n, f_n))$ is isomorphic to $l^1$, by corollary 8.9 c)).

It is known*) that a Banach space $E$ has the approximation property if and only if for every representation (8.110), (8.111) of $u \equiv 0 \in N(E, E)$

---

*) See [139], Ch. I, pp. 164–165, proposition 35, implication (B) $\Rightarrow$ (A).

we have

$$\sum_{i=1}^{\infty} g_i(y_i) = 0. \tag{8.155}$$

**Theorem 8.7.** *Every Banach space E with a series summable M-basis has the approximation property.*

*Proof.* Let $\{x_n\}$ be a series summable $M$-basis of $E$, with the a.s.f. $\{f_n\} \subset E^*$ and let (8.110), (8.111) be an arbitrary representation of $u \equiv 0 \in N(E, E)$, hence

$$u(x) = \sum_{i=1}^{\infty} g_i(x) y_i = 0 \qquad (x \in E).$$

Then the series $\sum_{i=1}^{\infty} \{g_i(x_j) f_j(y_i)\}_{j=1}^{\infty}$ in $\mathscr{S}(E, (x_n, f_n))$ converges to $\left\{\sum_{i=1}^{\infty} g_i(x_j) f_j(y_i)\right\}_{j=1}^{\infty} = 0$, since by (8.115) and (8.111) we have

$$\left\| \left\{ \sum_{i=1}^{\infty} g_i(x_j) f_j(y_i) \right\}_{j=1}^{\infty} - \sum_{i=1}^{n} \{g_i(x_j) f_j(y_i)\}_{j=1}^{\infty} \right\| =$$

$$= \left\| \left\{ \sum_{i=n+1}^{\infty} g_i(x_j) f_j(y_i) \right\}_{j=1}^{\infty} \right\| \leqslant \sum_{i=n+1}^{\infty} \|g_i\| \, \|y_i\| \to 0 \text{ as } n \to \infty.$$

Consequently, for $h_0 \in \mathscr{S}(E, (x_n, f_n))^*$ as in definition 8.6, we obtain

$$\sum_{i=1}^{\infty} g_i(y_i) = \sum_{i=1}^{\infty} h_0(\{g_i(x_j) f_j(y_i)\}_{j=1}^{\infty}) = h_0\left( \sum_{i=1}^{\infty} \{g_i(x_j) f_j(y_i)\}_{j=1}^{\infty} \right) = h_0(0) = 0,$$

i.e. (8.155), and thus $E$ has the approximation property. This completes the proof.

*Problem 8.4.* Does every separable Banach space with the approximation property have a series summable $M$-basis?

## § 9. Approximative bases. Quasi-bases. Finite-dimensional expansions of the identity. Commuting approximative bases

In the present and the next few sections we shall give some generalizations of bases in a different direction, related to the pointwise convergence of the sequence of partial sum operators $\{s_n\}$ associated with a basis $\{x_n\}$ to the identity operator. These generalizations will be either sequences of elements $\{x_n\} \subset E$ or sequences of operators $\{u_n\} \subset L(E, E)$.

*Definition 9.1.* Let $E$ be a Banach space.

a) A sequence $\{x_n\} \subset E$ is called an *approximative basis of elements* of $E$, if there exists a row-finite matrix of functionals $\{h_{ni}\}_{\substack{i=1,\dots,m_n\\ n=1,2,\dots}} \subset E^*$ such that

$$x = \lim_{n\to\infty} \sum_{i=1}^{m_n} h_{ni}(x)x_i \qquad (x \in E). \tag{9.1}$$

b) A sequence of finite rank*) endomorphisms $\{u_n\} \subset L(E, E)$ is called an *approximative basis of operators* of $E$, if

$$x = \lim_{n\to\infty} u_n(x) \qquad (x \in E); \tag{9.2}$$

if here $\sup_{1\leqslant n<\infty} \|u_n\| \leqslant \lambda$, we shall say that $\{u_n\}$ is a *$\lambda$-approximative basis (of operators)* of $E$.

c) An approximative basis of elements $\{x_n\} \subset E$ of $E$ or an approximative basis of operators $\{u_n\} \subset L(E, E)$ of $E$ is called an *approximative basis* of $E$.

Definition 9.1. c) leads to no contradiction, since we have

**Proposition 9.1.** *A Banach space $E$ has an approximative basis of elements if and only if it has an approximative basis of operators.*

*Proof.* If $\{x_n\} \subset E$ and $\{h_{ni}\} \subset E^*$ satisfy (9.1), then it is enough to put

$$u_n(x) = \sum_{i=1}^{m_n} h_{ni}(x)x_i \qquad (x \in E,\ n = 1,2,\dots). \tag{9.3}$$

Conversely, assume now that there exists a sequence of finite rank operators $\{u_n\} \subset L(E, E)$ satisfying (9.2). Then, since each $u_n$ is of finite rank, we can write

$$u_n(x) = \sum_{i=m_{n-1}+1}^{m_n} g_{ni}(x)y_{ni} \qquad (x \in E,\ n = 1,2,\dots;\ m_0 = 0), \tag{9.4}$$

where $\{y_{ni}\}_{i=m_{n-1}+1}^{m_n}$ is a basis of $u_n(E)$ $(n = 1,2, \dots)$ and $\{g_{ni}\} \subset E^*$ a suitable row-finite matrix of functionals. Define a sequence $\{x_n\} \subset E$

*) I.e., of finite-dimensional range $u_n(E)$.

and a row-finite matrix of functionals $\{h_{ni}\}_{\substack{i=1,\ldots,m_n\\ n=1,2,\ldots}} \subset E^*$ by

$$x_i = y_{ni} \qquad (i = m_{n-1}+1, \ldots, m_n;\ n = 1,2, \ldots), \tag{9.5}$$

$$h_{ni} = \begin{cases} 0 & \text{for } i = 1, \ldots, m_{n-1} \\ g_{ni} & \text{for } i = m_{n-1}+1, \ldots, m_n \end{cases} \qquad (n = 1,2, \ldots). \tag{9.6}$$

Then $x_n \neq 0$ $(n = 1,2, \ldots)$ and the sequence of finite rank endomorphisms (9.3) is nothing else than (9.4), so it satisfies (9.2). Therefore, by (9.3) and (9.2), we have (9.1) which completes the proof of proposition 9.1.

In the sequel it will be convenient to use, briefly, the term "approximative basis" of definition 9.1 c), which will lead to no confusion. In § 10 and § 11 we shall study some special classes of approximative bases of elements, and in definition 9.9 below and §§ 12—14 some special classes of approximative bases of operators, while in § 18 we shall study some generalizations of approximative bases of operators and of some special classes of approximative bases of operators.

Every basis $\{x_n\}$ of a Banach space $E$ is an approximative basis of $E$, since for $h_{ni} = f_i$ $(i = 1, \ldots, n;\ n = 1,2, \ldots)$, where $\{f_n\} \subset E^*$ is the a.s.c.f. to the basis $\{x_n\}$, we have

$$\left\| x - \sum_{i=1}^{n} h_{ni}(x)x_i \right\| = \left\| x - \sum_{i=1}^{n} f_i(x)x_i \right\| \to 0 \text{ as } n \to \infty;$$

in this case the finite rank operators $u_n$ defined by (9.3) are nothing else than

$$u_n(x) = \sum_{i=1}^{n} h_{ni}(x)x_i = \sum_{i=1}^{n} f_i(x)x_i = s_n(x) \qquad (x \in E,\ n = 1,2, \ldots),$$

i.e. the partial sum operators associated to the basis $\{x_n\}$. The converse is not true, since in the sequel we shall see many examples of approximative bases which are not bases.

*Problem 9.1.**) Does every Banach space $E$ with an approximative basis have a basis?

*Remark 9.1.* From Ch. I, § 20, theorem 20.1, equivalence 2° ⇔ 11°, it follows that *a Banach space E has a basis if and only if there exists a sequence of endomorphisms* $\{v_n\} \subset L(E, E)$ *such that*
a) $\dim v_n(E) = n$ $(n = 1,2, \ldots)$; b) $v_n(E) \subset v_{n+1}(E)$ $(n = 1,2, \ldots)$; c) $x = \lim_{n\to\infty} v_n(x)$ $(x \in E)$; d) $v_i v_n = v_i$ $(i = 1, \ldots, n;\ n = 1,2, \ldots)$ (hence, in particular, each $v_n$ is a projection); clearly, b) and d) together are equivalent to b′) $v_n v_m = v_m v_n = v_{\min(n,m)}$ $(n, m = 1,2, \ldots)$. Thus,

---

*) For a related problem see Ch. I, § 20, problem 20.1. Note that (9.2) is nothing else then condition c) of Ch. I, § 20, theorem 20.1.

problem 9.1 is equivalent to the following: If a Banach space $E$ admits a sequence of finite rank operators $\{u_n\} \subset L(E, E)$ satisfying (9.2), does it admit a sequence $\{v_n\} \subset L(E, E)$ satisfying a), b') and c)? In § 14, theorem 14.3 we shall see that if $E$ is the conjugate space $B^*$ of a Banach space $B$ with a basis, then the answer is affirmative.

Some results on bases can be extended to approximative bases, e.g. we mention the following duality results:

**Proposition 9.2.** *Let $\{x_n\}$ be an approximative basis of a Banach space $E$ and let $\{h_{ni}\} \subset E^*$, $\{u_n\} \subset L(E, E)$ be as in* (9.1)–(9.3). *Then*

$$[h_{ni}]_{\substack{i=1,\ldots,m_n \\ n=1,2,\ldots}} = \left[\bigcup_{n=1}^{\infty} u_n^*(E^*)\right], \tag{9.7}$$

$$f = w^*\text{-}\lim_{n\to\infty} u_n^*(f) \qquad (f \in E^*), \tag{9.8}$$

$$r\left(\left[\bigcup_{n=1}^{\infty} u_n^*(E^*)\right]\right) \geqslant \frac{1}{\sup\limits_{1\leqslant n<\infty} \|u_n\|} > 0, \tag{9.9}$$

*where $r(V)$ denotes the characteristic of $V \subset E^*$.*

*Proof.* a) By (9.3),

$$(u_n^*(f))(x) = f(u_n(x)) = \left(\sum_{i=1}^{m_n} f(x_i)h_{ni}\right)(x) \quad (x \in E, f \in E^*, n = 1,2,\ldots),$$

that is,

$$u_n^*(f) = \sum_{i=1}^{m_n} f(x_i)h_{ni} \qquad (f \in E^*, n = 1,2, \ldots), \tag{9.10}$$

whence (9.7) follows.

b) By (9.2) we have

$$f(x) = \lim_{n\to\infty} f(u_n(x)) = \lim_{n\to\infty} u_n^*(f)(x) \qquad (x \in E, f \in E^*),$$

that is, (9.8).

c) The proof of c) is*) similar to that of Ch. I, § 12, theorem 12.2 a). This completes the proof of proposition 9.2.

Formula (9.8) says that $\{u_n^*\}$ is a "weak* approximative basis" of $E^*$. However, in general $\{u_n^*|_V\}$ is not an approximative basis of $V = \left[\bigcup_{n=1}^{\infty} u_n^*(E^*)\right]$ for the norm topology, even when the $u_n$ are projections (see § 12, example 12.1).

---

*) Alternatively, for a simpler proof, see § 15, proof of theorem 15.7.

Obviously, every approximative basis $\{x_n\}$ of a Banach space $E$ is a complete sequence in $E$ and hence every Banach space $E$ with an approximative basis is separable. The converse is not true, since every Banach space with an approximative basis has the approximation property (see theorem 9.3 below), but there exist separable Banach spaces without the approximation property (see § 0).

We shall give some characterizations of Banach spaces having an approximative basis in theorem 9.3 below, for which we shall need some preparation. Firstly, we shall consider now a special class of approximative bases, called quasi-bases.

*Definition 9.2.* A sequence $\{x_n\}$ in a Banach space $E$, with $x_n \neq 0$ $(n = 1,2, \ldots)$, is called a *quasi-basis* of $E$ if there exists a sequence of functionals $\{f_n\} \subset E^*$ such that

$$x = \sum_{i=1}^{\infty} f_i(x)x_i \qquad (x \in E); \tag{9.11}$$

any such sequence $\{f_n\} \subset E^*$ is called an *admissible sequence* (for the quasi-basis $\{x_n\}$).

In general, the sequence $\{f_n\} \subset E^*$ is not uniquely determined; for example, if $\{y_n\}$ is a basis of $E$, with the a.s.c.f. $\{g_n\} \subset E^*$, then the sequence $\{x_n\} \subset E$ defined by

$$x_1 = x_2 = y_1, \quad x_n = y_{n-1} \qquad (n = 3,4, \ldots) \tag{9.12}$$

is a quasi-basis of $E$, satisfying (9.11) with both

$$f_1 = 0, \quad f_n = g_{n-1} \qquad (n = 2,3, \ldots) \tag{9.13}$$

and

$$f_1 = g_1, \quad f_2 = 0, \quad f_n = g_{n-1} \qquad (n = 3,4, \ldots). \tag{9.14}$$

Obviously, every basis $\{x_n\}$ of a Banach space $E$ is a quasi-basis of $E$. Actually, *a sequence $\{x_n\} \subset E$ is a basis of $E$ if and only if it is both a quasi-basis of $E$ and $\omega$-linearly independent.*

It is also immediate that every quasi-basis $\{x_n\}$ of $E$ is complete in $E$ and every admissible sequence $\{f_n\} \subset E^*$ for $\{x_n\}$ is total on $E$; moreover, as in Ch. I, § 12, theorem 12.2 a), we also have

$$r([f_n]) > 0. \tag{9.15}$$

*Every quasi-basis $\{x_n\}$ of $E$ is an approximative basis of $E$,* since for any admissible sequence $\{f_n\} \subset E^*$ we can take $h_{ni} = f_i$ $(i = 1, \ldots, n;$ $n = 1,2, \ldots)$. Clearly, the converse is not true, but we shall see in theorem 9.3 below (implication $1° \Rightarrow 7°$) that every Banach space with an approximative basis also has a quasi-basis and hence, by the remark made after proposition 9.2, the answer to the problem of existence of quasi-bases is negative. In this connection, the following observation may have some interest:

*Remark 9.2.* By § 5, theorem 5.2 c), every separable Banach space $E$ has a "*non-linear quasi-basis*", i.e., a sequence $\{x_n\}$ such that there exists a sequence of *continuous* (generally, non-linear) functionals $\{f_n\}$ on $E$ satisfying (9.11); in other words, $\{x_n\}$ is a pseudo-basis*) of $E$ and the set-valued mapping $x \to \left\{\{\alpha_n\} \subset K \,\middle|\, \sum_{i=1}^{\infty} \alpha_i x_i = x\right\}$ of $E$ into $2^{A_1}$ admits a continuous selection (see § 5, the proof of theorem 5.2).

Similarly, $\{x_n\}$ is a ("linear") quasi-basis of $E$ if and only if it is a pseudo-basis of $E$ and the above mapping admits a *continuous linear selection*. Some characterizations of ("linear") quasi-bases are given in

**Theorem 9.1.** *Let $\{x_n\}$ be a sequence in a Banach space $E$. The following statements are equivalent:*

1°. *$\{x_n\}$ is a quasi-basis of $E$.*

2°. *$\{x_n\}$ is a pseudo-basis of $E$ and*

$$G = \left\{\{\alpha_n\} \subset K \,\middle|\, \sum_{i=1}^{\infty} \alpha_i x_i = 0\right\} \tag{9.16}$$

*is a complemented subspace of the Banach space of sequences of scalars*

$$A_1 = \left\{\{\alpha_n\} \subset K \,\middle|\, \sum_{i=1}^{\infty} \alpha_i x_i \ \textit{converges}\right\} \tag{9.17}$$

$\left(\textit{with the norm } \|\{\alpha_n\}\| = \sup_{1 \leq n < \infty} \left\|\sum_{i=1}^{n} \alpha_i x_i\right\|\right)$, *introduced in Ch.* I, § 3, *proposition* 3.1.

3°. *$\{x_n\}$ is a pseudo-basis of $E$ and there exists an isomorphism $u$ of $E$ into $A_1$ such that*

$$A_1 = u(E) \oplus G, \tag{9.18}$$

*where $A_1$ and $G$ are as in* 2°.

4°. *There exist an isomorphism $u$ of $E$ into $A_1$ and a projection $v$ of $A_1$ onto $u(E)$ such that*

$$x_k = u^{-1} v(e_k) \qquad (k = 1,2, \ldots), \tag{9.19}$$

*where $\{e_n\}$ is the sequence of unit vectors in $A_1$.*

---

*) See § 5, definition 5.1.

5°. *There exist a Banach space $B$ with a basis $\{e_n\}$, a linear isometry $u$ of $E$ into $B$ and a projection $v$ of $B$ onto $u(E)$ such that we have* (9.19).

6°. *Same as 5°, with "linear isometry" replaced by "isomorphism".*

7°. *There exist a Banach space $D$ with a quasi-basis $\{e_n\}$ such that $E$ is a complemented subspace of $D$, and a projection $v$ of $D$ onto $E$ such that*

$$x_n = v(e_n) \qquad (n = 1,2, \ldots). \tag{9.20}$$

*Proof.* Let $\{x_n\}$ be a quasi-basis of $E$ and $\{f_n\} \subset E^*$ an admissible sequence for $\{x_n\}$. Then the mapping $u: E \to A_1$ defined by

$$u(x) = \{f_n(x)\} \qquad (x \in E) \tag{9.21}$$

is an isomorphism of $E$ into $A_1$, since $\sum_{i=1}^{\infty} f_i(x)x_i$ converges to $x$ by (9.11) and since

$$\|x\| = \left\| \sum_{i=1}^{\infty} f_i(x)x_i \right\| \leqslant \sup_{1 \leqslant n < \infty} \left\| \sum_{i=1}^{n} f_i(x)x_i \right\| = \|\{f_n(x)\}\| = \|u(x)\| \leqslant C\|x\| \qquad (x \in E),$$

where $C = \sup_{1 \leqslant n < \infty} \|s_n\| < \infty$ by the principle of uniform boundedness. Hence $u(E)$ is closed in $A_1$. Since

$$w: \{\alpha_n\} \to \sum_{i=1}^{\infty} \alpha_i x_i \qquad (\{\alpha_n\} \in A_1) \tag{9.22}$$

is a continuous linear mapping of $A_1$ onto $E$ (because $\left\| \sum_{i=1}^{\infty} \alpha_i x_i \right\| \leqslant \|\{\alpha_n\}\|$ for all $\{\alpha_n\} \in A_1$), and since $G = \operatorname{Ker} w$, $G$ is also closed in $A_1$. Furthermore, if $\{f_n(x)\} \in G$ for some $x \in E$, then, by (9.11), $x = 0$, whence $\{f_n(x)\} = 0$; thus, $u(E) \cap G = \{0\}$. Now, let $\{\alpha_n\} \in A_1$ be arbitrary. Then $\sum_{i=1}^{\infty} \alpha_i x_i = x \in E$ and we have

$$\{\alpha_n\} = \{f_n(x)\} + \{\alpha_n - f_n(x)\},$$

where $\{f_n(x)\} \in u(E)$ and $\{\alpha_n - f_n(x)\} \in G$ $\left(\text{because } \sum_{i=1}^{\infty} (\alpha_i - f_i(x))x_i = \right.$ $\left. = \sum_{i=1}^{\infty} \alpha_i x_i - \sum_{i=1}^{\infty} f_i(x)x_i = x - x = 0\right)$, which proves (9.18). Thus, 1° ⇒ 3°.

The implication 3° ⇒ 2° is obvious.

Assume now that we have 2°, say $G \oplus F = A_1$. Then $w|_F$, where $w$ is defined by (9.22), is an isomorphism of $F$ onto $E$. Indeed, if $w(\{\beta_n\}) = 0$

for some $\{\beta_n\} \in F$, then $\sum_{i=1}^{\infty} \beta_i x_i = 0$, whence $\{\beta_n\} \in G \cap F = \{0\}$, which proves that $w|_F$ is one-to-one. Furthermore, if $x \in E$, then, since $\{x_n\}$ is a pseudo-basis of $E$, there exists a sequence $\{\alpha_n\} \in A_1$ such that $x = \sum_{i=1}^{\infty} \alpha_i x_i = w(\{\alpha_n\})$. Writing $\{\alpha_n\} = \{\gamma_n\} + \{\beta_n\}$, where $\{\gamma_n\} \in G$, $\{\beta_n\} \in F$, we obtain $x = w(\{\gamma_n\}) + w(\{\beta_n\}) = w(\{\beta_n\})$, which proves that $w|_F$ maps $F$ onto $E$. Hence, by the inversion theorem of Banach, $w|_F$ is an isomorphism of $F$ onto $E$. Now let $x \in E$ be arbitrary and let $\{\beta_n(x)\} = (w|_F)^{-1}(x) \in F$. Then

$$x = w(\{\beta_n(x)\}) = \sum_{i=1}^{\infty} \beta_i(x) x_i \tag{9.23}$$

and each $\beta_n$ is linear. Also,

$$|\beta_n(x)| = \frac{1}{\|x_n\|} \|\beta_n(x) x_n\| \leqslant \frac{2}{\|x_n\|} \sup_{1 \leqslant k < \infty} \left\| \sum_{i=1}^{k} \beta_i(x) x_i \right\| =$$

$$= \frac{2}{\|x_n\|} \|\{\beta_n(x)\}\| \leqslant \frac{2}{\|x_n\|} \|(w|_F)^{-1}\| \, \|x\| \quad (x \in E,\ n = 1, 2, \ldots), \tag{9.24}$$

whence $\{x_n\}$ is a quasi-basis of $E$. Thus, $2° \Rightarrow 1°$.

Assume now again that $\{x_n\}$ is a quasi-basis of $E$ and $\{f_n\} \subset E^*$ an admissible sequence for $\{x_n\}$. By the above proof of the implication $1° \Rightarrow 3°$, let $u$ be the isomorphism (9.21) of $E$ into $A_1$ and let $v$ be the projection of $A_1$ onto $u(E)$ along $G$. Then

$$v(\{\alpha_n\}) = \left\{ f_n \left( \sum_{i=1}^{\infty} \alpha_i x_i \right) \right\} \qquad (\{\alpha_n\} \in A_1), \tag{9.25}$$

since for every $\{f_n(x)\} \in u(E)$ we have $v(\{f_n(x)\}) = \{f_n(x)\} = \left\{ f_n \left( \sum_{i=1}^{\infty} f_i(x) x_i \right) \right\}$ and since for every $\{\gamma_n\} \in G$ we have $v(\{\gamma_n\}) = 0 = \left\{ f_n \left( \sum_{i=1}^{\infty} \gamma_i x_i \right) \right\}$. By (9.25) we have, in particular,

$$v(e_k) = \left\{ f_n \left( \sum_{i=1}^{\infty} \delta_{ik} x_i \right) \right\} = \{f_n(x_k)\} = u(x_k) \qquad (k = 1,2, \ldots),$$

whence we infer (9.19). Thus, $1° \Rightarrow 4°$.

Assume that we have 4°. Then, by Ch. II, § 13, lemma 13.6, there exists an equivalent norm $\|\{\alpha_n\}\|_1$ on $A_1$ such that the subspace $u(E) \subset A_1$

endowed with this new norm is linearly isometric to $E$ and by Ch. I, § 8, proposition 8.1, $\{e_n\}$ is a basis of $A_1$, so we have 5° for $B = A_1$ endowed with the norm $\|\{\alpha_n\}\|_1$. Thus, 4° $\Rightarrow$ 5°.

The implication 5° $\Rightarrow$ 6° is obvious.

Assume now that we have 6°. Let $\{h_n\} \subset B^*$ be the a.s.c.f. to the basis $\{e_n\}$ of $B$ and let

$$g_n = h_n|_{v(B)} \qquad (n = 1,2, \ldots). \tag{9.26}$$

Then for every $y \in v(B)$ we have

$$y = v(y) = \sum_{i=1}^{\infty} h_i(y)v(e_i) = \sum_{i=1}^{\infty} g_i(y)v(e_i)$$

and therefore $\{v(e_n)\}$ is a quasi-basis of $v(B) = u(E)$. Hence, since $u$ is an isomorphism of $E$ onto $u(E)$, by (9.19) it follows that $\{x_n\}$ is a quasi-basis of $E$. Thus, 6° $\Rightarrow$ 1°.

Finally, the implication 1° $\Rightarrow$ 7° is obvious (with $D = E$, $\{e_n\} = \{x_n\}$ and $v = I_E$) and the implication 7° $\Rightarrow$ 1° follows with the argument of the above proof of the implication 6° $\Rightarrow$ 1°. This completes the proof of theorem 9.1.

We shall use the following extension of Ch. I, § 4, definition 4.4:

*Definition 9.3*. Let $E$ be a Banach space and $(x_n, f_n)$ a pair of sequences such that $\{x_n\} \subset E$, $\{f_n\} \subset E^*$ (not necessarily a biorthogonal system). The sequence of continuous linear operators $\{s_n\} \subset L(E, E)$, where

$$s_n(x) = \sum_{i=1}^{n} f_i(x)x_i \qquad (x \in E,\ n = 1,2, \ldots), \tag{9.27}$$

is called *the sequence of partial sum operators associated to the pair* $(x_n, f_n)$. If $\{x_n\}$ is a quasi-basis of $E$ and $\{f_n\} \subset E^*$ an admissible sequence for $\{x_n\}$, then $\{s_n\}$ is called *the sequence of partial sum operators associated to the quasi-basis* $\{x_n\}$ *with the admissible sequence* $\{f_n\}$.

*Remark 9.3*. a) For the isomorphism $u: E \to A_1$ occurring in 4° of theorem 9.1 we have $\|u^{-1}\| \leqslant 1$, $\|u\| \leqslant C = \sup_{1 \leqslant n < \infty} \|s_n\|$ (see (9.27)) and for the projection $v: A_1 \to u(E)$ of 4° we have $\|v\| \leqslant \|u\|$, since

$$\|v(\{\alpha_n\})\| = \left\|\left\{f_n\left(\sum_{i=1}^{\infty} \alpha_i x_i\right)\right\}\right\| = \left\|u\left(\sum_{i=1}^{\infty} \alpha_i x_i\right)\right\| \leqslant \|u\| \left\|\sum_{i=1}^{\infty} \alpha_i x_i\right\| \leqslant \|u\| \, \|\{\alpha_n\}\|$$
$$(\{\alpha_n\} \in A_1).$$

Consequently, for the projection $v: B \to u(E)$ occurring in 5° of theorem 9.1 we have $\|v\| = 1$. Furthermore, since by Ch. I, § 8, proposition 8.1, $\{e_n\}$ is a monotone basis of $A_1$ and since by Ch. II, § 13, proof of lemma 13.6 we have $\|\{\alpha_n\}\| \leqslant \|\{\alpha_n\}\|_1 \leqslant C\|\{\alpha_n\}\|$ $(\{\alpha_n\} \in A_1)$, where $\|\{\alpha_n\}\|_1$ denotes the norm on $B$, it follows that the norm of $\{e_n\}$ as a basis of $B$ is $\leqslant C$ (because for any scalars $\alpha_1, \ldots, \alpha_{n+m}$ we have $\left\|\sum_{i=1}^{n} \alpha_i e_i\right\|_1 \leqslant C \left\|\sum_{i=1}^{n} \alpha_i e_i\right\| \leqslant C \left\|\sum_{i=1}^{n+m} \alpha_i e_i\right\| \leqslant C \left\|\sum_{i=1}^{n+m} \alpha_i e_i\right\|_1$). Thus, we have the following sharpening of the implication $1° \Rightarrow 5°$ of theorem 9.1: *If $\{x_n\}$ is a quasi-basis of a Banach space $E$, with an admissible sequence $\{f_n\} \subset E^*$, then there exist a Banach space $B$ having a basis $\{e_n\}$ with $v_{\{e_n\}} \leqslant C = \sup_{1 \leqslant n < \infty} \|s_n\|$* (see (9.27)), *a linear isometry $u$ of $E$ into $B$ and a projection $v$ of $B$ onto $u(E)$, of norm $\|v\| = 1$, such that we have* (9.19). Note also that if $\dim E = n < \infty$ and $\{x_i\}_{i=1}^{N}$ is a finite quasi-basis of $E$, then, obviously, $\dim B = N$. b) The equivalence $1° \Leftrightarrow 5°$ of theorem 9.1 gives the "general form" of a quasi-basis as being, up to a linear isometry, *the image, by a projection, of a basis.* c) The implication $7° \Rightarrow 1°$ shows that *every complemented subspace of a space with a quasi-basis has a quasi-basis.* It is not known*) whether a similar result holds for bases. d) The implication $1° \Rightarrow 3°$ of theorem 9.1 may be regarded as a generalization of Ch. I, § 3, proposition 3.2 a) (which concerns the particular case when $\{x_n\}$ is a basis); see also § 5, remark 5.2, concerning the more general case of pseudo-bases.

The assumption in 3° (and hence in 2°) that $\{x_n\}$ is a pseudo-basis of $E$ is essential, as shown by

*Example 9.1.* Let $E$ be a Banach space with a basis $\{z_n\}$ and let $\{x_n\} \subset E$ be an $\omega$-linearly independent sequence in $E$, which is not a basis of $E$, such that $\{x_n\} \sim \{z_n\}$. For instance, one can take $E = c_0$, with the basis

$$z_n = \sum_{i=1}^{n} e_i = \{\underbrace{1, \ldots, 1}_{n}, 0, 0, \ldots\} \qquad (n = 1,2, \ldots) \tag{9.28}$$

and with

$$x_n = e_1 + e_{n+1} \qquad (n = 1,2, \ldots); \tag{9.29}$$

in this latter case $\{x_n\}$ is also complete in $E$ and minimal**) (see § 6, example 6.7 a)) and we have $\{x_n\} \sim \{z_n\}$ since both $\sum_{i=1}^{\infty} \alpha_i x_i$ and $\sum_{i=1}^{\infty} \alpha_i z_i$ converge if and only if $\sum_{i=1}^{\infty} \alpha_i$ converges (for the second property, see Ch. II, § 14, example 14.1). Then $\{x_n\}$ satisfies the second statement in 3°, but it is not a pseudo-basis (and hence not a quasi-basis) of $E$. Indeed,

*) By theorem 9.3 below, this problem is equivalent to problem 9.1.

**) We can even choose $\{x_n\}$ to be a basic sequence, e.g. any proper subsequence of (9.28).

since $\{x_n\}$ is $\omega$-linearly independent, the subspace $G$ of $A_1$ defined by (9.16) reduces to $\{0\}$ and hence (9.18) reduces to $A_1 = u(E)$. Now let $\{h_n\} \subset E^*$ be the a.s.c.f. to the basis $\{z_n\}$ of $E$ and let

$$u(x) = \{h_n(x)\} \qquad (x \in E). \tag{9.30}$$

Then $u(E) = A_1(\{z_n\}) = \left\{\{\alpha_n\} \subset K \,\middle|\, \sum_{i=1}^{\infty} \alpha_i z_i \text{ converges}\right\} = A_1(\{x_n\})$ (the last equality holds by $\{z_n\} \sim \{x_n\}$) and $u: E \to A_1(\{x_n\})$ is an isomorphism. However, $\{x_n\}$ is not a pseudo-basis of $E$, since it is $\omega$-linearly independent but not a basis of $E$.

In Ch. I, § 7, definition 7.1 the norm of a complete sequence $\{x_n\} \subset E$ with $x_n \neq 0$ $(n = 1,2, \ldots)$ was defined as $v_{\{x_n\}} = \sup_{1 \leqslant n < \infty} \|s_n\|$ whenever $\{x_n\}$ is a basis of $E$ and $\{s_n\}$ the associated sequence of partial sum operators and $v_{\{x_n\}} = \infty$ whenever $\{x_n\}$ is not a basis. However, for quasi-bases $\{x_n\}$ it is convenient to introduce another "norm", which coincides with $v_{\{x_n\}}$ when $\{x_n\}$ is a basis:

*Definition 9.4.* Let $\{x_n\}$ be a quasi-basis of a Banach space $E$. The number

$$v_{\{x_n\}} = \inf \sup_{1 \leqslant n < \infty} \|s_n\|, \tag{9.31}$$

where $\{s_n\}$ is the sequence of partial sum operators associated to $\{x_n\}$ paired with an admissible sequence $\{f_n\} \subset E^*$ and where the inf is taken over all admissible sequences $\{f_n\} \subset E^*$ for $\{x_n\}$, is called *the norm of the quasi-basis* $\{x_n\}$.

Since for every quasi-basis $\{x_n\}$ with any admissible sequence $\{f_n\} \subset E^*$ we have, by (9.11) and the principle of uniform boundedness, $1 \leqslant \sup_{1 \leqslant n < \infty} \|s_n\| < \infty$, it follows that

$$1 \leqslant v_{\{x_n\}} < \infty. \tag{9.32}$$

Let us consider now quasi-bases of finite-dimensional Banach spaces.

**Proposition 9.3.** *Let $E$ be a finite-dimensional Banach space, and let $\{y_i\}_{i=1}^n$ be a finite quasi-basis of $E$ (hence* dim $E \leqslant n$*). Then $E$ has a quasi-basis $\{x_i\}_{i=1}^{n^2}$ consisting of $n^2$ elements, such that*

$$v_{\{x_i\}_{i=1}^{n^2}} \leqslant \frac{n-1}{n} + \frac{v_{\{y_i\}_{i=1}^n}}{n} = 1 + \frac{v_{\{y_i\}_{i=1}^n} - 1}{n}. \tag{9.33}$$

*Proof.* By definition 9.4, for each $m$ there is an admissible sequence $\{g_i^{(m)}\}_{i=1}^n$ for $\{y_i\}_{i=1}^n$, such that the associated partial sum operators satisfy $\sup_{1 \leqslant k \leqslant n} \|s_k^{(m)}\| \leqslant v_{\{y_i\}_{i=1}^n} + \frac{1}{m}$. Then for every $x \in E$, $k = 1, \ldots, n$

and $m = 1,2, \ldots$ we have

$$|g_k^{(m)}(x)| = \frac{1}{\|y_k\|} \|g_k^{(m)}(x)y_k\| = \frac{1}{\|y_k\|} \|s_k^{(m)}(x) - s_{k-1}^{(m)}(x)\| \leqslant$$

$$\leqslant \frac{2}{\|y_k\|} \sup_{1 \leqslant k \leqslant n} \|s_k^{(m)}\| \, \|x\| \leqslant \frac{2}{\|y_k\|} \left( v_{\{y_i\}_{i=1}^n} + \frac{1}{m} \right) \|x\|$$

(where $s_0^{(m)}(x) = 0$), whence $\sup\limits_{1 \leqslant m < \infty} \|g_k^{(m)}\| < \infty$ $(k = 1, \ldots, n)$ and hence, since $\dim E^* < \infty$, each $\{g_k^{(m)}\}_{m=1}^{\infty}$ contains a subsequence converging to some $g_k \in E^*$ $(k = 1, \ldots, n)$. Then

$$\sum_{i=1}^{n} g_i(x) y_i = x \qquad (x \in E),$$

i.e., $\{g_i\}_{i=1}^n$ is an admissible sequence for $\{y_i\}_{i=1}^n$, and for $s_k(x) = \sum\limits_{i=1}^{k} g_i(x) y_i$ we have $\|s_k\| \leqslant v_{\{y_i\}_{i=1}^n}$, whence

$$\sup_{1 \leqslant k \leqslant n} \|s_k\| = v_{\{y_i\}_{i=1}^n} .$$

We claim that the sequence $\{x_i\}_{i=1}^{n^2}$ defined by

$$x_{rn+j} = \frac{1}{n} y_j \qquad (r = 0,1, \ldots, n-1; \; j = 1, \ldots, n) \tag{9.34}$$

has the required properties. Indeed, for $\{f_i\}_{i=1}^{n^2} \subset E^*$ defined by

$$f_{rn+j} = g_j \qquad (r = 0,1, \ldots, n-1; \; j = 1, \ldots, n) \tag{9.35}$$

we have

$$\sum_{i=1}^{n^2} f_i(x) x_i = n \sum_{j=1}^{n} \frac{1}{n} g_j(x) y_j = x \qquad (x \in E)$$

and thus $\{x_i\}_{i=1}^{n^2}$ is a quasi-basis of $E$, having $\{f_i\}_{i=1}^{n^2}$ as an admissible sequence. Furthermore, for any pair of integers $(r, k)$ with $0 \leqslant r \leqslant n-1$

and $1 \leqslant k \leqslant n$ we have

$$\left\| \sum_{i=1}^{rn+k} f_i(x) x_i \right\| = \left\| \frac{r}{n} \sum_{j=1}^{n} g_j(x) y_j + \frac{1}{n} \sum_{j=1}^{k} g_j(x) y_j \right\| =$$

$$= \left\| \frac{r}{n} x + \frac{1}{n} \sum_{j=1}^{k} g_j(x) y_j \right\| \leqslant \frac{r}{n} \|x\| + \frac{1}{n} \left\| \sum_{j=1}^{k} g_j(x) y_j \right\| \leqslant$$

$$\leqslant \left( \frac{n-1}{n} + \frac{\nu_{\{y_i\}_{i=1}^{n}}}{n} \right) \|x\| \qquad (x \in E),$$

whence we infer (9.33), which completes the proof of proposition 9.3.

Now we can prove

**Theorem 9.2.** *Let $E$ be an n-dimensional Banach space, where $2 \leqslant n < \infty$. Then*

a) *$E$ has a quasi-basis $\{x_i\}_{i=1}^{n^2}$ consisting of $n^2$ elements, such that*

$$\nu_{\{x_i\}_{i=1}^{n^2}} \leqslant \frac{2n-2}{n} < 2. \tag{9.36}$$

b) *For every $\varepsilon > 0$ the space $E$ has a finite quasi-basis $\{x_i\}_{i=1}^{N_\varepsilon}$ such that*

$$\nu_{\{x_i\}_{i=1}^{N_\varepsilon}} \leqslant 1 + \varepsilon. \tag{9.37}$$

c) *$E$ has an infinite quasi-basis $\{x_i\}_{i=1}^{\infty}$ such that*

$$\nu_{\{x_i\}_{i=1}^{\infty}} = 1. \tag{9.38}$$

*Proof.* a) By Ch. II, § 2, theorem 2.2, $E$ has a normal basis, i.e. a basis $\{y_j\}_{j=1}^{n}$ such that

$$\|y_j\| = \|g_j\| = 1 \qquad (j = 1, \ldots, n),$$

where $\{g_j\}_{j=1}^{n} \subset E^*$ is the a.s.c.f. to $\{y_j\}_{j=1}^{n}$. Then for $1 \leqslant k \leqslant n-1$ we have

$$\left\| \sum_{j=1}^{k} g_j(x) y_j \right\| \leqslant \sum_{j=1}^{k} \|g_j\| \, \|y_j\| \, \|x\| = k\|x\| \leqslant (n-1)\|x\| \qquad (x \in E)$$

and for $k = n$ we have

$$\left\| \sum_{j=1}^{n} g_j(x) y_j \right\| = \|x\| \leqslant (n-1)\|x\| \qquad (x \in E).$$

Therefore the basis $\{y_j\}_{j=1}^{n}$ is of norm

$$\nu_{\{y_j\}_{j=1}^{n}} \leqslant n - 1, \tag{9.39}$$

whence, by proposition 9.3, we infer that $E$ has a quasi-basis $\{x_i\}_{i=1}^{n^2}$ satisfying (9.36).

b) By part a) and proposition 9.3 above $E$ has a quasi-basis $\{x_i^{(1)}\}_{i=1}^{n^4}$ such that

$$\nu_{\{x_i^{(1)}\}_{i=1}^{n^4}} \leqslant \frac{n^2-1}{n^2} + \frac{\frac{2n-2}{n}}{n^2} = 1 + \frac{n-2}{n^3} < 1 + \frac{1}{n^2}.$$

Hence, applying again proposition 9.3, $E$ has a quasi-basis $\{x_i^{(2)}\}_{i=1}^{n^8}$ such that

$$\nu_{\{x_i^{(2)}\}_{i=1}^{n^8}} \leqslant \frac{n^4-1}{n^4} + \frac{1+\frac{1}{n^2}}{n^4} = 1 + \frac{1}{n^6}.$$

Continuing in this way, in a finite number of steps we obtain a finite quasi-basis $\{x_i\}_{i=1}^{N_\varepsilon}$ of $E$ satisfying (9.37).

c) As in the above proof of part a), let $\{y_j\}_{j=1}^{n}$ be a normal basis of $E$, with the a.s.c.f. $\{g_j\}_{j=1}^{n} \subset E^*$. We claim that the sequence $\{x_i\}_{i=1}^{\infty}$ defined by

$$x_{pn^2+rn+j} = \frac{1}{2^{p+1}n} y_j \quad (p = 0,1,\ldots; r = 0,1,\ldots,n-1; j = 1,\ldots,n) \tag{9.40}$$

has the required properties. Indeed, for $\{f_i\}_{i=1}^{\infty} \subset E^*$ defined by

$$f_{pn^2+rn+j} = g_j \qquad (p = 0,1,\ldots; r = 0,1,\ldots,n-1; j = 1,\ldots,n) \tag{9.41}$$

we have

$$\sum_{i=1}^{\infty} f_i(x)x_i = \sum_{p=0}^{\infty}\sum_{r=0}^{n-1}\sum_{j=1}^{n} f_{pn^2+rn+j}(x)x_{pn^2+rn+j} = \sum_{p=0}^{\infty} n \sum_{j=1}^{n} \frac{1}{2^{p+1}n} g_j(x)y_j =$$

$$= \sum_{p=0}^{\infty} \frac{1}{2^{p+1}} \sum_{j=1}^{n} g_j(x)y_j = \sum_{j=1}^{n} g_j(x)y_j = x \qquad (x \in E),$$

and thus $\{x_i\}_{i=1}^{\infty}$ is a quasi-basis of $E$, having $\{f_i\}_{i=1}^{\infty}$ as an admissible sequence. Furthermore, for any triple of integers $(p, r, k)$ with $0 \leqslant p < \infty$, $0 \leqslant r \leqslant n-1$ and $1 \leqslant k \leqslant n$ we have, taking into account (9.39),

$$\left\| \sum_{i=1}^{pn^2+rn+k} f_i(x)x_i \right\| = \left\| \sum_{l=1}^{pn+r} \sum_{i=(l-1)n+1}^{ln} f_i(x)x_i + \sum_{i=pn^2+rn+1}^{pn^2+rn+k} f_i(x)x_i \right\| =$$

$$= \left\| \sum_{l=1}^{p} n \sum_{j=1}^{n} \frac{1}{2^l n} g_j(x)y_j + r\frac{1}{2^{p+1}n} \sum_{j=1}^{n} g_j(x)y_j + \frac{1}{2^{p+1}n} \sum_{j=1}^{k} g_j(x)y_j \right\| =$$

$$= \left\| \sum_{l=1}^{p} \left( \frac{1}{2^l} + \frac{r}{2^{p+1}n} \right) x + \frac{1}{2^{p+1}n} \sum_{j=1}^{k} g_j(x)y_j \right\| \leqslant$$

$$\leqslant \left( \sum_{l=1}^{p} \frac{1}{2^l} + \frac{r}{2^{p+1}n} + \frac{n-1}{2^{p+1}n} \right) \|x\| \leqslant \|x\| \qquad (x \in E),$$

whence we infer (9.38), which completes the proof of theorem 9.2.

Combining theorem 9.2 b), c) with remark 9.3 a), we obtain

**Corollary 9.1.** a) *For every finite-dimensional Banach space $E$ and every $\varepsilon > 0$ there is a finite-dimensional Banach space $B$ having a basis $\{e_n\}$ with $v_{\{e_n\}} \leqslant 1 + \varepsilon$ such that there exist a linear isometry $u$ of $E$ into $B$ and a projection $v$ of $B$ onto $u(E)$, of norm* 1.

b) *For every finite-dimensional Banach space $E$ there is an infinite dimensional Banach space $B$ having a monotone basis $\{e_n\}$ such that there exist a linear isometry $u$ of $E$ into $B$ and a projection $v$ of $B$ onto $u(E)$, of norm* 1.

In connection with theorem 9.2 b), c) it is natural to ask whether every finite-dimensional Banach space $E$ has a finite quasi-basis $\{x_i\}_{i=1}^{N}$ such that $v_{\{x_i\}_{i=1}^{N}} = 1$ (in this case, by remark 9.3 a) it would follow that for every finite-dimensional Banach space $E$ there is a finite-dimensional Banach space $B$ having a monotone basis $\{e_n\}_{n=1}^{N}$ such that there exist a linear isometry $u$ of $E$ into $B$ and a projection $v$ of $B$ onto $u(E)$, of norm 1). However, we shall see later (in corollary 9.3) that *the answer is negative*.

**Corollary 9.2.** *For every finite-dimensional Banach space $E$ there exists a finite-dimensional Banach space $F$ such that $(E\times F)_{l^\infty}$ has a basis $\{x_n\}$ with $v_{\{x_n\}}\leqslant 5$.*

*Proof.* Let $\varepsilon>0$ be arbitrary and let $B=B(E,\varepsilon)$ be a finite-dimensional space as in corollary 9.1 a), hence $B$ has a basis $\{e_n\}$ with $v_{\{e_n\}}\leqslant 1+\varepsilon$ and $B=u(E)\oplus F$, where $u$ is a linear isometry of $E$ into $B$ and where $F=(I-v)(B)$, with $v$ a projection of norm 1 of $B$ onto $u(E)$. Then

$$\frac{1}{2}\max(\|x\|,\|y\|)=\frac{1}{2}\max(\|u(x)\|,\|y\|)=$$

$$=\frac{1}{2}\max(\|v(u(x)+y)\|,\|(I-v)(u(x)+y)\|)\leqslant$$

$$\leqslant\|u(x)+y\|\leqslant\|u(x)\|+\|y\|=\|x\|+\|y\|\leqslant$$

$$\leqslant 2\max(\|x\|,\|y\|)\qquad(u(x)+y\in u(E)\oplus F=B),$$

so the natural isomorphism $w$ of $B=u(E)\oplus F$ onto $(E\times F)_{l^\infty}$ satisfies $\|w\|\leqslant 2$, $\|w^{-1}\|\leqslant 2$. Consequently, by Ch. I, § 7, theorem 7.1, $\{x_n\}=\{w(e_n)\}$ is a basis of $(E\times F)_{l^\infty}$, with $v_{\{x_n\}}\leqslant 4(1+\varepsilon)\leqslant 5$ $\left(\text{for } \varepsilon\leqslant\frac{1}{4}\right)$, which completes the proof of corollary 9.2.

It is an unsolved problem whether a result similar to theorem 9.2 a) holds for bases, even with 2 replaced by a constant $C$:

*Problem 9.2.* Does there exist a constant $C$ such that every finite-dimensional Banach space $E$ has a basis of norm $\leqslant C$?

An affirmative answer to problem 9.2 would imply, among other things, that for every finite-dimensional Banach space $E$ and all Bohnenblust functions $\varphi_j$ (see Ch. I, § 11, formula (11.2)) we have $\varphi_j(E)\leqslant C$ $(j=1,\ldots,\dim E)$ and hence the conditions of Ch. I, § 11, theorem 11.1 (giving a possible way to construct a separable Banach space having no basis) would be void. For another consequence of an affirmative answer to problem 9.2 see § 13, the remark to problem 13.3. In any case, if such a constant $C$ exists, then it must be $>1$ (strictly), by any one of theorems 1.1—1.3 of Ch. II, § 1.

A related notion, which will also be used for characterizing Banach spaces with an approximative basis, is given in

*Definition 9.5.* Let $E$ be a Banach space. A sequence of non-zero endomorphisms of finite rank $\{v_n\}\subset L(E,E)$ is called *a finite dimensional expansion of the identity of $E$*, or shortly, *a finite dimensional expansion of $I_E$*, if

$$x=\sum_{i=1}^{\infty}v_i(x)\qquad(x\in E);\tag{9.42}$$

if here $\sup_{1 \leqslant n < \infty} \left\| \sum_{i=1}^{n} v_i \right\| \leqslant \lambda$, we shall say that $\{v_n\}$ is a *finite-dimensional $\lambda$-expansion of $I_E$*.

Clearly, $E$ has a quasi-basis if and only if $E$ has a one-dimensional expansion of $I_E$, since we have (9.11) if and only if we have (9.42) with $\{v_n\} \subset L(E, E)$ defined by

$$v_n(x) = f_n(x)x_n \qquad (x \in E,\ n = 1,2, \ldots). \tag{9.43}$$

It is also obvious that $E$ has a $\lambda$-approximative basis if and only if it has a finite-dimensional $\lambda$-expansion of $I_E$. Indeed, if $\{u_n\}$ satisfies (9.2) and $\sup_{1 \leqslant n < \infty} \|u_n\| \leqslant \lambda$ and if $u_1 \neq 0$, $u_n \neq u_{n+1}$ for $n = 1,2, \ldots$ (which we may assume), then the sequence

$$v_1 = u_1, \ v_{2n} = v_{2n+1} = \frac{1}{2}(u_{n+1} - u_n) \qquad (n = 1,2,\ldots) \tag{9.44}$$

satisfies (9.42) and $\sup_{1 \leqslant n < \infty} \left\| \sum_{i=1}^{n} v_i \right\| \leqslant \lambda$, $\sup_{1 \leqslant n < \infty} \|v_n\| \leqslant \lambda$ (note that for

$$v_1' = u_1, \ v_n' = u_n - u_{n-1} \qquad (n = 2,3, \ldots)$$

we have (9.42) and $\sup_{1 \leqslant n < \infty} \|v_n'\| \leqslant 2\lambda$); conversely, if $\{v_n\}$ satisfies (9.42) and $\sup_{1 \leqslant n < \infty} \left\| \sum_{i=1}^{n} v_i \right\| \leqslant \lambda$, then the sequence

$$u_n = \sum_{i=1}^{n} v_i \qquad (n = 1,2, \ldots) \tag{9.45}$$

satisfies (9.2) and $\sup_{1 \leqslant n < \infty} \|u_n\| \leqslant \lambda$.

**Proposition 9.4.** *Let $\{v_n\} \subset L(E, E)$ be a finite-dimensional expansion of the identity of a Banach space $E$ and for each $n$ let $\{x_1^{(n)}, \ldots, x_{m_n}^{(n)}\}$ be a quasi-basis of $v_n(E)$, such that $\sup_{1 \leqslant n < \infty} v_{\{x_i^{(n)}\}_{i=1}^{m_n}} = C_0 < \infty$. Then the sequence $\{x_n\} \subset E$ defined by*

$$x_{m_0+m_1+\ldots+m_n+i} = x_i^{(n+1)} \qquad (i = 1, \ldots, m_{n+1};\ n = 0,1,2, \ldots;\ m_0 = 0) \tag{9.46}$$

*(i.e., the sequence $x_1^{(1)}, \ldots, x_{m_1}^{(1)}, x_1^{(2)}, \ldots, x_{m_2}^{(2)}, \ldots$) is a quasi-basis of $E$.*

*Proof.* For each $n$ let $\{\varphi_i^{(n)}\}_{i=1}^{m_n} \subset v_n(E)^*$ be an admissible sequence for $\{x_i^{(n)}\}_{i=1}^{m_n}$ $\left(\text{hence } \sum_{i=1}^{m_n} \varphi_i^{(n)}(y)x_i^{(n)} = y \text{ for all } y \in v_n(E)\right)$ such that for the associated partial sum operators we have $\sup_{\substack{1 \leqslant i \leqslant m_n \\ 1 \leqslant n < \infty}} \|s_i^{(n)}\| \leqslant C_0$, and let

$$f_{m_0+m_1+\ldots+m_n+i} = \varphi_i^{(n+1)} \circ v_{n+1} \quad (i = 1, \ldots, m_{n+1};\ n = 0,1,2,\ldots;\ m_0 = 0). \tag{9.47}$$

For convenience of notation, put

$$r_n = m_0 + m_1 + \ldots + m_n \qquad (n = 0,1,2, \ldots). \tag{9.48}$$

Then for every $x \in E$ we have

$$\begin{aligned} s_{r_n+i_n}(x) &= \sum_{j=1}^{r_n+i_n} f_j(x)x_j = \sum_{k=1}^{n} \sum_{j=r_{k-1}+1}^{r_k} f_j(x)x_j + \sum_{j=r_n+1}^{r_n+i_n} f_j(x)x_j = \\ &= \sum_{k=1}^{n} \sum_{j=1}^{m_k} \varphi_j^{(k)}(v_k(x))x_j^{(k)} + \sum_{j=1}^{i_n} \varphi_j^{(n+1)}(v_{n+1}(x))x_j^{(n+1)} = \\ &= \sum_{k=1}^{n} v_k(x) + s_{i_n}^{(n+1)}(v_{n+1}(x)) \quad (i_n = 1, \ldots, m_{n+1};\ n = 0,1,2,\ldots). \end{aligned} \tag{9.48'}$$

Hence, since by our hypothesis $\sum_{k=1}^{n} v_k(x) \to x$ and $\|s_{i_n}^{(n+1)}(v_{n+1}(x))\| \leqslant \leqslant C_0\|v_{n+1}(x)\| \to 0$ as $n \to \infty$, it follows that $s_{r_n+i_n}(x) \to x$ as $n \to \infty$. Thus, $\{x_n\}$ is a quasi-basis of $E$, which completes the proof of proposition 9.4.

We recall that a Banach space $E$ is said to have the *bounded approximation property* if there exists a constant $\lambda \geqslant 1$ such that the identity operator $I_E: E \to E$ can be approximated, uniformly on every compact subset of $E$, by linear operators of finite rank, of norm $\leqslant \lambda$, that is, if there exists a constant $\lambda \geqslant 1$ with the following property: for every compact subset $Q \subset E$ and every $\varepsilon > 0$ there exists an endomorphism $u = u_{Q,\varepsilon} \in L(E, E)$ of finite rank, of norm $\|u\| \leqslant \lambda$, such that

$$\|u(x) - x\| < \varepsilon \qquad (x \in Q); \tag{9.49}$$

in this case we shall also say that $E$ has the *$\lambda$-approximation property*. If the property holds with $\lambda = 1$, then $E$ is said to have the *metric approximation property*. Obviously, the bounded approximation property (and hence, in particular, the metric approximation property) implies the approximation property (see § 0, formula (0.1)).

Let us also recall the following lemma, which we shall frequently use in the sequel:

**Lemma 9.1.** *Let $E$ be a Banach space, $G$ a $k$-dimensional subspace of $E$ (where $k < \infty$), $0 < \delta < 1$ and $u \in L(E, E)$ an endomorphism of* *finite rank such that*

$$\|I_G - u|_G\| < \delta < 1. \tag{9.50}$$

*Then there exists an endomorphism* $v \in L(E, E)$ *of finite rank, such that*

$$v|_G = I_G, \tag{9.51}$$

$$\|v - u\| < \frac{\delta k}{1-\delta}\|u\|. \tag{9.52}$$

*Proof.* From Ch. II, § 2, theorem 2.2 it follows that there always exists a projection of norm $\leqslant k$ from any Banach space onto any $k$-dimensional subspace. Let $v_1$ be a projection of norm $\leqslant k$ of $E$ onto $G$ and put

$$v = v_1 + u - uv_1. \tag{9.53}$$

Then, by $v_1|_G = I_G$, (9.50) and $\|v_1\| \leqslant k$, we obtain

$$v(x) = v_1(x) + u(x) - uv_1(x) = x + u(x) - u(x) = x \qquad (x \in G),$$

$$\|v - u\| = \|(I - u)v_1\| = \|(I_G - u|_G)v_1\| \leqslant \|I_G - u|_G\| \, \|v_1\| < \delta k.$$

Now, by (9.50), we have $|\,\|x\| - \|u(x)\|\,| \leqslant \|x - u(x)\| < \delta\|x\|$ for $x \in G$, whence $\|u(x)\| > \|x\| - \delta\|x\| = (1-\delta)\,\|x\|$ for $x \in G$, so $\|u\| \geqslant \|u|_G\| \geqslant 1 - \delta$. Consequently, $1 \leqslant \frac{\|u\|}{1-\delta}$, whence

$$\|v - u\| < \delta k \leqslant \frac{\delta k}{1-\delta}\|u\|,$$

which completes the proof of lemma 9.1. Note also that by (9.53) we have $v(E) \subset [G \cup u(E)]$ and by (9.51) we have $v(E) \supset v(G) = G$.

*Remark 9.4.* With a slightly more complicated construction, one can obtain a finite rank endomorphism $v \in L(E, E)$ as in lemma 9.1 (i.e., satisfying (9.51), (9.52)) and having some useful additional properties. Indeed, by (9.50), $u|_G$ is an isomorphism from $G$ onto*) $u(G)$, with $\|(u|_G)^{-1}\| < \frac{1}{1-\delta}$ (and, clearly, $\|u|_G\| \leqslant 1 + \delta$), whence, again by (9.50),

$$\|(u|_G)^{-1} - I_{u(G)}\| = \|(I_G - u|_G)(u|_G)^{-1}\| < \frac{\delta}{1-\delta}. \tag{9.54}$$

---

*) Hence, we must have $\dim u(E) \geqslant \dim u(G) = \dim G$.

Let $v_0$ be a projection of norm $\leqslant k$ of $u(E)$ onto $u(G)$ and put

$$w = I_{u(E)} - v_0 + (u|_G)^{-1}v_0, \tag{9.55}$$

$$v = wu = u - v_0u + (u|_G)^{-1}v_0u. \tag{9.56}$$

Then $\dim v(E) < \infty$ and for every $x \in G$ we have $v_0u(x) = u(x)$, whence

$$v(x) = u(x) - v_0u(x) + (u|_G)^{-1}v_0u(x) = (u|_G)^{-1}u(x) = x \quad (x \in G),$$

which proves (9.51). Furthermore, by (9.54),

$$\|w - I_{u(E)}\| = \|(u|_G)^{-1}v_0 - v_0\| = \|((u|_G)^{-1} - I_{u(G)})v_0\| < \frac{\delta k}{1-\delta}, \tag{9.57}$$

whence we infer

$$\|v - u\| = \|wu - u\| = \|(w - I_{u(E)})u\| < \frac{\delta k}{1-\delta}\|u\|,$$

i.e. (9.52), which proves again lemma 9.1. Also, by (9.56) we have $v(E) \subset [G \cup u(E)]$ and by (9.51) we have $v(E) \supset v(G) = G$. Furthermore, if $\delta > 0$ is so small that

$$\frac{\delta k}{1-\delta} < 1, \tag{9.58}$$

then, by (9.57), $w$ is an isomorphism of $u(E)$ onto $wu(E) = v(E)$, whence

$$\dim v(E) = \dim u(E). \tag{9.59}$$

Also, in this case, by (9.56) and since $w$ is an isomorphism, we have

$$\operatorname{Ker} v = \operatorname{Ker} u, \tag{9.60}$$

whence, since $v(E)$, $u(E)$ are closed (because they are of finite dimension), it follows*) that

$$v^*(E^*) = (\operatorname{Ker} v)^\perp = (\operatorname{Ker} u)^\perp = u^*(E^*). \tag{9.61}$$

---

*) See e.g. [87], p. 487, theorem 2.

Now we shall return to approximative bases. Some characterizations of Banach spaces having an approximative basis are given in

**Theorem 9.3.** *Let E be a Banach space. The following statements are equivalent:*

1°. *E has an approximative basis.*

2°. *E contains a sequence of subspaces $\{G_n\}$ with the following properties:*

a) $\dim G_n < \infty \quad (n = 1,2, \ldots)$;

b) $G_n \subset G_{n+1} \quad (n = 1,2, \ldots)$;

c) $\bigcup_{n=1}^{\infty} G_n$ *is dense in E;*

d) *there exists a finite rank endomorphism $v_n \in L(E, E)$ satisfying $v_n|_{G_n} = I_{G_n}$ $(n = 1,2, \ldots)$ and such that $\sup_{1 \leqslant n < \infty} \|v_n\| < \infty$.*

3°. *There exists a finite-dimensional expansion of $I_E$.*

4°. *E is separable and it has the bounded approximation property.*

5°. *E is separable and there exists a constant $\lambda \geqslant 1$ with the property that for every finite-dimensional subspace G of E and every $\delta > 0$ there exists an endomorphism $u = u_{G,\delta} \in L(E, E)$ of finite rank such that*

$$\|u(x) - x\| < \delta \|x\| \quad (x \in G), \tag{9.62}$$

$$\|u\| \leqslant \lambda. \tag{9.63}$$

6°. *E is separable and there exists a constant $\lambda \geqslant 1$ with the property that for every finite-dimensional subspace G of E and every $\varepsilon > 0$ there exists an endomorphism $v = v_{G,\varepsilon} \in L(E, E)$ of finite rank such that*

$$v(x) = x \qquad (x \in G), \tag{9.64}$$

$$\|v\| \leqslant \lambda + \varepsilon \tag{9.65}$$

*(or, in other words, the identity operator $I_G$ can be extended to an endomorphism $v \in L(E, E)$ of finite rank, of norm $\|v\| \leqslant \lambda + \varepsilon$).*

7°. *E has a quasi-basis.*

8°. *E is linearly isometric to a complemented subspace of a Banach space B with a basis.*

9°. *Same as 8°, with "linearly isometric" replaced by "isomorphic".*

10°. *E is isomorphic to a complemented subspace of the space $E_b$ of Ch. II, § 13.*

*These statements imply — and if E is isomorphic to a separable conjugate Banach space, they are equivalent to — the following statement:*

11°. *E has the approximation property.*

*Proof.* Assume that we have 1°, say $\{u_n\} \subset L(E, E)$ satisfies $\dim u_n(E) < \infty$ $(n = 1,2, \ldots)$ and (9.2). Then $E$ is separable and by Ch. I, § 17, proof of theorem 17.3, with $\{s_n\}$ and $v_{\{x_n\}} = \sup\limits_{1 \leqslant n < \infty} \|s_n\|$ replaced by $\{u_n\}$ and $\lambda = \sup\limits_{1 \leqslant n < \infty} \|u_n\|$ respectively, it follows that for every compact subset $Q$ of $E$ we have

$$\lim_{n\to\infty} \sup_{x\in Q} \|x - u_n(x)\| = 0, \tag{9.66}$$

whence we infer (9.49) for suitable $u = u_n$ (so $\|u\| \leqslant \lambda$). Thus, $1° \Rightarrow 4°$.

The implication $4° \Rightarrow 5°$ is obvious, since for every finite-dimensional subspace $G$ of $E$ the unit ball $S_G = \{x \in G |\ \|x\| \leqslant 1\}$ is compact.

Assume now that we have 5° and let $G_n = [y_1, \ldots, y_n]$ $(n = 1,2, \ldots)$, where $\{y_n\}$ is a dense sequence in $E$. Then, by 5° $\left(\text{for } G = G_n \text{ and } \varepsilon = \frac{1}{n}\right)$, for each $n$ there exists an endomorphism $u_n \in L(E, E)$ of finite rank, such that

$$\|u_n(x) - x\| < \frac{1}{n} \|x\| \qquad (x \in G_n;\ n = 1,2, \ldots), \tag{9.67}$$

$$\|u_n\| \leqslant \lambda \qquad (n = 1,2, \ldots). \tag{9.68}$$

Since $G_1 \subset G_2 \subset \ldots$, from (9.67) it follows that

$$\lim_{n\to\infty} u_n(x) = x \qquad \left(x \in \bigcup_{n=1}^{\infty} G_n\right),$$

which, by (9.68) and since $\bigcup\limits_{n=1}^{\infty} G_n$ is dense in $E$, implies (9.2). Thus, $5° \Rightarrow 1°$.

The implication $5° \Rightarrow 6°$ follows from lemma 9.1, by taking $\delta > 0$ so small that $\frac{\delta k \lambda}{1 - \delta} < \varepsilon$, where $k = \dim G$, since then by (9.52) and (9.63) we have (9.65).

Conversely, assume that we have 6° and let $G \subset E$ with $\dim G < \infty$ and $\varepsilon > 0$ be arbitrary. Then, if for $v$ as in 6° we have $\|v\| \leqslant \lambda$, then we have (9.62), (9.63) for $u = v$. On the other hand, if for $v$ as in 6° we have $\|v\| > \lambda$, then for $u = \frac{\lambda v}{\|v\|}$ we have (9.63) and

$$\|u(x) - x\| = \left\|\frac{\lambda v(x)}{\|v\|} - x\right\| = \left|\frac{\lambda}{\|v\|} - 1\right| \|x\| = \frac{\|v\| - \lambda}{\|v\|} \|x\| \leqslant$$

$$\leqslant \frac{\lambda + \varepsilon - \lambda}{\|v\|} \|x\| = \frac{\varepsilon}{\|v\|} \|x\| \quad (x \in G),$$

whence (9.62) whenever $\varepsilon > 0$ is chosen so small that $\frac{\varepsilon}{\|v\|} < \delta$. Thus, $6° \Rightarrow 5°$.

Assume now that we have 1°. Take any sequence of subspaces $\{G_n\}$ satisfying a), b) and c) of 2°. Then, by the implication $1° \Rightarrow 6°$ proved above, $\{G_n\}$ satisfies d) of 2°. Thus, $1° \Rightarrow 2°$.

Conversely, if we have 2°, then by b) and $v_n|_{G_n} = I_{G_n}$ $(n = 1,2, \dots)$ we have $\lim_{n\to\infty} v_n(x) = x$ $(x \in \bigcup_{n=1}^{\infty} G_n)$, whence, by c) and $\sup_{1\leqslant n<\infty} \|v_n\| < \infty$, we obtain $\lim_{n\to\infty} v_n(x) = x$ for all $x \in E$. Therefore, by a), $\{v_n\}$ is an approximative basis of $E$. Thus, $2° \Rightarrow 1°$.

The equivalence $1° \Leftrightarrow 3°$ and the implication $7° \Rightarrow 3°$ have been observed after definition 9.5. The implication $3° \Rightarrow 7°$ is a consequence of theorem 9.2 a) or b) and proposition 9.4.

The equivalences $7° \Leftrightarrow 8° \Leftrightarrow 9°$ follow from theorem 9.1, equivalences $1° \Leftrightarrow 5° \Leftrightarrow 6°$.

The equivalence $9° \Leftrightarrow 10°$ is obvious, since by the definition of $E_b$ every Banach space with a basis is isomorphic to a complemented subspace of $E_b$ and $E_b$ has a basis.

The implication $4° \Rightarrow 11°$ is obvious.

Finally, assume that $E$ is isomorphic to a separable conjugate space, say $B_1^*$, and that we have 11°. Then $B_1^*$ has also the metric approximation property*) and hence, since $E$ is isomorphic to $B_1^*$, $E$ has the bounded approximation property. Thus, in this case $11° \Rightarrow 4°$, which completes the proof of theorem 9.3.

*Remark 9.5.* a) The implication $1° \Rightarrow 10°$ of theorem 9.3 may be regarded as a sharpening of the implication $1° \Rightarrow 8°$, showing that the space $E_b$ is "complementably universal" for the family of all Banach spaces with an approximative basis (or, equivalently, of all separable Banach spaces with the bounded approximation property), in the sense that every such Banach space is isomorphic to a complemented subspace of $E_b$. It is natural to ask "how many" separable Banach spaces exist which are complementably universal in this sense. From the argument of Ch. II, § 13, remark 13.3 (with $E_s$ replaced by $E_b$) it follows that a separable Banach space with this property is unique up to an isomorphism and hence it must be isomorphic to $E_b$. b) For a possible sharpening of the implication $1° \Rightarrow 8°$ in a different direction, see § 14, problem 14.2. c) For any $\varepsilon > 0$, the above proof yields, using (9.44), (9.37) and (9.48′), a quasi-basis $\{x_n\}$ of norm $\nu_{\{x_n\}} \leqslant (2 + \varepsilon)\lambda$.

*Remark 9.6.* Other characterizations of Banach spaces having an approximative basis will be given, respectively can be deduced from

*) Indeed, combining [139], Ch. I, p. 122, theorem 8 with [139], Ch. I, p. 164, proposition 35, condition $(B_2)$ and [139], Ch. I, p. 180, proposition 40, it follows that *every separable conjugate Banach space with the approximation property has the metric approximation property* (and hence every isomorph of it has the bounded approximation property).

some results given, in the sequel. Thus, in § 10, theorem 10.1, such spaces are characterized by the existence of "operational bases". The proofs of § 12, theorem 12.1 $\gamma$), implications $1^\circ \Rightarrow 5^\circ \Rightarrow 4^\circ$ also yield, by taking linear operators instead of projections*), that *E has an approximative basis if and only if there exists a sequence of finite rank endomorphisms* $\{u_n\} \subset L(E, E)$ *satisfying* (9.2) *and* $u_m u_n = u_n$ *for all* $n < m$ (hence, in particular, $u_1(E) \subset u_2(E) \subset \ldots$). The proof of proposition 9.7 below shows that *condition* 2° d) *of theorem* 9.3 *can be improved to* $G_1 \subset v_1(E) \subset G_2 \subset v_2(E) \subset \ldots, \sup_{1 \leqslant n < \infty} \|v_n\| < \infty$. The proof of § 18, theorem 18.2 (necessity part) yields that *E has an approximative basis if and only if there exists a sequence of finite rank endomorphisms* $\{u_n\} \subset L(E, E)$ *satisfying* (9.2) *and* $\|u_n(x) - x\| < \frac{1}{n}$ $(x \in S_{\left[\bigcup_{i=1}^{n-1} u_i(E)\right]},$ $n = 1,2, \ldots)$. Some further equivalent conditions when $E$ is a separable conjugate space will be given in Vol. III, Ch. IV.

It is natural to ask whether the implication $11^\circ \Rightarrow 1^\circ$ of theorem 9.3 remains valid in the general case, i.e., without the assumption that $E$ is isomorphic to a (separable) conjugate space. We shall show in example 9.2 below that the answer is negative. In order to construct that example, we shall use

**Proposition 9.5.** *If* $\{E_n\}$ *is a sequence of Banach spaces such that for each n,* $E_n$ *has the approximation property but fails the n-approximation property, then the space* $E = (E_1 \times E_2 \times \ldots)_{l^2}$ *has the approximation property but fails the bounded approximation property.*

*Proof.* Let $Q$ be an arbitrary compact set in $E$ and let $\varepsilon > 0$. By the definition of $E$, the sequence of "section" operators

$$s_n(x) = \{x_1, \ldots, x_n, 0, 0, \ldots\} \quad (x = \{x_j\} \in E, \ n = 1,2, \ldots) \tag{9.69}$$

converges pointwise to $I_E$, whence also uniformly on $Q$ (by the argument of the proof of Ch. I, § 17, theorem 17.3), and thus there exists an index $N = N(\varepsilon)$ such that

$$\|x - s_N(x)\| < \frac{\varepsilon}{2} \qquad (x \in Q).$$

Since each $E_1, \ldots, E_N$ has the approximation property, there exists on $E_1 \times \ldots \times E_N \times \{0\} \times \{0\} \times \ldots (\subset E)$ an operator of finite rank $v_N$

---

*) Similarly, some other methods of $\pi$-bases and dual $\pi$-bases can be also applied to yield results on approximative bases.

such that

$$\|v_N(z) - z\| < \frac{\varepsilon}{2} \qquad (z \in p_1(Q) \times \ldots \times p_N(Q) \times \{0\} \times \{0\} \times \ldots),$$

where $p_n$ denotes the $n$-th coordinate projection $E \to E_n$ (i.e., $p_n(x) = x_n$ for $x = \{x_j\} \in E$). Put

$$u = v_N s_N. \tag{9.70}$$

Then, since $s_N(x) \in p_1(Q) \times \ldots \times p_N(Q) \times \{0\} \times \{0\} \times \ldots$ $(x \in Q)$, we obtain

$$\|u(x) - x\| \leqslant \|v_N(s_N(x)) - s_N(x)\| + \|s_N(x) - x\| < \varepsilon \quad (x \in Q),$$

which proves that $E$ has the approximation property.

On the other hand, suppose, a contrario, that $E$ has the $\lambda$-approximation property for some $\lambda$. We shall show that in this case each $E_n$ has the $\lambda$-approximation property, in contradiction with our assumption. Indeed, let $Q_n$ be an arbitrary compact set in $E_n$ and let $\varepsilon > 0$. Then $Q = \{0\} \times \ldots \times \{0\} \times Q_n \times \{0\} \times \ldots \subset E$ is compact and hence, since $E$ has the $\lambda$-approximation property, there exists an endomorphism $u \in L(E, E)$ of finite rank, with $\|u\| \leqslant \lambda$, such that $\|u(x) - x\| < \varepsilon$ $(x \in Q)$. Put

$$w_n = p_n u \tau_n, \tag{9.71}$$

where $p_n$ denotes the $n$-th coordinate projection $E \to E_n$ and $\tau_n$ the canonical linearly isometrical embedding $E_n \to \{0\} \times \ldots \times \{0\} \times E_n \times \{0\} \times \ldots$ Then $w_n$ is an endomorphism of finite rank in $L(E_n, E_n)$, with $\|w_n\| \leqslant \lambda$, and

$$\|w_n(x_n) - x_n\| = \|p_n u \tau_n(x_n) - p_n \tau_n(x_n)\| \leqslant$$

$$\leqslant \|p_n\| \, \|u(\tau_n(x_n)) - \tau_n(x_n)\| < \varepsilon \qquad (x_n \in Q_n), \tag{9.72}$$

so $E_n$ has the $\lambda$-approximation property, which completes the proof of proposition 9.5.*)

In order to show that there exists a sequence of Banach spaces $\{E_n\}$ satisfying the conditions of proposition 9.5 (and having, actually, some stronger properties), we shall use the following result, which may have interest also for other applications.

---

*) The proof of the second part also shows that if *$E$ has the $\lambda$-approximation property and if $G$ is a subspace of $E$ admitting a projection of norm $\leqslant \mu$, then $G$ has the $\lambda\mu$-approximation property.*

**Theorem 9.4.** *Let $E$ be a Banach space. If there exists a constant $\lambda \geqslant 1$ such that $(E, |.|)$ has the $\lambda$-approximation property for each equivalent norm $|.|$ on $E$, then $E^*$ has the $2\lambda(1+4\lambda)$-approximation property.*

*Proof.* Let $\Gamma$ be an arbitrary finite-dimensional subspace of $E^*$. We shall first show that for each $\beta > \lambda$ and $\varepsilon > 0$ there exists an operator of finite rank $u \in L(E, E)$ such that

$$\|u^*(h) - h\| \leqslant \left(\frac{1}{2} + \varepsilon\right) \|h\| \qquad (h \in \Gamma), \tag{9.73}$$

$$\|u^*\| = \|u\| \leqslant \beta(1 + 4\beta). \tag{9.74}$$

Define an equivalent norm $|f|$ on $E^*$ by

$$|f| = \|f\| + 4\beta \operatorname{dist}(f, \Gamma) \qquad (f \in E^*). \tag{9.75}$$

We claim that there exists an equivalent norm $|.|$ on $E$ such that $|f|$ is the dual norm of $|x|$ (i.e., $|f| = \sup_{\substack{x \in E \\ |x| \leqslant 1}} |f(x)|$ for all $f \in E^*$). Indeed, by Ch. II, § 5, lemma 5.1, it is sufficient to show that the set $A = \{f \in E^* \mid |f| \leqslant 1\}$ is closed for the weak* topology $\sigma(E^*, E)$. Let $f_0$ be any element in $\tilde{A}$, the weak* closure of $A$. Since $\|f_0\| = \sup_{\substack{x \in E \\ \|x\| \leqslant 1}} |f_0(x)|$ and*) $\operatorname{dist}(f_0, \Gamma) = \sup_{\substack{z \in \Gamma_\perp \\ \|z\| \leqslant 1}} |f_0(z)|$, for any $n$ there exist elements $x_n \in E$ and $z_n \in \Gamma_\perp$ with $\|x_n\|, \|z_n\| \leqslant 1$, such that

$$|f_0(x_n)| > \|f_0\| - \frac{1}{2n}, \qquad |f_0(z_n)| > \operatorname{dist}(f_0, \Gamma) - \frac{1}{2n}.$$

Since $f_0 \in \tilde{A}$, there exists an $f \in A \cap W_{x_n, z_n; \frac{1}{2n}}(f_0)$, that is, an $f \in A$ such that

$$|f(x_n) - f_0(x_n)| < \frac{1}{2n}, \qquad |f(z_n) - f_0(z_n)| < \frac{1}{2n}.$$

Then, by the above inequalities and since $f \in A$, we obtain

$$|f_0| = \|f_0\| + 4\beta \operatorname{dist}(f_0, \Gamma) < |f_0(x_n)| + \frac{1}{2n} + 4\beta\left(|f_0(z_n)| + \frac{1}{2n}\right) <$$

$$< |f(x_n)| + \frac{1}{n} + 4\beta\left(|f(z_n)| + \frac{1}{n}\right) \leqslant \|f\| + \frac{1}{n} + 4\beta\left(\operatorname{dist}(f, \Gamma) + \frac{1}{n}\right) =$$

$$= \|f\| + 4\beta \operatorname{dist}(f, \Gamma) + \frac{1}{n}(1 + 4\beta) \leqslant 1 + \frac{1}{n}(1 + 4\beta),$$

*) See e.g. [371], Ch. I, § 1.

whence, since $n$ was arbitrary, we infer that $f_0 \in A$, which proves our claim that $|f|$ is the dual norm of some equivalent norm $|x|$ on $E$.

Since $\dim \Gamma < \infty$, there exists a finite-dimensional subspace $G$ of $E$ such that

$$\sup_{\substack{y \in G \\ \|y\| \leqslant 1}} |h(y)| \geqslant \frac{1}{1+\varepsilon} \|h\| \qquad (h \in \Gamma); \tag{9.76}$$

indeed, picking a finite $\delta$-net $h_1, \ldots, h_m$ of $\sigma_\Gamma = \{h \in \Gamma \mid \|h\| = 1\}$ and then elements $y_i \in E$ with $\|y_i\| = 1$, $|h_i(y_i)| \geqslant \frac{1}{1+\delta}$ $(i = 1, \ldots, m)$, for $\delta$ sufficiently small the subspace $G = [y_1, \ldots, y_m]$ will satisfy (9.76).

Now, since by our assumption $(E, |.|)$ has the $\lambda$-approximation property and $\beta > \lambda$, there exists on $E$, by lemma 9.1, an operator $u$ of finite rank such that

$$u(y) = y \qquad (y \in G), \tag{9.77}$$

$$|u| \leqslant \beta. \tag{9.78}$$

By (9.75), for any $f \in E^*$ with $\|f\| \leqslant 1$ we have $|f| = \|f\| + 4\beta \operatorname{dist}(f, \Gamma) \leqslant 1 + 4\beta$, whence, by (9.78),

$$\|u^*(f)\| \leqslant |u^*(f)| \leqslant \beta |f| \leqslant \beta(1 + 4\beta) \quad (f \in E^*, \|f\| \leqslant 1),$$

which proves (9.74).

Furthermore, again by (9.75) and (9.78),

$$4\beta \operatorname{dist}(u^*(h), \Gamma) = |u^*(h)| - \|u^*(h)\| \leqslant$$

$$\leqslant |u^*(h)| \leqslant \beta |h| = \beta \|h\| \qquad (h \in \Gamma),$$

whence $\operatorname{dist}(u^*(h), \Gamma) \leqslant \frac{1}{4} \|h\|$ for all $h \in \Gamma$. Since $\operatorname{dist}(u^*(h), \Gamma)$ is attained (because $\dim \Gamma < \infty$), it follows that for each $h \in \Gamma$ there exists an $h_0 \in \Gamma$ satisfying

$$\|u^*(h) - h_0\| \leqslant \frac{1}{4} \|h\|. \tag{9.79}$$

But, by (9.77),

$$u^*(h)(y) = h(u(y)) = h(y) \quad (y \in G), \tag{9.80}$$

whence, by (9.79),

$$\sup_{\substack{y \in G \\ \|y\| \leqslant 1}} |h(y) - h_0(y)| = \sup_{\substack{y \in G \\ \|y\| \leqslant 1}} |u^*(h)(y) - h_0(y)| \leqslant$$

$$\leqslant \|u^*(h) - h_0\| \leqslant \frac{1}{4} \|h\|.$$

Therefore, by (9.76),

$$\|h - h_0\| \leqslant (1 + \varepsilon) \sup_{\substack{y \in G \\ \|y\| \leqslant 1}} |h(y) - h_0(y)| \leqslant \frac{1 + \varepsilon}{4} \|h\|,$$

whence, taking again into account (9.79), we obtain

$$\|u^*(h) - h\| \leqslant \|u^*(h) - h_0\| + \|h_0 - h\| \leqslant$$

$$\leqslant \frac{1}{4} \|h\| + \frac{1 + \varepsilon}{4} \|h\| = \left(\frac{1}{2} + \frac{\varepsilon}{4}\right) \|h\| \qquad (h \in \Gamma),$$

which proves (9.73). Thus, we have proved that for each $\beta > \lambda$ and $\varepsilon > 0$ there is, indeed, a finite rank operator $u \in L(E, E)$ satisfying (9.73), (9.74).

Now let $0 < \varepsilon < \frac{1}{2}$ be arbitrary and put, for simplicity, $\delta = \frac{1}{2} + \varepsilon$ (hence $\frac{1}{2} < \delta < 1$). By the above, one can construct, by induction, operators of finite rank $u_n$ on $E$ $(n = 1, 2, \ldots)$ such that

$$\|u_1^*(h) - h\| \leqslant \delta \|h\| \qquad (h \in \Gamma), \tag{9.81}$$

$$\|u_{n+1}^*(f) - f\| \leqslant \delta \|f\| \qquad \left(f \in \left[\Gamma \cup \bigcup_{i=1}^{n} u_i^*(E^*)\right]\right). \tag{9.82}$$

$$\|u_n^*\| = \|u_n\| \leqslant \beta(1 + 4\beta). \tag{9.83}$$

Put

$$v_n = I - (I - u_1)(I - u_2) \ldots (I - u_n), \tag{9.84}$$

hence

$$v_n^* = I - (I - u_n^*)(I - u_{n-1}^*) \ldots (I - u_1^*). \tag{9.85}$$

Then $v_n^*$ is an operator of finite rank on $E^*$ and by (9.82) and (9.81) we have

$$\|v_n^*(h) - h\| = \|(I - u_n^*)(I - u_{n-1}^*) \dots (I - u_1^*)(h)\| \leqslant$$

$$\leqslant \delta \|(I - u_{n-1}^*)(I - u_{n-2}^*) \dots (I - u_1^*)(h)\| \leqslant \dots$$

$$\dots \leqslant \delta^n \|h\| \qquad (h \in \Gamma). \tag{9.86}$$

Observe now that

$$v_n^* = I - (I - u_n^*)(I - v_{n-1}^*) = v_{n-1}^* + u_n^*(I - v_{n-1}^*)$$

$$(n = 2,3, \dots), \tag{9.87}$$

whence

$$v_n^*(f) = v_1^*(f) + \sum_{=2}^{n} (v_i^*(f) - v_{i-1}^*(f)) = u_1^*(f) + \sum_{i=2}^{n} u_i^*(I - v_{i-1}^*)(f)$$

$$(f \in E^*),$$

and thus

$$v_n^*(E^*) \subset \left[ \bigcup_{i=1}^{n} u_i^*(E^*) \right]. \tag{9.88}$$

Therefore, by (9.82),

$$\|(I - u_{n+1}^*)v_n^*(f)\| \leqslant \delta \|v_n^*(f)\| \leqslant \delta \|v_n^*\| \, \|f\| \qquad (f \in E^*),$$

whence

$$\|(I - u_{n+1}^*)v_n^*\| \leqslant \delta \|v_n^*\| \quad (n = 1,2, \dots). \tag{9.89}$$

We claim that for all $n = 1,2, \dots$ we have

$$\|v_n^*\| \leqslant (\delta^{n-1} + \dots + \delta + 1)\beta(1 + 4\beta) < \frac{\beta(1 + 4\beta)}{1 - \delta} =$$

$$= \frac{2\beta(1 + 4\beta)}{1 - 2\varepsilon}. \tag{9.90}$$

Indeed, for $n = 1$ this amounts to $\|v_1^*\| \leqslant \beta(1 + 4\beta)$, which holds by (9.83). Assuming that (9.90) holds for some $n$, from (9.87), (9.89) and

(9.83) we obtain

$$\|v_{n+1}^*\| = \|(I - u_{n+1}^*)v_n^* + u_{n+1}^*\| \leqslant \|(I - u_{n+1}^*)v_n^*\| +$$

$$+ \|u_{n+1}^*\| \leqslant \delta\|v_n^*\| + \beta(1 + 4\beta) \leqslant \delta(\delta^{n-1} + \ldots + \delta + 1)\beta(1 + 4\beta) +$$

$$+ \beta(1 + 4\beta) = (\delta^n + \ldots + \delta + 1)\beta(1 + 4\beta),$$

which proves (9.90). The inequalities (9.86) and (9.90) already show that $E^*$ has the $\frac{2\beta(1 + 4\beta)}{1 - 2\varepsilon}$-approximation property for all $\beta > \lambda$ and $0 < \varepsilon < \frac{1}{2}$. Finally, put

$$w_n^* = \begin{cases} v_n^* & \text{if } \|v_n^*\| \leqslant 2\lambda(1 + 4\lambda), \\ 2\lambda(1 + 4\lambda)\dfrac{v_n^*}{\|v_n^*\|} & \text{if } 2\lambda(1 + 4\lambda) < \|v_n^*\| \leqslant \dfrac{2\beta(1 + 4\beta)}{1 - 2\varepsilon}. \end{cases} \tag{9.91}$$

Then, in the first case, (9.86) works. In the second case we have $\|w_n^*\| = 2\lambda\,(1 + 4\lambda)$ and

$$\|w_n^*(h) - h\| \leqslant \|w_n^*(h) - v_n^*(h)\| + \|v_n^*(h) - h\| \leqslant$$

$$\leqslant |2\lambda(1 + 4\lambda) - \|v_n^*\||\frac{\|v_n^*(h)\|}{\|v_n^*\|} + \delta^n\|h\| \leqslant$$

$$\leqslant \left\{\left(\frac{2\beta(1 + 4\beta)}{1 - 2\varepsilon} - 2\lambda(1 + 4\lambda)\right) + \delta^n\right\}\|h\| \qquad (h \in \Gamma),$$

which, since $\delta^n \to 0$ as $n \to \infty$ and since $\beta > \lambda$ and $0 < \varepsilon < \frac{1}{2}$ have been arbitrary, shows that $E^*$ has the $2\lambda(1 + 4\lambda)$-approximation property*). This completes the proof of theorem 9.4.

---

*) Indeed, note that the implication 5° ⇒ 4° of theorem 9.3 remains valid also for arbitrary (not necessarily separable) Banach spaces $E$. For, if we have 5° and if $Q \subset E$ is compact, then for any $\varepsilon > 0$, taking a finite $\delta$-net $\{x_1, \ldots, x_m\}$ for $Q$ and applying 5° to $G = [x_1, \ldots, x_m]$ and $\delta$, we obtain a finite rank operator $u$ on $E$ with $\|u\| \leqslant \lambda$ and $\|u(x) - x\| \leqslant \|u(x - x_{i_0})\| + \|u(x_{i_0}) - x_{i_0}\| + \|x_{i_0} - x\| \leqslant (\lambda + \sup_{x' \in Q}\|x'\| + 1)\delta < \varepsilon$ ($\delta$ sufficiently small) for all $x \in Q$ (see also § 18, theorem 18.1, implication 4° ⇒ 3°).

Before giving the example announced above, let us make some remarks to theorem 9.4.

*Remark 9.7.* a) The inequalities (9.73) and (9.74) suggest the introduction of the following notion: For $\alpha$, $\lambda$ positive constants, a Banach space $E$ is said to have the *$(\alpha, \lambda)$-approximation property* if for every finite-dimensional subspace $G$ of $E$ and every $\varepsilon > 0$ there exists an operator of finite rank $u \in L(E, E)$ such that $\|u(y) - y\| \leqslant (\alpha + \varepsilon)\|y\|$ for all $y \in G$ and that $\|u\| \leqslant \lambda + \varepsilon$. In this terminology, the first part of the above proof shows that if $(E, |.|)$ has the $\lambda$-approximation property for each equivalent norm $|\cdot|$ on $E$, then $E^*$ has the $\left(\frac{1}{2}, \lambda(1+4\lambda)\right)$-approximation property and the second part shows that this already implies that $E^*$ has the $2\lambda(1 + 4\lambda)$-approximation property. With the same argument as in the second part, one can prove that if a Banach space $E$ has the $(\alpha, \lambda)$-approximation property, where $0 < \alpha < 1$, then $E$ has the $\frac{\lambda}{1-\alpha}$-approximation property. With the same argument as in the first part, (considering the norm $|f| = \|f\| + \frac{2\beta}{\alpha}\operatorname{dist}(f, \Gamma)$, where $\beta > \lambda$, instead of (9.75)), one can show that if $(E, |\cdot|)$ has the $\lambda$-approximation property for each equivalent norm $|\cdot|$ on $E$, then $E^*$ has the $\left(\alpha, \lambda\left(1 + \frac{2\lambda}{\alpha}\right)\right)$-approximation property for each $\alpha$ with $0 < \alpha < 1$*). Actually, the above proof shows that *the same conclusion, hence also the conclusion of theorem* 9.4, *holds under the weaker assumption that $(E, |.|)$ has the $\lambda$-approximation property for each norm $|\cdot|$ belonging to the family* $\mathfrak{A}$ of all equivalent norms $|\cdot|$ on $E$ whose dual norms on $E^*$ are of the form (9.75), where $\Gamma$ ranges over all finite dimensional subspaces of $E^*$ and $\beta$ ranges over all positive constants **). b) We have the following converse of this latter sharpening of theorem 9.4: *If $E^*$ has the $\lambda$-approximation property, then for each norm $|\cdot|$ of the form* (9.75) (*where $\Gamma \subset E^*$, $\dim \Gamma < \infty$ and $\beta > 0$*), *$(E^*, |\cdot|)$ has the $\lambda$-approximation property and hence $(E, |\cdot|)$ also has the $\lambda$-approximation property for each $|\cdot| \in \mathfrak{A}$.* Indeed, let $\Gamma_0$ be a finite-dimensional subspace

---

*) Hence, by the preceding observation, $E^*$ has the $\dfrac{\lambda\left(1 + \frac{2\lambda}{\alpha}\right)}{1-\alpha}$-approximation property for each $0 < \alpha < 1$. Note that for $\alpha \to 0$ or $\alpha \to 1$ this converges to $\infty$; in the above proof it is applied with $\alpha = \frac{1}{2}$.

**) For the proof of theorem 9.4 it has been sufficient to take only the constants $\beta > \lambda$, but for the sequel it will be more convenient to include all positive constants $\beta$ in the definition of $|\cdot|$.

of $E^*$ and let $\varepsilon>0$. Then, by our assumption and by lemma 9.1, for any finite-dimensional subspace $\Gamma \subset E^*$ there exists an operator $v$ of finite rank on $E^*$ such that

$$v(f)=f \qquad (f\in[\Gamma \cup \Gamma_0]), \tag{9.92}$$

$$\|v\| \leqslant \lambda + \varepsilon. \tag{9.93}$$

Then, by (9.92) for $h \in \Gamma$,

$$\operatorname{dist}(v(f), \Gamma) = \inf_{h\in\Gamma} \|v(f) - h\| = \inf_{h\in\Gamma} \|v(f-h)\| \leqslant$$

$$\leqslant \|v\| \inf_{h\in\Gamma} \|f-h\| = \|v\| \operatorname{dist}(f, \Gamma) \qquad (f\in E^*),$$

whence, for $|\cdot| \in \mathfrak{A}$ associated to $\Gamma$ and to any $\beta>0$,

$$|v(f)| = \|v(f)\| + 4\beta \operatorname{dist}(v(f), \Gamma) \leqslant$$

$$\leqslant \|v\|(\|f\| + 4\beta \operatorname{dist}(f, \Gamma)) = \|v\| \, |f| \qquad (f\in E^*).$$

Therefore, by (9.93),

$$|v| = \sup_{\substack{f\in E^* \\ |f|\leqslant 1}} |v(f)| \leqslant \|v\| \leqslant \lambda + \varepsilon,$$

which, together with (9.92) for $f\in\Gamma_0$, implies (see the end of the proof of theorem 9.4) that $(E^*, |\cdot|)$ has the $\lambda$-approximation property. Hence, by proposition 9.8 below, $(E, |\cdot|)$ also has the $\lambda$-approximation property, which completes the proof of our assertion.

*Remark 9.8.* a) In the particular case when $\lambda = 1$, the constant $2\lambda(1 + 4\lambda) = 10$ in the above sharpened version of theorem 9.4 can be replaced by 1, that is, *if* $(E, |\cdot|)$ *has the metric approximation property for each**) $|\cdot| \in \mathfrak{A}$, *then* $E^*$ *has the metric approximation property*. Indeed, let $\Gamma$ be an arbitrary finite-dimensional subspace of $E^*$ and let $\varepsilon > 0$. Define $|\cdot| \in \mathfrak{A}$ to be the equivalent norm on $E$ whose dual is

$$|f| = \|f\| + \frac{\varepsilon}{2} \operatorname{dist}(f, \Gamma) \qquad (f\in E^*). \tag{9.75'}$$

---

*) Note that such spaces do exist, e.g. every reflexive Banach space with the approximation property has the metric approximation property (see e.g. [139], Ch. I, p. 181, corollary 2).

Then, as above, for any $\delta>0$ there exists a finite-dimensional subspace $G$ of $E$ such that

$$\sup_{\substack{y\in G\\ \|y\|\leqslant 1}} |h(y)| \geqslant \frac{1}{1+\delta} \|h\| \qquad (h\in \Gamma). \tag{9.76'}$$

Since by our assumption $(E, |.|)$ has the metric approximation property, there exists, by lemma 9.1, an operator $u$ on $E$ of finite rank, such that we have (9.77) and

$$|u| \leqslant 1+\delta. \tag{9.78'}$$

Then, by (9.75'), for any $f\in E^*$ with $\|f\|\leqslant 1$ we have $|f|=\|f\|+ + \frac{\varepsilon}{2}\operatorname{dist}(f, \Gamma) \leqslant 1+\frac{\varepsilon}{2}$, whence, by (9.78'),

$$\|u^*(f)\| \leqslant |u^*(f)| \leqslant (1+\delta)|f| \leqslant (1+\delta)\left(1+\frac{\varepsilon}{2}\right) \quad (f\in E^*,\ \|f\|\leqslant 1),$$

and thus, if $\delta<\frac{\varepsilon}{2+\varepsilon}$ $\left(\text{hence } \delta\left(1+\frac{\varepsilon}{2}\right)<\frac{\varepsilon}{2+\varepsilon}\,\frac{2+\varepsilon}{2}=\frac{\varepsilon}{2}\right)$, then

$$\|u^*\| \leqslant (1+\delta)\left(1+\frac{\varepsilon}{2}\right)<1+\varepsilon. \tag{9.74'}$$

But, by (9.77) we have (9.80), whence, by (9.76'),

$$\frac{1}{1+\delta}\|h\| \leqslant \sup_{\substack{y\in G\\ \|y\|\leqslant 1}} |h(y)| = \sup_{\substack{y\in G\\ \|y\|\leqslant 1}} |u^*(h)(y)| \leqslant \|u^*(h)\| \quad (h\in\Gamma).$$

Consequently, by (9.75') and (9.78'),

$$\|u^*(h)\| + \frac{\varepsilon}{2}\operatorname{dist}(u^*(h), \Gamma) = |u^*(h)| \leqslant (1+\delta)\,|h| =$$

$$= (1+\delta)\|h\| \leqslant (1+\delta)^2\|u^*(h)\| \qquad (h\in\Gamma),$$

whence, by (9.74′),

$$\operatorname{dist}(u^*(h), \Gamma) \leqslant \frac{2(2\delta + \delta^2)}{\varepsilon} \|u^*(h)\| \leqslant \frac{2(2\delta + \delta^2)(1+\varepsilon)}{\varepsilon} \|h\| \quad (h \in \Gamma). \tag{9.79'}$$

Hence, as in the above proof of theorem 9.4, we obtain

$$\|u^*(h) - h\| \leqslant \|u^*(h) - h_0\| + \|h - h_0\| \leqslant (2+\delta)\|u^*(h) - h_0\| \leqslant$$

$$\leqslant \frac{2(2\delta + \delta^2)(2+\delta)(1+\varepsilon)}{\varepsilon} \|h\| \quad (h \in \Gamma),$$

which, together with (9.74′), implies (see the end of the proof of theorem 9.4) that $E^*$ has the metric approximation property. This completes the proof of our assertion. b) Even when $\lambda > 1$, it may happen that the constant $2\lambda(1 + 4\lambda)$ above can be replaced by 1. Namely, *if* $(E, |\cdot|)$ *has the* $\lambda$*-approximation property for each* $|\cdot| \in \mathfrak{A}$ (*where* $\lambda \geqslant 1$), *and if* $E^*$ *is separable, then* $E^*$, *and hence also* $E$, *has the metric approximation property*. Indeed, by the above, $E^*$ has the bounded approximation property, whence, since $E^*$ is separable, it also has the metric approximation property (as was observed in the proof of theorem 9.3, implication $11° \Rightarrow 4°$). Consequently, by proposition 9.8 below, $E$ also has the metric approximation property.

*Problem 9.3*. a) If $(E, |\cdot|)$ has the $\lambda$-approximation property for each $|\cdot| \in \mathfrak{A}$ (hence $E^*$ has the $2\lambda(1 + 4\lambda)$-approximation property), does $E^*$ have the $\lambda$-approximation property? b) If $E^*$ has the bounded approximation property, does it have also the metric approximation property?

We have seen in remark 9.8 that the answer to a) is affirmative for $\lambda = 1$[*)] and, if $E^*$ is separable, then for any $\lambda \geqslant 1$. Also, in the proof of theorem 9.3, implication $11° \Rightarrow 4°$, it was observed that the answer to

*) However, note that for $\lambda > 1$ the method of the proof of remark 9.8 a) does not give essentially more than the original proof of theorem 9.4. Indeed, for $\lambda > 1$ we obtain $|u| \leqslant \lambda + \delta$ instead of (9.78′), whence $\|u^*\| \leqslant (\lambda + \delta)\left(1 + \frac{\varepsilon}{2}\right) < \lambda + \varepsilon$ instead of (9.74′), but the inequality (9.79′) will be replaced by $\operatorname{dist}(u^*(h), \Gamma) \leqslant$ $\leqslant \frac{2}{\varepsilon}((\lambda + \delta)(1 + \delta) - 1)(\lambda + \varepsilon)\|h\| \quad (h \in \Gamma)$, which gives $\|u^*(h) - h\| \leqslant$ $\leqslant (2 + \delta)\|u^*(h) - h_0\| \leqslant \frac{(2+\delta)2}{\varepsilon}((\lambda + \delta)(1 + \delta) - 1)(\lambda + \varepsilon)\|h\|$ for all $h \in \Gamma$.

b) is affirmative if $E^*$ is separable. Clearly, an affirmative answer to b) would imply an affirmative answer to a).

Now we shall give the example announced before proposition 9.5.

*Example 9.2.* By § 0, example 0.1, there exists a conjugate Banach space $F^*$ with a monotone basis (hence $(F^*, |\cdot|)$ has the bounded approximation property for any equivalent norm $|\cdot|$ on $F^*$) such that $F^{**}$ is separable and fails the approximation property. By theorem 9.4, there exists a sequence $\{|\cdot|_n\}_{n=1}^{\infty}$ of equivalent norms on $F^*$, so that $(F^*, |\cdot|_n)$ fails the $n$-approximation property ($n = 1,2, \ldots$). Then the sequence of Banach spaces $\{E_n\} = \{(F^*, |\cdot|_n)\}$ satisfies the assumptions of proposition 9.5 and hence $E = (E_1 \times E_2 \times \ldots)_{l^2}$ is an example of a separable Banach space with the approximation property, which fails the bounded approximation property; moreover, $E^*$ is separable.

Some additional remarks to this example are collected in

*Remark 9.9.* a) Since $(F^*, |\cdot|_n)$ has the approximation property, but fails the $n$-approximation property, the norm $|\cdot|_n$ cannot be the dual of any norm on $F$ (and, moreover, $(F^*, |\cdot|_n)$ *is not isometric to any conjugate Banach space*), since otherwise, as was observed in the proof of theorem 9.3, implication $11° \Rightarrow 4°$, $(F^*, |\cdot|_n)$ would have the metric approximation property, whence also the $n$-approximation property. b) Similarly, since the space $E$ of example 9.2 is separable and has the approximation property, but fails the bounded approximation property, it follows that *$E$ is not isomorphic to any conjugate Banach space.* c) The space $E_n = (F^*, |\cdot|_n)$ above has a basis, but fails the $n$-approximation property and $E_n^* = (F^{**}, |\cdot|_n)$ is separable. d) The spaces $(F^*, |\cdot|_n)$ above are *examples of infinite dimensional Banach spaces $E$ with a basis, such that**) $0 < \Gamma(E) < 1$, giving thus an affirmative solution to Ch. II, § 1, problem 1.1; moreover, they also show that *for each $\varepsilon > 0$ there exists an infinite dimensional Banach space $E$ with a basis, such that* $0 < \Gamma(E) < \varepsilon$.

We have seen in the preceding that some results on bases and some concepts related to bases $\{x_n\}$ and their a.s.c.f. $\{f_n\}$, in which the biorthogonality of $(x_n, f_n)$ does not enter in an essential way, can be carried over to quasi-bases $\{x_n\}$ and their admissible sequences $\{f_n\}$ (see e.g. formula (9.15) or definitions 9.3, 9.4). Let us consider now two more ideas in this direction, namely, unconditional quasi-bases and monotone quasi-bases.

*Definition 9.6.* A quasi-basis $\{x_n\}$ of a Banach space $E$ is said to be *unconditional* if it has an admissible sequence $\{f_n\} \subset E^*$ such that all series (9.11) are unconditionally convergent.

By § 5, theorem 5.3 c), we have the following sharpening of remark 9.2 above: *every separable Banach space $E$ has a "non-linear unconditional quasi-basis"*. Furthermore, a sequence $\{x_n\} \subset E$ is a ("linear") unconditional quasi-basis of $E$ if and only if it is an unconditional

*) See Ch. I, § 7, definition 7.2.

pseudo-basis of $E$ and the set-valued mapping $x \to \left\{\{\alpha_n\} \subset K \,\middle|\, \sum_{i=1}^{\infty} \alpha_i x_i = x \text{ unconditionally}\right\}$ of $E$ into $2^{A_1^{(u)}}$ admits a continuous linear selection. *The unconditional analogue of theorem* 9.1 (i.e., theorem 9.1 with "unconditional" inserted before quasi-basis, pseudo-basis and basis and with $A_1$ replaced by $A_1^{(u)}$ of § 5, theorem 5.3) *is also valid,* with a similar proof, since we have observed in Ch. II, § 18, after the proof of proposition 18.5, that $\{e_n\}$ is an unconditional basis of $A_1^{(u)}$. In particular, let us point out that *if* $\{x_n\}$ *is an unconditional quasi-basis of E, then there exist a Banach space B with an unconditional basis* $\{e_n\}$, *a linear isometry u of E into B and a projection v of B onto u(E) such that we have* (9.19). Consequently, *if E cannot be embedded into any Banach space with an unconditional basis (e.g.**), *if* $E = C([0, 1])$, *or* $E = L^1([0, 1])$, *or* $E = J$), *then E has no unconditional quasi-basis.*

Similarly to definition 9.4, one can also define the *unconditional norm* $v_{\{x_n\}}^{(u)}$ *of an unconditional quasi-basis* $\{x_n\}$, extending the unconditional norm of an unconditional basis introduced in Ch. II, § 17, definition 17.1. However, if $E$ is an $n$-dimensional Banach space, where $n < \infty$, then the unconditional**) quasi-basis $\{x_i\}_{i=1}^{n^2}$ of $E$ constructed in formula (9.34) need not satisfy $v_{\{x_i\}_{i=1}^{n^2}}^{(u)} \leqslant 2$. Moreover, we shall see in Vol. III, Ch. IV, in connection with "local unconditional structures", that *the unconditional analogue of theorem* 9.2 (replacing the norms $v$ by unconditional norms $v^{(u)}$) *is not valid, even with* 2, $1 + \varepsilon$ *and* 1 *respectively replaced by a constant C.* Hence, in particular, the answer to the unconditional analogue of problem 9.2 is also negative.

*Definition 9.7.* A finite-dimensional expansion $\{v_n\}$ of the identity $I_E$ of a Banach space $E$ is said to be *unconditional* if all series (9.42) are unconditionally convergent.

In this case, by Ch. II, § 15, corollary 15.1 (implication 1° ⇒ 3°) and the principle of uniform boundedness, we have

$$\sup_{\substack{I \subset \mathcal{N} \\ I \text{ finite}}} \|\sum_{i \in I} v_i\| < \infty. \tag{9.94}$$

Clearly, $E$ has an unconditional quasi-basis if and only if $E$ has an unconditional one-dimensional expansion of $I_E$ (see formula (9.43)).

*The unconditional analogue of theorem* 9.3, *implication* 3° ⇒ 7° (or, equivalently, 3° ⇒ 9°) *is not valid.* Indeed, we have already mentioned

---

*) See Ch. II, § 15, theorems 15.1, 15.2 and 15.4. For an example which is, in addition, reflexive, see Ch. II, § 17, theorem 17.6.

**) Note that if $\dim E < \infty$, then, obviously, every *finite* quasi-basis of $E$ is unconditional.

that for each $n$ there is a finite-dimensional space $E_n$ such that the unconditional norm of every (finite or infinite) quasi-basis of $E_n$ is $\geqslant C_n^{(u)}$, where $\lim_{n\to\infty} C_n^{(u)} = \infty$. Then $E = (E_1 \times E_2 \times \ldots)_{l^2}$ has an unconditional finite-dimensional expansion of $I_E$ (given by the natural projections $v_n: E \to \{0\} \times \ldots \times \{0\} \times E_n \times \{0\} \times \ldots$), but $E$ has no unconditional quasi-basis $\{x_j\}$ (since otherwise $\{v_n(x_j)\}_{j=1}^{\infty}$ would be infinite unconditional quasi-bases of $\{0\} \times \ldots \times \{0\} \times E_n \times \{0\} \times \ldots \equiv E_n$ with uniformly bounded unconditional norms). However, a somewhat weaker result is still valid for spaces with an unconditional finite-dimensional expansion of $I_E$, as we shall see in § 15, corollary 15.7.

The notion of "monotone basis" can be also carried over to quasi-bases in the following natural way:

*Definition 9.8.* A quasi-basis $\{x_n\}$ of a Banach space $E$ is said to be *monotone* if it has an admissible sequence $\{f_n\} \subset E^*$ such that $\|s_n\| \leqslant 1$ $(n = 1,2, \ldots)$, where $s_n$ are the associated partial sum operators.

For a monotone quasi-basis $\{x_n\}$ we have, by (9.31) and (9.32),

$$\nu_{\{x_n\}} = 1; \tag{9.95}$$

it is also obvious that every Banach space $E$ with a monotone quasi-basis has the metric approximation property, but we do not know whether the converse is true for $\dim E = \infty$ (the proof of theorem 9.3, implication $4° \Rightarrow 7°$, yields*) only quasi-bases of norm $\leqslant 2 + \varepsilon$, for any $\varepsilon > 0$). The proof of theorem 9.2 c) (formulae (9.40), (9.41)) shows that *every finite-dimensional Banach space $E$ has an infinite monotone quasi-basis.* Now we shall prove (see corollary 9.3 below) the result announced after corollary 9.1, that there exist finite-dimensional Banach spaces $E$ which have no finite monotone quasi-basis. To this end, let us first prove the following theorem, which has interest also for other applications (see e.g. remark 9.10 a)).

**Theorem 9.5.** *Let $B$ be a finite-dimensional Banach space with a monotone basis $\{e_i\}_{i=1}^N$ and let $F$ be a subspace of $B$ such that there exists a projection $v$ of norm $\|v\| = 1$ of $B$ onto $F$. Then $F$ has a monotone basis.*

*Proof.* Let $n_0 \leqslant N$ be the last index for which $v(e_{n_0}) \neq 0$. Define on $F = v(B)$ the operators

$$w(x) = h_{n_0}(x) v(e_{n_0}) \qquad (x \in F), \tag{9.96}$$

$$u = I_F - w, \tag{9.97}$$

where $\{h_i\}_{i=1}^N \subset B^*$ is the a.s.c.f. to $\{e_i\}_{i=1}^N$. Then, since $x = v(x)$ for

---

*) See remark 9.5 c).

all $x \in F$ and since $v(e_{n_0+1}) = \ldots = v(e_N) = 0$, we have

$$u(x) = x - w(x) = v(x) - h_{n_0}(x)v(e_{n_0}) = v\left(\sum_{i=1}^{N} h_i(x)e_i\right) - h_{n_0}(x)v(e_{n_0}) =$$

$$= \sum_{i=1}^{n_0-1} h_i(x)v(e_i) = vs_{n_0-1}|_F(x) \qquad (x \in F),$$

where $s_{n_0-1}(x) = \sum_{i=1}^{n_0-1} h_i(x)e_i$ $(x \in B)$, whence, since $\{e_i\}_{i=1}^N$ is monotone,

$$\|u\| \leqslant \|v\| \, \|s_{n_0-1}\| = 1. \tag{9.98}$$

Consequently, by a well known version of the ergodic theorem*), there exists a projection $p_1$ of norm $\|p_1\|=1$ of $F$ onto $F_1 = \{y \in F | u(y) = y\}$. Since by (9.97) and (9.96) $F_1 = \operatorname{Ker} w = \operatorname{Ker} h_{n_0}$, $F_1$ is a subspace of codimension 1 of $F$. Furthermore, $p_1 v$ is a projection of norm 1 of $B$ onto $F_1$. Hence, repeating the above procedure $n$ times, where $n = \dim F$, we obtain projections of norm 1 of $B$, whence also of $F$, onto $F_1, F_2, \ldots, F_n = \{0\}$, where $F \supset F_1 \supset F_2 \supset \ldots \supset F_n = \{0\}$, $\dim F_j = n - j$ $(j = 1, \ldots, n)$. Consequently, by Ch. II, § 1, proposition 1.3, $F$ has a monotone basis, which completes the proof of theorem 9.5.

*Remark 9.10.* a) In Ch. II, § 1, theorem 1.2, we have seen that for each $n \geqslant 3$ there exists a real Banach space $E_n$ with $\dim E_n = n$, which has no monotone basis. Theorem 9.5 above can be applied to obtain a short proof of the following stronger result: *For each $n \geqslant 3$ there exists a real Banach space $E_n$ with* $\dim E_n = n$, *such that $E_n$ has no monotone basis, $E_n \subset E_{n+1}$ and $E_n$ has a basis $\{x_j^{(n)}\}_{j=1}^n$ with unconditional norm* $v^{(u)}_{\{x_j^{(n)}\}_{j=1}^n} \leqslant 2$ $(n = 1,2, \ldots)$. Indeed, by Ch. II, § 1, theorem 1.1, the statement is true for $n = 3$, so we have $E_3$. Put

$$E_n = (E_3 \times l^\infty_{n-3})_1. \tag{9.99}$$

Then $\dim E_n = n$ and the natural projection $v$ of $E_n$ onto $E_3 \times \{0\} \equiv E_3$ has norm $\|v\| = 1$, whence, by theorem 9.5, $E_n$ has no monotone basis $(n = 4,5, \ldots)$. Clearly, the sequence $\{E_n\}$ is increasing. Finally, since any basis of $E_3$ has unconditional norm $\leqslant 2$ (by $\dim E_3 = 3$) and since $l^\infty_{n-3}$ has a basis with unconditional norm 1, from Ch. II, § 17, corollary 17.4 it follows that $E_n$ has a basis with unconditional norm $\leqslant 2$. b) Since there exist finite-dimensional Banach spaces $E$ which have no monotone basis and since by corollary 9.1 b), for any such space there is an infinite dimensional Banach space $B$ having a monotone basis $\{e_n\}$

*) Let us recall this result (see e.g. [87], p. 662): If $F$ is a Banach space and $u$ a finite rank operator on $F$ with $\|u\| \leqslant 1$, then there exists a projection of norm 1 of $F$ onto the finite-dimensional subspace $\{x \in F | u(x) = x\}$.

such that there exist a linear isometry $u$ of $E$ into $B$ and a projection $v$ of $B$ onto $u(E) = F$ of norm 1, it follows that theorem 9.5 is no longer valid for $\dim B = \infty$, $\dim F < \infty$. We do not know whether theorem 9.5 remains valid for*) $\dim B = \dim F = \infty$ (see also the related problem 9.1 and the problem mentioned after definition 9.8).

Now we give the result announced above.

**Corollary 9.3.** *If $E$ is an $n$-dimensional Banach space, where $n < \infty$, and if $E$ has a finite monotone quasi-basis $\{x_i\}_{i=1}^N$, then $E$ has a monotone basis. Consequently, there exists a finite-dimensional Banach space $E_0$ which has no finite monotone quasi-basis and which cannot be embedded isometrically as a subspace admitting a projection of norm 1, into a finite-dimensional Banach space with a finite monotone quasi-basis.*

*Proof.* If $E$ has a finite monotone quasi-basis $\{x_i\}_{i=1}^N$, then by remark 9.3 a), $E$ can be embedded isometrically into an $N$-dimensional Banach space with a monotone basis, as a subspace admitting a projection of norm 1, and hence, by theorem 9.5, $E$ has a monotone basis. Consequently, if $E_0$ is a finite-dimensional Banach space having no monotone basis, then $E_0$ has no finite monotone quasi-basis. Also, the last statement holds since the image of a finite monotone quasi-basis by a norm 1 projection $v$ is a monotone quasi-basis of its span (indeed, by the proof of proposition 9.3 above, every finite monotone quasi-basis has an admissible sequence such that the associated partial sum operators $s_i$ have norm $\leqslant 1$ and hence $\|vs_i\| \leqslant 1$). This completes the proof of corollary 9.3.

In contrast with the situation for bases (see Ch. II, § 1, the remark made after proposition 1.3), it is not known whether every quasi-basis of a Banach space $E$ can be "monotonized" by replacing the norm of $E$ with a suitable equivalent norm. More generally**), the following problem is also open:

*Problem 9.4.* Is every Banach space $E$ with a quasi-basis (or, equivalently, every separable Banach space $E$ with the bounded approximation property) isomorphic to a Banach space having the metric approximation property?

Now we shall show (in theorem 9.6 below) that if $E^*$ has the bounded approximation property (hence $E$ also has this property, by proposition 9.8 below), then the answer is affirmative. To this end, we shall first show that if a Banach space $E$ has an approximative basis $\{u_n\}$ satisfying

*) However, the answer to the more general problem, whether every infinite dimensional Banach space $F$ with the metric approximation property has a monotone basis, is negative (see e.g. § 12, example 12.2 or the Notes and remarks to § 9, remark 9.9 d)).

**) This question seems indeed to be somewhat more general, since we do not know whether every infinite dimensional Banach space with the metric approximation property has a monotone quasi-basis.

a certain condition, then this forces $E$ to be isomorphic to a Banach space with the metric approximation property.

*Definition 9.9.* An approximative basis $\{u_n\}$ of a Banach space $E$ is said to be *commuting*, if

$$u_m u_n = u_n u_m = u_{\min(m,n)} \qquad (m \neq n;\, m, n = 1,2, \ldots). \qquad (9.100)$$

**Proposition 9.6.** *Every Banach space E with a commuting approximative basis is isomorphic to a Banach space having the metric approximation property.*

*Proof.* Let $\{u_n\} \subset L(E, E)$ be a sequence of endomorphisms of finite rank satisfying (9.2) and (9.100). Put

$$|||x||| = \sup_{1 \leqslant n < \infty} \|u_n(x)\| \qquad (x \in E). \qquad (9.101)$$

Then $|||\cdot|||$ is a norm on $E$, equivalent to the initial norm $\|\cdot\|$, since

$$\|x\| = \lim_{n\to\infty} \|u_n(x)\| \leqslant |||x||| \leqslant (\sup_{1\leqslant n<\infty} \|u_n\|)\, \|x\| \qquad (x \in E).$$

Define now a sequence $\{v_n\} \subset L(E, E)$ of endomorphisms of finite rank by taking the arithmetical means of $\{u_n\}$, that is,

$$v_n = \frac{1}{n} \sum_{i=1}^{n} u_i \qquad (n = 1, 2, \ldots). \qquad (9.102)$$

Then, by (9.2),

$$\lim_{n\to\infty} v_n(x) = x \qquad (x \in E). \qquad (9.103)$$

Furthermore, by (9.101), (9.100) and $\sup_{1\leqslant m<\infty} \|u_m\| < \infty$,

$$|||v_n||| = \sup_{\substack{x\in E \\ |||x|||\leqslant 1}} |||v_n(x)||| = \sup_{\substack{x\in E \\ \|u_1(x)\|,\, \|u_2(x)\|, \ldots \leqslant 1}} \sup_{1\leqslant m<\infty} \left\| \frac{1}{n} \sum_{i=1}^{n} u_m u_i(x) \right\| =$$

$$= \sup_{\substack{x\in E \\ \|u_1(x)\|,\, \|u_2(x)\|, \ldots \leqslant 1}} \max \left\{ \sup_{1\leqslant m\leqslant n} \left\| \frac{1}{n} \left( \sum_{i=1}^{m-1} u_i(x) + u_m^2(x) + \right.\right.\right.$$

$$\left.\left.\left. + \sum_{i=m+1}^{n} u_m(x) \right) \right\|, \left\| \frac{1}{n} \sum_{i=1}^{n} u_i(x) \right\| \right\} \leqslant$$

$$\leqslant \max \left\{ \sup_{1\leqslant m\leqslant n} \left( \frac{\|u_m\|}{n} + \frac{n-1}{n} \right),\, 1 \right\} = 1 + \varepsilon \qquad (n > N_1(\varepsilon)),$$

whence, by (9.103), $1 = |||I_E||| \leqslant \underline{\lim}_{n\to\infty} |||v_n||| \leqslant \overline{\lim}_{n\to\infty} |||v_n||| \leqslant 1$, so

$\lim_{n\to\infty} |||v_n||| = 1$. Consequently, again by (9.103),

$$\lim_{n\to\infty} \frac{v_n}{|||v_n|||}(x) = x \qquad (x \in E), \tag{9.104}$$

and thus $E$ endowed with the equivalent norm $|||\cdot|||$ has the metric approximation property, which completes the proof of proposition 9.6.

*Remark 9.11.* a) The final part of the proof of proposition 9.6 also shows that if a Banach space $E$ has an approximative basis $\{u_n\}$ such that $\lim_{n\to\infty} \|u_n\| = 1$ (for example, this happens when $\|u_n\| \leqslant 1 + \varepsilon_n$ for $n = 1, 2, \ldots$ where $\varepsilon_n > 0$, $\lim_{n\to\infty} \varepsilon_n = 0$), then $E$ already has the metric approximation property*). More generally, the same holds also for $\lim_{n\to\infty} \|u_n\| = \lambda$ and the $\lambda$-approximation property $\left(\text{taking } \frac{\lambda u_n}{\|u_n\|}\right)$. A similar remark was also used at the end of the proof of theorem 9.4. b) The operators $v_n$ defined by (9.102) do not satisfy the conditions (9.100). One can also obtain operators of finite rank $v_n$ satisfying (9.100) and (9.103), (9.104), by putting

$$v_n = \frac{1}{n} \sum_{i=\frac{(n-1)n}{2}+1}^{\frac{n(n+1)}{2}} u_i \qquad (n = 1, 2, \ldots). \tag{9.105}$$

It is an open problem whether the converse of proposition 9.6 holds:

*Problem 9.5.* a) Does every separable Banach space $E$ with the metric approximation property have a commuting approximative basis? b) What about every Banach space $E$ with an approximative basis?

The next step towards the proof of theorem 9.6 will be to give a sufficient condition for a Banach space $E$ to have a commuting approximative basis. To this end, let us recall that a Banach space $E$ is said to have the *$\lambda$-duality approximation property* (where $\lambda \geqslant 1$), if for every $\varepsilon > 0$ and every pair of finite-dimensional subspaces $G$ of $E$ and $\Gamma$ of $E^*$, there exists an endomorphism $u = u_{G,\Gamma,\varepsilon} \in L(E, E)$ of finite rank such that

$$\|u(y) - y\| < \varepsilon \|y\| \qquad (y \in G), \tag{9.106}$$

$$\|u^*(h) - h\| < \varepsilon \|h\| \qquad (h \in \Gamma), \tag{9.107}$$

$$\|u\| \leqslant \lambda. \tag{9.108}$$

*) This observation, together with § 1, remark 1.1, yields that every Banach space $E$ contains a subspace with a basis, having the metric approximation property.

Obviously, if $E$ has the $\lambda$-duality approximation property, then both $E$ and $E^*$ have the $\lambda$-approximation property. From proposition 9.8 below it follows that the converse is also true.

We recall the following characterization of such spaces, which we shall use in the sequel:

**Lemma 9.2.** *Let $E$ be a Banach space and $\lambda \geqslant 1$. The following statements are equivalent:*

*1°. $E$ has the $\lambda$-duality approximation property.*

*2°. For every pair of finite-dimensional subspaces $G$ of $E$ and $\Gamma$ of $E^*$ and every $\varepsilon > 0$ there exists an endomorphism $v = v_{G,\Gamma,\varepsilon} \in L(E, E)$ of finite rank such that*

$$v(y) = y \qquad (y \in G), \tag{9.109}$$

$$v^*(h) = h \qquad (h \in \Gamma), \tag{9.110}$$

$$\|v\| \leqslant \lambda + \varepsilon. \tag{9.111}$$

*Proof.* The implication 2° ⇒ 1° is immediate, by the argument of the proof of theorem 9.3, implication 6° ⇒ 5°.

Conversely, assume now that we have 1° and let $G \subset E$, $\Gamma \subset E^*$ be finite-dimensional and $\varepsilon > 0$. Choose $0 < \delta < 1$ so small that $\delta + \dfrac{\delta l(\lambda + \delta)}{1 - \delta} \leqslant \varepsilon$, where $l = \dim \Gamma$ and choose $0 < \gamma < 1$ small enough so that $\dfrac{\gamma k}{1 - \gamma} \leqslant \dfrac{\delta}{2\lambda}$, where $k = \dim G$. Let $u$ be an operator of finite rank on $E$ satisfying (9.106), (9.108) and

$$\|u^*(h) - h\| < \frac{\delta}{2} \|h\| \qquad (h \in \Gamma). \tag{9.107'}$$

Then, by lemma 9.1, there exists an operator $v_1$ of finite rank on $E$, such that

$$v_1|_G = I_G, \tag{9.112}$$

$$\|v_1 - u\| < \frac{\gamma k}{1 - \gamma} \|u\| \leqslant \frac{\gamma k \lambda}{1 - \gamma} \leqslant \frac{\delta}{2}\cdot \tag{9.113}$$

By (9.113) we have $\|v_1^* - u^*\| < \dfrac{\delta}{2}$, whence we obtain, taking into account (9.108) and (9.107′), the inequalities

$$\|v_1^*\| < \|u^*\| + \frac{\delta}{2} \leqslant \lambda + \frac{\delta}{2}, \tag{9.114}$$

$$\|v_1^*(h) - h\| \leqslant \|v_1^*(h) - u^*(h)\| + \|u^*(h) - h\| < \delta \|h\| \qquad (h \in \Gamma). \tag{9.115}$$

Let us apply now lemma 9.1 and its proof given in remark 9.4 to $E^*, \Gamma$ ($\dim \Gamma = l$) and $v_1^*$ instead of $E$, $G$ ($\dim G = k$) and $u$ respectively. Thus we get that if $v_0'$ is a projection of norm $\leqslant l$ of $v_1^*(E^*)$ onto $v_1^*(\Gamma)$ and if

$$v^* = (I_{v_1^*(E^*)} - v_0' + (v_1^*|_\Gamma)^{-1} v_0') v_1^*, \tag{9.116}$$

then $v^*$ satisfies (9.110) and $\|v^* - v_1^*\| < \dfrac{\delta l}{1-\delta}\|v_1^*\|$, whence also

$$\|v^*\| < \left(1 + \frac{\delta l}{1-\delta}\right)\|v_1^*\| < \left(1 + \frac{\delta l}{1-\delta}\right)(\lambda + \delta) =$$

$$= \lambda + \delta + \frac{\delta l(\lambda + \delta)}{1-\delta} \leqslant \lambda + \varepsilon,$$

that is, (9.111). Furthermore, $v^*$ is $\sigma(E^*, E)$-continuous (because $v_1^*$ is $\sigma(E^*, E)$-continuous and has finite rank), so $v^*$ is indeed the adjoint of some finite rank operator $v$ on $E$. Finally, let $y \in G$ and $f \in E^*$. Then by (9.112) we have $v_1(y) = y$, whence

$$f(v(y)) = v^*(f)(y) = v^*(f)(v_1(y)) =$$

$$= (v_1^* - v_0' v_1^* + (v_1^*|_\Gamma)^{-1} v_0' v_1^*)(f)(v_1(y)) =$$

$$= f(v_1(v_1(y))) - v_0' v_1^*(f)(v_1(y)) + (v_1^*(v_1^*|_\Gamma)^{-1} v_0' v_1^*(f))(y) = f(y).$$

Hence, since $E^*$ is total over $E$, we obtain (9.109), which completes the proof of lemma 9.2.

**Proposition 9.7.** *Let $E$ be a separable Banach space which has the $\lambda$-duality approximation property for some $\lambda \geqslant 1$. Then $E$ has a commuting approximative basis.*

*Proof.* Let $0 < \varepsilon_n < 1$, $\lim_{n\to\infty} \varepsilon_n = 0$ and let $\{y_n\} \subset E$ be a dense sequence in $E$. For the subspaces $G_1 = [y_1]$ of $E$, $\Gamma_1 = [f]$ of $E^*$ (where $f$ is any element of $E^*$) and for $\varepsilon_1$ there exists, by lemma 9.2, an operator $v_1 \in L(E, E)$ of finite rank, such that

$$v_1|_{G_1} = I_{G_1}, \quad v_1^*|_{\Gamma_1} = I_{\Gamma_1}, \quad \|v_1\| \leqslant \lambda + \varepsilon_1. \tag{9.117}$$

Furthermore, for $G_2 = [v_1(E) \cup \{y_2\}]$, $\Gamma_2 = v_1^*(E^*)$ and $\varepsilon_2$ there exists $v_2 \in L(E, E)$ of finite rank, such that

$$v_2|_{G_2} = I_{G_2}, \quad v_2^*|_{\Gamma_2} = I_{\Gamma_2}, \quad \|v_2\| \leqslant \lambda + \varepsilon_2. \tag{9.118}$$

Taking $G_3 = [v_2(E) \cup \{y_3\}]$, $\Gamma_3 = v_2^*(E^*)$ and $\varepsilon_3$ and continuing in this way indefinitely, we obtain two sequences of subspaces $\{G_n\}$, $\{\Gamma_n\}$ and a sequence of endomorphisms $\{v_n\} \subset L(E, E)$ of finite rank. We shall show that $\{v_n\}$ satisfies (9.100) (for $\{v_n\}$) and (9.103), which will complete the proof. Observe first that by (9.117), (9.118), ..., for each $y \in G_n$ we have $y = v_n(y) \in v_n(E)$, whence

$$G_1 \subset v_1(E) \subset G_2 \subset v_2(E) \subset G_3 \subset v_3(E) \subset \dots \tag{9.119}$$

and therefore $v_n(E) \subset G_m$ for all $n < m$. Hence, again by (9.117), (9.118), ... we obtain $v_m v_n(x) = v_n(x)$ $(x \in E,\ n < m)$, that is,

$$v_m v_n = v_n \qquad (n < m). \tag{9.120}$$

Similarly, we have $\Gamma_n \subset v_n^*(E^*) = \Gamma_{n+1}$ $(n = 1,2, \dots)$, whence

$$v_m^* v_n^* = v_n^* \qquad (n < m).$$

Consequently,

$$f(v_n v_m(x)) = v_m^* v_n^*(f)(x) = v_n^*(f)(x) = f(v_n(x)) \qquad (x \in E,\ f \in E^*,\ n < m),$$

whence $v_n v_m = v_n$ for all $n < m$, which, together with (9.120), gives (9.100) for $\{v_n\}$.

Finally, observe that $y_n \in G_n \subset v_n(E)$ $(n = 1, 2, \dots)$, whence $\{y_n\} \subset \subset \bigcup_{j=1}^{\infty} v_j(E)$. Therefore, since $[y_n] = E$, we obtain

$$\overline{\bigcup_{j=1}^{\infty} v_j(E)} = E. \tag{9.121}$$

Now let $y \in \bigcup_{j=1}^{\infty} v_j(E)$ be arbitrary, say $y \in v_N(E)$. Then, by (9.120), $v_m(y) = y$ for all $m > N$, and thus $\lim_{n\to\infty} v_n(y) = y$ for all $y \in \bigcup_{j=1}^{\infty} v_j(E)$. Consequently, by (9.121) and $\sup_{1 \leqslant n < \infty} \|v_n\| \leqslant \lambda + 1$ we obtain (see Ch. I, p. 26, footnote) that $\{v_n\}$ satisfies (9.103), which completes the proof of proposition 9.7.

*Remark 9.12.* a) The above proof shows that under the assumptions of proposition 9.7, for any $0 < \varepsilon_n < 1$, $\lim_{n\to\infty} \varepsilon_n = 0$, one can find $v_n$ of finite rank with $\|v_n\| \leqslant \lambda + \varepsilon_n$ $(n = 1, 2, \dots)$, satisfying (9.100) (for

$\{v_n\}$) and (9.103). b) The converse of proposition 9.7 is not valid, even if we assume that $E^*$ is separable, since we have seen in § 0, example 0.1, that there exists a Banach space with a basis (hence with a commuting approximative basis), whose dual is separable and fails the approximation property. c) The above method of proof also yields that *if $E^*$ is separable and $E$ has the $\lambda$-duality approximation property for some $\lambda \geqslant 1$, then $E$ has a "shrinking" commuting approximative basis.* Indeed, we only have to put $\Gamma_1' = [f_1]$, $\Gamma_2' = [v_1^*(E^*) \cup \{f_2\}]$, $\Gamma_3' = [v_2^*(E^*) \cup \cup \{f_3\}]$, ... where $\{f_n\}$ is a dense sequence in $E^*$ and then we shall also have $\lim_{n\to\infty} v_n^*(f) = f$ for all $f \in E^*$.

Let us also recall now

**Lemma 9.3.** a) *Let $E$ and $F$ be Banach spaces with* $\dim F < \infty$, *and let $V$ be a finite-dimensional subspace of $E^*$. Furthermore, let $u$ be a continuous linear mapping of $E^*$ into $F$ and let $\varepsilon > 0$. Then there exists a $w^*$-continuous*) linear mapping of $E^*$ into $F$ with the following properties:*

$$v|_V = u|_V, \tag{9.122}$$

$$\|v\| \leqslant \|u\| + \varepsilon. \tag{9.123}$$

b) *If, in addition, $F \subset E^*$, then there exists a continuous linear mapping $t: E \to E$ of finite rank, such that*

$$t^*(E^*) \subset F, \tag{9.124}$$

$$t^*|_V = u|_V, \tag{9.125}$$

$$\|t\| \leqslant \|u\| + \varepsilon. \tag{9.126}$$

*Proof.* We shall use some well known**) linear isometries between spaces of operators and tensor products or their conjugates, combined with Helly's theorem. Firstly, since $\dim F < \infty$, $L(E^*,F)$ is linearly isometric to $(E^* \otimes_\gamma F^*)^*$, by the mapping which carries each operator

$$u(f) = \sum_{i=1}^{n} \Phi_i(f) y_i \qquad (f \in E^*), \tag{9.127}$$

where $y_1, \ldots, y_n$ is a basis of $F$ and $\Phi_1, \ldots, \Phi_n \in E^{**}$, into $\Psi_u \in \in (E^* \otimes_\gamma F^*)^*$ defined by

$$\Psi_u\left(\sum_{j=1}^{n} f_j \otimes g_j\right) = \sum_{j=1}^{n} g_j(u(f_j)) = \sum_{i=1}^{n} \sum_{j=1}^{n} \Phi_i(f_j) g_j(y_i). \tag{9.128}$$

*) I.e., from $\sigma(E^*, E)$ to the norm topology of $F$ (since $\dim F < \infty$).
**) See e.g. [356].

But, again since $\dim F < \infty$, $E^* \otimes_\gamma F^*$ is linearly isometric to $(E \otimes_\lambda F)^*$, by the mapping which carries each $\sum_{j=1}^{n} f_j \otimes g_j \in E^* \otimes F^*$ into $\varphi_{\sum f_j \otimes g_j} \in (E \otimes_\lambda F)^*$ defined by

$$\varphi_{\sum f_j \otimes g_j}\left(\sum_{i=1}^{n} x_i \otimes y_i\right) = \sum_{i=1}^{n} \sum_{j=1}^{n} f_j(x_i) g_j(y_i), \tag{9.129}$$

and hence $(E^* \otimes_\gamma F^*)^*$ is linearly isometric to $(E \otimes_\lambda F)^{**}$, by the mapping which carries each $\Psi \in (E^* \otimes_\gamma F^*)^*$ into $\overline{\Psi} \in (E \otimes_\lambda F)^{**}$ defined by

$$\overline{\Psi}(\varphi_{\sum f_j \otimes g_j}) = \Psi\left(\sum_{j=1}^{n} f_j \otimes g_j\right) \left(\sum_{j=1}^{n} f_j \otimes g_j \in E^* \otimes F^*\right). \tag{9.130}$$

Thus, the composition of these mappings gives a linear isometry of $L(E^*, F)$ onto $(E \otimes_\lambda F)^{**}$, which carries each $u \in L(E^*, F)$ defined by (9.127) into $\overline{\Psi}_u \in (E \otimes_\lambda F)^{**}$ defined by

$$\overline{\Psi}_u(\varphi_{\sum f_j \otimes g_j}) = \sum_{j=1}^{n} g_j(u(f_j)) = \sum_{i=1}^{n} \sum_{j=1}^{n} \Phi_i(f_j) g_j(y_i). \tag{9.131}$$

Clearly, this latter mapping gives, by restriction, a linear isometry of the subspace of all $w^*$-continuous linear mappings

$$v(f) = \sum_{i=1}^{n} f(x_i) y_i \qquad (f \in E^*) \tag{9.132}$$

of $E^*$ into $F$ with $\pi(E \otimes_\lambda F)$, the canonical image of $E \otimes_\lambda F$ in $(E \otimes_\lambda F)^{**}$, carrying $v$ into $\overline{\Psi}_v \in \pi(E \otimes_\lambda F)$ defined by

$$\overline{\Psi}_v(\varphi_{\sum f_j \otimes g_j}) = \sum_{j=1}^{n} g_j(v(f_j)) = \sum_{i=1}^{n} \sum_{j=1}^{n} f_j(x_i) g_j(y_i) \tag{9.133}$$

$\left(\text{i.e., } \overline{\Psi}_v = \pi\left(\sum_{i=1}^{n} x_i \otimes y_i\right)\right)$.

Now let $u \in L(E^*, F)$ and $\varepsilon > 0$. Then, since $\dim V$, $\dim F < \infty$, for $\overline{\Psi}_u \in (E \otimes_\lambda F)^{**}$ there exists, by Helly's theorem*), an element $\sum_{i=1}^{n} x_i \otimes$

---

*) See e.g. [70], Ch. II, § 5, theorem 3.

$\otimes y_i \in E \otimes_\lambda F$ (where $\{y_1, \ldots, y_n\}$ is a basis of $F$ and $x_1, \ldots, x_n \in E$), such that

$$\varphi_{\sum f_j \otimes g_j}\left(\sum_{i=1}^{n} x_i \otimes y_i\right) = \overline{\Psi}_u(\varphi_{\sum f_j \otimes g_j}) \qquad \left(\sum_{j=1}^{n} f_j \otimes g_j \in V \otimes F^*\right)$$

$$\left\| \sum_{i=1}^{n} x_i \otimes y_i \right\|_\lambda \leqslant \|\overline{\Psi}_u\| + \varepsilon.$$

Define a $w^*$-continuous linear mapping of $E^*$ into $F$ by (9.132). Then, by the above,

$$\sum_{j=1}^{n} g_j(v(f_j)) = \sum_{i=1}^{n} \sum_{j=1}^{n} f_j(x_i) g_j(y_i) = \varphi_{\sum f_j \otimes g_j}\left(\sum_{i=1}^{n} x_i \otimes y_i\right) =$$

$$= \overline{\Psi}_u(\varphi_{\sum f_j \otimes g_j}) = \sum_{j=1}^{n} g_j(u(f_j)) \qquad \left(\sum_{j=1}^{n} f_j \otimes g_j \in V \otimes F^*\right),$$

$$\|v\| = \|\overline{\Psi}_v\| = \left\| \pi\left(\sum_{i=1}^{n} x_i \otimes y_i\right)\right\|_\lambda \leqslant \|\overline{\Psi}_u\| + \varepsilon = \|u\| + \varepsilon.$$

Therefore, in particular, $g(v(h)) = g(u(h))$ $(h \in V \subset E^*,\ g \in F^*)$, whence $v(h) = u(h)$ $(h \in V)$. Thus, $v$ satisfies (9.122) and (9.123).

b) Assume, in addition, that $F \subset E^*$. Let $v$ be as in part a) above, i.e., of the form

$$v(f) = \sum_{i=1}^{n} f(x_i) h_i \qquad (f \in E^*), \tag{9.134}$$

(where $x_1, \ldots, x_n \in E$ and where $\{h_1, \ldots, h_n\}$ is a basis of $F$) and satisfying (9.122), (9.123). Define a continuous linear mapping $t\colon E \to E$ by

$$t(x) = \sum_{i=1}^{n} h_i(x) x_i \qquad (x \in E). \tag{9.135}$$

Then

$$t^*(f)(x) = f(t(x)) = \sum_{i=1}^{n} h_i(x) f(x_i) = v(f)(x) \qquad (x \in E,\ f \in E^*),$$

whence $v = t^*$. Consequently, $t^*(E^*) = v(E^*) \subset [h_i]_{i=1}^n = F$ and, by (9.122), (9.123) we have (9.125), (9.126), which completes the proof of lemma 9.3.

Now we are ready to prove

**Proposition 9.8.** *Let $E$ be a Banach space and let $\lambda \geqslant 1$. If $E^*$ has the $\lambda$-approximation property, then $E$ has the $\lambda$-duality approximation property.*

*Proof.* By our assumption, for every finite-dimensional subspace $\Gamma$ of $E^*$ and every $\varepsilon > 0$ there exists a finite rank operator $u = u_{\Gamma,\varepsilon}$ on $E^*$ such that

$$\|u(h) - h\| \leqslant \frac{\varepsilon}{2}\|h\| \qquad (h \in \Gamma),$$

$$\|u\| \leqslant \lambda.$$

Then, by lemma 9.3 b) with $V = \Gamma \subset E^*$, there exists a continuous linear mapping $t$ of $E$ into $E$, of finite rank, such that

$$\|t^*(h) - h\| \leqslant \varepsilon\|h\| \qquad (h \in \Gamma),$$

$$\|t\| \leqslant \|u\| \leqslant \lambda;$$

indeed, if $t_0 \colon E \to E$ is a finite rank operator such that $t_0^*|_\Gamma = u|_\Gamma$, $\|t_0\| \leqslant \leqslant \|u\| + \frac{\varepsilon}{2}\|u\|$, and if $\|t_0\| > \|u\|$, then for $t = \frac{\|u\|}{\|t_0\|} t_0$ we have $\|t\| = \|u\| \leqslant \lambda$ and

$$\|t^*(h) - h\| \leqslant \left\| \frac{\|u\|}{\|t_0\|}(u(h) - h) \right\| + \left| \frac{\|u\|}{\|t_0\|} - 1 \right| \|h\| \leqslant$$

$$\leqslant \frac{\varepsilon}{2}\|h\| + \frac{\varepsilon\|u\|}{2\|t_0\|}\|h\| \leqslant \varepsilon\|h\| \qquad (h \in \Gamma),$$

while if $\|t_0\| \leqslant \|u\|$, then, clearly, one can take $t = t_0$.

Now let $\mathscr{D}$ be the directed set of all pairs $(\Gamma, \varepsilon)$, where $\Gamma$ is a finite-dimensional subspace of $E^*$ and $\varepsilon > 0$, and where $(\Gamma_1, \varepsilon_1) \geqslant (\Gamma_2, \varepsilon_2)$ if and only if $\Gamma_1 \supset \Gamma_2$ and $\varepsilon_1 \leqslant \varepsilon_2$. Furthermore, by the above, for each $d = (\Gamma, \varepsilon) \in \mathscr{D}$ let $t_d \in L(E, E)$ be a finite rank endomorphism such that $\|t_d^*(h) - h\| \leqslant \varepsilon\|h\|$ $(h \in \Gamma)$ and $\|t_d\| \leqslant \lambda$. If $f \in E^*$ and $\varepsilon > 0$, then putting $d_0 = ([f], \varepsilon)$, it follows that $\|t_d^*(f) - f\| \leqslant \varepsilon\|f\|$ $(d \geqslant d_0)$ and hence the net $\{t_d\}_{d \in \mathscr{D}}$ of finite rank operators on $E$ has the following properties:

$$\lim_{d \in \mathscr{D}} t_d^*(f) = f \qquad (f \in E^*), \tag{9.136}$$

$$\|t_d\| \leqslant \lambda \qquad (d \in \mathscr{D}). \tag{9.137}$$

Then, by (9.136),

$$\lim_{d\in\mathscr{D}} f(t_d(x)) = \lim_{d\in\mathscr{D}} t_d^*(f)(x) = f(x) \qquad (x\in E,\ f\in E^*), \quad (9.138)$$

and thus

$$t_d(x) \xrightarrow{w} x \qquad (x\in E). \quad (9.139)$$

Let $G\subset E$, $\Gamma\subset E^*$ be finite-dimensional subspaces and let $\varepsilon>0$. Then, since the convergence (9.136) is uniform on the compact set $\sigma_\Gamma = \{h\in\Gamma \mid \|h\|=1\}$ (by Ch. I, § 17, proof of theorem 17.3), there exists a $d_0 = d_0(\Gamma,\varepsilon)\in\mathscr{D}$ such that

$$\|t_d^*(h) - h\| \leqslant (\lambda+2)\,\varepsilon\|h\| \qquad (h\in\Gamma,\ d\geqslant d_0). \quad (9.140)$$

Let*) $\{y_1,\dots,y_k\}$ be an $\varepsilon$-net for $\sigma_G = \{y\in G \mid \|y\|=1\}$. Since by (9.139) the net $\{t_d(y_1),\dots,t_d(y_k)\}_{d\geqslant d_0}$ converges weakly to $\{y_1,\dots,y_k\}$ in $(\underbrace{E\times\dots\times E}_{k})_{l^\infty}$, there exists**) a convex combination

$\sum_{i=1}^{n}\alpha_i\{t_{d_i}(y_1),\dots,t_{d_i}(y_k)\}$ with $d_1,\dots,d_n\geqslant d_0$, $\alpha_1,\dots,\alpha_n\geqslant 0$, $\sum_{i=1}^{n}\alpha_i = 1$, such that

$$\left\|\sum_{i=1}^{n}\alpha_i\{t_{d_i}(y_1),\dots,t_{d_i}(y_k)\} - \{y_1,\dots,y_k\}\right\|_{l^\infty} < \varepsilon,$$

whence

$$\left\|\sum_{i=1}^{n}\alpha_i t_{d_i}(y_j) - y_j\right\| < \varepsilon \qquad (j=1,\dots,k).$$

Since $\{y_1,\dots,y_k\}$ is an $\varepsilon$-net for $\sigma_G$, it follows that for any $y\in\sigma_G$ we have $\|y-y_j\|<\varepsilon$ with a suitable $j\leqslant k$, whence, since $\left\|\sum_{i=1}^{n}\alpha_i t_{d_i}\right\| \leqslant \sum_{i=1}^{n}\alpha_i\|t_{d_i}\| \leqslant \lambda$, we get

$$\left\|\sum_{i=1}^{n}\alpha_i t_{d_i}(y) - y\right\| \leqslant \left\|\sum_{i=1}^{n}\alpha_i t_{d_i}(y-y_j)\right\| +$$

$$+ \left\|\sum_{i=1}^{n}\alpha_i t_{d_i}(y_j) - y_j\right\| + \|y_j - y\| < (\lambda+2)\varepsilon.$$

*) Alternatively, since $\{t_d(x)\}_{d\geqslant d_0} \xrightarrow{w} x$ $(x\in E)$, there exists a net of convex combinations $\{v_e\}_{e\in\mathscr{E}}$ of $\{t_d\}_{d\geqslant d_0}$ such that $v_e(x)\to x$ $(x\in E)$, hence uniformly on $\sigma_G$ (by Ch. I, § 17, proof of theorem 17.3).

**) See e.g. [355], Ch. II, § 9, corollary 2 of theorem 9.2 or [87], p. 477, corollary 5.

Hence we infer

$$\left\| \sum_{i=1}^{n} \alpha_i t_{d_i}(y) - y \right\| \leqslant (\lambda + 2)\varepsilon \|y\| \qquad (y \in G). \qquad (9.141)$$

On the other hand, by (9.140) and since $d_1, \ldots, d_n \geqslant d_0$,

$$\left\| \sum_{i=1}^{n} \alpha_i t_{d_i}^*(h) - h \right\| = \left\| \sum_{i=1}^{n} \alpha_i (t_{d_i}^*(h) - h) \right\| \leqslant$$

$$\leqslant \sum_{i=1}^{n} \alpha_i \| t_{d_i}^*(h) - h \| \leqslant (\lambda + 2)\varepsilon \|h\| \qquad (h \in \Gamma). \qquad (9.142)$$

Since $s = \sum_{i=1}^{n} \alpha_i t_{d_i}$ is an operator of finite rank on $E$, (9.141), (9.142) and $\|s\| \leqslant \lambda$ show that $E$ has the $\lambda$-duality approximation property, which completes the proof of proposition 9.8.

*Remark 9.13*. From proposition 9.8 and remark 9.12 c) it follows that for separable conjugate spaces the answer to problem 9.5 is affirmative, namely, *for every separable conjugate space $E^*$ with the* (*metric*) *approximation property*, *$E$ has a shrinking commuting approximative basis $\{v_n\}$, so $E^*$ has a commuting approximative basis $\{v_n^*\}$* (even with $\|v_n^*\| \leqslant 1 + \varepsilon_n$, where $\varepsilon_n > 0$, $\lim_{n\to\infty} \varepsilon_n = 0$).

Combining propositions 9.8, 9.7, and 9.6 we obtain

**Theorem 9.6.** *Let $E$ be a separable Banach space such that the conjugate space $E^*$ has the bounded approximation property. Then $E$ is isomorphic to a Banach space having the metric approximation property.*

From theorems 9.6 and 9.4 it follows

**Corollary 9.4.** *Let $E$ be a separable Banach space. If there exists a constant $\lambda \geqslant 1$ such that $(E, |.|)$ has the $\lambda$-approximation property for each equivalent norm $|.|$ on $E$, then $E$ is isomorphic to a Banach space having the metric approximation property.*

In the particular case when $E^*$ is separable, a stronger result was given in remark 9.8 b).

The notion of an approximative basis admits a natural extension to general topological linear spaces, namely, definition 9.1 makes sense in any such space $U$, and proposition 9.1 remains also valid in this situation. In the particular cases when $U$ is a Banach space $E$ endowed with the weak topology $\sigma(E, E^*)$ or a conjugate space $E^*$ with the weak* topology $\sigma(E^*, E)$, we thus arrive at the notions of *weak approximative basis* of $E$ and *weak* approximative basis* of $E^*$, respectively; the latter one was also considered in the preceding, in a remark to formula (9.8).

By the principle of uniform boundedness, for weak and weak* approximative bases we still have $\sup_{1 \leqslant n < \infty} \|u_n\| = \lambda < \infty$.

The methods used above permit also to obtain some properties of weak duality, collected in

**Theorem 9.7.** *Let $E$ be a separable Banach space. The following statements are equivalent:*

1°. *$E$ has an approximative basis.*

2°. *$E$ has a weak approximative basis.*

3°. *$E^*$ has a weak* approximative basis.*

*Proof.* The implications 1° $\Rightarrow$ 2° $\Rightarrow$ 3° are obvious (see proposition 9.2, formula (9.8)).

Assume now that we have 3°, so let $\{v_n\}$ be a weak* approximative basis of $E^*$. Then $\sup_{1 \leqslant n < \infty} \|v_n\| \leqslant \lambda < \infty$ and for each finite-dimensional subspace $\Gamma$ of $E^*$ and each $n \in \mathcal{N}$ there exists, by lemma 9.3 b), a finite rank operator $t_{\Gamma, \frac{1}{n}}$ on $E$ such that

$$t^*_{\Gamma, \frac{1}{n}}(h) = v_n(h) \qquad (h \in \Gamma), \tag{9.143}$$

$$\|t_{\Gamma, \frac{1}{n}}\| \leqslant \lambda + 1. \tag{9.144}$$

Now let $\mathcal{D}$ be the directed set of all pairs $\left(\Gamma, \frac{1}{n}\right)$, where $\Gamma$ is a finite-dimensional subspace of $E^*$ and $n \in \mathcal{N}$ and where $\left(\Gamma_1, \frac{1}{n_1}\right) \geqslant \left(\Gamma_2, \frac{1}{n_2}\right)$ if and only if $\Gamma_1 \supset \Gamma_2$ and $\frac{1}{n_1} \leqslant \frac{1}{n_2}$. Then, since $v_n(f) \xrightarrow{w^*} f$ $(f \in E^*)$, we obtain, similarly to the above proof of proposition 9.8, $t_d^*(f) \xrightarrow{w^*} f$ $(f \in E^*)$, whence

$$t_d(x) \xrightarrow{w} x \qquad (x \in E). \tag{9.145}$$

Let $\{y_n\}$ be a dense sequence in $E$. Then, by the above proof of proposition 9.8, we can find sequences $\{d_n\} \subset \mathcal{D}$ and $\{m_n\} \subset \mathcal{N}$ with $m_1 < m_2 < \ldots$ and, for each $n$, non-negative numbers $\alpha_{m_{n-1}+1}, \ldots, \alpha_{m_n}$ with $\sum_{i=m_{n-1}+1}^{m_n} \alpha_i = 1$, such that

$$\left\| \sum_{i=m_{n-1}+1}^{m_n} \alpha_i t_{d_i}(y_j) - y_j \right\| < \frac{1}{n} \qquad (j = 1, \ldots, n;\ n = 1,2, \ldots). \tag{9.146}$$

Then for the finite rank operators $u_n = \sum_{i=m_{n-1}+1}^{m_n} \alpha_i t_{d_i}$ we have $\lim_{n\to\infty} u_n(y_j) = y_j$ $(j = 1,2, \ldots)$ and $\|u_n\| \leqslant \sum_{i=m_{n-1}+1}^{m_n} \alpha_i \| t_{d_i} \| \leqslant \lambda + 1$ $(n = 1, 2, \ldots)$, whence (9.2), so $\{u_n\}$ is an approximative basis of $E$. Thus, $3^\circ \Rightarrow 1^\circ$, which completes the proof of theorem 9.7.

*Remark 9.14*. By (part of) proposition 9.8, if $E^*$ has an approximative basis, then so does $E$. The implication $3^\circ \Rightarrow 1^\circ$ of theorem 9.7 also yields this result, since if $E^*$ has an approximative basis $\{v_n\}$, then $E$ is separable and $\{v_n\}$ is a weak* approximative basis of $E^*$.

## § 10. Operational bases. Generalized summation bases

*Definition 10.1*. A complete minimal sequence $\{x_n\}$ in a Banach space $E$ is said to be an *operational basis* of $E$ if there exists a sequence of endomorphisms $v_n \in L(P_{(n)}, P_{(n)})$ $(n = 1,2, \ldots)$, where $P_{(n)} = [x_1, \ldots, x_n]$ $(n = 1,2, \ldots)$, such that

$$x = \lim_{n\to\infty} v_n s_n(x) \qquad (x \in E), \qquad (10.1)$$

where $\{s_n\}$ is the sequence of partial sum operators associated to the complete minimal sequence $\{x_n\}$. In this case we shall also call $\{x_n\}$ an *operational basis of E with respect to the sequence of endomorphisms* $\{v_n\}$.

If $\{f_n\} \subset E^*$ is the a.s.f. to $\{x_n\}$, we can write (10.1) in the form

$$x = \lim_{n\to\infty} v_n\left(\sum_{i=1}^{n} f_i(x) x_i\right) = \lim_{n\to\infty} \sum_{i=1}^{n} f_i(x) v_n(x_i) \qquad (x \in E). \qquad (10.2)$$

In particular, for $x = x_j$ we obtain, by biorthogonality,

$$\lim_{n\to\infty} v_n(x_j) = x_j \qquad (j = 1,2, \ldots). \qquad (10.3)$$

From definition 10.1 it follows that *every operational basis* $\{x_n\}$ *of E is an approximative basis of E* (since we have (9.2) with $u_n = v_n s_n$ and $\dim u_n(E) = \dim v_n s_n(E) \leqslant n < \infty$). Furthermore, we have

**Proposition 10.1.** *Every operational basis* $\{x_n\}$ *of a Banach space E is a norming**) *M-basis of E.*

*Proof*. Let $\{f_n\} \subset E^*$, $f_i(x_j) = \delta_{ij}$ $(i, j = 1,2, \ldots)$ and let $v_n$ be as in definition 10.1. If $x \in E$, $f_i(x) = 0$ $(i = 1,2, \ldots)$, then, by (10.2),

*) See § 8, definition 8.2.

$x = 0$. Thus, $\{x_n\}$ is an $M$-basis of $E$. Since $v_n \in L(P_{(n)}, P_{(n)})$, we can write

$$v_n(x_i) = \sum_{j=1}^{n} a_{ij}^{(n)} x_j \qquad (i = 1, \dots, n; n = 1, 2, \dots), \tag{10.4}$$

where

$$a_{ij}^{(n)} = f_j(v_n(x_i)) \qquad (i, j = 1, \dots, n;\ n = 1,2, \dots).^{*)} \tag{10.5}$$

Then

$$v_n s_n(x) = v_n\left(\sum_{i=1}^{n} f_i(x) x_i\right) = \sum_{j=1}^{n}\left(\sum_{i=1}^{n} a_{ij}^{(n)} f_i(x)\right) x_j \quad (x \in E,\ n = 1, 2, \dots), \tag{10.6}$$

whence for the $h_{ni}$ of § 9, proposition 9.2, associated to $u_n = v_n s_n$, we can take

$$h_{nj} = \sum_{i=1}^{n} a_{ij}^{(n)} f_i \qquad (j = 1, \dots, n;\ n = 1, 2, \dots). \tag{10.7}$$

Thus, $h_{nj} \in [f_i]_{i=1}^{n}$ $(j = 1, \dots, n;\ n = 1,2, \dots)$, whence, by § 9, proposition 9.2,

$$r([f_i]_{i=1}^{\infty}) \geqslant r([h_{nj}]_{\substack{j=1,\dots,n \\ n=1,2,\dots}}) > 0, \tag{10.8}$$

i.e., $\{x_n\}$ is a norming $M$-basis, which completes the proof.**)

The converse of proposition 10.1 is not valid, since there exist separable Banach spaces which have no approximative basis and hence no operational basis (see § 0), but every separable Banach space has a norming $M$-basis (by § 8, corollary 8.1). Of course, from proposition 10.1 it follows that *if* $\{x_n\}$ *is a non-norming M-basis* (we shall see in Vol. III that such $M$-bases exist in every separable non-quasi-reflexive***) Banach space $E$), *then* $\{x_n\}$ *is not an operational basis.*

We have the following property of weak duality:

**Proposition 10.2.** *Let* $\{x_n\}$ *be an operational basis of a Banach space* $E$, *with the a.s.f.* $\{f_n\} \subset E^*$. *Then* $\{f_n\}$ *is a "w*-operational basis" of*

---

*) Note also that by (10.5) and (10.3) we have $a_{ij}^{(n)} \to \delta_{ij}$ as $n \to \infty$ $(i, j = 1, 2, \dots)$.

**) We have also $r([f_i]) \geqslant \dfrac{1}{\sup\limits_{1 \leqslant n < \infty} \|v_n s_n\|} > 0$ (by $(v_n s_n)^*(P_{(n)}^*) \subset [f_i]$ and the proof of § 15, theorem 15.7).

***) I.e., such that $\dim E^{**}/\pi(E) = \infty$, where $\pi$ is the canonical embedding of $E$ into $E^{**}$.

*$E^*$, i.e. there exists a sequence of endomorphisms $w_n \in L(\Gamma_{(n)}, \Gamma_{(n)})$ $(n = 1,2, \ldots)$, where $\Gamma_{(n)} = [f_1, \ldots, f_n]$ $(n = 1,2, \ldots)$, such that*

$$f(x) = \lim_{n\to\infty} (w_n s_n^*(f))(x) \qquad (x \in E, f \in E^*), \tag{10.9}$$

*where $s_n(x) = \sum_{i=1}^{n} f_i(x)x_i$ $(x \in E, \ n = 1, 2, \ldots)$. Hence, in particular,*

$$\lim_{n\to\infty} (w_n(f_j))(x) = \lim_{n\to\infty} (w_n s_n^*(f_j))(x) = f_j(x) \qquad (x \in E, j = 1,2, \ldots). \tag{10.10}$$

*Proof.* Put

$$w_n(f_j) = \sum_{i=1}^{n} a_{ij}^{(n)} f_i = \sum_{i=1}^{n} f_j(v_n(x_i))f_i \quad (j=1, \ldots, n;\ n=1,2, \ldots), \tag{10.11}$$

and extend each $w_n$ by linearity to $[f_1, \ldots, f_n]$. Then, by (10.6) and (10.1),

$$(w_n s_n^*(f))(x) = \left(w_n\left(\sum_{j=1}^{n} f(x_j)f_j\right)\right)(x) = \sum_{j=1}^{n} (w_n(f_j)(x))f(x_j) =$$

$$= \sum_{j=1}^{n} \left(\sum_{i=1}^{n} a_{ij}^{(n)} f_i(x)\right) f(x_j) = f(v_n s_n(x)) \to f(x) \text{ as } n\to\infty \quad (x \in E, f \in E^*),$$

which completes the proof of proposition 10.2.

Note that for each $n$ the operator $w_n$ defined by (10.11) satisfies

$$(w_n(f_j))(x_k) = \begin{cases} f_j(v_n(x_k)) = (v_n^*(f_j|_{[x_1,\ldots,x_n]}))(x_k) & (j, k=1, \ldots, n) \\ 0 \quad (j = 1, \ldots, n;\ k = n+1,\ n+2, \ldots). \end{cases} \tag{10.12}$$

From the above computations it also follows the formula

$$w_n s_n^*(f) = (v_n s_n)^*(f|_{[x_1, \ldots, x_n]}) \quad (f \in E^*, n = 1,2, \ldots). \tag{10.13}$$

However, we shall see in example 10.1 below that in general operational bases $\{x_n\}$ do not have the natural property of strong duality, i.e., the a.s.f. $\{f_n\}$ to $\{x_n\}$ need not be an operational basis of $[f_n]$. Let us first prove

**Theorem 10.1.** *A Banach space $E$ has an operational basis if and only if it has an approximative basis.*

*Proof.* The condition is obviously necessary, since every operational basis is an approximative basis, as was observed before proposition 10.1.

Conversely, assume that there exists a sequence of endomorphisms $\{u_n\} \subset L(E, E)$ of finite rank such that

$$x = \lim_{n\to\infty} u_n(x) \qquad (x \in E). \tag{10.14}$$

Then, by § 8, theorem 8.1, $E$ has an $M$-basis $\{x_n\}$ such that

$$\bigcup_{n=1}^{\infty} u_n^*(E^*) \subset [f_n], \tag{10.15}$$

where $\{f_n\} \subset E^*$, $f_i(x_j) = \delta_{ij}$ $(i, j = 1,2, \ldots)$. We shall show that $\{x_n\}$ is an operational basis of $E$, which will complete the proof.

Since $\dim u_n(E) < \infty$ $(n = 1,2, \ldots)$, we can write

$$u_n(x) = \sum_{i=1}^{p_n} h_{ni}(x) y_{ni} \qquad (x \in E, n = 1, 2, \ldots), \tag{10.16}$$

where $\{y_{ni}\}_{i=1}^{p_n}$ is a basis of $u_n(E)$ and $\{h_{ni}\}_{i=1}^{p_n} \subset E^*$ for each $n$. Let $\{g_{nj}\}_{j=1}^{p_n} \subset E^*$, $g_{nj}(y_{ni}) = \delta_{ij}$ $(i, j = 1, \ldots, p_n)$. Then

$$u_n^*(g_{nj}) = \sum_{i=1}^{p_n} g_{nj}(y_{ni}) h_{ni} = h_{nj} \qquad (j = 1, \ldots, p_n)$$

and hence, by (10.15),

$$h_{nj} \in u_n^*(E^*) \subset [f_k] \qquad (j = 1, \ldots, p_n;\ n = 1, 2, \ldots). \tag{10.17}$$

Let $n$ be arbitrary. Then for any $\varepsilon > 0$ there is a positive integer $m_n = m_n(\varepsilon)$ such that for each $i = 1, \ldots, p_n$ there exist $\bar{y}_{ni} \in [x_1, \ldots, x_{m_n}]$ and $\bar{h}_{ni} \in [f_1, \ldots, f_{m_n}]$ satisfying

$$\|y_{ni} - \bar{y}_{ni}\| < \varepsilon, \ \|h_{ni} - \bar{h}_{ni}\| < \varepsilon \qquad (i = 1, \ldots, p_n). \tag{10.18}$$

Put

$$\bar{v}_{m_n}(x) = \sum_{i=1}^{p_n} \bar{h}_{ni}(x) \bar{y}_{ni} \qquad (x \in E). \tag{10.19}$$

Then we have

$$\|\bar{v}_{m_n}(x) - u_n(x)\| = \left\| \sum_{i=1}^{p_n} (\bar{h}_{ni}(x) - h_{ni}(x))\bar{y}_{ni} + \sum_{i=1}^{p_n} h_{ni}(x)(\bar{y}_{ni} - y_{ni}) \right\| \leqslant$$

$$\leqslant \left( \sum_{i=1}^{p_n} \|\bar{h}_{ni} - h_{ni}\| \, \|\bar{y}_{ni}\| + \sum_{i=1}^{p_n} \|h_{ni}\| \, \|\bar{y}_{ni} - y_{ni}\| \right) \|x\| \qquad (x \in E),$$

and hence, by (10.18), we can assure (taking $\varepsilon > 0$ sufficiently small) that

$$\|\bar{v}_{m_n} - u_n\| < \frac{1}{n}; \tag{10.20}$$

obviously, we may also take the sequence $\{m_n\}$ to be increasing.

Observe now that by $\bar{h}_{ni} \in [f_1, \ldots, f_{m_n}]$ and $f_i(x - s_k(x)) = 0$ $(x \in E,\ i = 1, \ldots, m_n;\ k = m_n,\ m_n + 1, \ldots)$ we have

$$\bar{v}_{m_n} s_k(x) = \sum_{i=1}^{p_n} \bar{h}_{ni}(s_k(x))\bar{y}_{ni} = \sum_{i=1}^{p_n} \bar{h}_{ni}(x)\bar{y}_{ni} = \bar{v}_{m_n}(x) \quad (x \in E, k \geqslant m_n), \tag{10.21}$$

whence, by (10.20) and (10.14), we obtain

$$\lim_{n\to\infty} \bar{v}_{m_n} s_{m_n}(x) = \lim_{n\to\infty} \bar{v}_{m_n}(x) = \lim_{n\to\infty} u_n(x) = x \qquad (x \in E). \tag{10.22}$$

Let us put

$$v_k = s_k|_{[x_1, \ldots, x_k]} \qquad (k=1, \ldots, m_1-1), \tag{10.23}$$

$$v_k = \bar{v}_{m_n}|_{[x_1, \ldots, x_k]} \quad (k=m_n, m_n+1, \ldots, m_{n+1}-1;\ n=1,2, \ldots). \tag{10.24}$$

Then each $v_k$ is a continuous linear operator defined on $[x_1, \ldots, x_k]$, with the range

$$v_k([x_1, \ldots, x_k]) = [x_1, \ldots, x_k] \qquad (k = 1, \ldots, m_1-1),$$

$$v_k([x_1, \ldots, x_k]) = \bar{v}_{m_n}([x_1, \ldots, x_k]) \subset [\bar{y}_{n1}, \ldots, \bar{y}_{np_n}] \subset [x_1, \ldots, x_{m_n}] \subset$$

$$\subset [x_1, \ldots, x_k] \qquad (k = m_n,\ m_n + 1, \ldots, m_{n+1}-1;\ n = 1, 2, \ldots,),$$

and by (10.21), (10.22) we have (10.1), which completes the proof of theorem 10.1.

Actually, the above argument shows that *if* $\{u_n\} \subset L(E, E)$ *is a sequence of endomorphisms of finite rank satisfying* (10.14), *then every* $M$*-basis* $\{x_n\}$ *of* $E$ *with the a.s.f.* $\{f_n\} \subset E^*$ *satisfying* (10.15) (*in particular, every shrinking* $M$*-basis of* $E$*, if it exists*), *is an operational basis of* $E$. Hence, since the conjugate space $E^*$ of a reflexive Banach space $E$ contains no proper total subspace, we obtain

**Corollary 10.1.** *Let* $E$ *be a reflexive Banach space which has an approximative basis. Then every* $M$*-basis of* $E$ *is an operational basis of* $E$.

In contrast with corollary 10.1, we have observed after proposition 10.1 that every separable non-quasi-reflexive Banach space $E$ has an $M$-basis which is not an operational basis of $E$. For the particular case when $E$ is a separable Hilbert space, we shall give a sharpening of corollary 10.1 in theorem 10.5 below.

Now we can give the example, announced above, that in general operational bases do not have the natural property of strong duality.

*Example 10.1.* By § 0, example 0.1, let $E$ be a Banach space with a basis, such that $E^*$ is separable but fails to have the approximation property and, by § 8, theorem 8.1, let $\{x_n\}$ be a shrinking $M$-basis of $E$, with the a.s.f. $\{f_n\} \subset E^*$. Then $E$ has an approximative basis and any approximative basis $\{u_n\}$ of $E$ satisfies (10.15) (since $[f_n] = E^*$), so $\{x_n\}$ is an operational basis of $E$ (by the remark made after theorem 10.1). However, $[f_n] = E^*$ has no operational basis whatsoever.

In § 9, remark 9.2, we have observed that by § 5, theorem 5.2 c), every separable Banach space has a "non-linear quasi-basis". Similarly, we shall consider now "non-linear operational bases", i.e. sequences $\{x_n\}$ as in definition 10.1, with $v_n\colon P_{(n)} \to P_{(n)}$ $(n = 1,2, \ldots)$ continuous (but not necessarily linear) operators and we shall prove that they exist in every separable Banach space. To this end, we need

**Lemma 10.1.** *Let* $E$ *be a separable normed linear space and let* $\{f_n\}$ *be a sequence in* $E^*$ *such that* $\frac{f_n}{\|f_n\|} \xrightarrow{w^*} 0$ *and that for the linear subspace* $V$ *of* $E^*$ *spanned by* $\{f_n\}$ *we have* $r(V) > 0$. *Then there exists a norm* $|.|$ *on* $E$*, equivalent to the initial norm* $\|.\|$ *on* $E$*, such that* $(E, |.|)$ *is strictly convex and has the following two properties:*

(K₁) *If* $\lim_{n\to\infty} f_k(x_n) = f_k(x_0)$ $(k = 1, 2, \ldots)$, *then* $\varliminf_{n\to\infty} |x_n| \geqslant |x_0|$.

(K₂) *If* $\lim_{n\to\infty} f_k(x_n) = f_k(x_0)$ $(k = 1,2, \ldots)$ *and* $\lim_{n\to\infty} |x_n| = |x_0|$,

*then* $\lim_{n\to\infty} |x_n - x_0| = 0$.

*Proof.* By Ch. II, § 16, lemma 16.3, there exists a norm $|||\,.\,|||$ on $E$, equivalent to the initial norm $\|\,.\,\|$ on $E$ and such that $(E, |||\,.\,|||)$ has the properties $(K_1)$ and $(K_2)$. Put

$$|x| = \left(|||x|||^2 + \sum_{k=1}^{\infty} \frac{|f_k(x)|^2}{2^k |||f_k|||^2}\right)^{\frac{1}{2}} \qquad (x \in E), \tag{10.25}$$

where $|||f_k||| = \sup\limits_{\substack{x \in E \\ |||x||| \leqslant 1}} |f_k(x)|$. Then $|\,.\,|$ is a norm on $E$, satisfying

$$|||x||| \leqslant |x| \leqslant \left(|||x|||^2 + \sum_{k=1}^{\infty} \frac{|||f_k|||^2\,|||x|||^2}{2^k |||f_k|||^2}\right)^{\frac{1}{2}} = \sqrt{2}\,|||x||| \qquad (x \in E),$$

so $|\,.\,|$ is equivalent to the norm $|||\,.\,|||$, whence also to the initial norm $\|\,.\,\|$ on $E$. Assume now that $x, y \in E \setminus \{0\}$, $|x + y| = |x| + |y|$. Then, by the triangle inequality in $l^2$,

$$\left(|||x|||^2 + \sum_{k=1}^{\infty} \frac{|f_k(x)|^2}{2^k |||f_k|||^2}\right)^{\frac{1}{2}} + \left(|||y|||^2 + \sum_{k=1}^{\infty} \frac{|f_k(y)|^2}{2^k |||f_k|||^2}\right)^{\frac{1}{2}} = |x| + |y| =$$

$$= |x + y| = \left(|||x + y|||^2 + \sum_{k=1}^{\infty} \frac{|f_k(x) + f_k(y)|^2}{2^k |||f_k|||^2}\right)^{\frac{1}{2}} \leqslant$$

$$\leqslant \left((|||x||| + |||y|||)^2 + \sum_{k=1}^{\infty} \frac{|f_k(x) + f_k(y)|^2}{2^k |||f_k|||^2}\right)^{\frac{1}{2}} \leqslant$$

$$\leqslant \left(|||x|||^2 + \sum_{k=1}^{\infty} \frac{|f_k(x)|^2}{2^k |||f_k|||^2}\right)^{\frac{1}{2}} + \left(|||y|||^2 + \sum_{k=1}^{\infty} \frac{|f_k(y)|^2}{2^k |||f_k|||^2}\right)^{\frac{1}{2}},$$

whence

$$\left((|||x||| + |||y|||)^2 + \sum_{k=1}^{\infty} \frac{|f_k(x) + f_k(y)|^2}{2^k |||f_k|||^2}\right)^{\frac{1}{2}} = \left(|||x|||^2 + \sum_{k=1}^{\infty} \frac{|f_k(x)|^2}{2^k |||f_k|||^2}\right)^{\frac{1}{2}} +$$

$$+ \left(|||y|||^2 + \sum_{k=1}^{\infty} \frac{|f_k(y)|^2}{2^k |||f_k|||^2}\right)^{\frac{1}{2}}.$$

Consequently, by the strict convexity of $l^2$, there exists a constant $c > 0$ such that

$$cf_k(x) = f_k(y) \qquad (k = 1, 2, \ldots),$$

whence, since $\{f_n\}$ is total on $E$, $cx = y$, which proves that $(E, |.|)$ is strictly convex.

Now let $\lim_{n\to\infty} f_k(x_n) = f_k(x_0)$ $(k = 1,2, \ldots)$. Then, since the norm $|||.|||$ has property $(K_1)$, $\varliminf_{n\to\infty} |||x_n||| \geqslant |||x_0|||$ and hence for any $\varepsilon > 0$ there exists a positive integer $N_1 = N_1(\varepsilon)$ such that

$$|||x_n|||^2 > |||x_0|||^2 - \frac{\varepsilon}{2} \qquad (n > N_1). \tag{10.26}$$

Furthermore, since $\frac{f_n}{\|f_n\|} \xrightarrow{w^*} 0$, we have also $\frac{f_n}{|||f_n|||} \xrightarrow{w^*} 0$, and hence there exists a positive integer $p = p(\varepsilon)$ such that

$$\frac{|f_k(x_0)|^2}{|||f_k|||^2} < \frac{\varepsilon}{2} \qquad (k > p).$$

Also, since $\lim_{n\to\infty} \frac{f_k}{|||f_k|||}(x_n) = \frac{f_k}{|||f_k|||}(x_0)$ $(k = 1, \ldots, p)$, there exists a positive integer $N_2 = N_2(\varepsilon, p(\varepsilon))$ such that

$$\frac{|f_k(x_n)|^2}{|||f_k|||^2} > \frac{|f_k(x_0)|^2}{|||f_k|||^2} - \frac{\varepsilon}{2} \qquad (n > N_2;\ k = 1, \ldots, p).$$

Then for all $n > N_2$ we have

$$\sum_{k=1}^{\infty} \frac{|f_k(x_n)|^2}{2^k|||f_k|||^2} \geqslant \sum_{k=1}^{p} \frac{|f_k(x_n)|^2}{2^k|||f_k|||^2} > \sum_{k=1}^{p} \frac{1}{2^k}\left(\frac{|f_k(x_0)|^2}{|||f_k|||^2} - \frac{\varepsilon}{2}\right) >$$

$$> \sum_{k=1}^{\infty} \frac{1}{2^k}\frac{|f_k(x_0)|^2}{|||f_k|||^2} - \sum_{k=p+1}^{\infty} \frac{1}{2^k}\frac{\varepsilon}{2} - \sum_{k=1}^{p} \frac{\varepsilon}{2^{k+1}} = \sum_{k=1}^{\infty} \frac{1}{2^k}\frac{|f_k(x_0)|^2}{|||f_k|||^2} - \frac{\varepsilon}{2}, \tag{10.27}$$

and hence, by (10.25) and (10.26),

$$|x_n|^2 = |||x_n|||^2 + \sum_{k=1}^{\infty} \frac{|f_k(x_n)|^2}{2^k|||f_k|||^2} > |||x_0|||^2 + \sum_{k=1}^{\infty} \frac{1}{2^k}\frac{|f_k(x_0)|^2}{|||f_k|||^2} - \varepsilon =$$

$$= |x_0|^2 - \varepsilon \qquad (n > N = \max(N_1, N_2)).$$

Consequently, $\underline{\lim}_{n\to\infty} |x_n| \geqslant |x_0|$, which proves that the norm $|.|$ has property $(K_1)$.

Finally, let $\lim_{n\to\infty} f_k(x_n) = f_k(x_0)$ $(k = 1,2, \ldots)$ and $\lim_{n\to\infty} |x_n| = |x_0|$, hence

$$\lim_{n\to\infty} \left( |||x_n|||^2 + \sum_{k=1}^{\infty} \frac{|f_k(x_n)|^2}{2^k |||f_k|||^2} \right) = |||x_0|||^2 + \sum_{k=1}^{\infty} \frac{|f_k(x_0)|^2}{2^k |||f_k|||^2}. \tag{10.28}$$

Let $\varepsilon > 0$ be arbitrary and let $N_3 = N_3(\varepsilon)$ be such that

$$|||x_0|||^2 + \sum_{k=1}^{\infty} \frac{|f_k(x_0)|^2}{2^k |||f_k|||^2} + \frac{\varepsilon}{2} \geqslant |||x_n|||^2 + \sum_{k=1}^{\infty} \frac{|f_k(x_n)|^2}{2^k |||f_k|||^2} \qquad (n > N_3).$$

Then, by (10.26), we obtain

$$|||x_0|||^2 + \sum_{k=1}^{\infty} \frac{|f_k(x_0)|^2}{2^k |||f_k|||^2} + \frac{\varepsilon}{2} \geqslant |||x_0|||^2 - \frac{\varepsilon}{2} + \sum_{k=1}^{\infty} \frac{|f_k(x_n)|^2}{2^k |||f_k|||^2}$$

$$(n > \max\ (N_1, N_3)),$$

whence

$$\sum_{k=1}^{\infty} \frac{|f_k(x_0)|^2}{2^k |||f_k|||^2} + \varepsilon \geqslant \sum_{k=1}^{\infty} \frac{|f_k(x_n)|^2}{2^k |||f_k|||^2} \qquad (n > \max\ (N_1, N_3)),$$

which, together with (10.27), implies

$$\lim_{n\to\infty} \sum_{k=1}^{\infty} \frac{|f_k(x_n)|^2}{2^k |||f_k|||^2} = \sum_{k=1}^{\infty} \frac{|f_k(x_0)|^2}{2^k |||f_k|||^2}. \tag{10.29}$$

Therefore, by (10.28), $\lim_{n\to\infty} |||x_n||| = |||x_0|||$, whence, since the norm $|||.|||$ has property $(K_2)$, $\lim_{n\to\infty} |||x_n - x_0||| = 0$ and hence, since the norms $|||.|||$ and $|.|$ are equivalent, we infer $\lim_{n\to\infty} |x_n - x_0| = 0$. Thus, the norm $|.|$ has property $(K_2)$, which completes the proof of lemma 10.1.

Now we can prove

**Theorem 10.2.** *Let $E$ be a separable Banach space and let $\{x_n\}$ be a norming M-basis of $E$. Then $\{x_n\}$ is a non-linear operational basis of $E$.*

*Proof.* Let $\{f_n\} \subset E^*$ be the a.s.f. to $\{x_n\}$ and let $|.|$ be an equivalent norm on $E$ as in lemma 10.1 (observe that $\frac{f_n}{\|f_n\|} \xrightarrow{w^*} 0$ by § 6, proposition 6.5, implication 1° ⇒ 2°). Then $(E, |.|)$ is strictly convex and hence for every finite-dimensional subspace $G$ of $E$ and every $x \in E \setminus G$ there exists a unique*) element $\pi_G(x) \in G$ such that

$$|x - \pi_G(x)| = \operatorname{dist}(x, G) = \min_{g \in G} |x - g| \tag{10.30}$$

and the mapping $\pi_G \colon E \to G$ is continuous; in general, $\pi_G$ is non-linear.

By § 6, proposition 6.10, there exists an increasing sequence of positive integers

$$m_1 = N(1,1),\ m_2 = N\left(m_1, \frac{1}{2}\right),\ m_3 = N\left(m_2, \frac{1}{3}\right), \ldots \tag{10.31}$$

such that

$$\operatorname{dist}(y, [x_i]_{m_{n-1}+1}^{m_n}) \leqslant \left(1 + \frac{1}{n}\right) \operatorname{dist}(y, [x_i]_{m_{n-1}+1}^{\infty})$$

$$(y \in [x_i]_1^{m_{n-1}};\ n = 1, 2, \ldots;\ m_0 = 1). \tag{10.32}$$

Let us put

$$v_k = s_k|_{[x_1, \ldots, x_k]} \qquad (k = 1, \ldots, m_1 - 1), \tag{10.33}$$

$$v_k\left(\sum_{i=1}^{k} \alpha_i x_i\right) = \sum_{i=1}^{m_{n-1}} \alpha_i x_i - \pi_{[x_i]_{m_{n-1}+1}^{m_n}}\left(\sum_{i=1}^{m_{n-1}} \alpha_i x_i\right)$$

$$\left(\sum_{i=1}^{k} \alpha_i x_i \in [x_i]_1^k;\ k = m_n, m_n + 1, \ldots, m_{n+1} - 1;\ n = 1, 2, \ldots\right). \tag{10.34}$$

Then each $v_k$ is a continuous (in general, non-linear) operator defined on $[x_1, \ldots, x_k]$, with the range

$$v_k([x_1, \ldots, x_k]) = [x_1, \ldots, x_k] \qquad (k = 1, \ldots, m_1 - 1),$$

$$v_k([x_1, \ldots, x_k]) \subset [x_1, \ldots, x_{m_n}] \subset [x_1, \ldots, x_k]$$

$$(k = m_n, m_n + 1, \ldots, m_{n+1} - 1;\ n = 1, 2, \ldots),$$

and it remains to prove that we have (10.1). Let $x \in E$ be arbitrary. Then, since

$$v_k s_k(x) - s_k(x) \in [x_i]_{m_{n-1}+1}^{k} \qquad (k = m_n, m_n + 1, \ldots, m_{n+1} - 1;\ n = 1, 2, \ldots),$$

*) See e.g. [371], Ch. I, § 3, corollary 3.3 and Ch. II, § 5, theorem 5.4.

we have, by biorthogonality,

$$f_i(v_k s_k(x)) = f_i(s_k(x)) = f_i(x) \qquad (i = 1, \ldots, m_{n-1};$$
$$k = m_n, m_n + 1, \ldots, m_{n+1} - 1;\ n = 1, 2, \ldots),$$

whence

$$\lim_{k\to\infty} f_i(v_k s_k(x)) = f_i(x) \qquad (i = 1, 2, \ldots), \tag{10.35}$$

and therefore, since the norm $|.|$ has property $(K_1)$,

$$\varliminf_{k\to\infty} |v_k s_k(x)| \geqslant |x|. \tag{10.36}$$

On the other hand, by (10.32) we have

$$|v_k s_k(x)| = |s_{m_{n-1}}(x) - \pi_{[x_i]_{m_{n-1}+1}^{m_n}}(s_{m_{n-1}}(x))| = \operatorname{dist}(s_{m_{n-1}}(x), [x_i]_{m_{n-1}+1}^{m_n}) \leqslant$$
$$\leqslant \left(1 + \frac{1}{n}\right) \operatorname{dist}(s_{m_{n-1}}(x), [x_i]_{m_{n-1}+1}^{\infty}) \leqslant$$
$$\leqslant \left(1 + \frac{1}{n}\right) |s_{m_{n-1}}(x) + (x - s_{m_{n-1}}(x))| = \left(1 + \frac{1}{n}\right) |x|$$
$$(k = m_n, m_n + 1, \ldots, m_{n+1} - 1;\ n = 1, 2, \ldots),$$

which, together with (10.36), implies

$$\lim_{k\to\infty} |v_k s_k(x)| = |x|. \tag{10.37}$$

Since the norm $|.|$ has property $(K_2)$, from (10.35) and (10.37) it follows that $\lim_{k\to\infty} |v_k s_k(x) - x| = 0$, whence, since the norm $|.|$ is equivalent to the initial norm $\|.\|$ on $E$, we infer (10.1), which completes the proof of theorem 10.2.

From theorem 10.2 and § 8, corollary 8.1, it follows

**Corollary 10.2.** *Every separable Banach space $E$ has a non-linear operational basis $\{x_n\}$.*

Let us return now to (linear) operational bases. The following extension of the notion of unconditional basis is suggested by Ch. II, § 17, theorem 17.1 (equivalence $1° \Leftrightarrow 2°$):

*Definition 10.2.* An operational basis $\{x_n\}$ of a Banach space $E$ with respect to the sequence of endomorphisms $v_n \in L(P_{(n)}, P_{(n)})$ $(n = 1,2, \ldots)$ is said to be an *unconditional operational basis* of $E$ if every permutation $\{x_{\sigma(n)}\}$ of $\{x_n\}$ is an operational basis of $E$ with respect to the sequence of those endomorphisms $v_n^\sigma \in L([x_{\sigma(1)}, \ldots, x_{\sigma(n)}], [x_{\sigma(1)}, \ldots, x_{\sigma(n)}])$ $(n = 1,2, \ldots)$ whose matrices in the bases $\{x_{\sigma(i)}\}_{i=1}^n$ coincide with $(a_{ij}^{(n)}) = (f_j(v_n(x_i)))$ $(n = 1,2, \ldots)$, i.e.

$$x = \lim_{n\to\infty} v_n^\sigma s_{\sigma, n}(x) \qquad (x \in E), \tag{10.38}$$

where*)

$$v_n^\sigma(x_{\sigma(i)}) = \sum_{j=1}^{n} a_{ij}^{(n)} x_{\sigma(j)} = \sum_{j=1}^{n} f_j(v_n(x_i)) x_{\sigma(j)} \quad (i=1, \ldots, n;\ n = 1, 2, \ldots), \tag{10.39}$$

$$s_{\sigma, n}(x) = \sum_{=1}^{n} f_{\sigma(i)}(x) x_{\sigma(i)} \qquad (x \in E,\ n = 1, 2, \ldots). \tag{10.40}$$

**Theorem 10.3.** *Every unconditional operational basis $\{x_n\}$ of a Banach space $E$ is an unconditional basis of $E$.*

*Proof.* We may assume, without loss of generality, that $\|x_n\| = 1$ $(n = 1,2, \ldots)$. Let us first prove that $\{x_n\}$ is a basis of $E$. Assume, a contrario, that $\{x_n\}$ is not a basis of $E$. Then, as in Ch. I, § 5, proof of theorem 5.1 (implication 3° $\Rightarrow$ 1°), there exist an increasing sequence of positive integers $\{m_n\}$ with $m_{n+1} \geqslant m_n + m_{n-1} + 1$ $(n = 1,2, \ldots;$ $m_0 = 0)$ and a sequence $\{y_n\} \subset E$, such that

$$y_n = \sum_{i=m_{n-1}+1}^{m_n} f_i(y_n) x_i \qquad (n = 1,2, \ldots; m_0 = 0), \tag{10.41}$$

$$\|y_n\| \leqslant \frac{1}{2^n} \qquad (n = 1, 2, \ldots), \tag{10.42}$$

$$\max_{m_{n-1}+1 \leqslant k \leqslant m_n} \left\| \sum_{i=m_{n-1}+1}^{k} f_i(y_n) x_i \right\| = \left\| \sum_{i=m_{n-1}+1}^{k_n} f_i(y_n) x_i \right\| = 2^n$$
$$(n = 1, 2, \ldots), \tag{10.43}$$

where $k_n$ are suitable integers with $m_{n-1} + 1 \leqslant k_n \leqslant m_n$ $(n = 1,2, \ldots;$ $m_0 = 0)$. Put

$$z_n = \sum_{i=m_{n-1}+1}^{k_n} f_i(y_n) x_i \qquad (n = 1, 2, \ldots); \tag{10.44}$$

*) For the operators $s_{\sigma, n}$ see Ch. II, § 16, formula (16.8).

then, by (10.43), $\|z_n\| = 2^n$ $(n = 1,2, \ldots)$. Since $\{x_n\}$ is an operational basis with respect to $\{v_n\}$, we have (10.3), whence, by (10.5), also

$$\lim_{n\to\infty} a_{ij}^{(n)} = \lim_{n\to\infty} f_j(v_n(x_i)) = \delta_{ij} \qquad (i,j = 1, 2, \ldots). \tag{10.45}$$

Consequently, since $\|z_1\| = 2$, there exists a positive integer $m_{n_1} = l_1$, where $n_1 > 1$, such that

$$\|v_{l_1}(z_1)\| > \frac{3}{2}, \tag{10.46}$$

$$\sum_{i=1}^{k_1} |f_i(z_1)| \sum_{j=k_1+1}^{m_1} |a_{ij}^{(l_1)}| < \frac{1}{4}\cdot \tag{10.47}$$

Define a permutation $\sigma_1$ of $\mathcal{N} = \{1,2,3, \ldots\}$ by

$$\sigma_1(i) = \begin{cases} i \text{ for } i=1, \ldots, k_1;\ i=m_1+1, \ldots, m_{n_1};\ i=m_{n_1}+m_1-k_1+1, \ldots \\ m_{n_1}+(i-k_1) \text{ for } i=k_1+1, \ldots, m_1 \\ k_1+(i-m_{n_1}) \text{ for } i=m_{n_1}+1, \ldots, m_{n_1}+m_1-k_1. \end{cases} \tag{10.48}$$

Then, by biorthogonality and by $m_{n_1} + m_1 - k_1 + 1 \leqslant m_{n_1} + m_1 + 1 \leqslant m_{n_1+1}$,

$$\sum_{i=1}^{l_1} |f_{\sigma_1(i)}(y_1)| = \sum_{i=1}^{k_1} |f_i(y_1)| = \sum_{i=1}^{k_1} |f_i(z_1)|, \tag{10.49}$$

$$s_{\sigma_1, l_1}(y_1) = \sum_{i=1}^{l_1} f_{\sigma_1(i)}(y_1)x_{\sigma_1(i)} = \sum_{i=1}^{k_1} f_i(y_1)x_i = z_1, \tag{10.50}$$

$$s_{\sigma_1, m_{n_1+1}}(y_1) = \sum_{i=1}^{m_{n_1+1}} f_{\sigma_1(i)}(y_1)x_{\sigma_1(i)} = \sum_{i=1}^{k_1} f_i(y_1)x_i +$$

$$+ \sum_{i=m_{n_1}+1}^{m_{n_1}+m_1-k_1} f_{k_1+(i-m_{n_1})}(y_1)x_{k_1+(i-m_{n_1})} = \sum_{i=1}^{m_1} f_i(y_1)x_i = y_1. \tag{10.51}$$

By (10.49), $\|x_n\| = 1$ $(n = 1,2, \ldots)$ and (10.47), we have

$$\|v_{l_1}^{\sigma_1}s_{\sigma_1, l_1}(y_1) - v_{l_1}s_{\sigma_1, l_1}(y_1)\| = \left\| \sum_{i=1}^{l_1} f_{\sigma_1(i)}(y_1)\,(v_{l_1}^{\sigma_1}(x_{\sigma_1(i)}) - v_{l_1}(x_{\sigma_1(i)})) \right\| =$$

$$= \left\| \sum_{i=1}^{l_1} f_{\sigma_1(i)}(y_1) \sum_{j=k_1+1}^{l_1} a_{ij}^{(l_1)}(x_{\sigma_1(j)} - x_j) \right\| \leqslant$$

$$\leqslant \sum_{i=1}^{l_1} |f_{\sigma_1(i)}(y_1)| \sum_{j=k_1+1}^{m_1} |a_{ij}^{(l_1)}|\, \|x_{\sigma_1(j)} - x_j\| < \frac{1}{2},$$

whence, by (10.50) and (10.46), we obtain

$$\|v_{l_1}^{\sigma_1} s_{\sigma_1, l_1}(y_1)\| > 1. \tag{10.52}$$

Similarly, since by our assumption $\{x_{\sigma_1(n)}\}$ is an operational basis with respect to $\{v_n^{\sigma_1}\}$, and since $\|y_1+z_{n_1+2}\| \geqslant \|z_{n_1+2}\| - \|y_1\| > 2^{n_1+2} - \frac{1}{2}$, there exists a positive integer $m_{n_2} = l_2$, where $n_2 > n_1 + 2$, such that

$$\|v_{l_2}^{\sigma_1}(y_1 + z_{n_1+2})\| > 2^{n_1+2} - \frac{3}{4}, \tag{10.53}$$

$$\sum_{i=1}^{k_{n_1}+2} |f_i(y_1 + z_{n_1+2})| \sum_{j=k_{n_1}+2+1}^{m_{n_1}+2} |a_{ij}^{(l_2)}| < \frac{1}{8}. \tag{10.54}$$

Define a permutation $\sigma_2$ of $\mathcal{N} = \{1, 2, 3, \ldots\}$ by

$$\sigma_2(i) = \begin{cases} \sigma_1(i) \text{ for } i=1,\ldots,k_{n_1+2};\ i=m_{n_1+2}+1,\ldots,m_{n_2};\ i=m_{n_2}+m_{n_1+2}-k_{n_1+2}+1,\ldots \\ m_{n_2}+(i-k_{n_1+2}) \text{ for } i=k_{n_1+2}+1, \ldots, m_{n_1+2} \\ k_{n_1+2}+(i-m_{n_2}) \text{ for } i=m_{n_2}+1, \ldots, m_{n_2}+m_{n_1+2}-k_{n_1+2}. \end{cases} \tag{10.55}$$

Then, by biorthogonality, (10.51) and $m_{n_2} + m_{n_1+2} - k_{n_1+2} + 1 \leqslant \leqslant m_{n_2} + m_{n_2-1} + 1 \leqslant m_{n_2+1}$,

$$\sum_{i=1}^{l_2} |f_{\sigma_2(i)}(y_1 + y_{n_1+2})| = \sum_{i=1}^{k_{n_1}+2} |f_{\sigma_1(i)}(y_1 + y_{n_1+2})| = \sum_{i=1}^{k_{n_1}+2} |f_i(y_1 + z_{n_1+2})|, \tag{10.56}$$

$$s_{\sigma_2, l_2}(y_1 + y_{n_1+2}) = \sum_{i=1}^{l_2} f_{\sigma_2(i)}(y_1 + y_{n_1+2}) x_{\sigma_2(i)} = \sum_{i=1}^{k_{n_1}+2} f_{\sigma_1(i)}(y_1 + y_{n_1+2}) x_{\sigma_1(i)} = y_1 + z_{n_1+2}, \tag{10.57}$$

$$s_{\sigma_2, m_{n_2}+1}(y_1 + y_{n_1+2}) = \sum_{i=1}^{m_{n_2}+1} f_{\sigma_2(i)}(y_1 + y_{n_1+2}) x_{\sigma_2(i)} = \sum_{i=1}^{m_1} f_i(y_1) x_i + \sum_{i=m_{n_1+1}+1}^{k_{n_1}+2} f_i(y_{n_1+2}) x_i + \sum_{i=m_{n_2}+1}^{m_{n_2}+m_{n_1}+2-k_{n_1}+2} f_{k_{n_1+2}+(i-m_{n_2})}(y_{n_1+2}) x_{k_{n_1+2}+(i-m_{n_2})} = y_1 + y_{n_1+2}. \tag{10.58}$$

By (10.56), $\|x_n\| = 1$ $(n = 1,2, \dots)$ and (10.54) we have

$$\|v_{l_2}^{\sigma_2} s_{\sigma_2, l_2}(y_1 + y_{n_1+2}) - v_{l_2}^{\sigma_1} s_{\sigma_2, l_2}(y_1 + y_{n_1+2})\| =$$

$$= \left\| \sum_{i=1}^{l_2} f_{\sigma_2(i)}(y_1 + y_{n_1+2}) \sum_{j=k_{n_1+2}+1}^{m_{n_1+2}} a_{ij}^{(l_2)}(x_{\sigma_2(j)} - x_{\sigma_1(j)}) \right\| < \frac{1}{4},$$

whence, by (10.57) and (10.53), we obtain

$$\|v_{l_2}^{\sigma_2} s_{\sigma_2, l_2}(y_1 + y_{n_1+2})\| > 2^{n_1+2} - 1. \tag{10.59}$$

Continuing in this way indefinitely, we obtain a subsequence $\{m_{n_p}\} = \{l_p\}$ of $\{m_n\}$, where $n_1 > 1$, $n_{p+1} > n_p + 2$ $(p = 1,2, \dots)$ and a sequence of permutations $\{\sigma_p\}$ of $\mathcal{N} = \{1,2,3, \dots\}$, such that

$$\sigma_{p+1}(i) =$$

$$= \begin{cases} \sigma_p(i) \text{ for } i=1,\dots,k_{n_p+2};\ i=m_{n_p+2}+1,\dots,m_{n_{p+1}};\ i=m_{n_{p+1}}+m_{n_p+2}-k_{n_p+2}+1,\dots \\ m_{n_{p+1}}+(i-k_{n_p+2}) \text{ for } i=k_{n_p+2}+1, \dots, m_{n_p+2} \\ k_{n_p+2}+(i-m_{n_{p+1}}) \text{ for } i=m_{n_{p+1}}+1, \dots, m_{n_{p+1}}+m_{n_p+2}-k_{n_p+2} \end{cases} \tag{10.60}$$

$(p = 0,1,2, \dots;\ \sigma_0(i) = i$ for $i = 1,2, \dots;\ n_0 = -1)$ and that

$$\left\| v_{l_p}^{\sigma_p} s_{\sigma_p, l_p}\left(\sum_{j=0}^{p-1} y_{n_j+2}\right) \right\| > 2^{n_{p-1}+2} - 1 \quad (p=1,2,\dots;\ n_0=-1). \tag{10.61}$$

By (10.42) and (10.60) we can define an element $y \in E$ and a permutation $\sigma$ of $\mathcal{N} = \{1,2,3, \dots\}$ by

$$y = \sum_{j=0}^{\infty} y_{n_j+2} \qquad (n_0 = -1), \tag{10.62}$$

$$\sigma(i) = \sigma_p(i) \text{ for } i = m_{n_{p-1}+1} + 1, \dots, m_{n_p+1} \quad (p = 1,2, \dots). \tag{10.63}$$

Then, by (10.41) and biorthogonality,

$$v_{l_1}^{\sigma} s_{\sigma, l_1}(y) = \sum_{i=1}^{l_1} f_{\sigma(i)}\left(\sum_{j=0}^{\infty} y_{n_j+2}\right) v_{l_1}^{\sigma}(x_{\sigma(i)}) = \sum_{=1}^{l_1} f_{\sigma_1(i)}(y_1)\, v_{l_1}^{\sigma_1}(x_{\sigma_1(i)}) =$$

$$= v_{l_1}^{\sigma_1} s_{\sigma_1, l_1}(y_1),$$

$$v_{l_2}^{\sigma} s_{\sigma, l_2}(y) = \sum_{i=1}^{l_2} f_{\sigma(i)}\left(\sum_{j=0}^{\infty} y_{n_j+2}\right) v_{l_2}^{\sigma}(x_{\sigma(i)}) =$$

$$= \sum_{i=1}^{m_{n_1}+1} f_{\sigma_1(i)}(y_1)\, v_{l_2}^{\sigma_1}(x_{\sigma_1(i)}) + \sum_{i=m_{n_1}+1+1}^{l_2} f_{\sigma_2(i)}(y_{n_1+2})\, v_{l_2}^{\sigma_2}(x_{\sigma_2(i)}) =$$

$$= \sum_{i=1}^{l_2} f_{\sigma_2(i)}(y_1 + y_{n_1+2})\, v_{l_2}^{\sigma_2}(x_{\sigma_2(i)}) = v_{l_2}^{\sigma_2} s_{\sigma_2, l_2}(y_1 + y_{n_1+2}),$$

and similarly, by induction,

$$v_{l_p}^{\sigma} s_{\sigma, l_p}(y) = v_{l_p}^{\sigma_p} s_{\sigma_p, l_p}\left(\sum_{j=0}^{p-1} y_{n_j+2}\right) \quad (p = 1,2, \ldots). \qquad (10.64)$$

Hence, by (10.61), we obtain

$$\lim_{p\to\infty} \|v_{l_p}^{\sigma} s_{\sigma, l_p}(y)\| = \infty, \qquad (10.65)$$

in contradiction with the assumption that $\{x_{\sigma(n)}\}$ is an operational basis with respect to $\{v_n^{\sigma}\}$. This proves that $\{x_n\}$ is a basis of $E$.

Now let $\{x_{\tau(n)}\}$ be an arbitrary permutation of $\{x_n\}$. Then, by our assumption, $\{x_{\tau(n)}\}$ is also an unconditional operational basis of $E$ and hence, by the above, $\{x_{\tau(n)}\}$ is a basis of $E$. Consequently, $\{x_n\}$ is an unconditional basis of $E$, which completes the proof of theorem 10.3.

*Remark 10.1.* If we do not require in the definition 10.2 of an unconditional operational basis $\{x_n\}$ that the sequence of matrices $(a_{ij}^{(n)})_{i,j=1}^{n}$ $(n = 1,2, \ldots)$ be the same for all sequences of endomorphisms $v_n^{\sigma}: [x_{\sigma(1)}, \ldots, x_{\sigma(n)}] \to [x_{\sigma(1)}, \ldots, x_{\sigma(n)}]$ $(n = 1,2, \ldots)$, then theorem 10.3 is no longer true, as shown by corollary 10.1.

Now we shall introduce generalized summation bases, which will turn out to constitute a subfamily of the family of all operational bases, but we shall see below that the question whether this is a proper subfamily is still open.

*Definition 10.3.* A sequence $\{x_n\}$ in a Banach space $E$ is said to be a *generalized summation basis* of $E$ if there exists a sequence of endomorphisms $v_n \in L(P_{(n)}, P_{(n)})$ $(n = 1,2, \ldots)$, where $P_{(n)} = [x_1, \ldots, x_n]$ $(n = 1,2, \ldots)$, with the following two properties:

a) We have

$$\lim_{n\to\infty} v_n(x_i) = x_i \quad (i = 1,2, \ldots). \tag{10.66}$$

b) For every $x \in E$ there exists a *unique* sequence of scalars $\{\alpha_n\} \subset K$ such that

$$x = \lim_{n\to\infty} v_n\left(\sum_{i=1}^{n} \alpha_i x_i\right). \tag{10.67}$$

In this case we shall also call $\{x_n\}$ a *generalized summation basis of E with respect to the sequence of endomorphisms* $\{v_n\}$.

From definition 10.3 it follows immediately that *every generalized summation basis* $\{x_n\}$ *is finitely linearly independent*. Indeed, if $\sum_{i=1}^{k} \beta_i x_i = 0$, then for $n \geqslant k$ and $\beta_{k+1} = \ldots = \beta_n = 0$ we have $v_n\left(\sum_{i=1}^{n} \beta_i x_i\right) = v_n(0) = 0 = v_n\left(\sum_{i=1}^{n} 0.x_i\right) \to 0$ as $n \to \infty$, whence, by the uniqueness condition in b) (for $x = 0$) it follows that $\beta_i = 0$ $(i = 1, \ldots, k)$. Using this remark, we shall prove now a stronger result, namely

**Theorem 10.4.** *Every generalized summation basis* $\{x_n\}$ *of a Banach space E is an operational basis of E, with the a.s.f.*

$$f_n(x) = \alpha_n \quad (x \in E,\ n = 1,2, \ldots), \tag{10.68}$$

*where* $\alpha_n$ *are as in condition* b) *of definition* 10.3, *and with respect to the same sequence of endomorphisms* $\{v_n\}$.

*Proof.* Let $A_1^{(v)}$ be the linear space of sequences of scalars

$$A_1^{(v)} = \left\{ \{\alpha_n\} \subset K \;\middle|\; \lim_{n\to\infty} v_n\left(\sum_{i=1}^{n} \alpha_i x_i\right) \text{ exists} \right\}, \tag{10.69}$$

endowed with the topology generated by the sequence of semi-norms

$$\|\{\alpha_n\}\|_{(0)} = \sup_{1 \leqslant n < \infty} \left\| v_n\left(\sum_{i=1}^{n} \alpha_i x_i\right)\right\|, \tag{10.70}$$

$$\|\{\alpha_n\}\|_{(m)} = \max_{1 \leqslant n \leqslant m} \left\| v_m\left(\sum_{i=1}^{n} \alpha_i x_i\right)\right\| \quad (m = 1,2, \ldots). \tag{10.71}$$

Then $A_1^{(v)}$ is a Hausdorff space, since $\|\cdot\|_{(0)}$ is a norm. Indeed, if $\{\alpha_n\} \in A_1^{(v)}$, $\|\{\alpha_n\}\|_{(0)} = 0$, then $v_n\left(\sum_{i=1}^{n} \alpha_i x_i\right) = 0$ $(n = 1,2, \ldots)$, whence $x = \lim_{n\to\infty} v_n\left(\sum_{i=1}^{n} \alpha_i x_i\right) = 0$, so $\alpha_n = f_n(x) = f_n(0) = 0$ $(n = 1,2, \ldots)$.

Now we shall prove that $A_1^{(v)}$ is complete, and hence a Fréchet space. Let $\{\alpha_n^{(k)}\}$ $(k = 1,2, \ldots)$ be a Cauchy sequence in $A_1^{(v)}$. Then for every $\varepsilon > 0$ and $m = 0,1,2, \ldots$ there exists a positive integer $N(\varepsilon, m)$ such that

$$\|\{\alpha_n^{(k)}\} - \{\alpha_n^{(l)}\}\|_{(0)} = \sup_{1 \leqslant n < \infty} \left\| v_n\left(\sum_{i=1}^{n} (\alpha_i^{(k)} - \alpha_i^{(l)}) x_i\right)\right\| < \varepsilon$$

$$(k, l > N(\varepsilon, 0)) \tag{10.72}$$

$$\|\{\alpha_n^{(k)}\} - \{\alpha_n^{(l)}\}\|_{(m)} = \max_{1 \leqslant n \leqslant m} \left\| v_m\left(\sum_{i=1}^{n} (\alpha_i^{(k)} - \alpha_i^{(l)}) x_i\right)\right\| < \varepsilon$$

$$(k, l > N(\varepsilon, m);\ m = 1,2, \ldots). \tag{10.73}$$

Since for each $n$ the elements $x_1, \ldots, x_n$ are linearly independent, by (10.66) and Ch. II, § 1, lemma 1.4 there exists a positive integer $m_n \geqslant n$ such that $v_{m_n}(x_1), \ldots, v_{m_n}(x_n)$ are linearly independent. Hence, by (10.73) and since all $n$-dimensional Banach spaces are isomorphic*), we have, with a suitable constant $C_n > 0$,

$$|\alpha_n^{(k)} - \alpha_n^{(l)}| \leqslant \sum_{i=1}^{n} |\alpha_i^{(k)} - \alpha_i^{(l)}| \leqslant C_n \left\| \sum_{i=1}^{n} (\alpha_i^{(k)} - \alpha_i^{(l)})\, v_{m_n}(x_i)\right\| \leqslant$$
$$\leqslant C_n \|\{\alpha_n^{(k)}\} - \{\alpha_n^{(l)}\}\|_{(m_n)} < C_n \varepsilon \qquad (k, l > N(\varepsilon, m_n)).$$

Consequently, there exists a sequence of scalars $\{\alpha_n\}$ such that

$$\lim_{l\to\infty} \alpha_n^{(l)} = \alpha_n \qquad (n = 1,2, \ldots). \tag{10.74}$$

Let us show that $\{\alpha_n\} \in A_1^{(v)}$. Since $\{\alpha_n^{(k)}\} \in A_1^{(v)}$ $(k = 1,2, \ldots)$, we can put

$$y_k = \lim_{n\to\infty} v_n\left(\sum_{i=1}^{n} \alpha_i^{(k)} x_i\right) \in E \qquad (k = 1,2, \ldots). \tag{10.75}$$

*) See e.g. [87], p. 245, corollary 3.

Then, by (10.72),

$$\|y_k - y_l\| = \lim_{n\to\infty}\left\|v_n\left(\sum_{i=1}^{n}(\alpha_i^{(k)} - \alpha_i^{(l)})x_i\right)\right\| \leqslant \|\{\alpha_n^{(k)}\} - \{\alpha_n^{(l)}\}\|_{(0)} < \varepsilon$$

$$(k, l > N(\varepsilon, 0)),$$

and hence there exists an $x \in E$ such that

$$\lim_{k\to\infty} y_k = x. \tag{10.76}$$

We shall show that $x = \lim_{n\to\infty} v_n\left(\sum_{i=1}^{n}\alpha_i x_i\right)$, which will prove that $\{\alpha_n\} \in A_1^{(v)}$. We have the inequalities

$$\left\|x - v_n\left(\sum_{i=1}^{n}\alpha_i x_i\right)\right\| \leqslant \|x - y_k\| + \left\|y_k - v_n\left(\sum_{i=1}^{n}\alpha_i^{(k)}x_i\right)\right\| +$$

$$+\left\|v_n\left(\sum_{i=1}^{n}(\alpha_i^{(k)} - \alpha_i)x_i\right)\right\| \qquad (n, k = 1,2, \ldots). \tag{10.77}$$

By (10.72), $\left\|v_n\left(\sum_{i=1}^{n}(\alpha_i^{(k)} - \alpha_i^{(l)})x_i\right)\right\| < \varepsilon$ $(k, l > N(\varepsilon, 0);\ n = 1,2,\ldots)$, whence, for $l \to \infty$ we obtain, by (10.74),

$$\left\|v_n\left(\sum_{i=1}^{n}(\alpha_i^{(k)} - \alpha_i)x_i\right)\right\| \leqslant \varepsilon \qquad (k > N(\varepsilon, 0);\ n = 1,2, \ldots). \tag{10.78}$$

Furthermore, by (10.76) there exists a $k = k(\varepsilon) > N(\varepsilon, 0)$ such that

$$\|x - y_k\| < \varepsilon, \tag{10.79}$$

and by (10.75) there exists a positive integer $N_k(\varepsilon)$ such that

$$\left\|y_k - v_n\left(\sum_{i=1}^{n}\alpha_i^{(k)}x_i\right)\right\| < \varepsilon \qquad (n > N_k(\varepsilon)). \tag{10.80}$$

Thus, by (10.77)—(10.80),

$$\left\|x - v_n\left(\sum_{i=1}^{n}\alpha_i x_i\right)\right\| < 3\varepsilon \qquad (n > N_k(\varepsilon)),$$

which proves that $x = \lim_{n\to\infty} v_n\left(\sum_{i=1}^{n}\alpha_i x_i\right)$ and that $\{\alpha_n\} \in A_1^{(v)}$.

Finally by (10.78) and (10.70) we have

$$\|\{\alpha_n^{(k)}\} - \{\alpha_n\}\|_{(0)} \leqslant \varepsilon \qquad (k > N(\varepsilon, 0)),$$

and similarly, by (10.73) for $l \to \infty$ and (10.74), we obtain

$$\|\{\alpha_n^{(k)}\} - \{\alpha_n\}\|_{(m)} \leqslant \varepsilon \qquad (k > N(\varepsilon, m);\ m = 1,2, \ldots),$$

whence $\lim_{k\to\infty} \{\alpha_n^{(k)}\} = \{\alpha_n\}$ in $A_1^{(v)}$, which proves that $A_1^{(v)}$ is a Fréchet space.

Define now a mapping $w: A_1^{(v)} \to E$ by

$$w(\{\alpha_n\}) = \lim_{n\to\infty} v_n\left(\sum_{i=1}^{n} \alpha_i x_i\right) \qquad (\{\alpha_n\} \in A_1^{(v)}). \tag{10.81}$$

The mapping $w$ is obviously linear and it satisfies

$$\|w(\{\alpha_n\})\| \leqslant \sup_{1\leqslant n<\infty} \left\| v_n\left(\sum_{i=1}^{n} \alpha_i x_i\right)\right\| = \|\{\alpha_n\}\|_{(0)} \quad (\{\alpha_n\} \in A_1^{(v)}), \tag{10.82}$$

whence $w$ is continuous. Furthermore, by condition b) of definition 10.1, $w$ is one to one and maps $A_1^{(v)}$ onto $E$. Hence, by the inversion theorem of Banach*), $w$ is an isomorphism of $A_1^{(v)}$ onto $E$ and thus there exist constants $M_m > 0$ $(m = 0,1,2, \ldots)$ such that

$$\|\{\alpha_n\}\|_{(m)} \leqslant M_m \|w(\{\alpha_n\})\| \quad (\{\alpha_n\} \in A_1^{(v)},\ m = 0,1,2, \ldots). \tag{10.83}$$

Consequently, using again that $v_{m_n}(x_1), \ldots, v_{m_n}(x_n)$ are linearly independent and that all $n$-dimensional Banach spaces are isomorphic, it follows that for every $x = w(\{\alpha_n\}) \in E$ we have, with suitable constants $C_n > 0$ $(n = 1,2, \ldots)$,

$$|\alpha_n| \leqslant \sum_{i=1}^{n} |\alpha_i| \leqslant C_n \left\| \sum_{i=1}^{n} \alpha_i v_{m_n}(x_i) \right\| \leqslant C_n \|\{\alpha_n\}\|_{(m_n)} \leqslant$$

$$\leqslant C_n M_{m_n} \|w(\{\alpha_n\})\| = C_n M_{m_n} \|x\| \qquad (n = 1,2, \ldots),$$

which proves that the linear functionals $f_n$ defined by (10.68) are continuous on $E$, that is,

$$f_n \in E^* \quad (n = 1,2, \ldots). \tag{10.84}$$

*) See e.g. [11], p. 41, theorem 5.

Now, by (10.66) we have

$$\lim_{n\to\infty} v_n\left(\sum_{i=1}^{n} \delta_{ij}x_i\right) = \lim_{n\to\infty} v_n(x_j) = x_j \qquad (j = 1,2, \ldots),$$

whence, by (10.67) and (10.68) for $x = x_j$,

$$f_n(x_j) = \delta_{nj} \qquad (n, j = 1,2, \ldots), \tag{10.85}$$

i.e., $(x_n, f_n)$ is a biorthogonal system. Consequently, by (10.68), we can write (10.67) in the form (10.1), where $\{s_n\}$ is the sequence of partial sum operators associated to the biorthogonal system $(x_n, f_n)$, which completes the proof of theorem 10.4.

Let us observe that condition a) of definition 10.3 is essential for the validity of theorem 10.4, since a sequence which satisfies only b) of definition 10.3 need not even be $\omega$-linearly independent, as shown by

*Example 10.2.* Let $E = l^2$ and let $\{x_n\}$ be*) a finitely linearly independent complete sequence in $E$ which is not $\omega$-linearly independent. Furthermore, let $\{z_n\}$ be the sequence in $E$ obtained from $\{x_n\}$ by the classical Gram-Schmidt orthonormalization procedure $\Big($hence $z_n = \sum_{i=1}^{n} \alpha_i^{(n)} x_i$ for suitable scalars $\alpha_i^{(n)}$ with $\alpha_n^{(n)} \neq 0$ for $n = 1,2, \ldots$ and $(z_i, z_j) = \delta_{ij}$ for $i, j = 1,2, \ldots\Big)$ and define $v_n \in L(P_{(n)}, P_{(n)})$ by

$$v_n(x_i) = z_i \qquad (i = 1, \ldots, n;\ n = 1,2, \ldots). \tag{10.86}$$

Then $\{x_n\}$ and $\{v_n\}$ satisfy b) of definition 10.3. Indeed, for $x \in E$ and $\alpha_n = (x, z_n)$ $(n = 1,2, \ldots)$ we have

$$v_n\left(\sum_{i=1}^{n} \alpha_i x_i\right) = \sum_{i=1}^{n} \alpha_i v_n(x_i) = \sum_{i=1}^{n} (x, z_i) z_i \to x \text{ as } n \to \infty,$$

and if $v_n\left(\sum_{i=1}^{n} \beta_i x_i\right) \to x$ as $n \to \infty$, then $\sum_{i=1}^{n} \beta_i z_i = \sum_{i=1}^{n} \beta_i v_n(x_i) = v_n\left(\sum_{i=1}^{n} \beta_i x_i\right) \to x$, i.e., $x = \sum_{i=1}^{\infty} \beta_i z_i$, whence $\beta_n = (x, z_n) = \alpha_n$ $(n = 1,2, \ldots)$, which proves our assertion.

---

*) See e.g. Ch. I, § 6, example 6.2

*Remark 10.2.* In some cases, for example if each matrix $(a_{ij}^{(n)})$ is diagonal (i.e.*), $v_n(x_i) = a_{ii}^{(n)} x_i$ for $i = 1, \ldots, n$; $n = 1,2, \ldots$), a certain "strong" converse of theorem 10.4 is also true, namely, every operational basis of a Banach space $E$ with respect to such a sequence of endomorphisms $\{v_n\}$ is a generalized summation basis of $E$ *with respect to the same* sequence of endomorphisms $\{v_n\}$, since $\lim_{n\to\infty} v_n\left(\sum_{i=1}^{n} \alpha_i x_i\right) = \lim_{n\to\infty} \sum_{i=1}^{n} \alpha_i a_{ii}^{(n)} x_i = 0$ implies $\alpha_j \lim_{n\to\infty} a_{jj}^{(n)} = \lim_{n\to\infty} f_j\left(\sum_{i=1}^{n} \alpha_i a_{ii}^{(n)} x_i\right) = 0$ $(j = 1,2, \ldots)$, whence by (10.45) for $i = j$, we obtain $\alpha_j = 0$ $(j = 1,2,\ldots)$. But, for arbitrary $\{v_n\}$ such a "strong" converse is no longer true, as shown by

*Example 10.3.* a) Let $E = c_0$, let $\{x_n\}$ be the unit vector basis of $E$, and let $v_1 = s_1|_{[x_1]}$. For each $n \geqslant 2$ put

$$v_n(x_i) = x_i \quad (i = 1, \ldots, n-1), \quad v_n(x_n) = -\sum_{i=1}^{n-1} x_i, \tag{10.87}$$

and extend $v_n$ by linearity to $[x_1, \ldots, x_n]$. Then $\{x_n\}$ is an operational basis of $E$ with respect to $\{v_n\}$, since by $\left\|\sum_{i=1}^{n-1} x_i\right\| = 1$ $(n \geqslant 2)$ we have

$$v_n s_n(x) = \sum_{i=1}^{n-1} f_i(x) x_i - f_n(x) \sum_{i=1}^{n-1} x_i \to x \text{ as } n \to \infty \qquad (x \in E),$$

where $\{f_n\} \subset E^*$, $f_i(x_j) = \delta_{ij}$. However, $\{x_n\}$ is not a generalized summation basis of $E$ with respect to $\{v_n\}$, since $v_n\left(\sum_{i=1}^{n} x_i\right) = 0$ $(n = 2,3,\ldots)$, whence $\lim_{n\to\infty} v_n\left(\sum_{i=1}^{n} x_i\right) = 0$, and thus the uniqueness condition in b) of definition 10.3 is not satisfied.

b) Let $E = l^2$, let $\{x_n\}$ be the unit vector basis of $E$ and let $v_1 = s_1|_{[x_1]}$. For each $n \geqslant 1$ put

$$v_{2n}\left(\sum_{i=1}^{2n} \alpha_i x_i\right) = v_{2n+1}\left(\sum_{i=1}^{2n+1} \alpha_i x_i\right) = \sum_{i=1}^{n} (\alpha_i - \alpha_{n+i}) x_i, \tag{10.88}$$

---

*) Note that this happens if and only if $\{x_n\}$ is a strongly series summable $M$-basis in the sense of § 11, definition 11.4.

i.e. $v_{2n}(x_i) = v_{2n+1}(x_i) = x_i$ for $i = 1, \ldots, n$ and $-x_{i-n}$ for $i = n+1, \ldots$ $\ldots, 2n$ and $v_{2n+1}(x_{2n+1}) = 0$. Then for every $x \in E$ we have

$$\|x - v_{2n}s_{2n}(x)\| = \|x - v_{2n+1}s_{2n+1}(x)\| = \left\| \sum_{i=1}^{\infty} f_i(x)x_i - \sum_{i=1}^{n} f_i(x)x_i + \right.$$

$$\left. + \sum_{i=1}^{n} f_{n+i}(x)x_i \right\| = \left( \sum_{i=n+1}^{\infty} |f_i(x)|^2 + \sum_{i=1}^{n} |f_{n+i}(x)|^2 \right)^{\frac{1}{2}} \leqslant$$

$$\leqslant 2 \left( \sum_{i=n+1}^{\infty} |f_i(x)|^2 \right)^{\frac{1}{2}} \to 0 \text{ as } n \to \infty$$

(where $\{f_n\} \subset E^*$, $f_i(x_j) = \delta_{ij}$), but $\lim_{n\to\infty} v_n \left( \sum_{i=1}^{n} x_i \right) = 0$.

However, the following problem is open:

*Problem 10.1.* Is every operational basis $\{x_n\}$ of a Banach space $E$ a generalized summation basis of $E$ (with respect to some other suitable sequence of endomorphisms $\{v'_n\}$)?

For separable Hilbert spaces the answer to problem 10.1 is affirmative, since for such spaces we have even the following sharpening of corollary 10.1:

**Theorem 10.5.** *Every M-basis $\{x_n\}$ of a separable Hilbert space $E$ is a generalized summation basis of $E$.*

*Proof.* As in the proof of theorem 10.2, define an increasing sequence of positive integers $\{m_n\}$ by (10.31), (10.32) and the endomorphisms $v_k$ by (10.33), (10.34), but with respect to the original norm of $E$ (instead of $|.|$); note that this is possible, since for $\{f_n\} \subset E^*$ with $f_i(x_j) = \delta_{ij}$ we have now $[f_n] = E^*$, and hence the original norm of $E$ has all properties of the norm $|.|$ of lemma 10.1. Then, since $E$ is a Hilbert space, for each subspace $G$ of $E$ the mapping $\pi_G$ defined by (10.30) (for the original norm of $E$) is linear (it is nothing else than the orthogonal projection of $E$ onto $G$) and hence all $v_k$ are continuous linear mappings. Consequently, by the proof of theorem 10.2, $\{x_n\}$ is a (linear) operational basis of $E$ and it remains to prove that the uniqueness condition in b) of definition 10.3 is also satisfied. However, this is immediate, since if $\lim_{n\to\infty} v_k \left( \sum_{i=1}^{k} \alpha_i x_i \right) = 0$, then

$$0 = \lim_{n\to\infty} \left( \sum_{i=1}^{m_{n-1}} \alpha_i x_i - \pi_{[x_i]_{m_{n-1}+1}^{m_n}} \left( \sum_{i=1}^{m_{n-1}} \alpha_i x_i \right) \right) =$$

$$= \lim_{n\to\infty} \left( \sum_{i=1}^{m_{n-1}} \alpha_i x_i - \sum_{i=m_{n-1}+1}^{m_n} \beta_i x_i \right)$$

for suitable $\beta_i$, whence

$$\alpha_j = \lim_{n\to\infty} f_j\left(\sum_{i=1}^{m_{n-1}} \alpha_i x_i - \sum_{i=m_{n-1}+1}^{m_n} \beta_i x_i\right) = 0 \quad (j = 1,2, \ldots),$$

which completes the proof of theorem 10.5.

From theorem 10.5 and § 8, proposition 8.12 it follows that there exist generalized summation bases (hence operational bases) which are not strong $M$-bases (and hence not series summable).

The family of all generalized summation bases (and hence also that of all operational bases) contains, in particular, all $T$-bases (=summation bases) with respect to*) triangular Toeplitz matrices $T = (t_{nm})$ (i.e., such that $t_{nm} = 0$ for $m > n$). One can extend further these families, by introducing double (and multiple) generalized summation bases (and such operational bases), so as to include all $T$-bases, with respect to arbitrary Toeplitz matrices $T = (t_{nm})$, as follows:

*Definition 10.4*. A sequence $\{x_n\}$ in a Banach space $E$ is said to be a *double generalized summation basis* of $E$ if there exists a double sequence of endomorphisms $\{v_{nm}\}$ with $v_{nm} \in L(P_{(m)}, P_{(m)})$ $(n, m = 1,2, \ldots)$, where $P_{(m)} = [x_1, \ldots, x_m]$ $(m = 1,2, \ldots)$, such that

a) We have

$$\lim_{n\to\infty} \lim_{m\to\infty} v_{nm}(x_i) = x_i \quad (i = 1,2, \ldots). \tag{10.89}$$

b) For every $x \in E$ there exists a *unique* sequence of scalars $\{\alpha_n\} \subset K$ such that

$$x = \lim_{n\to\infty} \lim_{m\to\infty} v_{nm}\left(\sum_{i=1}^{m} \alpha_i x_i\right). \tag{10.90}$$

Clearly, every generalized summation basis is a double generalized summation basis, with $v_{nm} = v_m$ $(n, m = 1,2, \ldots)$.

The properties of double (and multiple) generalized summation bases (and of such operational**) bases) are analogous to those of generalized summation bases (respectively, operational bases), with similar proofs (see also the proof of § 11, theorem 11.1), except that they need not be approximative bases.

In the next section and in § 13, definition 13.4 we shall consider some special classes of generalized summation bases and of multiple generalized summation bases.

---

*) See § 11, definition 11.1.

**) Naturally, a *double operational basis* $\{x_n\}$ is defined as a complete minimal sequence for which there exist $v_{nm} \in L(P_{(m)}, P_{(m)})$ $(n, m = 1, 2, \ldots)$ such that

$$x = \lim_{n\to\infty} \lim_{m\to\infty} v_{nm}\, s_m(x) \quad (x \in E).$$

## § 11. $T$-bases (summation bases). Strongly series summable $M$-bases. $\gamma$-bases

We recall that if $E$ is a Banach space and $T = (t_{nm})$ is a matrix of scalars, a sequence $\{z_n\} \subset E$ is called *$T$-limitable* to $x \in E$ if the elements $y_n = \sum_{j=1}^{\infty} t_{nj} z_j$ $(n = 1,2, \ldots)$ exist and $\lim_{n\to\infty} y_n = x$. A series $\sum_{i=1}^{\infty} x_i$ in $E$ is said to be *$T$-summable* to $x \in E$ if the sequence $\{z_n\} = \left\{\sum_{i=1}^{n} x_i\right\}$ of its partial sums is $T$-limitable to $x$. If $E$ is a Banach space, a matrix of scalars $T = (t_{nm})$ is called a *consistent matrix* (or a *Toeplitz matrix*, or a *permanent summation method*), if every convergent sequence $\{z_n\} \subset E$ is $T$-limitable to its limit. This notion does not depend on the space $E$, as shown by

**Lemma 11.1.** *Let $E$ be a Banach space. A matrix of scalars $T = (t_{nm})$ is consistent (for $E$) if and only if the following three conditions are satisfied:*

$$\sum_{j=1}^{\infty} |t_{nj}| \leqslant M < \infty \qquad (n = 1,2, \ldots), \tag{11.1}$$

$$\lim_{n\to\infty} t_{nj} = 0 \qquad (j = 1,2, \ldots), \tag{11.2}$$

$$\lim_{n\to\infty} \sum_{j=1}^{\infty} t_{nj} = 1. \tag{11.3}$$

*Proof.* Assume that $T = (t_{nm})$ is consistent (for $E$) and let $\{\alpha_n\} \in c$, $\lim_{n\to\infty} \alpha_n = \alpha$. Then for any $x \in E$ we have $\alpha_n x \to \alpha x$, whence, since $T$ is consistent for $E$, the elements $y_n = \sum_{i=1}^{\infty} t_{ni} \alpha_i x$ exist and converge to $\alpha x$. Therefore, taking $x \neq 0$, the scalars $\sum_{i=1}^{\infty} t_{ni} \alpha_i$ exist and converge to $\alpha$, so $T$ is also consistent for the field $K$ of the scalars and hence, by the necessity part of a well-known theorem of Silverman and Toeplitz,*) $T = (t_{nm})$ satisfies (11.1)–(11.3).

Conversely, assume now that $T = (t_{nm})$ satisfies (11.1)–(11.3), and consider the Banach space $c_E$ of all convergent sequences of elements of $E$ (with the norm $\|\{z_k\}\| = \sup_{1 \leqslant k < \infty} \|z_k\|$). By (11.1), the linear operators

$$u_n(\{z_k\}) = \sum_{j=1}^{\infty} t_{nj} z_j \qquad (\{z_k\} \in c_E) \tag{11.4}$$

---

*) See e.g. [11], p. 90, theorem 10, or [50], theorem (4.1.II), or [410], § 32, theorem I.

are well defined on $c_E$ and satisfy

$$\|u_n\| = \sup_{\substack{\{z_k\} \in c_E \\ \|z_1\|, \|z_2\|, \ldots \leqslant 1}} \|u_n(\{z_k\})\| = \sum_{j=1}^{\infty} |t_{nj}| \leqslant M \quad (n = 1,2, \ldots).$$

Furthermore, by (11.2), for any $\{x_1, \ldots, x_m, 0, 0, \ldots\} \in c_E$ we have

$$\lim_{n\to\infty} u_n(\{x_1, \ldots, x_m, 0, 0, \ldots\}) = \lim_{n\to\infty} \sum_{j=1}^{m} t_{nj}x_j = \sum_{j=1}^{m} (\lim_{n\to\infty} t_{nj})x_j =$$

$$= 0 = \lim \{x_1, \ldots, x_m, 0, 0, \ldots\}$$

and, by (11.3), for any $\{x, x, x, \ldots\} \in c_E$ we have

$$\lim_{n\to\infty} u_n(\{x, x, x, \ldots\}) = \lim_{n\to\infty} \sum_{j=1}^{\infty} t_{nj}x = x = \lim \{x, x, x, \ldots\}.$$

Hence, since the set $\mathscr{S}$ of all elements of the form $\{x_1, \ldots, x_m, 0,0,\ldots\}$ and $\{x, x, x, \ldots\}$ (where $x_1, \ldots, x_m \in E$, $1 \leqslant m < \infty$ and $x \in E$) is complete in $c_E$, it follows*) that

$$\sum_{j=1}^{\infty} t_{nj}z_j = u_n(\{z_k\}) \to \lim_{k\to\infty} z_k \text{ as } n \to \infty \quad (\{z_k\} \in c_E), \tag{11.5}$$

i.e., $T$ is consistent for $E$, which completes the proof of lemma 11.1.

*Definition 11.1.* Let $T = (t_{nm})$ be a consistent matrix. A sequence $\{x_n\}$ in a Banach space $E$ is said to be a *T-basis* (or a *summation basis*) of $E$ if for every $x \in E$ there exists a *unique* sequence of scalars $\{\alpha_n\} \subset K$ such that the series $\sum_{i=1}^{\infty} \alpha_i x_i$ is $T$-summable to $x$. A $T$-basis with respec to a triangular consistent matrix $T = (t_{nm})$ (i.e., such that**) $t_{nm} = 0$ for all $m>n$; $n = 1,2, \ldots$) is called a *triangular T-basis*. In particular, a $T$-basis of $E$ with respect to the triangular matrix $T = (t_{nm})$, where

$$t_{nm} = \frac{1}{n} \text{ for } m = 1, \ldots, n; \; t_{nm} = 0 \text{ for } m>n \quad (n = 1,2, \ldots), \tag{11.6}$$

is called a *Cesàro basis* of $E$.

Thus, in other words, $\{x_n\}$ is a $T$-basis of $E$ if for every $x \in E$ there exists a *unique* sequence $\{\alpha_n\} \subset K$ such that if we denote

$$s_n(x) = \sum_{i=1}^{n} \alpha_i x_i \qquad (n = 1,2, \ldots), \tag{11.7}$$

$$\sigma_{nm}(x) = \sum_{j=1}^{m} t_{nj}s_j(x) \qquad (n, m = 1,2, \ldots), \tag{11.8}$$

*) We also use that $u \in L(c_E, E)$, where $u(\{z_k\}) = \lim_{k\to\infty} z_k$ $(\{z_k\} \in c_E)$.

**) Throughout the sequel, by "triangular" we shall mean: lower triangular.

then the limits

$$\sigma_n(x) = \lim_{m\to\infty} \sigma_{nm}(x) = \sum_{j=1}^{\infty} t_{nj}s_j(x) \qquad (n = 1,2, \ldots) \tag{11.9}$$

exist and satisfy

$$x = \lim_{n\to\infty} \sigma_n(x). \tag{11.10}$$

Let us observe that in general (when $T = (t_{nm})$ is not row-finite) *the operators $\sigma_n$ are no longer of finite rank.*

In the sequel, whenever not specified otherwise, we shall assume that $T = (t_{nm})$ is a consistent matrix.

From definition 11.1 it follows immediately that *every $T$-basis $\{x_n\}$ is finitely linearly independent.* Indeed, if $\sum_{i=1}^{k} \beta_i x_i = 0$, then $\sigma_n\left(\sum_{i=1}^{k} \beta_i x_i\right) = = \sigma_n(0) = 0 = \sigma_n\left(\sum_{i=1}^{k} 0.\, x_i\right) \to 0$ as $n\to\infty$, whence, by the uniaueness condition in definition 11.1 (for $x = 0$) it follows that $\beta_i = 0$ ($i = 1, \ldots, k$). Using this remark, one can prove the following stronger result (which motivates the terms "double generalized summation basis" and "generalized summation basis" introduced in § 10, definitions 10.4 and 10.3):

**Theorem 11.1.** a) *Every $T$-basis $\{x_n\}$ of a Banach space $E$ is an $M$-basis of $E$, with the a.s.f.*

$$f_n(x) = \alpha_n \qquad (x \in E, n = 1,2, \ldots), \tag{11.11}$$

*where the $\alpha_n$ are as in definition* 11.1. *Moreover:*

b) *Every $T$-basis (in particular, every triangular $T$-basis) of a Banach space $E$ is a double generalized summation basis (respectively, a generalized summation basis) of $E$.*

*Proof.* a) Let $A_1^{(T)}$ be the linear space of sequences of scalars

$$A_1^{(T)} = \{\alpha = \{\alpha_n\} \subset K / f_n(x_\alpha) = \alpha_n \quad (n = 1,2, \ldots) \text{ for some (unique) } x_\alpha \in E\}, \tag{11.12}$$

endowed with the topology generated by the semi-norms

$$\|\{\alpha_j\}\|_{(0)} = \sup_{1\leqslant n<\infty} \|\sigma_n(x_\alpha)\|, \tag{11.13}$$

$$\|\{\alpha_j\}\|_{(n)} = \sup_{1\leqslant m<\infty} \|\sigma_{nm}(x_\alpha)\| \quad (n = 1,2, \ldots). \tag{11.14}$$

Then $A_1^{(T)}$ is a Hausdorff space, since $\|.\|_{(0)}$ is a norm. Indeed, if $\{\alpha_n\} \in A_1^{(T)}$, $\|\{\alpha_n\}\|_{(0)} = 0$, then $\sigma_n(x_\alpha) = 0$ $(n = 1,2, \dots)$, whence $x_\alpha = \lim_{n\to\infty} \sigma_n(x_\alpha) = 0$, so $\alpha_n = f_n(x_\alpha) = f_n(0) = 0$ $(n = 1,2, \dots)$.

Next, we shall prove that $A_1^{(T)}$ is complete, and hence a Fréchet space. Let $\alpha^{(k)} = \{\alpha_n^{(k)}\}$ $(k = 1,2, \dots)$ be a Cauchy sequence in $A_1^{(T)}$. Then for every $\varepsilon > 0$ and every $n = 0,1,2, \dots$ there exists a positive integer $N(\varepsilon, n)$ such that

$$\|\{\alpha_j^{(k)}\} - \{\alpha_j^{(l)}\}\|_{(0)} = \sup_{1 \leqslant n < \infty} \|\sigma_n(x_{\alpha^{(k)}} - x_{\alpha^{(l)}})\| < \varepsilon \quad (k, l > N(\varepsilon, 0)), \tag{11.15}$$

$$\|\{\alpha_j^{(k)}\} - \{\alpha_j^{(l)}\}\|_{(n)} = \sup_{1 \leqslant m < \infty} \|\sigma_{nm}(x_{\alpha^{(k)}} - x_{\alpha^{(l)}})\| < \varepsilon$$

$$(k, l > N(\varepsilon, n);\ n = 1,2, \dots). \tag{11.16}$$

By (11.3) and (11.2), for each $n$ there exists a positive integer $p_n \geqslant n$ such that

$$\left|\sum_{j=1}^{\infty} t_{p_n j} - 1\right| < \frac{1}{3}, \quad \sum_{j=1}^{n-1} |t_{p_n j}| < \frac{1}{3},$$

and then we can find $m_n \geqslant p_n$ such that $\left|\sum_{j=1}^{m_n} t_{p_n j} - 1\right| < \frac{2}{3}$, whence

$$\left|\sum_{j=i}^{m_n} t_{p_n j}\right| \geqslant \left|\sum_{j=1}^{m_n} t_{p_n j}\right| - \left|\sum_{j=1}^{i-1} t_{p_n j}\right| \geqslant$$

$$\geqslant 1 - \left|1 - \sum_{j=1}^{m_n} t_{p_n j}\right| - \left|\sum_{j=1}^{i-1} t_{p_n j}\right| > 1 - \frac{2}{3} - \frac{1}{3} = 0 \quad (i = 1, \dots, n).$$

Hence, since for each $n$ the elements $x_1, \dots, x_n$ are linearly independent, so are the elements $\left(\sum_{j=1}^{m_n} t_{p_n j}\right) x_1, \dots, \left(\sum_{j=n}^{m_n} t_{p_n j}\right) x_n$. Consequently, since all $n$-dimensional Banach spaces are isomorphic and since every linear operator defined on a finite-dimensional Banach space is continuous*),

---

*) We apply this to the operator $\sum_{i=1}^{m_n} \alpha_i y_i \to \sum_{i=1}^{n} \alpha_i y_i$, where $y_i = \left(\sum_{j=i}^{m_n} t_{p_n j}\right) x_i$ for $i = 1, \dots, m_n$. This operator is well defined, since $x_1, \dots, x_{m_n}$ are linearly independent (hence so are those $y_i$ which $\neq 0$) and since $y_i \neq 0$ for $i = 1, \dots, n$.

we obtain, with suitable constants $C_n, C'_n > 0$,

$$|\alpha_n^{(k)} - \alpha_n^{(l)}| \leqslant \sum_{i=1}^{n} |\alpha_i^{(k)} - \alpha_i^{(l)}| \leqslant C_n \left\| \sum_{i=1}^{n} (\alpha_i^{(k)} - \alpha_i^{(l)}) \left( \sum_{j=i}^{m_n} t_{p_n j} \right) x_i \right\| \leqslant$$

$$\leqslant C'_n \left\| \sum_{i=1}^{m_n} (\alpha_i^{(k)} - \alpha_i^{(l)}) \left( \sum_{j=i}^{m_n} t_{p_n j} \right) x_i \right\| = C'_n \|\sigma_{p_n m_n}(x_{\alpha^{(k)}} - x_{\alpha^{(l)}})\| \leqslant$$

$$\leqslant C'_n \|\{\alpha_j^{(k)}\} - \{\alpha_j^{(l)}\}\|_{(p_n)} < C'_n \varepsilon \quad (k, l > N(\varepsilon, n)),$$

and therefore the limits

$$\lim_{k \to \infty} \alpha_n^{(k)} = \alpha_n \quad (n = 1,2, \ldots)$$

exist. Let us show that $\{\alpha_n\} \in A_1^{(T)}$. By (11.10), $\{\alpha_n^{(k)}\} \in A_1^{(T)}$ $(k = 1,2, \ldots)$ and (11.15), we have

$$\|x_{\alpha^{(k)}} - x_{\alpha^{(l)}}\| = \lim_{n \to \infty} \|\sigma_n(x_{\alpha^{(k)}} - x_{\alpha^{(l)}})\| \leqslant$$

$$\leqslant \|\{\alpha_n^{(k)}\} - \{\alpha_n^{(l)}\}\|_{(0)} < \varepsilon \quad (k, l > N(\varepsilon, 0)),$$

and hence there exists an $x \in E$ such that

$$\lim_{k \to \infty} x_{\alpha^{(k)}} = x.$$

Since $\lim_{m \to \infty} \sigma_{nm}(x_{\alpha^{(k)}}) = \sigma_n(x_{\alpha^{(k)}})$, we have $\{\sigma_{nm}(x_{\alpha^{(k)}})\}_{m=1}^{\infty} \in c_E$ $(n, k = 1,2, \ldots)$. Furthermore, by (11.16), $\{\{\sigma_{nm}(x_{\alpha^{(k)}})\}_{m=1}^{\infty}\}_{k=1}^{\infty}$ is a Cauchy sequence in $c_E$ $(n = 1,2, \ldots)$, whence, since $c_E$ is complete, there exists $\{y_{nm}\}_{m=1}^{\infty} \in c_E$ $(n = 1,2, \ldots)$ such that

$$\lim_{k \to \infty} \sup_{1 \leqslant m < \infty} \|\sigma_{nm}(x_{\alpha^{(k)}}) - y_{nm}\| = 0 \quad (n = 1,2, \ldots).$$

But, since $\alpha_i^{(k)} \to \alpha_i$ as $k \to \infty$ $(i = 1,2, \ldots)$, we have

$$\sigma_{nm}(x_{\alpha^{(k)}}) = \sum_{i=1}^{m} \alpha_i^{(k)} \left( \sum_{j=i}^{m} t_{nj} \right) x_i \to \sum_{i=1}^{m} \alpha_i \left( \sum_{j=i}^{m} t_{nj} \right) x_i \text{ as } k \to \infty,$$

whence, by the above,

$$y_{nm} = \sum_{i=1}^{m} \alpha_i \left( \sum_{j=i}^{m} t_{nj} \right) x_i \quad (n, m = 1,2, \ldots).$$

Now, by $\{y_{nm}\} \in c_E$, let $y_n = \lim_{m \to \infty} y_{nm} \in E$ $(n = 1,2, \ldots)$. Then, since $z(.) \to \lim_{m \to \infty} z(m)$ is a continuous linear operator from $c_E$ into $E$ and since $\{\sigma_{nm}(x_{\alpha^{(k)}})\}_{m=1}^{\infty} \to \{y_{nm}\}_{m=1}^{\infty}$ in $c_E$ as $k \to \infty$ $(n = 1,2, \ldots)$, we obtain

$$\sigma_n(x_{\alpha^{(k)}}) = \lim_{m \to \infty} \sigma_{nm}(x_{\alpha^{(k)}}) \to \lim_{m \to \infty} y_{nm} = y_n \text{ as } k \to \infty \quad (n = 1,2, \ldots).$$

Moreover, since $\lim_{n\to\infty}\sigma_n(x_{\alpha(k)})=x_{\alpha(k)}$, we have $\{\sigma_n(x_{\alpha(k)})\}_{n=1}^{\infty}\in c_E$ $(k=1,2,\ldots)$ and, by (11.15), $\{\{\sigma_n(x_{\alpha(k)})\}_{n=1}^{\infty}\}_{k=1}^{\infty}$ is a Cauchy sequence in $c_E$. Hence, by the above,

$$\lim_{k\to\infty}\sup_{1\leqslant n<\infty}\|\sigma_n(x_{\alpha(k)})-y_n\|=0.$$

Thus, for each $\varepsilon>0$ there exists a $k=k(\varepsilon)$ such that

$$\|x-x_{\alpha(k)}\|<\varepsilon,\ \|\sigma_n(x_{\alpha(k)})-y_n\|<\varepsilon\quad(n=1,2,\ldots),$$

and, by $\lim_{n\to\infty}\sigma_n(x_{\alpha(k)})=x_{\alpha(k)}$, there exists a positive integer $N_k(\varepsilon)$ such that

$$\|x_{\alpha(k)}-\sigma_n(x_{\alpha(k)})\|<\varepsilon\quad(n>N_k(\varepsilon)).$$

Consequently,

$$\|x-y_n\|\leqslant\|x-x_{\alpha(k)}\|+\|x_{\alpha(k)}-\sigma_n(x_{\alpha(k)})\|+$$
$$+\|\sigma_n(x_{\alpha(k)})-y_n\|<3\varepsilon\quad(n>N_k(\varepsilon)),$$

whence $x=\lim_{n\to\infty}y_n$. This, together with $y_n=\lim_{m\to\infty}y_{nm}=\lim_{m\to\infty}\sum_{i=1}^{m}\alpha_i\left(\sum_{j=i}^{m}t_{nj}\right)x_i$ and the uniqueness condition in definition 11.1, implies $x=x_\alpha$ and $f_n(x)=f_n(x_\alpha)=\alpha_n$ $(n=1,2,\ldots)$. Thus, $\alpha=\{\alpha_n\}\in A^{(T)}$; also, $\sigma_{nm}(x)=$ $=\sigma_{nm}(x_\alpha)=y_{nm}$ $(n,m=1,2,\ldots)$ and $\sigma_n(x)=\sigma_n(x_\alpha)=y_n$ $(n=1,2,\ldots)$, whence, by the above, we obtain

$$\|\{\alpha_j^{(k)}\}-\{\alpha_j\}\|_{(0)}=\sup_{1\leqslant n<\infty}\|\sigma_n(x_{\alpha(k)})-\sigma_n(x_\alpha)\|=$$
$$=\sup_{1\leqslant n<\infty}\|\sigma_n(x_{\alpha(k)})-y_n\|\to 0\text{ as }k\to\infty,$$

$$\|\{\alpha_j^{(k)}\}-\{\alpha_j\}\|_{(n)}=\sup_{1\leqslant m<\infty}\|\sigma_{nm}(x_{\alpha(k)})-y_{nm}\|\to 0\text{ as }k\to\infty\quad(n=1,2,\ldots).$$

Thus, $\lim_{k\to\infty}\{\alpha_n^{(k)}\}=\{\alpha_n\}$ in $A_1^{(T)}$, which proves that $A_1^{(T)}$ is a Fréchet space.

Now, similarly to the proof of § 10, theorem 10.4, it follows that the mapping $w\colon A_1^{(T)}\to E$ defined by

$$w(\{\alpha_n\})=x_\alpha\quad(\{\alpha_n\}\in A_1^{(T)})\tag{11.17}$$

is an isomorphism of $A_1^{(T)}$ onto $E$ and hence, by the argument of that same proof $\left(\text{replacing again } v_{m_n}(x_i)\text{ by }\left(\sum_{j=i}^{m_n}t_{pnj}\right)x_i\text{ as above}\right)$,

$$f_n\in E^*\quad(n=1,2,\ldots).$$

Furthermore, since $x_j = \sum_{i=1}^{\infty} \delta_{ij} x_i$ and since $T$ is consistent, $\sum_{i=1}^{\infty} \delta_{ij} x_i$ is $T$-summable to $x_j$, whence, by (11.11),

$$f_i(x_j) = \delta_{ij} \qquad (i, j = 1,2, \ldots),$$

i.e., $(x_n, f_n)$ is a biorthogonal system. From definition 11.1 it is clear that $[x_n] = E$ and that $\{f_n\}$ is total on $E$. Thus, $\{x_n\}$ is an $M$-basis of $E$, with the a.s.f. (11.11).

b) Let $P_{(m)} = [x_1, \ldots, x_m]$ $(m = 1,2, \ldots)$ and let

$$v_{nm} = \sigma_{nm}|_{P_{(m)}} \qquad (n, m = 1,2, \ldots). \tag{11.18}$$

Then $v_{nm} \in L(P_{(m)}, P_{(m)})$ $(n, m = 1,2, \ldots)$. Furthermore, by (11.9) and (11.10) (for $x = x_i$), we have

$$\lim_{n\to\infty} \lim_{m\to\infty} v_{nm}(x_i) = \lim_{n\to\infty} \sigma_n(x_i) = x_i \qquad (i = 1,2, \ldots).$$

Now let $x \in E$ and let $\alpha_n = f_n(x)$ $(n = 1,2, \ldots)$. Then

$$s_j\left(\sum_{i=1}^{m} \alpha_i x_i\right) = \sum_{i=1}^{j} \alpha_i x_i = s_j(x) \quad (j = 1, \ldots, m;\ m = 1,2, \ldots), \tag{11.19}$$

whence, by (11.8),

$$v_{nm}\left(\sum_{i=1}^{m} \alpha_i x_i\right) = \sum_{j=1}^{m} t_{nj} s_j\left(\sum_{i=1}^{m} \alpha_i x_i\right) = \sum_{j=1}^{m} t_{nj} s_j(x) = \sigma_{nm}(x)$$

and thus, by (11.9), (11.10),

$$\lim_{n\to\infty} \lim_{m\to\infty} v_{nm}\left(\sum_{i=1}^{m} \alpha_i x_i\right) = \lim_{n\to\infty} \lim_{m\to\infty} \sigma_{nm}(x) = \lim_{n\to\infty} \sigma_n(x) = x.$$

The sequence $\{\alpha_n\} \subset K$ with this latter property is unique. Indeed, if $\lim_{n\to\infty} \lim_{m\to\infty} v_{nm}\left(\sum_{i=1}^{m} \alpha_i x_i\right) = \lim_{n\to\infty} \lim_{m\to\infty} \sigma_{nm}\left(\sum_{i=1}^{m} \alpha_i x_i\right) = 0$, then by (11.3) we have

$$0 = f_k\left(\lim_{n\to\infty} \lim_{m\to\infty} \sum_{j=1}^{m} t_{nj} \sum_{i=1}^{j} \alpha_i x_i\right) = \lim_{n\to\infty} \lim_{m\to\infty} \sum_{j=1}^{m} t_{nj} \sum_{i=1}^{j} \alpha_i \delta_{ik} =$$

$$= \lim_{n\to\infty} \lim_{m\to\infty} \sum_{j=1}^{m} t_{nj}\, \alpha_k = \lim_{n\to\infty} \sum_{j=1}^{\infty} t_{nj} \alpha_k = \alpha_k \qquad (k = 1,2, \ldots),$$

which proves that $\{x_n\}$ is a double generalized summation basis of $E$.

Finally, let $T = (t_{nm})$ be a triangular matrix and let

$$v_n = \sigma_n|_{P_{(n)}} \qquad (n = 1,2, \ldots). \tag{11.20}$$

Then, since $T$ is triangular, we have

$$\sigma_{nm}(x) = \sum_{j=1}^{m} t_{nj}s_j(x) = \sum_{j=1}^{n} t_{nj}s_j(x) \qquad (x \in E, \ m \geqslant n),$$

whence, by (11.9),

$$\sigma_n(x) = \lim_{m\to\infty} \sigma_{nm}(x) = \sum_{j=1}^{n} t_{nj}s_j(x) \qquad (x \in E, \ n = 1,2, \ldots), \quad (11.21)$$

and thus $v_n$ is well defined and $v_n \in L(P_{(n)}, P_{(n)})$. Also, by (11.10) (for $x = x_i$),

$$\lim_{n\to\infty} v_n(x_i) = x_i \qquad (i = 1,2, \ldots).$$

Now, if $x \in E$ and $\alpha_n = f_n(x)$ $(n = 1,2, \ldots)$, then, by (11.19) and (11.21),

$$v_n\left(\sum_{i=1}^{n} \alpha_i x_i\right) = \sigma_n\left(\sum_{i=1}^{n} \alpha_i x_i\right) = \sum_{j=1}^{n} t_{nj}s_j\left(\sum_{i=1}^{n} \alpha_i x_i\right) = \sum_{j=1}^{n} t_{nj}s_j(x) = \sigma_n(x),$$

and hence, by (11.10), $\lim_{n\to\infty} v_n\left(\sum_{i=1}^{n} \alpha_i x_i\right) = x$. As above, the sequence $\{\alpha_n\} \subset K$ with this latter property is unique, so $\{x_n\}$ is a generalized summation basis of $E$, which completes the proof of theorem 11.1.

Let us also note that by (11.6) and (11.21), for a Cesàro basis $\{x_n\}$ with the a.s.f. $\{f_n\} \subset E^*$ we have

$$\sigma_n(x) = \sum_{j=1}^{n} \frac{s_j(x)}{n} = \sum_{i=1}^{n} \frac{n-i+1}{n} f_i(x)x_i \quad (x \in E, \ n=1,2, \ldots). \ (11.22)$$

From theorem 11.1 and § 10, theorem 10.4 and proposition 10.1 (and their extensions to double generalized summation bases and double operational bases) it follows

**Corollary 11.1.** *Every (in particular, every triangular) $T$-basis $\{x_n\}$ of a Banach space $E$ is a double (respectively, an) operational basis of $E$ and $r([f_n]) > 0$, where $\{f_n\} \subset E^*$ is the a.s.f. to $\{x_n\}$.*

*Problem 11.1.* a) Does there exist a (triangular) $T$-basis in every Banach space $E$ with an operational basis (or, equivalently, in every separable space with the bounded approximation property)? b) What about every Banach space $E$ with a double operational basis or a (double) generalized summation basis?

**Corollary 11.2.** a) *Every T-basis* $\{x_n\}$ *of a Banach space E is a strong M-basis of E. Consequently:*

b) *Every separable infinite dimensional Banach space E has an M-basis which is not a T-basis of E for any consistent matrix T.*

*Proof.* a) By theorem 11.1 a), $\{x_n\}$ is an $M$-basis of $E$. Let $\{f_n\} \subset E^*$ be the a.s.f. to $\{x_n\}$. Then, by (11.7)–(11.10), we have

$$x = \lim_{n\to\infty} \lim_{m\to\infty} \sum_{j=1}^{m} t_{nj} \sum_{i=1}^{j} f_i(x) x_i \qquad (x \in E). \tag{11.10'}$$

Hence, if $\{n_k\} \subset \mathcal{N} = \{1,2, \ldots\}$ and $x \in [f_{n_k}]_\perp$, then $x \in [x_j]_{j \in \mathcal{N} \setminus \{n_k\}}$, so $\{x_n\}$ is a strong $M$-basis of $E$.

Finally, part b) is a consequence of part a) and of § 8, proposition 8.12. This completes the proof of corollary 11.2.

Another instance of $M$-bases which are not a $T$-basis for any consistent matrix $T$, is given in

**Proposition 11.1.** *If E is an infinite dimensional Banach space with a basis, then every conditional basis* $\{x_n\}$ *of E has a permutation* $\{x_{p(n)}\}$ *which is not a T-basis of E for any consistent matrix T.*

*Proof.* Note that by Ch. II, § 23, theorem 23.2, $E$ has a conditional basis. If $\{x_n\}$ is any conditional basis of $E$, then by Ch. II, § 16, theorem 16.1 a), implication $22° \Rightarrow 1°$, there exist an element $x = \sum_{i=1}^{\infty} \alpha_i x_i \in E$ and a functional $f \in E^*$ such that the numerical series $f(x) = \sum_{i=1}^{\infty} \alpha_i f(x_i)$ is not absolutely convergent. Take a permutation of this series, say $\sum_{i=1}^{\infty} \alpha_{p(i)} f(x_{p(i)})$, which converges to a number $\beta \neq f(x)$. Then $\{x_{p(n)}\}$ has the required property. Indeed, since $\{x_n\}$ is a basis, $\{x_{p(n)}\}$ is an $M$-basis of $E$. But, since for the "$T$-means" $\sigma_{p,n}(x)$ defined by (11.9) with $\{x_n\}$ replaced by $\{x_{p(n)}\}$ (where $T$ is any consistent matrix) we have $\lim_{n\to\infty} f(\sigma_{p,n}(x)) = \beta \neq f(x)$, $\{x_{p(n)}\}$ is not a $T$-basis of $E$. This completes the proof of proposition 11.1.

Every basis $\{x_n\}$ of a Banach space $E$ is obviously a $T$-basis of $E$, for any consistent matrix $T = (t_{nm})$. A $T$-basis for the identity matrix $T = (\delta_{nm})$ is a basis of $E$, with $\sigma_n = s_n$ $(n = 1,2, \ldots)$. However, there exist $T$-bases which are not bases, such as

*Example 11.1.* By Ch. I, § 4, example 4.1, the sequence

$$x_0(t) \equiv \frac{1}{2}, \quad x_{2n-1}(t) = \sin nt, \quad x_{2n}(t) = \cos nt$$

$$(t \in (-\infty, \infty), \; n = 1,2, \ldots) \tag{11.23}$$

is not a basis of $E = C_{2\pi}$. However, by a theorem of L. Fejér*), $\{x_n\}_0^\infty$ is a Cesàro basis of $E = C_{2\pi}$. Moreover, by the extension of Fejér's theorem due to M. Riesz**) $\{x_n\}_0^\infty$ is also a $T$-basis of $E$ for the Cesàro summation method***) $T = (C, \alpha)$ of order $\alpha$, for any $\alpha > 0$.

*Example 11.2.* In the space $E = A$ the sequence

$$x_n(\zeta) = \zeta^n \qquad (n = 0,1,2, \ldots) \tag{11.24}$$

is not a basis of $E$. Indeed, for the polynomials of Fejér

$$q_n(\zeta) = \frac{\zeta^n}{n} + \frac{\zeta^{n+1}}{n-1} + \ldots + \frac{\zeta^{2n-1}}{1} - \left(\frac{\zeta^{2n+1}}{1} + \frac{\zeta^{2n+2}}{2} + \ldots + \frac{\zeta^{3n}}{n}\right) \tag{11.5}$$

we have****)

$$\|q_n\| = \max_{|\zeta| \leqslant 1} |q_n(\zeta)| \leqslant C \qquad (n = 1,2, \ldots),$$

where $C$ is an absolute constant, but for the partial sum operators $s_n$ associated to the complete minimal sequence $\{x_n\}_0^\infty$ we have, clearly,

$$\|s_{2n-1}(q_n)\| = \max_{|\zeta| \leqslant 1} |s_{2n-1}(q_n)(\zeta)| \geqslant |s_{2n-1}(q_n)(1)| =$$

$$= \frac{1}{n} + \frac{1}{n-1} + \ldots + \frac{1}{1} > \ln n \to \infty \text{ as } n \to \infty.$$

Hence, $\sup\limits_{0 \leqslant n < \infty} \|s_n\| = \infty$ and thus $\{x_n\}_0$ is not a basis of $E$. However, by Fejér's theorem used in example 2.1 above, $\{x_n\}$ is a Cesàro basis of $E = A$.

---

*) See e.g. [411], Ch. III, theorem (3.4) or [13], Ch. I, § 47.

**) See e.g. [411], Ch. III, theorem (5.1) or [13], Ch. VII, § 4.

***) Such $T$-bases are sometimes also called "$(C, \alpha)$-*bases*". We recall that a series $\sum_{i=1}^{\infty} z_i$ is $(C, \alpha)$-*summable* ($\alpha \neq -1, -2, \ldots$) to $z$, if the sequence $\{\sigma_n^{(\alpha)}\}$ defined by

$$\sigma_n^{(\alpha)} = \frac{1}{\binom{\alpha+n-1}{n-1}} \sum_{i=1}^{n} \binom{\alpha+n-i}{n-i} z_i \quad \left(\text{where } \binom{\alpha+n}{n} = \frac{(\alpha+1)(\alpha+2)\ldots(\alpha+n)}{n!}\right)$$

converges to $z$ as $n \to \infty$.

****) See e.g. [411], Ch. VIII, § 1 or [13], Ch. I, § 43.

Let us also give the following example in Hilbert space:
*Example 11.3*. In the real space $E = l^2$ the sequence

$$x_n = e_n - e_{n+1} \qquad (n = 1,2, \ldots) \qquad (11.26)$$

is not a basis of $E$ (where $e_n$ is the $n$-th unit vector). Indeed, for

$$f_n(x) = (e_1 + \ldots + e_n, x) \quad (x \in E, n = 1,2, \ldots) \, (11.27)$$

we have $f_i(x_k) = \left(\sum_{j=1}^{i} e_j, e_k - e_{k+1}\right) = \delta_{ik}$ $(i, k=1,2, \ldots)$. Also, $[x_n]=E$ (since the relations $f \in E^*$, $f(x_n) = f(e_n) - f(e_{n+1}) = 0$ for $n = 1,2, \ldots$ imply $f(e_1) = f(e_2) = \ldots$ whence, by $\lim_{n\to\infty} f(e_n) = 0$ and $[e_n] = E$, we obtain $f = 0$). But, for the $E$-complete biorthogonal system $(x_n, f_n)$ we have

$$\|x_n\| = \sqrt{2}, \ \|f_n\| = \|e_1 + \ldots + e_n\| = \sqrt{n} \quad (n = 1,2, \ldots), \qquad (11.28)$$

and thus, by Ch. I, § 3, theorem 3.1, $\{x_n\}$ is not a basis of $E$. On the other hand, $\{x_n\}$ is a Cesàro basis of $E$. Indeed, by (11.22),

$$\sigma_n(x) = \sum_{i=1}^{n} \frac{n-i+1}{n} f_i(x) x_i = \sum_{i=1}^{n} \frac{n-i+1}{n} \left(\sum_{j=1}^{i} e_j, x\right)(e_i - e_{i+1}) =$$

$$= (e_1, x)e_1 + \sum_{i=2}^{n} \left(\frac{n-i+1}{n} \left(\sum_{j=1}^{i} e_j, x\right) - \frac{n-i+2}{n} \left(\sum_{j=1}^{i-1} e_j, x\right)\right) e_i -$$

$$- \frac{1}{n} \left(\sum_{j=1}^{n} e_j, x\right) e_{n+1} = \sum_{i=1}^{n} \frac{n-i+1}{n} (e_i, x) e_i -$$

$$- \frac{1}{n} \sum_{i=2}^{n+1} \left(\sum_{j=1}^{i-1} e_j, x\right) e_i \qquad (x \in E, n = 1,2, \ldots).$$

Since $\sum_{i=1}^{n} (e_i, x) x_i \to x$, we have $\sum_{i=1}^{n} \frac{n-i+1}{n} (e_i, x) e_i \to x$ as $n \to \infty$, in the norm topology of $E$. Furthermore, let us consider the continuous linear operators $u_n \colon E \to E$ defined by

$$u_n(x) = \frac{1}{n} \sum_{i=1}^{n} \left(\sum_{j=1}^{i} e_j, x\right) e_i \qquad (x \in E, n = 1,2, \ldots). \quad (11.29)$$

Clearly, for each $n, k = 1,2, \ldots$ we have

$$\|u_n(e_k)\|^2 = \frac{1}{n^2} \sum_{i=1}^{n} \left| \left( \sum_{j=1}^{i} e_j, e_k \right) \right|^2 = \frac{1}{n^2} \sum_{i=k}^{n} (e_k, e_k)^2 =$$

$$= \frac{n-k+1}{n^2} \to 0 \text{ as } n \to \infty,$$

hence $u_n(x) \to 0$ for all $x$ in the linear span of $\{e_n\}$, which is dense in $E$. But

$$\|u_n(x)\|^2 = \frac{1}{n^2} \sum_{i=1}^{n} \left| \left( \sum_{j=1}^{i} e_j, x \right) \right|^2 \leqslant \frac{1}{n^2} \sum_{i=1}^{n} \left\| \sum_{j=1}^{i} e_j \right\|^2 \|x\|^2 =$$

$$= \frac{1}{n^2} \sum_{i=1}^{n} i \|x\|^2 = \frac{n(n+1)}{2n^2} \|x\|^2 \leqslant \|x\|^2 \quad (x \in E, n = 1,2, \ldots),$$

so $\sup_{1 \leqslant n < \infty} \|u_n\| < \infty$, and hence $u_n(x) \to 0$ for all $x \in E$. Consequently, $\sigma_n(x) \to x$ for all $x \in E$, which proves that $\{x_n\}$ is a Cesàro basis of $E$.

*Problem 11.2.* Does every Banach space $E$ with a (triangular) $T$-basis have a basis?

As shown by example 11.2, one of the advantages of considering $T$-bases is that it is often much easier to show that a Banach space $E$ has a $T$-basis than to show that it has a basis (indeed, the problem of existence of a basis in $E = A$ was open for a long time*) and has been solved only recently**)). Another advantage of $T$-bases is that quite many of the known results on bases can be extended to $T$-bases. Thus, theorem 11.1 a) and corollary 11.1 above may be also regarded as such extensions of Ch. I, § 3, theorem 3.1 (first part) and Ch. I, § 12, theorem 12.2 a) (inequality $r([f_n]) > 0$) respectively. We shall give now some more results of this type.

Using formula (11.10′) above, the proof of Ch. I, § 4, proposition 4.1 can be adapted to show that *if* $\{x_n\}$ *is a $T$-basis of $E$ and* $\{i_n\} \subset \mathcal{N}$, *then* $\{x_{i_n}\}$ *and* $\{\omega(x_j)\}_{j \in \mathcal{N} \setminus \{i_n\}}$ *are* $T^{(i)}$*-bases* ($i = 1,2$) *of* $[x_{i_n}]$ *and* $E/[x_j]_{j \in \mathcal{N} \setminus \{i_n\}}$ *respectively, where* $\omega \colon E \to E/[x_{i_n}]$ *is the canonical mapping* (it is enough to define $T^{(1)}$ and $T^{(2)}$ by omitting from $T$ all $j$-th columns, where $j \in \mathcal{N} \setminus \{i_n\}$, respectively $j \in \{i_n\}$).

Let us give now some characterization theorems for $T$-bases. We have the following extension to $T$-bases of Ch. I, § 4, theorem 4.1:

---

*) See Ch. I, § 2, problem 2.3 and the corresponding Notes and remarks.
**) See [30].

**Theorem 11.2.** *Let $E$ be a Banach space and $(x_n, f_n)$ $(\{x_n\} \subset E, \{f_n\} \subset E^*)$ a biorthogonal system. Then the following statements are equivalent (where $T$ is a consistent matrix):*

1°. *$\{x_n\}$ is a $T$-basis of $E$.*

2°. *$\{x_n\}$ is a complete in $E$, and the continuous linear operators $\sigma_n$ defined by* (11.9) *exist and satisfy*

$$\sup_{1 \leqslant n < \infty} \|\sigma_n(x)\| < \infty \qquad (x \in E). \tag{11.30}$$

3°. *$\{x_n\}$ is complete in $E$, and the continuous linear operators $\sigma_n$ defined by* (11.9) *exist and satisfy*

$$\sup_{1 \leqslant n < \infty} \|\sigma_n\| < \infty. \tag{11.31}$$

*Proof.* The implication 1° $\Rightarrow$ 2° is obvious and the implication 2° $\Rightarrow$ 3° is a consequence of the principle of uniform boundedness. Finally, if $(x_n, f_n)$ is a biorthogonal system, then for every finite linear combination $p = \sum_{i=1}^{l} \beta_i x_i \in P$ we have $s_j(p) = p$ for $j \geqslant l$, whence $\lim_{j \to \infty} s_j(p) = p$ and therefore $\lim_{n \to \infty} \sigma_n(p) = p$. Hence it follows that 3° $\Rightarrow$ 1°, which completes the proof of theorem 11.2.

Next, let us give characterizations of triangular $T$-bases among $E$-complete biorthogonal systems, which are extensions of some results of Ch. I, § 5. Let $E$ be a Banach space and $(x_n, f_n)$ $(\{x_n\} \subset E, \{f_n\} \subset E^*)$ a biorthogonal system and let $T = (t_{nm})$ be a *triangular* consistent matrix, whence, by (11.21), all $\sigma_n(x)$ exist. We shall denote

$$|||x|||_{(T)} = \sup_{1 \leqslant n < \infty} \|\sigma_n(x)\| \qquad (x \in E), \tag{11.32}$$

where $|||x|||_{(T)} = \infty$ is also possible. Let

$$\mathscr{E}_0^{(T)} = \{x \in E \mid \lim_{n \to \infty} \sigma_n(x) = x\}, \tag{11.33}$$

$$\mathscr{E}_1^{(T)} = \{x \in E \mid \lim_{n \to \infty} \sigma_n(x) \text{ exists}\}, \tag{11.34}$$

$$\mathscr{E}_2^{(T)} = \{x \in E \mid \sup_{1 \leqslant n < \infty} \|\sigma_n(x)\| < \infty\} = \{x \in E \mid |||x|||_{(T)} < \infty\}, \tag{11.35}$$

$$\mathscr{E}_3^{(T)} = \{x \in E \mid \lim_{n \to \infty} \|\sigma_n(x)\| \text{ exists and } < \infty\} \cup$$

$$\cup \{x \in E \mid \lim_{n \to \infty} \|\sigma_n(x)\| \text{ does not exist}\}. \tag{11.36}$$

Then we have the inclusions $\mathscr{E}_0^{(T)} \subset \mathscr{E}_1^{(T)} \subset \mathscr{E}_2^{(T)} \subset \mathscr{E}_3^{(T)}$. Also, if $\{x_n\}$ is complete in $E$, the sets $\mathscr{E}_i^{(T)}$ $(i=0,1,2,3)$ are dense in $E$; indeed, we have seen in the above proof of theorem 11.2, implication $3° \Rightarrow 1°$, that

$$\lim_{n\to\infty} \sigma_n(p) = p \qquad (p \in P), \qquad (11.37)$$

where $P$ is the linear span of $\{x_n\}$. If $\{x_n\}$ is a $T$-basis of $E$, we have

$$\mathscr{E}_0^{(T)} = \mathscr{E}_1^{(T)} = \mathscr{E}_2^{(T)} = \mathscr{E}_3^{(T)} = E.$$

We have the following extension of Ch. I, § 5, proposition 5.1:

**Proposition 11.2.** *Let $E$ be a Banach space, $(x_n, f_n)$ $(\{x_n\} \subset E$, $\{f_n\} \subset E^*)$ a biorthogonal system and $T$ a triangular consistent matrix. We have $\mathscr{E}_0^{(T)} = \mathscr{E}_1^{(T)}$ if and only if $\{f_n\}$ is total on $E$.*

*Proof.* Assume that $\mathscr{E}_0^{(T)} = \mathscr{E}_1^{(T)}$ and let $x \in E$, $f_n(x) = 0$ $(n=1,2,\ldots)$. Then $s_n(x) = 0$ $(n=1,2,\ldots)$, whence $\sigma_n(x) = 0$ $(n=1,2,\ldots)$ and hence, by $\mathscr{E}_0^{(T)} = \mathscr{E}_1^{(T)}$, it follows that $x=0$.

Conversely, assume that $\{f_n\}$ is total on $E$, and let $x \in \mathscr{E}_1^{(T)}$. Then, by biorthogonality,

$$f_m(\sigma_n(x)) = f_m\left(\sum_{j=1}^{\infty} t_{nj} s_j(x)\right) = \left(\sum_{j=m}^{\infty} t_{nj}\right) f_m(x) \qquad (m, n = 1,2,\ldots),$$

whence, by (11.3), (11.2),

$$f_m(x - \lim_{n\to\infty} \sigma_n(x)) = f_m(x) - \lim_{n\to\infty}\left(\sum_{j=m}^{\infty} t_{nj}\right) f_m(x) =$$

$$= f_m(x)\left(1 - \lim_{n\to\infty}\left(\sum_{j=1}^{\infty} t_{nj}\right) + \lim_{n\to\infty}\left(\sum_{j=1}^{m-1} t_{nj}\right)\right) = 0 \qquad (m = 1,2,\ldots).$$

Consequently, since $\{f_n\}$ is total on $E$, $x - \lim_{n\to\infty} \sigma_n(x) = 0$, i.e., $x \in \mathscr{E}_0^{(T)}$, which completes the proof of proposition 11.2.

**Proposition 11.3.** *Let $E$ be a Banach space, $T$ a triangular consistent matrix and $(x_n, f_n)$ $(\{x_n\} \subset E$, $\{f_n\} \subset E^*)$ an $E$-complete biorthogonal system such that $\{x_n\}$ is not a $T$-basis of $E$. Then*

a) *The set $\mathscr{E}_2^{(T)}$ is of the first category.*

b) *The set $E \setminus \mathscr{E}_3^{(T)} = \{x \in E \mid \lim_{n\to\infty} \|\sigma_n(x)\| = \infty\}$ is of the first category.*

c) *If $A$ is a subset of $E$, such that every $x \in A$ is the limit of a sequence $\{y_n\} \subset E$ satisfying $\sup_{1 \leqslant n < \infty} |||y_n|||_{(T)} < \infty$, then $A$ is of the first category.*

Indeed, this is an extension of Ch. I, § 5, proposition 5.2, with an entirely similar proof (see also Ch. I, § 5, remark 5.1), replacing now $s_n$ by $\sigma_n$.

The next theorem is an extension of Ch. I, § 5, theorem 5.1, with a similar, but slightly more computational proof:

**Theorem 11.3.** *Let $E$ be a Banach space, $T$ a triangular consistent matrix and $(x_n, f_n)$ ($\{x_n\} \subset E$, $\{f_n\} \subset E^*$) an $E$-complete biorthogonal system. The following statements are equivalent:*

1°. *$\{x_n\}$ is a $T$-basis of $E$.*

2°. $\mathscr{E}_2^{(T)} = E$.

3°. $\mathscr{E}_1^{(T)} = \mathscr{E}_2^{(T)}$.

4°. $\mathscr{E}_2^{(T)} = \mathscr{E}_3^{(T)}$.

*For every $x \in E$ let $\{p_n(x)\}$ be a sequence of finite linear combinations of the elements $x_n$, such that $\lim_{n\to\infty} p_n(x) = x$. Then the above statements are equivalent to the following:*

5°. $\sup_{1\leqslant n<\infty} |||p_n(x)|||_{(T)} < \infty$.

*Proof.* 3° $\Rightarrow$ 1°. Assume that $\{x_n\}$ is not a $T$-basis of $E$. Then, by the implication 2° $\Rightarrow$ 1° of theorem 11.2, there exists an element $z_0 \in E$ such that

$$\sup_{1\leqslant k<\infty} \|\sigma_k(z_0)\| = \infty. \tag{11.38}$$

Hence, since $[x_n] = E$, there exists $y_1 = \sum_{i=1}^{m_1} f_i(y_1)x_i \in E$ with $\|y_1\| < \frac{1}{4}$, such that $\sup_{1\leqslant k<\infty} \|\sigma_k(y_1)\| = M_1 \geqslant 1$. Then, by $\lim_{k\to\infty} \sigma_k(y_1) = y_1$, there exists $q_1 > m_1$ such that $\|\sigma_k(y_1)\| < \frac{1}{4}$ for $k \geqslant q_1$.

Consider now $G_1 = [x_n]_{n=q_1+1}^{\infty}$ and let $z_1 = z_0 - s_{q_1}(z_0)$. Then $f_n(z_1) = 0$ for all $n \leqslant q_1$, hence $z_1 \in G_1$. Also, $\|\sigma_k(z_0)\| \leqslant \|\sigma_k(s_{q_1}(z_0))\| + \|\sigma_k(z_1)\|$ ($k = 1,2,\ldots$), whence, by (11.37) for $p = s_{q_1}(z_0)$ and by (11.38), $\sup_{1\leqslant k<\infty} \|\sigma_k(z_1)\| = \infty$. Therefore one can repeat the above argument for $G_1$, $\{x_n\}_{n=q_1+1}^{\infty}$ and $z_1$. Continuing in this way, we obtain two increasing sequences of positive integers $\{m_n\}$, $\{q_n\}$ with $1 < m_1 < q_1 < m_2 < q_2 < \ldots$ and a sequence $\{y_n\} \subset E$ with the following properties:

$$y_n = \sum_{i=q_{n-1}+1}^{m_n} f_i(y_n)x_i, \quad \|y_n\| < \frac{1}{4^n} \qquad (n = 1,2, \ldots; \ q_0 = 0), \tag{11.39}$$

$$\sup_{1\leqslant k<\infty} \|\sigma_k(y_n)\| = M_n \geqslant 1 \qquad (n = 1,2, \ldots), \tag{11.40}$$

$$\|\sigma_k(y_n)\| < \frac{1}{4^n} \qquad (k \geqslant q_n; \ n = 1,2, \ldots). \tag{11.41}$$

We shall show that the element

$$x = \sum_{j=1}^{\infty} \frac{1}{M_j} y_j \in E \tag{11.42}$$

satisfies $x \in \mathscr{E}_2^{(T)} \setminus \mathscr{E}_1^{(T)}$, whence $\mathscr{E}_1^{(T)} \neq \mathscr{E}_2^{(T)}$, which will prove that $3^\circ \Rightarrow 1^\circ$. In fact, let $k$ be an arbitrary index and let $n = n(k)$ be such that $q_{n-1} + 1 \leqslant k \leqslant q_n$. Then, by (11.21) and (11.39), $\sigma_k(y_j) = \sum_{l=1}^{k} t_{kl} s_l(y_j) = 0$ for $\geqslant n+1$, whence, by (11.40) and (11.41),

$$\|\sigma_k(x)\| = \left\| \sum_{j=1}^{\infty} \frac{\sigma_k(y_j)}{M_j} \right\| \leqslant \left\| \sum_{j=1}^{n-1} \frac{\sigma_k(y_j)}{M_j} \right\| + \left\| \frac{\sigma_k(y_n)}{M_n} \right\| < \sum_{j=1}^{n-1} \frac{1}{4^j} + 1 < 2,$$

and therefore, since $k$ was arbitrary, $x \in \mathscr{E}_2^{(T)}$. On the other hand, for suitable integers $k_n$ we have, by $\lim_{k\to\infty} \sigma_k(y_n) = y_n$ $(n = 1,2, \ldots)$,

$$\|\sigma_{k_n}(y_n)\| = \max_{1 \leqslant k < \infty} \|\sigma_k(y_n)\| = M_n \geqslant 1 \qquad (n = 1,2, \ldots);$$

here $q_{n-1} + 1 \leqslant k_n \leqslant q_n - 1$ $(n = 1,2, \ldots)$, since by (11.21) and (11.39) $\sigma_k(y_n) = 0$ for $k \leqslant q_{n-1}$ and since by (11.41) $\|\sigma_k(y_n)\| < \frac{1}{4^n}$ for $k \geqslant q_n$. Consequently, since $\sigma_{k_n}(y_j) = 0$ for $j \geqslant n+1$ and $\sigma_{q_{n-1}}(y_j) = \sum_{l=1}^{q_{n-1}} t_{q_{n-1}l} s_l(y_j) = 0$ for $j \geqslant n$, we obtain, using again (11.40) and (11.41),

$$\|\sigma_{k_n}(x) - \sigma_{q_{n-1}}(x)\| = \left\| \sum_{j=1}^{n-1} \left( \frac{\sigma_{k_n}(y_j)}{M_j} - \frac{\sigma_{q_{n-1}}(y_j)}{M_j} \right) + \frac{\sigma_{k_n}(y_n)}{M_n} \right\| \geqslant$$

$$\geqslant 1 - \left\{ \left\| \sum_{j=1}^{n-1} \frac{\sigma_{k_n}(y_j)}{M_j} \right\| + \left\| \sum_{j=1}^{n-1} \frac{\sigma_{q_{n-1}}(y_j)}{M_j} \right\| \right\} > 1 - 2 \sum_{j=1}^{n-1} \frac{1}{4^j} > \frac{1}{3}$$

$$(n = 2,3, \ldots),$$

which proves that $x \notin \mathscr{E}_1^{(T)}$. Thus, $3^\circ \Rightarrow 1^\circ$.

The proof of the other implications is similar to that of the corresponding implications of Ch. I, § 5, theorem 5.1, replacing now $s_n$ by $\sigma_n$. This completes the proof of theorem 11.3.

*Remark 11.1.* The above proof shows that a similar result remains valid in the more general case when $\{\sigma_n\}$ is replaced by a sequence of endomorphisms of finite rank $\{u_n\}$ such that $\lim\limits_{n\to\infty} u_n(p) = p$ for all finite linear combinations $p = \sum\limits_{i=1}^{l} \beta_i x_i$ and $u_k(y) = 0$ for all $y \in [x_n]_{n=k+1}^{\infty}$ $(k = 1,2, \ldots)$ and 1° is replaced by (9.2); for example, one can take $\{u_n\} = \{v_n s_n\}$, where $v_n \in L(P_{(n)}, P_{(n)})$ satisfy (10.66).

We shall now give characterizations of Cesàro bases among $M$-bases, in terms of properties of the set of multipliers, corresponding to Ch. I, § 5, theorem 5.2. To this end, we need some preparation.

For a sequence of scalars $\{\gamma_n\}$ we shall use the notations

$$\Delta\gamma_n = \gamma_n - \gamma_{n+1},\ \Delta^2\gamma_n = \Delta\gamma_n - \Delta\gamma_{n+1} = \gamma_n - 2\gamma_{n+1} + \gamma_{n+2}. \tag{11.43}$$

Thus, $\{\gamma_n\}$ is non-decreasing if and only if all $\gamma_n$ are real and $\Delta\gamma_n \geqslant 0$ $(n = 1,2, \ldots)$; furthermore, $\{\gamma_n\} \in bv$ if and only if*) $\|\{\gamma_n\}\|_{bv} = \sum\limits_{j=1}^{\infty} |\Delta\gamma_j| + \lim\limits_{n\to\infty} |\gamma_n| < \infty$. We recall that a sequence $\{\gamma_n\}$ is called *convex* if all $\gamma_n$ are real and $\Delta^2\gamma_n \geqslant 0$ $(n = 1,2, \ldots)$. A sequence $\{\gamma_n\}$ is said to be *of bounded 2-variation* if $\sum\limits_{j=1}^{\infty} j\,|\Delta^2\gamma_j| < \infty$. We recall

**Lemma 11.2.** *Every bounded sequence $\{\gamma_n\}$ of bounded 2-variation is in $bv$ (and hence convergent).*

*Proof.* We have

$$\Delta\gamma_k - \Delta\gamma_n = \sum_{j=k}^{n-1} (\Delta\gamma_j - \Delta\gamma_{j+1}) = \sum_{j=k}^{n-1} \Delta^2\gamma_j \qquad (k = 1, \ldots, n-1),$$

whence, by summing and adding $\Delta\gamma_n - \Delta\gamma_n = 0$,

$$\sum_{k=1}^{n} \Delta\gamma_k - n\Delta\gamma_n = \sum_{k=1}^{n-1} \sum_{j=k}^{n-1} \Delta^2\gamma_j = \sum_{j=1}^{n-1} j\Delta^2\gamma_j \tag{11.44}$$

and therefore

$$\sum_{j=1}^{n} |\Delta\gamma_j| \leqslant \sum_{j=1}^{n-1} j\,|\Delta^2\gamma_j| + n|\Delta\gamma_n| \qquad (n = 2,3, \ldots). \tag{11.45}$$

---

*) See Ch. I, § 5, formula (5.26).

From $\gamma_1 - \gamma_{n+1} = \sum_{k=1}^{n} \Delta\gamma_k$ and (11.44) we obtain

$$n\Delta\gamma_n = \gamma_1 - \gamma_{n+1} - \sum_{j=1}^{n} j\Delta^2\gamma_j \qquad (n = 2,3, \ldots), \tag{11.46}$$

whence, by our assumptions on $\{\gamma_n\}$,

$$\sup_{1 \leqslant n < \infty} n\,|\Delta\gamma_n| < \infty. \tag{11.47}$$

By (11.45), (11.47) and again $\sum_{j=1}^{\infty} j\,|\Delta^2\gamma_j| < \infty$, it follows that

$$\sum_{j=1}^{\infty} |\Delta\gamma_j| < \infty,$$

i.e., $\{\gamma_n\} \in bv$, which completes the proof of lemma 11.2.

The set $bv^2$ of all bounded sequences $\{\gamma_n\}$ of bounded 2-variation, endowed with the usual vector operations and with the norm

$$\|\{\gamma_n\}\|_{bv^2} = \sum_{j=1}^{\infty} j\,|\Delta^2\gamma_j| + \lim_{n\to\infty} |\gamma_n|, \tag{11.48}$$

is a Banach space. Note also that since $\{\gamma_n\}$ converges, by (11.46) $\{n\Delta\gamma_n\}$ converges. Moreover, we claim that

$$\lim_{n\to\infty} n\Delta\gamma_n = 0 \qquad (\{\gamma_n\} \in bv^2), \tag{11.49}$$

whence, by (11.45) for $n \to \infty$,

$$\sum_{j=1}^{\infty} |\Delta\gamma_j| \leqslant \sum_{j=1}^{\infty} j\,|\Delta^2\gamma_j| \qquad (\{\gamma_n\} \in bv^2), \tag{11.50}$$

and hence $\|\{\gamma_n\}\|_{bv} \leqslant \|\{\gamma_n\}\|_{bv^2}$, i.e., *the inclusion mapping* $bv^2 \to bv$ *is of norm* 1. Indeed, to prove (11.49), assume that $\lim_{n\to\infty} n\Delta\gamma_n = \beta \neq 0$. Then there exists $N$ such that $n|\Delta\gamma_n| > \frac{|\beta|}{2}$ for all $n > N$, whence

$$\sum_{n=1}^{\infty} |\Delta\gamma_n| > \sum_{n=N+1}^{\infty} |\Delta\gamma_n| > \sum_{n=N+1}^{\infty} \frac{|\beta|}{2n} = \infty,$$

in contradiction with lemma 11.2. This proves the claim (11.49) and hence (11.50). Finally, note that by (11.49) and (11.46),

$$\sum_{j=1}^{\infty} j\Delta^2\gamma_j = \gamma_1 - \lim_{n\to\infty} \gamma_n \qquad (\{\gamma_n\} \in bv^2). \tag{11.51}$$

Let us also recall

**Lemma 11.3.** *Every real sequence* $\{\gamma_n\} \in bv^2$ *converging to* 0 *can be written as the difference of two convex sequences converging to* 0.

*Proof.* Put

$$\alpha_n = \sum_{j=1}^{\infty} j\,|\Delta^2\gamma_{j+n-1}| \qquad (n = 1,2, \dots), \tag{11.52}$$

$$\beta_n = \gamma_n + \alpha_n \qquad (n = 1,2, \dots). \tag{11.53}$$

Then

$$\Delta\alpha_n = \alpha_n - \alpha_{n+1} = \sum_{j=1}^{\infty} j|\Delta^2\gamma_{j+n-1}| - \sum_{j=1}^{\infty} j\,|\Delta^2\gamma_{j+n}| =$$

$$= \sum_{j=1}^{\infty} j\,|\Delta^2\gamma_{j+n-1}| - \sum_{j=2}^{\infty} (j-1)\,|\Delta^2\gamma_{j+n-1}| =$$

$$= \Delta^2\gamma_n + \sum_{j=2}^{\infty} |\Delta^2\gamma_{j+n-1}| = \sum_{j=1}^{\infty} |\Delta^2\gamma_{j+n-1}|,$$

whence

$$\Delta^2\alpha_n = \Delta\alpha_n - \Delta\alpha_{n+1} = \sum_{j=1}^{\infty} |\Delta^2\gamma_{j+n-1}| - \sum_{j=1}^{\infty} |\Delta^2\gamma_{j+n}| =$$

$$= \sum_{j=1}^{\infty} |\Delta^2\gamma_{j+n-1}| - \sum_{j=2}^{\infty} |\Delta^2\gamma_{j+n-1}| = |\Delta^2\gamma_n| \geqslant 0 \qquad (n = 1,2, \dots),$$

so $\{\alpha_n\}$ is convex. Furthermore,

$$\alpha_n = \sum_{j=1}^{\infty} j|\Delta^2\gamma_{j+n-1}| = \sum_{j=n}^{\infty} (j-n+1)\,|\Delta^2\gamma_j| \leqslant \sum_{j=n}^{\infty} j\,|\Delta^2\gamma_j|$$

$$(n = 1,2, \dots),$$

whence $\lim_{n\to\infty} \alpha_n = 0$. On the other hand, by the above,

$$\Delta^2\beta_n = \Delta^2\gamma_n + \Delta^2\alpha_n = \Delta^2\gamma_n + |\Delta^2\gamma_n| \geqslant 0 \qquad (n = 1,2,\ldots),$$

so $\{\beta_n\}$ is convex, and $\lim_{n\to\infty} \beta_n = \lim_{n\to\infty} \gamma_n + \lim_{n\to\infty} \alpha_n = 0$. Thus, since by (11.53) $\{\gamma_n\} = \{\beta_n\} - \{\alpha_n\}$, the proof of lemma 11.3 is complete.

*Remark 11.2.* a) The above proof also shows that *every real sequence $\{\gamma_n\} \in bv^2$ can be written as the difference of two bounded convex sequences.* b) Let us observe that *the converse is also true*. Indeed, it is enough to show that *every bounded convex sequence $\{\gamma_n\}$ belongs to $bv^2$*. The convexity of $\{\gamma_n\}$ means that $\Delta(\Delta\gamma_n) = \Delta\gamma_n - \Delta\gamma_{n+1} \geqslant 0$ $(n = 1,2,\ldots)$, whence $\Delta\gamma_1 \geqslant \Delta\gamma_2 \geqslant \ldots$ Now, if there existed $N$ such that $\Delta\gamma_N < 0$, then we would have $\gamma_n - \gamma_N = \sum_{j=N}^{n-1} (-\Delta\gamma_j) \geqslant (n-N)(-\Delta\gamma_N) \to \infty$ as $n \to \infty$, in contradiction with $\sup_{1\leqslant n<\infty} |\gamma_n| < \infty$. Consequently, $\Delta\gamma_n \geqslant 0$ $(n = 1,2,\ldots)$, whence $\gamma_1 \geqslant \gamma_2 \geqslant \ldots$. Therefore $\{\gamma_n\}$ converges and hence $\lim_{n\to\infty} \Delta\gamma_n = 0$. But then $\sum_{k=j}^{\infty} \Delta^2\gamma_k = \sum_{k=j}^{\infty} (\Delta\gamma_k - \Delta\gamma_{k+1}) = \Delta\gamma_j$, whence, since $\Delta^2\gamma_j \geqslant 0$,

$$\sum_{j=1}^{\infty} j|\Delta^2\gamma_j| = \sum_{j=1}^{\infty} j\Delta^2\gamma_j = \sum_{j=1}^{\infty}\sum_{k=j}^{\infty} \Delta^2\gamma_k = \sum_{j=1}^{\infty} \Delta\gamma_j =$$

$$= \sum_{j=1}^{\infty} (\gamma_j - \gamma_{j+1}) = \gamma_1 - \lim_{n\to\infty} \gamma_n, \tag{11.54}$$

so $\{\gamma_n\} \in bv^2$, which completes the proof of our assertion.

Now we are ready to prove the following theorem, which corresponds, in a certain sense*⁾, to Ch. I, § 5, theorem 5.2:

**Theorem 11.4.** *Let $E$ be a Banach space and $\{x_n\}$ an $M$-basis of $E$, with the a.s.f. $\{f_n\} \subset E^*$. The following statements are equivalent:*

1°. *$\{x_n\}$ is a Cesàro basis of $E$.*

2°. *$M(E, (x_n, f_n)) \supset bv^2$.*

3°. *$M(E, (x_n, f_n))$ contains every convex sequence converging to 0.*

---

*⁾ Namely, they are the particular cases $p = 0$ and $p = 1$, respectively, of a general theorem on $(C, p)$-bases and $bv^{p+1}$ (see the Notes and remarks). We recall that $M(E,(x_n, f_n))$ denotes the set of all multipliers of $E$ (see Ch. I, § 5, definition 5.1).

*Proof.* $1^\circ \Rightarrow 2^\circ$. Assume that we have $1^\circ$ and let $\{\gamma_n\} \in bv^2$. Put

$$x_{\{\gamma_n\}} = \sum_{j=1}^{\infty} (j\Delta^2\gamma_j)\,\sigma_j(x) + (\lim_{n\to\infty} \gamma_n)\,x \qquad (x \in E). \qquad (11.55)$$

Note that $x_{\{\gamma_n\}}$ exists, since by $1^\circ$ and theorem 11.2,

$$\|x_{\{\gamma_n\}}\| \leqslant \sum_{j=1}^{\infty} j|\Delta^2\gamma_j|\,\|\sigma_j(x)\| + (\lim_{n\to\infty} |\gamma_n|)\|x\| \leqslant$$

$$\leqslant \left\{\sum_{j=1}^{\infty} j\,|\Delta^2\gamma_j| \sup_{1\leqslant n<\infty} \|\sigma_n\| + \lim_{n\to\infty} |\gamma_n|\right\} \|x\| \qquad (x \in E).$$

Furthermore, by (11.22) and biorthogonality,

$$f_k(\sigma_j(x)) = \begin{cases} f_k\left(\sum_{l=1}^{j} \frac{j-l+1}{j} f_l(x)x_l\right) = \frac{j-k+1}{j} f_k(x) \text{ for } j \geqslant k \\ 0 \text{ for } j < k, \end{cases}$$

whence, by (11.51),

$$f_k(x_{\{\gamma_n\}}) = \sum_{j=k}^{\infty} (j-k+1)\Delta^2\gamma_j f_k(x) + (\lim_{n\to\infty} \gamma_n) f_k(x) =$$

$$= \sum_{j=1}^{\infty} j\Delta^2\gamma_{k+j-1} f_k(x) + (\lim_{n\to\infty} \gamma_n) f_k(x) =$$

$$= (\gamma_k - \lim_{n\to\infty} \gamma_n)\, f_k(x) + (\lim_{n\to\infty} \gamma_n) f_k(x) = \gamma_k f_k(x) \qquad (x \in E, k = 1,2, \ldots),$$

that is, $\{\gamma_n\} \in M(E, (x_n, f_n))$.

$2^\circ \Rightarrow 1^\circ$. Assume that we have $2^\circ$. Then for each $x \in E$ we can define a mapping $u_x\colon bv^2 \to E$, by

$$u_x(\{\gamma_n\}) = x_{\{\gamma_n\}} \sim \sum_{k=1}^{\infty} \gamma_k f_k(x) x_k \qquad (\{\gamma_n\} \in bv^2). \qquad (11.56)$$

Then, as in Ch. I, § 5, proof of theorem 5.2, implication $2^\circ \Rightarrow 1^\circ$, it follows that each $u_x$ is linear and closed, whence also continuous on $bv^2$ (we use now that $\sup_{1\leqslant n<\infty} |\gamma_n| \leqslant \|\{\gamma_n\}\|_{bv} \leqslant \|\{\gamma_n\}\|_{bv^2}$ by (11.50)).

Observe now that for $\sum_{i=1}^{m} \frac{m-i+1}{m} e_i = \left\{1, \frac{m-1}{m}, \frac{m-2}{m}, \ldots \right.$ $\left. \ldots, \frac{1}{m}, 0,0, \ldots \right\} = \{\gamma_n\}$ we have $\Delta\gamma_j = \frac{1}{m}$ for $j = 1, \ldots, m$ and $\Delta\gamma_j = 0$ for $j > m$, whence $\Delta^2\gamma_j = \delta_{jm}\frac{1}{m}$ $(j=1,2, \ldots)$, so $\left\| \sum_{i=1}^{m} \frac{m-i+1}{m} e_i \right\|_{bv^2} =$ $= \sum_{j=1}^{\infty} j|\Delta^2\gamma_j| = 1$. Consequently, by (11.22) and $u_x(e_i) = f_i(x)x_i$ (Ch. I, § 5, formula (5.28)),

$$\|\sigma_m(x)\| = \left\| \sum_{i=1}^{m} \frac{m-i+1}{m} f_i(x)\, x_i \right\| = \left\| u_x \left( \sum_{i=1}^{m} \frac{m-i+1}{m} e_i \right) \right\| \leqslant$$

$$\leqslant \|u_x\| \left\| \sum_{i=1}^{m} \frac{m-i+1}{m} e_i \right\|_{bv^2} = \|u_x\| \qquad (x \in E,\ m = 1,2, \ldots), \tag{11.57}$$

whence, by theorem 11.2 (implication 2° ⇒ 1°), $\{x_n\}$ is a Cesàro basis of $E$.

The implication 2° ⇒ 3° is obvious by remark 11.2 b). Finally, the proof of the implication 3° ⇒ 2° is similar to that of Ch. I, § 5, theorem 5.2, implication 3° ⇒ 2°, using now lemma 11.3 above. This completes the proof of theorem 11.4.

Finally, we shall give now intrinsic characterizations of triangular $T$-bases which are extensions of part of Ch. I, § 7, theorem 7.1. We shall use, for a sequence $\{x_n\} \subset E$ and an arbitrary consistent matrix $T = (t_{nm})$, the notations

$$T_{nm} = \begin{cases} \sum_{j=m}^{n} t_{nj} \text{ if } m \leqslant n \\ 0 \text{ if } m > n \end{cases} \qquad (m, n = 1,2, \ldots), \tag{11.58}$$

$$\Sigma_n \left( \sum_{i=1}^{l} \beta_i x_i \right) = \sum_{i=1}^{l} T_{ni}\beta_i x_i \qquad (\beta_1, \ldots, \beta_l \in K;\ l,n = 1,2, \ldots) \tag{11.59}$$

(note that if $\{x_n\}$ is not finitely linearly independent, then for various representations of $\sum_{i=1}^{l} \beta_i x_i$ as a finite linear combination of the elements $x_n$ (11.59) yields different values for $\Sigma_n \left( \sum_{i=1}^{l} \beta_i x_i \right)$).

**Theorem 11.5.** *Let $E$ be a Banach space, $\{x_n\}$ a complete sequence in $E$ such that $x_n \neq 0$ $(n = 1,2, \ldots)$ and $T$ a triangular consistent matrix. Then the following statements are equivalent:*

*1°. $\{x_n\}$ is a $T$-basis of $E$.*

*2°. $\{x_n\}$ is finitely linearly independent and there exists a constant $C$ with $1 \leqslant C < \infty$, such that we have*

$$\left\| \sum_{i=1}^{l} T_{ni} \beta_i x_i \right\| \leqslant C \left\| \sum_{i=1}^{l} \beta_i x_i \right\| \tag{11.60}$$

*for all positive integers $l$, $n$ and all $\beta_1, \ldots, \beta_l \in K$.*

*3°. $\{x_n\}$ is finitely linearly independent and we have (where $P$ denotes the linear span of $\{x_n\}$)*

$$C' = \sup_{1 \leqslant n < \infty} \sup_{\substack{p \in P \\ \|p\| \leqslant 1}} \|\textstyle\sum_n(p)\| < \infty. \tag{11.61}$$

*If $T_{nn} \neq 0$ $(n = 1,2, \ldots)$, the finitely linearly independence of $\{x_n\}$ in 2° and 3° can be omitted.*

*Proof.* 1° $\Rightarrow$ 2°. If $\{x_n\}$ is a $T$-basis of $E$, then by theorem 11.2 we have $\sup_{1 \leqslant n < \infty} \|\sigma_n\| = C < \infty$, where $\sigma_n$ are the endomorphisms (11.21). But, by (11.58),

$$\begin{aligned} \sigma_n(x) &= \sum_{j=1}^{n} t_{nj} \sum_{i=1}^{j} f_i(x) x_i = \sum_{i=1}^{n} \left( \sum_{j=i}^{n} t_{nj} \right) f_i(x) x_i = \\ &= \sum_{i=1}^{n} T_{ni} f_i(x) x_i \qquad (x \in E,\ n = 1,2, \ldots), \end{aligned} \tag{11.62}$$

whence, by biorthogonality and since $T_{nm} = 0$ for $m > n$,

$$\sigma_n \left( \sum_{i=1}^{l} \beta_i x_i \right) = \sum_{i=1}^{l} T_{ni} \beta_i x_i \qquad (\beta_1, \ldots, \beta_l \in K;\ l, n = 1,2, \ldots). \tag{11.63}$$

Consequently,

$$\left\| \sum_{i=1}^{l} T_{ni} \beta_i x_i \right\| = \left\| \sigma_n \left( \sum_{i=1}^{l} \beta_i x_i \right) \right\| \leqslant C \left\| \sum_{i=1}^{l} \beta_i x_i \right\|,$$

so 1° $\Rightarrow$ 2°. Since $\lim_{n \to \infty} \|\sigma_n(x)\| = \|x\|$ $(x \in E)$, we have $C \geqslant 1$.

The implication 2° ⇒ 3° is obvious by (11.59).

3° ⇒ 1°. Let us first observe that by $t_{nj}=0$ for $j>n$ and (11.3), (11.2),

$$T_{ni}=\sum_{j=i}^{n}t_{nj}=\sum_{j=i}^{\infty}t_{nj}=\sum_{j=1}^{\infty}t_{nj}-\sum_{j=1}^{i-1}t_{nj}\to 1 \text{ as } n\to\infty \quad (i=1,2,\ldots), \tag{11.64}$$

hence for each $i$ there exists $n=n(i)$ such that $T_{ni}\neq 0$.

Assume now that we have 3°. We claim that in this case $\{x_n\}$ is minimal. Indeed, if not, then there exists a positive integer $i$ such that $x_i\in\in[x_1,\ldots,x_{i-1},x_{i+1},\ldots]$ and hence for each $\varepsilon>0$ there exists a finite sequence of scalars $\beta_1^{(\varepsilon)},\ldots,\beta_{i-1}^{(\varepsilon)},\beta_{i+1}^{(\varepsilon)},\ldots,\beta_{m(\varepsilon)}^{(\varepsilon)}$ such that

$$\left\|x_i-\sum_{j=1}^{i-1}\beta_j^{(\varepsilon)}x_j-\sum_{j=i+1}^{m(\varepsilon)}\beta_j^{(\varepsilon)}x_j\right\|<\frac{\varepsilon}{C'}|T_{ni}|,$$

where $n=n(i)>i$ is such that $T_{ni}\neq 0$; clearly, we may assume that $m(\varepsilon)$ is so large that $m(\varepsilon)\geqslant n$. Then, by 3° and $T_{nj}=0$ for $j>n$,

$$\left\|T_{ni}x_i-\sum_{j=1}^{i-1}T_{nj}\beta_j^{(\varepsilon)}x_j-\sum_{j=i+1}^{n}T_{nj}\beta_j^{(\varepsilon)}x_j\right\|=$$

$$=\left\|\Sigma_n\left(x_i-\sum_{j=1}^{i-1}\beta_j^{(\varepsilon)}x_j-\sum_{j=i+1}^{m(\varepsilon)}\beta_j^{(\varepsilon)}x_j\right)\right\|<\varepsilon|T_{ni}|,$$

and therefore

$$\left\|x_i-\sum_{j=1}^{i-1}\frac{T_{nj}}{T_{ni}}\beta_j^{(\varepsilon)}x_j-\sum_{j=i+1}^{n}\frac{T_{nj}}{T_{ni}}\beta_j^{(\varepsilon)}x_i\right\|<\varepsilon.$$

Since $\varepsilon>0$ was arbitrary and since the finite-dimensional subspace $[\{x_j\}_{j=1}^{i-1}\cup\{x_j\}_{j=i+1}^{n}]$ is closed, it follows that $x_i\in[\{x_j\}_{j=1}^{i-1}\cup\{x_j\}_{j=i+1}^{n}]$, in contradiction with the assumption that $\{x_n\}$ is finitely linearly independent. This proves the claim that $\{x_n\}$ is minimal. If*) $T_{nn}\neq 0$ $(n=1,2,\ldots)$, then 3° (or 2°) already implies that $\{x_n\}$ is finitely linearly independent (and hence we obtain, by the above, that $\{x_n\}$ is minimal), as follows: Assume that $\sum_{i=1}^{n}\gamma_ix_i=0$. Then, since $T_{ki}=0$ for $i>k$, from 3° (or 2°) it follows that

$$\left\|\sum_{i=1}^{k}T_{ki}\gamma_ix_i\right\|=\left\|\sum_{i=1}^{n}T_{ki}\gamma_ix_i\right\|=\left\|\Sigma_k\left(\sum_{i=1}^{n}\gamma_ix_i\right)\right\|\leqslant$$

$$\leqslant C'\left\|\sum_{i=1}^{n}\gamma_ix_i\right\|=0 \qquad (k=1,\ldots,n),$$

*) Note that in this case even the preceding part of the proof becomes simpler, since one can take there $n=n(i)=i$.

so $\sum_{i=1}^{k} T_{ki}\gamma_i x_i = 0$ $(k = 1, \ldots, n)$, whence, by $T_{nn} \neq 0$ $(n = 1,2, \ldots)$, we obtain successively that $\gamma_1 x_1 = \gamma_2 x_2 = \ldots = \gamma_n x_n = 0$. Hence, since $x_k \neq 0$ $(k = 1,2, \ldots)$, we get $\gamma_1 = \ldots = \gamma_n = 0$, which proves our assertion.

Define $\sigma_n$ by (11.21). Then, by (11.63), $\sigma_n|_P = \sum_n$, and hence, by 3°, $\sup_{1 \leqslant n < \infty} \|\sigma_n\| < \infty$. Consequently, by theorem 11.2, $\{x_n\}$ is a $T$-basis of $E$, which completes the proof of theorem 11.5.

In general, the assumption of finitely linearly independence of $\{x_n\}$ in 2° and 3° cannot be omitted, as shown by

*Example 11.4*. Let $\{y_n\}$ be a basis of a Banach space $E$, let

$$x_1 = y_1, \; x_n = y_{n-1} \qquad (n = 2,3, \ldots), \tag{11.65}$$

and let $T = (t_{nm})$, where

$$t_{nm} = \begin{cases} 0 & \text{for } n = 1;\ m = 1,2, \ldots \\ \delta_{nm} & \text{for } n = 2,3, \ldots;\ m = 1,2, \ldots \end{cases} \tag{11.66}$$

Then $T$ is a triangular consistent matrix and

$$T_{nm} = \sum_{j=m}^{\infty} t_{nj} = \begin{cases} 0 \text{ for } n = 1;\ m = 1,2, \ldots \\ \sum_{j=m}^{n} \delta_{nj} = 1 \text{ for } m = 1, \ldots, n;\ n = 2,3, \ldots \\ 0 \text{ for } m = n+1,\ n+2, \ldots;\ n = 2,3, \ldots \end{cases} \tag{11.67}$$

whence, for any scalars $\beta_1, \ldots, \beta_l \in K$,

$$\left\| \sum_{i=1}^{l} T_{1i}\beta_i x_i \right\| = 0 \leqslant \left\| \sum_{i=1}^{l} \beta_i x_i \right\|,$$

$$\left\| \sum_{i=1}^{l} T_{ni}\beta_i x_i \right\| = \left\| T_{n1}\beta_1 y_1 + \sum_{i=2}^{l} T_{ni}\beta_i y_{i-1} \right\| = \left\| (\beta_1 + \beta_2) y_1 + \right.$$

$$\left. + \sum_{i=3}^{n} \beta_i y_{i-1} \right\| \leqslant v_{\{y_j\}} \left\| (\beta_1 + \beta_2) y_1 + \sum_{i=3}^{l} \beta_i y_{i-1} \right\| =$$

$$= v_{\{y_j\}} \left\| \sum_{i=1}^{l} \beta_i x_i \right\| \qquad (2 \leqslant n \leqslant l),$$

$$\left\| \sum_{i=1}^{l} T_{ni}\beta_i x_i \right\| = \left\| \sum_{i=1}^{l} \beta_i x_i \right\| \qquad (n > l),$$

so $\{x_n\}$ satisfies (11.60) with $C = v_{\{y_n\}}$, whence also (11.61). However, $\{x_n\}$ is not finitely linearly independent (and hence not a $T$-basis of $E$), since $x_1 = x_2$. Note also that by (11.67) $T_{11} = 0$.

Since the image $\{y_n\}$ of a $T$-basis $\{x_n\}$ under an isomorphism is a $T$-basic sequence (i.e., a $T$-basis of $[y_n]$), the results of Ch. I, §§ 9–10 yield also various stability theorems for $T$-bases.

We have the following property of strong duality for $T$-bases, which is an extension of Ch. I, § 12, theorem 12.1:

**Theorem 11.6.** *Let $\{x_n\}$ be a T-basis of Banach space E, where $T=(t_{nm})$ is a consistent matrix and let $\{f_n\}\subset E^*$ be the a.s.f. to $\{x_n\}$. Then $\{f_n\}$ is a T-basic sequence and we have, in the norm topology of $E^*$,*

$$f=\lim_{n\to\infty}\sigma_n^*(f)=\lim_{n\to\infty}\sum_{j=1}^{\infty}t_{nj}s_j^*(f)\qquad (f\in[f_n]).\tag{11.68}$$

*Proof.* By (11.8) we have $\sigma_{nm}=\sum_{j=1}^{m}t_{nj}s_j$, whence

$$\sigma_{nm}^*=\sum_{j=1}^{m}t_{nj}s_j^*\qquad (n,\ m=1,2,\ldots).\tag{11.69}$$

Hence, by biorthogonality and (11.1) we obtain, in the norm topology of $E^*$,

$$\sigma_{nm}^*(f_k)=\left(\sum_{j=k}^{m}t_{nj}\right)f_k\to\left(\sum_{j=k}^{\infty}t_{nj}\right)f_k \text{ as } m\to\infty\qquad (n,k=1,2,\ldots).$$

On the other hand, by (11.9),

$$\sigma_n^*(f_k)=w^*\text{-}\lim_{m\to\infty}\sigma_{nm}^*(f_k)\qquad (n,k=1,2,\ldots).\tag{11.70}$$

Therefore, in the norm topology, $\sigma_{nm}^*(f_k)\to\sigma_n^*(f_k)=\left(\sum_{j=k}^{\infty}t_{nj}\right)f_k$ as $m\to\infty$, for each $n,k=1,2,\ldots$. Hence, since $\sup_{1\leqslant m<\infty}\|\sigma_{nm}^*\|<\infty$ for $n=1,2,\ldots$ (by (11.9) and the principle of uniform boundedness), we obtain, in the norm topology,

$$\sigma_n^*(f)=\lim_{m\to\infty}\sigma_{nm}^*(f)=\sum_{j=1}^{\infty}t_{nj}s_j^*(f)\qquad (f\in[f_k],\ n=1,2,\ldots).\tag{11.71}$$

Finally, from (11.3), (11.2) we get, in the norm topology,

$$\sigma_n^*(f_k)=\left(\sum_{j=1}^{\infty}t_{nj}-\sum_{j=1}^{k-1}t_{nj}\right)f_k\to f_k \text{ as } n\to\infty\qquad (k=1,2,\ldots),$$

whence, since $\sup_{1\leqslant n<\infty}\|\sigma_n^*\|<\infty$, we obtain (11.68), which completes the proof of theorem 11.6.

We have the following extensions of Ch. I, § 12, corollaries 12.1, 12.2 to $T$-bases:

**Corollary 11.3.** *Let $E$ be a Banach space and $(x_n, f_n)$ ($\{x_n\} \subset E$, $\{f_n\} \subset E^*$) a biorthogonal system. If $\{f_n\}$ is a $T$-basis of $E^*$, then $\{x_n\}$ is a $T$-basis of $E$.*

**Corollary 11.4.** *Let $\{x_n\}$ be a $T$-basis of a reflexive Banach space $E$ and let $\{f_n\} \subset E^*$ be the a.s.f. Then $\{f_n\}$ is a $T$-basis of $E^*$.*

Similarly to Ch. I, § 13, theorem 13.1, one can prove the following extension to $T$-bases of "the weak basis theorem" (with the obvious definition of the notion of a weak $T$-basis):

**Theorem 11.7.** *A sequence $\{x_n\}$ in a Banach space $E$ is a $T$-basis of $E$ if and only if it is a weak $T$-basis of $E$.*

Similarly to Ch. I, § 14, theorem 14.1, one can also prove the following extension to $T$-bases of that theorem of weak duality (with the obvious definition of the notion of a $w^*$-Schauder $T$-basis):

**Theorem 11.8.** *A sequence $\{f_n\}$ in a conjugate Banach space $E^*$ is a $w^*$-Schauder $T$-basis of $E^*$ if and only if $E$ has a $T$-basis $\{x_n\}$ whose a.s.f. is $\{f_n\}$.*

Since every $T$-basis is a (complete) minimal sequence and a generalized basis, the special classes of minimal sequences and generalized bases introduced in § 6 and § 7 define, naturally, corresponding special classes of $T$-bases (e.g. shrinking $T$-bases, which we shall use in Vol. III, Ch. IV). Besides these, let us introduce now some other special classes of $T$-bases, using the matrix $T = (t_{nm})$.

*Definition 11.2.* A $T$-basis $\{x_n\}$ of a Banach space $E$, where $T = (t_{nm})$, is said to be *$T$-boundedly complete*, if the relations $\{\alpha_n\} \subset K$, $\sup_{1 \leqslant n < \infty} \left\| \sum_{j=1}^{\infty} t_{nj} \sum_{i=1}^{j} \alpha_i x_i \right\| < \infty$ imply that $\sum_{i=1}^{\infty} \alpha_i x_i$ is $T$-summable.

In particular, if $T$ is a triangular matrix and $T_{nm}$ are the numbers defined by (11.58), a triangular $T$-basis $\{x_n\}$ is $T$-boundedly complete if and only if the relations $\{\alpha_n\} \subset K$, $\sup_{1 \leqslant n < \infty} \left\| \sum_{i=1}^{n} T_{ni} \alpha_i x_i \right\| < \infty$ imply that the sequence $\left\{ \sum_{i=1}^{n} T_{ni} \alpha_i x_i \right\}$ converges. The following extension of § 7, corollary 7.2 b) to triangular $T$-bases shows that this property is actually independent of the matrix $T = (t_{nm})$:

**Proposition 11.4.** *Let $\{x_n\}$ be a triangular $T$-basis of a Banach space $E$. The following statements are equivalent:*

1°. *$\{x_n\}$ is $T$-boundedly complete.*

2°. *$\{x_n\}$ is boundedly complete as an $M$-basis*) of $E$.*

---

*) See § 7, definition 7.8.

*Proof.* Let $\{f_n\} \subset E^*$ be the a.s.f. to $\{x_n\}$. Assumet hat we have 1° and let $\{y_j\} \subset E$ be a sequence such that $\sup\limits_{1 \leqslant j < \infty} \|y_j\| = M < \infty$ and that the limits $\lim\limits_{j \to \infty} f_n(y_j) = \alpha_n$ $(n = 1,2, \ldots)$ exist, whence $\sum\limits_{k=1}^{n} T_{nk}\alpha_k x_k = \lim\limits_{j \to \infty} \sum\limits_{k=1}^{n} T_{nk} f_k(y_j) x_k$ $(n = 1,2, \ldots)$. Then, by (11.62) and since $\{x_n\}$ is a $T$-basis, we have

$$\left\| \sum_{k=1}^{n} T_{nk} f_k(y_j) x_k \right\| = \|\sigma_n(y_j)\| \leqslant v_{\{x_k\}}^{(T)} M < \infty \qquad (n = 1,2, \ldots),$$

where $v_{\{x_k\}}^{(T)} = \sup\limits_{1 \leqslant n < \infty} \|\sigma_n\| < \infty$. Consequently, $\sup\limits_{1 \leqslant n < \infty} \left\| \sum\limits_{k=1}^{n} T_{nk}\alpha_k x_k \right\| < \infty$, and therefore, by 1°, the sequence $\left\{ \sum\limits_{k=1}^{n} T_{nk}\alpha_k x_k \right\}_{n=1}^{\infty}$ converges to an element $x \in E$. Then, by biorthogonality and (11.64), $f_i(x) = \lim\limits_{n \to \infty} T_{ni}\alpha_i = \alpha_i$ $(i = 1,2, \ldots)$. Thus, $1° \Rightarrow 2°$.

Conversely, assume that we have 2° and let $\{\alpha_n\} \subset K$, $\sup\limits_{1 \leqslant n < \infty} \left\| \sum\limits_{k=1}^{n} T_{nk}\alpha_k x_k \right\| < \infty$. Then for $y_j = \sum\limits_{k=1}^{j} T_{jk}\alpha_k x_k$ $(j = 1,2, \ldots)$

we have $\sup\limits_{1 \leqslant j < \infty} \|y_j\| < \infty$ and, by biorthogonality and (11.64), $f_n(y_j) = T_{jn}\alpha_n \to \alpha_n$ as $j \to \infty$ $(n = 1,2, \ldots)$. Hence, by 2°, there exists $x \in E$ such that $f_n(x) = \alpha_n$ $(n = 1,2, \ldots)$. Therefore, since $\{x_n\}$ is a $T$-basis, $\sum\limits_{k=1}^{n} T_{nk}\alpha_k x_k = \sum\limits_{k=1}^{n} T_{nk} f_k(x) x_k \to x$. Thus, $2° \Rightarrow 1°$, which completes the proof.

*Definition 11.3*. A $T$-basis $\{x_n\}$ of a Banach space $E$ is said to be *$T$-unconditional*, if every permutation $\{x_{p(n)}\}$ of $\{x_n\}$ is a $T$-basis of $E$ (for that given $T$).

By corollary 11.1, every $T$-unconditional $T$-basis is an unconditional double operational basis. Hence, by § 10, theorem 10.3 (extended to unconditional double operational bases) we have

**Theorem 11.9.** *Every $T$-unconditional $T$-basis of a Banach space $E$ is an unconditional basis of $E$.*

Using theorem 11.9, we obtain the following result, related to corollary 11.2 b) and proposition 11.1:

**Proposition 11.5.** *Let $E$ be an infinite dimensional separable Banach space and let $T$ be a consistent matrix. Then every $M$-basis $\{x_n\}$ of $E$ which is not an unconditional basis of $E$ has a permutation which is not a $T$-basis of $E$ (for that given $T$).*

*Proof*. Note first that such an $M$-basis $\{x_n\}$ exists, since by § 8, theorem 8.1 every separable Banach space $E$ has an $M$-basis and since by Ch. II, § 23, theorem 23.2 every infinite dimensional space $E$ with a basis has a conditional basis. Now, by theorem 11.9, $\{x_n\}$ is not a $T$-unconditional $T$-basis of $E$, so $\{x_n\}$ has a permutation $\{x_{p(n)}\}$ which is not a $T$-basis of $E$, which completes the proof of proposition 11.5.

We have seen in the above that various results on bases can be extended to $T$-bases. Let us give now some properties of $T$-bases which are *not* possessed by bases.

By Ch. I, § 3, corollary 3.1 a), for the a.s.c.f. $\{f_n\} \subset E^*$ to a basis $\{x_n\}$ of a Banach space $E$, with $\inf\limits_{1\leqslant n<\infty} \|x_n\| > 0$, we have $\sup\limits_{1\leqslant n<\infty} \|f_n\| < \infty$. This property is also possessed by some $T$-bases which are not bases, as shown by example 11.1, but not by all $T$-bases, as shown by example 11.3, in which $E = l^2$ and $\{x_n\}$ is a Cesàro basis with $\|x_n\| = \sqrt{2}$, $\|f_n\| = \sqrt{n}$ $(n = 1,2, \ldots)$. Thus, there arises naturally the problem of the order of growth of the sequences $\{\|f_n\|\}$.

**Proposition 11.6.** *Let* $\{x_n\}$ *be a Cesàro basis of a Banach space* $E$, *with the a.s.f.* $\{f_n\} \subset E^*$. *If* $\inf\limits_{1\leqslant n<\infty} \|x_n\| > 0$, *then there exists a constant* $C > 0$ *such that*

$$\|f_n\| \leqslant Cn \qquad (n = 1,2, \ldots). \tag{11.72}$$

*Proof*. Clearly, $|f_1(x)| = \dfrac{1}{\|x_1\|} \|s_1(x)\| = \dfrac{1}{\|x_1\|} \|\sigma_1(x)\| \leqslant \dfrac{\|\sigma_1\|}{\|x_1\|} \|x\|$ $(x \in E)$. Furthermore,

$$n\sigma_n(x) - (n-1)\sigma_{n-1}(x) = \sum_{j=1}^{n} s_j(x) - \sum_{j=1}^{n-1} s_j(x) = s_n(x) \qquad (x \in E,\ n = 1,2, \ldots), \tag{11.73}$$

whence

$$|f_n(x)| = \frac{1}{\|x_n\|} \|s_n(x) - s_{n-1}(x)\| = \frac{1}{\|x_n\|} \|n\sigma_n(x) - 2(n-1)\sigma_{n-1}(x) + $$

$$+ (n-2)\sigma_{n-2}(x)\| \leqslant \frac{\sup\limits_{1\leqslant j<\infty} \|\sigma_j\|}{\inf\limits_{1\leqslant j<\infty} \|x_j\|} (4n-4)\|x\| \quad (x \in E,\ n = 2,3, \ldots).$$

Consequently, we have (11.72) with $C = \dfrac{4 \sup\limits_{1\leqslant j<\infty} \|\sigma_j\|}{\inf\limits_{1\leqslant j<\infty} \|x_j\|}$, which completes the proof of proposition 11.6.

In general, the estimate (11.72) cannot be improved, since there exist Cesàro bases for which it is exact, as shown by

*Example 11.5.* In the real space $E = l^2$ let

$$x_{2n-1} = e_{2n-1} + \alpha_n e_{2n}, \; x_{2n} = e_{2n-1} + \beta_n e_{2n} \qquad (n = 1,2, \ldots), \quad (11.74)$$

where $\{e_n\}$ is the unit vector basis of $E$ and where

$$|\alpha_n| \leqslant 1, \; |\beta_n| \leqslant 1, \; |\alpha_n - \beta_n| = \frac{A}{n} \qquad (n = 1,2, \ldots) \quad (11.75)$$

for some constant $A > 0$. Then $\{x_n\}$ is a Cesàro basis of $E$, with the a.s.f. $\{f_n\} \subset E^*$ satisfying

$$1 \leqslant \|x_n\| \leqslant \sqrt{2}, \; \|f_n\| \geqslant C'n \quad (n = 1,2, \ldots) \qquad (11.76)$$

for some constant $C' > 0$. Indeed, to show first that $\{x_n\}$ is minimal, let (with the usual identification $E \equiv E^*$)

$$f_{2n-1} = \frac{1}{\beta_n - \alpha_n}(\beta_n e_{2n-1} - e_{2n}), \; f_{2n} = \frac{1}{\beta_n - \alpha_n}(-\alpha_n e_{2n-1} + e_{2n}) \quad (11.77)$$
$$(n = 1,2, \ldots).$$

Then obviously, $(x_n, f_n)$ is a biorthogonal system, so $\{x_n\}$ is minimal. Furthermore, putting $\delta_i = \beta_i - \alpha_i$, we have, for any $x \in E$ and $n = 1,2, \ldots$

$$s_{2n}(x) = \sum_{i=1}^{2n} f_i(x) x_i = \sum_{i=1}^{n} \frac{1}{\delta_i}(\beta_i e_{2i-1} - e_{2i}, x)(e_{2i-1} + \alpha_i e_{2i}) +$$
$$+ \sum_{i=1}^{n} \frac{1}{\delta_i}(-\alpha_i e_{2i-1} + e_{2i}, x)(e_{2i-1} + \beta_i e_{2i}) = \quad (11.78)$$
$$= \sum_{i=1}^{n} (e_{2i-1}, x) e_{2i-1} + \sum_{i=1}^{n} (e_{2i}, x) e_{2i} = \sum_{i=1}^{2n} (e_i, x) e_i = \mathscr{S}_{2n}(x),$$

$$s_{2n-1}(x) = s_{2n}(x) - f_{2n}(x) x_{2n} =$$
$$= \mathscr{S}_{2n}(x) - \frac{1}{\delta_n}\{-\alpha_n(e_{2n-1}, x) + (e_{2n}, x)\}(e_{2n-1} + \beta_n e_{2n}),$$

whence

$$\sigma_{2n}(x) = \frac{1}{2n}\sum_{j=1}^{2n} s_j(x) = \frac{1}{2n}\sum_{j=1}^{n} s_{2j}(x) + \frac{1}{2n}\sum_{j=1}^{n} s_{2j-1}(x) =$$

$$= \frac{1}{n}\sum_{j=1}^{n} \mathscr{S}_{2j}(x) + \frac{1}{2n}\sum_{j=1}^{n}\frac{1}{\delta_j}\{\alpha_j(e_{2j-1}, x) - (e_{2j}, x)\}(e_{2j-1} + \beta_j e_{2j}),$$

and, taking also into account (11.73),

$$\sigma_{2n-1}(x) = \frac{2n}{2n-1}\sigma_{2n}(x) - \frac{1}{2n-1}s_{2n}(x) =$$

$$= \frac{1}{2n-1}\left\{2\sum_{j=1}^{n-1}\mathscr{S}_{2j}(x) + \mathscr{S}_{2n}(x)\right\} +$$

$$+ \frac{1}{2n-1}\sum_{j=1}^{n}\frac{1}{\delta_j}\{\alpha_j(e_{2j-1}, x) - (e_{2j}, x)\}(e_{2j-1} + \beta_j e_{2j}).$$

Since $\mathscr{S}_{2n}(x) \to x$ $\left(\text{whence } \frac{1}{n}\sum_{j=1}^{n}\mathscr{S}_{2j}(x) \to x\right)$, we have $\frac{2}{2n-1}\sum_{j=1}^{n-1}\mathscr{S}_{2j}(x) \to x$ and $\frac{\mathscr{S}_{2n}(x)}{2n-1} \to 0$ as $n \to \infty$. Finally, let us consider the continuous linear operators $u_n\colon E \to E$ defined by

$$u_n(x) = \frac{1}{n}\sum_{j=1}^{n}\frac{1}{\delta_j}\{\alpha_j(e_{2j-1}, x) - (e_{2j}, x)\}(e_{2j-1} + \beta_j e_{2j}) \tag{11.79}$$

$$(x \in E,\ n = 1,2, \ldots).$$

By (11.75), for each $n$, $k = 1,2, \ldots$ we have

$$\|u_n(e_k)\|^2 = \frac{1}{n^2}\sum_{j=1}^{n}\frac{1}{|\delta_j|^2}|\alpha_j(e_{2j-1}, e_k) - (e_{2j}, e_k)|^2(1 + |\beta_j|^2) \leqslant$$

$$\leqslant \frac{1}{n^2}\frac{1}{|\delta_{\left[\frac{k+1}{2}\right]}|^2}(1 + |\beta_j|^2) \leqslant \frac{2}{n^2|\delta_{\left[\frac{k+1}{2}\right]}|^2} \to 0 \text{ as } n \to \infty,$$

whence $u_n(x) \to 0$ for all $x$ in the linear span of $\{e_n\}$, which is dense in $E$. But, by (11.75) and $\sum_{j=1}^{\infty}|(e_j, x)|^2 \leqslant \|x\|^2$,

$$\|u_n(x)\|^2 = \frac{1}{n^2}\sum_{j=1}^{n}\frac{1}{|\delta_j|^2}|\alpha_j(e_{2j-1}, x) - (e_{2j}, x)|^2(1 + |\beta_j|^2) \leqslant$$

$$\leqslant \frac{4}{n^2}\sum_{j=1}^{n}\frac{j^2}{A^2}(|(e_{2j-1}, x)|^2 + |(e_{2j}, x)|^2) \leqslant \frac{4}{A^2}\|x\|^2 \ (x \in E,\ n = 1,2, \ldots),$$

so $\sup_{1 \leqslant n < \infty} \|u_n\| < \infty$ and hence $u_n(x) \to 0$ for all $x \in E$. Consequently, $\sigma_n(x) \to x$ for all $x \in E$, which proves that $\{x_n\}$ is a Cesàro basis of $E$. Also, by (11.74), (11.75) we have $1 \leqslant \|x_n\| \leqslant \sqrt{2}$ $(n = 1,2, \dots)$, and by (11.77), (11.75) we have

$$\|f_{2n-1}\| = \frac{n}{A}(|\beta_n|^2 + 1) \geqslant \frac{1}{A} n, \; \|f_{2n}\| = \frac{n}{A}(|\alpha_n|^2 + 1) \geqslant \frac{1}{A} n$$

$$(n = 1,2, \dots),$$

that is, (11.76) with $C' = \dfrac{1}{A}$, which completes the proof of the assertions of example 11.5.

It is natural to ask whether for every consistent matrix $T = (t_{nm})$ one can give a "limiting" order of growth of $\{\|f_n\|\}$. The answer is negative, as shown by the following slight modification of example 11.5 above:

*Example 11.6.* Let $T = (t_{nm})$ be the consistent matrix corresponding to the convergence with respect to the subsequence of indices $\{i_n\} = \{2n\} \subset \mathcal{N}$, that is, the triangular matrix

$$t_{2n-1,m} = \delta_{2n-2,m}, \; t_{2n,m} = \delta_{2n,m} \qquad (n, m = 1,2, \dots). \quad (11.80)$$

Let*) $E = l^2$ and define $\{x_n\}$ by (11.74), where

$$0 < \alpha_n < \beta_n < 1 \qquad (n = 1,2, \dots). \qquad (11.81)$$

Then for $\{f_n\} \subset E^*$ defined by (11.77) we have $f_i(x_j) = \delta_{ij}$ $(i, j = 1,2, \dots)$, so $\{x_n\}$ is minimal. Furthermore, by (11.80) and (11.78),

$$\sigma_{2n}(x) = \sum_{j=1}^{2n} t_{2n,j} s_j(x) = s_{2n}(x) = \mathscr{S}_{2n}(x) \qquad (x \in E, \; n = 1,2, \dots),$$

$$\sigma_{2n-1}(x) = \sum_{j=1}^{2n-1} t_{2n-1,j} s_j(x) = s_{2n-2}(x) = \mathscr{S}_{2n-2}(x) \quad (x \in E, \; n=1,2,\dots).$$

Hence, since $\mathscr{S}_n(x) \to x$, we have $\sigma_n(x) \to x$ for all $x \in E$, so $\{x_n\}$ is a $T$-basis of $E$. Also, $1 \leqslant \|x_n\| \leqslant 2$ $(n = 1,2, \dots)$. However, by a suitable choice of $\alpha_n$ and $\beta_n$ (it is enough to take $\alpha_n$ very near to $\beta_n$), we can make $\|f_n\|$ arbitrarily large.

In Ch. I, § 19, corollary 19.1 and remark 19.2 we have seen that a complete minimal sequence $\{x_n\}$ in a Banach space $E$ is a basis of $E$ if and only if the approximations $\|x - s_n(x)\|$ by the associated partial sum operators are of the same order as the best approximations $\mu_n(x) = \inf_{y \in P_{(n)}} \|x - y\|$ (where $P_{(n)} = [x_1, \dots, x_n]$), i.e. if and only if there

*) More generally, this example works, with the same argument, if $E$ is any Banach space with a basis $\{e_n\}$ and if we replace in (11.77) $\{e_n\}$ by the a.s.c.f. $\{h_n\}$ to $\{e_n\}$.

exists a constant $C \geqslant 1$ depending only on the sequence $\{x_n\}$, such that

$$\|x - s_n(x)\| \leqslant C\mu_n(x) \qquad (x \in E,\ n = 1,2, \ldots). \tag{11.82}$$

Now we shall show that if $\{x_n\}$ is a *T*-basis of $E$, inequalities of type (11.82) still hold for those elements $x \in E$ for which the sequence $\{n_k\} = \{n \in \mathcal{N} \mid f_n(x) \neq 0\}$ is "sufficiently rare". We recall*) that a sequence $\{n_k\} \subset \mathcal{N}$ is called *lacunary* if $\inf\limits_{1 \leqslant k < \infty} \dfrac{n_{k+1}}{n_k} = \lambda > 1$. We shall denote by $\Lambda$ the class of all lacunary sequences and by $\Lambda_\sigma$ the class of all sequences which are the union of a finite number of lacunary sequences. We recall

**Lemma 11.4.** *If* $\{n_k\} \subset \Lambda_\sigma$, *then*

$$\sup_{1 \leqslant n < \infty} (N'(2n) - N'(n)) < \infty, \tag{11.83}$$

*where* $N'(n) = \sum\limits_{n_k \leqslant n} 1$ *(i.e., $N'(n)$ is the number of terms of the sequence $\{n_k\}$ which do not exceed $n$).*

*Proof.* Since $\{n_k\} \in \Lambda_\sigma$, we can write $\{n_k\} = \bigcup\limits_{j=1}^{r} \{n_k^{(j)}\}$, where $\{n_k^{(j)}\}_{k=1}^{\infty} \in \Lambda$ $(j = 1, \ldots, r)$. Let $N_j'(n) = \sum\limits_{n_k^{(j)} \leqslant n} 1$ $(j = 1, \ldots, r)$. Then

$$N'(2n) - N'(n) = \sum_{j=1}^{r} (N_j'(2n) - N_j'(n)) \leqslant r \max_{1 \leqslant j \leqslant r} (N_j'(2n) - N_j'(n)),$$

and hence it will be enough to prove (11.83) for lacunary sequences. Thus, let $\{n_k\} \in \Lambda$, say $\dfrac{n_{k+1}}{n_k} \geqslant \lambda > 1$ $(k = 1,2, \ldots)$. Fix $n$ and let $p = \max\limits_{n_k \leqslant n} k$. Then either $\{n_k\} \cap (n, 2n] = \varnothing$ or $\{n_k\} \cap (n, 2n] = \{n_{p+1}, n_{p+2}, \ldots, n_{p+N'(2n)-N'(n)}\}$. Hence, in the first case $N'(2n) - N'(n) = 0$, while in the second case

$$2 = \frac{2n}{n} > \frac{n_{p+N'(2n)-N'(n)}}{n_{p+1}} = \frac{n_{p+N'(2n)-N'(n)}}{n_{p+N'(2n)-N'(n)-1}} \cdots \frac{n_{p+2}}{n_{p+1}} \geqslant \lambda^{N'(2n)-N'(n)-1}.$$

Consequently, since $\log \lambda > 0$ (because $\lambda > 1$), we obtain

$$N'(2n) - N'(n) < 1 + \frac{\log 2}{\log \lambda},$$

which, since $n$ was arbitrary, completes the proof of lemma 11.4.

We recall that a series $\sum\limits_{j=1}^{\infty} y_j$ in a Banach space $E$ is said to have *the gap* $(p, r)$, if $y_j = 0$ for $p + 1 \leqslant j \leqslant r - 1$. We shall need

*) See § 4, formula (4.1).

**Lemma 11.5.** *If a series* $\sum_{j=1}^{\infty} y_j$ *in a Banach space* $E$ $\left(\text{with partial sums } s_n = \sum_{j=1}^{n} y_j \text{ and with } \sigma_n = \sum_{j=1}^{n} \frac{s_j}{n}\right)$ *has the gap* $(p, r)$ *and if* $\frac{r}{p} \geqslant \lambda > 1$, *then*

$$\|s_p\| \leqslant \frac{\lambda+1}{\lambda-1} \sup_{1 \leqslant n < \infty} \|\sigma_n\|. \tag{11.84}$$

*Proof.* Since $y_{p+1} = \ldots = y_{r-1} = 0$ and since $\sum_{j=1}^{n} s_j = n\sigma_n$, we have

$$(r-p)s_p = \sum_{j=p}^{r-1} s_j = \sum_{j=1}^{r-1} s_j - \sum_{j=1}^{p-1} s_j = (r-1)\sigma_{r-1} - (p-1)\sigma_{p-1},$$

whence

$$\|s_p\| = \left\| \frac{(r-1)\sigma_{r-1} - (p-1)\sigma_{p-1}}{r-p} \right\| \leqslant \frac{r+p-2}{r-p} \sup_{1 \leqslant n < \infty} \|\sigma_n\|.$$

But, by $2\lambda p \leqslant 2r$ we have $(\lambda-1)(r+p) \leqslant (\lambda+1)(r-p)$, whence, by $\lambda > 1$, $\frac{r+p-2}{r-p} \leqslant \frac{r+p}{r-p} \leqslant \frac{\lambda+1}{\lambda-1}$, which, together with the above inequality, gives (11.84). This completes the proof of lemma 11.5.

Now we are ready to prove

**Theorem 11.10.** *Let* $\{x_n\}$ *be a Cesàro basis of a Banach space* $E$, *with* $\sup_{1 \leqslant n < \infty} \|x_n\| \leqslant M < \infty$ *and with the a.s.f.* $\{f_n\} \subset E^*$ *and let* $x \in E$ *and* $\{n_k\} = \{n \in \mathcal{N} \mid f_n(x) \neq 0\}$.

a) *If* $\{n_k\} \in \Lambda_\sigma$ *and* $\sup_{1 \leqslant k < \infty} \|f_{n_k}\| \leqslant M_1 < \infty$, *then we have*

$$\|x - s_n(x)\| \leqslant C\mu_n(x) \qquad (n = 1, 2, \ldots), \tag{11.85}$$

*where* $C \geqslant 1$ *does not depend on* $n$ *or* $x$.

b) *If* $\{n_k\} \in \Lambda$, *then we have*

$$\|x - s_{n_k}(x)\| \leqslant C'\mu_{n_k}(x) \qquad (k = 1, 2, \ldots), \tag{11.86}$$

*where* $C' \geqslant 1$ *does not depend on* $k$ *or* $x$.

*Proof.* a) Let us consider the de la Vallée-Poussin means

$$\rho_n(x) = \frac{1}{n+1} \sum_{j=n}^{2n} s_j(x) \qquad (x \in E, n = 1, 2, \ldots). \tag{11.87}$$

We have

$$\rho_n(x) = \frac{\sum_{j=1}^{2n} s_j(x) - \sum_{j=1}^{n-1} s_j(x)}{n+1} = \frac{2n\sigma_{2n}(x) - (n-1)\sigma_{n-1}(x)}{n+1}, \tag{11.88}$$

whence, since $\{x_n\}$ is a $T$-basis,

$$\|\rho_n(x)\| \leqslant \left(\frac{3n-1}{n+1} \sup_{1 \leqslant j < \infty} \|\sigma_j\|\right) \|x\| = C_1 \|x\|$$

$$(x \in E, n = 1,2, \ldots). \tag{11.89}$$

Now let $\pi_n(x) \in P_{(n)} = [x_1, \ldots, x_n]$ be a polynomial of best approximation of $x$ (that is, $\|x - \pi_n(x)\| = \mu_n(x)$). Then, by (11.87) and (11.89),

$$\|\rho_n(x) - \pi_n(x)\| = \|\rho_n(x - \pi_n(x))\| \leqslant C_1 \|x - \pi_n(x)\| = C_1 \mu_n(x),$$

whence

$$\|\rho_n(x) - x\| \leqslant \|\rho_n(x) - \pi_n(x)\| + \|\pi_n(x) - x\| \leqslant (C_1 + 1)\mu_n(x). \tag{11.90}$$

Furthermore, since $f_j(x) = 0$ for $j \notin \{n_k\}$ and $f_{n_k}(\pi_n(x)) = 0$ for $n < n_k$, we have

$$\rho_n(x) - s_n(x) = \frac{1}{n+1} \sum_{j=n}^{2n} (s_j(x) - s_n(x)) = \frac{1}{n+1} \sum_{j=n+1}^{2n} \sum_{i=n+1}^{j} f_i(x) x_i =$$

$$= \frac{1}{n+1} \sum_{j=n+1}^{2n} (2n + 1 - j) f_j(x) x_j =$$

$$= \frac{1}{n+1} \sum_{n < n_k \leqslant 2n} (2n + 1 - n_k) f_{n_k}(x - \pi_n(x)) x_{n_k}.$$

But, since $\{n_k\} \in \Lambda_\sigma$, by lemma 11.4 the number of terms of this sum is $N'(2n) - N'(n) \leqslant C_2$ for some constant $C_2$ and each term of this sum has norm $\leqslant \dfrac{2n+1-n_k}{n+1} \|f_{n_k}\| \, \|x - \pi_n(x)\| \, \|x_{n_k}\| \leqslant 2MM_1\mu_n(x)$, whence, taking into account (11.90), we obtain

$$\|x - s_n(x)\| \leqslant \|x - \rho_n(x)\| + \|\rho_n(x) - s_n(x)\| \leqslant$$
$$\leqslant (C_1 + 1)\mu_n(x) + 2MM_1C_2\mu_n(x),$$

i.e. (11.85) with $C = C_1 + 1 + 2MM_1C_2$.

b) Let $\frac{n_{k+1}}{n_k} \geqslant \lambda > 1$ $(k = 1,2, \ldots)$. Since $f_j(x) = 0$ for $j \notin \{n_k\}$ and $f_j(\pi_{n_k}(x)) = 0$ for $n_k + 1 \leqslant j)$, we have $f_j(x - \pi_{n_k}(x)) = 0$ for $n_k + 1 \leqslant j \leqslant n_{k+1} - 1$. Thus, for each $k \in \mathcal{N}$, the series $\sum_{j=1}^{\infty} f_j(x - \pi_{n_k}(x))x_j$ has the gap $(n_k, n_{k+1})$, whence, by lemma 11.5 (with $y_j = f_j(x - \pi_{n_k}(x))x_j$, $p = n_k$, $r = n_{k+1}$), and since $\sup_{1 \leqslant n < \infty} \|\sigma_n\| \leqslant C_3 < \infty$ (because $\{x_n\}$ is a Cesàro basis),

$$\|s_{n_k}(x) - \pi_{n_k}(x)\| = \|s_{n_k}(x - \pi_{n_k}(x))\| \leqslant \frac{\lambda + 1}{\lambda - 1} \sup_{1 \leqslant n < \infty} \|\sigma_n(x - \pi_{n_k}(x))\| \leqslant$$

$$\leqslant \frac{\lambda + 1}{\lambda - 1} C_3 \|x - \pi_{n_k}(x)\| = C_4 \mu_{n_k}(x) \qquad (k = 1,2, \ldots).$$

Consequently, we have

$$\|x - s_{n_k}(x)\| \leqslant \|x - \pi_{n_k}(x)\| + \|\pi_{n_k}(x) - s_{n_k}(x)\| \leqslant (1 + C_4)\mu_{n_k}(x)$$

$$(k = 1, 2, \ldots),$$

i.e. (11.86) with $C' = 1 + C_4$, which completes the proof of theorem 11.10.

The assumption $\sup_{1 \leqslant k < \infty} \|f_{n_k}\| < \infty$ in theorem 11.10 a) cannot be omitted, even if we assume $\{n_k\} \in \Lambda$, as shown by

*Example 11.7.* In the real space $E = l^2$ define $\{x_n\}$ as in example 11.5, with $\alpha_n = 0$, $\beta_n = \frac{1}{n}$ $\left(\text{hence } \delta_n = \beta_n - \alpha_n = \frac{1}{n} \text{ for } n = 1,2, \ldots\right)$ and let $n_k = 2^k$ $(k = 1,2, \ldots)$, so $\{n_k\} \in \Lambda$. Put

$$x^{(n)} = \sum_{k=1}^{n+1} x_{2^k} \qquad (n = 1,2, \ldots).$$

Then, by biorthogonality, we have

$$s_{2^{n+1}-1}(x^{(n)}) = \sum_{j=1}^{2^{n+1}-1} f_j\left(\sum_{k=1}^{n+1} x_{2^k}\right)x_j = \sum_{k=1}^{n} x_{2^k}$$

(where $\{f_n\}$ is the a.s.f. to $\{x_n\}$), whence

$$\|x^{(n)} - s_{2^{n+1}-1}(x^{(n)})\| = \|x_{2^{n+1}}\| > 1 \qquad (n = 1,2, \ldots). \tag{11.91}$$

On the other hand,

$$\mu_{2^{n+1}-1}(x^{(n)}) \leqslant \|x^{(n)} - \sum_{k=1}^{n} x_{2^k} - x_{2^{n+1}-1}\| = \|x_{2^{n+1}} - x_{2^{n+1}-1}\| =$$

$$= \|(e_{2^{n+1}-1} + \beta_{2^n} e_{2^{n+1}}) - e_{2^{n+1}-1}\| = |\beta_{2^n}| = \frac{1}{2^n} \quad (n = 1,2, \dots),$$

which, together with (11.91) shows that we do not have (11.85) with $C$ independent of $x$ and $n$. The norms $\|x_n\|$, $\|f_n\|$ satisfy now (11.76).

The assumption $\{n_k\} \in \Lambda$ in theorem 11.10 b) cannot be replaced by the weaker assumption $\{n_k\} \in \Lambda_\sigma$, as shown by

*Example 11.8.* Define $\{x_n\}$ as in example 11.7 above and let $\{n_k\} = \{2^k - 1\}_{k=1}^{\infty} \cup \{2^k\}_{k=1}^{\infty}$, so $\{n_k\} \in \Lambda_\sigma$. Let $x \in E$ be such that $\{n_k\} = \{n \in \mathcal{N} | f_n(x) \neq 0\}$. Then, by (11.78),

$$s_{2^k}(x) = \mathcal{S}_{2^k}(x) \qquad (k = 1,2, \dots),$$

$$s_{2^k-1}(x) = \mathcal{S}_{2^k}(x) - 2^{k-1}(e_{2^k}, x)\left(e_{2^k-1} + \frac{1}{2^{k-1}} e_{2^k}\right) \qquad (k = 1,2, \dots),$$

whence

$$\|s_{2^k}(x) - s_{2^k-1}(x)\| = \left\| 2^{k-1}(e_{2^k}, x)\left(e_{2^k-1} + \frac{1}{2^{k-1}} e_{2^k}\right)\right\| =$$

$$= 2^{k-1}|(e_{2^k}, x)| \left(1 + \frac{1}{2^{2k-2}}\right)^{\frac{1}{2}} \geqslant 2^{k-1}|(e_{2^k}, x)| \qquad (k = 1,2, \dots).$$

Now let, in particular,

$$x = \sum_{k=1}^{\infty} \frac{1}{2^{\frac{k}{2}}} (e_{2^k-1} + e_{2^k});$$

then, by (11.77), $x$ satisfies $\{n_k\} = \{n \in \mathcal{N} | f_n(x) \neq 0\}$. However, $(e_{2^k}, x) = \frac{1}{2^{\frac{k}{2}}}$, whence

$$2^{\frac{k}{2}-1} = 2^{k-1}|(e_{2^k}, x)| \leqslant \|s_{2^k}(x) - s_{2^k-1}(x)\| \leqslant$$

$$\leqslant \|x - s_{2^k-1}(x)\| + \|x - s_{2^k}(x)\| \qquad (k = 1,2, \dots).$$

Since $\{e_n\}$ is a basis of $E$, we have $s_{2^k}(x) = \mathscr{S}_{2^k}(x) \to x$, whence we obtain

$$\lim_{k\to\infty} \|x - s_{2^k-1}(x)\| = \infty,$$

which shows that we do not have (11.86) with $C'$ independent of $k$ and $x$.

We have the following corollary of theorem 11.10 b):

**Corollary 11.5.** *Every Cesàro basis $\{x_n\}$ of a Banach space $E$ is a basis with respect to the class $\Lambda$ of all lacunary sequences of positive integers.**)

*Proof.* Clearly, $[x_n] = E$. Now let $\{n_k\} \in \Lambda$ and let $x \in [x_{n_k}]$. Then, by biorthogonality, $f_j(x) = 0$ $(j \in \mathscr{N} \setminus \{n_k\})$, whence for $\{n'_k\} = \{n \in \mathscr{N} | f_n(x) \neq 0\}$ we have $\{n'_k\} \subset \{n_k\}$, so $\{n'_k\} \in \Lambda$ (where $\{f_n\} \subset E^*$ is the a.s.f. to $\{x_n\}$). Hence, by theorem 11.10 b),

$$\|x - s_{n'_k}(x)\| \leqslant C'\mu_{n'_k}(x) \qquad (k = 1,2, \ldots), \qquad (11.86')$$

where $C'$ does not depend on $k$ or $x$. But, for the partial sum operators $S_m(x) = \sum_{i=1}^{m} f_{n_i}(x)x_{n_i}$ with respect to $\{x_{n_k}\}$ we have, by the definition of $\{n'_k\}$,

$$S_m(x) = s_{n'_k}(x) \quad (n'_k \leqslant n_m < n'_{k+1};\ m = 1,2, \ldots)$$

and hence, by (11.86'), $x = \lim_{m\to\infty} S_m(x)$. Thus, $\{x_{n_k}\}$ is a basic sequence, which completes the proof of corollary 11.5.

The converse of corollary 11.5 is not valid, as shown e.g. by § 4, theorem 4.1 (see also § 4, corollary 4.1).

*Definition 11.4.* A sequence $\{x_n\}$ in a Banach space $E$ is called a *strongly series summable M-basis* of $E$ if it is an $M$-basis of $E$ and if there exists a triangular matrix of scalars $(\lambda_{nm})$ such that

$$x = \lim_{n\to\infty} \sum_{i=1}^{n} \lambda_{ni} f_i(x) x_i \qquad (x \in E), \qquad (11.92)$$

where $\{f_n\} \subset E^*$ is the a.s.f. to $\{x_n\}$. We shall say that $(\lambda_{nm})$ is a *summability matrix* for $\{x_n\}$ and we shall also use the notations

$$\tau_n(x) = \sum_{i=1}^{n} \lambda_{ni} f_i(x) x_i \qquad (x \in E, n = 1,2, \ldots). \qquad (11.93)$$

---

*) See § 4, definition 4.1.

In particular, by (11.93) and biorthogonality we have

$$\tau_n(x_i) = \lambda_{ni} x_i \quad (i = 1, \ldots, n; n = 1,2, \ldots), \tag{11.94}$$

whence

$$\tau_n(x) = \sum_{i=1}^{n} f_i(x)\tau_n(x_i) = \tau_n s_n(x) \quad (x \in E, n = 1,2, \ldots), \tag{11.95}$$

and thus *every strongly series summable M-basis* $\{x_n\}$ *of E is an operational basis* of $E$; moreover, by (11.94) and § 10, remark 10.2, it is *even a generalized summation basis* of $E$. Also, by (11.94) and (11.92),

$$\lim_{n\to\infty} \lambda_{ni} x_i = \lim_{n\to\infty} \tau_n(x_i) = x_i \quad (i = 1,2, \ldots),$$

whence, since $x_i \neq 0$ $(i = 1,2, \ldots)$, we obtain

$$\lim_{n\to\infty} \lambda_{ni} = 1 \quad (i = 1,2, \ldots). \tag{11.96}$$

*Problem 11.3.* a) Does there exist a strongly series summable $M$-basis in every Banach space $E$ with an operational basis (or, equivalently, in every separable space with the bounded approximation property)? b) What about every Banach space $E$ with a generalized summation basis?

Since every strongly series summable $M$-basis $\{x_n\}$ is an operational basis, from § 10, proposition 10.1 it follows that *for every such* $\{x_n\}$ *we have* $r([f_n]) > 0$, where $\{f_n\} \subset E^*$ is the a.s.f. to $\{x_n\}$. Furthermore, since every strongly series summable $M$-basis is obviously a strong $M$-basis, from § 8, proposition 8.12 it follows that *every separable infinite dimensional Banach space E has an M-basis which is not strongly series summable.*

Every basis $\{x_n\}$ of a Banach space $E$ is obviously a strongly series summable $M$-basis of $E$, with respect to the summability matrix $(\lambda_{nm}) = (\delta_{nm})$ (but not with respect to an arbitrary summability matrix satisfying (11.96)) and conversely, a strongly series summable $M$-basis $\{x_n\}$ for $(\lambda_{nm}) = (\delta_{nm})$ is a basis of $E$, with $\tau_n = s_n$ $(n = 1,2, \ldots)$.

*Problem 11.4.* Does every Banach space $E$ with a strongly series summable $M$-basis have a basis?

Strongly series summable $M$-bases are a generalization of triangular $T$-bases, as shown by

**Proposition 11.7.** *Every triangular T-basis* $\{x_n\}$ *of a Banach space E is a strongly series summable M-basis of E with respect to the summability*

*matrix* $(\lambda_{nm}) = (T_{nm})$ *(where* $T_{nm}$ *are the numbers defined by* (11.58)), *which satisfies*

$$\sum_{i=1}^{\infty} |\lambda_{ni} - \lambda_{n,i+1}| \leqslant M < \infty \qquad (n = 1,2, \ldots). \qquad (11.97)$$

*Conversely, if* $\{x_n\}$ *is a strongly series summable M-basis of E with respect to a summability matrix* $(\lambda_{nm})$ *satisfying* (11.97), *then* $\{x_n\}$ *is a T-basis of E with respect to the triangular consistent matrix*

$$t_{nm} = \lambda_{nm} - \lambda_{n,m+1} \qquad (n, m = 1,2, \ldots). \qquad (11.98)$$

*Proof.* If $\{x_n\}$ is a triangular $T$-basis, then by (11.62) we have $\sum_{i=1}^{n} T_{ni} f_i(x) x_i = \sigma_n(x) \to x$ for all $x \in E$, so $\{x_n\}$ is a strongly series summable $M$-basis with respect to the matrix $(\lambda_{nm}) = (T_{nm})$ and we have $\tau_n = \sigma_n$ $(n = 1,2, \ldots)$. Also, by (11.1), $\sum_{i=1}^{\infty} |T_{ni} - T_{n,i+1}| = \sum_{i=1}^{\infty} |t_{ni}| \leqslant$ $\leqslant M < \infty$ $(n = 1,2, \ldots)$.

Conversely, assume now that $\{x_n\}$ is a strongly series summable $M$-basis of $E$ with respect to a summability matrix $(\lambda_{nm})$ satisfying (11.97) and define $T = (t_{nm})$ by (11.98). Then by (11.97) we have (11.1). Also, by (11.96) and since $(\lambda_{nm})$ is triangular, $\sum_{j=i}^{\infty} t_{nj} = \sum_{j=i}^{\infty} (\lambda_{nj} - \lambda_{n,j+1}) =$ $= \lambda_{ni} \to 1$ as $n \to \infty$ $(i = 1,2, \ldots)$, so we have (11.3) and $t_{ni} = \sum_{j=i+1}^{\infty} t_{nj} -$ $- \sum_{j=i}^{\infty} t_{nj} \to 1 - 1 = 0$ as $n \to \infty$ $(i = 1,2, \ldots)$, that is, (11.2). Thus, $T = (t_{nm})$ is a consistent matrix. Furthermore, since $(\lambda_{nm})$ is triangular, so is $T$. Finally, the numbers $T_{nm}$ defined by (11.58) are now

$$T_{nm} = \sum_{j=m}^{\infty} t_{nj} = \sum_{j=m}^{\infty} (\lambda_{nj} - \lambda_{n,j+1}) = \lambda_{nm} \quad (n, m = 1,2, \ldots),$$

whence, by (11.62) and (11.93), (11.92), $\sigma_n(x) = \sum_{i=1}^{n} T_{ni} f_i(x) x_i = \tau_n(x) \to x$ for all $x \in E$, so $\{x_n\}$ is a $T$-basis of $E$, which completes the proof.

The assumption (11.97) in the second part of proposition 11.7 cannot be omitted, as shown by

*Example 11.9.* Let $\{x_n\}$ be an unconditional basis of a Banach space $E$, with the a.s.c.f. $\{f_n\} \subset E^*$, let

$$t_{nm} = \frac{1}{\ln 2} \frac{(-1)^{n-m}}{n-m+1} \text{ for } m = 1, \ldots, n; \; t_{nm} = 0 \text{ for } m > n$$

$$(n = 1,2, \ldots) \qquad (11.99)$$

and let $(\lambda_{nm}) = (T_{nm})$, where $T_{nm}$ are the numbers defined by (11.58). Then $t_{nj} = \frac{1}{\ln 2} \frac{(-1)^{n-j}}{n-j+1} \to 0$ as $n \to \infty$ $(j = 1,2, \ldots)$ and

$$\sum_{j=1}^{\infty} t_{nj} = \frac{1}{\ln 2} \sum_{k=1}^{n} \frac{(-1)^{k-1}}{k} \to 1 \text{ as } n \to \infty,$$

so we have (11.2) and (11.3), whence also

$$\lambda_{ni} = T_{ni} = \sum_{j=i}^{\infty} t_{nj} = \sum_{j=1}^{\infty} t_{nj} - \sum_{j=1}^{i-1} t_{nj} \to 1 \text{ as } n \to \infty \quad (i = 1,2, \ldots),$$

i.e. (11.96). Consequently,

$$\tau_n(x_j) = \sum_{i=1}^{n} \lambda_{ni} f_i(x_j) x_i = \lambda_{nj} x_j \to x_j \text{ as } n \to \infty \quad (j = 1,2, \ldots).$$

Now, since $\{x_n\}$ is an unconditional basis of $E$, we have

$$\|\tau_n(x)\| = \left\| \sum_{i=1}^{n} \frac{1}{\ln 2} \left( \sum_{j=i}^{n} \frac{(-1)^{n-j}}{n-j+1} \right) f_i(x) x_i \right\| \leqslant v_{\{x_j\}}^{(u)} C \|x\|$$

$$(x \in E, n = 1,2, \ldots)$$

where $C = \frac{1}{\ln 2} \max_{\substack{1 \leqslant i \leqslant n \\ 1 \leqslant n < \infty}} \left| \sum_{j=i}^{n} \frac{(-1)^{n-j}}{n-j+1} \right| < \infty$. Hence $\tau_n(x) \to x$ for all $x \in E$ and thus $\{x_n\}$ is a strongly series summable $M$-basis of $E$ with the summability matrix $(\lambda_{nm})$. Also, the triangular matrix $T = (t_{nm})$ defined by (11.99) satisfies

$$t_{nm} = \sum_{j=m}^{\infty} t_{nj} - \sum_{j=m+1}^{\infty} t_{nj} = T_{nm} - T_{n,m+1} = \lambda_{nm} - \lambda_{n,m+1}$$

$$(n, m = 1,2, \ldots)$$

i.e. (11.98). However, $T = (t_{nm})$ is non-consistent, since

$$\sum_{j=1}^{\infty} |t_{nj}| = \frac{1}{\ln 2} \sum_{j=1}^{n} \frac{1}{n-j+1} = \frac{1}{\ln 2} \sum_{k=1}^{n} \frac{1}{k} \to \infty \text{ as } n \to \infty$$

(that is, (11.1) is violated); this latter relation, together with (11.98), also shows that $(\lambda_{nm})$ does not satisfy (11.97).

Moreover, *there exist strongly series summable M-bases which are not a T-basis for any consistent matrix* $T=(t_{nm})$, e.g. a suitable permutation of any conditional basis of a Banach space. Indeed, this follows from proposition 11.1 combined with

**Proposition 11.8.** *Every permutation* $\{x_{\rho(n)}\}$ *of a strongly series summable M-basis* $\{x_n\}$ *of a Banach space E is a strongly series summable M-basis of E.*

*Proof.* For each $m\in\mathcal{N}$ let $k_m>k_{m-1}$ $(k_0=0)$ be such that $\{1,\ldots,m\}\subset \subset\{\rho(i)\}_{i=1}^{k_m}$ and, if $(\lambda_{ni})$ is a summability matrix for $\{x_n\}$, let

$$\lambda_{ni}^{(\rho)}=\begin{cases}0 \text{ for } i=1,2,\ldots & (n=1,\ldots,k_1-1)\\ \lambda_{m,\rho(i)} \text{ for } i\in\{\rho^{-1}(j)\}_{j=1}^{m} & (n=k_m,\ldots,k_{m+1}-1;\ m=1,2,\ldots)\\ 0 \text{ for } i\in\mathcal{N}\setminus\{\rho^{-1}(j)\}_{j=1}^{m} & (n=k_m,\ldots,k_{m+1}-1;\ m=1,2,\ldots).\end{cases}$$

Then, by the definition of $k_m$,

$$\{\rho^{-1}(j)\}_{j=1}^{m}=\rho^{-1}(\{1,\ldots,m\})\subset\rho^{-1}(\{\rho(i)\}_{i=1}^{k_m})=\{1,\ldots,k_m\}$$

$$(m=1,2,\ldots),$$

and hence $(\lambda_{ni}^{(\rho)})$ is a triangular matrix. Furthermore, if $\{f_n\}\subset E^*$ is the a.s.f. to $\{x_n\}$,

$$\tau_n^{(\rho)}(x)=\sum_{i=1}^{n}\lambda_{ni}^{(\rho)}f_{\rho(i)}(x)x_{\rho(i)}=\sum_{i\in\{\rho^{-1}(j)\}_{j=1}^{m}}\lambda_{m,\rho(i)}f_{\rho(i)}(x)x_{\rho(i)}=$$

$$=\sum_{j=1}^{m}\lambda_{mj}f_j(x)x_j=\tau_m(x)\quad(x\in E,\ n=k_m,\ldots,k_{m+1}-1;\ m=1,2,\ldots),$$

whence $\lim_{n\to\infty}\tau_n^{(\rho)}(x)=\lim_{m\to\infty}\tau_m(x)=x$ for all $x\in E$. Thus, $\{x_{\rho(n)}\}$ is a strongly series summable $M$-basis of $E$, which completes the proof.

*Remark 11.3.* A similar result also holds for operational bases (by the proof of § 10, theorem 10.1), but not for $T$-bases (see proposition 11.1).

*Problem 11.5.* Does every Banach space $E$ with a strongly series summable $M$-basis $\{x_n\}$ have a triangular $T$-basis?

Returning now to problem 11.3 a), we shall show (in theorem 11.11 below) that if $E$ is a separable *complex* Banach space such that $E^*$ has the bounded approximation property, then the answer is affirmative. To this end, let us first prove

**Proposition 11.9.** *Let $E$ be a complex Banach space, $(x_i, f_i)_{i=1}^n$ ($\{x_i\}_{i=1}^n \subset E$, $\{f_i\}_{i=1}^n \subset E^*$) a finite biorthogonal system, $u$ an operator of finite rank on $E$ such that*

$$u(x_i) = x_i \qquad (i = 1, \ldots, n), \tag{11.100}$$

$$u^*(f_i) = f_i \qquad (i = 1, \ldots, n), \tag{11.101}$$

*and let $\varepsilon > 0$. Then there exist a finite biorthogonal system $(x_i, f_i)_{i=n+1}^{n+m}$ ($\{x_i\}_{i=n+1}^{n+m} \subset E$, $\{f_i\}_{i=n+1}^{n+m} \subset E^*$) and a finite set of complex numbers $\{\lambda_i\}_{i=1}^m$ such that $(x_i, f_i)_{i=1}^{n+m}$ is a biorthogonal system and that*

$$\|u - v\| < \varepsilon, \tag{11.102}$$

*where $v$ is the operator on $E$ defined by*

$$v(x) = \sum_{i=1}^n f_i(x)x_i + \sum_{i=n+1}^{n+m} \lambda_{i-n} f_i(x)x_i \quad (x \in E). \tag{11.103}$$

*Proof.* Put

$$s_n(x) = \sum_{i=1}^n f_i(x)x_i \qquad (x \in E), \tag{11.104}$$

$$E_0 = (I - s_n)(E). \tag{11.105}$$

If $u = s_n$, then one can take any $(x_i, f_i)_{i=n+1}^{n+m}$ such that $(x_i, f_i)_{i=1}^{n+m}$ is a biorthogonal system and any $\lambda_1, \ldots, \lambda_m$ such that $\sum_{i=1}^m |\lambda_i| < \dfrac{\varepsilon}{\max\limits_{n+1 \leqslant i \leqslant n+m} \|f_i\| \|x_i\|}$.

Assume now that $u \neq s_n$. By (11.100) and (11.101) we have

$$us_n(x) = \sum_{i=1}^n f_i(x)u(x_i) = s_n(x) \qquad (x \in E), \tag{11.106}$$

$$s_n u(x) = \sum_{i=1}^n f_i(u(x))x_i = \sum_{i=1}^n u^*(f_i)(x)x_i = s_n(x) \quad (x \in E). \tag{11.107}$$

Hence, if $x \in E_0$, say $x = (I - s_n)(y)$, then $u(x) = u(y) - us_n(y) = u(y) - s_n u(y) \in (I - s_n)(E) = E_0$ and thus

$$u(E_0) \subset E_0; \tag{11.108}$$

also, if $y \in E$ is such that $u(y) \neq s_n(y)$, then for $x = (I - s_n)(y) \in E_0$ we have $u(x) = u(y) - us_n(y) = u(y) - s_n(y) \neq 0$, whence $u(E_0) \neq \{0\}$. Furthermore, if $x \in \operatorname{Ker} u$, then $s_n(x) = s_n u(x) = 0$, whence $x = x - s_n(x) \in (I - s_n)(E) = E_0$ and thus

$$\operatorname{Ker} u \subset E_0. \tag{11.109}$$

Since $\dim (u(E) \cap \operatorname{Ker} u) \leqslant \dim u(E) < \infty$, let $G$ be a complement of $u(E) \cap \operatorname{Ker} u$ in $\operatorname{Ker} u$, that is,

$$\operatorname{Ker} u = (u(E) \cap \operatorname{Ker} u) \oplus G. \tag{11.110}$$

Then, by (11.109), $G \subset \operatorname{Ker} u \subset E_0$, so

$$(u(E) \cap E_0) \cap G = u(E) \cap G = (u(E) \cap \operatorname{Ker} u) \cap G = \{0\},$$

whence, since $\operatorname{codim}_{E_0} \operatorname{Ker} u = \dim E_0/\operatorname{Ker} u = \dim u(E_0) \leqslant \dim u(E) < \infty$ and $\operatorname{codim}_{\operatorname{Ker} u} G < \infty$, there exists a complement $G_0$ of $G$ in $E_0$ containing $u(E) \cap E_0$, that is, we can write

$$E_0 = G \oplus G_0, \tag{11.111}$$

$$G_0 \supset u(E) \cap E_0, \ \dim G_0 = m < \infty. \tag{11.112}$$

Then, by (11.108) and (11.112),

$$u(G_0) \subset u(E_0) \subset u(E) \cap E_0 \subset G_0; \tag{11.113}$$

also, $\{0\} \neq u(E_0) \subset G_0$, whence $\dim G_0 \geqslant 1$.

Choose*) a basis $\{z_1, \ldots, z_m\}$ of $G_0$ such that the matrix representation $(\alpha_{ij})_{i,j=1}^m$ for $u|_{G_0}$ with respect to $\{z_1, \ldots, z_m\}$ is lower triangular (i.e., $\alpha_{ij} = 0$ for $j > i$), so

$$u(z_i) = \sum_{j=1}^{i} \alpha_{ij} z_j \qquad (i = 1, \ldots, m). \tag{11.114}$$

By (11.111), let $p$ be the projection of $E_0$ onto $G_0$ along $G$ (i.e., with $p(E_0) = G_0$, $\operatorname{Ker} p = G$). Choose a sequence $\lambda_1, \ldots, \lambda_m$ of distinct

*) Since the scalars are complex, such a basis exists (see e.g. [155], Ch. II, § 56, theorem 2).

complex numbers, sufficiently close to $\alpha_{11}, \ldots, \alpha_{mm}$ respectively, such that the operator $q$ on $G_0$ defined by

$$q(z_i) = \sum_{j=1}^{i-1} \alpha_{ij} z_j + \lambda_i z_i \qquad (i = 1, \ldots, m) \tag{11.115}$$

satisfies

$$\|u|_{G_0} - q\| < \frac{\varepsilon}{\|I - s_n\| \|p\|}. \tag{11.116}$$

Since the matrix representation (11.115) for $q$ is lower triangular, $\{\lambda_i\}_{i=1}^m$ is the set of eigenvalues of $q$, whence, since the $\lambda_i$'s are distinct, there exists a basis $\{x_i\}_{i=n+1}^{n+m}$ of $G_0$ such that

$$q(x_i) = \lambda_{i-n} x_i \qquad (i = n+1, \ldots, n+m). \tag{11.117}$$

Then, since $x_{n+1}, \ldots, x_{n+m} \in E_0 = (I - s_n)(E) = \{x \in E | f_i(x) = 0$ $(i = 1, \ldots, n)\}$ (by (11.105), (11.104) and the biorthogonality of $(x_i, f_i)_{i=1}^n$), we have $f_i(x_{n+j}) = 0$ $(i = 1, \ldots, n;\ j = 1, \ldots, m)$. Furthermore, since by the above

$$E = s_n(E) \oplus E_0 = s_n(E) \oplus G_0 \oplus G = [x_i]_{i=1}^n \oplus [x_i]_{i=n+1}^m \oplus G, \tag{11.118}$$

there exist functionals $f_{n+1}, \ldots, f_{n+m} \in ([x_i]_{i=1}^n \oplus G)^\perp$ such that $f_{n+i}(x_{n+j}) = \delta_{ij}$ $(i, j = 1, \ldots, m)$. Clearly, $(x_i, f_i)_{i=1}^{n+m}$ is a biorthogonal system and hence, by (11.106), for $v$ defined by (11.103) we have $v|_{s_n(E)} = s_n|_{s_n(E)} = u|_{s_n(E)}$. Also, since $f_1, \ldots, f_n \in E_0^\perp \subset G^\perp$ and $f_{n+1}, \ldots, f_{n+m} \in G^\perp$ and since $G \subset \operatorname{Ker} u$, we have $v|_G = 0 = u|_G$. Finally, if $x \in G_0 = [x_{n+1}, \ldots, x_{n+m}]$, then $s_n(x) = 0$, whence, by (11.117), $v(x) = q\left(\sum_{i=n+1}^{n+m} f_i(x) x_i\right) = q(x)$, so $v|_{G_0} = q$. Consequently*), by (11.118)

---

*) Indeed, we have

$$\|(u - v)(x)\| \leqslant \|(u - v)(s_n(x))\| + \|(u - v)(I - s_n)(x)\| \leqslant$$
$$\leqslant \|(u - v)|_{s_n(E)}\| \|s_n\| \|x\| + \|(u - v)|_{E_0}\| \|I - s_n\| \|x\| \qquad (x \in E),$$

whence we obtain the first inequality of (11.119). The second one follows similarly, using (11.111) and the definition of $p$.

and (11.116),

$$\|u-v\| \leqslant \|s_n\| \, \|(u-v)|_{s_n(E)}\| + \|I-s_n\| \, \|(u-v)|_{E_0}\| \leqslant$$
$$\leqslant \|I-s_n\|(\|p\| \, \|(u-v)|_{G_0}\| + \|I-v\| \, \|(u-v)|_G\|) =$$
$$= \|I-s_n\| \, \|p\| \, \|u|_{G_0} - q\| < \varepsilon, \tag{11.119}$$

which completes the proof of proposition 11.9.

**Theorem 11.11.** *Let $E$ be a separable complex Banach space such that the conjugate space $E^*$ has the bounded approximation property and let $V$ be a separable subspace of $E^*$. Then $E$ has a strongly series summable M-basis $\{x_n\}$ such that $[f_n] \supset V$, where $\{f_n\} \subset E^*$ is the a.s.f. to $\{x_n\}$.*

*Proof.* Let $\{y_m\} \subset E$ be a dense sequence in $E$ and let $\{g_m\} \subset E^*$ be a dense sequence in $V$; we may assume, without loss of generality, that $g_1(y_1) = 1$. Put $x_1 = y_1, f_1 = g_1, k_1 = 1$ and $\lambda_{1i} = \delta_{1i}$ $(i = 1,2, \ldots)$. Assume now that $k_m$, $(x_i, f_i)_{i=1}^{k_m}$ and $\lambda_{ni}$ $(n = 1, \ldots, k_m;\ i = 1,2, \ldots)$ have been defined. We claim that $(x_i, f_i)_{i=1}^{k_m}$ can be extended to a biorthogonal system $(x_i, f_i)_{i=1}^{j_m}$ where $j_m = k_m$ or $k_m + 1$ or $k_m + 2$, so that $y_{m+1} \in [x_i]_{i=1}^{j_m}$, $g_{m+1} \in [f_i]_{i=1}^{j_m}$. Indeed, if $y_{m+1} \in [x_i]_{i=1}^{k_m}$ and $g_{m+1} \in [f_i]_{i=1}^{k_m}$, then we are done, with $j_m = k_m$. If one of these relations does not hold, say*) $y_{m+1} \notin [x_i]_{i=1}^{k_m}$, put $x_{k_m+1} = y_{m+1} - \sum_{i=1}^{k_m} f_i(y_{m+1})x_i$ (hence $x_{k_m+1} \notin$ $\notin [x_i]_{i=1}^{k_m}$) and take $f_{k_m+1} \in E^*$ such that $f_{k_m+1}(x) = 0$ $(x \in [x_i]_{i=1}^{k_m})$, $f_{k_m+1}(x_{k_m+1}) = 1$. Then $y_{m+1} \in [x_i]_{i=1}^{k_m+1}$ and $f_i(x_l) = \delta_{il}$ $(i, l = 1, \ldots, k_m+1)$, so if $g_{m+1} \in [f_i]_{i=1}^{k_m+1}$, we are done, with $j_m = k_m + 1$. If $g_{m+1} \notin [f_i]_{i=1}^{k_m+1}$, then putting $f_{k_m+2} = g_{m+1} - \sum_{i=1}^{k_m+1} g_{m+1}(x_i)f_i$ and taking $x_{k_m+2} \in E$ such that $f(x_{k_m+2}) = 0$ $(f \in [f_i]_{i=1}^{k_m+1})$, $f_{k_m+2}(x_{k_m+2}) = 1$, we obtain a biorthogonal system $(x_i, f_i)_{i=1}^{k_m+2}$ with $g_{m+1} \in [f_i]_{i=1}^{k_m+2}$, so we are done, with $j_m = k_m+2$, which proves our claim.

Now, by our assumption on $E^*$ and by § 9, proposition 9.8, there exists a $\lambda \geqslant 1$ such that $E$ has the $\lambda$-duality approximation property and hence, by § 9, lemma 9.2 $\left(\text{with } G = [x_i]_{i=1}^{j_m}, \ \Gamma = [f_i]_{i=1}^{j_m}, \ \varepsilon = \frac{1}{m}\right)$, there exists an operator $u_m$ of finite rank on $E$ satisfying $u_m(x_i) = x_i$, $u_m^*(f_i) = f_i$ $(i = 1, \ldots, j_m)$, $\|u_m\| \leqslant \lambda + \frac{1}{m}$. Therefore, by proposition 11.9 above, there exist a positive integer $k_{m+1} \geqslant j_m + 1$, a biorthogonal system $(x_i, f_i)_{i=j_m+1}^{k_{m+1}}$ $(\{x_i\}_{i=j_m+1}^{k_{m+1}} \subset E$, $\{f_i\}_{i=j_m+1}^{k_{m+1}} \subset E^*)$, and a set of complex

*) If $g_{m+1} \notin [f_i]_{i=1}^{k_m}$, the argument is similar (actually, if $y_{m+1} \in [x_i]_{i=1}^{k_m}$, then $j_m = k_m + 1$ will suffice).

numbers $\{\alpha_i\}_{i=j_m+1}^{k_{m+1}}$ such that $(x_i, f_i)_{i=1}^{k_{m+1}}$ is biorthogonal and that $\|u_m - v_m\| < \frac{1}{m}$, where $v_m$ is the operator on $E$ defined by

$$v_m(x) = \sum_{i=1}^{j_m} f_i(x)x_i + \sum_{i=j_m+1}^{k_{m+1}} \alpha_i f_i(x)x_i \qquad (x \in E). \qquad (11.120)$$

Put

$$\lambda_{ni} = \begin{cases} \lambda_{k_m,i} & \text{for } i = 1, \ldots, k_m \\ 0 & \text{for } i = k_m + 1, k_m + 2, \ldots \end{cases} \qquad (n = k_m + 1, \ldots, k_{m+1} - 1), \qquad (11.121)$$

$$\lambda_{k_{m+1},i} = \begin{cases} 1 & \text{for } i = 1, \ldots, j_m \\ \alpha_i & \text{for } i = j_m + 1, \ldots, k_{m+1} \\ 0 & \text{for } i = k_{m+1} + 1, k_{m+1} + 2, \ldots \end{cases} \qquad (11.122)$$

Continuing in this way indefinitely, we obtain an infinite biorthogonal system $(x_i, f_i)$ and an infinite triangular matrix $(\lambda_{ni})$. We shall show that $\{x_n\}$ is a strongly series summable $M$-basis of $E$ with the a.s.f. $\{f_n\}$ satisfying $V \subset [f_n]$, which will complete the proof.

By (11.121) and (11.122) we have

$$\tau_n(x) = \sum_{i=1}^{n} \lambda_{ni} f_i(x)x_i = \sum_{i=1}^{k_m} \lambda_{k_m,i} f_i(x)x_i =$$

$$= \sum_{i=1}^{j_{m-1}} f_i(x)x_i + \sum_{i=j_{m-1}+1}^{k_m} \alpha_i f_i(x)x_i = v_{m-1}(x)$$

$$(x \in E,\ n = k_m, \ldots, k_{m+1} - 1;\ m = 1,2, \ldots).$$

Consequently,

$$\|x_i - \tau_n(x_i)\| = \|u_m(x_i) - v_m(x_i)\| \leqslant \|u_m - v_m\| \, \|x_i\| < \frac{1}{m} \|x_i\|$$

$$(i = 1, \ldots, j_m;\ n = k_{m+1}, k_{m+1} + 1, \ldots, k_{m+2} - 1;\ m = 1,2, \ldots),$$

whence, since $[x_n] = E$ (because $x_1 = y_1$, $[x_1, \ldots, x_{j_m}] \ni y_{m+1}$ for $m = 1, 2, \ldots$ and $\{y_m\}$ is dense in $E$) and since $\|\tau_n\| = \|v_m\| < \|u_m\| +$

$+\frac{1}{m} \leqslant \lambda + \frac{2}{m}$ $(n = k_{m+1}, \ldots, k_{m+2} - 1;\ m = 1,2, \ldots)$, it follows that $x = \lim_{n\to\infty} \tau_n(x)$ for all $x \in E$, so $\{x_n\}$ is a strongly series summable $M$-basis of $E$ (with the summability matrix $(\lambda_{ni})$ defined by (11.121), (11.122)).

Finally, since $g_1 = f_1$ and $g_{m+1} \in [f_i]_{i=1}^{j_m}$ $(m = 1,2, \ldots)$ and since $\{g_m\}$ is dense in $V$, we have $V \subset [f_n]$, which completes the proof of theorem 11.11.

Many results on bases and triangular $T$-bases can be extended to strongly series summable $M$-bases $\{x_n\}$. For example, we have seen above that they are generalized summation bases and hence operational bases, so $r([f_n])>0$, where $\{f_n\} \subset E^*$ is the a.s.f. to $\{x_n\}$. It is also clear that if $\{x_n\}$ is a strongly series summable $M$-basis of $E$ and $\{i_n\} \subset \mathcal{N}$, then $\{x_{i_n}\}$ and $\{\omega(x_j)\}_{j\in\mathcal{N}\setminus\{i_n\}}$ are strongly series summable $M$-bases of $[x_{i_n}]$ and $E/[x_j]_{j\in\mathcal{N}\setminus\{i_n\}}$ respectively, where $\omega\colon E\to E/[x_{i_n}]$ is the canonical mapping. The characterization theorem 11.3 can be extended to strongly series summable $M$-bases, replacing $\sigma_n$ by $\tau_n$, because of remark 11.1. The intrinsic characterization theorem 11.5 can be also extended, with $\lambda_{ni}$ instead of $T_{ni}$, since its proof has not used (11.1). Clearly, the subsequent results on strong and weak duality also extend to strongly series summable $M$-bases (with $\sigma_n$ replaced by $\tau_n$).

The special classes of minimal sequences and generalized bases introduced in § 6 and § 7 define, naturally, corresponding special classes of strongly series summable $M$-bases (e.g., shrinking). We have the following corollary of theorem 11.11:

**Corollary 11.6.** *Let $E$ be a complex Banach space such that the conjugate space $E^*$ is separable and has the approximation property. Then $E$ has a shrinking strongly series summable $M$-basis.*

*Proof.* Since $E^*$ is separable and has the bounded approximation property*), one can apply theorem 11.11 above with $V = E^*$, which completes the proof.

It is also natural to give

*Definition 11.5.* A strongly series summable $M$-basis $\{x_n\}$ of a Banach space $E$ is said to be *boundedly complete* if there is a summability matrix $(\lambda_{ni})$ for $\{x_n\}$ such that the relations $\{\alpha_n\} \subset K$, $\sup_{1\leqslant n<\infty} \left\| \sum_{i=1}^{n} \lambda_{ni}\alpha_i x_i \right\| < \infty$ imply that $\left\{ \sum_{i=1}^{n} \lambda_{ni}\alpha_i x_i \right\}_{n=1}^{\infty}$ converges.

---

*) See § 9, theorem 9.3, implication $11° \Rightarrow 4°$.

The proof of proposition 11.4 (with $T_{ni}$ replaced by $\lambda_{ni}$) shows that *a strongly series summable M-basis* $\{x_n\}$ *is boundedly complete if and only if it is boundedly complete as an M-basis of E*. Hence, in definition 11.5 "there is a summability matrix" can be replaced by "for each summability matrix". Thus, using corollary 11.6, we obtain the following result, which gives an affirmative answer to problem 11.3 a) if $E$ is a complex conjugate space:

**Theorem 11.12.** *A complex Banach space E admits a boundedly complete strongly series summable M-basis if and only if E is isomorphic to a separable conjugate space $B^*$ and E has the bounded approximation property.*

*Proof.* If $E$ admits a boundedly complete strongly series summable $M$-basis $\{x_n\}$, with the a.s.f. $\{f_n\} \subset E^*$, then $E$ is separable and by § 7, theorem 7.9, $E$ is isomorphic to $B^*$, where $B = [f_n]$. Also, it is clear that $E$ has the bounded approximation property.

Conversely, if $E$ satisfies the condition of theorem 11.2, then, by corollary 11.6, $B$ has a shrinking strongly series summable $M$-basis $\{b_n\}$. Hence, by the extension of theorem 11.6 mentioned in the preceding, the a.s.f. to $\{b_n\}$ is a strongly series summable $M$-basis of $B^* \cong E$ and it is boundedly complete by § 8, proposition 8.17. This completes the proof of theorem 11.12.

Some characterizations of strongly series summable $M$-bases in terms of the set of multipliers, related to the results of the final part o § 8, are given in

**Proposition 11.10.** *Let $\{x_n\}$ be an M-basis of a Banach space E, wit the a.s.f. $\{f_n\} \subset E^*$. The following statements are equivalent:*

1°. *$\{x_n\}$ is a strongly series summable M-basis.*

2°. *We have*

$$([\varphi_n], \{\varphi_n\})^d = M(E, (x_n, f_n)), \tag{11.123}$$

*where $\{\varphi_n\} \subset [e_n]^*$ is the sequence of coordinate functionals on $[e_n] \subset$ $\subset M(E, (x_n, f_n))$.*

3°. *We have*

$$e \in ([\varphi_n], \{\varphi_n\})^d, \tag{11.124}$$

*i.e., there exists a functional* $\Psi \in [\varphi_n]^*$ *such that*

$$\Psi(\varphi_n) = 1 \qquad (n = 1,2, \ldots). \tag{11.125}$$

4°. *There exists a sequence of almost zero sequences (i.e., having a finite number of non-zero coordinates)* $\{\tilde{\gamma}_n^{(j)}\} \in M(E, (x_n, f_n))$ $(j = 1,2,\ldots)$ *such that*

$$\lim_{j\to\infty} \tilde{\gamma}_n^{(j)} = 1 \qquad (n = 1,2, \ldots), \tag{11.126}$$

$$\sup_{1 \leqslant j < \infty} \|\{\tilde{\gamma}_n^{(j)}\}\| < \infty. \tag{11.127}$$

*Proof.* Assume that we have 1°, so there is a triangular matrix $(\lambda_{ji})$ such that

$$\tau_j(x) = \sum_{i=1}^{j} \lambda_{ji} f_i(x) x_i \to x \text{ as } j \to \infty \qquad (x \in E), \tag{11.128}$$

and let $\{\gamma_n\} \in M(E, (x_n, f_n))$ be arbitrary. Put

$$\gamma_i^{(j)} = \lambda_{ji}\gamma_i \qquad (i, j = 1,2, \ldots), \tag{11.129}$$

$$v_j(x) = \sum_{i=1}^{j} \gamma_i^{(j)} f_i(x) x_i \qquad (x \in E, j = 1,2, \ldots). \tag{11.130}$$

Then, since $\{\gamma_n\} \in M(E, (x_n, f_n))$, we have $\gamma_i f_i(x) = f_i(x_{\{\gamma_n\}})$ $(x \in E,$ $i = 1,2, \ldots)$, whence, by (11.128),

$$v_j(x) = \sum_{i=1}^{j} \lambda_{ji}\gamma_i f_i(x) x_i = \sum_{i=1}^{j} \lambda_{ji} f_i(x_{\{\gamma_n\}}) x_i \to x_{\{\gamma_n\}} \text{ as } j \to \infty \quad (x \in E);$$

hence, by the principle of uniform boundedness, $\sup_{1 \leqslant j < \infty} \|v_j\| < \infty$ and therefore, by Ch. I, § 5, proposition 5.4 (since $v_j = v_{\{\gamma_n^{(j)}\}}$ of that proposition),

$$\sup_{1 \leqslant j < \infty} \|\{\gamma_n^{(j)}\}\| = \sup_{1 \leqslant j < \infty} \|v_j\| < \infty. \tag{11.131}$$

Furthermore, by (11.129) we have $\{\gamma_n^{(j)}\} \in [e_n] \subset M(E, (x_n, f_n))$ (because $(\lambda_{ji})$ is triangular), whence, by (11.96),

$$\varphi_k(\{\gamma_n^{(j)}\}) = \gamma_k^{(j)} = \lambda_{jk}\gamma_k \to \gamma_k = \varphi_k(\{\gamma_n\}) \text{ as } j \to \infty \qquad (k = 1,2, \ldots).$$

Define now a functional $\Psi$ on the linear span of $\{\varphi_n\}$ by

$$\Psi\left(\sum_{k=1}^{m} \beta_k \varphi_k\right) = \sum_{k=1}^{m} \beta_k \gamma_k. \tag{11.132}$$

Then for any scalars $\beta_1, \ldots, \beta_m$ we have

$$\left|\Psi\left(\sum_{k=1}^{m} \beta_k \varphi_k\right)\right| = \left|\sum_{k=1}^{m} \beta_k \lim_{j\to\infty} \varphi_k(\{\gamma_n^{(j)}\})\right| =$$

$$= \lim_{j\to\infty}\left|\sum_{k=1}^{m} \beta_k \varphi_k(\{\gamma_n^{(j)}\})\right| \leqslant \left(\sup_{1\leqslant j<\infty} \|\{\gamma_n^{(j)}\}\|\right)\left\|\sum_{k=1}^{m} \beta_k \varphi_k\right\|,$$

and hence, by (11.131), $\Psi$ is continuous, so it can be extended to a $\tilde{\Psi} \in [\varphi_n]^*$. Clearly,

$$\tilde{\Psi}(\varphi_n) = \Psi(\varphi_n) = \gamma_n \qquad (n = 1,2, \ldots),$$

so $\{\gamma_n\} \in ([\varphi_n], \{\varphi_n\})^d$. Consequently,

$$M(E, (x_n, f_n)) \subset ([\varphi_n], \{\varphi_n\})^d,$$

which, together with § 8, proposition 8.22 a), proves that $1° \Rightarrow 2°$.

The implication $2° \Rightarrow 3°$ is obvious, since $e \in M(E, (x_n, f_n))$.

Assume now that we have 3° and let $\Psi \in [\varphi_n]^*$ satisfy (11.125). Then for any scalars $\beta_1, \ldots, \beta_m$ we have $\left|\sum_{i=1}^{m} \beta_i \Psi(\varphi_i)\right| \leqslant \|\Psi\| \left\|\sum_{i=1}^{m} \beta_i \varphi_i\right\|$ and hence, by Helly's theorem*), we can find a sequence $\{\tilde{\gamma}_n^{(j)}\} \in [e_n] \subset M(E, (x_n, f_n))$ $(j = 1,2, \ldots)$ such that

$$1 = \Psi(\varphi_k) = \lim_{j\to\infty} \varphi_k(\{\tilde{\gamma}_n^{(j)}\}) = \lim_{j\to\infty} \tilde{\gamma}_k^{(j)} \qquad (k = 1,2, \ldots),$$

$$\|\{\tilde{\gamma}_n^{(j)}\}\| \leqslant \|\Psi\| + 1 \qquad (j = 1,2, \ldots).$$

Since $\{\tilde{\gamma}_n^{(j)}\} \subset [e_n]$, we may assume that each $\{\tilde{\gamma}_n^{(j)}\}$ $(j = 1,2, \ldots)$ is an almost zero sequence (replacing each $\{\tilde{\gamma}_n^{(j)}\}$ by a sufficiently near almost zero sequence, if necessary). Thus, $3° \Rightarrow 4°$.

Finally, assume that we have 4°. Let

$$\tilde{\tau}_j(x) = \sum_{i=1}^{\infty} \tilde{\gamma}_i^{(j)} f_i(x) x_i \qquad (x \in E, j = 1,2, \ldots). \tag{11.133}$$

---

*) See e.g. [70], Ch. II, § 4, theorem 3.

Then, by (11.126),

$$\tilde{\tau}_j(x_k) = \tilde{\gamma}_k^{(j)} x_k \to x_k \text{ as } j \to \infty \quad (k = 1,2, \ldots)$$

and, by (11.127) and Ch. I, § 5, proposition 5.4 (since $\tilde{\tau}_j = v_{\{\gamma_n^{(j)}\}}$ of that proposition),

$$\sup_{1 \leqslant j < \infty} \|\tilde{\tau}_j\| = \sup_{1 \leqslant j < \infty} \|\{\tilde{\gamma}_n^{(j)}\}\| < \infty$$

whence $\tilde{\tau}_j(x) \to x$ for all $x \in E$. Thus, $\{x_n\}$ is "strongly series summable with respect to the row-finite matrix $(\tilde{\lambda}_{ji})$" defined by

$$\tilde{\lambda}_{ji} = \tilde{\gamma}_i^{(j)} \quad (i, j = 1,2, \ldots) \tag{11.134}$$

and it remains to show that in this case there exists also a triangular summability matrix $(\lambda_{ji})$ with the same property. Let

$$k_m = \max_{\tilde{\lambda}_{mi} \neq 0} i \quad (m = 1,2, \ldots). \tag{11.135}$$

We may assume (by passing to a subsequence, if necessary) that $\{k_m\}$ is increasing. Let

$$\lambda_{ji} =$$

$$= \begin{cases} 0 & \text{for } i = 1,2, \ldots \ (j = 1, \ldots, k_1 - 1) \\ \tilde{\lambda}_{mi} & \text{for } i=1, \ldots, k_m \ (j=k_m, \ldots, k_{m+1}-1; m=1,2,\ldots) \\ 0 & \text{for } i = k_m + 1, k_m+2, \ldots \ (j=k_m, \ldots, k_{m+1}-1; m=1,2, \ldots). \end{cases} \tag{11.136}$$

Then $(\lambda_{ji})$ is triangular and

$$\tau_j(x) = \sum_{i=1}^{j} \lambda_{ji} f_i(x) x_i = \sum_{i=1}^{k_m} \tilde{\lambda}_{mi} f_i(x) x_i = \sum_{i=1}^{k_m} \tilde{\gamma}_i^{(m)} f_i(x) x_i = \tilde{\tau}_m(x)$$

$$(x \in E, \ j = k_m, \ldots, k_{m+1} - 1; \ m = 1,2,\ldots),$$

whence, by the above, $\tau_j(x) \to x$ for all $x \in E$, so $\{x_n\}$ is a strongly series summable $M$-basis of $E$. Thus, $4° \Rightarrow 1°$, which completes the proof of proposition 11.10.

*Remark 11.4.* One can also show that if we have 1° and $\{h_n\}$ is the sequence of coordinate functionals on $M(E, (x_n, f_n))$, then

$$([h_n], \{h_n\})^d = M(E, (x_n, f_n)) \tag{11.137}$$

(and hence $e \in ([h_n], \{h_n\})^d$); indeed, the proof is similar to that of the implication $1° \Rightarrow 2°$ above.

**Proposition 11.11.** *Every strongly series summable M-basis $\{x_n\}$ of a Banach space E is series summable.*

*Proof.* Let $\{f_n\} \subset E^*$ be the a.s.f. to $\{x_n\}$, let $(\lambda_{ni})$ be a triangular matrix such that we have (11.92) and let $\{e_n\}$ and $\{h_n\}$ be respectively the sequence of unit vectors and coordinate functionals for $\mathscr{S}(E, (x_n, f_n))$. Put

$$\tilde{h}_n = \sum_{i=1}^{n} \lambda_{ni} h_i \qquad (n = 1,2, \ldots).$$

Then $\tilde{h}_n \in \mathscr{S}(E, (x_n, f_n))^*$ and, by (11.96),

$$\tilde{h}_n(e_i) = \lambda_{ni} \to 1 \text{ as } n \to \infty \quad (i = 1,2, \ldots). \tag{11.138}$$

Let $\{\alpha_j\} \in \mathscr{S}(E, (x_n, f_n))$ be arbitrary, let $u \in N(E, E)$ be such that $\{\alpha_j\} = \{f_j(u(x_j))\}$ and let $\{y_n\} \subset E$, $\{g_n\} \subset E^*$ be such that

$$u(x) = \sum_{j=1}^{\infty} g_j(x) y_j \qquad (x \in E), \tag{11.139}$$

$$\sum_{j=1}^{\infty} \|g_j\| \, \|y_j\| < \infty. \tag{11.140}$$

Then, if $\tau_n$ is defined by (11.93),

$$|\tilde{h}_n(\{\alpha_j\})| = \left| \sum_{i=1}^{n} \lambda_{ni} h_i(\{f_j(u(x_j))\}) \right| = \left| \sum_{i=1}^{n} \lambda_{ni} f_i(u(x_i)) \right| =$$

$$= \left| \sum_{i=1}^{n} \lambda_{ni} \sum_{j=1}^{\infty} g_j(x_i) f_i(y_j) \right| = \left| \sum_{j=1}^{\infty} g_j \left( \sum_{i=1}^{\infty} \lambda_{ni} f_i(y_j) x_i \right) \right| =$$

$$= \left| \sum_{j=1}^{\infty} g_j(\tau_n(y_j)) \right| \leqslant \left( \sup_{1 \leqslant n < \infty} \|\tau_n\| \right) \sum_{j=1}^{\infty} \|g_j\| \, \|y_j\|,$$

whence, taking the inf over all representations (11.139), (11.140) of $u$, and then the inf over all $u \in N(E, E)$ such that $\{\alpha_j\}=\{f_j(u(x_j))\}$, we obtain

$$|\tilde{h}_n(\{\alpha_j\})| \leqslant (\sup_{1 \leqslant n < \infty} \|\tau_n\|)\|\{\alpha_j\}\|.$$

Thus, since $\{\alpha_j\} \in \mathscr{S}(E, (x_n, f_n))$ and $n \in \mathscr{N}$ were arbitrary, we have $\sup_{1 \leqslant n<\infty} \|\tilde{h}_n\| < \infty$, which, together with $[e_n] = \mathscr{S}(E, (x_n, f_n))$ and (11.138), implies*) that there exists $h_0 \in \mathscr{S}(E, (x_n, f_n))^*$ such that $\tilde{h}_n(\alpha) \rightarrow h_0(\alpha)$ for all $\alpha \in \mathscr{S}(E, (x_n, f_n))$. Then, by (11.138), $h_0(e_i) = 1$ $(i = 1,2, \ldots)$ and therefore, by § 8, proposition 8.24, $\{x_n\}$ is a series summable $M$-basis, which completes the proof.

The converse of proposition 11.11 is not valid, i.e., there exist series summable $M$-bases which are not strongly series summable, as shown by

*Example 11.10.* For each $j = 1,2, \ldots$ let $E_j$ be the space $l^1$ endowed with the equivalent norm

$$\|x\|_{(j)} = \frac{1}{j} \sum_{i=1}^{\infty} |\xi_i| + \left|\sum_{i=1}^{\infty} \xi_i\right| \qquad (x = \{\xi_n\} \in l^1) \tag{11.141}$$

and let

$$E = (E_1 \times E_2 \times \ldots)_{l^1}. \tag{11.142}$$

Furthermore, for each $j$ let $\{e_n^{(j)}\}$ be the unit vector basis of $E_j = (l^1, \|.\|_{(j)})$, with the a.s.f. $\{h_n^{(j)}\} \subset E_j^*$ and let $\{e_n\}$ be an arbitrary numbering of the set $\bigcup_{j=1}^{\infty} \bigcup_{n=1}^{\infty} \{\underbrace{0, \ldots, 0}_{j-1}, e_n^{(j)}, 0, 0, \ldots\} \subset E$ and $\{h_n\}$ the corresponding numbering of $\bigcup_{j=1}^{\infty} \bigcup_{n=1}^{\infty} \{\underbrace{0, \ldots, 0}_{j-1}, h_n^{(j)}, 0, 0, \ldots\} \subset E^*$. Then $\{e_n\}$ is a series summable $M$-basis of $E$, which is not strongly series summable. Indeed, it is obvious that $\{e_n\}$ is an $M$-basis of $E$, with the a.s.f. $\{h_n\} \subset E^*$. In order to prove that $\{e_n\}$ is series summable it will be sufficient, by § 8, proposition 8.26, to show that $(E, \{e_n\})^d$ is a $BK$-algebra containing $e$. To this end, observe first that for any $x = \{\xi_n\} \in E_j$ and $f \in E_j^*$ we have

$$\left\|\sum_{i=1}^{\infty} \xi_i f(e_i^{(j)}) e_i^{(j)}\right\|_{(j)} = \frac{1}{j} \sum_{i=1}^{\infty} |\xi_i f(e_i^{(j)})| +$$

$$+\left|f\left(\sum_{i=1}^{\infty} \xi_i e_i^{(j)}\right)\right| \leqslant \frac{1}{j} \sum_{i=1}^{\infty} |\xi_i| \|f\|_{(j)} \|e_i^{(j)}\|_{(j)} + \|f\|_{(j)} \|x\|_{(j)} \leqslant$$

$$\leqslant \left(2 + \frac{1}{j}\right) \|f\|_{(j)} \|x\|_{(j)}. \tag{11.143}$$

*) See e.g. [87], p. 55, theorem 18.

Now let $\{f_j\}, \{g_j\} \in (E_1^* \times E_2^* \times \ldots)_{l^\infty} \equiv E^*$ be arbitrary. Then, by (11.143), for each $j$ we have, in $(E_j, \{e_n^{(j)}\})^d$,

$$\|\{f_j(e_n^{(j)})g_j(e_n^{(j)})\}_{n=1}^\infty\|_{(j)} = \sup_{\substack{x=\{\xi_n\}\in E_j \\ \|x\|_{(j)}\leqslant 1}} \left|\sum_{i=1}^\infty \xi_i f_j(e_i^{(j)})g_j(e_i^{(j)})\right| \leqslant$$

$$\leqslant \|g_j\|_{(j)} \sup_{\substack{x=\{\xi_n\}\in E_j \\ \|x\|_{(j)}\leqslant 1}} \left\|\sum_{i=1}^\infty \xi_i f_j(e_i^{(j)})e_i^{(j)}\right\|_{(j)} \leqslant 3\|f_j\|_{(j)}\|g_j\|_{(j)},$$

and hence, in $(E, \{e_n\})^d$,

$$\|\{\{f_1(e_n^{(1)})g_1(e_n^{(1)})\}_{n=1}^\infty, \{f_2(e_n^{(2)})g_2(e_n^{(2)})\}_{n=1}^\infty, \ldots\}\| =$$

$$= \sup_{1\leqslant j<\infty} \|\{f_j(e_n^{(j)})g_j(e_n^{(j)})\}_{n=1}^\infty\|_{(j)} \leqslant 3 \sup_{1\leqslant j<\infty} \|f_j\|_{(j)} \sup_{1\leqslant j<\infty} \|g_j\|_{(j)} =$$

$$= 3\|\{\{f_1(e_n^{(1)})\}_{n=1}^\infty, \ \{f_2(e_n^{(2)})\}_{n=1}^\infty, \ldots\}\| \ \ \|\{\{g_1(e_n^{(1)})\}_{n=1}^\infty,$$

$$\{g_2(e_n^{(2)})\}_{n=1}^\infty, \ldots\}\|,$$

so $(E, \{e_n\})^d$ is a $BK$-algebra. Furthermore, for each $j$ we have, in $(E_j, \{e_n^{(j)}\})^d$,

$$\|\{e(e_n^{(j)})\}_{n=1}^\infty\|_{(j)} = \|\{1,1, \ldots\}\|_{(j)} = \sup_{\substack{x=\{\xi_n\}\in E_j \\ \|x\|_{(j)}\leqslant 1}} \left|\sum_{i=1}^\infty \xi_i\right| \leqslant 1,$$

whence, in $(E, \{e_n\})^d$,

$$\|\{\{e(e_n^{(1)})\}_{n=1}^\infty, \ \{e(e_n^{(2)})\}_{n=1}^\infty, \ldots\}\| = \sup_{1\leqslant j<\infty} \|\{e(e_n^{(j)})\}_{n=1}^\infty\|_{(j)} \leqslant 1,$$

which proves that $e = \{1,1, \ldots\} \in (E, \{e_n\})^d$. Thus, $\{e_n\}$ is series summable. Finally, in order to prove that $\{e_n\}$ is not strongly series summable, it will be sufficient, by the remark made after problem 11.3, to show that $r([h_n]) = 0$. Clearly, to this end it will be sufficient to prove that in each $E_j^*$ we have $r([h_n^{(j)}]) \leqslant \dfrac{2}{j+1}$ $(j = 1,2, \ldots)$. Now, for $y_n^{(j)} \in E_j$ defined by

$$y_n^{(j)} = \frac{j}{2}(e_1^{(j)} - e_{n+1}^{(j)}) \qquad (n = 1,2, \ldots) \tag{11.144}$$

we have $\|y_n^{(j)}\|_{(j)} = 1$ $(n = 1,2, \ldots)$ and for any $f = \sum_{i=1}^{\infty} \alpha_i h_i^{(j)} \in [h_m^{(j)}]$ we have $f(e_{n+1}^{(j)}) = \alpha_{n+1} \to 0$ as $n \to \infty$, whence $y_n^{(j)} \to \frac{j}{2} e_1^{(j)}$ for $\sigma(E_j, [h_m^{(j)}])$, as $n \to \infty$. Hence, since $\left\| \frac{j}{2} e_1^{(j)} \right\|_{(j)} = \frac{j}{2} \left( \frac{1}{j} + 1 \right) = \frac{j+1}{2}$, we obtain $\sup_{x \in \Sigma_{(j)}} \|x\|_{(j)} \geqslant \frac{j+1}{2}$, where $\Sigma_{(j)}$ is the closure of $S_{E_j} = \{x \in E_j | \, \|x\|_{(j)} \leqslant 1\}$ for $\sigma(E, [h_n^{(j)}])$. Consequently, by Ch. I, § 12, formula (12.15), $r([h_n^{(j)}]) = \frac{1}{\sup_{x \in \Sigma_{(j)}} \|x\|_{(j)}} \leqslant \frac{2}{j+1}$, which completes the proof of the assertions of example 11.10.

Finally, for the case when the summability matrix $(\lambda_{nm})$ is not necessarily triangular, let us mention

*Definition 11.6.* A sequence $\{x_n\}$ in a Banach space $E$ is called a *$\gamma$-basis* of $E$ if it is an $M$-basis of $E$ and if there exists a matrix of scalars $(\lambda_{nm})$ satisfying (11.96) *and* (11.97)*⁾, such that for each $x \in E$ the sequence $\{f_n(x)x_n\}$ is "$(\lambda_{nm})$-limitable" to $x$, i.e., the elements

$$\tau_n(x) = \sum_{i=1}^{\infty} \lambda_{ni} f_i(x) x_i \qquad (n = 1,2, \ldots) \tag{11.145}$$

exist (where $\{f_n\} \subset E^*$ is the a.s.f. to $\{x_n\}$) and

$$x = \lim_{n \to \infty} \tau_n(x); \tag{11.146}$$

in this case we shall also say that $\{x_n\}$ is a *$\gamma$-basis with respect to the matrix* $(\lambda_{nm})$.

We do not know whether the first part of proposition 11.7 can be extended to the non-triangular case, i.e., whether every $T$-basis $\{x_n\}$ of a Banach space $E$ is a $\gamma$-basis of $E$ with respect to the matrix

$$\lambda_{nm} = \sum_{j=m}^{\infty} t_{nj} \qquad (n, m = 1,2, \ldots). \tag{11.147}$$

---

*⁾ Similarly to lemma 11.1, one can show (see e.g. [50], theorem (4.2.II)) that these conditions are necessary and sufficient in order that for every convergent series $\sum_{i=1}^{\infty} z_i$ in $E$ with $\sum_{i=1}^{\infty} z_i = x$, the sequence $\{z_n\}$ be $(\lambda_{nm})$-limitable to $x$.

Note that, since $T = (t_{nm})$ is consistent and since $\lambda_{ni} - \lambda_{n,i+1} = t_{ni}$ ($n, i = 1,2, \ldots$), the matrix $(\lambda_{nm})$ satisfies (11.96) and (11.97). Also, if $\{f_n\} \subset E^*$ is the a.s.f. to $\{x_n\}$, then

$$\sum_{i=1}^{m} \lambda_{ni} f_i(x) x_i = \sum_{i=1}^{m} \left( \sum_{j=i}^{m} t_{nj} \right) f_i(x) x_i + \lambda_{n,m+1} \sum_{i=1}^{m} f_i(x) x_i =$$

$$= \sigma_{nm}(x) + \lambda_{n,m+1} \sum_{i=1}^{m} f_i(x) x_i$$

$$(x \in E;\ n, m = 1,2, \ldots).$$

The second part of proposition 11.7 does not extend to this more general case, as shown by

*Example 11.11.* Let $\{x_n\}$ be a $\gamma$-basis of a Banach space $E$ with respect to the matrix $(\lambda_{nm})$ defined by

$$\lambda_{2n-1,m} = \begin{cases} 1 \text{ for } m = 1, \ldots, 2n-1 \\ 0 \text{ for } m = 2n, 2n+1, \ldots \end{cases} \quad (n = 1,2, \ldots), \tag{11.148}$$

$$\lambda_{2n,m} = \begin{cases} 1 \text{ for } m = 1, \ldots, 2n \\ \dfrac{1}{2} \text{ for } m = 2n+1,\ 2n+2, \ldots \end{cases} \quad (n = 1,2, \ldots). \tag{11.149}$$

Then $(\lambda_{nm})$ satisfies indeed (11.96) and (11.97) with $M = 1$. However, for the matrix $T = (t_{nm})$ defined by (11.98) we have

$$t_{2n-1,m} = \delta_{2n-1,m},\ t_{2n,m} = \frac{1}{2} \delta_{2n,m} \qquad (n, m = 1,2, \ldots),$$

whence

$$\sum_{m=1}^{\infty} t_{2n-1,m} = 1,\ \sum_{m=1}^{\infty} t_{2n,m} = \frac{1}{2} \qquad (n = 1,2, \ldots),$$

so $\lim\limits_{n \to \infty} \sum\limits_{m=1}^{\infty} t_{nm}$ does not exist. Thus, $T = (t_{nm})$ does not satisfy (11.3) (hence it is not consistent) and therefore, by definition 11.1, $\{x_n\}$ is not a $T$-basis of $E$.

*Problem 11. 6.* Does every Banach space with a $T$-basis have a $\gamma$-basis?

*Remark 11.5.* Observe that the matrix $(\lambda_{nm})$ of formula (11.147) satisfies, in addition,

$$\lim_{j\to\infty} \lambda_{nj} = 0 \qquad (n = 1,2, \ldots), \qquad (11.150)$$

and conversely, if $(\lambda_{nm})$ satisfies (11.96), (11.97) and (11.150), then the matrix $T = (t_{nm})$ defined by (11.98) is consistent, since $\sum_{i=1}^{\infty} t_{ni} = \sum_{i=1}^{\infty} (\lambda_{ni} - \lambda_{n,i+1}) = \lambda_{n1} - \lim_{j\to\infty} \lambda_{nj} = \lambda_{n1} \to 1$ as $n \to \infty$. Also, if $\{x_n\}$ is a $\gamma$-basis with respect to such a matrix $(\lambda_{nm})$, then, writing

$$\sigma_{nm}(x) = \sum_{i=1}^{m} \left( \sum_{j=i}^{m} t_{nj} \right) f_i(x) x_i = \sum_{i=1}^{m} (\lambda_{ni} - \lambda_{n,m+1}) f_i(x) x_i =$$

$$= \sum_{i=1}^{m} \lambda_{ni} f_i(x) x_i - \lambda_{n,m+1} \sum_{i=1}^{m} f_i(x) x_i \qquad (x \in E;\ n, m = 1,2,\ldots),$$

we see that $\sigma_{nm}(x) \to \tau_n(x)$ for all $x \in P$ (the linear span of $\{x_n\}$) and all $n = 1,2, \ldots$, and that

$$\|\sigma_{nm}(x)\| \leqslant \sum_{i=1}^{m} |\lambda_{ni} - \lambda_{n,m+1}| \max_{1\leqslant j\leqslant m} \|f_j(x) x_j\|$$

$$(x \in E;\ n,m = 1,2, \ldots);$$

however, we do not know whether $\{x_n\}$ is a $T$-basis of $E$ for $T = (t_{nm})$ defined by (11.98).

*Problem 11.7.* Does every Banach space $E$ with a $\gamma$-basis have a $T$-basis?

Some of the properties of $T$-bases (and, in particular, of strongly series summable $M$-bases) can be carried over to $\gamma$-bases (see the Notes and remarks).

## § 12. $\Pi$-bases. $\pi$-bases. $\pi_1^\infty$-bases. The universal complements $C_p$

The finite rank endomorphisms $u_n$ in the definition of approximative bases (and of the other generalizations of bases, introduced in §§ 9—11) do not share with the partial sum operators $s_n$ associated to bases the property of being projections. Now we shall impose on the endomorphisms $u_n$ in the definition 9.1 of an approximative basis and in part of the definition 9.9 of a commuting approximative basis, to be also projections (i.e., $u_n^2 = u_n$).

*Definition 12.1.* Let $E$ be a Banach space. A sequence of finite rank projections $\{u_n\} \subset L(E, E)$ (or a sequence of elements $\{x_n\} \subset E$ for which there exists a sequence of finite rank projections $\{u_n\} \subset L(E, E)$ of the form (9.3), with the properties listed below) is called

a) a *$\Pi$-basis* of $E$, if

$$x = \lim_{n\to\infty} u_n(x) \qquad (x \in E); \qquad (12.1)$$

b) a *$\pi$-basis* of $E$, if we have (12.1) and

$$u_m u_n = u_n \qquad (n \leqslant m). \qquad (12.2)$$

If, in addition,

$$\|u_n\| \leqslant \lambda \qquad (n = 1,2, \ldots), \qquad (12.3)$$

then $\{u_n\}$ (respectively, $\{x_n\}$) is called a *$\Pi_\lambda$-basis* or a *$\pi_\lambda$-basis*, respectively.

In particular, if we have

$$\dim u_n(E) = n \qquad (n = 1,2, \ldots), \qquad (12.4)$$

then we shall use the terms *strict $\Pi$-basis, strict $\pi$-basis, strict $\Pi_\lambda$-basis* and *strict $\pi_\lambda$-basis* respectively.

By the above definition, every $\Pi_\lambda$-basis ($\pi_\lambda$-basis) is a $\Pi$-basis (respectively, a $\pi$-basis). Conversely, by the principle of uniform boundedness, every $\Pi$-basis ($\pi$-basis) is a $\Pi_\lambda$-basis (respectively, a $\pi_\lambda$-basis) for some suitable $\lambda$ (and hence for all $\lambda' \geqslant \lambda$).

Obviously, every $\pi$-basis is a $\Pi$-basis and hence an approximative basis.

*Problem 12.1.* a) Does every Banach space with an approximative basis have a $\Pi$-basis (or, equivalently*⁾, a $\pi$-basis)? b) How about a Banach space with a commuting approximative basis?

Let us recall the following observations:

**Lemma 12.1.** *Let $E$ be a Banach space and let $u$, $v \in L(E, E)$.*

a) *If $vu = u$, then $u(E) \subset v(E)$.*

b) *If $v$ is a projection and $u(E) \subset v(E)$, then $vu = u$.*

---

*⁾ By theorem 12.1 $\gamma$) below.

c) *If $u$ is a projection and $u(E) \subset v(E)$, then*

$$v(E) = u(E) \oplus (v(E) \cap \operatorname{Ker} u).$$

*Proof.* If $vu = u$, then $u(E) = vu(E) \subset v(E)$.

b) If $u(E) \subset v(E)$, then $u(x) \in u(E) \subset v(E)\, (x \in E)$, whence, since $v$ is a projection, $vu(x) = u(x)\, (x \in E)$.

c) Assume that $u(E) \subset v(E)$ and that $u$ is a projection. Then $v(x) = uv(x) + (I - u)v(x)\ \ (x \in E)$, where $(I - u)v(x) \in v(E) \cap \operatorname{Ker} u$ (since $(I - u)v(x) \in v(E) - u(E) = v(E)$ and $u(I - u)v(x) = 0$); also, $u(E) \cap (v(E) \cap \operatorname{Ker}\ u) \subset u(E) \cap \operatorname{Ker}\ u = \{0\}$. Conversely, $u(E) \oplus \oplus (v(E) \cap \operatorname{Ker}\ u) \subset u(E) + v(E) = v(E)$, which completes the proof.

Some remarks on the conditions occurring in definition 12.1 are collected in

**Proposition 12.1.** *Let $E$ be a Banach space and let $\{u_n\} \subset L(E, E)$.*

a) *If we have* (12.1), *then*

$$\overline{\bigcup_{n=1}^{\infty} u_n(E)} = E, \tag{12.5}$$

$$\sup_{1 \leqslant n < \infty} \|u_n\| < \infty. \tag{12.6}$$

b) *If we have* (12.2), (12.5) *and* (12.6), *then* (12.1) *holds.*

c) *We have* (12.2) *if and only if the $u_n$ are projections satisfying*

$$u_1(E) \subset u_2(E) \subset \ldots \tag{12.7}$$

*Proof.* a) Assume that (12.1) holds. Then, by the principle of uniform boundedness, we have (12.6). Also, by (12.1), every $x \in E$ is the limit of a sequence of elements of $\bigcup_{n=1}^{\infty} u_n(E)$, which proves (12.5).

Furthermore, the proof of b) is contained in the final part of the proof of § 9, proposition 9.7. Finally, c) is an immediate consequence of lemma 12.1 a), b) above. This completes the proof of proposition 12.1.

*Remark 12.1.* If we assume only (12.5) and (12.6) and that the $u_n$ are projections, then for every $x \in E$ and $\varepsilon > 0$ there exists an index $N = N(x, \varepsilon)$ such that $\|x - u_N(x)\| < \varepsilon$. Indeed, by (12.5) and (12.6) there exists an element $y \in \bigcup_{n=1}^{\infty} u_n(E)$ such that $\|x - y\| < \dfrac{\varepsilon}{2 \sup_{1 \leqslant n < \infty} \|u_n\|}$.

Then, if $N$ is such that $y \in u_N(E)$, we have (since $u_N$ is a projection)

$$\|u_N(x) - y\| = \|u_N(x) - u_N(y)\| \leqslant \sup_{1 \leqslant n < \infty} \|u_n\| \, \|x - y\| < \frac{\varepsilon}{2},$$

whence, by $\|u_n\| \geqslant 1$ $(n = 1,2, \ldots)$, we obtain

$$\|x - u_N(x)\| \leqslant \|x - y\| + \|y - u_N(x)\| < \frac{\varepsilon}{2 \sup\limits_{1 \leqslant n < \infty} \|u_n\|} + \frac{\varepsilon}{2} \leqslant \varepsilon,$$

which proves our assertion. However, it may also happen that the index $N = N(x, \varepsilon)$ above is unique, and hence $x = u_N(x)$ (see e.g. $\{u_n^*\}$ and $f^{(k)}$ in example 12.1 below).

From proposition 12.1 and its proof we obtain the following characterization of $\pi$-bases and $\pi_\lambda$-bases respectively:

**Corollary 12.1.** a) *Let $E$ be a Banach space. A sequence of finite rank projections $\{u_n\} \subset L(E, E)$ is a $\pi$-basis of $E$ if and only if it satisfies* (12.5), (12.6) *and* (12.7).

b) *A similar characterization holds for $\pi_\lambda$-bases, replacing* (12.6) *by* (12.3).

Furthermore, from proposition 12.1 c) and lemma 12.1 c) we obtain

**Corollary 12.2.** *Let $\{u_n\} \subset L(E, E)$ be a $\pi$-basis of a Banach space $E$. Then*

$$u_n(E) = u_{n-1}(E) \oplus (u_n(E) \cap \operatorname{Ker} u_{n-1}) \qquad (n = 2,3, \ldots), \tag{12.8}$$

*and hence*

$$u_n(E) = u_1(E) \oplus \sum_{i=2}^{n} \oplus \, u_i(E) \cap \operatorname{Ker} u_{i-1} \qquad (n = 2,3, \ldots). \tag{12.9}$$

One can also introduce *weak $\Pi$- ($\Pi_\lambda$-, $\pi$-, $\pi_\lambda$-) bases*, replacing in (12.1) norm-convergence by weak convergence. The *weak* $\Pi$- ($\Pi_\lambda$-, $\pi$-, $\pi_\lambda$-) bases* of a conjugate space $E^*$ are defined similarly, using weak* convergence. Then, from § 9, proposition 9.2, formula (9.8), it follows that *if $\{u_n\}$ is a $\Pi$-basis of $E$, then $\{u_n^*\}$ is a weak* $\Pi$-basis of $E^*$*. However, such a result of weak duality is *no longer true for $\pi$-bases*. Indeed, the equalities (12.2) are equivalent to $u_n^* u_m^* = u_n^*$ $(n \leqslant m)$, while we would need the "reverse" equalities $u_m^* u_n^* = u_n^*$ $(n \leqslant m)$, which are equivalent to $u_n u_m = u_n$ $(n \leqslant m)$, i.e., to § 13, formula (13.2); thus, if $\{u_n\}$ is a $\pi$-basis of $E$ which is not a dual $\pi$-basis of $E$ in the sense of § 13, definition 13.1 (see e.g. example 12.1 below), then $\{u_n^*\}$ is not a weak* $\pi$-basis of $E^*$ (but it is a "weak* dual $\pi$-basis" of $E^*$ in the sense of § 13, definition 13.3).

*Problem 12.2.* If a Banach space $E$ has a $\pi$-basis, does $E^*$ have a weak* $\pi$-basis?

Let us consider now properties of strong duality. If $\{u_n\}$ is a $\Pi$-basis (or even a $\pi$-basis) of $E$, then $\{u_n^*|_V\}$ need not be an approximative basis (or, what is now equivalent, a $\Pi$-basis) of $V=\left[\bigcup_{n=1}^{\infty} u_n^*(E^*)\right]$, as shown by the following slight modification of Ch. I, § 20, example 20.1 (of a strict $\pi$-basis which is not a dual $\pi$-basis):

*Example 12.1.* Let $\{x_n\}$ be a normalized basis of a Banach space $E$, with the a.s.c.f. $\{f_n\} \subset E^*$. Put

$$u_n(x) = s_n(x) + f_{n+1}(x) \sum_{i=1}^{n} \frac{1}{2^i} x_i \qquad (x \in E,\ n = 1,2, \ldots). \tag{12.10}$$

Then $u_n$ is a projection of $E$ onto $P_{(n)} = [x_1, \ldots, x_n]$, satisfying (12.1) (e.g. by Ch. II, § 7, proposition 7.4), and hence $\{u_n\}$ is a strict $\pi$-basis of $E$; clearly, $u_n^*(E^*) \subset [f_1, \ldots, f_{n+1}]$ $(n = 1,2, \ldots)$. However, for any $f \in [f_n]$ with $f\left(\sum_{i=1}^{\infty} \frac{1}{2^i} x_i\right) \neq 0$ we have

$$\varliminf_{n\to\infty} \|f - u_n^*(f)\| > 0,$$

and hence $\{u_n^*|_V\}$ is not an approximative basis of $V = \left[\bigcup_{n=1}^{\infty} u_n^*(E^*)\right]$. Indeed, for every $f \in [f_n]$ we have $\lim_{n\to\infty} f(x_n) = 0$ (since this holds for each $f = f_j$, $j = 1,2, \ldots$ and since $\sup_{1 \leqslant n < \infty} \|x_n\| = 1$), whence

$$\|f - u_n^*(f)\| \geqslant |f(x_{n+1}) - (u_n^*(f))(x_{n+1})| =$$

$$= \left| f(x_{n+1}) - f\left(\sum_{i=1}^{n} \frac{1}{2^i} x_i\right)\right| \to \left| f\left(\sum_{i=1}^{\infty} \frac{1}{2^i} x_i\right)\right| \quad \text{as } n \to \infty,$$

which will prove our assertions, provided that we shall produce an $f \in \left[\bigcup_{n=1}^{\infty} u_n^*(E^*)\right] \subset [f_n]$ with $f\left(\sum_{i=1}^{\infty} \frac{1}{2^i} x_i\right) \neq 0$. Put

$$f^{(k)} = k f_{k+1} + \sum_{j=1}^{k} 2^j f_j \qquad (k = 1,2, \ldots).$$

Then

$$u_k^*(f^{(k)})(x) = f^{(k)}(u_k(x)) = \left(k f_{k+1} + \sum_{j=1}^{k} 2^j f_j\right)\left(\sum_{i=1}^{k} f_i(x) x_i + \right.$$

$$\left. + f_{k+1}(x) \sum_{i=1}^{k} \frac{1}{2^i} x_i\right) = \sum_{i=1}^{k} 2^i f_i(x) + k f_{k+1}(x) = f^{(k)}(x) \qquad (x \in E),$$

so $f^{(k)} = u_k^*(f^{(k)}) \in u_k^*(E^*)$. Furthermore,

$$f^{(k)}\left(\sum_{i=1}^{\infty} \frac{1}{2^i} x_i\right) = \left(k f_{k+1} + \sum_{j=1}^{k} 2^j f_j\right)\left(\sum_{i=1}^{\infty} \frac{1}{2^i} x_i\right) = \frac{k}{2^{k+1}} + k \neq 0,$$

so $f^{(k)}$ has the required properties, which completes the proof of our assertions. Let us also note that

$$\|f^{(k)} - u_n^*(f^{(k)})\| \geqslant \left| f^{(k)}(x_{n+1}) - f^{(k)}\left(\sum_{i=1}^{n} \frac{1}{2^i} x_i\right)\right| =$$

$$= \left| \left(k f_{k+1} + \sum_{j=1}^{k} 2^j f_j\right)\left(x_{n+1} - \sum_{i=1}^{n} \frac{1}{2^i} x_i\right)\right| =$$

$$= \begin{cases} 2^{n+1} - n & \text{for } n = 1, \ldots, k-1 \\ \dfrac{k}{2^{k+1}} + k & \text{for } n = k+1,\ k+2, \ldots \end{cases}$$

whence $\|f^{(k)} - u_n^*(f^{(k)})\| \geqslant \frac{5}{4}$ for all $n \neq k$ $(n, k = 1, 2, \ldots)$, which proves the last assertion of remark 12.1.

We shall give some other results of duality for $\pi$-bases in § 13 and § 14. Now we shall give (in theorem 12.1 below) some characterizations of Banach spaces having a $\Pi$-basis or a $\pi$-basis, corresponding to some of those given in § 9, theorem 9.3 for approximative bases. To this end, we shall need a sharpening, for projections of finite rank, of § 9, lemma 9.1 and remark 9.4.

We recall that if $G_1$, $G_2$ are subspaces of a Banach space $E$ and if $\varepsilon > 0$, $G_2$ is said to be *$\varepsilon$-close* to $G_1$ if there exists an isomorphism $w$ of $G_1$ *onto* $G_2$ such that

$$\|w(x) - x\| \leqslant \varepsilon \|x\| \qquad (x \in G_1).$$

**Lemma 12.2.** *Let $E$ be a Banach space, $G$ a $k$-dimensional subspace of $E$ (where $k < \infty$) and $u \in L(E, E)$ a projection of finite rank such that*

$$\|I_G - u|_G\| < \delta < 1. \tag{12.11}$$

*Then there exists a projection $v \in L(E, E)$ of finite rank, with $v(E) \subset [G, u(E)]$, such that*

$$v|_G = I_G \text{ (or, equivalently, } v(E) \supset G), \tag{12.12}$$

$$\|v - u\| < \frac{\delta k}{1-\delta}\|u\|, \tag{12.13}$$

$$\|v|_{u(E)} - I_{u(E)}\| < \frac{\delta k}{1-\delta}. \tag{12.14}$$

*Moreover, if* $\frac{\delta k}{1-\delta} < \varepsilon < 1$, *then* $v$ *can be chosen so that, in addition,*

$$\operatorname{Ker} v = \operatorname{Ker} u, \tag{12.15}$$

$$v^*(E^*) = u^*(E^*), \tag{12.16}$$

$$v(E) \text{ is } \varepsilon\text{-close to } u(E). \tag{12.17}$$

*Proof.* We shall prove that the mapping $v$ defined in § 9, remark 9.4, formula (9.56), has the required properties. Clearly, it will be sufficient to prove that $v$ is a projection satisfying (12.17).

Observe first that with the notations of § 9, remark 9.4, we have

$$u(u|_G)^{-1}v_0 = I_{u(G)}v_0 = v_0,$$

and, since $u$ is now a projection on $E$, $uI_{u(E)} = I_{u(E)}$, $uv_0 = v_0$, whence

$$u(I_{u(E)} - v_0) = I_{u(E)} - v_0.$$

Consequently, by the definition (9.56) of $v$,

$$v^2 = ((u|_G)^{-1}v_0u + (I_{u(E)} - v_0)u)((u|_G)^{-1}v_0u + (I_{u(E)} - v_0)u) =$$

$$= (u|_G)^{-1}v_0u(u|_G)^{-1}v_0u + (I_{u(E)} - v_0)u(u|_G)^{-1}v_0u +$$

$$+ (u|_G)^{-1}v_0u(I_{u(E)} - v_0)u + (I_{u(E)} - v_0)u(I_{u(E)} - v_0)u =$$

$$= (u|_G)^{-1}v_0^2u + (I_{u(E)} - v_0)v_0u +$$

$$+ (u|_G)^{-1}v_0(I_{u(E)} - v_0)u + (I_{u(E)} - v_0)^2u,$$

whence, since $v_0^2 = v_0$, $(I_{u(E)} - v_0)v_0 = v_0(I_{u(E)} - v_0) = 0$ and $(I_{u(E)} - v_0)^2 = I_{u(E)} - v_0$ (because $v_0$ is a projection on $u(E)$), we obtain

$$v^2 = (u|_G)^{-1}v_0u + (I_{u(E)} - v_0)u = v,$$

which proves that $v$ is a projection of $E$.

Finally, by (9.56), (9.55) and (9.57) and since $u$ is a projection, we have

$$\|v|_{u(E)} - I_{u(E)}\| = \|wu|_{u(E)} - I_{u(E)}\| = \|w - I_{u(E)}\| < \frac{\delta k}{1-\delta},$$

i.e. (12.17), which completes the proof of lemma 12.2. Note that if $\frac{\delta k}{1-\delta} < \varepsilon < 1$, then also $v(u(E)) = v(E)$ (since $v(u(E)) \subset v(E)$ and $\dim v(u(E)) = \dim u(E) = \dim v(E)$).

It is natural to say that a Banach space $E$ has *the projection approximation property* if the identity operator $I_E\colon E \to E$ can be approximated, uniformly on every compact subset of $E$, by projections of finite rank, that is, if for every compact set $Q \subset E$ and every $\varepsilon > 0$ there exists a projection $u = u_{Q,\varepsilon} \in L(E, E)$ of finite rank satisfying (9.49). Similarly, a Banach space $E$ is said to have the *bounded projection approximation property* if there exists a constant $\lambda \geqslant 1$ such that the above projections of finite rank can be taken of norm $\|u\| = \|u_{Q,\varepsilon}\| \leqslant \lambda$; in this case we shall also say that $E$ has the *$\lambda$-projection approximation property*. Note that, using a similar terminology, a $\Pi$-basis ($\Pi_\lambda$-basis) might be also called a *projection approximative basis* (respectively, a *$\lambda$-projection approximative basis*).

Now we shall prove

**Theorem 12.1** *Let $E$ be a Banach space and let $\lambda \geqslant 1$.*

*$\alpha$) The following statements are equivalent:*

*1°. $E$ has a $\Pi_\lambda$-basis.*

*2°. $E$ is separable and has the $\lambda$-projection approximation property.*

*3°. $E$ is separable and for every finite-dimensional subspace $G$ of $E$ and every $\delta > 0$ there exists a projection $u = u_{G,\delta} \in L(E, E)$ of finite rank, such that*

$$\|u(x) - x\| < \delta \|x\| \qquad (x \in G), \tag{12.18}$$

$$\|u\| \leqslant \lambda. \tag{12.19}$$

*$\beta$) The following statements are equivalent:*

*1°. $E$ has a $\pi_\lambda$-basis.*

*2°. $E$ contains a sequence of subspaces $\{G_n\}$ with the following properties:*

a) $\dim G_n < \infty \qquad (n = 1,2, \ldots);$

b) $G_n \subset G_{n+1} \qquad (n = 1,2, \ldots);$

c) $\bigcup_{n=1}^{\infty} G_n$ *is dense in* $E$;

d) *there exists a projection $u_n$ of $E$ onto $G_n$ $(n = 1,2, \ldots)$, satisfying* (12.3).

$\gamma$) *The following conditions are equivalent to each other and they are satisfied when E has a $\Pi_\lambda$-basis:*

1°. *For every $\varepsilon > 0$, E has a $\Pi_{\lambda+\varepsilon}$-basis.*

2°. *E contains a sequence of subspaces $\{G_n\}$ satisfying* a), b), c) *above and*

d') *there exists a finite rank projection $v_n$ on E such that $v_n|_{G_n} = I_{G_n}$ and $\|v_n\| \leqslant \lambda + \varepsilon_n$, where $0 < \varepsilon_n < 1$ $(n = 1,2, \ldots)$, $\lim\limits_{n\to\infty} \varepsilon_n = 0$.*

3°. *For every $\varepsilon > 0$, E has a $\pi_{\lambda+\varepsilon}$-basis.*

4°. *There exists a (or, equivalently, for every) sequence $\{\varepsilon_n\}$ with $0 < \varepsilon_n < 1$, $\lim\limits_{n\to\infty} \varepsilon_n = 0$, and there exists a $\pi$-basis $\{v_n\} \subset L(E, E)$ of E satisfying*

$$\|v_n\| \leqslant \lambda + \varepsilon_n \qquad (n = 1,2, \ldots). \tag{12.20}$$

5°. *E is separable and for every finite-dimensional subspace G of E and every $\varepsilon > 0$ there exists a projection $v = v_{G,\varepsilon} \in L(E,E)$ of finite rank, satisfying*

$$v|_G = I_G \text{ (or, equivalently, } v(E) \supset G), \tag{12.21}$$

$$\|v\| \leqslant \lambda + \varepsilon. \tag{12.22}$$

$\delta$) *Consequently, a Banach space E has a $\pi$-basis if (and only if) it has a $\Pi$-basis.*

*Proof.* $\alpha$) The proofs of the implications $1° \Rightarrow 2° \Rightarrow 3° \Rightarrow 1°$ are similar to those of § 9, theorem 9.3, implications $1° \Rightarrow 4° \Rightarrow 5° \Rightarrow 1°$.

$\beta$) is a consequence of corollary 12.1 b).

$\gamma$) The proof of the equivalence $1° \Leftrightarrow 2°$ is similar to that of § 9, theorem 9.3, equivalence $1° \Leftrightarrow 2$.

Assume now that we have 1° and let $G \subset E$, $\dim G = k < \infty$, $\varepsilon > 0$. Choose $\delta > 0$ so small that $\delta + \frac{\delta k \lambda}{1-\delta} \leqslant \varepsilon$ and, by 1°, let $\{u_n\} \subset L(E, E)$ be a $\Pi_{\lambda+\delta}$-basis of $E$. Then, since the convergence (12.1) is uniform on the compact set $S_G = \{x \in G \mid \|x\| \leqslant 1\}$ (by Ch. I, § 17, proof of theorem 17.3, with $s_n$ replaced by $u_n$), there exists a positive integer $N = N(G, \delta)$ such that

$$\|I_G - u_N|_G\| < \delta. \tag{12.23}$$

Hence, by lemma 12.2, there exists a projection $v \in L(E, E)$ of finite rank satisfying (12.21) and

$$\|v\| < \|u_N\| + \frac{\delta k}{1-\delta} \|u_N\| \leqslant \lambda + \delta + \frac{\delta k \lambda}{1-\delta} \leqslant \lambda + \varepsilon,$$

i.e., (12.22). Thus, $1° \Rightarrow 5°$ (clearly, $1°$ implies that $E$ is separable).

The proof of the implication $5° \Rightarrow 4°$ is contained in § 9, proof of proposition 9.7, using now merely the assumption $5°$ instead of § 9, lemma 9.2.

Assume now that we have $4°$ and let $\varepsilon > 0$. Then, since $\lim_{n\to\infty} \varepsilon_n = 0$, we have $\varepsilon_n < \varepsilon$ for all $n > N(\varepsilon) = N$, whence, by (12.20), $\|v_n\| \leqslant \lambda + \varepsilon$ $(n = N+1, N+2, \ldots)$. Consequently, since $\{v_n\}$ satisfies (12.1) and (12.2), $\{v_n\}_{n=N+1}^{\infty}$ is a $\pi_{\lambda+\varepsilon}$-basis of $E$. Thus, $4° \Rightarrow 3°$.

Finally, the implication $3° \Rightarrow 1°$ is obvious (hence $1° \Leftrightarrow \ldots \Leftrightarrow 5°$) and it is also obvious that $1°$ is satisfied when $E$ has a $\Pi_\lambda$-basis.

$\delta$) If $E$ has a $\Pi$-basis, then this is a $\Pi_\lambda$-basis for a suitable $\lambda \geqslant 1$ and hence, by part $\gamma$), $E$ has a $\pi_{\lambda+\varepsilon}$-basis, which completes the proof of theorem 12.1.

*Remark 12.2.* Condition $\gamma$) $5°$ says, in other words, that the identity operator $I_G$ can be extended to a projection $v \in L(E, E)$ of finite rank (onto a suitable larger finite-dimensional subspace $F = F(G, \varepsilon') \supset G$), of norm $\leqslant \lambda + \varepsilon'$. The above proof of the implication $\gamma$) $1° \Rightarrow 5°$ shows that one can also obtain $v$ with the additional properties

$$\operatorname{Ker} v = \operatorname{Ker} u_N,\ v^*(E^*) = u_N^*(E^*),\ \|v|_{u_N(E)} - I_{u_N(E)}\| < \varepsilon \qquad (12.24)$$

and such that $v(E)$ is $\varepsilon$-close to $u_N(E)$. Hence, a similar remark holds for $4°$ and $3°$ as well.

For Banach spaces with a $\Pi_{1+\varepsilon}$-basis for each $\varepsilon > 0$ and for spaces with a $\pi_1$-basis we shall give a stronger result in § 13, proposition 13.7.

In contrast to the situation for approximative bases, the following problem is open:

*Problem 12.3.* a) If a Banach space $E$ has a $\Pi_{\lambda+\varepsilon}$-basis for every $\varepsilon > 0$, does $E$ have a $\Pi_\lambda$-basis? b) How about a $\pi_\lambda$-basis? c) In particular, does every Banach space $E$ with a $\Pi_\lambda$-basis have a $\pi_\lambda$-basis?

In connection with problem 12.3 b) let us observe that *every Banach space $E$ contains a subspace $G$ which has a $\pi_{1+\varepsilon}$-basis for each* $\varepsilon > 0$. Indeed, by § 1, remark 1.1, $E$ contains a basic sequence $\{x_n\}$ such that $\|s_n\| \leqslant 1 + \dfrac{1}{n}$ $(n = 1,2, \ldots)$ and then $G = [x_n]$ has the required property (by theorem 12.1 $\gamma$), implication $4° \Rightarrow 3°$).

Let us also observe that *there exist Banach spaces with bases* (hence with $\Pi$-bases) *which have no $\pi_1$-basis*, as shown by § 9, remark 9.9 c). Moreover, there are such spaces *even with an unconditional basis*, as shown by

*Example 12.2.* Let $E_0 = [t^{n_k}]$, the subspace of $C([0,1])$ constructed in Ch. II, § 1, corollary 1.1 and theorem 1.4. Then $E_0$ has no $\pi_1$-basis. Indeed, assume, a contrario, that $\{u_n\} \subset L(E_0, E_0)$ is a $\pi_1$-basis of $E_0$.

Choose a basis $\{x_i\}_{i=1}^{m_1}$ of $u_1(E_0)$. Next, by corollary 12.2, choose linearly independent elements $\{x_i\}_{i=m_1+1}^{m_2} \subset u_2(E_0) \cap \operatorname{Ker} u_1$ such that $\{x_i\}_{i=1}^{m_2}$ is a basis of $u_2(E_0)$. Continuing in this way indefinitely we obtain, by (12.5), a finitely linearly independent complete sequence $\{x_n\}$ in $E_0$ and an increasing sequence $\{m_n\}$ such that for each $n$ the projection of $[x_1, \ldots, x_{m_{n+1}}]$ onto $[x_1, \ldots, x_{m_n}]$ along $[x_{m_n+1}, \ldots, x_{m_{n+1}}]$ has norm 1. However, the proof of Ch. II, § 1, theorem 1.4 actually shows that if $\{x_n\}$ is an arbitrary finitely linearly independent complete sequence in $E_0$, then for every sufficiently large $n$ and every $m > n$, $(\widehat{[x_1, \ldots, x_n]; [x_{n+1}, \ldots, x_m]}) < 1$, in contradiction with the preceding construction of $\{x_n\}$. This proves that $E_0$ has no $\pi_1$-basis (however, we do not know whether $E_0$ has a $\pi_{1+\varepsilon}$-basis for each $\varepsilon > 0$). On the other hand, by § 4, theorem 4.1, $E_0$ is isomorphic to $c_0$ (so it has an unconditional basis). Thus, *there exists an equivalent norm* $|\cdot|$ *on* $c_0$ *such that* $(c_0, |\cdot|)$ *has no* $\pi_1$*-basis* (however, we note that $(c_0, |.|)$ has the metric approximation property, since the separable conjugate space $(c_0, |.|)^*$ has the approximation property, whence also the metric approximation property*); thus, *a separable Banach space* $E$ *with the metric approximation property need not have a* $\pi_1$*-basis*).

Obviously, every basis $\{x_n\}$ is a strict $\pi$-basis, since one can take $\{u_n\} = \{s_n\}$, the sequence of partial sum operators associated to the basis $\{x_n\}$. The converse is not true, as shown by

*Example 12.3.* In the real space $E = l^2$ let

$$x_n = e_n - e_{n+1} \qquad (n = 1,2, \ldots), \tag{12.25}$$

$$f_n(x) = (e_1 + \ldots. + e_n, x) \qquad (x \in E,\ n = 1,2, \ldots). \tag{12.26}$$

Then, by § 11, example 11.3, $\{x_n\}$ is a Cesàro basis of $E$, with the a.s.f. $\{f_n\}$, but $\{x_n\}$ is not a basis of $E$. Furthermore, since $E = l^2$, there exist projections $u_n$ of norm 1 of $E$ onto $[x_1, \ldots, x_n]$ $(n = 1,2, \ldots)$ and hence, by corollary 12.1 b), $\{x_n\}$ is a strict $\pi_1$-basis of $E$.

*Problem 12.4.* Does every Banach space $E$ with a $\Pi$-basis (or, equivalently, with a $\pi$-basis) have a basis?

From Ch. I, § 20, theorem 20.1, implication $5° \Rightarrow 11°$ (or, alternatively, from Ch. II, § 1, proposition 1.3) it follows that *if there exists an equivalent norm* $|||\cdot|||$ *on* $E$, *in which* $E$ *has a strict* $\pi_1$*-basis, then* $E$ *has a basis.* Also, Ch. I, § 20, problem 20.1 suggests the following problem: If $E$ has a strict $\pi$-basis, does $E$ have a basis, or, equivalently, a strict $\pi_1$-basis in some suitable equivalent norm $|||\cdot|||$ on $E$? Omitting here the condition "strict", we arrive at

---

*) See § 9, proof of theorem 9.3, implication $11° \Rightarrow 4°$ and § 9, proposition 9.8.

*Problem 12.5.* If a Banach space $E$ has a $\pi$-basis (or, equivalently, a $\Pi$-basis), does $E$ have a $\pi_1$-basis in some suitable equivalent norm on $E$?

Now we pass to a special class of $\pi_1$-bases $\{u_n\}$, obtained by imposing an additional condition on the ranges $u_n(E)$ $(n = 1,2, \ldots)$.

*Definition 12.2.* A $\pi_1$-basis $\{u_n\} \subset L(E, E)$ of a Banach space $E$ is called a $\pi_1^\infty$*-basis* of $E$ if $u_n(E)$ is linearly isometric to $l^\infty_{m_n}$ for some $m_n$ $(n = 1,2, \ldots)$.

We shall give some characterizations of Banach spaces with a $\pi_1^\infty$-basis (in theorem 12.2 below), from which it will follow that every such space has a monotone basis (theorem 12.3 below). To this end, we shall need some preparation.

Let us recall the following result, similar to Ch. II, § 3, lemma 3.1:

**Lemma 12.3.** *A linear mapping $T\colon l^\infty_m \to l^\infty_n$, with $T(e_j^{(m)}) = \{a_{ij}\}_{i=1}^n \in l^\infty_n$ $(j = 1, \ldots, m$; $\{e_j^{(m)}\}$ = the unit vector basis of $l^\infty_m)$ is an isometry if and only if*

$$\max_{1 \leqslant i \leqslant n} |a_{ij}| = 1 \qquad (j = 1, \ldots, m), \tag{12.27}$$

$$\sum_{j=1}^{m} |a_{ij}| \leqslant 1 \qquad (i = 1, \ldots, n). \tag{12.28}$$

*Proof.* If $T\colon l^\infty_m \to l^\infty_n$ is a linear isometry, we have

$$\max_{1 \leqslant i \leqslant n} |a_{ij}| = \|T(e_j^{(m)})\| = \|e_j^{(m)}\| = 1 \qquad (j = 1, \ldots, m),$$

and, putting $\varepsilon_{ij} = \operatorname{sign} a_{ij}$, we obtain

$$\sum_{j=1}^{m} |a_{ij}| = \left|\sum_{j=1}^{m} a_{ij}\varepsilon_{ij}\right| \leqslant \max_{1 \leqslant k \leqslant n} \left|\sum_{j=1}^{m} a_{kj}\varepsilon_{ij}\right| = \left\|\sum_{j=1}^{m} \varepsilon_{ij}\{a_{kj}\}_{k=1}^{n}\right\| =$$

$$= \left\|\sum_{j=1}^{m} \varepsilon_{ij} T(e_j^{(m)})\right\| = \left\|T\left(\sum_{j=1}^{m} \varepsilon_{ij} e_j^{(m)}\right)\right\| = \left\|\sum_{j=1}^{m} \varepsilon_{ij} e_j^{(m)}\right\| \leqslant 1$$

$(i = 1, \ldots, n)$, so the conditions are necessary.

Conversely, if $T\colon l^\infty_m \to l^\infty_n$ satisfies (12.27), (12.28), then, by (12.27), for each $j \leqslant m$ there exists an index $i_j \leqslant n$ such that

$$|a_{i_j j}| = 1 \qquad (j = 1, \ldots, m). \tag{12.29}$$

Hence, by (12.28), for each $j \leqslant m$ we have

$$1 \geqslant \sum_{s=1}^{m} |a_{i_j s}| = 1 + \sum_{\substack{s=1 \\ s \neq j}}^{m} |a_{i_j s}|,$$

and therefore

$$a_{i_j s} = 0 \qquad (s = 1, \ldots, j-1,\ j+1, \ldots, m).$$

Consequently, for any $x = \sum_{j=1}^{m} \xi_j e_j^{(m)} \in l_m^\infty$ we have

$$\|T(x)\| = \left\| \sum_{s=1}^{m} \xi_s T(e_s^{(m)}) \right\| = \max_{1 \leqslant i \leqslant n} \left| \sum_{s=1}^{m} a_{is}\xi_s \right| \geqslant$$

$$\geqslant \left| \sum_{s=1}^{m} a_{i_j s}\xi_s \right| = |\xi_j| \qquad (j = 1, \ldots, m),$$

whence

$$\|T(x)\| \geqslant \max_{1 \leqslant j \leqslant m} |\xi_j| = \|x\| \qquad \left(x = \sum_{j=1}^{m} \xi_j e_j^{(m)} \in l_m^\infty\right).$$

On the other hand, by (12.28) we have

$$\|T(x)\| = \left\| \sum_{j=1}^{m} \xi_j T(e_j^{(m)}) \right\| = \max_{1 \leqslant i \leqslant n} \left| \sum_{j=1}^{m} a_{ij}\xi_j \right| \leqslant$$

$$\leqslant \max_{1 \leqslant j \leqslant m} |\xi_j| \max_{1 \leqslant i \leqslant n} \sum_{j=1}^{m} |a_{ij}| \leqslant \|x\| \qquad \left(x = \sum_{j=1}^{m} \xi_j e_j^{(m)} \in l_m^\infty\right),$$

which, together with the preceding inequality, gives

$$\|T(x)\| = \|x\| \qquad (x \in l_m^\infty),$$

and hence $T$ is a linear isometry. This completes the proof.

**Lemma 12.4.** *Let $G$ be a subspace of $l_n^\infty$, which is linearly isometric to $l_m^\infty$, where $m < n$. Then there exists a subspace $F$ of $l_n^\infty$, such that $G \subset F$ and that $F$ is linearly isometric to $l_{m+1}^\infty$.*

*Proof.* Let $T$ be a linear isometry of $l_m^\infty$ onto $G$, with $T(e_j^{(m)}) = \{a_{ij}\}_{i=1}^{n} \in l_n^\infty$ $(j = 1, \ldots, m)$. Then, by lemma 12.3, we have (12.27) and (12.28) and hence, by (12.27), for each $j \leqslant m$ there exists an index $i_j \leqslant n$ satisfying (12.29). We shall consider two cases:

*Case 1°.* For each $j \leqslant m$ there exists exactly one such index $i_j \leqslant n$. Then, since $m < n$, there exists an index $i_0 \in \{1, \ldots, n\} \setminus \{i_1, \ldots, i_m\}$.

*Case 2°*. There exists a $j \leqslant m$ for which there are two such indices $i_j', i_j'' \leqslant n$. In this case, put $i_0 = i_j''$.

Then, in both cases, $i_0$ has the property that for each $j \leqslant m$ there exists an index $i_j \neq i_0$ satisfying (12.29), whence

$$\max_{\substack{1 \leqslant i \leqslant n \\ i \neq i_0}} |a_{ij}| = 1. \tag{12.30}$$

Define elements $z_j \in l_n^\infty$ ($j = 1, \ldots, m+1$) by

$$z_j = T(e_j^{(m)}) - a_{i_0 j} e_{i_0}^{(n)} \qquad (j = 1, \ldots, m), \tag{12.31}$$

$$z_{m+1} = e_{i_0}^{(n)}, \tag{12.32}$$

where $\{e_i^{(n)}\}$ is the unit vector basis of $l_n^\infty$, and let

$$F = [z_1, \ldots, z_{m+1}] \in l_m^\infty. \tag{12.33}$$

We shall show that $F$ has the required properties. Indeed, define $u: l_{m+1}^\infty \to l_n^\infty$ by

$$u\left(\sum_{j=1}^{m+1} \xi_j e_j^{(m+1)}\right) = \sum_{j=1}^{m+1} \xi_j z_j \tag{12.34}$$

and let

$$u(e_j^{(m+1)}) = \{b_{ij}\}_{i=1}^n \qquad (j = 1, \ldots, m+1). \tag{12.35}$$

Then, by (12.35), (12.34) and (12.31), (12.32),

$$\{b_{ij}\}_{i=1}^n = u(e_j^{(m+1)}) = z_j = \begin{cases} \{a_{ij}\}_{i=1}^n - a_{i_0 j} e_{i_0}^{(n)} & \text{for } j = 1, \ldots, m \\ e_{i_0}^{(n)} & \text{for } j = m+1, \end{cases}$$

whence

$$b_{ij} = \begin{cases} a_{ij} & \text{for } i \neq i_0;\ j = 1, \ldots, m \\ 0 & \text{for } i = i_0;\ j = 1, \ldots, m \\ \delta_{i_0 i} & \text{for } j = m+1. \end{cases} \tag{12.36}$$

Consequently, by (12.30) and (12.28),

$$\max_{1 \leq i \leq n} |b_{ij}| = 1 \qquad (j = 1, \ldots, m+1), \tag{12.37}$$

$$\sum_{j=1}^{m+1} |b_{ij}| = \begin{cases} \sum_{j=1}^{m} |a_{ij}| + 0 \leq 1 & \text{for } i \neq i_0 \\ \sum_{j=1}^{m} 0 + 1 = 1 & \text{for } i = i_0, \end{cases} \tag{12.38}$$

and hence, by lemma 12.3, $u: l_{m+1}^{\infty} \to l_n^{\infty}$ is a linear isometry. Since $u(l_{m+1}^{\infty}) \subset F = [z_1, \ldots, z_{m+1}]$, we have, clearly, $u(l_{m+1}^{\infty}) = F$, so $F$ is linearly isometric to $l_{m+1}^{\infty}$. Finally, since

$$T(e_j^{(m)}) = z_j + a_{i_0 j} e_{i_0}^{(n)} = z_j + a_{i_0 j} z_{m+1} \in F \qquad (j = 1, \ldots, m)$$

and since $G = [T(e_1^{(m)}), \ldots, T(e_m^{(m)})]$, it follows that $G \subset F$, which completes the proof of lemma 12.4.

*Remark 12.3.* Since the linearly isometric images of the unit balls of $l_n^{\infty}$ spaces (i.e., of $n$-dimensional hypercubes) are the $n$-dimensional hyperparallelepipeds*⁾, lemma 12.4 has the following geometric interpretation: *Let $W$ be an $n$-dimensional hyperparallelepiped in the $n$-dimensional (real or complex) euclidean space $E_n$, and let the origin be the center of symmetry of $W$. Let $G$ be an $m$-dimensional subspace of $E_n$, such that $G \cap W$ is an $m$-dimensional hyperparallelepiped, where $m < n$. Then there exists an $(m+1)$-dimensional subspace $F$ of $E_n$ such that $G \subset F$ and that $F \cap W$ is an $(m+1)$-dimensional hyperparallelepiped.*

**Lemma 12.5.** *Let $G$ be a subspace of a Banach space $E$, such that $G$ is linearly isometric to $l_n^{\infty}$. Then there exists a projection $v$ of $E$ onto $G$, of norm $\|v\| = 1$.*

*Proof.* Let $u$ be a linear isometry of $G$ onto $l_n^{\infty}$, let

$$x_i = u^{-1}(e_i^{(n)}) \in G \qquad (i = 1, \ldots, n), \tag{12.39}$$

where $\{e_i^{(n)}\}$ is the unit vector basis of $l_n^{\infty}$, and let $\{\varphi_i\}_{i=1}^n \subset G^*$ be the a.s.c.f. to the basis $\{x_i\}_{i=1}^n$ of $G$. Then

$$\max_{1 \leq i \leq n} |\varphi_i(x)| = \left\| \sum_{i=1}^{n} \varphi_i(x) e_i^{(n)} \right\| = \left\| \sum_{i=1}^{n} \varphi_i(x) x_i \right\| = \|x\| \qquad (x \in G),$$

*⁾ In Ch. II, § 2, theorem 2.1, condition 4°, the word "hyperparallelepiped" was used in a more general sense, but here (in remark 12.3) it is used in the ordinary sense of Euclidean geometry.

whence $\|\varphi_i\| \leqslant 1$ $(i = 1, \ldots, n)$. By the Hahn-Banach theorem, let $f_i \in E^*$, $f_i|_G = \varphi_i$, $\|f_i\| = \|\varphi_i\| \leqslant 1$ $(i = 1, \ldots, n)$. Put

$$v(x) = \sum_{i=1}^{n} f_i(x)x_i \qquad (x \in E). \tag{12.40}$$

Then $v$ has the required properties. Indeed, $v(E) \subset G$ and

$$v(x) = \sum_{i=1}^{n} f_i(x)x_i = \sum_{i=1}^{n} \varphi_i(x)x_i = x \qquad (x \in G),$$

so $v$ is a projection of $E$ onto $G$. Also,

$$\|v(x)\| = \left\| \sum_{i=1}^{n} f_i(x)x_i \right\| = \left\| \sum_{i=1}^{n} f_i(x)e_i^{(n)} \right\| = \max_{1 \leqslant i \leqslant n} |f_i(x)| \leqslant$$

$$\leqslant \max_{1 \leqslant i \leqslant n} \|f_i\| \, \|x\| \leqslant \|x\| \qquad (x \in E),$$

so $\|v\| \leqslant 1$, whence $\|v\| = 1$, which completes the proof.

Now we are ready to give the following characterizations of Banach spaces with a $\pi_1^\infty$-basis:

**Theorem 12.2.** *Let $E$ be a Banach space. The following statements are equivalent:*

1°. *$E$ has a $\pi_1^\infty$-basis.*

2°. *$E$ contains a sequence of subspaces $\{G_n'\}$ with the following properties:*

a') $G_n' \equiv l_{m_n}^\infty$ *for some* $m_n$ $(n = 1,2, \ldots)$;

b') $G_n' \subset G_{n+1}'$ $(n = 1,2, \ldots)$;

c') $\bigcup_{n=1}^{\infty} G_n'$ *is dense in* $E$.

3°. *$E$ contains a sequence of subspaces $\{G_n\}$ with the following properties:*

a) $G_n \equiv l_n^\infty$ $(n = 1,2, \ldots)$;

b) $G_n \subset G_{n+1}$ $(n = 1,2, \ldots)$;

c) $\bigcup_{n=1}^{\infty} G_n$ *is dense in* $E$.

*Proof.* If $\{u_n\}$ is a $\pi_1^\infty$-basis of $E$, then, by proposition 12.1, the subspaces $G_n' = u_n(E)$ $(n = 1,2, \ldots)$ satisfy a'), b') and c'). Thus, 1° $\Rightarrow$ 2°.

Assume now that we have 2°. Then, by lemma 12.4, if $n \in \mathcal{N}$ and $m_{n+1} - m_n > 1$, then there is a chain of subspaces $G'_n = F_0 \subset F_1 \subset \ldots \ldots \subset F_{m_{n+1}-m_n} = G'_{n+1}$ such that $F_j \equiv l^\infty_{m_n+j}$ for all $j = 0,1,2, \ldots \ldots, m_{n+1} - m_n$. Thus, 2° ⇒ 3°.

The implication 3° ⇒ 2° is obvious.

Assume, finally, that we have 2°. Then, by lemma 12.5, there exists a projection $u_n$ of $E$ onto $G'_n$, of norm $\|u_n\| = 1$ $(n = 1,2, \ldots)$. By proposition 12.1, $\{u_n\}$ satisfies (12.2) and (12.1), so $\{u_n\}$ is a $\pi_1^\infty$-basis of $E$. Thus, 2° ⇒ 1°, which completes the proof of theorem 12.2.

The following result shows that for Banach spaces with a $\pi_1^\infty$-basis the answer to problem 12.4 is affirmative:

**Theorem 12.3.** *Every Banach space $E$ with a $\pi_1^\infty$-basis has a monotone basis $\{x_n\}$ such that $[x_i]_{i=1}^n \equiv l_n^\infty$ $(n = 1,2, \ldots)$.*

*Proof.* By theorem 12.2, implication 1° ⇒ 3°, $E$ contains a sequence of subspaces $\{G_n\}$ with properties a), b), c) of 3°. By a) and lemma 12.5, each $G_n$ admits a projection $u_n$ of $E$ onto $G_n$, with $\|u_n\|=1$ $(n = 1,2, \ldots)$. Consequently, by Ch. II, § 1, proposition 1.3 and its proof, $E$ has a monotone basis $\{x_n\}$ such that $[x_i]_{i=1}^n = G_n \equiv l_n^\infty$ $(n = 1,2, \ldots)$, which completes the proof of theorem 2.3.

As an application we shall see, in § 18, corollary 18.3, that *every $C(Q)$ space ($Q$ compact metric) has a monotone basis* with the above property.

In § 9, theorem 9.3, implication 1° ⇒ 9°, we have seen that every Banach space $E$ with an approximative basis is isomorphic to a complemented subspace of a Banach space $E$ with a basis. Now we shall show that if $B$ is required to have only a $\pi$-basis, then one can even find a "universal complement" $C_p$ (i.e., $B = E \times C_p$ has a $\pi$-basis). In order to introduce the spaces $C_p$, we need some preparation.

We recall that for two isomorphic Banach spaces $A$, $B$ the *distance coefficient* $d(A, B)$ is defined by

$$d(A, B) = \inf \|w\| \, \|w^{-1}\|, \tag{12.41}$$

where the inf is taken over all isomorphisms $w$ of $A$ onto $B$. It is well known*) that $\rho(A, B) = \ln d(A, B)$ satisfies the axioms of a pseudometric in the collection $\mathcal{F}$ of all Banach spaces isomorphic to a given space (i.e., $\rho(A, B) = 0$ may happen also when $A \neq B$); thus, a metric space is obtained by considering in $\mathcal{F}$ the equivalence classes of nearly isometric spaces (two Banach spaces $A$, $B$ are called *nearly isometric* if $\rho(A, B) = 0$). This motivates the term "distance coefficient" and the use of the terminology of metric spaces when working, in the sequel, with $d(A, B)$ (rather than with $\rho(A, B)$). In particular, we shall say that $\mathcal{B} \subset \mathcal{F}$ is *dense* in $\mathcal{F}$ if for every $E \in \mathcal{F}$ and $\varepsilon > 0$ there exists a space

---

*) See e.g. [11], p. 242.

$B \in \mathscr{B}$ such that $d(E, B) \leqslant 1 + \varepsilon$; we shall say that $\mathscr{F}$ is *separable*, if it contains a countable dense subcollection $\mathscr{B}$. We recall

**Lemma 12.6.** *For each $n < \infty$, the collection $\mathscr{F}_n$ of all n-dimensional Banach spaces is separable.*

*Proof.* Let $E \in \mathscr{F}_n$ and let $0 < \varepsilon < 1$. Let $\{y_1, \ldots, y_k\} \subset \sigma_E$ be a finite $\varepsilon$-net for $\sigma_E = \{x \in E \mid \|x\| = 1\}$ and take $f_i \in E^*$ such that $\|f_i\| = 1$, $f_i(y_i) = 1$ $(i = 1, \ldots, k)$. Finally, let $B$ be the subspace of $c_0$ defined by

$$B = \{\{f_1(x), \ldots, f_k(x), 0, 0, \ldots\} \mid x \in E\}. \tag{12.42}$$

We claim that the mapping

$$w: x \to \{f_1(x), \ldots, f_k(x), 0, 0, \ldots\} \qquad (x \in E) \tag{12.43}$$

is an isomorphism of $E$ onto $B$, satisfying $\|w\| \leqslant 1$, $\|w^{-1}\| \leqslant \frac{1}{1-\varepsilon} = 1 + \frac{\varepsilon}{1-\varepsilon}$, whence $B \in \mathscr{F}_n$ and $d(E, B) \leqslant 1 + \frac{\varepsilon}{1-\varepsilon}$. Indeed, $w$ is linear and $\|w(x)\| = \max_{1 \leqslant i \leqslant k} |f_i(x)| \leqslant \|x\|$ for all $x \in E$, so $\|w\| \leqslant 1$. On the other hand, if $x \in \sigma_E$, then $\|x - y_{i_0}\| < \varepsilon$ for a suitable $i_0 \leqslant k$ whence

$$||f_{i_0}(x)| - 1| = ||f_{i_0}(x)| - |f_{i_0}(y_{i_0})|| \leqslant |f_{i_0}(x) - f_{i_0}(y_{i_0})| \leqslant$$

$$\leqslant \|f_{i_0}\| \, \|x - y_{i_0}\| < \varepsilon.$$

Consequently,

$$\|w(x)\| = \max_{1 \leqslant i \leqslant k} |f_i(x)| \geqslant |f_{i_0}(x)| \geqslant 1 - \varepsilon \qquad (x \in \sigma_E),$$

whence $\|w^{-1}\| \leqslant \frac{1}{1-\varepsilon}$, which proves our claim.

Now let

$$z_j = \{\zeta_1^{(j)}, \ldots, \zeta_k^{(j)}, 0, 0, \ldots\} \qquad (j = 1, \ldots, n)$$

be any basis of $B$. Take rational (complex or real) scalars $\tilde{\zeta}_i^{(j)}$ such that

$$|\zeta_i^{(j)} - \tilde{\zeta}_i^{(j)}| < \frac{\varepsilon}{2^{j+1}\|h_j\|} \qquad (i = 1, \ldots, k;\ j = 1, \ldots, n),$$

where $\{h_j\}_{j=1}^n \subset B^*$ is the a.s.c.f. to $\{z_j\}_{j=1}^n$, and let $\tilde{B} = [\tilde{z}_j]_{j=1}^n$, the subspace of $c_0$ spanned by

$$\tilde{z}_j = \{\tilde{\zeta}_1^{(j)}, \ldots, \tilde{\zeta}_k^{(j)}, 0,0, \ldots\} \qquad (j = 1, \ldots, n)$$

(with the norm induced by $c_0$). Then, clearly, $\dim \tilde{B} \leqslant n$. We shall show that $d(B, \tilde{B}) \leqslant \frac{1+\varepsilon}{1-\varepsilon}$, whence $\tilde{B} \in \mathscr{F}_n$ and

$$d(E, \tilde{B}) \leqslant d(E, B) d(B, \tilde{B}) \leqslant \frac{1+\varepsilon}{(1-\varepsilon)^2} = 1 + \frac{3\varepsilon - \varepsilon^2}{(1-\varepsilon)^2},$$

which, since the collection $\tilde{\mathscr{B}}_n$ of all subspaces of $c_0$ of the form $\tilde{B}$ is countable, will complete the proof.

Observe that, by the above choice of $\tilde{\zeta}_i^{(j)}$, we have $\|z_j - \tilde{z}_j\| < \frac{\varepsilon}{2^{j+1}\|h_j\|}$ $(j = 1, \ldots, n)$. Hence, for any scalars $\alpha_1, \ldots, \alpha_n$ we obtain, putting $x = \sum_{j=1}^n \alpha_j z_j = \sum_{j=1}^n h_j(x) z_j$,

$$\left| \left\| \sum_{j=1}^n \alpha_j z_j \right\| - \left\| \sum_{j=1}^n \alpha_j \tilde{z}_j \right\| \right| \leqslant \left\| \sum_{j=1}^n h_j(x)(z_j - \tilde{z}_j) \right\| \leqslant$$

$$\leqslant \sum_{j=1}^n \|h_j\| \, \|x\| \, \|z_j - \tilde{z}_j\| \leqslant \sum_{j=1}^n \frac{\varepsilon}{2^{j+1}} \|x\| < \varepsilon \left\| \sum_{j=1}^n \alpha_j z_j \right\|,$$

whence

$$(1-\varepsilon) \left\| \sum_{j=1}^n \alpha_j z_j \right\| \leqslant \left\| \sum_{j=1}^n \alpha_j \tilde{z}_j \right\| \leqslant (1+\varepsilon) \left\| \sum_{j=1}^n \alpha_j z_j \right\|.$$

Thus, $\tilde{w}: \sum_{j=1}^n \alpha_j z_j \to \sum_{j=1}^n \alpha_j \tilde{z}_j$ is an isomorphism of $B$ onto $\tilde{B}$, with $\|\tilde{w}\| \leqslant 1 + \varepsilon$, $\|\tilde{w}^{-1}\| \leqslant \frac{1}{1-\varepsilon}$, whence $d(B, \tilde{B}) \leqslant \frac{1+\varepsilon}{1-\varepsilon}$, which completes the proof of lemma 12.6.

*Remark 12.4.* The argument in the final part of the proof is essentially the same as that used in Ch. I, § 10, remark 10.5 (to theorem 10.3) combined with Ch. I, § 9, proof of theorem 9.1. This argument also shows the following relation between $\varepsilon$-close subspaces and their distance coefficients: *If* $B, \tilde{B}$, *are subspaces of a Banach space* $Z$ *and if* $\tilde{B}$ *is* $\varepsilon$-*close to* $B$, *where* $0 < \varepsilon < 1$, *then* $d(B, \tilde{B}) \leqslant \frac{1+\varepsilon}{1-\varepsilon}$. Hence, for example, in lemma 12.2 above we also have $d(u(E), v(E)) \leqslant \frac{1+\varepsilon}{1-\varepsilon}$ (by (12.17)).

Although $d(A, B)$ is defined in the collection $\bigcup_{n=1}^{\infty} \mathscr{F}_n$ of all finite-dimensional Banach spaces only for pairs $A, B$ with $\dim A = \dim B$, it will be convenient to say that a collection $\mathscr{B} \subset \bigcup_{n=1}^{\infty} \mathscr{F}_n$ is *dense* in $\bigcup_{n=1}^{\infty} \mathscr{F}_n$ if for every $E \in \bigcup_{n=1}^{\infty} \mathscr{F}_n$ and $\varepsilon > 0$ there exists a space $B \in \mathscr{B}$ such that $d(E, B) \leqslant 1 + \varepsilon$ (or, equivalently, if $\mathscr{B}_n = \mathscr{B} \cap \mathscr{F}_n$ is dense in $\mathscr{F}_n$ for $n = 1,2, \ldots$). Then, by lemma 12.6, $\bigcup_{n=1}^{\infty} \mathscr{F}_n$ contains a countable dense subcollection $\mathscr{B}$. This permits to introduce the spaces $C_p$ by

*Definition 12.3*. Let $\{G_n\}$ be a sequence of finite-dimensional Banach spaces such that

i) $\{G_n\}$ is dense in the collection $\bigcup_{n=1}^{\infty} \mathscr{F}_n$ of all finite-dimensional Banach spaces (in the above sense);

ii) each $G_i$ "occurs an infinity of times in $\{G_n\}$", that is, for each $i \in \mathscr{N}$ there exists an infinite subset $J_i$ of $\mathscr{N} = \{1,2,3, \ldots\}$ such that $G_i \equiv G_j$ for all $j \in J_i$.

Then, for each $p$ with $1 \leqslant p < \infty$, let $C_p$ be the separable Banach space*)

$$C_p = (G_1 \times G_2 \times \ldots)_{l^p} \tag{12.44}$$

and for $p = \infty$ let $C_\infty$ be the separable Banach space

$$C_\infty = (G_1 \times G_2 \times \ldots)_{c_0}. \tag{12.45}$$

Let us observe that the above definition of $C_p$ *does not depend on the sequence* $\{G_n\}$ (up to a nearly isometry), i.e., if $\{G_n'\}$ is another sequence in $\bigcup_{n=1}^{\infty} \mathscr{F}_n$, satisfying conditions i) and ii) and**) if

*) For the definition and some properties of such product spaces see Ch. II, § 18, especially lemmas 18.5 and 18.7 and Ch. II, § 8, the footnote on p. 303. Note that, by this definition, the spaces $C_p$ for $1 < p < \infty$ are reflexive, while $C_1$ and $C_\infty$ are non-reflexive.

**) We write the statement and its proof only for $1 \leqslant p < \infty$. In the case $p = \infty$, the only change is that $l^\infty$ must be replaced by $c_0$.

$C'_p=(G'_1 \times G'_2 \times \ldots)_{l^p}$, then $d(C_p, C'_p)=1$. Indeed, let $\varepsilon>0$. Then, by i), for each $n$ there exists an index $i_n$ such that $d(G_n, G'_{i_n}) \leqslant 1+\varepsilon$ and by ii) we may assume that here $i_j \neq i_k$ for $j \neq k$ $(j, k=1,2, \ldots)$. Hence, by Ch. II, § 18, lemma 18.7 a), $d(C_p, D'_1) \leqslant 1+\varepsilon$, where $D'_1=(G'_{i_1} \times G'_{i_2} \times \ldots)_{l^p}$. Similarly, for $\{m_n\}=\mathcal{N} \setminus \{i_n\}$ there exists a sequence $\{s_n\} \subset \mathcal{N}$ with $s_j \neq s_k$ for $j \neq k$ $(j, k=1,2, \ldots)$ such that $d(G'_{m_n}, G_{s_n}) \leqslant 1+\varepsilon$ $(n=1,2, \ldots)$, whence $d(D'_2, D) \leqslant 1+\varepsilon$, where $D'_2=(G'_{m_1} \times G'_{m_2} \times \ldots)_{l^p}$, $D=(G_{s_1} \times G_{s_2} \times \ldots)_{l^p}$. Hence, using again Ch. II, § 18, lemma 18.7 a), $d((C_p \times D)_{l^p}, (D'_1 \times D'_2)_{l^p}) \leqslant 1+\varepsilon$. But, by ii) and Ch. II, § 18, lemma 18.5, $(C_p \times D)_{l^p} \equiv C_p$ and, clearly, $(D'_1 \times D'_2)_{l^p} \equiv C'_p$. Consequently,

$$1 \leqslant d(C_p, C'_p)=d((C_p \times D)_{l^p}, (D'_1 \times D'_2)_{l^p}) \leqslant 1+\varepsilon,$$

whence, since $\varepsilon>0$ was arbitrary, $d(C_p, C'_p)=1$, which completes the proof of our assertion.

Now we can prove the property of $C_p$ of being a "universal complement", announced above.

**Theorem 12.4.** *Let $1 \leqslant p \leqslant \infty$. If a Banach space $E$ has an approximative basis, then $E \times C_p$ has a $\pi$-basis.*

*Proof.* Let $1 \leqslant p<\infty$ (in the case when $p=\infty$, the only change is that $l^\infty$ must be replaced by $c_0$). Let $\{F_n\}$ be any sequence of finite-dimensional subspaces of $E$, such that $F_1 \subset F_2 \subset \ldots$ and $\bigcup_{n=1}^{\infty} F_n=E$ and let $\{G_n\}$ be as in definition 12.3. Then

$$H_n=F_n \times(G_1 \times \ldots \times\{G_n\} \times\{0\} \times\{0\} \times \ldots)_{l^p} \quad (n=1,2, \ldots) \tag{12.46}$$

is a sequence of finite-dimensional subspaces of *) $E \times C_p$, such that $H_1 \subset H_2 \subset \ldots$ and $\bigcup_{n=1}^{\infty} H_n=E \times C_p$. We shall show that for each $n$ there exists a finite rank projection $v_n$ on $E \times C_p$ satisfying $v_n|_{H_n}=I_{H_n}$ and such that $\sup_{1 \leqslant n<\infty}\|v_n\|<\infty$, which, by theorem 12.1 $\beta$), implication $2^\circ \Rightarrow 1^\circ$, will complete the proof.

By our assumption on $E$ and § 9, theorem 9.3, there exists a finite rank operator $u=u_n$ on $E$ such that

$$u(x)=x \qquad (x \in F_n), \tag{12.47}$$

$$\|u\| \leqslant \lambda+1, \tag{12.48}$$

*) We recall (see Ch. II, § 18) that by $E \times F$ we denote the product endowed with an arbitrary norm equivalent to the norm of $(E \times F)_{l^p}$.

where $\lambda \geqslant 1$ is a constant which does not depend on $n$. Let

$$G = (I - u)u(E). \tag{12.49}$$

Observe that $G \subset u(E) - u^2(E) = u(E)$ (actually, one can show that $G = u(E) \cap (I - u)(E)$). Also, $\dim G < \infty$ and hence, by the definition of $C_p$, we can choose $m > n$ such that there exists an isomorphism $\tau$ of $G$ onto $G_m$, with $\|\tau\| = 1$, $\|\tau^{-1}\| < 2$. Let us denote by $q_m$ the natural projection of $C_p$ onto $G_m$ and by $q$ the natural projection of $C_p$ onto $(G_1 \times \dots \times G_n \times \{0\} \times \{0\} \times \dots)_{l^p}$. Define $v = v_n : E \times C_p \to E \times C_p$ by

$$v(\{x, y\}) = \{u(x) + \tau^{-1}q_m(y),\ \tau(I - u)(u(x) + \tau^{-1}q_m(y)) + q(y)\}$$
$$(\{x, y\} \in E \times C_p). \tag{12.50}$$

We shall show that $v = v_n$ is a finite rank projection on $E \times C_p$, having the required properties, which will complete the proof.

Clearly, $v$ is linear and of finite rank. If $x \in F_n$, then, by (12.47), $u(x) = x$ and $(I - u)u(x) = 0$, whence $v(\{x, 0\}) = \{x, 0\}$. If $y \in (G_1 \times \dots \times G_n \times \{0\} \times \{0\} \times \dots)_{l^p}$, then $q_m(y) = 0$ (since $m > n$) and $q(y) = y$, whence $v(\{0, y\}) = \{0, y\}$. Consequently,

$$v(\{x, y\}) = \{x, y\} \qquad (\{x, y\} \in H_n). \tag{12.51}$$

Furthermore, by $q_m\tau = \tau$ (since $\tau(G) = G_m = q_m(C_p)$, $q_m^2 = q_m$) and by $q\tau = 0$ (since $\tau(G) = G_m$ and $m > n$) we have

$$\begin{aligned} v^2(\{x, 0\}) &= v(\{u(x),\ \tau(I - u)\, u(x)\}) = \\ &= \{u^2(x) + \tau^{-1}q_m\tau(I - u)\, u(x), \\ &\qquad \tau(I - u)\,(u^2(x) + \tau^{-1}q_m\tau(I - u)\, u(x)) + q(\tau(I - u)\, u(x))\} = \\ &= \{u^2(x) + (I - u)u(x),\ \tau(I - u)(u^2(x) + (I - u)u(x)) + 0\} = \\ &= \{u(x),\ \tau(I - u)u(x)\} = v(\{x, 0\}) \quad (x \in E). \end{aligned}$$

Also, by $q_m\tau = \tau$, $q_mq = 0$ (since $m > n$) and $q\tau = 0$, $q^2 = q$, we have

$$\begin{aligned} v^2(\{0, y\}) &= v(\{\tau^{-1}q_m(y),\ \tau(I - u)\,\tau^{-1}q_m(y) + q(y)\}) = \\ &= \{u\tau^{-1}q_m(y) + \tau^{-1}q_m(\tau(I - u)\,\tau^{-1}q_m(y) + q(y)), \\ &\qquad \tau(I - u)\,[u\tau^{-1}q_m(y) + \tau^{-1}q_m(\tau(I - u)\,\tau^{-1}q_m(y) + q(y))] + \\ &\quad + q(\tau(I - u)\,\tau^{-1}q_m(y) + q(y))\} = \\ &= \{u\tau^{-1}q_m(y) + (I - u)\,\tau^{-1}q_m(y) + 0, \\ &\qquad \tau(I - u)(u\tau^{-1}q_m(y) + (I - u)\tau^{-1}q_m(y) + 0) + 0 + q^2(y)\} = \\ &= \{\tau^{-1}q_m(y),\ \tau(I - u)\tau^{-1}q_m(y) + q(y)\} = v(\{0, y\}) \qquad (y \in C_p). \end{aligned}$$

Consequently,

$$v^2(\{x, y\}) = v(\{x, y\}) \qquad (\{x, y\} \in E \times C_p), \tag{12.52}$$

so $v$ is a projection on $E \times C_p$.

Finally, if $\|\{x, y\}\|_{l^\infty} = \max (\|x\|, \|y\|) \leqslant 1$, then by $\|q_m\| = \|q\| = \|\tau\| = 1$, $\|\tau^{-1}\| < 2$ and (12.48) (whence $\|I - u\| \leqslant \lambda + 2$), we obtain

$$\|v(\{x, y\})\|_{l^\infty} = \max \{\|u(x) + \tau^{-1} q_m(y)\|,$$

$$\|\tau(I - u)(u(x) + \tau^{-1} q_m(y)) + q(y)\|\} \leqslant$$

$$\leqslant \max \{(\lambda + 1) + 2, \ (\lambda + 2)((\lambda + 1) + 2 + 1)\} = \lambda',$$

and thus, in $(E \times C_p)_{l^\infty}$,

$$\|v\| \leqslant \lambda', \tag{12.53}$$

which completes the proof of theorem 12.4.

*Problem 12.6.* If $E$ is a separable Banach space with the metric approximation property, does $E \times C_p$ have a $\pi_{1+\varepsilon}$-basis for every $\varepsilon > 0$? *)

We shall see in § 13 and § 14 that the spaces $C_p$ have bases and that $C_\infty$ has a shrinking basis. Also, some further universal complement properties of the spaces $C_p$ will be proved in the second part of § 14 and an extension of the universal complement property of $C_p$ given by theorem 12.4, to non-separable Banach spaces, will be proved in § 18, theorem 18.7. For some other properties of the spaces $C_p$, see also the Notes and remarks to § 0 and to § 9.

## § 13. Dual $\pi$-bases. Commuting $\pi$-bases. Bases with parentheses. Finite-dimensional decompositions

As we noted in § 12, the definition 12.1 of a $\pi$-basis was obtained by imposing on the finite rank endomorphisms $u_n$ in "part of" the definition 9.9 of a commuting approximative basis to be also projections, namely, we have required $u_m u_n = u_{\min(m, n)}$ to hold only for $n \leqslant m$. Now we shall consider the "dual" notion, i.e., we shall require that $u_m u_n = u_{\min(m, n)}$ hold for all $m \leqslant n$. In the sequel it will turn out that this is a much stronger property (see e.g. proposition 13.3 and theorem 13.1 below).

*Definition 13.1.* Let $E$ be a Banach space. A sequence of finite rank projections $\{u_n\} \subset L(E, E)$ (or a sequence of elements $\{x_n\} \subset E$ for which there exists a sequence of finite rank projections $\{u_n\} \subset L(E, E)$

---

*) Note that in this case $E \times C_p$ has also a basis (see the remark made after § 14, problem 14.2).

of the form (9.3), with the properties listed below) is called a *dual $\pi$-basis* of $E$, if

$$x = \lim_{n\to\infty} u_n(x) \qquad (x \in E), \tag{13.1}$$

$$u_m u_n = u_m \qquad (m \leqslant n). \tag{13.2}$$

If, in addition,

$$\|u_n\| \leqslant \lambda \qquad (n = 1,2, \ldots), \tag{13.3}$$

then $\{u_n\}$ (respectively, $\{x_n\}$) is called a *dual $\pi_\lambda$-basis* of $E$.

In particular, if we have

$$\dim u_n(E) = n \qquad (n = 1,2, \ldots), \tag{13.4}$$

then we shall use the terms *strict dual $\pi$-basis* and *strict dual $\pi_\lambda$-basis* respectively.

Finally, a dual $\pi_1$-basis $\{u_n\} \subset L(E, E)$ of Banach space $E$ is called a *dual $\pi_1^\infty$-basis* of $E$ if $u_n(E)$ is linearly isometric to $l_{m_n}^\infty$ for some $m_n$ $(n = 1,2, \ldots)$.

Clearly, every dual $\pi_\lambda$-basis is a dual $\pi$-basis and conversely, by the principle of uniform boundedness, every dual $\pi$-basis is a dual $\pi_\lambda$-basis for some $\lambda$ (and hence for all $\lambda' \geqslant \lambda$). It is also obvious that every dual $\pi$-basis is a $\Pi$-basis.

*Problem 13.1.* Does every Banach space with an approximative basis have a dual $\pi$-basis? How about a Banach space with a commuting approximative basis?

Some remarks on the conditions occurring in definition 13.1 are collected in the following proposition (see also § 12, proposition 12.1):

**Proposition 13.1.** *Let $E$ be a Banach space and let $\{u_n\} \subset L(E, E)$ be a sequence of projections on $E$.*

a) *If we have* (13.1), *then*

$$\bigcap_{n=1}^{\infty} \operatorname{Ker} u_n = \bigcap_{n=1}^{\infty} (I - u_n)(E) = \{0\}. \tag{13.5}$$

b) *If we have* (13.2) *and* (13.5) *and if* $\lim_{n\to\infty} u_n(x)$ *exists, then* $x = \lim_{n\to\infty} u_n(x)$.

c) *We have* (13.2) *if and only if*

$$\operatorname{Ker} u_1 \supset \operatorname{Ker} u_2 \supset \ldots \tag{13.6}$$

*Proof.* a) Assume that we have (13.1) and let $x \in \bigcap_{n=1}^{\infty} \operatorname{Ker} u_n$, hence $u_n(x) = 0$ $(n = 1,2,\ldots)$. Then, by (13.1), $x = \lim_{n\to\infty} u_n(x) = 0$, which proves (13.5).

b) Assume that we have (13.2) and (13.5) and let $x \in E$ be such that $y = \lim_{n\to\infty} u_n(x)$ exists. Then, by (13.2) for $n \to \infty$,

$$u_m(y) = \lim_{n\to\infty} u_m u_n(x) = u_m(x) \qquad (m = 1,2, \ldots),$$

whence, by (13.5), $x - y \in \bigcap_{m=1}^{\infty} \operatorname{Ker} u_m = \{0\}$, so $x = y$.

c) By § 12, lemma 12.1 a), b), we have $\operatorname{Ker} u_n = (I - u_n)(E) \subset \subset (I - u_m)(E) = \operatorname{Ker} u_m$ if and only if $(I - u_m)(I - u_n) = I - u_n$. But, this latter equality holds if and only if $u_m u_n = u_m$ (because $(I - u_m)(I - u_n) = I - u_n - u_m + u_m u_n$), which completes the proof of proposition 13.1.

From proposition 13.1 c) and § 12, lemma 12.1 c) (applied to $v = I - u_{n-1}$, $u = I - u_n$), we obtain

**Corollary 13.1.** *Let $\{u_n\} \subset L(E, E)$ be a dual $\pi$-basis of a Banach space E. Then*

$$\operatorname{Ker} u_{n-1} = \operatorname{Ker} u_n \oplus (u_n(E) \cap \operatorname{Ker} u_{n-1}) \qquad (n = 2,3, \ldots), \quad (13.7)$$

*and hence*

$$\operatorname{Ker} u_1 = \operatorname{Ker} u_n \oplus \sum_{i=2}^{n} \oplus\, u_i(E) \cap \operatorname{Ker} u_{i-1} \qquad (n = 2,3, \ldots). \quad (13.8)$$

In § 12 we have seen that if $\{u_n\}$ is a $\pi$-basis of $E$, then $\{u_n^*\}$ is a weak* dual $\pi$-basis of $E^*$. A similar argument also shows that *if $\{u_n\}$ is a dual $\pi$-basis of E, then $\{u_n^*\}$ is a weak* $\pi$-basis of $E^*$* (however, a stronger conclusion follows from proposition 13.3 below, which implies, in particular, that the "dual" of § 12, problem 12.2 has an affirmative answer).

In contrast to the situation for $\pi$-bases (see § 12, example 12.2), for dual $\pi$-bases we have the natural result of strong duality, namely:

**Proposition 13.2.** *If $\{u_n\}$ is a dual $\pi$-basis of a Banach space E, then $\{u_n^*|_V\}$ is a $\pi$-basis of* $V = \left[\bigcup_{n=1}^{\infty} u_n^*(E^*)\right]$.

*Proof.* By (13.2) and (13.3) we have $u_m^* u_n^* = u_n^*$ $(n \leqslant m)$ and $\|u_n^*\| \leqslant \lambda$ $(n = 1,2, \ldots)$, whence $u_1^*(V) \subset u_2^*(V) \subset \ldots$ (by § 12, proposition 12.1) and thus also*) $\bigcup_{n=1}^{\infty} u_n^*(V) = \left[\bigcup_{n=1}^{\infty} u_n^*(V)\right] = V$. Consequently, by § 12, corollary 12.1 a), $\{u_n^*|_V\}$ is a $\pi$-basis of $V$, which completes the proof.

*) Indeed, since $u_n^*$ is a projection, $u_n^*(E^*) = u_n^*(u_n^*(E^*)) \subset u_n^*(V) \subset u_n^*(E^*)$, whence $V = \left[\bigcup_{n=1}^{\infty} u_n^*(E^*)\right] = \left[\bigcup_{n=1}^{\infty} u_n^*(V)\right]$.

*Definition 13.2.* A $\pi$-basis $\{u_n\} \subset L(E, E)$ (or $\{x_n\} \subset E$) of a Banach space $E$ is said to be *commuting*, if it is also a dual $\pi$-basis of $E$, i.e., if

$$u_m u_n = u_n u_m = u_{\min(m,n)} \qquad (m, n = 1,2, \ldots). \tag{13.9}$$

As above, we shall also use the terms *commuting $\pi_\lambda$-basis*, *strict commuting $\pi$-basis* (or *$\pi_\lambda$-basis*) and *commuting $\pi_1^\infty$-basis*, *strict commuting $\pi_1^\infty$-basis*, according to the case.

We shall give some characterizations of Banach spaces which have a commuting $\pi$-basis, in theorem 13.1 below. The first main fact in this direction is

**Proposition 13.3.** *Let $E$ be a Banach space and let $\lambda \geqslant 1$, $0 < \varepsilon < 1$. If $E$ has a dual $\pi_\lambda$-basis, then $E$ has a commuting $\pi_{\lambda+\varepsilon}$-basis.*

We shall give here three different proofs of proposition 13.3. Although the first one of them is quite short, the second and third one are also of interest, since the second one will yield a sharpening of it for Banach spaces with dual $\pi_1$-bases (proposition 13.4 below), while in the third one the commuting $\pi_{\lambda+\varepsilon}$-basis $\{v_n\}$ and the subspaces $(v_n - v_{n-1})(E)$ will be of a particularly simple form (see formulae (13.30) and (13.35) below).

*First proof.* Let $\{u_n\} \subset L(E, E)$ be a dual $\pi_\lambda$-basis of $E$. Then, by the proof of § 12, theorem 12.1 $\gamma$), implication $1° \Rightarrow 3°$ and remark 12.2, there exist a sequence of finite rank projections $\{v_n\} \subset L(E, E)$ and an increasing sequence of indices $\{N(n)\}$ such that*)

$$x = \lim_{n\to\infty} v_n(x) \qquad (x \in E), \tag{13.10}$$

$$v_m v_n = v_n \qquad (n \leqslant m), \tag{13.11}$$

$$\|v_n\| \leqslant \lambda + \varepsilon \qquad (n = 1,2, \ldots), \tag{13.12}$$

$$\operatorname{Ker} v_n = \operatorname{Ker} u_{N(n)}, \ v_n^*(E^*) = u_{N(n)}^*(E^*), \ \|v_n|_{u_{N(n)}(E)} - I_{u_{N(n)}(E)}\| < \varepsilon \qquad (n = 1,2, \ldots) \tag{13.13}$$

and such that $v_n(E)$ is $\varepsilon$-close to $u_{N(n)}(E)$ $(n = 1,2, \ldots)$

By (13.6) and the first equality of (13.13), we have

$$\operatorname{Ker} v_1 \supset \operatorname{Ker} v_2 \supset \ldots \tag{13.14}$$

---

*) Indeed, for each $n$ take $G_n = [v_{n-1}(E) \cup y_n]$, where $v_0 = 0$, $\overline{\{y_j\}} = E$, then $\delta_n > 0$ with $\dfrac{\delta_n \dim G_n}{1 - \delta_n} < \varepsilon$, then $N(n) > N(n-1)$ with $\|I_{G_n} - u_{N(n)}|_{G_n}\| < \delta_n$ and, finally, $v_n$ as in § 12, lemma 12.2.

and hence, by proposition 13.1 c) (for $v_n$ instead of $u_n$),

$$v_m v_n = v_m \qquad (m \leqslant n), \tag{13.15}$$

which, together with (13.11), gives

$$v_m v_n = v_n v_m = v_{\min(m,n)} \qquad (m, n = 1,2, \ldots). \tag{13.16}$$

Thus, $\{v_n\}$ is a commuting $\pi_{\lambda+\varepsilon}$-basis of $E$, which completes the proof.

*Second proof.* Let $\{u_n\} \subset L(E, E)$ be a dual $\pi_\lambda$-basis of $E$. Then by § 12, formula (12.23), applied successively to $G = G_n = \left[\bigcup_{i=1}^{n} u_i(E)\right]$ $(n = 1,2, \ldots)$, we may assume (replacing $\{u_n\}$ with a suitable *subsequence* $\{u_{N(n)}\}$, which we denote again by $\{u_n\}$) that

$$\|x - u_{n+1}(x)\| < \frac{\varepsilon}{2^n} \quad \left(x \in \left[\bigcup_{i=1}^{n} u_i(E)\right], \|x\| \leqslant \lambda + \varepsilon;\ n = 1,2,\ldots\right). \tag{13.17}$$

Put

$$v_n^j = u_j u_{j-1} \cdots u_n \qquad (j \geqslant n;\ n = 1,2, \ldots). \tag{13.18}$$

We claim that

$$\|v_n^j\| \leqslant \sum_{i=n}^{j-1} \frac{\varepsilon}{2^i} + \lambda \qquad (j \geqslant n + 1;\ n = 1,2, \ldots). \tag{13.19}$$

Indeed, if $j = n + 1$, then by (13.3) and (13.17) we have

$$\|v_n^{n+1}(x)\| \leqslant \|u_{n+1}u_n(x) - u_n(x)\| + \|u_n(x)\| \leqslant \frac{\varepsilon}{2^n} + \lambda \quad (x \in E,\ \|x\| \leqslant 1).$$

In general, if (13.19) holds for some $j \geqslant n + 1$, then, by (13.17) we have

$$\|v_n^{j+1}(x)\| \leqslant \|u_{j+1}v_n^j(x) - v_n^j(x)\| + \|v_n^j(x)\| \leqslant$$

$$\leqslant \frac{\varepsilon}{2^j} + \sum_{i=n}^{j-1} \frac{\varepsilon}{2^i} + \lambda = \sum_{i=n}^{j} \frac{\varepsilon}{2^i} + \lambda \qquad (x \in E,\ \|x\| \leqslant 1),$$

so that (13.19) also holds for $j$ replaced by $j + 1$. With a similar argument, using (13.19) and again (13.17), it follows also that

$$\|v_n^j - v_n^l\| \leqslant \sum_{i=l}^{j-1} \frac{\varepsilon}{2^i} \qquad (j > l \geqslant n;\ n = 1,2, \ldots) \tag{13.20}$$

(indeed, it is enough to write $\|v_n^{j+1}(x) - v_n^l(x)\| \leqslant \|u_{j+1}v_n^j(x) - v_n^j(x)\| + \|v_n^j(x) - v_n^l(x)\|$) and hence, since $L(E, E)$ is complete, the limits

$$v_n = \lim_{j\to\infty} v_n^j \qquad (n = 1,2, \ldots) \tag{13.21}$$

exist and $v_n \in L(E, E)$ $(n = 1,2, \ldots)$. We shall show that $\{v_n\}$ is a sequence of projections of finite rank satisfying (13.10), (13.16) and (13.12), which will complete the proof.

By (13.2) we have, for $m \leqslant n$ and $m \leqslant j$,

$$\begin{aligned} v_m^j v_n &= u_j u_{j-1} \ldots u_m \lim_{i\to\infty} u_i u_{i-1} \ldots u_n = \\ &= \lim_{i\to\infty} u_j u_{j-1} \ldots u_m u_i u_{i-1} \ldots u_n = \\ &= \lim_{i\to\infty} u_j u_{j-1} \ldots u_m = v_m^j, \end{aligned}$$

whence (13.15). Also, again by (13.2) we have, for $j \geqslant m \geqslant n$,

$$\begin{aligned} v_m^j v_n &= u_j u_{j-1} \ldots u_m \lim_{i\to\infty} u_i u_{i-1} \ldots u_n = \\ &= \lim_{i\to\infty} u_j u_{j-1} \ldots u_m u_i u_{i-1} \ldots u_n = \\ &= u_j u_{j-1} \ldots u_m u_{m-1} \ldots u_n = v_n^j, \end{aligned}$$

whence (13.11), which together with (13.15), gives (13.16) and, in particular, that the $v_n$ are projections. Furthermore,

$$v_n u_n = \lim_{i\to\infty} u_i u_{i-1} \ldots u_n u_n = v_n \qquad (n = 1,2, \ldots) \tag{13.22}$$

and, using again (13.2), we also obtain

$$u_n v_n = \lim_{i\to\infty} u_n u_i u_{i-1} \ldots u_n = u_n \qquad (n = 1,2, \ldots), \tag{13.23}$$

whence

$$\operatorname{Ker} v_n = \operatorname{Ker} u_n, \ \dim v_n(E) = \dim u_n(E), \ v_n^*(E^*) = u_n^*(E^*) \qquad (n=1,2,\ldots), \tag{13.24}$$

and thus, in particular, each $v_n$ is of finite rank.

Now, since by (13.19) we have (13.12), it remains to prove (13.10). By (13.12) and $\bigcup_{n=1}^{\infty} u_n(E) = E$ (§ 12, proposition 12.1), it is sufficient to show that $\{v_n\}$ converges pointwise to $I$ on $\bigcup_{n=1}^{\infty} u_n(E)$. Let $x \in \bigcup_{n=1}^{\infty} u_n(E)$, say $x \in u_k(E)$ and assume that $\|x\| \leqslant 1$. If $j > n \geqslant k + 1$, then $x \in u_k(E) \subset \bigcup_{i=1}^{n-1} u_i(E)$ and hence, by (13.17) and (13.20) for $l = n$,

$$\|x - v_n^j(x)\| \leqslant \|x - u_n(x)\| + \|v_n^n(x) - v_n^j(x)\| < \frac{\varepsilon}{2^{n-1}} + \sum_{i=n}^{j-1} \frac{\varepsilon}{2^i}.$$

Hence, taking $j \to \infty$, it follows that

$$\|x - v_n(x)\| \leqslant \sum_{i=n-1}^{\infty} \frac{\varepsilon}{2^i} = \frac{\varepsilon}{2^{n-2}} \quad (x \in u_k(E), \|x\| \leqslant 1, n \geqslant k+1), \quad (13.25)$$

and thus we have (13.10) for all $x \in \bigcup_{n=1}^{\infty} u_n(E)$ with $\|x\| \leqslant 1$, whence also for all $x \in \bigcup_{n=1}^{\infty} u_n(E)$, which completes the proof.

*Remark 13.1.* By (13.22) and (13.23), $v_n|_{u_n(E)}$ is an isomorphism of $u_n(E)$ onto $v_n(E)$, with inverse $u_n|_{v_n(E)}$ and, by (13.22) and (13.20) for $l = n$ and $j \to \infty$, we have

$$\|v_n u_n - u_n\| = \|v_n - v_n^n\| \leqslant \sum_{i=n}^{\infty} \frac{\varepsilon}{2^i} = \frac{\varepsilon}{2^{n-1}} \quad (n = 1,2, \dots). \quad (13.26)$$

Thus, $v_n(E) = v_n(u_n(E))$ is $\frac{\varepsilon}{2^{n-1}}$-close to $u_n(E)$ and

$$\|v_n|_{u_n(E)} - I_{u_n(E)}\| \leqslant \frac{\varepsilon}{2^{n-1}} \quad (n = 1,2, \dots), \quad (13.27)$$

which corresponds now to the last inequality in formula (13.13) (because when we assumed (13.17), we replaced the initial $\{u_n\}$ by a subsequence $\{u_{N(n)}\}$, which we denoted again by $\{u_n\}$, for convenience). Let us also show that the subspaces $v_n(E)$ are now of the form

$$v_n(E) = \bigcap_{k=n}^{\infty} (u_k(E) \oplus \operatorname{Ker} u_{k+1}) \quad (n = 1,2, \dots), \quad (13.28)$$

and hence

$$(v_n - v_{n-1})(E) = \operatorname{Ker} u_{n-1} \cap \bigcap_{k=n}^{\infty} (u_k(E) \oplus \operatorname{Ker} u_{k+1}) \qquad (n = 2,3,\ldots). \tag{13.29}$$

Indeed, for any $x \in E$ and $k \geqslant n$ we have, by (13.21), (13.18) and (13.24),

$$v_n(x) = v_{k+1} u_k u_{k-1} \ldots u_n(x) = u_k u_{k-1} \ldots u_n(x) -$$

$$- (I - v_{k+1}) u_k u_{k-1} \ldots u_n(x) \in u_k(E) + \operatorname{Ker} v_{k+1} = u_k(E) \oplus \operatorname{Ker} u_{k+1}$$

(where the last sum is a direct sum since $u_k(E) \cap \operatorname{Ker} u_{k+1} \subset u_k(E) \cap \cap \operatorname{Ker} u_k = \{0\}$), whence $v_n(E) \subset \bigcap_{k=n}^{\infty} (u_k(E) \oplus \operatorname{Ker} u_{k+1})$. Conversely, let $x \in \bigcap_{k=n}^{\infty} (u_k(E) \oplus \operatorname{Ker} u_{k+1})$. Then $x = y + z$, where $y \in u_n(E)$, $z \in \operatorname{Ker} u_{n+1} \subset \operatorname{Ker} u_n$, whence $(I - u_n)(x) = z \in \operatorname{Ker} u_{n+1}$ and thus $u_{n+1}(I - u_n)(x) = 0$, so $u_{n+1} u_n(x) = u_{n+1}(x)$. Consequently,

$$v_n(x) = \lim_{j\to\infty} u_j u_{j-1} \ldots u_{n+1} u_n(x) = \lim_{j\to\infty} u_j u_{j-1} \ldots u_{n+1}(x) = v_{n+1}(x).$$

Continuing in this way indefinitely, we obtain

$$v_n(x) = v_{n+1}(x) = v_{n+2}(x) = \ldots$$

whence, by (13.10), $x = \lim_{j\to\infty} v_j(x) = v_n(x) \in v_n(E)$. Thus, $\bigcap_{k=n}^{\infty} (u_k(E) \oplus \oplus \operatorname{Ker} u_{k+1}) \subset v_n(E)$, whence we obtain (13.28). Now, by (13.16) and (13.24) we have

$$(v_n - v_{n-1})(E) = v_n(E) \cap \operatorname{Ker} v_{n-1} = v_n(E) \cap \operatorname{Ker} u_{n-1},$$

whence, by (13.28), we obtain (13.29). Finally, let us also observe that by (13.2) the mapping $v_n^j$ defined by (13.18) is a projection.

*Remark 13.2.* The second proof of proposition 13.3 shows, in particular, that *if* $\{u_n\} \subset L(E, E)$ *is a strict dual* $\pi_\lambda$*-basis of* $E$*, satisfying* (13.17), *then* $E$ *has a strict commuting* $\pi_{\lambda+\varepsilon}$*-basis, and hence also a basis* (by § 9, remark 9.1).

*Third proof.* Let $\{u_n\} \subset L(E, E)$ be a dual $\pi_\lambda$-basis of $E$. Then, as in the above, second proof, we may assume (by passing to a suitable subsequence $\{u_{N(n)}\}$ of $\{u_n\}$, which we denote again by $\{u_n\}$) that we have also (13.17). Put

$$v_n = I - (I - u_n)(I - u_{n-1}) \dots (I - u_1) \qquad (n = 1,2, \dots). \tag{13.30}$$

We shall show that $\{v_n\}$ is a sequence of projections of finite rank satisfying (13.10), (13.16) and (13.12), which will complete the proof.

Observe first (as in § 9, formulae (9.85), (9.87)) that

$$v_n = I - (I - u_n)(I - v_{n-1}) = v_{n-1} + u_n(I - v_{n-1}) =$$

$$= v_{n-1} + u_n - u_n v_{n-1} = u_n + (I - u_n)v_{n-1} \qquad (n = 2,3, \dots). \tag{13.31}$$

Next, observe that by (13.31) and (13.2) we have

$$u_k(I - v_n) = u_k(I - u_n)(I - v_{n-1}) = 0 \qquad (k \leqslant n),$$

whence

$$u_k v_n = u_k \qquad (k = 1, \dots, n;\ n = 1,2, \dots). \tag{13.32}$$

Now we shall prove (13.16) by induction. Let $n \geqslant 1$. By (13.32) we have

$$v_1 v_n = u_1 v_n = u_1 = v_1,$$

and, if we assume that $v_k v_n = v_k$ for some $k < n$, then

$$v_{k+1} v_n = (v_k + u_{k+1}(I - v_k))v_n =$$

$$= v_k v_n + u_{k+1} v_n - u_{k+1} v_k v_n = v_k + u_{k+1} - u_{k+1} v_k = v_{k+1},$$

which proves (13.15); therefore, in particular, each $v_n$ is a projection. Furthermore, if we assume that $v_k v_n = v_n$ for some $k \geqslant n$, then

$$v_{k+1} v_n = (v_k + u_{k+1}(I - v_k))v_n = v_k v_n + u_{k+1}(v_n - v_k v_n) = v_n,$$

which proves (13.11) and hence (13.16).

Observe now (as in § 9, formula (9.88)) that

$$v_n(E) \subset \left[\bigcup_{i=1}^{n} u_i(E)\right] \qquad (n = 1,2, \dots) \tag{13.33}$$

and thus each $v_n$ is of finite rank.

Clearly, $\|v_1\| = \|u_1\| \leqslant \lambda < \lambda + \varepsilon$ and, if we assume that $\|v_{n-1}\| \leqslant \leqslant \lambda + \varepsilon$, then by (13.31), (13.3), (13.33) and (13.17) we obtain

$$\|v_n(x)\| \leqslant \|u_n(x)\| + \|(I-u_n)v_{n-1}(x)\| \leqslant \lambda + \frac{\varepsilon}{2^{n-1}} < \lambda + \varepsilon \quad (x \in E, \ \|x\| \leqslant 1),$$

which proves (13.12).

Finally, by (13.31), (13.33), (13.12), (13.17) and (13.1) we get

$$\|(I - v_n)(x)\| \leqslant \|(I - u_n)(x)\| + \|(I - u_n)v_{n-1}(x)\| < \|(I - u_n)(x)\| +$$

$$+ \frac{\varepsilon}{2^{n-1}} \to 0 \text{ as } n \to \infty \qquad (x \in E, \ \|x\| \leqslant 1),$$

whence (13.10), which completes the proof.

*Remark 13.3*. For the projections $v_n$ constructed in the above third proof of proposition 13.3 the properties (13.22)—(13.24) and (13.26), (13.27) are still conserved. Indeed, let $n \geqslant 1$. Then, clearly, $v_1 u_n = u_1 u_n = u_1 = v_1$ and, if we assume $v_k u_n = v_k$ for some $k < n$, then, by (13.31) and (13.2),

$$v_{k+1}u_n = v_k u_n + u_{k+1}u_n - u_{k+1}v_k u_n = v_k + u_{k+1} - u_{k+1}v_k = v_{k+1},$$

which proves that

$$v_k u_n = v_k \qquad (k = 1, \ldots, n; \ n = 1,2, \ldots), \qquad (13.34)$$

whence also (13.22). Furthermore, (13.23) is a particular case of (13.32) and (13.24) follows from (13.22) and (13.23). Finally, by (13.22), (13.31), (13.33), (13.12) and (13.17) we obtain

$$\|v_n u_n - u_n\| = \|v_n - u_n\| = \|(I - u_n)v_{n-1}\| \leqslant \frac{\varepsilon}{2^{n-1}} \qquad (n = 2,3, \ldots),$$

whence (13.26) and (13.27). Let us also observe that now the subspaces $(v_n - v_{n-1})(E)$ are of the simple form

$$(v_n - v_{n-1})(E) = u_n(E) \cap \operatorname{Ker} u_{n-1} \qquad (n = 2,3, \ldots), \quad (13.35)$$

whence

$$v_n(E) = u_1(E) \oplus \sum_{i=2}^{n} \oplus \, u_i(E) \cap \operatorname{Ker} u_{i-1} \qquad (n = 1,2, \ldots) \quad (13.36)$$

(where the sum is a direct sum by (13.16)). Indeed, for every $x \in E$ and $n \geqslant 2$ we have, by (13.31),

$$(v_n - v_{n-1})(x) = u_n(I - v_{n-1})(x) \in u_n(E),$$

and, by (13.31), (13.2) and (13.23),

$$u_{n-1}(v_n - v_{n-1})(x) = u_{n-1}u_n(I - v_{n-1})(x) = u_{n-1}(I - v_{n-1})(x) = 0,$$

so $(v_n - v_{n-1})(E) \subset u_n(E) \cap \operatorname{Ker} u_{n-1}$. Conversely, let $x \in u_n(E) \cap \cap \operatorname{Ker} u_{n-1}$. Then, by (13.24) we have $x \in \operatorname{Ker} v_{n-1}$, whence, by $x \in u_n(E)$ and (13.31),

$$x = u_n(x) = u_n(I - v_{n-1})(x) = (v_n - v_{n-1})(x).$$

Therefore $u_n(E) \cap \operatorname{Ker} u_{n-1} \subset (v_n - v_{n-1})(E)$, whence we obtain (13.35) and (13.36).

The main advantage of the above second proof of proposition 13.3 is that it yields the following sharpening of proposition 13.3 for $\lambda = 1$:

**Proposition 13.4.** a) *Every Banach space $E$ with a dual $\pi_1$-basis has a commuting $\pi_1$-basis.*

b) *Every Banach space $E$ with a dual $\pi_1^\infty$-basis has a commuting $\pi_1^\infty$-basis.*

*Proof.* a) If in the second proof of proposition 13.3 we start with

$$\|u_n\| = 1 \qquad (n = 1,2, \ldots), \tag{13.37}$$

then for $v_n^j$ defined by (13.18) we have $\|v_n^j\| \leqslant 1$, whence, for $v_n$ defined by (13.21), $\|v_n\| \leqslant 1$ (and hence, since the $v_n$ are projections, $\|v_n\| = 1$), so $\{v_n\}$ is a commuting $\pi_1$-basis of $E$.

b) Let $\{u_n\}$ be a dual $\pi_1^\infty$-basis of $E$. Then, as was observed in remark 13.1, from (13.22) and (13.23) it follows that $v_n|_{u_n(E)}$ is an isomorphism of $u_n(E)$ onto $v_n(E)$, with inverse $u_n|_{v_n(E)}$ (where $v_n$ is defined by (13.21), (13.18)). But now, by the proof of part a), $\|v_n|_{u_n(E)}\| = \|u_n|_{v_n(E)}\| = 1$, whence $v_n(E) \equiv u_n(E) \equiv l_{m_n}^\infty$ $(n = 1,2, \ldots)$, so $\{v_n\}$ is a commuting $\pi_1^\infty$-basis of $E$, which completes the proof.

From proposition 13.4 a) and § 12, example 12.2, it follows

*Example 13.1.* The space $E_0$ (isomorphic to $c_0$) of § 12, example 12.2, has no dual $\pi_1$-basis.

*Problem 13.2.* Does every Banach space with a $\pi$-basis have a commuting $\pi$-basis?

Now we shall prove that for Banach spaces with $\pi$-bases satisfying an additional condition the answer is affirmative.

**Proposition 13.5.** *Let $E$ be a Banach space with a $\pi_\lambda$-basis $\{u_n\} \subset L(E, E)$ such that there exists a sequence of projections $p_n\colon u_{n+1}(E) \to u_n(E)$ $(n = 1,2, \ldots)$ satisfying*

$$\sup_{1 \leqslant n \leqslant k < \infty} \|p_n p_{n+1} \ldots p_k\| = \mu < \infty. \tag{13.38}$$

*Then $E$ has a commuting $\pi_\mu$-basis.*

*Proof.* Put

$$v_n^j = p_n p_{n+1} \ldots p_{j-1} u_j \qquad (j \geqslant n+1;\ n = 1,2, \ldots). \tag{13.39}$$

Then, by (13.38) and (12.3),

$$\|v_n^j\| \leqslant \|p_n p_{n+1} \cdots p_{j-1}\| \, \|u_j\| \leqslant \mu\lambda \quad (j \geqslant n+1;\ n = 1,2,\ldots). \tag{13.40}$$

Let $x \in u_k(E) \subset u_{k+1}(E) \subset \ldots$ (§ 12, proposition 12.1). Then

$$x = p_k(x) = p_{k+1}(x) = \ldots = u_k(x) = u_{k+1}(x) = \ldots$$

whence, for each $n \geqslant k$ and $j \geqslant n+1$,

$$x = p_n(x) = p_n u_{n+1}(x) = \ldots = p_n p_{n+1} \ldots p_{j-1} u_j(x) = v_n^j(x)$$

and therefore

$$x = \lim_{j \to \infty} v_n^j(x) \qquad (x \in u_k(E),\ n \geqslant k). \tag{13.41}$$

On the other hand, for each $n < k < j$ we have, by (13.39),

$$v_n^j = p_n p_{n+1} \ldots p_{k-1} v_k^j,$$

and hence, by (13.41),

$$\lim_{j \to \infty} v_n^j(x) = p_n p_{n+1} \ldots p_{k-1} \lim_{j \to \infty} v_k^j(x) =$$

$$= p_n p_{n+1} \ldots p_{k-1}(x) \qquad (x \in u_k(E),\ n < k). \tag{13.42}$$

Consequently, $\lim_{j \to \infty} v_n^j(x)$ exists for each $x \in \bigcup_{k=1}^{\infty} u_k(E)$, whence, by (13.40) and $\overline{\bigcup_{k=1}^{\infty} u_k(E)} = E$, also for each $x \in E$ $(n = 1, 2, \ldots)$. Put

$$v_n(x) = \lim_{j \to \infty} v_n^j(x) \qquad (x \in E,\ n = 1,2, \ldots). \tag{13.43}$$

Then, by the Banach-Steinhaus theorem, $v_n \in L(E, E)$ $(n = 1,2,\ldots)$. Also, by (13.41) and (13.42), we have

$$v_n(x) = x \qquad (x \in u_k(E),\ n \geqslant k), \tag{13.44}$$

$$v_n(x) = p_n p_{n+1} \cdots p_{k-1}(x) \qquad (x \in u_k(E),\ n < k). \tag{13.45}$$

We shall show that $\{v_n\}$ is a sequence of projections of finite rank satisfying (13.10), (13.16) and

$$\|v_n\| \leqslant \mu \qquad (n = 1,2, \ldots), \tag{13.46}$$

which will complete the proof.

Since $v_k(x) = \lim_{j\to\infty} p_k p_{k+1} \cdots p_{j-1} u_j(x) \in u_k(E)$ $(x \in E)$, by (13.44) and (13.45) we have

$$v_n v_k(x) = v_k(x) \qquad (x \in E,\ n \geqslant k),$$

$$v_n v_k(x) = p_n p_{n+1} \cdots p_{k-1} v_k(x) = p_n p_{n+1} \cdots p_{k-1} \lim_{j\to\infty} p_k p_{k+1} \cdots p_{j-1} u_j(x) =$$

$$= \lim_{j\to\infty} p_n p_{n+1} \cdots p_{j-1} u_j(x) = v_n(x) \qquad (x \in E,\ n < k),$$

i.e., (13.16). Furthermore, by (13.44), (13.45) and (13.38),

$$\|v_n(x)\| \leqslant \max\,(\|x\|,\ \sup_{n<k<\infty} \|p_n p_{n+1} \cdots p_{k-1}\|\,\|x\|) \leqslant \mu\|x\|$$

$$\left(x \in \bigcup_{k=1}^{\infty} u_k(E),\ n = 1,2, \ldots\right),$$

whence we infer (13.46). Finally, by (13.44) we have $\lim_{n\to\infty} v_n(x) = x$ for all $x \in \bigcup_{k=1}^{\infty} u_k(E)$, whence, by (13.46), also for all $x \in E$, which completes the proof.

*Remark 13.4*. For the projections $v_n$ constructed in the proof of proposition 13.5 we have

$$v_n u_n = u_n,\ u_n v_n = v_n,\ v_n(E) = u_n(E) \qquad (n = 1,2, \ldots). \tag{13.47}$$

Indeed, the first relation is obvious from (13.44). Furthermore, since $v_n(x) = \lim_{j\to\infty} p_n p_{n+1} \cdots p_{j-1} u_j(x) \in u_n(E)$, we have $u_n v_n(x) = v_n(x)$ $(x \in E)$. Consequently, $u_n(E) = v_n u_n(E) \subset v_n(E) = u_n v_n(E) \subset u_n(E)$, which proves (13.47). Let us also observe that now the subspaces $(v_n - v_{n-1})(E)$ are of the simple form

$$(v_n - v_{n-1})(E) = u_n(E) \cap \operatorname{Ker} p_{n-1} \qquad (n = 2,3, \ldots), \quad (13.48)$$

whence

$$\operatorname{Ker} v_n = \sum_{k=n+1}^{\infty} \oplus\, u_k(E) \cap \operatorname{Ker} p_{k-1} \qquad (n = 1,2,\ldots), \quad (13.49)$$

in the sense of definition 13.5 below. Indeed, for any $x \in E$ we have, by (13.43), (13.39), (13.47) and (12.7),

$$(v_n - v_{n-1})(x) = v_n(x) - \lim_{j\to\infty} p_{n-1}p_n \ldots p_{j-1}u_j(x) =$$

$$= v_n(x) - p_{n-1}v_n(x) \in (u_n(E) - u_{n-1}(E)) \cap \operatorname{Ker} p_{n-1} = u_n(E) \cap \operatorname{Ker} p_{n-1}.$$

Conversely, if $x \in u_n(E) \cap \operatorname{Ker} p_{n-1} = v_n(E) \cap \operatorname{Ker} p_{n-1}$, then $v_n(x) = x$ and $v_{n-1}(x) = p_{n-1}v_n(x) = p_{n-1}(x) = 0$, whence $x = v_n(x) - v_{n-1}(x) \in (v_n - v_{n-1})(E)$, proving (13.48). Finally, from (13.10) it follows that

$$x = v_n(x) + \sum_{k=n+1}^{\infty} (v_k - v_{k-1})(x) \qquad (x \in E), \qquad (13.50)$$

whence, by (13.16) and (13.48), we obtain

$$\operatorname{Ker} v_n = \sum_{k=n+1}^{\infty} \oplus\, (v_k - v_{k-1})(E) = \sum_{k=n+1}^{\infty} \oplus\, u_k(E) \cap \operatorname{Ker} p_{k-1},$$

i.e. (13.49). Note also, that by (12.7) the mapping $v_n^j$ defined by (13.39) is a projection.

Obviously, condition (13.38) of proposition 13.5 is satisfied (and hence $E$ has a commuting $\pi$-basis), in particular, whenever one of the following three conditions is satisfied:

$$\prod_{n=1}^{\infty} \|p_n\| < \infty, \qquad (13.51)$$

$$\sup_{1 \leqslant n \leqslant k < \infty} \|u_n u_{n+1} \ldots u_k\| < \infty, \qquad (13.52)$$

$$\prod_{n=1}^{\infty} \|u_n\| < \infty. \qquad (13.53)$$

If we have (13.52) or (13.53), the above proof of proposition 13.5 becomes dual to the second proof of proposition 13.3, in the following sense: If $v_n^j$, $v_n$ are defined by (13.18) and (13.21), then

$$(v_n^j)^* = u_n^* u_{n+1}^* \dots u_{j-1}^* u_j^* \qquad (j \geqslant n,\ n = 1,2, \dots). \quad (13.54)$$

$$v_n^* = \lim_{j\to\infty} (v_n^j)^* \qquad (n = 1,2, \dots), \quad (13.55)$$

which are, essentially, (13.39) (for $j \geqslant n$) and (13.43) for $u_n^*$ (note that the limit (13.55) exists now even in the uniform operator topology). It is natural to ask whether we can also "dualize" in this way the third proof of proposition 13.3, to obtain other commuting $\pi$-bases in Banach spaces with $\pi$-bases satisfying an additional condition. The following proposition and its proof show that the answer is affirmative:

**Proposition 13.6.** *Let $E$ be a Banach space with a $\pi_\lambda$-basis $\{u_n\} \subset L(E, E)$ satisfying*

$$\sup_{1\leqslant n<\infty} \|(I - u_1)(I - u_2) \dots (I - u_n)\| = \bar{\mu} < \infty. \quad (13.56)$$

*Then $E$ has a commuting $\pi_{\bar{\mu}+1}$-basis.*

*Proof.* Put

$$v_n = I - (I - u_1)(I - u_2) \dots (I - u_n) \qquad (n = 1,2, \dots). \quad (13.57)$$

We shall show that $\{v_n\}$ is a sequence of projections of finite rank satisfying (13.10), (13.16) and (13.46) with $\bar{\mu} + 1$ instead of $\mu$, which will complete the proof. By (13.57) we have

$$v_n^* = I - (I - u_n^*)(I - u_{n-1}^*) \dots (I - u_1^*) \qquad (n = 1,2, \dots), \quad (13.58)$$

whence, since by (12.2) $u_n^* u_m^* = u_n^*$ for $n \leqslant m$, from the third proof of proposition 13.3 it follows that

$$v_n^* v_m^* = v_{\min(n,m)}^* \qquad (m, n = 1,2, \dots).$$

Consequently,

$$f(v_m v_n(x)) = (v_m v_n)^*(f)(x) = v_n^* v_m^*(f)(x) = v_{\min(n,m)}^*(f)(x) =$$

$$= f(v_{\min(n,m)}(x)) \qquad (x \in E,\ f \in E^*;\ m, n = 1,2, \dots),$$

whence we obtain (13.16). Furthermore, by (13.56) and (13.57) we have

$$\|v_n\| \leqslant \|I - v_n\| + 1 = \|(I - u_1)(I - u_2) \dots (I - u_n)\| + 1 \leqslant \bar{\mu} + 1$$
$$(n = 1,2, \dots). \quad (13.59)$$

Finally, by (13.57) and (12.2) we have (13.44), whence $\lim_{n\to\infty} v_n(x) = x$ for all $x \in \bigcup_{k=1}^{\infty} u_k(E)$. Consequently, by (13.59) and (12.5), we obtain (13.10), which completes the proof of proposition 13.6.

*Remark 13.5.* For the projections $v_n$ constructed above the properties (13.47) are still conserved. Indeed, by (13.58) and the proofs of (13.32), (13.34), we have

$$u_k^* v_n^* = u_k^*, \; v_k^* u_n^* = v_k^* \qquad (k = 1, \ldots, n;\; n = 1,2, \ldots).$$

Hence, as in the above proof of proposition 13.6, we obtain

$$v_n u_k = u_k, \; u_n v_k = v_k \qquad (k = 1, \ldots, n;\; n = 1,2, \ldots), \tag{13.60}$$

whence, in particular, the first two equalities of (13.47), which, in their turn, imply the last equality of (13.47). Let us also observe that now the subspaces $\operatorname{Ker} v_n$ are of the form

$$\operatorname{Ker} v_n = \operatorname{Ker} u_1 \cap \bigcap_{k=2}^{n} (u_{k-1}(E) \oplus \operatorname{Ker} u_k) \qquad (n = 1,2, \ldots) \tag{13.61}$$

and hence

$$(v_n - v_{n-1})(E) = u_n(E) \cap \operatorname{Ker} u_1 \cap \bigcap_{k=2}^{n-1} (u_{k-1}(E) \oplus \operatorname{Ker} u_k) \qquad (n = 2,3, \ldots). \tag{13.62}$$

Indeed, for any $x \in \operatorname{Ker} v_n$ we have $u_1(x) = v_1(x) = v_1 v_n(x) = 0$, whence $\operatorname{Ker} v_n \subset \operatorname{Ker} u_1$. Furthermore, for any $x \in \operatorname{Ker} v_n$ and $k = 2,3, \ldots, n$ we have, by (13.57) and (13.47),

$$\begin{aligned} x = x - v_n(x) &= (I - v_{k-1})(I - u_k)(I - u_{k+1}) \ldots (I - u_n)(x) = \\ &= (I - u_k)(I - u_{k+1}) \ldots (I - u_n)(x) - v_{k-1}(I - u_k)(I - u_{k+1}) \ldots \\ &\ldots (I - u_n)(x) \in \operatorname{Ker} u_k + v_{k-1}(E) = \operatorname{Ker} u_k \oplus u_{k-1}(E) \end{aligned}$$

(where the last sum is a direct sum since $u_{k-1}(E) \cap \operatorname{Ker} u_k \subset u_k(E) \cap \operatorname{Ker} u_k = \{0\}$), whence $\operatorname{Ker} v_n \subset \operatorname{Ker} u_1 \cap \bigcap_{k=2}^{n} (u_{k-1}(E) \oplus \operatorname{Ker} u_k)$. Conversely, let $x \in \operatorname{Ker} u_1 \cap \bigcap_{k=2}^{n} (u_{k-1}(E) \oplus \operatorname{Ker} u_k)$. Then $x = y + z$, where

$y \in u_{k-1}(E) \subset u_k(E)$, $z \in \operatorname{Ker} u_k$, whence $u_k(x) = y \in u_{k-1}(E)$ and thus $(I - u_{k-1})\, u_k(x) = 0$, so $(I - u_{k-1})\,(I - u_k)\,(x) = (I - u_{k-1})\,(x)$ $(k = 2,3, \ldots, n)$. Consequently,

$$(I - v_n)(x) = (I - u_1)(I - u_2) \ldots (I - u_n)(x) =$$

$$= (I - u_1)(I - u_2) \ldots (I - u_{n-1})(x) = \ldots = (I - u_1)(x),$$

whence, since $x \in \operatorname{Ker} u_1$, we infer $x \in \operatorname{Ker} v_n$. Thus, $\operatorname{Ker} u_1 \cap \bigcap_{k=2}^{n} (u_{k-1}(E) \oplus \oplus \operatorname{Ker} u_k) \subset \operatorname{Ker} v_n$, whence we obtain (13.61). Finally, by (13.16) and (13.47) we have

$$(v_n - v_{n-1})(E) = v_n(E) \cap \operatorname{Ker} v_{n-1} = u_n(E) \cap \operatorname{Ker} v_{n-1},$$

whence, by (13.61), we obtain (13.62).

*Remark 13.6.* a) Obviously, condition (13.56) of proposition 13.6 is satisfied (and hence $E$ has a commuting $\pi$-basis), in particular, whenever we have

$$\prod_{n=1}^{\infty} \|I - u_n\| < \infty; \tag{13.63}$$

however, there exist Banach spaces which admit no infinite sequence of projections of finite rank $\{u_n\} \subset L(E, E)$ satisfying (13.63). Indeed, it is known*) that for $E = C([0, 1])$, every projection of finite rank $u \in L(E, E)$ satisfies

$$\|I - u\| = 1 + \|u\| \geqslant 2, \tag{13.64}$$

whence the assertion follows. It is also known**) that in the space $E = L^p([0, 1])$, where $1 < p < \infty$, $p \neq 2$, for every $\pi$-basis $\{u_n\} \subset L(E, E)$ we have $\underline{\lim}_{n \to \infty} \|I - u_n\| > 1$, and thus (13.63) cannot hold.

b) If there exists an equivalent norm on $E$ in which (13.56) holds, then it still follows that $E$ has a commuting $\pi$-basis. In the particular case when $\{u_n\} \subset L(E, E)$ is a strict $\pi$-basis satisfying $\sup_{1 \leqslant n_1 < \ldots < n_m < \infty} \|(I - u_{n_1})(I - u_{n_2}) \ldots (I - u_{n_m})\| < \infty$ in some equivalent norm on $E$, it has been proved in Ch. I, § 20, theorem 20.1, implication $4° \Rightarrow 11°$, that $E$ has a basis. An examination of that proof shows that it also works if we assume only (13.56) (in some equivalent norm on $E$), and that it leads to the commuting $\pi$-basis (13.57).

*) See [55] and [115].
**) See [230], p. 25, corollary and proof of theorem 2.2.3.

*Remark 13.7.* a) For any $\pi_\lambda$-basis $\{u_n\} \subset L(E, E)$ satisfying (13.52) (or, in particular, (13.53)) and for any dual $\pi_\lambda$-basis $\{u_n\} \subset L(E, E)$ satisfying (13.17), we have obtained above a commuting $\pi$-basis $\{v_n\} \subset L(E, E)$ of $E$ such that

$$v_1(E)=u_1(E),\ (v_n-v_{n-1})(E)=u_n(E) \cap \operatorname{Ker} u_{n-1} \qquad (n=2,3,\ldots). \quad (13.65)$$

Indeed, in the first case one can define projections $p_n\colon u_{n+1}(E)\to u_n(E)$ satisfying (13.38), by

$$p_n = u_n|_{u_{n+1}(E)} \qquad (n = 1,2, \ldots), \qquad (13.66)$$

and then formulae (13.47) for $n = 1$ and (13.48) give (13.65), while in the second case formulae (13.30) for $n = 1$ and (13.35) give again (13.65). Obviously, the commuting $\pi$-basis $\{v_n\}$ is uniquely determined by the sequence of subspaces $v_1(E) \cup \{(v_n - v_{n-1})(E)\}_2^\infty$ (indeed, $v_1$ and each $v_n - v_{n-1}$ are uniquely determined, since so are $v_1(E)$, $\operatorname{Ker} v_1$ and $(v_n - v_{n-1})(E)$, $\operatorname{Ker}(v_n - v_{n-1})$), so in the first case $v_n(x) = \lim_{j\to\infty} u_n u_{n+1} \ldots u_{j-1} u_j(x)$ $(x \in E,\ n = 1,2, \ldots)$, while in the second case $v_n = I - (I - u_n)(I - u_{n-1}) \ldots (I - u_1)$ $(n = 1,2, \ldots)$.

b) It is natural to raise the following question: If $\{u_n\} \subset L(E, E)$ is a dual $\pi_\lambda$-basis of $E$, satisfying (13.17) (respectively, a $\pi_\lambda$-basis of $E$ satisfying (13.52) or (13.53)), *and if* the sequence of finite-dimensional subspaces defined by $v_1(E) = u_1(E)$ and (13.62) (respectively, by $v_1(E)=\bigcap_{k=1}^{\infty}(u_k(E)\oplus\operatorname{Ker} u_{k+1})$ and (13.29)) determines a commuting $\pi$-basis $\{v_n\}$ of $E$, is this a new commuting $\pi$-basis or does it coincide with one of those obtained above? We shall now show that in both cases we have (13.65), so if $\{v_n\}$ is a commuting $\pi$-basis of $E$, then it is not a new one (namely, in the first case, $v_n = I-(I-u_n)(I-u_{n-1})\ldots(I-u_1)$, while in the second case $v_n(x) = \lim_{j\to\infty} u_n u_{n+1} \ldots u_{j-1} u_j(x)$).

Indeed, if $\{u_n\}$ is a dual $\pi_\lambda$-basis of $E$ and if we have $v_1(E) = u_1(E)$ and (13.62), then, by (13.6) and $0 \in u_{k-1}(E)$ $(k = 2, \ldots, n - 1)$, we obtain

$$v_1(E)\supset u_1(E),\ (v_n-v_{n-1})(E)\supset u_n(E) \cap \operatorname{Ker} u_{n-1} \qquad (n=2,3,\ldots); \quad (13.67)$$

on the other hand, if $\{u_n\}$ is a $\pi_\lambda$-basis of $E$ and if we have $v_1(E)= \bigcap_{k=1}^{\infty} (u_k(E)\oplus\operatorname{Ker} u_{k+1})$ and (13.29), then, by (12.7) and $0\in\operatorname{Ker} u_{k+1}$ $(k = n, n + 1,\ldots)$, we obtain again (13.67). Thus, it remains to show that in both cases, if $\{v_n\}$ is a commuting $\pi$-basis of $E$, then (13.67) implies (13.65). To this end, let us use in (13.65) the notation $v_n^0$ instead of $v_n$ and let us put $v_0 = v_0^0 = 0$; thus, we have to show that if $\{v_n\}$

is a commuting $\pi$-basis of $E$, the relations

$$(v_n - v_{n-1})(E) \supset (v_n^0 - v_{n-1}^0)(E) \qquad (n = 1,2, \ldots) \qquad (13.68)$$

imply

$$(v_n - v_{n-1})(E) = (v_n^0 - v_{n-1}^0)(E) \qquad (n = 1,2, \ldots). \qquad (13.69)$$

Now, for every $x \in E$ and $k \geqslant 1$ we have, since we know by part a) that $\{v_n^0\}$ is a commuting $\pi$-basis of $E$,

$$x = \sum_{\substack{i=1 \\ i \neq k}}^{\infty}{}' (v_i^0 - v_{i-1}^0)(x) + (v_k^0 - v_{k-1}^0)(x) = z + (v_k^0 - v_{k-1}^0)(x),$$

where, by (13.68) and since $\{v_n\}$ is a commuting $\pi$-basis of $E$,

$$z = \sum_{\substack{i=1 \\ i \neq k}}^{\infty}{}' (v_i^0 - v_{i-1}^0)(x) = \sum_{\substack{i=1 \\ i \neq k}}^{\infty}{}' (v_i - v_{i-1})(v_i^0 - v_{i-1}^0)(x) \in \operatorname{Ker}(v_k - v_{k-1}).$$

Consequently, if $x \in (v_k - v_{k-1})(E)$, then $z = 0$, whence

$$x = (v_k^0 - v_{k-1}^0)(x) \in (v_k^0 - v_{k-1}^0)(E).$$

Thus, $(v_k - v_{k-1})(E) \subset (v_k^0 - v_{k-1}^0)(E)$ $(k = 1,2, \ldots)$, which, together with (13.68), gives (13.69), completing the proof of our assertion.

Similarly to proposition 13.4, in the case when $\lambda$ is small we have a better result, namely

**Proposition 13.7.** a) *Every Banach space $E$ which has a $\Pi_{1+\varepsilon}$-basis for each $\varepsilon > 0$, has a commuting $\pi_{1+\varepsilon}$-basis for each $\varepsilon > 0$.*

b) *Every Banach space $E$ with a $\pi_1$-basis has a commuting $\pi_1$-basis.*

*Proof.* a) If $E$ has a $\Pi_{1+\varepsilon}$-basis for each $\varepsilon > 0$, then, by § 12, theorem 12.1 $\gamma$), implication $1° \Rightarrow 4°$, for every sequence $\{\varepsilon_n\}$ with $0 < \varepsilon_n < 1$, $\sum_{n=1}^{\infty} \varepsilon_n < \infty$, there exists a $\pi$-basis $\{u_n\} \subset L(E, E)$ of $E$ satisfying

$$\|u_n\| \leqslant 1 + \varepsilon_n \qquad (n = 1,2, \ldots). \qquad (13.70)$$

Now let $\varepsilon > 0$ be arbitrary and let $\varepsilon_n > 0$ be such that $\prod_{n=1}^{\infty} (1 + \varepsilon_n) = 1 + \varepsilon$. Define $p_n$ by (13.66). Then $p_n$ is a projection of $u_{n+1}(E)$ onto

$u_n(E)$ and, by (13.70),

$$\sup_{1\leqslant n\leqslant k<\infty} \|p_n p_{n+1} \cdots p_k\| \leqslant \prod_{n=1}^{\infty} \|p_n\| \leqslant \prod_{n=1}^{\infty} \|u_n\| \leqslant \prod_{n=1}^{\infty} (1+\varepsilon_n)=1+\varepsilon<\infty.$$

Consequently, by proposition 13.5, $E$ has a commuting $\pi_{1+\varepsilon}$-basis.
b) If in the above proof of part a) we start with

$$\|u_n\| = 1 \qquad (n = 1,2, \ldots), \tag{13.71}$$

then for $p_n\colon u_{n+1}(E) \to u_n(E)$ defined by (13.66) we have

$$\sup_{1\leqslant n\leqslant k<\infty} \|p_n p_{n+1} \cdots p_k\| \leqslant \prod_{n=1}^{\infty} \|p_n\| \leqslant \prod_{n=1}^{\infty} \|u_n\| = 1 < \infty,$$

whence, by proposition 13.5, $E$ has a commuting $\pi_1$-basis, which completes the proof of proposition 13.7.

**Corollary 13.2.** *A Banach space $E$ has a commuting $\pi$-basis if and only if there exists an equivalent norm $|||\cdot|||$ on $E$ such that $(E, |||\cdot|||)$ has a $\pi_1$-basis.*

*Proof.* If $\{u_n\}$ is a commuting $\pi$-basis of $E$, then for $|||x||| = \sup_{1\leqslant n<\infty} \|u_n(x)\|$ $(x \in E)$, $\{u_n\}$ becomes a $\pi_1$-basis of $(E, |||\cdot|||)$.
Conversely, if $(E, |||\cdot|||)$ has a $\pi_1$-basis, where $|||\cdot|||$ is an equivalent norm on $E$, then by proposition 13.7 b) $(E, |||\cdot|||)$, whence also $E$, has a commuting $\pi$-basis, which completes the proof.
From corollary 13.2 it follows that problem 13.2 is equivalent to § 12, problem 12.5.

*Definition 13.3.* Let $E$ be a Banach space. A sequence of finite rank projections $\{u_n\} \subset L(E, E)$ is called a *weak dual $\pi$- ($\pi_\lambda$-) basis* of $E$, if it satisfies the conditions of definition 13.1 with weak convergence in (13.1) instead of norm-convergence. A *weak* dual $\pi$- ($\pi_\lambda$-) basis* of a conjugate space $E^*$ is defined similarly, using weak* convergence.
Clearly, every dual $\pi_\lambda$-basis is a weak dual $\pi_\lambda$-basis. The converse is not true, as shown by

*Example 13.2.* Let $E$ be a reflexive Banach space with a basis $\{x_n\}$ and define $\{u_n\} \subset L(E, E)$ as in § 12, example 12.1, i.e., by (12.10). Then $\{u_n\}$ is a strict $\pi_\lambda$-basis of $E$, whence, as was observed in*) § 12, $\{u_n^*\}$ is a weak* dual $\pi_\lambda$-basis of $E^*$, which, since $E$ is reflexive, is also a weak dual $\pi_\lambda$-basis of $E^*$. However, as we have seen in § 12, example 12.1, there are many $f \in E^*$ for which $\lim_{n\to\infty} u_n^*(f) \neq f$.

*) See § 12, the part before problem 12.2.

Nevertheless, we have

**Proposition 13.8.** *Let $E$ be a Banach space with a weak dual $\pi_\lambda$-basis $\{u_n\}$. Then $E$ has a dual $\pi_{\lambda^2+2\lambda}$-basis $\{v_n\}$ satisfying, for a suitable increasing sequence of indices $\{m_n\}$,*

$$v_n^*(E^*) = u_{m_n}^*(E^*) \qquad (n = 1,2, \ldots). \tag{13.72}$$

*Proof.* Let $\{y_n\}$ be a dense sequence in $E$. Then we can find, similarly to the proof of § 9, proposition 9.8, an increasing sequence of positive integers $\{m_n\}$ and, for each $n$, non-negative numbers $\alpha_{m_{n-1}+1}, \ldots, \alpha_{m_n}$ ($m_0 = 0$) with $\sum_{i=m_{n-1}+1}^{m_n} \alpha_i = 1$, such that

$$\left\| \sum_{i=m_{n-1}+1}^{m_n} \alpha_i u_i(y_j) - y_j \right\| < \frac{1}{n} \qquad (j = 1, \ldots, n;\ n = 1,2, \ldots). \tag{13.73}$$

Indeed, by $u_n(y_1) \xrightarrow{w} y_1$, there exists*) a convex combination $\sum_{i=1}^{m_1} \alpha_i u_i(y_1)$ such that we have (13.73) for $n = 1$. Assume now that we have found $\{m_n\}_{n=1}^{k-1}$ and $\alpha_i$ such that (13.73) holds for $n = 1, \ldots, k-1$. Then, since $\{\{u_n(y_1), \ldots, u_n(y_k)\}\}_{n=m_{k-1}+1}^{\infty}$ converges weakly to $\{y_1, \ldots, y_k\} \in (\underbrace{E \times E \times \ldots \times E}_{k})_{l^\infty}$ as $n \to \infty$, there exists a convex combination $\sum_{i=m_{k-1}+1}^{m_k} \alpha_i \{u_i(y_1), \ldots, u_i(y_k)\}$ such that we have (13.73) for $n = k$.

Now put

$$w_n = \sum_{i=m_{n-1}+1}^{m_n} \alpha_i u_i \qquad (n = 1,2, \ldots); \tag{13.74}$$

$$v_n = w_n + u_{m_n} - u_{m_n} w_n \qquad (n = 1,2, \ldots). \tag{13.75}$$

We shall show that $\{v_n\}$ has the required properties, which will complete the proof. From (13.74) and $\alpha_i \geqslant 0$, $\sum_{i=m_{n-1}+1}^{m_n} \alpha_i = 1$, (13.3), we infer

$$\|w_n\| \leqslant \sum_{i=m_{n-1}+1}^{m_n} \alpha_i \|u_i\| \leqslant \lambda \qquad (n = 1,2, \ldots), \tag{13.76}$$

*) See e.g. [87], p. 422, corollary 14.

whence, by (13.75) and (13.3), $\|v_n\| \leqslant \lambda^2 + 2\lambda$ $(n = 1,2, \ldots)$. Also, (13.76) and (13.74), (13.73) imply

$$\lim_{n\to\infty} w_n(x) = x \qquad (x \in E), \tag{13.77}$$

whence, since by (13.75) and (13.3)

$$\|x - v_n(x)\| \leqslant \|x - w_n(x)\| + \|u_{m_n}\| \, \|x - w_n(x)\| \leqslant$$

$$\leqslant (1 + \lambda)\|x - w_n(x)\| \qquad (x \in E,\ n = 1,2, \ldots),$$

it follows that we have (13.10). Furthermore, by (13.74) and (13.2) we have $w_n u_{m_n} = w_n$, whence by (13.75) and $u_{m_n}^2 = u_{m_n}$,

$$v_n^2 = (w_n + u_{m_n}(I - w_n))(w_n + u_{m_n}(I - w_n)) = w_n^2 + u_{m_n}(w_n - w_n^2) +$$

$$+ w_n u_{m_n}(I - w_n) + u_{m_n}(I - w_n) - u_{m_n} w_n u_{m_n}(I - w_n) =$$

$$= w_n^2 + u_{m_n}(w_n - w_n^2) + w_n - w_n^2 + u_{m_n}(I - w_n) - u_{m_n}(w_n - w_n^2) =$$

$$= w_n + u_{m_n}(I - w_n) = v_n \qquad (n = 1,2, \ldots),$$

so each $v_n$ is a projection, obviously of finite rank (since so is each $u_n$). Also, for any $k \geqslant n + 1$ we have $w_n w_k = w_n$, $u_{m_n} w_k = u_{m_n}$, $w_n u_{m_k} = w_n$, whence

$$v_n v_k = (w_n + u_{m_n}(I - w_n))(w_k + u_{m_k}(I - w_k)) = w_n w_k + u_{m_n} w_k - u_{m_n} w_n w_k +$$

$$+ w_n u_{m_k}(I - w_k) + u_{m_n} u_{m_k}(I - w_k) - u_{m_n} w_n u_{m_k}(I - w_k) = w_n + u_{m_n} -$$

$$- u_{m_n} w_n + w_n(I - w_k) + u_{m_n}(I - w_k) - u_{m_n} w_n(I - w_k) = v_n$$

and thus we have (13.15), which proves that $\{v_n\}$ is a dual $\pi_{\lambda^2+2\lambda}$-basis of $E$. Finally,

$$v_n^* u_{m_n}^* = w_n^* u_{m_n}^* + u_{m_n}^* - w_n^* u_{m_n}^* = u_{m_n}^*,$$

so $v_n^*(E^*) \supset v_n^*(u_{m_n}^*(E^*)) = u_{m_n}^*(E^*)$ and, on the other hand,

$$v_n^*(E^*) \subset w_n^*(E^*) + u_{m_n}^*(E^*) - w_n^*(u_{m_n}^*(E^*)) \subset [u_i^*(E^*)]_{i=m_{n-1}+1}^{m_n} \subset u_{m_n}^*(E^*)$$

(by (13.74) and since $\{u_n^*\}$ is a weak* $\pi$-basis of $E^*$). Consequently, we have (13.72), which completes the proof.

*Definition 13.4.* A sequence $\{x_n\}$ in a Banach space $E$ is said to be a *basis with parentheses* of $E$ if $\{x_n\}$ is complete minimal and there exists an increasing sequence of positive integers $\{m_n\}$ such that

$$x = \lim_{n\to\infty} s_{m_n}(x) \qquad (x \in E), \tag{13.78}$$

where $\{s_n\}$ is the sequence of partial sum operators associated to $\{x_n\}$. In this case we shall also say that $\{x_n\}$ is a basis with parentheses of $E$ *with respect to* $\{m_n\}$.

If $\{f_n\} \subset E^*$ is the a.s.f. to $\{x_n\}$, we can write (13.78) in the form

$$\lim_{n\to\infty} \left\| x - \sum_{i=1}^{m_n} f_i(x)x_i \right\| = 0 \qquad (x \in E). \tag{13.79}$$

Obviously, every basis $\{x_n\}$ of $E$ is a basis with parentheses of $E$ (with respect to any increasing $\{m_n\} \subset \mathcal{N}$). The converse is not true, as shown by

*Example 13.3.* Let $E = c_0$ and let

$$x_{2n-1} = \frac{1}{2^{n-1}} e_{2n-1} - e_{2n},\ x_{2n} = e_{2n} \qquad (n = 1,2,\ldots), \tag{13.80}$$

$$f_{2n-1} = 2^{n-1}h_{2n-1},\ f_{2n} = 2^{n-1}h_{2n-1} + h_{2n} \qquad (n = 1,2,\ldots), \tag{13.81}$$

where $\{e_n\}$ is the unit vector basis of $E$ and $\{h_n\}$ the sequence of coordinate functionals on $E$. Then, as was observed in § 7, example 7.1, $\{x_n\}$ is a complete minimal sequence in $E$, with the a.s.f. $\{f_n\} \subset E^*$ total on $E$, and $\{x_n\}$ is not a basis of $E$. However, $\{x_n\}$ is a basis with parentheses of $E$, with respect to $\{m_n\} = \{2n\}$, since

$$s_{2n}(x) = \sum_{i=1}^{2n} f_i(x)x_i = \sum_{i=1}^{n} 2^{i-1}h_{2i-1}(x)\left(\frac{1}{2^{i-1}}e_{2i-1} - e_{2i}\right) +$$

$$+ \sum_{i=1}^{n} (2^{i-1}h_{2i-1} + h_{2i})(x)e_{2i} =$$

$$= \sum_{i=1}^{n} h_{2i-1}(x)e_{2i-1} + \sum_{i=1}^{n} h_{2i}(x)e_{2i} \to x \text{ as } n \to \infty \quad (x \in E).$$

*Problem 13.3.* Does every Banach space with a basis with parentheses have a basis?

If the answer to § 9, problem 9.2 were affirmative, then so would be the answer to problem 13.3, by proposition 13.11 and theorem 13.4

below. We also mention here that proposition 13.11 below permits to construct easily other examples of bases with parentheses which are not bases (namely, it is enough to take a finite-dimensional decomposition $\{G_n\}$ of $E$ with $\dim G_n \geqslant 2$ for $n=1,2, \ldots$ and in each $G_n$ a basis $\{x_j^{(n)}\}$ of $G_n$ such that $\lim_{n\to\infty} v_{\{x_j^{(n)}\}} = \infty$ and then to put $\{x_j\} = \bigcup_{n=1}^{\infty} \{x_j^{(n)}\}$).

Bases with parentheses are a special class of triangular $T$-bases. Namely, we have

**Proposition 13.9.** *A sequence $\{x_n\}$ in a Banach space $E$ is a basis with parentheses of $E$, with respect to $\{m_n\}$, if and only if $\{x_n\}$ is a $T$-basis of $E$ for the consistent matrix $T = (t_{ij})$ corresponding to the convergence with respect to the subsequence of indices $\{m_n\} \subset \mathcal{N}$, that is, for the triangular matrix**)

$$t_{ij} = \delta_{m_{n-1},j} \quad (i = m_{n-1}+1,\ldots,m_n-1;\ n,\ j = 1,2,\ldots;\ m_0 = 0), \tag{13.82}$$

$$t_{m_n,j} = \delta_{m_n,j} \quad (n,\ j = 1,2,\ \ldots). \tag{13.83}$$

*Proof.* It is enough to use § 11, theorem 11.1 a) and to observe that now we have

$$\sigma_i(x) = \sum_{j=1}^{i} t_{ij} s_j(x) = \sum_{j=1}^{i} \delta_{m_{n-1},j} s_j(x) =$$

$$= s_{m_{n-1}}(x) \qquad (x \in E,\ i = m_{n-1}+1,\ldots,m_n-1;\ n = 1,2,\ldots),$$

$$\sigma_{m_n}(x) = \sum_{j=1}^{m_n} t_{m_n,j} s_j(x) = \sum_{j=1}^{m_n} \delta_{m_n,j} s_j(x) = s_{m_n}(x) \quad (x \in E,\ n=1,2,\ldots),$$

and thus (13.78) holds if and only if $\sigma_n(x) \to x$ for all $x \in E$, which completes the proof.

By proposition 13.9, the results of § 11 on triangular $T$-bases can be applied to bases with parentheses. Thus, from proposition 13.9 and § 11, theorem 11.5 we deduce the following intrinsic characterization of bases with parentheses:

**Corollary 13.3.** *A finitely linearly independent complete sequence $\{x_n\}$ in a Banach space $E$ is a basis with parentheses of $E$, with respect to $\{m_n\}$, if and only if there exists a constant $C$ with $1 \leqslant C < \infty$ such that we have*

$$\left\| \sum_{j=1}^{m_n} \beta_j x_j \right\| \leqslant C \left\| \sum_{j=1}^{m_{n+p}} \beta_j x_j \right\| \tag{13.84}$$

*for all positive integers $n$, $p$ and all $\beta_1, \ldots, \beta_{m_{n+p}} \in K$.*

---

*) In particular, for $m_n = 2n$ this is nothing else than the matrix $T$ used in § 11, example 11.6 (formula (11.80)).

*Proof.* For the necessity part it is enough to observe that for the triangular matrix $T=(t_{ij})$ defined by (13.82), (13.83) the numbers $T_{ij}$ defined by § 11, formula (11.58) are

$$T_{ij}=\begin{cases}0 & (i=1,\ \ldots, m_1-1; \quad j=1,2,\ldots)\\ 1 & (i=m_n,\ldots,m_{n+1}-1;\ j=1,\ldots,m_n;\ n=1,2,\ldots)\\ 0 & (i=m_n,\ldots,m_{n+1}-1;\ j=m_n+1, m_n+2,\ldots;\ n=1,2,\ldots),\end{cases} \tag{13.85}$$

whence, for any $n, p=1,2,\ldots$ and $\beta_1,\ldots,\beta_{m_{n+p}}\in K$,

$$\sum_{j=1}^{m_{n+p}} T_{ij}\beta_j x_j=\sum_{j=1}^{m_n}\beta_j x_j \qquad (i=m_n,\ldots,m_{n+1}-1),$$

so we can apply proposition 13.9 and the implication $1^\circ \Rightarrow 2^\circ$ of § 11, theorem 11.5 $\Bigg($note that the necessity part also follows directly from definition 13.4 and the principle of uniform boundedness, observing that $\sum_{j=1}^{m_n}\beta_j x_j = s_{m_n}\left(\sum_{j=1}^{m_{n+p}}\beta_j x_j\right)\Bigg)$.

For the sufficiency part we have to observe, in addition, that if $l\geqslant 1$ and if $n$ is such that $m_{n-1}+1\leqslant l\leqslant m_n$, then by (13.85) we have, for any $\beta_1,\ldots,\beta_l\in K$,

$$\sum_{j=1}^{l} T_{ij}\beta_j x_j=\begin{cases}0 & (i=1,\ldots,m_1-1)\\ \sum_{j=1}^{m_k}\beta_j x_j & (i=m_k,\ldots,m_{k+1}-1;\ k=1,\ldots,n-1)\\ \sum_{j=1}^{l}\beta_j x_j & (i=m_n, m_n+1,\ldots),\end{cases}$$

Consequently, if we have (13.84), then, putting $\beta_{l+1}=\ldots=\beta_{m_n}=0$, we obtain

$$\left\|\sum_{j=1}^{l} T_{ij}\beta_j x_j\right\|\leqslant C\left\|\sum_{j=1}^{m_n}\beta_j x_j\right\|=C\left\|\sum_{j=1}^{l}\beta_j x_j\right\| \qquad (i=1,2,\ldots),$$

and therefore, by proposition 13.9 and § 11, theorem 11.5, implication $2^\circ\Rightarrow 1^\circ$, $\{x_n\}$ is a basis with parentheses of $E$, with respect to $\{m_n\}$, which completes the proof of corollary 13.3.

Note that corollary 13.3 also follows from proposition 13.11 below and § 15, theorem 15.5.

From proposition 13.9 and § 11, theorem 11.1 b) it follows that *every basis with parentheses* $\{x_n\}$ *of a Banach space* $E$ *is a generalized summation basis (and hence an operational basis) of* $E$ with respect to the sequence of endomorphisms $v_n \in L(P_{(n)}, P_{(n)})$ $(n = 1,2, \ldots)$ defined by $v_1 = \ldots = v_{m_1-1} = 0$ and

$$v_k\left(\sum_{i=1}^{k} \alpha_i x_i\right) = \sum_{i=1}^{m_n} \alpha_i x_i$$

$$\left(\sum_{i=1}^{k} \alpha_i x_i \in [x_i]_{i=1}^{k};\ k = m_n,\ m_n + 1, \ldots, m_{n+1} - 1;\ n = 1,2, \ldots\right).$$

Of course, it is also easy to see this directly from definition 13.4 and § 10, definition 10.3. Indeed, it is obvious that we have (10.66). Furthermore, if $x \in E$, then for $\alpha_i = f_i(x)$ $(i = 1,2, \ldots)$ we have (10.67) (by (13.79)). Conversely, if we have (10.67) for some $x$ and $\{\alpha_n\}$, then $\lim_{n\to\infty} \sum_{i=1}^{m_n} \alpha_i x_i = \lim_{k\to\infty} v_k\left(\sum_{i=1}^{k} \alpha_i x_i\right) = x$ whence, by biorthogonality,

$$\alpha_j = \lim_{n\to\infty} \sum_{i=1}^{m_n} \alpha_i f_j(x_i) = f_j\left(\lim_{n\to\infty} \sum_{i=1}^{m_n} \alpha_i x_i\right) = f_j(x) \qquad (j = 1,2, \ldots),$$

which proves the assertion.

Consequently, by the results of § 10, *every basis with parentheses* $\{x_n\}$ *of a Banach space* $E$ *is an approximative basis and a norming* $M$*-basis of* $E$. Moreover, from (13.78) it is clear that *every basis with parentheses* $\{x_n\}$ *of* $E$ *is a* $\pi$*-basis of* $E$. The converse statements are not valid, as shown by

*Example 13.4*. In the real space $E = l^2$ let

$$x_n = \sum_{i=1}^{n} e_i \qquad (n = 1,2, \ldots), \tag{13.86}$$

$$f_n = h_n - h_{n+1} \qquad (n = 1,2, \ldots), \tag{13.87}$$

where $\{e_n\}$ is the unit vector basis of $E$ and $\{h_n\}$ is the sequence of coordinate functionals on $E$. Then, by the argument of § 12, example 12.3, $\{x_n\}$ is a strict $\pi_1$-basis of $E$. Furthermore, by § 11, example 11.3 and corollary 11.1, $\{x_n\}$ is a Cesàro basis of $E$, with norming a.s.f. $\{f_n\}$, whence also a generalized summation basis of $E$. However, $\{x_n\}$ is not a basis with parentheses of $E$, since by Ch. I, § 4, formula (4.12), we have

$$s_n(x) = \sum_{i=1}^{n} f_i(x)x_i = \sum_{i=1}^{n} h_i(x)e_i - h_{n+1}(x) \sum_{i=1}^{n} e_i \qquad (x \in E,\ n=1,2,\ldots),$$

whence, by $h_i(e_j) = \delta_{ij}$ $(i, j = 1,2, \ldots)$,

$$\|s_n\| \geqslant \|s_n(e_{n+1})\| = \left\| \sum_{i=1}^{n} e_i \right\| = \sqrt{n} \qquad (n = 1,2, \ldots), \quad (13.88)$$

while for any basis with parentheses $\{x_n\}$ with respect to some $\{m_n\}$ we must have, by (13.78) and the principle of uniform boundedness,

$$\sup_{1 \leqslant n < \infty} \|s_{m_n}\| < \infty. \quad (13.89)$$

**Proposition 13.10.** *For a sequence $\{x_n\}$ in a Banach space $E$ the following statements are equivalent:*

1°. *$\{x_n\}$ is a basis with parentheses of $E$.*

2°. *$\{x_n\}$ is a finitely linearly independent commuting $\pi$-basis of $E$, for which there exist a sequence of projections of finite rank $\{u_n\} \subset L(E,E)$ satisfying* (13.1), (13.2) *and an increasing sequence of positive integers $\{m_n\}$ such that*

$$u_n(E) = [x_1, \ldots, x_{m_n}] \qquad (n = 1,2, \ldots), \quad (13.90)$$

$$x_j \in (u_n - u_{n-1})(E) \qquad (j = m_{n-1}+1,\ldots,m_n;\ m_0 = 0;\ u_0 = 0). \quad (13.91)$$

*Proof.* The implication 1°⇒2° is obvious, with $u_n = s_{m_n}$ $(n = 1,2,\ldots)$.

Conversely, assume now that we have 2°. Then clearly, $[x_n]=E$ and, by (13.90), we can write

$$u_n(x) = \sum_{i=1}^{m_n} f_i^{(n)}(x)x_i \qquad (x \in E,\ n = 1,2,\ldots), \quad (13.92)$$

with suitable $f_i^{(n)} \in E^*$ $(i = 1, \ldots, m_n;\ n = 1,2, \ldots)$. Since each $u_n$ is a projection, we have

$$\sum_{i=1}^{m_n} \delta_{ij}x_i = x_j = u_n(x_j) = \sum_{i=1}^{m_n} f_i^{(n)}(x_j)x_i \qquad (j = 1,\ldots, m_n;\ n=1,2,\ldots),$$

whence, since $\{x_n\}$ is finitely linearly independent,

$$f_i^{(n)}(x_j) = \delta_{ij} \qquad (i,j = 1, \ldots, m_n;\ n = 1,2, \ldots). \quad (13.93)$$

Furthermore, by (13.92), (13.91) and (13.2) we have

$$\sum_{i=1}^{m_n} f_i^{(n)}(x_j)x_i = u_n(x_j) \in u_n(u_{k+1} - u_k)(E) = \{0\}$$

$$(j = m_k + 1, \ldots, m_{k+1};\ k = n, n+1, n+2, \ldots;\ n=1,2,\ldots),$$

and hence, since $\{x_n\}$ is finitely linearly independent,

$$f_i^{(n)}(x_j) = 0 \quad (i = 1, \ldots, m_n;\ j = m_n + 1,\ m_n + 2, \ldots;\ n = 1,2,\ldots). \tag{13.94}$$

By (13.93), (13.94) and $[x_j] = E$, we can define functionals $f_i \in E^*$ satisfying $f_i(x_j) = \delta_{ij}$ $(i, j = 1,2, \ldots)$, by putting

$$f_i = f_i^{(n)} = f_i^{(n+1)} = f_i^{(n+2)} = \ldots \quad (i = m_{n-1}+1, \ldots, m_n;\ n = 1,2,\ldots). \tag{13.95}$$

Then, by (13.92), $u_n = s_{m_n}$ $(n = 1,2, \ldots)$ and hence, by (13.1), $x = \lim_{n\to\infty} s_{m_n}(x)$ $(x \in E)$. Thus, $2° \Rightarrow 1°$, which completes the proof.

**Corollary 13.4.** *A Banach space E has a basis with parentheses if and only if E has a commuting $\pi$-basis.*

*Proof.* The necessity is obvious by proposition 13.10, implication $1° \Rightarrow 2°$.

Conversely, assume now that $E$ has a commuting $\pi$-basis, say $\{u_n\}$. Choose, for each $n = 1,2,\ldots$ a basis $\{x_{m_{n-1}+1}, \ldots, x_{m_n}\}$ of $(u_n - u_{n-1})(E)$ (where $m_0 = 0$ and $u_0 = 0$). Then $\{x_n\}$ satisfies 2° of proposition 13.10 and hence $\{x_n\}$ is a basis with parentheses of $E$, which completes the proof.

Now we shall introduce a generalization of the sequence of one-dimensional subspaces spanned by the elements of a basis, which is closely related to the other notions considered in this section.

*Definition 13.5.* A sequence $\{G_n\}$ of finite-dimensional subspaces of a Banach space $E$, such that $G_n \neq \{0\}$ $(n = 1,2, \ldots)$, is called a *finite-dimensional decomposition* of $E$ if for every $x \in E$ there exists a *unique* sequence $\{y_n\} \subset E$ with $y_n \in G_n$ $(n = 1,2, \ldots)$ such that

$$x = \sum_{i=1}^{\infty} y_i = \lim_{n\to\infty} \sum_{i=1}^{n} y_i. \tag{13.96}$$

Clearly, every basis $\{x_n\}$ of $E$ generates a one-dimensional decomposition $G_1 = [x_1] = s_1(E)$, $G_2 = [x_2] = (s_2 - s_1)(E)$, ... of $E$. Conversely, every one-dimensional decomposition generates a family of bases,

namely, every sequence $\{x_n\} \subset E$ with*) $x_n \in G_n \setminus \{0\}$ $(n = 1,2, \ldots)$ is a basis of $E$ and all bases of this family have the same sequence of partial sum operators $\{s_n\}$ $\Big($indeed, if $x'_n = \alpha_n x_n$, then $f'_n = \frac{1}{\alpha_n} f_n$, where $\alpha_n \neq 0$ for $n = 1,2,\ldots$**) and hence $s'_n(x) = \sum_{i=1}^{n} f'_i(x)x'_i = \sum_{i=1}^{n} f_i(x)x_i\Big)$.
By abuse of language, we shall say that finite-dimensional decompositions are a "generalization" of bases.

Finite-dimensional decompositions are a special case of decompositions and therefore we shall freely use in this section, for finite-dimensional decompositions, the general results of § 15 on decompositions. In particular, from § 15, theorem 15.3 it follows that if $\{G_n\}$ is a finite-dimensional decomposition of $E$, then the finite rank projections $u_n : x \to \sum_{i=1}^{n} y_i$ (where $\{y_n\}$ corresponds to $x$ as in definition 13.5) of $E$ onto $\sum_{i=1}^{n} \oplus G_i$ (are continuous and) constitute a commuting $\pi$-basis of $E$ and conversely, if $\{u_n\}$ is a commuting $\pi$-basis of $E$, with $u_1 \neq 0$, $u_n \neq u_{n-1}$ $(n = 2,3, \ldots)$, then $G_1 = u_1(E)$, $G_n = (u_n - u_{n-1})(E) = u_n(E) \cap \operatorname{Ker} u_{n-1}$ $(n = 2,3, \ldots)$ is a finite-dimensional decomposition of $E$; these $\{G_n\}$ and $\{u_n\}$ are uniquely *determined* by each other (see also remark 13.7 a)). Thus, *a Banach space E has a finite-dimensional decomposition if and only if E has a commuting $\pi$-basis, or*, what is equivalent (by corollary 13.4), *E has a basis with parentheses.* The connection between finite-dimensional decompositions and bases with parentheses is given by

**Proposition 13.11.** *For a sequence $\{x_n\}$ in a Banach space E the following statements are equivalent:*

1°. *$\{x_n\}$ is a basis with parentheses of E with respect to $\{m_n\}$.*

2°. *$\{x_n\}$ is finitely linearly independent and $\{G_n\} = \{[x_i]_{i=m_{n-1}+1}^{m_n}\}$ (where $n_0 = 0$) is a finite-dimensional decomposition of E.*

*Proof.* If we have 1°, and $x \in E$ is arbitrary, then for $y_n = s_{m_n}(x) - s_{m_{n-1}}(x) \in G_n$ $(n = 1,2, \ldots; s_0 = 0)$ we have (13.96). This expansion is unique, since if $x = \sum_{i=1}^{\infty} y'_i$, where $y'_n \in G_n$ $(n = 1,2, \ldots)$, then, by biorthogonality,

$$f_k(y'_n) = \begin{cases} f_k(x) & \text{for } k = m_{n-1}+1, \ldots, m_n \\ 0 & \text{for } k \in \mathcal{N} \setminus \{m_{n-1}+1, \ldots, m_n\} \end{cases} \qquad (n = 1,2, \ldots),$$

whence, since $\{f_n\}$ is total on $E$, $y'_n = y_n$ $(n = 1,2, \ldots)$. Thus, 1° ⇒ 2°.

---

*) This is the reason why we have to assume in definition 13.5 that $G_n \neq \{0\}$ $(n = 1,2, \ldots)$.

**) Actually, this shows that $\{x_n\}$ and $\{x'_n\}$ are "related bases" in the sense of V. Ya. Kozlov (see Ch. I, Notes and remarks, p. 207).

Conversely, assume now that we have 2°. Then, as was observed above, the projections $u_n: x \to \sum_{i=1}^{n} y_i$ of $E$ onto $\sum_{i=1}^{n} \oplus G_i$ constitute a commuting $\pi$-basis of $E$, satisfying, obviously, (13.90) and (13.91). Therefore, by proposition 13.10, $\{x_n\}$ is a basis with parentheses of $E$, so 2° $\Rightarrow$ 1°. Alternatively, one can observe that if we have 2°, then, since $\{x_n\}$ is finitely linearly independent, each $\{x_i\}_{i=m_{n-1}+1}^{m_n}$ is a basis of $G_n$, say with the a.s.c.f. $\{\varphi_i\}_{i=m_{n-1}+1}^{m_n} \subset G_n^*$ $(n = 1,2, \ldots)$. Furthermore, since $\{G_n\} = \{[x_i]_{i=m_{n-1}+1}^{m_n}\}$ is a finite-dimensional decomposition of $E$, the sequence $\{v_n\} = \{u_n - u_{n-1}\}$ $(u_0 = 0)$, that is, the sequence of finite rank projections $v_n: x \to y_n$ of $E$ onto $G_n$ $(n = 1,2, \ldots)$, constitutes a finite-dimensional expansion*) of $I_E$. Then, putting

$$f_i = \varphi_i \circ v_n \qquad (i = m_{n-1} + 1, \ldots, m_n;\ n = 1,2, \ldots), \tag{13.97}$$

we have $\{f_n\} \subset E^*$ and $(x_n, f_n)$ is a biorthogonal system. Finally, as in the proof of § 9, proposition 9.4**), we obtain that $s_{m_n}(x) = \sum_{k=1}^{n} v_k(x) \to x$ for all $x \in E$, so 2° $\Rightarrow$ 1°, which completes the proof of proposition 13.11.

Now we are ready to give

**Theorem 13.1.** *For a Banach space E the following statements are equivalent:*

1°. *E has a dual $\pi$-basis.*

2°. *E has a weak dual $\pi$-basis.*

3°. *E is isomorphic to a space which has a $\Pi_{1+\varepsilon}$-basis for each $\varepsilon > 0$.*

4°. *E is isomorphic to a space with a $\pi_1$-basis.*

5°. *E has a commuting $\pi$-basis.*

6°. *E has a basis with parentheses.*

7°. *E has a finite-dimensional decomposition.*

*Proof.* The implications 5° $\Rightarrow$ 1° $\Rightarrow$ 2° and 4° $\Rightarrow$ 3° are obvious from the definitions. Furthermore, 1° $\Rightarrow$ 5° and 2° $\Rightarrow$ 1° by propositions 13.3 and 13.8, respectively. The implication 3° $\Rightarrow$ 5° follows from proposition 13.7 and the equivalence 5° $\Leftrightarrow$ 4° is nothing else than corollary 13.2. Thus, 1° $\Leftrightarrow \ldots \Leftrightarrow$ 5°. Finally, 5° $\Leftrightarrow$ 6° by corollary 13.4 and the equivalence 6° $\Leftrightarrow$ 7° was observed before proposition 13.11. This completes the proof of theorem 13.1.

Example 13.4 above shows, in particular, that a norming $M$-basis need not be a basis with parentheses. However, we have the following

---

*) See § 9, definition 9.5.

**) In our case it is not assumed that $\sup_{1 \leqslant n < \infty} \nu_{\{x_i\}_{i=m_{n-1}+1}^{m_n}} < \infty$ and therefore we do not get that $s_n(x) \to x$ for all $x \in E$.

positive result in this direction, which is a sharpening of § 6, theorem 6.8 for norming $M$-bases:

**Theorem 13.2.** *Let $\{x_n\}$ be a norming $M$-basis of a Banach space $E$. Then there exists a partition of $\{x_n\}$ into two subsequences $\{x_{n_k}\}$ and $\{x_j\}_{j\in\mathcal{N}\setminus\{n_k\}}$ such that each of these subsequences is a basis with parentheses of its closed linear span.*

*Proof.* By Ch. II, § 16, proposition 16.3, there exist two sequences of integers $\{m_n\}$, $\{l_n\}$ with

$$0 = m_0 < l_1 < m_1 < l_2 < m_2 < \dots \tag{13.98}$$

such that for every $x \in E$ satisfying $f_i(x) = 0$ for $i = l_n + 1, \dots, m_n$; $n = 1,2, \dots$ (or, respectively, $f_i(x) = 0$ for $i = m_{n-1} + 1, \dots, l_n$; $n = 1,2, \dots$), we have

$$x = \sum_{n=1}^{\infty} \left( \sum_{i=m_{n-1}+1}^{l_n} f_i(x)x_i \right) \tag{13.99}$$

$\left(\text{respectively, } x = \sum_{n=1}^{\infty} \left( \sum_{i=l_n+1}^{m_n} f_i(x)x_i \right)\right)$, where $\{f_n\} \subset E^*$ is the a.s.f. to $\{x_n\}$. Now let $\{x_{n_k}\}$ and $\{x_j\}_{j\in\mathcal{N}\setminus\{n_k\}}$ be the subsequences

$$x_1, \dots, x_{l_1}, x_{m_1+1}, \dots, x_{l_2}, x_{m_2+1}, \dots, x_{l_3}, \dots \tag{13.100}$$

and

$$x_{l_1+1}, \dots, x_{m_1}, x_{l_2+1}, \dots, x_{m_2}, x_{l_3+1}, \dots, x_{m_3}, \dots \tag{13.101}$$

respectively. Then for every $x \in [x_{n_k}]$ we have, by biorthogonality, $f_i(x) = 0$ ($i = l_{n+1}, \dots, m_n$; $n = 1,2, \dots$), whence (13.99), and thus $\{x_{n_k}\}$ is a basis with parentheses of $[x_{n_k}]$. Similarly, $\{x_j\}_{j\in\mathcal{N}\setminus\{n_k\}}$ is a basis with parentheses of $[x_j]_{j\in\mathcal{N}\setminus\{n_k\}}$, which completes the proof.

Since every $M$-basis of a reflexive Banach space is norming, from theorem 13.2 it follows

**Corollary 13.5.** *Every $M$-basis $\{x_n\}$ of a reflexive Banach space $E$ admits a partition into two subsequences $\{x_{n_k}\}$ and $\{x_j\}_{j\in\mathcal{N}\setminus\{n_k\}}$ such that each of these subsequences is a basis with parentheses of its closed linear span.*

*Problem 13.4.* Does theorem 13.2 remain valid for every $M$-basis of a Banach space $E$?

Note that in the above proof of theorem 13.2 the assumption that $\{x_n\}$ is norming has entered only when we used Ch. II, § 16, proposition 16.3, but in the proof of this latter the assumption that $\{x_n\}$ is norming has entered when using Ch. II, § 16, lemma 16.3, in which the normingness of $\{x_n\}$ cannot be omitted (see § 7, first footnote to remark 7.5).

Let us also mention that by § 1, corollary 1.3, every $M$-basis $\{x_n\}$ admits a (finite or infinite) partition into subsequences which are basic sequences.

By proposition 13.11, we can also interpret theorem 13.2 and its proof as follows: *If* $\{x_n\}$ *is a norming M-basis of a Banach space E, then there exists a partition of* $\mathcal{N} = \{1,2,3,\ldots\}$ *into two disjoint infinite subsets* $\sigma$ *and* $\Delta$ *with* $\sigma = \bigcup_{n=1}^{\infty} \sigma_n$, $\Delta = \bigcup_{n=1}^{\infty} \Delta_n$, *where* $\sigma_n$ *and* $\Delta_n$ $(n = 1,2,\ldots)$ *are disjoint and finite, such that* $\{\{x_i\}_{i\in\sigma_n}\}_{n=1}^{\infty}$ *and* $\{\{x_i\}_{i\in\Delta_n}\}_{n=1}^{\infty}$ *are finite-dimensional decompositions of* $[x_i]_{i\in\sigma} = ([f_i]_{i\in\Delta})_\perp$ *and of* $[x_i]_{i\in\Delta} = ([f_i]_{i\in\sigma})_\perp$ respectively, *where* $\{f_n\} \subset E^*$ *is the a.s.f. to* $\{x_n\}$. A related result will be given in theorem 13.3 below and its proof (see also corollary 13.6).

The special classes of bases and basic sequences admit natural extensions to bases with parentheses and basic sequences with parentheses, or equivalently, to finite-dimensional decompositions and finite-dimensional decompositions of subspaces, as we shall see in greater detail in § 15. Here we shall only give an extension of § 1, definition 1.3 and of part of § 1, proposition 1.8, which will be used in the proof of theorem 13.3 below.

*Definition 13.6.* A sequence $\{f_n\}$ in a conjugate Banach space $E^*$ is called a *w*-Schauder basic sequence with parentheses* if it is a Schauder basis with parentheses of $\widetilde{[f_n]}$ for the weak* topology $\sigma(E^*, E)$, i.e., if there exist a sequence $\{x_n\} \subset E$ and an increasing sequence of positive integers $\{m_n\}$ such that $(x_n, f_n)$ is a biorthogonal system and that

$$\mathcal{S}_{m_n}(f) = \sum_{i=1}^{m_n} f(x_i) f_i \xrightarrow{w^*} f \qquad (f \in \widetilde{[f_k]}). \tag{13.102}$$

**Proposition 13.12.** *Let E be a Banach space and* $(x_n, f_n)$ $(\{x_n\} \subset E$, $\{f_n\} \subset E^*)$ *a biorthogonal system and let* $\omega$ *denote the canonical mapping of E onto* $E/[f_n]_\perp$. *Then*

a) $\{f_n\}$ *is a w*-Schauder basic sequence with parentheses if and only if* $\{\omega(x_n)\}$ *is a basis with parentheses of* $E/[f_n]_\perp$, *with respect to the same sequence* $\{m_n\}$.

b) *If* $\{f_n\}$ *is a w*-Schauder basic sequence with parentheses, then* $\{f_n\}$ *is a basic sequence with parentheses, with respect to the same* $\{m_n\}$.

c) $\{f_n\}$ *is a basic sequence with parentheses, satisfying* $\widetilde{[f_n]} = [f_n]$, *if and only if* $\{\omega(x_n)\}$ *is a shrinking basis with parentheses of* $E/[f_n]_\perp$, *with respect to the same* $\{m_n\}$.

The proof of this proposition is analogous to that of the corresponding parts of § 1, proposition 1.8 (the other parts of § 1, proposition 1.8 can be also extended, but we shall not need them here) and we omit it.

In § 1 we have seen that every separable Banach space $E$ has a subspace $G_1$ with a basis and a quotient space $E/G_2$ with a basis. While it is not known whether one can always get this with $G_1 = G_2$, i.e., whether every separable Banach space $E$ has a subspace $G$ with a basis and such that $E/G$ has a basis, one can show that such a result is valid for bases with parantheses (or, equivalently, finite-dimensional decompositions). Namely, we have

**Theorem 13.3.** *Let $\{x_n\}$ be a norming M-basis of a Banach space E. Then there exists a partition of $\{x_n\}$ into two subsequences $\{x_{n_k}\}$ and $\{x_{l_k}\}=\{x_j\}_{j\in\mathcal{N}\setminus\{n_k\}}$ such that $\{x_{n_k}\}$ is a basis with parentheses of $[x_{n_k}]$ and $\{\omega(x_{l_k})\}$ is a basis with parentheses of $E/[x_{n_k}]$, where $\omega$ is the canonical mapping of E onto $E/[x_{n_k}]$.*

*Proof.* We may assume, without loss of generality, that $\|x_n\|=1$ $(n=1,2,\ldots)$ and that $r([f_n])=1$, where $\{f_n\}\subset E^*$ is the a.s.f. to $\{x_n\}$. We shall construct two sequences of finite sets $\sigma_1\subset\sigma_2\subset\ldots\subset\mathcal{N}$ and $\Delta_1\subset\Delta_2\subset\ldots\subset\mathcal{N}$ such that $\sigma=\bigcup_{n=1}^{\infty}\sigma_n$ and $\Delta=\bigcup_{n=1}^{\infty}\Delta_n$ are complementary infinite subsets of $\mathcal{N}=\{1,2,3,\ldots\}$ and that, for each $n=1,2,\ldots$ the following two conditions are satisfied:

(i) if $x\in[x_i]_{i\in\sigma_n}$, then there exists $f\in[f_i]_{i\in\sigma_n\cup\Delta_n}$ with $\|f\|=1$, $|f(x)|\geqslant\left(1-\dfrac{1}{n+1}\right)\|x\|$;

(ii) if $f\in[f_i]_{i\in\Delta_n}$, then there exists $x\in[x_i]_{i\in\Delta_n\cup\sigma_{n+1}}$ with $\|x\|=1$, $|f(x)|\geqslant\left(1-\dfrac{1}{n+1}\right)\|x\|$.

Namely, let $\sigma_1=\{1\}$. Assume that we have constructed finite sets $\Delta_1,\sigma_2,\ldots,\Delta_{n-1},\sigma_n$ with $\sigma_1\subset\ldots\subset\sigma_n$, $\Delta_1\subset\ldots\subset\Delta_{n-1}$, $\sigma_n\cap\Delta_{n-1}=\varnothing$ and $\sigma_n\cup\Delta_{n-1}\supset\{1,\ldots,2n-1\}$, satisfying (i) and (ii) for the indices involved. Let $y_1,\ldots,y_l$ be a $\dfrac{1}{2n+2}$- net for $\{x\in[x_i]_{i\in\sigma_n}\mid\|x\|=1\}$. Then, since $r([f_j])=1$, for each $y_k$ there exists a finite linear combination $g_k=\sum_{j\in J_k}\beta_j^{(k)}f_j\in[f_j]$ such that $\|g_k\|=1$, $|g_k(y_k)|\geqslant 1-\dfrac{1}{2n+2}$ $(k=1,\ldots,l)$. Put

$$\Delta_n=\begin{cases}\Delta_{n-1}\cup\{2n\}\cup\left(\bigcup_{k=1}^{l}J_k\right)\setminus\sigma_n & \text{if } 2n\notin\sigma_n\\[2ex] \Delta_{n-1}\cup\left(\bigcup_{k=1}^{l}J_k\right)\setminus\sigma_n & \text{if } 2n\in\sigma_n,\end{cases}$$

where $\{2n\}$ denotes the singleton containing $2n$. Then $\Delta_n\supset\Delta_{n-1}$, $\sigma_n\cap\Delta_n=\varnothing$ and $\sigma_n\cup\Delta_n\supset\{1,\ldots,2n\}$. Also, if $x\in[x_i]_{i\in\sigma_n}\setminus\{0\}$, then

$\left\| \frac{x}{\|x\|} - y_k \right\| < \frac{1}{2n+2}$ for a suitable $k \leqslant l$, whence

$$\left| g_k\left(\frac{x}{\|x\|}\right) \right| \geqslant |g_k(y_k)| - \frac{1}{2n+2} \geqslant 1 - \frac{1}{n+1},$$

where $g_k \in [f_j]_{j\in\sigma_n\cup\Delta_n}$ $\left(\text{because } \bigcup_{k=1}^{l} J_k \subset \sigma_n \cup \Delta_n\right)$, so (i) holds with $f = g_k$. Similarly (dually), using that $[x_n] = E$, one can construct a finite set $\sigma_{n+1} \subset \mathcal{N}$ with $\sigma_{n+1} \supset \sigma_n$, $\sigma_{n+1} \cap \Delta_n = \varnothing$ and $\sigma_{n+1} \cup \Delta_n \supset \{1,\ldots,2n+1\}$, satisfying (ii). Thus we obtain the two sequences of finite sets $\{\sigma_n\}$, $\{\Delta_n\}$ in $\mathcal{N}$, with the required properties.

Now put

$$u_n(x) = \sum_{i\in\sigma_n} f_i(x)x_i \qquad (x\in E,\ n = 1,2,\ldots), \tag{13.103}$$

$$v_n(x) = \sum_{i\in\Delta_n} f_i(x)x_i \qquad (x\in E,\ n = 1,2,\ldots). \tag{13.104}$$

We claim that

$$\|v_n^*|_{([x_i]_{i\in\sigma_{n+1}})^\perp}\| \leqslant 1 + \frac{1}{n} \qquad (n = 1,2,\ \ldots), \tag{13.105}$$

$$\|u_n|_{([f_i]_{i\in\Delta_n})_\perp}\| \leqslant 1 + \frac{1}{n} \qquad (n = 1,2,\ \ldots). \tag{13.106}$$

Indeed, let $f \in ([x_i]_{i\in\sigma_{n+1}})^\perp$. Since $v_n^*(f) = \sum_{i\in\Delta_n} f(x_i)f_i \in [f_i]_{i\in\Delta_n}$, by (ii) there exists $x \in [x_i]_{i\in\Delta_n\cup\sigma_{n+1}}$ with $\|x\| = 1$, such that

$$|v_n^*(f)(x)| \geqslant \left(1 - \frac{1}{n+1}\right)\|v_n^*(f)\|.$$

But then $x = \sum_{i\in\Delta_n\cup\sigma_{n+1}} f_i(x)x_i$, whence, since $f\in([x_i]_{i\in\sigma_{n+1}})^\perp$, we obtain $f(x) = \sum_{i\in\Delta_n} f_i(x)f(x_i) = v_n^*(f)(x)$. Consequently,

$$\|f\| \geqslant |f(x)| = |v_n^*(f)(x)| \geqslant \left(1 - \frac{1}{n+1}\right)\|v_n^*(f)\|,$$

whence

$$\|v_n^*(f)\| \leqslant \frac{1}{1-\frac{1}{n+1}}\|f\| = \frac{n+1}{n}\|f\|,$$

which proves (13.105). The inequalities (13.106) follow from (i) in a similar manner.

Now we shall show that

$$v_n^*(f) \overset{w^*}{\to} f \qquad (f \in ([x_i]_{i\in\sigma})^{\perp}). \tag{13.107}$$

Indeed, if $f \in ([x_i]_{i\in\sigma})^{\perp} \subset ([x_i]_{i\in\sigma_{n+1}})^{\perp}$, $\|f\| \leqslant 1$, then, by (13.105), $\sup_{1\leqslant n<\infty} \|v_n^*(f)\| \leqslant 2 < \infty$. Also, by biorthogonality, for each $n = 1,2,\ldots$ we have

$$v_n^*(f)(x_k) = \sum_{i\in\Delta_n} f(x_i)f_i(x_k) = \begin{cases} f(x_k) & (k \in \Delta_n) \\ 0 = f(x_k) & (k \in \sigma). \end{cases}$$

But, if $k \in \mathcal{N}$, then by $\Delta_1 \subset \Delta_2 \subset \ldots \subset \Delta = \bigcup_{n=1}^{\infty} \Delta_n$ and $\Delta \cup \sigma = \mathcal{N}$ we have $k \in \Delta_n \cup \sigma$ for all sufficiently large $n$. Consequently,

$$\lim_{n\to\infty} v_n^*(f)(x_k) = f(x_k) \qquad (k = 1,2,\ldots),$$

and hence, using also that $[x_k] = E$, we get (13.107).

Now, by (13.107) and biorthogonality, $([x_i]_{i\in\sigma})^{\perp} \subset \widetilde{[f_i]_{i\in\Delta}} \subset ([x_i]_{i\in\sigma})^{\perp}$, so

$$([x_i]_{i\in\sigma})^{\perp} = \widetilde{[f_i]_{i\in\Delta}}, \tag{13.108}$$

$$[x_i]_{i\in\sigma} = (([x_i]_{i\in\sigma})^{\perp})_{\perp} = (\widetilde{[f_i]_{i\in\Delta}})_{\perp} = ([f_i]_{i\in\Delta})_{\perp}. \tag{13.109}$$

Thus, by (13.107) and (13.108), $\{f_i\}_{i\in\Delta}$ is a $w^*$-Schauder basic sequence with parentheses and hence, by proposition 13.12 a) and (13.109), $\{\omega(x_i)\}_{i\in\Delta}$ is a basis with parentheses of $E/([f_i]_{i\in\Delta})_{\perp} = E/[x_i]_{i\in\sigma}$.

Finally, if $x \in [x_i]_{i\in\sigma} = ([f_i]_{i\in\Delta})_{\perp} \subset ([f_i]_{i\in\Delta_n})_{\perp}$, $\|x\| \leqslant 1$, then, by (13.106), $\sup_{1\leqslant n<\infty} \|u_n(x)\| \leqslant 2 < \infty$. Also, by biorthogonality,

$$u_n(x_k) = \sum_{i\in\sigma_n} f_i(x_k)x_i = x_k \qquad (k \in \sigma_n;\ n = 1,2,\ldots).$$

But, if $k \in \sigma$, then by $\sigma_1 \subset \sigma_2 \subset \ldots \subset \sigma = \bigcup_{n=1}^{\infty} \sigma_n$ we have $k \in \sigma_n$ for all sufficiently large $n$. Consequently,

$$\lim_{n\to\infty} u_n(x_k) = x_k \qquad (k \in \sigma),$$

and hence $\lim_{n\to\infty} u_n(x) = x$ for all $x \in [x_i]_{i\in\sigma}$. Thus, $\{x_i\}_{i\in\sigma}$ is a basis with parentheses of $[x_i]_{i\in\sigma}$, which completes the proof of theorem 13.3 (putting $\{n_k\} = \sigma$, $\{l_k\} = \Delta$).

*Remark 13.8.* If $E^*$ is separable and $\{x_n\}$ is a shrinking $M$-basis of $E$ (i.e., $[f_n] = E^*$), then $\{x_{n_k}\}$ above is clearly a shrinking basis with parentheses of $[x_{n_k}]$ and we can adapt the above construction in such a way that $\{\omega(x_{l_k})\}$ be a shrinking basis with parentheses of $E/[x_{n_k}]$. Indeed, since $E^*$ is separable there exists, by lemma 13.1 below, an equivalent norm $|||\cdot|||$ on $E$ such that for any sequence $\{g_n\} \subset E^*$ and any $g \in E^*$, if $g_n \xrightarrow{w^*} g$ and $|||g_n||| \to |||g|||$, then $g_n \to g$; thus, we may assume that the initial norm on $E$ has this property. Now let $f \in ([x_i]_{i\in\sigma})^\perp$. Then, by (13.107) and by

$$\|v_n^*|_{([x_i]_{i\in\sigma})^\perp}\| \leqslant \|v_n^*|_{([x_i]_{i\in\sigma_{n+1}})^\perp}\| \leqslant 1 + \frac{1}{n} \qquad (n = 1,2, \ldots)$$

(see (13.105)), we obtain

$$\|f\| \leqslant \underline{\lim}_{n\to\infty} \|v_n^*(f)\| \leqslant \overline{\lim}_{n\to\infty} \|v_n^*(f)\| \leqslant \overline{\lim}_{n\to\infty} \|v_n^*|_{([x_i]_{i\in\sigma})^\perp}\| \, \|f\| = \|f\|,$$

whence $\lim_{n\to\infty} \|v_n^*(f)\| = \|f\|$. Therefore, by (13.107) and by our assumption on the norm, $\lim_{n\to\infty} \|v_n^*(f) - f\| = 0$, so $f \in [f_i]_{i\in\Delta}$. Consequently, by (13.108),

$$\widetilde{[f_i]_{i\in\Delta}} = ([x_i]_{i\in\sigma})^\perp = [f_i]_{i\in\Delta},$$

and hence, by proposition 13.12 b), c) and (13.109), $\{\omega(x_i)\}_{i\in\Delta}$ is a shrinking basis with parentheses of $E/([f_i]_{i\in\Delta})_\perp = E/[x_i]_{i\in\sigma}$, which will prove our assertion, provided that we also prove

**Lemma 13.1.** *Let $E$ be a Banach space with separable conjugate space $E^*$. Then there exists a norm $|||\cdot|||$ on $E$, equivalent to the initial norm on $E$, such that the dual norm $|||f||| = \sup_{\substack{x\in E\\ |||x|||\leqslant 1}} |f(x)|$ has the following property:*

$$\textit{If } f_n \xrightarrow{w^*} f_0 \in E^* \textit{ and } \lim_{n\to\infty} |||f_n||| = |||f_0|||, \textit{ then } \lim_{n\to\infty} |||f_n - f_0||| = 0. \qquad (13.110)$$

*Proof.* By Ch. II, § 16, lemma 16.3, applied to $E^*$ and to the subspace $V = \pi(E)$ of $E^{**}$ (clearly, $r(V) = 1$), there exists a norm $|||\cdot|||$ on $E^*$, equivalent to the initial norm on $E^*$, such that we have (13.110) and the following property:

$$\text{If } f_n \xrightarrow{w^*} f_0 \in E^*, \text{ then } \underline{\lim}_{n\to\infty} |||f_n||| \geqslant |||f_0|||. \qquad (13.111)$$

We shall show that there exists a norm $|||\cdot|||$ on $E$, equivalent to the initial norm on $E$, such that $|||f||| = \sup_{\substack{x\in E\\ |||x|||\leqslant 1}} |f(x)|$ ($f\in E^*$), which will complete the proof. To this end*⁾, by Ch. II, § 5, lemma 5.1, it will be sufficient to show that the set $A=\{f\in E^* \mid |||f|||\leqslant 1\}$ is $w^*$-closed. Since $E^*$ is separable, so is $E$, whence $A$ is $w^*$-metrizable ($A$ is bounded, since $|||.|||$ is equivalent to the initial norm on $E^*$) and hence it will be sufficient to show that $A$ is sequentially $w^*$-closed, i.e., that the relations $f_n\in A$ ($n=1,2,\ldots$), $f_n \xrightarrow{w^*} f_0\in E^*$ imply $f_0\in A$. However, this is an obvious consequence of (13.111), which completes the proof of lemma 13.1.

From § 8, corollary 8.1 and proposition 13.11, theorem 13.3 and remark 13.8 it follows

**Corollary 13.6.** *Every separable Banach space E has a subspace G such that both G and E/G have finite-dimensional decompositions. Every Banach space E with separable conjugate space E* has a subspace G such that both G and E/G have shrinking finite-dimensional decompositions.*

The following version of § 9, proposition 9.4 for commuting $\pi$-bases gives a method of constructing bases in Banach spaces:

**Theorem 13.4.** *Let $\{G_n\}$ be a finite-dimensional decomposition of a Banach space E and for each n let $G_n$ have a basis $\{z_i\}_{i=m_{n-1}+1}^{m_n}$ (where $m_0=0$) such that the norms of these bases are uniformly bounded, say*

$$v_{\{z_i\}_{i=m_{n-1}+1}^{m_n}} \leqslant M < \infty \qquad (n=1,2,\ldots). \tag{13.112}$$

*Then $\{z_n\}$ is a basis of E. If $\{G_n\}$ is an unconditional finite-dimensional decomposition of E and if*

$$v^{(u)}_{\{z_i\}_{i=m_{n-1}+1}^{m_n}} \leqslant M' < \infty \qquad (n=1,2,\ldots), \tag{13.113}$$

*then $\{z_n\}$ is an unconditional basis of E.*

*Proof.* In the particular case when $G_n=[x_i]_{i=m_{n-1}+1}^{m_n}$ ($n=1,2,\ldots$), where $\{x_n\}$ is a basis of $E$, the first statement reduces to Ch. I, § 7, corollary 7.3. In the general case the proof is the same (using § 15, theorem 15.5). Finally, a slight modification of the argument (considering finite sums containing multipliers $\gamma_i$ with $|\gamma_i|\leqslant 1$ and applying § 15, theorem 15.18 and Ch. II, § 17, theorem 17.1, implication 7° ⇒ 1°) also proves the second statement and completes the proof of theorem 13.4 (see also Ch. II, § 8, example 8.1).

Now we shall apply theorem 13.4 to show the following property of the spaces $C_p$, which will imply (see corollary 13.7 below) that these

---

*⁾ For a similar argument, see § 7, the first footnote to remark 7.5.

spaces have bases (as was announced at the end of § 12) and which will be used in § 14 as well:

**Proposition 13.13.** *Let $1 \leqslant p \leqslant \infty$. If $E$ is a Banach space such that $E \times C_p$ has a finite-dimensional decomposition, then $E \times C_p$ has a basis.*

*Proof.* Let $1 \leqslant p < \infty$ (in the case when $p = \infty$, the only change is that $l^\infty$ must be replaced by $c_0$). Let $\{E_n\}$ be a finite-dimensional decomposition of $E \times C_p$. Then, by § 9, corollary 9.2, for each $E_n$ there exists a finite-dimensional Banach space $F_n$ such that $(E_n \times F_n)_{l^\infty}$ has a basis of norm $\leqslant 5$. By the definition of $C_p = (G_1 \times G_2 \times \ldots)_{l^p}$, for each $F_n$ there exists $G_{i_n}$ such that $d(F_n, G_{i_n}) \leqslant 2$. Then, by Ch. II, § 18, lemma 18.7 a) (which remains valid, with a similar proof, for $p = \infty$ as well), $d((E_n \times F_n)_{l^\infty}, (E_n \times G_{i_n})_{l^\infty}) \leqslant 2$ and hence (e.g. by Ch. I, § 7, theorem 7.1) $(E_n \times G_{i_n})_{l^\infty}$ has a basis of norm $\leqslant 10$ $(n = 1, 2, \ldots)$.

Now since $\{E_n\}$ and $\{G_{i_n}\}$ are finite-dimensional decompositions of $E \times C_p$ and $(G_{i_1} \times G_{i_2} \times \ldots)_{l^p}$ respectively*), it follows $\left(\text{writing } \left\{\sum_{k=1}^{\infty} y_k, \sum_{k=1}^{\infty} z_k\right\} = \sum_{k=1}^{\infty} \{y_k, z_k\}, \text{ where } y_k \in E_k, z_k \in G_{i_k}\right)$ that $\{E_n \times G_{i_n}\}$ is a finite-dimensional decomposition of $(E \times C_p)_{l^\infty} \times (G_{i_1} \times G_{i_2} \times \ldots)_{l^p}$. Hence, since each $(E_n \times G_{i_n})_{l^\infty}$ has a basis of norm $\leqslant 10$, from theorem 13.4 we infer that $(E \times C_p)_{l^\infty} \times (G_{i_1} \times G_{i_2} \times \ldots)_{l^p}$ has a basis. But, by Ch. II, § 18, lemma 18.5, we have the isomorphisms

$$(E \times C_p)_{l^\infty} \times (G_{i_1} \times G_{i_2} \times \ldots)_{l^p} \cong (E \times C_p)_{l^p} \times (G_{i_1} \times G_{i_2} \times \ldots)_{l^p} =$$

$$= (E \times (G_1 \times G_2 \times \ldots)_{l^p})_{l^p} \times (G_{i_1} \times G_{i_2} \times \ldots)_{l^p} \cong E \times C_p,$$

whence $E \times C_p$ has a basis, which completes the proof.

**Corollary 13.7.** *For each $p$ with $1 \leqslant p \leqslant \infty$, the space $C_p$ has a basis.*

*Proof.* Since $C_p$ has a natural finite-dimensional decomposition, so does $C_p \times C_p$ and hence, by proposition 13.13 (for $E = C_p$), $C_p \times C_p$ has a basis. But, by Ch. II, § 18, lemma 18.5, $C_p \times C_p \cong C_p$, so $C_p$ has a basis, which completes the proof.

---

*) Here, for notational convenience, we identify $G_{i_n}$ with $\underbrace{\{0\} \times \ldots \times \{0\}}_{n-1} \times G_{i_n} \times \{0\} \times \{0\} \times \ldots$ Note the contrast with § 1, proposition 1.13 and Ch. I, § 4, proposition 4.2.

## § 14. Duality theorems. Further universal complement properties of the spaces $C_p$

In § 9, propositions 9.8, 9.7 and remark 9.13 we have seen that if $E$ is a separable Banach space such that $E^*$ has the bounded approximation property, then $E$ has a commuting approximative basis $\{v_n\}$ and if, in addition, $E^*$ is separable, then $\{v_n\}$ can be chosen so that $\{v_n^*\}$ is a commuting approximative basis of $E^*$. In the first part of this section we shall prove some related results of (strong and weak) duality for $\pi$-bases, commuting $\pi$-bases and bases, which, in particular, will imply the equivalence of certain problems raised in § 12 and § 13 and will solve most of the problems of Ch. I, §§ 12—14. In the second part these duality results, applied to the spaces $E \times C_p$, will yield some further universal complement properties of the spaces $C_p$.

One of the main tools for obtaining the results of duality, will be the following version for projections of § 9, lemma 9.3 b):

**Lemma 14.1.** *Let $E$ be a Banach space, $\Gamma$ a finite-dimensional subspace of $E^*$, $w$ a finite rank continuous linear projection on $E^*$ and $\varepsilon > 0$. Then there exists a finite rank continuous linear projection $t$ on $E$ such that*

$$t^*(E^*) = w(E^*), \tag{14.1}$$

$$t^*|_\Gamma = w|_\Gamma, \tag{14.2}$$

$$\|t\| \leqslant \|w\| + \varepsilon. \tag{14.3}$$

*Proof.* By § 9, lemma 9.3 b) applied to $V = [\Gamma \cup w(E^*)]$, $F = w(E^*)$, there exists a continuous linear mapping $t: E \to E$ of finite rank satisfying $t^*(E^*) \subset w(E^*)$, $t^*|_V = w|_V$ and (14.3). Then, since $t^*(E^*) \subset w(E^*)$ and $t^*|_{w(E^*)} = w|_{w(E^*)} = I_{w(E^*)}$, it follows that $t^*$ is a projection of $E^*$ onto $w(E^*)$. Consequently, we have (14.1)—(14.3) and $t: E \to E$ is a projection *), which completes the proof.

Each of the next two propositions will play now the role of § 9, proposition 9.8.

**Proposition 14.1.** a) *Let $E$ be a Banach space with the $\lambda$-projection approximation property **) and such that $E^*$ has the $\mu$-approximation property. Then for every pair of finite-dimensional subspaces $G \subset E$ and $\Gamma \subset E^*$ there exists a finite rank projection $q = q_{G,\Gamma}$ on $E$ such that*

$$q(y) = y \qquad (y \in G), \tag{14.4}$$

$$q^*(h) = h \qquad (h \in \Gamma), \tag{14.5}$$

$$\|q\| \leqslant 2\lambda + 2\mu + 4\lambda\mu. \tag{14.6}$$

---

*) Indeed, applying twice the definition of $t^*$ and then the assumption $(t^*)^2 = t^*$, we obtain

$$f(t(t(x))) = t^*(f)(t(x)) = (t^*(t^*(f)))(x) = t^*(f)(x) = f(t(x)) \qquad (x \in E, f \in E^*),$$

whence $t^2 = t$.

**) See § 12, the remarks made before theorem 12.1.

b) *If, in addition, E is separable and* $\{u_n\}$ *is a* $\Pi_\lambda$*-basis of E and if* $0 < \varepsilon < 1$, *then* $q = q_{G,\Gamma,\varepsilon}$ *can be chosen to satisfy also*

$$q(E) \text{ is } \varepsilon\text{-close}^{*)} \text{ to } u_n(E), \tag{14.7}$$

*for a suitable positive integer n.*

*Proof*. a) Let $\delta > 0$ be so small that $\frac{\delta \dim \Gamma}{1-\delta} < \frac{1}{3}$. Then, by our assumption on $E^*$, for the pair $(\Gamma, \delta)$ there exists a finite rank operator $v$ on $E^*$ such that

$$\|v(h) - h\| < \delta\|h\| \qquad (h \in \Gamma),$$

$$\|v\| \leqslant \mu.$$

By § 9, lemma 9.1, there exists a finite rank operator $w$ on $E^*$ such that

$$w(h) = h \qquad (h \in \Gamma),$$

$$\|w\| \leqslant \|v\| + \frac{\delta \dim \Gamma}{1-\delta}\|v\| \leqslant \frac{4}{3}\mu.$$

Then, by § 9, lemma 9.3 b), there exists a finite rank operator $t: E \to E$ such that

$$t^*(h) = w(h) = h \qquad (h \in \Gamma), \tag{14.8}$$

$$\|t\| \leqslant \|w\| + \frac{\mu}{3} < 2\mu. \tag{14.9}$$

Now let $G' = [G \cup t(E)]$ and, for $0<\varepsilon<1$, choose $\gamma>0$ so small that $\frac{\gamma \dim G'}{1-\gamma} < \varepsilon$. Then, by our assumption on $E$, there exists a finite rank projection $u$ on $E$ such that

$$\|u(x) - x\| < \gamma\|x\| \qquad (x \in G'),$$

$$\|u\| \leqslant \lambda.$$

By § 12, lemma 12.2, there exists a finite rank projection $p$ on $E$ such that

$$p(x) = x \qquad (x \in G'), \tag{14.10}$$

$$\|p\| \leqslant \|u\| + \frac{\gamma \dim G'}{1-\gamma}\|u\| < 2\lambda. \tag{14.11}$$

---

*) See § 12, the part before lemma 12.2.

We shall show that the finite rank operator $q$ on $E$ defined by

$$q = t + p - tp \tag{14.12}$$

has all the required properties. Observe first that, by $t(E) \subset G'$ and (14.10) we have $pt = t$, whence, since $p^2 = p$,

$$pq = pt + p^2 - ptp = t + p - tp = q.$$

Consequently,

$$q^2 = (t + p - tp)q = tq + pq - tpq = tq + q - tq = q,$$

so $q$ is a projection. Furthermore, by $G \subset G'$ and (14.10) we have (14.4). Also, by (14.8),

$$q^*(h) = t^*(h) + p^*(h) - p^*t^*(h) = h + p^*(h) - p^*(h) = h \qquad (h \in \Gamma),$$

so we have (14.5). Finally, by (14.9) and (14.11) we obtain

$$\|q\| \leqslant \|t\| + \|p\| + \|t\| \, \|p\| \leqslant 2\lambda + 2\mu + 4\lambda\mu,$$

i.e. (14.6), which proves part a). Clearly, the bound in (14.6) could be replaced by $\lambda + \mu + \lambda\mu + \varepsilon'$, where $\varepsilon' > 0$. Let us also observe that

$$q(E) = p(E); \tag{14.13}$$

indeed, by the above, $pq = q$ and $qp = tp + p^2 - tp^2 = p$, whence (14.13) follows.

b) If $\{u_n\}$ is a $\Pi_\lambda$-basis of $E$, then we can choose, in the above proof of part a), $u = u_n$ for a suitable index $n$. Furthermore, for this $u = u_n$ we can choose, by § 12, lemma 12.2, a finite rank projection $p$ on $E$ satisfying (14.10), (14.11) and such that $p(E)$ is $\varepsilon$-close to $u_n(E)$ $\left(\text{since } \gamma \text{ was chosen so small that } \frac{\gamma \dim G'}{1-\gamma} < \varepsilon\right)$. Hence, by (14.13), we obtain (14.7), which completes the proof of proposition 14.1.

*Remark 14.1.* The argument of the proof of part b) also yields the following more general result: *If $E$ is a Banach space with a $\Pi_\lambda$-basis $\{u_n\}$ and if there exist a separable subspace $F$ of $E^*$ and a sequence of finite rank operators $\{v_n\}$ on $E^*$ such that*

$$f = \lim_{n \to \infty} v_n(f) \qquad (f \in F), \tag{14.14}$$

$$\sup_{1 \leqslant n < \infty} \|v_n\| < \infty, \tag{14.15}$$

*then for every pair of finite-dimensional subspaces $G \subset E$ and $\Gamma \subset F$ and every $\varepsilon$ with $0 < \varepsilon < 1$, there exists a finite rank projection $q = q_{G,\Gamma,\varepsilon}$ on $E$ satisfying* (14.4)–(14.7). *Moreover, if $u_n^*(E^*) \subset F$, $v_n(E^*) \subset F$ $(n = 1,2,\dots)$, then $q$ can be chosen to satisfy also $q(E^*) \subset F$.* Indeed, if $\Gamma \subset F$, then by (14.14), (14.15) we can choose, in the above proof of part b), $v = v_m$ for a suitable index $m$ and the rest of the proof remains the same. Finally, if $u_n^*(E^*) \subset F$, $v_n(E^*) \subset F$ $(n = 1,2,\dots)$, then in the above proof we can arrange also $t^*(E^*) = w(E^*) \subset [\Gamma \cup v_m(E^*)] \subset F$ and $p^*(E^*) = u_n^*(E^*) \subset F$ (by the same lemmas of § 9 and § 12), whence $q^*(E^*) \subset F$.

**Proposition 14.2.** *Let $E$ be a Banach space such that $E^*$ has the $\mu$-projection approximation property. Then for every pair of finite-dimensional subspaces $G \subset E$ and $\Gamma \subset E^*$ there exists a finite rank projection $q = q_{G,\Gamma}$ on $E$, satisfying* (14.4), (14.5) *and*

$$\|q\| \leqslant 4\mu + 4\mu^2. \tag{14.16}$$

*Proof.* By § 9, proposition 9.8, $E$ has the $\mu$-approximation property. Hence, for the pair $(G, \delta)$, where we choose $\delta > 0$ so small that $\frac{\delta \dim G}{1-\delta} < \frac{1}{3}$, there exists a finite rank operator $u$ on $E$ such that

$$\|u(y) - y\| < \delta\|y\| \qquad (y \in G),$$

$$\|u\| \leqslant \mu.$$

By § 9, lemma 9.1, there exists a finite rank operator $r$ on $E$ such that

$$r(y) = y \qquad (y \in G), \tag{14.17}$$

$$\|r\| \leqslant \|u\| + \frac{\delta \dim G}{1-\delta}\|u\| < 2\mu. \tag{14.18}$$

Now put $\Gamma' = [\Gamma \cup r^*(E^*)]$. Then, by our assumption on $E^*$, there exists a finite rank projection $v$ on $E^*$ such that

$$\|v(f) - f\| < \gamma\|f\| \qquad (f \in \Gamma'),$$

$$\|v\| \leqslant \mu,$$

where we choose $\gamma > 0$ so small that $\frac{\gamma \dim \Gamma'}{1-\gamma} < \frac{1}{3}$. By § 12, lemma 12.2, there exists a finite rank projection $w$ on $E^*$ such that

$$w(f) = f \qquad (f \in \Gamma'),$$

$$\|w\| \leqslant \|v\| + \frac{\gamma \dim \Gamma'}{1-\gamma}\|v\| < \frac{4}{3}\mu.$$

Finally, by lemma 14.1, there exists a finite rank projection $p$ on $E$ such that

$$p^*(f) = w(f) = f \qquad (f \in \Gamma'), \tag{14.19}$$

$$\|p\| \leqslant \|w\| + \frac{\mu}{3} < 2\mu. \tag{14.20}$$

We shall show that the finite rank operator $q$ on $E$ defined by

$$q = r + p - pr \tag{14.21}$$

has all the required properties. Observe first that, by $r^*(E^*) \subset \Gamma'$ and (14.19), we have $p^*r^* = r^*$, whence, since $(p^*)^2 = p^*$, it follows, as in the proof of proposition 14.1, that $q^* = r^* + p^* - r^*p^*$ is a projection, so $q$ is a projection*). Furthermore, by $\Gamma \subset \Gamma'$ and (14.19) we have (14.5). Also, by (14.17) we have (14.4). Finally, by (14.18) and (14.20) we obtain (14.16), which completes the proof of proposition 14.2. Clearly, the bound in (14.16) could be also replaced by $2\mu + \mu^2 + \varepsilon$, where $\varepsilon > 0$. Let us also observe that $q^*(E^*) = p^*(E^*)$ (since $p^*q^* = p^*r^* + (p^*)^2 - p^*r^*p^* = r^* + p^* - r^*p^* = q^*$ and $q^*p^* = r^*p^* + (p^*)^2 - r^*(p^*)^2 = p^*$).

*Remark 14.2.* The argument of the above proof also yields the following result (corresponding to remark 14.1): *If $E$ is a Banach space such that there exist a separable subspace $F$ of $E^*$ and sequences of finite rank operators $\{u_n\} \subset L(E, E)$, $\{v_n\} \subset L(E^*, E^*)$ satisfying* (14.14), (14.15) *and*

$$x = \lim_{n\to\infty} u_n(x) \qquad (x \in E), \tag{14.22}$$

$$u_n^*(E^*) \subset F \qquad (n = 1,2, \ldots), \tag{14.23}$$

$$v_n^2 = v_n \qquad (n = 1,2, \ldots), \tag{14.24}$$

*then for every pair of finite-dimensional subspaces $G \subset E$ and $\Gamma \subset F$ there exists a finite rank projection $q = q_{G,\Gamma}$ on $E$, satisfying* (14.4)–(14.6) *with $\lambda = \sup_{1 \leqslant n < \infty} \|u_n\|$. Moreover, if $v_n(E^*) \subset F$ $(n = 1,2, \ldots)$, then $q$ can be chosen to satisfy also $q^*(E^*) \subset F$.* Indeed, by (14.22) we can choose, in the above proof of proposition 14.2, $u = u_n$ for a suitable index $n$. Then, by (14.23) and § 9, remark 9.4, one can take $r: E \to E$ to satisfy also $r^*(E^*) = u_n^*(E^*) \subset F$, whence $\Gamma' = [\Gamma \cup r^*(E^*)] \subset F$. Hence, by (14.14), (14.15) and (14.24), we can choose,

*) Alternatively, using that $rp = r$, one can also show directly that $q^2 = q$, similarly to the equality $v_n^2 = v_n$ in § 13, proof of proposition 13.8.

in the above proof of proposition 14.2, $v = v_m$ for a suitable index $m$ and the rest of the proof remains the same. Moreover, if $v_n(E^*) \subset F$, $(n = 1,2, \ldots)$, then by § 12, lemma 12.2 we can take $w: E^* \to E^*$ to satisfy also $w(E^*) \subset [\Gamma' \cup v_m(E^*)] \subset F$ and, finally, by lemma 14.1, $p: E \to E$ to satisfy also $p^*(E^*) = w(E^*) \subset F$. Then, by (14.21), $q^*(E^*) \subset F$.

Now we can prove the following theorem, which gives some sufficient conditions for a separable Banach space $E$ to have a commuting $\pi$-basis and provides some partial answers to § 12, problems 12.2 and 12.5 and § 13, problem 13.2:

**Theorem 14.1.** *Let $E$ be a separable Banach space.*

a) *If $E$ has a $\Pi$-basis $\{u_n\}$ and $E^*$ has the bounded approximation property, then for any separable subspace $V$ of $E^*$ and any $\varepsilon_n$ with $0 < \varepsilon_n < 1$, $E$ has a commuting $\pi$-basis $\{q_n\}$ such that*

$$\left[\bigcup_{n=1}^{\infty} q_n^*(E^*)\right] \supset V, \tag{14.25}$$

$$q_n(E) \text{ is } \varepsilon_n\text{-close to } u_{m_n}(E) \qquad (n = 1,2, \ldots), \tag{14.26}$$

*for a suitable increasing sequence of indices $\{m_n\}$. Hence, in particular, if $E^*$ is also separable, then $E$ has a shrinking**) *commuting $\pi$-basis $\{q_n\}$ satisfying* (14.26) *for a suitable $\{m_n\}$.*

b) *If $E^*$ has the bounded projection approximation property, then $E$ has a commuting $\pi$-basis. If, in addition, $E^*$ is separable and $\{v_n\}$ is a $\pi$-basis of $E^*$, then $E$ has a shrinking commuting $\pi$-basis $\{q_n\}$ such that*

$$q_n^*(E^*) = v_{m_n}(E^*) \qquad (n = 1,2, \ldots) \tag{14.27}$$

*for a suitable increasing sequence of indices $\{m_n\}$.*

c) *If $E^*$ has a weak* $\pi$-basis $\{v_n\}$, then $E$ has a commuting $\pi$-basis $\{q_n\}$ satisfying* (14.27).

d) *If $E$ has a $\pi$-basis $\{u_n\}$ and $E$ is isomorphic to a conjugate Banach space, then $E$ has a commuting $\pi$-basis $\{q_n\}$ such that*

$$q_n(E) = u_{m_n}(E) \qquad (n = 1,2, \ldots) \tag{14.28}$$

*for a suitable increasing sequence of indices $\{m_n\}$.*

*Proof.* The proof of part a) is similar to the proof of § 9, remark 9.12 c), using now proposition 14.1 (instead of § 9, lemma 9.2) and taking $\{f_n\}$ to be a dense sequence in $V$; the sequence $\{\lambda + \varepsilon_n\}$ of that

---

*) I.e. (see § 13, the remark made before definition 13.6 and § 15, definition 15.15), such that $\{q_n^*\}$ is a commuting $\pi$-basis of $E^*$.

remark will be replaced by the constant $2\lambda + 2\mu + 4\lambda\mu$, where $\lambda, \mu$ are as in proposition 14.1 and then the finite rank operators $v_n$ obtained as in § 9, remark 9.12 c) will be the required projections $q_n$. The second statement is obtained by taking $V = E^*$.

The proof of the first statement of part b) is similar to the proof of § 9, proposition 9.7, using now proposition 14.2 (instead of § 9, lemma 9.2) and replacing $\{\lambda + \varepsilon_n\}$ by $4\mu + 4\mu^2$. However, in order to get the second statement of part b), specifically (14.27), we need a different construction*).

Thus, let us prove now the second statement in b). If $\{v_n\}$ is a $\pi_\mu$-basis of $E^*$, then for each $n$ there exists, by lemma 14.1, a finite rank projection $t_n$ on $E$ such that

$$t_n^*(E^*) = v_n(E^*), \tag{14.29}$$

$$t_n^*|_{v_n(E^*)} = v_n|_{v_n(E^*)} = I_{v_n(E^*)}\,, \tag{14.30}$$

$$\|t_n\| \leqslant \mu + \varepsilon, \tag{14.31}$$

where $\varepsilon > 0$ is arbitrary.

Now let $f \in \bigcup_{n=1}^{\infty} v_n(E^*)$, say $f \in v_k(E^*) \subset v_{k+1}(E^*) \subset \ldots$ (by § 12, proposition 12.1 c)). Then, by (14.30), $t_n^*(f) = f$ for all $n \geqslant k$, so $\lim_{n\to\infty} t_n^*(f) = f$. Hence by (14.31) and since $\overline{\bigcup_{n=1}^{\infty} v_n(E^*)} = E^*$, we obtain

$$\lim_{n\to\infty} t_n^*(f) = f \qquad (f \in E^*), \tag{14.32}$$

and therefore

$$t_n(x) \xrightarrow{w} x \qquad (x \in E). \tag{14.33}$$

Furthermore, by (14.29) and again since $\{v_n\}$ is a $\pi$-basis,

$$t_n^*(E^*) = v_n(E^*) \subset v_m(E^*) = t_m^*(E^*) \qquad (n \leqslant m),$$

whence

$$t_m^* t_n^*(f) = t_n^*(f) \qquad (f \in E^*,\ n \leqslant m)$$

and consequently

$$t_n t_m = t_n \qquad (n \leqslant m). \tag{14.34}$$

---

*) This construction also yields that if in proposition 14.2 $E^*$ is separable and $\{v_n\}$ is a $\pi_\mu$-basis of $E^*$, then $q$ can be chosen to satisfy also $q^*(E^*) = v_n(E^*)$ for a suitable positive integer $n$; however, we shall not need here this fact.

By (14.33), (14.34) and (14.31), $\{t_n\}$ is a weak dual $\pi_{\mu+\varepsilon}$-basis of $E$ and hence, by § 13, proposition 13.8, $E$ has a commuting $\pi$-basis $\{q_n\}$ such that

$$q_n^*(E^*) = t_{m_n}^*(E^*) \qquad (n = 1,2, \ldots) \tag{14.35}$$

for a suitable increasing sequence of indices $\{m_n\}$.

Finally, by (14.29) and (14.35) we have (14.27), which also implies that $\{q_n\}$ is shrinking (since $\{v_n\}$ was a $\pi_\mu$-basis of $E^*$). Thus, b) is now proved.

The proof of part c) is similar to that of part b), with the difference that if $\{v_n\}$ is a weak* $\pi_\mu$-basis of $E^*$, then in general $\overline{\bigcup_{n=1}^{\infty} v_n^*(E^*)} \neq E^*$, so the sequence $\{t_n\}$ above will not satisfy (14.32), (14.33), which makes it necessary to replace the sequence $\{t_n\}$ by a net $\{t_d\}_{d\in\mathscr{D}}$ (similarly to § 9, proof of theorem 9.7). Namely, for each finite-dimensional subspace $\Gamma$ of $E^*$ and each $n\in\mathscr{N}$ there exists, by lemma 14.1, a finite rank projection $t_{\Gamma,\frac{1}{n}}$ on $E$ such that

$$t^*_{\Gamma,\frac{1}{n}}(E^*) = v_n(E^*), \tag{14.29'}$$

$$t^*_{\Gamma,\frac{1}{n}}(h) = v_n(h) \qquad (h \in \Gamma), \tag{14.30'}$$

$$\|t_{\Gamma,\frac{1}{n}}\| \leqslant \mu + 1. \tag{14.31'}$$

Now let $\mathscr{D}$ be the directed set of all pairs $\left(\Gamma, \frac{1}{n}\right)$, where $\Gamma$ is a finite-dimensional subspace of $E^*$ and $n\in\mathscr{N}$ and where $\left(\Gamma_1, \frac{1}{n_1}\right) \geqslant \left(\Gamma_2, \frac{1}{n_2}\right)$ if and only if $\Gamma_1 \supset \Gamma_2$ and $\frac{1}{n_1} \leqslant \frac{1}{n_2}$. Then, by § 9, proof of proposition 9.8,

$$t_d(x) \xrightarrow{w} x \qquad (x \in E). \tag{14.33'}$$

Furthermore, if $d = \left(\Gamma, \frac{1}{n}\right) \leqslant \left(\Gamma', \frac{1}{n'}\right) = d'$, then, by (14.29') and since $\{v_n\}$ is a weak* $\pi$-basis, $t_d^*(E^*) = v_n(E^*) \subset v_{n'}(E^*) = t_{d'}^*(E^*)$, whence

$$t_{d'}^* t_d^*(f) = t_d^*(f) \qquad (f \in E^*).$$

Consequently,

$$t_d t_{d'} = t_d \qquad (d \leqslant d'). \tag{14.34'}$$

But, since $E$ is separable, from (14.33'), (14.34') and (14.31') it follows, as in § 13, proof of proposition 13.8, that $E$ has a dual $\pi$-basis $\{s_n\}$ satisfying $s_n^*(E^*) = t_{d_n}^*(E^*)$ $(n = 1,2, \ldots)$ for a suitable increasing sequence $\{d_n\} \subset \mathscr{D}$, say $d_n = \left(\Gamma_n, \frac{1}{k_n}\right)$ $(n = 1,2, \ldots)$. Hence, by § 13, (either) proof of proposition 13.3, $E$ has a commuting $\pi$-basis $\{q_n\}$ satisfying $q_n^*(E^*) = s_{l_n}^*(E^*) = t_{d_{l_n}}^*(E^*)$ for a suitable increasing sequence $\{l_n\} \subset \mathscr{N}$. Then by $d_{l_n} = \left(\Gamma_{l_n}, \frac{1}{k_{l_n}}\right)$ and (14.29') we obtain (14.27) with $\{m_n\} = \{k_{l_n}\}$, which proves c).

Finally, assume that $E$ has a $\pi$-basis $\{u_n\}$ and that there exist a separable Banach space $B$ and an isomorphism $\tau$ of $B^*$ onto $E$. Then $\{\tau^{-1}u_n\tau\}$ is a $\pi$-basis of $B^*$ and hence, by part b) above, $B$ has a shrinking commuting $\pi$-basis $\{p_n\}$ such that $p_n^*(B^*) = (\tau^{-1}u_{m_n}\tau)(B^*)$ $(n = 1,2, \ldots)$ for a suitable increasing sequence of indices $\{m_n\}$. Consequently, $\{q_n\} = \{\tau p_n^* \tau^{-1}\}$ is a commuting $\pi$-basis of $E$, satisfying

$$q_n(E) = \tau p_n^* \tau^{-1}(E) = \tau p_n^*(B^*) = \tau\tau^{-1}u_{m_n}\tau(B^*) = u_{m_n}(E) \quad (n = 1,2,\ldots),$$

i.e. (14.28), which completes the proof of theorem 14.1.

*Remark 14.3.* a) Theorem 14.1 c) implies that § 12, problem 12.2 is equivalent to § 13, problem 13.2 (which, in turn, is equivalent to § 12, problem 12.5 — as was observed in § 13, after corollary 13.2).

b) From the above proof of theorem 14.1 it follows that *if a Banach space $E$ satisfies either the assumptions of remark* 14.1 *or the assumptions of remark* 14.2, *with $u_n^*(E^*) \subset F$, $v_n(E^*) \subset F$ $(n = 1,2, \ldots)$, then $E$ has a commuting $\pi$-basis $\{q_n\}$ such that $\{q_n^*|_F\}$ is a commuting $\pi$-basis of $F$.* Indeed, to prove the last assertion, we take $\{f_n\}$ to be a dense sequence in $F$ and then, since by the construction of the above proof $F \supset q_n^*(E^*) \supset \{f_1, \ldots, f_n\}$ $(n = 1,2, \ldots)$, we obtain $\left[\bigcup_{n=1}^{\infty} q_n^*(E^*)\right] = F$; hence, since $q_n^* q_m^* = q_m^* q_n^* = q_{\min(n,m)}^*$ $(n, m = 1,2, \ldots)$ and $\|q_n^*\| \leqslant \leqslant 2\lambda + 2\mu + 4\lambda\mu$ $(n = 1,2, \ldots)$, the conclusion follows. Note that in the case of remark 14.1 we get also (14.26), while in the case of remark 14.2 $\{q_n\}$ can be chosen so that, in addition, *$q_n^*(E^*)$ is $\varepsilon_n$-close to $v_{m_n}(E^*)$* $(n = 1,2, \ldots)$ for a suitable sequence $\{m_n\}$.

In the second statement of theorem 14.1 a) one cannot omit the assumption that $E^*$ has the bounded approximation property and in part c) for separable $E^*$ one cannot always get a shrinking commuting $\pi$-basis, as shown by

*Example 14.1.* Let $E$ be a Banach space with a basis, such that $E^*$ is separable and does not have the approximation property (such a space exists, as we have seen in § 0, example 0.1). Then, of course, $E$ has a $\Pi$-basis, $E^*$ is separable and has a weak* $\pi$-basis, but $E$ has no shrinking commuting $\pi$-basis.

Now we shall apply the above results on the existence of commuting $\pi$-bases to obtain some similar duality theorems for bases. Firstly, we shall show (in theorem 14.2 below) that the answers to Ch. I, § 12, problem 12.1 (whence also to Ch. I, § 13, problem 13.1 c)) and even to Ch. II, § 5, problem 5.1 a), as well as to Ch. I, § 13, problem 13.2 a) for separable $E$, are affirmative. To this end, let us recall

**Lemma 14.2.** *Let $E$ be a Banach space, $G$ a complemented subspace of $E$ and $p_1, p_2$ two projections of $E$ onto $G$. Then there exists an isomorphism $\tau$ of* $\operatorname{Ker} p_1$ *onto* $\operatorname{Ker} p_2$, *such that*

$$\|\tau\| \leqslant 1 + \|p_2\|, \quad \|\tau^{-1}\| \leqslant 1 + \|p_1\|. \tag{14.36}$$

*Proof.* Let $\omega$ denote the canonical mapping $E \to E/G$. Then $u_i = \omega|_{\operatorname{Ker} p_i}$ is an isomorphism of $\operatorname{Ker} p_i$ onto $E/G$ (namely, $u_i^{-1}(x + G) = (I - p_i)(x)$ for $x + G \in E/G$), with $\|u_i\| \leqslant 1$ and

$$\|u_i^{-1}(x + G)\| = \|(I - p_i)(x)\| = \inf_{g \in G} \|(I - p_i)(x + g)\| \leqslant$$

$$\leqslant \|I - p_i\| \inf_{g \in G} \|x + g\| = \|I - p_i\| \, \|x + G\| \qquad (x + G \in E/G),$$

whence $\|u_i^{-1}\| \leqslant \|I - p_i\| \leqslant 1 + \|p_i\|$ $(i = 1,2)$. Consequently,

$$\tau = u_2^{-1} u_1 \tag{14.37}$$

is an isomorphism of $\operatorname{Ker} p_1$ onto $\operatorname{Ker} p_2$, satisfying (14.36), which completes the proof of lemma 14.2. Clearly, $\tau = (I - p_2)|_{\operatorname{Ker} p_1}$.

**Lemma 14.3.** *Let $p$ be a continuous linear projection on a Banach space $E$. Then the restriction mapping $\rho : f \to f|_{p(E)}$ from $p^*(E^*)$ to $p(E)^*$ is an isomorphism of $p^*(E^*)$ onto $p(E)^*$, satisfying*

$$\|\rho\| \leqslant 1, \quad \|\rho^{-1}\| \leqslant \|p\|. \tag{14.38}$$

*Proof.* Clearly, $\rho$ is a continuous linear mapping of $p^*(E^*)$ into $p(E)^*$, with $\|\rho\| \leqslant 1$. Also, $\rho$ is one-to-one, since $\rho(p^*(f)) = p^*(f)|_{p(E)} = 0$ implies $p^*(f) = 0$. Furthermore, if $\varphi \in p(E)^*$, then for any $f \in E^*$ with $f|_{p(E)} = \varphi$ we have $\rho(p^*(f)) = p^*(f)|_{p(E)} = f|_{p(E)} = \varphi$, so maps $p^*(E^*)$ *onto* $p(E)^*$. Since

$$\|p^*(f)\| = \sup_{\substack{x \in E \\ \|x\| \leqslant 1}} |f(p(x))| \leqslant \sup_{\substack{p(x) \in p(E) \\ \|p(x)\| \leqslant \|p\|}} |p^*(f)(p(x))| \leqslant$$

$$\leqslant \|p^*(f)|_{p(E)}\| \, \|p\| = \|\rho(p^*(f))\| \, \|p\| \qquad (f \in E^*),$$

we have $\|\rho^{-1}\| \leqslant \|p\|$ and $\rho$ is an isomorphism, which completes the proof of lemma 14.3. One can also show that $\|\rho^{-1}\| = \|p\|$ and $\rho$ is a $w^*$-$w^*$ isomorphism, since if $\bar{p}$ denotes the "astriction" of $p$

to $p(E)$ (that is, $\bar{p} \in L(E, p(E))$, $\bar{p}(x) = p(x)$ for all $x \in E$, i.e., $\bar{p}$ is $p$ regarded as an operator on $E$ with values in $p(E)$), then

$$\bar{p}^*\rho(p^*(f))(x) = \bar{p}^*(f|_{p(E)})(x) = f(p(x)) = p^*(f)(x) \qquad (x \in E,\ f \in E^*),$$

whence *) $u\rho^{-1} = \bar{p}^*$, where $u\colon p^*(E^*) \to E^*$ is the identical embedding, and thus $\rho^{-1}$ is $w^*$-$w^*$ continuous and $\|\rho^{-1}\| = \|\bar{p}^*\| = \|\bar{p}\| = \|p\|$.

Let us mention that a part of lemma 14.3 is also a consequence of § 1, lemma 1.3, implication $2° \Rightarrow 3°$, with $G = p(E)$, $V = p^*(E^*)$ (since $G^\perp = p(E)^\perp = \operatorname{Ker} p^*$, whence $G^\perp + V = E^*$).

Now we are ready to prove the duality results announced above:

**Theorem 14.2.** *Let $E$ be a Banach space.*

a) *If $E^*$ has a basis $\{h_n\}$, then $E$ has a shrinking basis $\{x_n\}$ with the a.s.c.f. $\{f_n\} \subset E^*$ such that*

$$f_i = h_i \qquad (i = 1, \ldots, m_1), \tag{14.39}$$

$$[f_i]_{i=1}^{m_n} = [h_i]_{i=1}^{m_n} \qquad (n = 1,2, \ldots), \tag{14.40}$$

*where $\{m_n\}$ is a suitable increasing sequence of positive integers, in which $m_1$ can be taken arbitrarily large.*

b) *If $E$ is separable and $E^*$ has a weak* basis $\{h_n\}$, then $E$ has a basis $\{x_n\}$ with the a.s.c.f. $\{f_n\} \subset E^*$ satisfying* (14.39), (14.40), *where $\{m_n\}$ is as above.*

*Proof.* In both cases, let $\{s_n\}$ denote the sequence of partial sum operators associated to $\{h_n\}$. Then, in case a), $\{s_n\}$ is a $\pi$-basis of $E^*$ and in case b) $\{s_n\}$ is a weak* $\pi$-basis of $E^*$. Hence, by theorem 14.1 b), c), in both cases $E$ has a commuting $\pi$-basis $\{q_n\}$ such that

$$q_n^*(E^*) = s_{m_n}(E^*) = [h_i]_{i=1}^{m_n} \qquad (n = 1,2, \ldots), \tag{14.41}$$

where $\{m_n\}$ is a suitable increasing sequence of indices; clearly, one can take $m_1$ arbitrarily large. Furthermore, $\{q_n^*|_{[h_j]}\}$ is a commuting $\pi$-basis of $[h_n] = \overline{\bigcup_{n=1}^{\infty} s_n(E^*)}$ (see e.g. § 13, proposition 13.2).

Now, since $s_{m_{n-1}}|_{q_n^*(E^*)}$ and $q_{n-1}^*|_{q_n^*(E^*)}$ are projections of $q_n^*(E^*)$ onto $q_{n-1}^*(E^*)$, there exists, by lemma 14.2 applied in $q_n^*(E^*)$, an isomorphism $\tau_{n-1}$ of $[h_i]_{i=m_{n-1}+1}^{m_n} = \operatorname{Ker} s_{m_{n-1}}|_{q_n^*(E^*)}$ onto $(q_n^* - q_{n-1}^*)(E^*) = \operatorname{Ker} q_{n-1}^*|_{q_n^*(E^*)}$, such that

$$\|\tau_{n-1}\| \leqslant \sup_{1 \leqslant j < \infty} \|q_j\| + 1, \quad \|\tau_{n-1}^{-1}\| \leqslant \nu_{\{h_j\}} + 1 \qquad (n = 2,3, \ldots).$$

Consequently, the finite sequence

$$f_i = \tau_{n-1}(h_i) \qquad (i = m_{n-1} + 1, \ldots, m_n;\ m_0 = 0;\ \tau_0 = I_{q_1^*(E^*)}) \tag{14.42}$$

*) Thus, in other words, $\rho^{-1}(\varphi) = \varphi \circ p \ (\varphi \in p(E)^*)$.

is a basis of $(q_n^* - q_{n-1}^*)(E^*)$ $(n = 1,2, \ldots; q_0 = 0)$ and the norms of these bases are uniformly bounded, namely,

$$\nu_{\{f_i\}_{i=m_{n-1}+1}^{m_n}} \leqslant (\sup_{1 \leqslant j < \infty} \|g_j\| + 1)(\nu_{\{h_j\}} + 1)\nu_{\{h_j\}} = C_0 \qquad (n = 1,2, \ldots).$$

But then, by lemma 14.3, $\{\rho_n(f_i)\}_{i=m_{n-1}+1}^{m_n} = \{f_i|_{(q_n-q_{n-1})(E)}\}_{i=m_{n-1}+1}^{m_n}$ is a basis of $(q_n - q_{n-1})(E)^*$, of norm

$$\nu_{\{\rho_n(f_i)\}_{i=m_{n-1}+1}^{m_n}} \leqslant \|\rho_n\| \, \|\rho_n^{-1}\| \, C_0 \leqslant \sup_{1 \leqslant j < \infty} \|q_j - q_{j-1}\| C_0 \quad (n=1,2,\ldots)$$

and hence, since $\dim (q_n - q_{n-1})(E) < \infty$, $(q_n - q_{n-1})(E)$ has a basis*) $\{x_i\}_{i=m_{n-1}+1}^{m_n}$ such that

$$f_i(x_j) = \rho_n(f_i)(x_j) = \delta_{ij} \quad (i, j = m_{n-1}+1,\ldots,m_n;\ n = 1,2,\ldots) \quad (14.43)$$

and that the norms of these bases are uniformly bounded (by $\sup_{1 \leqslant j < \infty} \|q_j - q_{j-1}\| C_0$). Hence, since $\{(q_n - q_{n-1})(E)\}$ is a finite-dimensional decomposition of $E$ (by § 13, the remarks made after definition 13.5), it follows from § 13, theorem 13.4, that $\{x_n\}$ is a basis of $E$. Also, by (14.43) and since $\{q_n\}$ is a commuting $\pi$-basis of $E$, it follows**) that $f_i(x_j) = \delta_{ij}$ $(i, j = 1,2, \ldots)$, so $\{f_n\}$ is the a.s.c.f. to $\{x_n\}$. Finally, since $[f_i]_{i=1}^{m_n} = q_n^*(E^*)$ $(n = 1,2, \ldots)$, from (14.41) we obtain (14.40), which completes the proof of theorem 14.2.

From theorem 14.2 b) and Ch. I, § 14, theorem 14.1 it follows that *for a separable Banach space $E$, if $E^*$ has a weak* basis, then $E^*$ has a weak* Schauder basis.* It is not known whether here the assumption of separability of $E$ can be removed (this question amounts to Ch. I, § 13, problem 13.2 b), which, in turn, is equivalent to Ch. I, § 13, problem 13.2 a), by virtue of theorem 14.2 above). Similarly, it is not known whether in theorem 14.1 c) the assumption of separability of $E$ can be removed. More generally, the following questions related to § 1, problem 1.5, arise naturally:

*Problem 14.1.* a) Does every conjugate Banach space $E^*$ contain a weak* basic sequence $\{h_n\}$ (i.e., such that $\{h_n\}$ is a weak* basis of $\widetilde{[h_n]}$, the weak* closure of $[h_n]$)? b) Does every locally convex space $U$ have a basic sequence? What about a Schauder basic sequence? c) The same questions, for "basic sequence" replaced by: basic sequence with parentheses.

Now we shall consider the "dual" problem raised in Ch. II, § 5, problem 5.1 b), whether a shrinking basis exists in every Banach space $E$ with a basis and with separable conjugate space $E^*$. We shall show (in theorem 14.3 below) that the answer is affirmative if $E^*$ has the

*) Namely, $\{x_i\}_{i=m_{n-1}+1}^{m_n}$ is the image of the a.s.c.f. to $\{\rho_n(f_i)\}_{i=m_{n-1}+1}^{m_n}$ by the canonical linear isometry $(q_n - q_{n-1})(E)^{**} \equiv (q_n - q_{n-1})(E)$.

**) By § 15, the "orthogonality relations" (15.6).

approximation property*). To this end, instead of theorem 14.1 b), c) we shall use now theorem 14.1 a) and therefore, instead of lemma 14.2, we shall need

**Lemma 14.4.** *Let $E$ be a Banach space, $G_1, G_2, F_1$ and $F_2$ finite-dimensional subspaces of $E$ with $F_i \subset G_i$ $(i = 1,2)$ and $p_i$ a projection of $G_i$ onto $F_i$ $(i = 1,2)$. If $0 < \varepsilon \leqslant \dfrac{1}{2(1 + 2\|p_1\|)}$ and if $G_2$ is $\varepsilon$-close to $G_1$ and $F_2$ is $\varepsilon$-close to $F_1$, then there exists an isomorphism $\tau$ of* Ker $p_1$ *onto* Ker $p_2$, *such that*

$$\|\tau\| \leqslant \frac{3}{2}(1 + \|p_2\|), \quad \|\tau^{-1}\| \leqslant 2(1 + \|p_1\|). \tag{14.44}$$

*Proof.* Let $u: G_1 \to G_2$ and $v: F_1 \to F_2$ be isomorphisms onto the second spaces, such that $\|I_{G_1} - u\| \leqslant \varepsilon$, $\|I_{F_1} - v\| \leqslant \varepsilon$. Define $t: G_1 \to G_2$ by

$$t(x) = vp_1(x) + u(x - p_1(x)) \qquad (x \in G_1). \tag{14.45}$$

Then, by our assumption on $\varepsilon$,

$$\|x - t(x)\| \leqslant \|x - p_1(x) - u(x - p_1(x))\| + \|p_1(x) - vp_1(x)\| \leqslant$$

$$\leqslant \varepsilon\|x - p_1(x)\| + \varepsilon\|p_1(x)\| \leqslant \varepsilon(1 + 2\|p_1\|)\|x\| \leqslant \frac{1}{2}\|x\| \qquad (x \in G_1),$$

whence

$$\frac{1}{2}\|x\| \leqslant \|t(x)\| \leqslant \frac{3}{2}\|x\| \qquad (x \in G_1), \tag{14.46}$$

so $t$ is an isomorphism of $G_1$ onto $t(G_1) \subset G_2$. But then $\dim t(G_1) = \dim G_1 = \dim u(G_1) = \dim G_2 < \infty$ (since $t$ and $u$ are isomorphisms and $u(G_1) = G_2$), whence $t(G_1) = G_2$, so $t$ maps $G_1$ *onto* $G_2$. Also, $t|_{F_1}$ maps $F_1$ *onto* $F_2$ (since by (14.45) $t(x) = v(x)$ for all $x \in F_1 = p_1(G_1)$). Consequently, $t$ induces an isomorphism $\varkappa$ of $G_1/F_1$ onto $G_2/F_2$, namely,

$$\varkappa(x + F_1) = t(x) + F_2 \qquad (x + F_1 \in G_1/F_1). \tag{14.47}$$

Then, since $\|t(x) + F_2\| = \inf_{z\in F_2} \|t(x) + z\| = \inf_{y\in F_1} \|t(x + y)\|$, from (14.46) we obtain

$$\frac{1}{2}\|x+F_1\| = \frac{1}{2}\inf_{y\in F_1}\|x+y\| \leqslant \|t(x)+F_2\| \leqslant \frac{3}{2}\|x+F_1\| \quad (x+F_1 \in G_1/F_1),$$

and thus $\|\varkappa\| \leqslant \dfrac{3}{2}$, $\|\varkappa^{-1}\| \leqslant 2$.

---

*) This is equivalent to the assumption that $E^*$ has the metric approximation property (see § 9, proof of theorem 9.3, implication 11° ⇒ 4°).

Now let $u_i$ denote the canonical isomorphism of $\operatorname{Ker} p_i$ onto $G_i/F_i$, with*) $\|u_i\| \leqslant 1$ and $\|u_i^{-1}\| \leqslant 1 + \|p_i\|$ $(i = 1,2)$. Then

$$\tau = u_2^{-1} \varkappa u_1 \tag{14.48}$$

is an isomorphism of $\operatorname{Ker} p_1$ onto $\operatorname{Ker} p_2$, satisfying (14.44), which completes the proof of lemma 14.4.

Now we can prove

**Theorem 14.3.** *If $E$ is a Banach space with a basis $\{z_n\}$ and if $E^*$ has the bounded approximation property, then for any separable subspace $V$ of $E^*$ and any $\varepsilon_n$ with $0 < \varepsilon_n \leqslant \dfrac{1}{2(1+2\nu_{\{z_j\}})}$, $E$ has a basis $\{x_n\}$, with the a.s.c.f. $\{f_n\}$, such that*

$$[f_n] \supset V, \tag{14.49}$$

$$[x_i]_{i=1}^{m_n} \text{ is } \varepsilon_n\text{-close to } [z_i]_{i=1}^{m_n} \qquad (n = 1,2, \ldots), \tag{14.50}$$

*where $\{m_n\}$ is a suitable increasing sequence of positive integers. Hence, in particular, if $E^*$ is also separable, then $E$ has a shrinking basis $\{x_n\}$ satisfying* (14.50) *for a suitable $\{m_n\}$.*

*Proof.* Let $\{s_n\}$ denote the sequence of partial sum operators associated to the basis $\{z_n\}$. Then $\{s_n\}$ is a $\Pi$-basis of $E$ and hence for any separable subspace $V$ of $E^*$ and any $\varepsilon_n$ with $0 < \varepsilon_n < 1$ there exists, by theorem 14.1 a), a commuting $\pi$-basis $\{q_n\}$ of $E$ satisfying (14.25) and such that

$$q_n(E) \text{ is } \varepsilon_n\text{-close to } s_{m_n}(E) = [z_i]_{i=1}^{m_n} \qquad (n = 1,2, \ldots), \tag{14.51}$$

for a suitable increasing sequence of indices $\{m_n\}$.

Now, since $s_{m_{n-1}}|_{s_{m_n}(E)}$ and $q_{n-1}|_{q_n(E)}$ are projections of $s_{m_n}(E)$ onto $s_{m_{n-1}}(E)$ and of $q_n(E)$ onto $q_{n-1}(E)$ respectively and since we have (14.51) and $0 < \varepsilon_n \leqslant \dfrac{1}{2(1+2\nu_{\{z_j\}})} \leqslant \dfrac{1}{2(1+2\|s_{m_{n-1}}|_{s_n(E)}\|)}$, there exists, by lemma 14.4, an isomorphism $\tau_{n-1}$ of $[z_i]_{i=m_{n-1}+1}^{m_n} = \operatorname{Ker} s_{m_{n-1}}|_{s_{m_n}(E)}$ onto $(q_n - q_{n-1})(E) = \operatorname{Ker} q_{n-1}|_{q_n(E)}$, such that

$$\|\tau_{n-1}\| \leqslant \frac{3}{2}\Big(\sup_{1 \leqslant j < \infty} \|q_j\| + 1\Big), \quad \|\tau_{n-1}^{-1}\| \leqslant 2(\nu_{\{z_j\}} + 1) \qquad (n = 1,2, \ldots),$$

where we have put $m_0 = 0$, $s_0 = q_0 = 0$**). Consequently, the finite sequence

$$x_i = \tau_{n-1}(z_i) \qquad (i = m_{n-1} + 1, \ldots, m_n) \tag{14.52}$$

---

*) See the proof of lemma 14.2.

**) Actually, by (14.51) for $n = 1$ and § 12, remark 12.4, we can choose $\tau_0$ with $\|\tau_0\| \, \|\tau_0^{-1}\| \leqslant \dfrac{1+\varepsilon}{1-\varepsilon}$.

is a basis of $(q_n - q_{n-1})(E)$ $(n = 1,2, \ldots)$ and the norms of these bases are uniformly bounded, namely,

$$v_{\{x_i\}_{i=m_{n-1}+1}^{m_n}} \leqslant 3\,(\sup_{1\leqslant j<\infty} \|q_j\| + 1)(v_{\{z_j\}} + 1)\, v_{\{z_j\}} = C \quad (n = 1,2, \ldots).$$

Hence, since $\{(q_n - q_{n-1})(E)\}$ is a finite dimensional decomposition of $E$, it follows, by § 13, theorem 13.4, that $\{x_n\}$ is a basis of $E$. Also, if $\{f_n\} \subset E^*$ is the a.s.c.f. to $\{x_n\}$, then, since $\sum_{\substack{i=1\\ i\neq n}}^{\infty}{}' \oplus (q_i - q_{i-1})(E) =$ $= \operatorname{Ker}(q_n - q_{n-1})$, we have

$$f_i \in \{\operatorname{Ker}(q_n - q_{n-1})\}^\perp = (q_n^* - q_{n-1}^*)(E^*)$$
$$(i = m_{n-1} + 1, \ldots, m_n;\ n = 1,2, \ldots).$$

But, by lemma 14.3,

$$\dim (q_n^* - q_{n-1}^*)(E^*) = \dim (q_n - q_{n-1})(E)^* = \dim (q_n - q_{n-1})(E) =$$
$$= \dim [x_i]_{i=m_{n-1}+1}^{m_n} = \dim [f_i]_{i=m_{n-1}+1}^{m_n} \qquad (n = 1,2, \ldots),$$

whence, by the above,

$$[f_i]_{i=m_{n-1}+1}^{m_n} = (q_n^* - q_{n-1}^*)(E^*) \qquad (n = 1,2, \ldots),$$

which, together with (14.25), gives (14.49). Finally, since $[x_i]_{i=1}^{m_n} =$ $= q_n(E)$ $(n = 1,2, \ldots)$, from (14.51) we get (14.50), which completes the proof of theorem 14.3.

The assumption that $E^*$ has the bounded approximation property cannot be omitted in the second statement of theorem 14.3, as shown by example 14.1.

**Corollary 14.1.** *The space $C_\infty$ has a shrinking basis.*

*Proof.* By § 13, corollary 13.7, $C_\infty$ has a basis. Furthermore, $C_\infty^* \equiv (G_1^* \times G_2^* \times \ldots)_{l^1}$ is separable and has the bounded approximation property. Hence, by theorem 14.3, $C_\infty$ has a shrinking basis, which completes the proof. Alternatively, one can also observe that $C_\infty^* \cong C_1$ (see the proof of theorem 14.4 below), whence, by § 13, corollary 13.7, $C_\infty^*$ has a basis and thus, by theorem 14.2, $C_\infty$ has a shrinking basis.

*Remark 14.4*. The above methods also yield the following more general result, which implies both theorem 14.2, without (14.39), (14.40), and theorem 14.3: *If $E$ is a Banach space such that there exist a subspace $F$ of $E^*$, with a basis $\{h_n\}$, and sequences of finite rank operators $\{u_n\} \subset L(E, E)$, $\{v_n\} \subset L(E^*, E^*)$ satisfying* (14.22), (14.14), (14.15) *and $u_n^*(E^*) \subset F$, $v_n(E^*) \subset F$ $(n=1,2, \ldots)$, and if $0 < \varepsilon_n \leqslant \dfrac{1}{2(1+2v_{\{h_j\}})}$, then $E$ has a basis $\{x_n\}$, with the a.s.c.f. $\{f_n\}$, such that $[f_n] = F$ and that $[f_i]_{i=1}^{m_n}$ is $\varepsilon_n$-close to $[h_i]_{i=1}^{m_n}$ for a suitable $m_n$ $(n = 1,2, \ldots)$.*

Indeed, let $\{s_n\} \subset L(F, F)$ be the sequence of partial sum operators associated to the basis $\{h_n\}$ of $F$. Then, as in remark 14.1 (with $\Gamma = s_n(F)$), for each $n$ there exists a finite rank operator $w_n: E^* \to F$ such that $w_n(h) = h$ $(h \in s_n(F))$ and $\|w_n\| \leqslant \frac{4}{3}\mu$. Then $v_n' = s_n w_n$ is a projection of $E^*$ onto $s_n(F)$, with $\|v_n'\| \leqslant \frac{4}{3}\mu \nu_{\{h_j\}} = \mu'$ $(n = 1,2, \ldots)$ and $\lim_{n\to\infty} v_n'(f) = f$ for all $f \in F$ (since this holds for all $f \in \bigcup_{m=1}^{\infty} s_m(F)$ and $\sup_{1\leqslant n<\infty} \|v_n'\| < \infty$). Thus, the conditions of remark 14.2 are satisfied for $E, F, \{u_n\}$ and $\{v_n'\}$ and hence, by remark 14.3 b), $E$ has a commuting $\pi$-basis $\{q_n\}$ such that $\{q_n^*|_F\}$ is a commuting $\pi$-basis of $F$ and that, given any $\varepsilon_n$ with $0 < \varepsilon_n < 1$, $q_n^*(E^*)$ is $\varepsilon_n$-close to $v_{m_n}'(E^*) = s_{m_n}(F) = [h_i]_{i=1}^{m_n}$ for a suitable $m_n$ $(n = 1,2, \ldots)$. Then, since $s_{m_{n-1}}|_{s_{m_n}(F)}$ and $q_{n-1}^*|_{q_n^*(E^*)}$ are projections of $s_{m_n}(F)$ onto $s_{m_{n-1}}(F)$ and of $q_n^*(E^*)$ onto $q_{n-1}^*(E^*)$ respectively, with $q_k^*(E^*)$ $\varepsilon_k$-close to $v_{m_k}'(E^*) = s_{m_k}(F)$ $(k = n, n-1)$, from lemma 14.4 it follows $\left(\text{assuming that } 0<\varepsilon_k\leqslant\frac{1}{2(1+2\nu_{\{h_j\}})}\right)$ that the spaces $(q_n^* - q_{n-1}^*)(E^*)$ $(n = 1,2, \ldots;\ q_0 = 0)$ have bases $\{f_i\}_{i=m_{n-1}+1}^{m_n}$ with uniformly bounded norms. Hence, as in the proof of theorem 14.2, the space $E$ has a basis $\{x_n\}$, with the a.s.c.f. $\{f_n\}$, such that $[f_n] = \left[\bigcup_{n=1}^{\infty} (q_n^* - q_{n-1}^*)(E^*)\right] = F$ and $[f_i]_{i=1}^{m_n} = q_n^*(E^*)$ $(n = 1,2, \ldots)$, which completes the proof of our assertion. Note that theorem 14.2 a), without (14.39), (14.40), is the particular case $F = E^*$ of the above result (by § 9, proposition 9.8 and remark 9.12 c)), while theorem 14.2 b), without (14.39), (14.40), is the particular case $F=[h_n]$, $v_n=s_n$ of the above result (by § 9, theorem 9.7 and its proof, refined by using also (9.124), and by Ch.I, § 13, theorem 13.1 b)). Furthermore, the second statement of theorem 14.3 is obtained by applying the above result for $E^*$, $\pi(E) \subset E^{**}$, $\{t_n^*\}$ and $\{s_n^{**}\}$ instead of $E$, $F \subset E^*$, $\{u_n\}$ and $\{v_n\}$ respectively, where (by § 9, proposition 9.8 and remark 9.12 c) $\{t_n\}$ is a sequence of finite rank operators on $E$, such that $\{t_n^*\}$ is an approximative basis of $E^*$. Similarly, the first statement of theorem 14.3 can be also deduced applying the above result for $V_0$, $u(E) \subset V_0^*$, $\{t_n^*|_{V_0}\}$ and $\{(s_n^*|_{V_0})^*\}$ instead of $E$, $F \subset E^*$, $\{u_n\}$ and $\{v_n\}$ respectively, where $V_0$ is a suitable subspace of $E^*$ with $V_0 \supset V$, $r(V_0) > 0$, $u$ is the canonical mapping of $E$ into $V_0^*$ and $\{t_n\}$ is a suitable sequence of finite rank operators on $E$. We omit the details.

Now we shall apply remark 14.3 to obtain some further "universal complement" properties of the spaces $C_p$, as was announced at the end of § 12.

**Theorem 14.4.** *Let $1 \leqslant p \leqslant \infty$. If $E$ is a Banach space with a commuting approximative basis $\{u_n\}$, then $E \times C_p$ has a basis.*

*Proof.* Let $\{G_n\}$ be a sequence of finite-dimensional Banach spaces as in § 12, definition 12.3 (of $C_p$). Then $C_p^* \equiv (G_1^* \times G_2^* \times \ldots)_{l^{p'}}$, canonically, where

$$p' = \begin{cases} 1 & \text{if } p = \infty \\ \dfrac{p}{p-1} & \text{if } 1 < p < \infty \\ \infty & \text{if } p = 1, \end{cases} \tag{14.53}$$

so we shall identify these two spaces. Clearly, $\{G_n^*\}$ also satisfies conditions i) and ii) of § 12, definition 12.3 (taking the adjoints of the isomorphisms and isometries occurring there for $\{G_n\}$) and hence, by the uniqueness property proved in § 12, $C_{p'} \sim (G_1^* \times G_2^* \times \ldots)_{l^{p'}} \equiv$ $\equiv C_p^*$ for $1 < p \leqslant \infty$, where $\sim$ means: nearly isometric. Also, with a similar argument, $C_\infty \equiv (G_1 \times G_2 \times \ldots)_{c_0}$ is nearly isometric to a separable subspace of $(G_1^* \times G_2^* \times \ldots)_{l^\infty} \equiv C_1^*$. We shall identify these nearly isometric spaces, so $V \times C_{p'}$ below will be a separable subspace of $E^* \times C_p^* \cong (E \times C_p)^*$ $(1 \leqslant p \leqslant \infty)$. Now let

$$H_n = \left[\bigcup_{i=1}^n u_i^*(E^*)\right] \times (G_1^* \times \ldots \times G_n^* \times \{0\} \times \{0\} \times \ldots)_{l^{p'}} \quad (n = 1, 2, \ldots). \tag{14.54}$$

By our assumption on $\{u_n\}$, we have

$$u_{n+1}^*(f) = f \qquad \left(f \in \left[\bigcup_{i=1}^n u_i^*(E^*)\right]\right),$$

$$\|u_{n+1}^*\| \leqslant \lambda = \sup_{1 \leqslant j < \infty} \|u_j\| < \infty.$$

Let

$$\Gamma = (I - u_{n+1}^*) u_{n+1}^*(E^*). \tag{14.55}$$

Then $\dim \Gamma < \infty$ and hence, by the above, we can choose $m > n$ such that there exists an isomorphism $\tau$ of $\Gamma$ onto $G_m^*$, with $\|\tau\| = 1$, $\|\tau^{-1}\| < 2$. Let us denote by $q_m$ the natural projection of $C_p^*$ onto $G_m^*$ and by $q$ the natural projection of $C_p^*$ onto $(G_1^* \times \ldots \times G_n^* \times \{0\} \times \{0\} \times$ $\times \ldots)_{l^{p'}}$. Define $v = v_n \colon E^* \times C_p^* \to E^* \times C_p^*$ by

$$v_n(\{f, h\}) = \{u_{n+1}^*(f) + \tau^{-1} q_m(h),$$

$$\tau(I - u_{n+1}^*)(u_{n+1}^*(f) + \tau^{-1} q_m(h)) + q(h)\} \quad (\{f, h\} \in E^* \times C_p^*). \tag{14.56}$$

Then, as in the proof of § 12, theorem 12.4, $\{v_n\}$ is a sequence of finite rank projections on $E^* \times C_p^*$, such that

$$v_n(\{f, h\}) = \{f, h\} \qquad (\{f, h\} \in H_n), \tag{14.57}$$

$$\sup_{1 \leqslant n < \infty} \|v_n\| < \infty. \tag{14.58}$$

Now let

$$V = \left[\bigcup_{n=1}^{\infty} u_n^*(E^*)\right] \subset E^*. \tag{14.59}$$

Then $H_1 \subset H_2 \subset \ldots \subset V \times C_{p'}$ and $\overline{\bigcup_{n=1}^{\infty} H_n} = V \times C_{p'}$, whence, by (14.57), (14.58), we obtain

$$\lim_{n\to\infty} v_n(\{f, h\}) = \{f, h\} \qquad (\{f, h\} \in V \times C_{p'}). \tag{14.60}$$

Furthermore, if $\{p_n\}$ denotes the natural commuting $\pi$-basis of $C_p$, then, clearly,

$$\lim_{n\to\infty} (u_n \times p_n)(\{x, y\}) = \lim_{n\to\infty} \{u_n(x), p_n(y)\} = \{x, y\} \quad (\{x, y\} \in E \times C_p). \tag{14.61}$$

Also, for any $\{x, y\} \in E \times C_p$ and $\{f, h\} \in E^* \times C_p^* \cong (E \times C_p)^*$ we have

$$(u_n \times p_n)^*(\{f, h\})(\{x, y\}) = \{f, h\}(\{u_n(x), p_n(y)\}) =$$
$$= f(u_n(x)) + h(p_n(y)) = \{u_n^*(f), p_n^*(h)\}(\{x, y\}),$$

whence, since $\{u_n^*(f), p_n^*(h)\} \in V \times C_{p'}$, we obtain

$$(u_n \times p_n)^*((E \times C_p)^*) \subset V \times C_{p'} \qquad (n = 1,2, \ldots).$$

Thus, since $v_n((E \times C_p)^*) \subset V \times C_{p'}$ $(n = 1,2, \ldots)$, the assumptions of remark 14.2 are satisfied for $E \times C_p$, $\{u_n \times p_n\}$, $V \times C_{p'} \subset E^* \times C_p^* \cong$ $\cong (E \times C_p)^*$ and $\{v_n\}$ and hence, by remark 14.3 b), $E \times C_p$ has a finite-dimensional decomposition. Consequently, by § 13, proposition 13.13, $E \times C_p$ has a basis, which completes the proof of theorem 14.4.

**Corollary 14.2.** *Let $1 \leqslant p \leqslant \infty$. If $E$ is a separable Banach space such that the conjugate space $E^*$ has the bounded approximation property, then $E \times C_p$ has a basis. If, in addition, $E^*$ is separable and if $1 < p \leqslant \infty$, then $E \times C_p$ has a shrinking basis.*

*Proof.* If $E$ is separable and $E^*$ has the bounded approximation property, then, by § 9, propositions 9.8 and 9.7, $E$ has a commuting approximative basis. Hence, by theorem 14.4, $E \times C_p$ has a basis. If, in addition, $E^*$ is separable and*) $1 < p \leqslant \infty$, then $(E \times C_p)^* \cong$ $\cong E^* \times C_{p'}$ is separable and has the bounded approximation property, whence, by theorem 14.3, $E \times C_p$ has a shrinking basis. This completes the proof.

**Corollary 14.3.** a) *If* $1 < p < \infty$, *then* $C_p$ *is a reflexive Banach space with a basis such that if* $E$ *is any reflexive Banach space with an approximative basis, then* $E \times C_p$ *is a reflexive space with a basis.*

b) $C_\infty$ *is (nearly isometric to) a subspace of* $c_0$ *with a shrinking basis and such that if* $E$ *is any Banach space for which* $E^*$ *has an approximative basis, then* $E \times C_\infty$ *has a shrinking basis.*

*Proof.* If $E$ is reflexive and has an approximative basis, then $E^*$ has the bounded approximation property (e.g. by § 9, proposition 9.8) and hence, by corollary 14.2, $E \times C_p$ has a basis, which proves a). Furthermore, by § 12, proof of lemma 12.6 and the uniqueness property given after definition 12.3, $C_\infty$ is (nearly isometric to) a subspace of $(c_0 \times c_0 \times \times \ldots)_{c_0} \equiv c_0$. Finally, $C_\infty$ has a shrinking basis by corollary 14.1 above and the last statement of part b) is simply the particular case $p = \infty$ of the last statement of corollary 14.2, which completes the proof.

*Remark 14.5.* Corollary 14.3 a) shows that *every reflexive Banach space with an approximative basis is isomorphic to a complemented subspace of a reflexive space with a basis*, while corollary 14.3 b) shows that *every Banach space* $E$ *for which* $E^*$ *has an approximative basis, is isomorphic to a complemented subspace of a Banach space with a shrinking basis;* moreover, in the first case any $C_p$ with $1 < p < \infty$ is a reflexive universal complement with a basis, while in the second case $C_\infty$ is a universal complement with a shrinking basis.

*Problem 14.2.* Let $1 \leqslant p \leqslant \infty$. If $E$ is a Banach space with an approximative basis, does $E \times C_p$ have a basis?

If the answer to § 12, problem 12.6 were affirmative, then by § 13, theorem 13.1 it would follow that for every separable Banach space $E$ with the metric approximation property $E \times C_p$ has a finite-dimensional decomposition and hence, by § 13, proposition 13.13, a basis. Thus, in this case we would have an affirmative answer to a particular case of problem 14.2 above.

## § 15. Decompositions (bases of subspaces). Schauder decompositions. Resolutions of the identity. Integral bases

If we drop the assumption of finite-dimensionality of the subspaces occurring in the definition of a finite-dimensional decomposition (§ 13, definition 13.5), we arrive naturally at

---

*) If $p = 1$ then $C_1^*$, whence also $E^* \times C_1^*$, are non-separable and do not have the bounded approximation property (see the Notes and remarks to § 0).

*Definition 15.1.* An infinite sequence $\{G_n\}$ of (not necessarily closed) linear subspaces of a Banach space $E$, such that $G_n \neq \{0\}$ $(n = 1,2, \ldots)$, is called a *decomposition* (or a *basis of subspaces*) of $E$ if for every $x \in E$ there exists a *unique* sequence $\{y_n\} \subset E$ with $y_n \in G_n$ $(n = 1,2, \ldots)$ such that

$$x = \sum_{i=1}^{\infty} y_i. \tag{15.1}$$

In other words, $\{G_n\}$ is a decomposition of $E$ if and only if $E$ is the "infinite direct sum of the subspaces $G_n$":

$$E = \sum_{i=1}^{\infty} \oplus\, G_i. \tag{15.2}$$

*Remark 15.1.* The assumption $G_n \neq \{0\}$ $(n = 1,2, \ldots)$ ensures that bases are a "particular case" of decompositions (see § 13, the remarks made after definition 13.5). However, we want to emphasize that, in contrast with the case of bases (see Ch. I, § 1, definition 1.1), in definition 15.1 it is essential that the sequence $\{G_n\}$ (whence also $\dim E$) is now assumed to be *infinite*. Indeed, although some results (e.g., some characterization theorems) can be formulated so as to include both the case of infinite direct sums and the case of finite direct sums, the above convention will allow us to avoid a number of trivial statements on *existence* of decompositions into a direct sum of *two* subspaces*) and to avoid to mention separately, in the majority of the subsequent results, that $\{G_n\}$ is infinite (e.g., in propositions 15.1, 15.4 and theorem 15.2 below).

Even with the convention that $\{G_n\}$ is infinite, the notion of a decomposition is too general, as shown by

**Proposition 15.1.** *Every Banach space**) $E$ has a decomposition.*

*Proof.* By § 1, theorem 1.1, let $\{x_n\}$ be an infinite basic sequence in $E$, such that $[x_n] \neq E$. Then there exists***) a linear (in general, discontinuous) projection $u$ of $E$ onto $[x_n]$. Put

$$G_1 = \operatorname{Ker} u,\ G_2 = [x_1],\ G_3 = [x_2], \ldots \tag{15.3}$$

Then $G_1 = \operatorname{Ker} u \neq \{0\}$ (since $[x_n] \neq E$). Also, every $x \in E$ has a unique decomposition $x = y + y_1$ with $y \in u(E) = [x_n]$ and $y_1 \in \operatorname{Ker} u = G_1$, and for every $y \in [x_n]$ there exists a unique sequence of scalars $\{\alpha_n\}$

*) Moreover, let us observe, that in this way we also avoid the trivial decomposition $G_1 = E$ (into one subspace).

**) Naturally, here and in the sequel we shall assume, without any special mention, that $\dim E = \infty$.

***) See e.g. [70], Ch. I, § 2, statement (1) (c).

such that $y = \sum_{i=1}^{\infty} \alpha_i x_i = \sum_{i=2}^{\infty} y_i$, where $y_n = \alpha_{n-1} x_{n-1} \in G_n$ $(n=2,3, \ldots)$. Thus, $\{G_n\}$ is a decomposition of $E$, which completes the proof.

Proposition 15.1 shows that it is necessary to restrict ourselves to a special class of decompositions, which will be more useful for applications. This class will be obtained as a natural generalization of the notion of a Schauder basis (Ch. I, § 14, definition 14.2). To this end, we need first to introduce a notion corresponding to that of coefficient functionals (Ch. I, § 3, definition 3.1). Rather than defining this concept as a sequence of subspaces*) of $E^*$, it will be more convenient to define it as a sequence of projections on $E$, as follows:

*Definition 15.2.* Let $\{G_n\}$ be a decomposition of a Banach space $E$. The sequence of linear projections $\{v_n\}$ on $E$, defined by

$$v_n(x) = y_n \qquad \left(x = \sum_{i=1}^{\infty} y_i \in E,\ n = 1,2, \ldots\right), \tag{15.4}$$

where $y_n \in G_n$ $(n = 1,2, \ldots)$, is called *the sequence of coordinate projections associated to the decomposition* $\{G_n\}$, or, shortly, *the associated sequence of coordinate projections* (we shall write: a.s.c.p.).

Thus, if $\{G_n\}$ is a decomposition of $E$, with the a.s.c.p. $\{v_n\}$, then every $x \in E$ has a unique expansion of the form

$$x = \sum_{i=1}^{\infty} v_i(x). \tag{15.5}$$

In particular, by (15.5) we have

$$\sum_{i=1}^{\infty} v_i(v_j(x)) = v_j(x) = \sum_{i=1}^{\infty} \delta_{ij} v_i(x) \qquad (x \in E,\ j = 1,2, \ldots),$$

whence, by the uniqueness condition of definition 15.1, we obtain

$$v_i v_j = \delta_{ij} v_i = \delta_{ij} v_j \qquad (i,j = 1,2, \ldots); \tag{15.6}$$

thus, the coordinate projections associated to a decomposition $\{G_n\}$ of a Banach space $E$ are mutually orthogonal.

In the particular case when $\{x_n\}$ is a basis of $E$ with the a.s.c.f. $\{f_n\} \subset E^*$ and $G_1 = [x_1]$, $G_2 = [x_2], \ldots$ is the one-dimensional decomposition of $E$ generated by this basis**), we have, clearly,

$$v_n(x) = f_n(x) x_n \qquad (x \in E,\ n = 1,2, \ldots); \tag{15.7}$$

*) See the remark made before definition 15.4.

**) See § 13, the remark made after definition 13.5.

in this case, from (15.6) one obtains again the biorthogonality relations $f_i(x_j) = \delta_{ij}$ $(i, j = 1,2, \ldots)$.

*Definition 15.3.* A decomposition $\{G_n\}$ of a Banach space $E$ is said to be a *Schauder decomposition* (or a *Schauder basis of subspaces*) of $E$, if all coordinate projections $v_n$ are continuous on $E$ (i.e., if $v_n \in L(E, E)$ for $n = 1,2, \ldots$).

In the particular case when $\{x_n\}$ is a basis of a Banach space $E$, the one-dimensional decomposition $G_1 = [x_1], G_2 = [x_2], \ldots$ of $E$ generated by $\{x_n\}$ is a Schauder decomposition of $E$ (by (15.7) and Ch. I, § 3, theorem 3.1). However, in the general case there are many decompositions of Banach spaces which are not Schauder decompositions; moreover, while every Banach space $E$ has a decomposition (by proposition 15.1), there are some frequently used Banach spaces $E$ which have no Schauder decomposition whatsoever (e.g., $E = l^\infty$, by theorem 15.2 below).

Now we shall give a characterization of Schauder decompositions (theorem 15.1 below), the sufficiency part of which is a generalization of Ch. I, § 3, theorem 3.1. To this end, let us first give the following version of Ch. I, § 3, proposition 3.1:

**Proposition 15.2.** *Let $\{G_n\}$ be a sequence of closed linear subspaces*) of a Banach space E and let $D_1$ be the linear space of sequences of elements*

$$D_1 = \left\{ \{y_n\} \subset E \,\middle|\, y_n \in G_n \quad (n = 1,2, \ldots), \sum_{i=1}^{\infty} y_i \text{ converges} \right\}, \tag{15.8}$$

*endowed with the norm*

$$\|\{y_n\}\| = \sup_{1 \leqslant n < \infty} \left\| \sum_{i=1}^{n} y_i \right\|. \tag{15.9}$$

*Then $D_1$ is a Banach space.*

*Proof.* Clearly, (15.9) is a norm on $D_1$. Now let $\{y_n^{(k)}\}$ $(k = 1,2, \ldots)$ be a Cauchy sequence in $D_1$, so

$$\|\{y_n^{(k)}\} - \{y_n^{(m)}\}\| = \sup_{1 \leqslant n < \infty} \left\| \sum_{i=1}^{n} (y_i^{(k)} - y_i^{(m)}) \right\| < \varepsilon \quad (k, m > N(\varepsilon)). \tag{15.10}$$

---

*) It is not necessary to assume that $G_n \neq \{0\}$ $(n = 1, 2, \ldots)$. In the particular case of Ch. I, § 3, the assumption $x_n \neq 0$ $(n = 1, 2, \ldots)$ was necessary in order to get a norm on $A_1$; note that $A_1$ was a Banach space of sequences of scalars, while $D_1$ is a Banach space of sequences of elements.

Then

$$\|y_n^{(k)} - y_n^{(m)}\| \leqslant \left\|\sum_{i=1}^{n} (y_i^{(k)} - y_i^{(m)})\right\| + \left\|\sum_{i=1}^{n-1} (y_i^{(k)} - y_i^{(m)})\right\| < 2\varepsilon$$

$$(k, m > N(\varepsilon);\ n = 1,2, \ldots),$$

whence, since by our assumption each $G_n$ is complete, $\lim_{k\to\infty} y_n^{(k)} = y_n \in G_n$ $(n = 1,2, \ldots)$. Then, from (15.10) for $m \to \infty$, we obtain

$$\left\|\sum_{i=1}^{n} (y_i^{(k)} - y_i)\right\| \leqslant \varepsilon \qquad (k > N(\varepsilon),\ n = 1,2, \ldots),$$

whence

$$\left\|\sum_{i=n+1}^{n+l} y_i\right\| \leqslant 2\varepsilon + \left\|\sum_{i=n+1}^{n+l} y_i^{(k)}\right\| \qquad (k > N(\varepsilon);\ n, l = 1,2, \ldots).$$

Consequently, since each series $\sum_{i=1}^{\infty} y_i^{(k)}$ converges and since $E$ is complete, it follows that $\sum_{i=1}^{\infty} y_i$ converges, i.e. $\{y_n\} \in D_1$. Also,

$$\|\{y_n^{(k)}\} - \{y_n\}\| = \sup_{1\leqslant n<\infty} \left\|\sum_{i=1}^{n} (y_i^{(k)} - y_i)\right\| \leqslant \varepsilon \qquad (k > N(\varepsilon)),$$

which completes the proof of proposition 15.2.

Next, we have the following version of Ch. I, § 3, proposition 3.2:

**Proposition 15.3.** *Let $\{G_n\}$ be a decomposition of a Banach space $E$, such that each $G_n$ $(n = 1,2, \ldots)$ is closed, and let $\{v_n\}$ be the a.s.c.p. to $\{G_n\}$. Then*

a) *The Banach space $D_1$ introduced in proposition* 15.2 *is isomorphic to $E$, by the mapping*

$$w\colon \{y_n\} \to \sum_{i=1}^{\infty} y_i\,. \tag{15.11}$$

b) *The numbers*

$$|||x||| = \sup_{1\leqslant n<\infty} \left\|\sum_{i=1}^{n} v_i(x)\right\| \qquad (x \in E) \tag{15.12}$$

*define a norm on the space $E$, equivalent to the initial norm of $E$.*

The proof is entirely similar to that of Ch. I, § 3, proposition 3.2 (replacing $A_1$ by $D_1$ and $\alpha_i x_i$ by $y_i$ of definition 15.1).

Now we are ready to prove

**Theorem 15.1.** *A decomposition $\{G_n\}$ of a Banach space $E$ is a Schauder decomposition of $E$ if and only if each $G_n$ $(n = 1,2, \ldots)$ is closed.*

*Proof.* If $\{G_n\}$ is a Schauder decomposition of $E$, then $v_n$ is a continuous linear projection of $E$ onto $G_n$ and hence $G_n = v_n(E)$ is closed $(n = 1,2, \ldots)$; indeed, if $x_k \in v_n(E) = G_n$ $(k = 1,2, \ldots)$ and $x_k \to x$, then $v_n(x_k) = x_k \to x$ and, on the other hand, since $v_n$ is continuous, $v_n(x_k) \to v_n(x)$ as $k \to \infty$, whence $x = v_n(x) \in G_n$. Thus, the condition is necessary.

For the sufficiency we shall give here three different proofs. Although the first one of them is quite short, the second and third one are also of interest.

1) Assume that $\{G_n\}$ is a decomposition of $E$ such that each $G_n$ $(n = 1,2, \ldots)$ is closed. Let $|||.|||$ be the norm on $E$ defined by (15.12). Then there exists, by proposition 15.3 b), a constant $C \geqslant 1$ such that $|||x||| \leqslant C\|x\|$ for all $x \in E$. Consequently,

$$\|v_n(x)\| \leqslant \left\|\sum_{i=1}^{n} v_i(x)\right\| + \left\|\sum_{i=1}^{n-1} v_i(x)\right\| \leqslant 2|||x||| \leqslant 2C\|x\|$$

$$(x \in E,\ n = 1,2, \ldots), \tag{15.13}$$

and thus $\{G_n\}$ is a Schauder decomposition of $E$, which completes the proof.

2) Assume again that $\{G_n\}$ is a decomposition of $E$ such that each $G_n$ $(n = 1,2, \ldots)$ is closed, and let

$$G^n = \operatorname{Ker} v_n = \{x \in E \mid v_n(x) = 0\} \quad (n = 1,2, \ldots). \tag{15.14}$$

We claim that *each $G^n$ $(n = 1,2, \ldots)$ is closed*. Indeed, fix $n$ and let $x_k \in G^n$ $(k = 1,2, \ldots)$, $x_k \to x \in E$. Then we can write

$$x_k = \sum_{i=1}^{\infty} y_i^{(k)} \ (k = 1,2, \ldots),\ x = \sum_{i=1}^{\infty} y_i,$$

where $y_i^{(k)}, y_i \in G_i$ $(i, k = 1,2, \ldots)$ and $y_n^{(k)} = v_n(x_k) = 0$ $(k = 1,2, \ldots)$. Then, by proposition 15.3 a),

$$\{y_i\} = w^{-1}\left(\sum_{i=1}^{\infty} y_i\right) = w^{-1}(x) = \lim_{k\to\infty} w^{-1}(x_k) = \lim_{k\to\infty} \{y_i^{(k)}\}$$

in $D_1$, that is,

$$\sup_{1 \leqslant n < \infty} \left\| \sum_{i=1}^{n} (y_i^{(k)} - y_i) \right\| \to 0 \qquad \text{as } k \to \infty.$$

Consequently,

$$\|y_n\| = \|y_n^{(k)} - y_n\| \leqslant \left\| \sum_{i=1}^{n} (y_i^{(k)} - y_i) \right\| + \left\| \sum_{i=1}^{n-1} (y_i^{(k)} - y_i) \right\| \to 0 \text{ as } k \to \infty,$$

whence $v_n(x) = y_n = 0$ and so $x \in G^n$, which proves the claim.

Now, since $G_n = v_n(E)$ and $G^n = \operatorname{Ker} v_n$ are closed, the projection $v_n$ is*) closed and hence, by the closed graph theorem, continuous ($n = 1,2, \ldots$), so $\{G_n\}$ is a Schauder decomposition of $E$, which completes the proof**). Thus, we see***) that *if $\{G_n\}$ is a Schauder decomposition of $E$*, then each $G_n$ has a closed complement $G^n$ in $E$, i.e., $E = G_n \oplus G^n$ ($n = 1,2, \ldots$); moreover, by formula (15.13) of the first proof, the norms of the projections $v_n$ of $E$ onto $G_n$ along $G^n$ ($n = 1,2, \ldots$) are uniformly bounded. This latter observation is a generalization of Ch. I, § 3, theorem 3.1, formula (3.7).

3) Let us first make the following observation, corresponding to Ch. I, § 8, proposition 8.1 a): *If $\{G_n\}$ is as in proposition* 15.2, *and each $G_n \neq \{0\}$, then the sequence of subspaces $\{F_n\}$ of $D_1$ defined by*

$$F_n = \underbrace{\{0\} \times \ldots \times \{0\}}_{n-1} \times G_n \times \{0\} \times \ldots \qquad (n = 1,2, \ldots) \quad (15.15)$$

is a Schauder decomposition of $D_1$. Indeed, since $G_n \neq \{0\}$, we have $F_n \neq \{0\}$ ($n = 1,2, \ldots$). Also if $\{y_n\} \in D_1$, then

$$\left\| \{y_n\} - \sum_{i=1}^{m} \{\underbrace{0, \ldots, 0}_{i-1}, y_i, 0, \ldots\} \right\| = \sup_{m+1 \leqslant k < \infty} \left\| \sum_{i=m+1}^{k} y_i \right\| \to \infty$$

as $m \to \infty$,

---

*) See e.g. [87], p. 480.

**) In the particular case of bases, this proof amounts to showing that if $\{x_n\}$ is a basis of a Banach space $E$, then for each $n$ the set $G^n = \left\{ x \in E \,\middle|\, x = \sum_{i=1}^{n-1} \alpha_i x_i + \sum_{i=n+1}^{\infty} \alpha_i x_i \right\} = \left\{ x \in E \,\middle|\, f_n(x) x_n = 0 \right\}$ is closed and hence the projection $\sum_{i=1}^{\infty} \alpha_i x_i \to \alpha_n x_n$ (i.e., the projection $x \to f_n(x) x_n$), with kernel $G^n$, is continuous.

***) Alternatively, this follows also from definition 15.2.

whence $\{y_n\} = \sum_{i=1}^{\infty} \{\underbrace{0, \ldots, 0}_{i-1}, y_i, 0, \ldots\}$. On the other hand, if

$$\sum_{i=1}^{\infty} \{\underbrace{0, \ldots, 0}_{-1},y_i,0, \ldots\} = \{0,0, \ldots\},$$

where $y_n \in G_n$ $(n = 1,2, \ldots)$, then, by (15.9),

$$\sup_{1 \leqslant n < \infty} \left\| \sum_{i=1}^{n} y_i \right\| = \lim_{n \to \infty} \sup_{1 \leqslant k \leqslant n} \left\| \sum_{i=1}^{k} y_i \right\| =$$

$$= \lim_{n \to \infty} \left\| \sum_{i=1}^{n} \{\underbrace{0, \ldots, 0}_{i-1}, y_i, 0, \ldots\} \right\| = \left\| \sum_{i=1}^{\infty} \{\underbrace{0, \ldots, 0}_{i-1}, y_i, 0, \ldots\} \right\| = 0,$$

whence, successively, $y_n = 0$ $(n = 1,2, \ldots)$. Thus, (15.15) is a decomposition of $D_1$. Finally, for the associated coordinate projections $\{y_i\} \to \{\underbrace{0, \ldots, 0}_{n-1}, y_n, 0, \ldots\}$ we have

$$\|(\{\underbrace{0, \ldots, 0}_{n-1}, y_n, 0, \ldots\}\| = \|y_n\| \leqslant \left\| \sum_{i=1}^{n} y_i \right\| + \left\| \sum_{i=1}^{n-1} y_i \right\| \leqslant 2\|\{y_j\}\|$$

$$(\{y_j\} \in D_1,\ n = 1,2, \ldots.),$$

which proves the assertion that (15.15) is a Schauder decomposition of $D_1$. Now, if $\{G_n\}$ is a decomposition of $E$, such that each $G_n$ $(n = 1,2, \ldots)$ is closed, then for the isomorphism $w$ of $D_1$ onto $E$ defined in proposition 15.3 a) we have

$$w(\{\underbrace{0, \ldots, 0}_{n-1}, y, 0, \ldots\}) = y \qquad (y \in G_n;\ n = 1,2, \ldots)$$

and hence, by the above, $\{w(F_n)\} = \{G_n\}$ is a Schauder decomposition of $E$, which completes the proof of theorem 15.1.

From now on we shall consider only decompositions which are Schauder decompositions (so each $G_n$ will be closed) and, as in the preceding sections, unless specified otherwise, by "subspace" we shall mean: closed linear subspace. Let us also mention separately that, since every finite-dimensional linear subspace is closed, from theorem 15.1 it follows

**Corollary 15.1.** *Every finite-dimensional decomposition of a Banach space $E$ is a Schauder decomposition of $E$.*

Let us consider now the problem of existence of Schauder decompositions. In addition to the results of § 14 on the existence of finite-dimensional decompositions, we have

**Proposition 15.4.** *If a Banach space $E$ contains a complemented subspace with a basis, then $E$ has a Schauder decomposition.*

The proof of proposition 15.4 is similar to that of proposition 15.1, with the only difference that now there exists, by the assumption of proposition 15.4, a continuous linear projection $u$ of $E$ onto $[x_n]$ and hence $G_1 = \operatorname{Ker} u$ is closed.

**Corollary 15.2.** a) *If $E$ is a Banach space with a boundedly complete basis, then $E^{**}$ has a Schauder decomposition.*

b) *If $E$ is a Banach space with a shrinking basis, then $E^{***}$ has a Schauder decomposition.*

*Proof.* a) If $E$ has a boundedly complete basis $\{x_n\}$, then, by Ch. II, § 6, theorem 6.2 we have $E^{**}=\pi(E)\oplus [f_n]^{\perp}$, where $\pi\colon E\to E^{**}$ is the canonical embedding, so $E^{**}$ satisfies the assumption of proposition 15.4. This proves a).

Finally, part b) is a consequence of part a) (since if $E$ has a shrinking basis, then $E^*$ has a boundedly complete basis), which completes the proof of corollary 15.2.

In order to prove one more corollary, let us first recall

**Lemma 15.1.** *If $G$ is a subspace of a separable Banach space $E$, such that $G$ is isomorphic to $c_0$, then $G$ is complemented in $E$.*

*Proof.* We shall show that if $v_0\colon G\to c_0$ is a continuous linear mapping of norm $\|v_0\|\leqslant\lambda$, then there exists an extension $v\colon E\to c_0$ of $v_0$, of norm $\|v\|\leqslant 2\lambda$. This will imply the statement of the lemma, since if $v_0$ is an isomorphism of $G\subset E$ onto $c_0$ and $v\colon E\to c_0$ an extension of $v_0$, then $w=v_0^{-1}v$ is a projection of $E$ onto $G$ (because then $w(E)=v_0^{-1}v(E)=$ $= G$ and $w^2 = v_0^{-1}v|_G v_0^{-1}v = v_0^{-1}v_0v_0^{-1}v = v_0^{-1}v = w$).

By Ch. I, § 17, corollary 17.6, there exists a sequence of functionals $\{\varphi_n\}\subset G^*$ with $\varphi_n \overset{w^*}{\to} 0$, such that

$$v_0(y) = \{\varphi_n(y)\} \qquad (y\in G),$$

$$\|v_0\| = \sup_{1\leqslant n<\infty} \|\varphi_n\| = \lambda.$$

By the Hahn-Banach theorem, let $\{f_n\}\subset E^*$, $f_n|_G = \varphi_n$, $\|f_n\| =$ $= \|\varphi_n\|\leqslant\lambda$ $(n = 1,2,\ldots)$. Thus $f_n\in\lambda S_{E^*}$ $(n = 1,2,\ldots)$ and, since $f_n|_G = \varphi_n \overset{w^*}{\to} 0$, every $w^*$-cluster point of $\{f_n\}$ is in the set

$$A = \lambda S_{E^*} \cap G^{\perp} \subset E^*.$$

Now, since $E$ is separable, there exists*) on $\lambda S_{E^*}$ a metric $\rho(.,.)$ generating a topology which coincides with the $w^*$-topology and such that $\rho(f-h, 0) = \rho(f, h)$ for all $f, h \in \lambda S_{E^*}$. Hence, by the above, $\lim_{n\to\infty} \rho(f_n, A) = 0$ (indeed, if $\rho(f_{n_k}, A) > \varepsilon > 0$ for $k = 1,2, \ldots$ then, since $\lambda S_{E^*}$ is $w^*$-compact, $\{f_{n_k}\}$ has a subsequence $\{f_{n_{k_m}}\}$ converging to an $f_0 \in A$, whence $\rho(f_{n_{k_m}}, A) \leqslant \rho(f_{n_{k_m}}, f_0) \to 0$, a contradiction). Let $\{h_n\} \subset A$ be an arbitrary sequence such that

$$\rho(f_n, h_n) \leqslant \rho(f_n, A) + \frac{1}{n} \qquad (n = 1,2, \ldots),$$

and put

$$v(x) = \{f_n(x) - h_n(x)\}_{n=1}^{\infty} \qquad (x \in E).$$

Then $\rho(f_n - h_n, 0) = \rho(f_n, h_n) \to 0$ as $n \to \infty$, whence $f_n - h_n \xrightarrow{w^*} 0$ and hence, by Ch. I, § 17, corollary 17.6, $v$ is a continuous linear mapping of $E$ into $c_0$, of norm

$$\|v\| = \sup_{1 \leqslant n < \infty} \|f_n - h_n\| \leqslant 2\lambda.$$

Finally, since $\{h_n\} \subset A \subset G^{\perp}$, we have

$$v(y) = \{f_n(y) - h_n(y)\} = \{f_n(y)\} = \{\varphi_n(y)\} = v_0(y) \qquad (y \in G),$$

that is, $v$ is an extension of $v_0$, which completes the proof of lemma 15.1.

Now we can give

**Corollary 15.3.** *If $E$ is a separable Banach space containing a subspace $G$ isomorphic to $c_0$, then $E$ has a Schauder decomposition. In particular, every subspace $E$ of $c_0$ has a Schauder decomposition.*

*Proof.* The first statement is a consequence of proposition 15.4 and lemma 15.1 (since $c_0$ has a basis). The second statement follows from the first one (since every subspace $E$ of $c_0$ contains**) a subspace isomorphic to $c_0$), which completes the proof. We shall give a sharpening of corollary 15.3 in proposition 15.10 below.

The assumption of separability of $E$ in corollary 15.3 is essential, as shown by theorem 15.2 below.

*Remark 15.2.* a) Proposition 15.4 and corollary 15.2 show that there exist many *non-separable* Banach spaces $E$ which have Schauder decom-

*) See e.g. [87], p. 426, theorem 1.

**) See e.g. [11], p. 194, theorem 1. Note that this also follows from § 1, corollary 1.2 and the perfect homogeneity of the unit vector basis of $c_0$.

positions. b) Since $c_0$ has a subspace $E$ which does not have the approximation property (see §0, theorems 0.1, 0.2), from corollary 15.3 it follows that *there exist Banach spaces with a Schauder decomposition, which do not have the approximation property**).

*Problem 15.1.* Does every separable Banach space $E$ have a Schauder decomposition?

The assumption of separability of $E$ in problem 15.1 is essential, as shown by the following negative result:

**Theorem 15.2.** *The space $E = l^\infty$ has no Schauder decomposition.*

*Proof.* Assume, a contrario, that $E = l^\infty$ has a Schauder decomposition $\{G_n\}$ and let $\{v_n\} \subset L(E, E)$ be the a.s.c.p. to $\{G_n\}$. We shall first show that in this case $\{v_n^*(E^*)\}$ is a Schauder decomposition of $E^* = (l^\infty)^*$, with the a.s.c.p. $\{v_n^*\} \subset L(E^*, E^*)$. Indeed, let $f \in E^*$. Then, by (15.5),

$$f(x) = \sum_{i=1}^{\infty} f(v_i(x)) = \sum_{i=1}^{\infty} v_i^*(f)(x) \qquad (x \in E),$$

and thus $\lim_{n\to\infty} (f - \sum_{i=1}^{n} v_i^*(f))(x) = 0$ for all $x \in E = l^\infty$. Hence, by a result of Grothendieck**),

$$\lim_{n\to\infty} \Phi\left(f - \sum_{i=1}^{n} v_i^*(f)\right) = 0 \qquad (\Phi \in E^{**}),$$

that is, $\sum_{i=1}^{n} v_i^*(f) \xrightarrow{w} f$. Consequently, if $\varepsilon > 0$, there exists***) a finite convex combination $g_\varepsilon = \sum_{n=1}^{k(\varepsilon)} \lambda_n \sum_{i=1}^{n} v_i^*(f) = \sum_{j=1}^{k(\varepsilon)} \mu_j v_j^*(f)$ such that $\|f - g_\varepsilon\| < \dfrac{\varepsilon}{1 + \sup_{1 \leqslant n < \infty} \left\| \sum_{i=1}^{n} v_i \right\|}$. But, if $m \geqslant k(\varepsilon)$, we have $\sum_{i=1}^{m} v_i^*(g_\varepsilon) = g_\varepsilon$ (by

---

*) However, if $E = \sum_{n=1}^{\infty} \oplus G_n$ and if each $G_n$ $(n = 1, 2, \ldots)$ has the approximation property, then $E$ has the approximation property (indeed, the proof is similar to that of the first statement of § 9, proposition 9.5).

**) See Ch. I, § 13, p. 151, first footnote.

***) See e.g. [87], p. 422, corollary 14. We have $\sup_{1 \leqslant n < \infty} \left\| \sum_{i=1}^{n} v_i \right\| < \infty$ by (15.5) and the principle of uniform boundedness.

(15.6)) and hence

$$\left\| f - \sum_{i=1}^{m} v_i^*(f) \right\| \leqslant \|f - g_\varepsilon\| + \left\| \sum_{i=1}^{m} v_i^*(g_\varepsilon) - \sum_{i=1}^{m} v_i^*(f) \right\| \leqslant$$

$$\leqslant \|f - g_\varepsilon\| \left( 1 + \sup_{1 \leqslant n < \infty} \left\| \sum_{i=1}^{n} v_i^* \right\| \right) < \varepsilon$$

which proves that, in the norm topology of $E^*$,

$$f = \sum_{i=1}^{\infty} v_i^*(f) \qquad (f \in E^*). \tag{15.16}$$

To show the uniqueness of these expansions, assume that $f = \sum_{i=1}^{\infty} g_i$, where $g_n \in v_n^*(E^*)$ $(n = 1,2, \ldots)$. Then, by (15.6),

$$v_j^*(g_i) = v_j^* v_i^*(g_i) = \delta_{ji} v_i^*(g_i) = \delta_{ij} g_i \qquad (i,\ j = 1,2, \ldots),$$

whence

$$v_j^*(f) = v_j^*\left( \sum_{i=1}^{\infty} g_i \right) = \sum_{i=1}^{\infty} v_j^*(g_i) = \sum_{i=1}^{\infty} \delta_{ij} g_i = g_j \qquad (j = 1,2, \ldots),$$

which proves the assertion that $\{v_n^*(E^*)\}$ is a Schauder decomposition of $E^*$, with the a.s.c.p. $\{v_n^*\}$.

Now for each $n = 1,2, \ldots$ choose $x_n \in G_n$ with $\|x_n\| = 1$ and then $\varphi_n \in G_n^*$ with $\|\varphi_n\| = 1$, $\varphi_n(x_n) = 1$ and let $f_n \in E^*$ be a Hahn-Banach extension of $\varphi_n$ (i.e., $f_n|_{G_n} = \varphi_n$, $\|f_n\| = \|\varphi_n\| = 1$). Then $v_i(x_j) = \delta_{ij} x_i$ (by the definition of $v_i$), whence

$$v_i^*(f_i)\,(x_j) = f_i(v_i(x_j)) = \delta_{ij} f_i(x_i) = \delta_{ij} \qquad (i, j = 1,2, \ldots). \tag{15.17}$$

Let $V = [v_n^*(f_n)] \subset E^*$. We claim that $V$ is reflexive. Indeed, since $\{v_n^*(E^*)\}$ is a Schauder decomposition of $E^*$, $\{v_n^*(f_n)\}$ is a basis of $V$ (by theorem 15.21 a) below). Hence*), if $\{h_n\}$ is any sequence in $V$ with $\|h_n\| \leqslant 1$ $(n = 1,2, \ldots)$, we can write, in the norm topology,

$$h_n = \sum_{i=1}^{\infty} \alpha_i^{(n)} v_i^*(f_i) \qquad (n = 1,2, \ldots), \tag{15.18}$$

---

*) Alternatively, instead of this argument one can also show that the basis $\{v_n^*(f_n)\}$ is shrinking and boundedly complete, whence $V$ is reflexive.

where the scalars $\alpha_i^{(n)}$ are uniquely determined. Also, by Ch. I, § 3, theorem 3.1, there exists a constant $M$ such that

$$|\alpha_i^{(n)}| \leqslant \frac{M}{\|v_i^*(f_i)\|} \|h_n\| \leqslant \frac{M}{\|v_i^*(f_i)\|} \qquad (i, n = 1,2, \ldots),$$

and hence, by the usual diagonal procedure, we can find an increasing sequence $\{n_j\} \subset \mathcal{N}$ such that the limits

$$\lim_{j \to \infty} \alpha_i^{(n_j)} = \alpha_i \qquad (i = 1,2, \ldots) \tag{15.19}$$

exist.

Let $x \in E$ and $\varepsilon > 0$ and, by (15.5), let $k = k(x, \varepsilon)$ be such that $\left\| \sum_{j=k}^{\infty} v_j(x) \right\| < \varepsilon$. Then, by (15.5), (15.18), $\|h_{n_p}\|$, $\|h_{n_q}\| \leqslant 1$, (15.6) and (15.19), we have

$$|h_{n_p}(x) - h_{n_q}(x)| \leqslant \left| (h_{n_p} - h_{n_q}) \left( \sum_{j=1}^{k-1} v_j(x) \right) \right| + \left| (h_{n_p} - h_{n_q}) \left( \sum_{j=k}^{\infty} v_j(x) \right) \right| \leqslant$$

$$\leqslant \left| \sum_{i=1}^{\infty} (\alpha_i^{(n_p)} - \alpha_i^{(n_q)}) v_i^*(f_i) \left( \sum_{j=1}^{k-1} v_j(x) \right) \right| + 2 \left\| \sum_{j=k}^{\infty} v_j(x) \right\| <$$

$$< \left| \sum_{i=1}^{k-1} (\alpha_i^{(n_p)} - \alpha_i^{(n_q)}) v_i^*(f_i)(x) \right| + 2\varepsilon < 3\varepsilon \quad (p, q > N(x, \varepsilon)),$$

so $\{h_{n_j}(x)\}$ is a Cauchy sequence for each $x \in E$ and hence*) there exists $h \in E^*$ such that $h_{n_j} \xrightarrow{w^*} h$. But then, by the result of Grothendieck used also above, $h_{n_j} \xrightarrow{w} h$, i.e., for $\sigma(E^*, E^{**})$, whence also for $\sigma(V, V^*)$. Thus, the unit ball $S_V$ of $V$ is sequentially weakly compact and hence**) $V$ is reflexive, as was claimed above. Therefore, $V^*$ is also reflexive.

Observe now that by $\|x_n\| = 1$, $v_i(x_n) = 0$ $(i = 1, \ldots, n-1)$ and (15.16) we have

$$|f(x_n)| \leqslant \left| \sum_{i=1}^{n-1} v_i^*(f)(x_n) \right| + \left\| f - \sum_{i=1}^{n-1} v_i^*(f) \right\| \to 0 \text{ as } n \to \infty \quad (f \in E^*),$$

---

*) See e.g. [87], p. 55, theorem 18.

**) See e.g. [87], p. 430, theorem 1 and p. 425, theorem 7.

so $x_n \xrightarrow{w} 0$. But, since $V^*$ is reflexive, the canonical mapping $u$ of $E$ into $V^*$ is weakly compact and hence*) it maps the weakly convergent sequence $\{x_n\} \subset E = l^\infty$ into a sequence $\{u(x_n)\}$ which is convergent in the norm topology, say to $\Psi \in V^*$. Then, since $u(x_n)(f) = f(x_n) \to 0$ $(f \in V)$, i.e. $u(x_n) \xrightarrow{w^*} 0$, we have $\Psi = 0$, so $\lim_{n\to\infty} \|u(x_n)\| = 0$. On the other hand, by (15.17) (for $i = j$), $\|f_n\| = 1$ and (15.13),

$$\|u(x_n)\| = \sup_{\substack{f \in V \\ \|f\| \leqslant 1}} |f(x_n)| \geqslant \frac{|v_n^*(f_n)(x_n)|}{\|v_i^*(f_n)\|} \geqslant \frac{1}{\sup\limits_{1 \leqslant j < \infty} \|v_j\|} > 0 \quad (n = 1,2, \ldots),$$

a contradiction which completes the proof of theorem 15.2.

*Remark 15.3.* a) The above argument shows that *the conclusion of theorem* 15.2 *remains valid, more generally*, for any Banach space $E$ with the following two properties: (i) $E$ is a "Grothendieck space", i.e., in the conjugate space $E^*$ weak* and weak convergence of sequences are equivalent. (ii) $E$ has the "Dunford-Pettis property", i.e. every weakly compact mapping $u$ of $E$ into any Banach space $B$ maps weakly convergent sequences into norm convergent sequences. For example, it is known **) that the spaces $E = C(Q)$, where $Q$ is a compact extremally disconnected space, whence also their complemented subspaces (the $\mathscr{P}_\lambda$ spaces), have properties (i) and (ii).

b) The first part of the above argument also shows that if $E$ is a Banach space with property (i), and if $\{G_n\}$ is a Schauder decomposition of $E$, with the a.s.c.p. $\{v_n\}$, then $\{v_n^*(E^*)\}$ is a Schauder decomposition***) of $E^*$, with the a.s.c.p. $\{v_n^*\}$.

Now we return to the process of carrying over the notions and results related with bases and with their generalizations, to Schauder decompositions. We have already seen that a natural "generalization" of coefficient functionals are the coordinate projections, for which one has suitable generalizations of the results of Ch. I, § 3, culminating in the sufficiency part of theorem 15.1. A natural generalization of the notion of a biorthogonal system could be a pair of sequences $(G_n, \Gamma_n)$ of non-zero subspaces $G_n \subset E$, $\Gamma_n \subset E^*$ $(n = 1,2, \ldots)$ such that $h_i(y_j) = 0$ $(h_i \in \Gamma_i, y_j \in G_j, i \neq j)$. However, it will be more convenient to restrict ourselves to pairs for which there exist projections $v_n \in L(E, E)$ with $v_n(E) = G_n$, $v_n^*(E^*) = \Gamma_n$ $(n = 1,2, \ldots)$ and to replace the second sequence of subspaces $\{\Gamma_n\}$ by the sequence of projections $\{v_n\}$ (as we also did in definition 15.2 above). Thus, we arrive at

*Definition 15.4.* Let $E$ be a Banach space. A pair of sequences $(G_n, v_n)$, where $\{G_n\}$ is a sequence of (closed linear) subspaces of $E$ with $G_n \neq \{0\}$ $(n = 1,2, \ldots)$ and $\{v_n\} \subset L(E, E)$ is a sequence of projec-

*) See e.g. [87], p. 494, theorem 4.
**) See e.g. [70], Ch. VI, § 4, statements 4 (b), (c) and [87], p. 494, theorem 4.
***) In other words, every Schauder decomposition of a Grothendieck space is shrinking (see definition 15.15 and theorem 15.13).

tions with $v_n(E) = G_n$ $(n = 1,2, \ldots)$, will be called a *generalized*[*]) *biorthogonal system*, if it satisfies (15.6). The generalized biorthogonal system $(G_n, v_n)$ is said to be *E-complete*, if[**]) $\left[\bigcup_{n=1}^{\infty} G_n\right] = E$. The *formal expansion* of an element $x \in E$ is $x \sim \sum_{i=1}^{\infty} v_i(x)$.

Clearly, if $\{G_n\}$ is a Schauder decomposition of $E$ with the a.s.c.p. $\{v_n\}$, then $(G_n, v_n)$ is an $E$-complete generalized biorthogonal system, but the converse is not true.

In view of (15.5) (and (15.7)), the natural generalization of the sequence of partial sum operators $\{s_n\}$ is given by

*Definition 15.5.* Let $E$ be a Banach space and $(G_n, v_n)$ $(G_n \subset E, v_n \in L(E, E), n = 1,2, \ldots)$ a generalized biorthogonal system. The sequence $\{u_n\} \subset L(E, E)$ defined by

$$u_n = \sum_{i=1}^{n} v_i \qquad (n = 1,2, \ldots) \tag{15.20}$$

will be called *the sequence of partial sum operators associated to* $(G_n, v_n)$ $\left(\text{or to } \{G_n\}, \text{ if } \left[\bigcup_{n=1}^{\infty} G_n\right] = E\right)$.

From the orthogonality relations (15.6) it is clear that $\{u_n\}$ satisfies the commutativity relations

$$u_n u_m = u_m u_n = u_{\min(n,m)} \qquad (n, m = 1,2, \ldots); \tag{15.21}$$

in particular, for $n = m$ we obtain that each $u_n$ is a projection of $E$ onto $\sum_{i=1}^{n} \oplus G_i$. Conversely, if $\{G_n\}$ is a sequence of linear subspaces of $E$ with $G_n \neq \{0\}$ $(n = 1,2, \ldots)$ and $\{u_n\} \subset L(E, E)$ a sequence of operators with $u_n(E) = \sum_{i=1}^{n} \oplus G_i$ $(n = 1,2, \ldots)$, satisfying (15.21), then the sequence $\{v_n\} \subset L(E, E)$ defined by

$$v_1 = u_1, \; v_n = u_n - u_{n-1} \qquad (n = 2,3, \ldots) \tag{15.22}$$

satisfies $v_n(E) = G_n$ $(n = 1,2, \ldots)$ and (15.6), so $(G_n, v_n)$ is a generalized biorthogonal system (each $G_n$ is closed, since $v_n$ is a continuous linear projection). Note also that $G_n \neq \{0\}$ is equivalent to $v_n \neq 0$ $(n = 1,2, \ldots)$, and to $u_1 \neq 0$, $u_n \neq u_{n-1}$ $(n = 2,3, \ldots)$.

---

*) Here the adjective "generalized" is used in a different sense than in the case of "generalized bases" (§ 7, definition 7.1), but this will cause no confusion.

**) Clearly, in this latter case the sequence $\{v_n\}$ is uniquely determined $\left(\text{since } v_n(E) = G_n \text{ and } \operatorname{Ker} v_n = \left[\bigcup_{i=1}^{n-1} G_i \cup \bigcup_{i=n+1}^{\infty} G_i\right] \text{ if } \left[\bigcup_{n=1}^{\infty} G_n\right] = E\right)$.

We have already observed (see the proof of theorem 15.1 or 15.2) that if $\{G_n\}$ is a Schauder decomposition of $E$, with the a.s.c.p. $\{v_n\}$, then $\sup_{1 \leqslant n < \infty} \|u_n\| = \sup_{1 \leqslant n < \infty} \left\| \sum_{i=1}^{n} v_i \right\| < \infty$. Also, from (15.21) it is clear that if $(G_n, v_n)$ is any generalized biorthogonal system, then

$$x = \lim_{n \to \infty} u_n(x) \qquad \left(x \quad \bigcup_{j=1}^{\infty} G_j\right).$$

Consequently, we have the following generalization of Ch. I, § 4, theorem 4.1 (we write it in a slightly different form, as a theorem on characterization*) of Banach spaces which admit Schauder decompositions):

**Theorem 15.3.** *For a Banach space $E$ the following statements are equivalent:*

1°. *$E$ has a Schauder decomposition $\{G_n\}$.*

2°. *There exists a sequence $\{v_n\} \subset L(E, E)$ with $v_n \neq 0$ $(n = 1, 2, \ldots)$, satisfying* (15.6) *and* (15.5) *for all $x \in E$.*

3°. *There exists a sequence $\{v_n\} \subset L(E, E)$ with $v_n \neq 0$ $(n = 1, 2, \ldots)$, satisfying* (15.6), $\left[\bigcup_{n=1}^{\infty} v_n(E)\right] = E$ *and* $\sup_{1 \leqslant n < \infty} \left\| \sum_{i=1}^{n} v_i \right\| < \infty$.

4°. *There exists a sequence $\{u_n\} \subset L(E, E)$ with $u_1 \neq 0$, $u_n \neq u_{n-1}$ $(n = 2, 3, \ldots)$, satisfying* (15.21) *and $x = \lim_{n \to \infty} u_n(x)$ for all $x \in E$.*

5°. *There exists a sequence $\{u_n\} \subset L(E, E)$ with $u_1 \neq 0$, $u_n \neq u_{n-1}$ $(n = 2, 3, \ldots)$, satisfying* (15.21), $\overline{\bigcup_{n=1}^{\infty} u_n(E)} = E$ *and* $\sup_{1 \leqslant n < \infty} \|u_n\| < \infty$.

*The subspaces $G_n$ and the projections $v_n$ and $u_n$ are related by $v_n(E) = G_n$ $(n = 1, 2, \ldots)$ and* (15.20) (*or* (15.22)).

In some cases, e.g. for the purpose of further extension (see §§ 18, 19), it will be convenient to consider the above sequence of projections $\{u_n\}$ instead of the associated Schauder decomposition $\{G_n\}$. Therefore, let us give

*Definition 15.6.* Let $E$ be a Banach space. A sequence of projections $\{u_n\} \subset L(E, E)$ is called *a resolution of the identity of $E$*, or shortly, *a resolution of $I_E$*, if it satisfies condition 4° (or, equivalently, 5°) of theorem 15.3.

For example, every commuting $\pi$-basis $\{u_n\}$ of a Banach space $E$, with $u_1 \neq 0$, $u_n \neq u_{n-1}$ $(n = 2, 3, \ldots)$, is a resolution of $I_E$.

Let us return to Schauder decompositions. As an application of theorem 15.3 we mention that Ch. I, § 4, propositions 4.1 and 4.2 have obvious generalizations to Schauder decompositions (however, see also § 13, proof of proposition 13.13).

---

*) Therefore, in connection with the equivalences $1° \Leftrightarrow 4° \Leftrightarrow 5°$ see also Ch. I, § 20, theorem 20.1, equivalence $11° \Leftrightarrow 2°$ and Ch. II, § 1, corollary 1.2.

We have the following generalization of Ch. I, § 5, remark 5.4: *Let $\{G_n\}$ be a sequence of subspaces of a Banach space $E$ and let $D_2$ be the linear space of sequences of elements*

$$D_2 = \left\{ \{y_n\} \subset E \;\middle|\; y_n \in G_n \ (n = 1,2, \ldots), \sup_{1 \leqslant n < \infty} \left\| \sum_{i=1}^{n} y_i \right\| < \infty \right\}, \tag{15.23}$$

*endowed with the norm* (15.9). *Then $D_2$ is a Banach space.* Indeed, the proof is similar to that of proposition 15.2 above.

One can extend the results of Ch. I, § 5 to obtain characterizations of Schauder decompositions among *total* generalized biorthogonal systems $(G_n, v_n)$ (i.e., such that the relations $x \in E$, $v_n(x) = 0$ for $n = 1,2, \ldots$ imply $x = 0$) in terms of multipliers, which are defined as follows.

*Definition 15.7.* Let $E$ be a Banach space and $(G_n, v_n)$ $(G_n \subset E$, $v_n \in L(E, E)$, $n = 1,2, \ldots)$ a total generalized biorthogonal system. A sequence of scalars $\{\gamma_n\}$ is called a *multiplier of an element* $x \in E$ if there exists an element $x_{\{\gamma_n\}} \in E$ such that

$$v_k(x_{\{\gamma_n\}}) = \gamma_k v_k(x) \qquad (x \in E,\ k = 1,2, \ldots); \tag{15.24}$$

since $\{v_k\}$ is total on $E$, this $x_{\{\gamma_n\}}$ is uniquely determined. The set of all multipliers $\{\gamma_n\}$ of an element $x \in E$ will be denoted by $M(x, (G_n, v_n))$. Also, we shall denote

$$M(E, (G_n, v_n)) = \bigcap_{x \in E} M(x, (G_n, v_n)). \tag{15.25}$$

**Theorem 15.4.** *Let $E$ be a Banach space and $(G_n, v_n)$ $(G_n \subset E$, $v_n \in L(E, E)$, $n = 1,2, \ldots)$ a total generalized biorthogonal system. The following statements are equivalent:*

1°. *$\{G_n\}$ is a Schauder decomposition of $E$.*

2°. *$M(E, (G_n, v_n)) \supset bv$.*

3°. *$M(E, (G_n, v_n))$ contains every non-increasing sequence $\{\gamma_n\}$ tending to zero.*

The proof is similar to that of Ch. I, § 5, theorem 5.2 and the proof of the implication 1° ⇒ 2° can be simplified similarly to § 11, theorem 11.4, proof of the implication 1° ⇒ 2°. In this connection let us mention that § 11, theorem 11.4 can be also extended to the more general setting of the present section, by using the following generalization of the notion of a Cesàro basis.

*Definition 15.8.* Let $E$ be a Banach space and $(G_n, v_n)$ $(G_n \subset E$, $v_n \in L(E, E)$, $n=1,2,\ldots)$ a generalized biorthogonal system. The sequence

$\{G_n\}$ is said to be a *Cesàro decomposition* of $E$, if for the operators $\sigma_n \in L(E, E)$ defined by

$$\sigma_n = \frac{1}{n} \sum_{i=1}^{n} u_i = \sum_{i=1}^{n} \frac{n-i+1}{n} v_i \qquad (n = 1,2, \ldots) \quad (15.26)$$

we have

$$x = \lim_{n\to\infty} \sigma_n(x) \qquad (x \in E). \qquad (15.27)$$

Clearly, every Schauder decomposition is a Cesàro decomposition, but the converse is not true. Note also that condition (15.27) is equivalent to $\left[\bigcup_{n=1}^{\infty} G_n\right] = E$ and $\sup_{1 \leqslant n < \infty} \|\sigma_n\| < \infty$. With this definition, one has a natural extension of § 11, theorem 11.4 to Cesàro decompositions.

One can also extend the other results of Ch. I, § 5 on multipliers, if one defines, as in Ch. I, § 5, proposition 5.4, a mapping $\{\gamma_n\} \to w_{\{\gamma_n\}}$ of $M(E, (G_n, v_n))$ into $L(E, E)$, by

$$w_{\{\gamma_n\}}(x) = x_{\{\gamma_n\}} \sim \sum_{k=1}^{\infty} \gamma_k v_k(x) \qquad (x \in E). \qquad (15.28)$$

The three types of linear independence of sequences considered in Ch. I, § 6, have the following natural generalizations:

*Definition 15.9.* A sequence of subspaces $\{G_n\}$ of a Banach space $E$ is said to be

a) *finitely linearly independent*, if $G_n \neq \{0\}$ $(n = 1,2, \ldots)$ and for every finite subsequence $\{G_n\}_{n=1}^{k}$ the relations $y_n \in G_n$ $(n = 1, \ldots, k)$, $\sum_{i=1}^{k} y_i = 0$ imply $y_n = 0$ $(n = 1, \ldots, k)$;

b) *ω-linearly independent*, if $G_n \neq \{0\}$ $(n = 1,2, \ldots)$ and the relations $y_n \in G_n$ $(n = 1,2, \ldots)$, $\sum_{i=1}^{\infty} y_i = 0$ imply $y_n = 0$ $(n = 1,2, \ldots)$;

c) *minimal*, if $G_n \neq \{0\}$ $(n = 1,2, \ldots)$ and

$$G_n \cap [\bigcup_{i \neq n} G_i] = \{0\} \qquad (n = 1,2, \ldots). \qquad (15.29)$$

Obviously, $\{G_n\}$ satisfies a) if and only if $G_n \neq \{0\}$ $(n = 1,2, \ldots)$ and for every finite subsequence $\{G_n\}_{n=1}^{k} \subset \{G_n\}$ we have $G_n \cap \left[\bigcup_{\substack{i=1 \\ i \neq n}}^{k} G_i\right] = \{0\}$ $(n = 1, \ldots, k)$. It is also obvious that every minimal sequence $\{G_n\}$ is $\omega$-linearly independent and every $\omega$-linearly independent sequence $\{G_n\}$ is finitely linearly independent. However, while if there exists

$\{v_n\} \subset L(E, E)$ such that $(G_n, v_n)$ is a generalized biorthogonal system, then $\{G_n\}$ is minimal, the converse is no longer true, since $G_n + [\bigcup_{i \neq n} G_i]$ need not be closed when $\{G_n\}$ is minimal. Thus, some parts of Ch. I, § 6. theorem 6.1 extend only to characterizations of sequences $\{G_n\}$ which admit $\{v_n\} \subset L(E, E)$ such that $(G_n, v_n)$ is an $E$-complete generalized biorthogonal system (rather than to characterizations of minimal sequences $\{G_n\}$), e.g. we get the following necessary and sufficient condition: $E = \left[\bigcup_{i=1}^{n} G_i\right] \oplus \left[\bigcup_{i=n+1}^{\infty} G_i\right]$ $(n = 1,2, \ldots)$, or, what is equivalent*): $\left[\bigcup_{n=1}^{\infty} G_n\right] = E$ and for each $n \in \mathcal{N}$ there exists a constant $C_n$ with $1 \leqslant C_n < \infty$, such that $\left\| \sum_{i=1}^{n} y_i \right\| \leqslant C_n \left\| \sum_{i=1}^{n+m} y_i \right\|$ for all positive integers $m$ and all $y_i \in G_i$ $(i = 1, \ldots, n+m)$.

We have the following intrinsic characterizations of Schauder decompositions, which generalize part of Ch. I, § 7, theorem 7.1 (the other parts can be also easily generalized):

**Theorem 15.5.** *Let $E$ be a Banach space and $\{G_n\}$ a sequence of subspaces of $E$, with $G_n \neq \{0\}$ $(n = 1,2, \ldots)$, $\left[\bigcup_{i=1}^{\infty} G_i\right] = E$. The following statements are equivalent:*

1°. *$\{G_n\}$ is a Schauder decomposition of $E$.*

2°. *There exists a constant $C$ with $1 \leqslant C < \infty$, such that*

$$\left\| \sum_{i=1}^{n} y_i \right\| \leqslant C \left\| \sum_{i=1}^{n+m} y_i \right\| \tag{15.30}$$

*for all positive integers $n$, $m$ and all $y_i \in G_i$ $(i = 1, \ldots, n+m)$.*

3°. *We have*

$$C' = \inf_{1 \leqslant n < \infty} \operatorname{dist}\left(\sigma_{(n)}, \left[\bigcup_{i=n+1}^{\infty} G_i\right]\right) = \inf_{1 \leqslant n < \infty} \widehat{\left(\left[\bigcup_{i=1}^{n} G_i\right]; \left[\bigcup_{i=n+1}^{\infty} G_i\right]\right)} > 0, \tag{15.31}$$

*where* $\sigma_{(n)} = \left\{ x \in \left[\bigcup_{i=1}^{n} G_i\right] \,\middle|\, \|x\| = 1 \right\}$ $(n = 1,2, \ldots)$.

---

*) Other equivalent conditions can be obtained by applying any necessary and sufficient condition in order that $E = Y \oplus Z$, to $Y = \left[\bigcup_{i=1}^{n} G_i\right]$, $Z = \left[\bigcup_{i=n+1}^{\infty} G_i\right]$ $(n = 1, 2, \ldots)$ or to $Y = G_n$, $Z = [\bigcup_{i \neq n} G_i]$ $(n = 1, 2, \ldots)$.

*Proof.* 1° ⇒ 2°. Assume that we have 1° and let $\{u_n\} \subset L(E, E)$ be the associated sequence of partial sum operators (defined by (15.20)). Then, by 1°, $\sup_{1 \leqslant j < \infty} \|u_j\| < \infty$, and hence for any $y_i \in G_i$ $(i = 1, \ldots \ldots, n+m)$ we have

$$\left\| \sum_{i=1}^{n} y_i \right\| = \left\| u_n \left( \sum_{i=1}^{n+m} y_i \right) \right\| \leqslant \sup_{1 \leqslant j < \infty} \|u_j\| \left\| \sum_{i=1}^{n+m} y_i \right\|.$$

2° ⇒ 1°. Assume that we have 2°. Then $\{G_n\}$ is finitely linearly independent, since if $y_n \in G_n$ $(n = 1, \ldots, k)$, $\sum_{i=1}^{k} y_i = 0$, then from (15.30) with $n + m = k$ we obtain, successively (for $n = 1, \ldots, k-1$ and $m = k - n$), $y_1 = 0$, $y_2 = 0, \ldots, y_{k-1} = 0$, whence also $y_k = \sum_{i=1}^{k} y_i - \sum_{i=1}^{k-1} y_i = 0$. Thus, we can define for each $n \in \mathcal{N}$ a linear operator $u_n$ on $\operatorname{lin}\left(\bigcup_{i=1}^{\infty} G_i\right)$, the linear span of $\bigcup_{i=1}^{\infty} G_i$, by

$$u_n\left(\sum_{i=1}^{k} y_i\right) = \begin{cases} 0 & \text{if } k < n \\ \sum_{i=1}^{n} y_i & \text{if } k \geqslant n \end{cases} \qquad (y_i \in G_i;\ i = 1, \ldots, k). \tag{15.32}$$

Then, again by (15.30), $\|u_n\| \leqslant C$ $(n = 1, 2, \ldots)$, so each $u_n$ can be extended to the whole space $E$, conserving $\sup_{1 \leqslant n < \infty} \|u_n\| \leqslant C$. Furthermore, by $G_n \neq \{0\}$ $(n = 1, 2, \ldots)$ we have $u_1 \neq 0$ and $u_n \neq u_{n-1}$ $(n = 2, 3, \ldots)$. Since by (15.32) and $\left[\bigcup_{i=1}^{\infty} G_i\right] = E$ we have (15.21) and $\bigcup_{n=1}^{\infty} u_n(E) = E$, from theorem 15.3, implication 5° ⇒ 1° it follows that $\{G_n\}$ is a Schauder decomposition of $E$.

Finally, the equivalence 2° ⇔ 3° is obvious, which completes the proof of theorem 15.5.

**Corollary 15.4.** *Let $\{G_n\}$ be a sequence of subspaces of a Banach space $E$ and let*

$$\sum_{i=1}^{\infty} G_i = \left\{ x \in E \;\middle|\; x = \sum_{i=1}^{\infty} y_i,\ y_n \in G_n\ (n = 1, 2, \ldots) \right\}. \tag{15.33}$$

*In order that $\sum_{i=1}^{\infty} G_i$ be closed it is sufficient and, if $\{G_n\}$ is $\omega$-linearly independent, then also necessary, that we have 2° of theorem 15.5.*

*Proof.* Assume that 2° of theorem 15.5 holds; we may assume (by omitting those $G_n$ which are $=\{0\}$) that $G_n \neq \{0\}$ $(n = 1,2, \ldots)$. Then, by theorem 15.5, $\{G_n\}$ is a Schauder decomposition of $\overline{\sum_{i=1}^{\infty} G_i}$. Hence, for any $x \in \overline{\sum_{i=1}^{\infty} G_i}$ we can find a (unique) sequence $\{y_n\}$ with $y_n \in G_n$ $(n = 1,2, \ldots)$ such that $x = \sum_{i=1}^{\infty} y_i \in \sum_{i=1}^{\infty} G_i$. Thus, $\sum_{i=1}^{\infty} G_i$ is closed.

Conversely, assume now that $\sum_{i=1}^{\infty} G_i$ is closed and that $\{G_n\}$ is $\omega$-linearly independent. Then, by (15.33), for each $x \in \overline{\sum_{i=1}^{\infty} G_i} = \sum_{i=1}^{\infty} G_i$ there exists a sequence $\{y_n\}$ with $y_n \in G_n$ $(n = 1,2, \ldots)$ such that $x = \sum_{i=1}^{\infty} y_i$ and, by the $\omega$-linearly independence of $\{G_n\}$, this sequence $\{y_n\}$ is unique. Thus, $\{G_n\}$ is a Schauder decomposition of $\overline{\sum_{i=1}^{\infty} G_i}$ and hence, by theorem 15.5, we have 2° of that theorem, which completes the proof of corollary 15.4.

In this connection, let us also mention

**Proposition 15.5.** *Let $\{G_n\}$ be a sequence of subspaces of a Banach space $E$, let $D_1$ be the Banach space defined in proposition* 15.2 *and let $D_0$ be the subspace of $D_1$ defined by*

$$D_0 = \left\{ \{y_n\} \subset E \,\middle|\, y_n \in G_n \ (n = 1,2, \ldots), \sum_{i=1}^{\infty} y_i = 0 \right\}. \tag{15.34}$$

*Furthermore, define a mapping $u\colon D_1/D_0 \to E$ by*

$$u(\{y_n\} + D_0) = \sum_{i=1}^{\infty} y_i \qquad (\{y_n\} + D_0 \in D_1/D_0). \tag{15.35}$$

*Then $u$ is a one-to-one continuous linear mapping of $D_1/D_0$ onto $\sum_{i=1}^{\infty} G_i$ (defined by* (15.33)*). Consequently, $\sum_{i=1}^{\infty} G_i$ is closed if and only if $u^{-1}$ is continuous.*

*Proof.* Clearly, $D_0$ is a closed linear subspace of $D_1$, since it is the kernel of the continuous linear mapping $w: D_1 \to E$ defined by $w(\{y_n\}) = \sum_{i=1}^{\infty} y_i$ $(\{y_n\} \in D_1)$; this latter is continuous since

$$\|w(\{y_n\})\| = \left\| \sum_{i=1}^{\infty} y_i \right\| \leqslant \sup_{1 \leqslant n < \infty} \left\| \sum_{i=1}^{n} y_i \right\| = \|\{y_n\}\| \text{ for all } \{y_n\} \in D_1.$$

Now, $u$ is linear and if $u(\{y_n\} + D_0) = \sum_{i=1}^{\infty} y_i = 0$, then (by (15.34)) $\{y_n\} \in D_0$, so $u$ is one-to-one. Furthermore, $u(D_1/D_0) = \sum_{i=1}^{\infty} G_i$ and

$$\|u(\{y_n\} + D_0)\| = \left\| \sum_{i=1}^{\infty} y_i \right\| = \left\| \sum_{i=1}^{\infty} (y_i + y_i^{(0)}) \right\| \leqslant$$

$$\leqslant \sup_{1 \leqslant n < \infty} \left\| \sum_{i=1}^{n} (y_i + y_i^{(0)}) \right\| = \|\{y_n\} + \{y_n^{(0)}\}\| \qquad (\{y_n\} \in D_1,\ \{y_n^{(0)}\} \in D_0),$$

whence

$$\|u(\{y_n\} + D_0)\| \leqslant \inf_{\{y_n^{(0)}\} \in D_0} \|\{y_n\} + \{y_n^{(0)}\}\| = \|\{y_n\} + D_0\|$$

$$(\{y_n\} + D_0 \in D_1/D_0),$$

so $u$ is continuous. Consequently, if $u^{-1}$ is continuous, then $u$ is an isomorphism and hence, since $D_1/D_0$ is complete, $\sum_{i=1}^{\infty} G_i = u(D_1/D_0)$ is closed. Conversely, if $\sum_{i=1}^{\infty} G_i = u(D_1/D_0)$ is closed, then, by the inversion theorem of Banach*), $u^{-1}$ is continuous. This completes the proof of proposition 15.5.

*Remark 15.4.* a) One can also express the last statement of proposition 15.5 in the following form: $\sum_{i=1}^{\infty} G_i$ is closed if and only if there exists a constant $C''$ such that

$$\inf_{\substack{\sum_{i=1}^{\infty} y_i = x \\ y_i \in G_i}} \sup_{1 \leqslant n < \infty} \left\| \sum_{i=1}^{n} y_i \right\| \leqslant C'' \|x\| \qquad \left(x \in \sum_{i=1}^{\infty} G_i\right); \quad (15.30')$$

*) See e.g. [11], p. 41, theorem 5.

indeed, the left side of (15.30′) is nothing else than $\|u^{-1}(x)\|$. b) Let us also show how proposition 15.5 implies again corollary 15.4. If we have 2° of theorem 15.5 and $x=\sum_{i=1}^{\infty} y_i \in \sum_{i=1}^{\infty} G_i$, $y_n \in G_n$ $(n=1,2,\ldots)$, then from (15.30) for $m\to\infty$ we obtain $\left\|\sum_{i=1}^{n} y_i\right\| \leqslant C\|x\|$ $(n=1,2,\ldots)$, whence (15.30′) with any $C''\geqslant C$. Conversely, if we have (15.30′) and if $\{G_n\}$ is $\omega$-linearly independent, then for any $y_i \in G_i$ $(i=1,\ldots$ $\ldots, n+m)$ we have, putting $x=\sum_{i=1}^{n+m} y_i$ and observing that by*) the $\omega$-linearly independence of $\{G_n\}$ this decomposition of $x$ is unique, $\sup_{1\leqslant n\leqslant m}\left\|\sum_{i=1}^{n} y_i\right\| \leqslant C''\left\|\sum_{i=1}^{n+m} y_i\right\|$, i.e. (15.30) with any $C\geqslant C''$, which completes the proof. c) In the particular case when $\{G_n\}$ is a Schauder decomposition of $E$, we have $D_0=\{0\}$ (hence $D_1/D_0=D_1$) and $\sum_{i=1}^{\infty} G_i = E$ is closed, whence, by the necessity part of proposition 15.5, we get again proposition 15.3 a).

Corresponding to block basic sequences with respect to bases (see Ch. I, § 7, definition 7.3), let us give now

*Definition 15.10.* Let $\{G_n\}$ be a Schauder decomposition of a Banach space $E$. A sequence of subspaces $\{F_n\}$ of $E$ is said to be a *block Schauder decomposition of E (with respect to* $\{G_n\}$), or shortly, a *blocking of* $\{G_n\}$, if it is of the form

$$F_n=\left[\bigcup_{i=m_{n-1}+1}^{m_n} G_i\right]=\sum_{i=m_{n-1}+1}^{m_n}\oplus G_i \qquad (n=1,2,\ldots), \quad (15.36)$$

where $\{m_n\}$ is an increasing sequence of positive integers and $m_0=0$.

Clearly, if $\{G_n\}$ is a Schauder decomposition of $E$, then every sequence $\{F_n\}$ of the form (15.36) is a Schauder decomposition *of the whole space E*. In the particular case when $\{G_n\}$ is the one-dimensional decomposition associated to a basis $\{x_n\}$ of $E$, the block basic sequences $\{z_n\}$ with respect to $\{x_n\}$ can be obtained by picking one non-zero element $z_n$ from each $F_n$, where $\{F_n\}$ is any blocking (15.36) with respect to $\{G_n\}$; in this case, Ch. I, § 7, theorem 7.2 amounts to the fact that the subspaces $F_n$ have bases containing $z_n$ and with uniformly bounded norms (which yield a basis of $E$, by § 13, theorem 13.4).

---

*) It is not enough to assume here that $\{G_n\}$ is finitely linearly independent, since then it may happen that $x=\sum_{i=1}^{n+m} y_i = \sum_{i=n+m+1}^{\infty} y_i$ with $y_i \in G_i$ $(i=1,2,\ldots)$ and that the inf in (15.30′) is attained for the latter $y_i$'s, but not for $y_1,\ldots,y_{n+m}$.

One can give the following generalization of part of Ch. I, § 8, definition 8.1, to sequences of subspaces:

*Definition 15.11.* Let $E$, $F$ be two Banach spaces. A sequence of subspaces $\{G_n\}$ of $E$ is said to *dominate strictly* a sequence of subspaces $\{F_n\}$ of $F$, and we write $\{G_n\} \gg \{F_n\}$, provided that there exists a continuous linear mapping $u \in L\left(\left[\bigcup_{n=1}^{\infty} G_n\right], \left[\bigcup_{n=1}^{\infty} F_n\right]\right)$ such that

$$u(G_n) = F_n \qquad (n = 1,2, \ldots). \tag{15.37}$$

The sequences $\{G_n\}$ and $\{F_n\}$ are said to be *strictly equivalent* (respectively, *fully equivalent*), and we write $\{G_n\} \approx \{F_n\}$ (respectively, $\{G_n\} \approxeq \{F_n\}$), if there exists an isomorphism $u$ of $\left[\bigcup_{n=1}^{\infty} G_n\right]$ onto $\left[\bigcup_{n=1}^{\infty} F_n\right]$ (respectively, of $E$ onto $F$) satisfying (15.37).

In contrast with the particular case of sequences of elements, the relations $\{G_n\} \gg \{F_n\} \gg \{G_n\}$ do not imply $\{G_n\} \approx \{F_n\}$, as shown e.g. by $E = (l^1 \times l^1 \times \ldots)_{l^1}$, $F = (L^1([0,1]) \times L^1([0,1]) \times \ldots)_{l^1}$ and $G_n = l^1$, $F_n = L^1([0,1])$ for $n = 1,2, \ldots$ (embedded in the natural way in $E$ and $F$ respectively).

Let us consider now, in greater detail, theorems of (strong and weak) duality for Schauder decompositions. Concerning theorems of strong duality, we have the following generalization of Ch. I, § 12, theorem 12.1:

**Theorem 15.6.** *Let $\{G_n\}$ be a Schauder decomposition of a Banach space $E$ and let $\{v_n\} \subset L(E, E)$ be the a.s.c.p. Then $\{v_n^*(E^*)\}$ is a Schauder decomposition of $\left[\bigcup_{n=1}^{\infty} v_n^*(E^*)\right]$ and we have, in the norm topology of $E^*$,*

$$f = \sum_{i=1}^{\infty} v_i^*(f) \qquad \left(f \in \left[\bigcup_{n=1}^{\infty} v_n^*(E^*)\right]\right). \tag{15.38}$$

*Proof.* By (15.6) we have $\sum_{i=1}^{n} v_i^*(v_j^*(f)) = v_j^*(f)$ for all $f \in E^*$ and $n \geqslant j$, whence $\lim_{n\to\infty} \sum_{i=1}^{n} v_i^*(v_j^*(f)) = v_j^*(f)$ $(f \in E^*)$. Hence, since $\sup_{1 \leqslant n < \infty} \left\| \sum_{i=1}^{n} v_i^* \right\| < \infty$, we obtain (15.38), which, by theorem 15.3, completes the proof. This argument also gives a simpler form to Ch. I, § 12, proof of theorem 12.1.

We have seen in remark 15.3 b) that if in the conjugate space $E^*$ weak* and weak convergence of sequences are equivalent, then in theorem 15.6 above we can replace $\left[\bigcup_{n=1}^{\infty} v_n^*(E^*)\right]$ by $E^*$. Hence, since reflexive spaces do have this property, one obtains, in particular, a generalization of Ch. I, § 12, corollary 12.2.

In § 14 we have seen that if $E^*$ has a basis, then so does $E$. The situation for Schauder decompositions is different, as shown by

*Example 15.1.* Let $E = l^\infty \equiv c_0^{**}$. Then, since $c_0$ has a shrinking basis, from corollary 15.2 it follows that $E^* \equiv c_0^{***}$ has a Schauder decomposition. However, by theorem 15.2, $E$ has no Schauder decomposition. Moreover, one can also show that for each $n = 1,2, \ldots$ the conjugate space $E^{(2n+1)} = (E^{(2n)})^*$ has a Schauder decomposition, but $E^{(2n)} = (E^{(2n-1)})^*$ has no Schauder decomposition.

We have the following generalization of Ch. I, § 12, theorem 12.2 a):

**Theorem 15.7.** *Let $\{G_n\}$ be a Schauder decomposition of a Banach space $E$, with the a.s.c.p. $\{v_n\} \subset L(E, E)$. Then*

$$r\left(\left[\bigcup_{n=1}^{\infty} v_n^*(E^*)\right]\right) \geqslant \frac{1}{\sup_{1 \leqslant n < \infty} \left\|\sum_{i=1}^{n} v_i\right\|} > 0. \tag{15.39}$$

*Proof.* By (15.5) we have

$$\sum_{i=1}^{n} v_i^*(f) = \left(\sum_{i=1}^{n} v_i\right)^*(f) \overset{w^*}{\to} f \qquad (f \in E^*).$$

But, clearly,

$$\sum_{i=1}^{n} v_i^*(f) \in \left[\bigcup_{m=1}^{\infty} v_m^*(E^*)\right] \qquad (f \in E^*,\ n = 1,2, \ldots),$$

$$\left\|\sum_{i=1}^{n} v_i^*(f)\right\| \leqslant 1 \qquad \left(f \in E^*,\ \|f\| \leqslant \frac{1}{\sup_{1 \leqslant m < \infty} \left\|\sum_{i=1}^{m} v_i\right\|},\ n = 1,2, \ldots\right).$$

Thus, the unit ball of $\left[\bigcup_{m=1}^{\infty} v_m^*(E^*)\right]$ is $\sigma(E^*, E)$-dense in the $\dfrac{1}{\sup_{1 \leqslant m < \infty} \left\|\sum_{i=1}^{m} v_i\right\|}$-ball of $E^*$, which, by the definition of $r\left(\left[\bigcup_{m=1}^{\infty} v_m^*(E^*)\right]\right)$, proves (15.39). Note that, in the particular case of bases, this argument yields a simpler proof for Ch. I, § 12, theorem 12.2 a).

Now we shall prove the following generalization of Ch. I, § 12, proposition 12.2:

**Proposition 15.6.** *Let* $\{G_n\}$ *be a Schauder decomposition of a Banach space* $E$, *with the a.s.c.p.* $\{v_n\} \subset L(E, E)$, *and let* $v = \sup_{1 \leqslant n < \infty} \left\| \sum_{i=1}^{n} v_i \right\|$ $(< \infty)$. *Then*

a) *We have*

$$\|f\| \leqslant \sup_{1 \leqslant n < \infty} \left\| \sum_{i=1}^{n} v_i^*(f) \right\| \leqslant v\|f\| \qquad (f \in E^*). \tag{15.40}$$

*Conversely, if for a sequence* $\{f_n\} \subset E^*$ *with* $f_n \in v_n^*(E^*)$ $(n = 1,2, \ldots)$ *we have* $\sup_{1 \leqslant n < \infty} \left\| \sum_{i=1}^{n} f_i \right\| < \infty$, *then there exists an* $f \in E^*$ *such that*

$$v_n^*(f) = f_n \qquad (n = 1,2, \ldots). \tag{15.41}$$

b) *We have*

$$\|\Psi\| \leqslant \sup_{1 \leqslant n < \infty} \left\| \sum_{i=1}^{n} (v_i^*|_V)^*(\Psi) \right\| \leqslant v\|\Psi\| \qquad (\Psi \in V^*), \tag{15.42}$$

*where* $V = \left[ \bigcup_{n=1}^{\infty} v_n^*(E^*) \right]$. *Conversely, if for a sequence* $\{\Psi_n\} \subset V^*$ *with* $\Psi_n \in (v_n^*|_V)^*(V^*)$ $(n = 1,2, \ldots)$ *we have* $\sup_{1 \leqslant n < \infty} \left\| \sum_{i=1}^{n} \Psi_i \right\| < \infty$, *then there exists a* $\Psi \in V^*$ *such that*

$$(v_n^*|_V)^*(\Psi) = \Psi_n \qquad (n = 1,2, \ldots). \tag{15.43}$$

c) *We have*

$$\sup_{1 \leqslant n < \infty} \left\| \sum_{i=1}^{n} v_i^{**}(\Phi) \right\| \leqslant v\|\Phi\| \qquad (\Phi \in E^{**}). \tag{15.44}$$

*Conversely, if for a sequence* $\{\Phi_n\} \in E^{**}$ *with* $\Phi_n \in v_n^{**}(E^{**})$ $(n = 1,2, \ldots)$ *we have* $\sup_{1 \leqslant n < \infty} \left\| \sum_{i=1}^{n} \Phi_i \right\| < \infty$, *then there exists a* $\Phi \in E^{**}$ *such that*

$$v_n^{**}(\Phi) = \Phi_n \qquad (n = 1,2, \ldots). \tag{15.45}$$

*Proof.* a) By (15.5) we have

$$f(x)=\sum_{i=1}^{\infty} f(v_i(x)) \qquad (x\in E, f\in E^*), \tag{15.46}$$

so

$$f=w^*\text{-}\lim_{n\to\infty}\sum_{i=1}^{n} v_i^*(f) \qquad (f\in E^*), \tag{15.47}$$

whence we obtain $\|f\|\leqslant \varliminf_{n\to\infty}\left\|\sum_{i=1}^{n} v_i^*(f)\right\|\leqslant \sup_{1\leqslant n<\infty}\left\|\sum_{i=1}^{n} v_i^*(f)\right\|$ for all $f\in E^*$, and the other inequality of (15.40) is obvious.

Conversely, assume now that $\sup_{1\leqslant n<\infty}\left\|\sum_{i=1}^{n} f_i\right\|<\infty$, where $f_n\in v_n^*(E^*)$ $(n=1,2,\ldots)$. Then $\left\{\sum_{i=1}^{n} f_i\right\}_{n\in\mathcal{N}}$ has a $w^*$-cluster point $f$, and hence, by (15.6), $\left\{v_k^*\left(\sum_{i=1}^{n} f_i\right)\right\}_{n\in\mathcal{N}}=\{\underbrace{0,\ldots,0}_{k-1},f_k,f_k,\ldots\}$ has the $w^*$-cluster point $v_k^*(f)$ $(k=1,2,\ldots)$, so we have (15.41).

b) and c). The first inequality in (15.42) follows from part a) applied to the Schauder decomposition $\{v_n^*|_V(V)\}=\{v_n^*(E^*)\}$ of $V=\left[\bigcup_{n=1}^{\infty} v_n^*(E^*)\right]$, since it has the a.s.c.p. $\{v_n^*|_V\}$. The second inequality in (15.42) and the inequality (15.44) are obvious, since $\left\|\sum_{i=1}^{n} v_i^*|_V\right\|\leqslant\left\|\sum_{i=1}^{n} v_i^*\right\|=\left\|\sum_{i=1}^{n} v_i\right\|$, $\left\|\sum_{i=1}^{n} v_i^{**}\right\|=\left\|\sum_{i=1}^{n} v_i\right\|$.

Conversely, assume now that $\sup_{1\leqslant n<\infty}\left\|\sum_{i=1}^{n}\Psi_i\right\|<\infty$, where $\Psi_n\in(v_n^*|_V)^*(V^*)$ $(n=1,2,\ldots)$. Then, by part a) applied to the Schauder decomposition $\{v_n^*|_V(V)\}$ of $V$, there exists a $\Psi\in V^*$ such that we have (15.43). Finally, if $\sup_{1\leqslant n<\infty}\left\|\sum_{i=1}^{n}\Phi_i\right\|<\infty$, where $\Phi_n\in v_n^{**}(E^{**})$ $(n=1,2,\ldots)$, then, taking a $w^*$-cluster point $\Phi$ of $\left\{\sum_{i=1}^{n}\Phi_i\right\}_{n\in\mathcal{N}}$ and using (15.6), we obtain, as in the proof of part a), that $\Phi$ satisfies (15.45), which completes the proof of proposition 15.6.

**Proposition 15.7.** *Let $E$ be a Banach space and $(G_n, v_n)$ $(G_n\subset E$, $v_n\in L(E,E)$, $n=1,2,\ldots)$ a generalized biorthogonal system. Then*

a) *$v_n^*(E^*)$ is isomorphic to $G_n^*$ ($n = 1,2,\dots$), by the mapping*

$$v_n^*(f) \to f|_{G_n} = v_n^*(f)|_{G_n} \qquad (f \in E^*). \tag{15.48}$$

b) *We have the isomorphisms $(v_n^*|_V)^*(V^*) \cong (v_n^*(E^*))^* \cong G_n^{**}$ and $v_n^{**}(E^{**}) \cong (v_n^*(E^*))^* \cong G_n^{**}$ ($n = 1,2,\dots$), by the mappings*

$$(v_n^*|_V)^*(\Psi) \to \Psi|_{v_n^*(E^*)} \to \Xi_n \qquad (\Psi \in V^*), \tag{15.49}$$

$$v_n^{**}(\Phi) \to \Phi|_{v_n^*(E^*)} \to \Xi_n \qquad (\Phi \in E^{**}), \tag{15.50}$$

*where $V = \left[\bigcup_{i=1}^{\infty} v_i^*(E^*)\right]$ and where $\Xi_n \in G^{**}$ is defined by*

$$\Xi_n(f|_{G_n}) = \Psi(v_n^*(f)) = \Phi(v_n^*(f)) \qquad (f \in E^*). \tag{15.51}$$

c) *If, in addition, each subspace $G_n = v_n(E)$ ($n = 1,2,\dots$) is reflexive, then for each $n = 1,2,\dots$*

$$(v_n^*|_V)^*(V^*) = u_E(G_n), \tag{15.52}$$

$$v_n^{**}(E^{**}) = \pi_E(G_n), \tag{15.53}$$

*where $u_E$ and $\pi_E$ are the canonical mappings of $E$ into $V^*$ and $E^{**}$ respectively. Hence, if also $(G_n, v_n)$ is $E$-complete and $r(V) > 0$, then*

$$u_E(E) = \left[\bigcup_{n=1}^{\infty} (v_n^*|_V)^*(V^*)\right], \tag{15.54}$$

$$\pi_E(E) = \left[\bigcup_{n=1}^{\infty} v_n^{**}(E^{**})\right]. \tag{15.55}$$

*Proof.* By § 14, lemma 14.3, we have a), with

$$\|f|_{G_n}\| \leqslant \|v_n^*(f)\| \leqslant \|v_n\| \, \|f|_{G_n}\| \qquad (f \in E^*). \tag{15.56}$$

Clearly, b) is a consequence of a) applied to the generalized biorthogonal systems $(v_n^*(E^*), v_n^*|_V)$ and $(v_n^*(E^*), v_n^*)$ respectively. Note that $\Xi_n \in G_n^{**}$, since by (15.56)

$$|\Xi_n(f|_{G_n})| = |\Psi(v_n^*(f))| \leqslant \|\Psi\| \, \|v_n^*(f)\| \leqslant \|\Psi\| \, \|v_n\| \, \|f|_{G_n}\| \qquad (f \in E^*)$$

and, similarly, $|\Xi_n(f|_{G_n})| \leqslant \|\Phi\| \, \|v_n\| \, \|f|_{G_n}\|$ $(f \in E^*)$ and since, by the Hahn-Banach theorem, $\{f|_{G_n} | f \in E^*\} = G_n^*$.

The inclusions

$$u_E(G_n) = u_E v_n(E) = (v_n^*|_V)^* u_E(E) \subset (v_n^*|_V)^*(V^*),$$

$$\pi_E(G_n) = \pi_E v_n(E) = v_n^{**} \pi_E(E) \subset v_n^{**}(E^{**})$$

are obvious. Conversely, assume now that each $G_n = v_n(E)$ $(n = 1,2,\ldots)$ is reflexive and let $\Psi \in (v_n^*|_V)^*(V^*)$. Then, for $\Xi_n$ as in (15.49), (15.51) and for $y_n = \pi_{G_n}^{-1}(\Xi_n) \in G_n$ (where $\pi_{G_n}$ is the canonical mapping of $G_n$ *onto* $G_n^{**}$, since $G_n$ is reflexive), we have

$$\Psi(f) = (v_n^*|_V)^*(\Psi)(f) = \Psi(v_n^*(f)) = \Xi_n(f|_{G_n}) = \\ = f|_{G_n}(y_n) = f(y_n) \qquad (f \in V),$$

and thus $\Psi = u_E(y_n) \in u_E(G_n)$. The proof of the inclusion $v_n^{**}(E^{**}) \subset \subset \pi_E(G_n)$ is similar, which proves (15.52), (15.53).

Consequently, if also $(G_n, v_n)$ is $E$-complete and $r(V) > 0$, then, since $u_E$ is an isomorphism (by $r(V) > 0$) and since $\pi_E$ is a linear isometry, we have (15.54), (15.55) (thus, for (15.55) the assumption $r(V) > 0$ is not needed), which completes the proof of proposition 15.7.

Applying propositions 15.6 and 15.7 we obtain the following generalization of Ch. I, § 12, theorem 12.5:

**Theorem 15.8.** *Let $\{G_n\}$ be a Schauder decomposition of a Banach space $E$, with the a.s.c.p. $\{v_n\} \subset L(E, E)$, let $V = \left[\bigcup_{n=1}^{\infty} v_n^*(E^*)\right]$ and let $u$, $\pi$ and $\pi_{G_n}$ be the canonical mappings of $E$ into $V^*$ and $E^{**}$ and of $G_n$ into $G_n^{**}$ respectively. Then*

a) *The conjugate space $E^*$ is isomorphic, by the mapping*

$$\eta : f \to \{v_n^*(f)\} \qquad (f \in E^*), \tag{15.57}$$

*to the Banach space of sequences of functionals*

$$D_4 = \left\{ \{f_n\} \subset E^* \middle| f_n \in v_n^*(E^*) \ (n = 1,2,\ldots), \ \sup_{1 \leqslant n < \infty} \left\| \sum_{i=1}^{n} f_i \right\| < \infty \right\}, \tag{15.58}$$

*in which the norm is defined by*

$$\|\{f_n\}\| = \sup_{1 \leqslant n < \infty} \left\| \sum_{i=1}^{n} f_i \right\|. \tag{15.59}$$

b) *The restriction* $\eta|_V$ *of the isomorphism* (15.57) *to* $V$ *maps* $V$ *onto the Banach space*

$$D_3 = \left\{\{f_n\} \subset E^* \,\middle|\, f_n \in v_n^*(E^*) \ (n = 1,2,\dots),\ \sum_{i=1}^{\infty} f_i \ \textit{converges}\right\}, \tag{15.60}$$

*in which the norm is defined by* (15.59).

c) *The space* $V^*(\equiv E^{**}/V^{\perp})$ *is isomorphic, by the mapping*

$$\tau\colon \Psi \to (v_n^*|_V)^*(\Psi) \qquad (\Psi \in V^*) \tag{15.61}$$

*to the Banach space of sequences of functionals*

$$D_6 = \left\{\{\Psi_n\} \subset V^* \,\middle|\, \Psi_n \in (v_n^*|_V)^*(V^*) \ (n = 1,2,\dots),\ \sup_{1 \leqslant n < \infty} \left\| \sum_{i=1}^{n} \Psi_i \right\| < \infty\right\}, \tag{15.62}$$

*in which the norm is defined by*

$$\|\{\Psi_n\}\| = \sup_{1 \leqslant n < \infty} \left\| \sum_{i=1}^{n} \Psi_i \right\|. \tag{15.63}$$

d) *The restriction* $\tau|_{u(E)}$ *of the isomorphism* (15.61) *to* $u(E)$ *maps* $u(E)$ *onto the Banach space*

$$D_5 = \left\{\{\Psi_n\} \subset V^* \,\middle|\, \Psi_n \in (v_n^*|_V)^*(V^*) \ (n = 1,2,\dots),\ \sum_{i=1}^{\infty} \Psi_i \ \textit{converges}\right\}, \tag{15.64}$$

*in which the norm is defined by* (15.63).

*If, in addition, each* $G_n = v_n(E)$ $(n = 1,2,\dots)$ *is reflexive, then*

c′) *The space* $V^*(\equiv E^{**}/V^{\perp})$ *is isomorphic, by the mapping*

$$\tau_0\colon \Psi \to \{\pi_{G_n}^{-1}(\Xi_n)\} \qquad (\Psi \in V^*), \tag{15.65}$$

*where* $\Xi_n$ *is as in* (15.49), (15.51), *to the space* $D_2$ *defined by* (15.23).

d′) *The restriction* $\tau_0|_{u(E)}$ *of the isomorphism* (15.65) *to* $u(E)$ *maps* $u(E)$ *onto the space* $D_1$ *defined by* (15.8).

e) *We have*

$$u(E) = \left\{\Psi \in V^* \,\middle|\, \sum_{i=1}^{\infty} (v_i^*|_V)^*(\Psi) \ \textit{converges}\right\}. \tag{15.66}$$

f) *We have*

$$\pi(E) \oplus V^{\perp} = \left\{ \Phi \in E^{**} \,\middle|\, \sum_{i=1}^{\infty} v_i^{**}(\Phi) \text{ converges} \right\}, \tag{15.67}$$

*and* $\pi(E) \oplus V^{\perp}$ *is norm-closed in* $E^{**}$.

*Proof.* c′) If each $G_n = v_n(E)$ is reflexive and $\Psi \in V^*$, then for $\Xi_n$ defined by (15.49), (15.51) and for $y_n = \pi_{G_n}^{-1}(\Xi_n) \in G_n$ we have

$$\sup_{1 \leqslant n < \infty} \left\| \sum_{i=1}^{n} (v_i^*|_V)^*(\Psi) \right\| = \sup_{1 \leqslant n < \infty} \sup_{\substack{f \in V \\ \|f\| \leqslant 1}} \left| \sum_{i=1}^{n} \Psi(v_i^*(f)) \right| =$$

$$= \sup_{1 \leqslant n < \infty} \sup_{\substack{f \in V \\ \|f\| \leqslant 1}} \left| \sum_{i=1}^{n} \Xi_i(f|G_i) \right| = \sup_{1 \leqslant n < \infty} \sup_{\substack{f \in V \\ \|f\| \leqslant 1}} \left| \sum_{i=1}^{n} f(y_i) \right| =$$

$$= \sup_{1 \leqslant n < \infty} \left\| \sum_{i=1}^{n} y_i \right\|_V .$$

Consequently, by part c) (which is a consequence of proposition 15.6 b)) and since by theorem 15.7 $\left\| \sum_{i=1}^{n} y_i \right\|_V \leqslant \left\| \sum_{i=1}^{n} y_i \right\| \leqslant \frac{1}{r(V)} \left\| \sum_{i=1}^{n} y_i \right\|_V$, we obtain c′).

d′) can be proved similarly, considering $\sum_{i=n+1}^{n+l}$ instead of $\sum_{i=1}^{n}$. Finally, the proofs of the other statements are straightforward.

The theory of spaces of sequences of scalars also admits a natural generalization to vector sequence spaces. Namely, if $\{G_n\}$ is a sequence of Banach spaces, a "sequence space" $S = S(\{G_n\})$ is a linear subspace of $s(\{G_n\}) = \prod_{n=1}^{\infty} G_n$ (the collection of all sequences $\{y_n\}$ with $y_n \in G_n$ for $n = 1,2,\ldots$, endowed with the natural linear operations). The *γ-dual* and *β-dual* of $S = S(\{G_n\})$ are defined by

$$S^{\gamma} = S^{\gamma}(\{G_n^*\}) = \left\{ \{\varphi_n\} \in s(\{G_n^*\}) \,\middle|\, \sup_{1 \leqslant n < \infty} \left| \sum_{i=1}^{n} \varphi_i(y_i) \right| < \infty \text{ for all } \{y_n\} \in S \right\}, \tag{15.68}$$

and, respectively, by

$$S^{\beta} = S^{\beta}(\{G_n^*\}) = \left\{ \{\varphi_n\} \in s(\{G_n^*\}) \,\middle|\, \sum_{i=1}^{\infty} \varphi_i(y_i) \text{ converges for all } \{y_n\} \in S \right\}. \tag{15.69}$$

In order to generalize $\gamma$-perfect and $\beta$-perfect spaces, let us introduce also the *$\gamma$-predual* and *$\beta$-predual* of $S = S(\{G_n^*\})$ by

$$S_\gamma = S_\gamma(\{G_n\}) = \left\{ \{y_n\} \in s(\{G_n\}) \;\middle|\; \sup_{1 \leqslant n < \infty} \left| \sum_{i=1}^{n} \varphi_i(y_i) \right| < \infty \text{ for all } \{\varphi_n\} \in S \right\}, \tag{15.70}$$

and, respectively, by

$$S_\beta = S_\beta(\{G_n\}) = \left\{ \{y_n\} \in s(\{G_n\}) \;\middle|\; \sum_{i=1}^{\infty} \varphi_i(y_i) \text{ converges for all } \{\varphi_n\} \in S \right\}; \tag{15.71}$$

then, $S = S(\{G_n\})$ is said to be *$\gamma$-perfect* (respectively, *$\beta$-perfect*) if $(S^\gamma)_\gamma = S$ (respectively, if $(S^\beta)_\beta = S$).

In order to generalize $BK$-spaces, let us observe that the natural generalization of the coordinate functionals on scalar sequence spaces to the setting of vector sequence spaces $S(\{G_j\})$ are the "*coordinate operators*" $\chi_n \colon S(\{G_j\}) \to G_n$ defined by $\chi_n(\{y_j\}) = y_n$. Then, $S(\{G_j\})$ is said to be a *BK-space* if the following conditions are satisfied: a) $S(\{G_j\})$ is a Banach space; b) the "coordinate operators" $\chi_n$ $(n = 1,2,\ldots)$ are continuous on $S(\{G_j\})$ (or, what is equivalent, the inclusion map of $S(\{G_j\})$ into $s(\{G_j\})$ is continuous, when $s(\{G_j\}) = \prod_{j=1}^{\infty} G_j$ is endowed with the product*) topology). The scalar $BK$-spaces containing all unit vectors $e_n$ are generalized by the spaces $S(\{G_j\})$ containing all $F_n = \underbrace{\{0\} \times \ldots \times \{0\}}_{n-1} \times G_n \times \{0\} \times \ldots$ (which are then closed linear subspaces of $S(\{G_j\})$, by b)).

We have the following generalization of part of Ch. I, § 12, lemmas 12.1 and 12.2 (with a similar proof):

**Lemma 15.2.** *If $\{G_j\}$ is a sequence of Banach spaces and $T = T(\{G_j\})$ is a BK-space, then the $\gamma$-dual $T^\gamma$ and the $\beta$-dual $T^\beta$ can be normed so as to become a Banach space, by*

$$\|\{\varphi_j\}\| = \sup_{\substack{\{y_j\} \in T \\ \|\{y_j\}\| \leqslant 1}} \sup_{1 \leqslant n < \infty} \left| \sum_{i=1}^{n} \varphi_i(y_i) \right| \qquad (\{\varphi_j\} \in T^\gamma \text{ or } T^\beta). \tag{15.72}$$

Using this norm, one can give the following supplement to theorem 15.8 a) above:

**Theorem 15.9.** *Let $\{G_n\}$ be a Schauder decomposition of a Banach space $E$. Then the conjugate space $E^*$ is isomorphic, by the mapping*

$$f \to \{f|_{G_n}\} \qquad (f \in E^*), \tag{15.73}$$

*) With this topology, $s(\{G_j\})$ is a Fréchet space (i.e., a locally convex complete metrizable space).

*to the $\gamma$-dual $D_1^\gamma$, where $D_1 = D_1(\{G_n\})$ is the sequence space defined by* (15.8).

*Proof.* By theorem 15.8 a), it is enough to show that if $\{v_n\} \subset L(E, E)$ is the a.s.c.p. to $\{G_n\}$, the mapping $\{v_n^*(f)\} \to \{f|_{G_n}\}$ $(f \in E^*)$ is an isomorphism of $D_4$ onto $D_1^\gamma$. Now, if $\{v_n^*(f)\} \in D_4$ and $\varphi_n = f|_{G_n} (n = 1,2,\ldots)$, then

$$\|\{v_n^*(f)\}\| = \sup_{1 \leqslant n < \infty} \left\| \sum_{i=1}^n v_i^*(f) \right\| = \sup_{1 \leqslant n < \infty} \sup_{\substack{x \in E \\ \|x\| \leqslant 1}} \left| \sum_{i=1}^n v_i^*(f)(x) \right| =$$

$$= \sup_{1 \leqslant n < \infty} \sup_{\substack{\sum_{i=1}^\infty y_i \in E \\ \left\|\sum_{i=1}^\infty y_i\right\| \leqslant 1}} \left| \sum_{i=1}^n \varphi_i(y_i) \right|. \tag{15.74}$$

But, by proposition 15.3, there exists a constant $C \geqslant 1$ such that

$$\left\| \sum_{i=1}^\infty y_i \right\| \leqslant \|\{y_n\}\| \leqslant C \left\| \sum_{i=1}^\infty y_i \right\| \qquad \left( \sum_{i=1}^\infty y_i \in E \right),$$

and hence, by (15.72) for $T = D_1$ and (15.74),

$$\|\{\varphi_n\}\| \leqslant \|\{v_n^*(f)\}\| \leqslant C \|\{\varphi_n\}\|, \tag{15.75}$$

which proves that $\{v_n^*(f)\} \to \{\varphi_n\} = \{f|_{G_n}\}$ is an isomorphism of $D_4$ into $D_1^\gamma$. Finally, to prove that this isomorphism maps $D_4$ *onto* $D_1^\gamma$, let $\{\varphi_n\} \in D_1^\gamma$, that is, $\varphi_n \in G_n^*$ $(n = 1,2,\ldots)$ and $\sup_{1 \leqslant n < \infty} \left| \sum_{i=1}^n \varphi_i(y_i) \right| < \infty$ for all $\sum_{i=1}^\infty y_i \in E$. Define $f_n$ on $E$ by

$$f_n\left( \sum_{i=1}^\infty y_i \right) = \varphi_n(y_n) \qquad \left( \sum_{i=1}^\infty y_i \in E, n = 1,2,\ldots \right). \tag{15.76}$$

Then $\left| f_n\left( \sum_{i=1}^\infty y_i \right) \right| \leqslant \|\varphi_n\| \left\| v_n\left( \sum_{i=1}^\infty y_i \right) \right\| \leqslant \|\varphi_n\| \|v_n\| \left\| \sum_{i=1}^\infty y_i \right\|$, so $f_n \in$ $\in E^*$. Moreover, since

$$f_n\left( \sum_{i=1}^\infty y_i \right) = \varphi_n(y_n) = f_n(y_n) = f_n\left( v_n\left( \sum_{i=1}^\infty y_i \right) \right) = v_n^*(f_n)\left( \sum_{i=1}^\infty y_i \right),$$

we have $f_n = v_n^*(f_n) \in v_n^*(E^*)$ $(n = 1,2,\ldots)$. Furthermore,

$$\sup_{1\leqslant n<\infty}\left|\sum_{j=1}^{n} f_j\left(\sum_{i=1}^{\infty} y_i\right)\right| = \sup_{1\leqslant n<\infty}\left|\sum_{i=1}^{n}\varphi_i(y_i)\right| < \infty \qquad \left(\sum_{i=1}^{\infty} y_i \in E\right),$$

whence, by the principle of uniform boundedness, $\sup_{1\leqslant n<\infty}\left\|\sum_{j=1}^{n} f_j\right\| < \infty$, and hence, by proposition 15.6 a), there exists $f \in E^*$ such that

$$v_n^*(f) = f_n \qquad (n = 1,2,\ldots).$$

Then

$$\begin{aligned} f\left(\sum_{i=1}^{\infty} y_i\right) &= \sum_{i=1}^{\infty} f(y_i) = \sum_{i=1}^{\infty} f(v_i(y_i)) = \sum_{i=1}^{\infty} v_i^*(f)(y_i) = \\ &= \sum_{i=1}^{\infty} f_i(y_i) = \sum_{i=1}^{\infty}\varphi_i(y_i) \qquad \left(\sum_{i=1}^{\infty} y_i \in E\right), \end{aligned} \tag{15.77}$$

whence, in particular, $f|_{G_n} = \varphi_n$ $(n = 1,2,\ldots)$. Thus, $\{\varphi_n\}$ is the image of $\{v_n^*(f)\}$ under the above isomorphism of $D_4$ into $D_1^\gamma$, which completes the proof of theorem 15.9. Note that $\{\varphi_n\} \to f$ $(\{\varphi_n\} \in D_1^\gamma)$, where $f \in E^*$ is defined by (15.77), is the inverse mapping of (15.73).

We mention also the following related isomorphic representation of $E^*$, which yields a generalization of Ch. I, § 12, proposition 12.3 b):

**Proposition 15.8.** *Let $\{G_n\}$ be a Schauder decomposition of a Banach space $E$. Then for $D_1 = D_1(\{G_n\})$ defined by* (15.8) *we have*

$$D_1^\beta = D_1^\gamma, \tag{15.78}$$

*and hence $E^*$ is isomorphic, by the mapping* (15.73), *to $D_1^\beta$ and the inverse mapping $\{\varphi_n\} \to f$ is given by* (15.77).

*Proof.* The inclusion

$$S^\beta \subset S^\gamma \tag{15.79}$$

is obvious for any sequence space $S = S(\{G_n\})$. The proof of the inclusion $D_1^\gamma \subset D_1^\beta$ is contained in the final part of the above proof of theorem 15.9. This completes the proof.

The obvious generalization of scalar sequence spaces associated to a basis (Ch. I, § 12, definition 12.1) does not give much in this setting, since now we already start with a given sequence of Banach spaces $\{G_n\}$ and therefore we can arrive (isomorphically) only at one space $E$ such that $E = \sum_{i=1}^{\infty} \oplus G_i$ and at $S(\{G_n\}) = D_1$ defined by (15.8). However, we have seen in theorems 15.8 and 15.9 (see also proposition 15.7) above that even for a given Schauder decomposition $\{G_n\}$ of a Banach space $E$, one has convenient generalizations of the scalar sequence space $(E, \{x_n\})^d = \{\{f(x_n)\} \mid f \in E^*\}$ (introduced in § 8, formula (8.129) for any complete minimal sequence $\{x_n\}$). Let us mention here that one can give another generalization (which is still a vector sequence space) of this space and of its subspaces of the form $\{\{f(x_n)\} \mid f \in V\}$, where $V \subset E^*$ and where $\{x_n\}$ is a complete sequence in $E$. Namely, one can put

$$a(E) = \{\{u(x_n)\} \mid u \in A(E, E)\}, \tag{15.80}$$

endowed with some natural norm, where $A(E, E)$ is a subspace of $L(E, E)$; for example, one can take $A(E, E) = L(E, E)$ itself, $A(E, E) = N(E, E)$ with the nuclear norm, $A(E, E) = \mathscr{C}(E, E)$ (the compact operators), with the norm induced by $L(E, E)$, etc. Then, it is natural to consider the mappings $A(E, E) \to a(E)$ defined by $u \to \{u(x_n)\}$, which relate the subspaces of $L(E, E)$ (in particular, the ideals of operators) to vector sequence spaces. Conversely, one can also start with vector sequence spaces and arrive at some operator spaces, as follows: Let $E, F$ be Banach spaces, $\{G_n\}$, $\{F_n\}$ sequences of subspaces of $E$ and $F$ respectively, and $S_1 = S_1(\{G_n\})$, $S_2 = S_2(\{G_n\})$ sequence spaces and let

$$\mathscr{M}(S_1, S_2) = \{u \in L(E, F) \mid \{u(y_n)\} \in S_2 \text{ for all } \{y_n\} \in S_1\}; \tag{15.81}$$

then, for example, one can give conditions on $S_1$ and $S_2$ in order that $\mathscr{M}(S_1, S_2)$ be an ideal of operators and a Banach space.

Returning to decompositions and Schauder decompositions, let us note that these notions can be introduced in any topological linear space $U$ as in definitions 15.1—15.3 above. Hence, similarly to Ch. I, § 13, definition 13.2, we have the notions of weak, bounded weak, weak* and bounded weak* decompositions and Schauder decompositions. Let us give here only

*Definition 15.12.* A sequence $\{G_n\}$ of linear subspaces of a Banach space $E$ is called a *weak decomposition* of $E$ if for every $x \in E$ there exists a *unique* sequence $\{y_n\} \subset E$ with $y_n \in G_n$ $(n = 1,2, \ldots)$ such that

$$x = w\text{-}\lim_{n\to\infty} \sum_{i=1}^{n} y_i; \tag{15.82}$$

a weak decomposition $\{G_n\}$ of $E$ is said to be a *weak Schauder decomposition* of $E$, if the coordinate projections $v_n$, defined similarly to (15.4), are weakly continuous on $E$ (i.e. continuous from $E$ endowed with $\sigma(E, E^*)$ into $E$ endowed with $\sigma(E, E^*)$). The notions of *weak* decomposition* and *weak* Schauder decomposition* of a conjugate Banach space $E^*$ are defined dually.

*Remark 15.5.* In the definition of a weak Schauder decomposition of a Banach space $E$, *one can replace the condition of weak continuity of each* $v_n$ $(n = 1,2, \ldots)$ *by*

$$v_n \in L(E, E) \qquad (n = 1,2, \ldots); \tag{15.83}$$

indeed, it is well known*) that $v_n$ is weakly continuous if and only if it is norm continuous. Consequently, *if* $\{G_n\}$ *is a weak Schauder decomposition of a Banach space* $E$, *then each* $G_n$ $(n = 1,2, \ldots)$ *is closed.* One can show that the converse is also true, that is, *if* $\{G_n\}$ *is a weak decomposition of a Banach space* $E$ *such that each* $G_n$ $(n = 1,2, \ldots)$ *is closed, then* $\{G_n\}$ *is a weak Schauder decomposition of* $E$. Similarly to theorem 15.1, the main part of the proof consists in showing (essentially with the argument of the proof of Ch. I, § 13, proposition 13.1) that if $\{G_n\}$ is a weak decomposition and each $G_n$ is closed, then the sequence space

$$D_1^{(w)} = D_1^{(w)}(\{G_n\}) = \left\{ \{y_n\} \subset E \,\middle|\, y_n \in G_n \ (n = 1,2, \ldots), \sum_{i=1}^{\infty} y_i \text{ is weakly convergent to an element of } E \right\}, \tag{15.84}$$

endowed with the norm (15.9), is a Banach space.

Let us give now the following partial**) generalization of the "weak basis theorem" (see Ch. I, § 13, theorem 13.1 a), implication $1° \Rightarrow 3°$):

**Theorem 15.10.** *Every weak Schauder decomposition* $\{G_n\}$ *of a Banach space* $E$ *is a Schauder decomposition of* $E$.

*Proof.* By (15.82) we have

$$f(x) = \lim_{n\to\infty} f\left(\sum_{i=1}^{n} v_i(x)\right) \qquad (x \in E, f \in E^*), \tag{15.85}$$

where $\{v_n\}$ is the a.s.c.p. to the weak decomposition $\{G_n\}$, and hence

$$\sup_{1 \leqslant n < \infty} \left| f\left(\sum_{i=1}^{n} v_i(x)\right)\right| < \infty \qquad (x \in E, f \in E^*). \tag{15.86}$$

---

*) See e.g. [87], p. 422, theorem 15.

**) This is only a partial generalization, since now we assume that the coordinate projections are weakly continuous. However, such an additional assumption is necessary, since proposition 15.1 and theorem 15.2 show that there exist weak decompositions which are not Schauder decompositions (indeed, every decomposition is a weak decomposition).

But, by remark 15.5, $\{v_n\} \subset L(E, E)$ $(n = 1,2, \ldots)$, whence, by (15.86) and the principle of uniform boundedness*),

$$\sup_{1 \leqslant n < \infty} \left\| \sum_{i=1}^{n} v_i \right\| < \infty. \tag{15.87}$$

Furthermore, $\{v_n\}$ obviously satisfies (15.6). Finally, by (15.82), the weakly closed linear span of $\bigcup_{n=1}^{\infty} v_n(E) = \bigcup_{n=1}^{\infty} G_n$ is $E$, whence also in the norm topology **) we have $\left[\bigcup_{n=1}^{\infty} v_n(E)\right] = E$. Consequently, by theorem 15.3 (implication $3° \Rightarrow 1°$), $\{G_n\}$ is a Schauder decomposition of $E$, which completes the proof.

*Remark 15.6.* The converse of theorem 15.10 is also valid, i.e., *every Schauder decomposition* $\{G_n\}$ *of a Banach space* $E$ *is a weak Schauder decomposition of* $E$. Indeed, if $\{v_n\} \subset L(E, E)$ is the a.s.c.p. to $\{G_n\}$ and $x \in E$, then $x = \sum_{i=1}^{\infty} v_i(x)$ and hence also $x = w\text{-}\lim_{n\to\infty} \sum_{i=1}^{n} v_i(x)$. In order to prove the uniqueness of this $w$-representation, assume that for a sequence $\{y_n\} \subset E$ with $y_n = v_n(y_n) \in v_n(E) = G_n$ $(n = 1,2, \ldots)$ we have $w\text{-}\lim_{n\to\infty} \sum_{i=1}^{n} y_i = 0$. Then, since $y_j = \sum_{i=1}^{n} v_j(y_i)$ $(n \geqslant j)$, we obtain

$$f(y_j) = \lim_{n\to\infty} \sum_{i=1}^{n} f(v_j(y_i)) = \lim_{n\to\infty} v_j^*(f)\left(\sum_{i=1}^{n} y_i\right) = 0 \quad (f \in E^*,\ j = 1,2,\ldots),$$

so $y_j = 0$ $(j = 1,2, \ldots)$, which proves that $\{G_n\}$ is a weak decomposition of $E$. Finally, since $v_n \in L(E,E)$, $v_n$ is also weakly continuous***), which completes the proof.

We have the following theorem of weak duality, which is a generalization of Ch. I, § 14, theorem 14.1:

**Theorem 15.11.** *A sequence* $\{\Gamma_n\}$ *of subspaces of a conjugate Banach space* $E^*$, *is a weak* Schauder decomposition of* $E^*$, *with the a.s.c.p.* $\{w_n\}$, *if and only if* $E$ *has a Schauder decomposition* $\{G_n\}$ *with the a.s.c.p.* $\{v_n\}$ *such that*

$$v_n^* = w_n \qquad (n = 1,2, \ldots). \tag{15.88}$$

*Proof.* Assume that $\{\Gamma_n\}$ is weak* Schauder decomposition of $E^*$, with the a.s.c.p. $\{w_n\}$. Then each $w_n$ $(n = 1,2, \ldots)$ is $w^*$-continuous on $E^*$ and hence there exists****) $\{v_n\} \subset L(E, E)$ such that we have (15.88).

*) See e.g. [11], p. 80, theorem 5.

**) See e.g. [355], Ch. II, § 9, corollary 2 of theorem 9.2.

***) See remark 15.5.

****) See e.g. [87], p. 513, exercise 13 (proof: since $w_n$ is $w^*$-continuous on $E^*$, we have $w_n^*(\pi_E(E)) \subset \pi_F(F)$ and we put $v_n = \pi_F^{-1} w_n^* \pi_E$, where $\pi_E: E \to E^{**}$, $\pi_F: F \to F^{**}$ are the canonical embeddings; in our case, $F = E$).

Then, since $\{v_n^*(E^*)\}$ is a weak* Schauder decomposition of $E^*$, we have

$$f(x) = \lim_{n\to\infty} \sum_{i=1}^{n} v_i^*(f)(x) = \lim_{n\to\infty} f\left(\sum_{i=1}^{n} v_i(x)\right) \qquad (x \in E,\ f \in E^*),$$

i.e. (15.85), whence, as in the above proof of theorem 15.10, we infer (15.87) and $\left[\bigcup_{n=1}^{\infty} v_n(E)\right] = E$. Furthermore, since

$$(v_j v_i)^* = v_i^* v_j^* = w_i w_j = \delta_{ij} w_j = \delta_{ij} v_j^* \quad (i, j = 1,2, \ldots),$$

we have $v_j v_i = \delta_{ij} v_j$ $(i, j = 1,2, \ldots)$, that is, (15.6). Consequently, by theorem 15.3, $\{G_n\} = \{v_n(E)\}$ is a Schauder decomposition of $E$, with the a.s.c.p. $\{v_n\}$.

Conversely, assume now that $\{G_n\}$ is a Schauder decomposition of $E$, with the a.s.c.p. $\{v_n\} \subset L(E, E)$. Then, by (15.5), we have (15.85) and hence every $f \in E^*$ has a $w^*$-representation $f = w^*\text{-}\lim_{n\to\infty} \sum_{i=1}^{n} v_i^*(f)$. In order to prove the uniqueness of this $w^*$-representation, assume that for a sequence $\{f_n\} \subset E^*$ with $f_n = v_n^*(f_n) \in v_n^*(E^*) = \Gamma_n$ $(n = 1,2, \ldots)$ we have $w^*\text{-}\lim_{n\to\infty} \sum_{i=1}^{n} f_i = 0$. Then, since $f_j = \sum_{i=1}^{n} v_j^*(f_i)$ $(n \geqslant j)$, we obtain

$$f_j(x) = \lim_{n\to\infty} \sum_{i=1}^{n} v_j^*(f_i)(x) = \lim_{n\to\infty} \left(\sum_{i=1}^{n} f_i\right)(v_j(x)) = 0 \quad (x \in E,\ j=1,2,\ldots),$$

so $f_j = 0$ $(j = 1,2, \ldots)$, which proves that $\{v_n^*(E^*)\}$ is a weak* decomposition of $E^*$, with the a.s.c.p. $\{v_n^*\}$. Finally, since $v_n \in L(E, E)$, $v_n^*$ is $w^*$-continuous $(n = 1,2, \ldots)$, so $\{w_n(E^*)\} = \{v_n^*(E^*)\}$ is a weak* Schauder decomposition of $E^*$, which completes the proof.

While the answer to Ch. I, § 13, problem 13.2 a) (if $E^*$ has a $w^*$-basis, does $E$ have a basis?) is still unknown (we have seen in § 14, theorem 14.2 b), that for separable $E$ the answer is affirmative), the corresponding problem for Schauder decompositions has a negative answer, as shown by

*Example* 15.2. Let $E = l^\infty$. Then, as we have seen in example 15.1, $E^*$ has a Schauder decomposition. We claim that *any Schauder decomposition* $\{\Gamma_n\}$ *of* $E^*$ *is a weak* decomposition of* $E^*$. Indeed, if $\{w_n\} \subset L(E^*, E^*)$ is the a.s.c.p. to $\{\Gamma_n\}$ and $f \in E^*$, then $f = \sum_{i=1}^{\infty} w_i(f)$ and hence also $f = w^*\text{-}\lim_{n\to\infty} \sum_{i=1}^{n} w_i(f)$. In order to prove the uniqueness of this $w^*$-representation, assume that for a sequence $\{f_n\} \subset E^*$ with $f_n \in \Gamma_n$ $(n = 1,2,\ldots)$ we have $w^*\text{-}\lim_{n\to\infty} \sum_{i=1}^{n} f_i = 0$. Then, by the theorem of Grothendieck used also in the proof of theorem 15.2, $w\text{-}\lim_{n\to\infty}\sum_{i=1}^{n} f_i=0$, whence, since $\{\Gamma_n\}$ is a weak Schauder decomposition of $E^*$ (by remark 15.6), we obtain $f_j = 0$ $(j = 1,2, \ldots)$, which proves the claim that $\{\Gamma_n\} = \{w_n(E^*)\}$ is a weak* decomposition of $E^*$. Thus, $E^*$ has a weak* decomposition, but $E = l^\infty$ has no Schauder decomposition (by theorem 15.2). Moreover, as in the case of example 15.1, one can also show that for each $n = 1,2, \ldots$ the conjugate space $E^{(2n+1)}$ has a weak* decomposition, but $E^{(2n)}$ has no Schauder decomposition.

In connection with problem 15.1 let us observe that in this problem the assumption that $E$ is a Banach space is essential, since already for separable locally convex spaces the answer is negative, as shown by

*Example 15.3.* Let $U = (l^\infty)^*$, endowed with the weak* topology $\sigma((l^\infty)^*, l^\infty)$; then, as we have seen in Ch. I, § 13, example 13.4, $U$ is a separable locally convex space which has no basis (and hence no Schauder basis). By theorems 15.2 and 15.11, $U$ has no Schauder decomposition either (although it does have a decomposition, as we have seen in example 15.2).

Let us also mention that the (non-locally convex) separable complete metric linear spaces $U = S([0, 1])$ and $U = L^p([0, 1])$ with $0 < p < 1$, which have no basis (see Ch. I, § 13, example 13.1), do have Schauder decompositions. For example, $\{G_n\} = \{\chi_{\left[\frac{n-1}{n}, \frac{n}{n+1}\right)} S([0, 1])\}$ is a Schauder decomposition of $U = S([0, 1])$, with the a.s.c.p. $\{v_n\} \subset L(U, U)$ given by $v_n(x) = \chi_{\left[\frac{n-1}{n}, \frac{n}{n+1}\right)} x$ $(x \in U,\ n = 1,2, \ldots)$, where $\chi_A$ denotes the characteristic function of $A$.

Let us show now that some special classes of bases of Banach spaces considered in Ch. II admit natural generalizations to special classes of Schauder decompositions.

*Definition 15.13.* A Schauder decomposition $\{G_n\}$ of a Banach space $E$ is said to be *monotone* if we have

$$\left\| \sum_{i=1}^{n} y_i \right\| \leqslant \left\| \sum_{i=1}^{n+m} y_i \right\| \tag{15.89}$$

for all positive integers $n, m$ and all $y_i \in G_i$ $(i = 1, \ldots, n + m)$.

From theorem 15.5 and its proof it follows that a Schauder decomposition $\{G_n\}$ of a Banach space $E$ is monotone if and only if $\sup_{1\leqslant n<\infty} \left\| \sum_{i=1}^{n} v_i \right\| = 1$, where $\{v_n\} \subset L(E, E)$ is the a.s.c.p. to $\{G_n\}$. Similarly to the particular case of bases, every Schauder decomposition $\{G_n\}$ of a Banach space $E$ can be "monotonized" by replacing the norm of $E$ with the equivalent norm (15.12). Also, similarly to Ch. II, § 1, proposition 1.4 a), if $\{G_n\}$ is a monotone Schauder decomposition of $E$, with the a.s.c.p. $\{v_n\}$, then $\{v_n^*(E^*)\}$ is a monotone Schauder decomposition of $\left[\bigcup_{n=1}^{\infty} v_n^*(E^*)\right]$.

*Definition 15.14.* A Schauder decomposition $\{G_n\}$ of a Banach space $E$ is said to be *normal*, if for the a.s.c.p. $\{v_n\} \subset L(E, E)$ we have

$$\|v_n\| = 1 \qquad (n = 1,2, \ldots). \tag{15.90}$$

If $\{x_n\}$ is a normal basis of a Banach space $E$, with the a.s.c.f. $\{f_n\} \subset E^*$, then by (15.7) the associated one-dimensional decomposition $G_1 = [x_1]$, $G_2 = [x_2]$, ... is normal. Conversely, if $\{G_n\}$ is a normal one-dimensional decomposition of $E$, then every sequence $x_n \in G_n$ with $\|x_n\| = 1$ $(n = 1,2, \ldots)$ is a normal basis of $E$.

*Problem 15.2.* Does every Banach space with a Schauder decomposition possess a normal Schauder decomposition?

Similarly to the particular case of bounded bases, every Schauder decomposition $\{G_n\}$ of a Banach space $E$ "can be made normal" by replacing the norm of $E$ with a suitable equivalent norm, that is, we have the following generalization of Ch. II, § 2, theorem 2.3:

**Theorem 15.12.** *Let $\{G_n\}$ be a Schauder decomposition of a Banach space $E$. Then there exists an equivalent norm on $E$, in which the Schauder decomposition $\{G_n\}$ is normal.*

*Proof.* We shall extend the argument of § 8, proof of proposition 8.15. Put

$$|||x||| = \max(\|x\|, \sup_{1 \leqslant n < \infty} \|v_n(x)\|) \qquad (x \in E), \tag{15.91}$$

where $\{v_n\} \subset L(E, E)$ is the a.s.c.p. to $\{G_n\}$. Then $|||\cdot|||$ is a norm on $E$ and by (15.13) we have

$$\|x\| \leqslant |||x||| \leqslant (\sup_{1 \leqslant n < \infty} \|v_n\|)\, \|x\| \qquad (x \in E),$$

so $|||\cdot|||$ is equivalent to the initial norm on $E$. Finally, by (15.6) we have

$$|||v_n(x)||| = \sup_{1 \leqslant j < \infty} \|v_j v_n(x)\| = \|v_n(x)\| \leqslant \sup_{1 \leqslant j < \infty} \|v_j(x)\| \leqslant |||x|||$$
$$(x \in E, \quad n = 1,2, \ldots),$$

whence $|||v_n||| \leqslant 1$ $(n = 1,2, \ldots)$. Consequently, since each $v_n$ is a projection, $|||v_n||| = 1$ $(n = 1,2, \ldots)$, which completes the proof.

Similarly to Ch. II, § 2, proposition 2.2, if $\{G_n\}$ is a normal Schauder decomposition of a Banach space $E$, with the a.s.c.p. $\{v_n\} \subset L(E, E)$, then $\{v_n^*(E^*)\}$ is a normal Schauder decomposition of $\left[\bigcup_{n=1}^{\infty} v_n^*(E^*)\right]$.

For any integer $k \geqslant 0$ one can introduce $k$-shrinking and $k$-boundedly complete Schauder decompositions, but we shall consider here only the case when $k = 0$.

*Definition 15.15.* A Schauder decomposition $\{G_n\}$ of a Banach space $E$ is called *shrinking* if for the a.s.c.p. $\{v_n\} \subset L(E, E)$ we have

$$\left[\bigcup_{n=1}^{\infty} v_n^*(E^*)\right] = E^*. \tag{15.92}$$

If $E$ is a Banach space and $(G_n, v_n)$ $(G_n \subset E, v_n \in L(E, E), n = 1,2, \ldots)$ is a generalized biorthogonal system, we shall use the notation

$$\|f\|_n = \|f|_{\left[\bigcup_{i=n+1}^{\infty} G_i\right]}\| \qquad (f \in E^*,\ n = 1,2, \ldots). \tag{15.93}$$

We have the following generalization of Ch. II, § 4, proposition 4.1:

**Proposition 15.9.** *Let $E$ be a Banach space and $(G_n, v_n)$ $(G_n \subset E$, $v_n \in L(E, E)$, $n = 1,2, \ldots)$ an $E$-complete generalized biorthogonal system. Then*

$$\|f\|_n = \operatorname{dist}\left(f, \left[\bigcup_{i=1}^{n} v_i^*(E^*)\right]\right) \qquad (f \in E^*,\ n = 1,2, \ldots). \tag{15.94}$$

The proof is similar to that of Ch. II, § 4, proposition 4.1, observing that $\left[\bigcup_{i=1}^{n} v_i^*(E^*)\right] = \sum_{i=1}^{n} \oplus\, v_i^*(E^*)$ is $w^*$-closed $\left(\text{because } \sum_{i=1}^{n} v_i^* \text{ is a } w^*\text{-continuous projection}\right)$ and that $E = \left[\bigcup_{i=1}^{n} G_i\right] \oplus \left[\bigcup_{i=n+1}^{\infty} G_i\right]$ (by the remarks made after definition 15.9).

Now we can give the following characterizations of shrinking Schauder decompositions, which generalize part of Ch. II, § 4, theorem 4.2:

**Theorem 15.13.** *Let $\{G_n\}$ be a Schauder decomposition of a Banach space $E$, with the a.s.c.p. $\{v_n\} \subset L(E, E)$. The following statements are equivalent:*

1°. *$\{G_n\}$ is shrinking.*

2°. $\lim_{n\to\infty} \|f\|_n = 0$ *$(f \in E^*)$.*

3°. *$\{v_n^*(E^*)\}$ is a Schauder decomposition of $E^*$.*

4°. *$E^{**}$ is isomorphic, by the mapping*

$$\tau: \Phi \to \{v_n^{**}(\Phi)\} \qquad (\Phi \in E^{**}), \tag{15.95}$$

*to the Banach space of sequences of scalars*

$$D_8 = \left\{ \{\Phi_n\} \subset E^{**} \,\middle|\, \Phi_n \in v_n^{**}(E^{**}) \; (n = 1,2, \ldots), \sup_{1 \leqslant n < \infty} \left\| \sum_{i=1}^{n} \Phi_i \right\| < \infty \right\}, \tag{15.96}$$

*in which the norm is defined by*

$$\|\{\Phi_n\}\| = \sup_{1 \leqslant n < \infty} \left\| \sum_{i=1}^{n} \Phi_i \right\|. \tag{15.97}$$

*If each subspace $G_n = v_n(E)$ $(n = 1,2, \ldots)$ is reflexive, these statements are equivalent to the following statements:*

5°. *$E^{**}$ is isomorphic, by the mapping*

$$\tau_1 : \Phi \to \{\pi_{G_n}^{-1}(\Xi_n)\} \qquad (\Phi \in E^{**}), \tag{15.98}$$

*where $\Xi_n$ is as in* (15.50), (15.51), *to the Banach space $D_2$ defined by* (15.23).

6°. *We have*

$$\pi(E) = \left\{ \Phi \in E^{**} \,\middle|\, \sum_{i=1}^{\infty} v_i^{**}(\Phi) \text{ converges} \right\} \tag{15.99}$$

*(where $\pi$ denotes the canonical mapping of $E$ into $E^{**}$), and the image of $\pi(E)$ in $D_2$ by the isomorphism* (15.98) *is the space $D_1$ defined by* (15.8).

*Proof.* By proposition 15.9 we have (even for any $E$-complete generalized biorthogonal system)

$$\left[ \bigcup_{n=1}^{\infty} v_n^*(E^*) \right] = \{f \in E^* \mid \lim_{n \to \infty} \|f\|_n = 0\}, \tag{15.100}$$

and thus 1° ⇔ 2°. Furthermore, the implication 3°⇒1° is obvious and the implications 1°⇒3° and 1°⇒4° are consequences of theorem 15.6 and theorem 15.8 c) respectively, with $V = E^*$. Conversely, if we do not have 1°, then, by a corollary of the Hahn-Banach theorem, there exists a $\Phi \in E^{**}$, $\Phi \neq 0$, such that $\Phi(v_n^*(f)) = 0$ $(f \in E^*, n = 1,2, \ldots)$, whence $v_n^{**}(\Phi) = 0$ $(n = 1,2, \ldots)$, so $\tau$ is not one-to-one. Thus, 4°⇒1°, which proves that 1° ⇔ ... ⇔ 4°.

Assume now that each $G_n = v_n(E)$ $(n = 1,2, \ldots)$ is reflexive. Then the implications 1° ⇒ 5° and 1° ⇒ 6° are consequences of theorem 15.8 c′), e) (or f)) and d′), with $V = E^*$. Conversely, if we do not have 1°, then, as above, there exists $\Phi \in E^{**}$, $\Phi \neq 0$, such that $v_n^{**}(\Phi) = 0$ $(n = 1,2, \ldots)$. But then, as was observed in the proof of proposition 15.7 c) (inclusion $v_n^{**}(E^{**}) \subset \pi_E(G_n)$), for $\Xi_n$ corresponding to this $\Phi$ by (15.50), (15.51) and for $y_n = \pi_{G_n}^{-1}(\Xi_n) \in G_n$ we have

$$f(\pi_{G_n}^{-1}(\Xi_n)) = f(y_n) = v_n^{**}(\Phi)(f) = 0 \quad (f \in E^*, \; n = 1,2, \ldots),$$

so $\tau_1$ is not one-to-one. Thus, 5°⇒1°. Finally, if we have 6°, then by theorem 15.8 f), $\left[\bigcup_{n=1}^{\infty} v_n^*(E^*)\right]^{\perp} = \{0\}$, whence $\left[\bigcup_{n=1}^{\infty} v_n^*(E^*)\right] = E^*$. Thus, 6°⇒1°, which completes the proof.

In particular, from the implication 1° ⇒ 3° and theorem 15.2 it follows that $l^1$ has no shrinking Schauder decomposition.

We have the following supplement to proposition 15.4:

**Proposition 15.10.** *If a Banach space $E$ contains a complemented subspace $G$ with a shrinking basis $\{x_n\}$, then $E$ has a shrinking Schauder decomposition.*

*Proof.* Let $u$ be a continuous linear projection of $E$ onto $G = [x_n]$ and define $\{G_n\}$ by (15.3), so $\{G_n\}$ is a Schauder decomposition of $E$. Then, since $\{x_n\}$ is shrinking, by Ch. II, § 4, theorem 4.2 we have

$$\lim_{n\to\infty} \|f\|_n = \lim_{n\to\infty} \left\| f\Big|_{\left[\bigcup_{i=n+1}^{\infty} G_i\right]} \right\| = \lim_{n\to\infty} \|f|_{[x_n,\, x_{n+1},\, \dots]}\| = 0 \qquad (f \in E^*),$$

whence, by theorem 15.13, implication 2° ⇒ 1°, $\{G_n\}$ is shrinking, which completes the proof. Note that the condition of proposition 15.10 is satisfied, in particular, if $E$ is a separable Banach space containing a subspace isomorphic to $c_0$ (see lemma 15.1), so we obtain a sharpening of corollary 15.3. Hence, for example, $E = C([0, 1])$ has a shrinking Schauder decomposition, although it has no shrinking basis.

*Definition 15.16.* A Schauder decomposition $\{G_n\}$ of a Banach space $E$ is said to be *boundedly complete*, if the relations $y_n \in G_n$ $(n = 1,2, \dots)$, $\sup_{1 \leq n < \infty} \left\| \sum_{i=1}^{n} y_i \right\| < \infty$ imply that $\sum_{i=1}^{\infty} y_i$ converges.

We have the following characterizations of boundedly complete Schauder decompositions, which generalize part of Ch. II, § 6, theorem 6.2:

**Theorem 15.14.** *Let $\{G_n\}$ be a Schauder decomposition of a Banach space $E$, with the a.s.c.p. $\{v_n\} \subset L(E, E)$, such that each $G_n = v_n(E)$ $(n = 1,2, \dots)$ is reflexive**) *and let $V = \left[\bigcup_{n=1}^{\infty} v_n^*(E^*)\right]$. The following statements are equivalent:*

1°. *$\{G_n\}$ is boundedly complete.*

2°. *$\pi(E) \oplus V^{\perp} = E^{**}$, where $\pi$ denotes the canonical mapping of $E$ into $E^{**}$.*

3°. *$u(E) = V^*$, where $u$ denotes the canonical mapping of $E$ into $V^*$ (hence, in this case $E$ is canonically isomorphic — and, if $\{G_n\}$ is a*

*) Actually, 3° implies this condition (see the Notes and remarks to § 18, theorem 18.10).

*monotone Schauder decomposition, E is canonically linearly isometric — to $V^*$).*

4°. *For every $\Phi \in E^{**}$ the series $\sum_{i=1}^{\infty} v_i^{**}(\Phi)$ converges.*

5°. *For every $\Psi \in V^*$ the series $\sum_{i=1}^{\infty} (v_i^*|_V)^*(\Psi)$ converges.*

*Proof.* The equivalences 2° ⇔ 4° and 3° ⇔ 5° follow from theorem 15.8 f), e). The implication 2° ⇒ 3° is obvious by restriction of 2° to $V$.

Assume now that we have 1° and let $\Phi \in E^{**}$. Since by proposition 15.7 c), $v_i^{**}(\Phi) \in \pi(G_i)$, let $y_i = \pi^{-1}(v_i^{**}(\Phi)) \in G_i$. Then, by proposition 15.6 c), $\sup_{1 \leqslant n < \infty} \left\| \sum_{i=1}^{n} y_i \right\| = \sup_{1 \leqslant n < \infty} \left\| \sum_{i=1}^{n} v_i^{**}(\Phi) \right\| < \infty$, whence, by 1°, $\sum_{i=1}^{\infty} y_i = x \in E$, and hence $\sum_{i=1}^{\infty} v_i^{**}(\Phi) = \sum_{i=1}^{\infty} \pi(y_i) = \pi(x)$. Thus, 1° ⇒ 4°.

Finally, assume that we have 5° and let $y_n \in G_n$ $(n = 1,2,\dots)$, $\sup_{1 \leqslant n < \infty} \left\| \sum_{i=1}^{n} y_i \right\| < \infty$, whence $\sup_{1 \leqslant n < \infty} \left\| \sum_{i=1}^{n} u(y_i) \right\| < \infty$. Then, by proposition 15.6 b), there exists $\Psi \in V^*$ such that

$$(v_n^*|_V)^*(\Psi) = u(y_n) \qquad (n = 1,2, \dots).$$

But then, by 5°, $\sum_{i=1}^{\infty} (v_i^*|_V)^*(\Psi) = \sum_{i=1}^{\infty} u(y_i)$ converges, whence, since $u$ is an isomorphism (by theorem 15.7), $\sum_{i=1}^{\infty} y_i$ converges. Thus, 5° ⇒ 1°, which completes the proof of theorem 15.14.

**Corollary 15.5.** *Let $\{G_n\}$ be a Schauder decomposition of a Banach space E, with the a.s.c.p. $\{v_n\} \subset L(E, E)$, such that each $G_n = v_n(E)$ $(n = 1,2, \dots)$ is reflexive. Then*

a) *$\{G_n\}$ is boundedly complete if and only if $\{v_n^*(E)^*\}$ is a shrinking Schauder decomposition of $V = \left[\bigcup_{n=1}^{\infty} v_n^*(E^*)\right]$.*

b) *$\{G_n\}$ is shrinking if and only if $\{v_n^*(E^*)\}$ is a boundedly complete Schauder decomposition of $V = \left[\bigcup_{n=1}^{\infty} v_n^*(E^*)\right]$.*

*Proof.* a) By theorem 15.7 and proposition 15.7 c), we have $\left[\bigcup_{n=1}^{\infty} (v_n^*|_V)^*(V^*)\right] = u(E)$. Hence, by theorem 15.14, equivalence 1°⇔3°, $\{G_n\}$ is boundedly complete if and only if $\left[\bigcup_{n=1}^{\infty} (v_n^*|_V)^*(V^*)\right] = V^*$. But, by theorem 15.6 and definition 15.15, this condition means that $\{v_n^*(E^*)\}$ is a shrinking Schauder decomposition of $V$.

b) Since the canonical mapping $u: E \to V^*$ is an isomorphism (by theorem 15.7), from proposition 15.7 c) it follows that $\{G_n\}$ is shrinking if and only if $\{u(G_n)\} = \{(v_n^*|_V)^*(V^*)\}$ is a shrinking Schauder decomposition of $u(E) = \left[\bigcup_{n=1}^{\infty} (v_n^*|_V)^*(V^*)\right]$. However, by theorem 15.6 and part a) above applied to the Schauder decomposition $\{v_n^*(E^*)\} = \{v_n^*|_V(V)\}$ of $V$, this happens if and only if $\{v_n^*(E^*)\}$ is a boundedly complete Schauder decomposition of $V$. This completes the proof of corollary 15.5.

Similarly to proposition 15.10, we have

**Proposition 15.11.** *If a Banach space $E$ contains a complemented subspace $G$ with a boundedly complete basis $\{x_n\}$, then $E$ has a boundedly complete Schauder decomposition.*

The assumption of reflexivity of each subspace $G_n = v_n(E)$ $(n = 1,2, \ldots)$ in theorem 15.14 is essential, as shown by

*Example 15.4.* a) Let $E = L^1([0, 1])$. Then $E$ contains*) a complemented subspace isomorphic to $l^1$ and hence, by proposition 15.11, $E$ has a boundedly complete Schauder decomposition. However, $E = L^1([0, 1])$ is not isomorphic to any conjugate Banach space**). b) *If $G$ is any Banach space, then $E = (G \times G \times \ldots)_{l^1}$ has a boundedly complete Schauder decomposition* $\{G_n\}$, namely, $G_n = \underbrace{\{0\} \times \ldots \times \{0\}}_{n-1} \times G \times \{0\} \times \ldots$ $(n=1,2, \ldots)$. However, for $G = c_0$, $E = (c_0 \times c_0 \times \ldots)_{l^1}$ is not isomorphic to any conjugate Banach space***). Note also that for $G = L^1([0, 1])$ we have $E = (L^1([0, 1]) \times L^1([0, 1]) \times \ldots)_{l^1} \equiv L^1([0, 1])$, which is not isomorphic to any conjugate Banach space. Indeed, $L^1([0, 1])$ has a (boundedly complete) Schauder $l^1$-decomposition****) similar to that of $U = S([0, 1])$ given after example 15.3, namely, $\{G_n'\} = \{\chi_{\left[\frac{n-1}{n}, \frac{n}{n+1}\right)} L^1([0, 1])\}$, where $\chi_A$ denotes the characteristic function of $A$; clearly, $\chi_{\left[\frac{n-1}{n}, \frac{n}{n+1}\right)} L^1([0, 1]) \equiv L^1([0, 1])$, whence, by Ch. II, § 18, lemma 18.7 a), we obtain the isometry $L^1([0, 1]) \equiv E$. Note also that these spaces $E$ have no boundedly complete basis (by Ch. II, § 6, theorem 6.2, implication 1° ⇒ 3°).

One can also give other characterizations of boundedly complete Schauder decompositions which generalize the corresponding results of § 7 and § 8. For example, we mention the following generalization of § 7, corollary 7.2 b) and theorem 7.9 (which we consider here only for Schauder decompositions of $E$):

*) See e.g. [205], theorem 6.
**) See e.g. Ch. II, § 15, theorem 15.3.
***) See e.g. [20], corollary 4.
****) See definition 15.18.

**Theorem 15.15.** *Let $\{G_n\}$ be a Schauder decomposition of a Banach space $E$, with the a.s.c.p. $\{v_n\} \subset L(E, E)$. The following statements are equivalent:*

*1°. $\{G_n\}$ is boundedly complete.*

*2°. For every bounded sequence $\{z_j\} \subset E$ such that the limits*

$$\lim_{j\to\infty} v_n(z_j) = y_n \qquad (n = 1,2, \dots) \tag{15.101}$$

*exist, there is an element $x \in E$ satisfying*

$$v_n(x) = y_n \qquad (n = 1,2, \dots). \tag{15.102}$$

*These statements are implied by — and, if each $G_n = v_n(E)$ $(n = 1,2, \dots)$ is reflexive*), they are equivalent to — the following statements (which are equivalent to each other for any separable total subspace $V$ of $E^*$):*

*3°. Every bounded $\sigma(E, V)$-Cauchy sequence $\{z_j\} \subset E$ is $\sigma(E, V)$-convergent to an element $x \in E$, where $V = \left[\bigcup_{n=1}^{\infty} v_n^*(E^*)\right]$.*

*4°. $E$ is canonically isomorphic to $V^*$, where $V = \left[\bigcup_{n=1}^{\infty} v_n^*(E^*)\right]$.*

*Proof.* The equivalence 3° ⇔ 4° holds for any separable total subspace $V$ of $E^*$, with the proof similar to that of § 7, theorem 7.9, equivalence 2° ⇔ 3° (taking now any $\{\varphi_n\} \subset V$ with $[\varphi_n] = V$).

Assume now 3° and let $\{z_j\} \subset E$ be a bounded sequence such that the limits (15.101) exist. We claim that $\{z_j\}$ is $\sigma(E, V)$-Cauchy. Indeed, if $f \in \bigcup_{n=1}^{\infty} v_n^*(E^*)$, say $f = v_n^*(f)$, then

$$|f(z_j) - f(z_{j+p})| = |v_n^*(f)(z_j) - v_n^*(f)(z_{j+p})| = |f(v_n(z_j) - v_n(z_{j+p}))| \leqslant$$
$$\leqslant \|f\| \, \|v_n(z_j) - v_n(z_{j+p})\| \qquad (j, p = 1,2, \dots),$$

whence $\{f(z_j)\}$ is Cauchy. Hence, since $\sup_{1\leqslant j<\infty} \|z_j\| < \infty$, it follows that the same holds for any $f \in V = \left[\bigcup_{n=1}^{\infty} v_n^*(E^*)\right]$, which proves the claim that $\{z_j\}$ is $\sigma(E, V)$-Cauchy. Consequently, by 3°, $\{z_j\}$ is $\sigma(E, V)$-convergent to an element $x \in E$. Then

$$f(v_n(x)) = v_n^*(f)(x) = \lim_{j\to\infty} v_n^*(f)(z_j) = \lim_{j\to\infty} f(v_n(z_j)) = f(y_n)$$
$$(f \in E^*, \; n = 1,2, \dots),$$

whence we obtain (15.102). Thus, 3° ⇒ 2°.

---

*) Actually, 4° implies this condition (see the Notes and remarks to § 18, theorem 18.10).

Assume now 2° and let $\{y_n\} \subset E$, $y_n \in G_n$ $(n=1,2,\dots)$, $\sup_{1 \leqslant n < \infty} \left\| \sum_{i=1}^{n} y_i \right\| < \infty$. Then $\lim_{j\to\infty} v_n\left(\sum_{i=1}^{j} y_i\right) = y_n$ $(n = 1,2, \dots)$, whence, by 2° $\left(\text{with } z_j = \sum_{i=1}^{j} y_i\right)$, there is an element $x \in E$ satisfying (15.102). But then, since $\{G_n\}$ is a Schauder decomposition, $\sum_{i=1}^{\infty} y_i = \sum_{i=1}^{\infty} v_i(x) = x$. Thus, $2° \Rightarrow 1°$.

Finally, assume 1° and that each $G_n = v_n(E)$ is reflexive. Then, by theorem 15.8 d'), c'), $\tau_0(u(E)) = D_1 = D_2 = \tau_0(V^*)$, whence, since $\tau_0$ is an isomorphism, $u(E) = V^*$. Thus, in this case, $1° \Rightarrow 4°$ (alternatively, one can observe that the equivalence $1° \Leftrightarrow 4°$ is nothing else than theorem 15.14, equivalence $1° \Leftrightarrow 3°$), which completes the proof of theorem 15.15.

*Definition 15.17.* A Schauder decomposition $\{G_n\}$ of a Banach space $E$ is said to be *of type* $wc_0$, if

$$y_n \in G_n \ (n = 1,2, \dots), \quad \sup_{1 \leqslant n < \infty} \|y_n\| < \infty \Rightarrow y_n \xrightarrow{w} 0. \qquad (15.103)$$

By theorem 15.13, implication $1° \Rightarrow 2°$, *every shrinking Schauder decomposition* $\{G_n\}$ *is of type* $wc_0$. In this connection, let us also mention the following generalization of § 6, proposition 6.5, implication $1° \Rightarrow 2°$ (and of Ch. II, § 7, proposition 7.4): *If* $(G_n, v_n)$ *is an E-complete generalized biorthogonal system, then for every sequence* $\{f_n\} \subset E^*$ *with* $f_n \in v_n^*(E^*)$ $(n = 1,2, \dots)$, $\sup_{1 \leqslant n < \infty} \|f_n\| < \infty$, *we have* $f_n \xrightarrow{w^*} 0$. Indeed, for any finite sum $\sum_{i=1}^{m} y_i$ with $y_i \in G_i$ $(i = 1, \dots, m)$ we have $\lim_{n\to\infty} f_n\left(\sum_{i=1}^{m} y_i\right) = \lim_{n\to\infty} v_n^*(f_n)\left(\sum_{i=1}^{m} y_i\right) = \lim_{n\to\infty} f_n\left(v_n\left(\sum_{i=1}^{m} y_i\right)\right) = \lim_{n\to\infty} f_n(0) = 0$, whence the assertion follows.

*Definition 15.18.* Let $F$ be a Banach space with an unconditional basis $\{e_n\}$. A Schauder decomposition $\{G_n\}$ of a Banach space $E$ is said to be

a) $(F, \{e_n\})$-*Besselian*, if

$$y_n \in G_n \ (n = 1,2, \dots), \ \sum_{i=1}^{\infty} y_i \text{ converges} \Rightarrow \sum_{i=1}^{\infty} \|y_i\| e_i \text{ converges}, \qquad (15.104)$$

i.e., if $\sum_{i=1}^{\infty} \|v_i(x)\| e_i$ converges for all $x \in E$, where $\{v_n\} \subset L(E, E)$ is the a.s.c.p. to $\{G_n\}$;

b) $(F, \{e_n\})$-*Hilbertian*, if

$$y_n \in G_n \ (n = 1,2, \ldots), \ \sum_{i=1}^{\infty} \|y_i\| e_i \text{ converges} \Rightarrow \sum_{i=1}^{\infty} y_i \text{ converges}, \quad (15.105)$$

i.e., if for every $\{y_n\} \subset E$ as in the left hand side of (15.105) there exists an $x \in E$ such that $v_n(x) = y_n$ $(n = 1,2, \ldots)$.

In the particular case when $F = l^p$ $(1 \leqslant p < \infty)$ or $c_0$ and $\{e_n\}$ = the unit vector basis of $F$, we use also the terms *p-Besselian*, *p-Hilbertian*, $\infty$-*Besselian* and $\infty$-*Hilbertian Schauder decompositions* respectively and, in particular when $p = 2$, the terms *Besselian* and *Hilbertian Schauder decompositions* respectively. A Schauder decomposition which is both $p$-Besselian and $p$-Hilbertian, where $1 \leqslant p \leqslant \infty$, is called an $l^p$-*decomposition* if $p < \infty$ and a $c_0$-*decomposition* if $p = \infty$.

Obviously, every Schauder decomposition $\{G_n\}$ of a Banach space $E$ is both $\infty$-Besselian and 1-Hilbertian. Also, every $c_0$-decomposition is shrinking, every $l^1$-decomposition is boundedly complete and every $l^p$-decomposition, where $1 < p < \infty$, is both shrinking and boundedly complete.

One can give various characterizations of $(F, \{e_n\})$-Besselian and $(F, \{e_n\})$-Hilbertian Schauder decompositions, which generalize the corresponding results of Ch. II, § 11, e.g. the following:

**Theorem 15.16.** *Let $F$ be a Banach space with an unconditional basis $\{e_n\}$ and let $\{G_n\}$ be a Schauder decomposition of a Banach space $E$. Then*

a) *$\{G_n\}$ is $(F, \{e_n\})$-Besselian if and only if there exists a constant $c > 0$ such that we have*

$$c \left\| \sum_{i=1}^{n} \|y_i\| e_i \right\| \leqslant \left\| \sum_{i=1}^{n} y_i \right\| \quad (15.106)$$

*for all finite sequences $y_i \in G_i$ $(i = 1, \ldots, n)$.*

b) *$\{G_n\}$ is $(F, \{e_n\})$-Hilbertian if and only if there exists a constant $C > 0$ such that we have*

$$\left\| \sum_{i=1}^{n} y_i \right\| \leqslant C \left\| \sum_{i=1}^{n} \|y_i\| e_i \right\| \quad (15.107)$$

*for all finite sequences $y_i \in G_i$ $(i = 1, \ldots, n)$.*

*Proof.* Consider the sequence space

$$D_F = \left\{ \{y_n\} \subset E \,\middle|\, y_n \in G_n \ (n = 1,2, \ldots), \ \sum_{i=1}^{\infty} \|y_i\| e_i \text{ converges} \right\}, \quad (15.108)$$

endowed with the norm

$$\|\{y_n\}\| = \sup_{1 \leqslant n < \infty} \left\| \sum_{i=1}^{n} \|y_i\| e_i \right\|. \quad (15.109)$$

Then, similarly to the proof of proposition 15.2, it follows that $D_F$ is a Banach space. Moreover, $D_F$ is a $BK$-space; indeed, since $\{e_n\}$ is a basis of $F$, the relations $\{y_n^{(j)}\} \in D_F$ $(j = 1,2, \ldots)$, $\{y_n\} \in D_F$, $\lim\limits_{j\to\infty} \sup\limits_{1\leqslant n<\infty} \left\| \sum\limits_{i=1}^{n} \|y_i^{(j)} - y_i\| e_i \right\| = 0$ clearly imply $\lim\limits_{j\to\infty} y_i^{(j)} = y_i$ $(i=1,2,\ldots)$.

Now, if $\{G_n\}$ is $(F, \{e_n\})$-Besselian, then we can define a linear mapping $v\colon E \to D_F$ by

$$v(x) = \{v_n(x)\} \qquad (x \in E), \tag{15.110}$$

where $\{v_n\} \subset L(E, E)$ is the a.s.c.p. to $\{G_n\}$.

We claim that $v$ is closed, and hence continuous. Indeed, let $\{x_j\} \subset E$, $x \in E$, $\lim\limits_{j\to\infty} x_j = x$ and $\lim\limits_{j\to\infty} v(x_j) = y = \{y_n\} \in D_F$. Then, since $D_F$ is a $BK$-space, $\lim\limits_{j\to\infty} v_n(x_j) = y_n$ $(n = 1,2, \ldots)$ and, on the other hand, $\lim\limits_{j\to\infty} v_n(x_j) = v_n(x)$ $(n = 1,2,\ldots)$. Therefore, $y = \{y_n\} = \{v_n(x)\} = v(x)$, which proves our claim. Consequently, there exists $c > 0$ such that we have (15.106).

Conversely, if there exists $c > 0$ such that we have (15.106), then, by Ch. I, § 7, theorem 7.1 for $\{e_n\}$, the linear mapping $v\colon \bigcup\limits_{n=1}^{\infty} G_n \to D_F$ defined by (15.110) is continuous and hence it can be extended to a mapping $v\colon E \to D_F$. Consequently, $\sum\limits_{i=1}^{\infty} \|v_i(x)\| e_i$ converges for all $x \in E$, so $\{G_n\}$ is $(F, \{e_n\})$-Besselian. This proves a).

The proof of part b) is similar, considering instead of $v$ the mapping $w\colon D_F \to E$ defined by

$$w(\{y_n\}) = \sum_{i=1}^{\infty} y_i \qquad (\{y_n\} \in D_F). \tag{15.111}$$

If $\{G_n\}$ is $(F, \{e_n\})$-Hilbertian, this mapping is well defined and closed. Indeed, let $\{y_n^{(j)}\} \in D_F$ $(j = 1,2, \ldots)$, $\{y_n\} \in D_F$, $\lim\limits_{j\to\infty} \{y_n^{(j)}\} = \{y_n\}$ and $\lim\limits_{j\to\infty} w(\{y_n^{(j)}\}) = \lim\limits_{j\to\infty} \sum\limits_{i=1}^{\infty} y_i^{(j)} = x \in E$. Then for $x_j = \sum\limits_{i=1}^{\infty} y_i^{(j)} \in E$ we have, since $D_F$ is a $BK$-space, $v_n(x) = \lim\limits_{j\to\infty} v_n(x_j) = \lim\limits_{j\to\infty} y_n^{(j)} = y_n$ $(n = 1,2, \ldots)$, whence $x = \sum\limits_{i=1}^{\infty} v_i(x) = \sum\limits_{i=1}^{\infty} y_i = w(\{y_n\})$, which proves that $w$ is closed. In the converse part, $w$ is defined and continuous on $\bigcup\limits_{n=1}^{\infty} G_1 \times \ldots \times G_n \times \{0\} \times \ldots$ endowed with the norm of $D_F$, so it can be extended to $w\colon D_F \to E$, which completes the proof of theorem 11.56.

We have the following generalization of Ch. II, § 11 part of corollary 11.2 and of Ch. II, § 25, theorem 25.1:

**Corollary 15.6.** *A Banach space $E$ has an $l^p$-decomposition, where $1 \leqslant p < \infty$, if and only if there exists a sequence of Banach spaces $\{E_n\}$ such that $E \cong (E_1 \times E_2 \times \ldots)_{l^p}$. A similar result also holds for $c_0$-decompositions.*

*Proof.* If $\{G_n\}$ is an $l^p$-decomposition of $E$, with a.s.c.p. $\{v_n\} \subset \subset L(E, E)$, then, by theorem 15.16, the mapping $v$ defined by (15.110) is an isomorphism of $E$ *onto* $D_F$ defined by (15.108), (15.109), with the inverse $w$ given by (15.111) (alternatively, one can also prove this similarly to the proof of proposition 15.3 a)). But, since $\{G_n\}$ is an $l^p$-decomposition of $E$, we have $F = l^p$, whence $D_F = (G_1 \times G_2 \times \ldots)_{l^p}$.

Conversely, if $u$ is an isomorphism of $E$ onto $(E_1 \times E_2 \times \ldots)_{l^p}$, then clearly $E = \sum_{n=1}^{\infty} \oplus G_n$, where $G_n = u^{-1}(\{0\} \times \ldots \times \{0\} \times E_n \times \{0\} \times \ldots)$ $(n=1,2, \ldots)$ and $\{y_n\} \in l^p$ for each $x = \sum_{i=1}^{\infty} y_i \in E$ with $y_n \in G_n$ $(n=1,2,\ldots)$. Finally, the argument for $c_0$ is similar, which completes the proof.

One can give various stability properties for $(F, \{e_n\})$-Besselian and $(F, \{e_n\})$-Hilbertian Schauder decompositions which generalize the corresponding results of Ch. II, § 11 (and, in particular, of Ch. I, § 10), e.g. the following:

**Theorem 15.17.** *Let $1 \leqslant p \leqslant \infty$ and let $\{G_n\}$ be a $p$-Besselian Schauder decomposition of a Banach space $E$. Then*

a) *There exists a constant $\lambda$ with $0 < \lambda < 1$ such that every sequence of subspaces $\{F_n\}$ of $E$, satisfying*

$$\left( \sum_{i=1}^{\infty} \theta(G_i, F_i)^q \right)^{\frac{1}{q}} \leqslant \lambda \tag{15.112}$$

*(when $q = \infty$, hence $p = 1$, by this we mean $\sup_{1 \leqslant n < \infty} \theta(G_n, F_n) \leqslant \lambda$), where $\theta(G_i, F_i)$ is the opening*) of the subspaces $G_i$, $F_i$ and where $\dfrac{1}{p} + \dfrac{1}{q} = 1$, is a Schauder decomposition of $E$, strictly equivalent**) to $\{G_n\}$.*

b) *Every sequence of subspaces $\{F_n\}$ of $E$, such that*

$$\sum_{i=1}^{\infty} \theta(G_i, F_i)^q < \infty \tag{15.113}$$

*(when $q = \infty$, hence $p = 1$, by this we mean $\lim_{n \to \infty} \theta(G_n, F_n) = 0$), and*

*) We recall (see Ch. II, § 1, definition 1.4) that $\theta(G, F) = \max \left( \sup_{\substack{x \in G \\ \|x\|=1}} \operatorname{dist}(x, F), \sup_{\substack{y \in F \\ \|y\|=1}} \operatorname{dist}(y, G) \right)$. Thus, if $\dim G = \dim F = 1$, $y \in G$, $z \in F$, $\|y\| = \|z\| = 1$, then $\theta(G, F) \leqslant \|y - z\|$.

**) See definition 15.11.

*admitting a sequence* $\{w_n\} \subset L(E, E)$ *such that* $(F_n, w_n)$ *is an E-complete generalized biorthogonal system, is a p-Besselian Schauder decomposition of E. If* $\dim G_n < \infty$ $(n = 1,2, \ldots)$, *i.e., if* $\{G_n\}$ *is a finite-dimensional decomposition of E, then the same conclusion holds for every ω-linearly independent sequence of subspaces* $\{F_n\}$ *satisfying* (15.113).

Since every Schauder decomposition is ∞-Besselian, for $p = \infty$, $q = 1$ we obtain a generalization of Ch. I, § 10, theorem 10.1 and part of theorem 10.2.

*Definition 15.19.* A Schauder decomposition $\{G_n\}$ of a Banach space $E$ is said to be *unconditional*, if every convergent series of the form $\sum_{i=1}^{\infty} y_i$, where $y_n \in G_n$ $(n = 1,2, \ldots)$, is unconditionally convergent, i.e., if for every $x \in E$ the series $\sum_{i=1}^{\infty} v_i(x)$ (where $\{v_n\} \subset L(E, E)$ is the a.s.c.p. to $\{G_n\}$) converges unconditionally. The Schauder decomposition $\{G_n\}$ is said to be *conditional* if it is non-unconditional.

The characterizations of unconditional bases, given in Ch. II, § 16 and § 17, admit natural generalizations to Schauder decompositions. As an example, we mention here only the following intrinsic characterizations of unconditional Schauder decompositions, which are generalizations of part of Ch. II, § 17, theorem 17.1:

**Theorem 15.18.** *Let E be a Banach space and* $\{G_n\}$ *a sequence of subspaces of E, with* $G_n \neq \{0\}$ $(n = 1,2, \ldots)$, $\left[\bigcup_{n=1}^{\infty} G_n\right] = E$. *The following statements are equivalent:*

1°. $\{G_n\}$ *is an unconditional Schauder decomposition of E.*

2°. *Every permutation* $\{G_{\sigma(n)}\}$ *of* $\{G_n\}$ *is a Schauder decomposition of E.*

3°. *There exists a constant M with* $1 \leqslant M < \infty$, *such that*

$$\left\| \sum_{j=1}^{n} y_{i_j} \right\| \leqslant M \left\| \sum_{j=1}^{n} y_{i_j} + \sum_{j=1}^{m} y_{l_j} \right\| \tag{15.114}$$

*for any* $n, m \in \mathcal{N}$ *and any* $y_{i_j} \in G_{i_j}$ $(j = 1, \ldots, n)$, $y_{l_j} \in G_{l_j}$ $(j = 1, \ldots, m)$ *which satisfy* $\{i_1, \ldots, i_n\} \cap \{l_1, \ldots, l_m\} = \varnothing$.

4°. *There exists a constant M′ with* $1 \leqslant M' < \infty$, *such that*

$$\left\| \sum_{i=1}^{n} \varepsilon_i y_i \right\| \leqslant M' \left\| \sum_{i=1}^{n} y_i \right\| \tag{15.115}$$

*for any* $n \in \mathcal{N}$, $y_i \in G_i$ *and* $\varepsilon_i \in K$ *with* $\varepsilon_i = \pm 1$ $(i = 1, \ldots, n)$.

5°. *There exists a constant $M''$ with $1 \leqslant M'' < \infty$, such that*

$$\left\| \sum_{i=1}^{n} \beta_i y_i \right\| \leqslant M'' \left\| \sum_{i=1}^{n} y_i \right\| \tag{15.116}$$

*for any $n \in \mathcal{N}$, $y_i \in G_i$ and $\beta_i \in K$ with $|\beta_i| \leqslant 1$ $(i = 1, \ldots, n)$.*

Similarly to propositions 15.10, 15.11 we have

**Proposition 15.12.** *If a Banach space $E$ contains a complemented subspace $G$ with an unconditional basis $\{x_n\}$, then $E$ has an unconditional Schauder decomposition.*

Hence, for example, $E = C([0, 1])$ and $E = L^1([0, 1])$ have unconditional Schauder decompositions, although they have no unconditional basis (see Ch. II, § 15). Also, taking into account lemma 15.1 and the proof of proposition 15.10, we obtain the following sharpening of corollary 15.3: *Every separable Banach space $E$ containing a subspace $G$ isomorphic to $c_0$ has an unconditional shrinking Schauder decomposition.* In particular, *every subspace of $c_0$ has an unconditional shrinking Schauder decomposition.*

In Ch. II, § 15 and § 17, we have seen that there exist Banach spaces (e.g. $C([0, 1])$, $L^1([0, 1])$, $J$) which cannot be embedded (isomorphically) into any Banach space with an unconditional basis. The same argument can be adapted (using unconditional bases with parentheses instead of unconditional bases) to show, more generally, that *these spaces cannot be embedded into any Banach space with an unconditional finite-dimensional decomposition.* However, for unconditional Schauder decompositions the situation is different, namely, *every Banach space $E$ can be embedded, as a complemented subspace, into a Banach space with an unconditional Schauder decomposition*, for example, into $E \times c_0$, or into $E \times l^1$ (clearly, a similar result also holds for the other special classes of Schauder decompositions considered above, e.g. monotone, normal, shrinking, boundedly complete).

We have the following generalization of Ch. II, § 18, theorem 18.2:

**Theorem 15.19.** a) *Every unconditional Schauder decomposition $\{G_n\}$ of $E = l^1$ is an $l^1$-decomposition.*

b) *Every unconditional Schauder decomposition $\{G_n\}$ of $E = c_0$ is a $c_0$-decomposition.*

*Proof.* a) Since by our assumption each expansion $x = \sum_{i=1}^{\infty} y_i \in E = l^1$ (where $y_n \in G_n$, $n = 1, 2, \ldots$) is unconditionally convergent, from Ch. II, § 14, lemma 14.10 it follows that $\sum_{i=1}^{\infty} \|y_i\|^2 < \infty$ for each $x = \sum_{i=1}^{\infty} y_i \in E$, so we can define a continuous linear mapping $u: E = l^1 \to l^2$ by

$$u(y_n) = \|y_n\| e_n \qquad \left(\sum_{i=1}^{\infty} y_i \in E;\ y_n \in G_n,\ n = 1,2,\dots\right), \tag{15.117}$$

where $\{e_n\}$ is the unit vector basis of $l^2$. Then, by Ch. II, § 18, lemma 18.3, $u$ is absolutely summing, whence, by (15.117) and the unconditional convergence of the expansion $x = \sum_{i=1}^{\infty} y_i$, we obtain

$$\sum_{i=1}^{\infty} \|y_i\| = \sum_{i=1}^{\infty} \|u(y_i)\| < \infty \qquad \left(x = \sum_{i=1}^{\infty} y_i \in E\right),$$

so $\{G_n\}$ is 1-Besselian. Hence, since every Schauder decomposition is 1-Hilbertian, $\{G_n\}$ is an $l^1$-decomposition of $E$.

b) If $\{G_n\}$ is an unconditional Schauder decomposition of $E = c_0$, then, by theorem 15.11, $\{v_n^*(E^*)\}$ is a weak*-unconditional Schauder decomposition of $E^* \equiv l^1$ (where $\{v_n\} \subset L(E, E)$ is the a.s.c.p. to $\{G_n\}$), whence, by Ch. II, § 15, lemmas 15.1 and 15.8, $f = \sum_{i=1}^{\infty} v_i^*(f)$ unconditionally, for all $f \in E^*$, in the norm topology of $E^*$, so $\{v_n^*(E^*)\}$ is an unconditional Schauder decomposition of $E^* \equiv l^1$. Therefore, by part a) proved above, $\{v_n^*(E^*)\}$ is an $l^1$-decomposition of $E^*$ and hence, by theorem 15.16 a), there exists a constant $c > 0$ such that

$$c \sum_{i=1}^{n} \|v_i^*(f)\| \leqslant \left\| \sum_{i=1}^{n} v_i^*(f) \right\| \qquad (f \in E^*,\ n = 1,2,\dots).$$

But then, taking into account (15.6),

$$\left\| \sum_{i=1}^{n} y_i \right\| = \sup_{\left\| \sum_{j=1}^{\infty} v_j^*(f) \right\| \leqslant 1} \left| \sum_{j=1}^{\infty} v_j^*(f) \left( \sum_{i=1}^{n} y_i \right) \right| \leqslant$$

$$\leqslant \frac{1}{c} \sup_{\sum_{j=1}^{\infty} \|v_j^*(f)\| \leqslant 1} \left| \sum_{j=1}^{n} v_j^*(f)(y_j) \right| \leqslant \frac{1}{c} \max_{1 \leqslant j \leqslant n} \|y_j\|$$

$$(y_i \in G_i;\ i = 1,\dots,n;\ n = 1,2,\dots),$$

and hence $\{G_n\}$ is $\infty$-Hilbertian. Thus, since every Schauder decomposition is $\infty$-Besselian, $\{G_n\}$ is a $c_0$-decomposition of $E$, which completes the proof of theorem 15.19.

In Vol. III we shall see that a result similar to a) holds for every "$\mathscr{L}_1$ space" $E$ (hence, e.g., for $E = L^1([0, 1])$) and a result similar to b) holds for every "$\mathscr{L}_\infty$ space" $E$ (hence, e.g., for $E = C([0, 1])$).

One can also define *orthogonal* and *hyperorthogonal* Schauder decompositions, by properties 3° and 5° of theorem 15.18 with $M = 1$ and $M'' = 1$, respectively. Furthermore, using the notion of strict equivalence of Schauder decompositions (definition 15.11), one can introduce Schauder decompositions which are called *subsymmetric* (unconditional and strictly equivalent to every subsequence) or *symmetric* (strictly equivalent to every permutation). Finally, we mention that one can define *absolutely convergent* Schauder decompositions $\{G_n\}$ by the condition that every convergent series of the form $\sum_{i=1}^{\infty} y_i$, where $y_n \in G_n$ $(n = 1,2, \ldots)$, is absolutely convergent $\left(\text{i.e., } \sum_{i=1}^{\infty} \|y_i\| < \infty\right)$, or, in other words, that $\sum_{i=1}^{\infty} \|v_i(x)\| < \infty$ for all $x \in E$, where $\{v_n\}$ is the a.s.c.p. to $\{G_n\}$. However, this latter one coincides with the notion of *$l^1$-decomposition*, introduced in definition 15.18 above. Hence, by corollary 15.6, $E$ has such a decomposition if and only if there exists a sequence of Banach spaces $\{E_n\}$ such that $E \cong (E_1 \times E_2 \times \ldots)_{l^1}$; in particular, by example 15.4 b), $E = L^1([0, 1])$ has such a decomposition, although it has no absolutely convergent basis (by Ch. II, § 25, theorem 25.1).

Some concepts and results of the previous sections of this Chapter can be also generalized to the setting of sequences of subspaces, as we have seen e.g. in theorem 15.15, definition 15.8 and the remark made after definition 15.8. Let us give now some other generalizations in this direction.

*Definition 15.20.* A sequence $\{G_n\}$ of subspaces of a Banach space $E$ is called an *$M$-decomposition* (or *Markuševič decomposition*) of $E$, if there exists a sequence *) $\{v_n\} \subset L(E, E)$ such that $(G_n, v_n)$ is an $E$-complete total generalized biorthogonal system**).

Clearly, every Schauder decomposition $\{G_n\}$ of a Banach space $E$ is an $M$-decomposition of $E$, but the converse is not true.

We have the following characterization of $M$-decompositions, which is a generalization of § 8, proposition 8.7':

**Proposition 15.13.** *A sequence $\{G_n\}$ of subspaces of a Banach space $E$, with $G_n \neq \{0\}$ $(n = 1,2, \ldots)$ and $\left[\bigcup_{n=1}^{\infty} G_n\right] = E$, is an $M$-decomposition of $E$ if and only if both of the following conditions hold:*

(i) *For every (finite or infinite) sequence of positive integers $\{n_k\}$ such that $\mathcal{N} \setminus \{n_k\}$ is infinite, we have* $\left[\bigcup_{k=1}^{\infty} G_{n_k}\right] \cap \left[\bigcup_{j \in \mathcal{N} \setminus \{n_k\}} G_j\right] = \{0\}$.

(ii) $G_n + [\bigcup_{j \neq n} G_j] = E$ $(n = 1,2, \ldots)$.

*Proof.* The proof of the necessity of (i) is immediate (similarly to that of § 8, proposition 8.6). Also, if $\{G_n\}$ is an $M$-decomposition of $E$,

---

*) Clearly, the sequence $\{v_n\}$ is uniquely determined.

**) See the remark before definition 15.7.

then, as was observed after definition 15.9, $E = G_n \oplus [\bigcup_{j \neq n} G_j]$ $(n = 1,2, \ldots)$, so (ii) is necessary.

Conversely, assume now that (i) and (ii) hold. Then, applying successively (i) to the singletons $\{n_k\} = \{1\}, \{2\}, \ldots$ and taking into account (ii), we obtain $E = G_n \oplus [\bigcup_{j \neq n} G_j]$ $(n = 1,2, \ldots)$ and hence, as was observed after definition 15.9, there exists a unique sequence $\{v_n\} \subset L(E, E)$ such that $(G_n, v_n)$ is an $E$-complete generalized biorthogonal system. Assume now that $\{v_n\}$ is not total on $E$, say $x \in E \setminus \{0\}$, $v_n^*(f)(x) = 0$ $(f \in E^*,\ n = 1,2, \ldots)$. Then, by the proof of proposition 15.9, we have

$$x \in \bigcap_{n=1}^{\infty} \bigcap_{j=1}^{n} v_j^*(E^*)_\perp = \bigcap_{n=1}^{\infty} \left[\bigcup_{j=1}^{n} v_j^*(E^*)\right]_\perp = \bigcap_{n=1}^{\infty} \left[\bigcup_{j=n+1}^{\infty} G_j\right].$$

Now, since $x \in \left[\bigcup_{j=1}^{\infty} G_j\right] = E$, there is an index $m_1$ such that $\operatorname{dist}\left(x, \left[\bigcup_{j=1}^{m_1} G_j\right]\right) < 1$. Next, since $x \in \left[\bigcup_{j=m_1+1}^{\infty} G_j\right]$, there is an index $m_2 > m_1$ such that $\operatorname{dist}\left(x, \left[\bigcup_{j=m_1+1}^{m_2} G_j\right]\right) < \frac{1}{2}$. Continuing in this way indefinitely, and then putting $\{n_k\} = \bigcup_{n=1}^{\infty} I_{2n-1}$ where $I_n = \{m_{n-1} + 1, \ldots, m_n\}$ $(n = 1,2, \ldots)$, we obtain, as in § 8, proof of proposition 8.7, that $\mathcal{N} \setminus \{n_k\} = \bigcup_{n=1}^{\infty} I_{2n}$ is infinite and $x \in \left(\left[\bigcup_{k=1}^{\infty} G_{n_k}\right] \cap [\bigcup_{j \in \mathcal{N} \setminus \{n_k\}} G_j]\right) \setminus \{0\}$, n contradiction with (i), which completes the proof of proposiion 15.13.

*Remark 15.7.* If $\dim G_n < \infty$ $(n = 1,2, \ldots)$ and $\left[\bigcup_{n=1}^{\infty} G_n\right] = E$, then condition (ii) is automatically satisfied (which shows, in particular, that proposition 15.13 yields indeed § 8, proposition 8.7′ as a particular case), since then $G_n + [\bigcup_{j \neq n} G_j]$ is*) closed $(n = 1,2, \ldots)$.

*Definition 15.21.* a) A sequence $\{G_n\}$ of (closed linear) subspaces of a Banach space $E$, with $G_n \neq \{0\}$ $(n = 1,2, \ldots)$, is called a *pseudo-decomposition***) of $E$, if for every $x \in E$ there exists a sequence $\{y_n\} \subset E$ with $y_n \in G_n$ $(n = 1,2, \ldots)$ such that we have (15.1) $\left(\text{i.e., if } E = \sum_{i=1}^{\infty} G_i,\right.$ where $\sum_{i=1}^{\infty} G_i$ is defined by $\left.(15.33)\right)$.

---

*) See e.g. [33], Ch. I, § 2, corollary 4 or [70], Ch. I, § 4, theorem 1.

**) We omit "Schauder", which will lead to no confusion.

b) A pseudo-decomposition $\{G_n\}$ of $E$ is said to be *unconditional* if every $x \in E$ has an unconditionally convergent expansion (15.1) with $y_n \in G_n$ $(n = 1,2, \dots)$.

*Definition 15.22.* a) A sequence $\{G_n\}$ of subspaces of a Banach space $E$, with $G_n \neq \{0\}$ $(n = 1,2, \dots)$, is called a *quasi-decomposition* of $E$, if there exists a sequence of operators (not necessarily projections) $v_n \in L(E, G_n)$ $(n = 1,2, \dots)$ such that we have (15.5) for every $x \in E$; any such sequence $\{v_n\}$ is called an *admissible sequence* (for $\{G_n\}$).

b) A quasi-decomposition $\{G_n\}$ of a Banach space $E$ is said to be *unconditional*, if it has an admissible sequence $\{v_n\}$ such that all series (15.5) are unconditionally convergent.

*Remark 15.8.* In contrast with the case of decompositions (see definition 15.1 and remark 15.1), the above definitions permit some trivial cases, too. For example, any Banach space $E$ admits the trivial quasi-decomposition $\{G_n\}$, where $G_1 = E$ and where $G_n \neq \{0\}$ $(n = 2,3, \dots)$ are arbitrary, with the admissible sequence $v_1 = I_E$, $v_n = 0$ $(n = 2,3, \dots)$. This is due, in part, to the fact that in § 9, definition 9.2 (of quasi-bases), admissible sequences $\{f_n\} \subset E^*$ with some $f_n = 0$ were also allowed and, in part, to the fact that theorem 15.20 below, implication $5° \Rightarrow 1°$, may lead indeed to such a trivial quasi-decomposition of $E$ (when $u_0(E) = F_1$ and $v(B) = F_1 = v(F_1)$, $v(F_2) = v(F_3) = \dots = 0$, i.e., $v = w_1$, where $\{w_n\}$ is the a.s.c.p. to $\{F_n\}$); however, the implications $1° \Rightarrow 2°, 3°, 4°, 5°$ of theorem 15.20 below are meaningful for any (trivial or non-trivial) quasi-decomposition $\{G_n\}$. One can exclude trivial quasi-decompositions assuming e.g. that $\dim G_n < \infty$ for all $n = 1,2, \dots$ (and, as before, $\dim E = \infty$).

Corresponding to § 9, theorem 9.1, let us prove now

**Theorem 15.20.** *Let $\{G_n\}$ be a sequence of subspaces of a Banach space E. The following statements are equivalent:*

1°. *$\{G_n\}$ is a quasi-decomposition of E.*

2°. *$\{G_n\}$ is a pseudo-decomposition of E and $D_0$ (defined by* (15.34)) *is a complemented subspace of the Banach space $D_1$ (defined by* (15.8), (15.9)).

3°. *$\{G_n\}$ is a pseudo-decomposition of E and there exists an isomorphism $u_0$ of E into $D_1$ such that*

$$D_1 = u_0(E) \oplus D_0, \tag{15.118}$$

*where $D_1$ and $D_0$ are as in* 2°.

4°. *There exist an isomorphism $u_0$ of E into $D_1$ and a projection v of $D_1$ onto $u_0(E)$ such that*

$$G_k = u_0^{-1} v(F_k) \qquad (k = 1,2, \dots), \tag{15.119}$$

*where $\{F_n\}$ is the Schauder decomposition of $D_1$ defined by* (15.15).

5°. *There exist a Banach space $B$ with a Schauder decomposition $\{F_n\}$, an isomorphism $u_0$ of $E$ into $B$ and a projection $v$ of $B$ onto $u_0(E)$ such that we have* (15.119).

*Proof.* The proof is similar to that of § 9, theorem 9.1, with the following adjustments: If we have 1°, the isomorphism $u_0$ of $E$ into $D_1$ is defined by

$$u_0(x) = \{v_n(x)\} \qquad (x \in E), \tag{15.120}$$

and to show (15.118) one can also use, alternatively, proposition 15.5. Thus, 1° ⇒ 3°. The implication 3° ⇒ 2° is obvious. If we have 2°, say $D_1 = D_0 \oplus F$ and if $w$ is the continuous linear mapping of $D_1$ onto $E$ defined by

$$w(\{y_n\}) = \sum_{i=1}^{\infty} y_i \qquad (\{y_n\} \in D_1), \tag{15.121}$$

then $w|_F$ is an isomorphism of $F$ onto $E$ and hence, for every $x \in E$, we can define $\{v_n(x)\} = (w|_F)^{-1}(x) \in F$. Then, since $F \subset D_1$, we have $v_n(x) \in G_n$ $(x \in E,\ n = 1,2, \dots)$. Also,

$$x = w(\{v_n(x)\}) = \sum_{i=1}^{\infty} v_i(x) \qquad (x \in E)$$

and each $v_n$ is linear on $E$ and satisfies

$$\|v_n(x)\| \leqslant 2 \sup_{1 \leqslant k < \infty} \left\| \sum_{i=1}^{k} v_i(x) \right\| = 2\|\{v_n(x)\}\| \leqslant 2\|(w|_F)^{-1}\| \, \|x\|$$
$$(x \in E,\ n = 1,2, \dots),$$

so $\{G_n\}$ is a quasi-decomposition of $E$. Thus, 2° ⇒ 1°. Assuming again that we have 1° and that $\{v_n\}$ is an admissible sequence for $\{G_n\}$, let $u_0$ be the isomorphism (15.120) of $E$ into $D_1$ and, by 3°, let $v$ be the projection of $D_1$ onto $u_0(E)$ along $D_0$. Then

$$v(\{y_n\}) = \left\{ v_n \left( \sum_{i=1}^{\infty} y_i \right) \right\} \qquad (\{y_n\} \in D_1), \tag{15.122}$$

since for every $\{v_n(x)\} \in u_0(E)$ we have $v(\{v_n(x)\}) = \{v_n(x)\} = \left\{ v_n \left( \sum_{i=1}^{\infty} v_i(x) \right) \right\}$ and since for every $\{y_n\} \in D_0$ we have $v(\{y_n\}) = 0 = \{v_n(0)\} = \left\{ v_n \left( \sum_{i=1}^{\infty} y_i \right) \right\}$. By (15.122) we have, in particular,

$$v(\{\delta_{nk} y_n\}) = \left\{ v_n \left( \sum_{i=1}^{\infty} \delta_{ik} y_i \right) \right\} = \{v_n(y_k)\} = u_0(y_k)$$
$$(\{\delta_{nk} y_n\} \in F_k;\ k = 1,2, \dots),$$

whence we infer (15.119). Thus, $1° \Rightarrow 4°$. The implication $4° \Rightarrow 5°$ is obvious, with $B = D_1$. Finally, if we have 5° and if $\{w_n\} \subset L(B, B)$ is the a.s.c.p. to $\{F_n\}$, then

$$y = v(y) = v\left(\sum_{i=1}^{\infty} w_i(y)\right) = \sum_{i=1}^{\infty} v(w_i(y)) \qquad (y \in v(B)),$$

and therefore, since $vw_i|_{v(B)} \in L(v(B), v(F_i))$ $(i = 1,2, \ldots)$, $\{v(F_n)\}$ is a quasi-decomposition of $v(B) = u_0(E)$. Hence, since $u_0$ is an isomorphism of $E$ onto $u_0(E)$, by (15.119) it follows that $\{G_n\}$ is a quasi-decomposition of $E$. Thus, $5° \Rightarrow 1°$, which completes the proof of theorem 15.20. Clearly, the other equivalences of § 9, theorem 9.1 can be also extended.

*The unconditional analogue of theorem* 15.20 (i.e., theorem 15.20 with "unconditional" inserted before "quasi-decomposition", "pseudo-decomposition" and "decomposition" and with $D_1$ replaced by the Banach space

$$D_1^{(u)} = \left\{\{y_n\} \subset E \,\middle|\, y_n \in G_n \ (n = 1,2, \ldots), \ \sum_{i=1}^{\infty} y_i \text{ is unconditionally convergent}\right\}, \qquad (15.123)$$

where the norm is defined by $\|\{y_n\}\| = \sup\limits_{|\beta_1|, |\beta_2|, \ldots \leq 1} \sup\limits_{1 \leq n < \infty} \left\| \sum\limits_{i=1}^{\infty} \beta_i y_i \right\|$ ), *is also valid*, with an analogous proof, since one can show, similarly to the third proof of theorem 15.1, that $\{F_n\}$ (defined by (15.15)) is an unconditional Schauder decomposition of $D_1^{(u)}$.

Let us make now the following observation: *A Banach space $E$ has a finite-dimensional quasi-decomposition* $\{G_n\}$ (i.e., such that $\dim G_n < \infty$ for all $n = 1,2, \ldots$) *if and only if $E$ admits a finite-dimensional expansion $\{v_n\}$ of $I_E$* (see § 9, definition 9.5). Indeed, if $\{G_n\}$ is a finite-dimensional quasi-decomposition of $E$, then any admissible sequence $\{v_n\}$ for $\{G_n\}$ is a finite-dimensional expansion of $I_E$ and conversely, if $\{v_n\}$ is a finite-dimensional expansion of $I_E$, then $\{G_n\} = \{v_n(E)\}$ is a finite-dimensional quasi-decomposition of $E$. Clearly, *the unconditional analogue of this statement is also valid*. Hence, from the unconditional analogue of theorem 15.20, equivalence $1° \Leftrightarrow 5°$, we obtain

**Corollary 15.7.** *A Banach space $E$ admits an unconditional finite-dimensional expansion of $I_E$ if and only if $E$ is isomorphic to a complemented subspace of a Banach space $E$ with an unconditional finite-dimensional decomposition.*

In § 9 we have seen (after definition 9.7) that in the necessity part of corollary 15.7 above one cannot replace "unconditional finite-dimensional decomposition" by "unconditional basis" (although in the non-unconditional case such a sharpening is possible, by § 9, theorem 9.3, implication $3° \Rightarrow 9°$).

We have observed (after proposition 15.12) that there exist Banach spaces (e.g. $C([0, 1])$, $L^1([0, 1])$, $J$) which cannot be embedded into any Banach space with an unconditional finite-dimensional decomposition. Now, from corollary 15.7 it follows, more generally, that *these spaces cannot be embedded into any Banach space $E$ admitting an unconditional finite-dimensional expansion of $I_E$.*

On the other hand, we have seen in Ch. II, § 15, theorem 15.5 and its proof, that the space $l^1$ has a subspace $E_0$ (e.g., $E_0 = u^{-1}(0)$, where $u$ is any continuous linear mapping of $l^1$ onto $L^1([0, 1])$), which is not isomorphic to any complemented subspace of any conjugate Banach space and hence $E_0$ has no unconditional basis. Now we shall prove, more generally, *that this subspace $E_0$ of $l^1$ does not admit any unconditional finite-dimensional expansion of $I_{E_0}$.* Assume, a contrario, that $E_0$ admits an unconditional finite-dimensional expansion $\{v_n\}$ of $I_{E_0}$. Then, by the unconditional analogue of theorem 15.20, implication $1° \Rightarrow 4°$ (with $E_0 = E$ and $\{G_n\} = \{v_n(E_0)\}$), $E_0$ is isomorphic to a complemented subspace of $D_1^{(u)} = D_1^{(u)}(E_0)$. We shall show that $D_1^{(u)}$ has a boundedly complete finite-dimensional decomposition, namely $\{F_n\}$ (defined by (15.15) with $E = E_0$ and $\{G_n\} = \{v_n(E_0)\}$) whence, by theorem 15.14 (or 15.15), $D_1^{(u)}$ is isomorphic to a conjugate Banach space and thus $E_0$ is isomorphic to a complemented subspace of a conjugate Banach space, a contradiction which will complete the proof. We have already observed above that $\{F_n\}$ is a Schauder decomposition (even unconditional) of $D_1^{(u)}$ and, by (15.15), we have $\dim F_n = \dim G_n = \dim v_n(E_0) < \infty$ $(n = 1,2, \ldots)$. Thus, it remains to show that $\{F_n\}$ is boundedly complete. Let $z_n = \{\underbrace{0, \ldots, 0}_{n-1}, y_n, 0, \ldots\} \in F_n$ $(n = 1,2, \ldots)$, $\sup_{1 \leqslant n < \infty} \left\| \sum_{i=1}^{n} z_i \right\| < \infty$.

Then, by the definition of the norm in $D_1^{(u)}$, $\sup_{|\beta_1|, |\beta_2|, \ldots \leqslant 1} \sup_{1 \leqslant n < \infty} \left\| \sum_{i=1}^{n} \beta_i y_i \right\| < \infty$. Hence, since $l^1$ (and therefore $E_0$) contains no subspace isomorphic to $c_0$, from Ch. II, § 15, corollary 15.1 and lemma 15.8 it follows that $\sum_{i=1}^{\infty} y_i$ is unconditionally convergent in $E_0$. Consequently, $\sum_{i=1}^{\infty} z_i = \sum_{i=1}^{\infty} \{\underbrace{0, \ldots, 0}_{i-1}, y_i, 0, \ldots\}$ converges in $D_1^{(u)}$, which completes the proof of our assertion.

One can also define *absolutely convergent* quasi-decompositions $\{G_n\}$ (finite-dimensional expansions $\{v_n\}$ of $I_E$) by the condition that there exists an admissible sequence $\{v_n\}$ such that (respectively, for the given sequence $\{v_n\}$) all series (15.5) are absolutely convergent $\left(\text{i.e., } \sum_{i=1}^{\infty} \|v_i(x)\| < \infty \text{ for all } x \in E\right)$. Then, similarly to the unconditional case, *a Banach* *space $E$ has an absolutely convergent finite-dimensional quasi-decomposition $\{G_n\}$ if and only if $E$ admits an absolutely convergent finite-dimensional expansion of $I_E$.* However, in contrast with theorem 15.19 a), there exists in $E = l^1$ an unconditional two-dimensional expansion $\{v_n\}$ of $I_E$ (i.e., such that $\dim v_n(E) \leqslant 2$ for all $n = 1,2, \ldots$), which is not absolutely convergent, as shown by

*Example 15.5.* Let $\{x_n\}$ be an unconditionally summable sequence in $E = l^1$, such that $\sum_{i=1}^{\infty} x_i = 0$ and $\sum_{i=1}^{\infty} \|x_i\| = \infty$ (such a sequence exists by the Dvoretzky-Rogers theorem*)). Put

$$v_n(x) = \xi_n e_n + \left(\sum_{k=1}^{\infty} \xi_{2^{k-1}(2j-1)}\right) x_i \qquad (x = \{\xi_k\} \in E, n = 2^{i-1}(2j-1); \; i, j = 1,2, \ldots). \tag{15.124}$$

Then $v_n(E) \subset [e_n, x_i]$, whence $\dim v_n(E) \leqslant 2$ $(n = 1,2, \ldots)$. Also, $\sum_{n=1}^{\infty} v_n(x) = \sum_{j=1}^{\infty} \xi_j e_j + \sum_{i=1}^{\infty} \sum_{j=1}^{\infty} \left(\sum_{k=1}^{\infty} \xi_{2^{k-1}(2j-1)}\right) x_i = x$ unconditionally, for all $x \in E$. However, for any $x = \{\xi_k\} \in E$ such that $\sum_{k=1}^{\infty} \xi_{2^{k-1}} = C > 0$ we have

$$\|v_{2^{i-1}}(x)\| = \|\xi_{2^{i-1}} e_{2^{i-1}} + C x_i\| \geqslant C\|x_i\| - |\xi_{2^{i-1}}| \qquad (i = 1,2, \ldots),$$

whence

$$\sum_{n=1}^{\infty} \|v_n(x)\| \geqslant \sum_{i=1}^{\infty} \|v_{2^{i-1}}(x)\| \geqslant C \sum_{i=1}^{\infty} \|x_i\| - \|x\| = \infty,$$

so $\{v_n\}$ is an unconditional two-dimensional expansion of $I_E$ in $E = l^1$, which is not absolutely convergent.

Leaving to the reader the other possible generalizations of the previous sections to this setting, let us give now some relations between Schauder decompositions, basic sequences and bases in Banach spaces. The intrinsic characterization of Schauder decompositions, given in theorem 15.5, equivalence $1° \Leftrightarrow 2°$, suggests that Schauder decompositions are closely related to basic sequences which one can "extract from them". The following theorem shows that this is indeed the case:

**Theorem 15.21.** a) *If $\{G_n\}$ is a Schauder decomposition of a Banach space $E$, then every sequence $\{y_n\} \subset E$ with $y_n \in G_n \setminus \{0\}$ $(n = 1,2, \ldots)$ is a basic sequence.*

b) *Conversely, if $\{G_n\}$ is a sequence of subspaces of a Banach space $E$, with $G_n \neq \{0\}$ $(n = 1,2, \ldots)$, such that every sequence $\{y_n\} \subset E$ with $y_n \in G_n \setminus \{0\}$ $(n = 1,2, \ldots)$ is a basic sequence, then there exists***)*

*) See § 1, the remark made after problem 1.2.

**) Clearly, this property is a generalization of the notion of deficient basic sequence, but there exists no special term for it.

*a positive integer N such that* $\{G_n\}_{n=N}^{\infty}$ *is a Schauder decomposition of* $\left[\bigcup_{n=N}^{\infty} G_n\right]$.

*Proof*. Part a) is an obvious consequence of theorem 15.5 and Ch. I, § 7, theorem 7.1, but it can be seen also directly, as follows: Let $y_n \in G_n \setminus \{0\}$ $(n = 1,2, \ldots)$ and $z \in [y_n]$. Then we can write $z = \lim_{n\to\infty} \sum_{i=1}^{m_n} \alpha_i^{(n)} y_i$. Hence, if $\{v_n\} \subset L(E, E)$ is the a.s.c.p. to $\{G_n\}$,

$$\alpha_j^{(n)} y_j = v_j\left(\sum_{i=1}^{m_n} \alpha_i^{(n)} y_i\right) \to v_j(z) \text{ as } n \to \infty \qquad (j = 1,2, \ldots),$$

whence $v_j(z) = \alpha_j y_j$ $(j = 1,2,\ldots)$, and therefore $z = \sum_{j=1}^{\infty} v_j(z) = \sum_{j=1}^{\infty} \alpha_j y_j$. Since $\alpha_j y_j \in G_j$ $(j = 1,2, \ldots)$ and since $\{G_j\}$ is a decomposition of $E$, the sequence $\{\alpha_j y_j\}$ is uniquely determined, whence, since $y_j \neq 0$ $(j = 1,2, \ldots)$, the sequence $\{\alpha_j\}$ is uniquely determined, which proves that $\{y_n\}$ is a basic sequence.

b) Assume that $\{G_n\}$ is a sequence of subspaces of $E$, with $G_n \neq \{0\}$ $(n = 1,2, \ldots)$, such that every sequence $y_n \in G_n \setminus \{0\}$ $(n = 1,2,\ldots)$ is a basic sequence. We claim that there exist a positive integer $N$ and a constant $C \geqslant 1$ such that for every such sequence $\{y_n\}$ we have $v_{\{y_j\}_{j=N}^{\infty}} \leqslant C$, where $v$ is the norm*) of the basic sequence.

Indeed, if $N$ and $C$ do not exist, then for each positive integer $n$ and each $M \geqslant 1$ there exists a sequence $y_j \in G_j \setminus \{0\}$ $(j = 1,2,\ldots)$ with $v_{\{y_j\}_{j=N}^{\infty}} > M$. Choose $y_j^{(1)} \in G_j \setminus \{0\}$ $(j = 1,2, \ldots)$ such that $v_{\{y_j^{(1)}\}} > 2$. Then there exist $q_1 > p_1$ and $\alpha_1, \ldots, \alpha_{q_1} \in K$ such that

$$\left\|\sum_{j=1}^{p_1} \alpha_j y_j^{(1)}\right\| > 2\left\|\sum_{j=1}^{q_1} \alpha_j y_j^{(1)}\right\|.$$

Continuing in this way, for each $n$ choose $y_j^{(n)} \in G_j \setminus \{0\}$ $(j=1,2,\ldots)$ such that $v_{\{y_j^{(n)}\}_{j=q_{n-1}+1}^{\infty}} > 2^n$ (where $q_0 = 0$) and then $q_n > p_n \geqslant$ $\geqslant q_{n-1} + 1$ and $\alpha_{q_{n-1}+1}, \ldots, \alpha_{q_n} \in K$ such that

$$\left\|\sum_{j=q_{n-1}+1}^{p_n} \alpha_j y_j^{(n)}\right\| > 2^n \left\|\sum_{j=q_{n-1}+1}^{q_n} \alpha_j y_j^{(n)}\right\|.$$

Put

$$y_j = y_j^{(n)} \qquad (q_{n-1} + 1 \leqslant j \leqslant q_n;\ n = 1,2, \ldots). \qquad (15.125)$$

*) See Ch. I, § 7, definition 7.1.

Then $y_j \in G_j \setminus \{0\}$ $(j = 1,2,\ldots)$ and, by Ch. I, § 7, theorem 7.1, $\{y_j\}$ is not a basic sequence, in contradiction with our assumption. This proves the claim that there exist $N$ and $C \geqslant 1$ such that for every sequence $y_n \in G_n \setminus \{0\}$ $(n = 1,2,\ldots)$ we have $v_{\{y_j\}_{j=N}^{\infty}} \leqslant C$.

Consequently, by theorem 15.5, implication $2^\circ \Rightarrow 1^\circ$, $\{G_n\}_{n=N}^{\infty}$ is a Schauder decomposition of $\left[\bigcup_{n=N}^{\infty} G_n\right]$, which completes the proof of theorem 15.21.

*Remark 15.9.* From the definition and theorem 15.13 it follows that the basic sequences $y_n \in G_n \setminus \{0\}$ $(n = 1,2,\ldots)$ conserve the special type of the Schauder decomposition $\{G_n\}$ of $E$. Namely, if $\{G_n\}$ is monotone, or shrinking*⁾, or boundedly complete, or of type $wc_0$, or unconditional, then so is $\{y_n\}$. Also, if $\{G_n\}$ is normal, or $p$-Besselian, or $p$-Hilbertian $(1 \leqslant p \leqslant \infty)$, then so is every basic sequence $\{y_n\} \subset E$ with $y_n \in G_n$, $\|y_n\| = 1$ $(n = 1,2,\ldots)$.

In theorem 15.21 b) it may happen that $N \geqslant 2$, as shown by

*Example 15.6.* Let $E$ be a separable Banach space, $\{x_n\}$ a basic sequence in $E$ such that $\dim [x_n] = \operatorname{codim} [x_n] = \infty$, and $G_1$ a subspace of $E$, which is a quasi-complement but not a complement**⁾ of $[x_n]$ in $E$. Put

$$G_2 = [x_1], \ G_3 = [x_2], \ldots \tag{15.126}$$

Then each sequence $y_n \in G_n \setminus \{0\}$ $(n = 1,2,\ldots)$ is a basic sequence, but $\{G_n\}$ is not a Schauder decomposition of $E$, since $G_1 \oplus \left[\bigcup_{n=2}^{\infty} G_n\right] = G_1 \oplus [x_n] \neq E$. Clearly, $\{G_n\}_{n=2}^{\infty}$ is a Schauder decomposition of $\left[\bigcup_{n=2}^{\infty} G_n\right]$, so $N = 2$.

From example 15.6 it follows that in order to have $N = 1$, i.e., in order that $\{G_n\}$ be a Schauder decomposition of $E$ when $E = \left[\bigcup_{n=1}^{\infty} G_n\right]$, it is necessary that for each $N$ we keep $\left[\bigcup_{n=N}^{\infty} G_n\right]$ from being a quasi-complement, but not a complement, of $\left[\bigcup_{n=1}^{N-1} G_n\right]$. It turns out that

---

*⁾ Indeed, observe that if $y_n \in G_n \setminus \{0\}$, $(n = 1, 2, \ldots)$, then $\|f|_{[y_i]_{i=n+1}^{\infty}}\| \leqslant \leqslant \|f|_{\left[\bigcup_{i=n+1}^{\infty} G_i\right]}\|$ $(f \in E^*)$.

**⁾ Such a pair of subspaces occurs also in § 8, corollary 8.3.

the addition of this condition is also sufficient. In fact, we have the following (intrinsic) characterizations of Schauder decompositions:

**Theorem 15.22.** *Let $\{G_n\}$ be a sequence of subspaces of a Banach space $E$, with $G_n \neq \{0\}$ $(n = 1,2, \ldots)$. The following statements are equivalent:*

1°. *$\{G_n\}$ is a Schauder decomposition of $E$.*

2°. *We have*

$$E = \left[\bigcup_{i=1}^{n-1} G_i\right] \oplus \left[\bigcup_{i=n}^{\infty} G_i\right] \qquad (n = 2,3, \ldots) \tag{15.127}$$

*and every sequence $\{y_n\} \subset E$ with $y_n \in G_n \setminus \{0\}$ $(n = 1,2,\ldots)$ is a basic sequence.*

3°. *We have*

$$E = G_n \oplus [\bigcup_{i \neq n} G_i] \qquad (n = 1,2, \ldots) \tag{15.128}$$

*and every sequence $\{y_n\} \subset E$ with $y_n \in G_n \setminus \{0\}$ $(n = 1,2,\ldots)$ is a basic sequence.*

*If* $\dim G_n < \infty$ $(n = 1,2, \ldots)$, *these statements are equivalent to the following statement:*

4°. $\left[\bigcup_{n=1}^{\infty} G_n\right] = E$ *and every sequence $\{y_n\} \subset E$ with $y_n \in G_n \setminus \{0\}$ $(n = 1,2, \ldots)$ is a basic sequence.*

*Proof.* The implications $1° \Rightarrow 2°$, $1° \Rightarrow 3°$ and $1° \Rightarrow 4°$ are obvious, taking into account theorem 15.21 a).

Assume now that we have 2°. Then, by theorem 15.21 b), there exists $N \in \mathcal{N}$ such that $\{G_n\}_{n=N}^{\infty}$ is a Schauder decomposition of $\left[\bigcup_{n=N}^{\infty} G_n\right]$. Also, from (15.127) we infer, by induction on $n$, that $\left[\bigcup_{i=1}^{n} G_i\right] = \sum_{i=1}^{n} \oplus G_i$ for $n = 1,2, \ldots$ ( indeed, if $u_{n-1}$ is the projection of $E$ onto $\left[\bigcup_{i=1}^{n-1} G_i\right]$ along $\left[\bigcup_{=n}^{\infty} G_i\right]$, then $u_{n-1}|_{\left[\bigcup_{i=1}^{n} G_i\right]}$ is a continuous linear projection of $\left[\bigcup_{i=1}^{n} G_i\right]$ onto $\left[\bigcup_{i=1}^{n-1} G_i\right]$ along $G_n$). Hence by (15.127) for $n = N$ and since $\{G_n\}_{n=N}^{\infty}$ is a Schauder decomposition of $\left[\bigcup_{n=N}^{\infty} G_n\right]$,

$$E = \sum_{=1}^{N-1} \oplus G_n \oplus \sum_{n=N}^{\infty} \oplus G_n = \sum_{n=1}^{\infty} \oplus G_n, \tag{15.129}$$

so $\{G_n\}$ is a Schauder decomposition of $E$. Thus, $2° \Rightarrow 1°$.

The equivalence $2° \Leftrightarrow 3°$ follows from the remarks made after definition 15.9.

Finally, assume that $\dim G_n < \infty$ $(n = 1,2, \ldots)$ and that we have $4°$. Again, by theorem 15.21 b), let $N \in \mathcal{N}$ be such that $\{G_n\}_{n=N}^{\infty}$ is a Schauder decomposition of $\left[\bigcup_{n=N}^{\infty} G_n\right]$. Since every sequence $y_n \in G_n \setminus \{0\}$ $(n = 1,2, \ldots)$ is a basic sequence, $\{G_n\}$ is finitely linearly independent; also, $\sum_{i=1}^{N-1} G_i$ is closed (since $\dim G_i < \infty$), so $\left[\bigcup_{i=1}^{N-1} G_i\right] = \sum_{i=1}^{N-1} \oplus G_i$. We claim that

$$\left[\bigcup_{i=1}^{N-1} G_i\right] \cap \left[\bigcup_{i=N}^{\infty} G_i\right] = \{0\}. \tag{15.130}$$

Indeed, if $x \in \left(\left[\bigcup_{i=1}^{N-1} G_i\right] \cap \left[\bigcup_{i=N}^{\infty} G_i\right]\right) \setminus \{0\}$, then we can write $x = \sum_{i=1}^{N-1} y_i = \sum_{i=N}^{\infty} y_i \neq 0$, where $y_n \in G_n$ $(n = 1,2, \ldots)$. But then $\{y_n\}$ is not a basic sequence and hence, replacing those $y_n$ which are 0 by elements $y_n \in G_n \setminus \{0\}$, we arrive at a contradiction with our assumption. This proves the claim (15.130). Now, since $\dim \left[\bigcup_{i=1}^{N-1} G_i\right] < \infty$, the sum $\left[\bigcup_{i=1}^{N-1} G_i\right] + \left[\bigcup_{i=N}^{\infty} G_i\right]$ is closed*) and hence, by (15.130) and $\left[\bigcup_{n=1}^{\infty} G_n\right] = E$, we obtain (15.129). Thus, $4° \Rightarrow 1°$, which completes the proof of theorem 15.22.

*Definition 15.23*. Let $\{G_n\}$ be a Schauder decomposition of a Banach space $E$, with the a.s.c.p. $\{v_n\} \subset L(E, E)$. Then every pair of sequences $(y_n, h_n)$ satisfying

$$y_n \in G_n, \ h_n \in v_n^*(E^*), \ h_n(y_n) = 1 \quad (n = 1,2, \ldots), \tag{15.131}$$

is called a *cofinal bibasic system***) for $\{G_n\}$.

From theorems 15.21 a) and 15.6 and the orthogonality relations (15.6) it follows that any such pair of sequences is indeed a bibasic system in the sense of § 1, definition 1.5.

---

*) See e.g. [33], Ch. I, § 2, corollary 4 or [70], Ch. I, § 4, theorem 1.

**) Similarly, one can also define *cofinal biorthogonal systems* for a generalized biorthogonal system $(G_n, v_n)$.

**Proposition 15.14.** *Let $\{G_n\}$ be a Schauder decomposition of a Banach space E, with the a.s.c.p. $\{v_n\} \subset L(E, E)$, and let $(y_n, h_n)$ be a cofinal bibasic system for $\{G_n\}$. Then $[y_n]$ and $[h_n]_\perp$ are quasi-complementary.*

*Proof.* If $x \in [y_n] \cap [h_n]_\perp$, then, since $\{y_n\}$ is a basic sequence, we can write $x = \sum_{i=1}^{\infty} \alpha_i y_i$. But then, by biorthogonality, $\alpha_n = h_n(x) = 0$ $(n = 1,2, \ldots)$, whence $x = \sum_{i=1}^{\infty} \alpha_i y_i = 0$, which proves that $[y_n] \cap [h_n]_\perp = \{0\}$.*)

Now let $x \in G_n$. Then for any $i \neq n$ we have, by (15.6),

$$h_i(x - h_n(x)y_n) = v_i^*(h_i)\,(v_n(x - h_n(x)y_n)) = h_i(v_i v_n(x - h_n(x)y_n)) = 0,$$

and for $i = n$ we have, clearly, $h_n(x - h_n(x)y_n) = 0$, so $x - h_n(x)y_n \in [h_i]_\perp$. Consequently,

$$x = h_n(x)y_n + (x - h_n(x)y_n) \in [y_i] + [h_i]_\perp$$

whence $\bigcup_{n=1}^{\infty} G_n \subset [y_i] + [h_i]_\perp$. Therefore, $[y_i] + [h_i]_\perp$ is dense in $E$, which completes the proof of proposition 15.14.

In some cases $[y_n]$ and $[h_n]_\perp$ are even complementary subspaces, as shown by

**Proposition 15.15.** *Let $\{G_n\}$ be an $l^1$-decomposition of a Banach space E, with the a.s.c.p. $\{v_n\} \subset L(E, E)$. Then for every sequence $\{y_n\} \subset E$ with $y_n \in G_n \setminus \{0\}$ $(n=1,2, \ldots)$ there exists a sequence $\{h_n\} \subset E^*$ with $h_n \in v_n^*(E^*) \setminus \{0\}$ $(n=1,2, \ldots)$ such that $(y_n, h_n)$ is a cofinal bibasic system and that $E = [y_n] \oplus [h_n]_\perp$.*

*Proof.* We may assume, without loss of generality, that $\|y_n\| = 1$ $(n = 1,2, \ldots)$. By a corollary of the Hahn-Banach theorem, there exists $\varphi_n \in G_n^*$ with $\|\varphi_n\| = 1$, $\varphi_n(y_n) = 1$ $(n = 1,2, \ldots)$. Put

$$h_n = \varphi_n \circ v_n \qquad (n = 1,2, \ldots). \tag{15.132}$$

Then $h_n \in E^*$ and

$$v_n^*(h_n)\,(x) = h_n(v_n(x)) = \varphi_n(v_n(v_n(x))) = \varphi_n(v_n(x)) = h_n(x) \qquad (x \in E,\ \ n = 1,2, \ldots),$$

---

*) This part is valid for any biorthogonal system $(y_n, h_n)$ such that $\{y_n\}$ is a basic sequence (see also § 1, proposition 1.15 b)).

whence $h_n = v_n^*(h_n) \in v_n^*(E^*)$ $(n = 1,2, \ldots)$. Also,

$$h_n(y_n) = \varphi_n(v_n(y_n)) = \varphi_n(y_n) = 1 \qquad (n = 1,2, \ldots),$$

and thus $(y_n, h_n)$ is a cofinal bibasic system. Then, by proposition 15.14 above, $[y_n] \cap [h_n]_\perp = \{0\}$ and $[y_n] + [h_n]_\perp$ is dense in $E$. We shall show that $[y_n] + [h_n]_\perp$ is also closed in $E$, which will complete the proof. To this end, by § 1, theorem 1.10, implication $9° \Rightarrow 13°$, it will be enough to show that

$$\sup_{1 \leqslant n < \infty} \|s_n(x)\| < \infty \qquad (x \in E),$$

where $s_n(x) = \sum_{i=1}^{n} h_i(x) y_i$ $(x \in E,\ n = 1,2, \ldots)$. But, by $\|y_n\| = 1 = \|\varphi_n\|$ $(n = 1,2, \ldots)$, (15.132) and since $\{G_n\}$ is an $l^1$-decomposition of $E$, we have

$$\|s_n(x)\| = \left\| \sum_{i=1}^{n} h_i(x) y_i \right\| \leqslant \sum_{i=1}^{n} |h_i(x)| = \sum_{i=1}^{n} |\varphi_i(v_i(x))| \leqslant$$

$$\leqslant \sum_{i=1}^{\infty} \|v_i(x)\| = M_x < \infty \qquad (x \in E,\ n = 1,2, \ldots),$$

which completes the proof of proposition 15.15. From § 1, theorem 1.11 it follows that in this case $\{h_n\}$ is a $w^*$-Schauder basic sequence, equivalent to the a.s.c.f. $\{h_n|_{[y_j]}\}$ to the basis $\{y_n\}$ of $[y_n]$.

In § 13, theorem 13.4 we have seen that if the subspaces $\{G_n\}$ of a finite-dimensional decomposition (respectively, an unconditional finite-dimensional decomposition) of a Banach space $E$ have bases with uniformly bounded norms (respectively, unconditional norms), then one can "stick them together" to obtain a basis (respectively, an unconditional basis) of $E$. It is natural to ask whether such a result remains valid also for Schauder decompositions $\{G_n\}$, with a careful enumeration of the basis elements. In Ch. I, § 8, proposition 8.3 and Ch. II, § 17, corollary 17.4 we have seen that this is indeed the case for $l^2$-decompositions $\{G_n\}$. The same argument works also for $l^p$-decompositions $(1 \leqslant p < \infty)$ and $c_0$-decompositions and, more generally, for any "lattice type" Schauder decompositions $\{G_n\}$ (i.e., such that the relations $x = \sum_{i=1}^{\infty} y_i \in E$, $z = \sum_{i=1}^{\infty} z_i \in E$, $y_n, z_n \in G_n$, $\|y_n\| \leqslant \|z_n\|$ for $n = 1,2, \ldots$ imply $\|x\| \leqslant \|z\|$); however, it no longer works even for unconditional Schauder decompositions.

Finally, let us introduce another natural generalization of the notion of a basis, called integral basis, to which one can associate, in a natural way, a Schauder decomposition of the same space. In this generalization of bases, summation on the index set $\mathcal{N} = \{1, 2, 3, \ldots\}$ (discrete case) is replaced by integration on a locally compact*) space $T$ (continuous case). The "basis" $\{x_n\}$ will be replaced by a function $X: t \to X(t)$ from $T$ into $E$, and the unique representation of each $x \in E$ as $\sum_{i=1}^{\infty} \alpha_i x_i$, where $\{\alpha_n\} \subset K$, will be replaced now by the unique representation of each $x \in E$ as the limit, in $E$, of integrals**) $\int_{Q_n} X(t) d\mu_x(t)$, where $\{Q_n\}$ is a fixed increasing sequence of compact subsets of $T$ and where $\mu_x$ is a (Radon) measure on $T$, depending on $x$ (thus, $\{Q_n\}$ determines***) a method of "ordering" $T$ and the values $X(t)$ and $\{X|_{Q_n \setminus Q_{n-1}}\}$ is a *countable* generalization of $\{x_n\}$). We shall assume that $\mu_x$ lies in some Fréchet space of measures and if $\mu_x$ depends continuously on $x$, we shall use the term "Schauder integral basis".

After this rough description, let us pass now to the precise definitions. All integrals which we shall use will be in the weak (Gelfand-Pettis) sense. We recall that if $A \subset T$ is $\mu$-measurable and $Y: t \to Y(t)$ is a mapping of $A$ into $E$, the (weak) integral $\int_A Y(t) d\mu(t)$ is said to *exist* whenever $t \to f(Y(t))$ is in $L^1(A, \mu|_A)$ for each $f \in E^*$ and then the value of this integral is introduced as the (unique) element $\Phi \in (E^*)^\#$ (the algebraic dual of $E^*$) defined by

$$\Phi(f) = \int_A f(Y(t)) d\mu(t) \qquad (f \in E^*).$$

It is well known****) that, since $E^*$ is a Banach space, we have actually $\Phi \in E^{**}$. In general, $\Phi$ need not belong to $\pi(E)$, the canonical image

*) For another version, in which the index set is any measurable space, see the Notes and remarks.

**) The "partial integrals" $\int_{Q_n} X(t) d\mu_x(t)$ generalize the partial sums $\sum_{i=1}^{n} \alpha_i x_i$. It would be too restrictive to require the unique representation of each $x \in E$ as a weakly absolutely convergent integral $\int_T X(t) d\mu_x(t)$, since in the particular case when $T = \mathcal{N} = \{1, 2, 3, \ldots\}$, this would reduce us to (weak, hence strong) unconditional bases, as we shall see in § 17.

***) Therefore, integral bases may be regarded as "countable" generalizations of bases (see also theorem 15.25 below).

****) See e.g. [95], Ch. VIII, proposition 8.14.11(a). For some simple sufficient conditions in order that $\int_A Y(t) d\mu(t) \in E$, see e.g. [95], Ch. VIII, or [32], Ch. III, § 4.

of $E$ in $E^{**}$. For simplicity of notation, if the integral $\int_A Y(t)d\mu(t)$ exists (in the above sense) and if $\Phi \in \pi(E)$, it is customary to write $\int_A Y(t)d\mu(t) \in E$, and we shall also adopt this notation.

*Definition 15.24.* Let $E$ be a Banach space, $T$ a locally compact space, $\{Q_n\}$ an increasing sequence of compact subsets of $T$, $X: t \to X(t)$ a mapping of $T$ into $E$ and $\mathfrak{M}$ a Fréchet space of measures on $T$, such that

(i) $\int_{Q_n} X(t)d\mu(t) \in E \qquad (n = 1,2, \ldots; \mu \in \mathfrak{M})$;

(ii) there exists a set $\Gamma \subset E^*$, which is total on $E$, such that for each $f \in \Gamma$ and each $n = 1,2, \ldots$ the linear functional $\mu \to \int_{Q_n} f(X(t))d\mu(t)$ is continuous on $\mathfrak{M}$;

(iii) for each $n = 1,2, \ldots$ there exists a $\mu^{(n)} \in \mathfrak{M}$ such that $\int_{Q_n} X(t)d\mu^{(n)}(t) \neq \int_{Q_{n-1}} X(t)d\mu^{(n)}(t)$ $\left(\text{where} \int_{Q_0} = 0\right)$.

A system $(X, T, \{Q_n\}, \mathfrak{M}\}$ satisfying (i)—(iii) is called an *integral basis* of $E$, if for every $x \in E$ there exists a *unique* measure $\mu_x \in \mathfrak{M}$ such that, in the norm topology of $E$,

$$x = \lim_{n\to\infty} \int_{Q_n} X(t)d\mu_x(t). \tag{15.133}$$

In particular, if we take $T = \mathcal{N} = \{1,2,3, \ldots\}$ with the discrete topology, $Q_n = \{1, \ldots, n\}$ $(n = 1,2, \ldots)$, $\mathfrak{M} = s = K^{\mathcal{N}}$ (the Fréchet space [*]) of all scalar sequences $\mu = \{\mu(n)\}$, regarded as measures on $\mathcal{N}$) and $X(n) = x_n$ $(n = 1,2, \ldots)$, then $(X, T, \{Q_n\}, \mathfrak{M})$ is an integral basis of $E$ if and only if $\{x_n\}$ is a basis of $E$ (in the usual sense). Indeed, in this case integrals are replaced by sums, namely, if $A \subset \mathcal{N}$, $\{Y(n)\} = \{y_n\} \subset E$ and $\mu = \{\mu(n)\} \in s$, the integral $\int_A Y d\mu \in E^{**}$ exists if and only if $\sum_{n\in A} |f(\mu(n)y_n)| < \infty$ for all $f \in E^*$ and then $\left(\int_A Y d\mu\right)(f) =$

---

[*]) We recall that $s = K^{\mathcal{N}}$ is endowed with the usual product topology, which amounts to the topology of coordinatewise convergence.

$= \sum_{n \in A} f(\mu(n)y_n)$ $(f \in E^*)$. Thus, if $A$ is finite, $\int_A Y d\mu = \sum_{n \in A} \mu(n) y_n \in E$ $\left(\text{in particular, } \int_{Q_n} X d\mu = \sum_{i=1}^{n} \mu(i) x_i \in E, \text{ so we have (i)}\right)$. Furthermore, if $\mu_k = \{\mu_k(j)\} \to \mu = \{\mu(j)\}$ in $\mathfrak{M}$, then for each $f \in E^*$ and $n = 1,2, \ldots$ we have $\sum_{i=1}^{n} \mu_k(i) f(x_i) \to \sum_{i=1}^{n} \mu(i) f(x_i)$ as $k \to \infty$, so (ii) holds with $\Gamma = E^*$; clearly, (iii) is also satisfied. Finally, condition (15.133) means that $x = \sum_{i=1}^{\infty} \mu_x(i) x_i$ in the norm topology of $E$, and the uniqueness of $\mu_x$ is obvious from the definition of bases and integral bases.

More generally, if $\{m_n\} \subset \mathcal{N}$, $m_1 < m_2 < \ldots$, $T = \mathcal{N}$ with the discrete topology, $Q_n = \{1, \ldots, m_n\}$ $(n = 1,2, \ldots)$ and $\mathfrak{M} = s = K^{\mathcal{N}}$, a similar argument shows that $(X, T, \{Q_n\}, \mathfrak{M})$ is an integral basis of $E$ if and only if $\{x_n\}$ is a basis with parentheses of $E$ with respect to $\{m_n\}$ (in the sense of § 13, definition 13.4). Indeed, in this case $\int_{Q_n} X d\mu = \sum_{i=1}^{m_n} \mu(i) x_i$, so (15.133) means that $x = \lim_{n \to \infty} \sum_{i=1}^{m_n} \mu_x(i) x_i$. Thus, if $\{x_n\}$ is a basis with parentheses with respect to $\{m_n\}$, the uniqueness of $\mu_x$ follows from biorthogonality. Conversely, if $(X, T, \{Q_n\}, \mathfrak{M})$ is an integral basis of $E$, then, by § 13, proposition 13.9, $\{x_n\}$ is a basis with parentheses of $E$, with respect to $\{m_n\}$, which proves our assertion.

Condition (iii) in definition 15.24 has the role of excluding trivial cases; in particular, this condition implies that $Q_1 \neq \emptyset$, $Q_n \neq Q_{n-1}$ $(n = 2,3, \ldots)$.

Clearly, the mapping $x \to \mu_x$ of $E$ into $\mathfrak{M}$ is the natural extension to integral bases of the mapping $x \to \{f_n(x)\}$, where $\{f_n\} \subset E^*$ is the a.s.c.f. to a (usual) basis $\{x_n\}$. By the uniqueness condition in definition 15.24, the mapping $x \to \mu_x$ is linear.

*Definition 15.25.* An integral basis $(X, T, \{Q_n\}, \mathfrak{M})$ of a Banach space $E$ is called a *Schauder integral basis* of $E$, if the linear mapping $x \to \mu_x$ of $E$ into $\mathfrak{M}$ is continuous on $E$.

As a generalization of Ch. I, § 3, theorem 3.1, we shall show (in theorem 15.23 below) that every integral basis of a Banach space $E$ is a Schauder integral basis of $E$. To this end, let us first give the following result, corresponding to Ch. I, § 3, proposition 3.1:

**Proposition 15.16** *Let $E$ be a Banach space and $(X, T, \{Q_n\}, \mathfrak{M})$ a system satisfying the conditions* (i) *and* (ii) *of definition* 15.24 *and* *let* $A_1'$ *be the linear space of measures*

$$A_1' = \left\{ \mu \in \mathfrak{M} \,\middle|\, \int_{Q_n} X(t) d\mu(t) \in E \quad (n = 1,2, \ldots) \text{ and } \right.$$

$$\left. \lim_{n\to\infty} \int_{Q_n} X(t) d\mu(t) \text{ exists} \right\}, \qquad (15.134)$$

*endowed with the topology generated by the semi-norms*

$$q_j(\mu) = p_j(\mu) + \sup_{1 \leqslant n < \infty} \left\| \int_{Q_n} X(t) d\mu(t) \right\| \quad (\mu \in A_1', \ j = 1,2, \ldots), \quad (15.135)$$

*where* $\{p_n\}$ *is a sequence of semi-norms generating the topology of* $\mathfrak{M}$. *Then* $A_1'$ *is a Fréchet space.*

*Proof.* $A_1'$ is a Hausdorff space, since the relations $q_j(\mu) = 0$ ($j = 1,2, \ldots$) imply $p_j(\mu) = 0$ ($j = 1,2, \ldots$), whence $\mu = 0$ (because $\mathfrak{M}$ is a Hausdorff space). Thus, it remains to show that $A_1'$ is complete.

Let $\{\mu_k\}$ be a Cauchy sequence in $A_1'$. Then, by $\mu_k \in A_1'$, we can put

$$y_{k,n} = \int_{Q_n} X(t)\, d\mu_k(t) \in E \quad (k, n = 1,2, \ldots), \qquad (15.136)$$

$$y_k = \lim_{n\to\infty} y_{k,n} \in E \quad (k = 1,2, \ldots), \qquad (15.137)$$

and, since $\{\mu_k\}$ is Cauchy in $A_1'$, for every $\varepsilon > 0$ and $j = 1,2, \ldots$ there exists a positive integer $N(\varepsilon, j)$ such that

$$q_j(\mu_k - \mu_m) = p_j(\mu_k - \mu_m) + \sup_{1 \leqslant n < \infty} \|y_{k,n} - y_{m,n}\| < \varepsilon$$

$$(k, m > N(\varepsilon, j);\ j = 1,2, \ldots). \qquad (15.138)$$

By (15.137), for each $k, m > N(\varepsilon, 1)$ we can take $n = n(k, m)$ so large that $\|y_k - y_{k,n}\| < \varepsilon$, $\|y_m - y_{m,n}\| < \varepsilon$, whence, by (15.138),

$$\|y_k - y_m\| \leqslant \|y_k - y_{k,n}\| + \|y_{k,n} - y_{m,n}\| + \|y_{m,n} - y_m\| < 3\varepsilon$$

$$(k, m > N(\varepsilon, 1)).$$

Consequently, since $E$ is a Banach space,

$$\lim_{n\to\infty} y_k = x \in E. \tag{15.139}$$

On the other hand, again by (15.138), $\{\mu_k\}$ and each $\{y_{k,n}\}_{k=1}^{\infty}$ $(n = 1,2, \ldots)$ are Cauchy sequences in $\mathfrak{M}$ and $E$ respectively, whence, since $\mathfrak{M}$ is a Fréchet space and $E$ a Banach space,

$$\lim_{k\to\infty} \mu_k = \mu \in \mathfrak{M}, \tag{15.140}$$

$$\lim_{k\to\infty} y_{k,n} = z_n \in E \quad (n = 1,2, \ldots). \tag{15.141}$$

Then, for each $f \in \Gamma$ as in (ii) and each $n = 1,2, \ldots$ we have, by (ii) and (i),

$$f(z_n) = \lim_{k\to\infty} f(y_{k,n}) = \lim_{k\to\infty} \int_{\Omega_n} f(X(t))d\mu_k(t) =$$

$$= \int_{\Omega_n} f(X(t))d\mu(t) = f\left(\int_{\Omega_n} X(t)d\mu(t)\right),$$

whence, since $\Gamma$ is total on $E$,

$$z_n = \int_{\Omega_n} X(t)d\mu(t) \quad (n = 1,2, \ldots). \tag{15.142}$$

But, by (15.138),

$$\|y_{k,n} - y_{m,n}\| < \varepsilon \quad (k, m > N(\varepsilon, 1);\ n = 1,2, \ldots),$$

whence, for $m \to \infty$ we obtain, by (15.141),

$$\|y_{k,n} - z_n\| < \varepsilon \quad (k > N(\varepsilon, 1);\ n = 1,2, \ldots).$$

Thus, by (15.139) we can take $k > N(\varepsilon, 1)$ so large that $\|x - y_k\| < \varepsilon$ and then, by (15.137), $N_k(\varepsilon)$ so large that $\|y_k - y_{k,n}\| < \varepsilon$ $(n > N_k(\varepsilon))$, whence

$$\left\|x - \int_{\Omega_n} X(t)\,d\mu(t)\right\| = \|x - z_n\| \leqslant \|x - y_k\| + \|y_k - y_{k,n}\| +$$

$$+ \|y_{k,n} - z_n\| < 3\varepsilon \qquad (n > N_k(\varepsilon)),$$

which proves (15.133) and $\mu \in A_1'$. Finally, by (15.138),

$$p_j(\mu_k - \mu_m) + \|y_{k,n} - y_{m,n}\| < \varepsilon \ (k, m > N(\varepsilon, j);\ j, n = 1,2, \ldots),$$

whence, for $m \to \infty$ we obtain, by (15.140), (15.141),

$$p_j(\mu_k - \mu) + \|y_{k,n} - z_n\| \leqslant \varepsilon \quad (k > N(\varepsilon, j);\ j, n = 1,2, \ldots).$$

Consequently, by (15.135), (15.136) and (15.142),

$$q_j(\mu_k - \mu) = p_j(\mu_k - \mu) + \sup_{1 \leqslant n < \infty} \|y_{k,n} - z_n\| \leqslant \varepsilon$$

$$(k > N(\varepsilon, j);\ j = 1,2, \ldots),$$

so $\mu_k \to \mu$ in $A_1'$, which completes the proof of proposition 15.16.

Now we can prove

**Theorem 15.23.** *Every integral basis $(X, T, \{Q_n\}, \mathfrak{M})$ of a Banach space $E$ is a Schauder integral basis of $E$.*

*Proof.* We shall use the notations of proposition 15.16 and its proof. Define a mapping $w: A_1' \to E$ by

$$w(\mu) = \lim_{n \to \infty} \int_{Q_n} X(t) d\mu(t) \quad (\mu \in A_1'). \tag{15.143}$$

The mapping $w$ is obviously linear and, by (15.135), we have

$$\|w(\mu)\| = \|\lim_{n \to \infty} \int_{Q_n} X(t) d\mu(t)\| \leqslant \sup_{1 \leqslant n < \infty} \|\int_{Q_n} X(t) d\mu(t)\| \leqslant q_1(\mu) \quad (\mu \in A_1'),$$

whence $w$ is continuous. Furthermore, by the uniqueness and existence of $\mu = \mu_x$ in (15.133), $w$ is one-to-one and maps $A_1'$ onto $E$. Hence, by the inversion theorem of Banach*), $w$ is an isomorphism of $A_1'$ onto $E$ and thus there exist constants $M_j > 0$ $(j = 1,2, \ldots)$ such that

$$q_j(\mu) \leqslant M_j \|w(\mu)\| \quad (\mu \in A_1',\ j = 1,2, \ldots). \tag{15.144}$$

Hence, by (15.135) and (15.133) we obtain

$$p_j(\mu_x) \leqslant q_j(\mu_x) \leqslant M_j \|\lim_{n \to \infty} \int_{Q_n} X(t) d\mu_x(t)\| = M_j \|x\|$$

$$(x \in E,\quad j = 1,2, \ldots), \tag{15.145}$$

*) See e.g. [11], p. 41, theorem 5.

which completes the proof of theorem 15.23.

*Definition 15.26.* Let $(X, T, \{Q_n\}, \mathfrak{M})$ be an integral basis of a Banach space $E$. The sequence of linear operators $\{S_n\}$ on $E$ defined by

$$S_n(x) = \int_{Q_n} X(t)d\mu_x(t) \quad (x \in E, \ n = 1,2, \ldots) \tag{15.146}$$

is called the *sequence of partial integral operators associated to the integral basis* $(X, T, \{Q_n\}, \mathfrak{M})$.

In the particular case of usual bases this reduces, clearly, to the sequence of partial sum operators associated to the basis, while in the case of bases with parentheses it yields the sequence $\{s_{m_n}\}$.

We have the following generalization of Ch. I, § 4, theorem 4.1, implication $1° \Rightarrow 4°$, to integral bases:

**Theorem 15.24.** *For every integral basis $(X, T, \{Q_n\}, \mathfrak{M})$ of a Banach space $E$ the associated partial integral operators $S_n$ are uniformly bounded (and hence, in particular, continuous) on $E$.*

*Proof.* With the notations of the preceding proof (of theorem 15.23), we have

$$\sup_{1 \leqslant n < \infty} \|S_n(x)\| = \sup_{1 \leqslant n < \infty} \|\int_{Q_n} X(t)d\mu_x(t)\| \leqslant q_1(\mu_x) \leqslant$$

$$\leqslant M_1 \|w(\mu_x)\| = M_1 \|x\| \qquad (x \in E),$$

which completes the proof.

In § 13, propositions 13.10 and 13.11 we have seen that every basis with parentheses generates a resolution of the identity and hence a Schauder decomposition of the space, namely, $\{s_{m_n}\}$ and $\{(s_{m_n} - s_{m_{n-1}})(E)\}$ $(m_0 = 0)$ respectively. We have the following generalization of this result to integral bases:

**Theorem 15.25.** *For every integral basis $(X, T, \{Q_n\}, \mathfrak{M})$ of a Banach space $E$, the associated sequence of partial integral operators $\{S_n\}$ is a resolution of the identity and hence $\{(S_n - S_{n-1})(E)\}$ $(S_0 = 0)$ is a Schauder decomposition of $E$.*

*Proof.* For each $n$ and each measure $\mu$ on $T$ define a measure $\rho_n(\mu)$ on $T$ by putting, for any $\mu$-measurable set $A \subset T$,

$$\rho_n(\mu)(A) = \mu(A \cap Q_n). \tag{15.147}$$

Then, since $\{Q_n\}$ is increasing,

$$\int_{Q_n} X(t)d\mu(t) = \int_{Q_{n+k}} X(t)d\rho_n(\mu)(t) \quad (n = 1,2, \ldots; k = 0,1,2, \ldots). \tag{15.148}$$

Consequently, for the operators $S_n$ defined by (15.146),

$$\lim_{k\to\infty} \int_{Q_{n+k}} X(t)\, d\mu_{S_n(x)}(t) = S_n(x) = \int_{Q_n} X(t)d\mu_x(t) =$$

$$= \lim_{k\to\infty} \int_{Q_{n+k}} X(t)d\rho_n(\mu_x)(t) \quad (x \in E,\ n = 1,2, \ldots),$$

whence, by the uniqueness of the representing measures in the "expansions" (15.133), we obtain

$$\mu_{S_n(x)} = \rho_n(\mu_x) \quad (x \in E,\ n = 1,2, \ldots). \tag{15.149}$$

Also, again since $\{Q_n\}$ is increasing, for any $\mu$-measurable set $A \subset T$ we have

$$\rho_m(\rho_n(\mu))(A) = \rho_n(\mu)(A \cap Q_m) = \mu(A \cap Q_m \cap Q_n) =$$

$$= \mu(A \cap Q_{\min(n,m)}) = \rho_{\min(n,m)}(\mu)(A) \quad (n, m = 1,2, \ldots),$$

that is,

$$\rho_m(\rho_n(\mu)) = \rho_{\min(n,m)}(\mu) \quad (n, m = 1,2, \ldots). \tag{15.150}$$

Therefore,

$$\mu_{S_m(S_n(x))} = \rho_m(\mu_{S_n(x)}) = \rho_m(\rho_n(\mu_x)) =$$

$$= \rho_{\min(n,m)}(\mu_x) = \mu_{S_{\min(n,m)}(x)} \quad (x \in E;\ n, m = 1,2, \ldots),$$

whence, by (15.133),

$$S_m(S_n(x)) = \lim_{k\to\infty} \int_{Q_k} X(t)\, d\mu_{S_m(S_n(x))}(t) =$$

$$= \lim_{k\to\infty} \int_{Q_k} X(t)\, d\mu_{S_{\min(n,m)}(x)}(t) = S_{\min(n,m)}(x) \quad (x \in E;\, n, m = 1,2, \ldots)$$

so $\{S_n\}$ satisfies (15.21). Furthermore, by (15.133) we have $x = \lim_{n\to\infty} S_n(x)$ $(x \in E)$. Finally, if $\mu^{(n)} \in \mathfrak{M}$ is as in condition (iii) of definition 15.24, then, by (i) of the same definition, $\int_{Q_n} X(t) d\mu^{(n)}(t) = \int_{Q_n} X(t) d\rho_n(\mu^{(n)})(t) =$

$= y_n \in E$, whence, by (15.148), $y_n = \lim_{k\to\infty} \int_{Q_{n+k}} X(t) d\rho_n(\mu^{(n)})(t)$. Consequently, by the uniqueness condition in definition 15.24, $\mu_{y_n} = \rho_n(\mu^{(n)})$, whence, by (iii) and $Q_n \supset Q_{n-1}$,

$$S_n(y_n) = \int_{Q_n} X(t)\, d\rho_n(\mu^{(n)})\,(t) = \int_{Q_n} X(t)\, d\mu^{(n)}(t) \neq$$

$$\neq \int_{Q_{n-1}} X(t)\, d\mu^{(n)}(t) = \int_{Q_{n-1}} X(t)\, d\rho_n(\mu^{(n)})\,(t) = S_{n-1}(y_n),$$

which proves that $S_n \neq S_{n-1}$ $(n = 1,2, \ldots;\ S_0 = 0)$. Thus, $\{S_n\}$ is a resolution of $I_E$, whence, by theorem 15.3, $\{(S_n - S_{n-1})(E)\}$ $(S_0 = 0)$ is a Schauder decomposition of $E$, which completes the proof of theorem 15.25.

*Remark 15.10.* a) Since $\{Q_n\}$ is increasing, we have

$$(S_n - S_{n-1})(x) = \int_{Q_n \setminus Q_{n-1}} X(t) d\mu_x(t) \quad (x \in E, n = 1,2, \ldots). \quad (15.151)$$

b) From theorems 15.25 and 15.2 it follows that *the space $E = l^\infty$ has no integral basis.*

Theorem 15.25 permits us to introduce the following terminology:

*Definition 15.27.* Let $(X, T, \{Q_n\}, \mathfrak{M})$ be an integral basis of a Banach space $E$ and let $\{S_n\}$ be the associated sequence of partial integral operators. Then $\{(S_n - S_{n-1})(E)\}$ $(S_0 = 0)$ is called *the Schauder decomposition of $E$ associated to the integral basis* $(X, T, \{Q_n\}, \mathfrak{M})$.

Now we can define, in a natural way, special classes of integral bases, with the aid of the properties of the associated Schauder decompositions. For example, one can give

*Definition 15.28.* An integral basis $(X, T, \{Q_n\}, \mathfrak{M})$ of a Banach space $E$ is said to be *shrinking* (respectively, *boundedly complete*, respectively, *unconditional*, etc.), if so is the Schauder decomposition $\{(S_n - S_{n-1})(E)\}$ $(S_0 = 0)$ of $E$ associated to this integral basis.

In the particular cases of usual bases and bases with parentheses, this yields again the usual special classes.

One can also give another generalization of unconditional bases to the continuous case. Since this generalization does not involve countability, we shall give it in § 17.

## § 16. Bases of sets

Let us observe that a sequence $\{x_n\}$ in a Banach space $E$ is a basic sequence if and only if there exists a (unique) subspace $G$ of $E$ such that $\{x_n\} \subset G$ is a basis of the space $G$, i.e., such that for every $x \in G$ there exists a unique sequence of scalars $\{\alpha_n\} \subset K$ satisfying $x = \sum_{i=1}^{\infty} \alpha_i x_i$ (indeed, if $\{x_n\}$ is a basic sequence, then by definition $\{x_n\}$ is a basis of the subspace $G = [x_n]$ of $E$ and, conversely, if $\{x_n\}$ is a basis of some subspace $G$ of $E$, then $G = [x_n]$, whence $\{x_n\}$ is a basic sequence). Replacing here the subspace $G$ by an arbitrary subset $\mathscr{M}$ of $E$, we arrive naturally at

*Definition 16.1.* Let $\mathscr{M}$ be a set in a Banach space $E$. A sequence $\{x_n\} \subset \mathscr{M}$ is called a *basis* of $\mathscr{M}$ if for every $x \in \mathscr{M}$ there exists a *unique* sequence of scalars $\{\alpha_n\} \subset K$ such that $x = \sum_{i=1}^{\infty} \alpha_i x_i$ (actually, $\sum_{i=1}^{\infty}$ must be replaced by a finite sum if $\dim [\mathscr{M}] < \infty$, but we make the convention of writing $\sum_{i=1}^{\infty}$).

*Remark 16.1.* One can also define, more generally, a *basis for* $\mathscr{M}$, by omitting in definition 16.1 the requirement $\{x_n\} \subset \mathscr{M}$. Then, for example, every basis of $E$ will be a basis for any subset $\mathscr{M}$ of $E$. However, in the present section we shall not consider this more general notion.

**Proposition 16.1.** *Let $\{x_n\}$ be a basis of a set $\mathscr{M}$ in a Banach space $E$. Then $\{x_n\}$ is $\omega$-linearly independent.*

*Proof.* If $\sum_{i=1}^{\infty} \alpha_i x_i = 0$ and $\alpha_n \neq 0$, then

$$x_n = 1.x_n + \sum_{i \neq n} 0.x_i = 0.x_n + \sum_{i \neq n} \left(-\frac{\alpha_i}{\alpha_n}\right) x_i,$$

which, since $x_n \in \mathscr{M}$, contradicts the uniqueness condition in definition 16.1. This completes the proof.

Clearly, every basis $\{x_n\}$ of a Banach space $E$ is a basis of any set $\mathscr{M}$ containing $\{x_n\}$. One can also give some results in the converse direction, e.g. the following: *If $\{x_n\}$ is a basis of $\mathscr{M} = \{x \in E | f(x) = c\}$, where $f \in E^*$ and $0 = f(0) \neq c$* (i.e., $\mathscr{M}$ is a hyperplane not containing 0), *then $\{x_n\}$ is a basis of the whole space $E$.* Indeed, if $x \in E$ and $f(x) \neq 0$, then $\frac{c}{f(x)} x \in \mathscr{M}$, whence $\frac{c}{f(x)} x = \sum_{i=1}^{\infty} \alpha_i x_i$ and hence $x =$

$=\sum_{i=1}^{\infty}\left(\frac{f(x)}{c}\alpha_i\right)x_i$; on the other hand, if $x \in E$ and $f(x)=0$, then for any $x_0=\sum_{i=1}^{\infty}\alpha_i x_i \in \mathcal{M}$ we have $x+x_0 \in \mathcal{M}$, whence $x+x_0=\sum_{i=1}^{\infty}\beta_i x_i$ and hence $x=-x_0+\sum_{i=1}^{\infty}\beta_i x_i=\sum_{i=1}^{\infty}(\beta_i-\alpha_i)x_i$. Thus, every $x \in E$ admits an expansion $\sum_{i=1}^{\infty}\gamma_i x_i$. By proposition 16.1, this expansion is unique, so $\{x_n\}$ is a basis of $E$.

In the sequel we shall consider bases of certain cones $\mathcal{M}$ in real Banach spaces.

We recall that a set $\mathcal{M}$ in a real Banach space is said to be *acute angled* if

$$\|x+y\|-1 \geqslant \theta(t) > 0 \quad (x, y \in \mathcal{M},\ \|x\| \geqslant 1,\ \|y\| \geqslant t > 0) \quad (16.1)$$

and in this case the function

$$\theta_{\mathcal{M}}(t)=\inf_{\substack{x,y\in\mathcal{M}\\ \|x\|\geqslant 1,\ \|y\|\geqslant t}}(\|x+y\|-1) \quad (0 \leqslant t < \infty) \quad (16.2)$$

is called the *angular modulus* of $\mathcal{M}$. For example, the cone $\mathcal{K}_{\{e_n\}}$ associated to the unit vector basis*) $\{e_n\}$ of the real Hilbert space $l^2$ is acute angled, with the angular modulus

$$\theta_{\mathcal{K}_{\{e_n\}}}(t)=\sqrt{1+t^2}-1 \quad (0 \leqslant t < \infty). \quad (16.3)$$

Clearly, any subset of an acute angled set is acute angled. It is also obvious that for any set $\mathcal{M} \subset E$,

$$\theta_{\mathcal{M}}(t_1) \leqslant \theta_{\mathcal{M}}(t_2) \quad (t_1 \leqslant t_2). \quad (16.4)$$

**Lemma 16.1.** *Let $\mathcal{K}$ be an acute angled cone**) in a real Banach space $E$. If $\{x_n\} \subset \mathcal{K}$ and if there exists an increasing sequence of*

---

*) We recall (see Ch. II, § 10, definition 10.2) that $\mathcal{K}_{\{e_n\}}=\left\{\sum_{i=1}^{\infty}\alpha_i e_i \in l^2 \,\middle|\, \alpha_n \geqslant 0\ (n=1,2,\ldots)\right\}$.

**) We recall (see Ch. II, § 10) that by "cone" we mean "closed convex cone having the origin as extreme point", i.e. a closed set $\mathcal{K}$ such that $\mathcal{K}+\mathcal{K}\subset\mathcal{K}$, $\lambda\mathcal{K}\subset\mathcal{K}$ $(\lambda \geqslant 0)$ and $\mathcal{K}\cap(-\mathcal{K})=\{0\}$. In this section we shall only consider cones in *real* Banach spaces (see Ch. II, § 10, footnote to definition 10.2).

*positive integers* $\{m_n\}$ *such that*

$$\sup_{1 \leqslant n < \infty} \left\| \sum_{i=1}^{m_n} x_i \right\| \leqslant C < \infty , \tag{16.5}$$

*then the series* $\sum_{i=1}^{\infty} x_i$ *converges to an element* $x_0 \in \mathscr{K}$.

*Proof.* Since $\mathscr{K}$ is an acute angled cone, by (16.1) we have

$$\left\| \frac{x}{\|x\|} + \frac{y}{\|x\|} \right\| - 1 > 0 \qquad (x, y \in \mathscr{K} \setminus \{0\}),$$

whence

$$\|x + y\| \geqslant \|x\| \qquad (x, y \in \mathscr{K}). \tag{16.6}$$

Applying this to $x = \sum_{i=1}^{n} x_i \in \mathscr{K}$, $y = x_{n+1} \in \mathscr{K}$ $(n = 1,2, \ldots)$, it follows that $\left\{ \left\| \sum_{i=1}^{n} x_i \right\| \right\}$ is non-decreasing and hence, by (16.5), bounded, so there exists $c \leqslant C$ such that

$$\sup_{1 \leqslant n < \infty} \left\| \sum_{i=1}^{n} x_i \right\| = c = \lim_{n \to \infty} \left\| \sum_{i=1}^{n} x_i \right\| . \tag{16.7}$$

Assume now, a contrario, that $\sum_{i=1}^{\infty} x_i$ does not converge, so there exist a $\delta > 0$ and an increasing sequence $\{j_n\} \subset \mathscr{N}$ such that

$$\left\| \sum_{i=j_n+1}^{j_{n+1}} x_i \right\| \geqslant \delta \qquad (n = 1,2, \ldots).$$

By the assumption that $\sum_{i=1}^{\infty} x_i$ does not converge, we have $c \neq 0$ (indeed, otherwise from (16.7) we obtain $\sum_{i=1}^{\infty} x_i = 0$). Thus, by (16.7), for any $\varepsilon$ with $0 < \varepsilon < c$ there exists $N = N(\varepsilon)$ such that

$$\left\| \sum_{i=1}^{j_n} \right\| \geqslant c - \varepsilon \qquad (n \geqslant N).$$

Consequently, by (16.4) with $t_1 = \frac{\delta}{c} \leqslant \frac{\left\|\sum_{i=j_N+1}^{j_{N+1}} x_i\right\|}{\left\|\sum_{i=1}^{j_N} x_i\right\|} = t_2$, we obtain

$$\left\|\sum_{i=1}^{j_{N+1}} x_i\right\| = \left\|\sum_{i=1}^{j_N} x_i\right\| \left\| \frac{\sum_{i=1}^{j_N} x_i}{\left\|\sum_{i=1}^{j_N} x_i\right\|} + \frac{\sum_{i=j_N+1}^{j_{N+1}} x_i}{\left\|\sum_{i=1}^{j_N} x_i\right\|} \right\| \geqslant$$

$$\geqslant (c-\varepsilon)\left(1+\theta_{\mathscr{K}}\left(\frac{\left\|\sum_{i=j_N+1}^{j_{N+1}} x_i\right\|}{\left\|\sum_{i=1}^{j_N} x_i\right\|}\right)\right) \geqslant (c-\varepsilon)\left(1+\theta_{\mathscr{K}}\left(\frac{\delta}{c}\right)\right).$$

Since $\theta_{\mathscr{K}}\left(\frac{\delta}{c}\right) > 0$, for $\varepsilon > 0$ sufficiently small the last expression becomes greater than $c$, whence $\left\|\sum_{i=1}^{j_{N+1}} x_i\right\| > c$, in contradiction with (16.7), which proves that $\sum_{i=1}^{\infty} x_i$ converges to an element $x_0 \in E$. Since $\mathscr{K}$ is closed and $\sum_{i=1}^{n} x_i \in \mathscr{K}$ $(n = 1,2, \ldots)$, we have $x_0 \in \mathscr{K}$, which completes the proof of lemma 16.1.

**Lemma 16.2.** *Let $\mathscr{K}$ be an acute angled cone in a real Banach space E. If $\{x_n\} \subset \mathscr{K}$, $\{y_n\} \subset \mathscr{K}$ are such that*

$$\lim_{n\to\infty}\left(\sum_{i=1}^{n} x_i + y_n\right) = x, \tag{16.8}$$

*then the series $\sum_{i=1}^{\infty} x_i$ converges to an element $x_0 \in \mathscr{K}$.*

*Proof.* We claim that

$$\left\|\sum_{i=1}^{n} x_i\right\| \leqslant \|x\| \qquad (n = 1,2, \ldots). \tag{16.9}$$

Indeed, if for some $n_0$ we have $\left\|\sum_{i=1}^{n_0} x_i\right\| > \|x\|$, then by (16.6) for $\sum_{i=1}^{n_0} x_i, \sum_{i=n_0+1}^{n} x_i + y_n \in \mathscr{X}$, we obtain

$$\left\|\sum_{i=1}^{n} x_i + y_n\right\| \geqslant \left\|\sum_{i=1}^{n_0} x_i\right\| > \|x\| \qquad (n > n_0),$$

in contradiction with (16.8). This proves the claim (16.9).

By (16.9) and lemma 16.1, $\sum_{i=1}^{\infty} x_i$ converges to some element $x_0 \in \mathscr{X}$, which completes the proof.

*Remark 16.2.* Clearly, $\|x_0\| \leqslant \|x\|$. Moreover, if $\|x_0\| = \|x\|$, then $x_0 = x$. Indeed, let $\|x_0\| = \|x\|$. We shall show that in this case $\lim_{n\to\infty} y_n = 0$, whence, by (16.8), $x_0 = \sum_{i=1}^{\infty} x_i = x$. Assume, a contrario, that there exist a $\delta > 0$ and an increasing sequence $\{j_n\} \subset \mathscr{N}$ such that

$$\|y_{j_n}\| > \delta \qquad (n = 1,2, \ldots).$$

Then, with the same argument*) as in the proof of lemma 16.1 (replacing $\sum_{i=j_n+1}^{j_{n+1}} x_i$ by $y_{j_n}$ and $c$ by $\|x_0\|$) we obtain that for suitable $0 < \varepsilon < \|x_0\|$ and $N = N(\varepsilon)$,

$$\left\|\sum_{i=1}^{j_n} x_i + y_{j_n}\right\| \geqslant (\|x_0\| - \varepsilon)\left(1 + \theta_{\mathscr{X}}\left(\frac{\delta}{\|x_0\|}\right)\right) > \|x_0\| \qquad (n \geqslant N),$$

whence, by $\|x_0\| = \|x\|$ and (16.8),

$$\|x_0\| = \|x\| = \lim_{n\to\infty} \left\|\sum_{i=1}^{j_n} x_i + y_{j_n}\right\| > \|x_0\|,$$

which is impossible. This completes the proof of our assertion.

---

*) We mention that in lemma 16.1 the sequence $\left\{\left\|\sum_{i=1}^{j_n} x_i + \sum_{i=j_n+1}^{j_{n+1}} x_i\right\|\right\}$ was non-decreasing, while $\left\{\left\|\sum_{i=1}^{j_n} x_i + y_{j_n}\right\|\right\}$ need not be so. Observe that $x_0 \neq 0$, since otherwise, by (16.8), $\lim_{n\to\infty} \|y_n\| = \|x\| = \|x_0\| = 0$.

*Definition 16.2.* Let $\{x_n\}$ be a sequence in a real Banach space $E$. The set

$$\mathscr{C}_{\{x_n\}} = \left\{ \lim_{n\to\infty} \sum_{i=1}^{m_n} \alpha_i^{(n)} x_i \in E \,\middle|\, \alpha_i^{(n)} \geqslant 0 \ (i=1, \ldots, m_n;\ n=1,2, \ldots) \right\} \quad (16.10)$$

is called *the cone associated to the sequence* $\{x_n\}$.

In particular, if $\{x_n\}$ is a basis of $E$, then $\mathscr{C}_{\{x_n\}} = \mathscr{K}_{\{x_n\}}$, where $\mathscr{K}_{\{x_n\}}$ is the cone associated to the basis $\{x_n\}$, i.e.*)

$$\mathscr{K}_{\{x_n\}} = \left\{ \sum_{i=1}^{\infty} \alpha_i x_i \in E \,\middle|\, \alpha_n \geqslant 0 \quad (n = 1,2, \ldots) \right\}. \qquad (16.11)$$

Indeed, if $x = \sum_{i=1}^{\infty} \alpha_i x_i$, $\alpha_n \geqslant 0$ $(n = 1,2, \ldots)$, then $x = \lim_{n\to\infty} \sum_{i=1}^{n} \alpha_i x_i$, $\alpha_n \geqslant 0$ $(n = 1,2, \ldots)$, so $\mathscr{K}_{\{x_n\}} \subset \mathscr{C}_{\{x_n\}}$. Conversely, if $x = \lim_{n\to\infty} \sum_{i=1}^{m_n} \alpha_i^{(n)} x_i$, $\alpha_i^{(n)} \geqslant 0$ $(i = 1, \ldots, m_n;\ n = 1,2, \ldots)$ then $f_j(x) = \lim_{n\to\infty} \alpha_j^{(n)} \geqslant 0$ $(j = 1,2, \ldots)$, where $\{f_n\} \subset E^*$ is the a.s.c.f. to the basis $\{x_n\}$, whence $x = \sum_{j=1}^{\infty} f_j(x) x_j$, $f_j(x) \geqslant 0$ $(j = 1,2, \ldots)$, so $\mathscr{C}_{\{x_n\}} \subset \mathscr{K}_{\{x_n\}}$ and thus **) $\mathscr{C}_{\{x_n\}} = \mathscr{K}_{\{x_n\}}$.

The last part of the above argument also shows that *if* $(x_n, f_n)$ $(\{x_n\} \subset E,\ \{f_n\} \subset E^*)$ *is a total biorthogonal system, then* $\mathscr{C}_{\{x_n\}} \subset \mathscr{K}_{(x_n, f_n)}$, where $\mathscr{K}_{(x_n, f_n)}$ is the cone associated to the biorthogonal system $(x_n, f_n)$, i.e.***)

$$\mathscr{K}_{(x_n, f_n)} = \{x \in E \mid f_n(x) \geqslant 0 \quad (n = 1,2, \ldots)\}; \qquad (16.12)$$

in general, the inclusion is strict, since if $[x_n] \neq E$, then there exist an infinity of sequences $\{f_n\} \subset E^*$ such that $(x_n, f_n)$ is a biorthogonal system (see § 6, proposition 6.4), and each gives another cone $\mathscr{K}_{(x_n, f_n)}$.

We shall say that a sequence $\{x_n\}$ in a real Banach space $E$ is *acute angled* if the associated cone $\mathscr{C}_{\{x_n\}}$ is acute angled; in this case, the function

$$\theta_{\{x_n\}}(t) = \theta_{\mathscr{C}_{\{x_n\}}}(t) \quad (0 \leqslant t < \infty) \qquad (16.13)$$

is called the *angular modulus* of $\{x_n\}$.

---

*) See Ch. II, § 10, definition 10.2. Such a cone also occurs in formula (16.3).

**) With a slightly different argument, this was also proved in Ch. II, § 10, the remark after definition 10.2.

***) See Ch. II, § 16, definition 16.3.

**Theorem 16.1.** *For an acute angled sequence* $\{x_n\}$ *in a real Banach space* $E$ *the following statements are equivalent:*

1°. $\{x_n\}$ *is a basis of* $\mathscr{C}_{\{x_n\}}$.

2°. $\{x_n\}$ *is an M-basis of the linear subspace*)* $\mathscr{C}_{\{x_n\}} - \mathscr{C}_{\{x_n\}}$ *of* $E$.

3°. $\{x_n\}$ *is a basis of the linear subspace* $\mathscr{C}_{\{x_n\}} - \mathscr{C}_{\{x_n\}}$ *of* $E$.

*Proof.* Assume that we have 1° and let $x \in \mathscr{C}_{\{x_n\}} - \mathscr{C}_{\{x_n\}}$. Then $x = y - z$, where $y, z \in \mathscr{C}_{\{x_n\}}$ and hence, since $\{x_n\}$ is a basis of $\mathscr{C}_{\{x_n\}}$, there exist sequences of scalars $\{\alpha_n\}$, $\{\beta_n\}$ such that $x = \sum_{i=1}^{\infty} (\alpha_i - \beta_i) x_i$. Furthermore, by proposition 16.1, this expansion is unique. Thus, 1° $\Rightarrow$ 3°.

The implication 3° $\Rightarrow$ 2° is obvious.

Finally, assume that we have 2° and let $\{\varphi_n\} \subset (\mathscr{C}_{\{x_n\}} - \mathscr{C}_{\{x_n\}})^* \equiv [x_n]^*$ be the a.s.f. to $\{x_n\}$. Let $x \in \mathscr{C}_{\{x_n\}}$, say

$$x = \lim_{n \to \infty} \sum_{i=1}^{m_n} \alpha_i^{(n)} x_i, \tag{16.14}$$

where $\alpha_i^{(n)} \geqslant 0$ $(i = 1, \ldots, m_n;\ n = 1, 2, \ldots)$. Then

$$\varphi_j(x) = \lim_{n \to \infty} \alpha_j^{(n)} \geqslant 0 \qquad (j = 1, 2, \ldots), \tag{16.15}$$

and hence for each $k$ there exists a positive integer $n_k$ with $m_{n_k} > k$, such that

$$\left\| x - \sum_{i=1}^{m_{n_k}} \alpha_i^{(n_k)} x_i \right\| < \frac{1}{2k}, \quad \max_{1 \leqslant i \leqslant k} |\alpha_i^{(n_k)} - \varphi_i(x)| < \frac{1}{2k \sum_{i=1}^{k} \|x_i\|}.$$

Then

$$\left\| x - \sum_{i=1}^{k} \varphi_i(x) x_i - \sum_{i=k+1}^{m_{n_k}} \alpha_i^{(n_k)} x_i \right\| \leqslant \left\| x - \sum_{i=1}^{m_{n_k}} \alpha_i^{(n_k)} x_i \right\| + \left\| \sum_{i=1}^{k} (\alpha_i^{(n_k)} - \varphi_i(x))\, x_i \right\| < \frac{1}{k} \qquad (k = 1, 2, \ldots).$$

Hence, by lemma 16.2 (with $\{\varphi_k(x) x_k\} \subset \mathscr{C}_{\{x_n\}}$ and $\left\{ \sum_{i=k+1}^{m_{n_k}} \alpha_i^{(n_k)} x_i \right\} \subset \mathscr{C}_{\{x_n\}}$), we obtain that $\sum_{i=1}^{\infty} \varphi_i(x) x_i$ converges to an element $x_0 \in \mathscr{C}_{\{x_n\}}$.

---

*) $\mathscr{C}_{\{x_n\}} - \mathscr{C}_{\{x_n\}}$ need not be closed, but the notions of $M$-basis and basis are also well defined for incomplete normed linear spaces.

Then, by biorthogonality, $\varphi_j(x_0)=\varphi_j(x)$ $(j=1, 2, \ldots)$, whence, since $\{\varphi_n\}$ is total on $\mathscr{C}_{\{x_n\}} - \mathscr{C}_{\{x_n\}}$, $x_0 = x$, so $x = \sum_{i=1}^{\infty} \varphi_i(x)x_i$. Using again biorthogonality and that $\{\varphi_n\}$ is total on $\mathscr{C}_{\{x_n\}} - \mathscr{C}_{\{x_n\}}$, it follows that this expansion is unique. Thus, $2° \Rightarrow 1°$, which completes the proof of theorem 16.1.

**Corollary 16.1.** *If an acute angled minimal sequence $\{x_n\}$ in a real Banach space $E$ is a T-basis of $\mathscr{C}_{\{x_n\}}$, then $\{x_n\}$ is a basis of $\mathscr{C}_{\{x_n\}}$.*

*Proof.* Clearly, $\{x_n\}$ satisfies 2° of theorem 16.1.

Let us consider now some special classes of bases (defined in the obvious way) of the cones (16.12) associated to total biorthogonal systems. We have the following version of Ch. II, § 16, theorem 16.3 for such cones, which has a somewhat different proof:

**Theorem 16.2.** *Let $E$ be a real Banach space and let $(x_n, f_n)$ ($\{x_n\} \subset E$, $\{f_n\} \subset E^*$) be a biorthogonal system such that $\{f_n\}$ is total on $E$ and that $\{x_n\}$ is a basis of the cone $\mathscr{K}_{(x_n, f_n)}$. The following statements are equivalent:*

1°. *$\{x_n\}$ is an unconditional basis of $\mathscr{K}_{(x_n, f_n)}$.*

2°. *For every $x \in \mathscr{K}_{(x_n, f_n)}$ the series $\sum_{i=1}^{\infty} f_i(x)x_i$ is weakly unconditionally Cauchy* $\left(\text{i.e., } \sum_{i=1}^{\infty} |f_i(x)f(x_i)| < \infty \text{ for all } f \in E^*\right)$.

3°. *$\mathscr{K}_{(x_n, f_n)}$ is normal.*

4°. *For every $x \in \mathscr{K}_{(x_n, f_n)}$, the set*

$$\mathscr{P}_x = \mathscr{K}_{(x_n, f_n)} \cap (x - \mathscr{K}_{(x_n, f_n)}) = \{y \in E \mid 0 \leqslant y \leqslant x\} \quad (16.16)$$

*is bounded (in the norm).*

5°. *For every $x \in \mathscr{K}_{(x_n, f_n)}$ the set $\mathscr{P}_x$ above is linearly homeomorphic either to a finite dimensional cube or to the fundamental parallelotope of Hilbert (hence compact).*

*Moreover, if we have 1°, then $\mathscr{K}_{(x_n, f_n)}$ is minihedral.*

*Proof.* Assume that we have 1° and let $x \in \mathscr{K}_{(x_n, f_n)}$. Then there exists a (unique) sequence of scalars $\{\alpha_n\}$ such that $\sum_{i=1}^{\infty} \alpha_i x_i$ is unconditionally convergent to $x$. Then, by biorthogonality, $f_j(x) = \alpha_j$ $(j = 1, 2, \ldots)$, so $\sum_{j=1}^{\infty} f_j(x)x_j$ is unconditionally convergent to $x$. Thus, $1° \Rightarrow 2°$.

Assume now that we have 2°. Let $x, y \in \mathscr{K}_{(x_n, f_n)}$ be such that $y \leqslant x$, i.e. $0 \leqslant f_i(y) \leqslant f_i(x)$ $(i = 1,2, \ldots)$, and let $f \in E^*$ be arbitrary. Then, since $\{x_n\}$ is a basis of $\mathscr{K}_{(x_n, f_n)}$ and by 2°,

$$|f(y)| = \left| \sum_{i=1}^{\infty} f_i(y) f(x_i) \right| \leqslant \sum_{i=1}^{\infty} f_i(y)\, |f(x_i)| \leqslant \sum_{i=1}^{\infty} f_i(x)\, |f(x_i)| = M_{f,x} < \infty,$$

which shows that for every $x \in \mathscr{K}_{(x_n, f_n)}$, the set $\mathscr{P}_x = \{y \in E \mid 0 \leqslant y \leqslant x\}$ is weakly bounded, whence also strongly bounded. Thus, $2° \Rightarrow 4°$.

The equivalence $3° \Leftrightarrow 4°$ is well known.*)

Assume now that we have 3°. Let $x \in \mathscr{K}_{(x_n, f_n)}$ and $\varepsilon > 0$ be arbitrary. Since $\{x_n\}$ is a basis of $\mathscr{K}_{(x_n, f_n)}$, there exists a positive integer $N$ such that

$$\left\| \sum_{i=N}^{\infty} f_i(x) x_i \right\| < \frac{\varepsilon}{L}, \tag{16.17}$$

where $L > 0$ is a constant such that $0 \leqslant z \leqslant y$ implies $\|z\| \leqslant L\|y\|$ (such a constant $L$ exists, by the definition**) of normal cones). Now let $\sum_{i=1}^{\infty} f_{n_i}(x) x_{n_i}$ be an arbitrary subseries of $\sum_{i=1}^{\infty} f_i(x) x_i$. Choose $i_0$ such that $n_i \geqslant N$ whenever $i \geqslant i_0$. Then, for any $p \geqslant q \geqslant i_0$,

$$0 \leqslant \sum_{i=p}^{q} f_{n_i}(x) x_{n_i} \leqslant \sum_{i=N}^{\infty} f_i(x) x_i,$$

whence

$$\left\| \sum_{i=p}^{q} f_{n_i}(x) x_{n_i} \right\| \leqslant L \left\| \sum_{i=N}^{\infty} f_i(x) x_i \right\| < L \frac{\varepsilon}{L} = \varepsilon,$$

which proves (since $E$ is complete) that $\sum_{i=1}^{\infty} f_i(x) x_i$ is unconditionally convergent to an element of $E$. Since $\{f_n\}$ is total, $\sum_{i=1}^{\infty} f_i(x) x_i = x$. Thus, $3° \Rightarrow 1°$.

The implication $5° \Rightarrow 4°$ is obvious. Finally, the implications $1° \Rightarrow \mathscr{K}_{(x_n, f_n)}$ is minihedral and $1° \Rightarrow 5°$ follow by observing that in the proof of Ch. II, § 16, theorem 16.3, implication $1° \Rightarrow 5°$, only expansions of elements of $\mathscr{K}_{(x_n, f_n)}$ are used. This completes the proof of theorem 16.2.

---

*) See e.g. [5], p. 1165, lemma 2.

**) See Ch. II, § 10, formula (10.26).

A basis $\{x_n\}$ of the whole space $E$, with the a.s.c.f. $\{f_n\} \subset E^*$, may be an unconditional basis of $\mathscr{K}_{(x_n, f_n)}$ without being an unconditional basis of $E$, as shown by theorem 16.2 and Ch. II, § 16, example 16.3. Another natural example is the following:

*Example 16.1.* Let $\{x_n\}_0^\infty$ be the usual Schauder basis*) of $E = C([0, 1])$ and let $\{f_n\}_0^\infty \subset E^*$ be the a.s.c.f. to $\{x_n\}$. Then $x_n \geqslant 0$ $(n = 0,1,2, \ldots)$ and hence $\mathscr{K}_{(x_n, f_n)_0^\infty} = \mathscr{K}_{\{x_n\}_0^\infty}$ is contained in the natural positive cone of $E = C([0, 1])$, which is obviously normal**). Therefore $\mathscr{K}_{(x_n, f_n)_0^\infty}$ is normal, whence, by theorem 16.2, $\{x_n\}_0^\infty$ is an unconditional basis of $\mathscr{K}_{(x_n, f_n)_0^\infty}$, although $\{x_n\}_0^\infty$ is not an unconditional basis of $E$ (by Ch. II, § 15, theorem 15.1).

Since a cone $\mathscr{K} \subset E$ is normal if and only if***) there exists a constant $\delta > 0$ such that

$$\|x + y\| \geqslant \delta \qquad (x, y \in \mathscr{K}, \ \|x\| = \|y\| = 1), \tag{16.18}$$

it is obvious that *every acute angled cone is normal* (one can take $\delta = 1 + \theta(1)$). However, the converse is not true, even for cones associated to bases of Banach spaces, as shown by example 16.1 above (indeed, one can take two consecutive elements $x_k, x_{k+1}$ of the normalized Schauder basis $\{x_n\}_0^\infty$ of $E = C([0, 1])$ such that $\|x_k + x_{k+1}\| = 1$, so $\mathscr{K}_{(x_n, f_n)_0^\infty} = \mathscr{C}_{\{x_n\}_0^\infty}$ is not acute angled).

The converse of the last assertion of theorem 16.2 is not valid, as shown by

*Example 16.2.* Let $E = l^1$ and let

$$x_1 = e_1, \ x_n = e_{n-1} - e_n \qquad (n = 1, 2, \ldots), \tag{16.19}$$

where $\{e_n\}$ is the unit vector basis of $l^1$. Then, by Ch. II, § 16, example 16.1, $\{x_n\}$ is a conditional basis of $E$, such that $\sum_{i=1}^{\infty} |\alpha_i| x_i$ converges whenever $\sum_{i=1}^{\infty} \alpha_i x_i$ converges. Thus, for each $x \in E$ there exists the element $|x| \in E$ (in the partial order induced by $\mathscr{K}_{(x_n, f_n)}$, where $\{f_n\} \subset E^*$ is the a.s.c.f. to $\{x_n\}$), whence also the element $x_+ = \sup(x, 0) = \dfrac{x + |x|}{2}$, so $\mathscr{K}_{(x_n, f_n)}$ is minihedral and generating. However, $\mathscr{K}_{(x_n, f_n)}$ is not

*) See Ch. I, § 2, example 2.2.

**) See Ch. II, § 16, example 16.6.

***) See Ch. II, § 10, footnote to formula (10.26).

normal, since otherwise, by Ch. II, § 16, theorem 16.3, $\{x_n\}$ would be an unconditional basis of $E$. One can also see directly that $\mathscr{K}_{(x_n, f_n)}$ is not normal; indeed, for the sequences $\{y_n\} \subset E$, $\{z_n\} \subset E$ defined by

$$y_n = \frac{1}{n} \sum_{i=1}^{n} (e_{2i-1} - e_{2i}) = \frac{1}{n} \sum_{i=1}^{n} x_{2i}, \; z_n = \frac{1}{n}(e_1 - e_{2n}) =$$

$$= \frac{1}{n} \sum_{i=2}^{2n} x_i \qquad (n = 1, 2, \ldots) \tag{16.20}$$

we have $0 \leqslant y_n \leqslant z_n$ (in the partial order induced by $\mathscr{K}_{(x_n, f_n)}$), but $\|y_n\| = 2$, $\|z_n\| = \dfrac{2}{n}$ $(n = 1, 2, \ldots)$.

The assumption in theorem 16.2 that $\{x_n\}$ is a basis of $\mathscr{K}_{(x_n, f_n)}$ cannot be omitted, as shown by Ch. II, § 16, example 16.5, where $(x_n, f_n)$ is a total biorthogonal system for which we have $2° - 5°$, but not $1°$ (or by the example of the unit vectors $x_n$ and coordinate functionals $f_n$ on $E = m$, for which we have $2° - 4°$ but not $1°$ or $5°$). In the case when $\mathscr{K}_{(x_n, f_n)}$ is also sequentially weakly complete, we have the following result, in which it is no longer assumed that $\{x_n\}$ is a basis of $\mathscr{K}_{(x_n, f_n)}$:

**Theorem 16.3.** *Let $E$ be a real Banach space and let $(x_n, f_n)$ ($\{x_n\} \subset \subset E$, $\{f_n\} \subset E^*$) be a biorthogonal system such that $\mathscr{K}_{(x_n, f_n)}$ is normal and sequentially weakly complete. Then $\{x_n\}$ is an unconditional basis of $\mathscr{K}_{(x_n, f_n)}$, which is also boundedly complete on $\mathscr{K}_{(x_n, f_n)}$* (i.e., the relations $\alpha_i \geqslant 0$ for $i = 1,2, \ldots$ and $\sup_{1 \leqslant n < \infty} \left\| \sum_{i=1}^{n} \alpha_i x_i \right\| < \infty$ imply that $\sum_{i=1}^{\infty} \alpha_i x_i$ converges).

*Proof.* Let us first prove that $\{x_n\}$ is unconditionally boundedly complete on $\mathscr{K}_{(x_n, f_n)}$. Let $\{\alpha_n\}$ be a sequence of scalars such that $\alpha_n \geqslant 0$ $(n = 1,2, \ldots)$, $\sup_{1 \leqslant n < \infty} \left\| \sum_{i=1}^{n} \alpha_i x_i \right\| = M < \infty$, and let $J$ be an arbitrary finite set of positive integers. Choose $n$ such that $J \subset \{1, \ldots, n\}$. Then $0 \leqslant \sum_{i \in J} \alpha_i x_i \leqslant \sum_{i=1}^{n} \alpha_i x_i$, whence, since $\mathscr{K}_{(x_n, f_n)}$ is normal, we obtain, with a suitable constant $L > 0$,

$$\left\| \sum_{i \in J} \alpha_i x_i \right\| \leqslant L \left\| \sum_{i=1}^{n} \alpha_i x_i \right\| \leqslant LM.$$

Consequently, by Ch. II, § 15, corollary 15.1, $\sum_{i=1}^{\infty} \alpha_i x_i$ is weakly unconditionally Cauchy (i.e., $\sum_{i=1}^{\infty} |f(\alpha_i x_i)| < \infty$ for all $f \in E^*$), whence, since $\mathscr{K}_{(x_n, f_n)}$ is sequentially weakly complete, every subseries $\sum_{i=1}^{\infty} \alpha_{n_i} x_{n_i}$ converges weakly to an element of $\mathscr{K}_{(x_n, f_n)}$. Therefore, by the Orlicz-Pettis theorem*), the series $\sum_{i=1}^{\infty} \alpha_i x_i$ converges unconditionally in the norm topology, which proves that $\{x_n\}$ is unconditionally boundedly complete on $\mathscr{K}_{(x_n, f_n)}$.

Furthermore, since $(x_n, f_n)$ is a biorthogonal system, for every $x \in \mathscr{K}_{(x_n, f_n)}$ we have

$$0 \leqslant \sum_{i=1}^{n} f_i(x) x_i \leqslant x \qquad (n = 1,2, \ldots),$$

whence, since $\mathscr{K}_{(x_n, f_n)}$ is normal (with the constant $L$),

$$\sup_{1 \leqslant n < \infty} \left\| \sum_{i=1}^{n} f_i(x)\, x_i \right\| \leqslant L \|x\| < \infty,$$

and therefore, by the above, $\sum_{i=1}^{\infty} f_i(x) x_i$ converges unconditionally in the norm topology to an element $x_0 \in \mathscr{K}_{(x_n, f_n)}$. Since $\{f_n\}$ is total on $E$, we must have $x_0 = x$, so $\{x_n\}$ is an unconditional basis of $\mathscr{K}_{(x_n, f_n)}$, which completes the proof of theorem 16.3.

The converse of theorem 16.3 is not valid, as shown by

*Example 16.3*. Let $\{x_n\}$ be the standard conditional basis

$$x_n = \sum_{i=1}^{n} e_i \qquad (n = 1, 2, \ldots) \tag{16.21}$$

of $E = c_0$ and let $\{f_n\} \subset E^*$ be the a.s.c.f. to $\{x_n\}$. Then $\{x_n\} \subset \mathscr{K}_{(x_n, f_n)}$ is a weak Cauchy sequence which is not weakly convergent to any element of $E$, so $\mathscr{K}_{(x_n, f_n)}$ is not sequentially weakly complete. However, as was observed after theorem 16.2, $\{x_n\}$ is an unconditional basis

*) See e.g. [87], p. 318, theorem 1.

of $\mathscr{K}_{(x_n,f_n)}$. Finally, $\{x_n\}$ is also boundedly complete on $\mathscr{K}_{(x_n,f_n)}$, since the relations $\alpha_i \geqslant 0$ $(i = 1,2, \ldots)$ and

$$\sup_{1 \leqslant n < \infty} \left\| \sum_{i=1}^{n} \alpha_i x_i \right\| = \sup_{1 \leqslant n < \infty} \sup_{1 \leqslant j \leqslant n} \left| \sum_{i=j}^{n} \alpha_i \right| = \sup_{1 \leqslant n < \infty} \left| \sum_{i=1}^{n} \alpha_i \right| < \infty$$

imply that $\sum_{i=1}^{\infty} \alpha_i < \infty$ and hence, by Ch. II, § 14, example 14.1, $\sum_{i=1}^{\infty} \alpha_i x_i$ converges in the norm topology.

## II. Generalizations of Bases, Without Assuming Countability

The term "countable" in the title of part I referred to the cardinality of the set of objects (which can be elements of the space $E$, or operators, or subspaces) occurring in the definitions and was not related to the separability of the space $E$. Thus, we have seen that every (separable or non-separable) Banach space $E$ contains basic sequences and various other non-complete sequences of elements which generalize the notion of a basis and that some (separable or non-separable) Banach spaces $E$ have complete sequences (of course, not of elements, when $E$ is non-separable) which generalize bases, for example, Schauder decompositions.

In the present part, having in view future applications to the study of non-separable Banach spaces, we shall consider generalizations of bases which need not form a countable set of objects (but, for the particular case of separable spaces, most of them will turn out to be countable). Naturally, we shall introduce them by extending the countable generalizations of part I to the case of not necessarily countable sets of objects; in some natural cases we shall even use for them the same term, adding the prefix "extended".

### § 17. *ER*-sets. Extended unconditional bases. Transfinite bases. Strongly unconditional integral bases

The problem of extension of the notion of a basis $\{x_n\}$ of a separable Banach space $E$ to the case when $E$ is not assumed to be separable, replacing the sequence $\{x_n\}$ by a family $\{x_i\}_{i \in I}$ which is not assumed countable, presents the main difficulty of finding for this case a suitable generalization of conditional convergence of series. Since the concept of conditional convergence of a series $\sum_{i=1}^{\infty} y_i$ relies on the ordering of the underlying countable set of elements as a sequence $\{y_n\} = \{y_1, y_2, y_3, \ldots\}$, the generalizations of bases which we shall give in this section will involve some "ordering" of the family $\{x_i\}_{i \in I}$ or of its countable subfamilies. Of course, since for unconditional convergence such ordering problems do not arise, the extensions of unconditional bases will be "easier" to introduce.

The following definition, involving only the usual convergence of series, generalizes, in a natural way, the notion of a countable set having an ordering which is a basis (see corollary 17.1 below):

*Definition 17.1.* A family $\{x_i\}_{i \in I}$ of elements in a Banach space $E$ is called an *ER-set* (or *Enflo-Rosenthal set*) of $E$, if it is complete in $E$ (i.e., $[x_i]_{i \in I} = E$) and if every countable subfamily of $\{x_i\}_{i \in I}$ has an ordering which is a basic sequence.

By Ch. I, § 4, proposition 4.1 a), every basis of a separable Banach space $E$ is an $ER$-set of $E$, with $I = \mathcal{N}$ and the initial ordering. An example of a not necessarily separable Banach space $E$, having an $ER$-set, is the following:

*Example 17.1.* Let $(T, \nu)$ be a positive measure space and let $1 < p < \infty$. Then $E = L^p(T, \nu)$ has an $ER$-set.

The proof, which will be given in Vol. III, uses the known structure of $L^p(T, \nu)$ spaces [257] and the following result of R.E.A.C. Paley [298]: *The Walsh system**) $\{w_j\}$ *is a basis of* $L^p([0, 1])$. The proof of this latter result is a consequence of the machinery used in Ch. II, § 14, proof of theorem 14.1 and will be also given in Vol. III.

*Problem 17.1.* Let $(T, \nu)$ be a positive measure space such that $\nu(T) < \infty$ and that $L^1(T, \nu)$ is non-separable. Does $E = L^1(T, \nu)$ have an $ER$-set?

**Proposition 17.1.** *If $\{x_i\}_{i \in I}$ is an ER-set of a Banach space E, then for every separable subspace G of E there exists a separable subspace F of E with $F \supset G$, such that some countable subfamily $\{x_i\}_{i \in I_0}$ of $\{x_i\}_{i \in I}$ has an ordering which forms a basis of F.*

*Proof.* Let $\{y_k\}$ be a dense sequence in $G$. Then, since $[x_i]_{i \in I} = E$, for every $k = 1,2, \ldots$ there exist a sequence $\{I_k^n\}_{n=1}^{\infty}$ of finite subsets of $I$ and a sequence $\{\sum_{i \in I_k^n} \alpha_i^{(k,n)} x_i\}$ of finite linear combinations such that $\lim_{n \to \infty} \sum_{i \in I_k^n} \alpha_i^{(k,n)} x_i = y_k$ $(k = 1,2, \ldots)$. Let $F = [x_i]_{i \in I_0}$, where $I_0 =$

*) See Ch. II, § 14, formulae (14.6)÷(14.13).

$= \bigcup_{k=1}^{\infty} \bigcup_{n=1}^{\infty} I_k^n$. Then, since $I_0$ is countable, $\{x_i\}_{i \in I_0}$ has an ordering which is a basic sequence, hence a basis of $F$. Also, $F \supset \{y_k\}$, whence $F \supset G$, which completes the proof of proposition 17.1.

**Corollary 17.1.** *A family $\{x_i\}_{i \in I}$ is an ER-set of a separable Banach space $E$ if and only if $\{x_i\}_{i \in I}$ is countable and has an ordering which is a basis of $E$. Consequently, a separable Banach space $E$ has an ER-set if and only if it has a basis.*

*Proof.* If $\{x_i\}_{i \in I}$ is countable and has an ordering $\{x_{i_n}\}$ which is a basis of $E$, then Ch. I, § 4, proposition 4.1, every subsequence of $\{x_{i_n}\}$ is a basic sequence and thus $\{x_i\}_{i \in I}$ is an *ER*-set of $E$.

Conversely, assume now that $E$ is separable and that $\{x_i\}_{i \in I}$ is an *ER*-set of $E$. Then, by proposition 17.1 above with $G = E$, some countable subfamily $\{x_i\}_{i \in I_0}$ of $\{x_i\}_{i \in I}$ has an ordering which is a basis of $E$. Now, if there existed an $i_1 \in I \setminus I_0$, then $I_1 = I_0 \cup \{i_1\}$ would be countable, whence $\{x_i\}_{i \in I_1}$ would have an ordering which is a basic sequence, But then $x_{i_1} \notin [x_i]_{i \in I_0}$, in contradiction with $[x_i]_{i \in I_0} = E$. Consequently, $I = I_0$, which completes the proof.

It is natural to ask, to what extent the known properties of bases remain valid for *ER*-sets. Firstly, we shall show below that one can define coefficient functionals for *ER*-sets and that these functionals are still continuous; moreover, if the *ER*-set is normalized, then they are also uniformly bounded. The proof will not be an extension of the proof of Ch. I, § 3, theorem 3.1, but actually it will make use of that result.

Let $\{x_i\}_{i \in I}$ be an *ER*-set of a Banach space $E$. Then, clearly, $\{x_i\}_{i \in I}$ is finitely linearly independent and hence we can define a family of linear functionals $\{f_i\}_{i \in I}$ on $P = \text{lin}\,\{x_i\}_{i \in I}$ (the linear span of $\{x_i\}_{i \in I}$) by

$$f_i\left(\sum_{k=1}^{n} \alpha_{i_k} x_{i_k}\right) = \begin{cases} \alpha_i & \text{if } i \in \{i_k\}_{k=1}^n \\ 0 & \text{if } i \in I \setminus \{i_k\}_{k=1}^n \end{cases} \quad \left(\sum_{k=1}^{n} \alpha_{i_k} x_{i_k} \in P\right). \tag{17.1}$$

**Theorem 17.1.** *Let $\{x_i\}_{i \in I}$ be an ER-set of a Banach space $E$. Then the functionals $f_i$ defined by* (17.1) *are continuous on $P = \text{lin}\,\{x_i\}_{i \in I}$ and hence they admit unique extensions to $f_i \in E^*$ $(i \in I)$. Moreover, there exists a constant $M$ such that*

$$1 \leqslant \|x_i\|\,\|f_i\| \leqslant M \qquad (i \in I). \tag{17.2}$$

*Proof.* We may assume $\left(\text{replacing } (x_i, f_i) \text{ by } \left(\frac{x_i}{\|x_i\|}, \|x_i\| f_i\right)\right)$ that $\|x_i\| = 1$ $(i \in I)$. We shall show that there exists a constant $M$ such that

$$|f_i(x)| \leqslant M\|x\| \qquad (x \in P,\ i \in I), \tag{17.3}$$

which will complete the proof. Assume that (17.3) does not hold. Then there exist a sequence $\{i_n\} \subset I$ and a sequence $\{y_n\} \subset P$ with $\|y_n\| = 1$ $(n = 1,2, \ldots)$ such that

$$\lim_{n\to\infty} f_{i_n}(y_n) = \infty. \tag{17.4}$$

Since $[x_{i_n}, y_n]$ is a separable subspace of $E$, by proposition 17.1 there exists a countable subfamily $\{x_i\}_{i\in I_0}$ of $\{x_i\}_{i\in I}$ such that $[x_{i_n}, y_n] \subset \subset [x_i]_{i\in I_0}$ and that $\{x_i\}_{i\in I_0}$ has an ordering $\{x_{i_n^0}\}$ which is a basis of $[x_i]_{i\in I_0}$. Let $\{\varphi_{i_n^0}\} \subset ([x_i]_{i\in I_0})^*$ be the a.s.c.f. to $\{x_{i_n^0}\}$. Then, by $\{i_n\} \subset I_0$ and Ch. I, § 3, theorem 3.1,

$$\sup_{1\leqslant n<\infty} |f_{i_n}(y_n)| = \sup_{1\leqslant n<\infty} |\varphi_{i_n}(y_n)| \leqslant \sup_{1\leqslant n<\infty} \|\varphi_{i_n}\| \leqslant$$

$$\leqslant \sup_{i\in I_0} \|\varphi_i\| = \sup_{i\in I_0} \|x_i\| \, \|\varphi_i\| < \infty,$$

which contradicts (17.4). Thus, we have (17.3), which completes the proof.

Now we are ready to give

*Definition 17.2.* Let $\{x_i\}_{i\in I}$ be an *ER*-set of a Banach space $E$. The family $\{f_i\}_{i\in I} \subset E^*$ of theorem 17.1 above is called the family of *coefficient functionals* associated to the *ER*-set $\{x_i\}_{i\in I}$ (or, shortly, a.f.c.f.).

The term "coefficient functionals" is motivated by formula (17.1) and by

**Corollary 17.2.** *Let $\{x_i\}_{i\in I}$ be an ER-set of a Banach space E and let $\{f_i\}_{i\in I} \subset E^*$ be the a.f.c.f. to $\{x_i\}_{i\in I}$. Then for each $x \in E$ there exists a unique family of scalars $\{\alpha_i\}_{i\in I} \subset K$ such that the set $\mathscr{S}_x = \{i \in I \mid \alpha_i \neq 0\}$ is countable and has an ordering $\{i_n\}$ such that*

$$x = \sum_{n=1}^{\infty} \alpha_{i_n} x_{i_n}; \tag{17.5}$$

*namely,*

$$\alpha_i = f_i(x) \qquad (i \in I). \tag{17.6}$$

*Proof.* Let $x \in E$. Since $[x_i]_{i\in I} = E$, there exists a countable subset $I_0$ of $E$ such that $x \in [x_i]_{i\in I_0}$. Then $\{x_i\}_{i\in I_0}$ has an ordering $\{x_{i_n^0}\}$ which is a basis of $[x_i]_{i\in I_0}$, so there is a sequence of scalars $\{\alpha_{i_n^0}\}$ such that

$$x = \sum_{n=1}^{\infty} \alpha_{i_n^0} x_{i_n^0}. \tag{17.7}$$

Define now

$$\alpha_i = 0 \qquad (i \in I \setminus I_0). \tag{17.8}$$

We shall show that $\{\alpha_i\}_{i\in I}$ has the required properties. Indeed, by (17.8), $\mathcal{S}_x = \{i \in I \mid \alpha_i \neq 0\} \subset I_0$ (so $\mathcal{S}_x$ is countable) and hence, by (17.7), we have (17.5) for the ordering $\{i_n\}$ of $\mathcal{S}_x$ induced by $\{i_n^0\}$. Finally, by $f_i \in E^*$, (17.7), (17.8) and the biorthogonality of $(x_i, f_i)_{i\in I}$ we have (17.6), which completes the proof of corollary 17.2.

*Problem 17.2.* a) Is the converse of corollary 17.2 valid? b) If $\{x_n\}$ is a sequence in a separable Banach space $E$ such that for each $x \in E$ there exist a unique sequence of scalars $\{\alpha_n\}$ and a (clearly, non-unique) permutation $\sigma$ of $\mathcal{N}$ such that $x = \sum_{i=1}^{\infty} \alpha_{\sigma(i)} x_{\sigma(i)}$, then is some permutation $\{x_{\tau(n)}\}$ of $\{x_n\}$ a basis of $E$? If, in addition, $\{x_n\}$ is minimal, is some permutation $\{x_{\tau(n)}\}$ of $\{x_n\}$ a basis of $E$?

**Proposition 17.2.** *Let $\{x_i\}_{i\in I}$ be an ER-set of a Banach space E, with the a.f.c.f. $\{f_i\}_{i\in I} \subset E^*$. Then $\{f_i\}_{i\in I}$ is total on E.*

*Proof.* Let $x \in E$, $f_i(x) = 0$ $(i \in I)$. Then, for $\{\alpha_i\}_{i\in I} \subset K$ as in corollary 17.2, $x = \sum_{n=1}^{\infty} \alpha_{i_n} x_{i_n} = \sum_{n=1}^{\infty} f_{i_n}(x) x_{i_n} = 0$, which completes the proof.

As a consequence, we obtain that there are many usual non-separable Banach spaces which have no *ER*-set:

**Corollary 17.3.** *No non-reflexive Grothendieck space**) *E (hence, in particular, no $\mathcal{P}_\lambda$ space) admits an ER-set.*

*Proof.* By theorem 17.1 and proposition 17.2, *every ER-set $\{x_i\}_{i\in I}$ of a Banach space E is an extended M-basis of E* (in the sense of § 20, definition 20.4). Hence the conclusion follows from § 20, proposition 20.5.

*Definition 17.3.* Let $\{x_i\}_{i\in I}$ be an *ER*-set of a Banach space $E$, with the a.f.c.f. $\{f_i\}_{i\in I} \subset E^*$. Any family of finite rank operators $\{s_{\{i_k\},n}\}$ on $E$, of the form

$$s_{\{i_k\},n}(x) = \sum_{k=1}^{n} f_{i_k}(x) x_{i_k} \qquad (x \in E), \tag{17.9}$$

where $\{i_k\}$ is an ordering of the countable set $\mathcal{S}_x = \{i \in I \mid f_i(x) \neq 0\}$ as in corollary 17.2 and where $n = 1, 2, \ldots$, is called a family of *partial sum operators* associated to the *ER*-set $\{x_i\}_{i\in I}$.

*) See § 15, remark 15.3 a).

Thus, there are many families of partial sum operators associated to an $ER$-set, corresponding to the selections of an ordering $\{i_k\}$ of the set $\mathcal{S}_x$ for each $x \in E$. By theorem 17.1, these partial sum operators are continuous on $E$, so it is natural to ask whether there exists at least one associated family of partial sum operators which is uniformly bounded on $E$. In general, even the answer to the question whether the partial sum operators have the following considerably weaker property, is also negative: Is it possible to select, for each countable subset $I_0$ of $I$, an ordering $\{i_k^0\}$ of $I_0$ such that

$$\sup_{1 \leqslant n < \infty} \|s_{\{i_k^0\}, n}\| \leqslant M(\{i_k^0\}) < \infty \tag{17.10}$$

(where $\sup\limits_{I_0} \sup\limits_{\{i_k^0\}} M(\{i_k^0\}) = \infty$ is also permitted)? Indeed, if $I = \mathcal{N}$ and if $E$ is reflexive, then by (17.10), for any infinite subset $I_0$ of $I$, $\{x_{i_k^0}\}$ is a boundedly complete basic sequence, whence, by (17.10) and § 1, proposition 1.14 a), $E = [x_i]_{i \in I_0} \oplus ([f_i]_{i \in I_0})_{\perp}$. But, since $I_0 \subset I$ and some ordering $\{x_{i_n}\}$ of $\{x_i\}_{i \in I}$ is a basis of $E$, we have $([f_i]_{i \in I_0})_{\perp} = [x_j]_{j \in I \setminus I_0}$, whence $E = [x_i]_{i \in I_0} \oplus [x_j]_{j \in I \setminus I_0}$. Consequently, by § 7, theorem 7.3, $\{x_{i_n}\}$ is an unconditional basis of $E$ (which, of course, a reflexive space $E$ with a basis need not have).

However, we shall prove now the following positive result:

**Theorem 17.2.** *Let $\{x_i\}_{i \in I}$ be an ER-set of a Banach space E. Then for each countable subset $I_0$ of I one can select an ordering $\{i_k^0\}$ of $I_0$ such that the corresponding family $\bigcup\limits_{I_0} \{s_{\{i_k^0\}, n}|_{[x_i]_{i \in I_0}}\}_{n=1}^{\infty}$ is uniformly bounded, i.e. such that*

$$\sup_{I_0} \nu_{\{x_{i_k^0}\}} < \infty, \tag{17.11}$$

*where the* sup *is taken over all countable subsets $I_0$ of I.*

*Proof.* Choose, for each countable subset $I_0$ of $I$, an ordering $\{i_k^0\}$ of $I_0$ such that

$$\nu_{\{x_{i_k^0}\}} \leqslant \inf_{\{i_k\}} \nu_{\{x_{i_k}\}} + 1 = \nu(I_0) + 1, \tag{17.12}$$

where the inf is taken over all orderings $\{i_k\}$ of $I_0$ such that $\{x_{i_k}\}$ is a basic sequence. We shall show that this selection satisfies (17.11). Indeed, if not, then $\sup\limits_{I_0} \nu(I_0) = \infty$, which will lead to a contradiction. Observe that if $I_1, I_2$ are countable subsets of $I$ with $I_1 \subset I_2$, then $\nu(I_1) \leqslant \nu(I_2)$, since if $\{i_k^2\}$ is an ordering of $I_2$ such that $\{x_{i_k^2}\}$ is a basic sequence, then by Ch. I, § 4, proposition 4.1, $\{x_{i_k^2}\}_{i_k^2 \in I_1}$ is a basis of $[x_i]_{i \in I_1}$, satisfying, clearly, $\nu_{\{x_{i_k^2}\}_{i_k^2 \in I_1}} \leqslant \nu_{\{x_{i_k^2}\}}$. Now, by our assumption there exists a sequence $\{I_n\}$ of countable subsets of $I$ such that $\lim_{n\to\infty} \nu(I_n) = \infty$. Let $I_0 = \bigcup_{n=1}^{\infty} I_n$. Then $I_0 \supset I_n$, whence, by the preceding observation, $\nu(I_0) \geqslant \nu(I_n)$ $(n = 1,2, \ldots)$, so $\nu(I_0)=\infty$, which contradicts the assumption that $\{x_i\}_{i\in I}$ is an *ER*-set of $E$. This completes the proof of theorem 17.2.

One can define various special classes of *ER*-sets by extending the corresponding special classes of bases in a similar way to definition 17.1 above. For example, let us give

*Definition 17.4.* An *ER*-set $\{x_i\}_{i\in I}$ of a Banach space $E$ is said to be *shrinking* (respectively, *boundedly complete*), if every countable subfamily of $\{x_i\}_{i\in I}$ has an ordering which is a shrinking (respectively, a boundedly complete) basic sequence.

Similarly, one can define unconditional *ER*-sets, which we shall also call "extended unconditional bases" (since in this case the preceding complications with existence of suitable orderings do not arise), as follows:

*Definition 17.5.* A family $\{x_i\}_{i\in I}$ of elements in a Banach space $E$ is called an *extended unconditional basis* of $E$ (or, an *unconditional ER-set* of $E$), if it is complete in $E$ and if every countable subfamily of $\{x_i\}_{i\in I}$ is an unconditional basic sequence.

Clearly, a family of elements $\{x_i\}_{i\in I}$ in a *separable* Banach space $E$ is an extended unconditional basis of $E$ if and only if $\{x_i\}_{i\in I}$ is a countable unconditional basis of $E$.

There are many concrete Banach spaces which have no extended unconditional basis. For example, we have the following generalization of Ch. II, § 15, theorems 15.4 and 15.1:

**Theorem 17.3.** *If $E$ is a Banach space containing a subspace isomorphic to $J$ (in particular, if $E$ contains a subspace isomorphic to $C([0, 1])$ or to $m = l^\infty$) then $E$ has no extended unconditional basis.*

We recall that for any normed linear space $E$, the *density character* of $E$, denoted dens $E$, is defined as the smallest cardinal $\mathfrak{m}$ for which $E$ has a dense subset of cardinality $\mathfrak{m}$.

**Theorem 17.4.** *Let $1 < p < \infty$, $p \neq 2$ and let $\nu$ be a finite positive measure on some measurable space $T$, such that* dens $L^p(T, \nu) \geqslant \aleph_\omega$, *where $\omega$ is the first infinite ordinal number. If $E$ is a Banach space containing a subspace isomorphic to $L^p(T, \nu)$, then $E$ has no extended unconditional basis.*

The proof, which will be given in Vol. III, uses the known structure of $L^p(T, \nu)$ spaces [257] and a combinatorial lemma. Theorem 17.4, together with example 17.1, show that some frequently used non-separable Banach spaces have an *ER*-set, but no extended unconditional basis.

It is not known whether theorem 17.4 remains valid if one only assumes that $L^p(T, \nu)$ is non-separable. In particular:

*Problem 17.3.* Let $(T, \nu)$ be a positive measure space such that $\nu(T) < \infty$ and that $L^p(T, \nu)$ is non-separable, where $1 < p < \infty$, $p \neq 2$. Does $E = L^p(T, \nu)$ have an extended unconditional basis?

In general, the results on unconditional bases of separable Banach spaces admit natural generalizations to extended unconditional bases. For example, we mention here the following characterizations of extended unconditional bases:

**Theorem 17.5.** *For a family of elements $\{x_i\}_{i \in I}$ in a Banach space $E$ the following statements are equivalent:*

1°. *$\{x_i\}_{i \in I}$ is an extended unconditional basis of $E$.*

2°. *For every $x \in E$ there exists a unique family of scalars $\{\alpha_i\}_{i \in I} \subset K$ such that*

$$x = \lim_{d \in \mathscr{D}} \sum_{i \in d} \alpha_i x_i, \tag{17.13}$$

*where $\mathscr{D}$ is the collection of all finite subsets of the set $I$, directed by inclusion.*

3°. *For every $x \in E$ there exists a unique family of scalars $\{\alpha_i\}_{i \in I} \subset K$ such that the set $\mathscr{S}_x = \{i \in I | \alpha_i \neq 0\}$ is countable and that we have*

$$x = \sum_{i \in \mathscr{S}_x} \alpha_i x_i, \tag{17.14}$$

*where the series is unconditionally convergent.*

4°. *There exists a constant $M_1$ with $1 \leqslant M_1 < \infty$, such that*

$$\left\| \sum_{i \in d_1} \alpha_i x_i \right\| \leqslant M_1 \left\| \sum_{i \in d_2} \alpha_i x_i \right\| \tag{17.15}$$

*for any $d_1, d_2 \in \mathscr{D}$ with $d_1 \subset d_2$ (where $\mathscr{D}$ is as in 2°) and any $\{\alpha_i\}_{i \in d_2} \subset K$.*

5°. *There exists a constant $M_2$ with $1 \leqslant M_2 < \infty$ such that*

$$\left\| \sum_{i \in d} \varepsilon_i \alpha_i x_i \right\| \leqslant M_2 \left\| \sum_{i \in d} \alpha_i x_i \right\| \tag{17.16}$$

*for any $d \in \mathscr{D}$ and any $\{\varepsilon_i\}_{i \in d}$, $\{\alpha_i\}_{i \in d} \subset K$ with $\varepsilon_i = \pm 1$ $(i \in d)$.*

6°. *There exists a constant $M_3$ with $1 \leqslant M_3 < \infty$, such that*

$$\left\| \sum_{i \in d} \beta_i \alpha_i x_i \right\| \leqslant M_3 \left\| \sum_{i \in d} \alpha_i x_i \right\| \tag{17.17}$$

*for any $d \in \mathscr{D}$ and any $\{\beta_i\}_{i \in d}$, $\{\alpha_i\}_{i \in d} \subset K$ with $|\beta_i| \leqslant 1$ $(i \in d)$.*

We have the following generalization of Ch. II, § 18, theorems 18.1 and 18.2:

**Theorem 17.6.** *In each of the spaces**) $l^2(I)$, $l^1(I)$ *and* $c_0(I)$ *all bounded extended unconditional bases are equivalent.*

*Definition 17.6.* An extended unconditional basis $\{x_i\}_{i\in I}$ of a Banach space $E$ is said to be *orthogonal* (respectively, *strictly orthogonal*) if we have (17.15) with $M_1 = 1$ (respectively,

$$\left\|\sum_{i\in d_1} \alpha_i x_i\right\| < \left\|\sum_{i\in d_2} \alpha_i x_i\right\| \tag{17.18}$$

for any $d_1, d_2 \in \mathscr{D}$ with $d_1 \subset d_2$ and any $\{\alpha_i\}_{i\in d_2} \subset K$ such that $\sum_{i\in d_2 \setminus d_1} |\alpha_i| \neq 0$). The extended unconditional basis $\{x_i\}_{i\in I}$ is said to be *hyperorthogonal* (respectively, *strictly hyperorthogonal*), if we have (17.17) with $M_3 = 1$ (respectively,

$$\left\|\sum_{i\in d} \beta_i \alpha_i x_i\right\| < \left\|\sum_{i\in d} \alpha_i x_i\right\| \tag{17.19}$$

for any $d \in \mathscr{D}$ and any $\{\beta_i\}_{i\in d}, \{\alpha_i\}_{i\in d} \subset K$ such that $|\beta_{i_0}| < 1, \alpha_{i_0} \neq 0$ for some $i_0 \in d$).

For example, the natural unit vector basis $\{e_i\}_{i\in I}$ of $l^p(I)$ $(1 \leqslant p < \infty)$, defined by

$$e_i = \{\delta_{ij}\}_{j\in I} \qquad (i \in I), \tag{17.20}$$

is strictly hyperorthogonal, while the unit vector basis $\{e_i\}_{i\in I}$ of $c_0(I)$ is hyperorthogonal, but not strictly hyperorthogonal.

The partial sum operators

$$s_{\{\beta_n\},d}(x) = \sum_{i\in d} \beta_i f_i(x) x_i \qquad (x \in E,\ d \in \mathscr{D},\ |\beta_i| \leqslant 1 \text{ for } i \in d) \tag{17.21}$$

---

*) We recall that for $1 \leqslant p < \infty$, $l^p(I)$ is the set of all scalar-valued functions on $I$ whose $p$-th power is unconditionally summable, with the norm $\|x(.)\| = (\sum_{i\in I} |x(i)|^p)^{\frac{1}{p}}$; $l^\infty(I)$ is the set of all scalar-valued bounded functions on $I$, with the norm $\|x(.)\| = \sup_{i\in I} |x(i)|$; $c_0(I)$ is the closed linear subspace of $l^\infty(I)$ consisting of all $x(.) \in l^\infty(I)$ such that for each $\varepsilon > 0$ the set $\{i \in I \mid |x(i)| > \varepsilon\}$ is finite (see e.g. [70], Ch. II, § 2). Clearly, for $I = \mathscr{N} = \{1, 2, \ldots\}$ these spaces reduce to the usual spaces $l^p$, $l^\infty$ and $c_0$ respectively.

associated to an extended unconditional basis $\{x_i\}_{i\in I}$ are uniformly bounded. Consequently, every Banach space $E$ with an extended unconditional basis $\{x_i\}_{i\in I}$ admits an equivalent norm $|||\,.\,|||$ in which $\{x_i\}_{i\in I}$ is strictly hyperorthogonal, namely,

$$|||x||| = \sup_{d\in\mathscr{D}} \sup_{|\beta_i|\leqslant 1(i\in d)} \left\| \sum_{i\in d} \beta_i f_i(x)\, x_i \right\| + \sum_{n=1}^{\infty} \frac{1}{2^{i_n}} \| f_{i_n}(x) x_{i_n} \| \qquad (x\in E), \tag{17.22}$$

where $\{f_i\}_{i\in I}\subset E^*$ is the a.f.c.f. to $\{x_i\}_{i\in I}$, $\mathscr{D}$ is the collection of all finite subsets of $I$, and $\{i_n\}$ is an ordering of the countable set $\mathscr{S}_x = \{i\in I \mid f_i(x)\neq 0\}$ such that $|f_{i_1}(x)|\geqslant |f_{i_2}(x)|\geqslant \ldots$

Now we pass to another generalization of bases, namely, to the notion of a transfinite basis of a Banach space $E$, which uses, instead of a suitable ordering of every countable subset $I_0$ of the index set $I$, a well ordering of the whole index set $I$; or, what amounts to the same thing, we replace the index set $\mathscr{N} = \{1,2,3, \ldots\}$ in the definition of usual bases by a set of the form $[1, \vartheta)$ (the set of all*) ordinal numbers $<\vartheta$), where $\vartheta$ is an ordinal number. This generalization has the additional advantage that we may (and we shall) consider any set of ordinal numbers as a topological space in its order topology (i.e., the topology in which the open intervals $(\xi, \eta) = \{\lambda \mid \xi < \lambda < \eta\}$ form a base of neighbourhoods).

The main tool in the definition of transfinite bases will be a natural definition of the sum of a transfinite series, which is a suitable generalization of the notion of the sum of a usual (countable) series. We recall that a function defined on $[1, \vartheta)$, where $\vartheta$ is an ordinal, is called a *transfinite sequence*. If $\{y_\lambda\}_{\lambda<\vartheta}$ is**) a transfinite sequence of elements in a Banach space $E$, we say that the transfinite series $\sum_{\lambda<\vartheta} y_\lambda$ *converges* to an element $x\in E$, which is called *the sum of the transfinite series* $\sum_{\lambda<\vartheta} y_\lambda$, and we write

$$x = \sum_{\lambda<\vartheta} y_\lambda,$$

if there exists a (clearly, unique) continuous function $S$: $[1, \vartheta] \to E$ such that

$$S(1) = 0, \quad S(\vartheta) = x, \tag{17.23}$$

$$S(\lambda+1) = S(\lambda) + y_\lambda \qquad (\lambda < \vartheta); \tag{17.24}$$

*) For notational convenience, we exclude 0 from the ordinals.

**) Since we have excluded 0 from the ordinals, $\lambda < \vartheta$ means: $1 \leqslant \lambda < \vartheta$ (or $\lambda \in [1, \vartheta)$).

in other words, $S$ is defined inductively, by (17.23) and $S(\lambda)=S(\lambda-1)+y_{\lambda-1}$ if $\lambda-1$ exists and by the continuity of $S$ if $\lambda$ is a limit ordinal. In the particular case when $\vartheta=\omega$, the first infinite ordinal, this definition is equivalent to the usual definition of the sum of a series, since the space $C_E([1,\omega])$ (of all continuous functions on the compact*) space $[1,\omega]$, with values in $E$) is canonically linearly isometric to the space $c_E$ (of all convergent sequences of elements of $E$ — both spaces being endowed with the supremum norm); in this case $S(\lambda+1)$ is the $\lambda$-th partial sum of the series (this notational discrepancy is necessary because of the existence of limit ordinals $\lambda$ in the general case, but it will not lead to any confusion).

If $x=\sum_{\lambda<\vartheta} y_\lambda$, then since $[1,\vartheta]$ is compact and $S$ is continuous, $S([1,\vartheta])$ is a compact subset of $E$, whence separable. Furthermore, we recall the following lemma, which will be used in later sections as well:

**Lemma 17.1.** *If $x=\sum_{\lambda<\vartheta} y_\lambda$ in a Banach space $E$, then for each $\varepsilon>0$ the set $B_\varepsilon=\{\lambda<\vartheta \mid \|y_\lambda\|>\varepsilon\}$ is finite and hence $y_\lambda=0$ except for at most a countable number of values of $\lambda$.*

*Proof.* Let $S\colon[1,\vartheta]\to E$ be the continuous function occurring in the definition of $\sum_{\lambda<\vartheta} y_\lambda$ and assume, a contrario, that for some $\varepsilon>0$ the set $B_\varepsilon$ is infinite. Then there exists an infinite sequence of ordinals $\lambda_1<\lambda_2<\ldots<\vartheta$ such that $\|S(\lambda_i+1)-S(\lambda_i)\|=\|y_{\lambda_i}\|>\varepsilon$ $(i=1,2,\ldots)$. Let $\lambda_0$ be the smallest ordinal larger than all $\lambda_i$. Then $\lambda_0$ is a limit ordinal and $\lambda_0=\lim_{i\to\infty}\lambda'_i$, where $\lambda'_{2i-1}=\lambda_i$, $\lambda'_{2i}=\lambda_i+1$ $(i=1,2,\ldots)$. Hence, any neighbourhood of the form $(\mu,\lambda_0)$ (where $\mu<\lambda_0$) contains a closed interval $[\lambda'_{2i-1},\lambda'_{2i}]=[\lambda_i,\lambda_i+1]$ on which the oscillation of $S$ is $>\varepsilon$, contradicting the continuity of $S$ at $\lambda_0$. This proves the first statement of lemma 17.1. The second statement also follows, since $\{\lambda<\vartheta \mid y_\lambda\neq 0\}=\bigcup_{n=1}^{\infty} B_{\frac{1}{n}}$, which completes the proof of lemma 17.1.

*Definition 17.7.* A transfinite sequence of elements $\{x_\lambda\}_{\lambda<\vartheta}$ in a Banach space $E$ is called a *transfinite basis* (of type $\vartheta$) of $E$ if for every $x\in E$ there exists a *unique* transfinite sequence of scalars $\{\alpha_\lambda\}_{\lambda<\vartheta}\subset K$ such that

$$x=\sum_{\lambda<\vartheta}\alpha_\lambda x_\lambda. \tag{17.25}$$

---

*) See e.g. [128], Ch. V, proposition 5.11 (c) or [360], Ch. II, § 8, remark 8.6.1.

Obviously, for $\vartheta = \omega$ this reduces to the definition of usual bases.

*Definition 17.8.* Let $\{x_\lambda\}_{\lambda<\vartheta}$ be a transfinite basis of a Banach space $E$. The transfinite sequence of linear functionals $\{f_\lambda\}_{\lambda<\vartheta}$ defined by

$$f_\varkappa(x) = \alpha_\varkappa \qquad (x = \sum_{\lambda<\vartheta} \alpha_\lambda x_\lambda \in E, \varkappa < \vartheta) \tag{17.26}$$

is called the *associated transfinite sequence of coefficient functionals* (a.t.s.c.f.).

Thus, if $\{x_\lambda\}_{\lambda<\vartheta}$ is a transfinite basis of the space $E$ and $\{f_\lambda\}_{\lambda<\vartheta}$ the a.t.s.c.f., then every $x \in E$ has a unique transfinite expansion of the form

$$x = \sum_{\lambda<\vartheta} f_\lambda(x) x_\lambda. \tag{17.27}$$

As in the case of usual bases, it turns out that the coefficient functionals associated to a transfinite basis are continuous. The proof is similar to the one for usual bases, with some additional care for the way how convergence and sums of transfinite series have been defined. We shall give here the proof, in order to show the techniques for the transfinite case. We shall use

**Lemma 17.2.** *If* $x = \sum_{\lambda<\vartheta} y_\lambda$ *in a Banach space* $E$ *and if* $S\colon [1, \vartheta] \to E$ *is the continuous function satisfying* (17.23), (17.24), *then*

$$S(\lambda) = \sum_{\varkappa<\lambda} y_\varkappa \qquad (\lambda \leqslant \vartheta). \tag{17.28}$$

*Proof.* For any $\lambda \leqslant \vartheta$, let $S_\lambda\colon [1, \lambda] \to E$ be the continuous function occurring in the definition of $\sum_{\varkappa<\lambda} y_\varkappa$. We claim that

$$S_\lambda|_{[1,\eta]} = S_\eta \qquad (\eta < \lambda < \vartheta). \tag{17.29}$$

Indeed, fix $\eta < \lambda < \vartheta$. Then, clearly, $S_\lambda(1) = 0 = S_\eta(1)$. Assume now that for some $\nu \leqslant \eta$ we have

$$S_\lambda(\mu) = S_\eta(\mu) \qquad (\mu < \nu). \tag{17.30}$$

If $\nu$ is a nonlimit ordinal, then, by (17.24), (17.30) and again (17.24), we have

$$S_\lambda(\nu) = S_\lambda(\nu - 1) + y_{\nu-1} = S_\eta(\nu - 1) + y_{\nu-1} = S_\eta(\nu).$$

On the other hand, if $\nu$ is a limit ordinal, then, by the continuity of $S_\lambda$, $S_\eta$ and by (17.30) we obtain

$$S_\lambda(\nu) = \lim_{\mu<\nu} S_\lambda(\mu) = \lim_{\mu<\nu} S_\eta(\mu) = S_\eta(\nu),$$

which proves the claim (17.29).

Let us prove now (17.28), which, by (17.23) (for $S_\lambda$ and $\lambda$), amounts to

$$S(\lambda) = S_\lambda(\lambda) \qquad (\lambda \leqslant \vartheta). \tag{17.28'}$$

Clearly, $S(1) = 0 = S_1(1)$. Let $\lambda \leqslant \vartheta$ and assume that

$$S(\mu) = S_\mu(\mu) \qquad (\mu < \lambda). \tag{17.31}$$

If $\lambda$ is a nonlimit ordinal, then, by (17.24), (17.31), (17.29) and again (17.24), we have

$$S(\lambda) = S(\lambda - 1) + y_{\lambda-1} = S_{\lambda-1}(\lambda - 1) + y_{\lambda-1} = S_\lambda(\lambda - 1) +$$
$$+ y_{\lambda-1} = S_\lambda(\lambda).$$

On the other hand, if $\lambda$ is a limit ordinal, then, by the continuity of $S$, $S_\lambda$ and by (17.31), (17.29) we obtain

$$S(\lambda) = \lim_{\mu<\lambda} S(\mu) = \lim_{\mu<\lambda} S_\mu(\mu) = \lim_{\mu<\lambda} S_\lambda(\mu) = S_\lambda(\lambda),$$

which completes the proof of lemma 17.2.

**Proposition 17.3.** *Let $\{x_\lambda\}_{\lambda<\vartheta}$ be a transfinite sequence in a Banach space $E$, such that $x_\lambda \neq 0$ $(\lambda < \vartheta)$ and let $A_1^{(\vartheta)}$ be the linear space of transfinite sequences of scalars*

$$A_1^{(\vartheta)} = \{\{\alpha_\lambda\}_{\lambda<\vartheta} \subset K \mid \sum_{\lambda<\vartheta} \alpha_\lambda x_\lambda \ \textit{converges}\}, \tag{17.32}$$

*endowed with the norm*

$$\| \{\alpha_\lambda\}_{\lambda<\vartheta} \| = \sup_{\lambda\leqslant\vartheta} \| \sum_{\varkappa<\lambda} \alpha_\varkappa x_\varkappa \| . \tag{17.33}$$

*Then $A_1^{(\vartheta)}$ is a Banach space.*

*Proof.* If $\{\alpha_\lambda\}_{\lambda<\vartheta} \in A_1^{(\vartheta)}$ and if $S\colon [1, \vartheta] \to E$ is the continuous function occurring in the definition of $\sum_{\lambda<\vartheta} \alpha_\lambda x_\lambda$, then, by lemma 17.2, $S(\lambda) = \sum_{\varkappa<\lambda} \alpha_\varkappa x_\varkappa$ $(\lambda \leqslant \vartheta)$, so $\|\{\alpha_\lambda\}_{\lambda<\vartheta}\| = \sup_{\lambda\leqslant\vartheta} \|S(\lambda)\|$, which is finite since $[1, \vartheta]$ is compact and which is indeed a norm on $A_1^{(\vartheta)}$, by (17.24) and $x_\lambda \neq 0$ $(\lambda < \vartheta)$.

Let $\{\alpha_\lambda^{(n)}\}_{\lambda<\vartheta}$ be a Cauchy sequence in $A_1^{(\vartheta)}$ and let $S^{(n)}(\lambda) = \sum_{\varkappa<\lambda} \alpha_\varkappa^{(n)} x_\varkappa$ $(\lambda \leqslant \vartheta;\ n = 1,2, \ldots)$. Then for every $\varepsilon > 0$ there exists a positive integer $N(\varepsilon)$ such that

$$\|\{\alpha_\lambda^{(n)}\}_{\lambda<\vartheta} - \{\alpha_\lambda^{(m)}\}_{\lambda<\vartheta}\| = \sup_{\lambda\leqslant\vartheta} \|S^{(n)}(\lambda) - S^{(m)}(\lambda)\| < \varepsilon \quad (n, m > N(\varepsilon)), \tag{17.34}$$

so $\{S^{(n)}(.)\}_{n=1}^{\infty}$ is a Cauchy sequence in the Banach space $C_E([1,\vartheta])$, whence there exists a continuous function $S\colon [1, \vartheta] \to E$ such that

$$\lim_{n\to\infty} \sup_{\lambda\leqslant\vartheta} \|S^{(n)}(\lambda) - S(\lambda)\| = 0.$$

Then $S(1) = \lim_{n\to\infty} S^{(n)}(1) = 0$ and there exist the limits

$$S(\vartheta) = \lim_{n\to\infty} S^{(n)}(\vartheta) = \lim_{n\to\infty} \sum_{\lambda<\vartheta} \alpha_\lambda^{(n)} x_\lambda = x \in E,$$

$$S(\lambda + 1) - S(\lambda) = \lim_{n\to\infty} (S^{(n)}(\lambda + 1) - S^{(n)}(\lambda)) = \lim_{n\to\infty} \alpha_\lambda^{(n)} x_\lambda = y_\lambda \in E$$

$$(\lambda < \vartheta).$$

Since $x_\lambda \neq 0$ $(\lambda < \vartheta)$, from the latter it follows that the limits $\lim_{n\to\infty} \alpha_\lambda^{(n)} = \alpha_\lambda$ exist and hence

$$S(\lambda + 1) - S(\lambda) = \lim_{n\to\infty} \alpha_\lambda^{(n)} x_\lambda = \alpha_\lambda x_\lambda \qquad (\lambda < \vartheta).$$

Consequently, by the continuity of $S\colon [1, \vartheta] \to E$ and the definition of $\sum_{\lambda<\vartheta} \alpha_\lambda x_\lambda$, we have $\{\alpha_\lambda\}_{\lambda<\vartheta} \in A_1^{(\vartheta)}$ (and $\sum_{\lambda<\vartheta} \alpha_\lambda x_\lambda = S(\vartheta) = x$). Finally, from (17.34) for $m \to \infty$ we obtain

$$\sup_{\lambda\leqslant\vartheta} \|S^{(n)}(\lambda) - S(\lambda)\| \leqslant \varepsilon \qquad (n > N(\varepsilon)),$$

so $\lim_{n\to\infty} \{\alpha_\lambda^{(n)}\} = \{\alpha_\lambda\}$ in $A_1^{(\vartheta)}$, which completes the proof of proposition 17.3.

*Remark 17.1.* In the particular case when $\vartheta = \omega$, the above proof of proposition 17.3 gives a slightly different proof of Ch. I, § 3, proposition 3.1, using the Banach space $c_E$ (of all convergent sequence in $E$, with the supremum norm). Note that in this case $\|\{\alpha_\lambda\}_{\lambda<\omega}\| = \sup\limits_{\lambda\leqslant\omega} \|\sum\limits_{\varkappa<\lambda} \alpha_\varkappa x_\varkappa\| = \sup\limits_{\lambda<\omega} \|\sum\limits_{\varkappa<\lambda} \alpha_\varkappa x_\varkappa\|$ (since $\omega$ is a limit point of $[1, \omega]$ and since $S(.)$ is continuous), so $A_1^{(\omega)} = A_1$.

**Proposition 17.4.** *Let $E$ be a Banach space with a transfinite basis $\{x_\lambda\}_{\lambda<\vartheta}$ and let $\{f_\lambda\}_{\lambda<\vartheta}$ be the a.t.s.c.f. Then*

a) *The Banach space $A_1^{(\vartheta)}$ introduced in proposition 17.3 is isomorphic to $E$, by the mapping*

$$w\colon \{\alpha_\lambda\}_{\lambda<\vartheta} \to \sum_{\lambda<\vartheta} \alpha_\lambda x_\lambda. \tag{17.35}$$

b) *The numbers*

$$|||x||| = \sup_{\lambda\leqslant\vartheta} \|\sum_{\varkappa<\lambda} f_\varkappa(x) x_\varkappa\| \qquad (x \in E) \tag{17.36}$$

*define a norm on the space $E$, equivalent to the initial norm of $E$.*

*Proof.* The proof is similar to that of Ch. I, § 3, proposition 3.2, with the difference that the inequality $\left\|\sum\limits_{i=1}^{\infty} \alpha_i x_i\right\| \leqslant \sup\limits_{1\leqslant n<\infty} \left\|\sum\limits_{i=1}^{n} \alpha_i x_i\right\|$ is replaced by the trivial inequality $\|\sum\limits_{\lambda<\vartheta} \alpha_\lambda x_\lambda\| \leqslant \sup\limits_{\lambda\leqslant\vartheta} \|\sum\limits_{\varkappa<\lambda} \alpha_\varkappa x_\varkappa\|$.

**Theorem 17.7.** *Let $\{x_\lambda\}_{\lambda<\vartheta}$ be a transfinite basis of a Banach space $E$. Then the coefficient functionals $f_\lambda$ associated to $\{x_\lambda\}_{\lambda<\vartheta}$ are continuous on $E$, i.e. $f_\lambda \in E^*(\lambda < \vartheta)$. Moreover, there exists a constant $M$ such that*

$$1 \leqslant \|x_\lambda\| \|f_\lambda\| \leqslant M \qquad (\lambda < \vartheta). \tag{17.37}$$

*Proof.* The proof is analogous to that of Ch. I, § 3, theorem 3.1, with the difference that the inequality $\|f_n(x) x_n\| \leqslant \left\|\sum\limits_{i=1}^{n} f_i(x) x_i\right\| + \left\|\sum\limits_{i=1}^{n-1} f_i(x) x_i\right\|$ is replaced by

$$\|f_\lambda(x) x_\lambda\| \leqslant \|S(\lambda+1)\| + \|S(\lambda)\|,$$

where $S\colon [1, \vartheta] \to E$ is the continuous function occurring in the definition of $\sum\limits_{\lambda<\vartheta} f_\lambda(x) x_\lambda$.

By lemma 17.2, one can also define*) the transfinite sequence of partial sum operators $\{s_\lambda\}_{\lambda\leqslant\vartheta}$ associated to a transfinite basis $\{x_\lambda\}_{\lambda<\vartheta}$, as follows:

*Definition 17.9.* Let $\{x_\lambda\}_{\lambda<\vartheta}$ be a transfinite basis of a Banach space $E$, with the a.t.s.c.f. $\{f_\lambda\}_{\lambda<\vartheta}\subset E^*$. The transfinite sequence of linear operators $\{s_\lambda\}_{\lambda\leqslant\vartheta}$, where**)

$$s_\lambda(x)=\sum_{\varkappa<\lambda} f_\varkappa(x)x_\varkappa \qquad (x\in E,\ \lambda\leqslant\vartheta), \tag{17.38}$$

is called *the transfinite sequence of partial sum operators associated to* $\{x_\lambda\}_{\lambda<\vartheta}$.

Thus, by lemma 17.2,

$$s_1=0,\ s_\vartheta=I_E,\ s_{\lambda+1}(x)=s_\lambda(x)+f_\lambda(x)x_\lambda \qquad (x\in E,\ \lambda<\vartheta), \tag{17.39}$$

and the mapping $\lambda\to s_\lambda(x)$ (from $[1,\vartheta]$ into $E$) is continuous for each fixed $x\in E$. By (17.39) and $x_\lambda\neq 0$ we have

$$\dim\,(s_{\lambda+1}-s_\lambda)\,(E)=1 \qquad (\lambda<\vartheta) \tag{17.40}$$

(namely, $(s_{\lambda+1}-s_\lambda)\,(E)$ is the one-dimensional subspace of $E$ spanned by $x_\lambda$), whence $s_{\lambda+1}\neq s_\lambda$ $(\lambda<\vartheta)$.

As in the case of usual bases, from (17.38) and proposition 17.4 b) it follows that for any transfinite basis $\{x_\lambda\}_{\lambda<\vartheta}$ of $E$ *the partial sum operators* $s_\lambda$ *are uniformly bounded* (hence continuous) on $E$, that is,

$$\sup_{\lambda\leqslant\vartheta}\|s_\lambda\|<\infty. \tag{17.41}$$

Note also that

$$s_\lambda s_\varkappa=s_\varkappa s_\lambda=s_{\min(\lambda,\varkappa)} \qquad (\lambda,\varkappa\leqslant\vartheta), \tag{17.42}$$

so $\{s_\lambda\}_{\lambda\leqslant\vartheta}$ is a commuting transfinite sequence of projections.

---

*) However, in contrast with the particular case of usual bases, $s_\lambda$ is of finite rank if and only if $\lambda<\omega$ and thus one cannot use the same definition for general transfinite biorthogonal systems $(x_\lambda,f_\lambda)_{\lambda<\vartheta}$ with $\vartheta>\omega$ (since the transfinite series (17.38) occurring in the definition need not converge if $\{x_\lambda\}_{\lambda<\vartheta}$ is not a transfinite basis).

**) Again (see the remarks made after the definition of the sum of a transfinite series), (17.38) gives a slight discrepancy with the notations used in the particular case $\vartheta=\omega$ (of usual bases), since now $s_{\lambda+1}$ is the $\lambda$-th partial sum operator of that case; as we have noted, this cannot be avoided and will lead to no confusion.

Thus, if we consider the set $\mathscr{P}$ of all continuous linear projections on $E$, endowed with the natural order (i.e.*), $u \leqslant v$ if and only if $uv = vu = u$) and with the topology of pointwise convergence, then by the above, $\lambda \rightarrow s_\lambda$ is a strictly ascending continuous function from $[1, \vartheta]$ into $\mathscr{P}$.

As we have already mentioned, in contrast with the particular case of usual bases, for transfinite bases the partial sum operator $s_\lambda$ is of finite rank if and only if $\lambda < \omega$ (and, clearly, in this case $\dim s_\lambda(E) = \lambda - 1$ and $\operatorname{dens} s_\lambda(E) = \aleph_0$). In the case of transfinite bases, for the infinite ordinals $\lambda$ we have

$$\operatorname{dens} s_\lambda(E) = \operatorname{card} \lambda \qquad (\omega \leqslant \lambda \leqslant \vartheta); \tag{17.43}$$

indeed, this follows e.g. from § 20, proposition 20.1 (since $(x_\varkappa, f_\varkappa)_{\varkappa < \lambda}$ is a transfinite biorthogonal system with $[x_\varkappa]_{\varkappa < \lambda} = s_\lambda(E)$).

In order to give an intrinsic characterization of transfinite bases (theorem 17.8 below), let us prove first the following generalization of Ch. I, § 8, proposition 8.1:

**Proposition 17.5.** *Let $\{x_\lambda\}_{\lambda < \vartheta}$ be a transfinite sequence in a Banach space $E$, such that $x_\lambda \neq 0$ $(\lambda < \vartheta)$ and let $A_1^{(\vartheta)}$ be the space introduced in proposition 17.3. Then the unit vectors*

$$e_\lambda = \{\delta_{\lambda\varkappa}\}_{\varkappa < \vartheta} \qquad (\lambda < \vartheta) \tag{17.44}$$

*constitute a transfinite basis of $A_1^{(\vartheta)}$.*

*Proof.* Let $\{\alpha_\lambda\}_{\lambda < \vartheta} \in A_1^{(\vartheta)}$. We shall show that

$$\{\alpha_\lambda\}_{\lambda < \vartheta} = \sum_{\lambda < \vartheta} \alpha_\lambda e_\lambda \tag{17.45}$$

and that this expansion is unique, which will complete the proof.

Define a function $S_0\colon [1, \vartheta] \rightarrow A_1^{(\vartheta)}$ by

$$(S_0(\lambda))_\varkappa = \begin{cases} \alpha_\varkappa & \text{for } \varkappa < \lambda \\ 0 & \text{for } \lambda \leqslant \varkappa < \vartheta, \end{cases} \tag{17.46}$$

where $(S_0(\lambda))_\varkappa$ is the $\varkappa$-th coordinate of $S_0(\lambda)$. Then

$$S_0(1) = 0,\ S_0(\vartheta) = \{\alpha_\lambda\}_{\lambda < \vartheta},\ S_0(\lambda + 1) = S_0(\lambda) + \alpha_\lambda e_\lambda \qquad (\lambda < \vartheta),$$

---

*) See e.g. [87], p. 481. By § 12, lemma 12.1 and § 13, proof of proposition 13.1 c), we have $u \leqslant v$ if and only if both $u(E) \subset v(E)$ and $\operatorname{Ker} u \supset \operatorname{Ker} v$. One can show ([49], theorem 2.4) that $\mathscr{P}$ is a directed set, but we shall not use here this remark.

so in order to prove (17.45) it remains to show that $S_0$ is continuous. Since every nonlimit ordinal is an isolated point of $[1, \vartheta]$, and any function is continuous from the right at every point of $[1, \vartheta]$, it is enough to prove that $S_0$ is continuous from the left at every limit ordinal $\lambda_0 \in [1, \vartheta]$. Let $S\colon [1, \vartheta] \to E$ be the continuous function occurring in the definition of $\sum_{\lambda<\vartheta} \alpha_\lambda x_\lambda$ and let $\varepsilon > 0$. Then there exists $\mu_0 < \lambda_0$ such that

$$\| S(\lambda) - S(\lambda_1) \| < \varepsilon \qquad (\mu_0 < \lambda_1 < \lambda \leqslant \lambda_0).$$

Consequently, by (17.46) and the definition of the norm in $A_1^{(\vartheta)}$,

$$\| S_0(\lambda_0) - S_0(\lambda_1) \| = \sup_{\lambda_1 < \lambda \leqslant \lambda_0} \| \sum_{\lambda_1 \leqslant \varkappa < \lambda} \alpha_\varkappa x_\varkappa \| =$$

$$= \sup_{\lambda_1 < \lambda \leqslant \lambda_0} \| S(\lambda) - S(\lambda_1) \| \leqslant \varepsilon \qquad (\mu_0 < \lambda_1 < \lambda_0),$$

which proves the continuity of $S_0$ and thus (17.45) holds.

Assume now that $\{\alpha_\lambda\}_{\lambda<\vartheta} \subset K$, $\sum_{\lambda<\vartheta} \alpha_\lambda e_\lambda = 0$ in $A_1^{(\vartheta)}$. Let $S_0\colon [1, \vartheta] \to A_1^{(\vartheta)}$ be the continuous function occurring in the definition of $\sum_{\lambda<\vartheta} \alpha_\lambda e_\lambda$, hence

$$S_0(1) = 0, \ \ S_0(\vartheta) = \sum_{\lambda<\vartheta} \alpha_\lambda e_\lambda = 0, \ \ S_0(\lambda + 1) = S_0(\lambda) + \alpha_\lambda e_\lambda \ \ (\lambda < \vartheta).$$

We claim that $S_0$ satisfies (17.46). Indeed, for $\lambda = 1$ this is obvious. Assume now that (17.46) holds for all $\mu < \lambda$, where $\lambda$ is some ordinal number $\leqslant \vartheta$. If $\lambda$ is a nonlimit ordinal, then by $S_0(\lambda) = S_0(\lambda - 1) + \alpha_{\lambda-1} e_{\lambda-1}$ and by the induction hypothesis,

$$(S_0(\lambda))_\varkappa = (S_0(\lambda - 1) + \alpha_{\lambda-1} e_{\lambda-1})_\varkappa = \begin{cases} \alpha_\varkappa \text{ for } \varkappa < \lambda - 1 \\ \alpha_\varkappa \text{ for } \varkappa = \lambda - 1 \\ 0 \text{ for } \lambda \leqslant \varkappa < \vartheta, \end{cases}$$

so (17.46) holds for $\lambda$. On the other hand, if $\lambda$ is a limit ordinal, then by the continuity of $S_0$ and of the coordinate projections*) $\{\alpha_\lambda\}_{\lambda<\vartheta} \to$

---

*) Indeed, these projections are linear and we have

$$|\alpha_\varkappa| = \frac{1}{\|x_\varkappa\|} \|\alpha_\varkappa x_\varkappa\| \leqslant \frac{2}{\|x_\varkappa\|} \|\{\alpha_\lambda\}_{\lambda<\vartheta}\|.$$

$\to \alpha_\varkappa$ ($\varkappa < \vartheta$) in $A_1^{(\vartheta)}$ and by the induction hypothesis,

$$(S_0(\lambda))_\varkappa = \lim_{\mu<\lambda} (S_0(\mu))_\varkappa = \lim_{\mu<\lambda} \begin{cases} \alpha_\varkappa & \text{for } \varkappa < \mu \\ 0 & \text{for } \mu \leqslant \varkappa < \vartheta, \end{cases}$$

so again (17.46) holds for $\lambda$, which proves our claim.

By $S_0(\lambda) \in A_1^{(\vartheta)}$ and (17.46), the transfinite series $\sum_{\varkappa<\mu} \alpha_\varkappa x_\varkappa = \sum_{\varkappa<\mu} (S_0(\lambda))_\varkappa x_\varkappa$ converges for each $\mu \leqslant \lambda \leqslant \vartheta$ and we have

$$\| S_0(\lambda) \| = \sup_{\mu\leqslant\vartheta} \| \sum_{\varkappa<\mu} (S_0(\lambda))_\varkappa x_\varkappa \| = \sup_{\mu\leqslant\lambda} \| \sum_{\varkappa<\mu} \alpha_\varkappa x_\varkappa \| \qquad (\lambda \leqslant \vartheta).$$

Observe also that $\|S_0(.)\|: [1, \vartheta] \to K$ is an ascending continuous scalar-valued function, since for $\lambda_1 \leqslant \lambda_2 \leqslant \vartheta$ we have

$$\| S_0(\lambda_1) \| = \sup_{\mu\leqslant\lambda_1} \| \sum_{\varkappa<\mu} \alpha_\varkappa x_\varkappa \| \leqslant \sup_{\mu\leqslant\lambda_2} \| \sum_{\varkappa<\mu} \alpha_\varkappa x_\varkappa \| = \| S_0(\lambda_2) \|.$$

Consequently,

$$\| \{\alpha_\lambda\}_{\lambda<\vartheta} \| = \sup_{\lambda\leqslant\vartheta} \| \sum_{\varkappa<\lambda} \alpha_\varkappa x_\varkappa \| = \sup_{\lambda\leqslant\vartheta} \sup_{\mu\leqslant\lambda} \| \sum_{\varkappa<\mu} \alpha_\varkappa x_\varkappa \| =$$

$$= \sup_{\lambda\leqslant\vartheta} \| S_0(\lambda) \| = \lim_{\lambda\leqslant\vartheta} \| S_0(\lambda) \| = \| S_0(\vartheta) \| = \| \sum_{\lambda<\vartheta} \alpha_\lambda e_\lambda \| = 0,$$

so $\alpha_\lambda = 0$ ($\lambda < \vartheta$), which completes the proof of proposition 17.5.

Now we can prove the following intrinsic characterization of transfinite bases, generalizing part of Ch. I, § 7, theorem 7.1:

**Theorem 17.8.** *Let $E$ be a Banach space and $\{x_\lambda\}_{\lambda<\vartheta}$ a complete transfinite sequence in $E$ (i.e., $[x_\lambda]_{\lambda<\vartheta} = E$), such that $x_\lambda \neq 0$ ($\lambda < \vartheta$). The following statements are equivalent:*

1°. *$\{x_\lambda\}_{\lambda<\vartheta}$ is a transfinite basis of $E$.*

2°. *There exists a constant $C$ with $1 \leqslant C < \infty$, such that we have*

$$\|y\| \leqslant C \|y + z\| \tag{17.47}$$

*for all $y \in [x_\varkappa]_{\varkappa<\lambda}$, $z \in [x_\varkappa]_{\lambda\leqslant\varkappa<\vartheta}$ and all $\lambda < \vartheta$.*

3°. *There exists a constant $C$ with $1 \leqslant C < \infty$, such that we have* (17.47) *for all $y \in \operatorname{lin} \{x_\varkappa\}_{\varkappa<\lambda}$, $z \in \operatorname{lin} \{x_\varkappa\}_{\lambda\leqslant\varkappa<\vartheta}$ and all $\lambda < \vartheta$.*

*Proof.* The equivalence 2° ⇔ 3° is obvious.

If we have 1° and $\{s_\lambda\}_{\lambda\leqslant\vartheta}$ is the transfinite sequence of partial sum operators associated to $\{x_\lambda\}_{\lambda<\vartheta}$, then, by (17.41),

$$\|y\| = \|s_\lambda(y + z)\| \leqslant \|s_\lambda\| \, \|y + z\| \leqslant C\|y + z\|$$

for all $y \in [x_\varkappa]_{\varkappa<\lambda}$, $z \in [x_\varkappa]_{\lambda\leqslant\varkappa<\vartheta}$ and all $\lambda < \vartheta$. Thus, 1° ⇒ 2°.

Finally, assume that we have 2° and let us consider the linear mapping $w: A_1^{(\vartheta)} \to E$ defined by (17.35). Clearly,

$$w(e_\lambda) = x_\lambda \qquad (\lambda < \vartheta),$$

and, hence since $[x_\lambda]_{\lambda<\vartheta} = E$, $w(A_1^{(\vartheta)})$ is dense in $E$. We shall show that $w$ is an isomorphism, whence, by propositions 17.3 *) and 17.5, $w(A_1^{(\vartheta)}) = E$ and $\{w(e_\lambda)\}_{\lambda<\vartheta} = \{x_\lambda\}_{\lambda<\vartheta}$ is a transfinite basis of $E$, which will complete the proof.

Let $\{\alpha_\lambda\}_{\lambda<\vartheta} \in A_1^{(\vartheta)}$ and let $S: [1, \vartheta] \to E$ be the continuous function occurring in the definition of $\sum_{\lambda<\vartheta} \alpha_\lambda x_\lambda$. Then

$$\|w(\{\alpha_\lambda\}_{\lambda<\vartheta})\| = \|\sum_{\lambda<\vartheta} \alpha_\lambda x_\lambda\| = \|S(\vartheta)\| \leqslant \sup_{\lambda\leqslant\vartheta} \|S(\lambda)\| = \|\{\alpha_\lambda\}_{\lambda<\vartheta}\|,$$

so $w$ is continuous (we have $\|w\| \leqslant 1$). Also, by (17.47), for any $\lambda < \vartheta$ and for $S(\lambda) = \sum_{\varkappa<\lambda} \alpha_\varkappa x_\varkappa = y \in [x_\varkappa]_{\varkappa<\lambda}$ and $S(\vartheta) - S(\lambda) = \sum_{\lambda\leqslant\varkappa<\vartheta} \alpha_\varkappa x_\varkappa = z \in [x_\varkappa]_{\lambda\leqslant\varkappa<\vartheta}$ we have

$$\|S(\lambda)\| \leqslant C\|S(\lambda) + (S(\vartheta) - S(\lambda))\| = C\|S(\vartheta)\| = C\|\sum_{\varkappa<\vartheta} \alpha_\varkappa x_\varkappa\|,$$

whence

$$\|\{\alpha_\lambda\}_{\lambda<\vartheta}\| = \sup_{\lambda\leqslant\vartheta} \|S(\lambda)\| \leqslant C\|\sum_{\varkappa<\vartheta} \alpha_\varkappa x_\varkappa\| = C\|w(\{\alpha_\lambda\}_{\lambda<\vartheta})\|,$$

so $w$ is one-to-one and $w^{-1}$ is continuous. Thus, 2° ⇒ 1°, which completes the proof of theorem 17.8.

*Remark 17.2.* In the particular case when $\vartheta = \omega$, the above proof of the implications 3° ⇒ 2° ⇒ 1° gives a different proof of Ch. I, § 7, theorem 7.1, implication 4° ⇒ 1°.

Let us show now, using theorem 17.8, that for any ordinal $\vartheta$ the Banach space $C([1, \vartheta])$ has a standard transfinite basis of type $\vartheta + 1$.

**Proposition 17.6.** *Let $\vartheta$ be any ordinal number. Then the transfinite sequence $\{x_\lambda\}_{\lambda\leqslant\vartheta} = \{x_\lambda\}_{\lambda<\vartheta+1} \subset C([1, \vartheta])$ defined by*

$$x_\lambda(\varkappa) = \begin{cases} 1 \text{ for } \varkappa \leqslant \lambda \\ 0 \text{ for } \lambda + 1 \leqslant \varkappa \leqslant \vartheta \end{cases} \qquad (\lambda \leqslant \vartheta) \tag{17.48}$$

*is a transfinite basis of $E = C([1, \vartheta])$.*

---

*) We also use the (obvious) fact that the image of a Banach space by an isomorphism is a Banach space.

*Proof*. Clearly, $x_\lambda \neq 0$ $(\lambda < \vartheta)$. Furthermore, since for any finite sequence of scalars $\alpha_1, \ldots, \alpha_{n+m}$ and any $k \leqslant n$ we have $\left|\sum_{i=k}^{n} \alpha_i\right| \leqslant \left|\sum_{i=k}^{n+m} \alpha_i\right| + \left|\sum_{i=n+1}^{n+m} \alpha_i\right|$, whence*) $\left\|\sum_{i=1}^{n} \alpha_i x_{\lambda_i}\right\| = \max_{1\leqslant k\leqslant n} \left|\sum_{i=k}^{n} \alpha_i\right| \leqslant 2 \max_{1\leqslant k\leqslant n+m} \left|\sum_{i=k}^{n+m} \alpha_i\right| = 2\left\|\sum_{i=1}^{n+m} \alpha_i x_{\lambda_i}\right\|$ for any $\lambda_1 < \ldots < \lambda_{n+m}$, it follows that for all $y \in \operatorname{lin}\{x_\varkappa\}_{\varkappa<\lambda}$, $z \in \operatorname{lin}\{x_\varkappa\}_{\lambda\leqslant\varkappa\leqslant\vartheta}$ and all $\lambda \leqslant \vartheta$ we have $\|y\| \leqslant 2\|y + z\|$. Consequently, by theorem 17.8, it will be enough to show that $[x_\lambda]_{\lambda\leqslant\vartheta} = E$.

Assuming that $[x_\lambda]_{\lambda\leqslant\vartheta} \neq E$, for any $\lambda \leqslant \vartheta$ let

$$E_\lambda = \{y \in C([1, \vartheta]) \mid y(\varkappa) = 0 \text{ for all } \varkappa \in [\lambda + 1, \vartheta]\}$$

and let $\mu$ be the least ordinal for which $E_\mu \not\subset [x_\lambda]_{\lambda\leqslant\vartheta}$ (such a $\mu$ exists, since $E_\vartheta = E \not\subset [x_\lambda]_{\lambda\leqslant\vartheta}$). If $\mu$ is a nonlimit ordinal, then for any $y \in E_\mu$ we have $y - y(\mu)x_\mu \in E_{\mu-1}$ (since $y(\varkappa) = 0 = x_\mu(\varkappa)$ for $\varkappa \geqslant \mu + 1$ and $x_\mu(\mu) = 1$), whence, by the minimality of $\mu$, $y \in y(\mu)x_\mu + E_{\mu-1} \subset [x_\lambda]_{\lambda\leqslant\vartheta}$, in contradiction with $E_\mu \not\subset [x_\lambda]_{\lambda\leqslant\vartheta}$. On the other hand, if $\mu$ is a limit ordinal, then for any $y \in E_\mu$ and $\varepsilon > 0$ there exists $\lambda < \mu$ such that $|y(\varkappa) - y(\mu)| < \varepsilon$ $(\lambda < \varkappa \leqslant \mu)$, whence for $y' \in E_\lambda$ defined by

$$y'(\varkappa) = \begin{cases} y(\varkappa) - y(\mu) & \text{for } \varkappa \leqslant \lambda \\ 0 & \text{for } \lambda + 1 \leqslant \varkappa \leqslant \vartheta \end{cases}$$

we have $\|y - y(\mu)x_\mu - y'\| \leqslant \varepsilon$ (since $(y - y(\mu)x_\mu - y')(\varkappa) = 0$ for $\varkappa \leqslant \lambda$ and for $\mu + 1 \leqslant \varkappa \leqslant \vartheta$ and $|y(\varkappa) - y(\mu)x_\mu(\varkappa) - y'(\varkappa)| = |y(\varkappa) - y(\mu)| < \varepsilon$ for $\lambda + 1 \leqslant \varkappa \leqslant \mu$). Hence, since $\varepsilon > 0$ was arbitrary, $y - y(\mu)x_\mu \in E_\lambda$, which, as in the preceding case, leads to a contradiction (because $\lambda < \mu$). This completes the proof of proposition 17.6.

*Remark 17.3*. The Banach space $c_0([1, \vartheta])$, considered in theorem 17.6 (with $I = [1, \vartheta]$), is not a subspace of $C([1, \vartheta])$, even when $\vartheta = \omega$, since for example the characteristic function of the singleton $\{\omega\}$ (or of any limit ordinal) is in $c_0([1, \vartheta])$, but not in $C([1, \vartheta])$. One can slightly modify the definition of $c_0([1, \vartheta])$, to arrive at a subspace of $C([1, \vartheta])$, as follows: We shall call $c_0^\vartheta$ the space $c_0(\mathscr{M})$, on the set $\mathscr{M}$ of all nonlimit ordinals $\varkappa \leqslant \vartheta$. Then, clearly, $c_0^\vartheta$ is canonically isometric to the subspace $\{x(.) \in C([1, \vartheta]) \mid x(\lambda) = 0 \ (\lambda \text{ limit ordinal } \leqslant \vartheta)\}$ of $C([1, \vartheta])$; in particular, $c_0^\omega = c_0([1, \omega)) = c_0 \subset c \equiv C([1, \omega])$.

---

*) See Ch. II, § 17, example 17.2, for similar computations.

It is known*) that for $\omega \leqslant \vartheta_1 \leqslant \vartheta_2 < \omega_1$ (the first uncountable ordinal), $C([1, \vartheta_1])$ is isomorphic to $C([1, \vartheta_2])$ if and only if $\vartheta_2 < \vartheta_1^\omega$, so the numbers $\omega, \omega^\omega, (\omega^\omega)^\omega, \ldots$ determine the successive isomorphic types of these spaces (there are $\aleph_1$ different such types). Each of these spaces $C([1, \vartheta])$ (with $\vartheta < \omega_1$) is separable and its transfinite basis (17.48) is countable; however, this transfinite basis is different from the usual bases of $C([1, \vartheta])$ (we shall see in § 18, corollary 18.3, that if $Q$ is a compact space such that $C(Q)$ is separable, then $C(Q)$ has a monotone basis in the usual sense).

There exist non-separable $C(Q)$ spaces ($Q$ compact), which have no transfinite basis, as shown by

*Example 17.2.* Let $E = l^\infty \equiv C(\beta\mathcal{N})$, where $\beta\mathcal{N}$ is the Stone-Čech compactification of $\mathcal{N} = \{1,2,3, \ldots\}$. Then $E$ has no transfinite basis. Indeed, every non-separable space with a transfinite basis has a separable infinite dimensional complemented subspace (since $s_\omega$ is a projection), but it is known**) that $l^\infty$ has no such subspace.

The conclusion of example 17.2 also follows from § 20, corollary 20.2 (see also remark 20.3), since every transfinite basis is an extended $M$-basis. Alternatively, it also follows from § 15, theorem 15.2, combined with

**Proposition 17.7.** *Every infinite dimensional Banach space $E$ with a transfinite basis has a Schauder decomposition.*

*Proof.* Let $\{x_\lambda\}_{\lambda<\vartheta}$ be a transfinite basis of $E$, where $\vartheta \geqslant \omega$, and let $\{s_\lambda\}_{\lambda\leqslant\vartheta}$ be the associated transfinite sequence of partial sum operators. If $\vartheta = \omega$, then $\{x_\lambda\}_{\lambda<\vartheta}$ is a basis of $E$. If $\vartheta > \omega$, then the sequence $\{G_n\}$ defined by

$$G_1 = (I - s_\omega)(E),\ G_2 = [x_1],\ G_3 = [x_2], \ldots \tag{17.49}$$

is a Schauder decomposition of $E$, since $G_1 = (I - s_\omega)(E) \neq \{0\}$ (by $\vartheta > \omega$) and since every $x \in E$ has a unique expansion of the form $x = \sum_{\lambda<\vartheta} \alpha_\lambda x_\lambda = \sum_{\lambda<\omega} \alpha_\lambda x_\lambda + \sum_{\omega\leqslant\lambda<\vartheta} \alpha_\lambda x_\lambda$, with $\sum_{\omega\leqslant\lambda<\vartheta} \alpha_\lambda x_\lambda = (I - s_\omega)(x) \in G_1$. This completes the proof of proposition 17.7.

Various other results on bases can be also generalized to transfinite bases. For example, we have the following result of strong duality, which is a generalization of Ch. I, § 12, theorem 12.1:

**Theorem 17.9.** *Let $\{x_\lambda\}_{\lambda<\vartheta}$ be a transfinite basis of a Banach space $E$ and let $\{f_\lambda\}_{\lambda<\vartheta} \subset E^*$ be the a.t.s.c.f. Then $\{f_\lambda\}_{\lambda<\vartheta}$ is a transfinite basis of $V = [f_\lambda]_{\lambda<\vartheta}$ and we have*

$$f = \sum_{\lambda<\vartheta} f(x_\lambda) f_\lambda \qquad (f \in V = [f_\lambda]_{\lambda<\vartheta}). \tag{17.50}$$

*) See [24] or [360], p. 381.
**) See e.g. [299], corollary 5.

*Proof.* By (17.39) we have

$$s_1^* = 0,\ s_\vartheta^* = I_{E^*},\ s_{\lambda+1}^*(f) = s_\lambda^*(f) + f(x_\lambda)f_\lambda \quad (f \in E^*,\ \lambda < \vartheta).$$

Furthermore, by (17.38) and biorthogonality,

$$s_\lambda^*(f_\mu)(x) = f_\mu(\sum_{\varkappa<\lambda} f_\varkappa(x)x_\varkappa) = \begin{cases} f_\mu(x) & \text{if } \mu < \lambda \\ 0 & \text{if } \mu \geqslant \lambda \end{cases} \quad (x \in E;\ \lambda \leqslant \vartheta,\ \mu < \vartheta),$$

whence

$$s_\lambda^*(f_\mu) = \begin{cases} f_\mu & \text{if } \mu < \lambda \\ 0 & \text{if } \mu \geqslant \lambda \end{cases} \qquad (\lambda \leqslant \vartheta,\ \mu < \vartheta) \tag{17.51}$$

and hence, since each $s_\lambda^*$ is continuous, we get $s_\lambda^*(V) \subset V$ $(\lambda \leqslant \vartheta)$. Also, for any limit ordinal $\lambda_0 \leqslant \vartheta$, we obtain

$$\lim_{\lambda<\lambda_0} s_\lambda^*(f_\mu) = \begin{Bmatrix} f_\mu & \text{if } \mu < \lambda_0 \\ 0 & \text{if } \mu \geqslant \lambda_0 \end{Bmatrix} = s_{\lambda_0}^*(f_\mu) \qquad (\mu < \vartheta),$$

whence, since $\sup_{\lambda\leqslant\vartheta} \|s_\lambda^*\| < \infty$, it follows that $\lim_{\lambda<\lambda_0} s_\lambda^*(f) = s_{\lambda_0}^*(f)$ $(f \in V)$. Thus, for each $f \in V$, the function $\lambda \to s_\lambda^*(f)$ from $[1, \vartheta]$ into $V$ is continuous, which proves (17.50). Finally, the uniqueness of the coefficients $f(x_\lambda)$ in (17.50) is obvious (by biorthogonality), which completes the proof.

Furthermore, one can prove *the weak basis theorem* for transfinite bases of Banach spaces, by using a suitable generalization of Ch. I, § 13, proposition 13.1 (see the techniques of the above proof of proposition 17.3). Also, we have the following generalization of Ch. I, § 14, theorem 14.1: *A transfinite sequence $\{f_\lambda\}_{\lambda<\vartheta}$ in a conjugate Banach space $E^*$ is a "transfinite $w^*$-Schauder basis" of $E^*$ if and only if $E$ has a transfinite basis $\{x_\lambda\}_{\lambda<\vartheta}$ whose a.t.s.c.f. is $\{f_\lambda\}_{\lambda<\vartheta}$* (if the condition is satisfied, then from $s_{\lambda_0}(x) = \lim_{\lambda<\lambda_0} s_\lambda(x)$ it follows that $s_{\lambda_0}^*(f) = w^*\text{-}\lim_{\lambda<\lambda_0} s_\lambda^*(f)$ for all $f \in E^*$ and $\lambda_0 \leqslant \vartheta$, so for each fixed $f \in E^*$ the function $\lambda \to s_\lambda^*(f)$ from $[1, \vartheta]$ into $E^*$ is $w^*$-continuous).

One can also define some special classes of transfinite bases, which are generalizations of the corresponding special classes of (usual) bases, for example:

*Definition 17.10.* A transfinite basis $\{x_\lambda\}_{\lambda<\vartheta}$ of a Banach space $E$ is said to be *monotone*, if it satisfies condition 2° (or 3°) of theorem 17.8 with $C = 1$ (or, what is equivalent, if the associated partial sum operators satisfy $\|s_\lambda\| = 1$ for $2 \leqslant \lambda \leqslant \vartheta$).

By theorem 17.8, a transfinite sequence $\{x_\lambda\}_{\lambda<\vartheta}$ is a monotone transfinite basis of $E$ if and only if $x_\lambda \neq 0$ $(\lambda < \vartheta)$, $[x_\lambda]_{\lambda<\vartheta} = E$ and 2° (or 3°) of theorem 17.8 holds with $C = 1$.

For instance, the unit vectors (17.44) constitute a monotone transfinite basis of $l^p([1, \vartheta])$, $1 \leqslant p < \infty$, and of $c_0([1, \vartheta))$. Also, the unit vectors (17.44), where $\lambda$ runs over the set of all nonlimit ordinals $\leqslant \vartheta$, constitute a monotone transfinite basis of the space $c_0^\vartheta$ introduced in remark 17.3.

The following proposition, which generalizes Ch. II, § 1, proposition 1.3, gives a necessary and sufficient condition for a Banach space $E$ to have a monotone transfinite basis:

**Proposition 17.8.** *A Banach space $E$ has a monotone transfinite basis if and only if there exists a transfinite sequence of projections $\{u_\lambda\}_{\lambda\leqslant\vartheta}$ on $E$ with the following properties:*

a) $\dim (u_{\lambda+1} - u_\lambda)(E) = 1$ $(\lambda < \vartheta)$;

b) $u_\lambda(E) \subset u_\mu(E)$ $(\lambda \leqslant \mu \leqslant \vartheta)$;

c) $u_1 = 0$, $u_\vartheta = I_E$ *and for each* $x \in E$ *the mapping* $\lambda \to u_\lambda(x)$ *is continuous on* $[1, \vartheta]$;

d) $\|u_\lambda\| = 1$ $(2 \leqslant \lambda \leqslant \vartheta)$.

*Proof.* If $\{x_\lambda\}_{\lambda<\vartheta}$ is a monotone transfinite basis of $E$, then the associated partial sum operators $\{u_\lambda\}_{\lambda\leqslant\vartheta} = \{s_\lambda\}_{\lambda\leqslant\vartheta}$ constitute a transfinite sequence of projections on $E$, satisfying a)—d).

Conversely, assume that the condition is satisfied. Then, by a) and b), for each $\lambda < \vartheta$ we can pick an element $x_\lambda \in u_{\lambda+1}(E)\setminus\{0\}$ such that $u_\lambda(x_\lambda) = 0$. We claim that

$$u_\lambda(E) \subset [x_\varkappa]_{\varkappa<\lambda} \qquad (2 \leqslant \lambda \leqslant \vartheta), \tag{17.52}$$

whence, by part of c), $E = u_\vartheta(E) \subset [x_\varkappa]_{\varkappa<\vartheta}$. Indeed, by $u_1 = 0$ and a), we have $\dim u_2(E) = \dim (u_2(E) - u_1(E)) = 1$, whence, by $x_1 \in u_2(E)\setminus\{0\}$, we get $u_2(E) = [x_1]$, so (17.52) holds for $\lambda = 2$. Assume now that (17.52) is true for all $\mu < \lambda$, where $\lambda \leqslant \vartheta$ .If $\lambda$ is a nonlimit ordinal, then by a), $\dim (u_\lambda - u_{\lambda-1})(E) = 1$, whence, since $x_{\lambda-1} \in u_\lambda(E) \cap \operatorname{Ker} u_{\lambda-1} \subset (u_\lambda - u_{\lambda-1})(E)$, we get $(u_\lambda - u_{\lambda-1})(E) = [x_{\lambda-1}]$, the one-dimensional subspace of $E$ spanned by $x_{\lambda-1}$. Consequently, by the induction hypothesis, $u_\lambda(E) \subset u_{\lambda-1}(E) + [x_{\lambda-1}] \subset [x_\varkappa]_{\varkappa<\lambda-1} + [x_{\lambda-1}] \subset [x_\varkappa]_{\varkappa<\lambda}$, so (17.52) holds for $\lambda$. On the other hand, if $\lambda$ is a limit ordinal, then by c) and the induction hypothesis, for each $x \in E$ we have $u_\lambda(x) = \lim_{\mu<\lambda} u_\mu(x) \in [[x_\varkappa]_{\varkappa<\mu}]_{\mu<\lambda} \subset [x_\varkappa]_{\varkappa<\lambda}$, so (17.52) holds for $\lambda$. This proves (17.52) and hence $[x_\lambda]_{\lambda<\vartheta} = E$.

Finally, we shall show that $\{x_\lambda\}_{\lambda<\vartheta}$ satisfies 3° of theorem 17.8 with $C = 1$, which (by theorem 17.8) will complete the proof. Let $y \in \operatorname{lin} \{x_\varkappa\}_{\varkappa<\lambda}$ and $\alpha_\lambda \in K$ be arbitrary. Then $y$ is a finite sum of the form $\sum_i \alpha_{\varkappa_i} x_{\varkappa_i}$, with all $\varkappa_i < \lambda$, whence*) $y \in \sum_i u_{\varkappa_i+1}(E) \subset u_\lambda(E)$ (by $x_{\varkappa_i} \in u_{\varkappa_i+1}(E)$ and b)). Consequently, by $u_\lambda^2 = u_\lambda$, $u_\lambda(x_\lambda) = 0$ and d),

$$\|y\| = \|u_\lambda(y + \alpha_\lambda x_\lambda)\| \leqslant \|y + \alpha_\lambda x_\lambda\|.$$

Hence, by (finite) induction, we obtain that $\{x_\lambda\}_{\lambda<\vartheta}$ satisfies 3° of theorem 17.8 with $C = 1$, which completes the proof of proposition 17.8.

*Remark 17.4.* α) For the transfinite basis $\{x_\lambda\}_{\lambda<\vartheta}$ constructed in the proof of the sufficiency part of proposition 17.8, $(u_{\lambda+1} - u_\lambda)(E)$ is nothing else than the one-dimensional subspace of $E$ spanned by $x_\lambda$ $(\lambda < \vartheta)$. Indeed, this follows from a) and $x_\lambda \in u_{\lambda+1}(E) \cap \operatorname{Ker} u_\lambda \subset \subset (u_{\lambda+1} - u_\lambda)(E)$ $(\lambda < \vartheta)$.

β) In the particular case when $\vartheta = \omega$, condition c) of proposition 17.8 implies that $x = u_\omega(x) = \lim_{\lambda<\omega} u_\lambda(x)$ $(x \in E)$, whence

c') $\overline{\bigcup_{\lambda<\omega} u_\lambda(E)} = E$, with $u_1 = 0$.

Conversely, b), c') and d) imply $\lim_{\lambda<\omega} u_\lambda(x) = x$ for all $x \in E$ (see § 12, proposition 12.1). Also, c') and b) imply $E = \overline{\bigcup_{\lambda<\omega} u_\lambda(E)} \subset u_\omega(E)$, so $E = u_\omega(E)$, whence, since $u_\omega$ is a projection, we get $u_\omega = I_E$. Thus, b), c') and d) imply c). Finally, it is clear that $u_1 = 0$ and a), b) imply $\dim u_\lambda(E) = \lambda - 1$ $(\lambda < \omega)$ and conversely, $\dim u_\lambda(E) = \lambda - 1$ $(\lambda<\omega)$ and b) imply a), so proposition 17.8 is indeed a generalization of Ch. II, § 1, proposition 1.3 (with $G_\lambda = u_{\lambda+1}(E)$ for each $\lambda < \omega = \vartheta$).

γ) We shall see in § 19, proposition 19.5, that condition c) in proposition 17.8 above can be also replaced by

c'') $u_1 = 0, u_\vartheta = I_E, u_\lambda u_\mu = u_\lambda \; (\lambda \leqslant \mu \leqslant \vartheta)$ and $u_\lambda(E) \subset \overline{\bigcup_{\mu<\lambda} u_{\mu+1}(E)}$ $(\lambda \leqslant \vartheta)$.

Some other special classes of transfinite bases are introduced as follows:

*Definition 17.11.* A transfinite basis $\{x_\lambda\}_{\lambda<\vartheta}$ of a Banach space $E$ is said to be

a) *shrinking*, if $\{f_\lambda\}_{\lambda<\vartheta} \subset E^*$, the a.t.s.c.f. to $\{x_\lambda\}_{\lambda<\vartheta}$, is a transfinite basis of $E^*$;

b) *boundedly complete*, if for any $\lambda_1 < \lambda_2 < \ldots < \lambda \leqslant \vartheta$, with $\lim_{j\to\infty} \lambda_j = \lambda$ and any (countable) bounded sequence $\{y_j\} \subset E$ such that

$$s_{\lambda_n}(y_j) = y_n \qquad (n = 1, \ldots, j;\; j = 1,2, \ldots), \tag{17.53}$$

*) This argument also implies that $[x_\varkappa]_{\varkappa<\lambda} \subset u_\lambda(E)$, which, together with (17.52), gives $[x_\varkappa]_{\varkappa<\lambda} = u_\lambda(E)$ $(\lambda \leqslant \vartheta)$.

where $\{s_\lambda\}_{\lambda \leqslant \vartheta}$ is the transfinite sequence of partial sum operators associated to $\{x_\lambda\}_{\lambda < \vartheta}$, the limit

$$x = \lim_{j \to \infty} y_j$$

exists.

*Remark 17.5.* In the particular case when $\vartheta = \omega$, boundedly completeness of a basis $\{x_\lambda\}_{\lambda<\omega}$ in the sense of definition 17.11 b) is equivalent to usual boundedly completeness. Indeed, assume that $\{x_\lambda\}_{\lambda<\omega}$ is boundedly complete in the sense of definition 17.11 b) and let $\{\alpha_j\} \subset K$, $\sup_{1 \leqslant j < \infty} \left\| \sum_{i=1}^{j} \alpha_i x_i \right\| < \infty$. Then $\lambda_j = j$ $(j = 1,2,\ldots)$ satisfies $\lambda_1 < \lambda_2 < \ldots$ $\ldots < \omega = \vartheta$, $\lim_{j \to \infty} \lambda_j = \omega$ and for this $\{\lambda_j\}$ the bounded sequence $y_1 = 0$, $y_j = \sum_{i=1}^{j-1} \alpha_i x_i$ $(j = 2,3, \ldots)$ satisfies (17.53) (with $s_\lambda$ defined by (17.38), i.e., $s_\lambda$ is the $(\lambda - 1)$-th partial sum operator in the usual sense). Consequently, $\sum_{i=1}^{\infty} \alpha_i x_i = \lim_{j \to \infty} y_j$ exists.

Conversely, assume now that $\{x_\lambda\}_{\lambda<\omega}$ is boundedly complete in the usual sense and let $\lambda_1 < \lambda_2 < \ldots < \omega$, whence $\lim_{j \to \infty} \lambda_j = \omega$. Also, let $\{y_j\} \subset E$ be a bounded sequence satisfying (17.53). Then, since $\{x_\lambda\}_{\lambda<\omega}$ is a basis of $E$, we can write $y_j = \sum_{\lambda<\omega} \alpha_\lambda^{(j)} x_\lambda$ $(j = 1,2,\ldots)$, so (17.53) gives

$$\sum_{\lambda<\lambda_n} \alpha_\lambda^{(j)} x_\lambda = s_{\lambda_n}(y_j) = y_n = \sum_{\lambda<\omega} \alpha_\lambda^{(n)} x_\lambda \qquad (n = 1, \ldots, j;\ j = 1,2, \ldots).$$

Hence, using again that $\{x_\lambda\}_{\lambda<\omega}$ is a basis, we obtain

$$\alpha_\lambda^{(n)} = \begin{cases} \alpha_\lambda^{(j)} & \text{for } \lambda < \lambda_n \\ 0 & \text{for } \lambda \geqslant \lambda_n \end{cases} \qquad (n = 1, \ldots, j;\ j = 1,2, \ldots),$$

whence

$$y_j = \sum_{n=1}^{j} \sum_{\lambda_{n-1} \leqslant \lambda < \lambda_n} \alpha_\lambda^{(n)} x_\lambda \qquad (j = 1,2, \ldots;\ \lambda_0 = 1).$$

Define now $\{\alpha_\lambda\}_{\lambda<\omega} \subset K$ by

$$\alpha_\lambda = \alpha_\lambda^{(n)} \qquad (\lambda_{n-1} \leqslant \lambda < \lambda_n;\ n = 1,2, \ldots;\ \lambda_0 = 1).$$

Then, since $\{x_\lambda\}_{\lambda<\omega}$ is a basis, for any $\mu < \omega$ with $\mu < \lambda_j$ we have $\|\sum_{\lambda<\mu} \alpha_\lambda x_\lambda\| \leqslant v \|\sum_{\lambda<\lambda_j} \alpha_\lambda x_\lambda\| = v\|y_j\|$, where $v$ is the norm of the basis $\{x_\lambda\}_{\lambda<\omega}$. Hence, since $\{y_j\}$ is bounded, $\sup_{\mu<\omega} \|\sum_{\lambda<\mu} \alpha_\lambda x_\lambda\| \leqslant v \sup_{1 \leqslant j < \infty} \|y_j\| < \infty$, whence, by the boundedly completeness of $\{x_\lambda\}_{\lambda<\omega}$ in the usual sense, $\lim_{j \to \infty} y_j = \sum_{n=1}^{\infty} \sum_{\lambda_{n-1} \leqslant \lambda < \lambda_n} \alpha_\lambda^{(n)} x_\lambda = \sum_{\lambda<\omega} \alpha_\lambda x_\lambda$ exists. This proves our assertion.

In order to define unconditional transfinite bases, it seems natural to define the unconditional convergence of a transfinite series $\sum_{\lambda<\vartheta} y_\lambda$ in $E$, to an element $x \in E$, by the condition that for every one-to-one mapping $\sigma$ of $[1, \vartheta)$ onto itself and the order relation*) $\prec$ on $\sigma([1, \vartheta))$ defined by $\sigma(\lambda) \prec \sigma(\mu)$ if and only if $\lambda < \mu$ in $[1, \vartheta)$, the transfinite series $\sum_{\sigma(\lambda)\prec\sigma(\vartheta)} y_{\sigma(\lambda)}$ is convergent to $x$. However, with this definition, we have the following generalization of Ch. II, § 16, lemma 16.1, equivalence $1° \Leftrightarrow 2°$:

**Lemma 17.3.** *A transfinite series* $\sum_{\lambda<\vartheta} y_\lambda$ *in a Banach space* $E$ *is unconditionally convergent to* $x \in E$ *if and only if*

$$x = \lim_{d\in\mathscr{D}} \sum_{\lambda\in[1,\vartheta)} y_\lambda, \tag{17.54}$$

*where* $\mathscr{D}$ *is the collection of all finite subsets of the set* $[1, \vartheta)$, *ordered by inclusion.*

*Proof.* Assume that $\sum_{\lambda<\vartheta} y_\lambda$ is unconditionally convergent to $x$. Then, by lemma 17.1, we may assume that in the initial well ordering of $[1, \vartheta)$ we have $\{\lambda < \vartheta \mid y_\lambda \neq 0\} = [1, \omega)$, so $y_\omega = y_{\omega+1} = y_{\omega+2} = \ldots = y_\vartheta = 0$. Thus, for the continuous function $S: [1, \vartheta) \to E$ occurring in the definition of $\sum_{\lambda<\vartheta} y_\lambda$, we have

$$S(\omega) = S(\omega + 1) = \ldots = S(\vartheta) = x,$$

whence, by our assumption, $x = \sum_{\lambda<\omega} y_\lambda$ unconditionally (since every permutation of $[1, \omega)$ can be extended to a one-to-one mapping $\sigma$ of $[1, \vartheta)$ onto itself). Consequently, by Ch. II, § 16, lemma 16.1, implication $1° \Rightarrow 2°$, we have (17.54).

Conversely, assume that (17.54) holds and let us consider, for simplicity of notation, the initial well ordering of $[1, \vartheta)$ (for any well ordering of $[1, \vartheta)$ the same argument works). By (17.54), there exist the limits

$$S(\lambda) = \lim_{d\in\mathscr{D}_\lambda} \sum_{\varkappa\in[1,\lambda)} y_\varkappa \qquad (2 \leqslant \lambda \leqslant \vartheta), \tag{17.55}$$

where $\mathscr{D}_\lambda$ denotes the collection of all finite subsets of $[1, \lambda)$, ordered by inclusion. Let $S(1)=0$. We shall show that $S: [1, \vartheta) \to E$ is a continuous function satisfying (17.23), (17.24), which will complete the proof.

---

*) Clearly, with this ordering, $\sigma([1, \vartheta)) = [\sigma(1), \sigma(\vartheta))$ becomes well ordered and, conversely, to every well ordering $<$ of $[1, \vartheta)$ corresponds such a mapping $\sigma$, defined by $\sigma(1) = \min\, [1, \vartheta)$, $\sigma(\lambda) = \min\, [1, \vartheta)\setminus\{\sigma(\mu)\}_{\mu<\lambda}$ $(2 \leqslant \lambda < \vartheta)$, where "min" is taken with respect to $<$. Thus, in other words, the condition means that $\sum_{\lambda<\vartheta} y_\lambda$ converges to $x$ "under every well ordering of $[1, \vartheta)$".

Clearly, by (17.54) we have (17.23). In order to show (17.24), let $\varepsilon > 0$ and let $d_\lambda(\varepsilon) \in \mathscr{D}_\lambda$ and $d_{\lambda+1}(\varepsilon) \in \mathscr{D}_{\lambda+1}$ with $d_{\lambda+1}(\varepsilon) \supset d_\lambda(\varepsilon)$ be such that

$$\| S(\lambda+1) - \sum_{\varkappa \in d} y_\varkappa \| < \varepsilon \qquad (d \in \mathscr{D}_{\lambda+1},\ d \supset d_{\lambda+1}(\varepsilon)),$$

$$\| S(\lambda) - \sum_{\varkappa \in d} y_\varkappa \| < \varepsilon \qquad (d \in \mathscr{D}_\lambda,\ d \supset d_\lambda(\varepsilon)). \qquad (17.56)$$

If $d_{\lambda+1}(\varepsilon) \not\ni \lambda$, then $d_{\lambda+1}(\varepsilon) \in \mathscr{D}_\lambda$, whence

$$\| S(\lambda+1) - y_\lambda - \sum_{\varkappa \in d_{\lambda+1}(\varepsilon)} y_\varkappa \| = \| S(\lambda+1) - \sum_{\varkappa \in d_{\lambda+1}(\varepsilon) \cup \{\lambda\}} y_\varkappa \| < \varepsilon,$$

$$\| S(\lambda) - \sum_{\varkappa \in d_{\lambda+1}(\varepsilon)} y_\varkappa \| < \varepsilon.$$

On the other hand, if $d_{\lambda+1}(\varepsilon) \ni \lambda$, then for $d = d_{\lambda+1}(\varepsilon) \setminus \{\lambda\} \in \mathscr{D}_\lambda$ we have $d \supset d_\lambda(\varepsilon)$, whence

$$\| S(\lambda+1) - y_\lambda - \sum_{\varkappa \in d} y_\varkappa \| = \| S(\lambda+1) - \sum_{\varkappa \in d_{\lambda+1}(\varepsilon)} y_\varkappa \| < \varepsilon,$$

$$\| S(\lambda) - \sum_{\varkappa \in d} y_\varkappa \| < \varepsilon.$$

Consequently, $\| S(\lambda+1) - S(\lambda) - y_\lambda \| < 2\varepsilon$, whence, since $\varepsilon > 0$ was arbitrary, we get (17.24).

Finally, let us prove that $S$ is continuous from the left at any limit ordinal $\lambda \leqslant \vartheta$. Let $\varepsilon > 0$ and let $d_\lambda(\varepsilon) \subset \mathscr{D}_\lambda$ be such that we have (17.56). Put $\mu_\varepsilon = \max_{\varkappa \in d_\lambda(\varepsilon)} \varkappa$. Then for any $\mu \in (\mu_\varepsilon, \lambda]$ we have $d_\lambda(\varepsilon) \in \mathscr{D}_\mu$ and hence there exists $d_\mu(\varepsilon) \in \mathscr{D}_\mu$ with $d_\mu(\varepsilon) \supset d_\lambda(\varepsilon)$, such that

$$\| S(\mu) - \sum_{\varkappa \in d} y_\varkappa \| < \varepsilon \qquad (d \in \mathscr{D}_\mu,\ d \supset d_\mu(\varepsilon)).$$

Hence, for $d = d_\mu(\varepsilon) \in \mathscr{D}_\mu \subset \mathscr{D}_\lambda$ we obtain, taking also into account (17.56), that $\|S(\lambda) - S(\mu)\| < 2\varepsilon$, which, since $\mu \in (\mu_\varepsilon, \lambda]$ was arbitrary, completes the proof of lemma 17.3.

From lemma 17.3 and theorem 17.3, equivalence $1° \Leftrightarrow 2°$, it follows that the natural definition of unconditional transfinite bases leads us to the class of extended unconditional bases, which was considered already in the preceding (see e.g. theorem 17.3).

It is natural to introduce the following generalization of (usual) basic sequences:

*Definition 17.12.* A transfinite sequence $\{x_\lambda\}_{\lambda<\vartheta}$ in a Banach space $E$ is said to be a *transfinite basic sequence* (of type $\vartheta$) if $\{x_\lambda\}_{\lambda<\vartheta}$ is a transfinite basis of $[x_\lambda]_{\lambda<\vartheta}$.

The special classes of transfinite basic sequences are defined similarly.

In § 1, remark 1.1, we have seen that in every infinite dimensiona Banach space $E$ there exists an "asymptotically monotone" basic sequence $\{x_\lambda\}_{\lambda<\omega}$ (of type $\omega$). Now we shall show that in a wide class of non-separable Banach spaces $E$ (including all reflexive non-separable Banach spaces) there exists a monotone transfinite basic sequence of the largest possible*) cardinality, i.e., a monotone transfinite basic sequence $\{x_\lambda\}_{\lambda<\vartheta}$ of type $\vartheta$, where $\vartheta$ is the first ordinal number of cardinality dens $E$. We recall that for any normed linear space $E$, the *$w^*$-density character* of $E^*$, denoted $w^*$-dens $E^*$, is defined as the smallest cardinal $\mathfrak{m}$ for which $E^*$ has a $w^*$-dense (or, equivalently, a total) subset of cardinality $\mathfrak{m}$. For any normed linear space $E$ we have

$$w^*\text{-dens } E^* \leqslant \text{dens } E. \tag{17.57}$$

For, if $\{x_i\}_{i\in I}$ is a dense set in $\sigma_E = \{x\in E \mid \|x\| = 1\}$, with card $I =$ dens $E$, and if $f_i \in E^*$, $\|f_i\| = 1 = f_i(x_i)$ $(i\in I)$, then we have $r([f_i]_{i\in I}) = 1$ (whence, in particular, $\{f_i\}_{i\in I}$ is total on $E$ and therefore $w^*$-dens $E^* \leqslant$ card $I =$ dens $E$). Indeed, for any $x\in\sigma_E$ and $\varepsilon > 0$ there exists $i\in I$ such that $\|x-x_i\|<\varepsilon$, whence $|f_i(x)| \geqslant \big| |f_i(x_i)| - |f_i(x-x_i)| \big| \geqslant 1-\varepsilon$, which proves our assertion. There exist Banach spaces**) for which the inequality in (17.57) is strict, for example, $E = l^\infty$ (since the sequence of coordinate functionals is total on $E$).

**Theorem 17.10.** *If $E$ is a Banach space such that*

$$w^*\text{-dens } E^* = \text{dens } E > \aleph_0, \tag{17.58}$$

*then there exists in $E$ a monotone transfinite basic sequence $\{x_\lambda\}_{\lambda<\vartheta}$ of type $\vartheta =$ the first ordinal of cardinality* dens $E$.

*Proof.* We shall define a transfinite sequence $\{x_\lambda\}_{\lambda<\vartheta}$ with $x_\lambda \neq 0$ $(\lambda < \vartheta)$, such that for each $\lambda < \vartheta$ we have

$$\|y\| \leqslant \|y + \alpha_\lambda x_\lambda\| \qquad (y\in \text{lin}\,\{x_\varkappa\}_{\varkappa<\lambda},\ \alpha_\lambda\in K), \tag{17.59}$$

which will complete the proof (since then, by — finite — induction it follows that $\{x_\lambda\}_{\lambda<\vartheta}$ satisfies 3° of theorem 17.8 with $C = 1$).

Let $x_1 \in E\setminus\{0\}$ be arbitrary. Assume that for some $\mu < \vartheta$ we have defined $\{x_\lambda\}_{\lambda<\mu}$ with $x_\lambda \neq 0$ $(\lambda < \mu)$, satisfying (17.59) for each $\lambda < \mu$. Let $\mathscr{B}_\mu$ be the set of all finite linear combinations of $\{x_\varkappa\}_{\varkappa<\mu}$ with rational coefficients and, for each $x\in\mathscr{B}_\mu$, pick a functional $f_x\in E^*$ such that $\|f_x\| = 1$, $f_x(x) = \|x\|$. Then the set $\mathscr{C}_\mu = \{f_x \mid x\in\mathscr{B}_\mu\}$ has

$$\text{card } \mathscr{C}_\mu \leqslant \text{card } \mathscr{B}_\mu \leqslant \aleph_0 \text{card } \mu < \text{dens } E, \tag{17.60}$$

---

*) By § 20, proposition 20.1. For the definition of the density character dens $E$, see the part before theorem 17.3.

**) Obviously, any such space $E$ must be non-separable.

whence, by (17.58), there exists an element $x_\mu \in (\mathscr{C}_\mu)_\perp \setminus \{0\}$. Clearly,

$$\|x\| = |f_x(x)| = |f_x(x + \alpha_\mu x_\mu)| \leqslant \|x + \alpha_\mu x_\mu\| \qquad (x \in \mathscr{B}_\mu, \alpha \in K),$$

whence, since $\mathscr{B}_\mu$ is dense in lin $\{x_\varkappa\}_{\varkappa<\mu}$, it follows that we have (17.59) for $\lambda = \mu$. This completes the proof.

*Remark 17.6.* a) Every non-separable reflexive space $E$ satisfies (17.58). Indeed, for every reflexive $E$ we have $w^*$-dens $E^* = w$-dens $E^*$ and for every Banach space $E$ we have*) $w$-dens $E^* =$ dens $E^* \geqslant$ dens $E$, which together with (17.57), gives $w^*$-dens $E^* =$ dens $E$. More generally, every non-separable weakly compactly generated space $E$ also satisfies**) (17.58).

b) We shall see in § 19 that every non-separable Banach space $E$ having a (monotone) transfinite Schauder decomposition $\{G_\lambda\}_{\lambda<\vartheta}$ of type $\vartheta$, where $\vartheta$ is the first ordinal number of cardinality dens $E$ (for example, every weakly compactly generated non-separable space) contains a (monotone) transfinite basic sequence $\{x_\lambda\}_{\lambda<\vartheta}$ of type $\vartheta$, with $x_\lambda \in G_\lambda$ $(\lambda < \vartheta)$.

Finally, as announced at the end of § 15, let us introduce another natural generalization of unconditional bases (and, actually, of extended unconditional bases), to the continuous case. The "index set" will be an infinite locally compact space $T$, on which we shall not assume any ordering (neither by an increasing sequence $\{Q_n\}$ of compact subsets of $T$, nor a well ordering). We shall modify § 15, definition 15.24, to give the following definition (where the notation $\int_A Y(t)d\mu(t) \in E$ is used in the sense explained in § 15):

*Definition 17.13.* Let $E$ be an infinite dimensional Banach space, $T$ an infinite locally compact space, $X: t \to X(t)$ a mapping of $T$ into $E$ and $\mathfrak{M}$ a locally convex space of measures on $T$, such that the closed graph theorem is valid***) for mappings from any Banach space into $\mathfrak{M}$. Assume that

(i$_u$) $\int_T X(t)d\mu(t) \in E \qquad (\mu \in \mathfrak{M})$;

(ii$_u$) there exists a set $\Gamma \subset E^*$, which is total on $E$, such that for each $f \in \Gamma$ the linear functional $\mu \to \int_T f(X(t))d\mu(t)$ is continuous on $\mathfrak{M}$.

---

*) For the equality $w$-dens $E^*$=dens $E^*$, see e.g. [87], p. 422, theorem 13. The inequality dens $E^* \geqslant$ dens $E$ is proved in [11], p. 189, theorem 12, for separable $E$ and in the general case the proof is similar.

**) See [242], proposition 2.2.

***) I.e., every linear mapping from a Banach space $F$ into $\mathfrak{M}$, with closed graph, is continuous.

A system $(X, T, \mathfrak{M})$ satisfying $(\mathrm{i}_u)$, $(\mathrm{ii}_u)$ is called a *strongly unconditional integral basis* of $E$, if for every $x \in E$ there exists a *unique* measure $\mu_x \in \mathfrak{M}$ such that

$$x = \int_T X(t) d\mu_x(t). \tag{17.61}$$

In particular, if we take $T = I$ with the discrete topology, $X(i) = x_i$ $(i \in I)$ and $\mathfrak{M}$ = the linear subspace of $s(I) = K^I$ (the linear space of all families of scalars $\mu = \{\mu(i)\}_{i \in I}$, regarded as measures on $I$) consisting of all $\mu = \{\mu(i)\}_{i \in I}$ such that $\sum_{i \in I} |f(\mu(i)x_i)| < \infty$ $(f \in E^*)$ and that $\sum_{i \in I} \mu(i)x_i$ is $\sigma(E, E^*)$-convergent to an element $x$ of $E$, endowed*) with the norm $\|\mu\| = \|x\|$, then $(X, T, \mathfrak{M})$ is a strongly unconditional integral basis of $E$ if and only if $\{x_i\}_{i \in I}$ is an extended unconditional basis of $E$. Indeed, condition $(\mathrm{i}_u)$ is satisfied**) by our definition of $\mathfrak{M}$ and, if $\{x_i\}_{i \in I}$ is an extended unconditional basis of $E$, then $(\mathrm{ii}_u)$ is obviously satisfied for $\Gamma = \{f_i\}_{i \in I} \subset E^*$, the a.f.c.f. to $\{x_i\}_{i \in I}$; also, in this case each $x \in E$ admits a representation of the form (17.61), with a unique $\mu_x \in \mathfrak{M}$. Conversely, if $(X, T, \mathfrak{M})$ is a strongly unconditional integral basis of $E$, then for each $x \in E$ there exists a unique $\mu_x \in \mathfrak{M}$ such that $\sum_{i \in I} |f(\mu_x(i)x_i)| < \infty$ and $f(x) = \sum_{i \in I} \mu_x(i) f(x_i)$ $(f \in E^*)$, so $\sum_{i \in I} \mu_x(i)x_i$ is weakly unconditionally convergent to $x$. Thus, $\{x_i\}_{i \in I}$ is an extended unconditional basis of $E$ (e.g. by the Orlicz-Pettis theorem***), or by the generalized weak basis theorem), which proves our assertion.

*Remark 17.7.* The following observation may partially motivate the term introduced in definition 17.13: If $(X, T, \mathfrak{M})$ is a strongly unconditional integral basis of a Banach space $E$, with $\mathfrak{M}$ a Fréchet space, and if there exists an increasing sequence $\{Q_n\}$ of compact subsets of $T$, such that****) $\mu_x\left(T \setminus \bigcup_{n=1}^{\infty} Q_n\right) = 0$ $(x \in E)$ and satisfying (i)—(iii) of § 15, definition 15.24, then $(X, T, \{Q_n\}, \mathfrak{M})$ is an unconditional integral basis of $E$ in the sense of § 15, definition 15.28, with the same representing measures $\mu_x$. Indeed, let $\sigma$ be any permutation of $\mathcal{N} = \{1,2,3, \ldots\}$

---

*) Thus, $\mathfrak{M}$ is a Banach space, and therefore the closed graph theorem is valid for mappings from any Banach space into $\mathfrak{M}$.

**) See e.g. [32], Ch. III, § 2, No. 2.

***) See e.g. [70], Ch. IV, § 1, theorem 1.

****) This happens, for example, when $T = \bigcup_{n=1}^{\infty} Q_n$ (i.e., $T$ is "countable at the infinity" [31]).

and let

$$v_{\sigma(n)}(x) = (S_{\sigma(n)} - S_{\sigma(n)-1})(x) \qquad (x \in E,\ n = 1,2, \ldots), \quad (17.62)$$

where $S_n$ is defined by § 15, formula (15.146) and $S_0 = 0$. Then, since $\{Q_n\}$ is increasing, we have

$$\sum_{i=1}^{n} v_{\sigma(i)}(x) = \sum_{i=1}^{n} \int_{Q_{\sigma(i)} \setminus Q_{\sigma(i)-1}} X(t) d\mu_x(t) = \int_{B_n} X(t) d\mu_x(t),$$

where $B_n = \bigcup_{i=1}^{n} (Q_{\sigma(i)} \setminus Q_{\sigma(i)-1})$, $Q_0 = \emptyset$. But, $\{B_n\}$ is increasing and, by our assumption, $\mu_x\left(T \setminus \bigcup_{n=1}^{\infty} B_n\right) = \mu_x\left(T \setminus \bigcup_{n=1}^{\infty} Q_n\right) = 0 \quad (x \in E)$, whence $|1 - \chi_{B_n}(t)| \to 0$, $\mu_x$-almost everywhere on $T$ $(x \in E)$, where $\chi_A(\cdot)$ denotes the characteristic function of $A$; also, clearly, $|1 - \chi_{B_n}(t)| \leqslant 1$ $(t \in T,\ n = 1,2, \ldots)$ and, by $(\mathrm{i}_u)$, $\int_T |f(X(t))| d\mu_x(t) < \infty$ $(f \in E^*)$. Consequently, by (17.61) and the theorem of Lebesgue*),

$$\left| f\left(x - \sum_{i=1}^{n} v_{\sigma(i)}(x)\right) \right| = \left| \int_T (1 - \chi_{B_n}(t)) f(X(t)) d\mu_x(t) \right| \to 0 \qquad (f \in E^*),$$

so $\sum_{i=1}^{\infty} v_i(x)$ is weakly unconditionally convergent to $x$. Hence, by the Orlicz-Pettis theorem**), $\sum_{i=1}^{\infty} v_i(x)$ is unconditionally convergent to $x$ in the norm-topology, which, since the uniqueness of $\mu_x$ is obvious, completes the proof of our assertion.

For strongly unconditional integral bases we have the following result, similar to § 15, theorem 15.23, but with a simpler proof:

**Theorem 17.11.** *For every strongly unconditional integral basis* $(X, T, \mathfrak{M})$ *of a Banach space* $E$, *the linear mapping* $x \to \mu_x$ *is continuous on* $E$.

*Proof.* Since by our assumption on $\mathfrak{M}$, the closed graph theorem is valid for mappings from $E$ into $\mathfrak{M}$, it is sufficient to show that the mapping $x \to \mu_x$ is closed. Assume that $\{y_d\}_{d \in \Delta} \subset E$, $\lim_{d \in \Delta} y_d = y \in E$ and $\lim_{d \in \Delta} \mu_{y_d} = \mu \in \mathfrak{M}$, where $\Delta$ is a directed set of indices. Then, by (17.61)

---

*) See e.g. [32], Ch. IV, § 3, theorem 6.
**) See e.g. [70], Ch. IV, § 1, theorem 1.

and (ii$_u$), (i$_u$),

$$f(y) = \lim_{d \in \Delta} f(y_d) = \lim_{d \in \Delta} \int_T f(X(t)) d\mu_{y_d}(t) = \int_T f(X(t)) d\mu(t) =$$

$$= f\left(\int_T X(t) d\mu(t)\right) \qquad (f \in \Gamma),$$

whence, since $\Gamma$ is total on $E$, $y = \int_T X(t) d\mu(t)$. Consequently, by the uniqueness condition in definition 17.13, $\mu = \mu_y$, which completes the proof of theorem 17.11.

## § 18. Extended approximative bases. Extended $\Pi$-bases. Extended resolutions of the identity

One can obtain natural extensions of approximative bases $\{u_n\}$ and of other sequences of endomorphisms considered in some of the sections of part I, replacing these sequences $\{u_n\}$ by uniformly bounded nets $\{u_d\}_{d \in \mathscr{D}}$ (where $\mathscr{D}$ is a directed set), converging pointwise to the identity operator. We have already seen in part I (e.g. in § 9, proofs of proposition 9.8 and theorem 9.7 and in § 14, proof of theorem 14.1 c)) that such nets appear naturally in spaces having various approximation properties, with $\mathscr{D}$ related to the set of all finite-dimensional subspaces of $E$, directed by inclusion. The condition of uniform boundedness, which was a consequence of pointwise convergence in the separable case, must be required separately in the general case.

*Definition 18.1.* Let $E$ be a Banach space. A net of endomorphisms $\{u_d\}_{d \in \mathscr{D}} \subset L(E, E)$ of finite rank is called an *extended approximative basis* of $E$ if

$$x = \lim_{d \in \mathscr{D}} u_d(x) \qquad (x \in E), \tag{18.1}$$

$$\sup_{d \in \mathscr{D}} \|u_d\| < \infty. \tag{18.2}$$

If in (18.2) $\sup_{d \in \mathscr{D}} \|u_d\| \leqslant \lambda$, we shall say that $\{u_d\}_{d \in \mathscr{D}}$ is an *extended $\lambda$-approximative basis* of $E$.

For example, taking $\mathscr{D} = \mathscr{N} = \{1,2, \ldots\}$ (directed by $\geqslant$), we see that every approximative basis $\{u_n\}$ is an extended approximative basis. Also, by § 17, theorem 17.3, implication $1° \Rightarrow 2°$, the net of all partial sum operators $\{s_d\}_{d \in \mathscr{D}}$ associated to an extended unconditional basis

$\{x_i\}_{i\in I}$ of $E$ (where $\mathscr{D}$ is the collection of all finite subsets of the index set $I$, directed by inclusion), is an extended approximative basis of $E$.

We have the following generalization of part of § 9, theorem 9.3:

**Theorem 18.1.** *Let $E$ be a Banach space and let $\lambda \geqslant 1$. The following statements are equivalent:*

1°. *$E$ has an extended $\lambda$-approximative basis.*

2°. *$E$ contains a set of subspaces $\{G_d\}_{d\in\mathscr{D}}$ with the following properties:*

a) $\dim G_d < \infty$ $(d \in \mathscr{D})$;

b) $\{G_d\}_{d\in\mathscr{D}}$ *is directed by inclusion;*

c) $\bigcup_{d\in\mathscr{D}} G_d$ *is dense in $E$;*

d) *for every $d \in \mathscr{D}$ and $\varepsilon > 0$ there exists a finite rank endomorphism $v_{d,\varepsilon} \in L(E, E)$ satisfying $v_{d,\varepsilon}|_{G_d} = I_{G_d}$, $\|v_{d,\varepsilon}\| \leqslant \lambda + \varepsilon$.*

3°. *$E$ has the $\lambda$-approximation property.*

4°. *For every finite-dimensional subspace $G$ of $E$ and every $\delta > 0$ there exists a finite rank operator $u = u_{G,\delta} \in L(E, E)$ such that*

$$\|u(x) - x\| < \delta\|x\| \qquad (x \in G), \tag{18.3}$$

$$\|u\| \leqslant \lambda. \tag{18.4}$$

5°. *For every finite-dimensional subspace $G$ of $E$ and every $\varepsilon > 0$ there exists a finite rank operator $v = v_{G,\varepsilon} \in L(E, E)$ such that*

$$v(x) = x \qquad (x \in G), \tag{18.5}$$

$$\|v\| \leqslant \lambda + \varepsilon. \tag{18.6}$$

*Proof.* If we have 1°, then, by the argument of the proof of Ch. I, § 17, theorem 17.3 (using now also that $\mathscr{D}$ is a directed set), it follows that for every compact set $Q$ in $E$ we have

$$\lim_{d\in\mathscr{D}} \sup_{x\in Q} \|x - u_d(x)\| = 0,$$

whence we infer that § 9, formula (9.49) holds for suitable $u = u_d$ (hence $\|u\| \leqslant \lambda$). Thus, 1° ⇒ 3°.

The implication 3° ⇒ 4° is obvious, since for every finite-dimensional subspace $G$ of $E$ the unit ball $S_G = \{x \in G \mid \|x\| \leqslant 1\}$ is compact.

Assume now that we have 4°. Let $\mathscr{D}$ be the directed set of all pairs $(G, \delta)$, where $G$ is a finite-dimensional subspace of $E$ and $\delta > 0$, and where $(G_1, \delta_1) \geqslant (G_2, \delta_2)$ if and only if $G_1 \supset G_2$ and $\delta_1 \leqslant \delta_2$. Furthermore, by 4°, for each $d = (G, \delta) \in \mathscr{D}$, let $u = u_d \in L(E, E)$ be an endomorphism of finite rank satisfying (18.3), (18.4). If $x \in E$ and $\delta > 0$, then, putting

$d_0 = ([x], \delta)$, it follows that $\|u_d(x) - x\| < \delta\|x\|$ $(d \geqslant d_0)$, which proves (18.1). Thus*), $4° \Rightarrow 1°$.

The proof of the equivalence $4° \Leftrightarrow 5°$ is contained in § 9, proof of theorem 9.3, equivalence $5° \Leftrightarrow 6°$.

If we have 5°, then any set of subspaces $\{G_d\}_{d \in \mathscr{D}}$ satisfying a), b), c) of 2° (for example, the set of all finite-dimensional subspaces of $E$, indexed by itself) satisfies also d) of 2°. Thus, $5° \Rightarrow 2°$.

Finally, assume that $E$ satisfies 2°. Let $\mathscr{A}$ be the directed set of all pairs $(d, \varepsilon)$, where $d \in \mathscr{D}$ and $\varepsilon > 0$ and where $(d_1, \varepsilon_1) \geqslant (d_2, \varepsilon_2)$ if and only if $G_{d_1} \supset G_{d_2}$ and $\varepsilon_1 \leqslant \varepsilon_2$ ($\mathscr{A}$ is a directed set by b)). Let $x \in \bigcup_{d \in \mathscr{D}} G_d$, say $x \in G_{d_0}$, and let $\varepsilon_0 > 0$. Then, since $v_{d,\varepsilon}|_{G_d} = I_{G_d}$ $(d \in \mathscr{D}, \varepsilon > 0)$, we have $v_{d,\varepsilon}(x) = x$ $((d, \varepsilon) \geqslant (d_0, \varepsilon_0))$. Hence, by c) and $\sup_{(d,\varepsilon) \in \mathscr{A}} \|v_{d,\varepsilon}\| < \infty$, we get $x = \lim_{(d,\varepsilon) \in \mathscr{A}} v_{d,\varepsilon}(x)$ for all $x \in E$ and thus, by a) and $\|v_{d,\varepsilon}\| \leqslant \lambda + \varepsilon$ $(d \in \mathscr{D}, \varepsilon > 0)$, $\{v_{d,\varepsilon}\}_{(d,\varepsilon) \in \mathscr{A}}$ is an extended approximative basis of $E$. Consequently, putting, similarly to § 9, formula (9.91),

$$w_{d,\varepsilon} = \begin{cases} v_{d,\varepsilon} & \text{if } \|v_{d,\varepsilon}\| \leqslant \lambda \\ \dfrac{\lambda v_{d,\varepsilon}}{\|v_{d,\varepsilon}\|} & \text{if } \lambda < \|v_{d,\varepsilon}\| \leqslant \lambda + \varepsilon, \end{cases}$$

we obtain an extended $\lambda$-approximative basis $\{w_{d,\varepsilon}\}_{(d,\varepsilon) \in \mathscr{A}}$ of $E$. Thus, $2° \Rightarrow 1°$, which completes the proof of theorem 18.1.

*Remark 18.1.* a) If we only want to deduce from 2° (or, in particular, from 5°) the existence of an extended $(\lambda+\varepsilon)$-approximative basis of $E$, then it is enough to consider, instead of the above set $\mathscr{A}$, the set $\mathscr{D}$ of 2°, in which we put, by definition, $d_1 \geqslant d_2$ if and only if $G_{d_1} \supset G_{d_2}$ and to take, for each $d \in \mathscr{D}$, the finite rank endomorphism $v_{d,\varepsilon} = v_d$ of 2° d). Indeed, the above proof of the implication $2° \Rightarrow 1°$ shows that $\{v_d\}_{d \in \mathscr{D}}$ is an extended $(\lambda + \varepsilon)$-approximative basis of $E$.

b) With the notations of the above proof of the implication $4° \Rightarrow 1°$, the proof of § 9, theorem 9.3, implication $5° \Rightarrow 1°$ can be expressed as follows: If $G_n = [y_1, \ldots, y_n]$ $(n = 1,2, \ldots)$, where $\{y_n\}$ is a dense sequence in $E$ and if $u_n = u_{d_n}$, where $d_n = \left(G_n, \dfrac{1}{n}\right)$ $(n = 1,2, \ldots)$, then $x = \lim_{n \to \infty} u_n(x)$ $(x \in E)$. Thus, in this case, $\{u_d\}_{d \in \mathscr{D}}$ can be replaced by the sequence $\{u_n\} = \{u_{d_n}\}$.

By the equivalence $1° \Leftrightarrow 3°$ above and, by § 9, theorem 9.3, equivalence $1° \Leftrightarrow 4°$, a *separable* Banach space $E$ has an extended approxima-

---

*) Note that this argument has been also used in § 9, proof of proposition 9.8, formulae (9.136), (9.137).

tive basis if and only if it has an approximative basis. More generally, we shall prove now

**Theorem 18.2.** *Let $\lambda \geqslant 1$. A Banach space $E$ has an extended $\lambda$-approximative basis if and only if every separable subspace $G$ of $E$ is contained in a separable subspace $F$ of $E$ with a $\lambda$-approximative basis.*

*Proof.* Assume that $E$ has an extended $\lambda$-approximative basis $\{u_d\}_{d\in\mathscr{D}}$. Let $G$ be a separable subspace of $E$ and let $\{y_n\} \subset G$ be a dense sequence in $S_G = \{x \in G \mid \|x\| \leqslant 1\}$. Choose $d_1 \in \mathscr{D}$ such that $\|u_{d_1}(y_1) - y_1\| < 1$. If $d_{n-1} \in \mathscr{D}$ has been chosen, select $d_n > d_{n-1}$ in $\mathscr{D}$ such that

$$\|u_{d_n}(x) - x\| < \frac{1}{n} \qquad (x \in \{y_1, \dots, y_n\} \cup S_{\left[\bigcup_{i=1}^{n-1} u_{d_i}(E)\right]}); \tag{18.7}$$

this is possible, since the pointwise convergence of the uniformly bounded net $\{u_d\}_{d\in\mathscr{D}}$ to $I_E$ implies that the convergence is uniform on compact sets. Put

$$F = \{x \in E \mid \lim_{n\to\infty} u_{d_n}(x) = x\}. \tag{18.8}$$

Then $F$ is a closed linear subspace of $E$. Indeed, if $x_k \in F$, $\lim_{k\to\infty} x_k = x \in E$, then $\lim_{n\to\infty} u_{d_n}(x_k) = x_k$ $(k = 1,2, \dots)$, whence, by

$$\|x - u_{d_n}(x)\| \leqslant \|x - x_k\| + \|x_k - u_{d_n}(x_k)\| + \sup_{d\in\mathscr{D}} \|u_d\| \, \|x_k - x\|$$

and by (18.2), we obtain $\lim_{n\to\infty} u_{d_n}(x) = x$, so $x \in F$.

Furthermore, by (18.7), $\lim_{n\to\infty} u_{d_n}(y_i) = y_i$ $(i = 1,2, \dots)$, whence, by $[y_n] = G$ and $[F] = F$, it follows that $G \subset F$. Also, again by (18.7)

$$\|u_{d_m}u_{d_n}(x) - u_{d_n}(x)\| < \frac{1}{m}\|u_{d_n}(x)\| \qquad (x \in E,\ m \geqslant n),$$

whence

$$u_{d_n}(x) = \lim_{m\to\infty} u_{d_m}u_{d_n}(x) \qquad (x \in E,\ n = 1,2, \dots), \tag{18.9}$$

which proves that

$$u_{d_n}(E) \subset F \qquad (n = 1,2, \dots). \tag{18.10}$$

Consequently, in particular, $u_{d_n}|_F \in L(F, F)$ $(n = 1,2, \dots)$ so by (18.8), (18.2), $\{u_{d_n}|_F\}$ is a $\lambda$-approximative basis of $F$, which proves the necessity of the condition of theorem 18.2. Note that actually we have proved somewhat more, namely, that *if $E$ has an extended $\lambda$-approximative basis $\{u_d\}_{d\in\mathscr{D}}$, then for every separable subspace $G$ of $E$ there exists a sequence $d_1 < \ldots < d_n < \ldots$ in $\mathscr{D}$ (depending on $G$) such that* $G \subset F = \left[\bigcup_{n=1}^{\infty} u_{d_n}(E)\right]$ (this latter equality follows from (18.8) and (18.10)) *and that $\{u_{d_n}|_F\}$ is a $\lambda$-approximative basis of $F$.*

Conversely, assume now that every separable subspace $G$ of $E$ is contained in a separable subspace $F$ of $E$ with a $\lambda$-approximative basis. Let $B$ be a finite-dimensional subspace of $E$. Then, by a particular case of § 19, lemma 19.3, there exists a separable subspace $C$ of $E$ such that, given any finite-dimensional subspace $Z$ of $E$ with $Z \supset B$, there exists a linear operator $w_Z\colon Z \to C$ satisfying $w_Z|_B = I_B$ and $\|w_Z\| \leqslant 1 + \dfrac{1}{\dim Z}$.
By our assumption, $C$ is contained in a separable subspace $F$ of $E$ with a $\lambda$-approximative basis. Then, by § 9, theorem 9.3, for any $\varepsilon > 0$ there exists a finite rank linear operator $u\colon F \to F$ such that $\|u(x) - x\| < \varepsilon\|x\|$ for all $x \in B$ (hence $\|u(w_Z(x)) - x\| < \varepsilon\|x\|$ for all $x \in B$) and $\|u\| \leqslant \lambda$.

Consider now the net of functions $\tilde{u}_Z\colon E \to u(F)$ defined by

$$\tilde{u}_Z(x) = \begin{cases} u(w_Z(x)) & \text{for } x \in Z \\ 0 & \text{for } x \in E\setminus Z, \end{cases} \tag{18.11}$$

where the index set $\{Z \subset E \mid \dim Z < \infty,\ Z \supset B\}$ is directed by inclusion. Since $\dim u(F) < \infty$, a compactness argument similar to that used in the proof of § 19, lemma 19.4*) yields that $\{\tilde{u}_Z\}$ has a pointwise convergent subnet $\{\tilde{u}_{Z'}\}$, say $\tilde{u}_{Z'}(x) \to \tilde{u}(x)$ for all $x \in E$, where $\tilde{u}\colon E \to E$ has the following properties: $\tilde{u}$ is linear and of finite rank, $\|\tilde{u}(x) - x\| \leqslant \varepsilon\|x\|$ $(x \in B)$ and $\|\tilde{u}\| \leqslant \lambda$. Thus, by theorem 18.1, implication $4° \Rightarrow 1°$, $E$ has an extended $\lambda$-approximative basis, which completes the proof of theorem 18.2.

The natural existence problems are unsolved even for some usual concrete spaces, for example:

*Problem 18.1.***) a) Does the Banach space $L(l^2, l^2)$ (of all continuous linear mappings of $l^2$ into itself, with the usual operator norm) have an extended approximative basis (or, equivalently, the bounded approximation property)? b) What about the approximation property? c) What about the space $E = H^\infty$ (of all bounded analytic functions on the open unit disk $|\zeta| < 1$, with the norm $\|x\| = \sup\limits_{|\zeta|<1} |x(\zeta)|$)?

---

*) Namely, since $\tilde{u}_Z(S_E) \subset \alpha_Z S_{u(F)}$, where $\alpha_Z = \lambda\left(1 + \dfrac{1}{\dim Z}\right) \leqslant 2\lambda$ and since by Tychonov's theorem $(2\lambda S_{u(F)})^{S_E}$ is compact, $\{\tilde{u}_Z|_{S_E}\}$ has a subnet converging pointwise to some $u \in (2\lambda S_{u(F)})^{S_E}$, which can be extended by homogeneity to $\tilde{u}\colon E \to E$.

**) Recently, problems 18.1 a), b) have been solved in the negative (see the Notes and remarks).

The notion of $\lambda$-duality approximation property (see § 9) admits the following characterization in terms of extended $\lambda$-approximative bases:

**Proposition 18.1.** *Let $\lambda \geqslant 1$. A Banach space $E$ has the $\lambda$-duality approximation property if and only if $E$ has an extended $\lambda$-approximative basis $\{u_d\}_{d \in \mathscr{D}}$ satisfying*

$$f = \lim_{d \in \mathscr{D}} u_d^*(f) \qquad (f \in E^*). \tag{18.12}$$

*Proof.* The argument is similar to that of the above proof of theorem 18.1, equivalence $1^\circ \Leftrightarrow 3^\circ$, taking now $\mathscr{D}$ to be the directed set of all triples $d = (G, \Gamma, \varepsilon)$ (where $G$ and $\Gamma$ are finite-dimensional subspaces of $E$ and $E^*$ respectively, and $\varepsilon > 0$) and letting $u_d \in L(E, E)$ be an endomorphism of finite rank satisfying § 9, formulae (9.106) — (9.108).

In particular, when $E$ is separable, the net $\{u_d\}_{d \in \mathscr{D}}$ can be replaced by a sequence of endomorphisms $\{u_n\} \subset L(E, E)$ of finite rank satisfying $x = \lim_{n \to \infty} u_n(x)$ $(x \in E)$, $f = \lim_{n \to \infty} u_n^*(f)$ $(f \in E^*)$ and $\sup_{1 \leqslant n < \infty} \|u_n\| \leqslant \lambda$ (see remark 18.1 b) above).

*Definition 18.2.* Let $E$ be a Banach space. A net $\{u_d\}_{d \in \mathscr{D}} \subset L(E, E)$ is called an *extended commuting approximative ($\lambda$-approximative) basis* of $E$, if it is an extended approximative ($\lambda$-approximative) basis of $E$, satisfying

$$u_{d'} u_d = u_d u_{d'} = u_d \qquad (d < d'). \tag{18.13}$$

We have the following generalization of § 9, proposition 9.7 and remark 9.12 c):

**Proposition 18.2.** *Let $E$ be a Banach space which has the $\lambda$-duality approximation property for some $\lambda \geqslant 1$. Then $E$ has a shrinking extended commuting approximative basis $\{v_d\}_{d \in \mathscr{D}}$ (i.e., such that $\{v_d^*\}_{d \in \mathscr{D}}$ is an extended commuting approximative basis of $E^*$).*

*Proof.* Let $\mathscr{D}$ be the set of all pairs $d = (G, \Gamma)$, where $G$ and $\Gamma$ are finite-dimensional subspaces of $E$ and $E^*$ respectively, directed by inclusion of components (i.e., $(G_1, \Gamma_1) \geqslant (G_2, \Gamma_2)$ if and only if $G_1 \supset G_2$ and $\Gamma_1 \supset \Gamma_2$). For each $d = (G, \Gamma) \in \mathscr{D}$, let $v_d \in L(E, E)$ be a finite rank operator satisfying § 9, conditions (9.109) — (9.111) with $\varepsilon = 1$. Then, by an argument similar to that of remark 18.1 a) above, we obtain

$$x = \lim_{d \in \mathscr{D}} v_d(x) \qquad (x \in E), \tag{18.14}$$

$$f = \lim_{d \in \mathscr{D}} v_d^*(f) \qquad (f \in E^*), \tag{18.15}$$

$$\|v_d\| \leqslant \lambda + 1 \qquad (d \in \mathscr{D}). \tag{18.16}$$

Finally, let $d < d'$. Then, by § 9, formulae (9.109), (9.110) for $G = v_d(E)$, $\Gamma = v_d^*(E^*)$, we obtain

$$v_{d'}v_d(x) = v_d(x) \qquad (x \in E),$$

$$v_{d'}^*v_d^*(f) = v_d^*(f) \qquad (f \in E^*),$$

and the latter gives

$$v_d v_{d'}(x) = v_d(x) \qquad (x \in E),$$

so we have (18.13) for $\{v_d\}_{d \in \mathscr{D}}$. Thus, $\{v_d\}_{d \in \mathscr{D}}$ is a shrinking extended commuting approximative basis of $E$, which completes the proof.

Combining proposition 18.2 and § 9, proposition 9.8, we obtain

**Corollary 18.1.** *Let $E$ be a Banach space. If $E^*$ has the bounded approximation property, then $E$ has a shrinking extended commuting approximative basis.*

Since for subsets of $L(E, E)$ and $L(E^*, E^*)$ weak and weak* boundedness are equivalent to norm-boundedness (by the principle of uniform boundedness), it is natural to give

*Definition 18.3.* Let $E$ be a Banach space.

a) A net of finite rank operators $\{u_d\}_{d \in \mathscr{D}} \subset L(E, E)$ is called an *extended weak approximative basis* of $E$, if it satisfies

$$f(x) = \lim_{d \in \mathscr{D}} f(u_d(x)) \qquad (x \in E, f \in E^*) \tag{18.17}$$

and (18.2).

b) A net of finite rank operators $\{v_d\}_{d \in \mathscr{D}} \subset L(E^*, E^*)$ is called an *extended weak* approximative basis* of $E^*$, if

$$f(x) = \lim_{d \in \mathscr{D}} v_d(f)(x) \qquad (x \in E, f \in E^*), \tag{18.18}$$

$$\sup_{d \in \mathscr{D}} \|v_d\| < \infty. \tag{18.19}$$

If in (18.2) or (18.19) $\sup_{d \in \mathscr{D}} \|u_d\| \leqslant \lambda$ or $\sup_{d \in \mathscr{D}} \|v_d\| \leqslant \lambda$, we shall use the terms *extended weak $\lambda$-approximative basis* or *extended weak* $\lambda$-approximative basis*, respectively.

We have the following extension of § 9, theorem 9.7:

**Theorem 18.3.** *Let $E$ be a Banach space. The following statements are equivalent:*

1°. *$E$ has an extended approximative basis.*

2°. *$E$ has an extended weak approximative basis.*

3°. *$E^*$ has an extended weak* approximative basis.*

*Proof.* The implications $1° \Rightarrow 2° \Rightarrow 3°$ are obvious (with the same net $\{u_d\}_{d\in\mathcal{D}}$, respectively with $\{u_d^*\}_{d\in\mathcal{D}}$).

Assume now that we have 3°, say $\{v_d\}_{d\in\mathcal{D}}$ is an extended weak* $\lambda$-approximative basis of $E^*$. Then, by § 9, lemma 9.3 b), for each finite-dimensional subspace $\Gamma$ of $E^*$ and each $d\in\mathcal{D}$ there exists a finite rank operator $t_{\Gamma,d}\in L(E,E)$ such that

$$t_{\Gamma,d}^*(h) = v_d(h) \qquad (h\in\Gamma), \tag{18.20}$$

$$\|t_{\Gamma,d}\| \leqslant \lambda + 1. \tag{18.21}$$

Now let $\mathcal{A}$ be the directed set of all pairs $(\Gamma, d)$, where $\Gamma$ is a finite-dimensional subspace of $E^*$ and $d\in\mathcal{D}$, and where $(\Gamma_1, d_1)\geqslant(\Gamma_2, d_2)$ if and only if $\Gamma_1\supset\Gamma_2$ and $d_1\geqslant d_2$. For each $\alpha=(\Gamma, d)\in\mathcal{A}$ let $t_\alpha = t_{\Gamma,d}\in$ $\in L(E,E)$ be a finite rank operator satisfying (18.20), (18.21). We claim that

$$\lim_{\alpha\in\mathcal{A}} f(t_\alpha(x)) = f(x) \qquad (x\in E, f\in E^*). \tag{18.22}$$

Indeed, let $x\in E$, $f\in E^*$ and $\varepsilon>0$ be arbitrary. Then, by (18.18), there exists $d_0 = d_0(x, f, \varepsilon)\in\mathcal{D}$ such that

$$|v_d(f)(x) - f(x)| < \varepsilon \qquad (d\geqslant d_0);$$

hence, putting $\alpha_0 = ([f], d_0)\in\mathcal{A}$ and using (18.20), we obtain

$$|f(t_\alpha(x)) - f(x)| \leqslant |t_\alpha^*(f)(x) - v_d(f)(x)| + |v_d(f)(x) - f(x)| =$$

$$= |v_d(f)(x) - f(x)| < \varepsilon \qquad (\alpha = (\Gamma, d)\geqslant([f], d_0) = \alpha_0),$$

which proves the claim (18.22). Thus*), $3° \Rightarrow 2°$.

Finally, assume that we have 2°, say $\{t_d\}_{d\in\mathcal{D}}$ is an extended weak $\lambda$-approximative basis of $E$. Then, by § 9, proof**) of proposition 9.8, for each finite-dimensional subspace $G$ of $E$ and each $\varepsilon>0$ there exists a convex combination $s = s_{G,\varepsilon} = \sum_{i=1}^n \lambda_i t_{d_i}$, where $\lambda_1,\ldots,\lambda_n\geqslant 0$, $\sum_{i=1}^n \lambda_i = 1$, such that

$$\|s(y) - y\| \leqslant (\lambda+2)\varepsilon\|y\| \qquad (y\in G), \tag{18.3'}$$

$$\|s\| \leqslant \lambda. \tag{18.4'}$$

*) Alternatively, considering the directed set $\mathcal{B}$ of all triples $(\Gamma, d, \varepsilon)$, replacing (18.21) by $\|t_{\Gamma,d,\varepsilon}\|\leqslant\lambda+\varepsilon$ and then defining $w_{\Gamma,d,\varepsilon}$ similarly to $w_{d,\varepsilon}$ of the proof of theorem 18.1, one obtains even an extended weak $\lambda$-approximative basis $\{w_\beta\}_{\beta\in\mathcal{B}}$ of $E$.

**) Indeed, in that part of the proof, we have not used § 9, formula (9.136). Clearly, now $d\geqslant d_0$ and $d_1,\ldots,d_n\geqslant d_0$ of that proof can be replaced by $d\in\mathcal{D}$ and $d_1,\ldots,d_n\in\mathcal{D}$ respectively.

Hence, by theorem 18.1, implication $4° \Rightarrow 1°$, $E$ has an extended $\lambda$-approximative basis (namely, the net $\{s_{G,\varepsilon}\}$ above, with the natural directing of the pairs $(G, \varepsilon)$). Thus, $2° \Rightarrow 1°$, which completes the proof of theorem 18.3.

Let us pass now to the case of finite rank projections. Similarly to the above, it is natural to give

*Definition 18.4*. Let $E$ be a Banach space. A net of projections $\{u_d\}_{d\in\mathscr{D}}\subset L(E, E)$ of finite rank is called

a) an *extended $\Pi$-basis* of $E$, if it satisfies (18.1) and (18.2);

b) an *extended $\pi$-basis* of $E$ if, in addition,

$$u_{d'}u_d = u_d \qquad (d \leqslant d'); \tag{18.23}$$

c) an *extended dual $\pi$-basis* of $E$, if we have (18.1), (18.2) and

$$u_d u_{d'} = u_d \qquad (d \leqslant d'); \tag{18.24}$$

d) an *extended commuting $\pi$-basis* of $E$, if we have (18.1), (18.2) and

$$u_{d'}u_d = u_d u_{d'} = u_d \qquad (d \leqslant d'). \tag{18.25}$$

If in (18.2) $\sup_{d\in\mathscr{D}}\|u_d\|\leqslant\lambda$, we shall use the terms *extended $\Pi_\lambda$-, $\pi_\lambda$-, dual $\pi_\lambda$-*, and *commuting $\pi_\lambda$-basis* respectively. If $\{u_d\}_{d\in\mathscr{D}}$ is an extended $\pi_1$-*basis* of $E$, such that $u_d(E)$ is linearly isometric to $l^\infty_{m_d}$ for some $m_d\in\mathscr{N}$ $(d\in\mathscr{D})$, then $\{u_d\}_{d\in\mathscr{D}}$ is called an *extended $\pi_1^\infty$-basis* of $E$; the terms *extended dual $\pi_1^\infty$-basis* and *extended commuting $\pi_1^\infty$-basis* are defined similarly.

For example, by § 17, theorem 17.3, implication $1° \Rightarrow 2°$, the net of all partial sum operators $\{s_d\}_{d\in\mathscr{D}}$ associated to an extended unconditional basis $\{x_i\}_{i\in I}$ of $E$ (where $\mathscr{D}$ is the collection of all finite subsets of the index set $I$, directed by inclusion), is an extended commuting $\pi$-basis of $E$.

Clearly, every extended $\Pi$- ($\pi$-, etc.) basis of $E$ is an extended approximative basis of $E$.

By § 12, lemma 12.1 a), b) and since all $u_d$ $(d\in\mathscr{D})$ are projections, condition (18.23) is equivalent to

$$u_d(E) \subset u_d'(E) \qquad (d \leqslant d') \tag{18.26}$$

and condition (18.24) is equivalent to

$$\operatorname{Ker} u_d \supset \operatorname{Ker} u_{d'} \qquad (d \leqslant d'). \tag{18.27}$$

We have the following generalization of § 12, theorem 12.1:

**Theorem 18.4.** *Let $E$ be a Banach space and let $\lambda \geqslant 1$.*

*α) The following statements are equivalent:*

*1°. $E$ has an extended $\Pi_\lambda$-basis.*

*2°. $E$ has the $\lambda$-projection approximation property.*

*3°. For every finite-dimensional subspace $G$ of $E$ and every $\delta > 0$ there exists a projection $u = u_{G,\delta} \in L(E, E)$ of finite rank, such that*

$$\|u(x) - x\| < \delta \|x\| \qquad (x \in G), \tag{18.28}$$

$$\|u\| \leqslant \lambda. \tag{18.29}$$

*β) The following conditions are equivalent:*

*1°. $E$ has an extended $\pi_\lambda$-basis.*

*2°. $E$ contains a set of subspaces $\{G_d\}_{d \in \mathscr{D}}$ with the following properties:*

a) $\dim G_d < \infty$ $(d \in \mathscr{D})$;

b) $\{G_d\}_{d \in \mathscr{D}}$ *is directed by inclusion;*

c) $\bigcup_{d \in \mathscr{D}} G_d$ *is dense in $E$;*

d) *for every $d \in \mathscr{D}$ there exists a projection $u_d$ of $E$ onto $G_d$, satisfying $\|u_d\| \leqslant \lambda$.*

*γ) The following conditions are equivalent to each other and they are satisfied when $E$ has an extended $\Pi_\lambda$-basis:*

*1°. For every $\varepsilon > 0$, $E$ has an extended $\Pi_{\lambda+\varepsilon}$-basis.*

*2°. $E$ contains a set of subspaces $\{G_d\}_{d \in \mathscr{D}}$ with the properties* a), b), c) *above and*

d') *for every $d \in \mathscr{D}$ and $\varepsilon > 0$ there exists a finite rank projection $v_{d,\varepsilon}$ on $E$ satisfying $v_{d,\varepsilon}|G_d = I_{G_d}$, $\|v_{d,\varepsilon}\| \leqslant \lambda + \varepsilon$.*

*3°. For every $\varepsilon > 0$, $E$ has an extended $\pi_{\lambda+\varepsilon}$-basis.*

*4°. For every finite-dimensional subspace $G$ of $E$ and every $\varepsilon > 0$ there exists a projection $v = v_{G,\varepsilon} \in L(E, E)$ of finite rank, such that*

$$v|_G = I_G \qquad (\textit{or, equivalently, } v(E) \supset G), \tag{18.30}$$

$$\|v\| \leqslant \lambda + \varepsilon. \tag{18.31}$$

*δ) Consequently, a Banach space $E$ has an extended $\pi$-basis if (and only if) it has an extended $\Pi$-basis.*

*Proof.* α) The proofs of the implications $1° \Rightarrow 2° \Rightarrow 3° \Rightarrow 1°$ are similar to those of theorem 18.1, implications $1° \Rightarrow 3° \Rightarrow 4° \Rightarrow 1°$.

$\beta$) If $\{u_d\}_{d\in\mathscr{D}}$ is an extended $\pi_\lambda$-basis of $E$, then the set $\{G_d\}_{d\in\mathscr{D}} = \{u_d(E)\}_{d\in\mathscr{D}}$ satisfies a) – d) of 2°. Thus, 1° $\Rightarrow$ 2°.

Conversely, assume now that $E$ satisfies 2°, whence $G_d = u_d(E)$ $(d\in\mathscr{D})$. Define a partial order relation in $\mathscr{D}$ by (18.26). Then, by b), $\mathscr{D}$ becomes directed by inclusion. Also, since each $u_d$ is a projection, we have $u_d(u_{d_0}(x)) = u_{d_0}(x)$ $(x\in E,\ d\geqslant d_0)$. Consequently, by c) and $\sup_{d\in\mathscr{D}}\|u_d\|\leqslant\lambda<\infty$, we obtain (18.1), and hence, by a), $\{u_d\}_{d\in\mathscr{D}}$ is an extended $\pi_\lambda$-basis of $E$. Thus, 2° $\Rightarrow$ 1°.

$\gamma$) If we have 1°, then, by the implication $\alpha$) 1° $\Rightarrow$ 3° proved above, we have $\alpha$) 3° with $\lambda$ replaced by $\lambda+\frac{\varepsilon}{2}$ in (18.29). Hence, by § 12, lemma 12.2, we obtain $\gamma$) 4°. Thus, 1° $\Rightarrow$ 4°.

Furthermore, if we have 4°, then any set of subspaces $\{G_d\}_{d\in\mathscr{D}}$ satisfying a), b), c) of 2° (for example, the set of all finite-dimensional subspaces of $E$, indexed by itself) satisfies also d') of 2°. Thus, 4° $\Rightarrow$ 2°.

The proof of the implication 2° $\Rightarrow$ 1° is similar to that of theorem 18.1, implication 2° $\Rightarrow$ 1° (see also remark 18. 1a)).

Assume now again that we have 4°. Fix $\varepsilon>0$. For each finite-dimensional subspace $G$ of $E$ choose a finite rank projection $v=v_G\in L(E,E)$ satisfying (18.30), (18.31) and let $E_G=v_G(E)$. Then the set $\{v_G(E)\}=\{E_G\}$ (where $G$ runs over the set $\mathscr{G}$ of all finite-dimensional subspaces of $E$), directed by inclusion, satisfies the conditions of $\beta$) 2°, with $\lambda$ replaced by $\lambda+\varepsilon$. Indeed, obviously $E=\bigcup_{G\in\mathscr{G}} v_G(E)$ (for any $x\in E$, take $G=[x]\in\mathscr{G}$, then $x=v_G(x)\in v_G(E)$ by (18.30)) and, if $G_1, G_2\in\mathscr{G}$ and $G=[v_{G_1}(E), v_{G_2}(E)]\in\mathscr{G}$, then $v_G(E)\supset G\supset v_{G_1}(E)$, $v_{G_2}(E)$ by (18.30), so $\{v_G(E)\}$ is directed by inclusion. Consequently, by $\beta$) 2° $\Rightarrow$ $\beta$) 1°, $E$ has an extended $\pi_{\lambda+\varepsilon}$-basis. Thus, 4° $\Rightarrow$ 3°.

Also, it is obvious that 3° $\Rightarrow$ 1° and that 1° is satisfied when $E$ has a $\Pi_\lambda$-basis, which proves $\gamma$).

Finally, $\delta$) is an immediate consequence of $\gamma$), which completes the proof of theorem 18.4.

**Theorem 18.5.** *Let $\lambda\geqslant 1$. If a Banach space $E$ has an extended $\Pi_\lambda$- (respectively, $\pi_\lambda$-, dual $\pi_\lambda$-, $\pi_1^\infty$-, or dual $\pi_1^\infty$-) basis $\{u_d\}_{d\in\mathscr{D}}$, then for every separable subspace $G$ of $E$ there exists a sequence $d_1<\ldots<d_n<\ldots$ in $\mathscr{D}$ such that $G\subset F=\left[\bigcup_{n=1}^{\infty} u_{d_n}(E)\right]$ and that $\{u_{d_n}|_F\}$ is a $\Pi_\lambda$- (respectively, $\pi_\lambda$-, dual $\pi_\lambda$-, $\pi_1^\infty$-, dual $\pi_1^\infty$-) basis of $F$. Consequently, if $\{u_d\}_{d\in\mathscr{D}}$ is*

a) *an extended dual* $\pi_\lambda$*-basis of* $E$*, then this* $F = \left[\bigcup_{n=1}^{\infty} u_{d_n}(E)\right]$ *has a commuting* $\pi_{\lambda+\varepsilon}$*-basis for any* $\varepsilon > 0$*, whence also a finite-dimensional decomposition;*

b) *an extended* $\pi_1$*- (or dual* $\pi_1$*-) basis of* $E$*, then* $F$ *has a commuting* $\pi_1$*-basis and hence a monotone finite-dimensional decomposition;*

c) *an extended* $\pi_1^\infty$*- (or dual* $\pi_1^\infty$*-) basis of* $E$*, then* $F$ *has a commuting* $\pi_1^\infty$*-basis and hence a monotone basis* $\{x_n\}$ *satisfying*

$$[x_i]_{i=1}^n \equiv l_n^\infty \qquad (n = 1,2, \ldots). \tag{18.32}$$

*Proof.* The proof of the first statement is similar to the proof of the necessity part of theorem 18.2. The statements a) and b) follow from § 13, propositions 13.3, 13.7 b) and 13.4 a), together with § 13, theorem 13.1, equivalence 5° ⇔ 7° and its proof (or, alternatively, see § 15, proof of theorem 15.3, equivalence 1° ⇔ 4°). Finally, c) follows from § 12, theorem 12.3 and § 13, proposition 13.4 b). This completes the proof.

In particular, obviously, we have

**Corollary 18.2.** *Let* $\lambda \geqslant 1$*. A separable Banach space* $E$ *has an extended* $\Pi_\lambda$*- (respectively,* $\pi_\lambda$*-, dual* $\pi_\lambda$*-,* $\pi_1^\infty$*-, or dual* $\pi_1^\infty$*-) basis if and only if* $E$ *has a* $\Pi_\lambda$*- (respectively,* $\pi_\lambda$*-, dual* $\pi_\lambda$*-,* $\pi_1^\infty$*-, dual* $\pi_1^\infty$*-) basis. In the last three cases* $E$ *has a finite-dimensional decomposition; moreover, in the last two cases* $E$ *has a monotone basis* $\{x_n\}$ *satisfying* (18.32).

Now we shall show that the $C(Q)$ spaces ($Q$ compact) have an extended dual $\pi_1^\infty$-basis (theorem 18.6 below) and the consequence of this fact, announced in § 12, that the separable $C(Q)$ spaces have a monotone basis $\{x_n\}$ satisfying (18.32) (corollary 18.3 below).

Let $Q$ be a compact space. We recall that a finite sequence $\{x_1, \ldots, x_n\} \subset C(Q)$ is called a *peaked partition of unity* (on $Q$) if it is a partition of unity $\left(\text{i.e., the } x_i \text{ are non-negative continuous functions on } Q \text{ such that } \sum_{i=1}^n x_i(q) = 1 \text{ for all } q \in Q\right)$ and $\|x_i\| = 1$ for $i = 1, \ldots, n$. The subspace $[x_1, \ldots, x_n] \subset C(Q)$ spanned by a peaked partition of unity is called a *peaked partition subspace*.

**Lemma 18.1.** *Every peaked partition subspace* $G$ *of* $E = C(Q)$ *(*$Q$ *compact) is linearly isometric to* $l_n^\infty$ *for some* $n$*.*

*Proof.* Let $G = [x_1, \ldots, x_n]$, where $\{x_1, \ldots, x_n\}$ is a peaked partition of unity on $Q$. Since $\|x_j\| = 1$, we can pick $q_j \in Q$ such that

$$x_j(q_j) = 1 \qquad (j = 1, \ldots, n); \tag{18.33}$$

then $1 = \sum_{k=1}^{n} x_k(q_j) \geqslant x_j(q_j) + x_i(q_j) = 1 + x_i(q_j) \geqslant 1 \quad (i \neq j)$, whence

$$x_i(q_j) = 0 \qquad (i \neq j). \tag{18.34}$$

We shall show that for any scalars $\alpha_1, \ldots, \alpha_n$ we have

$$\left\| \sum_{i=1}^{n} \alpha_i x_i \right\| = \max_{1 \leqslant i \leqslant n} |\alpha_i|, \tag{18.35}$$

which will complete the proof. Clearly,

$$\left| \sum_{i=1}^{n} \alpha_i x_i(q) \right| \leqslant \max_{1 \leqslant i \leqslant n} |\alpha_i| \sum_{j=1}^{n} |x_j(q)| = \max_{1 \leqslant i \leqslant n} |\alpha_i| \qquad (q \in Q),$$

whence $\left\| \sum_{i=1}^{n} \alpha_i x_i \right\| \leqslant \max_{1 \leqslant i \leqslant n} |\alpha_i|$. On the other hand, by (18.33), (18.34),

$$\left\| \sum_{i=1}^{n} \alpha_i x_i \right\| \geqslant \left| \sum_{i=1}^{n} \alpha_i x_i(q_j) \right| = |\alpha_j| \qquad (j = 1, \ldots, n),$$

whence $\left\| \sum_{i=1}^{n} \alpha_i x_i \right\| \geqslant \max_{1 \leqslant j \leqslant n} |\alpha_j|$. Consequently, we have (18.35), which completes the proof of lemma 18.1.*)

**Lemma 18.2.** *Let $\{U_1, \ldots, U_n\}$ be an open covering of a compact space $Q$ and let $q_i \in U_i$ $(i = 1, \ldots, n)$ be such that $q_i \neq q_j$ for $i \neq j$. Then there exists a peaked partition of unity $\{x_1, \ldots, x_n\}$ on $Q$ satisfying* (18.33) *and*

$$x_i(q) = 0 \qquad (q \in Q \setminus U_i;\ i = 1, \ldots, n). \tag{18.36}$$

*Proof.* Put

$$V_i = U_i \setminus \{q_1, \ldots, q_{i-1}, q_{i+1}, \ldots, q_n\} \qquad (i = 1, \ldots, n). \tag{18.37}$$

Then $\{V_1, \ldots, V_n\}$ is also an open covering of $Q$. Let $\{x_1, \ldots, x_n\}$ be any partition of unity on $Q$ such that

$$x_i(q) = 0 \qquad (q \in Q \setminus V_i;\ i = 1, \ldots, n); \tag{18.38}$$

---

*) It is also easy to show that, conversely, if $G \subset C(Q)$, $G \equiv l_n^\infty$ and $1 \in G$, then $G$ is a peaked partition subspace of $C(Q)$ ([282], proposition 5.1), but we shall not need this result.

it is well known*) that such a partition exists. Then, since $q_j \in Q \setminus V_i$ $(j \neq i)$, we have $x_i(q_j) = 0$ for $j \neq i$, whence $1 = \sum_{i=1}^{n} x_i(q_j) = x_j(q_j)$ $(j = 1, \ldots, n)$, so (18.33) is satisfied. Since $Q \setminus U_i \subset Q \setminus V_i$, (18.36) also holds, which completes the proof of lemma 18.2.

Now we are ready to prove

**Theorem 18.6.** *Let $Q$ be a compact space. Then $E = C(Q)$ has an extended dual $\pi_1^\infty$-basis.*

*Proof*. By lemma 18.1, it will be enough to prove that $E = C(Q)$ has an extended dual $\pi_1$-basis $\{u_d\}_{d \in \mathscr{D}}$ such that each $u_d(E)$ $(d \in \mathscr{D})$ is a peaked partition subspace. Let $\mathscr{D}$ be the collection of all ordered pairs $(\{U_i\}_{i=1}^n, \{q_i\}_{i=1}^n)$ such that $\{U_i\}_{i=1}^n$ is a minimal open covering (i.e., an open covering which has no proper open subcovering) of $Q$ and that $q_i \in U_i \setminus \bigcup_{j \neq i} U_j$ $(i = 1, \ldots, n)$; such points $q_i$ exist, since otherwise we would have $U_i \subset \bigcup_{j \neq i} U_j$ for some $i$, whence $Q = \bigcup_{j \neq i} U_j$, contradicting the minimality of $\{U_i\}_{i=1}^n$. Partially order $\mathscr{D}$ by $(\{U_i\}_{i=1}^n, \{q_i\}_{i=1}^n) \leqslant (\{V_j\}_{j=1}^m, \{t_j\}_{j=1}^m)$ if and only if $\{V_j\}_{j=1}^m$ refines $\{U_i\}_{i=1}^n$ and $\{q_i\}_{i=1}^n \subset \{t_j\}_{j=1}^m$; then $\mathscr{D}$ is directed by $\geqslant$.

By lemma 18.2, to each $d = (\{U_i\}_{i=1}^n, \{q_i\}_{i=1}^n) \in \mathscr{D}$ we can associate a peaked partition of unity $\{x_1, \ldots, x_n\}$ on $Q$ satisfying (18.33), (18.36) (and hence (18.34) too, so $x_i(q_j) = \delta_{ij}$). For each $d \in \mathscr{D}$ let us define a finite rank projection $u_d \in L(E, E)$ by

$$u_d(x) = \sum_{i=1}^{n} x(q_i) x_i \qquad (x \in E = C(Q)), \tag{18.39}$$

where $\{x_1, \ldots, x_n\}$ is the peaked partition of unity associated to $d = (\{U_i\}_{i=1}^n, \{q_i\}_{i=1}^n)$. Then, clearly,

$$\operatorname{Ker} u_d = \{x \in C(Q) \mid x(q_1) = \ldots = x(q_n) = 0\}, \tag{18.40}$$

whence $\{u_d\}_{d \in \mathscr{D}}$ satisfies (18.27) (or, equivalently, (18.24)). Also, by (18.35), $\|u_d(x)\| = \max_{1 \leqslant i \leqslant n} |x(q_i)| \leqslant \|x\|$ for all $x \in E = C(Q)$, whence $\|u_d\| = 1$ $(d \in \mathscr{D})$. Thus, in order to complete the proof it remains to show that $\{u_d\}_{d \in \mathscr{D}}$ satisfies (18.1).

Let $x \in C(Q)$ and $\varepsilon > 0$. Choose a minimal open covering $\{U_i\}_{i=1}^n$ of $Q$ such that if $\{q, t\} \subset U_i$, then $|x(q) - x(t)| < \varepsilon$, and choose $q_i \in U_i \setminus \bigcup_{j \neq i} U_j$ $(i = 1, \ldots, n)$. Let

$$d = (\{V_j\}_{j=1}^m, \{t_j\}_{j=1}^m) \geqslant (\{U_i\}_{i=1}^n, \{q_i\}_{i=1}^n) = d_0$$

*) See e.g. [360], pp. 352–353.

be arbitrary and let $\{x_1, \ldots, x_m\}$ be the peaked partition of unity associated to $d$. Then for all $q \in Q$,

$$|x(q) - u_d(x)(q)| = \left| x(q) - \sum_{j=1}^{m} x(t_j)x_j(q) \right| =$$

$$= \left| \sum_{j=1}^{m} x_j(q)(x(q) - x(t_j)) \right| \leqslant \sum_{j=1}^{m} x_j(q)\, |x(q) - x(t_j)|.$$

Now, if $q \in V_j$, then, since $t_j \in V_j$ and since $\{V_j\}_{j=1}^m$ refines $\{U_i\}_{i=1}^n$, we have $|x(q) - x(t_j)| < \varepsilon$. On the other hand, if $q \notin V_j$, then, by (18.36) (for $\{V_j\}_{j=1}^m$) we have $x_j(q) = 0$. Consequently,

$$|x(q) - u_d(x)(q)| \leqslant \sum_{j=1}^{m} x_j(q)\, |x(q) - x(t_j)| <$$

$$< \sum_{j=1}^{m} x_j(q)\varepsilon = \varepsilon \qquad (q \in Q,\ d \geqslant d_0),$$

so $\lim_{d \in \mathscr{D}} \|x - u_d(x)\| = 0$, which completes the proof of theorem 18.6.

**Corollary 18.3.** *Let $Q$ be a compact metric space. Then $E = C(Q)$ has a monotone basis $\{x_n\}$ satisfying* (18.32).

*Proof.* By theorem 18.6, $E = C(Q)$ has an extended dual $\pi_1^\infty$-basis. But, since $Q$ is compact metric, $E = C(Q)$ is separable*), and hence, by corollary 18.2, $E$ has a monotone basis $\{x_n\}$ satisfying (18.32), which completes the proof.

The spaces $C_p$ have the following "universal complement" property, which is an extension of § 12, theorem 12.4:

**Theorem 18.7.** *Let $1 \leqslant p \leqslant \infty$. If a Banach space $E$ has an extended approximative basis, then $E \times C_p$ has an extended $\pi$-basis.*

*Proof.* The proof is similar to that of § 12, theorem 12.4, with the following differences: Instead of the sequence of subspaces $\{H_n\}$ defined by (12.46) we consider now the set of all subspaces of $E \times C_p$ of the form

$$H = H_{F,n} = F \times (G_1 \times \ldots \times G_n \times \{0\} \times \{0\} \times \ldots)_{l^p}, \quad (18.41)$$

where $F$ is a finite-dimensional subspace of $E$ and $n = 1,2, \ldots$ (and where $\{G_n\}$ is as in § 12, definition 12.3), directed by inclusion. For each such subspace $H$ we obtain a finite rank projection $v = v_H$ on $(E \times C_p)_{l^\infty}$, satisfying $v|_H = I_H$ and $\|v\| \leqslant \lambda'$, where $\lambda' (> \lambda)$ does not depend on $H$. Hence, by theorem 18.4 $\gamma$), implication $2° \Rightarrow 1°$, $E$ has an extended $\pi$-basis, which completes the proof.

Similarly to definition 18.3, one can also introduce *extended weak $\Pi$- ($\Pi_\lambda$-, $\pi$-, $\pi_\lambda$-, dual $\pi$-, dual $\pi_\lambda$-,* etc.) *bases* and *extended weak**

*) See e.g. [140], Ch. I, § 9, proposition 16 or [360], Ch. II, § 7, proposition 7.6.2.

$\Pi$- ($\Pi_\lambda$-, etc.) *bases*, replacing in (18.1) norm-convergence by weak convergence, respectively weak* convergence.

We have the following partial extension of § 13, proposition 13.8:

**Proposition 18.3.** *Let $E$ be a Banach space with an extended weak dual $\pi_\lambda$-basis $\{u_d\}_{d\in\mathscr{D}}$. Then $E$ has an extended $\Pi_{\lambda^2+2\lambda}$-basis and hence, for each $\varepsilon > 0$, an extended $\pi_{\lambda^2+2\lambda+\varepsilon}$-basis.*

*Proof.* By the above proof of theorem 18.3, implication $2^\circ \Rightarrow 1^\circ$, for each finite-dimensional subspace $G$ of $E$ and each $\varepsilon > 0$ there exists a convex combination $w = \sum_{i=1}^{n} \alpha_i u_{d_i}$, where $\alpha_1, \ldots, \alpha_n \geqslant 0$, $\sum_{i=1}^{n} \alpha_i = 1$, such that

$$\|y - w(y)\| \leqslant (\lambda + 2)\varepsilon\|y\| \qquad (y \in G), \tag{18.42}$$

$$\|w\| \leqslant \lambda. \tag{18.43}$$

We may (and shall) assume, without loss of generality, that here $d_n \geqslant d_1, \ldots, d_{n-1}$ (indeed, since $\mathscr{D}$ is directed, we can add to any $\{d_1, \ldots, d_n\}$ which fails to have this property, an element $d_{n+1} \in \mathscr{D}$ with $d_{n+1} \geqslant d_1, \ldots, d_n$ and then write $w = \sum_{i=1}^{n} \alpha_i u_{d_i} + 0.u_{d_{n+1}}$). Let

$$v = w + u_{d_n} - u_{d_n} w. \tag{18.44}$$

Then, by (18.42), (18.43) and $\|u_{d_n}\| \leqslant \lambda$,

$$\|y - v(y)\| \leqslant \|y - w(y)\| + \|u_{d_n}\| \; \|y - w(y)\| \leqslant$$

$$\leqslant (1 + \lambda)\|y - w(y)\| \leqslant (1 + \lambda)(\lambda + 2)\varepsilon\|y\| \qquad (y \in G),$$

$$\|v\| \leqslant \lambda + \lambda + \lambda^2 = \lambda^2 + 2\lambda.$$

Moreover, by $d_n \geqslant d_1, \ldots, d_{n-1}$ and (18.24) we have $wu_{d_n} = w$, whence as in § 13, proof of proposition 13.8, we obtain $v^2 = v$, so $v$ is a projection. Consequently, by theorem 18.4 $\alpha$), implication $3^\circ \Rightarrow 1^\circ$, $E$ has an extended $\Pi_{\lambda^2+2\lambda}$-basis, whence, by theorem 18.4 $\gamma$), also an extended $\pi_{\lambda^2+2\lambda+\varepsilon}$-basis for each $\varepsilon > 0$, which completes the proof.

We have the following partial extension of § 14, theorem 14.1:

**Theorem 18.8.** *Let $E$ be a Banach space.*

a) *If $E$ has an extended $\Pi$-basis and $E^*$ has the bounded approximation property, then $E$ has a shrinking extended commuting $\pi$-basis $\{q_d\}_{d\in\mathscr{D}}$ (i.e., such that $\{q_d^*\}_{d\in\mathscr{D}}$ is an extended commuting $\pi$-basis of $E^*$).*

b) *If $E^*$ has an extended $\Pi$-basis, then $E$ has a shrinking extended commuting $\pi$-basis.*

c) *If $E^*$ has an extended weak* $\pi_\lambda$-basis, then $E$ has an extended weak dual $\pi_{\lambda+\varepsilon}$-basis, for each $\varepsilon > 0$. Consequently, $E$ has an extended $\pi_{\lambda^2+2\lambda+\varepsilon}$-basis, for each $\varepsilon > 0$.*

*Proof.* a) Let $\mathscr{D}$ be the set of all pairs $d = (G, \Gamma)$, where $G$ and $\Gamma$ are finite-dimensional subspaces of $E$ and $E^*$ respectively, ordered by inclusion of components.Then, by § 14, proposition 14.1 a), for each $d = (G, \Gamma) \in \mathscr{D}$, there exists a finite rank projection $q = q_d$ on $E$, satisfying § 14, conditions (14.4) — (14.6). Hence, by the above proof of proposition 18.2, $\{q_d\}_{d\in\mathscr{D}}$ and $\{q_d^*\}_{d\in\mathscr{D}}$ are extended commuting $\pi$-bases of $E$ and $E^*$ respectively.

b) The proof is similar to the above proof of part a), using now § 14, proposition 14.2.

c) Assume that $\{v_d\}_{d\in\mathscr{D}}$ is an extended weak* $\pi_\lambda$-basis of $E^*$ and let $\varepsilon > 0$. Then, by § 14, lemma 14.1, for each finite-dimensional subspace of $E^*$ and each $d \in \mathscr{D}$, there exists a finite rank projection $t_{\Gamma,d}$ on $E$ such that

$$t^*_{\Gamma,d}(E^*) = v_d(E^*), \tag{18.45}$$

$$t^*_{\Gamma,d}(h) = v_d(h) \qquad (h \in \Gamma), \tag{18.46}$$

$$\|t_{\Gamma,d}\| \leqslant \lambda + \varepsilon. \tag{18.47}$$

Now let $\mathscr{A}$ be the directed set of all pairs $(\Gamma, d)$, where $\Gamma$ is a finite-dimensional subspace of $E^*$ and $d \in \mathscr{D}$, and where $(\Gamma_1, d_1) \geqslant (\Gamma_2, d_2)$ if and only if $\Gamma_1 \supset \Gamma_2$ and $d_1 \geqslant d_2$. For each $\alpha = (\Gamma, d) \in \mathscr{A}$ let $t_\alpha = t_{\Gamma,d}$ be a finite rank projection on $E$, satisfying (18.45)—(18.47). Then, by the above proof of theorem 18.3, implication $3° \Rightarrow 2°$, we have (18.22). Also, if $\alpha = (\Gamma, d) \leqslant (\Gamma', d') = \alpha'$, then, by (18.45) and since $\{v_d\}_{d\in\mathscr{D}}$ is a weak* $\pi$-basis, $t^*_\alpha(E^*) = v_d(E^*) \subset v_{d'}(E^*) = t^*_{\alpha'}(E^*)$, whence

$$t^*_{\alpha'} t^*_\alpha(f) = t^*_\alpha(f) \qquad (f \in E^*).$$

Therefore,

$$t_\alpha t_{\alpha'} = t_\alpha \qquad (\alpha \leqslant \alpha'),$$

so $\{t_\alpha\}_{\alpha\in\mathscr{A}}$ is an extended weak dual $\pi_{\lambda+\varepsilon}$-basis of $E$. Consequently, by proposition 18.3, $E$ has an extended $\pi_{\lambda^2+2\lambda+\varepsilon}$-basis, for each $\varepsilon > 0$, which completes the proof of theorem 18.8.

*Remark 18.2.* From theorem 18.8 b) it follows that if the second conjugate space $E^{**}$ has an extended $\Pi$-basis, then $E$ has an extended commuting $\pi$-basis. One can also show*) that if $E^{**}$ has an extended $\Pi_\lambda$-basis, then $E$ has an extended $\pi_{\lambda+\varepsilon}$-basis, for each $\varepsilon > 0$.

---

*) See [187], corollary 3.4.

If we drop the assumption of finite-dimensionality of the ranges in the definition of an extended commuting $\pi$-basis, consisting of distinct non-zero projections, we arrive at

*Definition 18.5.* Let $E$ be a Banach space. An infinite net of projections $\{u_d\}_{d\in\mathscr{D}}\subset L(E, E)$, with $0\neq u_d\neq u_{d'}$ for $d\neq d'$, is called an *extended resolution of the identity of E* (or, briefly, *of* $I_E$), if it satisfies (18.1), (18.2) and (18.25) and if $u_d\neq I_E$ $(d\in\mathscr{D})$.

The last condition in definition 18.5 is necessary, since otherwise (18.1) would be trivial; in definitions 18.1—18.4 this condition was automatically satisfied, since there each $u_d$ was of finite rank.

The above proof of the necessity part of theorem 18.2 no longer yields a result similar to theorem 18.5 for extended resolutions of the identity, since $S_{\left[\bigcup_{i=1}^{n-1} u_{d_i}(E)\right]}$ need not be compact, whence (18.7) may fail, so (18.9), (18.10) and even $u_{d_n}(F)\subset F$ may fail. However, by using only the part of that argument involving $\{y_n\}$, we obtain that a result similar to corollary 18.2 holds for extended resolutions of the identity. Namely, we have

**Proposition 18.4.** *If $\{u_d\}_{d\in\mathscr{D}}$ is an extended resolution of the identity of a separable Banach space E, then there exists a sequence $d_1<\dots$ $\dots<d_n<\dots$ in $\mathscr{D}$ such that $\{u_{d_n}\}$ is a resolution of $I_E$. Hence, a separable Banach space E has an extended resolution of the identity if and only if it has a resolution of the identity.*

*Proof.* Let $\{y_n\}$ be a dense sequence in $S_E=\{x\in E\mid \|x\|\leqslant 1\}$. Choose $d_1\in\mathscr{D}$ such that $\|u_{d_1}(y_1)-y_1\|<1$. If $d_{n-1}\in\mathscr{D}$ has been chosen' select $d_n>d_{n-1}$ in $\mathscr{D}$ such that

$$\|u_{d_n}(y_i)-y_i\|<\frac{1}{n}\qquad (i=1,\dots,n). \tag{18.48}$$

Then $\lim\limits_{n\to\infty} u_{d_n}(y_i)=y_i$ $(i=1,2,\dots)$, whence, by (18.2) and $[y_n]=E$, we obtain $\lim\limits_{n\to\infty} u_{d_n}(x)=x$ $(x\in E)$. Hence, by (18.25) for $d, d'\in\{d_n\}$, it follows that $\{u_{d_n}\}$ is a resolution of the identity of $E$, which completes the proof of proposition 18.4.

Some duality results for resolutions of the identity can be generalized, with a similar proof, to extended resolutions of the identity. For example, we have the following generalization of § 15, theorem 15.6 and part of theorem 15.11 $\left(\text{using there } u_n=\sum_{i=1}^{n} v_i \text{ for } n=1,2,\dots\right)$, with the obvious definition of an extended weak* resolution of $I_{E^*}$:

**Theorem 18.9.** *Let $\{u_d\}_{d\in\mathscr{D}}$ be an extended resolution of the identity of a Banach space E. Then $\{u_d^*|_V\}_{d\in\mathscr{D}}$ is an extended resolution of the* *identity of* $V = \overline{\bigcup_{d \in \mathscr{D}} u_d^*(E^*)}$ *and* $\{u_d^*\}_{d \in \mathscr{D}}$ *is an extended weak* resolution of the identity of* $E^*$.

One can also introduce various special classes of extended resolutions of the identity. For example, one can give the following generalizations of § 15, definitions 15.13, 15.15 and 15.16:

*Definition 18.6*. An extended resolution $\{u_d\}_{d \in \mathscr{D}}$ of the identity of a Banach space $E$ is called

a) *monotone*, if $\|u_d\| = 1$ $(d \in \mathscr{D})$;

b) *shrinking*, if $\overline{\bigcup_{d \in \mathscr{D}} u_d^*(E^*)} = E^*$;

c) *boundedly complete*, if for every bounded net $\{z_d\}_{d \in \mathscr{D}} \subset E$ such that the limits

$$\lim_{d' \in \mathscr{D}} u_d(z_{d'}) = x_d \qquad (d \in \mathscr{D}) \tag{18.49}$$

exist, there is an element $x \in E$ satisfying

$$u_d(x) = x_d \qquad (d \in \mathscr{D}). \tag{18.50}$$

We note that in the separable case the latter condition amounts actually to that given in § 15, theorem 15.15, statement 2° $\left(\text{using there } u_n = \sum_{i=1}^{n} v_i \text{ and } x_n = \sum_{i=1}^{n} y_i \text{ for } n = 1, 2, \ldots\right)$. We have the following generalization of part of § 15, theorem 15.14:

**Theorem 18.10.***) *Let* $\{u_d\}_{d \in \mathscr{D}}$ *be an extended resolution of the identity of a Banach space* $E$. *If each* $u_d(E)$ $(d \in \mathscr{D})$ *is reflexive and if* $\{u_d\}_{d \in \mathscr{D}}$ *is boundedly complete, then* $E$ *is canonically isomorphic to* $V^*$, *where* $V = \overline{\bigcup_{d \in \mathscr{D}} u_d^*(E^*)}$.

## § 19. Transfinite decompositions. Transfinite Schauder decompositions. Ordinal resolutions of the identity

In this section we shall consider some concepts which may be regarded both as transfinite generalizations of decompositions, Schauder decompositions and resolutions of the identity and as generalizations of transfinite (whence also of usual) bases. We shall show that they exist in large classes of non-separable Banach spaces (containing,

---

*) The converse is also true (see the Notes and remarks).

in particular, all reflexive non-separable Banach spaces and the conjugate spaces of very smooth non-separable Banach spaces). Similarly to § 15, throughout the sequel we shall assume, without any special mention, that dim $E = \infty$.

*Definition 19.1.* A transfinite sequence $\{G_\lambda\}_{\lambda<\vartheta}$, with $\vartheta \geqslant \omega$, of linear subspaces of a Banach space $E$, such that $G_\lambda \neq \{0\}$ $(\lambda < \vartheta)$, is called a *transfinite decomposition*, or a *transfinite basis of subspaces* (of type $\vartheta$) of $E$, if for every $x \in E$ there exists a *unique* transfinite sequence $\{y_\lambda\}_{\lambda<\vartheta} \subset \subset E$ with $y_\lambda \in G_\lambda$ $(\lambda < \vartheta)$, such that*)

$$x = \sum_{\lambda<\vartheta} y_\lambda. \tag{19.1}$$

In other words, $\{G_\lambda\}_{\lambda<\vartheta}$ is a transfinite decomposition of $E$ if and only if $E$ is the "transfinite direct sum of the subspaces $G_\lambda$":

$$E = \sum_{\lambda<\vartheta} \oplus G_\lambda. \tag{19.2}$$

Clearly, in the particular case when $\vartheta = \omega$, this reduces to the definition of usual decompositions (§ 15, definition 15.1). It is also clear that every transfinite basis $\{x_\lambda\}_{\lambda<\vartheta}$ of $E$ (§ 17, definition 17.7) generates a (one-dimensional) transfinite decomposition $G_1 = [x_1]$, $G_2 = [x_2]$, ...

*Definition 19.2.* Let $\{G_\lambda\}_{\lambda<\vartheta}$ be a transfinite decomposition of a Banach space $E$. The transfinite sequence of linear projections $v_\lambda : E \to G_\lambda$ defined by

$$v_\varkappa(x) = y_\varkappa \qquad (x = \sum_{\lambda<\vartheta} y_\lambda \in E, \quad \varkappa < \vartheta), \tag{19.3}$$

where $y_\lambda \in G_\lambda$ $(\lambda < \vartheta)$, is called *the associated transfinite sequence of coordinate projections* (we shall write: a.t.s.c.p.).

Thus, if $\{G_\lambda\}_{\lambda<\vartheta}$ is a transfinite decomposition of $E$, with the a.t.s. c.p. $\{v_\lambda\}_{\lambda<\vartheta}$, then every $x \in E$ has a unique expansion of the form

$$x = \sum_{\lambda<\vartheta} v_\lambda(x). \tag{19.4}$$

Also, as in the case of usual decompositions, we have

$$v_\lambda v_\mu = \delta_{\lambda\mu} v_\lambda = \delta_{\lambda\mu} v_\mu \qquad (\lambda, \mu < \vartheta). \tag{19.5}$$

*) For the definition of convergence and sum of transfinite series, see § 17.

In the particular case when $\{x_\lambda\}_{\lambda<\vartheta}$ is a transfinite basis of $E$, with the a.t.s.c.f. $\{f_\lambda\}_{\lambda<\vartheta} \subset E^*$ (§ 17, definition 17.8) and $G_1 = [x_1]$, $G_2 = [x_2]$, ... is the (one-dimensional) transfinite decomposition of $E$ generated by $\{x_\lambda\}_{\lambda<\vartheta}$, we have, clearly,

$$v_\lambda(x) = f_\lambda(x)x_\lambda \qquad (x \in E, \lambda < \vartheta). \tag{19.6}$$

*Definition 19.3.* A transfinite decomposition $\{G_\lambda\}_{\lambda<\vartheta}$ of a Banach space $E$ is said to be a *transfinite Schauder decomposition* (or a *transfinite Schauder basis of subspaces*) of $E$, if all coordinate projections $v_\lambda$ are continuous on $E$ (i.e., if $v_\lambda \in L(E, E)$ for $\lambda < \vartheta$).

Combining the proofs of § 17, proposition 17.3 (using the compactness of $[1, \vartheta]$ and the continuity of the function $S$: $[1, \vartheta] \to E$ occurring in the definition of $\sum_{\lambda<\vartheta} y_\lambda$) and of § 15, proposition 15.2, we obtain

**Proposition 19.1.** *Let $\{G_\lambda\}_{\lambda<\vartheta}$ be a transfinite sequence of closed linear subspaces of a Banach space E and let $D_1^{(\vartheta)}$ be the linear space of transfinite sequences of scalars*

$$D_1^{(\vartheta)} = \{\{y_\lambda\}_{\lambda<\vartheta} \subset E \mid y_\lambda \in G_\lambda \ \ (\lambda < \vartheta), \sum_{\lambda<\vartheta} y_\lambda \text{ converges}\}, \tag{19.7}$$

*endowed with the norm*

$$\|\{y_\lambda\}_{\lambda<\vartheta}\| = \sup_{\lambda \leqslant \vartheta} \|\sum_{\varkappa<\lambda} y_\varkappa\|. \tag{19.8}$$

*Then $D_1^{(\vartheta)}$ is a Banach space.*

Similarly to § 17, proposition 17.4 and § 15, proposition 15.3, we have

**Proposition 19.2.** *Let $\{G_\lambda\}_{\lambda<\vartheta}$ be a transfinite decomposition of a Banach space E, such that each $G_\lambda$ $(\lambda < \vartheta)$ is closed, and let $\{v_\lambda\}_{\lambda<\vartheta}$ be the a.t.s.c.p. to $\{G_\lambda\}_{\lambda<\vartheta}$. Then*

a) *The Banach space $D_1^{(\vartheta)}$ introduced in proposition 19.1 is isomorphic to E, by the mapping*

$$w\colon \{y_\lambda\}_{\lambda<\vartheta} \to \sum_{\lambda<\vartheta} y_\lambda. \tag{19.9}$$

b) *The numbers*

$$|||x||| = \sup_{\lambda \leqslant \vartheta} \|\sum_{\varkappa<\lambda} v_\varkappa(x)\| \qquad (x \in E) \tag{19.10}$$

*define a norm on the space E, equivalent to the initial norm of E.*

Using these two propositions and e.g. the first proof of § 15, theorem 15.1 $\left(\text{with the difference that the inequality } \|v_n(x)\| \leqslant \left\|\sum_{i=1}^{n} v_i(x)\right\| + \right.$ $\left. + \left\|\sum_{i=1}^{n-1} v_i(x)\right\| \text{ is replaced by}\right.$

$$\|v_\lambda(x)\| \leqslant \|S(\lambda + 1)\| + \|S(\lambda)\|,$$

where $S\colon [1, \vartheta] \to E$ is the continuous function occurring in the definition of $\left.\sum_{\lambda<\vartheta} v_\lambda(x)\right)$, we obtain

**Theorem 19.1.** *A transfinite decomposition $\{G_\lambda\}_{\lambda<\vartheta}$ of a Banach space $E$ is a transfinite Schauder decomposition of $E$ if and only if each $G_\lambda$ $(\lambda < \vartheta)$ is closed.*

From now on we shall consider only transfinite Schauder decompositions (so each $G_\lambda$ will be closed) and, as in the preceding sections, unless specified otherwise, by "subspace" we shall mean: closed linear subspace.

By § 17, lemma 17.2, one can also define partial sum operators for transfinite Schauder decompositions, as follows:

*Definition 19.4.* Let $\{G_\lambda\}_{\lambda<\vartheta}$ be a transfinite Schauder decomposition of a Banach space $E$, with the a.t.s.c.p. $\{v_\lambda\}_{\lambda<\vartheta} \subset L(E, E)$. The transfinite sequence of linear operators $\{u_\lambda\}_{\lambda\leqslant\vartheta}$, where

$$u_\lambda(x) = \sum_{\varkappa<\lambda} v_\varkappa(x) \qquad (x \in E,\ \lambda \leqslant \vartheta), \tag{19.11}$$

is called the *transfinite sequence of partial sum operators associated to* $\{G_\lambda\}_{\lambda<\vartheta}$.

Thus, by § 17, lemma 17.2,

$$u_1 = 0,\ u_\vartheta = I_E,\ u_{\lambda+1}(x) = u_\lambda(x) + v_\lambda(x) \quad (x \in E,\ \lambda < \vartheta), \tag{19.12}$$

and the mapping $\lambda \to u_\lambda(x)$ (from $[1, \vartheta]$ into $E$) is continuous for each fixed $x \in E$. Furthermore, from (19.11) and proposition 19.2 b) it follows that for any transfinite Schauder decomposition $\{G_\lambda\}_{\lambda<\vartheta}$ of $E$ *the partial sum operators $u_\lambda$ are uniformly bounded* (hence continuous) on $E$, that is,

$$\sup_{\lambda\leqslant\vartheta} \|u_\lambda\| < \infty. \tag{19.13}$$

Note that $G_\lambda \neq \{0\}$ is equivalent to $v_\lambda \neq 0$ $(\lambda < \vartheta)$ which, by (19.12), s equivalent to $u_{\lambda+1} \neq u_\lambda$ $(\lambda < \vartheta)$. Also, from the orthogonality relations (19.5) it follows that $\{u_\lambda\}_{\lambda\leqslant\vartheta}$ satisfies the commutativity relations

$$u_\lambda u_\varkappa = u_\varkappa u_\lambda = u_{\min(\lambda,\varkappa)} \qquad (\lambda, \varkappa \leqslant \vartheta); \tag{19.14}$$

thus, $\lambda \to u_\lambda$ is a strictly ascending*) continuous function from $[1, \vartheta]$ into $\mathscr{P}$, the set of all continuous linear projections on $E$, endowed with the natural order and with the topology of pointwise convergence.

We shall use the following simple

**Lemma 19.1.** *If $x = \sum\limits_{\lambda<\vartheta} y_\lambda$ in a Banach space $E$ and if $v \in L(E, F)$, where $F$ is a Banach space, then*

$$v(x) = v(\sum_{\lambda<\vartheta} y_\lambda) = \sum_{\lambda<\vartheta} v(y_\lambda). \tag{19.15}$$

*Proof.* By $x = \sum\limits_{\lambda<\vartheta} y_\lambda$, there exists a continuous function $S\colon [1, \vartheta] \to E$ such that $S(1) = 0$, $S(\vartheta) = x$ and $S(\lambda+1) = S(\lambda) + y_\lambda$ for $\lambda < \vartheta$. Define $S'\colon [1, \vartheta] \to F$ by $S' = v \circ S$. Then $S'$ is continuous and

$$S'(1) = v(S(1)) = v(0) = 0,$$

$$S'(\vartheta) = v(S(\vartheta)) = v(x),$$

$$S'(\lambda+1) = v(S(\lambda+1)) = v(S(\lambda) + y_\lambda) =$$

$$= v(S(\lambda)) + v(y_\lambda) = S'(\lambda) + v(y_\lambda) \qquad (\lambda < \vartheta),$$

so we have (19.15), which completes the proof of lemma 19.1.

Now we are ready to prove the following characterization of transfinite Schauder decompositions and of Banach spaces admitting transfinite Schauder decompositions, which generalizes part of § 15, theorem 15.3 (see also § 17, proposition 17.8 for a related result):

**Theorem 19.2.** *For a Banach space $E$ the following statements are equivalent:*

1°. *$E$ has a transfinite Schauder decomposition $\{G_\lambda\}_{\lambda<\vartheta}$.*

2°. *There exists a transfinite sequence of continuous linear projections $\{u_\lambda\}_{\lambda\leq\vartheta}$ on $E$, with the following properties:*

a) $u_1 = 0$, $u_\vartheta = I_E$;

b) *we have* $u_{\lambda+1} \neq u_\lambda$ $(\lambda < \vartheta)$ *and* (19.14);

c) *for each fixed $x \in E$, the mapping $\lambda \to u_\lambda(x)$ of $[1, \vartheta]$ into $E$ is continuous***).

---

*) See § 17, the remark made after formula (17.42).

**) In other words, the mapping $\lambda \to u_\lambda$ is a strictly ascending continuous function from $[1, \vartheta]$ into $\mathscr{P}$, the set of all continuous linear projections on $E$, with the natural order and with the topology of pointwise convergence, such that $u_1 = 0$, $u_\vartheta = I_E$.

*The subspaces* $G_\lambda$ *and the projections* $u_\lambda$ *are related by* $u_\lambda(x) = \sum_{\varkappa<\lambda} y_\varkappa$ $(x = \sum_{\varkappa<\vartheta} y_\varkappa \in E, \lambda \leqslant \vartheta)$, *where* $y_\varkappa \in G_\varkappa$ $(\varkappa<\vartheta)$, *and by* $G_\lambda = (u_{\lambda+1} - u_\lambda)(E)$ $(\lambda < \vartheta)$.

*Proof.* The implication $1° \Rightarrow 2°$ was observed above.

Conversely, assume that we have 2° and let

$$G_\lambda = (u_{\lambda+1} - u_\lambda)(E) \qquad (\lambda < \vartheta). \quad (19.16)$$

Then, by $u_{\lambda+1} \neq u_\lambda$, we have $G_\lambda \neq \{0\}$ $(\lambda < \vartheta)$. We claim that

$$x = \sum_{\lambda<\vartheta} (u_{\lambda+1} - u_\lambda)(x) \qquad (x \in E). \quad (19.17)$$

Indeed, let $x \in E$ and define $S$: $[1, \vartheta] \to E$ by $S(\lambda) = u_\lambda(x)$. Then, by 2° a), $S(1) = 0$, $S(\vartheta) = x$. Also, by 2° c), $S$ is continuous and, clearly, $S(\lambda + 1) = u_{\lambda+1}(x) = u_\lambda(x) + (u_{\lambda+1} - u_\lambda)(x) = S(\lambda) + (u_{\lambda+1} - u_\lambda)(x)$, which proves the claim (19.17).

Finally, we show that the expansion (19.17) of each $x \in E$ is unique. Let $y_\lambda \in G_\lambda = (u_{\lambda+1} - u_\lambda)(E)$, $\sum_{\lambda<\vartheta} y_\lambda = 0$. Observe that if we put, for brevity, $v_\lambda = u_{\lambda+1} - u_\lambda$ $(\lambda < \vartheta)$, then, by (19.14) we have (19.5). Hence, using also lemma 19.1, we obtain

$$y_\varkappa = v_\varkappa(y_\varkappa) = \sum_{\lambda<\vartheta} v_\varkappa v_\lambda(y_\lambda) = v_\varkappa(\sum_{\lambda<\vartheta} v_\lambda(y_\lambda)) =$$

$$= v_\varkappa(\sum_{\lambda<\vartheta} y_\lambda) = v_\varkappa(0) = 0 \qquad (\varkappa < \vartheta),$$

so $\{G_\lambda\}_{\lambda<\vartheta}$ defined by (19.16) is a transfinite Schauder decomposition of $E$, which completes the proof of theorem 19.2.

Similarly to § 15, definition 15.6, it is natural to give

*Definition 19.5.* Let $E$ be a Banach space. A transfinite sequence of projections $\{u_\lambda\}_{\lambda\leqslant\vartheta} \subset L(E, E)$ is called an *ordinal resolution of the identity of* $E$, or shortly, an *ordinal resolution of* $I_E$ (of type $\vartheta + 1$) if it satisfies condition 2° of theorem 19.2.

In order to give an intrinsic characterization of transfinite Schauder decompositions, let us first prove the following generalization of Ch. I, § 8, proposition 8.1 a) (see also § 15, third proof of theorem 15.1 and § 17, proposition 17.5):

**Proposition 19.3.** *Let $\{G_\lambda\}_{\lambda<\vartheta}$ be a transfinite sequence of closed linear subspaces of a Banach space E and define $D_1^{(\vartheta)}$ as in proposition 19.1. Then the transfinite sequence of closed linear subspaces $\{F_\lambda\}_{\lambda<\vartheta}$ of $D_1^{(\vartheta)}$ defined by*

$$F_\lambda = \{\{y_\varkappa\}_{\varkappa<\vartheta} \in D_1^{(\vartheta)} \mid y_\varkappa = 0 \ (\varkappa \neq \lambda)\} \quad (\lambda < \vartheta) \tag{19.18}$$

*is a transfinite Schauder decomposition of $D_1^{(\vartheta)}$.*

*Proof.* Let $\{y_\lambda\}_{\lambda<\vartheta} \in D_1^{(\vartheta)}$ and for each $\lambda<\vartheta$ define $\{z_\varkappa^{(\lambda)}\}_{\varkappa<\vartheta} \in F_\lambda$ by

$$z_\varkappa^{(\lambda)} = \begin{cases} y_\lambda & \text{for } \varkappa = \lambda \\ 0 & \text{for } \varkappa \neq \lambda. \end{cases} \tag{19.19}$$

We shall show that

$$\{y_\lambda\}_{\lambda<\vartheta} = \sum_{\lambda<\vartheta} \{z_\varkappa^{(\lambda)}\}_{\varkappa<\vartheta} \tag{19.20}$$

and that this expansion is unique, which will complete the proof.

Define a function $S_0$: $[1, \vartheta] \to D_1^{(\vartheta)}$ by

$$(S_0(\lambda))_\varkappa = \begin{cases} y_\varkappa & \text{for } \varkappa < \lambda \\ 0 & \text{for } \lambda \leqslant \varkappa < \vartheta, \end{cases} \tag{19.21}$$

where $(S_0(\lambda))_\varkappa$ is the $\varkappa$-th component of $S_0(\lambda)$. Then

$$S_0(1) = 0, \ S_0(\vartheta) = \{y_\lambda\}_{\lambda<\vartheta}, \ S_0(\lambda+1) = S_0(\lambda) + \{z_\varkappa^{(\lambda)}\}_{\varkappa<\vartheta} \quad (\lambda < \vartheta),$$

so in order to prove (19.20) it remains to show that $S_0$ is continuous. The proof of this fact is similar to that of the corresponding statement in § 17, proof of proposition 17.5 (using now the continuous function $S$: $[1, \vartheta] \to E$ occurring in the definition of $\sum_{\lambda<\vartheta} y_\lambda$). Thus, (19.20) holds.

Assume now that $\sum_{\lambda<\vartheta} \{z_\varkappa^{(\lambda)}\}_{\varkappa<\vartheta} = 0$ in $D_1^{(\vartheta)}$, where $\{z_\varkappa^{(\lambda)}\}_{\varkappa<\vartheta} \in F_\lambda$ $(\lambda < \vartheta)$, so each $z_\varkappa^{(\lambda)}$ is of the form (19.19) with $y_\lambda \in G_\lambda$. Let $S_0$: $[1, \vartheta] \to D_1^{(\vartheta)}$ be the continuous function occurring in the definition of $\sum_{\lambda<\vartheta} \{z_\varkappa^{(\lambda)}\}_{\varkappa<\vartheta}$, hence

$$S_0(1) = 0, \ S_0(\vartheta) = \sum_{\lambda<\vartheta} \{z_\varkappa^{(\lambda)}\}_{\varkappa<\vartheta} = 0, \ S_0(\lambda+1) = S_0(\lambda) + \{z_\varkappa^{(\lambda)}\}_{\varkappa<\vartheta} \ (\lambda<\vartheta).$$

Then, similarly to § 17, proof of proposition 17.5, it follows that $S_0$ satisfies (19.21) and that, consequently, $\{z_\varkappa^{(\lambda)}\}_{\varkappa<\vartheta} = 0$ in $F_\lambda$ $(\lambda < \vartheta)$, which completes the proof of proposition 19.3.

Now we can prove the following intrinsic characterization of transfinite Schauder decompositions, which is a generalization of § 15, theorem 15.5 and § 17, theorem 17.8:

**Theorem 19.3.** *Let $E$ be a Banach space and $\{G_\lambda\}_{\lambda<\vartheta}$ a transfinite sequence of subspaces of $E$, with $G_\lambda \neq \{0\}$ $(\lambda < \vartheta)$, $[\bigcup_{\lambda<\vartheta} G_\lambda] = E$. The following statements are equivalent:*

1°. *$\{G_\lambda\}_{\lambda<\vartheta}$ is a transfinite Schauder decomposition of $E$.*

2°. *There exists a constant $C$ with $1 \leqslant C < \infty$, such that we have*

$$\|y\| \leqslant C\|y+z\| \tag{19.22}$$

*for all $y \in [\bigcup_{\varkappa<\lambda} G_\varkappa]$, $z \in [\bigcup_{\lambda\leqslant\varkappa<\vartheta} G_\varkappa]$ and all $\lambda < \vartheta$.*

3°. *There exists a constant $C$ with $1 \leqslant C < \infty$, such that we have* (19.22) *for all $y \in \operatorname{lin}\{\bigcup_{\varkappa<\lambda} G_\varkappa\}$, $z \in \operatorname{lin}\{\bigcup_{\lambda\leqslant\varkappa<\vartheta} G_\varkappa\}$ and all $\lambda < \vartheta$.*

*Proof.* The equivalence 2° $\Leftrightarrow$ 3° is obvious.

If we have 1° and $\{u_\lambda\}_{\lambda\leqslant\vartheta}$ is the transfinite sequence of partial sum operators associated to $\{G_\lambda\}_{\lambda<\vartheta}$, then, by (19.13),

$$\|y\| = \|u_\lambda(y+z)\| \leqslant \|u_\lambda\|\,\|y+z\| \leqslant C\|y+z\|$$

for all $y \in [\bigcup_{\varkappa<\lambda} G_\varkappa]$, $z \in [\bigcup_{\lambda\leqslant\varkappa<\vartheta} G_\varkappa]$ and all $\lambda < \vartheta$. Thus, 1° $\Rightarrow$ 2°.

Finally, assume that we have 2° and let us consider the mapping $w: D_1^{(\vartheta)} \to E$ defined by (19.9). Clearly, for any $\{z_\varkappa^{(\lambda)}\}_{\varkappa<\vartheta} \in F_\lambda$ and $y_\lambda \in G_\lambda$ related by (19.19) we have

$$w(\{z_\varkappa^{(\lambda)}\}_{\varkappa<\vartheta}) = y_\lambda,$$

and this gives a linear isometry of $F_\lambda$ onto $G_\lambda$, so

$$w(F_\lambda) = G_\lambda \qquad (\lambda < \vartheta). \tag{19.23}$$

Hence, since $[\bigcup_{\lambda<\vartheta} G_\lambda] = E$, $w(D_1^{(\vartheta)})$ is dense in $E$. Thus, if one proves that $w$ is an isomorphism, then $w(D_1^{(\vartheta)}) = E$ (by proposition 19.1 and since the image of a Banach space by an isomorphism is a Banach space) and $\{w(F_\lambda)\}_{\lambda<\vartheta} = \{G_\lambda\}_{\lambda<\vartheta}$ is a transfinite Schauder decomposition of $E$ (by proposition 19.3), so the proof of 2° $\Rightarrow$ 1° will be complete. But, the proof that $w: D_1^{(\vartheta)} \to E$ is one-to-one and bicontinuous is similar to that of the corresponding statement in § 17, proof of theorem 17.8. This concludes the proof of theorem 19.3.

We have the following obvious consequence of theorem 19.3 and § 17, theorem 17.8, which is a generalization of § 15, theorem 15.21 a) (see also § 17, theorem 17.10 and remark 17.6 b)):

**Corollary 19.1.** *If* $\{G_\lambda\}_{\lambda<\vartheta}$ *is a transfinite Schauder decomposition of a Banach space* $E$, *then every transfinite sequence* $\{y_\lambda\}_{\lambda<\vartheta}\in E$ *with* $y_\lambda\in G_\lambda\setminus\{0\}$ $(\lambda<\vartheta)$ *is a transfinite basic sequence of type* $\vartheta$. *If* $\{G_\lambda\}_{\lambda<\vartheta}$ *is monotone**⁾, *then so is* $\{y_\lambda\}_{\lambda<\vartheta}$.

Various other results on Schauder decompositions and on transfinite bases can be also generalized to transfinite Schauder decompositions. For example, one can generalize the results of duality of § 17 (theorem 17.9 and the results after it) and one can prove that *every transfinite weak Schauder decomposition of a Banach space E is a transfinite Schauder decomposition of E.*

One can also introduce some special classes of transfinite Schauder decompositions and of ordinal resolutions of the identity, for example:

*Definition 19.6.* A transfinite Schauder decomposition $\{G_\lambda\}_{\lambda<\vartheta}$ of a Banach space $E$ is said to be

a) *monotone*, if it satisfies condition 2° (or 3°) of theorem 19.3 with $C=1$ (or, what is equivalent, if the associated partial sum operators satisfy $\|u_\lambda\|=1$ for $2\leqslant\lambda\leqslant\vartheta$);

b) *shrinking*, if for the a.t.s.c.p. $\{v_\lambda\}_{\lambda<\vartheta}\subset L(E,E)$, $\{v_\lambda^*(E^*)\}_{\lambda<\vartheta}$ is a transfinite Schauder decomposition of $E^*$ (or, what is equivalent, if $\{u_\lambda^*\}_{\lambda\leqslant\vartheta}$ is an ordinal resolution of $I_{E^*}$, where the $u_\lambda$'s are the partial sum operators associated to $\{G_\lambda\}_{\lambda<\vartheta}$);

c) *boundedly complete*, if for any $\lambda_1<\lambda_2<\ldots<\lambda\leqslant\vartheta$, with $\lim_{j\to\infty}\lambda_j=\lambda$ and any bounded sequence $\{y_j\}\subset E$ such that

$$u_{\lambda_n}(y_j)=y_n \qquad (n=1,\ldots,j;\, j=1,2,\ldots), \tag{19.24}$$

where $\{u_\lambda\}_{\lambda\leqslant\vartheta}$ is the transfinite sequence of partial sum operators associated to $\{G_\lambda\}_{\lambda<\vartheta}$, there exists the limit

$$x=\lim_{j\to\infty} y_j. \tag{19.25}$$

The notions of *monotone, shrinking* and *boundedly complete ordinal resolutions of* $I_E$ are defined similarly (i.e., directly by the above conditions on $\{u_\lambda\}_{\lambda\leqslant\vartheta}$).

Let us pass now to the problem of existence of transfinite Schauder decompositions (or, equivalently, of ordinal resolutions of the identity).

---

*⁾ See definition 19.6 a).

There exist non-separable Banach spaces of type $C(Q)$ ($Q$ compact) which have no transfinite Schauder decomposition, for example $E = l^\infty \equiv C(\beta\mathcal{N})$. Indeed, this follows from § 15, theorem 15.2, combined with the following generalization and sharpening of the case $\vartheta > \omega$ of § 17, proposition 17.7:

**Proposition 19.4.** *If a Banach space $E$ has a transfinite Schauder decomposition $\{G_\lambda\}_{\lambda<\vartheta}$ of type $\vartheta > \omega$, then for each $\vartheta_0$ with $\omega \leqslant \vartheta_0 < \vartheta$ the space $E$ has a transfinite Schauder decomposition of type $\vartheta_0$. Hence, in particular, every Banach space with a transfinite Schauder decomposition has a Schauder decomposition.*

*Proof.* Let $\{G_\lambda\}_{\lambda<\vartheta}$ be a transfinite Schauder decomposition of $E$, where $\vartheta > \omega$, let $\{u_\lambda\}_{\lambda\leqslant\vartheta}$ be the associated transfinite sequence of partial sum operators and let $\omega \leqslant \vartheta_0 < \vartheta$. Then the transfinite sequence of subspaces $\{F_\lambda\}_{\lambda<\vartheta_0}$ defined by

$$F_1 = (I - u_{\vartheta_0})(E), \ F_{1+\lambda} = G_\lambda \qquad (\lambda < \vartheta_0), \tag{19.26}$$

is a transfinite Schauder decomposition of $E$, since $F_1 = (I - u_{\vartheta_0})(E) \neq \{0\}$ (by $\vartheta_0 < \vartheta$) and since every $x \in E$ has a unique expansion of the form

$$x = \sum_{\lambda<\vartheta} y_\lambda = \sum_{\lambda<\vartheta_0} y_\lambda + \sum_{\vartheta_0\leqslant\lambda<\vartheta} y_\lambda,$$

where $y_\lambda \in G_\lambda = F_{1+\lambda}$ $(\lambda < \vartheta_0)$, $\sum_{\vartheta_0\leqslant\lambda<\vartheta} y_\lambda \in (I - u_{\vartheta_0})(E) = F_1$. This completes the proof of proposition 19.4.

Now we shall prove, as announced in the introduction to this section, some positive results on the existence of ordinal resolutions of the identity (or, equivalently, of transfinite Schauder decompositions) in certain classes of non-separable Banach spaces. To this end, we shall need extensive preparation, which, however, will be also used later in this section and in part of § 20.

**Lemma 19.2.** *Let $B$, $G_1$ be finite-dimensional*) subspaces of a normed linear space $E$. Then there exists a continuous linear projection $v$ of $E$ onto $B$ such that*

$$v(G_1) \subset G_1; \tag{19.27}$$

*note that then for any subspace $F_1$ of $E$ with $B \subset F_1$, we have*

$$v(F_1) \subset v(E) = B \subset F_1. \tag{19.28}$$

---

*) Actually, the same result remains valid if we assume only that $\dim B < \infty$ (changing slightly the proof), but we shall not need here this remark.

*Proof.* Let $\{x_i\}$ be a basis of $B \cap G_1$. Choose $\{y_i\}$ and $\{z_i\}$ in $E$ such that $\{x_i\} \cup \{y_i\}$ is a basis of $B$ and $\{x_i\} \cup \{z_i\}$ is a basis of $G_1$. We claim that $\{x_i\} \cup \{y_i\} \cup \{z_i\}$ is linearly independent. Indeed, if $\sum \alpha_i x_i + \sum \beta_i y_i + \sum \gamma_i z_i = 0$, then

$$\sum \alpha_i x_i + \sum \beta_i y_i = - \sum \gamma_i z_i \in B \cap G_1 = \mathrm{lin}\,\{x_i\},$$

whence $\sum \beta_i y_i = 0$ and $\sum \alpha_i x_i + \sum \gamma_i z_i = 0$; consequently, by our assumptions, $\beta_i = 0$, $\alpha_i = 0$, $\gamma_i = 0$ for all $i$, which proves the claim. Hence, since obviously

$$\mathrm{lin}\,\{B \cup G_1\} = \mathrm{lin}\,(\{x_i\} \cup \{y_i\} \cup \{z_i\}),$$

we can define a linear projection $v$ of $\mathrm{lin}\,\{B \cup G_1\}$ onto $B$ by

$$v(\sum \alpha_i x_i + \sum \beta_i y_i + \sum \gamma_i z_i) = \sum \alpha_i x_i + \sum \beta_i y_i;$$

then, since $v(\sum \alpha_i x_i + \sum \gamma_i z_i) = \sum \alpha_i x_i \in G_1$, we have (19.27).

Now, we can extend $v$ to a bounded linear projection of $E$ onto $B$, satisfying (19.27), which we shall denote again by $v$ (indeed, by the Hahn-Banach theorem, extend the coefficient functionals $\alpha_i, \beta_i \in B^*$ associated to the basis $\{x_i\} \cup \{y_i\}$ of $B$, to $\alpha_i, \beta_i \in E^*$ satisfying $\alpha_i(z) = 0$, $\beta_i(z) = 0$ for all $z \in \mathrm{lin}\,\{z_i\}$ and then put $v(x) = \sum \alpha_i(x) x_i + \sum \beta_i(x) y_i$ for all $x \in E$). This completes the proof of lemma 19.2.

**Lemma 19.3.** *Let $E$ be a linear space with two norms $\|\cdot\|_1$, $\|\cdot\|_2$ such that*

$$\|x\|_1 \leqslant \|x\|_2 \qquad (x \in E), \tag{19.29}$$

*and with a third norm $\|\cdot\|_3$ defined on a linear subspace $F \subset E$, such that*

$$\|x\|_1 \leqslant \|x\|_3 \qquad (x \in F). \tag{19.30}$$

*Let $B$ be a finite-dimensional subspace of $F$, let $f_1, \ldots, f_m \in (E, \|\cdot\|_2)^*$ let $n \in \mathcal{N}$ and let $G$ be a linear subspace of $E$. Then there exists an $\aleph_0$-dimensional* *) *linear subspace $C$ of $E$, containing $B$, such that for every $\varepsilon > 0$, every pair of finite-dimensional subspaces $F_1 \subset F$, $G_1 \subset G$ with $B \subset F_1$, and every finite-dimensional subspace $Z$ of $E$ with $Z \supset B$, $\dim Z/B = n$, there is a linear operator $u\colon Z \to C$ satisfying*

$$u(Z \cap G_1) \subset G, \quad u(Z \cap F_1) \subset F, \tag{19.31}$$

$$\|u\|_1 \leqslant 1 + \varepsilon, \ \|u\|_2 \leqslant 1 + \varepsilon, \ \|u|_{Z \cap F_1}\|_3 \leqslant 1 + \varepsilon, \tag{19.32}$$

$$u(y) = y \qquad (y \in B), \tag{19.33}$$

$$|f_k(z) - f_k(u(z))| \leqslant \varepsilon \|z\|_2 \quad (z \in Z,\ k = 1, \ldots, m). \tag{19.34}$$

---

*) I.e., $C$ is the linear span of some countable subset of $E$.

*Proof.* Let $r \in \mathcal{N}$. Choose $y_1, \ldots, y_p \in B$ such that for each $y \in B$ with $\|y\|_\alpha \leqslant r$ there is some $y_h$, where $1 \leqslant h \leqslant p$, satisfying $\|y - y_h\|_\alpha < \frac{1}{r}$ ($\alpha = 1,2,3$); in other words, $\{y_1, \ldots, y_p\}$ is a $\frac{1}{r}$-net for the $r$-ball of $(B, \|\cdot\|_\alpha)$ ($\alpha = 1,2,3$). Clearly, $p = p(r)$.

Furthermore, choose $\lambda^1 = \{\lambda_i^1\}_{i=1}^n, \ldots, \lambda^q = \{\lambda_i^q\}_{i=1}^n \in l_n^1$ with $\|\lambda^j\|_{l_n^1} = \sum_{i=1}^n |\lambda_i^j| = 1$ $(j = 1, \ldots, q)$, such that for each $\lambda = \{\lambda_i\}_{i=1}^n \in l_n^1$ with $\lambda_{s+1} = \ldots = \lambda_n = 0$ for some $s \leqslant n$ (of course, when $s = n$, this condition means no restriction on $\lambda$) and with $\|\lambda\|_{l_n^1} = 1$, there is some $\lambda^j$, where $1 \leqslant j \leqslant q$, satisfying $\lambda_{s+1}^j = \ldots = \lambda_n^j = 0$ and $\|\lambda - \lambda^j\|_{l_n^1} < \frac{1}{r}$ $\left(\text{such a } \frac{1}{r}\text{-net can be constructed by induction on } s\right)$. Clearly, $q = q(r)$.

Define an auxiliary function $\rho_F$ on $E$ by

$$\rho_F(x) = \begin{cases} x \text{ for } x \in F \\ 0 \text{ for } x \in E \setminus F. \end{cases} \tag{19.35}$$

For each $r \in \mathcal{N}$ and each triple of natural numbers $a, b, c \in \{0,1, \ldots, n\}$ with $a + b + c \leqslant n$ define $3n + 3pq + mn = M = M(r)$ scalar-valued functions of

$$\{x_1, \ldots, x_a, x_{a+1}, \ldots, x_{a+b}, x_{a+b+1}, \ldots, x_{a+b+c}, \ldots, x_n\} \in$$

$$\in (F \cap G)^a \times F^b \times G^c \times E^{n-a-b-c} = D_{abc} \subset E^n, \tag{19.36}$$

as follows:

$$\|x_i\|_\alpha, \ \|\rho_F(x_i)\|_3, \ \left\| y_h + \sum_{i=1}^n \lambda_i^j x_i \right\|_\alpha, \ \left\| y_h + \sum_{i=1}^n \lambda_i^j \rho_F(x_i) \right\|_3, \ f_k(x_i)$$

$$(1 \leqslant i \leqslant n, \quad 1 \leqslant \alpha \leqslant 2, \quad 1 \leqslant h \leqslant p, \quad 1 \leqslant j \leqslant q, \quad 1 \leqslant k \leqslant m). \tag{19.37}$$

These functions can be regarded as a function $\varphi$ from $D_{abc}$ into $l_M^\infty$. Choose a sequence $\{x_1^s, \ldots, x_n^s\} \in D_{abc}$ $(s = 1,2, \ldots)$, such that $\{\varphi(\{x_1^s, \ldots, x_n^s\})\}_{s=1}^\infty$ is $\|\cdot\|_{l_M^\infty}$-dense in $\varphi(D_{abc})$. This sequence is constructed for the fixed $r, a, b, c$, so we can denote it by $\{x_1^s, \ldots, x_n^s\} = \{x_1^{srabc}, \ldots, x_n^{srabc}\}$ $(s = 1, 2, \ldots)$.

Define a linear subspace $C$ of $E$ by

$$C = \operatorname{lin} \{B \cup \{x_i^{srabc}\} \mid i=1, \ldots, n; \ s, r = 1,2, \ldots; \ a, b, c \in \{0,1, \ldots, n\}; \ a + b + c \leqslant n\}. \tag{19.38}$$

We shall show that the subspace $C$ has the required properties, which will complete the proof.

It is obvious that $C$ is $\aleph_0$-dimensional and that $C$ contains $B$. Let $\varepsilon>0$, let $F_1 \subset F$, $G_1 \subset G$ be finite-dimensional subspaces with $B \subset F_1$, and let $Z$ be a finite-dimensional subspace of $E$ with $Z \supset B$, $\dim Z/B = n$. Applying lemma 19.2 in $(E, \|\cdot\|_1)$, choose a projection $v$ of $E$ onto $B$ with $\|v\|_1 < \infty$, satisfying (19.27), (19.28). Then $v(Z) \subset v(E) = B = v(B) \subset v(Z)$, whence $v(Z) = B$ and $(I - v)(Z) \subset Z - B = Z$. Also, by (19.29), $F_1 \subset F$, (19.30) and since all norms on the finite-dimensional space $v(E) = B$ are equivalent, there are two constants $N_2$, $N_3 < \infty$ such that

$$\|v(x)\|_2 \leqslant N_2\|v(x)\|_1 \leqslant N_2\|v\|_1 \|x\|_1 \leqslant N_2\|v\|_1 \|x\|_2 \qquad (x \in E),$$

$$\|v(x)\|_3 \leqslant N_3\|v(x)\|_1 \leqslant N_3\|v\|_1 \|x\|_1 \leqslant N_3\|v\|_1 \|x\|_3 \qquad (x \in F_1),$$

so $\|v\|_2 \leqslant N_2\|v\|_1 < \infty$, $\|v|_{F_1}\|_3 \leqslant N_3\|v\|_1 < \infty$. Let

$$L = \max (\|v\|_1, \|v\|_2, \|v|_{F_1}\|_3), \tag{19.39}$$

$$N \geqslant \max\left(1, \frac{6(1+L)}{\varepsilon}\right). \tag{19.40}$$

Choose a basis $\{z_1, \ldots, z_n\}$ of $(I - v)(Z)$ such that
(i) $\{z_1, \ldots, z_a\}$ is a basis of $(I - v)(Z) \cap F_1 \cap G_1$;
(ii) $\{z_1, \ldots, z_{a+b}\}$ is a basis of $(I - v)(Z) \cap F_1$;
(iii) $\{z_1, \ldots, z_a, z_{a+b+1}, \ldots, z_{a+b+c}\}$ is a basis of $(I - v)(Z) \cap G_1$.
Then, by $F_1 \subset F$, $G_1 \subset G$ and the definition (19.36) of $D_{abc}$, we have

$$\{z_1, \ldots, z_n\} \in (F_1 \cap G_1)^a \times F_1^b \times G_1^c \times E^{n-a-b-c} \subset D_{abc}. \tag{19.41}$$

Furthermore, by multiplying all $z_i$ with a sufficiently large number, we may assume that

$$\left\|\sum_{i=1}^{n} \lambda_i z_i\right\|_\alpha \geqslant \sum_{i=1}^{n} |\lambda_i| \qquad (\lambda = (\lambda_1, \ldots, \lambda_n) \in l_n^1;\ \alpha = 1, 2), \tag{19.42}$$

$$\left\|\sum_{i=1}^{a+b} \lambda_i z_i\right\|_3 \geqslant \sum_{i=1}^{a+b} |\lambda_i| \qquad (\lambda = (\lambda_1, \ldots, \lambda_n) \in l_n^1); \tag{19.43}$$

indeed, since all norms on $l_n^1$ are equivalent, there are constants $M_\alpha$ such that $\left\|\sum_{i=1}^{n} \lambda_i z_i\right\|_\alpha \geqslant M_\alpha \sum_{i=1}^{n} |\lambda_i|$ for all $\lambda = (\lambda_1, \ldots, \lambda_n) \in l_n^1$ and

$\alpha=1,2$, whence $\left\|\sum_{i=1}^{n}\lambda_i c z_i\right\|_\alpha \geqslant cM_\alpha \sum_{i=1}^{n}|\lambda_i| \geqslant \sum_{i=1}^{n}|\lambda_i|$ for all $\lambda=(\lambda_1, \ldots$ $\ldots, \lambda_n)\in l_n^1$ and $\alpha=1,2$, provided that $cM_\alpha \geqslant 1$ ($\alpha=1,2$). For (19.43) the argument is similar.

Now let $t \geqslant 1$ be such that

$$\|z_i\|_\alpha \leqslant t, \ \ \|\rho_F(z_i)\|_3 \leqslant t \quad (i=1, \ldots, n; \alpha=1,2), \tag{19.44}$$

and let $r \in \mathcal{N}$ be such that

$$2t+1 < \varepsilon(r-t), \qquad \frac{t}{r} < \frac{1}{N}, \tag{19.45}$$

where $N$ is as in (19.40). Thus, we have chosen the natural numbers $a, b, c$ with $a+b+c \leqslant n$ and the integer $r \in \mathcal{N}$. *We shall keep these $r, a, b, c$ fixed throughout the rest of the proof.*

By (19.41), let $\{x_1, \ldots, x_n\} = \{x_1^{srabc}, \ldots, x_n^{srabc}\}$ be an element of the sequence in $D_{abc}$ defining $C$ (see formula (19.38)), such that

$$\|\varphi(\{z_1, \ldots, z_n\}) - \varphi(\{x_1, \ldots, x_n\})\|_{l_M^\infty} < \frac{1}{N}. \tag{19.46}$$

Define on $Z = B \oplus (I-v)(Z)$ a linear operator $u\colon Z \to C$ by

$$u\left(y+\sum_{i=1}^{n}\lambda_i z_i\right) = y + \sum_{i=1}^{n}\lambda_i x_i \qquad (y\in B; \lambda_1, \ldots, \lambda_n \in K). \tag{19.47}$$

We shall show that $u$ satisfies (19.31)—(19.34), which will complete the proof.

From (19.47) it is obvious that $u$ satisfies (19.33).

If $z = y + \sum_{i=1}^{n}\lambda_i z_i \in Z \cap G_1$, then, by (19.27), $y = v(z) \in v(G_1) \subset G_1$, whence $\sum_{i=1}^{n}\lambda_i z_i = (I-v)(z) = z - y \in (I-v)(Z) \cap G_1$. Therefore, by (iii), $\lambda_{a+1} = \ldots = \lambda_{a+b} = 0 = \lambda_{a+b+c+1} = \ldots = \lambda_n$, whence, by $y \in G_1 \subset G$ and (19.36),

$$u(z) = y + \sum_{i=1}^{a}\lambda_i x_i + \sum_{i=a+b+1}^{a+b+c}\lambda_i x_i \in G,$$

which proves that $u(Z\cap G_1) \subset G$. The inequality $u(Z \cap F_1) \subset F$ is proved similarly, using now (19.28) and (ii) (it turns out that if $z = y + \sum_{i=1}^{n}\lambda_i z_i \in Z \cap F_1$, then $\lambda_{a+b+1} = \ldots = \lambda_n = 0$, whence $u(z)\in F$), so (19.31) holds; in particular, by (ii), $x_i = u(z_i) \in F$ $(i=1, \ldots, a+b)$.

Now we shall prove the last inequality in (19.32). By the above and by the homogeneity of the norm, it will be enough to prove that we have

$$\left\| y + \sum_{i=1}^{a+b} \lambda_i x_i \right\|_3 \leqslant (1+\varepsilon) \left\| y + \sum_{i=1}^{a+b} \lambda_i z_i \right\|_3$$

$$\left( y + \sum_{i=1}^{a+b} \lambda_i z_i \in Z \cap F_1, \sum_{i=1}^{a+b} |\lambda_i| = 1 \right). \tag{19.48}$$

*Case 1°*: $\|y\|_3 \geqslant r$. Then, since by $\sum_{i=1}^{n} |\lambda_i| = 1$ and (19.44)

$$\left\| \sum_{i=1}^{a+b} \lambda_i z_i \right\|_3 = \left\| \sum_{i=1}^{a+b} \lambda_i \rho_F(z_i) \right\|_3 \leqslant \sum_{i=1}^{a+b} |\lambda_i| \max_{1 \leqslant j \leqslant a+b} \|\rho_F(z_j)\|_3 \leqslant t,$$

we obtain

$$\left\| y + \sum_{i=1}^{a+b} \lambda_i z_i \right\|_3 \geqslant \|y\|_3 - \left\| \sum_{i=1}^{a+b} \lambda_i z_i \right\|_3 \geqslant r - t. \tag{19.49}$$

But, since $z_i \in Z \cap F_1 \subset F$ and $x_i = u(z_i) \in u(Z \cap F_1) \subset F$ $(i = 1, \ldots, a+b)$, we have $\rho_F(z_i) = z_i$ and $\rho_F(x_i) = x_i$ $(i = 1, \ldots, a+b)$, whence, by the definition (19.37) of $\varphi$ and by (19.46), (19.40), we get

$$\|x_i\|_3 - \|z_i\|_3 = \|\rho_F(x_i)\|_3 - \|\rho_F(z_i)\|_3 \leqslant$$

$$\leqslant \|\varphi(\{x_1, \ldots, x_n\}) - \varphi(\{z_1, \ldots, z_n\})\|_{l_M^\infty} < \frac{1}{N} \leqslant 1 \quad (i=1, \ldots, a+b).$$

Therefore, by (19.44), $\|x_i\|_3 \leqslant \|z_i\|_3 + 1 = \|\rho_F(z_i)\|_3 + 1 \leqslant t+1$ $(i = 1, \ldots, a+b)$, whence, by $\sum_{i=1}^{a+b} |\lambda_i| = 1$, we obtain

$$\left\| \sum_{i=1}^{a+b} \lambda_i x_i \right\|_3 \leqslant \sum_{i=1}^{a+b} |\lambda_i| \max_{1 \leqslant j \leqslant a+b} \|x_j\|_3 \leqslant t+1.$$

Consequently, by $\|z_i\|_3 = \|\rho_F(z_i)\|_3 \leqslant t$ $(i = 1, \ldots, a+b)$, (19.45) and (19.49),

$$\left\| y + \sum_{i=1}^{a+b} \lambda_i x_i \right\|_3 \leqslant \left\| y + \sum_{i=1}^{a+b} \lambda_i z_i \right\|_3 + \left\| \sum_{i=1}^{a+b} \lambda_i z_i \right\|_3 + \left\| \sum_{i=1}^{a+b} \lambda_i x_i \right\|_3 \leqslant$$

$$\leqslant \left\| y + \sum_{i=1}^{a+b} \lambda_i z_i \right\|_3 + t + (t+1) \leqslant$$

$$\leqslant \left\| y + \sum_{i=1}^{a+b} \lambda_i z_i \right\|_3 + \varepsilon(r-t) \leqslant (1+\varepsilon) \left\| y + \sum_{i=1}^{a+b} \lambda_i z_i \right\|_3.$$

*Case 2°*: $\|y\|_3 \leqslant r$. Then, by the construction of $\{y_1, \ldots, y_p\}$ and $\{\lambda^1, \ldots, \lambda^q\}$, let $y_h \in B$ be a $\frac{1}{r}$-approximation to $y$ in $\|\cdot\|_\alpha$ $(\alpha = 1, 2, 3)$ and let $\lambda^j = \{\lambda_1^j, \ldots, \lambda_n^j\}$ with $\lambda_{a+b+1}^j = \ldots = \lambda_n^j = 0$, $\|\lambda^j\|_{l_n^1} = 1$ be a $\frac{1}{r}$-approximation to $\lambda = \{\lambda_1, \ldots, \lambda_n\}$ with $\lambda_{a+b+1} = \ldots = \lambda_n = 0$, $\|\lambda\|_{l_n^1} = 1$. Thus, $\|y - y_h\|_\alpha < \frac{1}{r}$ $(\alpha = 1, 2, 3)$, $\sum_{i=1}^{a+b} |\lambda_i| = \sum_{i=1}^{n} |\lambda_i| = 1$ (as above), $\sum_{i=1}^{a+b} |\lambda_i^j| = \sum_{i=1}^{n} |\lambda_i^j| = 1$ and $\sum_{i=1}^{a+b} |\lambda_i - \lambda_i^j| = \sum_{i=1}^{n} |\lambda_i - \lambda_i^j| < \frac{1}{r}$. Clearly,

$$\Big\|y + \sum_{i=1}^{a+b} \lambda_i x_i\Big\|_3 + \Big\|y_h + \sum_{i=1}^{a+b} \lambda_i^j z_i\Big\|_3 \leqslant \|y - y_h\|_3 + \\ + \Big\|y_h + \sum_{i=1}^{a+b} \lambda_i^j x_i\Big\|_3 + \Big\|\sum_{i=1}^{a+b} (\lambda_i^j - \lambda_i) x_i\Big\|_3 + \\ + \|y_h - y\|_3 + \Big\|y + \sum_{i=1}^{a+b} \lambda_i z_i\Big\|_3 + \Big\|\sum_{i=1}^{a+b} (\lambda_i - \lambda_i^j) z_i\Big\|_3 .$$

But, as was observed in case 1° above, $\rho_F(x_i) = x_i$, $\rho_F(z_i) = z_i$ $(i = 1, \ldots, a+b)$, whence, by the definition (19.37) of $\varphi$ and by (19.46),

$$\Big\|y_h + \sum_{i=1}^{a+b} \lambda_i^j x_i\Big\|_3 - \Big\|y_h + \sum_{i=1}^{a+b} \lambda_i^j z_i\Big\|_3 = \Big\|y_h + \sum_{i=1}^{n} \lambda_i^j \rho_F(x_i)\Big\|_3 - \\ - \Big\|y_h + \sum_{i=1}^{n} \lambda_i^j \rho_F(z_i)\Big\|_3 \leqslant \|\varphi(\{x_1, \ldots, x_n\}) - \varphi(\{z_1, \ldots, z_n\})\|_{l_M^\infty} < \frac{1}{N} .$$

Also, as was observed in case 1° above, we have $\|x_i\|_3 \leqslant t+1$, $\|z_i\|_3 \leqslant t$ $(i = 1, \ldots, a+b)$. Consequently, since by (19.45) and $t \geqslant 1$, $\frac{1}{r} < \frac{1}{Nt} \leqslant \frac{1}{N}$, we obtain

$$\Big\|y + \sum_{i=1}^{a+b} \lambda_i x_i\Big\|_3 - \Big\|y + \sum_{i=1}^{a+b} \lambda_i z_i\Big\|_3 \leqslant 2\|y - y_h\|_3 + \\ + \Big\|y_h + \sum_{i=1}^{a+b} \lambda_i^j x_i\Big\|_3 - \Big\|y_h + \sum_{i=1}^{a+b} \lambda_i^j z_i\Big\|_3 + \Big\|\sum_{i=1}^{a+b} (\lambda_i^j - \lambda_i) x_i\Big\|_3 + \\ + \Big\|\sum_{i=1}^{a+b} (\lambda_i^j - \lambda_i) z_i\Big\|_3 \leqslant \frac{2}{r} + \frac{1}{N} + \frac{t+1}{r} + \frac{t}{r} = \frac{3}{r} + \frac{2t}{r} + \frac{1}{N} \leqslant \frac{6}{N} . \tag{19.50}$$

On the other hand, $\sum_{i=1}^{a+b} \lambda_i z_i = (I - v|_{F_1}) \left( y + \sum_{i=1}^{a+b} \lambda_i z_i \right)$ (since $y \in B \subset \subset F_1$), whence, by (19.39),

$$\left\| \sum_{i=1}^{a+b} \lambda_i z_i \right\|_3 \leqslant \|I - v|_{F_1}\|_3 \left\| y + \sum_{i=1}^{a+b} \lambda_i z_i \right\|_3 \leqslant (1+L) \left\| y + \sum_{i=1}^{a+b} \lambda_i z_i \right\|_3 .$$

Consequently, using also (19.43), $\sum_{i=1}^{a+b} |\lambda_i| = 1$ and (19.40), we get

$$\varepsilon \left\| y + \sum_{i=1}^{a+b} \lambda_i z_i \right\|_3 \geqslant \frac{\varepsilon}{1+L} \left\| \sum_{i=1}^{a+b} \lambda_i z_i \right\|_3 \geqslant \frac{\varepsilon}{1+L} \geqslant \frac{6}{N} .$$

Thus, by (19.50),

$$\left\| y + \sum_{i=1}^{a+b} \lambda_i x_i \right\|_3 \leqslant \left\| y + \sum_{i=1}^{a+b} \lambda_i z_i \right\|_3 + \frac{6}{N} \leqslant (1+\varepsilon) \left\| y + \sum_{i=1}^{a+b} \lambda_i z_i \right\|_3 ,$$

which proves (19.48) and hence that $\|u|_{Z \cap F_1}\|_3 \leqslant 1 + \varepsilon$. The proof of the inequalities $\|u\|_1 \leqslant 1 + \varepsilon$, $\|u\|_2 \leqslant 1 + \varepsilon$ is similar (even simpler, since then we work directly in $E$ with $\sum_{i=1}^{n}$, instead of $F$, $\sum_{i=1}^{a+b}$ and $\rho_F$), so $u$ satisfies (19.32).

Finally, let us prove (19.34). Let $z = y + \sum_{i=1}^{n} \lambda_i z_i \in Z$. Then, by the definition (19.37) of $\varphi$ and by (19.46),

$$|f_k(z) - f_k(u(z))| = \left| f_k(y) + \sum_{i=1}^{n} \lambda_i f_k(z_i) - f_k \left( y + \sum_{i=1}^{n} \lambda_i x_i \right) \right| =$$

$$= \left| \sum_{i=1}^{n} \lambda_i (f_k(z_i) - f_k(x_i)) \right| \leqslant \sum_{i=1}^{n} |\lambda_i| \, \|\varphi(\{z_1, \ldots, z_n\}) -$$

$$- \varphi(\{x_1, \ldots, x_n\})\|_{l_M^\infty} < \frac{1}{N} \sum_{i=1}^{n} |\lambda_i| \qquad (k = 1, \ldots, m).$$

On the other hand, by (19.39), $\left\| \sum_{i=1}^{n} \lambda_i z_i \right\|_2 = \|(I - v)(z)\|_2 \leqslant \|I - - v\|_2 \|z\|_2 \leqslant (1 + L) \|z\|_2$, whence, by (19.42),

$$\|z\|_2 \geqslant \frac{1}{1+L} \left\| \sum_{i=1}^{n} \lambda_i z_i \right\|_2 \geqslant \frac{1}{1+L} \sum_{i=1}^{n} |\lambda_i| .$$

Consequently, by (19.40) we obtain

$$\frac{|f_k(z)-f_k(u(z))|}{\|z\|_2} \leqslant \frac{\sum_{i=1}^{n}|\lambda_i|}{N} \frac{1+L}{\sum_{i=1}^{n}|\lambda_i|} = \frac{1+L}{N} < \varepsilon \qquad (k=1,\ldots,m)$$

that is, (19.34), which completes the proof of lemma 19.3.

We recall that a set $\mathscr{M}$ in a Banach space $E$ is said to *generate* the space $E$ if $E=[\mathscr{M}]$. A Banach space $E$ is said to be *weakly compactly generated* if it contains a weakly compact set $\mathscr{C}$ which generates $E$; it is known*) that in this case $E$ also contains a weakly compact circled convex set $\mathscr{K}$ which generates $E$ (and hence $E=[\mathscr{K}]=\overline{\bigcup_{n=1}^{\infty} n\mathscr{K}}$).

For brevity, throughout the sequel in this section, unless specified otherwise, all topological terms (weak topology, density character, closure, etc.) will refer to the $\|\cdot\|$-norm.

**Lemma 19.4.** *Let $(E, \|\cdot\|)$ be a Banach space which is generated by a weakly compact circled convex set $\mathscr{K}$ and let $|\cdot|$ be another norm on $E$, such that*

$$|x| \leqslant \|x\| \qquad (x\in E). \tag{19.51}$$

*Then for every finite-dimensional subspace $B\subset \operatorname{lin}\mathscr{K}$, every closed linear subspace $G$ of $E$ and every sequence $\{f_k\}\subset (E, \|\cdot\|)^*$ there exists a linear operator $u\colon E\to E$ satisfying*

$$|u| = \|u\| = 1, \tag{19.52}$$

$$u(\mathscr{K})\subset\mathscr{K}, \tag{19.53}$$

$$u(y)=y \qquad (y\in B), \tag{19.54}$$

$$u^*(f_k)=f_k \qquad (k=1,2,\ldots), \tag{19.55}$$

$$u(G)\subset G, \tag{19.56}$$

$$u(E) \text{ is separable.} \tag{19.57}$$

*Proof.* Let $F=\operatorname{lin}\mathscr{K}$ and let $|||x||| = \inf_{\substack{\lambda>0\\ x\in\lambda\mathscr{K}}} \lambda$ $(x\in F)$ be the Minkowski functional of $\mathscr{K}$ on $F$, so $|||\cdot|||$ is the norm of $F$ whose unit ball is $\mathscr{K}$.

*) Indeed, the circled closed convex hull $\mathscr{K}$ of $\mathscr{C}$ is weakly compact (see e.g. [87], p. 434, theorem 4).

Since $\mathscr{K}$ is bounded, we may assume (by contracting $\mathscr{K}$ with a suitable homothety, if necessary) that $\mathscr{K} \subset S_{(F, \|\cdot\|)}$ or, what is equivalent, that

$$\|x\| \leqslant |||x||| \qquad (x \in F). \tag{19.58}$$

Then, by (19.51) and (19.58), $|x| \leqslant |||x|||$ for all $x \in F$, so we are in the situation of lemma 19.3, with $\|\cdot\|_1 = |\cdot|$, $\|\cdot\|_2 = \|\cdot\|$, $\|\cdot\|_3 = |||\cdot|||$. Let $C_n$ be the $\aleph_0$-dimensional linear subspace of $E$ given, according to lemma 19.3, by the subspaces $B$, $G$ of $E$, the integer $n$ and the $m = n$ functionals $f_1, \ldots, f_n \in (E, \|\cdot\|)^*$, and let

$$C = \left[\bigcup_{n=1}^{\infty} C_n\right] \tag{19.59}$$

(in the sense of the $\|\cdot\|$-norm). Furthermore, let $\mathscr{D}$ be the set of triples $(Z, F_1, G_1)$, where $Z, F_1, G_1$ are finite-dimensional subspaces of $E$ with $Z \supset B$, $B \subset F_1 \subset F$ and $G_1 \subset G$, directed by inclusion of components, i.e.,

$$d=(Z, F_1, G_1)\geqslant(Z', F_1', G_1')=d' \Leftrightarrow Z\supset Z',\ F_1\supset F_1',\ G_1\supset G_1'. \tag{19.60}$$

Then, by the above definition of $C_n$, for each $d = (Z, F_1, G_1) \in \mathscr{D}$ there exists a linear operator $u_{Z, F_1, G_1} = u_d\colon Z \to C_n \subset C$ (where $n = \dim Z/B$), such that

$$u_d(Z \cap G_1) \subset G, u_d(Z \cap F_1) \subset F, \tag{19.61}$$

$$\max(|u_d|, \|u_d\|, |||u_d|_{Z\cap F_1}|||) \leqslant 1 + \frac{1}{n}, \tag{19.62}$$

$$u_d(y) = y \qquad (y \in B), \tag{19.63}$$

$$|f_k(z) - f_k(u_d(z))| \leqslant \frac{1}{n} \|z\| \quad (z \in Z, k = 1, \ldots, n). \tag{19.64}$$

Put

$$u'_{Z, F_1, G_1}(x) = \begin{cases} u_{Z, F_1, G_1}(x) \text{ for } x \in Z \cap F_1 \\ 0 \quad \text{for } x \in E \setminus (Z \cap F_1). \end{cases} \tag{19.65}$$

Then for every $x \in \mathscr{K}$ (that is, for every $x \in F$ with $|||x||| \leqslant 1$) we have, by (19.61) and (19.62),

$$|||u'_{Z, F_1, G_1}(x)||| = \begin{cases} |||u_{Z, F_1, G_1}(x)||| \leqslant 1 + \dfrac{1}{n} \text{ for } x \in Z \cap F_1 \\ 0 \quad \text{for } x \in E \setminus (Z \cap F_1), \end{cases}$$

so $u'_{Z, F_1, G_1}|_{\mathscr{K}} \in (2\mathscr{K})^{\mathscr{K}}$, the space of all functions from $\mathscr{K}$ into $2\mathscr{K}$. If we take in $\mathscr{K}$ the weak topology *), then, by our assumption on $\mathscr{K}$ and by Tychonov's theorem, $(2\mathscr{K})^{\mathscr{K}}$ is compact, hence the net $\{u'_d|_{\mathscr{K}}\} = \{u'_{Z, F_1, G_1}|_{\mathscr{K}}\}$ has **) a subnet, which we shall denote again by $\{u'_d|_{\mathscr{K}}\}$, converging pointwise to some $u \in (2\mathscr{K})^{\mathscr{K}}$ in the weak topology, that is: $u'_d(x) \xrightarrow{w} u(x)$ for all $x \in \mathscr{K}$.

The operator $u$ is linear on $\mathscr{K}$. Indeed, if $x, y, \alpha x + \beta y \in Z \cap F_1 \cap \mathscr{K} = \{z \in Z \cap F_1 |\ |||z||| \leqslant 1\}$, then, by (19.65) and since $u_{Z, F_1, G_1}$ is linear, we have

$$f(u'_{Z, F_1, G_1}(\alpha x + \beta y) - \alpha u'_{Z, F_1, G_1}(x) - \beta u'_{Z, F_1, G_1}(y)) = 0 \quad (f \in (E, \|\cdot\|)^*),$$

hence the same holds for the weak limit $u$, which implies that

$$u(\alpha x + \beta y) - \alpha u(x) - \beta u(y) = 0 \quad (x, y, \alpha x + \beta y \in \mathscr{K}). \tag{19.66}$$

Therefore, we can extend $u$ homogeneously, defining

$$u(x) = \begin{cases} |||x|||\, u\left(\dfrac{x}{|||x|||}\right) & \text{for } x \in F \setminus \{0\} \\ 0 & \text{for } x = 0, \end{cases} \tag{19.67}$$

so we get a linear operator $u: F \to F$. Indeed, if $x, y, \alpha x + \beta y \in F \setminus \{0\}$, then, by (19.66),

$$u(\alpha x + \beta y) = |||\alpha x + \beta y|||\, u\left(\frac{\alpha x + \beta y}{|||\alpha x + \beta y|||}\right) =$$

$$= |||\alpha x + \beta y|||\, u\left(\frac{\alpha |||x|||}{|||\alpha x + \beta y|||}\frac{x}{|||x|||} + \frac{\beta |||y|||}{|||\alpha x + \beta y|||}\frac{y}{|||y|||}\right) =$$

$$= |||\alpha x + \beta y||| \left(\frac{\alpha |||x|||}{|||\alpha x + \beta y|||}\, u\left(\frac{x}{|||x|||}\right) + \frac{\beta |||y|||}{|||\alpha x + \beta y|||}\, u\left(\frac{y}{|||y|||}\right)\right) =$$

$$= \alpha u(x) + \beta u(y),$$

while for the other $x, y, \alpha x + \beta y \in F$, the relation $u(\alpha x + \beta y) = \alpha u(x) + \beta u(y)$ is again immediate.

Obviously (by homogeneity), $u'_d(x) \xrightarrow{w} u(x)$ for all $x \in F$. Conse-

---

*) We recall that, by our convention, this means the weak topology with respect to the norm $\|\cdot\|$.

**) See e.g. [215], p. 136, theorem 2.

quently, by (19.62), $\|u(x)\| \leqslant \varliminf_{d \in \mathscr{D}} \|u'_d(x)\| \leqslant 2\|x\|$ $(x \in F)$, so $\|u\| \leqslant 2$. Therefore, since by our assumption $F = \operatorname{lin} \mathscr{K}$ is dense in $(E, \|\cdot\|)$, there is a (unique) extension of $u$ to a $\|\cdot\|$-continuous linear mapping of $E$ into $E$, which we shall denote again by $u$. We shall show that this linear operator $u: E \to E$ has all the required properties, which will complete the proof. To this end, observe that, since $u'_d(x) \xrightarrow{w} u(x)$ for all $x$ in the dense subset $F$ of $E$ and since $\sup_{d \in \mathscr{D}} \|u'_d\| \leqslant 2 < \infty$, we have

$$u'_d(x) \xrightarrow{w} u(x) \qquad (x \in E).$$

Now, if $y \in B$ and $f \in (E, \|\cdot\|)^*$, then, by (19.63),

$$f(u(y)) = \lim_{d \in \mathscr{D}} f(u'_d(y)) = f(y),$$

which proves (19.54).

If $x \in E$ and $f \in (E, \|\cdot\|)^*$, then, by (19.62),

$$|f(u(x))| = \lim_{d \in \mathscr{D}} |f(u'_d(x))| \leqslant$$

$$\leqslant \|f\| \varlimsup_{d \in \mathscr{D}} \|u'_d(x)\| \leqslant \|f\| \varlimsup_{n \to \infty} \left(1 + \frac{1}{n}\right) \|x\| = \|f\| \, \|x\|,$$

whence $\|u\| \leqslant 1$. Similarly, if $x \in E$ and $f \in (E, |\cdot|)^*$, then again $|f(u(x))| = \lim_{d \in \mathscr{D}} |f(u'_d(x))|$ (since (19.51) implies that $(E, |\cdot|)^* \subset (E, \|\cdot\|)^*$) whence, by (19.62), we obtain $|u| \leqslant 1$. Thus, by (19.54), we have (19.52).

Furthermore, by (19.61), (19.62), $u'_d(x) = u'_{Z, F_1, G_1}(x) \in \left(1 + \frac{1}{n}\right)\mathscr{K}$ whenever $\dim Z/B \geqslant n$ and $x \in Z \cap F_1$. Hence, since $\left(1 + \frac{1}{n}\right)\mathscr{K}$ is weakly closed, we obtain $u(x) = w\text{-}\lim_{d \in \mathscr{D}} u'_d(x) \in \left(1 + \frac{1}{n}\right)\mathscr{K}$ $(x \in \mathscr{K})$, so $u(x) \in \bigcap_{n=1}^{\infty} \left(1 + \frac{1}{n}\right)\mathscr{K} = \mathscr{K}$ $(x \in \mathscr{K})$, which proves (19.53).

Also, by (19.61), $u'_d(x) = u'_{Z, F_1, G_1}(x) \in G$ whenever $x \in Z \cap G_1$, whence $u(x) = w\text{-}\lim_{d \in \mathscr{D}} u'_d(x) \in \overline{G} = G$ for each $x \in G$ (since the weak closure of any linear subspace coincides*) with its norm closure), which proves (19.56).

---

*) See e.g. [355], Ch. II, § 9, corollary 2 of theorem 9.2.

If $x \in E$, then, by (19.64),

$$|u^*(f_k)(x) - f_k(x)| = |f_k(u(x)) - u(x)| =$$

$$= \lim_{d \in \mathscr{D}} |f_k(u_d'(x)) - f_k(x)| = 0 \qquad (k = 1,2, \ldots),$$

which proves (19.55).

Finally, since $C$ is $\|\cdot\|$-closed, whence also weakly closed, and since $u_d'(Z) \subset C$ $(d \in \mathscr{D})$, we have $u(x) = w\text{-}\lim_{d \in \mathscr{D}} u_d'(x) \in C$ $(x \in E)$, so $u(E) \subset \subset C$. Hence, since $C$ is separable by its definition (19.59), we obtain (19.57), which completes the proof of lemma 19.4.

**Lemma 19.5.** *Let* $(E, \|\cdot\|)$ *be a Banach space which is generated by a weakly compact circled convex set* $\mathscr{K}$, *let* $|.|$ *be another norm on* $E$, *satisfying* (19.51), *and let* $\mathfrak{m}$ *be an infinite cardinal number. Furthermore, let* $G$ *be a closed linear subspace of* $E$, $B$ *a linear subspace of* $F = \operatorname{lin} \mathscr{K}$ *with dens* $B \leqslant \mathfrak{m}$ *and* $V$ *a linear subspace of* $(E, \|\cdot\|)^*$ *with**) $w^*$-dens $V \leqslant \leqslant \mathfrak{m}$. *Then there exists a linear projection* $p \colon E \to E$ *satisfying*

$$|p| = \|p\| = 1, \tag{19.68}$$

$$p(\mathscr{K}) \subset \mathscr{K}, \tag{19.69}$$

$$p(y) = y \qquad (y \in B), \tag{19.70}$$

$$p^*(f) = f \qquad (f \in V), \tag{19.71}$$

$$p(G) \subset G, \tag{19.72}$$

$$\operatorname{dens} p(E) \leqslant \mathfrak{m}. \tag{19.73}$$

*Proof.* We shall use transfinite induction on $\mathfrak{m}$.

Assume first that $\mathfrak{m} = \aleph_0$. Let $\{y_i^0\}$ be a dense sequence in $B_0 = B$ and $\{f_k\}$ a $w^*$-dense sequence in $V$. By lemma 19.4 (for the subspace $[y_i^0]$ of $F = \operatorname{lin} \mathscr{K}$), there is a linear operator $u_1$ from $E$ onto a separable subspace $u_1(E) = B_1 \subset E$, such that $|u_1| = \|u_1\| = 1$, $u_1(\mathscr{K}) \subset \mathscr{K}$, $u_1(y_i^0) = y_1^0$, $u_1^*(f_k) = f_k$ $(k = 1,2,\ldots)$ (whence $u_1^*(f) = f$ for all $f \in V$) and $u_1(G) \subset G$. Now, since $F = \operatorname{lin} \mathscr{K}$ is dense in $E$ and $u_1$ is continuous, $u_1(F)$ is dense in $u_1(E) = B_1$. But, by $u_1(\mathscr{K}) \subset \mathscr{K}$ we have $u_1(F) \subset F \cap B_1$, hence $F \cap B_1$ is dense in $B_1$. Let $\{y_i^1\}$ be a dense sequence in $F \cap B_1$, hence also in $B_1$. Then, by lemma 19.4 (for the subspace $[y_1^0, y_2^0, y_1^1, y_2^1]$ of $F = \operatorname{lin} \mathscr{K}$), there is a linear operator $u_2$ from $E$ onto a separable subspace $u_2(E) = B_2 \subset E$, such that $|u_2| = \|u_2\| = 1$, $u_2(\mathscr{K}) \subset \mathscr{K}$, $u_2(y_i^j) = y_i^j$ $(i = 1,2; j = 0,1)$, $u_2^*(f) = f$ $(f \in V)$ and $u_2(G) \subset \subset G$. Continuing in this way we can define, inductively, a sequence of

*) We recall that $w^*$-dens $V$ is defined similarly to $w^*$-dens $E^*$ (see § 17, the part before theorem 17.10).

separable subspaces $B_j$ of $E$ ($j = 0,1,2, \ldots$), with dense subsequences $\{y_i^j\}_{i=1}^{\infty}$ ($j = 0,1,2, \ldots$) and linear operators $u_n\colon E \to u_n(E) = B_n \subset E$ satisfying $|u_n| = \|u_n\| = 1$, $u_n(\mathcal{K}) \subset \mathcal{K}$, $u_n(y_i^j) = y_i^j$ ($i = 1, \ldots, n$; $j = 0,1, \ldots, n-1$), $u_n^*(f) = f$ ($f \in V$), and $u_n(G) \subset G$ ($n = 1,2, \ldots$).

Then, as in the proof of lemma 19.4 above (i.e., looking first at $\{u_n|_{\mathcal{K}}\}$, etc.) it follows that $\{u_n\}$ has a weak cluster point $p\colon E \to E$, which is linear and satisfies $|p| = \|p\| = 1$, $p(\mathcal{K}) \subset \mathcal{K}$, $p(y) = y$ $\left(y \in \left[\bigcup_{n=0}^{\infty} B_n\right]\right)$, $p^*(f) = f$ ($f \in V$) and $p(G) \subset G$. But, since $\left[\bigcup_{n=0}^{\infty} B_n\right]$ is also weakly closed, from the definition of $p$ it follows that $p(E) \subset \left[\bigcup_{n=0}^{\infty} B_n\right]$, and hence $p$ is a projection of $E$ onto its separable subspace $\left[\bigcup_{n=0}^{\infty} B_n\right]$. This completes the proof in the case when $\mathfrak{m} = \aleph_0$.

Now let $\mathfrak{m} > \aleph_0$ and assume that the lemma has been proved for all cardinal numbers $< \mathfrak{m}$. Let $\vartheta$ be the first ordinal number of cardinality $\mathfrak{m}$ (hence $\vartheta$ is a limit ordinal) and let*) $\{y_\lambda\}_{\lambda < \vartheta}$ be a dense transfinite sequence in $B$ and $\{f_\lambda\}_{\lambda < \vartheta}$ a $w^*$-dense transfinite sequence in $V$. We claim that there exists a "long sequence" of linear projections**) $\{p_\lambda\}_{\omega \leqslant \lambda < \vartheta}$ satisfying $|p_\lambda| = \|p_\lambda\| = 1$, $p_\lambda(\mathcal{K}) \subset \mathcal{K}$, $p_\lambda(y_\varkappa) = y_\varkappa$ ($\varkappa < \lambda$), $p_\lambda p_\varkappa = p_\varkappa$ ($\omega \leqslant \varkappa < \lambda$), $p_\lambda^*(f_\varkappa) = f_\varkappa$ ($\varkappa < \lambda$), $p_\lambda(G) \subset G$ and $\operatorname{dens} p_\lambda(E) \leqslant \operatorname{card} \lambda$, for all $\lambda$ with $\omega \leqslant \lambda < \vartheta$.

Indeed, $p_\omega$ exists by the first part of the proof. Let $\lambda < \vartheta$ and assume that we have constructed all $p_\varkappa$, $\omega \leqslant \varkappa < \lambda$, satisfying the above conditions. Let $B_\lambda = \operatorname{lin}\{\bigcup_{\omega \leqslant \varkappa < \lambda} p_\varkappa(E) \cup \{y_\varkappa\}_{\varkappa < \lambda}\}$, $V_\lambda = \operatorname{lin}\{f_\varkappa\}_{\varkappa < \lambda}$. Since $F$ is dense in $E$ and $p_\varkappa$ is continuous, $\bigcup_{\omega \leqslant \varkappa < \lambda} p_\varkappa(F)$ is dense in $\bigcup_{\omega \leqslant \varkappa < \lambda} p_\varkappa(E)$, whence $\operatorname{lin}\{\bigcup_{\omega \leqslant \varkappa < \lambda} p_\varkappa(F) \cup \{y_\varkappa\}_{\varkappa < \lambda}\}$ is dense in $B_\lambda$. But, by $\bigcup_{\omega \leqslant \varkappa < \lambda} p_\varkappa(\mathcal{K}) \subset \mathcal{K}$ and $\{y_\varkappa\}_{\varkappa < \lambda} \subset B \cap B_\lambda \subset F \cap B_\lambda$, we have $\operatorname{lin}\{\bigcup_{\omega \leqslant \varkappa < \lambda} p_\varkappa(F) \cup \{y_\varkappa\}_{\varkappa < \lambda}\} \subset F \cap B_\lambda$, whence $F \cap B_\lambda$ is dense in $B_\lambda$. Consequently, since by our construction assumption $\operatorname{dens} p_\varkappa(E) \leqslant \operatorname{card} \varkappa$ for $\omega \leqslant \varkappa < \lambda$, we obtain

$$\operatorname{dens}(F \cap B_\lambda) = \operatorname{dens} B_\lambda \leqslant \sum_{\varkappa \leqslant \lambda} \operatorname{card} \varkappa = \operatorname{card} \lambda < \mathfrak{m};$$

also, clearly, $w^*$-dens $V_\lambda \leqslant \operatorname{card} \lambda < \mathfrak{m}$. Hence, by our induction hypo thesis applied to the subspaces $F \cap B_\lambda$ and $V_\lambda$ of $F = \operatorname{lin} \mathcal{K}$ and

---

*) Some of the $y_\lambda$ (and of the $f_\lambda$) may also coincide (for example, when $\operatorname{dens} B < \operatorname{card} B < \mathfrak{m}$ and $w^*$-dens $V = \mathfrak{m}$.

**) The term "long sequence" is used for any function defined on $[\omega, \vartheta)$ (this is neither a sequence, nor a transfinite sequence, since it is not defined for $[1, \omega)$, but it would be easy to reindex it to become a transfinite sequence). In proposition 19.5 we shall use this term for a function defined on $[\omega, \vartheta] = [\omega, \vartheta + 1)$.

$(E, \|\cdot\|)^*$ respectively, there exists a linear projection $p_\lambda\colon E \to E$ satisfying $|p_\lambda| = \|p_\lambda\| = 1$, $p_\lambda(\mathscr{X}) \subset \mathscr{X}$, $p_\lambda(y) = y$ (for all $y \in F \cap B_\lambda$, whence also for all $y \in B_\lambda$), $p_\lambda^*(f) = f$ $(f \in V_\lambda)$, $p_\lambda(G) \subset G$ and $\operatorname{dens} p_\lambda(E) \leqslant \operatorname{card} \lambda$, which proves (by the above definitions of $B_\lambda$ and $V_\lambda$) our claim on the existence of $\{p_\lambda\}_{\omega \leqslant \lambda < \vartheta}$.

Now, as in the proof of lemma 19.4 above (i.e., looking first at $\{p_\lambda|_{\mathscr{X}}\}_{\omega \leqslant \lambda < \vartheta}$, etc.), it follows that the net $\{p_\lambda\}_{\omega \leqslant \lambda < \vartheta}$ has a subnet, which we shall denote again by $\{p_\lambda\}_{\omega \leqslant \lambda < \vartheta}$, converging pointwise in the weak topology to a $p\colon E \to E$, which is linear and satisfies $|p| = \|p\| = 1$, $p(\mathscr{X}) \subset \mathscr{X}$, $p(y) = y$ (for all*) $y \in \{y_\lambda\}_{\lambda < \vartheta}$, whence also for all $y \in B$), $p^*(f) = f$ (for all $f \in \overline{\{f_\lambda\}_{\lambda < \vartheta}}$, whence also for all $f \in V$), $p(G) \subset G$ and $p(E) \subset \overline{\bigcup_{\omega \leqslant \lambda < \vartheta} p_\lambda(E)}$, whence

$$\operatorname{dens} p(E) \leqslant \operatorname{dens} \overline{\bigcup_{\omega \leqslant \lambda < \vartheta} p_\lambda(E)} \leqslant \sum_{\omega \leqslant \lambda < \vartheta} \operatorname{card} \lambda \leqslant \mathfrak{m}.$$

Finally, if $x \in E$ and $f \in (E, \|\cdot\|)^*$, then by $p_\lambda p_\varkappa = p_\varkappa$ $(\omega \leqslant \varkappa < \lambda)$, we obtain

$$f(p^2(x)) = p^*(f)(p(x)) = \lim_{\varkappa < \vartheta} p^*(f)(p_\varkappa(x)) = \lim_{\varkappa < \vartheta} f(pp_\varkappa(x)) =$$

$$= \lim_{\varkappa < \vartheta} \lim_{\lambda < \vartheta} f(p_\lambda p_\varkappa(x)) = \lim_{\varkappa < \vartheta} f(p_\varkappa(x)) = f(p(x)),$$

whence $p$ is a projection, which completes the proof of lemma 19.5.

*Remark 19.1.* a) Note that only the case $\mathfrak{m} < \operatorname{dens} E$ of lemma 19.5 is of interest for applications, since if $\mathfrak{m} \geqslant \operatorname{dens} E$, then already $p = I_E$ has the required properties.

b) Lemma 19.2 remains valid, with a similar proof, if the assumption $B \subset F_1$ is replaced by the assumptions $\dim F_1 < \infty$, $G_1 \subset F_1$ and if the conclusions (19.27), (19.28) are kept unchanged (indeed, let $\{x_i\}$, $\{x_i\} \cup \{y_i\}$, $\{x_i\} \cup \{y_i\} \cup \{z_i\}$, $\{x_i\} \cup \{d_i\}$, $\{x_i\} \cup \{y_i\} \cup \{d_i\} \cup \{e_i\}$ and $\{x_i\} \cup \{y_i\} \cup \cup \{z_i\} \cup \{d_i\} \cup \{e_i\}$ be bases of $B \cap G_1$, $B \cap F_1$, $B$, $G_1$, $F_1$ and $\operatorname{lin}\{B \cup F_1\}$, let $v(\sum \alpha_i x_i + \sum \beta_i y_i + \sum \gamma_i z_i + \sum \delta_i d_i + \sum \varepsilon_i e_i) = \sum \alpha_i x_i + \sum \beta_i y_i + + \sum \gamma_i z_i$ and then use the Hahn-Banach theorem). From this fact it follows that lemma 19.3 remains valid if the assumption $B \subset F$ is replaced by the assumption that $B \subset E$ and if in the conclusion we replace the condition $B \subset F_1$ by $G_1 \subset F_1$ and keep $F_1 \subset F$, $G_1 \subset G$ unchanged (indeed, the proof is similar, with (i) being omitted and (19.41) being slightly simplified, because of $F_1 \cap G_1 = G_1$). Hence, lemma 19.4 remains valid if the assumptions that $B \subset \operatorname{lin} \mathscr{X}$ and $G \subset E$ are replaced by the assumption that $B$ is a finite-dimensional subspace

*) Here we use that $\vartheta$ is a limit ordinal.

of $E$ and $G$ is a linear subspace of $\operatorname{lin}\mathscr{K}$, such that $\bar{G}=G$ in $E$ (indeed, the proof is similar, with the difference that now $\mathscr{D}$ will be the set of all triples $(Z, F_1, G_1)$, where $Z, F_1, G_1$ are finite-dimensional subspaces of $E$ with*) $Z \supset B$, $G_1 \subset F_1 \subset F$ and $G_1 \subset G$, directed by (19.60)). Consequently, using this latter result, we obtain that *lemma* 19.5 *remains valid if the assumptions that $B$ is a linear subspace of $F=\operatorname{lin}\mathscr{K}$ with* $\operatorname{dens} B \leqslant \mathfrak{m}$ *and $G$ is a closed linear subspace of $E$, are replaced by the assumptions that $B$ is a linear subspace of $E$ with* $\operatorname{dens} B \leqslant \mathfrak{m}$ *and $G$ is a linear subspace of $F=\operatorname{lin}\mathscr{K}$, such that $\bar{G}=G$ in $E$* (indeed, the proof is similar, with the difference that now we do not need to work with $F \cap B_n$ and $F \cap B_\lambda$, but directly with $B_n$ and $B_\lambda$). Note that *we can apply this modified lemma for any weakly compactly generated subspace $G$* of $E$, since if $\mathscr{K}_0$ is a weakly compact set generating**) $G$, then $\mathscr{K}_1$, the closure of the circled convex hull of $\mathscr{K} \cup \mathscr{K}_0$, is a weakly compact***) circled convex set generating $E$ and $F_1=\operatorname{lin}\mathscr{K}_1 \supset G$.

Now we recall a lemma about density characters.

**Lemma 19.6.** *Let $E$ be a normed linear space and $p: E \to E$ a continuous linear projection. Then*

$$w^*\text{-dens}\, p^*(E^*) \leqslant \operatorname{dens} p(E). \tag{19.74}$$

*Proof.* By § 14, lemma 14.3 and the remark made after it, $w^*\text{-dens}\, p^*(E^*)=w^*\text{-dens}\, p(E)^*$. Hence, by § 17, formula (17.57) (applied to $p(E)$), we obtain (19.74), which completes the proof. Alternatively, one can also show, directly, that if $\{x_\lambda\}_{\lambda \in \Lambda}$ is a dense set in $p(E)$, with $\operatorname{card} \Lambda = \operatorname{dens} p(E)$, then the sets

$$\hat{W}_{x_{\lambda_1}, \ldots, x_{\lambda_n};\, r_1, \ldots, r_n;\, t}=\{f \in S_{p^*(E^*)} \mid |f(x_{\lambda_i})-r_i|<t \ (i=1, \ldots, n)\}, \tag{19.75}$$

where $\lambda_i \in \Lambda$, $r_i$ are rational numbers ($i=1, \ldots, n$) and $t$ is a positive rational number, form a basis of the $w^*$-topology in $S_{p^*(E^*)}= =\{f \in p^*(E^*) \mid \|f\| \leqslant 1\}$, so $S_{p^*(E^*)}$, whence also $p^*(E^*)$, has a $w^*$-dense subset of cardinality $\operatorname{dens} p(E)$, which proves (19.74).

**Proposition 19.5** *Let $(E, \|\cdot\|)$ be a Banach space which is generated by a weakly compact circled convex set $\mathscr{K}$, let $|.|$ be another norm on $E$, satisfying* (19.51), *and let $\vartheta$ be the first ordinal number of cardinality* $\operatorname{dens} E$. *Furthermore, let $G$ be a closed linear subspace of $E$, and*

*) Therefore, the assumption $G \subset F$ is now necessary (since $G_1 \subset F$ for all $d=(Z, F_1, G_1) \in \mathscr{D}$).

**) I.e., $G=[\mathscr{K}_0]$, so $G$ is closed.

***) See e.g. [87], p. 434, theorem 4.

*let* $\{x_\lambda\}_{\lambda<\vartheta}$ *be a complete**) *transfinite sequence in* $F = \text{lin}\,\mathscr{K}$. *Then there exists a long sequence of linear projections* $\{p_\lambda\}_{\omega\leqslant\lambda\leqslant\vartheta}$ *on* $E$, *satisfying*

$$|p_\lambda| = \|p_\lambda\| = 1 \qquad (\omega\leqslant\lambda\leqslant\vartheta), \tag{19.76}$$

$$p_\lambda(\mathscr{K}) \subset \mathscr{K} \qquad (\omega\leqslant\lambda\leqslant\vartheta), \tag{19.77}$$

$$x_\varkappa \in p_\lambda(E) \qquad (\varkappa<\lambda\leqslant\vartheta), \tag{19.78}$$

$$p_\lambda p_\varkappa = p_\varkappa p_\lambda = p_\varkappa \qquad (\omega\leqslant\varkappa\leqslant\lambda\leqslant\vartheta), \tag{19.79}$$

$$p_\lambda(G) \subset G \qquad (\omega\leqslant\lambda\leqslant\vartheta), \tag{19.80}$$

$$\text{dens}\, p_\lambda(E) \leqslant \text{card}\,\lambda \qquad (\omega\leqslant\lambda\leqslant\vartheta), \tag{19.81}$$

$$p_\lambda(E) \subset \overline{\bigcup_{\omega\leqslant\varkappa<\lambda} p_{\varkappa+1}(E)} \qquad (\omega < \lambda\leqslant\vartheta). \tag{19.82}$$

*Moreover, from* $[x_\lambda]_{\lambda<\vartheta} = \overline{F} = E$, (19.78) *and* $p_\vartheta p_\varkappa = p_\varkappa$ $(\omega\leqslant\varkappa\leqslant\vartheta)$ *it follows that*

$$p_\vartheta = I_E \tag{19.83}$$

*and from* $\|p_\lambda\| = 1$ $(\omega\leqslant\lambda\leqslant\vartheta)$, (19.79) *and* (19.82) *it follows that for each fixed* $x \in E$, *the mapping* $\lambda \to p_\lambda(x)$ *from* $[\omega, \vartheta]$ *into* $E$ *is continuous.*

*Proof.* If $\text{dens}\, E = \aleph_0$ (hence $\vartheta = \omega$), then $p_\omega = I_E$ satisfies the required conditions (in this case condition (19.82) is void).

Now let $\text{dens}\, E > \aleph_0$. By lemma 19.5 with $B = \text{lin}\,\{x_\varkappa\}_{\varkappa<\omega} \subset F$, there exists a linear projection $p_\omega\colon E \to E$ satisfying $|p_\omega| = \|p_\omega\| = 1$, $p_\omega(\mathscr{K}) \subset \mathscr{K}$, $p_\omega(x_\varkappa) = x_\varkappa$ for all $\varkappa < \omega$ (or, equivalently, $x_\varkappa \in p_\omega(E)$ for all $\varkappa < \omega$), $p_\omega(G) \subset G$ and $\text{dens}\, p_\omega(E) \leqslant \aleph_0$. Let $\omega < \lambda \leqslant \vartheta$ and assume that we have constructed $p_\varkappa$, $\omega\leqslant\varkappa < \lambda$, satisfying the required conditions.

*Case 1°.* $\lambda$ is a nonlimit ordinal, say $\lambda = \gamma + 1$. Let $B_\lambda = \text{lin}\,\{\bigcup_{\omega\leqslant\varkappa\leqslant\gamma} p_\varkappa(E) \cup \{x_\varkappa\}_{\varkappa\leqslant\gamma}\}$ (by our construction assumption, this coincides with $\text{lin}\,\{p_\gamma(E), x_\gamma\}$) and let $V_\lambda = p_\gamma^*(E^*)$. Then, as in the above proof of lemma 19.5, $F \cap B_\lambda$ is dense in $B_\lambda$, $\text{dens}\,(F \cap B_\lambda) \leqslant \text{card}\,(\gamma + 1)$ and, by lemma 19.6 and**) (19.81), $w^*\text{-dens}\, p_\gamma^*(E^*) \leqslant \text{dens}\, p_\gamma(E) \leqslant \text{card}\,(\gamma + 1)$. Hence, by lemma 19.5 (applied to the subspaces $F \cap B_\lambda$ and $V_\lambda$ of $F$ and $E^*$ respectively), there exists a linear projection

---

*) We take here a complete transfinite sequence rather than a dense one, since this will be used in the sequel (see remarks 19.2 b) and 19.4 a) and theorem 19.4).

**) By our construction assumption, (19.81) holds for $\gamma < \lambda = \gamma + 1$.

$p_{\gamma+1} = p_\lambda: E \to E$ satisfying $|p_{\gamma+1}| = \|p_{\gamma+1}\| = 1$, $p_{\gamma+1}(\mathscr{K}) \subset \mathscr{K}$, $p_{\gamma+1}(x) = x$ (for all $x \in F \cap B_\lambda$, whence also for all $x \in B_\lambda$), $p^*_{\gamma+1}(f) = f$ ($f \in p^*_\gamma(E^*)$), $p_{\gamma+1}(G) \subset G$ and $\operatorname{dens} p_{\gamma+1}(E) \leqslant \operatorname{card}(\gamma+1)$. Then, in particular $p_{\gamma+1}p_\varkappa = p_\varkappa\ (\omega \leqslant \varkappa \leqslant \gamma)$ and $x_\varkappa \in p_{\gamma+1}(E)\ (\varkappa \leqslant \gamma)$. Also, $(p_\gamma p_{\gamma+1})^* = p^*_{\gamma+1}p^*_\gamma = p^*_\gamma$, whence $p_\gamma p_{\gamma+1} = p_\gamma$, and hence for every $\omega \leqslant \varkappa < \gamma$ we have, using also our construction assumption, $p_\varkappa p_{\gamma+1} = (p_\varkappa p_\gamma)p_{\gamma+1} = p_\varkappa p_\gamma = p_\varkappa$. Finally, it is obvious that $p_{\gamma+1}(E) \subset \overline{\bigcup_{\omega \leqslant \varkappa \leqslant \gamma} p_{\varkappa+1}(E)}$.

*Case 2°*. $\lambda$ is a limit ordinal. Then, as in the preceding proofs, it follows that $\{p_\varkappa\}_{\omega \leqslant \varkappa < \lambda}$ has a weak cluster point $p_\lambda$, which is linear and satisfies $|p_\lambda| = \|p_\lambda\| = 1$, $p_\lambda(\mathscr{K}) \subset \mathscr{K}$, $p_\lambda(x_\varkappa) = x_\varkappa\ (\varkappa < \lambda)$, $p_\lambda(x) = x\ (x \in p_\varkappa(E); \omega \leqslant \varkappa < \lambda)$, $p_\lambda(G) \subset G$, $p_\lambda(E) \subset \overline{\bigcup_{\omega \leqslant \varkappa < \lambda} p_\varkappa(E)} = \overline{\bigcup_{\omega \leqslant \varkappa < \lambda} p_{\varkappa+1}(E)}$, whence $\operatorname{dens} p_\lambda(E) \leqslant \operatorname{dens} \overline{\bigcup_{\omega \leqslant \varkappa < \lambda} p_\varkappa(E)} \leqslant \sum_{\varkappa < \lambda} \operatorname{card} \varkappa \leqslant \operatorname{card} \lambda$. Also, as in the final part of the proof of lemma 19.5, $p_\lambda$ is a projection. Finally, since by our construction assumption $p^*_\varkappa(f)(p_\beta(x)) = f(p_\varkappa p_\beta(x)) = f(p_\varkappa(x))$ $(x \in E, f \in E^*; \omega \leqslant \varkappa \leqslant \beta < \lambda)$ and since $p_\lambda$ is a weak cluster point of $\{p_\beta\}_{\omega \leqslant \beta < \lambda}$, we get $f(p_\varkappa p_\lambda(x)) = p^*_\varkappa(f)(p_\lambda(x)) = f(p_\varkappa(x))$ $(x \in E, f \in E^*; \omega \leqslant \varkappa < \lambda)$, whence $p_\varkappa p_\lambda = p_\varkappa\ (\omega \leqslant \varkappa < \lambda)$. Thus, there exists $\{p_\lambda\}_{\omega \leqslant \lambda \leqslant \vartheta}$ satisfying (19.76)—(19.82).

Furthermore, by $p_\vartheta p_\varkappa = p_\varkappa\ (\omega \leqslant \varkappa < \vartheta)$, we have $p_\varkappa(E) \subset p_\vartheta(E)$ $(\omega \leqslant \varkappa < \vartheta)$, whence, by (19.78), $x_\varkappa \in p_\vartheta(E)\ (\varkappa < \vartheta)$. Consequently, by $[x_\varkappa]_{\varkappa < \vartheta} = \overline{F} = E$ we obtain $E = [x_\varkappa]_{\varkappa < \vartheta} \subset p_\vartheta(E) \subset E$, so $p_\vartheta(E) = E$, whence, by $p^2_\vartheta = p_\vartheta$, it follows (19.83).

Finally, in order to prove the last statement, it is enough to show that if $\lambda \leqslant \vartheta$ is a limit ordinal and $x \in E$, then $p_\lambda(x) = \lim_{\varkappa < \lambda} p_\varkappa(x)$. By (19.82), there exist $\mu < \lambda$ and $y \in p_\mu(E)$ such that $\|p_\lambda(x) - y\| < \frac{\varepsilon}{2}$. But, by $p_\varkappa p_\mu = p_\mu\ (\mu \leqslant \varkappa < \lambda)$ we have $p_\varkappa(y) = p_\varkappa(p_\mu(y)) = p_\mu(y) = y$ $(\mu \leqslant \varkappa < \lambda)$. Hence, by $p_\varkappa p_\lambda = p_\varkappa\ (\mu \leqslant \varkappa < \lambda)$ and $\|p_\varkappa\| = 1$, we obtain

$$\|p_\lambda(x) - p_\varkappa(x)\| = \|p_\lambda(x) - y - p_\varkappa(p_\lambda(x) - y)\| < \frac{\varepsilon}{2} + \frac{\varepsilon}{2} = \varepsilon$$
$$(\mu \leqslant \varkappa < \lambda),$$

which completes the proof of proposition 19.5.

*Remark 19.2*. a) From (19.82) and $p_{\varkappa+1}(E) \subset p_\lambda(E)\ (\omega \leqslant \varkappa < \lambda \leqslant \vartheta)$ it follows that we have actually the equalities

$$p_\lambda(E) = \overline{\bigcup_{\omega \leqslant \varkappa < \lambda} p_{\varkappa+1}(E)} \qquad (\omega < \lambda \leqslant \vartheta). \tag{19.82'}$$

Also, from the continuity of the mappings $\lambda \to p_\lambda(x)$ of $[\omega, \vartheta]$ into $E$ $(x \in E)$ it follows that

$$p_\lambda(x) \in [p_\omega(x) \cup \bigcup_{\omega \leqslant \varkappa < \lambda} (p_{\varkappa+1} - p_\varkappa)(x)] \quad (x \in E,\ \omega < \lambda \leqslant \vartheta) \quad (19.82'')$$

and that *for each* $x \in E$ *and* $\varepsilon > 0$ *the set* $B_{x,\varepsilon} = \{\lambda \in [\omega, \vartheta) \mid \|p_{\lambda+1}(x) - p_\lambda(x)\| \geqslant \varepsilon\}$ *is finite.*

Indeed, for $\lambda = \omega + 1$ we have $p_{\omega+1}(x) = p_\omega(x) + (p_{\omega+1} - p_\omega)(x)$ $(x \in E)$. Let $\omega + 1 < \mu \leqslant \vartheta$ and assume that (19.82″) holds for all ordinal numbers $\lambda$ with $\omega < \lambda < \mu$. Let $x \in E$. If $\mu$ is a nonlimit ordinal, then, by the induction hypothesis, $p_{\mu-1}(x) \in [p_\omega(x) \cup \bigcup_{\omega \leqslant \varkappa < \mu-1} (p_{\varkappa+1} - p_\varkappa)(x)]$, whence $p_\mu(x) = p_{\mu-1}(x) + (p_\mu - p_{\mu-1})(x) \in [p_\omega(x) \cup \bigcup_{\omega \leqslant \varkappa < \mu} (p_{\varkappa+1} - p_\varkappa)(x)]$. If $\mu$ is a limit ordinal, then by the continuity of $\lambda \to p_\lambda(x)$ and by the induction hypothesis, $p_\mu(x) = \lim_{\lambda < \mu} p_\lambda(x) \in$ $\in [\bigcup_{\omega \leqslant \lambda < \mu} [p_\omega(x) \cup \bigcup_{\omega \leqslant \varkappa < \lambda} (p_{\varkappa+1} - p_\varkappa)(x)]] \subset [p_\omega(x) \cup \bigcup_{\omega \leqslant \varkappa < \mu} (p_{\varkappa+1} - p_\varkappa)(x)]$, which completes the proof of (19.82″).

Finally, assume, a contrario, that for some $x \in E$ and $\varepsilon > 0$ the set $B_{x,\varepsilon}$ is infinite. Then, by the argument of the proof of § 17, lemma 17.1, we arrive at a contradiction with the continuity of $\lambda \to p_\lambda(x)$.

b) By choosing in a special way the transfinite sequence $\{x_\lambda\}_{\lambda < \vartheta}$, one can even construct the long sequence $\{p_\lambda\}_{\omega \leqslant \lambda \leqslant \vartheta}$ of proposition 19.5 so as to have (19.81) replaced by

$$\operatorname{dens} p_\lambda(E) = \operatorname{card} \lambda \qquad (\omega \leqslant \lambda \leqslant \vartheta). \qquad (19.81')$$

Indeed, by § 20, theorem 20.5 a), $E$ has an extended $M$-basis $\{x_i\}_{i \in I}$ such that $x_i \in \mathscr{K}$ $(i \in I)$. Then, by § 20, proposition 20.1, $\operatorname{card} I = \operatorname{dens} E$ and hence we may assume that $I = [1, \vartheta)$ (with the usual ordering), so $\{x_i\}_{i \in I} = \{x_\lambda\}_{\lambda < \vartheta}$. Since $x_\lambda \in \mathscr{K}$ $(\lambda < \vartheta)$, $\{x_\lambda\}_{\lambda < \vartheta}$ is a complete transfinite sequence in $F = \operatorname{lin} \mathscr{K}$, so we can apply proposition 19.5 for this $\{x_\lambda\}_{\lambda < \vartheta}$. Then, by (19.78), $p_\lambda(E) \supset [x_\varkappa]_{\varkappa < \lambda}$ $(\omega \leqslant \lambda \leqslant \vartheta)$. But, the transfinite subsequence $\{x_\varkappa\}_{\varkappa < \lambda}$ of $\{x_\varkappa\}_{\varkappa < \vartheta}$ is an extended $M$-basis of $[x_\varkappa]_{\varkappa < \lambda}$ (by the obvious extension of § 8, proposition 8.4) and hence, again by § 20, proposition 20.1, we obtain $\operatorname{dens} p_\lambda(E) \geqslant \operatorname{dens} [x_\varkappa]_{\varkappa < \lambda} = \operatorname{card} [1, \lambda) = \operatorname{card} \lambda$. This, together with (19.81), yields (19.81′).

c) If $\vartheta \geqslant \omega_1$ and (19.81′) holds and if $\omega_\alpha < \omega_{\alpha+1} \leqslant \vartheta$ denote the first ordinals of cardinality $\aleph_\alpha$ and $\aleph_{\alpha+1}$ respectively*) (where $\alpha \geqslant 0$), then

$$\operatorname{card} \{\varkappa \in [\omega_\alpha, \omega_{\alpha+1}) \mid p_{\varkappa+1} \neq p_\varkappa\} = \aleph_{\alpha+1}, \qquad (19.84)$$

*) See e.g. [1], Ch. III, § 6.

and hence

$$\operatorname{card}\{\varkappa \in [\omega, \lambda) \mid p_{\varkappa+1} \neq p_\varkappa\} = \operatorname{card} \lambda = \operatorname{dens} p_\lambda(E) \quad (\omega_1 \leqslant \lambda \leqslant \vartheta) \tag{19.85}$$

(thus, in particular, $\operatorname{card}\{\varkappa \in [\omega, \vartheta) \mid p_{\varkappa+1} \neq p_\varkappa\} = \operatorname{dens} E$).

Indeed, for any fixed $\alpha \geqslant 0$ with $\omega_{\alpha+1} \leqslant \vartheta$, from (19.82″) and (19.79) it follows that

$$p_{\omega_{\alpha+1}}(E) \subset [p_{\omega_\alpha}(E) \cup \bigcup_{\omega_\alpha \leqslant \varkappa < \omega_{\alpha+1}} (p_{\varkappa+1} - p_\varkappa)(E)],$$

where (by (19.81)) $\operatorname{dens} p_{\omega_\alpha}(E) \leqslant \operatorname{card} \omega_\alpha = \aleph_\alpha$ and $\operatorname{dens}(p_{\varkappa+1} - p_\varkappa)(E) \leqslant \leqslant \operatorname{card}(\varkappa + 1) = \operatorname{card} \omega_\sigma = \aleph_\alpha$ $(\omega_\sigma \leqslant \varkappa < \omega_{\alpha+1})$. Hence, if $\operatorname{card}\{\varkappa \in [\omega_\alpha, \omega_{\alpha+1}) \mid (p_{\varkappa+1} - p_\varkappa)(E) \neq \{0\}\} = \mathfrak{m}_\alpha \leqslant \aleph_\alpha$, then, since the union of any collection of cardinality $\leqslant \aleph_\alpha$ of sets of cardinality $\leqslant \aleph_\alpha$ is*) of cardinality $\leqslant \aleph_\alpha$, we obtain

$$\operatorname{dens} p_{\omega_{\alpha+1}}(E) \leqslant \mathfrak{m}_\alpha \cdot \aleph_\alpha \leqslant \aleph_\alpha < \aleph_{\alpha+1} = \operatorname{card} \omega_{\alpha+1},$$

in contradiction with (19.81′). This proves that $\operatorname{card}\{\varkappa \in [\omega_\alpha, \omega_{\alpha+1}) \mid p_{\varkappa+1} \neq \neq p_\varkappa\} \geqslant \aleph_{\alpha+1}$ and hence the equality (19.84).

Finally, (19.85) is a consequence of (19.84). Indeed, let $\omega_1 \leqslant \lambda \leqslant \vartheta$ and let $\tau \geqslant 1$ be the ordinal index for which $\omega_\tau \leqslant \lambda < \omega_{\tau+1}$, hence $\operatorname{card} \lambda = = \operatorname{card} \omega_\tau = \aleph_\tau \geqslant \aleph_1$. Then, by (19.84),

$$\operatorname{card}\{\varkappa \in [\omega, \lambda) \mid p_{\varkappa+1} \neq p_\varkappa\} \geqslant \operatorname{card}\{\varkappa \in [\omega_\alpha, \omega_{\alpha+1}) \mid p_{\varkappa+1} \neq p_\varkappa\} =$$
$$= \aleph_{\alpha+1} \qquad (0 \leqslant \alpha < \tau),$$

whence

$$\operatorname{card} \lambda \geqslant \operatorname{card}\{\varkappa \in [\omega, \lambda) \mid p_{\varkappa+1} \neq p_\varkappa\} \geqslant \aleph_\tau = \operatorname{card} \lambda,$$

which proves (19.85).

Now, the projections $p_\lambda$ of proposition 19.5 satisfy most of the conditions required in the definition of an ordinal resolution of the identity. However, they form a long sequence instead of a transfinite sequence, the first one is $\neq 0$ (these differences can be corrected easily) and they are ascending instead of being *strictly* ascending (i.e., they need not satisfy $p_{\lambda+1} \neq p_\lambda$ for $\lambda < \vartheta$). In order to overcome this latter difficulty, we shall use

**Lemma 19.7.** *Let $A$ be a partially ordered set, $\vartheta$ an ordinal number, and $\Phi$ an ascending function from $[1, \vartheta]$ into $A$. Then there exist an*

*) See e.g. [1], Ch. III, § 6, theorem 22.

*ordinal number* $\mu$ *and an ascending function* $\xi$ *from* $[1, \mu]$ *into* $[1, \vartheta]$, *such that:*

(i) $\xi(1) = 1$ *and* $\lambda \to \Phi(\xi(\lambda))$ *is strictly ascending for* $1 \leqslant \lambda \leqslant \mu$;

(ii) $\xi$ *is continuous (in the order topology).*

*If* $A$ *is endowed with a topology such that* $\Phi$ *is continuous, then* $\xi$ *may be chosen so as to satisfy in addition:*

(iii) *for any* $\lambda \leqslant \mu$, $\lambda$ *is a limit ordinal if and only if* $\xi(\lambda)$ *is a limit ordinal;*

(iv) *if for all* $\nu < \vartheta$, $\Phi(\nu) \neq \Phi(\vartheta)$ *and if* $\vartheta$ *is a limit ordinal, then* $\xi(\mu) = \vartheta$ *and* $\mu$ *is a limit ordinal.*

*Proof.* We define $\xi$ by transfinite induction. Let $\xi(1) = 1$. If $\xi(\varkappa)$ are defined for $\varkappa < \lambda$, where $\lambda \geqslant 2$, let

$$C_\lambda = \{\Phi(\xi(\varkappa)) \mid \varkappa < \lambda\} = \Phi\{\xi([1, \lambda))\} \subset A,$$

$$D_\lambda = \{\nu \leqslant \vartheta \mid \Phi(\nu) \notin C_\lambda\}.$$

If $D_\lambda \neq \emptyset$, we put

$$\xi(\lambda) = \min_{\nu \in D_\lambda} \nu, \tag{19.86}$$

while if $D_\lambda = \emptyset$, we stop the induction and put $\rho = \lambda$, hence $D_\rho = \emptyset$. We shall show that this $\xi$ determines $\mu$ for which all conditions are satisfied.

Clearly, $\xi(\lambda) \in [1, \vartheta]$ for all $\lambda$ in the domain of definition of $\xi$.

Let us observe the following properties, which follow from the above definition of $\xi(\lambda)$:

(v) $\xi(\lambda) \in D_\lambda$, hence $\Phi(\xi(\lambda)) \notin C_\lambda$;

(vi) if $\nu \in D_\lambda$, then $\nu \geqslant \xi(\lambda)$.

Now let $\alpha < \lambda$, in the domain of $\xi$. Then $C_\alpha \subset C_\lambda$, whence $D_\alpha \supset D_\lambda$. Therefore

$$\xi(\alpha) = \min_{\nu \in D_\alpha} \nu \leqslant \min_{\nu \in D_\lambda} \nu = \xi(\lambda)$$

(so $\xi$ is ascending on its domain of definition), whence, since $\Phi$ is ascending, $\Phi(\xi(\alpha)) \leqslant \Phi(\xi(\lambda))$. But, $\Phi(\xi(\alpha)) \in C_\lambda$ (by $\alpha < \lambda$) and $\Phi(\xi(\lambda)) \notin C_\lambda$, so we must have $\Phi(\xi(\alpha)) < \Phi(\xi(\lambda))$. Thus, $\Phi \circ \xi$ is strictly ascending on its domain of definition. This also shows that the induction must stop (indeed, after at most card $\vartheta$ steps $\Phi(\xi(\lambda))$ will exhaust $\Phi\{\xi([1, \vartheta))\} = C_\vartheta$, so either $D_\vartheta = \emptyset$ or $D_{\vartheta+1} = \emptyset$) and thus $\rho$ will be defined. Since $D_\rho = \emptyset$,

$$\{\Phi(\xi(\varkappa)) \mid \varkappa < \rho\} = C_\rho = \{\Phi(\nu) \mid \nu \leqslant \vartheta\}, \tag{19.87}$$

and hence there exists $\mu < \rho$ such that

$$\Phi(\xi(\mu)) = \Phi(\vartheta). \tag{19.88}$$

This $\mu$ is unique (since $\Phi \circ \xi$ is one-to-one) and $\rho = \mu + 1$ (indeed, if $\mu + 1 < \rho$, then $\xi(\mu + 1) \leqslant \vartheta$ is defined and hence, since $\Phi \circ \xi$ is strictly ascending on its domain and $\Phi$ is defined and ascending on $[1, \vartheta]$, we obtain $\Phi(\xi(\mu + 1)) > \Phi(\xi(\mu)) = \Phi(\vartheta) \geqslant \Phi(\xi(\mu + 1))$, which is absurd). Thus, the domain of definition of $\xi$ is $[1, \rho) = [1, \mu]$.

We have already proved that (i) is satisfied. In order to show (ii) it is enough to prove that if $\lambda \leqslant \mu$ is a limit ordinal, then for every $\nu < \xi(\lambda)$ there exists $\varkappa < \lambda$ such that the relations $\varkappa < \sigma < \lambda$ imply $\nu < \xi(\sigma) < \xi(\lambda)$. Thus, let $\lambda \leqslant \mu$ be a limit ordinal and let $\nu < \xi(\lambda)$. Then, since $\Phi$ is ascending, we have $\Phi(\nu) \leqslant \Phi(\xi(\lambda))$. We claim that $\Phi(\nu) < \Phi(\xi(\lambda))$. Indeed, otherwise, by (v), $\Phi(\nu) = \Phi(\xi(\lambda)) \notin C_\lambda$, whence $\nu \in D_\lambda$ and hence, by (vi), $\nu \geqslant \xi(\lambda)$, a contradiction which proves the claim. But then, by (19.87), there exists $\varkappa \leqslant \mu$ such that $\Phi(\xi(\varkappa)) = \Phi(\nu) < \Phi(\xi(\lambda))$, whence, by (i), $\varkappa < \lambda$. Also, if $\varkappa < \sigma < \lambda$, then again by (i) we have $\Phi(\nu) = \Phi(\xi(\varkappa)) < \Phi(\xi(\sigma)) < \Phi(\xi(\lambda))$, whence, since $\Phi$ is ascending, we obtain $\nu < \xi(\sigma) < \xi(\lambda)$. This proves (ii).

Assume now that $A$ is given a topology such that $\Phi$ is continuous. If $\lambda \leqslant \mu$ is a limit ordinal and $\nu < \xi(\lambda)$, then by the above there exists $\varkappa < \lambda$ such that the relations $\varkappa < \sigma < \lambda$ imply $\nu < \xi(\sigma) < \xi(\lambda)$. But, since $\lambda$ is a limit ordinal, such $\sigma$ exists, and hence $\xi(\lambda)$ is a limit ordinal. This proves one implication*) in (iii). To prove the converse implication in (iii) assume, a contrario, that there exists a nonlimit ordinal $\lambda + 1 \leqslant \mu$ such that $\xi(\lambda + 1)$ is a limit ordinal. Then, by $\lambda < \lambda + 1$ and (i) we have $\Phi(\xi(\lambda)) < \Phi(\xi(\lambda + 1))$, whence, since $\Phi$ is ascending, we obtain $\xi(\lambda) < \xi(\lambda + 1)$. Also, for any $\nu$ with $\xi(\lambda) < \nu < \xi(\lambda + 1)$ (there exist an infinity of such $\nu$, since $\xi(\lambda + 1)$ is a limit ordinal) we have $\nu \notin D_{\lambda+1}$ (since $\nu < \xi(\lambda + 1) = \min_{\nu' \in D_{\lambda+1}} \nu'$), whence there exists $\sigma < \lambda + 1$ such that $\Phi(\nu) = \Phi(\xi(\sigma))$. But, then $\sigma = \lambda$ (indeed, if $\sigma < \lambda$, then, by (i), $\Phi(\nu) = \Phi(\xi(\sigma)) < \Phi(\xi(\lambda))$, which, since $\Phi$ is ascending, contradicts $\xi(\lambda) < \nu$). Thus, the relations $\xi(\lambda) < \nu < \xi(\lambda + 1)$ imply $\Phi(\nu) = \Phi(\xi(\sigma)) = \Phi(\xi(\lambda))$. Since $\Phi$ is continuous, it follows that $\Phi(\xi(\lambda + 1)) = \lim_{\nu < \xi(\lambda+1)} \Phi(\nu) = \Phi(\xi(\lambda))$, which, since $\lambda < \lambda + 1$, contradicts (i). This proves the converse implication in (iii).

Finally, to prove (iv), assume that $\Phi(\nu) \neq \Phi(\vartheta)$ for all $\nu < \vartheta$. If $\xi(\mu) < \vartheta$, then, by our assumption, $\Phi(\xi(\mu)) \neq \Phi(\vartheta)$, in contradiction with (19.88), so we must have $\xi(\mu) = \vartheta$. Hence, if $\vartheta$ is a limit ordinal, then, by (iii), $\mu$ is a limit ordinal. Thus, (iv) holds, which completes the proof of lemma 19.7.

*Remark 19.3.* a) Actually, the function $\xi : [1, \mu] \to [1, \vartheta]$ of lemma 19.7 is even *strictly ascending* and hence we also have

$$\xi(\lambda) \geqslant \lambda \qquad (\lambda \leqslant \mu) \qquad (19.86')$$

---

*) In this part we have not used the continuity of $\Phi$.

(thus, in particular, $\vartheta = \xi(\mu) \geqslant \mu$). Indeed, if $\lambda_1 < \lambda_2 \leqslant \mu$, then, by (i), $\Phi(\xi(\lambda_1)) < \Phi(\xi(\lambda_2))$, whence, since $\Phi$ is ascending, $\xi(\lambda_1) < \xi(\lambda_2)$. Thus*⁾, $\xi$ is strictly ascending. Furthermore, (19.86′) holds for $\lambda = 1$. Now, let us assume that (19.86′) holds for all ordinals $\lambda < \lambda_0$, where $\lambda_0 \leqslant \mu$. Then, since $\xi$ is strictly ascending, $\xi(\lambda_0) > \xi(\lambda) \geqslant \lambda$ for all $\lambda < \lambda_0$, whence $\xi(\lambda_0) \geqslant \lambda_0$, which proves (19.86′).

b) If $\Phi(\nu+1) \neq \Phi(\nu)$ for some $\nu < \vartheta$, then for $\lambda' = \min_{\substack{\lambda \leqslant \vartheta \\ \Phi(\nu) \in C_\lambda}} \lambda \leqslant \mu$ we have $\nu + 1 = \xi(\lambda')$.

Now we are ready to prove

**Theorem 19.4.** *Let $E$ be a non-separable Banach space which is generated by a weakly compact circled convex set $\mathscr{K}$, let $\vartheta$ be the first ordinal number of cardinality* dens $E$ *and let $G$ be a closed linear subspace of $E$. Then $E$ has a monotone ordinal resolution of the identity $\{u_\lambda\}_{\lambda \leqslant \vartheta}$, of type $\vartheta + 1$, such that*

$$u_\lambda(\mathscr{K}) \subset \mathscr{K} \qquad (\lambda \leqslant \vartheta), \tag{19.89}$$

$$u_\lambda(G) \subset G \qquad (\lambda \leqslant \vartheta), \tag{19.90}$$

$$\operatorname{dens} u_\lambda(E) = \begin{cases} 1 & \text{for } \lambda = 1 \\ \aleph_0 & \text{for } 2 \leqslant \lambda < \omega \\ \operatorname{card} \lambda & \text{for } \omega \leqslant \lambda \leqslant \vartheta. \end{cases} \tag{19.91}$$

*Proof.* Let $\{p_\lambda\}_{\omega \leqslant \lambda \leqslant \vartheta}$ be the long sequence of projections on $E$, given by proposition 19.5 and remark 19.2 b). Put

$$u_1' = 0, \; u_2' = p_\omega, \; u_{2+\nu}' = p_{\omega+\nu} \qquad (\nu \leqslant \vartheta). \tag{19.92}$$

Then, in particular, since $\vartheta \geqslant \omega_1$,

$$u_\vartheta' = u_{2+\vartheta}' = p_{\omega+\vartheta} = p_\vartheta = I_E.$$

Furthermore, by (19.81′),

$$\operatorname{dens} u_\nu'(E) = \begin{cases} \operatorname{dens} \{0\} = 1 < \aleph_0 \text{ for } \nu = 1 \\ \operatorname{dens} p_{\omega+(\nu-2)}(E) = \operatorname{card}(\omega + (\nu - 2)) = \aleph_0 \text{ for } 2 \leqslant \nu < \omega \\ \operatorname{dens} u_{2+\nu}'(E) = \operatorname{dens} p_{\omega+\nu}(E) = \operatorname{card}(\omega + \nu) = \operatorname{card} \nu \\ \qquad\qquad\qquad\qquad\qquad \text{for } \omega \leqslant \nu \leqslant \vartheta; \end{cases}$$

also, by (19.76), $\|u_\nu'\| = 1$ for $2 \leqslant \nu \leqslant \vartheta$.

---

*⁾ Actually, we have also used this argument in the proof of (iii), inequality $\xi(\lambda) < \xi(\lambda + 1)$.

Define now a function $\Phi$ from $[1, \mu]$ into $\mathscr{P}$, the set of all continuous linear projections on $E$, endowed with the natural order and with the topology of pointwise convergence, by

$$\Phi(\nu) = u'_\nu \qquad (\nu \leqslant \vartheta). \tag{19.93}$$

Then, by (19.79) and the last statement of proposition 19.5, $\Phi$ is ascending and continuous. Let $\mu$ and $\xi: [1, \mu] \to [1, \vartheta]$ be the ordinal number and the (strictly) ascending continuous function given by lemma 19.7 with $A = \mathscr{P}$. Then, by the above, $\operatorname{dens} u'_\nu(E) < \operatorname{dens} E$ $(\nu < \vartheta)$, whence

$$\Phi(\nu) = u'_\nu \neq I_E = u'_\vartheta = \Phi(\vartheta) \qquad (\nu < \vartheta),$$

and therefore, since $\vartheta$ is a limit ordinal, by lemma 19.7 (iv) it follows that $\mu$ is a limit ordinal and $\xi(\mu) = \vartheta$. Put

$$u_\lambda = u'_{\xi(\lambda)} \qquad (\lambda \leqslant \mu). \tag{19.94}$$

Then $u_1 = u'_{\xi(1)} = u'_1 = 0$, $u_\mu = u'_{\xi(\mu)} = u'_\vartheta = I_E$ and, by lemma 19.7 (i), (ii), the function $\lambda \to u_\lambda = u'_{\xi(\lambda)} = \Phi(\xi(\lambda))$ from $[1, \mu]$ into $\mathscr{P}$ is strictly ascending and continuous. Hence, by theorem 19.2, $\{u_\lambda\}_{\lambda \leqslant \mu}$ is an ordinal resolution of $I_E$. Also by remark 19.3, a), $\xi(\lambda) \geqslant 2$ for $2 \leqslant \lambda \leqslant \mu$, whence

$$\|u_\lambda\| = \|u'_{\xi(\lambda)}\| = 1 \qquad (2 \leqslant \lambda \leqslant \mu),$$

so $\{u_\lambda\}_{\lambda \leqslant \mu}$ is monotone. The relations (19.89), (19.90) (for $\lambda \leqslant \mu$) are obvious from (19.77), (19.80). We shall show that $\mu = \vartheta$ and that we have (19.91), which will complete the proof.

We claim that *if* $\xi(\lambda_0) = \omega_\tau$ *for some* $\tau \geqslant 1$ *and* $\omega_1 \leqslant \lambda_0 \leqslant \mu$, *then* $\lambda_0 = \xi(\lambda_0) = \omega_\tau$; hence, in particular, since $\xi(\mu) = \vartheta$ is of the form $\omega_\sigma$ for some $\sigma \geqslant 1$ (by the definition of $\vartheta$), it follows that*) $\mu = \xi(\mu) = \vartheta$. Indeed, by $u_{\lambda+1} \neq u_\lambda$ $(\lambda < \mu)$, (19.93), remark 19.3b), (19.94), (19.92) and (19.85), we obtain

$$\operatorname{card} \lambda_0 = \operatorname{card} \{\lambda < \lambda_0 \mid u_{\lambda+1} \neq u_\lambda\} = \operatorname{card} \{\lambda < \lambda_0 \mid u'_{\xi(\lambda+1)} \neq u'_{\xi(\lambda)}\} =$$

$$= \operatorname{card} \{\omega \leqslant \nu < \xi(\lambda_0) \mid u'_{\nu+1} \neq u'_\nu\} =$$

$$= \operatorname{card} \{\omega \leqslant \varkappa < \xi(\lambda_0) \mid p_{\varkappa+1} \neq p_\varkappa\} = \operatorname{card} \xi(\lambda_0).$$

*) Note that although $\xi: [1, \vartheta] \to [1, \vartheta]$ is strictly ascending, $\xi$ need not be the identity.

Hence, since $\xi(\lambda_0) = \omega_\tau$ is the first ordinal number of cardinality $\aleph_\tau = \operatorname{card} \xi(\lambda_0)$, it follows that $\lambda_0 \geqslant \xi(\lambda_0)$. Consequently, by remark 19.3 a), $\lambda_0 = \xi(\lambda_0)$, which proves the claim.

From this observation it follows that

$$\xi(\omega_\tau) = \omega_\tau \qquad (\tau \geqslant 0,\ \omega_\tau \leqslant \vartheta). \tag{19.86''}$$

Indeed, by remark 19.3 a), $\xi(\omega_\tau) \geqslant \omega_\tau$. Now, if the strict inequality

$$\omega_\tau < \xi(\omega_\tau) = \min_{u'_\nu \notin \{u'_{\xi(\varkappa)}\}_{\varkappa < \omega_\tau}} \nu$$

holds (see the definition of $\xi$ in the proof of lemma 19.7), then $u'_{\omega_\tau} \in \{u'_{\xi(\varkappa)}\}_{\varkappa < \omega_\tau}$, so there exists $\varkappa_0 < \omega_\tau$ such that $\xi(\varkappa_0) = \omega_\tau$. Hence, if $\tau \geqslant 1$, then, by the preceding observation, $\varkappa_0 = \xi(\varkappa_0) = \omega_\tau$, in contradiction with $\varkappa_0 < \omega_\tau$, while if $\tau = 0$, then $\varkappa_0 < \omega_0$ is a non-limit ordinal and $\xi(\varkappa_0) = \omega_0$ is a limit ordinal, in contradiction with lemma 19.7 (iii). This proves (19.86'').

Now we can prove (19.91). Clearly, $\operatorname{dens} u_1(E) = \operatorname{dens} \{0\} = 1$. If $2 \leqslant \lambda < \omega = \omega_0$, then by (19.86'') for $\tau = 0$ and since $\xi$ is strictly ascending, we have $1 = \xi(1) < \xi(2) \leqslant \xi(\lambda) < \xi(\omega_0) = \omega_0$, whence, by $\operatorname{dens} u'_\nu(E) = \aleph_0$ $(2 \leqslant \nu < \omega)$, we get

$$\operatorname{dens} u_\lambda(E) = \operatorname{dens} u'_{\xi(\lambda)}(E) = \aleph_0.$$

Finally, let $\omega \leqslant \lambda < \vartheta$ and let $\tau \geqslant 0$ be the ordinal index for which $\omega_\tau \leqslant \lambda < \omega_{\tau+1} \leqslant \vartheta$. Then, by (19.86'') and since $\xi$ is strictly ascending $\omega_\tau = \xi(\omega_\tau) \leqslant \xi(\lambda) < \xi(\omega_{\tau+1}) = \omega_{\tau+1}$, whence, by $\operatorname{dens} u'_\nu(E) = \operatorname{card} \nu$, $(\omega \leqslant \nu \leqslant \vartheta)$, we obtain

$$\operatorname{dens} u_\lambda(E) = \operatorname{dens} u'_{\xi(\lambda)}(E) = \operatorname{card} \xi(\lambda) = \operatorname{card} \omega_\tau = \aleph_\tau = \operatorname{card} \lambda,$$

which proves*) (19.91) and completes the proof of theorem 19.4.

*Remark 19.4.* a) In the above we have constructed an ordinal resolution of the identity $\{u_\lambda\}_{\lambda \leqslant \mu}$ and then proved that $\mu = \vartheta$, by taking a long sequence $\{p_\lambda\}_{\omega \leqslant \lambda \leqslant \vartheta}$ as in proposition 19.5, which satisfies also (19.81') (using to this end extended $M$-bases), then "counting" the sets $\{\omega \leqslant \varkappa < \lambda \mid p_{\varkappa+1} \neq p_\varkappa\}$ for $\omega_1 \leqslant \lambda \leqslant \vartheta$ (formula (19.85)) and then, finally, taking a suitable transfinite subsequence $\{u'_{\xi(\lambda)}\}_{\lambda \leqslant \mu}$ of $\{p_\lambda\}_{\omega \leqslant \lambda \leqslant \vartheta}$ consisting of distinct elements**), to which we joined $u_1 = 0$ and which we reindexed (formulae (19.92), (19.94)). In the particular case when $\operatorname{dens} E$ is of the form $\aleph_{\tau+1}$ for some $\tau \geqslant 0$ (hence $\vartheta = \omega_{\tau+1}$), one can

*) Clearly, $\operatorname{dens} u_\vartheta(E) = \operatorname{dens} E = \operatorname{card} \vartheta$.

**) Actually, by our construction (see (19.86)), the *sets* $\{u'_{\xi(\lambda)}\}_{\lambda \leqslant \mu}$ and $\{p_\lambda\}_{\omega \leqslant \lambda \leqslant \vartheta}$ coincide.

also give the following simpler proof of the equality $\mu = \vartheta$ (similar to the above proof of the equality (19.84)), which makes use only of (19.81), but not of (19.81'): By remark 19.3 a) and lemma 19.7, $\mu \leqslant \xi(\mu) = \vartheta$. Assume now that $\mu < \vartheta$, so card $\mu \leqslant \aleph_\tau$. Then, by (19.81) and since $\xi(\lambda+1) < \xi(\mu) = \vartheta = \omega_{\tau+1}$ $(\lambda < \mu)$, where $\vartheta$ is the first ordinal number of cardinality dens $E$, we have

$$\operatorname{dens}(u_{\lambda+1} - u_\lambda)(E) = \operatorname{dens}(u'_{\xi(\lambda+1)} - u'_{\xi(\lambda)})(E) \leqslant$$

$$\leqslant \operatorname{dens} u'_{\xi(\lambda+1)}(E) \leqslant \operatorname{card} \xi(\lambda+1) \leqslant \aleph_\tau \qquad (\lambda < \mu).$$

But, since $\{u_\lambda\}_{\lambda \leqslant \mu}$ is an ordinal resolution of $I_E$, we have (see (19.17))

$$E = u_\mu(E) = [\bigcup_{\lambda<\mu} (u_{\lambda+1} - u_\lambda)(E)].$$

Consequently,

$$\aleph_{\tau+1} = \operatorname{dens} E \leqslant \sum_{\lambda<\mu} \operatorname{dens}(u_{\lambda+1} - u_\lambda)(E) \leqslant (\operatorname{card} \mu)\, \aleph_\tau \leqslant$$

$$\leqslant \aleph_\tau \cdot \aleph_\tau = \aleph_\tau,$$

which is impossible. This proves that $\mu = \vartheta$. In § 20, remark 20.6, we shall give, in the general case (i.e., for any dens $E$), another construction of an ordinal resolution of the identity $\{u_\lambda\}_{\lambda \leqslant \vartheta}$, of type $\vartheta+1$, by using extended $M$-bases to replace $\{p_\lambda\}_{\omega \leqslant \lambda \leqslant \vartheta}$ directly by a long sequence of the same type $\vartheta + 1$, consisting of distinct elements and then again joining 0 and reindexing.

b) Under the assumptions of theorem 19.4, we also have

$$\operatorname{dens}(u_{\lambda+1} - u_\lambda)(E) < \operatorname{dens} E \qquad (\lambda < \vartheta); \qquad (19.91')$$

indeed, by (19.91) and $(u_{\lambda+1} - u_\lambda)(E) \subset u_{\lambda+1}(E)$ $(\lambda < \vartheta)$,

$$\operatorname{dens}(u_{\lambda+1} - u_\lambda)(E) \leqslant \operatorname{dens} u_{\lambda+1}(E) \leqslant \operatorname{card}(\lambda+1) = \operatorname{card} \lambda <$$

$$< \operatorname{dens} E \ (\lambda < \vartheta).$$

c) The converse of theorem 19.4 is not valid, since for example the space $E = l^1(I)$, where $I$ is uncountable, is non-separable and has a monotone ordinal resolution of the identity (generated by the natural extended unconditional basis (17.20) of $l^1(I)$ and any well ordering of $I$), but $l^1(I)$, where $I$ is uncountable, is not even isomorphic to any subspace of a weakly compactly generated Banach space (see § 20, corollary 20.5).

The norm $|\cdot|$ on $E$ (satisfying (19.51)) and the equalities $|p_\lambda| = 1$ $(\omega \leqslant \lambda \leqslant \vartheta)$ occurring in proposition 19.5 are not necessary for the proof of theorem 19.4 and, in fact, have not yet been used at all. However, we shall use them below, combined with the following lemma, which enables us to apply the above results to conjugate Banach spaces so as to get even $w^*$-continuous projections:

**Lemma 19.8.** *Let $(E, \|\cdot\|)$ be a Banach space which is generated by a weakly compact circled convex set $\mathscr{K}$. Put*

$$|f| = \sup_{x \in \mathscr{K}} |f(x)| \qquad (f \in E^*). \tag{19.95}$$

*If $u: E^* \to E^*$ is a linear operator which is continuous in both norms $|\cdot|$ and $\|\cdot\|$, then $u$ is $w^*$-$w^*$ continuous.*

*Proof.* We claim that the identity mapping $I_{E^*}: E^* \to E^*$ is $w^*$-$w_{|\cdot|}$continuous, where $w_{|\cdot|}$ denotes the weak topology $\sigma((E^*, |\cdot|), (E^*, |\cdot|)^*)$. Indeed, let $\{f_d\}_{d \in \mathscr{D}} \subset E^*, f_0 \in E^*, f_d \overset{w^*}{\to} f_0$ and let $\Phi \in (E^*, |\cdot|)^*$. Observe that $(E^*|_{\mathscr{K}}, |\cdot|)$ is a subspace of $C(\mathscr{K})$, linearly isometric to $(E^*, |\cdot|)$ by the mapping $f|_{\mathscr{K}} \to f$ $(f \in E^*)$, since the relations $f', f'' \in E^*, f'|_{\mathscr{K}} = f''|_{\mathscr{K}}$ imply $f' = f''$ (by $[\mathscr{K}] = E$) and since, by (19.95), $\|f|_{\mathscr{K}}\|_{C(\mathscr{K})} = \sup_{x \in \mathscr{K}} |f(x)| = |f|$ $(f \in E^*)$. Define a functional $\varphi$ on $(E^*|_{\mathscr{K}}, |\cdot|)$ by

$$\varphi(f|_{\mathscr{K}}) = \Phi(f) \qquad (f \in E^*).$$

Then, since $\Phi \in (E^*, |\cdot|)^*$, we have

$$|\varphi(f|_{\mathscr{K}})| = |\Phi(f)| \leqslant |\Phi|\, |f| = |\Phi| \sup_{x \in \mathscr{K}} |f(x)| \qquad (f \in E^*),$$

so $\varphi \in (E^*|_{\mathscr{K}}, |\cdot|)^*$ and hence $\varphi$ can be extended to a Radon measure $\mu_0 \in C(\mathscr{K})^*$. Thus, in order to prove the claim that $\Phi(f_d) \to \Phi(f_0)$, it will be enough to prove that $f_d|_{\mathscr{K}} \to f_0|_{\mathscr{K}}$ in the weak topology $\sigma(C(\mathscr{K}), C(\mathscr{K})^*)$, i.e., that $\mu(f_d|_{\mathscr{K}}) \to \mu(f_0|_{\mathscr{K}})$ for every $\mu \in C(\mathscr{K})^*$, or, what is equivalent, for every probability measure $\mu$ (i.e., positive and of total mass 1). But, since $\mathscr{K}$ is a weakly compact convex subset of $(E, \|\cdot\|)$, for every probability measure $\mu$ on $\mathscr{K}$ there exists*) a (unique) $x_\mu \in \mathscr{K}$ such that $\mu(f|_{\mathscr{K}}) = \int_{\mathscr{K}} f(y)d\mu(y) = f(x_\mu)$ $(f \in E^*)$. Hence, since by our assumption $f_d \overset{w^*}{\to} f_0$, we obtain that $f_d|_{\mathscr{K}} \to f_0|_{\mathscr{K}}$ for $\sigma(C(\mathscr{K}), C(\mathscr{K})^*)$, which proves our claim. On the other hand, since $S_{(E^*, \|\cdot\|)}$ is $w^*$-compact, for every ball $rS_{(E^*, \|\cdot\|)}$ the restriction $I_{E^*}|_{rS_{(E^*, \|\cdot\|)}}$ is also $w_{|\cdot|}$-$w^*$ continuous. Consequently, the $w^*$ and $w_{|\cdot|}$ topologies coincide on each ball $rS_{(E^*, \|\cdot\|)}$.

*) See e.g. [316], p. 4, proposition 1.1.

Now, since by our assumption $u$ is $|.|$-continuous, it is also $w_{|.|}$-$w_{|.|}$ continuous*). Furthermore, since by our assumption $u$ is $\|.\|$-continuous, we have $u(S_{(E^*,\|.\|)}) \subset rS_{(E^*,\|.\|)}$ for some $r > 0$. Consequently, by the above, $u$ is $w^*$-$w^*$ continuous on $S_{(E^*,\|.\|)}$, whence**) also on $E^*$, which completes the proof of lemma 19.8.

**Proposition 19.6.** *Let $E$ be a Banach space which is generated by a weakly compact circled convex set $\mathscr{K}$ and such that $E^*$ is generated by a weakly compact circled convex set $\mathscr{L}$, let $\Gamma$ be a closed linear subspace of $E^*$, and let $\{f_\lambda\}_{\lambda<\vartheta}$ be a dense transfinite sequence in $V = \operatorname{lin} \mathscr{L}$, where $\vartheta$ is the first ordinal number of cardinality* dens $V =$ dens $E^*$. *Then there exists a long sequence of linear projections $\{p_\lambda\}_{\omega\leqslant\lambda\leqslant\vartheta}$ on $E$, satisfying* (19.79), (19.82), (19.83) *and*

$$\|p_\lambda\| = 1 \qquad (\omega \leqslant \lambda \leqslant \vartheta), \tag{19.96}$$

$$p_\lambda^*(\mathscr{L}) \subset \mathscr{L} \qquad (\omega \leqslant \lambda \leqslant \vartheta), \tag{19.97}$$

$$f_\varkappa \in p_\lambda^*(E^*) \qquad (\varkappa < \lambda \leqslant \vartheta), \tag{19.98}$$

$$p_\lambda^*(\Gamma) \subset \Gamma \qquad (\omega \leqslant \lambda \leqslant \vartheta), \tag{19.99}$$

$$\operatorname{dens} p_\lambda^*(E^*) \leqslant \operatorname{card} \lambda \qquad (\omega \leqslant \lambda \leqslant \vartheta), \tag{19.100}$$

$$p_\lambda^*(E^*) \subset \overline{\bigcup_{\omega\leqslant\varkappa<\lambda} p_{\varkappa+1}^*(E^*)} \qquad (\omega < \lambda \leqslant \vartheta), \tag{19.101}$$

*and such that for each fixed $x \in E$, $f \in E^*$ the mappings $\lambda \to p_\lambda(x)$ and $\lambda \to p_\lambda^*(f)$ from $[\omega, \vartheta]$ into $E$ and $E^*$ respectively, are continuous.*

*Proof.* Since $\mathscr{K}$ is bounded, we may assume (by contracting $\mathscr{K}$ with a suitable homothety, if necessary) that $\mathscr{K} \subset S_{(E,\|.\|)}$. Define a norm $|.|$ on $E^*$ by (19.95). Then, by our assumption,

$$|f| = \sup_{x\in\mathscr{K}} |f(x)| \leqslant \sup_{\substack{x\in E\\ \|x\|\leqslant 1}} |f(x)| = \|f\| \qquad (f \in E^*). \tag{19.102}$$

Hence, by proposition 19.5, there exists a long sequence of linear projections $\{q_\lambda\}_{\omega\leqslant\lambda\leqslant\vartheta}$ on $E^*$, satisfying $|q_\lambda| = \|q_\lambda\| = 1$, $q_\lambda(\mathscr{L}) \subset \mathscr{L}$ $(\omega\leqslant\lambda\leqslant\vartheta)$, $f_\varkappa \in q_\lambda(E^*)$ $(\varkappa < \lambda \leqslant \vartheta)$, $q_\lambda q_\varkappa = q_\varkappa q_\lambda = q_\varkappa$ $(\omega \leqslant \varkappa \leqslant \lambda \leqslant \vartheta)$, $q_\lambda(\Gamma)\subset\Gamma$, dens $q_\lambda(E^*) \leqslant$ card $\lambda$ $(\omega \leqslant \lambda\leqslant \vartheta)$, $q_\lambda(E^*) \subset \overline{\bigcup_{\omega\leqslant\varkappa<\lambda} q_{\varkappa+1}(E^*)}$

*) See e.g. [87], p. 422, theorem 5.

**) See e.g. [87], p. 428, theorem 6, applied to each $u^*(\pi(x)) \in E^{**}$ $(x \in E)$, where $\pi : E \to E^{**}$ is the canonical embedding.

$(\omega < \lambda \leqslant \vartheta)$, $q_\vartheta = I_{E^*}$ and such that for any fixed $f \in E^*$ the mapping $\lambda \to q_\lambda(f)$ of $[\omega, \vartheta]$ into $E^*$ is continuous. Then, by lemma 19.8, each $q_\lambda$: $E^* \to E^*$ is $w^*$-$w^*$ continuous and hence*) there exists $p_\lambda \in L(E, E)$ such that $p_\lambda^* = q_\lambda$ $(\omega \leqslant \lambda \leqslant \vartheta)$. Then, by the above, the long sequence of linear projections $\{p_\lambda\}_{\omega \leqslant \lambda \leqslant \vartheta}$ on $E$ will have all the required properties, provided that we prove (19.82) (since then, by proposition 19.5, for each fixed $x \in E$ the mapping $\lambda \to p_\lambda(x)$ from $[\omega, \vartheta]$ into $E$ will be continuous). Clearly, (19.82) holds for nonlimit ordinals $\lambda \in (\omega, \vartheta]$. Now let $x \in E$, $f \in E^*$ and let $\lambda \in (\omega, \vartheta]$ be a limit ordinal. Then, since the mapping $\lambda \to p_\lambda^*(f)$ of $[\omega, \vartheta]$ into $E^*$ is continuous, we have $\lim_{\varkappa < \lambda} f(p_\varkappa(x)) = \lim_{\varkappa < \lambda} p_\varkappa^*(f)(x) = p_\lambda^*(f)(x) = f(p_\lambda(x))$, whence, since $f \in E^*$ was arbitrary, $p_\lambda(x) = w\text{-}\lim_{\varkappa < \lambda} p_\varkappa(x)$. But then**) $p_\lambda(x) \in \overline{\bigcup_{\omega \leqslant \varkappa < \lambda} p_\varkappa(E)} = \overline{\bigcup_{\omega \leqslant \varkappa < \lambda} p_{\varkappa+1}(E)}$, so (19.82) holds, which completes the proof of proposition 19.6. Note also that, similarly to remark 19.2 b), we can replace in (19.100) the inequalities by equalities.

**Theorem 19.5.** *Let $E$ be a non-separable Banach space which is generated by a weakly compact circled convex set $\mathscr{K}$ and such that $E^*$ is generated by a weakly compact circled convex set $\mathscr{L}$, let $\vartheta$ be the first ordinal of cardinality****) dens $E^*$, *and let $\Gamma$ be a closed linear subspace of $E^*$. Then $E$ has a shrinking monotone ordinal resolution of the identity $\{u_\lambda\}_{\lambda \leqslant \vartheta}$, of type $\vartheta + 1$, such that*

$$u_\lambda^*(\mathscr{L}) \subset \mathscr{L} \qquad (\lambda \leqslant \vartheta), \tag{19.103}$$

$$u_\lambda^*(\Gamma) \subset \Gamma \qquad (\lambda \leqslant \vartheta), \tag{19.104}$$

$$\operatorname{dens} u_\lambda^*(E^*) = \begin{cases} 1 & \text{for } \lambda = 1 \\ \aleph_0 & \text{for } 2 \leqslant \lambda < \omega \\ \operatorname{card} \lambda & \text{for } \omega \leqslant \lambda \leqslant \vartheta. \end{cases} \tag{19.105}$$

The proof is similar to that of theorem 19.4, using now proposition 19.6 (instead of proposition 19.5).

Now we shall show, adapting the above methods so as to exploit that conjugate Banach spaces are "weakly* compactly generated" (because of the $w^*$-compactness of balls in $E^*$) that a certain geometric property of a non-separable space implies the existence of a monotone ordinal resolution of the identity in the conjugate space. We recall that

---

*) See the footnote to § 15, proof of theorem 15.11.

**) See e.g. [355], Ch. II, § 9, corollary 2 of theorem 9.2.

***) Actually (see the Notes and remarks to § 20), under these assumptions $E$ has a shrinking extended $M$-basis and hence dens $E^*$ = dens $E$.

a Banach space $E$ is said to be *smooth* if for each $x \in E$ with $\|x\| = 1$ there exists a *unique* $f_x \in E^*$ such that $f_x(x)=1$ and $\|f_x\|=1$. If $E$ is smooth, then, clearly (by homogeneity), for each $x \in E$ there exists a *unique* $h_x = \|x\| f_{\frac{x}{\|x\|}} \in E^*$ such that $h_x(x) = \|h_x\| \, \|x\|$ and $\|h_x\| = \|x\|$ and it is well known*) that the relations $x_n$, $x \in E$, $x_n \to x$ imply $h_{x_n} \xrightarrow{w^*} h_x$. Also, we recall that a Banach space $E$ is said to be *very smooth* if it is smooth and if the relations $x_n$, $x \in E$, $x_n \to x$ imply $h_{x_n} \xrightarrow{w} h_x$. Our next aim is to prove that the conjugate space of a very smooth Banach space admits a long sequence of projections with certain properties similar to the above ones and hence, if $E$ is a non-separable**) Banach space isomorphic to a very smooth space, then $E^*$ has a monotone ordinal resolution of the identity. To this end, we shall need some preparation.

If $C$ is a subspace of a Banach space $E$ and if $v \in L(C^*, E^*)$, we shall denote by $\tilde{v}$ the linear mapping of $E^*$ into $E^*$ defined by

$$\tilde{v}(f) = v(f|_C) \qquad (f \in E^*); \tag{19.106}$$

since $\|\tilde{v}(f)\| \leqslant \|v\| \, \|f|_C\| \leqslant \|v\| \, \|f\|$ $(f \in E^*)$, we have $\tilde{v} \in L(E^*, E^*)$ and

$$\|\tilde{v}\| \leqslant \|v\|. \tag{19.107}$$

We shall retain this notation until the end of the present section.

**Lemma 19.9.** *Let $E$ be a Banach space and $B$ a finite-dimensional subspace of $E$. Then there exist a separable subspace $C$ of $E$, containing $B$, and an operator $v \in L(C^*, E^*)$ such that*

$$\|\tilde{v}\| = \|v\| = 1, \tag{19.108}$$

$$\tilde{v}^*(\pi(y)) = \pi(y) \qquad (y \in B), \tag{19.109}$$

*where $\pi$ is the canonical embedding of $E$ into $E^{**}$.*

*Proof.* By a particular case of lemma 19.3, for every $n$ there exists a separable subspace $C_n$ of $E$ such that for every finite-dimensional subspace $Z$ of $E$ with $Z \supset B$, $\dim Z/B = n$, there is a linear operator $u_Z \colon Z \to C_n$ satisfying

$$\|u_Z\| \leqslant 1 + \frac{1}{n}, \tag{19.110}$$

$$u_Z(y) = y \qquad (y \in B). \tag{19.111}$$

*) See e.g. [389], § 1.

**) By lemma 19.13, this is equivalent to the assumption that $E^*$ is non-separable.

Let

$$C=\left[\bigcup_{n=1}^{\infty} C_n\right]. \tag{19.112}$$

Then $C$ is a separable subspace of $E$ and we may regard each $u_Z$ above as a mapping from $Z$ into $C$, with the properties (19.110), (19.111).

Let $\mathscr{D}$ be the set of all finite-dimensional subspaces of $E$ with $Z\supset B$, directed by inclusion. For each $Z\in\mathscr{D}$, extend $u_Z$ to a (non-linear) operator $u'_Z: E\to C$, by putting

$$u'_Z(x)=\begin{cases} u_Z(x) & \text{for } x\in Z\\ 0 & \text{for } x\in E\setminus Z.\end{cases} \tag{19.113}$$

Let us denote by $E^h$ the space of all homogeneous functionals on $E$, which are bounded on the unit ball $S_E$ of $E$, endowed with the norm $\|f\|=\sup_{\substack{x\in E\\ \|x\|\leqslant 1}}|f(x)|$ $(f\in E^h)$. Then for each $Z\in\mathscr{D}$ the adjoint $(u'_Z)^*$ of $u'_Z$, defined by

$$(u'_Z)^*(\varphi)\,(x)=\varphi(u'_Z(x))\qquad (x\in E,\ \varphi\in C^*), \tag{19.114}$$

maps $C^*$ into $E^h$ and, moreover, by (19.113), (19.110),

$$\|(u'_Z)^*(\varphi)\|=\sup_{\substack{x\in E\\ \|x\|\leqslant 1}}|(u'_Z)^*(\varphi)(x)|\leqslant \|u_Z\|\,\|\varphi\|\leqslant 2\|\varphi\|\qquad (\varphi\in C^*), \tag{19.115}$$

so $u'_Z|_{S_{C^*}}\in (2S_{E^h})^{S_{C^*}}$, the space of all functions from $S_{C^*}$ into $2S_{E^h}$. If we take on $S_{E^h}$ the $\sigma(E^h, E)$-topology, then $S_{E^h}$ is compact*) and hence, by Tychonov's theorem, $(2S_{E^h})^{S_{C^*}}$ is compact, so the net $\{u'_Z|_{S_{C^*}}\}_{Z\in\mathscr{D}}$ has a subnet, which we shall denote again by $\{u'_Z|_{S_{C^*}}\}_{Z\in\mathscr{D}}$, converging pointwise to some $v\in(2S_{E^h})^{S_{C^*}}$ in the $\sigma(E^h, E)$-topology, that is, $(u'_Z)^*(\varphi)\,(x)\to v(\varphi)\,(x)$ for all $x\in E$, $\varphi\in S_{C^*}$.

We claim that $v(\varphi)\in E^*$ $(\varphi\in S_{C^*})$. Indeed, for any $\varphi\in S_{C^*}$ and $x, y, \alpha x+\beta y\in E$ we can take a finite-dimensional subspace $Z$ of $E$ with $Z\supset[B, x, y, \alpha x+\beta y]$ and then, by (19.113) and the linearity of $u_Z: Z\to C$,

$$(u'_Z)^*(\varphi)\,(\alpha x+\beta y)-\alpha(u'_Z)^*(\varphi)\,(x)-\beta(u'_Z)^*(\varphi)\,(y)=$$

$$=\varphi(u_Z(\alpha x+\beta y))-\alpha\varphi(u_Z(x))-\beta\varphi u_Z(y)=0,$$

*) This follows e.g. from the argument of [87], p. 423, proof of lemma 1.

so the same holds also for the $\sigma(E^h, E)$-limit $v(\varphi)$, which proves the claim.

Consequently, we can extend $v$ homogeneously to an operator $v: C^* \to E^*$, defining

$$v(\varphi) = \begin{cases} \|\varphi\| \, v\left(\dfrac{\varphi}{\|\varphi\|}\right) & \text{for } \varphi \in C^* \setminus \{0\} \\ 0 & \text{for } \varphi = 0. \end{cases} \tag{19.116}$$

We shall show that $v$ satisfies (19.108), (19.109), which will complete the proof. Since each $(u'_Z)^*$ is linear, $v$ is linear on $S_{C^*}$, whence also on $C^*$. Obviously (by homogeneity), $(u'_Z)^*(\varphi) \to v(\varphi)$ for $\sigma(E^h, E)$, for all $\varphi \in C^*$.

If $x \in E$ and $\varphi \in C^*$, then, by (19.110),

$$|v(\varphi)(x)| = |\lim_{Z \in \mathscr{D}} (u'_Z)^*(\varphi)(x)| = \lim_{Z \in \mathscr{D}} |\varphi(u'_Z(x))| \leqslant$$

$$\leqslant \|\varphi\| \overline{\lim_{Z \in \mathscr{D}}} \|u'_Z(x)\| \leqslant \|\varphi\| \overline{\lim_{n \to \infty}} \left(1 + \frac{1}{n}\right) \|x\| = \|\varphi\| \, \|x\|,$$

whence, by (19.107), $\|\tilde{v}\| \leqslant \|v\| \leqslant 1$. Furthermore, by (19.106) and (19.111),

$$\tilde{v}^*(\pi(y))(f) = \pi(y)(\tilde{v}(f)) = \tilde{v}(f)(y) = v(f|_C)(y) = \lim_{Z \in \mathscr{D}} (u'_Z)^*(f|_C)(y) =$$

$$= \lim_{Z \in \mathscr{D}} f(u'_Z(y)) = f(y) = \pi(y)(f) \qquad (y \in B, \ f \in E^*),$$

which proves (19.109). This, together with $\|\tilde{v}\| \leqslant \|v\| \leqslant 1$, yields (19.108), completing the proof of lemma 19.9.

The following lemma will play the role of lemma 19.4:

**Lemma 19.10.** *Let $E$ be a smooth Banach space, let $x_1, \ldots, x_n \in E$, $f_1, \ldots, f_m \in E^*$ (where $n, m < \infty$), and let $\varepsilon > 0$. Then there exist a separable subspace $C$ of $E$, containing $x_1, \ldots, x_n$ and an operator $v \in L(C^*, E^*)$ such that*

$$\|\tilde{v}\| = \|v\| = 1, \tag{19.117}$$

$$\tilde{v}^*(\pi(x_i)) = \pi(x_i) \qquad (i = 1, \ldots, n), \tag{19.118}$$

$$\|\tilde{v}(f_j) - f_j\| < \varepsilon \qquad (j = 1, \ldots, m). \tag{19.119}$$

*Proof.* Since every Banach space $E$ is*) "subreflexive" (i.e. the set of all $f \in E^*$ which attain their norm on $S_E$ is norm dense in $E^*$), there exist $g_j \in E^*$ and $y_j \in E$ with $g_j(y_j) = \|g_j\|$, $\|y_j\| = 1$ $(j = 1, \dots, m)$, such that

$$\|f_j - g_j\| < \frac{\varepsilon}{2} \qquad (j = 1, \dots, m). \tag{19.120}$$

By lemma 19.9, there exist a separable subspace $C$ of $E$, containing $x_1, \dots, x_n, y_1, \dots, y_m$ and an operator $v \in L(C^*, E^*)$ satisfying (19.117), (19.118) and

$$\tilde{v}^*(\pi(y_j)) = \pi(y_j) \qquad (j = 1, \dots, m), \tag{19.121}$$

whence

$$\tilde{v}(f)(y_j) = \pi(y_j)\,(\tilde{v}(f)) = \tilde{v}^*(\pi(y_j))\,(f) = \pi(y_j)\,(f) = f(y_j)$$

$$(f \in E^*,\ j = 1, \dots, m).$$

In particular, for $f = g_j$ we obtain, by (19.117),

$$\tilde{v}(g_j)\,(y_j) = g_j(y_j) = \|g_j\| \geqslant \|\tilde{v}(g_j)\| \qquad (j = 1, \dots, m),$$

and therefore, since $\|y_j\| = 1$, we must have

$$\tilde{v}(g_j)\,(y_j) = \|\tilde{v}(g_j)\|, \quad \|\tilde{v}(g_j)\| = \|g_j\| \qquad (j = 1, \dots, m).$$

Hence, by the definition of $g_j$, $y_j$ $(j = 1, \dots, m)$ and since $E$ is smooth, we obtain

$$\tilde{v}(g_j) = g_j \qquad (j = 1, \dots, m). \tag{19.122}$$

Consequently, by (19.122), (19.120) and (19.117), we get

$$\|v(f_j) - f_j\| \leqslant \|v(f_j) - \tilde{v}(g_j)\| + \|g_j - f_j\| \leqslant (\|\tilde{v}\| + 1)\frac{\varepsilon}{2} < \varepsilon \quad (j=1, \dots, m),$$

i.e., (19.119), which completes the proof of lemma 19.10.

If $G$ is a (closed linear) subspace of a Banach space $E$, we shall denote by $D_{E^*}(G)$ the set of all functionals $f \in E^*$ which attain their norm on the unit ball $S_G$ of $G$. We shall retain this notation throughout the present section.

In order to prove a substitute of lemma 19.5, we shall need

*) See e.g. [70], Ch. II, § 2, statement (11) or [29].

**Lemma 19.11.** *Let $G$ be a closed linear subspace of a Banach space $E$. If $\overline{D_{E^*}(G)}$ is a closed linear subspace of $E^*$, then it is linearly isometric to $G^*$, by the restriction map.*

*Proof.* Let $\rho: \overline{D_{E^*}(G)} \to G^*$ be the restriction map, i.e.

$$\rho(f) = f|_G \qquad (f \in \overline{D_{E^*}(G)}). \tag{19.123}$$

For any $f \in D_{E^*}(G)$ there exists $y \in G$ with $\|y\|=1$, $f(y)=\|f\|$, whence

$$\|\rho(f)\| \geqslant |\rho(f)(y)| = |f|_G(y)| = |f(y)| = \|f\| \geqslant \|f|_G\| = \|\rho(f)\| \qquad (f \in D_{E^*}(G)),$$

and therefore

$$\|\rho(f)\| = \|f\| \qquad (f \in \overline{D_{E^*}(G)}). \tag{19.124}$$

Thus, $\rho$ is a linear norm preserving map, and hence a linear isometry, of $\overline{D_{E^*}(G)}$ into $G^*$. We shall show that $\rho$ maps $\overline{D_{E^*}(G)}$ *onto* $G^*$, which will complete the proof.

Let $\varphi \in D_{G^*}(G)$ ($\subset G^*$) and let $f \in E^*$ be a Hahn-Banach extension of $\varphi$, that is, $f|_G = \varphi$, $\|f\| = \|\varphi\|$. Then, clearly, $f \in D_{E^*}(G)$ and $\rho(f) = \varphi$. Thus, the linear isometry $\rho$ maps $\overline{D_{E^*}(G)}$ into a closed subset of $G^*$, containing $D_{G^*}(G)$. But $G$ is subreflexive*), i.e., $D_{G^*}(G)$ is (norm) dense in $G^*$, so $\rho(\overline{D_{E^*}(G)}) = G^*$, which completes the proof of lemma 19.11.

The following lemma corresponds now to lemma 19.5:

**Lemma 19.12.** *Let $E$ be a smooth Banach space, $\mathfrak{m}$ an infinite cardinal number, and $B \subset E$, $V \subset E^*$ subspaces of $E$, $E^*$ respectively, with* dens $B \leqslant \mathfrak{m}$. *Then there exist a subspace $C$ of $E$, containing $B$, with* dens $C \leqslant \mathfrak{m}$, *and an operator $v \in L(C^*, E^*)$ such that $q = \tilde{v}$ is a linear projection on $E^*$, satisfying*

$$\|q\| = 1, \tag{19.125}$$

$$q(f) = f \qquad (f \in V), \tag{19.126}$$

$$q^*(\pi(x)) = \pi(x) \qquad (x \in C), \tag{19.127}$$

$$q(E^*) = \overline{D_{E^*}(C)}; \tag{19.128}$$

*hence, in particular, $q(E^*)$ is linearly isometric to $C^*$.*

---

*) See the beginning of the proof of lemma 19.10.

*Proof.* We shall use transfinite induction on $\mathfrak{m}$.

Assume first that $\mathfrak{m} = \aleph_0$. Let $\{f_i\}$ be a dense sequence in $V$ and $\{x_i\}$ a dense sequence in $B$. By lemma 19.10 we can construct, inductively, a sequence $\{C_n\}$ of separable subspaces of $E$, with $C_n = \overline{\{x_i^n\}_{i=1}^{\infty}}$ containing $x_i$, $x_i^k$ $(i = 1, \ldots, n; k = 1, \ldots, n-1)$ for each $n = 1,2, \ldots$ and a sequence of operators $v_n \in L(C_n^*, E^*)$ $(n = 1,2, \ldots)$ such that

$$\|\tilde{v}_n\| = \|v_n\| = 1 \qquad (n = 1,2, \ldots), \tag{19.129}$$

$$\tilde{v}_n^*(\pi(x_i)) = \pi(x_i) \qquad (i = 1, \ldots, n;\ n = 1,2, \ldots), \tag{19.130}$$

$$\tilde{v}_n^*(\pi(x_i^k)) = \pi(x_i^k) \qquad (i=1, \ldots, n;\ k=1, \ldots, n-1;\ n=1,2, \ldots), \tag{19.131}$$

$$\|\tilde{v}_n(f_i) - f_i\| < \frac{1}{n} \qquad (i = 1, \ldots, n; n = 1,2, \ldots), \tag{19.132}$$

where, as before, $\tilde{v}_n \colon E^* \to E^*$ is defined by

$$\tilde{v}_n(f) = v_n(f|_{C_n}) \qquad (f \in E^*, n = 1,2, \ldots). \tag{19.133}$$

Put

$$C = \left[\bigcup_{n=1}^{\infty} C_n\right]. \tag{19.134}$$

Then $C$ is a separable subspace of $E$, containing $\{x_i\}$, whence also $C \supset B$. Let us define operators $\bar{v}_n \colon C^* \to E^*$ by

$$\bar{v}_n(\varphi) = v_n(\varphi|_{C_n}) \qquad (\varphi \in C^*,\ n = 1,2, \ldots); \tag{19.135}$$

hence, similarly to (19.107), we have $\bar{v}_n \in L(C^*, E^*)$ and

$$\|\bar{v}_n\| \leqslant \|v_n\| \qquad (n = 1,2, \ldots). \tag{19.136}$$

Then, using the techniques of the above proof of lemma 19.9 (i.e., looking first at $\{\bar{v}_n|_{S_{C^*}}\} \subset (S_{E^*})^{S_{C^*}}$, etc.) it follows that the net $\{\bar{v}_n\}_{n \in \mathcal{N}}$ has a subnet, which we shall denote again by $\{\bar{v}_n\}_{n \in \mathcal{N}}$, converging pointwise in the $\sigma(E^*, E)$-topology to some $v \in L(C^*, E^*)$, i.e., $\bar{v}_n(\varphi) \xrightarrow{w^*} v(\varphi)$ $(\varphi \in C^*)$. Put

$$q = \tilde{v}, \tag{19.137}$$

where $\tilde{v}$ is defined by (19.106).

Then, by (19.136) and (19.129),

$$\|v(\varphi)\| \leqslant \varliminf_{n \in \mathcal{N}} \|\bar{v}_n(\varphi)\| \leqslant \varlimsup_{n \in \mathcal{N}} \|\bar{v}_n\| \, \|\varphi\| \leqslant \varlimsup_{n \in \mathcal{N}} \|v_n\| \, \|\varphi\| = \|\varphi\| \qquad (\varphi \in C^*),$$

whence, by (19.107),

$$\|q\| = \|\tilde{v}\| \leqslant \|v\| \leqslant 1.$$

Furthermore, by (19.135) and (19.132),

$$q(f_i)(x) = \tilde{v}(f_i)(x) = v(f_i|_C)(x) = \lim_{n\in\mathcal{N}} \bar{v}_n(f_i|_C)(x) =$$

$$= \lim_{n\in\mathcal{N}} v_n(f_i|_{C_n})(x) = \lim_{n\in\mathcal{N}} \tilde{v}_n(f_i)(x) = f_i(x) \; (x\in E, i = 1,2, \ldots),$$

whence, since $V = \overline{\{f_i\}}$, we obtain (19.126). This, together with $\|q\| \leqslant 1$, yields (19.125).

Similarly, by (19.131),

$$q^*(\pi(x_i^k))(f) = \pi(x_i^k)(q(f)) = q(f)(x_i^k) = \lim_{n\in\mathcal{N}} \tilde{v}_n(f)(x_i^k) =$$

$$= \lim_{n\in\mathcal{N}} \pi(x_i^k)(\tilde{v}_n(f)) = \lim_{n\in\mathcal{N}} \tilde{v}_n^*(\pi(x_i^k))(f) = \pi(x_i^k)(f)$$

$$(f\in E^*;\; i, k = 1,2, \ldots),$$

whence, by $C_k = \overline{\{x_i^k\}}$ and (19.134), we obtain (19.127).

Finally, if $f\in D_{E^*}(C)$, $\|f\| = 1$, then there exists $x\in C$ with $\|x\| = 1$, $f(x) = 1$ (so $f = f_x$). On the other hand, by (19.127),

$$q(f)(x) = \pi(x)(q(f)) = q^*(\pi(x))(f) = \pi(x)(f) = f(x) = 1$$

and, by (19.125), $\|q(f)\| \leqslant \|f\| = 1$, whence, since $\|x\| = 1$, we obtain $\|q(f)\| = 1$. Consequently, since $E$ is smooth, we must have $q(f) = f$. Thus, since $f\in D_{E^*}(C)$ was arbitrary,

$$q(f) = f \qquad (f\in \overline{D_{E^*}(C)}). \tag{19.138}$$

Now let $f\in E^*$ be arbitrary. Then $f|_C \in C^*$, whence, since $C$ is subreflexive*), there exists a sequence $\{\varphi_n\} \subset D_{C^*}(C)$ such that $\varphi_n \to f|_C$ in the norm topology of $C^*$. Let $f_n\in E^*$, $f_n|_C = \varphi_n$, $\|f_n\| = \|\varphi_n\|$ $(n = 1,2, \ldots)$. Then, clearly, $f_n\in D_{E^*}(C)$ and, by (19.138),

$$f_n = q(f_n) = \tilde{v}(f_n) = v(f_n|_C) = v(\varphi_n) \to v(f|_C) = q(f),$$

so $q(f)\in \overline{D_{E^*}(C)}$. Thus, since $f\in E^*$ was arbitrary, $q(E^*)\subset\overline{D_{E^*}(C)}$. This, together with (19.138), yields (19.128) and that $q$ is a projection of $E^*$ onto $\overline{D_{E^*}(C)}$. Hence, in particular, by lemma 19.11, $q(E^*) \equiv C^*$. This completes the proof in the case when $\mathfrak{m} = \aleph_0$.

Assume now that $\mathfrak{m} > \aleph_0$ and that the lemma has been proved for all cardinal numbers $< \mathfrak{m}$ and let $\vartheta$ be the first ordinal number of cardinality $\mathfrak{m}$. Let $\{f_\lambda\}_{\lambda<\vartheta}$ be a dense transfinite sequence in $V$ and $\{x_\lambda\}_{\lambda<\vartheta}$ a dense transfinite sequence in $B$. By our hypothesis we can construct, inductively, for each $\lambda$ with $\omega \leqslant \lambda < \vartheta$, a subspace $C_\lambda$ of $E$ with dens $C_\lambda \leqslant$ card $\lambda$ and such that $C_\lambda \supset B_\lambda = [\bigcup_{\omega\leqslant\varkappa<\lambda} C_\varkappa \cup \{x_\varkappa\}_{\varkappa<\lambda}]$ (here

*) I.e., $C^* = \overline{D_{C^*}(C)}$ (see the beginning of the proof of lemma 19.10).

dens $B_\lambda \leqslant \sum_{\omega \leqslant \varkappa \leqslant \lambda}$ card $\varkappa \leqslant$ card $\lambda < \mathfrak{m}$) and an operator $v \in L(C^*, E^*)$ such that $q_\lambda = \tilde{v}_\lambda$ is a linear projection on $E$, satisfying $\|q_\lambda\| = 1$, $q_\lambda(f) = f$ $(f \in V_\lambda = [f_\varkappa]_{\varkappa \leqslant \lambda})$, $q_\lambda^*(\pi(x)) = \pi(x)$ $(x \in C_\lambda)$ and $q_\lambda(E^*) = \overline{D_{E^*}(C_\lambda)}$. Put

$$C = \overline{\bigcup_{\omega \leqslant \lambda < \vartheta} C_\lambda} \tag{19.139}$$

and define operators $\bar{v}_\lambda \colon C^* \to E^*$ $(\omega \leqslant \lambda < \vartheta)$ similarly to (19.135). Then, as in the first part of the proof, it follows that the net $\{\bar{v}_\lambda\}_{\omega \leqslant \lambda < \vartheta}$ has a subnet converging pointwise in the $\sigma(E^*, E)$-topology to some $v \in L(C^*, E^*)$ and that $C$ and $v$ satisfy the required conditions. This completes the proof of lemma 19.12.

Again, as in remark 19.1 a), only the case $\mathfrak{m} <$ dens $E$ of lemma 19.12 is of interest for applications.

In order to prove a substitute of proposition 19.5, we shall need two properties of very smooth spaces, given in the following two lemmas.

**Lemma 19.13.** *If $E$ is a very smooth Banach space, then*

$$\text{dens } E = \text{dens } E^*. \tag{19.140}$$

*Proof.* Since*) dens $E \leqslant$ dens $E^*$ (for any Banach space), we have to prove that dens $E^* \leqslant$ dens $E$. Let $\vartheta$ be the first ordinal number of cardinality dens $E$ and let $\{x_\lambda\}_{\lambda < \vartheta}$ be a dense transfinite sequence in $E$, The set $\mathfrak{R}$ consisting of all finite linear combinations with rational coefficients of the functionals $h_{x_\lambda} \in D_{E^*}(E)$ is a set of cardinality dens $E$, where $h_{x_\lambda} \in E^*$, $h_{x_\lambda}(x_\lambda) = \|h_{x_\lambda}\| \|x_\lambda\|$, $\|h_{x_\lambda}\| = \|x_\lambda\|$ for each $\lambda < \vartheta$. Furthermore, $\mathfrak{R}$ is dense in $D_{E^*}(E)$ (indeed, if $x \in E$, then there is a sequence $\{x_{\lambda_n}\}$ such that $x_{\lambda_n} \to x$, whence, since $E$ is very smooth, $h_{x_{\lambda_n}} \xrightarrow{w} h_x$ and therefore**) $h_x \in \overline{\mathfrak{R}}$, the norm closure of $\mathfrak{R}$ in $E^*$; but, if $g \in D_{E^*}(E)$ and $y \in E$, $\|y\| = 1$, $g(y) = \|g\|$, then for $x = \|g\| y$ we have $\|g\| = \|x\|$ and $g(x) = \|g\|^2 = \|g\| \|x\|$, so $g = h_x$, whence $g \in \overline{\mathfrak{R}}$). Consequently, since $E$ is subreflexive***), $\mathfrak{R}$ is dense in $E^*$, whence dens $E^* \leqslant$ card $\mathfrak{R}$ = dens $E$, which completes the proof of lemma 19.13.

In general, property (19.140) is not inherited by all subspaces $G$ of a Banach space $E$. However, from lemma 19.13 it follows

**Corollary 19.2.** *If $E$ is a very smooth Banach space, then for every subspace $G$ of $E$ we have*

$$\text{dens } G = \text{dens } G^*. \tag{19.140'}$$

*) See e.g. [11], p. 189, theorem 12 for dens $E^* = \aleph_0$. In the case when dens $E^*$ is arbitrary, the proof is similar.

**) See e.g. [87], p. 422, corollary 14.

***) I.e., $D_{E^*}(E)$ is dense in $E^*$ (see the beginning of the proof of lemma 19.10).

*Proof.* If $E$ is very smooth, then so is every subspace*) $G$ of $E$ and hence the conclusion follows from lemma 19.13 applied to $G$.

*Problem 19.1.* a) If we have (19.140′) for all subspaces $G$ of a Banach space $E$, does there exist an equivalent norm $|||\cdot|||$ on $E$ such that $(E, |||\cdot|||)$ is very smooth? b) Do the results of § 19 and § 20 on properties of Banach spaces $E$ isomorphic to very smooth spaces remain valid if we replace the assumption that $E$ is isomorphic to a very smooth space by the assumption that (19.140′) holds for all subspaces $G$ of $E$?

**Lemma 19.14.** *Let $E$ be a very smooth Banach space and let $\{C_\varkappa\}_{\omega\leqslant\varkappa<\lambda}$ be a long sequence of subspaces of $E$, such that $C_\varkappa\subset C_\mu$ for all $\omega\leqslant\varkappa<\mu<\lambda$. If $\overline{\bigcup_{\omega\leqslant\varkappa<\lambda} D_{E^*}(C_\varkappa)}$ is a subspace of $E^*$, then*

$$\overline{D_{E^*}(\overline{\bigcup_{\omega\leqslant\varkappa<\lambda} C_\varkappa})}=\overline{\bigcup_{\omega\leqslant\varkappa<\lambda} D_{E^*}(C_\varkappa)}. \tag{19.141}$$

*Proof.* It is clear from the definition that $\bigcup_{\omega\leqslant\varkappa<\lambda} D_{E^*}(C_\varkappa)\subset D_{E^*}(\overline{\bigcup_{\omega\leqslant\varkappa<\lambda} C_\varkappa})$, whence $\overline{\bigcup_{\omega\leqslant\varkappa<\lambda} D_{E^*}(C_\varkappa)}\subset \overline{D_{E^*}(\overline{\bigcup_{\omega\leqslant\varkappa<\lambda} C_\varkappa})}$, so it remains to show that the opposite inclusion holds. To this end, it is enough to prove that

$$D_{E^*}(\overline{\bigcup_{\omega\leqslant\varkappa<\lambda} C_\varkappa})\subset\overline{\bigcup_{\omega\leqslant\varkappa<\lambda} D_{E^*}(C_\varkappa)}.$$

Let**) $h_x\in D_{E^*}(\overline{\bigcup_{\omega\leqslant\varkappa<\lambda} C_\varkappa})$, so $x\in\overline{\bigcup_{\omega\leqslant\varkappa<\lambda} C_\varkappa}$. Then there exists a sequence $\{x_n\}\subset\bigcup_{\omega\leqslant\varkappa<\lambda} C_\varkappa$ such that $x_n\to x$, whence, since $E$ is very smooth, $h_{x_n}\xrightarrow{w} h_x$. Hence, since $\{h_{x_n}\}\subset\bigcup_{\omega\leqslant\varkappa<\lambda} D_{E^*}(C_\varkappa)$ and since $\overline{\bigcup_{\omega\leqslant\varkappa<\lambda} D_{E^*}(C_\varkappa)}$ is a subspace, it follows***) that $h_x\in\overline{\bigcup_{\omega\leqslant\varkappa<\lambda} D_{E^*}(C_\varkappa)}$, which completes the proof of lemma 19.14.

Now we can prove the following result, corresponding to proposition 19.5:

**Proposition 19.7.** *Let $E$ be a very smooth Banach space and let $\vartheta$ be the first ordinal number of cardinality* dens $E$. *Then there exist a*

---

*) Indeed, $G$ is smooth and if $x_n\to x$ in $G$ and $\Psi\in G^{**}$, then for $\Phi=\tilde{\Psi}\in E^{**}$ defined by $\Phi(f)=\Psi(f|_G)$ $(f\in E^*)$ we have $\Psi(\varphi_{x_n})=\Phi(f_{x_n})\to\Phi(f_x)=\Psi(\varphi_x)$, whence $\varphi_{x_n}\xrightarrow{w}\varphi_x$ (where $\varphi_x\in G^*$, $\varphi_x(x)=\|\varphi_x\|\,\|x\|$, $\|\varphi_x\|=\|x\|$).

**) See the above proof of lemma 19.13.

***) See e.g. [87], p. 422, corollary 14.

*long sequence of subspaces* $\{C_\lambda\}_{\omega\leqslant\lambda\leqslant\vartheta}$ *of E and a long sequence of operators* $v_\lambda \in L(C_\lambda^*, E^*)$ $(\omega\leqslant\lambda\leqslant\vartheta)$ *such that* $q_\lambda = \tilde{v}_\lambda$ *is a linear projection on* $E^*$, *satisfying*

$$\|q_\lambda\| = 1 \qquad (\omega\leqslant\lambda\leqslant\vartheta), \tag{19.142}$$

$$q_\lambda q_\varkappa = q_\varkappa q_\lambda = q_\varkappa \qquad (\omega\leqslant\varkappa\leqslant\lambda\leqslant\vartheta), \tag{19.143}$$

$$\text{dens } C_\lambda \leqslant \text{card } \lambda \qquad (\omega\leqslant\lambda\leqslant\vartheta), \tag{19.144}$$

$$q_\lambda(E^*) = \overline{D_{E^*}(C_\lambda)} \equiv C_\lambda^* \qquad (\omega\leqslant\lambda\leqslant\vartheta), \tag{19.145}$$

$$q_\lambda(E^*) \subset \overline{\bigcup_{\omega\leqslant\varkappa<\lambda} q_{\varkappa+1}(E^*)} \qquad (\omega<\lambda\leqslant\vartheta), \tag{19.146}$$

$$q_\vartheta = I_{E^*}, \tag{19.147}$$

*and such that for each fixed* $f\in E^*$, *the mapping* $\lambda\to q_\lambda(f)$ *from* $[\omega, \vartheta]$ *into* $E^*$ *is continuous.*

*Proof.* By lemma 19.13, let $\{f_\lambda\}_{\lambda<\vartheta}$ be a dense transfinite sequence in $E^*$. We shall construct, by transfinite induction, long sequences $\{C_\lambda\}_{\omega\leqslant\lambda\leqslant\vartheta}$ and $\{v_\lambda\}_{\omega\leqslant\lambda\leqslant\vartheta}$ satisfying the required conditions, together with the additional condition

$$f_\varkappa \in q_\lambda(E^*) \qquad (\varkappa < \lambda\leqslant\vartheta). \tag{19.148}$$

If dens $E = \aleph_0$, then $C_\omega = E$ and $v_\omega = I_{E^*}$ (hence $q_\omega = \tilde{v}_\omega = I_{E^*}$) have the required properties (in this case (19.145) is nothing else than the subreflexivity of $E$ and (19.146) is void).

Assume now that dens $E > \aleph_0$. By lemma 19.12 with $V= [f_\varkappa]_{\varkappa<\omega}$, there exist a separable subspace $C_\omega$ of $E$ and an operator $v_\omega \in L(C_\omega^*, E^*)$ such that $q_\omega = \tilde{v}_\omega$ is a linear projection of $E^*$, satisfying $\|q_\omega\| = 1$, $q_\omega(f_\varkappa) = f_\varkappa$ for all $\varkappa < \omega$ (or, equivalently, $f_\varkappa \in q_\omega(E^*)$ for all $\varkappa < \omega$), $q_\omega^*(\pi(x)) = \pi(x)(x\in C_\omega)$ and $q_\omega(E^*) = \overline{D_{E^*}(C_\omega)} \equiv C_\omega^*$. Assume that $\lambda < \vartheta$ and that we have constructed $C_\varkappa$, $v_\varkappa$ $(\omega\leqslant\varkappa < \lambda)$ satisfying the required conditions and (19.148).

*Case 1°.* $\lambda$ is a nonlimit ordinal, say $\lambda = \gamma + 1$. Let $B = C_\gamma$, $V = [\bigcup_{\omega\leqslant\varkappa\leqslant\gamma} q_\varkappa(E^*) \cup \{f_\varkappa\}_{\varkappa\leqslant\gamma}]$ (by our construction assumption, this coincides with $[q_\gamma(E^*), f_\gamma]$). Then, by our assumption, dens $B=$ dens $C_\gamma \leqslant$ card $(\gamma + 1)$ and dens $V=$ dens $q_\gamma(E^*) =$dens $C_\gamma^*$. Hence, since $E$ is very smooth, by corollary 19.2 we obtain dens $V=$ dens $C_\gamma^*=$ dens $C_\gamma\leqslant$ card $(\gamma + 1)$. Consequently, by lemma 19.12, there exist a subspace $C_{\gamma+1}$ of $E$, containing $B = C_\gamma$, with dens $C_{\gamma+1}\leqslant$card $(\gamma + 1)$, and an operator $v_{\gamma+1} \in L(C_{\gamma+1}^*, E^*)$ such that $q_{\gamma+1} = \tilde{v}_{\gamma+1}$ is a linear projection on $E^*$, satisfying $\|q_{\gamma+1}\| = 1$, $q_{\gamma+1}(f) = f\,(f\in V)$, $q_{\gamma+1}^*(\pi(x)) = \pi(x)$ $(x\in C_{\gamma+1})$ and $q_{\gamma+1}(E^*) = \overline{D_{E^*}(C_{\gamma+1})} \equiv C_{\gamma+1}^*$. Then, in particular,

$q_{\gamma+1}(f_\varkappa) = f_\varkappa$ $(\varkappa < \gamma+1)$, so $f_\varkappa \in q_{\gamma+1}(E^*)$ $(\varkappa < \gamma+1)$, and $q_{\gamma+1}q_\varkappa = q_\varkappa$ $(\omega \leqslant \varkappa \leqslant \gamma)$. Let us prove now that

$$(q_\gamma q_{\gamma+1})^* = q_{\gamma+1}^* q_\gamma^* = q_\gamma^*, \tag{19.149}$$

whence it will follow, using also our construction assumption, that $q_\varkappa q_{\gamma+1} = (q_\varkappa q_\gamma) q_{\gamma+1} = q_\varkappa q_\gamma = q_\varkappa$ $(\omega \leqslant \varkappa \leqslant \gamma)$. By lemma 19.11, for each $\Phi \in E^{**}$ one can define a $\Psi \in C_\gamma^{**}$ by

$$\Psi(f|_{C_\gamma}) = \Phi(f) \qquad (f \in \overline{D_{E^*}(C_\gamma)} = q_\gamma(E^*)). \tag{19.150}$$

Hence, by Goldstine's theorem*), there exists a bounded net $\{x_d\}_{d\in\mathscr{D}} \subset C_\gamma$ such that

$$\Phi(f) = \Psi(f|_{C_\gamma}) = \lim_{d\in\mathscr{D}} f(x_d) \qquad (f \in q_\gamma(E^*)). \tag{19.151}$$

In particular, if $\Phi \in q_\gamma^*(E^{**})$, so $\Phi = q_\gamma^*(\Phi)$, then for every $f \in E^*$ we have (using that $q_\gamma^*(\pi(x)) = \pi(x)$ for all $x \in C_\gamma$)

$$\Phi(f) = q_\gamma^*(\Phi)(f) = \Phi(q_\gamma(f)) = \lim_{d\in\mathscr{D}} q_\gamma(f)(x_d) = \lim_{d\in\mathscr{D}} \pi(x_d)(q_\gamma(f)) =$$

$$= \lim_{d\in\mathscr{D}} q_\gamma^*(\pi(x_d))(f) = \lim_{d\in\mathscr{D}} \pi(x_d)(f) = \lim_{d\in\mathscr{D}} f(x_d),$$

that is, $\Phi = w^*\text{-}\lim_{d\in\mathscr{D}} \pi(x_d)$. Consequently, using that $q_{\gamma+1}^*(\pi(x)) = \pi(x)$ for all $x \in C_\gamma$ $(\subset C_{\gamma+1})$ and the $w^*$-$w^*$ continuity of $q_{\gamma+1}^*$, we obtain

$$q_{\gamma+1}^*(\Phi) = q_{\gamma+1}^*(w^*\text{-}\lim_{d\in\mathscr{D}} \pi(x_d)) = w^*\text{-}\lim_{d\in\mathscr{D}} q_{\gamma+1}^*(\pi(x_d)) =$$

$$= w^*\text{-}\lim_{d\in\mathscr{D}} \pi(x_d) = \Phi \qquad (\Phi \in q_\gamma^*(E^{**})),$$

which proves (19.149). Finally, obviously, $q_{\gamma+1}(E^*) \subset \overline{\bigcup_{\omega\leqslant\varkappa\leqslant\gamma} q_{\varkappa+1}(E^*)}$.

*Case 2°.* $\lambda$ is a limit ordinal. Let $C_\lambda = \overline{\bigcup_{\omega\leqslant\varkappa<\lambda} C_\varkappa}$ and let us define operators $\bar{v}_\varkappa \colon C_\lambda^* \to E^*$ $(\omega \leqslant \varkappa < \lambda)$ similarly to (19.135), that is, by

$$\bar{v}_\varkappa(\varphi) = v_\varkappa(\varphi|_{C_\varkappa}) \qquad (\varphi \in C_\lambda^*;\ \omega \leqslant \varkappa < \lambda). \tag{19.152}$$

Then, using the techniques of the proof of lemma 19.9 above, it follows that the net $\{\bar{v}_\varkappa\}_{\omega\leqslant\varkappa<\lambda}$ has a subnet, which we shall denote again by $\{\bar{v}_\varkappa\}_{\omega\leqslant\varkappa<\lambda}$, converging pointwise in the $\sigma(E^*, E)$-topology to some $v_\lambda \in L(C_\lambda^*, E^*)$, i.e., $\bar{v}_\varkappa(\varphi) \xrightarrow{w^*} v_\lambda(\varphi)$ $(\varphi \in C_\lambda^*)$. Define $q_\lambda = \tilde{v}_\lambda \in L(E^*, E^*)$ by

$$q_\lambda(f) = \tilde{v}_\lambda(f) = v_\lambda(f|_{C_\lambda}) \qquad (f \in E^*). \tag{19.153}$$

---

*) See e.g. [87], p. 424, theorem 5.

Then, as in the above proof of lemma 19.12, we obtain $\|q_\lambda\| = \|\tilde{v}_\lambda\| \leqslant \|v_\lambda\| \leqslant 1$ and $q_\lambda(f_\varkappa)(x) = f_\varkappa(x)$ $(x \in E, \varkappa < \lambda)$, whence $q_\lambda(f_\varkappa) = f_\varkappa$ $(\varkappa < \lambda)$ and $\|q_\lambda\| = 1$. Also, by the definition of $C_\lambda$ and the induction hypothesis, dens $C_\lambda \leqslant \sum_{\omega \leqslant \varkappa < \lambda}$ card $\varkappa \leqslant$ card $\lambda$. Again, as in the above proof of lemma 19.12, the smoothness of $E$ yields $q_\lambda(f) = f$ $(f \in \overline{D_{E^*}(C_\lambda)})$ and then the subreflexivity of $C_\lambda$ yields $q_\lambda(f) \in \overline{D_{E^*}(C_\lambda)}$ $(f \in E^*)$, so $q_\lambda$ is a projection of $E^*$ onto $\overline{D_{E^*}(C_\lambda)}$ and hence, by lemma 19.11, $q_\lambda(E^*) \equiv C_\lambda^*$. This, together with the definition of $C_\lambda$, also implies $q_\lambda(E^*) = \overline{D_{E^*}(C_\lambda)} \supset \overline{D_{E^*}(C_\varkappa)} = q_\varkappa(E^*)$ $(\omega \leqslant \varkappa < \lambda)$, whence $q_\lambda q_\varkappa = q_\varkappa$ $(\omega \leqslant \varkappa < \lambda)$. Also, similarly to the above proof of (19.149), we obtain $(q_\varkappa q_\lambda)^* = q_\lambda^* q_\varkappa^* = q_\varkappa^*$, whence $q_\varkappa q_\lambda = q_\varkappa$ $(\omega \leqslant \varkappa < \lambda)$. Furthermore, by lemma 19.14 and the induction hypothesis,

$$q_\lambda(E^*) = \overline{D_{E^*}(C_\lambda)} = \overline{D_{E^*}(\overline{\bigcup_{\omega \leqslant \varkappa < \lambda} C_\varkappa})} = \overline{\bigcup_{\omega \leqslant \varkappa < \lambda} D_{E^*}(C_\varkappa)} \subset$$

$$\subset \overline{\bigcup_{\omega \leqslant \varkappa < \lambda} \overline{D_{E^*}(C_\varkappa)}} = \overline{\bigcup_{\omega \leqslant \varkappa < \lambda} q_\varkappa(E^*)} = \overline{\bigcup_{\omega \leqslant \varkappa < \lambda} q_{\varkappa+1}(E^*)}.$$

Finally, as was observed in proposition 19.5, the relations $\overline{\{f_\lambda\}_{\lambda < \vartheta}} = E^*$, (19.148), (19.143), (19.142) and (19.146) imply (19.147) and that for each fixed $f \in E^*$, the mapping $\lambda \to q_\lambda(f)$ from $[\omega, \vartheta]$ into $E^*$ is continuous. This completes the proof of proposition 19.7.

**Theorem 19.6.** *Let $E$ be a non-separable Banach space isomorphic to a very smooth space and let $\vartheta$ be the first ordinal number of cardinality* dens $E$. *Then the conjugate space $E^*$ has a monotone ordinal resolution of the identity $\{w_\lambda\}_{\lambda \leqslant \mu}$, with $\mu$ a limit ordinal satisfying $\mu \leqslant \vartheta$, such that*

$$\text{dens } w_\lambda(E^*) < \text{dens } E \qquad (\lambda < \mu). \tag{19.154}$$

*Proof.* We may assume, without loss of generality, that the space $E$ itself is very smooth. Then the proof is similar to that of theorem 19.4, using proposition 19.7.

*Remark 19.5.* Here only $\mu \leqslant \vartheta$, and the inequalities (19.154) are of weaker form than the corresponding equalities (19.91) and (19.105) of theorems 19.4 and 19.5 respectively. This is due to the fact that it is not known whether under the assumptions of theorem 19.6, $E^*$ has an extended $M$-basis*) and therefore we cannot apply the argument of remark 19.2 b) to replace the inequalities in (19.144) by equalities (which, in turn, would yield $\mu = \vartheta$ and the equalities of types (19.91), (19.105) for $\{w_\lambda\}_{\lambda \leqslant \vartheta}$).

---

*) Recently it has been shown that if $E^*$ is locally uniformly convex (which implies that $E$ is very smooth), then $E^*$ has an extended $M$-basis (see § 20 and the Notes and remarks to § 20).

Now we shall combine the assumptions of theorems 19.4 and 19.6, considering non-separable weakly compactly generated Banach spaces isomorphic to very smooth spaces and we shall get shrinking ordinal resolutions of the identity for such spaces. Let us first recall

**Lemma 19.15.** *Let $E$ be a smooth Banach space and $p$ a linear projection on $E$ with $\|p\| = 1$. Then*

$$p^*(E^*) = \overline{D_{E^*}(p(E))}. \tag{19.155}$$

*Proof.* Let $f \in D_{E^*}(p(E))$ and let $x = p(x) \in p(E)$, $f(x) = \|f\|$, $\|x\| = 1$. Then $p^*(f)(x) = f(p(x)) = f(x) = \|f\| = \|p^*\| \|f\| \geqslant \|p^*(f)\|$, whence $\|p^*(f)\| = \|f\|$. Hence, since $E$ is smooth, $f = p^*(f) \in p^*(E^*)$. Therefore, $\overline{D_{E^*}(p(E))} \subset p^*(E^*)$.

Conversely *), assume now that $f = p^*(f) \in p^*(E^*)$ and let $\varphi = f|_{p(E)} \in p(E)^*$. Then, since by $\|p\| = 1$ and § 14, lemma 14.3, the restriction mapping $\rho: p^*(E^*) \to p(E)^*$ is a linear isometry, we have $\|f\| = \|\varphi\|$. Now, since $p(E)$ is subreflexive, there exist sequences $\{\varphi_n\} \subset p(E)^*$ and $\{x_n\} \subset p(E)$ with $\varphi_n(x_n) = \|\varphi_n\|$, $\|x_n\| = 1$ $(n = 1,2, \ldots)$, such that $\varphi_n \to \varphi$. Let

$$f_n = \rho^{-1}(\varphi_n) = \varphi_n \circ p \qquad (n = 1,2, \ldots).$$

Then, by the above, $f_n \in E^*$, $f_n|_{p(E)} = \varphi_n$, $\|f_n\| = \|\varphi_n\|$, whence $f_n(x_n) = \|f_n\|$, so $f_n \in D_{E^*}(p(E))$ $(n = 1,2, \ldots)$. Also, $f_n = \rho^{-1}(\varphi_n) \to \rho^{-1}(\varphi) = f$, whence $f \in \overline{D_{E^*}(p(E))}$. Thus, $p^*(E^*) \subset \overline{D_{E^*}(p(E))}$, which, together with the opposite inclusion proved above, completes the proof of lemma 19.15.

**Proposition 19.8.** *Let $E$ be a very smooth Banach space and assume that there exists a long sequence of linear projections $\{p_\lambda\}_{\omega \leqslant \lambda \leqslant \vartheta}$ on $E$, satisfying*

$$\|p_\lambda\| = 1 \qquad (\omega \leqslant \lambda \leqslant \vartheta) \tag{19.76'}$$

*and* (19.79), (19.81), (19.82) *and* (19.83) *of proposition* 19.5, *where $\vartheta$ is the first ordinal of cardinality* dens $E$. *Then*

$$\|p_\lambda^*\| = 1 \qquad (\omega \leqslant \lambda \leqslant \vartheta), \tag{19.156}$$

$$p_\lambda^* p_\varkappa^* = p_\varkappa^* p_\lambda^* = p_\varkappa^* \qquad (\omega \leqslant \varkappa \leqslant \lambda \leqslant \vartheta), \tag{19.157}$$

$$\text{dens}\, p_\lambda^*(E^*) \leqslant \text{card}\, \lambda \qquad (\omega \leqslant \lambda \leqslant \vartheta), \tag{19.158}$$

$$p_\lambda^*(E^*) \subset \overline{\bigcup_{\omega \leqslant \varkappa < \lambda} p_{\varkappa+1}^*(E^*)} \qquad (\omega < \lambda \leqslant \vartheta), \tag{19.159}$$

$$p_\vartheta^* = I_{E^*}, \tag{19.160}$$

*and for each fixed $f \in E^*$, the mapping $\lambda \to p_\lambda^*(f)$ from $[\omega, \vartheta]$ into $E^*$ is continuous.*

---

*) In this part we do not use the smoothness of $E$.

*Proof.* (19.156), (19.157) and (19.160) are obvious from (19.76′), (19.79) and (19.83) respectively.

By § 14, lemma 14.3, $\operatorname{dens} p_\lambda^*(E^*) = \operatorname{dens} p_\lambda(E)^*$. Hence, since $E$ is very smooth, by corollary 19.2 and by (19.81) we obtain $\operatorname{dens} p_\lambda^*(E^*) = \operatorname{dens} p_\lambda(E)^* = \operatorname{dens} p_\lambda(E) \leqslant \operatorname{card} \lambda$ $(\omega \leqslant \lambda \leqslant \vartheta)$, so (19.158) holds.

Furthermore, by lemma 19.15, (19.82) and lemma 19.14,

$$p_\lambda^*(E^*) = \overline{D_{E^*}(p_\lambda(E))} \subset \overline{D_{E^*}(\overline{\bigcup_{\omega \leqslant \varkappa < \lambda} p_{\varkappa+1}(E)})} = \overline{\bigcup_{\omega \leqslant \varkappa < \lambda} D_{E^*}(p_{\varkappa+1}(E))} \subset \overline{\bigcup_{\omega \leqslant \varkappa < \lambda} p_{\varkappa+1}^*(E^*)},$$

so (19.159) holds.

Finally, as was observed in proposition 19.5, relations (19.156), (19.157) and (19.159) imply that for each fixed $f \in E^*$, the mapping $\lambda \to p_\lambda^*(f)$ of $[\omega, \vartheta]$ into $E^*$ is continuous. This completes the proof of proposition 19.8. Note also that if we replace the assumption (19.81) by (19.81′), then, by the above proof, we can replace the inequalities in (19.158) by equalities.

**Corollary 19.3.** *Every monotone ordinal resolution of the identity* $\{u_\lambda\}_{\lambda \leqslant \vartheta}$ *of a very smooth Banach space* $E$ *is shrinking.*

*Proof.* Put

$$p_\omega = u_2, \; p_{\omega+\lambda} = u_{2+\lambda} \qquad (\lambda \leqslant \vartheta). \tag{19.161}$$

Then $\{p_\lambda\}_{\omega \leqslant \lambda \leqslant \vartheta}$ satisfies the assumptions of proposition 19.8, except perhaps (19.81), and hence, by the proof of that proposition, $\{u_\lambda\}_{\lambda \leqslant \vartheta}$ is shrinking, which completes the proof of corollary 19.3.

**Theorem 19.7.** *Let* $E$ *be a non-separable space isomorphic to a very smooth Banach space and generated by a weakly compact circled convex set* $\mathscr{K}$, *let* $\vartheta$ *be the first ordinal of cardinality* $\operatorname{dens} E$ *and let* $G$ *be a closed linear subspace of* $E$. *Then* $E$ *has a shrinking ordinal resolution of the identity* $\{u_\lambda\}_{\lambda \leqslant \vartheta}$, *of type* $\vartheta + 1$, *satisfying* (19.89)–(19.91) *and* (19.105).

*Proof.* We may assume, without loss of generality, that the space $E$ itself is very smooth. Then, by theorem 19.4, $E$ has a monotone ordinal resolution of the identity $\{u_\lambda\}_{\lambda \leqslant \vartheta}$, of type $\vartheta + 1$, satisfying (19.89)–(19.91). By corollary 19.3, $\{u_\lambda\}_{\lambda \leqslant \vartheta}$ is shrinking. Finally, as in the above proof of (19.158), we have $\operatorname{dens} u_\lambda^*(E^*) = \operatorname{dens} u_\lambda(E)^* = \operatorname{dens} u_\lambda(E)$ $(\lambda \leqslant \vartheta)$, whence, by (19.91), we obtain (19.105), which completes the proof.

Since there exist*) very smooth Banach spaces which even cannot be embedded isomorphically into any weakly compactly generated Banach space, it is natural to raise

*) See [184].

*Problem 19.2**). Does theorem 19.7 remain valid for any (not necessarily weakly compactly generated) non-separable Banach space $E$, isomorphic to a very smooth Banach space?

## § 20. Extended biorthogonal systems. Extended $M$-bases

*Definition 20.1.* Let $E$ be a Banach space. A pair of families $(x_i, f_i)_{i \in I}$, where $\{x_i\}_{i \in I} \subset E$, $\{f_i\}_{i \in I} \subset E^*$, is called an *extended biorthogonal system* if

$$f_i(x_j) = \delta_{ij} \qquad (i, j \in I). \tag{20.1}$$

The extended biorthogonal system $(x_i, f_i)_{i \in I}$ is said to be *E-complete* if $[x_i]_{i \in I} = E$ and it is said to be *total* if $\{f_i\}_{i \in I}$ is total on $E$.

In order to avoid the terms "extended complete minimal family" and "extended generalized basis" we shall work with the notions of $E$-complete and total extended biorthogonal systems, as defined above (see Ch. I, § 6, theorem 6.1, equivalence $1° \Leftrightarrow 2°$, which remains valid for any set of indices, with the same proof).

**Proposition 20.1.** *Let $E$ be a Banach space. Then for every extended biorthogonal system $(x_i, f_i)_{i \in I}$ ($\{x_i\}_{i \in I} \subset E$, $\{f_i\}_{i \in I} \subset E^*$) we have*

$$\text{card } I \leqslant \text{dens } E \tag{20.2}$$

*(thus, in particular, in a separable Banach space $E$ every extended biorthogonal system $(x_i, f_i)_{i \in I}$ is at most countable). Consequently, for every infinite E-complete extended biorthogonal system $(x_i, f_i)_{i \in I}$ we have*

$$\text{card } I = \text{dens } E. \tag{20.2'}$$

*Proof.* By (20.1) we have $f_i \neq 0$ $(i \in I)$ and hence we may assume (by considering $\left(\|f_i\| x_i, \dfrac{f_i}{\|f_i\|}\right)_{i \in I}$ instead of $(x_i, f_i)_{i \in I}$, if necessary) that $\|f_i\| = 1$ $(i \in I)$. Then

$$1 = |f_i(x_i - x_j)| \leqslant \|x_i - x_j\| \qquad (i, j \in I;\ i \neq j). \tag{20.3}$$

Let $\{y_\lambda\}_{\lambda < \vartheta}$ be an arbitrary dense transfinite sequence in $E$. Then we can define

$$\lambda(i) = \min\left\{\lambda < \vartheta \,\middle|\, \|x_i - y_\lambda\| < \frac{1}{2}\right\} \qquad (i \in I). \tag{20.4}$$

---

*) See the Appendix.

We have $\lambda(i) \neq \lambda(j)$ for $i \neq j$, since otherwise

$$\|x_i - x_j\| \leqslant \|x_i - y_{\lambda(i)}\| + \|x_j - y_{\lambda(j)}\| < \frac{1}{2} + \frac{1}{2} = 1,$$

in contradiction with (20.3). Thus, $i \to \lambda(i)$ is a one-to-one mapping of $I$ into $[1, \vartheta)$, whence

$$\operatorname{card} I \leqslant \operatorname{card}[1, \vartheta) = \operatorname{card} \vartheta.$$

Hence, since $\{y_\lambda\}_{\lambda<\vartheta}$ was an arbitrary dense subset of $E$, we infer (20.2). Finally, if $(x_i, f_i)_{i \in I}$ is infinite and $E$-complete, then the set of all finite linear combinations $\sum_{n=1}^{m} r_{i_n} x_{i_n}$, where the $r_{i_n}$ are (complex or real) rational numbers and $m = 1,2, \ldots$, is a dense subset of $E$, of cardinality card $I$, whence dens $E \leqslant$ card $I$. This, together with (20.2), gives (20.2′), which completes the proof of proposition 20.1.

One cannot replace "$E$-complete" by "total" in proposition 20.1, since e.g. the non-separable space $E = l^\infty$ admits a natural countable total biorthogonal system. Also, one cannot replace dens $E$ by $w^*$-dens $E^*$ in (20.2), since $E = l^\infty$ admits an $E$-complete extended biorthogonal system $(x_i, f_i)_{i \in I}$ (see remark 20.3 below), whence, by proposition 20.1, card $I$ = dens $E = \aleph_1 > \aleph_0 = w^*$-dens $E^*$.

*Definition 20.2.* Let $E$ be a Banach space and let $(x_i, f_i)_{i \in I}, (y_l, g_l)_{l \in L}$ ($\{x_i\}_{i \in I}, \{y_l\}_{l \in L} \subset E, \{f_i\}_{i \in I}, \{g_l\}_{l \in L} \subset E^*$) be two extended biorthogonal systems. The system $(x_i, f_i)_{i \in I}$ is said to be an *extension**) of $(y_l, g_l)_{l \in L}$, if $L \subset I$ and $x_l = y_l$, $f_l = g_l$ for $l \in L$. An extended biorthogonal system $(x_i, f_i)_{i \in I}$ is said to be *maximal* if it has no proper extension (i.e., no extension different from $(x_i, f_i)_{i \in I}$ itself).

For example, every $E$-complete extended biorthogonal system $(x_i, f_i)_{i \in I}$ and every total extended biorthogonal system $(x_i, f_i)_{i \in I}$ (see definition 20.1) is maximal.

We have the following characterizations of maximal extended biorthogonal systems, which generalize § 6, proposition 6.9:

**Proposition 20.2.** *Let $E$ be a Banach space, $(x_i, f_i)_{i \in I}$ ($\{x_i\}_{i \in I} \subset E$, $\{f_i\}_{i \in I} \subset E^*$) an extended biorthogonal system and $W = \widetilde{[f_i]_{i \in I}}$, the $\sigma(E^*, E)$-closed linear subspace of $E^*$ spanned by $\{f_i\}_{i \in I}$. The following statements are equivalent:*

1°. *$(x_i, f_i)_{i \in I}$ is a maximal extended biorthogonal system.*

2°. *$W_\perp = ([f_i]_{i \in I})_\perp \subset [x_i]_{i \in I}$.*

3°. *$([x_i]_{i \in I})^\perp \subset W$.*

The proof is similar to that of § 6, proposition 6.9.

*) This abuse of language for the terms "extension" and "extended biorthogonal system" will lead to no confusion.

By proposition 20.1, the following result is a generalization of § 6, theorem 6.6:

**Theorem 20.1.** *Let $E$ be a Banach space and let $(y_l, g_l)_{l\in L}$ ($\{y_l\}_{l\in L}\subset E$, $\{g_l\}_{l\in L}\subset E^*$) be an extended biorthogonal system. Then $(y_l, g_l)_{l\in L}$ admits an extension to a maximal extended biorthogonal system.*

*Proof.* If $(y_l, g_l)_{l\in L}$ is not maximal, let $\mathcal{M}$ be the set of all extended biorthogonal systems $(x_i, f_i)_{i\in I}$ which are extensions of $(y_l, g_l)_{l\in L}$, partially ordered by extension (i.e., $(x_i, f_i)_{i\in I}\leqslant(z_j, h_j)_{j\in J}$ in $\mathcal{M}$ if and only if $(z_j, h_j)_{j\in J}$ is an extension of $(x_i, f_i)_{i\in I}$). Clearly, every totally ordered subset of $\mathcal{M}$ has an upper bound (namely, the union of its members, considered as an extended biorthogonal system). Hence, by Zorn's lemma*), $\mathcal{M}$ has a maximal element $(x_i, f_i)_{i\in I}$, which completes the proof of theorem 20.1.

*Problem 20.1***). Does every Banach space $E$ admit an $E$-complete extended biorthogonal system?

*Remark 20.1.* If $E$ admits an $E$-complete extended biorthogonal system $(x_i, f_i)_{i\in I}$, then $E$ has a separable quotient space and thus an affirmative answer to problem 20.1 would yield affirmative answers to § 1, problems 1.5′ and 1.5. Indeed, let $\{i_n\}$ be any countable infinite subset of $I$ and let $I_0=I\setminus\{i_n\}$. Then the sequence of subspaces $G_n=$ $=[x_i]_{i\in I_0\cup\{i_1,\dots,i_n\}}$ $(n=1,2,\dots)$ satisfies $G_1\subset G_2\subset G_3\subset\dots$, $G_1\neq$ $\neq G_2\neq G_3\neq\dots$ and $\overline{\bigcup_{n=1}^{\infty} G_n}=E$ and hence, by § 1, the remark made after problem 1.5′, $E$ has a separable quotient space, which proves our assertion.

Now we shall show (in theorem 20.2 below) that the answer to the similar question for total extended biorthogonal systems is affirmative. To this end, we shall need some preparation.

We recall the following obvious observation, which will be used repeatedly in the sequel:

**Lemma 20.1.** *Let $G$ be a subspace of a Banach space $E$ and let $x\in E$, $f\in E^*$ be such that*

$$\|f\|=f(x)=1,\qquad \operatorname{dist}(x, G)<1. \tag{20.5}$$

*Then*

$$G\not\subset \operatorname{Ker} f. \tag{20.6}$$

*Proof.* If $G\subset\operatorname{Ker} f$, then

$$\|x-y\|\geqslant|f(x-y)|=|f(x)|=1\qquad (y\in G),$$

whence $\operatorname{dist}(x, G)\geqslant 1$, in contradiction with (20.5).

*) See e.g. [215], Ch. 0, theorem 25.

**) See the Appendix.

**Proposition 20.3.** *Let $E$ be an infinite-dimensional Banach space, $G$ a subspace of $E$ and $V$ a subspace of $E^*$, with $V_\perp \neq \{0\}$, such that*

$$\widehat{(G;\ V_\perp)} > 0. \tag{20.7}$$

*Furthermore, let $(x_i, f_i)_{i\in I}$ be an extended biorthogonal system, such that*

$$\{x_i\}_{i\in I} \subset G, \qquad \{f_i\}_{i\in I} \subset V, \tag{20.8}$$

*and let $0 < \varepsilon < 1$. Then there exists a proper extension of $(x_i, f_i)_{i\in I}$ to an extended biorthogonal system $(x_i, f_i)_{i\in I\cup J}$, such that*

$$\widehat{([x_i]_{i\in I\cup J};\ [\{f_i\}_{i\in I\cup J} \cup V]_\perp)} \geqslant 1 - \varepsilon, \tag{20.9}$$

$$\{x_j\}_{j\in J} \subset V_\perp, \qquad \{f_j\}_{j\in J} \subset G^\perp, \tag{20.10}$$

$$\operatorname{card} J \begin{cases} < \aleph_0 \ \textit{if} \ \operatorname{card} I < \aleph_0 \\ \leqslant \operatorname{card} I \ \textit{if} \ \operatorname{card} I \geqslant \aleph_0. \end{cases} \tag{20.11}$$

*Proof.* We shall proceed by induction on card $I$.

Let card $I = \aleph_0$, so we may assume that $I = \mathscr{N} = \{1,2,3,\ldots\}$ (in the case when card $I < \aleph_0$, there will be some obvious simplifications). Also, we may assume, without loss of generality, that

$$0 < \widehat{(G;\ V_\perp)} \leqslant \widehat{([x_n];\ V_\perp)} < 1 - \varepsilon. \tag{20.12}$$

Indeed, since $V_\perp \neq \{0\}$, there exists $x_0 \in V_\perp$ with $\|x_0\| = 1$. Then, by $\widehat{(G;\ V_\perp)} > 0$, we have $x_0 \notin G$ and hence, by a corollary of the Hahn-Banach theorem, there exists a functional $f_0 \in G^\perp$ such that $f_0(x_0) = 1$. Thus, $(\{x_n\} \cup \{x_0\},\ \{f_n\} \cup \{f_0\})$ is a biorthogonal system, which is a proper extension of $(x_n, f_n)$. Let

$$G' = [G \cup \{x_0\}], \qquad V' = [V \cup \{f_0\}]. \tag{20.13}$$

We claim that $\widehat{(G';\ V'_\perp)} > 0$. Indeed, let $z + \beta x_0 \in G'$, where $z \in G$, and let $y \in V'_\perp$. Then $\beta x_0 - y \in V_\perp$, whence

$$\|z + \beta x_0 - y\| \geqslant \widehat{(G;\ V_\perp)}\, \|z\|.$$

On the other hand,

$$\|z+\beta x_0 - y\| \geqslant \frac{1}{\|f_0\|} |f_0(z+\beta x_0 - y)| =$$

$$= \frac{1}{\|f_0\|} |\beta| = \frac{1}{\|f_0\|} \|\beta x_0\|.$$

Consequently, for $\gamma = \min\left(\widehat{(G; V_\perp)}, \frac{1}{\|f_0\|}\right) > 0$ we have

$$\|z+\beta x_0 - y\| \geqslant \frac{1}{2}\left(\widehat{(G; V_\perp)}\|z\| + \frac{1}{\|f_0\|}\|\beta x_0\|\right) \geqslant \frac{\gamma}{2}\|z+\beta x_0\|,$$

which proves the claim $\widehat{(G'; V'_\perp)} > 0$.

Now, if $\widehat{([\{x_n\} \cup \{x_0\}]; V'_\perp)} \geqslant 1-\varepsilon$, then, since $[\{f_n\} \cup \{f_0\} \cup V]_\perp \subset \subset V'_\perp$, we have (20.9)—(20.11) for $J=\{0\}$, so the proof of proposition 20.3 is complete. On the other hand, if $0 < \widehat{(G'; V'_\perp)} \leqslant \leqslant \widehat{([\{x_n\} \cup \{x_0\}]; V'_\perp)} < 1-\varepsilon$ (hence $V'_\perp \neq \{0\}$), then we can consider the subspaces $G' \subset E$, $V' \subset E^*$ and the biorthogonal system $(\{x_n\} \cup \{x_0\}, \{f_n\} \cup \{f_0\})$ instead of $G$, $V$ and $(x_n, f_n)$ respectively. This proves that we may assume (20.12).

We shall construct non-void collections $\bigcup_{n=1}^{\infty} \{z_j^n\}_{j=1}^{k_n} \subset V_\perp$ and $\bigcup_{n=1}^{\infty} \{h_j^n\}_{j=1}^{k_n} \subset G^\perp$ such that for each $n=1,2,\ldots$

$$\widehat{\left(\left[\{x_j\}_{j=1}^{p} \cup \bigcup_{i=1}^{p-1} \{z_j^i\}_{j=1}^{k_i}\right]; \left[\bigcup_{i=1}^{p} \{h_j^i\}_{j=1}^{k_i} \cup V\right]_\perp\right)} \geqslant 1-\varepsilon \ (p=1,\ldots,n), \tag{20.14}$$

$$h_j^i(z_l^p) = \delta_{ij}^{pl} \quad (j=1,\ldots,k_i;\ l=1,\ldots,k_p;\ i,p=1,\ldots,n) \tag{20.15}$$

$\left(\text{where the term} \bigcup_{i=1}^{p-1} \text{in (20.14) is missing if } p=1\right)$, whenever $\bigcup_{i=1}^{p}\{h_j^i\}_{j=1}^{k_i} \neq \neq \emptyset$, $(z_l^p, h_l^p) \neq \emptyset$, $(z_j^i, h_j^i) \neq \emptyset$.

Let $0 < \varepsilon' < \frac{\varepsilon}{2}$. We shall proceed by induction on $n$.

Assume that for some $n\geqslant 1$ we have constructed $\bigcup_{i=1}^{n-1} \{z_j^i\}_{j=1}^{k_i} \subset V_\perp$ and $\bigcup_{i=1}^{n-1} \{h_j^i\}_{j=1}^{k_i} \subset G^\perp$ satisfying (20.14), (20.15) for $n$ replaced by $n-1$ (if $n=1$, we make no assumption). Let $\{y_j^n\}_{j=1}^{k_n}$ be a finite $\varepsilon'$-net of $\sigma_n = \{x \in [\{x_i\}_{i=1}^n \cup \bigcup_{i=1}^{n-1} \{z_j^i\}_{j=1}^{k_i}] \mid \|x\| = 1\}$ and let $\{g_j^n\}_{j=1}^{k_n} \subset E^*$ be such that

$$\|g_j^n\| = g_j^n(y_j^n) = 1 \quad (j=1,\ldots,k_n). \tag{20.16}$$

Furthermore, let $1\leqslant l\leqslant k_n$ and assume that we have constructed $\{z_m^n\}_{m=1}^{l-1} \subset V_\perp$ and $\{h_m^n\}_{m=1}^{l-1} \subset G^\perp$ (if $l=1$, we make no assumption; thus, our first step, when $n=1$, $l=1$, will yield the construction of $(z_1^1, h_1^1)$).

*Case 1°.* $\operatorname{dist}\left(y_l^n, \left[\bigcup_{i=1}^{n-1} \{h_j^i\}_{j=1}^{k_i} \cup \{h_m^n\}_{m=1}^{l-1} \cup V\right]_\perp\right) < 1-\varepsilon'$ (where the terms involving $n-1$ or $l-1$ are missing if*) $n=1$ or $l=1$ respectively). Then, by lemma 20.1, there exists an element

$$z_l^n \in \left[\bigcup_{i=1}^{n-1} \{h_j^i\}_{j=1}^{k_i} \cup \{h_m^n\}_{m=1}^{l-1} \cup V\right]_\perp, \quad z_l^n \notin \operatorname{Ker} g_l^n, \tag{20.17}$$

with $\|z_l^n\| = 1$. We claim that $z_l^n \notin [G \cup F]$, where

$$F = \left[\bigcup_{i=1}^{n-1} \{z_j^i\}_{j=1}^{k_i} \cup \{z_m^n\}_{m=1}^{l-1} \cup \left[\bigcup_{i=1}^{n-1} \{h_j^i\}_{j=1}^{k_i} \cup \{h_m^n\}_{m=1}^{l-1} \cup \{g_l^n\} \cup V\right]_\perp\right].$$

Indeed, otherwise we can write

$$z_l^n = \lim_{q\to\infty} (s_q + t_q),$$

where $s_q \in G$, $t_q \in F$ $(q=1,2,\ldots)$. But, by our construction assumption, $\bigcup_{i=1}^{n-1} \{z_j^i\}_{j=1}^{k_i} \cup \{z_m^n\}_{m=1}^{l-1} \subset V_\perp$, whence $F \subset V_\perp$. Also, by (20.17), $z_l^n \in V_\perp$, so $z_l^n - t_q \in V_\perp$ $(q=1,2,\ldots)$. Therefore

$$\|(z_l^n - t_q) - s_q\| \geqslant \widehat{(G; V_\perp)}\, \|s_q\| \quad (q=1,2,\ldots),$$

*) Thus, for $n=1$ and $l=1$ this means that $\operatorname{dist}(y_1^1, V_\perp) < 1-\varepsilon'$.

whence, since $\widehat{(G; V_\perp)} > 0$, we obtain $\lim_{q\to\infty} s_q = 0$. Thus, $z_l^n = \lim_{q\to\infty} t_q \in$ $\in F \subset \operatorname{Ker} g_l^n$, in contradiction with (20.17), which proves the claim.

Consequently, by a corollary of the Hahn-Banach theorem, we can choose a functional

$$h_l^n \in \left[G \cup \bigcup_{i=1}^{n-1}\{z_j^i\}_{j=1}^{k_i} \cup \{z_m^n\}_{m=1}^{l-1} \cup \left[\bigcup_{i=1}^{n-1}\{h_j^i\}_{j=1}^{k_i} \cup \{h_m^n\}_{m=1}^{l-1} \cup \{g_l^n\} \cup V\right]_\perp\right]^\perp, \tag{20.18}$$

satisfying $h_l^n(z_l^n) = 1$.

*Case 2°*. $\operatorname{dist}\left(y_l^n, \left[\bigcup_{i=1}^{n-1}\{h_j^i\}_{j=1}^{k_i} \cup \{h_m^n\}_{m=1}^{l-1} \cup V\right]_\perp\right) \geqslant 1 - \varepsilon'$. In this case, we take

$$(z_l^n, h_l^n) = \varnothing. \tag{20.19}$$

Observe now that at least for one $n \geqslant 1$ and $l \leqslant k_n$ we are in case 1°. Indeed, otherwise, since $\{y_j^n\}_{j=1}^{k_n}$ is an $\varepsilon'$-net for $\sigma_n$ $(n = 1,2, \ldots)$, we would obtain, by (20.19),

$$\operatorname{dist}(\sigma_n, V_\perp) \geqslant 1 - 2\varepsilon' > 1 - \varepsilon \quad (n = 1,2, \ldots),$$

whence, again by (20.19), $\widehat{([x_n]; V_\perp)} \geqslant 1 - \varepsilon$, in contradiction with (20.12).

We shall show that the non-void collections $\bigcup_{i=1}^{n}\{z_j^i\}_{j=1}^{k_i} \subset V_\perp$ and $\bigcup_{i=1}^{n}\{h_j^i\}_{j=1}^{k_i} \subset G^\perp$ constructed in this way satisfy (20.14), (20.15) whenever $\bigcup_{i=1}^{p}\{h_j^i\}_{j=1}^{k_i} \neq \varnothing$, $(z_l^p, h_l^p) \neq \varnothing$, $(z_j^i, h_j^i) \neq \varnothing$. Indeed, let $x \in \sigma_n$, $y \in$ $\in \left[\bigcup_{i=1}^{n}\{h_j^i\}_{j=1}^{k_i} \cup V\right]_\perp$. Since $\{y_j^n\}_{j=1}^{k_n}$ is an $\varepsilon'$-net of $\sigma_n$, there exists an $l \leqslant k_n$ such that $\|x - y_l^n\| < \varepsilon'$. If $y_l^n$ is in case 1° above, then by (20.18) and since $\dim\left[\bigcup_{i=1}^{n-1}\{h_j^i\}_{j=1}^{k_i} \cup \{h_m^n\}_{m=1}^{l-1} \cup \{g_l^n\}\right] < \infty$, we can write

$$h_l^n = \sum_{i=1}^{n-1}\sum_{j=1}^{k_i} \alpha_j^i h_j^i + \sum_{m=1}^{l-1} \beta_m^n h_m^n + \gamma_l^n g_l^n + g,$$

where $g \in (V_\perp)^\perp$. Then, since $y \in \left[\bigcup_{i=1}^{n}\{h_j^i\}_{j=1}^{k_i} \cup V\right]_\perp$, we have $h_j^i(y) = 0$

$(j=1,\ldots,k_i;\ i=1,\ldots,n)$, $g(y)=0$, whence $\gamma_l^n g_l^n(y)=0$. But, by $h_l^n(z_l^n)=1$ and (20.17), we have

$$1=h_l^n(z_l^n)=\left(\sum_{i=1}^{n-1}\sum_{j=1}^{k_i}\alpha_j^i h_j^i+\sum_{m=1}^{l-1}\beta_m^n h_m^n+\gamma_l^n g_l^n+g\right)(z_l^n)=\gamma_l^n g_l^n(z_l^n),$$

whence $\gamma_l^n\neq 0$ and therefore $g_l^n(y)=0$. Consequently, by (20.16),

$$\|x-y\|\geqslant\|y_l^n-y\|-\|x-y_l^n\|>|g_l^n(y_l^n-y)|-\varepsilon'=$$

$$=|g_l^n(y_l^n)|-\varepsilon'=1-\varepsilon'>1-2\varepsilon'>1-\varepsilon.$$

On the other hand, if $y_l^n$ is in case 2° above, then $\|y_l^n-y\|\geqslant 1-\varepsilon'$ $\left(\text{because } y\in\left[\bigcup_{i=1}^{n-1}\{h_j^i\}_{j=1}^{k_i}\cup\{h_m^n\}_{m=1}^{l-1}\cup V\right]_\perp\right)$, whence

$$\|x-y\|\geqslant\|y_l^n-y\|-\|x-y_l^n\|\geqslant 1-\varepsilon'-\varepsilon'=1-2\varepsilon'>1-\varepsilon.$$

Consequently, $\operatorname{dist}\left(\sigma_n,\left[\bigcup_{i=1}^{n}\{h_j^i\}_{j=1}^{k_i}\cup V\right]_\perp\right)\geqslant 1-\varepsilon$, which proves (20.14). Also, (20.15) for $(z_l^p,h_l^p)\neq\emptyset$, $(z_j^i,h_j^i)\neq\emptyset$, is obvious from the above construction.

Now put

$$(x_j,f_j)_{j\in J}=\bigcup_{n=1}^{\infty}{}'(z_j^n,h_j^n)_{j=1}^{k_n},\tag{20.20}$$

where $\bigcup'$ means that we include only the pairs $(z_j^n,h_j^n)\neq\emptyset$. We shall show that $(x_i,f_i)_{i\in I\cup J}$ is the required proper extension of $(x_i,f_i)_{i\in I}$. Indeed, by the above construction and (20.8) we have $\{f_j\}_{j\in J}=$ $=\bigcup_{n=1}^{\infty}{}'\{h_j^n\}_{j=1}^{k_n}\subset G^\perp\subset[\{x_i\}_{i\in I}]^\perp$ and $\{x_j\}_{j\in J}=\bigcup_{n=1}^{\infty}{}'\{z_j^n\}_{j=1}^{k_n}\subset V_\perp\subset$ $\subset[\{f_i\}_{i\in I}]_\perp$, which, together with (20.15), shows that $(x_i,f_i)_{i\in I\cup J}$ is an extended biorthogonal system, satisfying (20.10). Also, card $J\leqslant\aleph_0$ is obvious from (20.20). Finally, let $x\in\operatorname{lin}\{x_i\}_{i\in I\cup J}$, $\|x\|=1$ and let $y\in$ $\in[\{f_i\}_{i\in I\cup J}\cup V]_\perp$. Then there exists an integer $n$ such that $x\in\left[\{x_j\}_{j=1}^{n}\cup\bigcup_{i=1}^{n-1}\{z_j^i\}_{j=1}^{k_i}\right]$ and, clearly, we have also $y\in\left[\bigcup_{i=1}^{n}\{h_j^i\}_{j=1}^{k_i}\cup V\right]_\perp$. Hence, by (20.14), $\|x-y\|\geqslant 1-\varepsilon$ and therefore, since $\operatorname{lin}\{x_i\}_{i\in I\cup J}$ is dense in $[x_i]_{i\in I\cup J}$, we obtain (20.9). This proves proposition 20.3 for card $I\leqslant\aleph_0$.

Assume now that for each pair of subspaces $G \subset E$, $V \subset E^*$ with $V_\perp \neq \{0\}$, $\widehat{(G; V_\perp)} > 0$, every extended biorthogonal system $(x_i, f_i)_{i \in I}$ with $\aleph_0 \leqslant \operatorname{card} I < \aleph_\tau$ (where $\tau \geqslant 1$), satisfying (20.8), admits a proper extension to an extended biorthogonal system $(x_i, f_i)_{i \in I \cup J}$ satisfying (20.9)—(20.11). Let $G \subset E$, $V \subset E^*$ be a pair of subspaces with $V_\perp \neq \{0\}$, $\widehat{(G; V_\perp)} > 0$, and let $(x_i, f_i)_{i \in I}$ be an extended biorthogonal system with $\operatorname{card} I = \aleph_\tau$, satisfying (20.8). We shall show that $(x_i, f_i)_{i \in I}$ admits a proper extension to an extended biorthogonal system $(x_i, f_i)_{i \in I \cup J}$ satisfying (20.9)—(20.11), which will complete the proof.

Since $\operatorname{card} I = \aleph_\tau$, we may assume that $I = [1, \omega_\tau)$ and $(x_i, f_i)_{i \in I} = (x_\lambda, f_\lambda)_{\lambda < \omega_\tau}$. We shall first show, using the above induction hypothesis, that *for each* $\varkappa \in [\omega, \omega_\tau)$, *the extended biorthogonal system* $(x_\lambda, f_\lambda)_{\lambda < \varkappa}$ *admits a proper extension to an extended biorthogonal system of the form* $(x_\lambda, f_\lambda)_{\lambda < \varkappa} \cup (z_\lambda, h_\lambda)_{\lambda \in I_\varkappa}$, *where* $\operatorname{card} I_\varkappa \leqslant \operatorname{card} \varkappa$ *and where* $(z_\lambda, h_\lambda)_{\lambda \in I_\nu}$ *is a (not necessarily proper) extension of* $(z_\lambda, h_\lambda)_{\lambda \in I_\varkappa}$ *for* $\varkappa < \nu < \omega_\tau$, *such that*

$$\widehat{([\{x_\lambda\}_{\lambda < \varkappa} \cup \{z_\lambda\}_{\lambda \in I_\varkappa}]; [\{f_\lambda\}_{\lambda < \varkappa} \cup \{h_\lambda\}_{\lambda \in I_\varkappa} \cup V]_\perp)} \geqslant 1 - \varepsilon, \tag{20.21}$$

$$\{z_\lambda\}_{\lambda \in I_\varkappa} \subset V_\perp, \quad \{h_\lambda\}_{\lambda \in I_\varkappa} \subset G^\perp. \tag{20.22}$$

We shall proceed by transfinite induction on $\varkappa$. If $\varkappa = \omega$, the statement on the proper extension of $(x_\lambda, f_\lambda)_{\lambda < \omega}$ is true by our induction hypothesis above (since $\operatorname{card}[1, \omega) = \aleph_0 < \aleph_\tau$).

Now let $\nu \in [\omega + 1, \omega_\tau)$ and assume that our statement (on the proper extension of $(x_\lambda, f_\lambda)_{\lambda < \varkappa}$) is true for each $\varkappa < \nu$.

*Case 1°.* $\nu$ is a nonlimit ordinal. Let

$$G' = [G \cup \{z_\lambda\}_{\lambda \in I_{\nu-1}}], \quad V' = [V \cup \{h_\lambda\}_{\lambda \in I_{\nu-1}}]. \tag{20.23}$$

If $V'_\perp = \{0\}$, then, putting $(z_\lambda, h_\lambda)_{\lambda \in I_\nu} = (z_\lambda, h_\lambda)_{\lambda \in I_{\nu-1}}$, we obtain the required proper extension of $(x_\lambda, f_\lambda)_{\lambda < \nu}$, satisfying (20.21), (20.22) for $\varkappa = \nu$. Therefore, let us assume that $V'_\perp \neq \{0\}$.

We claim that $\widehat{(G'; V'_\perp)} > 0$. Indeed, we may assume, without loss of generality, that $\|z_\lambda\| = 1$ $(\lambda \in I_{\nu-1})$. Let $z + \sum_\lambda \beta_\lambda z_\lambda \in \operatorname{lin}(G \cup \{z_\lambda\}_{\lambda \in I_{\nu-1}})$, where $z \in G$, and let $y \in V'_\perp$. Then, by the assumption (20.22) (for $\varkappa = \nu - 1$), $\sum_\lambda \beta_\lambda z_\lambda - y \in V_\perp - V'_\perp = V_\perp$, whence

$$\|z + \sum_\lambda \beta_\lambda z_\lambda - y\| \geqslant \widehat{(G; V_\perp)} \|z\|.$$

Furthermore, again by (20.22) (for $\varkappa = \nu - 1$), $h = \sum_\lambda (\operatorname{sign} \beta_\lambda) h_\lambda \in$ $\in G^\perp \cap V'$, whence, by biorthogonality and $\|z_\lambda\| = 1$, we obtain

$$\|z + \sum_\lambda \beta_\lambda z_\lambda - y\| \geqslant \frac{1}{\|h\|} |h(\sum_\lambda \beta_\lambda z_\lambda)| = \frac{1}{\|h\|} \sum_\lambda |\beta_\lambda| \geqslant$$

$$\geqslant \frac{1}{\|h\|} \|\sum_\lambda \beta_\lambda z_\lambda\|,$$

Consequently, $\widehat{(\operatorname{lin}(G \cup \{z_\lambda\}_{\lambda \in I_{\nu-1}}); V'_\perp)} > 0$ (see the argument at the beginning of this proof, concerning (20.12)), whence the claim $\widehat{(G'; V'_\perp)} > 0$ follows.

Now, the extended biorthogonal system

$$(\tilde{x}_\lambda, \tilde{f}_\lambda)_{\lambda \in [1, \nu) \cup I_{\nu-1}} = (x_\lambda, f_\lambda)_{\lambda < \nu} \cup (z_\lambda, h_\lambda)_{\lambda \in I_{\nu-1}} \tag{20.24}$$

satisfies (20.8), for $G$, $V$ and $I$ replaced by $G'$, $V'$ and $[1, \nu) \cup I_{\nu-1}$ respectively, where $\operatorname{card}([1, \nu) \cup I_{\nu-1}) = \operatorname{card}(\nu - 1) < \aleph_\tau$. Hence, by our induction hypothesis, (20.24) admits a proper extension to an extended biorthogonal system

$$(\tilde{x}_\lambda, \tilde{f}_\lambda)_{\lambda \in [1, \nu) \cup I_{\nu-1} \cup J} = (x_\lambda, f_\lambda)_{\lambda < \nu} \cup (z_\lambda, h_\lambda)_{\lambda \in I_\nu}, \tag{20.25}$$

with $\operatorname{card} J \leqslant \operatorname{card}([1, \nu) \cup I_{\nu-1}) = \operatorname{card}(\nu - 1)$ (whence $\operatorname{card} I_\nu =$ $= \operatorname{card}(I_{\nu-1} \cup J) \leqslant \operatorname{card}(\nu - 1) = \operatorname{card} \nu$), satisfying (20.9)—(20.11) for $G$, $V$ and $I \cup J$ replaced by $G'$, $V'$ and $[1, \nu) \cup I_\nu$ respectively, whence also (20.21), (20.22) for $\varkappa = \nu$ (and the initial $G$, $V$), because $(G')^\perp \subset G^\perp$, $V'_\perp \subset V_\perp$.

*Case 2°.* $\nu$ is a limit ordinal. Let

$$(z_\lambda, h_\lambda)_{\lambda \in I_\nu} = \bigcup_{\omega \leqslant \varkappa < \nu} (z_\lambda, h_\lambda)_{\lambda \in I_\varkappa}. \tag{20.26}$$

Then $\operatorname{card} I_\nu = \operatorname{card}(\bigcup_{\omega \leqslant \varkappa < \nu} I_\varkappa) \leqslant \sum_{\omega \leqslant \varkappa < \nu} \operatorname{card} \varkappa = \operatorname{card} \nu$ and $(x_\lambda, f_\lambda)_{\lambda < \nu} \cup (z_\lambda, h_\lambda)_{\lambda \in I_\nu}$ is an extended biorthogonal system, which is a proper extension of $(x_\lambda, f_\lambda)_{\lambda < \nu}$, satisfying (20.21), (20.22) for $\varkappa = \nu$ (indeed, (20.21) follows looking first at $\operatorname{lin}(\{x_\lambda\}_{\lambda < \nu} \cup \{z_\lambda\}_{\lambda \in I_\nu})$). This proves our statement on the proper extension of $(x_\lambda, f_\lambda)_{\lambda < \varkappa}$ for each $\varkappa \in [\omega, \omega_\tau)$.

Now put

$$(x_i, f_i)_{i \in I \cup J} = (x_i, f_i)_{i \in I} \cup \bigcup_{\omega \leqslant \varkappa < \omega_\tau} (z_\lambda, h_\lambda)_{\lambda \in I_\varkappa} . \tag{20.27}$$

Then $(x_i, f_i)_{i \in I \cup J}$ has the required properties (indeed, card $J =$ $=$ card $(\bigcup_{\omega \leqslant \varkappa < \omega_\tau} I_\varkappa) \leqslant \sum_{\omega \leqslant \varkappa < \omega_\tau}$ card $\varkappa =$ card $\omega_\tau$ and (20.9) follows by looking first at lin $\{x_i\}_{i \in I \cup J}$), which completes the proof of proposition 20.3.

Now we are ready to prove

**Theorem 20.2.** *Every Banach space $E$ admits a total extended biorthogonal system $(x_i, f_i)_{i \in I}$ $(\{x_i\}_{i \in I} \subset E, \{f_i\}_{i \in I} \subset E^*)$.*

*Proof.* If $\dim E < \infty$, the statement is obviously true.

If $\dim E = \infty$, we shall proceed similarly to the proof of theorem 20.1 above. Let $0 < \varepsilon < 1$ and let $\mathscr{M}$ be the set of all extended biorthogonal systems $(x_i, f_i)_{i \in I}$, satisfying

$$\widehat{([x_i]_{i \in I}; [\{f_i\}_{i \in I}]_\perp)} \geqslant 1 - \varepsilon,$$

partially ordered by extension. Then $\mathscr{M} \neq \emptyset$, since for any $x_1 \in E$, $f_1 \in E^*$ with $f_1(x_1) = 1 = \|x_1\| = \|f_1\|$ we have $\widehat{([x_1]; [f_1]_\perp)} = 1$, so $(x_1, f_1) \in \mathscr{M}$. Also, $\mathscr{M}$ is inductive (the fact that the union of the members of any totally ordered subset of $\mathscr{M}$ belongs to $\mathscr{M}$ is shown by looking first at the linear span of the $x_i$'s occurring in the members of the union). Hence, by Zorn's lemma, $\mathscr{M}$ has a maximal element $(x_i, f_i)_{i \in I}$. Then, from proposition 20.3 (with $G = [x_i]_{i \in I}$, $V = [f_i]_{i \in I}$), it follows that $[\{f_i\}_{i \in I}]_\perp = \{0\}$, which completes the proof of theorem 20.2.

*Remark 20.2.* a) Taking any $(x_1, f_1) \in \mathscr{M}$ as above and applying the case card $I < \aleph_0$ of proposition 20.3 (with $G = [x_i]_{i \in I}$, $V = [f_i]_{i \in I}$) countably many times, it follows that *for each $\varepsilon$ with $0 < \varepsilon < 1$, every infinite-dimensional Banach space $E$ admits an infinite biorthogonal system $(x_n, f_n)$ $(\{x_n\} \subset E, \{f_n\} \subset E^*)$ such that*

$$\widehat{([x_n]; [f_n]_\perp)} \geqslant 1 - \varepsilon. \tag{20.28}$$

One can also give the following alternative proof of this result, which yields even an infinite biorthogonal system $(x_n, f_n)$ satisfying (20.28) and $\widehat{([x_n]^\perp; [f_n])} \geqslant 1 - \varepsilon$: Take $z_1 \in E$, $h_1 \in E^*$ with $\|z_1\| = \|h_1\| =$ $= h_1(z_1) = 1$ and put $Z_1 = [z_1]$, $V_1 = [h_1]$. Assume now that we have constructed finite-dimensional subspaces $Z_1 \subset \ldots \subset Z_n \subset E$ and

$V_1 \subset \ldots \subset V_n \subset E^*$ such that

$$\widehat{(Z_{k-1}; (V_k)_\perp)} \geqslant 1 - \varepsilon \qquad (k = 1, \ldots, n), \tag{20.29}$$

$$\widehat{(Z_k^\perp; V_{k-1})} \geqslant 1 - \varepsilon \qquad (k = 1, \ldots, n), \tag{20.30}$$

where $Z_0 = \{0\}$, $V_0 = \{0\}$. Then, similarly to § 1, proof of theorem 1.9, taking a finite $\frac{\varepsilon}{2}$-net $\{y_1, \ldots, y_p\}$ of $\sigma_{Z_n}$ and then functionals $g_i \in E^*$ with $\|g_i\| = 1$, $|g_i(y_i)| \geqslant 1 - \frac{\varepsilon}{2}$ $(i = 1, \ldots, p)$ and putting $V_{n+1} = [V_n \cup \{g_i\}_{i=1}^p] \supset V_n$, we get $\widehat{(Z_n; (V_{n+1})_\perp)} \geqslant 1 - \varepsilon$. Dually, starting with a finite $\frac{\varepsilon}{2}$-net of $\sigma_{V_n}$ in $E^*$, we obtain a finite-dimensional subspace $Z_{n+1}$ of $E$ with $Z_{n+1} \supset Z_n$, satisfying $\widehat{(Z_{n+1}^\perp; V_n)} \geqslant 1 - \varepsilon$. Thus, we have constructed two increasing infinite sequences of finite-dimensional subspaces $\{Z_n\}$, $\{V_n\}$ satisfying (20.29), (20.30) for all $n = 1,2, \ldots$ Then the $\aleph_0$-dimensional subspaces $Z = \bigcup_{n=1}^{\infty} Z_n \subset E$, $V = \bigcup_{n=1}^{\infty} V_n \subset E^*$ satisfy, clearly,

$$\widehat{(Z; V_\perp)} \geqslant 1 - \varepsilon, \quad \widehat{(Z^\perp; V)} \geqslant 1 - \varepsilon. \tag{20.31}$$

Hence, by § 8, propositions 8.1 and 8.2, there exists a biorthogonal system $(x_n, f_n)$ such that $Z = \operatorname{lin}\{x_n\}$, $V = \operatorname{lin}\{f_n\}$. Clearly, (20.31) remains still valid with $Z$, $V$ replaced by $\bar{Z} = [x_n]$ and $\bar{V} = [f_n]$ respectively, which completes the proof. However, the first method of proof (using lemma 20.1) has the advantages that it gives a step by step construction of the biorthogonal system and that it does not make use of finite $\varepsilon$-nets in the conjugate space $E^*$ and therefore it is suitable for further induction, involving even the infinite-dimensional subspaces of $E^*$ occurring in proposition 20.3.

b) For any biorthogonal system $(x_n, f_n)$ satisfying (20.28), $[x_n] \oplus [f_n]_\perp$ is closed in $E$ $\left(\text{the natural projection of } [x_n] \oplus [f_n]_\perp \text{ onto } [x_n] \text{ has norm } \leqslant 1 + \frac{\varepsilon}{1 - \varepsilon}\right)$. However, this does not solve the problem mentioned in § 1, remark 1.14, whether every Banach space contains a closed linear subspace which is the direct sum of two infinite dimensional subspaces, since in the above constructions it may happen that $[f_n]_\perp = \{0\}$ (i.e., that $\{f_n\}$ is total on $E$).

c) Theorem 20.2 permits us to regard every Banach space $E$ as an "extended $BK$-space", containing all unit vectors on the index set. Indeed, if $(x_i, f_i)_{i\in I}$ ($\{x_i\}_{i\in I}\subset E$, $\{f_i\}_{i\in I}\subset E^*$) is a total extended biorthogonal system, then, similarly to § 7, remark 7.1, $E$ is linearly isometric to the "extended $BK$-space"

$$S = S(E, \{f_i\}_{i\in I}) = \{\{f_i(x)\}_{i\in I} | \, x \in E\} \tag{20.32}$$

(where the norm is defined by $\|\{f_i(x)\}_{i\in I}\| = \|x\|$), by the mapping $v\colon x \to \{f_i(x)\}_{i\in I}$. Clearly, $v(x_i) = e_i$ $(i \in I)$, so $S$ contains all unit vectors on $I$.

*Definition 20.3.* Let $E$ be a Banach space. An extended biorthogonal system $(x_i, f_i)_{i\in I}$ ($\{x_i\}_{i\in I} \subset E$, $\{f_i\}_{i\in I} \subset E^*$) is said to be

a) *strictly bounded*, if $\sup_{i\in I} \|x_i\| < \infty$, $\sup_{i\in I} \|f_i\| < \infty$;

b) *normal*, if $\|x_i\| = \|f_i\| = 1$ $(i \in I)$.

*Problem 20.2.* a) Does every Banach space $E$ admit*) a strictly bounded $E$-complete extended biorthogonal system? b) Does every Banach space $E$ admit**) a strictly bounded total extended biorthogonal system? c) and d) Same questions, with "strictly bounded" replaced by "normal".

In §§ 6—8 we have seen that for all separable spaces the answers to a), b) are affirmative. Now we shall show that for some classes of non-separable spaces the answers to problems 20.2 a), b) are affirmative as well.

Let us first prove the following auxiliary result:

**Proposition 20.4.** *Let $E$ be a non-separable weakly compactly generated Banach space and let $\vartheta$ be the first ordinal number of cardinality* dens $E$. *Then $E$ has a quotient space $E/G$ with* dens $E/G =$ dens $E$, *which admits a strictly bounded $E/G$-complete extended biorthogonal system $\bigcup_{\lambda<\vartheta} (z_n^\lambda, \psi_n^\lambda)_{n=1}^\infty$ such that for each $\lambda < \vartheta$, $0$ is a weak limit point of $\{z_n^\lambda\}_{n=1}^\infty$.*

*Proof.* By § 19, theorem 19.4, $E$ admits a monotone ordinal resolution of the identity $\{u_\lambda\}_{\lambda\leq\vartheta}$, such that each subspace $(u_{\lambda+1} - u_\lambda)(E)$ of the associated transfinite Schauder decomposition has infinite dimension.

For each $\lambda < \vartheta$, write

$$\lambda = \mu_\lambda + \nu_\lambda, \tag{20.33}$$

---

*) See the Appendix.

**) Recently, problem 20.2 b) has been solved in the affirmative (see the Notes and remarks).

where $\mu_\lambda$ is a limit ordinal or zero, $\nu_\lambda$ is a non-negative integer and "+" denotes ordinal addition. By § 6, proposition 6.6, for each $\lambda < \vartheta$ we can choose a finite biorthogonal system $(z_i^\lambda, \varphi_i^\lambda)_{i=1}^{\nu_\lambda+1}$ ($\{z_i^\lambda\}_{i=1}^{\nu_\lambda+1} \subset (u_{\lambda+1} - u_\lambda)(E)$, $\{\varphi_i^\lambda\}_{i=1}^{\nu_\lambda+1} \subset (u_{\lambda+1} - u_\lambda)(E)^*$) with $\|z_i^\lambda\| = \|\varphi_i^\lambda\| = 1$ $(i=1, \ldots, \nu_\lambda + 1)$, such that $\{z_i^\lambda\}_{i=1}^{\nu_\lambda+1}$ is $\frac{1}{2}$-equivalent to the unit vector basis of $l^2_{\nu_\lambda+1}$. Then, clearly,

$$(u_{\lambda+1} - u_\lambda)(E) = [z_i^\lambda]_{i=1}^{\nu_\lambda+1} \oplus [\{\varphi_i^\lambda\}_{i=1}^{\nu_\lambda+1}]_\perp \qquad (\lambda < \vartheta). \tag{20.34}$$

Now put

$$h_i^\lambda = \varphi_i^\lambda \circ (u_{\lambda+1} - u_\lambda) \qquad (i = 1, \ldots, \nu_\lambda + 1;\ \lambda < \vartheta). \tag{20.35}$$

Then $h_i^\lambda \in E^*$ and by § 19, formulae (19.12) and (19.5),

$$h_i^\lambda(z_j^\varkappa) = \varphi_i^\lambda(u_{\lambda+1} - u_\lambda)(z_j^\varkappa) = \delta_{i\lambda}^{j\varkappa} \qquad (i = 1, \ldots, \nu_\lambda + 1;\ j = 1, \ldots \ldots, \nu_\varkappa + 1;\ \lambda, \varkappa < \vartheta),$$

so $\bigcup_{\lambda<\vartheta} (z_i^\lambda, h_i^\lambda)_{i=1}^{\nu_\lambda+1}$ is an extended biorthogonal system.

If $x \in E$, then, by § 19, formula (19.17) and § 17, lemma 17.1, for each $\varepsilon > 0$ there exists a finite sum $y = \sum_{i=1}^{n} (u_{\lambda_i+1} - u_{\lambda_i})(x)$ such that $\|x - y\| < \varepsilon$. Therefore, by (20.34), the set $\operatorname{lin}\{\bigcup_{\lambda<\vartheta} \{z_i^\lambda\}_{i=1}^{\nu_\lambda+1}\} + \operatorname{lin}\{\bigcup_{\lambda<\vartheta} [\{\varphi_i^\lambda\}_{i=1}^{\nu_\lambda+1}]_\perp\}$, whence also the larger set $[\bigcup_{\lambda<\vartheta} \{z_i^\lambda\}_{i=1}^{\nu_\lambda+1}] + [\bigcup_{\lambda<\vartheta} \{h_i^\lambda\}_{i=1}^{\nu_\lambda+1}]_\perp$, is dense in $E$. Thus, by reindexing $\bigcup_{\lambda<\vartheta} (z_i^\lambda, h_i^\lambda)_{i=1}^{\nu_\lambda+1}$, we obtain that $E$ admits a strictly bounded extended biorthogonal system*) $\bigcup_{\lambda<\vartheta} (\tilde{z}_n^\lambda, \tilde{h}_n^\lambda)_{n=1}^\infty$ such that

(i) $[\bigcup_{\lambda<\vartheta} \{\tilde{z}_n^\lambda\}_{n=1}^\infty] + [\bigcup_{\lambda<\vartheta} \{\tilde{h}_n^\lambda\}_{n=1}^\infty]_\perp$ is dense in $E$;

(ii) for each $\lambda < \vartheta$ and $n \in \mathcal{N} = \{1,2, \ldots\}$ there exists $m_n^\lambda \in \mathcal{N}$ such that $\{\tilde{z}^\lambda_{m_n^\lambda+k}\}_{k=1}^n$ is $\frac{1}{2}$-equivalent to the unit vector basis of $l_n^2$.

From (ii) it follows (by the argument of Ch. II, § 8, example 8.1), that for each $\lambda < \vartheta$, 0 is a weak limit point of $\{\tilde{z}_n^\lambda\}_{n=1}^\infty$. Now put

$$G = [\bigcup_{\lambda<\vartheta} \{\tilde{h}_n^\lambda\}_{n=1}^\infty]_\perp, \tag{20.36}$$

$$Z_n^\lambda = \omega_G(\tilde{z}_n^\lambda) \qquad (\lambda < \vartheta;\ n = 1,2, \ldots), \tag{20.37}$$

---

*) Namely, $\{\tilde{z}_n^1\}_{n=1}^\infty = \{z_1^1, z_2^1; z_1^2, z_2^2, z_3^2; \ldots\}$, $\{\tilde{z}_n^2\}_{n=1}^\infty = \{z_1^\omega; z_1^{\omega+1}, z_2^{\omega+1}; z_1^{\omega+2}, z_2^{\omega+2}, z_3^{\omega+2}; \ldots\}, \ldots$

where $\omega_G$ is the canonical mapping of $E$ onto $E/G$. Since $\bigcup_{\lambda<\vartheta}\{\tilde{h}_n^\lambda\}_{n=1}^\infty \subset G^\perp$, let $\psi_n^\lambda \in (E/G)^*$ be the image of $\tilde{h}_n^\lambda$ by the canonical linear isometry $G^\perp \equiv (E/G)^*$, i.e., $\psi_n^\lambda(\omega_G(x)) = \tilde{h}_n^\lambda(x)\ (x \in E)$. Then $\psi_n^\lambda(Z_m^\mu) = \psi_n^\lambda(\omega_G(\tilde{z}_m^\mu)) = \tilde{h}_n^\lambda(\tilde{z}_m^\mu) = \delta_{n\lambda}^{m\mu}$, i.e., $\bigcup_{\lambda<\vartheta}(Z_n^\lambda, \psi_n^\lambda)_{n=1}^\infty$ is an extended biorthogonal system, which is $E/G$-complete by (i). Furthermore, since $\bigcup_{\lambda<\vartheta}(\tilde{z}_n^\lambda, \tilde{h}_n^\lambda)_{n=1}^\infty$ is strictly bounded, so is $\bigcup_{\lambda<\vartheta}(Z_n^\lambda, \psi_n^\lambda)_{n=1}^\infty$. Finally, for each $\lambda < \vartheta$, since 0 is a weak limit point of $\{\tilde{z}_n^\lambda\}_{n=1}^\infty$ in $E$, so is 0 for $\{Z_n^\lambda\}_{n=1}^\infty$ in $E/G$. This completes the proof of proposition 20.4.

Now we can prove

**Theorem 20.3.** *Let $E$ be a non-separable Banach space which has a weakly compactly generated quotient space $E/F$ with* dens $E/F =$ dens $E$. *Then $E$ admits a strictly bounded $E$-complete extended biorthogonal system.*

*Proof.* By proposition 20.4, $E/F$, whence also $E$, has a quotient space, say $E/G$, with dens $E/G =$ dens $E/F =$ dens $E$, which admits a strictly bounded $E/G$-complete extended biorthogonal system $\bigcup_{\lambda<\vartheta}(Z_n^\lambda, \psi_n^\lambda)_{n=1}^\infty$ (where $\vartheta$ is the first ordinal number of cardinality dens $E$) such that for each $\lambda < \vartheta$, 0 is a weak limit point of $\{Z_n^\lambda\}_{n=1}^\infty$. Let $h_n^\lambda \in G^\perp$ be the image of $\psi_n^\lambda$ by the canonical linear isometry $(E/G)^* \equiv G^\perp$, i.e. $h_n^\lambda(x) = \psi_n^\lambda(x + G)\ (x \in E,\ n = 1,2,\ldots)$. Then for any $z_n^\lambda \in Z_n^\lambda$ with $\|z_n^\lambda\| \leqslant \|Z_n^\lambda\| + 1\ (\lambda < \vartheta;\ n = 1,2,\ldots)$ we have that $\bigcup_{\lambda<\vartheta}(z_n^\lambda, h_n^\lambda)_{n=1}^\infty$ is an extended biorthogonal system (because $h_n^\lambda(z_m^\mu) = \psi_n^\lambda(Z_m^\mu) = \delta_{n\lambda}^{m\mu}$) satisfying $\sup_{\lambda<\vartheta,\,1\leqslant n<\infty}\|z_n^\lambda\| < \infty$, $\sup_{\lambda<\vartheta,\,1\leqslant n<\infty}\|h_n^\lambda\| < \infty$. Furthermore, since $[\bigcup_{\lambda<\vartheta}\{Z_n^\lambda\}_{n=1}^\infty] = E/G$, we have that $[\bigcup_{\lambda<\vartheta}\{z_n^\lambda\}_{n=1}^\infty] + G$ is dense*) in $E$; indeed, if $x \in E$ and $\varepsilon > 0$, then there exist scalars $\alpha_1,\ldots,\alpha_m$ and ordinals $\lambda_1,\ldots,\lambda_m < \vartheta$, such that $\left\|\omega_G\left(x - \sum_{i=1}^m \alpha_i z_i^{\lambda_i}\right)\right\| = \left\|\omega_G(x) - \sum_{i=1}^m \alpha_i Z_i^{\lambda_i}\right\| < \varepsilon$ (where $\omega_G$ is the canonical mapping of $E$ onto $E/G$), whence also an element $y \in G$ such that $\left\|x - \sum_{i=1}^m \alpha_i z_i^{\lambda_i} - y\right\| \leqslant \left\|\omega_G\left(x - \sum_{i=1}^m \alpha_i z_i^{\lambda_i}\right)\right\| + \varepsilon < 2\varepsilon$.

Now let $\{y_\lambda\}_{\lambda<\vartheta}$ with $\|y_\lambda\| = 1\ (\lambda < \vartheta)$ be such that $[y_\lambda]_{\lambda<\vartheta} = G$; such a transfinite sequence $\{y_\lambda\}_{\lambda<\vartheta}$ exists, since $\vartheta$ is the first ordinal

*) Consequently, $[\bigcup_{\lambda<\vartheta}\{z_n^\lambda\}_{n=1}^\infty] + [\bigcup_{\lambda<\vartheta}\{h_n^\lambda\}_{n=1}^\infty]_\perp$ is also dense in $E$, but we shall not use here this remark.

of cardinality dens $E$ ($\geqslant$ dens $G$). Put

$$x_n^\lambda = z_n^\lambda - z_{n+1}^\lambda - y_\lambda \qquad (\lambda < \vartheta;\ n = 1,2, \dots), \tag{20.38}$$

$$f_1^\lambda = h_1^\lambda \qquad (\lambda < \vartheta), \tag{20.39}$$

$$f_n^\lambda = h_{n-1}^\lambda + h_n^\lambda \qquad (\lambda < \vartheta;\ n = 2,3, \dots). \tag{20.40}$$

Then $\bigcup_{\lambda<\vartheta} (x_n^\lambda, f_n^\lambda)_{n=1}^\infty$ is an extended biorthogonal system*) with $\sup_{\lambda<\vartheta, 1\leqslant n<\infty} \|x_n^\lambda\| < \infty$, $\sup_{\lambda<\vartheta, 1\leqslant n<\infty} \|f_n^\lambda\| < \infty$. We shall show that $[\bigcup_{\lambda<\vartheta} \{x_n^\lambda\}_{n=1}^\infty]^\perp = \{0\}$, which will complete the proof.

Let $f \in [\bigcup_{\lambda<\vartheta} \{x_n^\lambda\}_{n=1}^\infty]^\perp$. Then, adding the relations

$$f(y_\lambda) = f(z_n^\lambda) - f(z_{n+1}^\lambda) \qquad (\lambda < \vartheta;\ n = 1,2, \dots) \tag{20.41}$$

for fixed $\lambda < \vartheta$ and for $n = 1, \dots, m$, we obtain

$$mf(y_\lambda) = f(z_1^\lambda) - f(z_{m+1}^\lambda) \qquad (\lambda < \vartheta;\ m = 1,2, \dots),$$

whence, by the boundedness of $\bigcup_{\lambda<\vartheta} \{z_n^\lambda\}_{n=1}^\infty$,

$$f(y_\lambda) = 0 \qquad (\lambda < \vartheta). \tag{20.42}$$

Thus, $f \in G^\perp \equiv (E/G)^*$, whence, since for each $\lambda < \vartheta$, 0 is a weak limit point of $\{Z_n^\lambda\}_{n=1}^\infty \subset E/G$, and since $f(z_1^\lambda) = f(z_2^\lambda) = \dots$ (by (20.41) and (20.42)), it follows that $f \in [\bigcup_{\lambda<\vartheta} \{z_n^\lambda\}_{n=1}^\infty]^\perp$. Consequently, since $[\bigcup_{\lambda<\vartheta} \{z_n^\lambda\}_{n=1}^\infty] + G$ is dense in $E$, we obtain $f = 0$, which completes the proof of theorem 20.3.

For non-separable Grothendieck spaces the converse of theorem 20.3 is also true, as shown by

**Corollary 20.1.** *Let $E$ be a non-separable Grothendieck space***). *The following statements are equivalent:*

1°. *$E$ admits an $E$-complete extended biorthogonal system.*

2°. *$E$ admits a strictly bounded $E$-complete extended biorthogonal system.*

3°. *$E$ has a weakly compactly generated quotient space $E/F$ with* dens $E/F$ = dens $E$.

4°. *$E$ has a reflexive quotient space $E/F$ with* dens $E/F$ = dens $E$.

---

*) See the Appendix.

**) See § 15, remark 15.3 a). If $E$ is separable, then 1°–4° are satisfied (clearly, every separable Grothendieck space is reflexive).

*Proof.* The implications $4° \Rightarrow 3°$ and $2° \Rightarrow 1°$ are obvious and the implication $3° \Rightarrow 2°$ is nothing else than theorem 20.3.

Assume, finally, that we have $1°$ and let $(x_i, f_i)_{i \in I}$ ($\{x_i\}_{i \in I} \subset E$, $\{f_i\}_{i \in I} \subset E^*$) be an $E$-complete extended biorthogonal system. We claim that $[f_i]_{i \in I}$ *is reflexive*. Indeed, let $\{g_n\}_{n=1}^{\infty}$ be an arbitrary sequence in $S_{[f_i]_{i \in I}} = \{f \in [f_i]_{i \in I} \mid \|f\| \leqslant 1\}$. Since $f_j(x_i) = \delta_{ij}$ and since each $g_n$ is the norm limit of a sequence from lin $\{f_j\}_{j \in I}$, it follows that for each $n$ the set $I_n = \{i \in I \mid g_n(x_i)\} \neq 0$ is countable, and hence $\bigcup_{n=1}^{\infty} I_n$ is countable. By the classical diagonal procedure, there exists an increasing sequence of positive integers $\{m_n\}$ such that $\lim_{n \to \infty} g_{m_n}(x_i)$ exists for each $i \in \bigcup_{n=1}^{\infty} I_n$, whence also for each $i \in I$ $\left(\text{since } g_m(x_i) = 0 \text{ for all } i \in I \setminus \bigcup_{n=1}^{\infty} I_n \text{ and } m = 1,2, \dots\right)$. Hence, since $\|g_m\| \leqslant 1$ $(m = 1,2, \dots)$ and since $[x_i]_{i \in I} = E$, it follows that $\lim_{n \to \infty} g_{m_n}(x)$ exists for each $x \in E$. Consequently*), $\{g_{m_n}\}$ is $\sigma(E^*, E)$-convergent to some $g \in E^*$. But then, since $E$ is a Grothendieck space, $\{g_{m_n}\}$ is $\sigma(E^*, E^{**})$-convergent to $g$ and therefore $g \in [f_i]_{i \in I}$ (since the weak and norm closures of lin $\{f_i\}_{i \in I}$ in $E^*$ are**) the same). Consequently, $\{g_{m_n}\}$ is $\sigma([f_i]_{i \in I}, ([f_i]_{i \in I})^*)$-convergent to $g \in S_{[f_i]_{i \in I}}$ and thus, by the Eberlein-Šmulyan theorem ***), $[f_i]_{i \in I}$ is reflexive, which proves the claim.

Now, since $[f_i]_{i \in I}$ is reflexive, $S_{[f_i]_{i \in I}}$ is $\sigma(E^*, E^{**})$-compact, whence also $\sigma(E^*, E)$-compact and hence, by the Krein-Šmulyan theorem****), $[f_i]_{i \in I}$ *is* $\sigma(E^*, E)$*-closed*. Consequently, $(E/[\{f_i\}_{i \in I}]_{\perp})^* \equiv ([\{f_i\}_{i \in I}]_{\perp})^{\perp} = [f_i]_{i \in I}$, which is reflexive, whence so is $E/[\{f_i\}_{i \in I}]_{\perp}$. Hence, by proposition 20.1, dens $E/[\{f_i\}_{i \in I}]_{\perp} =$ dens $(E/[\{f_i\}_{i \in I}]_{\perp})^* =$ dens $[f_i]_{i \in I} =$ card $I =$ dens $E$. Thus, $1° \Rightarrow 4°$ (with $F = E/[\{f_i\}_{i \in I}]_{\perp}$), which completes the proof of corollary 20.1.

By theorem 20.3, every non-separable weakly compactly generated*****) Banach space $E$ and hence, in particular, every non-separable reflexive space, admits a strictly bounded $E$-complete extended biorthogonal system. Consequently, by duality (i.e., applying this observation to $E^*$), it follows that every non-separable reflexive space admits a

*) See e.g. [87], p. 494, theorem 4.

**) See e.g. [355], Ch. II, § 9, corollary 2 of theorem 9.2.

***) See e.g. [87], p. 430, theorem 1.

****) See e.g. [87], p. 429, theorem 7.

*****) In the meantime it has been proved that it has even a strictly bounded extended $M$-basis (see the Notes and remarks).

strictly bounded total extended biorthogonal system. This latter result can be also obtained as a consequence of

**Theorem 20.4.***) *Let $E$ be a Banach space which has a subspace $G$ with* dens $G =$ dens $E$ *admitting a strictly bounded extended M-basis***) $\{x_i\}_{i\in I}$. *Then there exists a family* $\{f_i\}_{i\in I}\subset E^*$ *such that* $(x_i, f_i)_{i\in I}$ *is a strictly bounded total extended biorthogonal system.*

*Proof.* We claim that for the a.f.f. $\{\varphi_i\}_{i\in I}\subset G^*$ to $\{x_i\}_{i\in I}$, we have***) $\{\varphi_i(x)\}_{i\in I}\in c_0(I)$ $(x\in G)$. Indeed, $\sup_{i\in I}|\varphi_i(x)|\leqslant\sup_{i\in I}\|\varphi_i\|\,\|x\|<\infty$ $(x\in G)$ and we shall show that for each $x\in G$ and $\varepsilon>0$ the set $\{i\in I\mid |\varphi_i(x)|>\varepsilon\}$ is finite. Assume, a contrario, that for some $x\in G$ and $\varepsilon>0$ there exists an infinite sequence $\{i_n\}\subset I$ such that $|\varphi_{i_n}(x)|>\varepsilon$ $(n=1,2,\ldots)$. Since $G=[x_i]_{i\in I}$, there exists $p=p_{x,\varepsilon}\in\operatorname{lin}\{x_i\}_{i\in I}$ such that $\|x-p\|<\dfrac{\varepsilon}{M}$, where $M=\sup_{i\in I}\|\varphi_i\|<\infty$. Then

$$\big||\varphi_i(x)|-|\varphi_i(p)|\big|\leqslant|\varphi_i(x)-\varphi_i(p)|\leqslant M\|x-p\|<\varepsilon \qquad (i\in I),$$

whence $|\varphi_{i_n}(p)|>|\varphi_{i_n}(x)|-\varepsilon>0$ for $n=1,2,\ldots$, which contradicts the fact that the set $\{i\in I\mid \varphi_i(p)\neq 0\}$ is finite. This proves the claim $\{\varphi_i(x)\}_{i\in I}\in c_0(I)$ $(x\in G)$.

We shall assume, without loss of generality, that the index set $I$ is a directed set. Then, by the above, $\lim_{i\in I}\varphi_i(x)=0$. Let $\{g_i\}_{i\in I}\subset E^*$, $g_i|_G=\varphi_i$, $\|g_i\|=\|\varphi_i\|$ $(i\in I)$, and let $i\to i_j^{(k)}$ (where $j\in I$, $k\in\mathcal{N}=\{1,2,\ldots\}$) be a one-to-one mapping of $I$ onto $I\times\mathcal{N}$, so $I\times\mathcal{N}$ is also directed by the induced ordering. Then, since $\sup_{i\in I}\|g_i\|=\sup_{i\in I}\|\varphi_i\|<\infty$, for each $k=1,2,\ldots$ there exists a subnet of $\{g_{i_j^{(k)}}\}$, $w^*$-converging to some $\bar{g}_k\in E^*$; obviously, $\|\bar{g}_k\|\leqslant\sup_{i\in I}\|\varphi_i\|$ $(k=1,2,\ldots)$.

Furthermore, since****) $w^*$-dens $(E/G)^*\leqslant$ dens $E/G\leqslant$ dens $E=$ dens $G$ and $G=[x_i]_{i\in I}$, $(E/G)^*$ contains a total family $\{\psi_i\}_{i\in I}$; we may assume that $\sup_{i\in I}\|\psi_i\|\leqslant\varepsilon$, where $\varepsilon$ is any positive number, given in advance. Let $\{h_i\}_{i\in I}\subset G^\perp$ be the image of $\{\psi_i\}_{i\in I}$ by the canonical linear isometry $(E/G)^*\equiv G^\perp$. Put

$$f_{i_j^{(k)}}=g_{i_j^{(k)}}-\bar{g}_k+h_k \qquad (j\in I;\ k=1,2,\ldots). \tag{20.43}$$

---

*) As we already mentioned, in the meantime it has been proved that every Banach space $E$ has a strictly bounded total extended biorthogonal system.

**) See definition 20.4.

***) For the definition of $c_0(I)$ see § 17, the remark made after definition 17.6. For separable $E$, the claim is nothig else than § 6, proposition 6.5, implication 1° ⇒ 2°.

****) See § 17, formula (17.57).

Then, as in § 7, proof of lemma 7.8, it follows that $f_{i_j^{(k)}}|_G = \varphi_{i_j^{(k)}}$ $(j \in I, k = 1,2, \ldots)$, $\sup_{i \in I} \|f_i\| < \infty$ and $[\{f_i\}_{i \in I}]_\perp = \{0\}$, so $(x_i, f_i)_{i \in I}$ is a strictly bounded total extended biorthogonal system. This completes the proof of theorem 20.4.

The above argument is a generalization of § 7, proof of theorem 7.8.

*Definition 20.4.* A family of elements $\{x_i\}_{i \in I}$ in a Banach space $E$ is called an *extended M-basis* of $E$, if $[x_i]_{i \in I} = E$ and if there exists a total family $\{f_i\}_{i \in I} \subset E^*$ such that $(x_i, f_i)_{i \in I}$ is an extended biorthogonal system. The (unique) family $\{f_i\}_{i \in I} \subset E^*$ is called *the family of functionals associated to the extended M-basis* $\{x_i\}_{i \in I}$, or, shortly, *the a.f.f. to* $\{x_i\}_{i \in I}$. Also, we make the convention that $\emptyset$ is the unique extended $M$-basis of $\{0\}$.

Extended $M$-bases have already occurred in the preceding. For example, we have observed in § 17, proof of corollary 17.3, that every $ER$-set $\{x_i\}_{i \in I}$ of a Banach space $E$ is an extended $M$-basis of $E$, and we have used extended $M$-bases in the proof of § 19, remark 19.2 b) (which was essential for the proof of § 19, theorem 19.4). Also, extended $M$-bases have occurred in the statement of theorem 20.4.

Let us consider the problem of existence of extended $M$-bases. We shall see that in this problem the geometric properties of $E$ play a great role.

We know from § 8 that every separable Banach space has an $M$-basis. Now we shall show (in corollary 20.2 below) that there exist some frequently used non-separable Banach spaces which have no extended $M$-basis. To this end, let us first prove

**Proposition 20.5.** *Let $E$ be a Grothendieck space. If $E$ admits an extended M-basis, then $E$ is reflexive.*

*Proof.* Since every separable Grothendieck space is reflexive, let us assume that $E$ is non-separable. Let $\{x_i\}_{i \in I}$ be an extended $M$-basis of $E$, with the a.f.f. $\{f_i\}_{i \in I} \subset E^*$. Then, by the proof of corollary 20.1, implication $1° \Rightarrow 4°$, $E/[\{f_i\}_{i \in I}]_\perp$ is reflexive, whence, since $[\{f_i\}]_\perp = \{0\}$, $E$ is reflexive. This completes the proof of proposition 20.5.

We shall see (in theorem 20.5 below) that every reflexive space $E$ admits indeed an extended $M$-basis (and thus the converse of proposition 20.5 also holds).

**Corollary 20.2.** *Let $E$ be an infinite dimensional $\mathscr{P}_\lambda$ space**⁾. *Then $E$ admits no extended M-basis.*

---

*⁾ See § 15, remark 15.3 a) for the notion of a $\mathscr{P}_\lambda$ space.

*Proof.* By § 15, remark 15.3 a), every $\mathscr{P}_\lambda$ space $E$ is a Grothendieck space, with the property that each weakly compact mapping $u$ of $E$ into any Banach space $B$ maps weakly convergent sequences into norm convergent sequences. Thus, if $E$ is reflexive (so $u = I_E: E \to E$ is weakly compact), then every bounded sequence in $E$ contains a subsequence which is weakly convergent, whence norm convergent, in contradiction with the assumption that $\dim E = \infty$. Consequently, $E$ is non-reflexive and hence, by proposition 20.5, $E$ admits no extended $M$-basis. This completes the proof of corollary 20.2.

*Remark 20.3.* Since for any infinite set $I$, the Banach space $l^\infty(I)$ has*) a quotient space isomorphic to a Hilbert space $H$ with dens $H = 2^{\operatorname{card} I} = \operatorname{dens} l^\infty(I)$, from theorem 20.3 it follows that $E = l^\infty(I)$ *($I$ infinite) admits a strictly bounded $E$-complete extended biorthogonal system.* Also, it is obvious that the unit vectors in $E = l^\infty(I)$, together with the coefficient functionals, constitute a *strictly bounded total extended biorthogonal system for* $E = l^\infty(I)$. However, by corollary 20.2, $E = l^\infty(I)$ *($I$ infinite) admits no extended $M$-basis.*

In particular (for $I = \mathscr{N} = \{1,2,3, \ldots\}$), *the space $l^\infty$ admits no extended $M$-basis.* Therefore, it is natural to raise

*Problem 20.3.* If a Banach space $E$ has an extended $M$-basis, can $E$ contain a subspace isomorphic to $l^\infty$?

We shall see in theorem 20.5 below that every weakly compactly generated Banach space has an extended $M$-basis; clearly, for such spaces the answer to problem 20.3 is negative (a weakly compactly generated space $E$ contains no subspace $G$ isomorphic**) to $l^\infty$). Also, in corollary 20.8 and in the Notes and remarks to § 20, on norming extended $M$-bases, we shall see that for Banach spaces with a boundedly complete or a norming extended $M$-basis the answer to problem 20.3 is negative. Now we shall show (in corollary 20.3 below) that if we replace in problem 20.3 $l^\infty$ by $l^\infty(I)$, where $I$ is any uncountable set, then the answer is negative. Let us first prove

**Proposition 20.6.** *Let $E$ be a Banach space with an extended $M$-basis. Then there exists a one-to-one continuous linear mapping of $E$ into some $c_0(I)$ and hence $E$ admits an equivalent strictly convex norm.*

*Proof.* Let $\{x_i\}_{i\in I}$ be an extended $M$-basis of $E$, with the a.f.f. $\{f_i\}_{i\in I} \subset E^*$. We may assume (replacing $(x_i, f_i)_{i\in I}$ by $\left(\|f_i\| x_i, \frac{f_i}{\|f_i\|}\right)_{i\in I}$, if necessary) that $\|f_i\| = 1$ $(i \in I)$. We claim that the mapping $u$ defined by

$$u(x) = \{f_i(x)\}_{i\in I} \qquad (x \in E) \tag{20.44}$$

is a one-to-one continuous linear mapping of $E$ into $c_0(I)$. Indeed, by the proof of theorem 20.4 above (with $G$ and $\{\varphi_i\}_{i\in I}$ replaced by $E$ and

*) See e.g. [339], p. 203, remark 2.

**) See e.g. corollary 20.5 below.

$\{f_i\}_{i\in I}$ respectively), we have $\{f_i(x)\}_{i\in I} \in c_0(I)$ $(x \in E)$, so $u$ maps $E$ into $c_0(I)$. Furthermore, by $\|f_i\| = 1$ $(i \in I)$, we have $\|u(x)\| = \sup_{i\in I} |f_i(x)| \leqslant \|x\|$ $(x \in E)$, so $u$ is continuous. Finally, since $\{f_i\}_{i\in I}$ is total on $E$, $u$ is one-to-one, which proves our claim.

Now, $c_0(I)$ admits*) an equivalent strictly convex norm, namely,

$$D(z(\cdot)) = \sup_{\substack{i_1, \ldots, i_n \in I \\ 1 \leqslant n < \infty}} \left( \sum_{k=1}^{n} \frac{1}{2^k} |z(i_k)|^2 \right)^{\frac{1}{2}} \qquad (z(\cdot) \in c_0(I)). \quad (20.45)$$

Also, it is well known**) that if there exists a one-to-one continuous linear operator of $E$ into a Banach space $F$ which admits an equivalent strictly convex norm, say $|||\cdot|||$, then $E$ itself admits an equivalent strictly convex norm, namely, $|x| = \|x\| + |||u(x)|||$, or, alternatively, $|x| = (\|x\|^2 + |||u(x)|||^2)^{\frac{1}{2}}$ $(x \in E)$. This completes the proof of proposition 20.6.

The converse of proposition 20.6 is not valid, since e.g. $E = l^\infty$ admits***) an equivalent strictly convex norm, but it has no extended $M$-basis (see remark 20.3 above).

**Corollary 20.3.** *If a Banach space $E$ admits an extended $M$-basis, then $E$ contains no subspace isomorphic to $l^\infty(I)$, for any uncountable set $I$.*

*Proof.* By proposition 20.6, $E$ admits an equivalent strictly convex norm, which induces an equivalent strictly convex norm onto any subspace $G$ of $E$. However, it is well known****) that $l^\infty(I)$, with $I$ uncountable, admits no equivalent strictly convex norm. Consequently, $E$ contains no subspace isomorphic to $l^\infty(I)$, with $I$ uncountable, which completes the proof of corollary 20.3.

One can prove positive results on the existence of extended $M$-bases when there exists a long sequence of projections $\{p_\lambda\}_{\omega \leqslant \lambda \leqslant \vartheta}$ on $E$ such that the subspaces $(p_{\lambda+1} - p_\lambda)(E)$ are suitable for transfinite induction. For example, this is the case for weakly compactly generated Banach spaces and their subspaces, as shown by

**Theorem 20.5.** a) *A Banach space $E$ is weakly compactly generated if and only if it has an extended $M$-basis $\{x_i\}_{i\in I}$ such that the set $\{x_i\}_{i\in I} \cup \{0\}$ is weakly compact. Moreover such an extended $M$-basis can be found in any weakly compact circled convex set $\mathcal{K}$ generating $E$.*

b) *Every subspace $G$ of a weakly compactly generated Banach space $E$ has an extended $M$-basis.*

---

*) See e.g. [70], Ch. VII, § 4, statement (3) (a).

**) See e.g. [70], Ch. VII, § 4, lemma 1, or [77], p. 100, theorem 1; actually, a particular case of this was shown in § 10, first part of the proof of lemma 10.1 and in the general case the proof is similar.

***) Indeed, this follows e.g. from corollary 20.5 below (applied to $l^1$).

****) See e.g. [70], Ch. VII, § 4, statement (1) (a).

*Proof.* a) The sufficiency part of the first statement is obvious, since if the condition is satisfied, then $E = [x_i]_{i \in I} = [\{x_i\}_{i \in I} \cup \{0\}]$.

Conversely, assume now that $E$ is weakly compactly generated and let $\mathscr{K}$ be any weakly compact circled convex set which generates $E$. We shall first show that there exists an extended $M$-basis $\{x_i\}_{i \in I}$ of $E$ such that $x_i \in \mathscr{K}$ $(i \in I)$.

We shall proceed by transfinite induction on dens $E$. If $E$ is separable, then $F = \operatorname{lin} \mathscr{K}$ is a separable normed linear space and hence, by § 8, remark 8.1 and theorem 8.1, $F$ has an $M$-basis $\{x_n\}$, which is also an $M$-basis of $E$. But, since $\mathscr{K}$ is a circled convex set, we have $F = \operatorname{lin} \mathscr{K} = \bigcup_{n=1}^{\infty} n\mathscr{K}$, and hence for each $n$ there exists a scalar $\lambda_n > 0$ such that $\lambda_n x_n \in \mathscr{K}$. Thus, $\{\lambda_n x_n\}$ is an $M$-basis of $E$ with the required property.

Assume now that dens $E > \aleph_0$ and that the statement holds for all Banach spaces $E_0$ with dens $E_0 <$ dens $E$. By § 19, proposition 19.5 (see also § 19, remark 19.2 a)), there exists a long sequence of projections $\{p_\lambda\}_{\omega \leqslant \lambda \leqslant \vartheta}$ satisfying (19.76) — (19.83) and (19.82″), where $\vartheta$ is the first ordinal of cardinality dens $E$. Then $E_\omega = p_\omega(E)$ is generated by $\mathscr{K}_\omega = p_\omega(\mathscr{K})$, which is a weakly compact circled convex set (since $\mathscr{K}$ is such a set) and dens $E_\omega \leqslant$ card $\omega = \aleph_0 <$ dens $E$. Similarly, for each $\lambda$ with $\omega \leqslant \lambda < \vartheta$, $E_{\lambda+1} = (p_{\lambda+1} - p_\lambda)(E)$ (possibly $E_{\lambda+1} = \{0\}$) is generated by the weakly compact circled convex set $\mathscr{K}_{\lambda+1} = \frac{1}{2}(p_{\lambda+1} - p_\lambda)(\mathscr{K})$ and has dens $E_{\lambda+1} \leqslant$ card $(\lambda + 1) <$ card $\vartheta =$ dens $E$. Hence, by the induction hypothesis, $E_\omega$ and each $E_{\lambda+1}$ have*) an extended $M$-basis $\{x_i^\omega\}_{i \in I_\omega} \subset \mathscr{K}_\omega$ and $\{x_i^{\lambda+1}\}_{i \in I_{\lambda+1}} \subset \mathscr{K}_{\lambda+1}$ $(\omega \leqslant \lambda < \vartheta)$ respectively. Then, by (19.82″) and (19.79), $\{x_i^\omega\}_{i \in I_\omega} \cup \bigcup_{\omega \leqslant \lambda < \vartheta} \{x_i^{\lambda+1}\}_{i \in I_{\lambda+1}}$ is an extended $M$-basis of $E$ (with the a.f.f. $\{f_i^\omega\}_{i \in I_\omega} \cup \bigcup_{\omega \leqslant \lambda < \vartheta} \{f_i^{\lambda+1}\}_{i \in I_{\lambda+1}} = \{\varphi_i^\omega \circ p_\omega\}_{i \in I_\omega} \cup \bigcup_{\omega \leqslant \lambda < \vartheta} \{\varphi_i^{\lambda+1} \circ (p_{\lambda+1} - p_\lambda)\}_{i \in I_{\lambda+1}} \subset E^*$, where $\{\varphi_i^\nu\}_{i \in I_\nu} \subset E_\nu^*$ is the a.f.f. to $\{x_i^\nu\}_{i \in I_\nu} \subset E_\nu$). Also, by (19.77) and since $\mathscr{K}$ is a circled convex set, $\{x_i^\omega\}_{i \in I_\omega} \cup \bigcup_{\omega \leqslant \lambda < \vartheta} \{x_i^{\lambda+1}\}_{i \in I_{\lambda+1}} \subset \mathscr{K}_\omega \cup \bigcup_{\omega \leqslant \lambda < \vartheta} \mathscr{K}_{\lambda+1} = p_\omega(\mathscr{K}) \cup \bigcup_{\omega \leqslant \lambda < \vartheta} \frac{1}{2}(p_{\lambda+1} - p_\lambda)(\mathscr{K}) \subset \mathscr{K}$. This proves that $E$ has an extended $M$-basis $\{x_i\}_{i \in I}$ such that $x_i \in \mathscr{K}$ $(i \in I)$.

Finally, in order to complete the proof of part a) it will be enough to show that *for any extended M-basis $\{x_i\}_{i \in I}$ of E with $x_i \in \mathscr{K}$ $(i \in I)$, the set $\{x_i\}_{i \in I} \cup \{0\}$ is weakly compact;* since $\mathscr{K}$ is weakly compact and $0 \in \mathscr{K}$, it will be sufficient to prove that *$\{x_i\}_{i \in I} \cup \{0\}$ is weakly closed.* Let $x \in E$ be a weak limit point of $\{x_i\}_{i \in I}$. We shall show that $x = 0$ $(\in \{x_i\}_{i \in I} \cup \{0\})$, which will complete the proof. For each $j \in I$ and

---

*) We recall that, by our convention, $\varnothing$ is the unique extended $M$-basis of $\{0\}$.

$n \in \mathcal{N}$, the weak neighbourhood $V_{f_j;\frac{1}{n}}(x)$ of $x$ contains some $x_{i_n(j)} \in \in \{x_i\}_{i \in I}$, whence, by biorthogonality, $f_j(x) = \lim_{n\to\infty} f_j(x_{i_n(j)}) = 0$ $(j \in I)$. Since $\{f_j\}_{j \in I}$ is total on $E$, it follows that $x = 0$, which completes the proof of part a).

b) Again, we shall proceed by transfinite induction on dens $E$. If $E$ is separable, then so is $G \subset E$ and hence, by § 8, theorem 8.1, $G$ has an $M$-basis.

Now let $G \subset E$, dens $E > \aleph_0$, and assume that the statement has been proved for all Banach spaces $E_0$ with dens $E_0 <$ dens $E$. Let $\{p_\lambda\}_{\omega \leqslant \lambda \leqslant \vartheta}$ be as in the above proof of part a). Then $G_\omega = p_\omega(G)$ is a subspace of the weakly compactly generated space $E_\omega = p_\omega(E)$, which has dens $E_\omega \leqslant \aleph_0 <$ dens $E$ and, similarly, each $G_{\lambda+1} = (p_{\lambda+1} - p_\lambda)(G)$ is a subspace (possibly $G_{\lambda+1} = \{0\}$) of the weakly compactly generated space $E_{\lambda+1} = (p_{\lambda+1} - p_\lambda)(E)$, which has dens $E_{\lambda+1} \leqslant$ dens $p_{\lambda+1}(E) \leqslant \leqslant$ card $(\lambda + 1) <$ card $\vartheta =$ dens $E$ $(\omega \leqslant \lambda < \vartheta)$. Hence, by the induction hypothesis, each $G_{\lambda+1}$ has an extended $M$-basis $\{x_i^{\lambda+1}\}_{i \in I_{\lambda+1}}$ $(\omega \leqslant \lambda < \vartheta)$. But, by (19.82″) (for $\lambda = \vartheta$) and (19.80) we have

$$G = [p_\omega(G) \cup \bigcup_{\omega \leqslant \lambda < \vartheta} (p_{\lambda+1} - p_\lambda)(G)], \tag{20.46}$$

whence, by (19.79), $\{x_i^\omega\}_{i \in I_\omega} \cup \bigcup_{\omega \leqslant \lambda < \vartheta} \{x_i^{\lambda+1}\}_{i \in I_{\lambda+1}}$ is an extended $M$-basis of $G$ (with the a.f.f. $\{f_i^\omega\}_{i \in I_\omega} \cup \bigcup_{\omega \leqslant \lambda < \vartheta} \{f_i^{\lambda+1}\}_{i \in I_{\lambda+1}} = \{\varphi_i^\omega \circ p_\omega|_G\}_{i \in I_\omega} \cup \cup \bigcup_{\omega \leqslant \lambda < \vartheta} \{\varphi_i^{\lambda+1} \circ (p_{\lambda+1}|_G - p_\lambda|_G)\}_{i \in I_{\lambda+1}}) \subset G^*$, where $\{\varphi_i^\nu\}_{i \in I_\nu} \subset G_\nu^*$ is the a.f.f. to $\{x_i^\nu\}_{i \in I_\nu} \subset G_\nu$). This completes the proof of theorem 20.5.

*Remark 20.4.* a) The final part of the above proof of a) is a generalization of Ch. II, § 7, proof of proposition 7.2, to extended $M$-bases. Note also that by the obvious generalization of Ch. II, proposition 8.1 a), *all points of any extended M-basis $\{x_i\}_{i \in I}$ of $E$ are $\sigma(E, E^*)$-isolated points* (and isolated points in the norm topology). Thus, we obtain that *every weakly compactly generated Banach space $E$ is generated by a subset which is in the weak topology the one point compactification of a discrete subset of $E$.*

b) The converse of theorem 20.5 b) is not valid, since for example the space $E = l^1(I)$, where $I$ is uncountable, has an extended $M$-basis (even an extended unconditional basis), but cannot be embedded isomorphically into any weakly compactly generated Banach space (by corollary 20.5 below).

**Corollary 20.4.** *If $E$ is a weakly compactly generated Banach space, then there exists a one-to-one continuous linear mapping $u$ of $E^*$ into* *some* $c_0(I)$, *which is also* $w^*$-$w$ *continuous* (*i.e., with respect to* $\sigma(E^*, E)$ *and* $\sigma(c_0(I), c_0(I)^*)$).

*Proof.* By theorem 20.5 a), $E$ has an extended $M$-basis $\{x_i\}_{i\in I}$ such that $\{x_i\}_{i\in I} \cup \{0\}$ is weakly compact. Put

$$u(f) = \{f(x_i)\}_{i\in I} \qquad (f \in E^*). \tag{20.47}$$

We claim that $u(f) \in c_0(I)$ $(f \in E^*)$. Indeed, $\sup_{i\in I} |f(x_i)| \leqslant \|f\| \sup_{i\in I} \|x_i\| < \infty$ $(f \in E^*)$. Assume now, a contrario, that for some $f \in E^*$ and $\varepsilon > 0$ there exists an infinite sequence $\{i_n\} \subset I$ such that $|f(x_{i_n})| > \varepsilon$ $(n = 1, 2, \ldots)$. Then, since $\{x_i\}_{i\in I} \cup \{0\}$ is weakly compact, by the Eberlein-Šmulyan theorem*) $\{x_{i_n}\}$ has a subsequence $\{x_{i_{n_m}}\}$ converging weakly to some $y \in \{x_i\}_{i\in I} \cup \{0\}$. But then, since $|f(y)| = \lim_{m\to\infty} |f(x_{i_{n_m}})| \geqslant \varepsilon$, we have $y \neq 0$, whence $w\text{-}\lim_{m\to\infty} x_{i_{n_m}} = y \in \{x_i\}_{i\in I}$, in contradiction with remark 20.4 a). Thus, $u$ maps $E^*$ into $c_0(I)$.

Clearly, $u$ is linear and $\|u\| \leqslant \sup_{i\in I} \|x_i\|$. Also, since $[x_i]_{i\in I} = E$, $u$ is one-to-one. Finally, let us show that $u$ is $w^*$-$w$ continuous. If $\{g_d\}_{d\in\mathscr{D}} \subset E^*$, $\|g_d\| \leqslant 1$ ($d \in \mathscr{D}$, where $\mathscr{D}$ is a directed set) and $g_d \xrightarrow{w^*} g \in E^*$, then $\lim_{d\in\mathscr{D}} g_d(x_i) = g(x_i)$ $(i \in I)$ and hence**) $u(g_d) = \{g_d(x_i)\}_{i\in I} \xrightarrow{w} \{g(x_i)\}_{i\in I} = u(g)$ in $c_0(I)$. Thus, $u$ is $w^*$-$w$ continuous on $S_{E^*} = \{f \in E^* \mid \|f\| \leqslant 1\}$, whence***) also on $E^*$, which completes the proof of corollary 20.4.

**Corollary 20.5.** *If $E$ is a weakly compactly generated Banach space, then the conjugate space $E^*$ admits an equivalent strictly convex norm which is the dual of some equivalent smooth norm on $E$. Consequently, $l^\infty(I)$ for any infinite set $I$, and $l^1(I)$ for any uncountable set $I$, cannot be embedded isomorphically into a weakly compactly generated Banach space.*

*Proof.* Let $u: E^* \to c_0(I)$ be as in corollary 20.4. Then, by the proof of proposition 20.6 above, $E^*$ admits an equivalent strictly convex norm, namely,

$$|f| = \|f\| + D(u(f)) \qquad (f \in E^*), \tag{20.48}$$

where $D(\cdot)$ is the equivalent strictly convex norm on $c_0(I)$, given by formula (20.45). We shall show that $|\cdot|$ is the dual of some equivalent norm $|\cdot|$ on $E$ (hence****) $(E, |\cdot|)$ is smooth), which will prove the first as-

---

*) See e.g. [87], p. 430, theorem 1.

**) See e.g. [70], Ch. II, § 2, theorem 1 and [11], p. 133, theorem 1 (which remains valid also for nets, with the same proof).

***) See e.g. [87], p. 428, theorem 6, applied to each $u^*(h) \in E^{**}$ $(h \in c_0(I)^*)$.

****) See e.g. [70], Ch. VII, § 2, statement (3) (a).

sertion. To this end it is sufficient, by Ch. II, § 5, lemma 5.1, to show that the set $A = \{f \in E^* \mid |f| \leqslant 1\}$ is $w^*$-closed, or, what is equivalent, that the functional $f \to |f|$ is $w^*$-lower semi-continuous. Since $u: E^* \to c_0(I)$ is $w^*$-$w$ continuous, each functional $f \to u(f)(i)$ is $w^*$-continuous and hence, for any $i_1, \ldots, i_n \in I$, the functional $f \to \left(\sum_{j=1}^{n} \frac{1}{2^j} |u(f)(i_j)|^2\right)^{\frac{1}{2}}$ is $w^*$-lower semi-continuous. Therefore, by (20.45), so is the functional $f \to D(u(f))$, whence also the functional $f \to |f|$, which proves the first assertion of corollary 20.5. Finally, it is known*) that $l^\infty(I)$ for any infinite set $I$ and $l^1(I)$ for any uncountable set $I$, do not admit an equivalent smooth norm. Hence, since the property of smoothness is inherited by subspaces, it follows that these spaces cannot be embedded isomorphically into any weakly compactly generated Banach space, which completes the proof of corollary 20.5.

*Remark 20.5.* Since $l^1(I)$ has an extended $M$-basis, the above proof of the second part of corollary 20.5 shows that the first part of corollary 20.5 does not remain valid for all spaces with extended $M$-bases.

*Definition 20.5.* An extended $M$-basis $\{x_i\}_{i \in I}$ of a Banach space $E$ is said to be *norming* if $r([f_i]_{i \in I}) > 0$, where $\{f_i\}_{i \in I} \subset E^*$ is the a.f.f. to $\{x_i\}_{i \in I}$ and $r([f_i]_{i \in I})$ is the characteristic of the subspace $[f_i]_{i \in I}$ of $E^*$.

In view of theorem 20.5, it is natural to ask

*Problem 20.4.* a) Does every weakly compactly generated Banach space, or every subspace thereof, have a norming extended $M$-basis? b) How about a normal extended $M$-basis?

In the above we have used § 19, proposition 19.5 to obtain the extended $M$-bases of theorem 20.5 a). On the other hand, for these extended $M$-bases, by slightly modifying the proof of § 19, proposition 19.5, we obtain the following result, which will be used in the sequel:

**Proposition 20.7.** *Let $E$ be an infinite dimensional Banach space with an extended $M$-basis $\{x_i\}_{i \in I}$ such that the set $\{x_i\}_{i \in I} \cup \{0\}$ is weakly compact**) and let $\vartheta$ be the first ordinal of cardinality* dens $E$. *Then there exist a long sequence $\{p_\lambda\}_{\omega \leqslant \lambda \leqslant \vartheta}$ of projections on $E$ and a long sequence $\{I_\lambda\}_{\omega \leqslant \lambda \leqslant \vartheta}$ of subsets of $I$, with the following properties (where $\{f_i\}_{i \in I} \subset E^*$ is the a.f.f. to $\{x_i\}_{i \in I}$):*

$$\|p_\lambda\| = 1 \qquad (\omega \leqslant \lambda \leqslant \vartheta), \qquad (20.49)$$

$$p_\lambda p_\varkappa = p_\varkappa p_\lambda = p_\varkappa \qquad (\omega \leqslant \varkappa \leqslant \lambda \leqslant \vartheta), \qquad (20.50)$$

$$\text{dens } p_\lambda(E) = \text{card } \lambda \qquad (\omega \leqslant \lambda \leqslant \vartheta), \qquad (20.51)$$

$$p_\vartheta = I_E, \qquad (20.52)$$

$$p_\lambda(E) = [x_i]_{i \in I_\lambda} \qquad (\omega \leqslant \lambda \leqslant \vartheta), \qquad (20.53)$$

*) See e.g. [70], Ch. VII, § 4, statements (1) (b) and (2) (g). For a short proof, see [199], corollary 1.

**) See theorem 20.5 a).

$$p_\lambda^*(E^*) \supset [f_i]_{i\in I_\lambda} \qquad (\omega \leqslant \lambda \leqslant \vartheta), \tag{20.54}$$

$$I_\lambda \subset I_{\lambda+1} \qquad (\omega \leqslant \lambda < \vartheta), \tag{20.55}$$

$$I_\lambda = \bigcup_{\omega\leqslant\varkappa<\lambda} I_{\varkappa+1} \qquad (\omega < \lambda \leqslant \vartheta). \tag{20.56}$$

*Proof.* Let $\mathscr{K}$ be the circled convex hull of $\{x_i\}_{i\in I} \cup \{0\}$, so $\mathscr{K}$ is weakly compact*). We shall construct $\{p_\lambda\}_{\omega\leqslant\lambda\leqslant\vartheta}$ and $\{I_\lambda\}_{\omega\leqslant\lambda\leqslant\vartheta}$ by transfinite induction.

If dens $E = \aleph_0$, then $p_\omega = I_E$ and $I_\omega = I$ satisfy the required conditions.

Assume now that dens $E > \aleph_0$. By proposition 20.1 we have card $I =$ $=$ dens $E$, so we can write $I = [1, \vartheta)$ (by considering a well ordering of $I$). Let $J_0 = [1, \omega)$. Then, by § 19, lemma 19.5 (for $B = \operatorname{lin}\{x_i\}_{i\in J_0}$, $V = \operatorname{lin}\{f_i\}_{i\in J_0}$), there exists a linear projection $q_\omega\colon E \to E$ satisfying $\|q_\omega\| = 1$, $q_\omega(\mathscr{K}) \subset \mathscr{K}$, $[x_i]_{i\in J_0} \subset q_\omega(E)$, $[f_i]_{i\in J_0} \subset q_\omega^*(E^*)$ and dens $q_\omega(E) =$ $= \aleph_0$. If $q_\omega(E) = [x_i]_{i\in J_0}$, let $p_\omega = q_\omega$ and $I_\omega = J_0$. If not, then, since $q_\omega(E)$ is separable, there exists a countable subset $J_1$ of $I$, with $J_1 \supset J_0$, such that $q_\omega(E) \subset [x_i]_{i\in J_1}$ (see e.g. the proof of § 17, proposition 17.1). Then, by § 19, lemma 19.5, we find a projection $q_{\omega+1}\colon E \to$ $\to E$ satisfying $\|q_{\omega+1}\| = 1$, $q_{\omega+1}(\mathscr{K}) \subset \mathscr{K}$, $[x_i]_{i\in J_1} \subset q_{\omega+1}(E)$, $[f_i]_{i\in J_1} \cup$ $\cup\ q_\omega^*(E^*) \subset q_{\omega+1}^*(E^*)$, and dens $q_{\omega+1}(E) = \aleph_0$. Then, as was observed in the proof of § 19, proposition 19.5, $q_\omega q_{\omega+1} = q_{\omega+1} q_\omega = q_\omega$. Continuing in this way, we obtain a sequence of projections $\{q_{\omega+n}\}_{n=0}^\infty$ on $E$ and an increasing sequence of countable subsets $\{J_n\}_{n=0}^\infty$ of $I$, with $[1, \omega) = J_0$, such that $\|q_{\omega+n}\| = 1$, $q_{\omega+n}(\mathscr{K}) \subset \mathscr{K}$ $(n = 0,1,2, \ldots)$, $q_{\omega+n}q_{\omega+m} = q_{\omega+m}q_{\omega+n} = q_{\min(\omega+n,\omega+m)}$ $(n, m = 0,1,2, \ldots)$, $[x_i]_{i\in J_0} \subset$ $\subset q_\omega(E) \subset [x_i]_{i\in J_1} \subset q_{\omega+1}(E) \subset [x_i]_{i\in J_2} \subset \ldots$, $[f_i]_{i\in J_n} \subset q_{\omega+n}^*(E^*)$ and dens $q_{\omega+n}(E) = \aleph_0$ $(n = 0,1,2, \ldots)$. Then, as in the proof of § 19, lemma 19.4, it follows that $\{q_{\omega+n}\}_{n=0}^\infty$ has a weak cluster point $p_\omega$ $\Big($hence $p_\omega(E) \subset \overline{\bigcup_{n=0}^\infty q_{\omega+n}(E)}\Big)$, which is a projection on $E$, satisfying $\|p_\omega\| = 1$, $p_\omega(\mathscr{K}) \subset \mathscr{K}$, $\quad p_\omega q_{\omega+m} = q_{\omega+m} p_\omega = q_{\omega+m} \quad (m = 0,1,2, \ldots)$ $\Big($hence $p_\omega(E) \supset \overline{\bigcup_{n=0}^\infty q_{\omega+n}(E)}\Big)$ and dens $p_\omega(E) = \aleph_0$. But, since $\bigcup_{n=0}^\infty q_{\omega+n}(E) =$ $= \bigcup_{n=0}^\infty [x_i]_{i\in J_n}$, we have $p_\omega(E) = \overline{\bigcup_{n=0}^\infty q_{\omega+n}(E)} = [x_i]_{i\in I_\omega}$, where $I_\omega =$ $= \bigcup_{n=0}^\infty J_n \subset I$. Also, $p_\omega^*(E^*) \ni f_i$ $(i \in I_\omega)$, whence $p_\omega^*(E^*) \supset [f_i]_{i\in I_\omega}$. Thus, $p_\omega$ and $I_\omega$ satisfy (20.49), (20.51), (20.53) and (20.54) for $\lambda = \omega$ and, in addition, $p_\omega(\mathscr{K}) \subset \mathscr{K}$, $[1, \omega) \subset I_\omega$.

*) See e.g. [87], p. 434, theorem 4.

Assume now that $\omega < \lambda \leqslant \vartheta$ and that we have constructed all $p_\varkappa$ and $I_\varkappa$, $\omega \leqslant \varkappa < \lambda$, satisfying the required conditions and $p_\varkappa(\mathscr{X}) \subset \mathscr{X}$, $[1, \varkappa) \subset \subset I_\varkappa$ $(\omega \leqslant \varkappa < \lambda)$. If $\lambda$ is a nonlimit ordinal, we construct $p_\lambda$ and $I_\lambda \supset I_{\lambda-1} \cup \cup \{\lambda - 1\}$ with card $I_\lambda =$ card $\lambda$, from $p_{\lambda-1}$ and $I_{\lambda-1}$, similarly to the above construction of $p_\omega$ and $I_\omega$, taking care to have, in addition, (20.50). Namely, if $J_\lambda = I_{\lambda-1} \cup \{\lambda - 1\}$, then, by § 19, lemma 19.5 (for $B = B_\lambda \cap \operatorname{lin} \mathscr{X}$, where $B_\lambda = \operatorname{lin} \{p_{\lambda-1}(E) \cup \{x_i\}_{i \in J_\lambda}\}$, and for $V = \operatorname{lin} \{p^*_{\lambda-1}(E^*) \cup \{f_i\}_{i \in J_\lambda}\}$), there exists a linear projection $q_\lambda \colon E \to E$ satisfying $\|q_\lambda\| = 1$, $q_\lambda(\mathscr{X}) \subset \mathscr{X}$, $p_{\lambda-1}(E) \cup [x_i]_{i \in J_\lambda} \subset q_\lambda(E)$, $p^*_{\lambda-1}(E^*) \cup \cup [f_i]_{i \in J_\lambda} \subset q^*_\lambda(E^*)$, and dens $q_\lambda(E) =$ card $\lambda$. If $q_\lambda(E) = [x_i]_{i \in J_\lambda}$, let $p_\lambda = q_\lambda$ and $I_\lambda = J_\lambda$. If not, take $J_{\lambda+1} \subset I$ with $J_{\lambda+1} \supset J_\lambda$, card $J_{\lambda+1} = =$ card $\lambda$, such that $q_\lambda(E) \subset [x_i]_{i \in J_{\lambda+1}}$. Then, constructing $\{q_{\lambda+n}\}_{n=0}^\infty$ and $\{J_{\lambda+n}\}_{n=0}^\infty$ similarly to the above, we take $p_\lambda$ to be a weak cluster point of $\{q_{\lambda+n}\}_{n=0}^\infty$ and $I_\lambda = \bigcup_{n=0}^\infty J_{\lambda+n} \subset I$. On the other hand, if $\lambda$ is a limit ordinal, we take a weak cluster point $p_\lambda$ of $\{p_\varkappa\}_{\omega \leqslant \varkappa < \lambda}$ and we put $I_\lambda = \bigcup_{\omega \leqslant \varkappa < \lambda} I_\varkappa$. Then, similarly to § 19, proof of proposition 19.5, we get (20.50), hence (20.54) and $p_\lambda(E) = \overline{\bigcup_{\omega \leqslant \varkappa < \lambda} p_\varkappa(E)} = [x_i]_{i \in I_\lambda}$, whence also dens $p_\lambda(E) \leqslant \sum_{\omega \leqslant \varkappa < \lambda}$ dens $p_\varkappa(E) = \sum_{\omega \leqslant \varkappa < \lambda}$ card $\varkappa =$ card $\lambda$. On the other hand, by lemma 20.1, dens $p_\lambda(E) =$ card $I_\lambda \geqslant$ card $I_\varkappa =$ dens $p_\varkappa(E) =$ card $\varkappa$ $(\omega \leqslant \varkappa < \lambda)$, whence, finally, dens $p_\lambda(E) =$ card $\lambda$. Also, $[1, \lambda) = = \bigcup_{\omega \leqslant \varkappa < \lambda} [1, \varkappa) \subset \bigcup_{\omega \leqslant \varkappa < \lambda} I_{\varkappa+1} = I_\lambda$. Thus, we have constructed $\{p_\lambda\}_{\omega \leqslant \lambda \leqslant \vartheta}$ and $\{I_\lambda\}_{\omega \leqslant \lambda \leqslant \vartheta}$ satisfying (20.49) – (20.51) and (20.53)–(20.56) and such that $I = [1, \vartheta) \subset I_\vartheta \subset I$, whence $I_\vartheta = I$. But then $p_\vartheta(E) = = [x_i]_{i \in I_\vartheta} = [x_i]_{i \in I} = E$, whence, by $p^2_\vartheta = p_\vartheta$, it follows that $p_\vartheta = I_E$, which completes the proof of proposition 20.7.

*Remark 20.6.* Since card $I_\lambda =$ card $\lambda <$ card $\vartheta =$ card $I$ for $\omega \leqslant \lambda < \vartheta$, we have $I_\lambda \neq I$ for $\omega \leqslant \lambda < \vartheta$, and hence the above construction can be made so that $I_{\lambda+1} \neq I_\lambda$ for all $\omega \leqslant \lambda < \vartheta$ (by omitting those $\lambda$ for which $I_{\lambda+1} = I_\lambda$ and reindexing). In this case, since $\{x_i\}_{i \in I}$ is an $M$-basis, we have $p_{\lambda+1}(E) = [x_i]_{i \in I_{\lambda+1}} \neq \neq [x_i]_{i \in I_\lambda} = p_\lambda(E)$, whence $p_{\lambda+1} \neq p_\lambda$, for all $\omega \leqslant \lambda < \vartheta$. Note also that, by the results of § 19 which have been used in the above proof, one can also add to (20.49)–(20.56) the property $p_\lambda(G) \subset G$ $(\omega \leqslant \lambda \leqslant \vartheta)$, where $G$ is any given closed linear subspace of $E$. Thus, reindexing $\{0\} \cup \{p_\lambda\}_{\omega \leqslant \lambda \leqslant \vartheta}$, we obtain a new proof of § 19, theorem 19.4 (we start with theorem 20.5 a)).

*Definition 20.6.* An extended $M$-basis $\{x_i\}_{i \in I}$ of a Banach space $E$ is said to be *shrinking*, if for the a.f.f. $\{f_i\}_{i \in I} \subset E^*$ to $\{x_i\}_{i \in I}$ we have

$$[f_i]_{i \in I} = E^*. \tag{20.57}$$

Also, we make the convention that $\emptyset$ is the only shrinking extended $M$-basis of $\{0\}$.

We have the following generalization of the first implication of Ch. II, § 12, theorem 12.1 a):

**Proposition 20.8.** *Let $E$ be a Banach space with a shrinking extended $M$-basis $\{x_i\}_{i\in I}$ such that* $\sup_{i\in I}\|x_i\| < \infty$. *Then* $\{f(x_i)\}_{i\in I}\in c_0(I)$ $(f\in E^*)$, *hence the set $\{x_i\}_{i\in I}\cup\{0\}$ is weakly compact (and hence $E$ is weakly compactly generated).*

*Proof.* Let $\{f_i\}_{i\in I}\subset E^*$ be the a.f.f. to $\{x_i\}_{i\in I}$. Since $\{x_i\}_{i\in I}$ is shrinking, we have $[f_i]_{i\in I} = E^*$ and hence, by the proof of proposition 20.6 (applied to the $M$-basis $\{f_i\}_{i\in I}$ of $E^*$, with the a.f.f. $\{\pi(x_i)\}_{i\in I}$, where $\pi$ is the canonical embedding of $E$ into $E^{**}$), we have $\{f(x_i)\}_{i\in I}\in c_0(I)$ for all $f\in E^*$. Then, by the finiteness of the sets $\{i\in I \mid |f(x_i)|\geqslant\varepsilon\}$ $(f\in E^*, \varepsilon>0)$, every infinite subsequence $\{x_{i_n}\}$ of $\{x_i\}_{i\in I}$ is weakly convergent to 0 and hence, by the Eberlein-Šmulyan theorem*), $\{x_i\}_{i\in I}\cup\{0\}$ is weakly compact, which completes the proof of proposition 20.8.

For shrinking extended $M$-bases we have the following existence theorem, similar to theorem 20.5 b) above:

**Theorem 20.6.** *If $G$ is a subspace of a weakly compactly generated Banach space $E$ and if $G$ is isomorphic to a very smooth Banach space, then $G$ has a shrinking extended $M$-basis.*

*Proof.* Let us first make the following additional observation to the proof of theorem 20.5 b): If $G$ is a very smooth subspace of $E$ and if $\{x_i^{\omega}\}_{i\in I_\omega}$ and $\{x_i^{\lambda+1}\}_{i\in I_{\lambda+1}}$ are shrinking extended $M$-bases of $G_\omega = p_\omega(G)$ and $G_{\lambda+1} = (p_{\lambda+1}-p_\lambda)(G)$ $(\omega\leqslant\lambda<\vartheta)$ respectively, where $\{p_\lambda\}_{\omega\leqslant\lambda\leqslant\vartheta}$ is as in § 19, proposition 19.5, then the extended $M$-basis $\{x_i^{\omega}\}_{i\in I_\omega}\cup\bigcup_{\omega\leqslant\lambda<\vartheta}\{x_i^{\lambda+1}\}_{i\in I_{\lambda+1}}$ of $G$ is shrinking, that is,

$$[\{f_i^{\omega}\}_{i\in I_\omega}\cup\bigcup_{\omega\leqslant\lambda<\vartheta}\{f_i^{\lambda+1}\}_{i\in I_{\lambda+1}}] =$$

$$= [\{\varphi_i^{\omega}\circ p_\omega|_G\}_{i\in I_\omega}\cup\bigcup_{\omega\leqslant\lambda<\vartheta}\{\varphi_i^{\lambda+1}\circ(p_{\lambda+1}|_G - p_\lambda|_G)\}_{i\in I_{\lambda+1}}] = G^*,$$

where $\{\varphi_i^{\nu}\}_{i\in I_\nu}\subset G_\nu^*$ is the a.f.f. to $\{x_i^{\nu}\}_{i\in I_\nu}\subset G_\nu$. Indeed, let $\rho_\omega$ and $\rho_{\lambda+1}$ be the canonical isomorphisms of $(p_\omega|_G)^*(G^*)$ and $((p_{\lambda+1}|_G)^* - (p_\lambda|_G)^*)(G^*)$ onto $p_\omega(G)^*$ and $(p_{\lambda+1}-p_\lambda)(G)^*$ $(\omega\leqslant\lambda<\vartheta)$ respectively, given by § 14, lemma 14.3. Then $f_i^{\omega} = \varphi_i^{\omega}\circ p_\omega|_G = \rho_\omega^{-1}(\varphi_i^{\omega})$ $(i\in I_\omega)$ and, similarly, $f_i^{\lambda+1} = \rho_{\lambda+1}^{-1}(\varphi_i^{\lambda+1})$ $(i\in I_{\lambda+1}, \omega\leqslant\lambda<\vartheta)$, whence, since $[\varphi_i^{\omega}]_{i\in I_\omega} = p_\omega(G)^*$ and $[\varphi_i^{\lambda+1}]_{i\in I_{\lambda+1}} = (p_{\lambda+1}-p_\lambda)(G)^*$, we obtain

$$[f_i^{\omega}]_{i\in I_\omega} = [\rho_\omega^{-1}(\varphi_i^{\omega})]_{i\in I_\omega} = \rho_\omega^{-1}([\varphi_i^{\omega}]_{i\in I_\omega}) = \rho_\omega^{-1}(p_\omega(G)^*) = (p_\omega|_G)^*(G^*)$$

*) See e.g. [87], p. 430, theorem 1.

and, similarly, $[f_i^{\lambda+1}]_{i\in I_{\lambda+1}} = ((p_{\lambda+1}|_G)^* - (p_\lambda|_G)^*)(G^*)$ $(\omega \leqslant \lambda < \vartheta)$. Hence, in order to show that $\{x_i^{(\omega)}\}_{i\in I_\omega} \cup \bigcup_{\omega\leqslant\lambda<\vartheta} \{x_i^{\lambda+1}\}_{i\in I_{\lambda+1}}$ is shrinking, it remains to prove that we have

$$G^* = [(p_\omega|_G)^*(G^*) \cup \bigcup_{\omega\leqslant\lambda<\vartheta} ((p_{\lambda+1}|_G)^* - (p_\lambda|_G)^*)(G^*)]. \quad (20.46')$$

To this end, observe first that, by § 19, proposition 19.5, $\{p_\lambda|_G\}_{\omega\leqslant\lambda\leqslant\vartheta}$ is a long sequence of projections on $G$, satisfying $\|p_\lambda|_G\| = 0$ or $1$ $(\omega\leqslant\lambda\leqslant\vartheta)$; moreover, since $p_\vartheta = I_E$ and since for each $x\in G$ the mapping $\lambda \to p_\lambda(x)$ from $[\omega, \vartheta]$ into $G$ is ascending and continuous, there exists an ordinal $\lambda_0$ with $\omega \leqslant \lambda_0 < \vartheta$, such that $p_\lambda|_G = 0$ for $\omega \leqslant \lambda < \lambda_0$ and $\|p_\lambda|_G\| = 1$ for $\lambda_0 \leqslant \lambda \leqslant \vartheta$. Hence, since $G$ is very smooth, by § 19, lemma 19.15, proposition 19.5 (formulae (19.80) and (19.82)) and lemma 19.14, we obtain*)

$$(p_\lambda|_G)^*(G^*) = \overline{D_{G^*}(p_\lambda(G))} \subset \overline{D_{G^*}(\overline{\bigcup_{\omega\leqslant\varkappa<\lambda} p_{\varkappa+1}(G)})} =$$

$$= \overline{\bigcup_{\omega\leqslant\varkappa<\lambda} D_{G^*}(p_{\varkappa+1}(G))} \subset \overline{\bigcup_{\omega\leqslant\varkappa<\lambda} (p_{\varkappa+1}|_G)^*(G^*)} \qquad (\omega<\lambda\leqslant\vartheta).$$

But then, by § 19, the last statement of proposition 19.5 (applied to $\{(p_\lambda|_G)^*\}_{\omega\leqslant\lambda\leqslant\vartheta}$ on $G^*$), for each fixed $f\in G^*$ the mapping $\lambda \to (p_\lambda|_G)^*(f)$ from $[\omega, \vartheta]$ into $G^*$ is continuous. Consequently, by § 19, remark 19.2 a) (applied to $\lambda \to (p_\lambda|_G)^*(f)$) and (19.83), we obtain (20.46′), which proves the assertion that $\{x_i^{(\omega)}\}_{i\in I_\omega} \cup \bigcup_{\omega\leqslant\lambda<\vartheta} \{x_i^{\lambda+1}\}_{i\in I_{\lambda+1}}$ is a shrinking extended $M$-basis of $G$.

Let us prove now theorem 20.6. By Ch. II, § 13, lemma 13.6, we may assume that $G$ itself is very smooth. If $E$ is separable, so is $G$ and hence by § 19, lemma 19.13, so is $G^*$. Therefore, by § 8, theorem 8.1, $G$ has a shrinking $M$-basis.

Assume now that dens $E > \aleph_0$ and that the statement holds for all Banach spaces $E_0$ with dens $E_0 <$ dens $E$. Let $\{p_\lambda\}_{\omega\leqslant\lambda\leqslant\vartheta}$ be a long sequence of projections on $E$ given by § 19, proposition 19.5. Then, since $G_\omega$ and $G_{\lambda+1}$ are very smooth subspaces of the weakly compactly generated spaces $E_\omega = p_\omega(E)$ and $E_{\lambda+1} = (p_{\lambda+1} - p_\lambda)(E)$ respectively, with dens $p_\omega(E)$, dens $(p_{\lambda+1} - p_\lambda)(E) <$ dens $E$, by our induction hypothesis $G_\omega$ and $G_{\lambda+1}$ $(\omega\leqslant\lambda < \vartheta)$ have shrinking extended $M$-bases. Hence, by the observation made above, $G$ has a shrinking extended $M$-basis, which completes the proof of theorem 20.6.

---

*) This part of the argument is similar to § 19, proof of proposition 19.8, formula (19.159).

**Corollary 20.6.** *If $G$ is a subspace of a weakly compactly generated Banach space and if $G$ is isomorphic to a very smooth Banach space, then $G$ is weakly compactly generated. Hence, in particular, every subspace $G$ of a weakly compactly generated very smooth Banach space $E$ is weakly compactly generated.*

*Proof.* By theorem 20.6, $G$ has a shrinking extended $M$-basis and hence, by proposition 20.8, the conclusion follows.

Now we shall prove (in theorem 20.7 below) that the converse of the particular case $G = E$ of theorem 20.6 (whence the converse of theorem 20.6 as well) is also valid and we shall give some other characterizations, too, of Banach spaces which have a shrinking extended $M$-basis. To this end, we shall need some preparation.

**Proposition 20.9.** *Let $\{x_i\}_{i\in I}$ be a shrinking extended M-basis of a Banach space E, such that $\sup_{i\in I}\|x_i\|<\infty$, and let $\{p_\lambda\}_{\omega\leqslant\lambda\leqslant\vartheta}$ and $\{I_\lambda\}_{\omega\leqslant\lambda\leqslant\vartheta}$ be as in proposition 20.7. Then*

a) *For each $\lambda$ with $\omega\leqslant\lambda<\vartheta$, $\{(p_{\lambda+1}-p_\lambda)(x_i)\}_{i\in I_{\lambda+1}\setminus I_\lambda}$ is a shrinking extended M-basis of $(p_{\lambda+1}-p_\lambda)(E)$.*

b) *We have*

$$p_\lambda^*(f)\in\overline{\{p_{\varkappa+1}^*(f)\}_{\omega\leqslant\varkappa<\lambda}}\qquad (f\in E^*,\ \omega\leqslant\lambda\leqslant\vartheta). \tag{20.58}$$

*Consequently, for each $f\in E^*$ and $\varepsilon>0$, the set*

$$B_{f,\varepsilon}=\{\lambda\in[\omega,\vartheta)\,|\ \|p_{\lambda+1}^*(f)-p_\lambda^*(f)\|>\varepsilon\} \tag{20.59}$$

*is finite.*

*Proof.* We note first that by proposition 20.8 the set $\{x_i\}_{i\in I}\cup\{0\}$ is weakly compact and thus we can indeed apply proposition 20.7 to get $\{p_\lambda\}_{\omega\leqslant\lambda\leqslant\vartheta}$ and $\{I_\lambda\}_{\omega\leqslant\lambda\leqslant\vartheta}$.

a) If $p_{\lambda+1}=p_\lambda$, then $(p_{\lambda+1}-p_\lambda)(E)=\{0\}$ and $[x_i]_{i\in I_{\lambda+1}}=p_{\lambda+1}(E)=$ $=p_\lambda(E)=[x_i]_{i\in I_\lambda}$, whence, since $\{x_i\}_{i\in I}$ is an extended $M$-basis, $I_{\lambda+1}=I_\lambda$, so the statement is true.

Assume now that $p_{\lambda+1}\neq p_\lambda$ and let $x\in(p_{\lambda+1}-p_\lambda)(E)\setminus\{0\}$. Then $x\notin p_\lambda(E)$ (since otherwise $x\in(p_{\lambda+1}-p_\lambda)(E)\cap p_\lambda(E)=\{0\}$). Take any $\varepsilon$ such that $0<\varepsilon<\operatorname{dist}(x,p_\lambda(E))$. Since $x\in p_{\lambda+1}(E)=[x_i]_{i\in I_{\lambda+1}}$, there exist a finite subset $\{i_n\}_{n=1}^m\subset I_{\lambda+1}$ and scalars $\alpha_{i_1},\ldots,\alpha_{i_m}\in K$ such that

$$\left\|x-\sum_{n=1}^m\alpha_{i_n}x_{i_n}\right\|<\varepsilon. \tag{20.60}$$

We have $\{i_n\}_{n=1}^m\cap(I_{\lambda+1}\setminus I_\lambda)\neq\varnothing$, since otherwise $\{i_n\}_{n=1}^m\subset I_\lambda$, whence, by (20.53) and (20.60), $\operatorname{dist}(x,p_\lambda(E))=\operatorname{dist}(x,[x_i]_{i\in I_\lambda})<\varepsilon$, in contradiction with our choice of $\varepsilon$. We may write (by reindexing $\{i_n\}_{n=1}^m$, if necessary) that $\{i_n\}_{n=1}^m \cap (I_{\lambda+1} \setminus I_\lambda) = \{i_n\}_{n=1}^k$. Then, since $(p_{\lambda+1} - p_\lambda)(x_i) = x_i - x_i = 0$ $(i \in I_\lambda)$, we obtain

$$\left\| x - \sum_{n=1}^{k} \alpha_{i_n}(p_{\lambda+1} - p_\lambda)(x_{i_n}) \right\| = \left\| (p_{\lambda+1} - p_\lambda)\left( x - \sum_{n=1}^{m} \alpha_{i_n} x_{i_n} \right) \right\| \leqslant$$

$$\leqslant \|p_{\lambda+1} - p_\lambda\| \left\| x - \sum_{n=1}^{m} \alpha_{i_n} x_{i_n} \right\| < 2\varepsilon,$$

which proves that $(p_{\lambda+1} - p_\lambda)(E) = [(p_{\lambda+1} - p_\lambda)(x_i)]_{i \in I_{\lambda+1} \setminus I_\lambda}$.

Furthermore, let us show that $\{(p_{\lambda+1} - p_\lambda)(x_i)\}_{i \in I_{\lambda+1} \setminus I_\lambda}$ admits the a.f.f. $\{f_i|_{(p_{\lambda+1} - p_\lambda)(E)}\}_{i \in I_{\lambda+1} \setminus I_\lambda}$, where $\{f_i\}_{i \in I} \subset E^*$ is the a.f.f. to $\{x_i\}_{i \in I}$. Indeed, since $p_\lambda(x) \in [x_j]_{j \in I_\lambda}$ $(x \in E)$, we have $f_i(p_\lambda(x)) = 0$ $(x \in E,$ $i \in I_{\lambda+1} \setminus I_\lambda)$. Consequently, since $p_{\lambda+1}(x_j) = x_j$ $(j \in I_{\lambda+1})$, we obtain

$$f_i((p_{\lambda+1} - p_\lambda)(x_j)) = f_i(p_{\lambda+1}(x_j)) - f_i(p_\lambda(x_j)) = \delta_{ij} \quad (i,\ j \in I_{\lambda+1} \setminus I_\lambda),$$

which proves our assertion.

Finally, we shall show that $[f_i|_{(p_{\lambda+1} - p_\lambda)(E)}]_{i \in I_{\lambda+1} \setminus I_\lambda} = (p_{\lambda+1} - p_\lambda)(E)^*$, which will prove a). Let $\varphi \in (p_{\lambda+1} - p_\lambda)(E)^* \setminus \{0\}$ and let $0 < \varepsilon < \|\varphi\|$. Take any extension $f \in E^*$ of $\varphi$ to the whole space $E$. Then, since $E^* = [f_i]_{i \in I}$, there exist a finite subset $\{i_n\}_{n=1}^m$ of $I$ and scalars $\alpha_{i_1}, \ldots, \alpha_{i_m} \in K$ such that

$$\left\| f - \sum_{n=1}^{m} \alpha_{i_n} f_{i_n} \right\| < \varepsilon. \tag{20.61}$$

Let us observe now that we have

$$f_i(x) = 0 \qquad (x \in (p_{\lambda+1} - p_\lambda)(E),\ i \in I \setminus (I_{\lambda+1} \setminus I_\lambda)). \tag{20.62}$$

Indeed, if $i \in I \setminus I_{\lambda+1}$, then, since $p_{\lambda+1}(E) = [x_j]_{j \in I_{\lambda+1}}$, we have $f_i(x) = 0$ $(x \in p_{\lambda+1}(E))$. On the other hand, if $i \in I_\lambda$, then, by (20.54), $f_i \in p_\lambda^*(E^*)$, whence, by (20.50), $f_i(p_{\lambda+1}(x)) = p_\lambda^*(f_i)(p_{\lambda+1}(x)) = f_i(p_\lambda p_{\lambda+1}(x)) = f_i(p_\lambda(x))$ $(x \in E)$, which proves (20.62). From this we infer that $\{i_n\}_{n=1}^m \cap (I_{\lambda+1} \setminus I_\lambda) \neq \emptyset$ (since otherwise $\{i_n\}_{n=1}^m \subset I \setminus (I_{\lambda+1} \setminus I_\lambda)$, whence, by (20.62) and (20.61), $|\varphi(x)| = \left| f(x) - \sum_{n=1}^{m} \alpha_{i_n} f_{i_n}(x) \right| \leqslant \left\| f - \sum_{n=1}^{m} \alpha_{i_n} f_{i_n} \right\| \|x\| < \varepsilon \|x\|$ for all $x \in (p_{\lambda+1} - p_\lambda)(E)$, so $\|\varphi\| \leqslant \varepsilon$, in contradiction with our choice of $\varepsilon$). We may write (by reindexing $\{i_n\}_{n=1}^m$, if necessary) that $\{i_n\}_{n=1}^m \cap (I_{\lambda+1} \setminus I_\lambda) = \{i_n\}_{n=1}^k$. Then, again by (20.62) and (20.61), we obtain

$$\left|\varphi(x) - \sum_{n=1}^{k} \alpha_{i_n} f_{i_n}(x)\right| = \left|f(x) - \sum_{n=1}^{m} \alpha_{i_n} f_{i_n}(x)\right| \leqslant$$

$$\leqslant \left\|f - \sum_{n=1}^{m} \alpha_{i_n} f_{i_n}\right\| \|x\| < \varepsilon \|x\| \qquad (x \in (p_{\lambda+1} - p_\lambda)(E)),$$

whence $\left\|\varphi - \sum_{n=1}^{k} \alpha_{i_n} f_{i_n}|_{(p_{\lambda+1}-p_\lambda)(E)}\right\| \leqslant \varepsilon$. This proves a).

b) Let us first prove the inclusions

$$p_\lambda^*(E^*) \subset \overline{\bigcup_{\omega \leqslant \varkappa < \lambda} p_{\varkappa+1}^*(E^*)} \qquad (\omega \leqslant \lambda \leqslant \vartheta). \tag{20.63}$$

Let $\omega \leqslant \lambda \leqslant \vartheta$, $f \in p_\lambda^*(E^*) \setminus \{0\}$ and let $0 < \varepsilon < \|f|_{p_\lambda(E)}\|$. Then, since $E^* = [f_i]_{i \in I}$, there exist a finite subset $\{i_n\}_{n=1}^m$ of $I$ and scalars $\alpha_{i_1}, \ldots, \alpha_{i_m} \in K$ such that we have (20.61). Observe that by $p_\lambda(E) = [x_j]_{j \in I_\lambda}$ we have $f_i(p_\lambda(x)) = 0$ for all $x \in E$ and $i \in I \setminus I_\lambda$. From this we infer that $\{i_n\}_{n=1}^m \cap I_\lambda \neq \emptyset$ (since otherwise $\{i_n\}_{n=1}^m \subset I \setminus I_\lambda$, whence, by (20.61), $|f(x)| = \left|f(x) - \sum_{n=1}^{m} \alpha_{i_n} f_{i_n}(x)\right| \leqslant \left\|f - \sum_{n=1}^{m} \alpha_{i_n} f_{i_n}\right\| \|x\| < \varepsilon \|x\|$ for all $x \in p_\lambda(E)$, in contradiction with our choice of $\varepsilon$). We may write (by reindexing $\{i_n\}_{n=1}^m$, if necessary) that $\{i_n\}_{n=1}^m \cap I_\lambda = \{i_n\}_{n=1}^k$. Then, using (20.54) and again that $f_i(p_\lambda(x)) = 0$ $(x \in E,\ i \in I \setminus I_\lambda)$, we obtain

$$\left\|f - \sum_{n=1}^{k} \alpha_{i_n} f_{i_n}\right\| = \left\|p_\lambda^*\left(f - \sum_{n=1}^{m} \alpha_{i_n} f_{i_n}\right)\right\| \leqslant \left\|f - \sum_{n=1}^{m} \alpha_{i_n} f_{i_n}\right\| < \varepsilon,$$

which proves (20.63).

Finally, (20.58) (which means, in other words, that for each fixed $f \in E^*$ the mapping $\lambda \to p_\lambda^*(f)$ from $[\omega, \vartheta]$ into $E^*$ is continuous) follows from (20.63), (20.49) and (20.50) (see § 19, the last statement of proposition 19.5), which completes the proof of proposition 20.9 (the last statement in b) follows from (20.58) by § 19, remark 19.2 a) ).

We recall that (the norm of) a Banach space $E$ is said to be *locally uniformly convex* if the relations $\{x_n\} \subset E$, $x \in E$, $\|x_n\| = \|x\| = 1$ $(n = 1,2, \ldots)$ and $\lim_{n\to\infty} \|x_n + x\| = 2$ imply $\lim_{n\to\infty} \|x_n - x\| = 0$. Clearly, every locally uniformly convex space is strictly convex. It is well known*) that if the conjugate space $E^*$ is locally uniformly convex, then $E$ is $(F)$ (i.e., the norm of $E$ is Fréchet differentiable at every nonzero point) and hence very smooth (the converses of these implications are not valid). We shall follow this way to obtain the desired converse of the particular case $G = E$ of theorem 20.6, announced above. For this purpose,

*) See [70], Ch. VII, § 2, statement (10) (d) or [253], theorem 2.3.

we shall first give, in lemma 20.3 below, a sufficient condition for a (conjugate) Banach space to admit an equivalent (dual) locally uniformly convex norm.

We recall that the Banach space $c_0(\Delta)$ (where $\Delta$ is any set) admits an equivalent locally uniformly convex norm, namely, the norm defined by*)

$$D(z(.)) = \sup_{\substack{\delta_1, \ldots, \delta_n \in \Delta \\ 1 \leqslant n < \infty}} \left( \sum_{j=1}^{n} \frac{1}{2^j} |z(\delta_j)|^2 \right)^{\frac{1}{2}} \qquad (z(.) \in c_0(\Delta)). \quad (20.64)$$

**Lemma 20.2.** *Let $\{\Psi_\delta\}_{\delta \in \Delta}$ be a transfinite sequence of functionals on a Banach space $E$, which are either sublinear and positive-homogeneous or linear on $E$ and such that $\{\Psi_\delta(x)\}_{\delta \in \Delta} \in c_0(\Delta)$ $(x \in E)$. Define an operator $Q\colon E \to c_0(\Delta)$ by*

$$Q(x) = \{\Psi_\delta(x)\}_{\delta \in \Delta} \qquad (x \in E) \quad (20.65)$$

*and let $D$ be the norm* (20.64) *on $c_0(\Delta)$. Then*

$$x \to D(Q(x)) \qquad (x \in E) \quad (20.66)$$

*is a sublinear and positive-homogeneous functional on $E$, such that the relations $\{x_k\} \subset E$, $x \in E$ and*

$$D(Q(x_k)) = D(Q(x)) = 1 \ (k = 1, 2, \ldots), \ \lim_{k \to \infty} D(Q(x_k + x)) = 2 \quad (20.67)$$

*imply*

$$\lim_{k \to \infty} \|Q(x_k) - Q(x)\|_{c_0(\Delta)} = 0. \quad (20.68)$$

*Proof.* By our assumption on the $\Psi_\delta$'s we have, in the natural partial ordering of $c_0(\Delta)$ (i.e., coordinatewise),

$$|Q(x+y)| = \{|\Psi_\delta(x+y)|\}_{\delta \in \Delta} \leqslant \{|\Psi_\delta(x) + \Psi_\delta(y)|\}_{\delta \in \Delta} = $$
$$= |Q(x) + Q(y)| \qquad (x, y \in E),$$

whence**)

$$D(Q(x+y)) \leqslant D(Q(x)+Q(y)) \leqslant D(Q(x)) + D(Q(y)) \quad (x, y \in E). \quad (20.69)$$

Similarly, again by our assumption on the $\Psi_\delta$'s and since $D$ is a norm on $c_0(\Delta)$, we obtain the relation of positive-homogeneity $D(Q(\alpha x)) = |\alpha| D(Q(x))$ $(x \in E,\ \alpha \in K)$.

---

*) See e.g. [70], Ch. VII, § 4, statement (3) (a) and [327], [328], [290]. Actually, we have used the strict convexity of this equivalent norm $D(.)$ on $c_0(\Delta)$ in the proofs of proposition 20.6 and corollary 20.5 above.

**) Indeed, by (20.64), the relations $z_1(.), z_2(.) \in c_0(\Delta)$, $|z_1(\delta)| \leqslant |z_2(\delta)|$ $(\delta \in \Delta)$ imply $D(z_1(.)) \leqslant D(z_2(.))$.

Finally, assume that we have (20.67). Then, by (20.69),

$$D(Q(x_k + x)) \leqslant D(Q(x_k) + Q(x)) \leqslant 2 \qquad (k = 1, 2, \ldots),$$

whence, by $\lim_{k\to\infty} D(Q(x_k + x)) = 2$, we obtain $\lim_{k\to\infty} D(Q(x_k) + Q(x)) = 2$. But, since $D$ is a locally uniformly convex norm on $c_0(\Delta)$, these relations, together with $D(Q(x_k)) = D(Q(x)) = 1$ $(k = 1,2, \ldots)$, imply $\lim_{k\to\infty} D(Q(x_k) - Q(x)) = 0$. Hence, since $D$ is equivalent to the initial norm on $c_0(\Delta)$, it follows that we have (20.68), which completes the proof of lemma 20.2.

**Lemma 20.3.** a) *Let $E$ be a Banach space such that there exists a continuous linear one-to-one mapping $u\colon E \to c_0(I)$ and assume that there exists a transfinite sequence of non-zero continuous linear operators $\{v_\lambda\}_{\lambda\leqslant\zeta}$ on $E$, with the following properties (where we add $v_0 = 0$ to $\{v_\lambda\}_{\lambda\leqslant\zeta}$, for notational convenience):*

(i) *For each $x \in E$ and $\varepsilon > 0$, the set*

$$\Lambda(x, \varepsilon) = \{\lambda \in [0, \zeta) \mid \|v_{\lambda+1}(x) - v_\lambda(x)\| > \varepsilon(\|v_{\lambda+1}\| + \|v_\lambda\|)\} \qquad (20.70)$$

*is finite (or empty);*

(ii) *for each $x \in E$ we have*

$$x \in Y_x = [\bigcup_{\lambda\in\Lambda(x)} (v_{\lambda+1} - v_\lambda)(E)], \qquad (20.71)$$

*where*

$$\Lambda(x) = \bigcup_{\varepsilon>0} \Lambda(x, \varepsilon) = \{\lambda \in [0, \zeta) \mid v_{\lambda+1}(x) \neq v_\lambda(x)\}; \qquad (20.72)$$

(iii) dens $(v_{\lambda+1} - v_\lambda)(E) \leqslant \aleph_0$ $(0 \leqslant \lambda < \zeta)$, *i.e., all $(v_{\lambda+1} - v_\lambda)(E)$ are separable.*

*Then $E$ admits an equivalent locally uniformly convex norm.*

b) *If there exists a one-to-one (norm) continuous linear mapping $u\colon E^* \to c_0(I)$ which is also $w^*$-$w$ continuous (i.e., with respect to $\sigma(E^*, E)$ and $\sigma(c_0(I), c_0(I)^*)$), and if there exists a transfinite sequence of non-zero continuous linear operators $\{v_\lambda\}_{\lambda\leqslant\zeta}$ on $E$, such that $\{v_\lambda^*\}_{\lambda\leqslant\zeta} \cup \{v_0^* = 0\}$ satisfies (i) — (iii) above (with $E$ replaced by $E^*$), then $E^*$ admits an equivalent locally uniformly convex norm, which is the dual of some equivalent norm on $E$.*

*Proof.* a) By (iii), let $\{e_n^\lambda\}_{n=1}^\infty$ be a complete sequence in $(v_{\lambda+1} - v_\lambda)(E)$ $(0 \leqslant \lambda < \zeta)$. Let us denote

$$\mathfrak{A}_n = \{A_n \subset [0, \zeta] \mid \text{card } A_n \leqslant n\} \qquad (n = 1, 2, \ldots), \qquad (20.73)$$

and let us introduce the functionals

$$\mathscr{E}_A^{(n)}(x) = \text{dist }(x, [\bigcup_{\lambda\in A} \{e_i^\lambda\}_{i=1}^n]) \quad \left(x \in E,\ A \in \bigcup_{j=1}^\infty \mathfrak{A}_j,\ n = 1,2, \ldots\right), \qquad (20.74)$$

$$t_\lambda(x) = \frac{\|v_{\lambda+1}(x) - v_\lambda(x)\|}{\|v_{\lambda+1}\| + \|v_\lambda\|} \qquad (x \in E,\ 0 \leqslant \lambda < \zeta), \tag{20.75}$$

$$\mathscr{F}_A(x) = \sum_{\lambda \in A} t_\lambda(x) \qquad \left(x \in E,\ A \in \bigcup_{j=1}^{\infty} \mathfrak{A}_j\right), \tag{20.76}$$

$$\mathscr{G}_1(x) = \|x\|,\ \mathscr{G}_{n+1}(x) = \sup_{A \in \mathfrak{A}_n} (\mathscr{E}_A^{(n)}(x) + n\mathscr{F}_A(x)) \quad (x \in E,\ n = 1,2, \dots). \tag{20.77}$$

Let us define the disjoint union

$$\varDelta = \{-1, -2, \dots\} \cup [0, \xi) \cup I \tag{20.78}$$

and the operator $Q : E \to c_0(\varDelta)$ by

$$Q(x)(\delta) = \begin{cases} \dfrac{1}{2^{-\delta}} \mathscr{G}_{-\delta}(x) & \text{for } \delta = -1,\ -2,\ \dots \\ t_\delta(x) & \text{for } \delta \in [0, \zeta) \\ u(x)(\delta) & \text{for } \delta \in I \end{cases} \qquad (x \in E). \tag{20.79}$$

We have indeed $Q(E) \subset c_0(\varDelta)$. For, the relations $\mathscr{E}_A^{(n)}(x) \leqslant \|x\|$ and $n\mathscr{F}_A(x) = n \sum_{\lambda \in A} t_\lambda(x) \leqslant n^2 \|x\|$ $(x \in E,\ A \in \mathfrak{A}_n;\ n = 1, 2, \dots)$ imply that $\frac{1}{2^{n+1}} \mathscr{G}_{n+1}(x) \leqslant \frac{1}{2^{n+1}} (1 + n^2)\|x\| \to 0$ as $n \to \infty$, for each $x \in E$; furthermore, by (i), $\{t_\lambda(x)\}_{0 \leqslant \lambda < \zeta} \in c_0([0, \zeta))$ $(x \in E)$ and, finally, $u(E) \subset c_0(I)$, whence the assertion $Q(E) \subset c_0(\varDelta)$ follows.

Let

$$|||x||| = D(Q(x)) \qquad (x \in E), \tag{20.80}$$

where $D$ is the norm on $c_0(\varDelta)$ defined by (20.64). We shall show that $|||\cdot|||$ is an equivalent locally uniformly convex norm on $E$, which will prove a).

Let $\Psi_\delta(x) = Q(x)(\delta)$ $(x \in E,\ \delta \in \varDelta)$, where $Q(x)(\delta)$ is defined by (20.79). Then, since $\frac{1}{2^{-\delta}} \mathscr{G}_{-\delta}$ $(\delta = -1, -2, \dots)$ and $t_\delta$ $(\delta \in [0, \zeta))$ are sublinear and positive-homogeneous and $x \to u(x)(\delta)$ $(\delta \in I)$ is linear, from lemma 20.2 it follows that $|||\cdot|||$ is a sublinear and positive-homogeneous functional on $E$. Also, it is clear that $|||0||| = D(Q(0)) = D(\{\Psi_\delta(0)\}_{\delta \in \varDelta}) = D(0) = 0$. Conversely, if $|||x||| = D(Q(x)) = D(\{\Psi_\delta(x)\}_{\delta \in \varDelta}) = 0$, then, since $D$ is a norm on $c_0(\varDelta)$, $\{\Psi_\delta(x)\}_{\delta \in \varDelta} = 0$, so in particular $\Psi_{-1}(x) = \mathscr{G}_1(x) = \|x\| = 0$, whence $x = 0$. Thus, $|||\cdot|||$ is a norm on $E$.

Furthermore, $\left|\frac{1}{2^{n+1}} \mathscr{G}_{n+1}(x)\right| \leqslant \frac{1}{2^{n+1}} (1 + n^2)\|x\| \leqslant 2\|x\|$ $(x \in E,\ n = 1, 2, \dots)$, $|t_\lambda(x)| \leqslant \|x\|$ $(x \in E,\ 0 \leqslant \lambda < \zeta)$ and $|u(x)(i)| \leqslant \|u(x)\|_{c_0(I)} \leqslant$

$\leqslant \|u\| \, \|x\| \; (x \in E, \; i \in I)$, whence, since by (20.64) $\frac{1}{\sqrt{2}} \sup_{\delta \in \Delta} |z(\delta)| \leqslant$ $\leqslant D(z(.)) \leqslant \sup_{\delta \in \Delta} |z(\delta)| \; (z(.) \in c_0(\Delta))$, we obtain

$$\frac{1}{2} \|x\| = \frac{1}{2} \mathscr{G}_1(x) = Q(x)(-1) \leqslant \|Q(x)\|_{c_0(\Delta)} \leqslant \sqrt{2}\, D(Q(x)) =$$

$$= \sqrt{2}\, |||x||| \leqslant \sqrt{2}\, \|Q(x)\|_{c_0(\Delta)} \leqslant \sqrt{2} \max(2, \|u\|) \|x\| \qquad (x \in E),$$

and thus $|||\cdot|||$ is equivalent to the initial norm $\|\cdot\|$ on $E$.

Finally, let us show that $|||\cdot|||$ is locally uniformly convex. Assume that (20.67) holds. We have to show that in this case

$$\lim_{k\to\infty} |||x_k - x||| = \lim_{k\to\infty} D(Q(x_k - x)) = 0; \tag{20.81}$$

since the only immediate consequence of (20.68) is that $\lim_{k\to\infty} D(Q(x_k) - Q(x)) = 0$, the proof of (20.81) needs further argument.

By (20.67) and lemma 20.2 we have (20.68), whence, by (20.79),

$$\lim_{k\to\infty} \mathscr{G}_n(x_k) = \mathscr{G}_n(x) \qquad (n = 1, 2, \ldots), \tag{20.82}$$

$$\lim_{k\to\infty} t_\lambda(x_k) = t_\lambda(x) \qquad (0 \leqslant \lambda < \zeta), \tag{20.83}$$

$$\lim_{k\to\infty} \|u(x_k) - u(x)\|_{c_0(I)} = 0. \tag{20.84}$$

Now, in order to prove (20.81) it will be sufficient to prove that $\{x_k\}_{k=1}^{\infty}$ is relatively compact. Indeed, if this is proved, then any subsequence $\{x_{k_n}\}$ of $\{x_k\}$ contains a subsequence $\{x_{k_{n_m}}\}$ converging to some $y \in E$. But then, since $u: E \to c_0(I)$ is continuous, we have $\lim_{m\to\infty} \|u(x_{k_{n_m}}) - u(y)\|_{c_0(I)} = 0$, which, together with (20.84), gives $u(y) = u(x)$. Hence, since $u: E \to c_0(I)$ is one-to-one, $y = x$. Thus, every subsequence of $\{x_k\}$ contains a subsequence converging to $x$, whence $\{x_k\}$ itself converges to $x$. Consequently, since $|||\cdot|||$ is equivalent to the initial norm on $E$, we obtain (20.81).

In order to prove that $\{x_k\}_{k=1}^{\infty}$ is relatively compact, take any $\varepsilon > 0$. Then, by (ii) and the definition of $e_n^\lambda$, there exist $B \subset \bigcup_{j=1}^{\infty} \mathfrak{A}_j$ with $B \subset \Lambda(x)$ and $m \in \mathscr{N}$ such that $\mathscr{E}_B^{(m)}(x) < \frac{\varepsilon}{4}$. Let

$$\Lambda_B(x) = \{0 \leqslant \lambda < \zeta \,|\, t_\lambda(x) < \min_{\mu \in B} t_\mu(x)\}, \tag{20.85}$$

$$j = \operatorname{card}(\Lambda(x) \setminus \Lambda_B(x)), \tag{20.86}$$

$$b = \inf_{\substack{\lambda \in \Lambda_B(x) \\ \mu \in B}} (t_\mu(x) - t_\lambda(x)). \tag{20.87}$$

We note that, since $B$ is a finite subset of $\Lambda(x)$, $\min_{\mu \in B} t_\mu(x) > 0$ in (20.85). Also, by (i) (with $\varepsilon = \min_{\mu \in B} t_\mu(x)$), we have $j < \infty$. Furthermore, since $\{t_\lambda(x)\}_{0 \leqslant \lambda < \zeta} \in c_0([0, \zeta))$, we have $b > 0$ in (20.87). Let

$$n = \max\left(m,\ j,\ \frac{\varepsilon + 4\|x\|}{4b}\right) + 1 \tag{20.88}$$

and, by (20.77), let $A \in \mathfrak{A}_n$ be such that

$$\mathscr{G}_{n+1}(x) - (\mathscr{E}_A^{(n)}(x) + n\mathscr{F}_A(x)) \leqslant \frac{\varepsilon}{4}. \tag{20.89}$$

We claim that $A \supset B$. Indeed, assume, a contrario, that there exists a $\mu_1 \in B \setminus A$. We shall consider two cases:

*Case 1°*. card $A < n$. Then, taking $C = A \cup \{\mu_1\}$, we have $C \in \mathfrak{A}_n$, $C \supset A$ and

$$\mathscr{F}_C(x) - \mathscr{F}_A(x) = \sum_{\mu \in C} t_\mu(x) - \sum_{\lambda \in A} t_\lambda(x) = t_{\mu_1}(x) \geqslant b.$$

*Case 2°*. card $A = n$. Then, by card $A = n > j = \mathrm{card}\,(\Lambda(x) \setminus \Lambda_B(x))$, there exists a $\lambda_1 \in A \setminus (\Lambda(x) \setminus \Lambda_B(x))$, so either $\lambda_1 \in A \cap \Lambda_B(x)$ or $\lambda_1 \in A$, $t_{\lambda_1}(x) = 0$. Hence, taking $C = (A \cup \{\mu_1\}) \setminus \{\lambda_1\}$, we have $C \in \mathfrak{A}_n$ and

$$\mathscr{F}_C(x) - \mathscr{F}_A(x) = \sum_{\mu \in C} t_\mu(x) - \sum_{\lambda \in A} t_\lambda(x) = t_{\mu_1}(x) - t_{\lambda_1}(x) \geqslant b.$$

Consequently, in either case,

$$\mathscr{G}_{n+1}(x) - (\mathscr{E}_A^{(n)}(x) + n\mathscr{F}_A(x)) \geqslant \mathscr{E}_C^{(n)}(x) + n\mathscr{F}_C(x) - (\mathscr{E}_A^{(n)}(x) + n\mathscr{F}_A(x)) =$$

$$= \mathscr{E}_C^{(n)}(x) - \mathscr{E}_A^{(n)}(x) + n(\mathscr{F}_C(x) - \mathscr{F}_A(x)) \geqslant 0 - \|x\| + nb >$$

$$> -\|x\| + \max\left(mb,\ jb,\ \frac{\varepsilon + 4\|x\|}{4}\right) \geqslant -\|x\| + \frac{\varepsilon}{4} + \|x\| = \frac{\varepsilon}{4},$$

in contradiction with (20.89). This proves the claim $A \supset B$.

From $A \supset B$ and $n > m$ we obtain

$$\mathscr{E}_A^{(n)}(x) \leqslant \mathscr{E}_B^{(m)}(x) < \frac{\varepsilon}{4}. \tag{20.90}$$

Furthermore, by (20.82), (20.83) there exists an $M = M(\varepsilon)$ such that

$$|\mathscr{G}_{n+1}(x) - \mathscr{G}_{n+1}(x_k)| < \frac{\varepsilon}{4}, \quad |t_\lambda(x) - t_\lambda(x_k)| < \frac{\varepsilon}{4n^2} \qquad (\lambda \in A,\ k > M).$$

Then $|\mathscr{F}_A(x)-\mathscr{F}_A(x_k)|=|\sum_{\lambda\in A}t_\lambda(x)-\sum_{\lambda\in A}t_\lambda(x_k)|<n\frac{\varepsilon}{4n^2}=\frac{\varepsilon}{4n}$ for $k>M$, whence, by (20.77) and (20.89), (20.90),

$$\mathscr{E}_A^{(n)}(x_k)\leqslant\mathscr{G}_{n+1}(x_k)-n\mathscr{F}_A(x_k)\leqslant\mathscr{G}_{n+1}(x)-n\mathscr{F}_A(x_k)+\frac{\varepsilon}{4}\leqslant$$

$$\leqslant\mathscr{G}_{n+1}(x)-n\mathscr{F}_A(x)+\frac{2\varepsilon}{4}\leqslant\mathscr{E}_A^{(n)}(x)+\frac{3\varepsilon}{4}<\varepsilon\qquad(k>M).$$

Consequently, for any element of best approximation $\pi_A^{(n)}(x_k)$ of $x_k$ in the finite-dimensional subspace $[\bigcup_{\lambda\in A}\{e_i^\lambda\}_{i=1}^n]$ we have

$$\|x_k-\pi_A^{(n)}(x_k)\|=\mathscr{E}_A^{(n)}(x_k)<\varepsilon\qquad(k>M),$$

and hence $\{x_k\}_{k=1}^M\cup\{\pi_A^{(n)}(x_k)\}_{k=M+1}^\infty$ is an $\varepsilon$-net for $\{x_k\}_{k=1}^\infty$. But, since $|||x_k|||=1$ $(k=1,2,\ldots)$, we have

$$\sup_{1\leqslant k<\infty}\|\pi_A^{(n)}(x_k)\|\leqslant 2\sup_{1\leqslant k<\infty}\|x_k\|\leqslant 2\sqrt{2}\sup_{1\leqslant k<\infty}|||x_k|||=2\sqrt{2}<\infty,$$

whence, since $\{x_k\}_{k=1}^M\cup\{\pi_A^{(n)}(x_k)\}_{k=M+1}^\infty\subset[\{x_k\}_{k=1}^M\cup\bigcup_{\lambda\in A}\{e_i^\lambda\}_{i=1}^n]$ (which is a finite-dimensional subspace of $E$), the set $\overline{\{x_k\}_{k=1}^M\cup\{\pi_A^{(n)}(x_k)\}_{k=M+1}^\infty}$ is compact. Thus, for each $\varepsilon>0$, $\{x_k\}_{k=1}^\infty$ has a compact $\varepsilon$-net, so $\{x_k\}_{k=1}^\infty$ is relatively compact, which proves a).

b) Under the assumptions of b) there exists, by the above proof of part a), an equivalent locally uniformly convex norm on $E^*$, defined by

$$|||f|||=D(Q(f))=D(\{Q(f)(\delta)\}_{\delta\in\Delta})\qquad(f\in E^*),\quad(20.80')$$

where $D(\cdot)$ is the norm (20.64) on $c_0(\Delta)$ and where

$$Q(f)(\delta)=\begin{cases}\frac{1}{2^{-\delta}}\mathscr{G}_{-\delta}(f) & \text{for } \delta=-1,-2,\ldots\\ t_\delta(f) & \text{for } \delta\in[0,\zeta)\\ u(f)(\delta) & \text{for } \delta\in I.\end{cases}\qquad(f\in E^*).\quad(20.79')$$

In order to show that $|||\cdot|||$ is the dual of some equivalent norm $|||\cdot|||$ on $E$, it will be sufficient, by Ch. II, § 5, lemma 5.1, to show that the set $S=\{f\in E^*|\ |||f|||\leqslant 1\}$ is $w^*$-closed, or, what is equivalent, that the functional $f\to|||f|||$ is $w^*$-lower semi-continuous. Now, since $\mathscr{E}_A^{(n)}(f)=$

$= \operatorname{dist}(f, \Gamma)$, where $\Gamma$ is a finite-dimensional subspace of $E^*$, it is $w^*$-lower semi-continuous by § 9, proof of theorem 9.4 (using the relation $\operatorname{dist}(f, \Gamma) = \sup_{\substack{z \in \Gamma_\perp \\ \|z\| \leq 1}} |f(z)|$). Furthermore, since each $v_\lambda^*$ is a $w^*$-continuous linear operator on $E^*$, each $t_\lambda$ corresponding to $v_\lambda^*$ is $w^*$-lower semi-continuous on $E^*$, so each $\mathscr{F}_\Delta$, whence also each $\mathscr{G}_n$, is $w^*$-lower semi-continuous. Finally, since $u: E^* \to c_0(I)$ is $w^*$-$w$ continuous, each functional $f \to u(f)(i)$ is $w^*$-continuous. Consequently, for each $\delta \in \Delta$, the functional $f \to Q(f)(\delta)$ is $w^*$-lower semi-continuous and hence so are the functionals $f \to \left( \sum_{j=1}^{n} \frac{1}{2^j} |Q(f)(\delta_j)|^2 \right)^{\frac{1}{2}}$, for all $\delta_1, \ldots, \delta_n \in \Delta$ and $n = 1, 2, \ldots$ Therefore, by (20.64), the functional $f \to |||f|||$ is $w^*$-lower semi-continuous, which completes the proof of lemma 20.3.

Now we are ready to prove

**Theorem 20.7.** *Let $E$ be a Banach space. The following statements are equivalent:*

1°. *$E$ has a shrinking extended $M$-basis.*

2°. *$E$ is weakly compactly generated and there exists an equivalent norm $|||\cdot|||$ on $E$ such that $(E, |||\cdot|||)^*$ is locally uniformly convex.*

3°. *$E$ is weakly compactly generated and admits an equivalent $(F)$-norm.*

4°. *$E$ is weakly compactly generated and admits an equivalent very smooth norm.*

2′—4′. *Same as 2°—4° respectively, with "$E$ is weakly compactly generated" replaced by: $E$ is isomorphic to a subspace of a weakly compactly generated Banach space.*

*Proof.* We have already mentioned above (before lemma 20.2) that 2° ⇒ 3° ⇒ 4° and 2′ ⇒ 3′ ⇒ 4′ (with the same equivalent norm $|||\cdot|||$ on $E$). The implication 4°⇒1° is nothing else than the particular case $G = E$ of theorem 20.6. The implication 2°⇒2′ is obvious and the implication 4′⇒4° follows from corollary 20.6.

Let us prove, finally, that 1°⇒2°. By proposition 20.8, it will be sufficient to prove that if we have 1°, then $E$ satisfies the conditions of lemma 20.3 b). Since $E$ is weakly compactly generated, there exists, by corollary 20.4*), a one-to-one continuous linear mapping $u$ of $E^*$ into some $c_0(I)$, which is also $w^*$-$w$ continuous. Let us show the existence of a transfinite sequence of non-zero continuous linear operators $\{v_\lambda\}_{\lambda \leq \zeta}$ on $E$ satisfying the assumptions of lemma 20.3 b) and $v_\zeta = I_E$ (this latter will be helpful in the transfinite induction process, to prove (20.92) below).

---

*) Actually, since $E$ has a shrinking extended $M$-basis, a somewhat simpler proof of $u(E^*) \subset c_0(I)$ is given by the argument of the proof of proposition 20.8.

We shall proceed by transfinite induction on dens $E$. If $E$ is separable, then $\{v_1^*\} \cup \{v_0^* = 0\}$, where $v_1 = I_E$, satisfies (i)—(iii) (indeed, by 1°, $v_1^*(E^*) = E^*$ is separable).

Assume now that dens $E > \aleph_0$ and that the statement (on the existence of $\{v_\lambda\}_{\lambda \leqslant \zeta}$) has been proved for all Banach spaces $E_0$ with dens $E_0 <$ < dens $E$. Let $\{p_\lambda\}_{\omega \leqslant \lambda \leqslant \vartheta}$ be a long sequence of projections on $E$ and $\{I_\lambda\}_{\omega \leqslant \lambda \leqslant \vartheta}$ a long sequence of subsets of $I$, given by proposition 20.7*). Let us reindex $\{p_\lambda\}_{\omega \leqslant \lambda \leqslant \vartheta}$ and $\{I_\lambda\}_{\omega \leqslant \lambda \leqslant \vartheta}$ as follows: put $p_1' = p_\omega$, $p_{1+\lambda}' = p_{\omega+\lambda}$ $(\lambda < \vartheta)$ and then denote $p_\lambda'$ by $p_\lambda$ $(\lambda \leqslant \vartheta)$ and proceed similarly for $I_\lambda$. In this way we obtain two transfinite sequences $\{p_\lambda\}_{\lambda \leqslant \vartheta}$ and $\{I_\lambda\}_{\lambda \leqslant \vartheta}$ satisfying (20.49)—(20.56) with $\omega$ replaced by 1 and with dens $p_\lambda(E) = \aleph_0$ $(\lambda < \omega)$ added to (20.51). Then $p_1(E) = [x_i]_{i \in I_1}$ has a shrinking extended $M$-basis (namely, $\{x_i\}_{i \in I_1}$) and, by proposition 20.9 a), each $(p_{\lambda+1} - p_\lambda)(E)$ $(\lambda < \vartheta)$ has a shrinking extended $M$-basis. Also, by (20.51), dens $(p_{\lambda+1} - p_\lambda)(E) <$ dens $E$ $(\lambda < \vartheta)$. Therefore, by our induction hypothesis, for each $(p_{\lambda+1} - p_\lambda)(E)$ $(0 \leqslant \lambda < \vartheta$, where $p_0 = 0)$ there exists a transfinite sequence of non-zero continuous linear operators $\{q_\beta^\lambda\}_{\beta \leqslant \zeta_\lambda}$ on $(p_{\lambda+1} - p_\lambda)(E)$, such that $\{(q_\beta^\lambda)^*\}_{\beta \leqslant \zeta_\lambda} \cup \{(q_0^\lambda)^* = 0\}$ satisfies (i)—(iii) of lemma 20.3 b) on $(p_{\lambda+1} - p_\lambda)(E)^*$ and $q_{\zeta_\lambda}^\lambda = I_{(p_{\lambda+1} - p_\lambda)(E)}$. Let**) $\mathscr{M}$ be the collection of all pairs $(\lambda, \beta)$, where $0 \leqslant \lambda \leqslant \vartheta$ and where $0 \leqslant \beta \leqslant \zeta_\lambda$ if $0 \leqslant \lambda < \vartheta$ and $\beta = 0$ if $\lambda = \vartheta$. Let us consider on $\mathscr{M}$ the lexicographical ordering, that is, $(\lambda_1, \beta_1) > (\lambda_2, \beta_2)$ if and only if either $\lambda_1 > \lambda_2$ or $\lambda_1 = \lambda_2$ and $\beta_1 > \beta_2$. Then $\mathscr{M}$ becomes a well ordered set, in which $(\lambda, \beta) + 1 = (\lambda, \beta + 1)$ if $\beta < \zeta_\lambda$ and $= (\lambda + 1, 0)$ if $\beta = \zeta_\lambda$ (where $(\lambda, \beta) + 1$ denotes the smallest element of the set $\{(\lambda', \beta') \in \mathscr{M} \mid (\lambda', \beta') > (\lambda, \beta)\}$) and thus $\mathscr{M}$ can be identified with some ordinal interval $[0, \zeta]$.

Define a "transfinite sequence" of continuous linear operators $\{v_{(\lambda,\beta)}\}_{(\lambda,\beta) \in \mathscr{M}} = \{v_{(\lambda,\beta)}\}_{(0,0) \leqslant (\lambda,\beta) \leqslant (\vartheta,0)}: E \to E$ by

$$v_{(\lambda,\beta)} = \begin{cases} q_\beta^\lambda(p_{\lambda+1} - p_\lambda) + p_\lambda & \text{if } 0 \leqslant \lambda < \vartheta,\ 0 \leqslant \beta \leqslant \zeta_\lambda \\ I_E & \text{if } \lambda = \vartheta \text{ (i.e., if } (\lambda, \beta) = (\vartheta, 0)). \end{cases} \tag{20.91}$$

We claim that $v_{(\lambda,\beta)} \neq 0$ for $(\lambda, \beta) \neq (0, 0)$. Indeed, otherwise there exists a pair $(\lambda, \beta) \neq (0, 0)$ such that $q_\beta^\lambda(p_{\lambda+1} - p_\lambda)(x) = -p_\lambda(x) \in (p_{\lambda+1} - p_\lambda)(E) \cap p_\lambda(E) = \{0\}$ $(x \in E)$. Then, by $\|p_\lambda\| = 1$ $(1 \leqslant \lambda < \vartheta)$, we obtain $\lambda = 0$, whence $\beta \neq 0$. But, $v_{(0,\beta)} = q_\beta^0 p_1$, where, by our induction hypothesis, $q_\beta^0 \neq 0$ (since $\beta \neq 0$), whence $v_{(0,\beta)} \neq 0$, in contradiction with our assumption. This proves the claim.

We shall prove that $\{v_{(\lambda,\beta)}^*\}_{(\lambda,\beta) \in \mathscr{M}}$ has the properties required in lemma 20.3 b). Let us observe first that, for $(0,0) \leqslant (\lambda, \beta) < (\vartheta, 0)$,

$$v_{(\lambda,\beta)+1} = \begin{cases} q_{\beta+1}^\lambda(p_{\lambda+1} - p_\lambda) + p_\lambda & \text{if } \beta < \zeta_\lambda \\ v_{(\lambda,\beta)} & \text{if } \beta = \zeta_\lambda. \end{cases} \tag{20.92}$$

*) We may apply proposition 20.7, by 1° and proposition 20.8.

**) Thus, $\mathscr{M} = \{(\lambda, \beta) \mid 0 \leqslant \lambda < \vartheta,\ 0 \leqslant \beta \leqslant \zeta_\lambda\} \cup \{(\vartheta, 0)\}$.

Indeed, if $\beta < \zeta_\lambda$, then $(\lambda, \beta) + 1 = (\lambda, \beta + 1)$ and hence (20.92) holds by (20.91). On the other hand, if $\beta = \zeta_\lambda$, then $(\lambda, \beta) + 1 = (\lambda + 1, 0)$, whence, by (20.91) and $q_0^{\lambda+1} = 0$, we obtain $v_{(\lambda,\beta)+1} = p_{\lambda+1}$; but, by our induction hypothesis, $q_\beta^\lambda = q_{\zeta_\lambda}^\lambda = I_{(p_{\lambda+1}-p_\lambda)(E)}$, whence $q_\beta^\lambda(p_{\lambda+1} - p_\lambda) + p_\lambda = p_{\lambda+1}$, so again (20.92) holds by (20.91).

Let us also observe that

$$v_{(\lambda,\beta)}(x) = (q_\beta^\lambda(p_{\lambda+1} - p_\lambda) + p_\lambda)(x) = q_\beta^\lambda(x)$$
$$(x \in (p_{\lambda+1} - p_\lambda)(E), \quad (0, 0) \leqslant (\lambda, \beta) < (\vartheta, 0)),$$

that is, $v_{(\lambda,\beta)}|_{(p_{\lambda+1}-p_\lambda)(E)} = q_\beta^\lambda$ $((0, 0) \leqslant (\lambda, \beta) < (\vartheta, 0))$, whence

$$\|v_{(\lambda,\beta)}\| \geqslant \|q_\beta^\lambda\| \qquad ((0, 0) \leqslant (\lambda, \beta) < (\vartheta, 0)). \tag{20.93}$$

Now, for each $\lambda$ with $0 \leqslant \lambda < \vartheta$, let $\rho_\lambda$ be the canonical isomorphism of $(p_{\lambda+1} - p_\lambda)^*(E^*)$ onto $(p_{\lambda+1} - p_\lambda)(E)^*$ given by § 14, lemma 14.3 and let $\iota_\lambda$ be the identical embedding of $(p_{\lambda+1} - p_\lambda)(E)$ into $E$. Thus, $\rho_\lambda = \iota_\lambda^*$ (the operator of restriction to $(p_{\lambda+1} - p_\lambda)(E)$) and $u_\lambda\rho_\lambda^{-1} = \overline{(p_{\lambda+1} - p_\lambda)}^*$, where $\overline{p_{\lambda+1} - p_\lambda}$ denotes the "astriction" of $p_{\lambda+1} - p_\lambda$ to $(p_{\lambda+1} - p_\lambda)(E)$ and $u_\lambda\colon (p_{\lambda+1} - p_\lambda)^*(E^*) \to E^*$ is the identical embedding (hence $\|u_\lambda\rho_\lambda^{-1}\| \leqslant 2$). Then by, (20.92) and (20.91), for each $f \in E^*$ and $(0, 0) \leqslant (\lambda, \beta) < (\vartheta, 0)$ with $\beta < \zeta_\lambda$ we have

$$\begin{aligned} v_{(\lambda,\beta)+1}^*(f) - v_{(\lambda,\beta)}^*(f) &= \overline{(p_{\lambda+1} - p_\lambda)}^* (q_{\beta+1}^\lambda - q_\beta^\lambda)^* \iota_\lambda^*(f) = \\ &= u_\lambda\rho_\lambda^{-1}(q_{\beta+1}^\lambda - q_\beta^\lambda)^*(f|_{(p_{\lambda+1}-p_\lambda)(E)}) = \\ &= u_\lambda\rho_\lambda^{-1}(q_{\beta+1}^\lambda - q_\beta^\lambda)^*(\rho_\lambda((p_{\lambda+1} - p_\lambda)^*(f))), \end{aligned} \tag{20.94}$$

whence, by (20.92) and (20.93), we obtain

$$\frac{\|v_{(\lambda,\beta)+1}^*(f) - v_{(\lambda,\beta)}^*(f)\|}{\|v_{(\lambda,\beta)+1}^*\| + \|v_{(\lambda,\beta)}^*\|} \leqslant \frac{2\|(q_{\beta+1}^\lambda - q_\beta^\lambda)^*(\rho_\lambda((p_{\lambda+1} - p_\lambda)^*(f)))\|}{\|q_{\beta+1}^\lambda\| + \|q_\beta^\lambda\|} \leqslant$$
$$\leqslant \frac{2\|q_{\beta+1}^\lambda - q_\beta^\lambda\| \, \|\rho_\lambda\| \, \|(p_{\lambda+1} - p_\lambda)^*(f)\|}{\|q_{\beta+1}^\lambda\| + \|q_\beta^\lambda\|} \leqslant 2\|p_{\lambda+1}^*(f) - p_\lambda^*(f)\|. \tag{20.95}$$

Now let $f \in E^*$ and $\varepsilon > 0$. Then, by proposition 20.9 b), the set $B'_{,\frac{\varepsilon}{2}} = \left\{0 \leqslant \lambda < \vartheta \mid \|p_{\lambda+1}^*(f) - p_\lambda^*(f)\| > \frac{\varepsilon}{2}\right\}$ is finite and, by our induction hypothesis, for each $\lambda \in B'_{f,\frac{\varepsilon}{2}}$ the set

$$C_{f,\frac{\varepsilon}{2},\lambda} = \Big\{0 \leqslant \beta < \zeta_\lambda \mid \|(q_{\beta+1}^\lambda - q_\beta^\lambda)^*(\rho_\lambda((p_{\lambda+1} - p_\lambda)^*(f)))\| >$$
$$> \frac{\varepsilon}{2}(\|q_{\beta+1}^\lambda\| + \|q_\beta^\lambda\|)\Big\} = \Big\{0 \leqslant \beta < \zeta_\lambda \mid \|(q_{\beta+1}^\lambda - q_\beta^\lambda)^*(f|_{(p_{\lambda+1}-p_\lambda)(E)})\| >$$
$$> \frac{\varepsilon}{2}(\|q_{\beta+1}^\lambda\| + \|q_\beta^\lambda\|)\Big\}$$

is finite. Hence, by (20.92) and (20.95), the set

$$\Lambda(f, \varepsilon) = \{(0, 0) \leqslant (\lambda, \beta) < (\vartheta, 0) |\ \|v^*_{(\lambda,\beta)+1}(f) - v^*_{(\lambda,\beta)}(f)\| >$$

$$> \varepsilon(\|v^*_{(\lambda,\beta)+1}\| + \|v^*_{(\lambda,\beta)}\|)\} \subset \{(0, 0) \leqslant (\lambda, \beta) < (\vartheta, 0) |\ \lambda \in B'_{f,\frac{\varepsilon}{2}},$$

$$\beta \in C_{f,\frac{\varepsilon}{2},\lambda}\}$$

is finite. Thus, $\{v^*_{(\lambda,\beta)}\}_{(\lambda,\beta)\in\mathcal{M}}$ satisfies (i) on $E^*$.

Now we shall show, by transfinite induction, that

$$p^*_\mu(f) \in Y_f = [\bigcup_{(\lambda,\beta)\in\Lambda(f)} (v^*_{(\lambda,\beta)+1} - v^*_{(\lambda,\beta)})(E^*)] \qquad (f \in E^*, 0 \leqslant \mu \leqslant \vartheta), \tag{20.96}$$

where $\Lambda(f) = \{(0, 0) \leqslant (\lambda, \beta) < (\vartheta, 0) |\ v^*_{(\lambda,\beta)+1}(f) \neq v^*_{(\lambda,\beta)}(f)\}$, which will imply, in particular, $f = p^*_\vartheta(f) \in Y_f$ $(f \in E^*)$, proving (ii).

Clearly $p^*_0(f) = 0 \in Y_f$ $(f \in E^*)$. Let $\lambda \leqslant \vartheta$ and assume that (20.96) holds for all ordinal numbers $\mu < \lambda$.

*Case 1°*: $\lambda$ is a nonlimit ordinal. Then, by our induction hypothesis on (20.96), $p^*_{\lambda-1}(f) \in Y_f$ and we shall show that $(p_\lambda - p_{\lambda-1})^*(f) \in Y_f$, whence $p^*_\lambda(f) \in Y_f$ $(f \in E^*)$. By our initial induction hypothesis,

$$f|_{(p_\lambda - p_{\lambda-1})(E)} \in [\bigcup_{\beta\in\Lambda(f|_{(p_\lambda - p_{\lambda-1})(E)})} (q^{\lambda-1}_{\beta+1} - q^{\lambda-1}_\beta)^*((p_\lambda - p_{\lambda-1})(E)^*)] =$$

$$= Y_{f|_{(p_\lambda - p_{\lambda-1})(E)}} \qquad (f \in E^*),$$

where $\Lambda(f|_{(p_\lambda - p_{\lambda-1})(E)}) = \{0 \leqslant \beta \leqslant \zeta_\lambda |\ (q^{\lambda-1}_{\beta+1} - q^{\lambda-1}_\beta)^*(f)|_{(p_\lambda - p_{\lambda-1})(E)} \neq$ $\neq 0\}$. Observe that if $\beta \in \Lambda(f|_{(p_\lambda - p_{\lambda-1})(E)})$, then, by (20.94) and since $u_{\lambda-1}\rho^{-1}_{\lambda-1}$ is one-to-one,

$$v^*_{(\lambda-1,\beta)+1}(f) - v^*_{(\lambda-1,\beta)}(f) = u_{\lambda-1}\rho^{-1}_{\lambda-1}(q^{\lambda-1}_{\beta+1} - q^{\lambda-1}_\beta)^*(f|_{(p_\lambda - p_{\lambda-1})(E)}) \neq 0,$$

whence $(\lambda - 1, \beta) \in \Lambda(f)$. Hence, since $u_{\lambda-1}\rho^{-1}_{\lambda-1}$ is an isomorphism, we obtain

$$p_\lambda - p_{\lambda-1})^*(f) = u_{\lambda-1}\rho^{-1}_{\lambda-1}(f|_{(p_\lambda - p_{\lambda-1})(E)}) \in u_{\lambda-1}\rho^{-1}_{\lambda-1}(Y_{f|(p_\lambda - p_{\lambda-1})(E)}) =$$

$$= [\bigcup_{\beta\in\Lambda(f|_{(p_\lambda - p_{\lambda-1})(E)})} u_{\lambda-1}\rho^{-1}_{\lambda-1}(q^{\lambda-1}_{\beta+1} - q^{\lambda-1}_\beta)^*((p_\lambda - p_{\lambda-1})(E)^*)] =$$

$$= [\bigcup_{\beta\in\Lambda(f|(_{p_\lambda - p_{\lambda-1}})(E))} (v^*_{(\lambda-1,\beta)+1} - v^*_{(\lambda-1,\beta)})(E^*)] \subset Y_f \qquad (f \in E^*).$$

*Case 2°*: $\lambda$ is a limit ordinal. Then, by our induction hypothesis on (20.96), $p_\mu^*(f) \in Y_f$ $(f \in E^*, \mu < \lambda)$, whence, by proposition 20.9 b), $p_\lambda^*(f) = \lim_{\mu<\lambda} p_\mu^*(f) \in Y_f$ $(f \in E^*)$. This proves (ii).

Finally, since by our induction hypothesis, $\operatorname{dens}(q_{\beta+1}^\lambda - q_\beta^\lambda)^*((p_{\lambda+1} - p_\lambda)(E)^*) \leqslant \aleph_0$ $((0,0) \leqslant (\lambda, \beta) < (\vartheta, 0))$, from (20.94) and since $u_\lambda \rho_\lambda^{-1}$ is an isomorphism it follows that $\operatorname{dens}(v_{(\lambda,\beta)+1}^* - v_{(\lambda,\beta)}^*)(E^*) \leqslant \aleph_0$ $((0,0) \leqslant (\lambda, \beta) < (\vartheta, 0))$, which proves (iii). Hence, by lemma 20.3 b), $E^*$ admits an equivalent locally uniformly convex norm, which is the dual of some equivalent norm on $E$. Thus, 1°⇒2°, which completes the proof of theorem 20.7.

*Remark 20.7.* a) Due to the special form of $v_{(\lambda,\beta)}$ in the above construction (20.91), one can prove more, namely, the following: *If $E$ is a Banach space with a shrinking extended M-basis, then there exists a transfinite sequence of non-zero continuous linear projections $\{v_\lambda\}_{\lambda \leqslant \zeta}$ on $E$, with the following properties:*

$$v_\lambda v_\varkappa = v_\varkappa v_\lambda = v_\varkappa \qquad (\varkappa \leqslant \lambda \leqslant \zeta), \tag{20.97}$$

$$v_\zeta = I_E, \tag{20.98}$$

$$v_\lambda^*(f) \in \overline{\{v_{\varkappa+1}^*(f)\}_{\varkappa<\lambda}} \qquad (f \in E^*, \lambda \leqslant \zeta), \tag{20.99}$$

$$\operatorname{dens}(v_{\lambda+1}^* - v_\lambda^*)(E^*) \leqslant \aleph_0 \qquad (0 \leqslant \lambda < \zeta;\ v_0 = 0) \tag{20.100}$$

(indeed, it is enough to assume, inductively, the corresponding properties for $\{q_\beta^\lambda\}_{\beta \leqslant \zeta_\lambda}$ on $(p_{\lambda+1} - p_\lambda)(E)$, where $p_0 = 0$, $q_0^\lambda = 0$). Next, one can prove*) that (20.97)—(20.99) already imply (i) and (ii) on $E^*$ (which constitutes another way of proving (i) and (ii)) and then one can apply lemma 20.3 b). However, the projections $v_\lambda$ above (and those of theorem 20.7) need not be uniformly bounded (we only know that $\|v_{(\lambda,\beta)}\| \leqslant \leqslant \|q_\beta^\lambda\| \, \|p_{\lambda+1} - p_\lambda\| + \|p_\lambda\| \leqslant 2\|q_\beta^\lambda\| + 1$, where $\|q_\beta^\lambda\| \geqslant 1$ whenever $\beta \neq 0$).

b) Similarly to the above proof of the implication 1°⇒2°, one can also show (using only lemma 20.3 a)) that every weakly compactly generated Banach space admits an equivalent locally uniformly convex norm.

**Corollary 20.7.** *Let $G$ be a subspace of a Banach space $E$ with a shrinking extended M-basis. Then $G$ has a shrinking extended M-basis (and hence $G$ is weakly compactly generated).*

*Proof.* By theorem 20.7, implication 1°⇒4°, $E$ is weakly compactly generated and admits an equivalent very smooth norm $|||\cdot|||$, whence

---

*) See § 19, remark 19.2 a).

$G(\subset E)$ is isomorphic to a very smooth space. Consequently, by theorem 20.6, $G$ has a shrinking extended $M$-basis, which completes the proof of corollary 20.7.

Similarly to theorem 20.6 one can prove (by transfinite induction on dens $E^*$ and using § 19, proposition 19.6 and § 14, lemma 14.3) that *if both E and $E^*$ are weakly compactly generated**), *then E has a shrinking extended M-basis* (and hence, by corollary 20.7, so does every subspace $G$ of $E$). Hence, by theorem 20.7, implication $1° \Rightarrow 2°$, $E$ admits an equivalent norm $|||\cdot|||$ such that $(E, |||\cdot|||)^*$ is locally uniformly convex. The converses of these statements are not valid, as shown by $E = c_0(I)$, with $I$ uncountable (which has a shrinking extended $M$-basis, but $E^* \equiv l^1(I)$ is not weakly compactly generated).

*Definition 20.7.* An extended $M$-basis $\{x_i\}_{i \in I}$ of a Banach space $E$, with the a.f.f. $\{f_i\}_{i \in I} \subset E^*$, is said to be *boundedly complete*, if every bounded $\sigma(E, [f_i]_{i \in I})$-Cauchy net $\{z_d\}_{d \in \mathscr{D}} \subset E$ (i.e., such that $\{f(z_d)\}_{d \in \mathscr{D}}$ converges for each $f \in [f_i]_{i \in I}$) is $\sigma(E, [f_i]_{i \in I})$-convergent to an element $x \in E$.

Similarly to § 7, theorem 7.9, implication $1° \Rightarrow 3°$ and § 8, proposition 8.17 a), it follows that *if $\{x_i\}_{i \in I}$ is a boundedly complete extended M-basis of a Banach space E, with the a.f.f. $\{f_i\}_{i \in I} \subset E^*$, then the canonical mapping u of E into $([f_i]_{i \in I})^*$ is an isomorphism onto $([f_i]_{i \in I})^*$ and $\{f_i\}_{i \in I}$ is a shrinking extended M-basis of $[f_i]_{i \in I}$, with the a.f.f. $\{u(x_i)\}_{i \in I}$.* Hence, by theorem 20.7, implication $1° \Rightarrow 2°$ (applied to $[f_i]_{i \in I}$), we obtain

**Corollary 20.8.** *Every Banach space E with a boundedly complete extended M-basis admits an equivalent locally uniformly convex norm.*

In particular**), such a space $E$ contains no subspace isomorphic to $l^\infty$ and thus in this case the answer to problem 20.3 is affirmative.

We have seen in § 19, proposition 19.7, that if $E$ is a non-separable very smooth Banach space, then $E^*$ admits a long sequence of projections $\{q_\lambda\}_{\omega \leqslant \lambda \leqslant \vartheta}$ with certain properties which ensure that there exists an ordinal resolution of $I_{E^*}$. However, the projections $q_\lambda$ need not be $w^*$-continuous, so the subspaces $(q_{\lambda+1} - q_\lambda)(E^*)$ are not suitable for transfinite induction, since they need not be duals of very smooth spaces. Therefore it is natural to raise

*Problem 20.5.****) If $E$ is a non-separable very smooth Banach space, does the conjugate space $E^*$ have an extended $M$-basis?

---

*) For various recent extensions of this result, see the Notes and remarks.

**) Indeed, obviously, local uniform convexity is inherited by subspaces and it is well known (see e.g. [70], Ch. VII, § 4, statement (1) (b) or [395], proposition 4) that $l^\infty$ admits no equivalent locally uniformly convex norm.

***) Recently it has been proved that if $E^*$ is locally uniformly convex, then the answer is affirmative (see the Notes and remarks).

# Notes and remarks

**§ 0.** The approximation property, in the form presented here, was defined by A. Grothendieck ([139], Ch. I, § 5, definition 9), within the more general framework of locally convex spaces $E$ (replacing, in the definition, "every $\varepsilon > 0$" and $\|v(x) - x\| < \varepsilon$ by "every neighbourhood $U$ of 0 in $E$" and $v(x) - x \in U$ respectively). A further generalization of the notion of a space with the approximation property, called an *admissible space*, was introduced by V. Klee ([219], [220]), within the framework of (Hausdorff) topological linear spaces $E$ (replacing, in the definition, the finite rank operator $v \in L(E, E)$ by a continuous finite rank operator $v: Q \to E$). V. Klee [220] has observed that, by a result of M. Nagumo [292], every locally convex space is admissible. Also, V. Klee has shown [220] that there exist admissible spaces which are not locally convex (e.g., the spaces $l^p(I)$ for every set $I$ and every $p$ with $0 < p < 1$) and has raised the problem, whether all topological linear spaces are admissible. In this direction, T. Riedrich ([336], [337]) has shown that the spaces*) $L^p([0, 1])$, where $0 < p < 1$, and the space $S([0, 1])$ are admissible.

One of A. Grothendieck's main motivations, in [139], for introducing the approximation property and for raising "*the approximation problem*" (i.e., the problem whether every locally convex space $E$ has the approximation property), was the following result ([139], Ch. I, part of proposition 35), which relates this problem to the considerably older problem**) of uniform approximability of compact linear operators on Banach spaces by finite rank linear operators: *A Banach space $E$ has the approximation property if and only if for every Banach space $F$, every compact linear operator from $F$ into $E$ can be approximated, in the norm-topology of $L(F, E)$, by linear operators of finite rank from $F$ into $E$.* In this connection, let us also mention the following "dual" result of A. Grothendieck ([139], Ch. I, proposition 36, part 2): *The conjugate space $E^*$ of a Banach space $E$ has the approximation property***) if and only if for every Banach space $F$, every compact linear operator from $E$ into $F$ can be approximated, in the norm topology of $L(E, F)$, by linear operators of finite rank from $E$ into $F$.* R. S. Phillips has shown ([317], § 6) that the usual concrete Banach spaces satisfy this latter condition; consequently, their conjugate spaces have the

---

*) Note that these spaces have no basis (see Ch. I, § 13, example 13.1).

**) Concerning this latter problem and its relation to the problem of existence of approximative bases, see the Notes and remarks to § 9. We mention here also the following related result of J. W. Brace ([35], theorem 1): *A linear operator $u: F \to E$ is compact if and only if it is a cluster point for the topology of "almost uniform convergence"* [34], *on the unit ball $S_F = \{x \in F \mid \|x\| \leqslant 1\}$ and using the norm topology on $E$, of a sequence of finite rank linear operators from $F$ into $E$).*

***) In this case, $E$ also has the approximation property ([139], part 1 of the same proposition).

approximation property. A. Grothendieck [139] has shown that the usual concrete spaces have the approximation property and even some stronger properties (see e.g. the Notes and remarks to § 18). For the spaces $E = C(Q)$ ($Q$ compact), some other proofs were given, subsequently, by C. T. Taam [388] (in terms of the uniform approximability of compact linear operators) and J. Gil de Lamadrid [127] (for the metric approximation property). Also, C. T. Taam [388] observed that this property of $C(Q)$, combined with the fact that for every Banach space $E$ there is a canonical linearly isometric embedding $j_E$ of $E$ into $C(S_{E^*})$ (where $S_{E^*}$ is compact for $\sigma(E^*, E)$), has the following obvious consequence: *An operator $u \in L(F, E)$ is compact if and only if the operator $j_E u \in L(F, C(S_{E^*}))$ can be approximated, in the norm topology, by linear operators of finite rank from $F$ into $C(S_{E^*})$.*

A. Grothendieck ([139], Ch. II, p. 135) called the approximation problem "undoubtedly the most important problem" occurring in [139] and conjectured*) that the answer will be negative. Also, he mentioned ([139], Ch. I, p. 164) that a negative answer would give a negative answer to various other interesting questions. In fact, A. Grothendieck has given various equivalent formulations of the approximation problem, some of which **) we mention now here. Namely ([139], Ch. I, part of proposition 37), *the following statements are equivalent:*

1°. *Every locally convex space $E$ has the approximation property.*

2°. *Every Banach space $E$ has the approximation property.*

3°. *Every subspace $E$ of $c_0$ has the approximation property.*

4°. *If $A = (a_{ij})$ is an infinite matrix satisfying $(a_{ij})_{j=1}^{\infty} \in c_0$ $(i = 1, 2, \dots$ $\dots)$, $\sum_{i=1}^{\infty} \sup_{1 \leqslant j < \infty} |a_{ij}| < \infty$ and $A^2 = 0$***), then* $\operatorname{Tr} A = \sum_{i=1}^{\infty} a_{ii} = 0$.

5°. *If $x(s, q)$ is a continuous function on* $[0, 1] \times [0, 1]$, *such that*

$$\int_0^1 x(s, t) x(t, q)\, dt = 0 \quad \text{for all } s, q \in [0, 1], \quad \text{then} \int_0^1 x(t, t)\, dt = 0.$$

6°. *For every continuous function $x(s, q)$ on* $[0, 1] \times [0, 1]$ *and every* $\varepsilon > 0$, *there exist* $s_1, \dots, s_n, q_1, \dots, q_n \in [0, 1]$ *and scalars* $\alpha_1, \dots, \alpha_n$ *such that*

$$\left| x(s, q) - \sum_{i=1}^{n} \alpha_i x(s, q_i) x(s_i, q) \right| < \varepsilon \qquad (s, q \in [0, 1]).$$

According to the survey paper of A. Pelczynski [305], the implication 6°⇒1° had been proved, before 1940, by S. Mazur (unpublished).

---

*) He used the words: "it seems probable to me".

**) We omit, among others, those which involve, in their statement, topological tensor products.

***) I.e., $\sum_{j=1}^{\infty} a_{ij} a_{jk} = 0$.

Since we have seen that the approximation property is transmitted from any conjugate Banach space to the initial space, *the above assertions are also equivalent to the following one:*

7°. *Every conjugate Banach space has the approximation property.* Without aiming at giving a history of the approximation problem (for such a history and connections with various other problems, see the recent survey paper of K.-D. Bierstedt [27]), let us mention now some other results on this problem, restricting ourselves only to Banach spaces. Even before the approximation problem was solved, T. Figiel had shown ([105], § 3) that the above assertions are equivalent to the following ones:

8°. *There exists p with* $1 \leqslant p \leqslant \infty$, *such that every subspace of**) $C_p$ *has the approximation property.*

9°. *There exists p with* $1 \leqslant p \leqslant \infty$, *such that every subspace of* $(l_1^\infty \times l_2^\infty \times \ldots)_{l^p}$ *has the approximation property.*

10°. *Every reflexive Banach space has the approximation property.* The implications 10°⇒8° and 10°⇒9° are obvious, since for every $p$ with $1 < p < \infty$, the space $C_p$ is reflexive. T. Figiel [105] observed that, since $C_\infty$ is**) isomorphic to a subspace of $c_0$, the implication 8° ⇒ 2° (with $p = \infty$) yields again the implication 3°⇒2° of A. Grothendieck. Also, T. Figiel [105] observed that the implication 10°⇒2° invalidates the statement***) of A. Grothendieck ([139], Ch. II, p. 135), according to which "... one can hope, even if the approximation problem were solved in the negative in the general case (which seems probable to me) that... every reflexive Banach space has the approximation property". The proof of the implication 8°⇒2° is based on the following factorization theorem, due to W. B. Johnson ([176], theorem 1): *For every p with* $1 \leqslant p \leqslant \infty$, *every pair of Banach spaces* $F$, $\tilde{E}$ *and every operator* $\tilde{u} \in L(F, \tilde{E})$ *which is the limit, in the norm topology, of a sequence of finite rank linear operators from F into* $\tilde{E}$, *there exist compact operators* $\tilde{v} \in L(F, C_p)$ *and* $\tilde{w} \in L(C_p, \tilde{E})$ *such that* $\tilde{u} = \tilde{w}\tilde{v}$. Combining this result with the above mentioned observation of C. T. Taam [388], T. Figiel obtained the following factorization theorem ([105], proposition 3.1): *For every p with* $1 \leqslant p \leqslant \infty$, *every pair of Banach spaces F, E and every compact* $u \in L(F, E)$ *there exist a subspace G of* $C_p$ *and compact operators* $v \in L(F, G)$ *and* $w \in L(G, E)$ *such that* $u = wv$. Indeed, by C. T. Taam's observation, $\tilde{u} = j_E u \in L(F, C(S_{E^*}))$ can be approximated, in the norm topology, by linear operators of finite rank from $F$ into $\tilde{E} = C(S_{E^*})$ and hence, by W. B. Johnson's factorization theorem, there

*) For the definition of $C_p$, $1 \leqslant p \leqslant \infty$, see § 12.

**) See § 14, corollary 14.3 b).

***) Actually, T. Figiel [105] (see also J. Lindenstrauss [239]) interprets this statement as containing two conjectures, although the words "one can hope" might mean slightly less than a conjecture.

exist compact operators $\tilde{v} \in L(F, C_p)$ and $\tilde{w} \in L(C_p, C(S_{E^*}))$, such that $j_E u = \tilde{w}\tilde{v}$; then, putting $G = \overline{\tilde{v}(F)}$, $v =$ the astriction of $\tilde{v}$ to $G$ (i.e., $v \in L(F, G)$, $v(y) = \tilde{v}(y)$ for all $y \in F$) and $w = (j_E)^{-1}\tilde{w}i_G$, where $i_G$ is the canonical embedding of $G$ into $C_p$ (so $\tilde{v} = i_G v$), one obtains the desired factorization $u = (j_E)^{-1}\tilde{w}\tilde{v} = (j_E)^{-1}\tilde{w}i_G v = wv$, with $w$ and $v$ compact. This proves Figiel's factorization theorem. Now the proof of the implication 8°⇒2° above is immediate. Indeed, if we have 8° and if $u \in L(F, E)$ is any compact operator and $G \subset C_p$, $v \in L(F, G)$ and $w \in L(G, E)$ are as above, then, by 8°, there exists a sequence $\{v_n\} \subset L(F, G)$ of finite rank operators such that $\lim_{n\to\infty} \|v - v_n\| = 0$, whence the operators $wv_n \in L(F, E)$ are of finite rank and

$$\|u - wv_n\| = \|wv - wv_n\| \leqslant \|w\| \, \|v - v_n\| \to 0 \text{ as } n \to \infty,$$

which completes the proof. For the implication 9°⇒2° the proof is similar [105].

W. B. Johnson showed ([180], theorem 3) that the above assertions are equivalent to the following one:

11°. *$C_1^*$ has the approximation property.*

Obviously, 2°⇒11°. The proof of the implication 11°⇒3° (⇒2°) is based on the following theorem of W. B. Johnson ([180], theorem 1): *For every separable Banach space $E$ there exists $u \in L(C_1, E)$ such that $u^*$ is an isometry and that there exists a projection**) *$v$ of $C_1^*$ onto $u^*(E^*)$* (thus, $C_1^*$ is "complementably universal" for the family of all Banach spaces which are conjugates of separable Banach spaces). Now the proof of the implication 11°⇒3° becomes immediate, using also that the approximation property is inherited by complemented subspaces and that it is transmitted from any conjugate Banach space to the initial space. W. B. Johnson has also proved ([180], remark 1) that $C_1^*$ is isomorphic to $E_b^*$, where $E_b$ is the space occurring in § 9, theorem 9.3, assertion 10°.

The approximation problem was completely solved, in the negative, by P. Enflo [99] (according to the survey paper of A. Pelczynski [305], this result was obtained by P. Enflo in May 1972). Actually, P. Enflo proved ([99], theorem 1) that *there exist a separable reflexive Banach space $E$, with a sequence $\{G_n\}$ of finite-dimensional subspaces satisfying* $\lim_{n\to\infty} \dim G_n = \infty$ *and a constant $C$, such that for every finite rank operator $v \in L(E, E)$,*

$$\|v|_{G_n} - I_{G_n}\| \geqslant 1 - \frac{C\,\|v\|}{\log \dim G_n} \qquad (n = 1, 2, \ldots); \qquad (0.121)$$

hence, since $E$ is separable and reflexive, $E$ does not have the approximation property (e.g., by § 9, theorem 9.3, implication 11°⇒5° or 11°⇒6°). Subsequently, various authors substantially simplified the method of P. Enflo [99] and, using still the same basic ideas of the

*) Moreover [180], $u^*$ is also a weak* isomorphism and $\|v\| = 1$.

paper of P. Enflo [99], they obtained other Banach spaces which do not have the approximation property. Without specifying here in detail these contributions to the simplification of Enflo's method, let us mention that subspaces $E$ of $c_0$ and of $l^p$, for each $p$ with $2 < p < \infty$, which do not have the approximation property, were obtained by A. Pelczynski and T. Figiel*) [308] and, respectively, T. Figiel [106], using a combinatorial lemma (simpler than that used by P. Enflo [99]). Furthermore, S. Kwapien [227] and A. M. Davie [56] have replaced the use of combinatorial arguments by the use of probabilistic inequalities and have obtained subspaces $E$ of $L^p$ (on the unit circumference, with the Haar measure), for each $p$ with $2 < p < \infty$, respectively, again of $c_0$ and of $l^p$, for each $p$ with $2 < p < \infty$, which do not have the approximation property. The proof of A. M. Davie is also presented in the book of J. Lindenstrauss and L. Tzafriri [247] and in the survey paper of D. Dacunha-Castelle [53a], while the method of T. Figiel is also exposed in the survey paper of W. B. Johnson [182]. Recently, using a simple combinatorial lemma, A. Szankowski has obtained ([387], proposition 1), for any $p, q$ with $1 \leqslant q < p \leqslant \infty$, a *Banach lattice* which can be linearly isometrically embedded into $(E_1 \times E_2 \times \dots)_{l^p}$, where $E_n = L^q([0, 1])$ $(n = 1,2, \dots)$ and which fails to have the approximation property. Consequently, by taking $1 < q < p < \infty$, it follows ([387], proposition 2) that *there exists a uniformly convex* (*hence reflexive*) *Banach lattice without the approximation property*. The existence of a reflexive Banach lattice which does not have the approximation property was also shown by A. V. Buhvalov [40], with a different method, using the existence of a non-reflexive Banach lattice without the approximation property, proved by A. Szankowski in an earlier version of [387] (which did not include the reflexive case) and a factorization of compact linear operators with values in Banach lattices. We shall present the example of A. Szankowski in Vol. III.

The above spaces $E$ without the approximation property give counterexamples to the assertions 1°—3° and 7°—10° above and hence they imply that 4°—6° and 11° are not valid. A direct proof of the fact that 4° and 5° are not valid was given, with similar methods, by A. M. Davie [57]. In connection with 4°, A. Grothendieck showed ([139], Ch. I, remark 14) that if $A = (a_{ij})$ is an infinite matrix satisfying $(a_{ij})_{j=1}^{\infty} \in c_0$ $(i = 1,2, \dots)$, $\sum_{i=1}^{\infty} (\sup_{1 \leqslant j < \infty} |a_{ij}|)^{\frac{2}{3}} < \infty$ and $A^2 = 0$, then $\operatorname{Tr} A = \sum_{i=1}^{\infty} a_{ii} = 0$. A. M. Davie proved ([57], remark (iv)) that in this result the exponent $\frac{2}{3}$ is sharp.

In § 0 we have presented the proofs of T. Figiel [106] (with slight changes in the final part) and A. M. Davie [56] of the existence of sub-

---

*) The authors are in this order, since the paper appeared in Russian (in the Russian alphabet the letter P precedes the letter F).

spaces of $c_0$ and $l^p$ $(2 < p < \infty)$ which do not have the approximation property. In contrast with the other existing presentations of these proofs, mentioned above (including the original ones of T. Figiel [106] and A. M. Davie [56]), where some of the details are only sketched or left to the reader, we have given here all details of these proofs, in order to make their understanding easier; also, we have attempted to explain intuitively the underlying ideas of these proofs, in a somewhat different way.

Lemma 0.1 b) is a version of a remark of T. Figiel ([106], § 2), who has shown directly that (0.2), (0.3) and (0.6)—(0.8) imply that $E$ does not have the approximation property. Remark 0.1 a) was communicated to us by T. Figiel in 1973, for the case of his version of lemma 0.1 b). Each of the other papers mentioned above also contains (sometimes only implicitly) some sufficient condition for a Banach space to fail the approximation property, expressed directly in terms of the jumps of the average trace (except in the survey paper of W. B. Johnson [182], where it is expressed in terms similar to those of remark 0.1 a)).

The notion of average trace, which plays a fundamental role in the proofs of the existence of Banach spaces without the approximation property, mentioned above, was introduced by P. Enflo [99]. Actually, P. Enflo [99] has introduced the average trace with respect to a finitely linearly independent sequence $\{x_n\} \subset E$ (hence every finite subsequence of $\{x_n\}$ admits a sequence of biorthogonal functionals), of any "finite expansion operator" $u \in L(E, E)$ $\Big($i.e., for which there exists an infinite matrix $(a_{ij})$ with each row having only a finite number of non-zero entries, such that $u(x_j) = \sum_i a_{ij} x_i$ for $j = 1,2, \ldots$; then, by definition, $\mathscr{T}(M; u) = \frac{1}{|M|} \sum_{i \in M} a_{ii}\Big)$. S. Kwapien [228] has defined the average trace of $u \in L(E, E)$ with respect to any pair of sequences $(x_n, f_n)$ $(\{x_n\} \subset E, \{f_n\} \subset E^*)$, by formula (0.15) and has given the following sufficient condition for a Banach space $E$ to fail the approximation property ([228], proposition 1): If there exist a sequence $\{M_n\}$ of finite subsets of $\mathscr{N}$ and a sequence $\{\alpha_n\}$ of positive numbers with $\sum_{n=1}^{\infty} \alpha_n < \infty$, such that

$$\lim_{n\to\infty} \mathscr{T}(M_n; I_E) \neq 0, \tag{0.122}$$

$$\lim_{n\to\infty} \mathscr{T}(M_n; v) = 0 \qquad (v \in L(E, E), \dim v(E) < \infty), \tag{0.123}$$

$$|\mathscr{T}(M_{n+1}; u) - \mathscr{T}(M_n; u)| \leqslant \alpha_n \|u\| \quad (u \in L(E, E),\ n = 1,2, \ldots), \tag{0.124}$$

then $E$ does not have the approximation property.*) Also, S. Kwapien observed [228] that (0.122), (0.123) are satisfied not only when $(x_n, f_n)$ is an $E$-complete biorthogonal system, but also when $f_n(x_n) = 1$ $(n = 1,2,\dots \dots)$, $\sup_{1 \leqslant n < \infty} \|x_n\| \|f_n\| < \infty$ and either $x_n \xrightarrow{w} 0$ or $f_n \xrightarrow{w^*} 0$. T. Figiel observed (verbal communication, 1973, reproduced also by W. B. Johnson and A. Szankowski [188], Appendix, lemma and W. B. Johnson [182], problem 4.2) that if the $M_n$ are pairwise disjoint, we have even

$$\lim_{n\to\infty} \mathscr{T}(M_n; v) = 0 \qquad (v \in L(E, E),\ v \text{ compact}); \qquad (0.125)$$

indeed, if $v$ is compact and $\sup_{1 \leqslant n < \infty} \|x_n\| < \infty, f_n \xrightarrow{w^*} 0$, then $Q = \overline{\{v(x_n)\}}$ is compact and hence

$$|\mathscr{T}(M_n; v)| = \frac{1}{|M_n|} \left| \sum_{i \in M_n} f_i(v(x_i)) \right| \leqslant \max_{i \in M_n} |f_i(v(x_i))| \leqslant$$

$$\leqslant \max_{i \in M_n} \sup_{x \in Q} |f_i(x)| \to 0 \text{ as } n \to \infty,$$

while if $v$ is compact and $\sup_{1 \leqslant n < \infty} \|f_n\| < \infty$, $x_n \xrightarrow{w} 0$, then $Q' = \overline{\{v^*(f_n)\}} \subset \subset E^*$ is compact and hence

$$|\mathscr{T}(M_n; v)| \leqslant \max_{i \in M_n} |v^*(f_i)(x_i)| \leqslant \max_{i \in M_n} \sup_{f \in Q'} |f(x_i)| \to 0 \text{ as } n \to \infty.$$

As a consequence of this latter fact, T. Figiel observed (see [188], [182]) that *the known Banach spaces which do not have the approximation property, fail to have even the compact approximation property* (a Banach

---

*) The proof uses the fact ([139], Ch. I, proposition 35) that $E$ has approximation property (if and) only if the canonical mapping of $E^* \otimes_\gamma E$ into $L(E, E)$ is one-to-one. Note that here (0.124) is weaker than (0.10) (or (0.4), (0.5)) and, in particular, it implies that in lemma 0.1 a) one can replace (0.5) by the weaker condition $\sup_{1 \leqslant n < \infty} \max_{x \in \mathscr{A}_n} \|x\| < \infty$ (one can also see this fact directly, using the observation made at the beginning of the proof of lemma 0.1 b); indeed, if we have (0.4) and the latter condition, then, taking any sequence $\{\alpha'_n\}$ of positive numbers such that $\sum_{n=1}^{\infty} \alpha'_n < \infty$, $\lim_{n\to\infty} \frac{\alpha_n}{\alpha'_n} = 0$ and putting $\mathscr{A}'_n = \frac{\alpha_n}{\alpha'_n} \mathscr{A}_n$, we shall have (0.4) and (0.5) for $\alpha'_n$, $\mathscr{A}'_n$, since $\alpha'_n \max_{x \in \mathscr{A}'_n} \|u(x)\| = \alpha'_n \max_{x \in \mathscr{A}_n} \left\| u\left( \frac{\alpha_n}{\alpha'_n} x \right) \right\| = \alpha_n \max_{x \in \mathscr{A}_n} \|u(x)\|$).

space $E$ is said to have the *compact approximation property* if for every compact subset $Q$ of $E$ and every $\varepsilon > 0$ there exists a compact*) operator $v \in L(E, E)$ satisfying (0.1)); indeed, this follows as in the proof of lemma 0.1 a), or by the argument of remark 0.1 a).

In the equivalent language of the Walsh functions on $2^{[1,n]}$ (see remark 0.2), lemma 0.2 was given, with a more complicated proof (using induction), by A. Pelczynski and T. Figiel ([308], proposition 2, 1°); the present form and proof (using the function $y_0$ defined by (0.22)) are due to T. Figiel ([106], lemma 1). Part of remark 0.2 can be found in the paper of A. Pelczynski and T. Figiel ([308], remark to proposition 2), who have also observed that one could replace lemma 0.2 by another version (sufficient for their construction of a subspace**) of $c_0$ without the approximation property), which does not involve any group structure, but uses, instead, the decomposition of $l^\infty_{2^n}$ into a direct sum of two subspaces with "bad" projection constants, due to A. Sobczyk ([381]; for a simplified proof, see G. Köthe [223], § 31, no. 3); in this connection, see also remark 0.5 and the comments to the first part of this remark, which will be made below. Lemma 0.3 is due to T. Figiel ([106], lemma 2). A more general version of this lemma can be found, with a similar proof, in the paper of A. Pelczynski and T. Figiel ([308], § 3, combinatorial lemma), with the mention that the construction used in the proof is not new (see e.g. M. Hall, Jr. [154], proof of theorem 15.2.3). Remark 0.3 a) was communicated to us by T. Figiel. Remark 0.3 b) is contained, implicitly, in the paper of T. Figiel ([106], p. 9). The definition (0.44) of the $x_w$'s was given by A. Pelczynski and T. Figiel ([308], § 4) with the sets $h^{-1}(w)$ having higher cardinalities, which are increasing in the process of construction; the present simple form of the $x_w$'s, with all $|h^{-1}(w)| = 4$, is due to T. Figiel [106]. Theorem 0.1 was proved by T. Figiel [106], who used a slightly different method to estimate $C_n$ in the case when $2 < p < \infty$ and obtained (0.68) with $B'_p = 6$ (independently on $p$). To this end, instead of (0.67), T. Figiel [106] observed that

$$\|x_w\|_2^2 = 2^n + 4.2^{n+1} = 9.2^n \qquad (w \in W^+_{n,k}), \tag{0.126}$$

whence

$$\Big\| \sum_{w \in W^+} w(g) x_w - \sum_{w \in W^-} w(g) x_{h(w)} \Big\|_2^2 =$$

$$= 2^{n-1}.9.2^n + 2^{n-1}.9.2^{n-1} = 27.2^{2n-2}, \tag{0.127}$$

which, together with the $l^\infty$-estimate (0.65), yields the above mentioned inequality for $C_n$ (since $\|\cdot\|_p \leqslant \|\cdot\|_2^{\frac{2}{p}} \|\cdot\|_\infty^{1-\frac{2}{p}}$). The possibility of obtaining

*) Since every finite rank operator $v \in L(E, E)$ is compact, it is obvious that the approximation property implies the compact approximation property. Similarly, one can also introduce the bounded compact approximation property (see [188], [182]).

**) S. Kwapien [228] has observed that a similar remark is also valid for the case of $l^p$, $2 < p < \infty$, using the Sobczyk decomposition of $l^p_{2^n}$.

an $l^p$-estimate of the type (0.68) by a direct computation was mentioned to us, in a conversation, by T. Figiel; such a computation is made in (0.67), (0.68).

Lemma 0.4 was given in the paper of A. M. Davie ([56], lemma), with the mention that it is taken from the theory of random trigonometrical series (see e.g. J.-P. Kahane and R. Salem [208]). Lemma 0.5 is contained, implicitly, in the paper of A. M. Davie ([56], p. 262). Theorem 0.2 is due to A. M. Davie [56], who also made remark 0.4 ([56], remark (i)). The first part (namely, the inequality (0.101)) of remark 0.5 is due to S. Kwapien [228], who also formulated [228] the following related conjecture: For every $n$-dimensional Banach space $X$ there exist an $X$-complete biorthogonal system $(x_i, h_i)_{i\in I}$ $(\{x_i\}\subset X, \{h_i\}\subset X^*)$ and a subset $A\subset I$ such that the average trace $\mathscr{T}$ with respect to $(x_i, h_i)_{i\in I}$ satisfies (0.101) with*)

$$\alpha = \frac{C}{d(X, l_n^2)}, \tag{0.128}$$

where $C$ is a universal constant. According to S. Kwapien [228], it has been observed by A. Pelczynski that the conjecture is true for $X=l_n^p$, with $C=C(p)$, where $1\leqslant p\leqslant\infty$ (in this case it is known that $d(l_n^p, l_n^2)=n^{\left|\frac{1}{p}-\frac{1}{2}\right|}$), namely, the Sobczyk decomposition of $l_n^p$ (mentioned above) gives the desired property; a different proof for this case was sketched by S. Kwapien [228] (using his methods of [227]).

In connection with problem 0.1 let us mention that, even before the approximation problem was solved, the following result had been proved by T. Figiel ([105], corollary 6.3): *If the approximation problem has a negative solution, then every Banach space Z with the "subspace factorization property" contains a subspace E which does not have the approximation property*. Here, a Banach space $Z$ is said to have the *subspace factorization property* [105], if for every pair of Banach spaces $F, E$ and every compact $u\in L(F, E)$ there exist a subspace $G$ of $Z$ and compact operators $v\in L(F, G)$ and $w\in L(G, E)$ such that**) $u = wv$. We have seen above that for every $p$ with $1\leqslant p\leqslant\infty$, $C_p$ has the subspace factorization property. T. Figiel gave some necessary and sufficient conditions for a Banach space $Z$ to have the subspace factorization property, e.g., the following one ([105], theorem 6.1):

$$\inf_{\substack{B\subset Z\\ \dim B=n}} d(B, l_n^\infty) = 1 \qquad (n = 1,2, \ldots); \tag{0.129}$$

*) For the definition of the distance coefficient $d(A,B)$ see § 12, formula (12.41).

**) If this condition is satisfied for $F = E = l^p$, $1\leqslant p\leqslant\infty$, where $l^\infty$ stands for $c_0$, then $Z$ is said to have [105] the *subspace factorization property for* $L^p$. T. Figiel has given some characterizations of such spaces, e.g. by (0.129), with $\infty$ replaced by $p$ and $= 1$ replaced by $\leqslant M$, for a constant $M$ which does not depend on $n$ ([105], theorem 7.7); for $p=\infty$, see also the subsequent remark 1.17.

T. Figiel observed (verbal communication, in 1973) that this criterion, together with P. Enflo's negative solution of the approximation problem, yields for example, that *the space $J$ of R. C. James**) *contains a subspace $E$ which does not have the approximation property* (indeed, it is known**) that some isomorph of $J$ satisfies (0.129)).

After the approximation problem had been solved in the negative, problem 0.1 was raised by T. Figiel, who conjectured that the answer is affirmative ([105], the note added in proof) and by S. Kwapien ([228], problem 1). However, W. J. Davis informed us that recently W. B. Johnson has solved problem 0.1 in the negative, by showing that *there exist Banach spaces $X$ which are not isomorphic to $l^2$* (namely, of the form $(l^{p_1}_{k_1} \times l^{p_2}_{k_2} \times \dots)_{l^2}$, with suitable $k_n \to \infty$ and $p_n \to 2$), *such that every subspace $E$ of $X$ has even a finite-dimensional decomposition and the uniform****) *approximation property* (whence all conjugates $E^*, E^{**}, \dots$ have the bounded approximation property); it is not known whether every such $E$ has a basis****). The particular case $X = l^p$, $1 \leqslant p < 2$, of problem 0.1 was raised simultaneously by P. Enflo [99] and A. M. Davie [56] (the latter paper has also suggested a possible way of giving an affirmative answer to it) and it can be also found in the subsequent survey papers mentioned above. The particular case $X = L^p$, $1 \leqslant p < 2$, of problem 0.1 was raised by S. Kwapien ([228], problem 2). Even before the approximation problem was solved, T. Figiel had shown ([105], corollary 7.5) that for any $p$ with $1 \leqslant p \leqslant \infty$, the particular cases $X = l^p$ and $X = L^p$ of problem 0.1 are equivalent and ([105], corollary 7.6) that if for some $p$ with $1 \leqslant p < 2$, $l^p$ has a subspace without the approximation property, then so does every $l^r$, where $p < r \leqslant 2$. W. J. Davis communicated us that recently A. Szankowski has solved the particular case $X = l^p$, $1 \leqslant p < 2$, of problem 0.1, in the affirmative; moreover, he has shown that *if $X$ is a Banach space "with $p(X) < 2$" or "with $q(X) > 2$", then $X$ contains a subspace which does not have the approximation property*. Here $p(X) = \sup\{p \mid X \text{ is of type } p\}$, $q(X) = \inf\{q \mid X \text{ is of cotype } q\}$. We recall that for any real number $p \geqslant 1$ (respectively, $q > 0$) a Banach space $E$ is said to be *of type $p$* (respectively, *of cotype $q$*), if there exist a constant $C$ and a real number $\alpha \in (0, \infty)$ such that for any finite sequence $\{x_1, \dots, x_n\} \subset E$ we have

$$\left(\int_0^1 \left\| \sum_{i=1}^n r_i(t) x_i \right\|^\alpha dt\right)^{\frac{1}{\alpha}} \leqslant C \left(\sum_{i=1}^n \|x_i\|^p\right)^{\frac{1}{p}} \tag{0.130}$$

*) See Ch. II, § 4, example 4.1.

**) See D. P. Giesy and R. C. James [126], theorem 4.

***) See the Notes and remarks to § 18.

****) See the Appendix.

(respectively,

$$\left(\sum_{i=1}^{n}\|x_i\|^q\right)^{\frac{1}{q}} \leqslant C\left(\int_0^1\left\|\sum_{i=1}^{n} r_i(t)x_i\right\|^{\alpha} dt\right)^{\frac{1}{\alpha}}\Bigg), \tag{0.131}$$

where $\{r_n(t)\}$ is the sequence of Rademacher functions*) on [0, 1]; it is known**) that here one can replace $\alpha$ by any other number $\beta \in \in (0, \infty)$ (modifying $C$). We shall consider such spaces in Vol. III. Let us mention that *there exist some concrete Banach spaces about which it is not yet known whether they have the approximation property* (see e.g. § 18, problem 18.1 c)).

In the particular case when $E^*$ has a monotone boundedly complete basis, lemma 0.6 was proved by R. C. James ([160], theorem 1); in the general case it was given by J. Lindenstrauss ([240], theorem), with a modification of James' proof. Also, J. Lindenstrauss proved the following corollary of this result ([240], corollary 3 and the remark to it): If there exists a Banach space which does not have the approximation property, then there exists a Banach space $E$ with a basis, whose conjugate space $E^*$ is separable and does not have the approximation property; of course, this result, together with the (subsequent) negative solution of the approximation problem, yields example 0.1. Besides the other applications of theorems 0.1, 0.2 given in the present volume (for example, to the proof of the existence of Banach spaces with the approximation property which do not have the bounded approximation property***) or of Banach spaces with bases which do not have the uniform approximation property****), etc.), we shall give in Vol. III further applications of these theorems.

**§ 1.** The notion of a basic sequence in a Banach space can be found in Banach's monograph [11], where theorem 1.1 is given without proof ([11], p. 238). The term "basic sequence" is due to C. Bessaga and A. Pelczynski [21]. Proposition 1.1 and its application to the proof of theorem 1.1 were given by C. Bessaga and A. Pelczynski ([21], theorem 3 and corollary 3); also, corollaries 1.1 and 1.2 were obtained in the same paper ([21], corollaries 1 and 2). The proof of theorem 1.2 (and hence of theorem 1.1), presented here, is due to B. R. Gelbaum [125], while the proof, given in remark 1.1, can be found in the paper [23] of C. Bessaga and A. Pelczynski, with the mention that it is due to S. Mazur. Lemma 1.1, which is used (implicitly) in the proofs of the above mentioned authors (and in the proof of theorem 1.4 and remark 1.5), was given explicitly in [61], remark 1. Let us also mention the following equivalent version of lemma 1.1, which was used (impli-

*) See Ch. II, § 14.
**) See e.g. B. Maurey and G. Pisier [269], remarks 1.1 and 2.1.
***) See § 9, example 9.2.
****) See the Notes and remarks to § 18.

citly) in the sequel, e.g. in §§ 1, 9, 13, 20, proofs of theorems 1.9, 9.4 and 13.3 and remark 20.2 a) respectively (the equivalence of these versions is clear from their proofs and it is quite natural in the light of § 1, lemma 1.3, equivalence $1° \Leftrightarrow 2°$ and its dual):

**Lemma 1.1′.** *Let $E$ be a Banach space and* $0 < \varepsilon < 1$.

a) *If $G$ is a finite-dimensional subspace of $E$, then there exists a finite-dimensional subspace $\Gamma$ of $E^*$ such that $r_\Gamma(G) > 1 - \varepsilon$. Moreover, we can find such a $\Gamma$ in any subspace $V$ of $E^*$ with $r(V) > 1 - \varepsilon'$, where* $0 < \varepsilon' < \varepsilon$.

b) *If $\Gamma$ is a finite-dimensional subspace of $E^*$, then there exists a finite-dimensional subspace $G$ of $E$ such that $r_G(\Gamma) > 1 - \varepsilon$.*

Further research on selection of basic sequences was stimulated by the problem (which we shall present in Vol. III, Ch. IV) of existence of basic sequences of types $l_+$, $P$ and $P^*$ in non-reflexive Banach spaces, raised in [363] and solved in the affirmative by A. Pelczynski [300]. Thus, proposition 1.3 and the second statement in theorem 1.3 were obtained by A. Pelczynski [301], while theorem 1.3 is due to M. I. Kadec and A. Pelczynski ([206], theorem 2). For the space $E = l^1$ (in which*⁾ weak convergence and norm convergence of sequences coincide), V. I. Gurariĭ and N. I. Gurariĭ have proved ([146], theorem 1) the following sharpening of part of theorem 1.3: *Every bounded sequence $\{x_n\}$ in $l^1$, which is not conditionally compact, contains a subsequence $\{x_{n_k}\}$ which is a basic sequence, equivalent to the unit vector basis of $l^1$ and such that $[x_{n_k}]$ is complemented in $l^1$.* H. P. Rosenthal has proved for real Banach spaces $E$ ([342], main theorem) and L. E. Dor has extended to complex Banach spaces [83] the following theorem, which implies the first statement of the above mentioned result: *If $\{x_n\}$ is a bounded sequence in a Banach space $E$, which has no weak Cauchy subsequence, then $\{x_n\}$ contains a subsequence $\{x_{n_k}\}$ which is equivalent to the unit vector basis of $l^1$.* We shall present this latter theorem and some of its consequences in Vol. III, Ch. IV. For basic sequences in $l^p$, $1 \leqslant p < \infty$, see V. I. Gurariĭ and N. I. Gurariĭ [146] and V. I. Gurariĭ, N. I. Gurariĭ and V. I. Liokumovič [146a].

For $r(V) > 0$, proposition 1.4 can be found in the paper [300] of A. Pelczynski, with the mention that it has been first proved by C. Bessaga in his thesis and that the idea of the proof belongs to S. Mazur; in the general case (i.e., for $V$ only total), proposition 1.4 was proved, independently, in the original version**⁾ of [62] and by V. D. Milman ([287], theorem 1.12′). Corollary 1.3 was proved in the original version of [62]. The equivalence $1° \Leftrightarrow 3°$ and for $r(V) > 0$, the implication $2° \Rightarrow 1°$,

---

*⁾ See e.g. [11], p. 137.

**⁾ Proposition 1.4 and corollary 1.3, as well as some related results, were deleted from the final version of [62], upon the suggestion of the referee of [62]. A similar remark (which we shall not mention again explicitly) is also valid for the subsequent references to the original versions of [86] and [376].

of corollary 1.4, as well as corollary 1.5, are due to M. I. Kadec and A. Pelczynski ([206], theorem 1 and corollaries 1 and 3). Examples of hypercomplete sequences*) in certain concrete spaces have been known for a long time; for example, G. Szegö proved (see [52], I, p. 86) that if $0 < \lambda_n \to \infty$, then the sequence $x_n(t) = \dfrac{1}{t+\lambda_n}$ $(t \in [0, 1]; n = 1,2, \ldots)$ is complete in $C([0, 1])$ and hence (as observed by P. Erdös and E. G. Straus [102]) hypercomplete in $C([0, 1])$. Proposition 1.5 is due to V. Klee [217]; the proof given here was communicated to us in a letter by M. I. Kadec and the constructive proof given in remark 1.3 is due to H. Brass [36] (we note that still another proof can be found in the survey paper of V. D. Milman ([287], § 1), with the mention that it is due to Yu. I. Lyubich. The observation that for a separable space $E$ the statements "$E$ has a basis" and "every infinite set $A$ in $E$ contains an infinite subset $B$ such that the space $[B]$ has a basis", are equivalent (leading to example 1.1), can be found in the paper [206] of M. I. Kadec and A. Pelczynski, with the mention that it was made by V. I. Gurariĭ.

Problem 1.1 was raised in [367], problem 3.1. In connection with this problem we mention**) here that *if $E^*$ is not $w^*$-separable, then $E$ contains a monotone basic sequence*, since in this case we can replace in the proof of theorem 1.2 the finite $\varepsilon$-nets of the unit spheres of $P_{(n)} = [x_1, \ldots, x_n]$ $(n = 1,2, \ldots)$ by countable dense subsets thereof and the finite-codimensional subspaces $Y_n$ $(n = 1,2, \ldots)$ by the corresponding countable codimensional subspaces (which cannot be $\{0\}$, since $E^*$ is not $w^*$-separable). Hence, by § 17, remark 17.6 a), *every non-separable weakly compactly generated Banach space $E$ contains a monotone basic sequence.* Furthermore by § 17, remark 17.6 b) and by the results of § 19, there are also other non-separable Banach spaces which contain a monotone basic sequence (e.g., the conjugate space $E^*$ of any non-separable very smooth Banach space $E$). Some other positive results on the existence of monotone basic sequences (e.g., that *every subspace of $c_0$ and every subspace of $L^{2n}([0, 1])$ contains a monotone basic sequence*) can be deduced from the paper [232] of A. Lazar and M. Zippin. Let us also observe that the following notion appears to be useful for the study of the problem of existence of monotone basic sequences in (infinite dimensional) Banach spaces: We shall say that a subspace $G$ of a Banach space $E$ is *strong*, if there exist a subspace $G_0 \subset E$ with $G_0 \supset G$ and a projection $u$ of $G_0$ onto $G$, of norm $\|u\| = 1$. One can show that *the following statements are equivalent*: 1°. *$G$ is strong.* 2°. *There exists an element $x \in E$ such that****) $G \perp x$.* 3°. *There exists a subspace $V$ of $E^*$, satisfying $r_G(V) = 1$, which is not $w^*$-dense in $E^*$.* From Ch. II, § 1, proposition 1.3 it is clear that *the following statements are equivalent*:

*) Such sequences are also called "overfilling sequences" (see e. g. [287], § 1).

**) This remark is due to M. M. Day ([70], p. 93), even with $\{x_n\}$ monotone and normal (using the proof of theorem 1.4).

***) I.e. (see Ch. II, p. 215), such that $\|g + \alpha x\| \geqslant \|g\|$ $(g \in G, \alpha \in K)$.

1°. *E contains a monotone basic sequence.* 2°. *E contains an increasing sequence of strong subspaces* $\{G_n\}$ *with* $\dim G_n = n$ $(n = 1,2, \ldots)$. 3°. *E contains an increasing sequence of strong subspaces* $\{G_n\}$. However, even the following problem is open: *Does every infinite dimensional Banach space E contain a strong subspace G with* $\dim G \geqslant 2$? We conjecture that the answer is negative and hence so is the answer to the problem of existence of monotone basic sequences in infinite dimensional Banach spaces (in this direction, see also our remarks made after problem 1.1).

Proposition 1.6 was given in [368], theorem 5.3. For various special classes of basic sequences (including non-monotone, non-strictly monotone and non-normal basic sequences) remark 1.4 was made in [367], Ch. III, § 2. Theorem 1.4 and remark 1.5 are due to M. M. Day (announced without proof in the 1958 edition of [70] and given with proof in [69]). The result, mentioned after remark 1.5, that in every Banach space with a basis there exist, for any integer $k > 0$, a non-$k$-shrinking and a non-$k$-boundedly complete basis, was given in [74].

Problem 1.2 was raised by C. Bessaga and A. Pelczynski ([22], problem 5.1), who have also observed that an affirmative answer would immediately reduce the proof of the Dvoretzky-Rogers theorem to the easy case of the space $l^1$. Proposition 1.7 and corollary 1.6 are due to C. Bessaga and A. Pelczynski ([21], corollaries 1 and 4). In connection with problem 1.2 and proposition 1.7, it is natural to ask the following problem, which can be found in the paper of C. Bessaga and A. Pelczynski ([22], problem 5.12; as has been communicated to us by C. Bessaga, this problem was raised by S. Mazur in 1955): If $\{y_n\}$ is a sequence in a Banach space $E$, satisfying (1.1) and $y_n \overset{w}{\to} 0$, does $\{y_n\}$ contain a subsequence $\{y_{p_n}\}$ which is an unconditional basic sequence? (Obviously, an affirmative answer to this problem would yield an affirmative answer to problem 1.2). Recently, B. Maurey and H. P. Rosenthal have shown [270] that the answer is negative. Let us give here their counterexample.

*Example 1.5.* Let $\mathscr{D}$ be the (countable) family of all finite sequences $\{F_1, \ldots, F_k\}$ of finite subsets of $\mathscr{N} = \{1,2,3, \ldots\}$, let

$$m_k = 4^{k^2} \qquad (k = 1,2, \ldots), \tag{1.106}$$

and let $\psi : \mathscr{D} \to \{m_k\}$ be a one-to-one mapping. Furthermore, let $\mathscr{F}$ be the family of all infinite sequences $\{F_j\}$ of finite subsets of $\mathscr{N}$, such that*)

$$|F_1| = 1, \tag{1.107}$$

$$\max F_{j-1} < \min F_j \qquad (j = 2,3, \ldots), \tag{1.108}$$

$$|F_j| = \psi(\{F_1, \ldots, F_{j-1}\}) \qquad (j = 2,3, \ldots). \tag{1.109}$$

*) We recall that for any finite set $A$, we denote by $|A|$ the cardinality of $A$.

Define a norm on the linear space $\mathscr{S}$ of all real-valued functions*) on $\mathscr{N}$ with finite support, by

$$\|x\| = \sup_{\{F_j\} \in \mathscr{F}} \left( x, \sum_{j=1}^{\infty} \frac{1}{\sqrt{|F_j|}} \chi_{F_j} \right), \tag{1.110}$$

where $\chi_{F_j}$ denotes the characteristic function of $F_j$ and $\sum_{j=1}^{\infty} \frac{1}{\sqrt{|F_j|}} \chi_{F_j}$ denotes the function $\varphi$ defined by

$$\varphi(n) = \begin{cases} \dfrac{1}{\sqrt{|F_j|}} & \text{if } n \in F_j \ (j = 1,2, \ldots) \\ 0 & \text{if } n \notin \bigcup_{j=1}^{\infty} F_j \end{cases}$$

and where $(x, y)$ denotes the usual scalar product $\sum_{n=1}^{\infty} x(n)y(n)$. Finally, let $E$ be the completion of $\mathscr{S}$ with respect to the norm (1.110). Then the sequence $\{y_n\} \subset E$ defined by**)

$$y_n(i) = \delta_{ni} \qquad (i, n = 1,2, \ldots) \tag{1.111}$$

satisfies (1.1) and $y_n \xrightarrow{w} 0$, but contains no subsequence which is an unconditional basic sequence.

In order to prove the assertions of this example, it will be convenient to summarize some properties of $\{m_k\}$ and $\mathscr{F}$ in the following lemma:

**Lemma 1.4.** *Let* $\{m_k\}$ *and* $\mathscr{F}$ *be as above. Then*

o) *We have****)

$$\sum_{\substack{i,j=0 \\ i \neq j}}^{\infty}{}' \min \left\{ \sqrt{\frac{m_j}{m_i}}, \sqrt{\frac{m_i}{m_j}} \right\} = c < 2. \tag{1.112}$$

i) *For each* $\{F_j\} \in \mathscr{F}$, *we have* (1.108) *and*

$$|F_1| = 1, \qquad |F_j| \in \{m_k\} \qquad (j = 2,3, \ldots), \tag{1.113}$$

$$|F_i| \neq |F_j| \qquad (i \neq j). \tag{1.113'}$$

*) In this example, we prefer to use the language of real-valued functions on $\mathscr{N}$, rather than that of sequences of real numbers.

**) In other words, for each $n \in \mathscr{N}$, $y_n = \chi_{\{n\}}$, the characteristic function of the singleton $\{n\}$.

***) Note that in this sum the number $m_0 = 4^{0^2} = 1$ also occurs.

ii) *For each* $\{F_j\} \in \mathscr{F}$ *and each pair of positive integers* $k, n$ *with* $k \leqslant n$, *there exists* $\{G_j\} \in \mathscr{F}$ *such that*

$$F_j = G_j \qquad (j = 1, \ldots, k), \tag{1.114}$$

$$\max F_n < \min G_{k+1}. \tag{1.115}$$

iii) *For each pair* $\{F_j\}$, $\{G_j\} \in \mathscr{F}$ *and each pair of positive integers* $k, n$, *the relation* $|F_{n+1}| = |G_{k+1}|$ *implies*

$$n = k \text{ and } F_j = G_j \qquad (j = 1, \ldots, k). \tag{1.116}$$

iv) *For each positive integer* $n$ *there exists* $\{F_j\} \in \mathscr{F}$ *such that* $F_1 = \{n\}$ (the singleton consisting of the element $n$).

v) *For each infinite subset* $\mathscr{N}'$ *of* $\mathscr{N} = \{1,2,3, \ldots\}$ *there exists* $\{F_j\} \in \mathscr{F}$ *such that*

$$F_j \subset \mathscr{N}' \qquad (j = 1,2, \ldots). \tag{1.117}$$

*Proof.* o) Since $\{m_k\}_{k=0}^{\infty}$ is increasing and since the summands in (1.112) are symmetric with respect to $i$ and $j$, we have

$$\sum_{\substack{i,j=0\\ i \neq j}}{}' \min\left\{\sqrt{\frac{m_j}{m_i}}, \sqrt{\frac{m_i}{m_j}}\right\} = 2\sum_{i=0}^{\infty}\sum_{j=0}^{i-1}\sqrt{\frac{m_j}{m_i}} =$$

$$= 2\sum_{i=1}^{\infty}\sum_{j=0}^{i-1}\sqrt{4^{j^2-i^2}} = 2\sum_{i=1}^{\infty}\sum_{j=0}^{i-1}\frac{1}{2^{i^2-j^2}} <$$

$$< 2\sum_{i=1}^{\infty}\frac{i}{2^{i^2-(i-1)^2}} < 2\sum_{i=1}^{\infty}\frac{2^{i-1}}{2^{2i-1}} = 2\sum_{i=1}^{\infty}\frac{1}{2^i} = 2,$$

i.e., (1.112). One can also show, similarly, that for each $\varepsilon > 0$ there exists a positive integer $n$ such that $m_k = n^{k^2}$ $(k = 0,1,2, \ldots)$ satisfies (1.112) with 2 replaced by $\varepsilon$.

Formulae (1.113), (1.113') in i) are obvious consequences of (1.107), (1.109) and the definition of $\psi$.

Property ii) is obvious from the definition of $\mathscr{F}$. In the particular case when $k = n$, one can even take $G_j = F_j$ (by (1.108)).

Now let $\{F_j\}$, $\{G_j\} \in \mathscr{F}$ and $k, n \in \mathscr{N}$ be such that $|F_{n+1}| = |G_{k+1}|$. Then, by (1.109),

$$\psi(\{F_1, \ldots, F_n\}) = |F_{n+1}| = |G_{k+1}| = \psi(\{G_1, \ldots, G_k\}),$$

whence, since $\psi$ is one-to-one,

$$\{F_1, \ldots, F_n\} = \{G_1, \ldots, G_k\},$$

which yields (1.116). This proves iii).

Property iv) is obvious from the definition of $\mathscr{F}$.

Finally, let $\mathscr{N}' \subset \mathscr{N}$. Define $F_1$ to be any singleton $\{j\}$, where $j \in \mathscr{N}'$. Assuming that $F_1, \ldots, F_k$ have been defined, let $F_{k+1}$ be any finite subset *of* $\mathscr{N}'$, with $|F_{k+1}| = \psi(\{F_1, \ldots, F_k\})$ and $\min F_{k+1} > \max F_k$. The infinite sequence $\{F_j\}$, defined in this way, belongs to $\mathscr{F}$ and satisfies (1.117). Thus, we have v), which completes the proof of lemma 1.4.

Let us make the following observation, which will be used repeatedly in the sequel: For any pair of finite subsets $F$, $G$ of $\mathscr{N}$ we have

$$\left(\frac{1}{\sqrt{|F|}}\chi_F, \frac{1}{\sqrt{|G|}}\chi_G\right) = \frac{1}{\sqrt{|F|}}\frac{1}{\sqrt{|G|}}\sum_{n=1}^{\infty}\chi_F(n)\chi_G(n) =$$

$$= \frac{1}{\sqrt{|F|}}\frac{1}{\sqrt{|G|}}|F \cap G| \leqslant \frac{1}{\sqrt{|F|}}\frac{1}{\sqrt{|G|}} \min\{|F|, |G|\} =$$

$$= \min\left\{\sqrt{\frac{|F|}{|G|}}, \sqrt{\frac{|G|}{|F|}}\right\}. \tag{1.118}$$

Now we shall prove another lemma, giving some estimates for the norms of characteristic functions and their linear combinations, in the space $E$ of example 1.5:

**Lemma 1.5.** a) *For any finite subset $F$ of $\mathscr{N}$ we have*

$$\|\chi_F\| \leqslant (2 + c)\sqrt{|F|}, \tag{1.119}$$

*where c is the number* (1.112).

b) *For any $\{F_j\} \in \mathscr{F}$ and any finite sequence of real numbers $\alpha_1, \ldots, \alpha_n$ we have*

$$\max_{1 \leqslant k \leqslant n}\left|\sum_{j=1}^{k}\alpha_j\right| \leqslant \left\|\sum_{i=1}^{n}\alpha_i \frac{1}{\sqrt{|F_i|}}\chi_{F_i}\right\| \leqslant (1 + 2c)\max_{1 \leqslant k \leqslant n}\left|\sum_{j=1}^{k}\alpha_j\right|, \tag{1.120}$$

*where c is as above.*

*Proof.* a) Let $\{G_j\} \in \mathscr{F}$ and let

$$\{j_1, \ldots, j_r\} = \{j \in \mathscr{N} \mid G_j \cap F \neq \varnothing\}, \tag{1.121}$$

$$s_i = |G_{j_i} \cap F|, \; a_i = |G_{j_i}| \qquad (i = 1, \ldots, r). \tag{1.122}$$

Then $\sum_{i=1}^{r} s_i \leqslant |F|$, $1 \leqslant s_i \leqslant a_i$ $(i = 1, \ldots, r)$ and, by lemma 1.4 i), we have $\{a_i\}_{i=1}^{r} \subset \{m_k\}_{k=0}^{\infty}$, $a_i \neq a_l$ $(i \neq l)$. Also, by the first part of (1.118),

$$\left(\chi_F, \sum_{j=1}^{\infty} \frac{1}{\sqrt{|G_j|}} \chi_{G_j}\right) = \sum_{i=1}^{r} \frac{1}{\sqrt{a_i}} s_i. \tag{1.123}$$

Let

$$A = \{1 \leqslant i \leqslant r \mid a_i \leqslant |F|\}. \tag{1.124}$$

If $A \neq \varnothing$, then, by (1.112), $\sum_{i \in A} \frac{\sqrt{a_i}}{\max_{k \in A} \sqrt{a_k}} \leqslant 1 + c$, whence, by the definition of $A$,

$$\sum_{i \in A} \frac{1}{\sqrt{a_i}} s_i \leqslant \sum_{i \in A} \sqrt{a_i} \leqslant (1 + c) \max_{i \in A} \sqrt{a_i} \leqslant (1 + c) \sqrt{|F|}.$$

If $\{1, \ldots, r\} \setminus A \neq \varnothing$, then for each $i \in \{1, \ldots, r\} \setminus A$ we have $a_i > |F|$, whence

$$\sum_{i \notin A} \frac{1}{\sqrt{a_i}} s_i \leqslant \max_{k \notin A} \frac{1}{\sqrt{a_k}} \sum_{i \notin A} s_i < \frac{1}{\sqrt{|F|}} |F| = \sqrt{|F|}.$$

Consequently,

$$\sum_{i=1}^{r} \frac{1}{\sqrt{a_i}} s_i \leqslant (1 + c) \sqrt{|F|} + \sqrt{|F|} = (2 + c) \sqrt{|F|},$$

which, together with (1.123) and the definition (1.110) of the norm in $E$, yields (1.119).

b) Let $k \leqslant n$ and choose, by lemma 1.4 ii), a $\{G_j\} \in \mathscr{F}$ satisfying (1.114), (1.115). Then, taking also into account (1.108), it follows that the sets $G_1 = F_1, \ldots, G_k = F_k, F_{k+1}, \ldots, F_n, G_{k+1}, G_{k+2}, \ldots$ are pairwise disjoint and hence, by (1.110) and the first part of (1.118), we obtain

$$\left\| \sum_{i=1}^{n} \alpha_i \frac{1}{\sqrt{|F_i|}} \chi_{F_i} \right\| \geqslant \left| \left( \sum_{i=1}^{n} \alpha_i \frac{1}{\sqrt{|F_i|}} \chi_{F_i}, \sum_{j=1}^{\infty} \frac{1}{\sqrt{|G_j|}} \chi_{G_j} \right) \right| = \left| \sum_{j=1}^{k} \alpha_j \right|,$$

which, since $k \leqslant n$ was arbitrary, proves the first inequality in (1.120).

In order to prove the second inequality in (1.120), let $\{G_j\}$ be an arbitrary element of $\mathscr{F}$ and let $s$ be the largest integer with $s \leqslant n$, such that $|F_s| = |G_s|$ (it is also possible that $s = 0$). If $s \geqslant 2$, then, by lemma 1.4 iii), $F_i = G_i$ $(i = 1, \ldots, s-1)$, whence, by (1.108), $F_i \cap G_j = \emptyset$ $(i = s, \ldots, n;\ j = 1, \ldots, s-1)$. Consequently, in any case we can write

$$\left(\sum_{i=1}^{n} \alpha_i \frac{1}{\sqrt{|F_i|}} \chi_{F_i}, \sum_{j=1}^{\infty} \frac{1}{\sqrt{|G_j|}} \chi_{G_j}\right) =$$

$$= \left(\sum_{i=1}^{s-1} \alpha_i \frac{1}{\sqrt{|F_i|}} \chi_{F_i}, \sum_{j=1}^{\infty} \frac{1}{\sqrt{|G_j|}} \chi_{G_j}\right) + \left(\alpha_s \frac{1}{\sqrt{|F_s|}} \chi_{F_s}, \frac{1}{\sqrt{|G_s|}} \chi_{G_s}\right) +$$

$$+ \left(\sum_{i=s}^{n} \alpha_i \frac{1}{\sqrt{|F_i|}} \chi_{F_i}, \sum_{j=s+1}^{\infty} \frac{1}{\sqrt{|G_j|}} \chi_{G_j}\right) = \eta_1 + \eta_2 + \eta_3$$

$\Bigg($where, if $s = 0$, then $\eta_1$, $\eta_2$ are missing and $\sum\limits_{i=0}^{n} = \sum\limits_{i=1}^{n}$ in $\eta_3$ and, if $s=1$, then $\eta_1$ is missing$\Bigg)$. Then, by the first part of (1.118),

$$\eta_1 = \sum_{i=1}^{s-1} \alpha_i, \quad \eta_2 = \alpha_s \tag{1.125}$$

and, by the second part of (1.118),

$$|\eta_3| \leqslant \max_{s \leqslant p \leqslant n} |\alpha_p| \sum_{i=s}^{n} \sum_{j=s+1}^{\infty} \left(\frac{1}{\sqrt{|F_i|}} \chi_{F_i}, \frac{1}{\sqrt{|G_j|}} \chi_{G_j}\right) \leqslant$$

$$\leqslant \max_{s \leqslant p \leqslant n} |\alpha_p| \sum_{i=s}^{n} \sum_{j=s+1}^{\infty} \min \left\{\sqrt{\frac{|F_i|}{|G_j|}}, \sqrt{\frac{|G_j|}{|F_i|}}\right\}. \tag{1.126}$$

We claim that $\{|F_i|\}_{i=s}^{n} \cup \{|G_j|\}_{=s+1}^{\infty}$ consists of distinct members of $\{m_k\}_{k=0}^{\infty}$ and hence the sum occurring in (1.126) is a partial sum of the series (1.112). Indeed, by lemma 1.4 i), $\{|F_i|\} \cup \{|G_j|\} \subset \{m_k\}_{k=0}^{\infty}$. Furthermore, if for some pair $i, j \in \mathscr{N}$ with $s \leqslant i \leqslant n$, $s+1 \leqslant j < \infty$ and $i, j \geqslant 2$ we had $|F_i| = |G_j|$, then, by lemma 1.4 iii) and by the definition of $s$, it would follow that $i = j \leqslant s$, in contradiction with $s + 1 \leqslant j$. On the other hand, if for $i, j \in \mathscr{N}$ with $s \leqslant i \leqslant n$, $s + 1 \leqslant j < \infty$ and either $i = 1$ or $j = 1$ we had $|F_i| = |G_j|$, then, by lemma 1.4 i) it would follow that both $i = j = 1$, which contradicts $s + 1 \leqslant j$ when $s = 1$ and contradicts the definition of $s$ when $s = 0$.

Thus, $|F_i| \neq |G_j|$ for all $i, j \in \mathcal{N}$ with $s \leqslant i \leqslant n$, $s+1 \leqslant j < \infty$, which proves the claim. Consequently, by lemma 1.4 o),

$$|\eta_3| \leqslant \max_{s \leqslant p \leqslant n} |\alpha_p| c \leqslant 2 \max_{1 \leqslant k \leqslant n} \left| \sum_{j=1}^{k} \alpha_j \right| c,$$

whence, by (1.125),

$$\left( \sum_{i=1}^{n} \alpha_i \frac{1}{\sqrt{|F_i|}} \chi_{F_i}, \sum_{j=1}^{\infty} \frac{1}{\sqrt{|G_j|}} \chi_{G_j} \right) \leqslant (1+2c) \max_{1 \leqslant k \leqslant n} \left| \sum_{j=1}^{k} \alpha_j \right|,$$

which, together with (1.110), proves the second inequality in (1.120). This completes the proof of lemma 1.5*).

Finally, we are now ready to prove the assertions of example 1.5. For each $n \in \mathcal{N}$ there exists, by lemma 1.4 iv), an $\{F_j\} \in \mathcal{F}$ such that $F_1 = \{n\}$, whence, by (1.108) and the definition (1.111) of $\{y_n\}$,

$$\|y_n\| \geqslant \left| \left( y_n, \sum_{j=1}^{\infty} \frac{1}{\sqrt{|F_j|}} \chi_{F_i} \right) \right| = 1 \qquad (n = 1,2, \ldots),$$

so $\{y_n\}$ satisfies (1.1) (actually, by lemma 1.4 i), (1.110) and (1.111) it is obvious that $\|y_n\| \leqslant 1$ for $n = 1,2, \ldots$ and thus we have $\|y_n\| = 1$ for $n = 1,2, \ldots$). Assume now that $\{y_n\}$ does not converge weakly to 0. Then there exist an $\varepsilon > 0$, an $f \in E^*$ with $\|f\| = 1$ and an increasing sequence $\{n_k\} \subset \mathcal{N}$ such that

$$f(y_{n_k}) > \varepsilon \qquad (k = 1,2, \ldots).$$

Hence, since $\chi_{\{n_1, \ldots, n_k\}} = \sum_{j=1}^{k} y_{n_j}$, we obtain

$$\|\chi_{\{n_1, \ldots, n_k\}}\| \geqslant f\left( \sum_{j=1}^{k} y_{n_j} \right) > \varepsilon k \qquad (k = 1,2, \ldots),$$

which, for sufficiently large $k$, contradicts lemma 1.5 a). This proves that $y_n \xrightarrow{w} 0$, so it remains to show that $\{y_n\}$ contains no unconditional basic subsequence.

Let $\{y_n\}_{n \in \mathcal{N}'}$ be an arbitrary subsequence of $\{y_n\}$. By lemma 1.4 v), choose $\{F_j\} \in \mathcal{F}$ satisfying (1.117). Put

$$z_n = \frac{1}{\sqrt{|F_n|}} \chi_{F_n} = \frac{1}{\sqrt{|F_n|}} \sum_{j \in F_n} y_j \qquad (n = 1,2, \ldots) \qquad (1.127)$$

*) Note that we have not yet used lemma 1.4 iv), v).

and let $\{x_n\} = \left\{\sum_{i=1}^{n} y_i\right\}$ be the standard conditional basis of $c_0$, given in Ch. II, § 14, example 14.1 $\left(\text{so } \left\|\sum_{i=1}^{n} \alpha_i x_i\right\| = \max_{1 \leqslant k \leqslant n} \left|\sum_{j=k}^{n} \alpha_j\right| \text{ for any } \alpha_1, \ldots, \alpha_n\right)$. Then, since

$$\frac{1}{2} \max_{1\leqslant k\leqslant n} \left|\sum_{j=1}^{k} \alpha_j\right| \leqslant \max_{1\leqslant k\leqslant n} \left|\sum_{j=k}^{n} \alpha_j\right| \leqslant 2 \max_{1\leqslant k\leqslant n} \left|\sum_{j=1}^{k} \alpha_j\right|,$$

from lemma 1.5 b) and Ch. I, § 8, theorem 8.1 d) it follows that $\{z_n\} \approx \{x_n\}$ and thus $\{z_n\}$ is a conditional basic sequence. But, by (1.117), (1.108) and (1.127), $\{z_n\}$ is a block sequence with respect to $\{y_n\}_{n\in\mathcal{N}'}$ and hence, by Ch. II, § 17, corollary 17.2, $\{y_n\}_{n\in\mathcal{N}'}$ is not an unconditional basic sequence. This completes the proof of the assertions of example 1.5.

In the same paper [270], B. Maurey and H. P. Rosenthal observed that the conjugate space $E^*$ of the space $E$ of example 1.5 is non-separable and they constructed other examples $E$ containing a sequence $\{y_n\}$ with the same properties and such that $E^*$ is separable, namely, a space $E$ which can be embedded isomorphically into*) $C([1, \omega^{\omega^2}])$ and a uniformly convex space $E$ (this latter is the first example of a uniformly convex space which cannot be embedded into a Banach space with an unconditional basis — by § 1, proposition 1.7). On the positive side they have shown [270] that in the spaces $E = C([1, \vartheta])$, where $\vartheta \leqslant \omega^\omega$, every sequence $\{y_n\} \subset E$ satisfying (1.1) and $y_n \xrightarrow{w} 0$, contains a subsequence which is an unconditional basic sequence. Also, B. Maurey and H. P. Rosenthal [270] have raised some related problems, e.g. the following ones: a) Does every sequence $\{y_n\} \subset L^1([0,1])$, satisfying (1.1) and $y_n \xrightarrow{w} 0$, contain an unconditional basic subsequence?**) They observed in [270] that, by a result of H. P. Rosenthal ([340], theorem 8), the answer is affirmative if $[y_n]$ is reflexive. b) Does there exist a reflexive Banach space $E$ such that *every* normalized sequence $\{y_n\} \subset E$ with $y_n \xrightarrow{w} 0$ has a block sequence $\{z_n\}$ equivalent to the standard conditional basis $\{x_n\}$ of $c_0$? They observed in [270] that such a space $E$ would solve problem 1.2 in the negative (since in a reflexive space every normalized basic sequence $\{y_n\}$ satisfies***) $y_n \xrightarrow{w} 0$).

M. I. Kadec and A. Pelczynski showed ([205], theorem 6) that every non-reflexive subspace of $L^1([0,1])$ contains an unconditional basic sequence (even equivalent to the unit vector basis of $l^1$ and spanning a complemented subspace of $L^1([0,1])$), while H. P. Rosenthal

---

*) For the spaces $C([1, \vartheta])$, where $\vartheta$ is an ordinal number, see § 17 (e.g., proposition 17.6).

**) We recall (see Ch. II, § 15, theorem 15.2) that $L^1([0,1])$ cannot be embedded into any Banach space with an unconditional basis.

***) See e.g. Ch. II, § 7, the remark made before theorem 7.2. Also, apply Ch. II, § 17, corollary 17.2 and Ch. I, § 8, theorem 8.1 d), implication 6°⇒1°.

showed ([340], corollary 12) that every subspace of $L^1([0,1])$ contains an unconditional basic sequence, so the answer to problem 1.2 for $E = L^1([0,1])$ is affirmative. We shall present the proofs of these results and further progress on problem 1.2 (e.g., results of T. Figiel, W. B. Johnson and L. Tzafriri [109]), in Vol. III.

The affirmative answer to the problem of existence of conditional basic sequences, mentioned after corollary 1.6, was obtained, independently, by V. I. Gurariĭ [141] and A. Pelczynski (unpublished), both using the theorem of Dvoretzky mentioned in Ch. II, § 8. However, this result has become a simple consequence of Ch. II, § 23, theorem 23.2 (due to [311]), the proof of which does not make use of Dvoretzky's theorem. Problem 1.3 was raised in [367], problem 3.4. The similar problems for symmetric*) and for perfectly homogeneous basic sequences were raised in [367], problem 3.5. By Ch. II, § 24, theorem 24.1, the latter problem (of the existence of perfectly homogeneous basic sequences) is equivalent to the problem whether every Banach space $E$ contains a subspace isomorphic to $c_0$ or $l^p$ for some $p$ with $1 \leqslant p < \infty$. V. D. Milman ([285 a], [287], Ch. III, § 6 and [288], Ch. V, § 1) has given some sufficient conditions, which he has called "hypotheses", in order that the answer to this problem be affirmative. J. Lindenstrauss and L. Tzafriri ([244], [246]) have shown that the answer is affirmative for Orlicz sequence spaces $E$ and for subspaces $E$ of separable Orlicz sequence spaces. In the negative direction, J. Lindenstrauss and L. Tzafriri [245] have constructed the first example of a Banach space (namely, a certain reflexive Orlicz sequence space) which contains no *complemented* subspace isomorphic to $c_0$ or $l^p$, $1 \leqslant p < \infty$. B. S. Tsirelson [397] solved the problem completely, in the negative, and Y. Beniaminy observed (in a Colloquium talk in Oberwolfach, 1973) that Tsirelson's counterexample also gives a negative solution to the problem of the existence of subsymmetric basic sequences. We shall present now this counterexample, as well as some other properties of Tsirelson's space, proved in [397].

*Definition 1.6.* A finite sequence $\{x_1 = \{\xi_k^{(1)}\}, \ldots, x_n = \{\xi_k^{(n)}\}\}$ in $c_0$ is said to be a *quasi-block* (with respect to the unit vector basis $\{e_k\}$ of $c_0$) if for every pair of integers $i, j$ with $1 \leqslant i < j \leqslant n$ and every pair of indices $l, p$ with $\xi_l^{(i)} \neq 0$, $\xi_p^{(j)} \neq 0$ we have $l < p$.

In other words, $\{x_1, \ldots, x_n\}$ is a quasi-block if and only if $x_1, \ldots, x_n$ have disjoint successive supports**) (and hence all $x_i$, except perhaps one, —namely, the last non-zero one —, have a finite number of non-zero coordinates).

*) The problem of the existence of subsymmetric basic sequences was not raised in [367]; this latter notion was introduced explicitly, as a special class of bases in its own right, only in Vol. I (although it occurred implicitly also in [364] and some other papers).

**) Regarded as functions on $\mathcal{N} = \{1, 2, 3, \ldots\}$.

For any $x = \{\xi_k\} \in l^\infty$ and any integer $n \geqslant 1$ we shall use the notations

$$r_n(x) = x - s_n(x) = \{0, \ldots, 0, \xi_{n+1}, \xi_{n+2}, \ldots\}. \tag{1.128}$$

**Lemma 1.6.** *There exists a weakly compact set $Q \subset c_0$ with the following properties:*

1°. $\|x\|_\infty \leqslant 1$ *for all* $x \in Q$ *and* $e_n \in Q$ $(n = 1,2, \ldots)$.

2°. *If* $x \in Q$, $y \in c_0$ *and* $|y| \leqslant |x|$ *(coordinatewise), then* $y \in Q$.

3°. *If* $\{x_1, \ldots, x_n\} \subset Q$ *is a quasi-block, then* $\frac{1}{2} r_n(x_1 + \ldots + x_n) \in Q$.

4°. *For each* $x \in Q$ *there exists* $N$ *such that* $2r_N(x) \in Q$.

*Proof.* Let

$$A_1 = \{\alpha e_n \mid |\alpha| \leqslant 1, \ n = 1,2, \ldots\}, \tag{1.129}$$

$$A_{j+1} = A_j \cup B_j \qquad (j = 1,2, \ldots), \tag{1.130}$$

where $B_j$ is the set of all elements of the form $\frac{1}{2} r_n(x_1 + \ldots + x_n)$, with $\{x_1, \ldots, x_n\} \subset A_j$ a quasi-block. Put*)

$$A = A_1 \cup A_2 \cup A_3 \cup \ldots = A_1 \cup B_1 \cup B_2 \cup \ldots \tag{1.131}$$

and let $Q$ be the closure of $A$ in $l^\infty$ in the topology of coordinatewise convergence. We shall show that $Q$ has the required properties.

Assume, for a moment, that 1° — 4° have been already proved and let $x = \{\xi_k\} \in Q$. Then, by 4°, there exists $N_1$ such that $2r_{N_1}(x) \in Q$. Applying 4° to $r_{N_1}(x)$, there exists $N_2$ such that $2r_{N_1}(2r_{N_2}(x)) \in Q$, whence, since $r_{N_1}r_{N_2} = r_{\max(N_1, N_2)}$, we obtain $4r_{\max(N_1, N_2)}(x) \in Q$. Continuing in this way indefinitely, it follows that for each integer $n \geqslant 1$ there exists $N = N(n)$ such that $2^n r_N(x) \in Q$, whence, by 1°, $\|r_N(x)\|_\infty = \sup_{N+1 \leqslant k < \infty} |\xi_k| \leqslant \frac{1}{2^n}$, so $x \in c_0$. Thus, $Q \subset c_0$. Therefore, by 1°, we have $Q \subset S_{c_0}$, whence, since on $S_{c_0}$ the topology of coordinatewise convergence coincides**) with the weak topology $\sigma(c_0, l^1)$, $Q$ is a $\sigma(c_0, l^1)$-closed subset of $S_{c_0}$. Consequently, $Q$ is a $\sigma(l^\infty, l^1)$-closed subset of $S_{l^\infty}$, so $Q$ is $\sigma(l^\infty, l^1)$-compact, whence also $\sigma(c_0, l^1)$-compact. Thus, it remains to prove that $Q$ satisfies 1°—4°.

---

*) Thus, $A$ is "the smallest" set having properties 1°—3°.

**) See e.g. [70], Ch. II, § 2, theorem 1 and [11], p. 133, theorem 1.

Clearly, $Q$ satisfies 1°—3° (since $A$ satisfies the similar conditions*) and since they are preserved when taking closure in $l^\infty$ in the topology of coordinatewise convergence). Let us prove 4°. Let $x = \{\xi_k\} \in Q$. If $x$ has only a finite number of non-zero coordinates, then $2r_N(x) = 0 \in A$ for sufficiently large $N$, so $x$ satisfies 4°. Assume now that $x$ has an infinity of non-zero coordinates. Take $x^{(s)} = \{\xi_k^{(s)}\} \in A$ such that $x^{(s)} \to x$ coordinatewise and let

$$k_0 = \min_{\xi_k \neq 0} k, \tag{1.132}$$

so $\xi_{k_0}$ is the first non-zero coordinate of $x$. Then, for large $s$, $\xi_{k_0}^{(s)} \neq 0$ and (by our assumption on $x$) $\xi_k^{(s)} \neq 0$ for at least one $k \neq k_0$, whence $x^{(s)} \notin A_1$. Therefore $x^{(s)}$ belongs to one of the sets $B_j$, that is,

$$x^{(s)} = \frac{1}{2} r_{n_s}(x_1^{(s)} + \ldots + x_{n_s}^{(s)}), \tag{1.133}$$

where $\{x_1^{(s)}, \ldots, x_{n_s}^{(s)}\} \subset A_j \subset A$ is a quasi-block. Since $\xi_{k_0}^{(s)} \neq 0$, we have

$$n_s \leqslant k_0 \tag{1.134}$$

(indeed, if $n_s > k_0$, then for any $y \in c_0$ the $k_0$-th coordinate of $\frac{1}{2} r_{n_s}(y)$ is 0, whence by (1.133), $\xi_{k_0}^{(s)} = 0$) and hence an infinity of the numbers $n_s$ must be equal to each other. Therefore we may assume (by passing to this subsequence of $\{n_s\}$) that

$$n_s = n \qquad (s = 1,2, \ldots). \tag{1.135}$$

Furthermore, since $\{x_i^{(s)}\} \subset A \subset S_{c_0}$, we may assume (by passing to subsequences) that each of the $n$ sequences $\{x_i^{(s)}\}_{s=1}^{\infty}$ $(i = 1, \ldots, n)$ occurring in (1.133) is coordinatewise convergent, say to $x_i \in Q$ $(i = 1, \ldots, n)$. Then, for $s \to \infty$ we obtain, by (1.133) and (1.135),

$$x = \frac{1}{2} r_n(x_1 + \ldots + x_n), \tag{1.136}$$

where $\{x_1, \ldots, x_n\} \subset Q$ is a quasi-block (since it is the coordinatewise limit of the sequence of quasi-blocks $\{x_1^{(s)}, \ldots, x_n^{(s)}\}$).

---

*) Actually each $A_j$ satisfies 1° and 2°.

Now let $x_l$ be the last non-zero one among*) $x_1, \ldots, x_n$ (that is, $l = \max\limits_{x_i \neq 0} i$) and let $p$ be the index of the last non-zero coordinate of $x_{l-1}$, hence

$$x_l = r_p(x_1 + \ldots + x_n). \tag{1.137}$$

Then, by (1.136), (1.137), $x_l \in Q$ and 2°,

$$2r_{\max(p,n)}(x) = 2r_p r_n(x) = r_p r_n(x_1 + \ldots + x_n) =$$

$$= r_n r_p(x_1 + \ldots + x_n) = r_n(x_l) \in Q,$$

so $x$ satisfies 4°, which completes the proof of lemma 1.6.

**Lemma 1.7.** *If $Q$ is a weakly compact set in $c_0$ with properties* 1°—4° *of lemma* 1.6, *then***) $S = \overline{\text{co}}\, Q$ *is a weakly compact circled convex set in $c_0$ having properties* 1°—4°.

*Proof.* Since $Q$ is weakly compact, so is***) $S$. Furthermore, since $Q$ satisfies 1° and 2°, so does $M = \text{co}\, Q$, whence also $S = \overline{M}$ (since the norm closure $\overline{M}$ of $M$ coincides****) with its weak closure and hence, by 1°, also with its closure in the topology of coordinatewise convergence, which preserves 2°) and thus, in particular, $S$ is circled (by 2°).

In order to prove 3° it is sufficient to show that $M = \text{co}\, Q$ satisfies 3° (by the above observation and since closure in the topology of coordinatewise convergence preserves 3°).

Let $\{x_1, \ldots, x_n\} \subset M$ be a quasi-block. Then we can write $x_i = \sum\limits_{s=1}^{p} \lambda_i^{(s)} x_i^{(s)}$, where $x_i^{(s)} \in Q$, $\lambda_i^{(s)} \geqslant 0$ $(s = 1, \ldots, p)$, $\sum\limits_{s=1}^{p} \lambda_i^{(s)} = 1$ $(i = 1, \ldots, n)$. Since $Q$ satisfies 2°, we may assume that the support of $x_i^{(s)}$ is contained in the support of $x_i$ $(s = 1, \ldots, p;\, i = 1, \ldots, n)$; indeed, if necessary, we can replace $x_i^{(s)}$ by $z_i^{(s)}$ which coincides with $x_i^{(s)}$ on supp $x_i$ and is 0 outside supp $x_i$ (then, by 2°, $z_i^{(s)} \in Q$ and, clearly, $x_i = \sum\limits_{s=1}^{n} \lambda_i^{(s)} z_i^{(s)}$). Then, since $x_1, \ldots, x_n$ is a quasi-block, it follows that for

---

*) Hence, by the definition of a quasi-block, $x_1, \ldots, x_{l-1}$ have a finite number of non-zero coordinates.

**) We recall that co and $\overline{\text{co}}$ stand for the convex hull and (norm-) closed convex hull respectively.

***) See e.g. [87], p. 434, theorem 4.

****) See e.g. [355], Ch. II, § 9, corollary 2 of theorem 9.2.

any $s_1, \ldots, s_n \in [1, p]$, $\{x_1^{(s_1)}, x_2^{(s_2)}, \ldots, x_n^{(s_n)}\}$ is a quasi-block and hence, since $Q$ satisfies 3°, we have

$$\frac{1}{2} r_n(x_1^{(s_1)} + x_2^{(s_2)} + \ldots + x_n^{(s_n)}) \in Q. \tag{1.138}$$

We claim that $x_1 + \ldots + x_n$ is a convex combination of elements of the form $x_1^{(s_1)} + x_2^{(s_2)} + \ldots + x_n^{(s_n)}$, where $s_1, \ldots, s_n \in [1, p]$. Indeed, for $n = 2$ we have

$$x_1 + x_2 = \sum_{s=1}^{p} \lambda_1^{(s)} x_1^{(s)} + \sum_{s=1}^{p} \lambda_2^{(s)} x_2^{(s)} = \sum_{s_1=1}^{p} \sum_{s_2=1}^{p} \lambda_1^{(s_1)} \lambda_2^{(s_2)} (x_1^{(s_1)} + x_2^{(s_2)}),$$

where the coefficients $\alpha_{s_1 s_2} = \lambda_1^{(s_1)} \lambda_2^{(s_2)}$ satisfy $\alpha_{s_1 s_2} \geqslant 0$, $\sum_{s_1=1}^{p} \sum_{s_2=1}^{p} \alpha_{s_1 s_2} = 1$, and the induction step from $n=l$ to $n=l+1$ is similar (with $x_1 + \ldots + x_l$ and $x_{l+1}$ playing the parts of the above $x_1$ and $x_2$ respectively), which proves our claim. Consequently, $\frac{1}{2} r_n(x_1 + \ldots + x_n)$ is a convex combination of elements of the form $\frac{1}{2} r_n(x_1^{(s_1)} + x_2^{(s_2)} + \ldots + x_n^{(s_n)})$ whence, by (1.138), $\frac{1}{2} r_n(x_1 + \ldots + x_n) \in M$. Thus $M$, whence also $S = \overline{M}$, satisfies 3°.

Finally, let us prove that $S$ satisfies 4°. Let $x_0 \in S$. Then there exists*) a (weakly Borel regular) probability measure $\mu$ on $Q$ such that

$$f(x_0) = \int_Q f(x) d\mu(x) \qquad (f \in c_0^*). \tag{1.139}$$

Put

$$\mathscr{D}_n = \{x \in Q \mid 4 r_n(x) \in Q\} \qquad (n = 1,2, \ldots). \tag{1.140}$$

Then $\{\mathscr{D}_n\}_{n=1}^{\infty}$ is an increasing sequence of weakly closed subsets of $Q$, such that $\bigcup_{n=1}^{\infty} \mathscr{D}_n = Q$. Indeed, since $4|r_{n+1}(x)| \leqslant 4|r_n(x)|$ $(x \in c_0)$ and since $Q$ satisfies 2°, we have $\mathscr{D}_n \subset \mathscr{D}_{n+1}$ $(n = 1,2 \ldots)$. Also, it is clear that $\mathscr{D}_n$ is weakly closed (since $r_n$ is weakly continuous and $Q$ is weakly closed). Finally, as was observed at the beginning of the proof of lemma 1.6, for any $x \in Q$ there exists $N$ such that $4 r_N(x) \in Q$, whence $x \in \mathscr{D}_N$.

---

*) See e.g. [316], p. 5, proposition 1.2.

Since $\mu(\mathscr{D}_n) \to 1$ (because $1 = \mu(Q) = \mu\left(\mathscr{D}_n \cup \bigcup_{i=n}^{\infty} (\mathscr{D}_{i+1} \setminus \mathscr{D}_i)\right) = \mu(\mathscr{D}_n) + \sum_{i=n}^{\infty} \mu(\mathscr{D}_{i+1} \setminus \mathscr{D}_i)$ for $n = 1,2, \ldots$), there exists $n_0$ such that

$$1 \geqslant \mu(\mathscr{D}_{n_0}) \geqslant \frac{3}{4}. \tag{1.141}$$

We shall show that $2r_{n_0}(x_0) \in S$, which will complete the proof (since $x_0 \in S$ was arbitrary). Assume, a contrario, that $2r_{n_0}(x_0) \notin S$. Then $x_0 \notin D$, where $D$ is the weakly compact circled convex set defined by

$$D = \{x \in 2S \mid 2r_{n_0}(x) \in S\}, \tag{1.142}$$

and hence there exists*) a functional $f \in c_0^*$ such that

$$|f(x_0)| > 1, \ \sup_{x \in D} |f(x)| \leqslant 1. \tag{1.143}$$

Observe that

$$\sup_{x \in \mathscr{D}_{n_0}} |f(x)| \leqslant \frac{1}{2}, \ \sup_{x \in Q} |f(x)| \leqslant 2. \tag{1.144}$$

Indeed, if $x \in \mathscr{D}_{n_0}$, then $2x \in 2Q \subset 2S$, $2r_{n_0}(2x) = 4r_{n_0}(x) \in Q \subset S$, whence $2x \in D$. On the other hand, if $x \in Q$, then $\frac{1}{2} x \in \frac{1}{2} Q \subset \frac{1}{2} S \subset \subset 2S$ and $2r_{n_0}\left(\frac{1}{2} x\right) = r_{n_0}(x) \in Q \subset S$ (by 2°), whence $\frac{1}{2} x \in D$. Consequently, by the second part of (1.143), we have (1.144).

From (1.143) (first part), (1.139), (1.144) and (1.141) we obtain

$$1 < |f(x_0)| = \left| \int_{\mathscr{D}_{n_0}} f(x) d\mu(x) + \int_{Q \setminus \mathscr{D}_{n_0}} f(x) d\mu(x) \right| \leqslant$$

$$\leqslant \frac{1}{2} \mu(\mathscr{D}_{n_0}) + 2\mu(Q \setminus \mathscr{D}_{n_0}) \leqslant \frac{1}{2} + 2 \cdot \frac{1}{4} = 1,$$

a contradiction which completes the proof of lemma 1.7.

**Lemma 1.8.** *Let $S$ be a weakly compact circled convex set in $c_0$ having properties* 1°—4° *above and let $E$ be the linear span of $S$ in $c_0$,*

*) See e.g. [409], p. 108, theorem 3.

*endowed with the norm**) $\|x\|_E = \inf_{\substack{\lambda>0 \\ x\in\lambda S}} \lambda$. *Then $E$ is a reflexive Banach space, the sequence of unit vectors $\{e_n\}$ is a hyperorthogonal***) *(hence unconditional) basis of $E$ and if $\{x_n\}$ is any normalized block basic sequence with respect to $\{e_n\}$, then*

$$\left\| r_n\left(\sum_{i=1}^{n} \alpha_i x_i\right)\right\|_E \leqslant 2 \max_{1\leqslant i\leqslant n} |\alpha_i| \tag{1.145}$$

*for all scalars $\alpha_1, \ldots, \alpha_n$ $(1 \leqslant n < \infty)$.*

*Proof.* Since $S_E = S \subset S_{c_0}$, we have

$$\|x\|_{c_0} \leqslant \|x\|_E \qquad (x \in E). \tag{1.146}$$

Let $\{z_n\} \subset E$ be a Cauchy sequence in $E$. Then for every $\varepsilon > 0$ there exists a positive integer $N(\varepsilon)$ such that

$$\|z_n - z_m\|_E < \varepsilon \qquad (n, m > N(\varepsilon)). \tag{1.147}$$

Hence, by (1.146), $\|z_n - z_m\|_{c_0} < \varepsilon$ for $n, m > N(\varepsilon)$ and therefore there exists an element $z \in c_0$ such that $\|z_n - z\|_{c_0} \to 0$ as $n \to \infty$. Since by (1.147) $\sup_{1\leqslant n<\infty} \|z_n\|_E = \lambda < \infty$, that is, $z_n \in \lambda S_E$ $(n = 1,2, \ldots)$ and since $S_E$ is norm-closed in $c_0$, we have $z \in \lambda S_E \subset E$. Since $z_n - z_m \to z_n - z$ in $c_0$ as $m \to \infty$ and $z_n - z_m \in \varepsilon S_E$ $(n, m > N(\varepsilon))$ (by (1.147)), and since $\varepsilon S_E$ is norm-closed in $c_0$, we have $z_n - z \in \varepsilon S_E$ $(n > N(\varepsilon))$, whence $\|z_n - z\|_E \to 0$, which proves that $E$ is complete.

Furthermore, as was observed at the beginning of the proof of lemma 1.6, by property 4° (applied now for $S$ instead of $Q$) we have $\|r_n(x)\|_E \to 0$ for each $x \in S$, whence also for each $x \in E$, and thus $\{e_n\}$ is a basis of $E$. By property 2° (for $S$) this basis is hyperorthogonal, and hence unconditional.

Let us show now that $\{e_n\}$ is shrinking. If not, then by Ch. II, § 12, theorem 12.2, there exist a normalized block basic sequence $\{x_n\}$ with respect to $\{e_n\}$, a functional $f \in E^*$ and $\varepsilon > 0$ such that $f(x_n) > \varepsilon$ $(n = 1,2, \ldots)$. Then $\{x_{n+1}, \ldots, x_{2n}\} \subset S$ is a quasi-block $(n = 1,2, \ldots)$ and hence, by 3°,

$$\frac{1}{2}(x_{n+1} + \ldots + x_{2n}) = \frac{1}{2} r_n(x_{n+1} + \ldots + x_{2n}) \in S,$$

*) I.e., $\|\cdot\|_E$ is the Minkowski functional of $S$ and the unit ball $S_E = \{x \in E \mid \|x\|_E \leqslant 1\}$ of $E$ is $S$.

**) See Ch. II, § 20, definition 20.2.

whence

$$2\|f\|_{E^*} \geqslant \|x_{n+1} + \ldots + x_{2n}\|_E \|f\|_{E^*} \geqslant f(x_{n+1} + \ldots + x_{2n}) > n\varepsilon \to \infty,$$

a contradiction which proves that $\{e_n\}$ is shrinking.

To show that $\{e_n\}$ is also boundedly complete, let $\sup_{1 \leqslant n < \infty} \left\| \sum_{i=1}^{n} \alpha_i e_i \right\|_E = \lambda < \infty$, that is, $\sum_{i=1}^{n} \alpha_i e_i \in \lambda S_E$ $(n=1,2,\ldots)$. Then, since $\lambda S_E$ is bounded in $c_0$ and $\sigma(c_0, c_0^*)$-closed, whence also closed under coordinatewise limits, and since $\sum_{i=1}^{n} \alpha_i e_i \to \{\alpha_1, \alpha_2, \ldots\}$ coordinatewise, it follows that $\{\alpha_1, \alpha_2, \ldots\} \in \lambda S_E \subset E$, whence, since $\{e_n\}$ is a basis of $E$, $\{\alpha_1, \alpha_2, \ldots\} = \sum_{i=1}^{\infty} \alpha_i e_i$ converges. This proves that $\{e_n\}$ is boundedly complete. Consequently, since $\{e_n\}$ is also shrinking, from Ch. II, § 23, remark 23.1 it follows that $E$ is reflexive.

Finally, let $\{x_n\}$ be an arbitrary normalized block basic sequence with respect to $\{e_n\}$. Then $\{x_1, \ldots, x_n\} \subset S$ is a quasi-block $(n = 1,2, \ldots)$, whence, by 3°, $\frac{1}{2} r_n(x_1 + \ldots + x_n) \in S$ and therefore, by 2°, for any scalars $\alpha_1, \ldots, \alpha_n$ we obtain

$$\frac{1}{2} r_n \left( \sum_{i=1}^{n} \frac{\alpha_i}{\max\limits_{1 \leqslant j \leqslant n} |\alpha_j|} \cdot x_i \right) \in S.$$

Thus, we have (1.145), which completes the proof of lemma 1.8.

*Definition 1.7.* A Banach space $G$ is said to be *finitely universal*, if there exists a constant $C \geqslant 1$ such that for each finite-dimensional Banach space $B$ there exist a subspace $F$ of $G$ with $\dim F = \dim B$ and an isomorphism $u$ of $B$ onto $F$ satisfying*) $\|u\| \, \|u^{-1}\| \leqslant C$.

*Remark 1.17. A Banach space G is finitely universal if ond only if G contains uniformly**)* $l_n^\infty$. Indeed, obviously every finitely universal space $G$ contains uniformly $l_n^\infty$. Conversely, assume that $G$ contains uniformly $l_n^\infty$ and let $B$ be an arbitrary finite-dimensional Banach space.

---

*) Or, equivalently, the distance coefficient (see § 12, formula (12.41)) satisfies $d(B, F) \leqslant C$.

**) I.e. (see [268]), there exists a constant $C \geqslant 1$ such that for each $n$ there is an $n$-dimensional subspace $F$ of $G$ with $d(l_n^\infty, F) \leqslant C$. From [340], corollary 10, it follows that this property is equivalent to (0.129); by [268], it is also equivalent to $q(G) = \infty$ (see (0.131)).

Then for any $\varepsilon > 0$ there exists an isomorphism*) $v$ of $B$ into a space $l_N^\infty$ of sufficiently high dimension $N = N(\dim B)$, with $\|v\|\,\|v^{-1}\| \leqslant 1 + \varepsilon$. Now, by our assumption, there is an isomorphism $w$ of $l_N^\infty$ into $G$ such that $\|w\|\,\|w^{-1}\| \leqslant C$, where $C$ is an absolute constant. Then, clearly, $u = wv$ is an isomorphism of $B$ into $G$, with $\|u\|\,\|u^{-1}\| \leqslant C(1+\varepsilon)$, which completes the proof of our assertion.

**Theorem 1.12.** *Let $E$ be the reflexive Banach space with an unconditional basis, constructed in lemma* 1.8. *Then*

a) *Every infinite dimensional subspace $G$ of $E$ is finitely universal.*

b) *$E$ contains no subsymmetric (and hence no symmetric) basic sequence. Hence, in particular, $E$ contains no subspace isomorphic to $c_0$ or $l^p$* $(1 \leqslant p < \infty)$.

*Proof.* a) By remark 1.2 b), for any $\varepsilon > 0$, every infinite dimensional subspace $G$ of $E$ contains a basic sequence $\{y_n\}$ which is $\varepsilon$-equivalent to a normalized block basic sequence $\{x_n\} = \left\{\sum_{k=m_{n-1}+1}^{m_n} \beta_k e_k\right\}$ (i.e., such that the mapping $v\colon y_n \to x_n$ yields an isomorphism of $[y_n]$ onto $[x_n]$ with $\max(\|v\|, \|v^{-1}\|) \leqslant \dfrac{1}{1-\varepsilon}$). Then, for any scalars $\alpha_{n+1}, \ldots$ $\ldots, \alpha_{2n}$ we have (coordinatewise)

$$|\alpha_j x_j| = \left|\sum_{k=m_{j-1}+1}^{m_j} \alpha_j \beta_k e_k\right| \leqslant \left|\sum_{i=n+1}^{2n} \sum_{k=m_{i-1}+1}^{m_i} \alpha_i \beta_k e_k\right| =$$
$$= \left|\sum_{i=n+1}^{2n} \alpha_i x_i\right| \qquad (j = n+1, \ldots, 2n),$$

whence, by hyperorthogonality and formula (1.145) of lemma 1.8, we obtain

$$\max_{n+1 \leqslant j \leqslant 2n} |\alpha_j| = \max_{n+1 \leqslant j \leqslant 2n} \|\alpha_j x_j\|_E \leqslant \left\|\sum_{i=n+1}^{2n} \alpha_i x_i\right\|_E =$$
$$= \left\|r_n\left(\sum_{i=n+1}^{2n} \alpha_i x_i\right)\right\|_E \leqslant 2 \max_{n+1 \leqslant j \leqslant 2n} |\alpha_j|. \tag{1.148}$$

Thus, $v_1\colon \{\alpha_{n+1}, \ldots, \alpha_{2n}\} \to \sum_{i=n+1}^{2n} \alpha_i x_i$ is an isomorphism of $l_n^\infty$ onto $F_n = [x_{n+1}, \ldots, x_{2n}]$, with $\|v_1\|\,\|v_1^{-1}\| \leqslant 2$. Also, $v_2\colon x_i \to y_i$ $(i = n+1, \ldots$

---

*) Indeed, using an $\dfrac{\varepsilon}{1+\varepsilon}$-net of $S_B$, take $h_i \in B^*$ with $\|h_i\| = 1$ $(i = 1, \ldots, N$ such that

$$\max_{1 \leqslant j \leqslant N} |h_j(x)| \leqslant \|x\| \leqslant (1+\varepsilon) \max_{1 \leqslant j \leqslant N} |h_j(x)| \qquad (x \in B)$$

and then put $v(x) = \{h_1(x), \ldots, h_N(x)\}$ $(x \in B)$. A similar argument was used in § 12, proof of lemma 12.6.

$\ldots, 2n)$ yields an isomorphism, with $\|v_2\|\,\|v_2^{-1}\| \leqslant \frac{1}{(1-\varepsilon)^2}$. Hence $u = v_2 v_1$ is an isomorphism of $l_n^\infty$ onto $[y_n] \subset G$, with $\|u\|\,\|u^{-1}\| \leqslant \frac{2}{(1-\varepsilon)^2}$, which, by remark 1.17, proves a).

b) Assume, a contrario, that $E$ contains a subsymmetric basic sequence $\{y_n\}$; then, by Ch. II, § 21, proposition 21.4, $0 < \inf_{1 \leqslant n < \infty} \|y_n\|_E \leqslant \sup_{1 \leqslant n < \infty} \|y_n\|_E < \infty$. Therefore, since $E$ is reflexive, we have*) $y_n \xrightarrow{w} 0$ and hence, by proposition 1.1, $\{y_n\}$ contains a basic subsequence $\{y_{k_n}\}$ which is equivalent to a block basic sequence $\{x_n\}$ with respect to $\{e_n\}$. Since $\{y_{k_n}\}$ is subsymmetric, so is $\{x_n\}$, and hence, by Ch. II, § 21, theorem 21.2, there exists a constant $C_1 \geqslant 1$ such that

$$\frac{1}{C_1}\left\|\sum_{i=1}^{n} \alpha_i x_i\right\|_E \leqslant \left\|\sum_{i=n+1}^{2n} \alpha_{i-n} x_i\right\|_E \leqslant C_1 \left\|\sum_{i=1}^{n} \alpha_i x_i\right\|_E \tag{1.149}$$

for all scalars $\alpha_1, \ldots, \alpha_n$. Hence, by the above proof of part a),

$$\max_{1 \leqslant j \leqslant n} |\alpha_j| \leqslant \frac{1}{\inf\limits_{1 \leqslant k \leqslant n} \|x_k\|_E} \max_{1 \leqslant j \leqslant n} \|\alpha_j x_j\|_E \leqslant \frac{1}{\inf\limits_{1 \leqslant k < \infty} \|x_k\|_E} \left\|\sum_{i=1}^{n} \alpha_i x_i\right\|_E \leqslant$$

$$\leqslant \frac{C_1}{\inf\limits_{1 \leqslant k < \infty} \|x_k\|_E} \left\|\sum_{i=n+1}^{2n} \alpha_{i-n} x_i\right\|_E \leqslant \frac{2C_1}{\inf\limits_{1 \leqslant k < \infty} \|x_k\|_E} \max_{n+1 \leqslant j \leqslant 2n} |\alpha_{j-n}| =$$

$$= \frac{2C_1}{\inf\limits_{1 \leqslant k < \infty} \|x_k\|_E} \max_{1 \leqslant j \leqslant n} |\alpha_j|.$$

Consequently**), $\{x_n\}$ is fully equivalent to the unit vector basis of $c_0$, whence the subspace $[x_n]$ of $E$ is isomorphic to $c_0$, in contradiction with the reflexivity of $E$. This completes the proof of theorem 1.12.

*Remark 1.18.* a) The above proof of part a) shows that for all infinite dimensional subspaces $G$ of $E$ the constant $C$ of definition 1.7 (of finitely universal spaces) is the same and can be taken arbitrarily near to 2. Moreover, if we replace in the construction of $Q$ (in property 3°) $\frac{1}{2} r_n(x_1 + \ldots + x_n)$ by $(1-\varepsilon) r_n(x_1 + \ldots + x_n)$, we obtain that *for each* $\varepsilon$ *with* $0 < \varepsilon < 1$ *there exists a reflexive Banach space* $E$, *with an unconditional basis, such that every infinite dimensional subspace* $G$ *of* $E$ *is finitely universal, with constant* $C \leqslant \frac{1}{1-\varepsilon}$.

b) In the above proof of part b) we have only used the first inequality in (1.149) and therefore $E$ contains no unconditional basic sequence

*) E.g., by Ch. II, § 7, the remark made before theorem 7.2.
**) See Ch. I, § 8, theorem 8.1 d), implication 2°⇒1°.

satisfying this first inequality (it is known*) that this family is strictly larger than that of all subsymmetric basic sequences). Let us also note that $\{e_n\}$ *is an example of an unconditional basis which contains no subsymmetric basic subsequence***).

Now we shall prove another property of the Banach space $E$ of theorem 1.12, namely, that it contains no subspace isomorphic to a uniformly convex space. We recall that a Banach space $E$ is said to be *uniformly convex* if for every $\varepsilon$ with $0 < \varepsilon \leqslant 2$ there exists a $\delta = \delta(\varepsilon)$, $0 < \delta < 1$, such that the relations $\|x\| = \|y\| = 1$, $\|x - y\| \geqslant \varepsilon$ imply $\left\|\frac{x+y}{2}\right\| \leqslant \delta$.

**Lemma 1.9.** *If $G$ is a uniformly convex Banach space, then $G$ is not finitely universal.*

*Proof.* Assume, a contrario, that $G$ is finitely universal with constant $C$ and uniformly convex. Then there exists $\delta$, $0 < \delta < 1$, such that the relations $x, y \in G$, $\|x\| = \|y\| = 1$, $\|x - y\| \geqslant \frac{2}{C}$ imply $\left\|\frac{x+y}{2}\right\| \leqslant \delta$. We shall show that for each $n$ and any isomorphism $u$ of $l_n^\infty$ into $G$ we have

$$\|u\| \, \|u^{-1}\| > \min\,(C, \delta^{2-n}), \tag{1.150}$$

which, for $n$ so large that $C \leqslant \delta^{2-n}$, contradicts the assumption that $G$ is finitely universal with constant $C$. For $n = 1$ we have (1.150) since $\|u\| \, \|u^{-1}\| \geqslant 1 > \delta = \delta^{2-n} = \min\,(C, \delta^{2-n})$. Assume now that (1.150) with $n$ replaced by $n - 1$ holds for any isomorphism of $l_{n-1}^\infty$ into $G$, but

$$\|u\| \, \|u^{-1}\| \leqslant \min\,(C, \delta^{2-n}). \tag{1.151}$$

Let $\iota$ be the natural embedding $\{\xi_1, \ldots, \xi_{n-1}\} \to \{\xi_1, \ldots, \xi_{n-1}, 0\}$ of $l_{n-1}^\infty$ into $l_n^\infty$ and let

$$v = u\iota. \tag{1.152}$$

Then for each $x \in l_{n-1}^\infty$ with $\|x\| \leqslant 1$ we have

$$\iota(x) = \{\xi_1, \ldots, \xi_{n-1}, 0\} = \frac{\{\xi_1, \ldots, \xi_{n-1}, 1\} + \{\xi_1, \ldots, \xi_{n-1}, -1\}}{2} = \frac{y+z}{2},$$

*) The question whether this family is strictly larger was raised in Ch. II, § 21, p. 573 and was solved by P. G. Casazza and Bor-Luh Lin [45].

**) This solves a problem raised by H. P. Rosenthal (in a conversation).

where $y,\ z \in l_n^\infty$, $\|y\|$, $\|z\| \leqslant 1$, $\|y-z\| = \|\{0, \ldots, 0,2\}\| = 2$. Hence, by the definition of $\delta$, for each $x \in l_{n-1}^\infty$ with $\|x\| \leqslant 1$ we have

$$\|v(x)\| = \|u\iota(x)\| = \left\| u\left(\frac{y+z}{2}\right)\right\| \leqslant \|u\| \left\|\frac{y+z}{2}\right\| \leqslant \|u\|\,\delta,$$

and thus, taking into account (1.151),

$$\|v\| \leqslant \delta \|u\| \leqslant \frac{\delta \min (C, \delta^{2-n})}{\|u^{-1}\|} = \frac{1}{\|u^{-1}\|} \min (\delta C, \delta^{2-(n-1)}). \quad (1.153)$$

On the other hand, since $v = u\iota$ is an isomorphism of $l_{n-1}^\infty$ into $G$, by our induction hyperthesis we have

$$\min (C, \delta^{2-(n-1)}) < \|v\|\, \|v^{-1}\|. \quad (1.154)$$

Observe now that by (1.152) $\|v^{-1}\| = \|pu^{-1}\| \leqslant \|u^{-1}\|$ (where $p$ is the natural projection $\{\xi_1, \ldots, \xi_n\} \to \{\xi_1, \ldots, \xi_{n-1}\}$ of $l_n^\infty$ onto $l_{n-1}^\infty$), whence, by (1.154) and (1.153),

$$\min (C, \delta^{2-(n-1)}) < \|v\|\, \|v^{-1}\| < \min (\delta C, \delta^{2-(n-1)}),$$

which is impossible (indeed, if $\delta C \geqslant \delta^{2-(n-1)}$, then $C \geqslant \delta^{2-(n-1)}$ and hence $\min (\delta C, \delta^{2-(n-1)}) = \delta^{2-(n-1)} = \min (C, \delta^{2-(n-1)})$, while if $\delta C < \delta^{2-(n-1)}$, then $\min (\delta C, \delta^{2-(n-1)}) = \delta C < \min (C, \delta^{2-(n-1)})$). This completes the proof of lemma 1.9.

From theorem 1.12 a) and lemma 1.9 we obtain

**Corollary 1.14.** *The Banach space $E$ of theorem* 1.12 *contains no subspace isomorphic to a uniformly convex space.*

From corollary 1.14 and the reflexivity of $E$ it follows again that $E$ contains no subspace isomorphic to $c_0$ or $l^p$ $(1 \leqslant p < \infty)$.

B. S. Tsirelson [397] has also made the following observation: One can show that for every infinite dimensional subspace $V$ of $E^*$ (where $E$ is the space of theorem 1.12) and every $n$ there exists an isomorphism $v$ of $l_n^1$ into $V$ with $\|v\|\|v^{-1}\| \leqslant 3$ (hence $V$ contains $l_n^1$ uniformly) and consequently $E^*$, *too, contains no subspace isomorphic to $c_0$ or $l^p$* $(1 \leqslant p < \infty)$. T. Figiel and W. B. Johnson [108] have given another proof of this latter observation, by using the following analytical description of the norm of $E^*$: If $A$, $B$ are finite subsets of $\mathcal{N} = \{1,2,3, \ldots\}$, let us write $A < B$ if $\max A < \min B$. Given finite subsets $A_1, \ldots, A_k$ of $\mathcal{N}$, let us say that $\{A_i\}_{i=1}^k$ is *admissible* if $\{k\} < A_1 < \ldots < A_k$. Define a sequence of norms on the linear space $X$ of all scalar sequences which have only finitely many non-zero terms, as follows: $\|x\|_0 = \|x\|_{c_0}$ and

$$\|x\|_{n+1} = \max \left\{ \|x\|_n, \frac{1}{2} \max_{\{A_i\}_{i=1}^k} \sum_{i=1}^k \|\chi_{A_i} \cdot x\|_n \right\}, \quad (1.155)$$

where max is taken over all admissible $\{A_i\}_{i=1}^k$ and where $\chi_{A_i}$ denotes the characteristic function of $A_i$. Then one can show that $\|x\| = \lim_{n\to\infty} \|x\|_n$ is a norm on $X$ and that *the completion of* $(X, \|\cdot\|)$ *is the conjugate space* $E^*$ *of the space* $E$ *of theorem* 1.12.

J. Lindenstrauss and L. Tzafriri conjectured ([245], problem 6 and the comments to it) that every Banach space *E with a symmetric basis* contains a subspace isomorphic to $c_0$ or $l^p$ for some $p$ with $1 \leqslant p < \infty$. However, T. Figiel and W. B. Johnson [108] have shown that *there exists* 1) *a reflexive Banach space with a symmetric basis which contains no subspace isomorphic to a uniformly convex space* (*and hence no subspace isomorphic to* $c_0$ *or* $l^p$, $1 \leqslant p < \infty$) *and* 2) *a uniformly convex* (*hence reflexive*) *Banach space with a symmetric basis which contains no subspace isomorphic to* $c_0$ *or* $l^p$ $(1 \leqslant p < \infty)$.

It is not known whether the following conjecture is true (see e.g. the survey paper of P. Enflo [100], p. 54): *Every Banach space contains an infinite dimensional subspace that is either reflexive or is isomorphic to* $c_0$ *or* $l^1$. We shall study this problem and related problems in Vol. III, Ch. IV.

Definition 1.3 of a $w^*$-Schauder basic sequence was given by W. B. Johnson and H. P. Rosenthal [186], who used the term "$w^*$-basic sequence"; however, we prefer the term "$w^*$-Schauder basic sequence", in concordance with the terminology of Ch. I, § 14. Remark 1.6 and proposition 1.8 are due, essentially, to W. B. Johnson and H. P. Rosenthal [186]. Proposition 1.9, theorem 1.5 and remarks 1.8, 1.9 too, are due to W. B. Johnson and H. P. Rosenthal [186]. Furthermore, they have also proved in [186] some results on selections of shrinking basic sequences and shrinking or boundedly complete $w^*$-Schauder basic sequences, which we shall give in Vol. III, Ch. IV. W. B. Johnson has noted ([181], the observation made before theorem 1), that the first part of theorem 1.5 (namely, that every $\{f_n\} \subset E^*$ satisfying (1.33) and (1.34) contains a $w^*$-Schauder basic subsequence $\{f_{k_n}\}$) remains also valid for weakly compactly generated Banach spaces $E$; indeed, by § 19, lemma 19.5 and § 14, lemma 14.3, the general case reduces to the separable case.

It is also natural to consider the problem of *uniqueness*, up to equivalence, of certain special classes of normalized unconditional basic sequences. From Ch. II, § 24, theorem 24.1 it follows*) that in the spaces $c_0$ and $l^p$ $(1 \leqslant p < \infty)$ all normalized perfectly homogeneous basic sequences are equivalent. Combining Ch. II, § 24, theorem 24.1 (or 24.3) with Ch. II, § 8, example 8.1 and remark 8.3 (which yield normalized unconditional basic sequences in $l^1$ and $c_0$, not equivalent to the unit vector basis) and with Ch. II, § 18, theorem 18.3, it follows that

*) One also uses e.g. [11], p. 194, theorem 1.

there exists no Banach space $E$, except $E \cong l^2$, in which all normalized unconditional basic sequences are equivalent. Recently*), Z. Altshuler [2] has shown that *a Banach space $E$ with a symmetric basis $\{x_n\}$ is isomorphic to $c_0$ or $l^p$ for some $p$ with $1 \leqslant p < \infty$ if and only if all normalized symmetric basic sequences in $E$ are equivalent and all normalized symmetric basic sequences in $[f_n] \subset E^*$ are equivalent to $\{f_n\}$, where $\{f_n\} \subset E^*$ is the a.s.c.f. to $\{x_n\}$*. Also, Z. Altshuler [3] has constructed *a Banach space $E$ with a symmetric basis, having a unique (up to equivalence) normalized symmetric basic sequence, which contains no subspace isomorphic to $c_0$ or $l^p$ $(1 \leqslant p < \infty)$.*

Block basic sequences with respect to an arbitrary (not necessarily basic) sequence were considered in [75]. Theorem 1.6 and corollaries 1.7—1.9 are due to [75]. The equivalence of problems 1.2 and 1.4 was observed in [75]. Some results on existence of shrinking block basic sequences have been also proved in [75]; we shall give them in Vol. III, Ch. IV.

Problem 1.5 was raised by A. Pelczynski ([302], problem 5). In [75] it has been shown that for a large class of conjugate Banach spaces, containing all separable conjugate spaces, the answer is affirmative. Theorem 1.7 was proved by W. B. Johnson and H. P. Rosenthal [186]. In connection with the remark made after theorem 1.7 let us mention that the problem whether for every Banach space the conjugate space $E^*$ contains a sequence $\{f_n\}$ satisfying (1.33) and (1.34), was raised by E. Thorp and R. Whitley ([390], p. 59), who also observed ([391], § 2, p. 118) that for every Banach space $E$ admitting a separable quotient space, $E^*$ contains such a sequence $\{f_n\}$. As we have mentioned, the problem was completely solved, in the affirmative, by A. Nissenzweig [294] and B. Josephson [192]. H. P. Rosenthal noted ([339], remark 2 to corollary 2.2) that problem 1.5′ is equivalent to the following problem: *Does every Banach space have a quasi-complemented separable subspace?* The observation made after problem 1.5′ is due to W. B. Johnson and H. P. Rosenthal ([186], remark III. 3). For some related results, see also H. E. Lacey [229] and S. A. Saxon and A. Wilansky [354a].

Theorem 1.8 and corollaries 1.10 and 1.11 were given in [75]. In the meantime, the general problem of extension of basic sequences $\{y_n\}$ to bases $\{x_n\}$ (Ch. I, § 4, problem 4.1) was solved in the negative by A. Pelczynski and H. P. Rosenthal, for $E = L^p([0,1])$ $(2 < p < \infty)$ and $E = L^1([0,1])$ [310]. Since in their counterexamples $\{y_n\}$ had some permutation $\{y_{\sigma(n)}\}$ which can be extended to a basis of $E$, they raised the problem whether there exists a basic sequence $\{y_n\}$ in some Banach space $E$ with a basis, such that no permutation $\{y_{\sigma(n)}\}$ of $\{y_n\}$ can be extended to a basis of $E$. In [378] it has been proved that even more is

---

*) Concerning bases, J. Lindenstrauss and L. Tzafriri [244] showed that there exist Banach spaces (namely, certain Orlicz sequence spaces) having a countably infinite number of mutually non-equivalent symmetric bases; this solves Ch. II, § 22, problem 22.1, in the negative.

true, namely, *there exists a subspace G with a basis, of some Banach space E with a basis, such that no basis of G can be extended to a basis of E.* Let us also mention the following related problems, raised in [378]: (1) In which Banach spaces $E$ with a basis does there exist (a) a basic sequence $\{y_n\}$ which cannot be extended to a basis of $E$? (b) a basic sequence $\{y_n\}$ such that no permutation $\{y_{\sigma(n)}\}$ of $\{y_n\}$ can be extended to a basis of $E$? (c) a subspace $G$ with a basis such that no basis of $G$ can be extended to a basis of $E$? (It is even conceivable that every Banach space $E$ with a basis, which is not isomorphic to $l^2$, contains such a basic sequence $\{y_n\}$ or such a subspace $G$). (2) If $u$ is a continuous linear mapping of $l^1$ onto a separable Banach space $F$, does Ker $u$ have a basis? (The answer is not known even for $F = l^2$ or $c_0$). In [378] it was observed that an affirmative answer to this latter question would yield another example of $G$ and $E$ of the above type.

For a basic sequence $\{x_n\}$, the subspace $V'$ defined by (1.69) was considered by W. J. Davis, D. W. Dean and B.-L. Lin [60] (for the particular case of bases see Ch. II, § 4, proposition 4.2 and the Notes and remarks to Ch. II, § 4). Propositions 1.10 and 1.11 for basic sequences are due to the same authors [60], who also observed a particular case of remark 1.10 (namely, that there exists a strictly bounded biorthogonal system $(x_n, f_n)$ in $E = l^2 \times R$, such that $\{x_n\}$ is basic, but $\{f_n\}$ is not basic). Problem 1.6 a) appears in the paper [303] of A. Pelczynski, with the mention that it is due to J. R. Retherford. Definition 1.5 of a bibasic system was given by W. J. Davis, D. W. Dean and B.-L. Lin [60], who used the term "bibasic sequence"; however, since this is actually a pair of sequences, we use the term "bibasic system" (in accordance with "biorthogonal system"). Theorem 1.9 is due to the same authors ([60], theorem 1). Remark 1.11 was made by W. J. Davis and W. B. Johnson ([63], remark 1). Proposition 1.13, examples 1.3 and 1.4 and lemma 1.3 were given by W. J. Davis, D. W. Dean and B.-L. Lin [60]. Corollary 1.12 a) is due to S. Karlin ([213], theorem 8 and the proof of theorem 9; for (i), see also F. J. Murray [291]). The equivalences $1° \Leftrightarrow \ldots \Leftrightarrow 8°$ and the implications $2° \Rightarrow 10°$, $2° \Rightarrow 11°$, $2° \Rightarrow 13°$ and (for shrinking $\{x_n\}$) $2° \Rightarrow 15°$, of theorem 1.10, are due to W. J. Davis, D. W. Dean and B.-L. Lin [60], who also observed that if we have 2°, then the subspace $B$ of $E^*$ defined in remark 1.13, formula (1.103), is closed. The first two statements in remark 1.14 (concerning 9° and 10° of theorem 1.10) and proposition 1.15 a) are also due to the same authors [60]. For recent progress related to the last problem mentioned in remark 1.14 (on 13° of theorem 1.10) see [375]. As we already mentioned above (in connection with corollary 1.12 a)), the equivalences $2° \Leftrightarrow 3°$ and $7° \Leftrightarrow 8°$ of theorem 1.11 were proved by S. Karlin [213]. Finally, the equivalence $8° \Leftrightarrow 9°$ of theorem 1.11 (for $\{x_n\}$ boundedly complete), corollary 1.13 and problem 1.7 are due, in equivalent formulations, to W. J. Davis, D. W. Dean and B.-L. Lin [60].

**§ 2.** Definition 2.1 of deficient basic sequences and of the number $\alpha_{\{x_n\}}$ was given by I. C. Gohberg and A. S. Markus [131], who also proved proposition 2.1 and corollary 2.1. The numbers $\beta_{\{x_n\}}$ and $\varkappa_{\{x_n\}}$ defined by (2.9) and (2.10) were introduced by the same authors, who have then also proved propositions 2.4—2.6, theorem 2.1 (using lemma 2.3) and corollary 2.2. The definition of a $\Phi_+$-operator, used here, is that of I. C. Gohberg and A. S. Markus [131] and it is slightly different from the definition of a $\Phi_+$-operator given previously by I.C. Gohberg and M. G. Kreĭn ([130], p, 89), who required also the additional condition $\dim u(E_1)^{\perp} = \infty$ (which is equivalent to $\beta_u = \dim E_2/u(E_1) = \infty$, since $(E_2/u(E_1))^* \equiv u(E_1)^{\perp}$). The $\Phi_+$-operators used here are semi-Fredholm (and hence Fredholm) operators in the sense used e.g. in the monograph of T. Kato ([214], p. 230); however, the term "Fredholm operator" is also used in a completely different sense, e.g. by A. Grothendieck ([139], Ch. I, p. 80, definition 4). Lemmas 2.4 and 2.5 are consequences of some results of I. C. Gohberg and M. G. Kreĭn [130], but the proofs given here are different. Theorems 2.2 and 2.3 are generalizations, to the framework introduced in [369], of some theorems of I. C. Gohberg and A. S. Markus [131], who considered only Besselian bases with respect to symmetric gauge functions (see the Notes and remarks to Ch. II, § 11).

Theorem 2.4 is due to V. I. Gurariĭ and M. I. Kadec ([147], theorem 3). Remark 2.7. was made by V. I. Gurariĭ and M. A. Meletidi ([149], theorem 1).

Theorem 2.5 and a result similar to corollary 2.3 were proved by V. D. Milman ([284], theorems 4 and 5). In the particular case when $\{x_n\}$, $\{y_n\}$ are basic sequences, lemma 2.7 is due to V. D. Milman ([284], lemma 2) and in the general case it has been given in Vol. I, p. 205. The notions of $\{\gamma_n\}$-stable and deficient $\{\gamma_n\}$-stable sequences were introduced by V. D. Milman [284], who used the terms "symmetrically stable" and "symmetrically deficient stable" sequences respectively. Theorem 2.6 and corollary 2.4 were proved in the same paper ([284], theorem 6 and its corollary).

Propositions 2.7 and 2.8 were given, the first of them without proof, by V. D. Milman ([285], lemma 1). Example 2.1 is due to V. D. Milman ([287], p. 159) and so is corollary 2.6 ([285], theorem 3a and [287], corollary 3.14).

Propositions 2.9 and 2.10 were proved in [369], which also contains remarks 2.9—2.12.

**§ 3.** Theorem 3.2 on the stability of completeness*) was announced, without proof, by H. Brass ([36], theorem 9).

Lemma 3.1 and theorems 3.1, 3.2 were proved by V. I. Gurariĭ and M.A. Meletidi ([149], lemma 1 and theorems 2 and 3). Actually, theorem 3.2 also appears, with a different proof, in the survey paper of V. D. Milman ([287], theorem 4.7), with the mention that the result is

*) Concerning the term "complete sequence", see Ch. I, p. 24, footnote.

due to V. I. Gurariĭ, who communicated it to him in 1967. V. I. Gurariĭ and M. A. Meletidi [150] considered the problem of characterizing the sequences $\{\lambda_i\}$ of theorem 3.1 and proved the following theorem ([150], theorem 2): *Let* $E=\{x\in C([0,1])\mid x(0)=0\}$, *with the norm induced by* $C([0,1])$ *and let* $x_n(t)=t^n$ $(n=1,2,\ldots)$. *A sequence of positive numbers* $\{\lambda_i\}$ *has the property of theorem* 3.1 (*for this particular* $\{x_n\}$) *if and only if for each* $a>1$ *we have* $\underline{\lim}_{n\to\infty}\frac{a^n}{\lambda_n}=0$.

Proposition 3.1 is due to V. D. Milman ([284], lemma 3). Remark 3.1 was made, essentially, by H. Brass ([36], the remark after theorem 10). In the paper [284] V. D. Milman proved the following complement to remark 3.1, on minimal sequences ([284], theorem 9): *Let* $\{x_n\}$ *be a minimal* (*complete or incomplete*) *sequence in a separable Banach space E, let* $\{f_n\}\subset E^*$, $f_i(x_j)=\delta_{ij}$ $(i,j=1,2,\ldots)$ *and let* $\{\gamma_n\}$ *be a sequence of positive numbers such that the necessary condition* (2.52) *of* § 2 *is not satisfied, i.e., such that*

$$b_n=\sup_{|\varepsilon_i|=1}\sup_{n\leqslant m<\infty}\left\|\sum_{i=n}^{m}\varepsilon_i\gamma_i f_i|_{[x_j]}\right\|>1\qquad(n=1,2,\ldots).$$

*Then there exists a complete minimal sequence* $\{y_n\}$ *in E satisfying* $\|x_n-y_n\|\leqslant\gamma_n$ $(n=1,2,\ldots)$, *such that* $\{x_n\}\not\approx\{y_n\}$. *If* $\{x_n\}$ *is a basic sequence, then we can choose* $\{y_n\}$ *with the above properties and non-basic.*

Let us also mention the following consequences of proposition 3.1 and of this result, due to V. D. Milman ([284], theorem 10 and its corollary): *Let* $\{x_n\}$ *be a sequence in a separable Banach space E, with* $\sup_{1\leqslant n<\infty}\|x_n\|<\infty$. *If there exists no complete sequence* $\{y_n\}$ *in E such that* $\lim_{n\to\infty}\|x_n-y_n\|=0$, *then one can omit from* $\{x_n\}$ *a finite number of elements in such a way that the remaining sequence be strictly equivalent to the unit vector basis* $\{e_n\}$ *of* $l^1$. *Consequently, a separable Banach space E contains no subspace isomorphic to* $l^1$ *if and only if for every sequence* $\{x_n\}\subset E$ *with* $\sup_{1\leqslant n<\infty}\|x_n\|<\infty$ *there exists a complete sequence* $\{y_n\}$ *in E such that* $\lim_{n\to\infty}\|x_n-y_n\|=0$. Independently, H. Brass has announced, without proof, the following related result ([36], theorem 10): *A separable Banach space E contains no subspace isomorphic to* $l^1$ *if and only if for every sequence* $\{x_n\}\subset E$ *with* $\sup_{1\leqslant n<\infty}\|x_n\|<\infty$ *and every* $\varepsilon>0$ *there exists a complete sequence* $\{y_n\}$ *in E such that* $\|x_n-y_n\|<\varepsilon$ $(n=1,2,\ldots)$. The sufficiency parts of both results follow from the observation that if $E=l^1$, $x_n=e_{n+1}$ $(n=1,2,\ldots)$ and $0\leqslant\varepsilon<1$, then every sequence $\{y_n\}\subset E=l^1$ with $\|x_n-y_n\|<\varepsilon$ $(n=1,2,\ldots)$ is $PW$-near to $\{x_n\}$ $\left(\text{since }\left\|\sum_{i=1}^{n}\alpha_i(x_i-y_i)\right\|\leqslant\sum_{i=1}^{n}|\alpha_i|\,\|x_i-y_i\|<\varepsilon\sum_{i=1}^{n}|\alpha_i|=\varepsilon\left\|\sum_{i=1}^{n}\alpha_i x_i\right\|\right)$

and hence, by Ch. I, § 9, theorem 9.2, $\{y_n\}$ is incomplete in $E = l^1$. In Vol. III, Ch. IV, we shall see some other characterizations of Banach spaces which contain no subspace isomorphic to $l^1$.

Following V. I. Gurariĭ [145], a sequence $\{x_n\}$ in a Banach space $E$ is said to have the "property of instability of incompleteness" with respect to a (given) sequence of positive numbers $\{\gamma_n\}$, if there exists a complete sequence $\{y_n\} \subset E$ such that $\|x_n - y_n\| \leqslant \gamma_n$ $(n = 1,2, \ldots)$. For Hilbert spaces, V. I. Gurariĭ obtained the following characterization of such sequences of scalars $\{\gamma_n\}$ ([145], theorem 4.4.2): *A normalized sequence $\{x_n\}$ in $E = l^2$ has the property of instability of incompleteness with respect to $\{\gamma_n\}$ (where $\gamma_n > 0$ for $n = 1,2, \ldots$) if and only if* $\sum_{n=1}^{\infty} |\gamma_n|^2 = \infty$.

Theorem 3.3 was proved by A. I. Markuševič ([261], theorem 2).

**§ 4.** The notion of a basis with respect to a class $\mathfrak{M}$ of sequences of indices was introduced by V. I. Gurariĭ [142]. Theorem 4.1 is due to V. I. Gurariĭ and V. I. Macaev ([148], theorem 2); they call the sequences $\{x_{i_n}\} \subset E$ with property (4.3) "non-approaching". The notion of a uniformly minimal sequence (in the sense of theorem 4.1, condition (4.4)) had been considered previously by M. M. Grinblium [135], who used for it the term "strictly minimal". Theorem 4.2 was proved by V. I. Gurariĭ and V. I. Macaev [148].

The classes of sequences $\mathfrak{M}_{\{i_n\}}$ were considered by V. I. Gurariĭ in the paper [142], where theorem 4.3 and corollary 4.1 were announced, with a sketch of a different proof for theorem 4.3. The inductive proof of theorem 4.3, given here, is an improved version of an argument communicated to us in a letter by V. I. Gurariĭ.

**§ 5.** The notion of a pseudo-basis was introduced by B. R. Gelbaum [125]. For pseudo-bases (instead of unconditional pseudo-bases) theorem 5.1 a) was proved by B. R. Gelbaum ([125], § 3, theorem 1), while theorem 5.1 b) was given in [367] ([367], proposition 3.3 b) and theorem 3.8). Pseudo-bases were considered, under different terminology (e.g., as "serially total sequences") by several authors, e.g. Ben-Ami Braun ([37], [38]).

The notion of a semi-basis was considered by V. Klee [217], who used for it the term "quasi-basis". However, following the terminology of [367], we call it semi-basis and reserve the term quasi-basis for a different notion (see § 9).

**§ 6.** The observation that for every complete finitely linearly independent sequence $\{y_n\} \subset E$ there exists a complete minimal sequence $\{x_n\} \subset E$ satisfying (6.1), can be found, in the more general framework of locally convex spaces, in a paper of V. Klee ([217], proposition 2.2), with the mention that it is implicit in a footnote of Mazur and Sternbach [271]; for stronger results see § 8 and the Notes and remarks to it.

The problem, whether every finitely linearly independent normalized sequence in a Banach space $E$ contains a subsequence which is linearly independent in some stronger sense, was raised by A. Dvoretzky (see [102]). A concrete version of example 6.1, in $E = C([0,1])$, was given by P. Erdös and E. G. Straus ([102], § 2). The notion of a $\{\lambda_i\}$-linearly independent sequence was introduced by P. Erdös and E. G. Straus [102] (without using any term for it), who also observed, without proof, proposition 6.1 and that the converse statements are not valid. Theorem 6.1 was given by P. Erdös and E. G. Straus ([102], § 3), but we modified here, in case 2° of the proof, their (generally, impossible) selection of $\{x_{l_k}\}$. Let us also mention the following related observation of S. Mazur (unpublished): *For every finitely linearly independent sequence $\{x_n\}$ in a Banach space $E$ there exists a sequence $\lambda_i > 0$ $(i = 1,2, \ldots)$ such that the relations*

$$|\alpha_i| \leqslant \lambda_i \quad (i = 1,2, \ldots), \quad \sum_{i=1}^{\infty} \alpha_i x_i = 0 \tag{6.95}$$

*imply* $\alpha_i = 0$ $(i = 1,2, \ldots)$; a simple proof of this result can be found in the monograph of C. Bessaga and A. Pelczynski ([26], p. 268).

Theorem 6.2 is due to V. I. Gurariĭ and M. I. Kadec (see [147], proof of theorem 1).

Under the stronger assumption that $\{x_n\}$ is an $M$-basis of $E$, remark 6.2 was made by W. H. Courage and W. J. Davis ([51], proposition).

Strictly bounded and normal biorthogonal systems were considered by S. Banach ([11] , Appendix, Remarks to Ch. VII, § 1). The equivalences 1° $\Leftrightarrow$ 2° $\Leftrightarrow$ 3° of proposition 6.5 are due to M. M. Grinblium [135] (who stated them under the additional—superfluous—assumption that $\sup_{1 \leqslant n < \infty} \|x_n\| < \infty$), while their equivalence to 4° was observed in [367], proposition 3.6. The problem whether in every separable Banach space $E$ there exists a strictly bounded complete minimal sequence $\{x_n\}$ (which is a particular case of Banach's problem [11] on the existence of strictly bounded $M$-bases) was raised in [367] ([367], problem 3.10) and then raised again in [373]. Proposition 6.6 and theorem 6.3, which give an affirmative solution to this problem, were proved by W. J. Davis and W. B. Johnson ([63], lemma 1 and theorem 1 (a)), with the mention ([63], remark 2) that the perturbation technique used in the proof was suggested by the proof of [372], proposition 1. For the particular case of reflexive spaces theorem 6.3, with $2 + \varepsilon$ instead of $1 + \varepsilon$ in (6.48), was also obtained, with a different proof, in [374] ([374], theorem 4). Let us note that if we replace in (6.49) $\varepsilon$ by $\varepsilon_{n_i(k)}$, where $\varepsilon_{n_i(k)} < \varepsilon$, $\varepsilon_{n_i(k)} \to 0$, then we obtain that *for every separable Banach space $E$ and every $\varepsilon > 0$ there exists a complete minimal sequence $\{x_n\} \subset E$ satisfying* (6.48) *and* $\lim_{n\to\infty} \|x_n\| = 1$. An equivalent form of this result $\Big($involving $\operatorname{dist}(x_n, [x_i]_{i\neq n})$, which, by Ch. II, § 2, corollary 2.1, coincides with $\dfrac{1}{\|f_n\|}\Big)$ was also obtained, with a similar proof, by A. N. Plička ([319], theorem 1). On the other hand, W. J. Davis and W. B. Johnson have observed (unpublished remark)*) that the existence of strictly bounded complete minimal sequences in every separable Banach space $E$ (i.e., theorem 6.3 with $\leqslant C$ instead of $\leqslant 1 + \varepsilon$ in (6.48)) can be proved without using the theorem of Dvoretzky and without using § 1, lemma 1.2 (which was needed in proposition 6.6), as follows: Firstly, it is known that *for every separable Banach space $E$ there exists a biorthogonal system* $(z_n, f_n)$ $(\{z_n\} \subset E,\ \{f_n\} \subset E^*)$ *such that* $\sup_{1 \leqslant n < \infty} \|z_n\| < \infty$, $\sup_{1 \leqslant n < \infty} \|f_n\| < \infty$ and that $[z_n] + [f_n]_\perp$ is dense in $E$ (see § 1, remark 1.11). Now, if $E$ contains no subspace isomorphic to $l^1$, then using $(z_n, f_n)$, one obtains, as in the proof of theorem 6.3, a strictly bounded complete minimal sequence $\{x_n\}$ in $E$, with the a.s.f. $\{f_n\}$. On the other hand, if $E$ does contain a subspace isomorphic to $l^1$, then $E$ has a quotient space isomorphic to $c_0$ (indeed, if $\omega$ is**) a continuous linear mapping of $l^1$ *onto* $c_0$ and $u: l^1 \to E$ is an isomorphic embedding, then $v_0 = \omega u^{-1}$ is a continuous linear mapping of $G = u(l^1) \subset E$ onto $c_0$ and hence, by § 15, proof of lemma 15.1, there exists a continuous linear mapping $v$ of $E$ onto $c_0$, extending $v_0$; then the quotient space $E/\operatorname{Ker} v$ is isomorphic to $c_0$). Hence, the technique of the proof***) of § 20, theorem 20.3, yields a strictly bounded complete minimal sequence in $E$, which completes the proof. Let us add the following remark: Using § 1, lemma 1.2 (but not the theorem of Dvoretzky), we obtain proposition 6.6 (i), (ii), (iv) and hence, if $E$ contains no subspace isomorphic to $l^1$, then we obtain theorem 6.3 for $E$, with the "good" bound $1 + \varepsilon$ in (6.48); thus, actually, Dvoretzky's theorem was needed only for the case when $E$ does contain a subspace isomorphic to $l^1$. In order to find a proof (for the bound $1 + \varepsilon$) which does not use Dvoretzky's theorem, it may be useful to mention here the following result of R. C. James ([161], lemma 2.1): If a Banach space $E$ contains a subspace isomorphic to $l^1$, then for every $\delta > 0$, $E$ has a subspace isomorphic to $l^1$, by an isomorphism $u$ with $\|u\| = 1$, $\|u^{-1}\| \leqslant 1 + \delta$.

Theorem 6.4 was proved in [374], theorem 3. Example 6.3 appeared in [374], with the mention that it is due to W. B. Johnson (communicated in a letter). Problem 6.3 was raised in [367], problem 3.10.

Shrinking minimal sequences were introduced in [367] ([367], definition 3.8).

Strongly boundedly complete minimal sequences were considered by A. Wilansky [406], who has stated that they must be basic sequences ([406], theorem 3); the proof sketched in [406] was not correct, but in [367] it has been observed that the result itself is valid ([367], Ch. III, § 5).

---

*) Of course, this happened before the discovery of the proof, by R. I. Ovsepian and A. Pelczynski [297], of § 8, theorem 8.5.

**) Such a mapping exists (see e.g. [140], p. 73, exercise 1 or [223], p. 283, theorem 1).

***) See the Appendix.

Quasi-boundedly complete minimal sequences were introduced in [367], where example 6.4 was also given ([367], definition 3.10 and example 3.2). Although the term "boundedly complete" is used in connection with minimal sequences (and with $M$-basic sequences, basic sequences, etc.), it should be noted that such sequences need not be "complete" minimal sequences, nor conversely.

Example 6.6, which disproves a claim of V. D. Milman ([287], proposition 3.4 b)), was given, in the context of boundedly complete $M$-bases, in [66] ([66], example 1).

Minimal sequences of types $l_+$, $l_+^0$, $P$, $P^*$, were considered by V. Pták [324], without using any term for them; these terms were introduced in [363]. V. Pták [324] has shown that minimal sequences of type $l_+$ exist in every non-reflexive Banach space and in the Hilbert space*) $L^2([0,1])$. Corollary 6.2 (with the proof given in remark 6.5) and the equivalence of theorem 6.5 with theorem 6.3 were proved in [373] ([373], theorems 2 and 1). Remarks 6.6 and 6.7 are due to V. Pták ([324], theorem 3, implication $3° \Rightarrow 1°$ and the necessity parts of theorems 1 and 2).

Maximal biorthogonal systems were considered by J. Dieudonné [81], who also gave, in a more general form (not assuming countability) proposition 6.9 and — with a different proof — theorem 6.6 ([81], propositions 3 and 2). Example 6.7 a) can be found in the monograph of S. Kaczmarz and H. Steinhaus ([194], Ch. VIII, § 1). Example 6.7 b) is due, essentially, to G. Julia [193] and example 6.7 c) to J. Dieudonné [81].

Lemma 6.1 and theorem 6.7 were proved in [372] ([372], proposition 1 and theorem 1). Proposition 6.10 is due to M. I. Kadec ([200], lemma).

In the particular case when $E = C_{2\pi}$ and $\{x_n\}$ is the trigonometric system $\left(\text{hence } \sum_{i=1}^{\infty} f_i(x)x_i \text{ is the Fourier series of } x\right)$, theorem 6.8 is due to D. E. Menšov ([279], theorem 4). In the general case, theorem 6.8 was given by A. I. Markuševič ([261], theorem 3 and [260]).

**§ 7.** The notions of generalized basis, admissible sequence and admissible mapping were introduced by M. G. Arsove and R. E. Edwards ([7], definition 1), who used the terms "coefficient functionals" and "coefficient mapping" for the last two concepts; the adjective "admissible" was preferred in [367], § 8.

The literature on the theory of sequence spaces is vast (see e.g. the monograph of K. Zeller [410] and the paper [123] of D. J. H. Garling and the references therein, for the notions and results which we mentioned in § 7). Apparently, the equivalence of the theory of generalized bases with the theory of $BK$- spaces, observed in remark 7.1, was not

*) Actually, a simpler example of this latter can be found already in the monograph of S. Kaczmarz and H. Steinhaus ([194], Ch. VIII, example 8.1.6 and remark 8.4.1), namely, $x_n = e_1 + e_{n+1}$ $(n = 1, 2, \ldots)$, where $\{e_n\}$ is a complete orthonormal system in $L^2([0,1])$.

exploited in the literature. A more complicated equivalent version of example 7.1 was communicated to us in a letter by D. J. H. Garling.

Lemma 7.1 can be found in the paper [358] of G. L. Seever ([358], lemma 3.1). Lemma 7.2 (and hence also lemma 7.3) is due to J. Dieudonné [80], but the proof given here follows the lines of the proof of a more general result due to G. L. Seever ([358], theorem 3.2). Lemma 7.4 was given by G. F. Bachelis and H. P. Rosenthal ([10], lemma 1), with the mention that it is due to G. L. Seever [358]. In the particular case when $[x_n] = E$, theorem 7.2 b) is due to E. R. Lorch ([251], lemma; see also G. F. Bachelis [9], theorem 3.4, implication $4° \Rightarrow 3°$). As has been already announced (in the Notes and remarks to Ch. II, § 16, problem 16.1), in the general case theorem 7.2 b) was proved in [62], theorem 1. Independently, G. F. Bachelis and H. P. Rosenthal ([10], theorem 1) have proved the stronger theorem 7.2 a) (even replacing the assumption that $E$ is separable by the weaker assumption that $E$ contains no subspace isomorphic to $l^\infty = m$). Example 7.2 is due to G. F. Bachelis and H. P. Rosenthal ([10], remark (IV)). Proposition 7.1 and remark 7.4 were given in [62], lemma 3 and the remark to it. The second proof of theorem 7.2 b) is the one of [62], mentioned above. Also, remark 7.5 was made in [62].

Theorem 7.3 is due to E. R. Lorch ([251], theorem A).

Proposition 7.2 was observed by W. J. Davis ([58], lemma 2 (i)).

Lemma 7.5, corollary 7.1 and theorem 7.5 were proved in [372] and remark 7.6 was also made there. Lemma 7.6 is due to M. Nakamura and S. Kakutani ([293], theorem 5), with a slightly different proof; the proof given here appeared in a paper of R. Whitley ([405], lemma), with the mention that it is due to A. Kruse. Lemma 7.7 is inherent in a paper of A. Pelczynski and V. N. Sudakov ([312], lemma 2); the proof given here is due to R. Whitley [405].

Similar generalized bases were considered by M. G. Arsove and R. E. Edwards ([7], definition 2), who also proved theorem 7.6 ([7], theorem 2), even for complete metric linear spaces; we note that they use a slightly different definition of similarity, for which the converse of theorem 7.6 is also valid (namely, two generalized bases $\{x_n\}$, $\{y_n\}$ of Banach spaces $E$ and $F$, respectively, are called similar [7], if there exist for them admissible sequences $\{f_n\} \subset E^*$ and $\{g_n\} \subset F^*$, respectively, satisfying (7.53)). Similar biorthogonal systems were considered by W. J. Davis [58], who has also given proposition 7.3 ([58], theorem 1).

Theorem 7.7 was proved by W. J. Davis and W. B. Johnson ([63], theorem 1(b), with the mention ([63], remark 2) that the perturbation technique used in the proof was suggested by the proof of [372], proposition 1 and that the theorem itself, with $1 + \varepsilon$ replaced by $2 + \varepsilon$, was also obtained in [372]. Lemma 7.8 and theorem 7.8 were proved in [372], theorem 1, corollary 1 and theorem 2. Remark 7.7 was made by D. van Dulst [85 a].

Definition 7.8 of boundedly complete generalized bases and the equivalent condition 2° of theorem 7.9 were suggested by the similar definitions of boundedly complete $M$-bases, given, independently, by V. D. Milman [286], [287] and W. B. Johnson [174] respectively. They have also proved, in the context of $M$-bases, the relations $1° \Leftrightarrow 3°$ and $2° \Rightarrow 3°$ of theorem 7.9 ([287], remark 1 and theorem 3.1 and [174], theorem II.5) and corollary 7.2 b) ([287], proposition 3.3 and [174], theorem II.3 and the remark before it).

**§ 8.** $M$-bases were considered by S. Banach ([11], Appendix, Remarks to Ch. VII, § 1), who called them*) "complete biorthogonal systems". He noted, without proof, that they exist in every separable Banach space $E$ and raised the problem whether strictly bounded $M$-bases exist in every such space. $M$-bases in Fréchet spaces were considered by A. I. Markuševič [259], [262], who called them "bases in the wide sense"; he has regarded $M$-bases as a natural generalization of the trigonometric system in $C_{2\pi}$ (see Ch. I, § 4, example 4.1) and hence as a natural replacement for bases and has studied biorthogonal expansions with respect to $M$-bases which are not bases. As was already mentioned in the Notes and remarks to Ch. I, § 4 and § 6, A. I. Markuševič [259] has proved the existence, in every separable Banach space, of an $M$-basis satisfying (8.1) and has given some characterizations of $M$-bases. The "biorthogonalization" of an arbitrary pair of finitely linearly independent sequences $\{y_n\} \subset E$, $\{g_n\} \subset E^*$ can be found, for $E = L^2([0,1])$, in the paper of S. Lewin [233] and for $E = L^p([0,1])$, where $1 \leqslant p < \infty$, in the monograph of S. Kaczmarz and H. Steinhaus ([194], Ch. VIII, § 2; the term "biorthogonalization" is used since it extends the classical "orthogonalization" procedure in $L^2([0,1])$, of Gram-Schmidt—see [357]) and it has been extended to Banach spaces and more general spaces $E$, by O. Frink ([116], § 10) and G. W. Mackey [254], [255]. This procedure was rediscovered later, in various equivalent forms, by several authors (e.g. V. F. Gapoškin and M. I. Kadec [121], theorem 4, W. B. Johnson [174], theorem III.1). Proposition 8.1 was given, without proof, by V. D. Milman ([287], lemma 1.6). Formulae (8.17), (8.18) of proposition 8.2 can be found in the paper [289] of D. P. Milman and V. D. Milman ([289], § 2, no. 2), who call them "biorthogonalization formulae". The remaining part of proposition 8.2 says, essentially, that these formulae yield necessarily the same biorthogonal system as those discovered previously by other authors (for example, G. W. Mackey), mentioned above; in particular, this observation is also valid for the $M$-bases of theorem 8.1 a). Theorem 8.1 b), in the equivalent form given in remark 8.2, can be found in the survey paper [287] of V. D. Milman ([287], theorem 1.6), with the mention that it was obtained jointly with Yu. B. Tumarkin. Remark 8.3 was given, without proof, by M. M. Grinblium ([135], theorem 1, 2°). Lemma 8.1 and corollary 8.1 were observed by W. H. Ruckle ([345], theorem 2.1),

*) This term was also conserved by some other authors, for example, in the papers of W. H. Ruckle.

with the mention that the argument of the proof of lemma 8.1 can be frequently encountered (for example, in Kadec-Pelczynski [206], proof of lemma 4). Proposition 8.3 was noted by W. H. Ruckle ([345], statement 7.1 (b)). Let us also mention here the following negative results in more general spaces: There exist separable locally convex spaces $U$ which do not admit any total sequence of continuous linear functionals and hence they have no $M$-basis, for example, $U = m^*$ endowed with the topology $\sigma(m^*, m)$; another such example (namely, $U =$ the product of $\aleph_1$ copies of the real line, with the product topology) was noted by W. B. Johnson [174]. The (non-locally convex) separable complete metric linear space $U = S([0,1])$ admits no non-zero continuous linear functional and hence it has no biorthogonal system whatsoever (see also Ch. I, § 13, example 13.1).

Example 8.1 and proposition 8.5 were given by W. H. Courage and W. J. Davis ([51], example and theorem 1, equivalence (1) $\Leftrightarrow$ (2)), as a solution of [372], problem 1. Proposition 8.6 is inherent in the paper of G. W. Mackey [255]. Proposition 8.7 was given by W. H. Courage and W. J. Davis ([51], theorem 1, equivalence (1)$\Leftrightarrow$(3)), as a solution of [372], problem 2. Corollary 8.2 was obtained, with a somewhat different proof, by G. W. Mackey ([255], theorem). The weaker form of theorem 8.2, in which $F$ does not occur (i.e., the fact that every $M$-basic sequence in a separable Banach space $E$ can be extended to an $M$-basis of $E$) was proved by V. I. Gurariĭ and M. I. Kadec ([147], theorem 1). Theorem 8.2 was stated, without proof, by V. D. Milman ([287], theorem 1.8). Theorem 8.3, remark 8.4 and corollary 8.3 were given in [372], theorem 3, remark 2 and corollary 3. Lemma 8.2 is due to A. Grothendieck ([138], footnote to the proof of theorem 2). Corollary 8.4 was announced, without proof, by V. G. Vinokurov ([402], theorem 1); the proof given here has appeared in [372]. Proposition 8.8 is due to V. I. Gurariĭ and M. I. Kadec ([147], theorem 2). Some applications of proposition 8.8 to quasi-complements and some other results on quasi-complements will be given in Vol. III, Ch. IV.

Proposition 8.9 slightly improves a result of B. R. Gelbaum ([125], § 4, theorem 1 and remark 1). Theorem 8.4 and corollary 8.5 were given in [66], theorem 3 and corollary. Let us observe that corollary 8.5, together with § 1, theorem 1.10, implication 1° $\Rightarrow$ 5°, yields the following result of J. R. Holub and J. R. Retherford [156], mentioned in the Notes and remarks to Ch. II, § 17, proposition 17.3 and example 17.2 (see Vol. I, p. 640): If $\{x_n\}$ is a basis of a Banach space $E$, with the a.s.c.f. $\{f_n\} \subset E^*$, such that for every infinite sequence $\{n_k\}$ of positive integers, $\{f_{n_k}\} \sim \{f_{n_k}|_{[x_{n_k}]}\}$, then $\{x_n\}$ is an unconditional basis of $E$. Example 8.3 was stated by W. J. Davis, D. W. Dean and Bor-Luh Lin ([60], example 5), with the mention that it had been pointed out to them by

A. Pelczynski. Some other results on complemented subspaces will be given in Vol. III, Ch. IV.

For $M$-bases, proposition 8.10 was observed by L. Sternbach ([385], proposition 4.3).

Strong $M$-bases were considered (without using any special term for them) by W. H. Ruckle [345], who observed that every series summable $M$-basis has both this property and the one occurring in proposition 8.11, condition 4° (but did not note their equivalence). In the same paper it has been remarked ([345], p. 524) that a certain $M$-basis in a separable Banach space, constructed by A. Wilansky [407], is not strong, — with the mention that this observation is due to A. K. Snyder. Strong $M$-bases were considered further (without using any special term for them) in [66], § 1, where problem 8.1 was also raised ([66], problem 2). Furthermore, in [66] and later, in a paper of W. H. Ruckle ([347], theorem 1.2), it has been observed that *an $M$-basis $\{x_n\}$, with the a.s.f. $\{f_n\}$, is strong if and only if for each $x \in E$ there exists a triangular matrix of scalars $(\lambda_{nm}(x))$ such that* $x = \lim_{n\to\infty} \sum_{i=1}^{n} \lambda_{ni}(x) f_i(x) x_i$. Indeed, it is obvious that if for each $x \in E$ such a matrix exists, then $\{x_n\}$ is a strong $M$-basis. Conversely, if $\{x_n\}$ is a strong $M$-basis and $x \in E$, let $\{n_k\} = \{j \in \mathscr{N} | f_j(x) = 0\}$ (possibly $\{n_k\} = \emptyset$). Then $x \in [f_{n_k}]_\perp = [x_j]_{j\in\mathscr{N}\setminus\{n_k\}}$, whence $x = \lim_{n\to\infty} \sum_{j=1}^{m_n}{}' \alpha_j^{(n)}(x) x_j$, where each $\sum'$ contains only elements $x_j$ with $j \in \mathscr{N}\setminus\{n_k\}$. Thus, $x = \lim_{n\to\infty} \sum_{j=1}^{m_n} \lambda'_{nj}(x) f_j(x) x_j$, where $\lambda'_{nj}(x) = \dfrac{\alpha_j^{(n)}(x)}{f_j(x)}$ for $j \in \mathscr{N}\setminus\{n_k\}$ and $\lambda'_{nj}(x) = 0$ for $j \in \{n_k\}$*). The matrix $\{\lambda'_{nj}(x)\}$ is row-finite, so it can be easily replaced by a triangular matrix $\{\lambda_{nj}(x)\}$ with the same property (see § 11), which completes the proof. This notion is somewhat more general than that of a strongly series summable $M$-basis, in which the multipliers $\lambda_{ni}$ do not depend on $x$ (see § 11 and the Notes and remarks to it for conditions under which strongly series summable $M$-bases exist). From the above result and a theorem of S. Banach ([11], p. 213, theorem 2) it follows that *every strong $M$-basis is a norming $M$-basis;* the converse is not true, as shown by § 6, example 6.6 (e.g. combined with proposition 8.16) or by proposition 8.12 applied to any norming $M$-basis. In the paper [347] W. H. Ruckle has called the strong $M$-bases "1-series summable" $M$-bases, because of the above observation. Let us also add to the con-

---

*) Thus, in other words, *the $M$-basis $\{x_n\}$ is strong if and only if*

$$x \in [f_j(x) x_j] = [s_n(x)] \qquad (x \in E).$$

ditions 1°—4° of proposition 8.11 the following equivalent condition, which gives an intrinsic characterization of strong $M$-bases:

5°. *For every sequence of positive integers* $\{n_k\}$ *we have*

$$\bigcap_{j\in\{n_k\}} [x_i]_{i\neq j} = [x_j]_{j\in\mathcal{N}\setminus\{n_k\}}\,. \tag{8.57'}$$

Indeed, for the proof of the equivalence 1° $\Leftrightarrow$ 5° it is enough to observe that for any complete minimal sequence we have $\operatorname{Ker} f_j = [x_i]_{i\neq j}$ $(j=1,2,\dots)$, whence $[f_{n_k}]_\perp = \bigcap_{j\in\{n_k\}} [x_i]_{i\neq j}$. W. H. Ruckle has given some other characterizations of strong $M$-bases ([347], theorem 1.2) and some more restrictive notions such as $k$-series summable, finitely series summable, series summable and strongly series summable $M$-bases (the last two of these are studied here in § 8 and § 11), as well as some other similar special classes of $M$-bases, classified "according to positivity". The observation (slightly weaker than proposition 8.12) that every separable Banach space has a non-strong $M$- basis can be found, without proof, in the paper [347] of W. H. Ruckle, with the mention that it was pointed out to him by W. B. Johnson.

As we already mentioned, the problem of the existence of strictly bounded $M$-bases in every separable Banach space was raised by S. Banach [11] and various partial results were obtained since 1973 (see §§ 6,7 and the Notes and remarks to them). The complete solution of this problem, in the affirmative (see theorem 8.5), was obtained by R. I. Ovsepian and A. Pelczynski ([297], theorem 1), "using a trick invented by A. M. Olevskiĭ" ([295], lemmas 3 and 4). Proposition 8.13 is implicitly contained in the paper of R. I. Ovsepian and A. Pelczynski ([297], proof of theorem 1 and remark 1), where proposition 8.14 is also given ([297], lemma 1), with the mention that it is a modification of a lemma of A. M. Olevskiĭ ([295], lemma 3). Proposition 8.15 can be found in the same paper ([297], corollary 1), with the mention that it is due to C. Bessaga; independently, it has been observed also by A. N. Plička ([322], corollary 1), with the same proof. Recently, A. Pelczynski has shown ([307], theorem 1 and its proof) that theorem 8.5 remains valid with the bounds $1+\sqrt{2}+\varepsilon$ in (8.89) replaced by $1+\varepsilon$, where $\varepsilon>0$ is arbitrary. The proof in [307] uses a consequence of the initial form of the theorem of Dvoretzky (which is stronger than the one applied in the proof of § 6, proposition 6.6, namely, in which 2 is replaced by $1+\varepsilon$), and the $2^n\times 2^n$ Walsh orthogonal matrix $w_{ij}^n = \frac{1}{2^{\frac{n}{2}}} w_i\left(\frac{2j-1}{2^{n+1}}\right)$, where $\{w_i\}$ is the Walsh orthonormal system (see Ch. II, § 14). A simpler proof of the existence, in every separable Banach space $E$, of an $M$-basis $\{x_n\}$ such that $\sup\limits_{1\leqslant n<\infty}\|x_n\|<1+\varepsilon$, $\sup\limits_{1\leqslant n<\infty}\|f_n\|<1+\varepsilon$, where $\varepsilon>0$ is arbitrary and $\{f_n\}$ is the a.s.f. to $\{x_n\}$, using only proposition 6.6 (hence only the weakened form of

Dvoretzky's theorem), was given by A. N. Pličko ([322], theorem 1). His construction proceeds as follows: Let $\{m_n\}$ be an increasing sequence of positive integers and let

$$q_0 = 0, \quad q_n = \sum_{i=1}^{n} m_i \qquad (n = 1,2, \ldots), \quad (8.156)$$

$$B_n = \{j \in \mathcal{N} \mid q_{n-1} + 1 \leqslant j \leqslant q_n\} \qquad (n = 1,2, \ldots). \quad (8.157)$$

By § 6, proposition 6.6 and its proof (see formula (6.47)), there exists a biorthogonal system $(y_n, g_n)$ ($\{y_n\} \subset E$, $\{g_n\} \subset E^*$), with $\{y_n\}$ a basic sequence, such that

a) $\|y_n\| = \|g_n\| = g_n(y_n) = 1 \qquad (n = 1,2, \ldots)$;

b) $\widehat{([y_j]_{j=1}^{q_n}; [y_j]_{j=q_n+1}^{\infty})} \geqslant \frac{1}{2} \qquad (n = 1,2, \ldots)$;

c) $\frac{1}{2}\left(\sum_{j \in B_n} |\alpha_j|^2\right)^{\frac{1}{2}} \leqslant \left\|\sum_{j \in B_n} \alpha_j y_j\right\| \leqslant \frac{3}{2}\left(\sum_{j \in B_n} |\alpha_j|^2\right)^{\frac{1}{2}}$ for all scalars $\alpha_{q_{n-1}+1}, \ldots, \alpha_{q_n} \in K$ $(n = 1,2, \ldots)$;

d) $[y_n] + [g_n]_\perp$ is dense in $E$.

Then, by theorem 8.3 (with $F = [g_n]_\perp$) and its proof, there exists a biorthogonal system $(z_n, h_n)$ ($\{z_n\} \subset [g_n]_\perp$, $\{h_n\} \subset [y_n]^\perp$) such that $\{y_n\} \cup \{z_n\}$is an $M$-basis of $E$, with the a.s.f. $\{g_n\} \cup \{h_n\}$, and $\|z_n\| = 1$ $(n = 1, 2, \ldots)$. Choose $\{n_i^{(k)}\} \subset \{m_n\}$ such that for all $k$, $n$, $i$ we have

$$\frac{6}{\sqrt{n_i^{(k)}}} \leqslant \varepsilon; \quad \frac{n_i^{(k)} - 1}{8} - \frac{2^9}{\varepsilon} > \begin{cases} \sqrt{n_i^{(1)}}\,\|h_i\| & \text{if } k = 1 \\ n_i^{(k-1)} & \text{if } k > 1, \end{cases} \quad (8.158)$$

and let $B_i^{(k)}$ be that $B_l$ for which $n_i^{(k)} = m_l$. Then the sequence $\{x_n\} \subset E$ defined by

$$x_n = \begin{cases} y_n & \text{for } n \notin \bigcup_{i,k} B_i^{(k)} \\ y_n + \dfrac{z_i}{\sqrt{n_i^{(1)}}} & \text{for } n \in B_i^{(1)} \\ y_n + \dfrac{\sum_{j \in B_i^{(k-1)}} y_j}{n_i^{(k-1)}} & \text{for } n \in B_i^{(k)} \quad (k > 1), \end{cases} \quad (8.159)$$

has all the required properties [322]. Problem 8.2 b) was raised by A. Pelczynski ([307], remark A)).

Proposition 8.16 gives a sufficient condition for a boundedly complete $M$-basis $\{x_n\}$ in order that every subsequence $\{x_{n_k}\}$ of $\{x_n\}$ be a boundedly complete $M$-basic sequence. In [66] the following problem was considered: What conditions on a biorthogonal system $(x_n, f_n)$ with $[x_n] = E$, $[f_n] = E^*$ (so $\{f_n\}$ is a boundedly complete $M$-basis of $E^*$, by proposition 8.17 b)) will guarantee that a given (or every) basic subsequence $\{f_{n_k}\}$ of $\{f_n\}$ is boundedly complete? In [66] it has been observed that a sufficient condition is that $[f_{n_k}]$ be $\sigma(E^*, E)$-closed or, in particular, that we have $[f_{n_k}]=([x_j]_{j\in\mathcal{N}\setminus\{n_k\}})^\perp$, and therefore the following problem was raised ([66], problem 1): If $E^*$ is separable, does there exist a biorthogonal system $(x_n, f_n)$ with $[x_n] = E$, $[f_n] = E^*$ and such that $[f_{n_k}]$ is $\sigma(E^*, E)$-closed for every $\{n_k\} \subset \mathcal{N}$?

Proposition 8.17 was given, independently, by V. D. Milman ([287], theorem 3.3) and W. B. Johnson ([174], theorems II.4 and II.5), with the mention that it is suggested by the corresponding results about bases (see Ch. II, § 6, corollary 6.1 and the Notes and remarks to it). As was observed in § 8, the equivalences $1° \Leftrightarrow 2° \Leftrightarrow 3°$ of theorem 8.6 are nothing else than the particular case $[x_n] = E$ of § 7, theorem 7.9. The equivalence $1 \Leftrightarrow 5°$ of theorem 8.6 (via $3° \Rightarrow 5° \Rightarrow 2°$) was proved in [66], theorem 2, where the $M$-bases $\{x_n\}$ with property 5° were called "norm-boundedly complete". Proposition 8.18 was announced, without proof, in the same paper [66]. Remark 8.9 was made by L. Sternbach ([385], definitions 4.2 and 4.5).

Proposition 8.19 is contained, implicitly, in the paper [278] of R. J. McGivney and W. H. Ruckle ([278], the remark after corollary 3.3). The series space of an $M$-basis was introduced by W. H. Ruckle [345], who has also given propositions 8.20 and 8.21 ([345], § 6). The $BK$-space $(E, \{x_n\})^d$ defined by (8.129) is also called the "dual associated sequence space" and some authors use instead of $d$ (which stands for "dual") other upper symbols, such as $\delta$ or $f$ (see e.g. [278], [345]). Proposition 8.22 is due to W. H. Ruckle ([345], 7.1 (c) and 6.3). Proposition 8.23 was given by R. J. McGivney and W. H. Ruckle ([278], proposition 3.5). Series summable $M$-bases were introduced by W. H. Ruckle [345], who has also given proposition 8.24, equivalences $1°\Leftrightarrow 2°\Leftrightarrow 3°$ ([345], theorem 6.4) and proposition 8.25 ([345], 5.5). The original proofs of the above mentioned results of R. J. McGivney and W. H. Ruckle [278] and W. H. Ruckle [345] use certain sequence spaces associated to circled convex subsets of $s$ and they are different from the direct proofs given here. Problem 8.3 was raised by W. H. Ruckle ([347], problem 5.1). Let us observe that if the answer to problem 8.1 is affirmative, or, at least, if the space $E$ of § 0, theorem 0.1 (or 0.2) has a strong $M$-basis, then the answer to problem 8.3 is negative (by theorem 8.7).

Proposition 8.26 and remark 8.10 b) were given, with a different proof, by L. Crone, D. J. Fleming and P. Jessup ([53], theorem 4.16);

the proof given here was communicated to us in a letter by W. H. Ruckle. Theorem 8.7 is due to W. H. Ruckle ([345], 6.6).

Let us also mention that some results on tensor products of $M$-bases, generalizing those of Ch. I, § 18, were obtained by W. H. Ruckle [346].

§ 9. Approximative bases consisting of compact operators (instead of finite rank operators) were considered, without using any special term for them, by S. Banach ([11], p. 237), who raised the problem, whether they exist in every separable Banach space $E$. Approximative bases of elements were considered by I. Maddaus [256], who observed that every compact linear operator from a Banach space $F$ into a space $E$ with an approximative basis (called, in [256], "space of type $A$") is the limit, in the norm-topology (of $L(F, E)$), of a sequence of finite-rank linear operators from $F$ into $E$; the latter observation, in the particular case when $E$ has a basis, was also made later in the book of R. Schatten ([356], Ch. III, remark 3.4). Approximative bases of (finite-rank) operators were considered by R. S. Phillips [317], who called them "generalized bases" and observed (implicitly) proposition 9.1 (the necessity part is inherent also in [256]). In the above papers these objects were used to obtain criteria for compactness of sets $A \subset E$. For approximative bases of elements, the term "approximative basis" was suggested in [367], definition 3.19, where it was also observed that every basis is an approximative basis and where problem 9.1 was raised ([367], problem 3.12); problem 9.1 is also mentioned in the more recent survey paper of J. Lindenstrauss [239], where it is conjectured that the answer is negative. The Banach spaces with an approximative basis of operators were called, by M. I. Kadec [203], "spaces with the Banach approximation property".

Let us also mention here (similarly to remark 9.2 and to other results on non-linear concepts, such as § 10, theorem 10.2 and corollary 10.2), the observation of N. Aronszajn and K. T. Smith [6] that every separable Banach space $E$ has a "*non-linear approximative basis*", i.e., a sequence $\{x_n\} \subset E$ or $\{u_n\} \subset L(E, E)$ as in definition 9.1, but with the $h_{ni}$ or, respectively, the $u_n$, only continuous (generally, non-linear). Indeed, it is enough to renorm $E$ with an equivalent strictly convex norm*) and then to take $\{u_n\}$ to be the best approximation operators with respect to any fixed increasing sequence of finite-dimensional subspaces $\{G_n\}$ of $E$. This latter observation was combined in [361] with a suitable choice of the sequence of subspaces $\{G_n\}$, to obtain, for each compact linear operator $u: F \to E$, a sequence of "nearly linear" uniformly continuous finite rank operators from $F$ into $E$, converging uniformly to $u$.

Quasi-bases were introduced by B. R. Gelbaum [125], who has also given theorem 9.1, equivalence $1° \Leftrightarrow 2°$ and, implicitly, theorem 9.1,

---

*) Such a norm exists (see e.g. § 10, lemma 10.1 or § 20, proposition 20.6).

implication 1° ⇒ 3° ([125], § 3, theorem 2 and its proof); the latter implication was also found, independently, by A. Pelczynski ([304], proof of theorem 1), who has been unaware of [125]. The fact*) that $\{e_n\}$ is a basis of $A_1$, was observed, independently, by B. R. Gelbaum ([125], § 3, proof of theorem 3) and C. Bessaga and A. Pelczynski ([23], lemma 1,2°); the latters have also observed that this basis is monotone ([23], lemma 1, 3°). The implications 1° ⇒ 4°, 1° ⇒ 5° and 1° ⇒ 6° of theorem 9.1 are contained, implicitly, in the paper of A. Pelczynski ([304], proof of theorem 1) and so are, essentially, proposition 9.3, theorem 9.2 and corollaries 9.1, 9.2 ([304], proof of theorem 1 and remarks 2,3 and 4); independently, corollary 9.2 was also obtained, with a larger bound for $\nu_{\{x_n\}}$, by W. B. Johnson, H. P. Rosenthal and M. Zippin ([187], corollary 4.12 (a)). Let us observe that for real scalars the inequality (9.36) of theorem 9.2 a) can be sharpend to

$$\nu_{\{x_i\}_{i=1}^{n^2}} \leqslant \frac{n-1}{n} + \frac{1}{\sqrt{n}}. \tag{9.147}$$

Indeed, F. John [162] proved that every $n$-dimensional real Banach space $E$ has a basis $\{y_i\}_{i=1}^n$ such that for all scalars $\alpha_1, \ldots, \alpha_n$ we have

$$\sqrt{\sum_{i=1}^n \alpha_i^2} \leqslant \left\| \sum_{i=1}^n \alpha_i y_i \right\| \leqslant \sqrt{n} \sqrt{\sum_{i=1}^n \alpha_i^2}, \tag{9.148}$$

whence, for any $k$ with $1 \leqslant k \leqslant n$,

$$\left\| \sum_{i=1}^k \alpha_i y_i \right\| \leqslant \sqrt{n} \sqrt{\sum_{i=1}^k \alpha_i^2} \leqslant \sqrt{n} \sqrt{\sum_{i=1}^n \alpha_i^2} \leqslant \sqrt{n} \left\| \sum_{i=1}^n \alpha_i y_i \right\|,$$

and thus $\nu_{\{y_i\}_{i=1}^n} \leqslant \sqrt{n}$; hence, by proposition 9.3, we infer that $E$ has a quasi-basis $\{x_i\}_{i=1}^{n^2}$ satisfying (9.147). The above observation, that every $n$-dimensional real Banach space $E$ has a basis $\{y_i\}_{i=1}^n$ with $\nu_{\{y_i\}_{i=1}^n} \leqslant \sqrt{n}$, was made (with the above proof), by D. Rutowicz ([351], theorem 4), with the mention that it seems likely that even if $\nu_{\{y_i\}_{i=1}^n} \to \infty$, one can obtain $\nu_{\{y_i\}_{i=1}^n}$ very much smaller than $\sqrt{n}$. Problem 9.2 was raised by D. Rutovicz ([351], § 4), who called it "the finite-dimensional counterpart of the basis problem". Actually, D. Rutowicz ([351], § 4) has

---

*) This fact is nothing else than Ch. I, § 8, proposition 8.1 a). Thus, in the Notes and remarks to that result, the reference to [21] should be replaced by [23] and the reference to B. R. Gelbaum [125] should be added.

considered the numbers*)

$$\beta_n = \sup_{\dim B_n = n} \inf_{\{x_j\}_{j=1}^n} \nu_{\{x_j\}_{j=1}^n} \qquad (n = 1,2, \ldots), \tag{9.149}$$

where the inf is taken over all bases $\{x_j\}_{j=1}^n$ of $B_n$ and the sup is taken over all $n$-dimensional Banach spaces $B_n$, and has raised the problem of the behaviour of $\beta_n$ as $n$ increases.

Let us also make the following observation: Corollary 9.1 shows that (with the exception of the situation of theorem 9.5) the existence of a norm 1 projection of a (finite or infinite dimensional) Banach space $B$ with a basis $\{e_n\}$ of norm $\nu_{\{e_n\}} \leqslant 1 + \varepsilon$, onto a finite-dimensional subspace $G$, has no influence on the basis constant $\beta(G) = \frac{1}{\Gamma(G)}$ of the subspace $G$. We do not know whether a similar remark is also valid for infinite dimensional subspaces.

Finite-dimensional expansions of the identity were considered by A. Pelczynski and P. Wojtaszczyk ([313], definition 1.1), who also made the observation following definition 9.5 ([313], proposition 1.1). Proposition 9.4 is contained, implicitly, in the paper of A. Pelczynski ([304], proof of theorem 1).

The bounded approximation property was introduced, essentially, by A. Grothendieck ([139], p. 182), who called it "a purely topological variant of the metric approximation property". The term "bounded approximation property" was used by W. B. Johnson, H. P. Rosenthal and M. Zippin [187] and A. Pelczynski [304] (the latter used it only for separable spaces, in lieu of the term "Banach approximation property" of M. I. Kadec, mentioned above). Some authors (e.g. W. B. Johnson, H. P. Rosenthal and M. Zippin [187], W. B. Johnson [178]) have used for the $\lambda$-approximation property the term "$\lambda$-metric approximation property". Lemma 9.1 and remark 9.4, with the proof given in remark 9.4 for both of them, are due to W. B. Johnson, H. P. Rosenthal and M. Zippin ([187], lemma 2.4 (a)). M. I. Kadec [203] has shown the existence of a separable Banach space $B_1$ such that every Banach space satisfying 1° of theorem 9.3 is isomorphic to a complemented subspace of $B_1$ and, subsequently, A. Pelczynski and P. Wojtaszczyk have shown ([313], proposition 4.1) that this space $B_1$ has a finite-dimensional decomposition and have also given a different construction ([313], theorem 3.2) of a space $B_2$ with the same two properties. Furthermore, they have shown ([313], theorem 3.3) that any two Banach spaces with these

---

*) Obviously, we have $\beta_n = \sup_{\dim B_n = n} \frac{1}{\Gamma(B_n)} = \frac{1}{\Gamma(n)}$ $(n = 1, 2, \ldots)$, where $\Gamma(B_n)$ is the index of the space $B_n$ (see Ch. I, § 7) and where $\Gamma(n)$ are the numbers considered in Ch. II, § 1, formula (1.58); thus, $\inf_{1 \leqslant n < \infty} \beta_n > 1$ (by Ch. II, § 1, theorem 1.3).

properties*) are isomorphic to each other (whence, in particular, $B_1 \cong B_2$) and have given also a direct proof ([313], theorem 1.1), similar to the proof of theorem 9.3, implication $3° \Rightarrow 9°$, of the fact that every Banach space satisfying 3° of theorem 9.3 is isomorphic to a complemented subspace of a Banach space with a finite-dimensional decomposition. The equivalences $1° \Leftrightarrow 3° \Leftrightarrow 4°$ of theorem 9.3 were observed by A. Pelczynski and P. Wojtaszczyk ([313], proposition 1.1), with the mention that they are well known. In the particular case of the metric approximation property and $\lambda = 1$, the equivalence $4° \Leftrightarrow 6°$ of theorem 9.3 was observed, without proof, in the survey paper of J. Lindenstrauss ([239], pp. 169—170). In the particular case when $E$ is a separable conjugate space, the implication $11° \Rightarrow 9°$ of theorem 9.3, even with $B$ being a conjugate space in 9°, was proved by J. Lindenstrauss ([240], corollary 4), using the implication $11° \Rightarrow 4°$, § 0, lemma 0.6 and § 14, theorem 14.3. For a separable conjugate space $E = E_1^*$, the implication $11° \Rightarrow$ $\Rightarrow 4°$ and the footnote used in its proof, were observed by W. B. Johnson, H. P. Rosenthal and M. Zippin ([187], remark 4.11). The equivalences $1° \Leftrightarrow 7° \Leftrightarrow 8° \Leftrightarrow 9° \Leftrightarrow 10°$ of theorem 9.3 were proved by A. Pelczynski ([304], theorem 1 and its proof and remark 6) and, independently, W. B. Johnson, H. P. Rosenthal and M. Zippin ([187], proposition 1.1, corollary 4.12 (b) and remark 4.13) have proved the equivalences $1° \Leftrightarrow 4° \Leftrightarrow 5° \Leftrightarrow 6°$ and $4° \Leftrightarrow 9° \Leftrightarrow 10°$ of theorem 9.3; here we presented the proof of A. Pelczynski of the equivalence $1° \Leftrightarrow 9°$ since it is simpler and gives better constants (see e.g. the above comments to corollary 9.2). Remark 9.5 a) was made, independently, by A. Pelczynski ([304], remark 6) and W. B. Johnson, H. P. Rosenthal and M. Zippin ([187], remark 4.13). In connection with theorem 9.3, condition 2° and remark 9.6, let us also mention the following result of M. I. Kadec ([203], lemma 1): *If $E$ has an approximative basis, then $E$ contains a sequence of subspaces $\{G_n\}$ such that* $\dim G_n = n$, $G_n \subset G_{n+1}$ $(n = 1,2, \ldots)$ *and that there exists a sequence of finite rank endomorphisms* $\{u_n\} \subset L(E,E)$ *satisfying* (9.2) *and* $u_n(E) \subset G_n$ $(n = 1,2, \ldots)$.

The second statement of proposition 9.5 is due, essentially, to A. Grotendieck ([139], Ch. I, p. 182); the footnote to its proof was observed by T. Figiel (verbal communication). Theorem 9.4 and remark 9.7 a), b) were given by T. Figiel and W. B. Johnson ([107], theorem 1, propositions 1, 2 and remark 1); the construction (9.85) and some of the related computations, used in the proof of the last part of the proof of theorem 9.4 and in the corresponding part of remark 9.7 a), are due to [376] (see [376], theorem 1 and its proof). Remark 9.8 a) was made by T. Figiel and W. B. Johnson ([107], remark 2), with the mention that the result given in this remark is due to W. B. Johnson ([180], theorem 4); the proof presented here is different, being a modification of that of

*) Actually, one can replace here the second property by the bounded approximation property.

theorem 9.4. Problem 9.3 a), b) was raised by T. Figiel and W. B. Johnson ([107], remark 2). Recently, J. Diestel has given a partial answer to problem 9.3 b), by showing ([78], theorem 1) that if $E^*$ has the bounded approximation property and the Radon-Nikodym property*), then $E^*$ has the metric approximation property. In view of remark 9.8, in our lectures given at the University of Amsterdam in 1973 we have raised the problem of replacing the bound $2\lambda(1 + 4\lambda)$ of the sharpened version of theorem 9.4 (given in the final part of remark 9.7 a)) by a bound $\varphi(\lambda)$ such that $\varphi(1) = 1$. Some partial results concerning this problem were obtained by W. Oostenbrink [296]. Example 9.2 of a Banach space with the approximation property, which fails to have the bounded approximation property**), is due to T. Figiel and W. B. Johnson ([107], example); remark 9.9 c) is also contained, implicity, in their paper [107]. Let us observe that if we drop the condition of separability of $E_n^*$, then another example as in remark 9.9 c) can be obtained replacing $F^*$ by $C_1$. Indeed, $C_1$ has a basis (by § 13, corollary 13.7) and for each $n = 1,2, \ldots$ there exists on $C_1$ an equivalent norm $|\cdot|_n$ such that $E_n' = (C_1, |\cdot|_n)$ fails the $n$-approximation property (since otherwise, by theorem 9.4, $C_1^*$ would have the bounded approximation property, in contradiction with the Notes and remarks to § 0). Hence, by proposition 9.5, we also obtain another example of a Banach space with the approximation property, which fails the bounded approximation property, namely, $E' = (E_1' \times E_2' \times \ldots)_{l^2}$. We mention that T. Figiel has communicated to us, at the Colloquium on Geometry of Banach spaces held in Oberwolfach in 1973, yet another way of obtaining such examples, as follows: Let us call *the bounded approximation constant* $\Lambda(E)$ of a Banach space $E$ the infimum of all numbers $\lambda$ such that $E$ has the $\lambda$-approximation property (by § 18, theorem 18.1 and its proof, for every Banach space $E$ with the bounded approximation property this inf is attained); this number $\Lambda(E)$ was introduced by A. Grothendieck ([139], Ch. I, p. 182). Now, T. Figiel has observed that for any pair of Banach spaces $E$, $F$ we have***) $\Lambda(E \otimes_\gamma F) = \Lambda(E)\Lambda(F)$ and hence, taking *any Banach space* $E_0$ *with* $1 < \Lambda(E_0) < \infty$ (which exists, by theorem 9.4 *or its version mentioned in remark* 9.8 a)), we can apply proposition 9.5, with

$$E_1 = E_0,\ E_2 = E_0 \otimes_\gamma E_0,\ E_3 = (E_0 \otimes_\gamma E_0) \otimes_\gamma E_0, \ldots \tag{9.150}$$

---

*) I.e., every $E^*$-valued countably additive vector measure (defined on a sigma-algebra), with finite variation, is differentiable (in the sense of Bochner integral) with respect to its variation. For example, it is known that *weakly compactly generated conjugate spaces $E^*$ have the Radon-Nikodym property*. For various characterizations of Banach spaces with the Radon-Nikodym property see e.g. the book of J. Diestel ([77], Ch. VI) and the survey paper of J. Diestel and J. J. Uhl, Jr. [79].

**) The problem whether the approximation property implies the metric approximation property, was raised by A. Grothendieck ([139], Ch. II, p. 135).

***) For the definition of $E \otimes_\gamma F$ see e.g. R. Schatten [356]. The space $E \otimes_\gamma F$ was also used e.g. in the proof of lemma 9.3.

In connection with remark 9.9 d), let us mention that the first example of a Banach space $E$ with $\Gamma(E) < 1$ was given by P. Enflo [98], even before his solution of the basis problem. Actually, P. Enflo has proved in [98] somewhat more, namely, that *there exists a separable infinite dimensional Banach space $E$ with a two-dimensional subspace $G_1$, having the following property: There is a $C > 1$ such that, if $G$ is any two-dimensional subspace of $E$, sufficiently close to $G_1$ and $F$ is any finite-dimensional subspace of $E$ with $F \supset G$, then there is no projection from $E$ onto $F$, of norm $\leqslant C$.* Thus, in particular, by § 12, theorem 12.1 $\gamma$)), there exists $\varepsilon > 0$ such that $E$ has no $\Pi_{1+\varepsilon}$-basis. In [98] it is not stated whether $E$ has a basis (i.e., whether $\Gamma(E) > 0$), but it is mentioned ([98], remark) that, as was shown by J. Lindenstrauss (unpublished), the same construction of $E$ can be modified so that one obtains a uniformly convex Banach space *$E$ isomorphic to $l^2$*, with $0 < \Gamma(E) < 1$ (thus, solving in the affirmative Ch. II, § 1, problem 1.1). In view of these results, the following problems were raised in the initial version of [86]: Does every Banach space $E$ with a basis admit: a) an equivalent norm $|||\cdot|||$ in which it has no monotone basis? b) an equivalent norm $|||\cdot|||$ in which it has no basis $\{x_n\}$ with the associated partial sum operators $s_n$ satisfying $\lim_{n\to\infty} |||s_n||| = 1$? (such bases are called, following V. D. Milman [287], asymptotically monotone); c) an equivalent norm $|||\cdot|||$ such that $0<\Gamma_{|||\cdot|||}(E) < 1$? d) for each $n$, an equivalent norm $|||\cdot|||_n$ such that $0 < \Gamma_{|||\cdot|||_n}(E) < \frac{1}{n}$?

Let us note that in proposition 9.5, theorem 9.4, remarks 9.7, 9.8 and problem 9.3, the space $E$ need not be separable, so the natural framework for these results would be § 18; however, we have given them in § 9, since they are a necessary preparation for example 9.2, in which $E$ (and even $E^*$) is separable.

The unconditional analogue of theorem 9.1, equivalence $1^\circ \Leftrightarrow 6^\circ$, is due to A. Pelczynski and P. Wojtaszczyk ([313], corollary 1.1), who also observed ([313], the remark after corollary 1.1), as a consequence, that the Banach spaces which cannot be embedded into any Banach space with an unconditional basis, cannot be subspaces of any space with an unconditional quasi-basis. The fact that the answer to the unconditional analogue of problem 9.2 is negative and, more generally, that the unconditional analogue of theorem 9.2, even with a constant $C$, is not valid, was shown by Y. Gordon and D. R. Lewis ([132], [133]). The problem of the validity of the unconditional analogue of theorem 9.3, implication $3^\circ \Rightarrow 7^\circ$ (or, equivalently $3^\circ \Rightarrow 9^\circ$), was raised by A. Pelczynski ([304], remark 5). We have seen in the remarks made after definition 9.7 that the answer to this problem is negative. Let us note that those remarks give actually an example of a Banach space with an unconditional finite-dimensional decomposition (and with a basis, since the $E_n$ are known to have uniformly bounded basis constants), which is not isomorphic*) to any complemented subspace of a Banach space with an unconditional basis, so it has no unconditional quasi-basis. The spaces $C_p$, where $1 \leqslant p < \infty$, also have these properties $\left(\text{since } \{G_n\} \text{ is dense in } \bigcup_{n=1}^{\infty} \mathscr{F}_n\right)$, which answers in the negative the question of W. B. Johnson ([180], p. 306, remark 2) as to whether $C_1$ has an unconditional basis.

I. Maddaus [256] has defined a concept which, actually, generalizes unconditional quasi-bases, replacing the sequence $\{x_n\}$ by a function $X: t \to X(t)$ from [0,1] into $E$, with $\|X(t)\| = 1$ $(t \in [0, 1])$, the sequences $\{f_n(x)\}$ by functions $\lambda_x: t \to \lambda_x(t)$, where, for each fixed $t \in [0,1]$, the functionals $x \to \lambda_x(t)$ belong to $E^*$, and the (unconditionally convergent) series expansions $x = \sum_{i=1}^{\infty} f_i(x)x_i$ by representations of the elements $x \in E$ as strong (i.e., norm convergent) Riemann-Stieltjes integrals $x = \int_0^1 X(t)d\lambda_x(t)$. I. Maddaus [256] has shown that *such a space $E$ has necessarily an approximative basis of elements* (hence, in particular, $E$ is separable), which can be obtained by taking a suitable sequence of subdivisions of [0,1] and then the values of $X$ at one point chosen from each of these subdivisions. Also, I. Maddaus [256] has observed, without proof, that *every Banach space with a basis has such an object* $(X, \{\lambda_x\})$. For a related generalization of unconditional bases, see § 17, definition 17.13 and remark 17.7.

The fact that every finite-dimensional Banach space has an infinite monotone quasi-basis, was shown by A. Pelczynski ([304], remark 4), who also raised ([304], remark 4) the problem whether one can find a finite monotone quasi-basis in every such space. Theorem 9.5 and the last statement of corollary 9.3 (which gives a negative answer to the preceding problem) were given by L. E. Dor [82]. Problem 9.4 was raised by W. B. Johnson [180]. Note that an affirmative answer to problem 9.1 or 9.5 b) would imply the same for problem 9.4. Furthermore, let us observe that in order to obtain an affirmative answer to problem 9.4, it would be sufficient, by the proof of theorem 9.1, implication $6° \Rightarrow 1°$, to show the following: If $B$ is a Banach space with a basis $\{x_n\}$ and $E$ is a complemented subspace of $B$, then there exists an equivalent norm $|||\cdot|||$ on $B$, in which a) $\{x_n\}$ is monotone and b) $E$ admits a projection of norm 1

*) However, according to W. B. Johnson and L. Tzafriri ([189], remark 1), recently it has been shown by J. Lindenstrauss that such spaces can be isomorphically embedded (as non-complemented subspaces) into Banach spaces with an unconditional basis; the fact that all known examples of Banach spaces which are not isomorphic to any subspace of a Banach space with an unconditional basis are also not isomorphic to any subspace of a Banach space with an unconditional finite-dimensional decomposition, had been noted previously by A. Pelczynski and P. Wojtaszczyk ([313], the remarks after corollary 1.1).

(by the argument of § 12, the remark made after problem 12.3, even $\{x_n\}$ asymptotically monotone would suffice). It is well known that one can achieve a) and b) separately, using two equivalent norms $|||\cdot|||_1$ and $|||\cdot|||_2$ on $B$ (for a), see the remark made before problem 9.4 and for b) take $|||x|||_2 = \|p(x)\| + \|x - p(x)\|$, where $p$ is any projection of $B$ onto $E$). so the problem might be, how to combine these two norms.

According to W. B. Johnson [180], the notion of a Banach space with a commuting $\lambda$-approximative basis was introduced by H. P. Rosenthal (unpublished), who used the term "Banach space with the commuting $\lambda$-metric approximation property". Proposition 9.6 and remark 9.11 b) are due to W. B. Johnson ([180], theorem 5). As was communicated to us by W. B. Johnson (in 1970), problem 9.5 a) was raised by H. P. Rosenthal, who conjectured that the answer is affirmative. Clearly, an affirmative answer to problem 9.5 b) would imply (by proposition 9.6) an affirmative answer to problem 9.4.

The $\lambda$-duality approximation property was considered by J. Lindenstrauss [236], who has not used any term for it, and by W. B. Johnson [178], who has used for it the term "$\lambda$-duality metric approximation property". Lemma 9.2 is due to W. B. Johnson ([178], lemma 3). Proposition 9.7 is contained, implicitly, in a paper of W. B. Johnson [180] (in the observation of [180], p. 309, according to which from lemma 9.2 it follows*) that if $E^*$ has the bounded approximation property, then $E$ has a commuting approximative basis; the method of proof is similar to that of J. Lindenstrauss [236], p. 203, proof of the theorem). Lemma 9.3 a) was proved by W. B. Johnson ([178], lemma 1) and, with a different method, using the "principle of local reflexivity", by W. B. Johnson, H. P. Rosenthal and M. Zippin ([187], lemma 3.1), who have also proved lemma 9.3 b) ([187], corollary 3.2); we presented here the method of W. B. Johnson [178], using the well-known isometries between spaces of operators and tensor products or their conjugates and then Helly's theorem (actually, as was shown by D. W. Dean [72], the principle of local reflexivity can be also proved with such methods and even replacing the use of tensor products by certain cartesian products). Proposition 9.8 was proved, for reflexive spaces**), by J. Lindenstrauss ([236], lemma 1) and in the general case by W. B. Johnson ([178], lemma 2). Theorem 9.6 is contained, implicitly, in a paper of W. B. Johnson [180] (see the above comments to propositions 9.7 and 9.6).

Let us give now some characterizations of Banach spaces with a commuting approximative basis. To this end, let us prove first the following generalization of a particular case***) of lemma 9.2, implication $1° \Rightarrow 2°$:

---

*) In fact, this follows from propositions 9.7 (which uses lemma 9.2) and 9.8.

**) Under the assumption (equivalent, for reflexive spaces, to that of proposition 9.8) that $E$ has the metric approximation property.

***) Namely, if $E^*$ is separable, then one can show (similarly to the proof of theorem 9.3, equivalence $1° \Leftrightarrow 5°$) that the assumption of lemma 9.4 for $F = E^*$ is equivalent to condition 1° of lemma 9.2.

**Lemma 9.4.** *Let $\{\tau_n\}$ be a $\lambda$-approximative basis of a Banach space $E$, satisfying*

$$f = \lim_{n\to\infty} \tau_n^*(f) \qquad \left(f \in F = \left[\bigcup_{n=1}^{\infty} \tau_n^*(E^*)\right]\right). \tag{9.151}$$

*Then for every pair of finite-dimensional subspaces $G$ of $E$ and $\Gamma$ of $F$ and every $\varepsilon > 0$ there exists a finite rank operator $v = v_{G,\Gamma,\varepsilon}$ on $E$ satisfying* (9.109)–(9.111) *and*

$$v^*(E^*) \subset F. \tag{9.152}$$

*Proof.* We shall proceed similarly to the proof of lemma 9.2, taking care to get also (9.152).

Choose $0 < \delta < 1$ and $0 < \gamma < 1$ as in the proof of lemma 9.2. Since $\Gamma \subset F$ and since the convergences $x = \lim_{n\to\infty} \tau_n(x)$ $(x \in E)$ and (9.151) are uniform on compact sets, there exists $n$ so large that

$$\|\tau_n(y) - y\| < \varepsilon\|y\| \qquad (y \in G), \tag{9.153}$$

$$\|\tau_n^*(h) - h\| < \frac{\delta}{2}\|h\| \qquad (h \in \Gamma). \tag{9.154}$$

Then, by (9.153) and remark 9.4, there exists a finite rank operator $v_1$ on $E$, satisfying (9.112), (9.113) (where $u = \tau_n$) and

$$v_1^*(E^*) = \tau_n^*(E^*). \tag{9.155}$$

Now let $p$ be a projection of norm $\leqslant l = \dim \Gamma$ of $v_1^*(E^*)$ onto $v_1^*(\Gamma)$ and let

$$v^* = (I_{v_1^*(E^*)} - p + (v_1^*|_\Gamma)^{-1}p)v_1^*. \tag{9.156}$$

Then, as was shown in the proof of lemma 9.2, $v^*$ is $\sigma(E^*, E)$-continuous, whence it is indeed the adjoint of some finite rank operator $v$ on $E$, and we have (9.109)–(9.111). Finally, since

$$pv_1^*(E^*) = v_1^*(\Gamma) \subset v_1^*(E^*) = \tau_n^*(E^*),$$

we also have $v^*(E^*) \subset \tau_n^*(E^*) \subset F$, which completes the proof of lemma 9.4.

Now we can give the following generalization of remark 9.12 c):

**Proposition 9.9.** *Under the assumptions of lemma* 9.4, *$E$ has a commuting approximative basis $\{u_n\}$ such that*

$$\left[\bigcup_{n=1}^{\infty} u_n^*(E^*)\right] = \left[\bigcup_{n=1}^{\infty} \tau_n^*(E^*)\right]. \tag{9.157}$$

*Proof.* The proof is obtained by combining the proofs of proposition 9.7 and remark 9.12 c) $\Big($where $\{f_n\}$ is now a dense sequence in $F = \left[\bigcup_{n=1}^{\infty} \tau_n^*(E^*)\right]\Big)$ and using lemma 9.4 instead of lemma 9.2.

Now we are ready to prove the following proposition, which gives some characterizations of Banach spaces with a commuting approximative basis, in terms of formally weaker properties (the equivalence $1° \Leftrightarrow 2°$ of this proposition was given, without proof, by W. B. Johnson [180], p. 309, with the mention that it is due to H. P. Rosenthal):

**Proposition 9.10.** *For a Banach space $E$ the following statements are equivalent:*

*1°. $E$ has a commuting approximative basis $\{u_n\}$.*

*2°. $E$ has an approximative basis $\{v_n\}$ satisfying*

$$v_m v_n = v_n v_m \qquad (m, n = 1,2, \ldots). \tag{9.158}$$

*3°. $E$ has an approximative basis $\{\tau_n\}$ satisfying* (9.151).

*Proof.* The implication $1° \Rightarrow 2°$ is obvious, with $v_n = u_n$ $(n = 1,2, \ldots)$ and the implication $3° \Rightarrow 1°$ follows from proposition 9.9.

Assume, finally, that we have 2° and let $\lambda = \sup_{1 \leqslant n < \infty} \|v_n\|$, hence $1 \leqslant \lambda < \infty$. Then, by (9.158),

$$\begin{aligned} &\|v_n^*(v_m^*(f)) - v_m^*(f)\| = \|v_m^*(v_n^*(f) - f)\| = \\ &= \sup_{\substack{x \in E \\ \|x\| \leqslant 1}} |(v_n^*(f) - f)(v_m(x))| \leqslant \sup_{\substack{v_m(x) \in v_m(E) \\ \|v_m(x)\| \leqslant \lambda}} |(v_n^*(f) - f)(v_m(x))| = \\ &= \lambda \|(v_n^*(f) - f)|_{v_m(E)}\| \qquad (f \in E^*;\, m, n = 1,2, \ldots). \end{aligned} \tag{9.159}$$

But, by proposition 9.2, we have $\lim_{n\to\infty} (v_n^*(f) - f)(x) = 0$ $(f \in E^*$, $x \in E)$, whence, since for each $f \in E^*$ this convergence is uniform on every $S_G = \{y \in G \mid \|y\| \leqslant 1\}$, where $\dim G < \infty$, we have $\lim_{n\to\infty} \|(v_n^*(f) - f)|_G\| = 0$ $(f \in E^*)$ whenever $\dim G < \infty$. Consequently, by (9.159),

$$\lim_{n\to\infty} \|v_n^*(v_m^*(f)) - v_m^*(f)\| = 0 \qquad (f \in E^*, m = 1,2, \ldots),$$

that is,

$$\lim_{n\to\infty} \|v_n^*(f) - f\| = 0 \qquad \left(f \in \bigcup_{m=1}^{\infty} v_m^*(E^*)\right),$$

whence, since $\sup_{1 \leqslant n < \infty} \|v_n\| < \infty$, we obtain (9.151) with $\tau_n = v_n$ $(n = 1,2, \ldots)$. Thus, $2° \Rightarrow 3°$, which completes the proof of proposition 9.10. Moreover, this proof also shows that if we have 3° (or 2°), then $E$ has a commuting approximative basis $\{u_n\}$ satisfying (9.157) $\left(\text{respectively, } \left[\bigcup_{n=1}^{\infty} u_n^*(E^*)\right] = \left[\bigcup_{n=1}^{\infty} v_n^*(E^*)\right]\right)$.

Weak* approximative bases were considered by D. W. Dean [73], who used the term "weak-star $\lambda$-metric approximation property", in the underlying conjugate spaces ([73], p. 727). Theorem 9.7 was proved by D. W. Dean ([73], theorem 1 and the remarks at the end of the paper).

**§ 10.** Operational bases were introduced by M. I. Kadec [196], who called them "generalized summation bases" (this term was also adopted by W. B. Johnson [174]). In a later paper, M. I. Kadec [200] introduced non-linear operational bases (and called them*) so); in order to unify terminology we have preferred to use here the term "operational basis" in the linear case as well (this term was also adopted e.g. by F. S. Vaher [399]). Proposition 10.1 was announced, without proof, by M. I. Kadec ([196], theorem 3). A proof for generalized summation bases was given by V. F. Gapoškin and M. I. Kadec ([121], theorem 2) and it also works, more generally, for operational bases; the proof given here is different. Also, V. F. Gapoškin and M. I. Kadec have observed ([121], theorem 5) that if $E$ is separable and $E^*$ contains a total subspace $V$ with $r(V) = 0$, then $E$ has an $M$-basis which is not a generalized summation basis of $E$ (and not an operational basis of $E$, as shown by the same argument). For generalized summation bases, proposition 10.2 was given in the same paper ([121], remark 5).

Theorem 10.1 is due to W. B. Johnson ([174], theorem IV.1). M. I. Kadec has announced, without proof ([196], theorem 4) that if a reflexive Banach space $E$ has an operational basis, then every $M$-basis of $E$ is an operational basis of $E$; by theorem 10.1, this result is equivalent to corollary 10.1, which was proved by W. B. Johnson ([174], corollary IV.2). Recently, in [380], the Banach spaces with the property of corollary 10.1 have been completely determined; namely, we have

**Theorem 10.6.** *For a separable Banach space E the following statements are equivalent:*

1°. *Every M-basis of E is an operational basis of E.*

2°. *E is quasi-reflexive**) and has an approximative basis.*

The implication 1° ⇒ 2° was observed in § 10, before propositions 10.2 and 10.1 (it makes use of proposition 10.1 and of the result of W. J. Davis and J. Lindenstrauss [65] according to which for every separable non-quasi-reflexive space $E$ the dual $E^*$ contains a separable total subspace $V$ of characteristic zero, combined with § 8, theorem 8.1) and for reflexive spaces the implication 2° ⇒ 1° is nothing else than corollary 10.1. Now we shall prove the implication 2° ⇒ 1° in the general case. To this end, let us first give

**Lemma 10.2.** *Let E be a quasi-reflexive Banach space, G a finite-dimensional subspace of E and V a total closed linear subspace of $E^*$. Then for every continuous linear operator $u: E \to G$ there exists a*

---

*) Actually, the Russian term used by him might be translated as (non-linear) "operatorial basis" or "operator basis", but we have preferred to use "operational basis", in order to avoid possible confusion (some authors, for example W. B. Johnson [177], [179], have used the term "operator basis" for certain nets of operators, which are called here extended $\pi$-bases; see the Notes and remarks to § 18).

**) I.e., dim $E^{**}/\pi(E) < \infty$, where $\pi$ is the canonical embedding of $E$ into $E^{**}$. For example, a non-reflexive quasi-reflexive space is the space $J$ considered in Ch. II, § 4, example 4.1.

*$\sigma(E, V)$-continuous linear operator $v: E \to G$ such that*

$$v|_G = u|_G, \tag{10.91}$$

$$\|v\| \leqslant C\|u\|, \tag{10.92}$$

*where $C$ is a constant which does not depend on* $\dim G$.

*Proof.* Since $E$ is quasi-reflexive, we have $\operatorname{codim}_{E^*} V = \dim E^*/V = \dim V^\perp < \infty$ (since otherwise, by Ch. II, § 4, lemma 4.1, we would have $V^\perp \cap \pi(E) \neq \{0\}$, in contradiction with the assumption that $V$ is total on $E$) and $\operatorname{codim}_{E^{**}}(\pi(E) \oplus V^\perp) \leqslant \operatorname{codim}_{E^{**}}\pi(E) < \infty$. The latter inequality implies that $\operatorname{codim}_{V^*}\varphi(E) < \infty$, where $\varphi$ denotes the canonical mapping of $E$ into $V^*$, defined by $\varphi(x) = \pi(x)|_V$ $(x \in E)$; note that since $\operatorname{codim}_{E^*} V < \infty$, the characteristic $r(V)$ of $V$ is*) $> 0$ and hence $\varphi$ is**) an isomorphism with $\|\varphi^{-1}\| \leqslant \dfrac{1}{r(V)}$. Let $p$ be any continuous linear projection of $V^*$ onto $\varphi(E)$. If $u: E \to G$ is a continuous linear operator, define $\tilde{u}: V^* \to G$ by

$$\tilde{u} = u\varphi^{-1}p. \tag{10.93}$$

Then, by § 9, lemma 9.3 a) (applied to $V, G$ and $\Gamma = \varphi(G) \subset V^*$ in the role of $E$, $F$ and $V \subset E^*$ of that lemma), for $0 < \varepsilon < \|\tilde{u}\|$ there exists a $w^*$-continuous linear operator $\tilde{v}: V^* \to G$ such that

$$\tilde{v}(\varphi(y)) = \tilde{u}(\varphi(y)) = u\varphi^{-1}p(\varphi(y)) = u\varphi^{-1}\varphi(y) = u(y) \quad (y \in G), \tag{10.94}$$

$$\|\tilde{v}\| \leqslant \|\tilde{u}\| + \varepsilon \leqslant 2\|\tilde{u}\| \leqslant 2\|\varphi^{-1}\|\,\|p\|\,\|u\| \leqslant \frac{2\|p\|}{r(V)}\|u\|. \tag{10.95}$$

Now define $v: E \to G$ by

$$v = \tilde{v}\varphi. \tag{10.96}$$

Then, since $\tilde{v}$ is $\sigma(V^*, V)$-continuous, $v$ is $\sigma(E, V)$-continuous. Also, by (10.94) we have (10.91) and by (10.95) and $\|\varphi\| \leqslant 1$ we have (10.92) with $C = \dfrac{2\|p\|}{r(V)}$, which completes the proof of lemma 10.2.

---

*) See e.g. Ch. I, § 12, the last term of the equalities (12.15).
**) See e.g. Ch. I, § 12, proof of theorem 12.2 e).

**Lemma 10.3.** *Let $E$ be a separable quasi-reflexive Banach space with an approximative basis $\{u_n\}$ and let $V$ be a total closed linear subspace of $E^*$. Then $E$ has an approximative basis $\{v_n\}$ such that*

$$\bigcup_{n=1}^{\infty} v_n^*(E^*) \subset V. \tag{10.97}$$

*Proof*. Since the convergence of $\{u_n\}$ to $I_E$ is uniform on compact subsets of $E$, we may assume (by passing to a subsequence of $\{u_n\}$) that

$$\|x - u_n(x)\| < \frac{1}{n} \|x\| \quad \left(x \in \left[\bigcup_{i=1}^{n-1} u_i(E)\right], \ n = 2,3, \ldots\right). \tag{10.98}$$

By lemma 10.2 above, since $u_n(E) \subset G_n = \left[\bigcup_{i=1}^{n} u_i(E)\right]$, for each $n = 1,2, \ldots$ there exists a $\sigma(E, V)$-continuous linear operator $v_n\colon E \to G_n$ such that

$$v_n|_{G_n} = u_n|_{G_n} \qquad (n = 1,2, \ldots), \tag{10.99}$$

$$\|v_n\| \leqslant C\|u_n\| \leqslant C\lambda < \infty \qquad (n = 1,2, \ldots), \tag{10.100}$$

where $C$ and $\lambda = \sup_{1 \leqslant n < \infty} \|u_n\|$ do not depend on $n$. Then, from (10.98) and (10.99) it follows that $x = \lim_{n\to\infty} v_n(x)$ $(x \in \bigcup_{m=1}^{\infty} u_m(E))$, whence, by (10.100) and since $\bigcup_{m=1}^{\infty} u_m(E)$ is dense in $E$, we obtain $x = \lim_{n\to\infty} v_n(x)$ $(x \in E)$, so $\{v_n\}$ is an approximative basis of $E$. Finally, since $v_n$ is $\sigma(E, V)$-continuous, we can write $v_n(x) = \sum_{i=1}^{m_n} h_i^{(n)}(x) y_i^{(n)}$, where $\{y_i^{(n)}\}_{i=1}^{m_n}$ is a basis of $G_n$ and $\{h_i^{(n)}\}_{i=1}^{m_n} \subset V$. But then $v_n^*(f) = \sum_{i=1}^{m_n} f(y_i^{(n)}) h_i^{(n)} \in V$ $(n = 1,2, \ldots)$, so (10.97) holds, which completes the proof of lemma 10.3.

Now we are ready to give the

*Proof of theorem 10.6*. As was observed above, it remains to prove only the implication $2° \Rightarrow 1°$. Assume that we have $2°$ and let $\{x_n\}$ be an arbitrary $M$-basis of $E$, with the a.s.f. $\{f_n\} \subset E^*$. Then, by lemma 10.3, with*) $V = [f_n]$, there exists an approximative basis $\{v_n\}$ of $E$ satisfying

---

*) $V = [f_n]$ is total on $E$, since $\{x_n\}$ is an $M$-basis of $E$.

$\bigcup_{n=1}^{\infty} v_n^*(E^*) \subset [f_n]$. Hence, by the remark made after the proof of theorem 10.1, $\{x_n\}$ is an operational basis of $E$, which completes the proof of theorem 10.6.

Let us also mention the following problems raised in [380]: We do not know any characterization a) of all Banach spaces $E$ for which lemma 10.2 or lemma 10.3 remains valid, with "$V$ total" replaced by $r(V) > 0$; b) of the Banach spaces $E$ such that every norming $M$-basis $\{x_n\}$ of $E$ is an operational basis of $E$. By the above, every quasi-reflexive space (respectively, having also an approximative basis) has these properties, but there might exist non-quasi reflexive ones, too. For example, we do not know whether the non-quasi reflexive spaces*) $E$ such that every subspace $V$ of $E^*$ with $r(V) > 0$ is of finite codimension in $E^*$, have these properties**).

If $\{x_n\}$ is an operational basis of a Banach space $E$, with the a.s.f. $\{f_n\} \subset E^*$, then the subspace $[f_n]$ of $E^*$ is called (following W. B. Johnson [178]) the *coefficient space* of $\{x_n\}$. The proof of theorem 10.1 shows that *if $E$ has an operational basis and if $V$ is any separable subspace of $E^*$, then $E$ has also an operational basis $\{x_n\}$ with coefficient space* $[f_n] \supset V$ (thus, $E$ has operational bases whose coefficient spaces are "arbitrarily large"); indeed, if $\{u_n\}$ is any approximative basis of $E$, it is enough to take an $M$-basis $\{x_n\}$ of $E$ with coefficient space $[f_n] \supset$ $\supset V \cup \bigcup_{n=1}^{\infty} u_n^*(E^*)$. W. B. Johnson observed ([178], remark 1) that *if $V$ is a separable subspace of $E^*$ and $V$ contains a subspace which is the coefficient space of some operational basis of $E$, then $V$ itself is the coefficient space of some operational basis of $E$*; indeed, it is enough to take any $M$-basis $\{x_n\}$ of $E$ with coefficient space $[f_n] = V$.

In the other direction, F. S. Vaher [399] considered the following problem***), which she called "the local problem of existence of operational bases": If $E$ has an operational basis and if $V$ is a closed linear subspace of $E^*$ with $r(V) > 0$, does $E$ have an operational basis $\{x_n\}$ with coefficient space $[f_n] \subset V$? (in other words, does $E$ have operational bases whose coefficient spaces are "arbitrarily small"?). Lemma 10.3 above shows that if $E$ is quasi-reflexive, the answer is affirmative. F. S. Vaher [399] has proved that the answer is negative for a suitable subspace $V$ of $A^*$ (the space $A$ has an operational basis, e.g. by § 11, example 11.2); of course, this also shows that lemma 10.3 (and hence lemma 10.2) does not remain valid for all separable Banach spaces with an approximative basis.

---

*) The existence of such spaces $E$ was shown in the paper of W. J. Davis and W. B. Johnson ([64], example 1), with the mention that it was discovered in conversation with D. W. Dean.

**) See the Appendix.

***) Actually, F. S. Vaher [399] omitted in the formulation of the problem the assumption that $E$ has an operational basis, but then any space with no operational basis gives a negative answer (such spaces exist by § 0, theorems 0.1, 0.2).

As we already mentioned above, non-linear operational bases were introduced by M. I. Kadec [200]. Lemma 10.1 and theorem 10.2 (and hence implicitly, also corollary 10.2) were proved in the same paper [200] of M. I. Kadec.

For unconditional generalized summation bases theorem 10.3 was proved by V. F. Gapoškin and M. I. Kadec ([121], theorem 6) and the same proof also works, more generally, for operational bases (as given here). Remark 10.1 was made by V. F. Gapoškin and M. I. Kadec [121].

Generalized summation bases were introduced by V. F. Gapoškin and M. I. Kadec ([121], definition 1), who used for them the term "operational basis". However, we preferred to use here a different terminology, since (as mentioned above) it unifies the terms used in the linear and non-linear cases and since the generalized summation bases (of definition 10.3), while indeed more general than the triangular summation bases (of § 11), are in fact particular cases of the operational bases (of definition 10.1); this is shown by theorem 10.4, due to V. F. Gapoškin and M. I. Kadec ([121], theorem 1), who have also given example 10.2 ([121], remark 2). For some time there seemed to exist (see e.g. [200], [367]) the belief that the two notions are equivalent (with respect to the same sequence of endomorphisms $\{v_n\}$); as a negative answer to this question, M. I. Kadec and W. B. Johnson have communicated us, in letters, example 10.3 a) and b) respectively.

Theorem 10.5 was given by V. F. Gapoškin and M. I. Kadec ([121], theorem 3), with the mention that it is essentially contained in the paper [234] of S. S. Lewin.

Double (and multiple) generalized summation bases were introduced by V. F. Gapoškin and M. I. Kadec ([121], remark 4), who also observed that their properties are analogous to those of generalized summation bases. However, there is at least one exception to this observation. Namely, although *every Banach space with a double operational basis has the approximation property*, even *a space with a double generalized summation basis need not have an approximative basis*. Indeed, let $\varepsilon > 0$ and let $Q$ be an arbitrary compact subset of a Banach space $E$ with a double operational basis $\{x_n\}$, say with respect to $v_{nm} \in L(P_{(m)}, P_{(m)})$ $(n, m = 1,2, \ldots)$, so the limits

$$v_n(x) = \lim_{m\to\infty} v_{nm}s_m(x) \qquad (x \in E, n = 1,2, \ldots) \tag{10.101}$$

exist*) and satisfy

$$x = \lim_{n\to\infty} v_n(x) \qquad (x \in E). \tag{10.102}$$

*) Note that each $v_n \in L(E, E)$, but it need not be of finite rank.

Then the convergences (10.102) and (10.101) are*) uniform on $Q$ and hence there exist positive integers $n_0 = n_0(\varepsilon)$ and $m_0 = m_0(n_0(\varepsilon))$ such that

$$\|x - v_{n_0}(x)\| < \frac{\varepsilon}{2}, \ \|v_{n_0}(x) - v_{n_0 m_0} s_{m_0}(x)\| < \frac{\varepsilon}{2} \quad (x \in Q).$$

Thus, $\|x - v_{n_0 m_0} s_{m_0}(x)\| < \varepsilon$ $(x \in Q)$, which, since $v_{n_0 m_0} s_{m_0}$ is a finite rank operator on $E$, proves that $E$ has the approximation property. Hence, by § 0, there exist separable Banach spaces which have no double operational basis.

On the other hand, T. Figiel has suggested, in a conversation, a possible method of constructing a Banach space with a $T$-basis which does not have an approximative basis. We shall show now that his idea yields, in fact, a space with a double generalized summation basis (but which is not even a $\gamma$-basis for the matrix $(\lambda_{nm})$ satisfying (10.106) below), which has no approximative basis.

*Example 10.4*. Let $E = (E_1 \times E_2 \times \ldots)_{l^2}$ be the Banach space of § 9, example 9.2, which has the approximation property but has no approximative basis. Then each $E_n$ has a basis $\{x_j^{(n)}\}_{j=1}^{\infty}$ and we shall show that the sequence

$$\{x_i\} = \bigcup_{n=1}^{\infty} \bigcup_{j=1}^{\infty} \{\underbrace{0, \ldots, 0}_{n-1}, x_j^{(n)}, 0, 0, \ldots\} \tag{10.103}$$

is a double generalized summation basis of $E$, under any ordering with the following property: for $x_{i_1} = \{\underbrace{0, \ldots, 0}_{n-1}, x_{j_1}^{(n)}, 0, \ldots\}$, $x_{i_2} = \{\underbrace{0, \ldots, 0}_{n-1}, x_{j_2}^{(n)}, 0, \ldots\}$, we have $i_1 < i_2$ if and only if $j_1 < j_2$. Indeed, clearly, $\{x_i\}$ is an $M$-basis of $E$, with the a.s.f.**)

$$\{f_i\} = \bigcup_{n=1}^{\infty} \bigcup_{j=1}^{\infty} \{\underbrace{0, \ldots, 0}_{n-1}, f_j^{(n)}, 0, 0, \ldots\} \tag{10.104}$$

(under the same ordering as $\{x_i\}$), where $\{f_j^{(n)}\}_{j=1}^{\infty} \subset E_n^*$ is the a.s.c.f. to the basis $\{x_j^{(n)}\}_{j=1}^{\infty}$ of $E_n$ $(n = 1,2, \ldots)$. Define an infinite matrix of

*) See e.g. Ch. I, § 17, proof of theorem 17.3; indeed, that argument works for any sequence of operators $\{s_n\}$ on $E$ (not necessarily of finite rank), such that $x = \lim_{n \to \infty} s_n(x)$ $(x \in E)$.

**) In other words,

$$f_i(x) = f_j^{(n)}(p_n(x)) \qquad (x = \{p_1(x), p_2(x), \ldots\} \in E).$$

scalars $(\lambda_{ni})$ by

$$\lambda_{ni} = \begin{cases} 1 \text{ if } x_i \in \bigcup_{k=1}^{n} \bigcup_{j=1}^{\infty} \{\underbrace{0, \ldots, 0}_{k-1}, x_j^{(k)}, 0, \ldots\} \\ 0 \text{ if } x_i \notin \bigcup_{k=1}^{n} \bigcup_{j=1}^{\infty} \{\underbrace{0, \ldots, 0}_{k-1}, x_j^{(k)}, 0, \ldots\} \end{cases} \tag{10.105}$$

and let

$$v_{nm}(x) = \sum_{i=1}^{m} \lambda_{ni} f_i(x) x_i \quad (x \in P_{(m)};\ n, m = 1,2, \ldots). \tag{10.106}$$

Then $v_{nm} \in L(P_{(m)}, P_{(m)})$, where $P_{(m)} = [x_1, \ldots, x_m]$ and, by the above ordering of $\{x_i\}$, we have

$$\lim_{m\to\infty} v_{nm} s_m(x) = \sum_{i=1}^{\infty} \lambda_{ni} f_i(x) x_i =$$

$$= \left\{ \sum_{j=1}^{\infty} f_j^{(1)}(p_1(x)) x_j^{(1)}, \ldots, \sum_{j=1}^{\infty} f_j^{(n)}(p_n(x)) x_j^{(n)}, 0, 0, \ldots \right\} =$$

$$= \{p_1(x), \ldots, p_n(x), 0, 0, \ldots\} \quad (x \in E,\ n = 1,2, \ldots),$$

where $p_n$ is the coordinate projection of $E$ onto $E_n$ (so $x = \{p_1(x), p_2(x), \ldots\}$ for each $x \in E$). Thus, the limits (10.101) exist and, by the definition of $E$, they satisfy (10.102), which proves that $\{x_i\}$ is a double operational basis of $E$. In particular, by (10.101) and (10.102) (for $x = x_i$), we have

$$\lim_{n\to\infty} \lim_{m\to\infty} v_{nm}(x_i) = \lim_{n\to\infty} v_n(x_i) = x_i \quad (i = 1,2, \ldots); \tag{10.107}$$

also, by (10.101) and (10.102), for each $x \in E$ we have, putting $\alpha_n = f_n(x)$ $(n = 1,2, \ldots)$,

$$\lim_{n\to\infty} \lim_{m\to\infty} v_{nm}\left(\sum_{i=1}^{m} \alpha_i x_i\right) = \lim_{n\to\infty} \lim_{m\to\infty} v_{nm}(s_m(x)) = \lim_{n\to\infty} v_n(x) = x.$$

Finally, the sequence $\{\alpha_n\} \subset K$ with this latter property is unique. Indeed, observe first that, by (10.106),

$$v_{nm}(x_k) = \sum_{i=1}^{m} \lambda_{ni} \delta_{ik} x_i = \lambda_{nk} x_k \quad (k = 1, \ldots, m;\ n, m = 1,2, \ldots),$$

whence, by (10.107), we obtain*)

$$\lim_{n\to\infty} \lambda_{nk} = 1 \quad (k = 1,2, \ldots). \tag{10.108}$$

Hence, if $\{\alpha_n\} \subset K$, $\lim_{n\to\infty}\lim_{m\to\infty} v_{nm}\left(\sum_{i=1}^{m} \alpha_i x_i\right) = 0$, then, by (10.106) and (10.108),

$$0 = f_k\left(\lim_{n\to\infty}\lim_{m\to\infty} \sum_{i=1}^{m} \lambda_{ni}\alpha_i x_i\right) = \lim_{n\to\infty}\lim_{m\to\infty} \sum_{i=1}^{m} \lambda_{ni}\alpha_i \delta_{ik} =$$

$$= \lim_{n\to\infty}\lim_{m\to\infty} \lambda_{nk}\alpha_k = \alpha_k \qquad (k = 1,2, \ldots),$$

which proves that $\{x_i\}$ is a generalized summation basis of $E$. However, $(\lambda_{nm})$ cannot be a summability matrix for any $\gamma$-basis**) of $E$, since it does not satisfy (11.97) (and hence $T = (t_{ni}) = (\lambda_{ni} - \lambda_{n,i+1})$ is not consistent, since it does not satisfy (11.1)). Indeed, by (10.105), for each fixed $n$ there exist an infinity of indices $i$ such that $\lambda_{ni} - \lambda_{n,i+1} = 1$, whence $\sum_{i=1}^{\infty} |\lambda_{ni} - \lambda_{n,i+1}| = \infty$ $(n = 1,2, \ldots)$.

§ **11.** The idea of applying summation methods to the usual Fourier series, and, more generally, to expansions with respect to orthonormal systems, in spaces of continuous functions, is very old (see e.g. [194], [411], [13] and the references therein). The application of summation methods to biorthogonal expansions in Banach spaces was suggested by O. Frink ([116], § 6), but he considered only row-finite and row-infinite methods of $(\lambda_{nm})$-limitability (see § 11, definitions 11.4 and 11.6), where $(\lambda_{nm})$ satisfies (11.96) (he called such methods "semi-regular"). Lemma 11.1 (the generalization of the Silverman-Toeplitz***) theorem from the case of numerical series to the case of series in Banach spaces) was proved by B. R. Gelbaum ([124], theorem 15); the proof of the sufficiency part is a generalization of Banach's techniques ([11], p. 90, theorem 10). $T$-bases of Banach spaces were introduced, independently, by B. R. Gelbaum [124] and V. Ya. Kozlov [224]; the latter has also considered, in particular, Cesàro bases [224]. Let us observe that if $\{x_n\}$ satisfies the conditions of definition 11.1 for some matrix $T = (t_{nm})$, then this matrix is necessarily "semi-consistent", in the sense that it satisfies (11.2) and (11.3). Indeed, by (11.8) and (11.7),

$$\sigma_{nm}(x_i) = \sum_{j=1}^{m} t_{nj} s_j(x_i) = \left(\sum_{j=i}^{m} t_{nj}\right) x_i \quad (m \geqslant i; n = 1,2, \ldots),$$

*) Alternatively, (10.108) follows also directly from (10.105).
**) See § 11, definition 11.6.
***) For references, see § 11, the footnote to the proof of lemma 11.1.

whence, by definition 11.1, the limits

$$\sigma_n(x_i) = \lim_{m\to\infty} \sigma_{nm}(x_i) = \left(\sum_{j=i}^{\infty} t_{nj}\right) x_i \qquad (i, n = 1,2, \ldots)$$

exist and satisfy

$$x_i = \lim_{n\to\infty} \sigma_n(x_i) = \lim_{n\to\infty} \left(\sum_{j=i}^{\infty} t_{nj}\right) x_i \qquad (i = 1,2, \ldots).$$

Consequently, since $x_i \neq 0$ (see the remark before theorem 11.1), we obtain

$$\lim_{n\to\infty} \sum_{j=i}^{\infty} t_{nj} = 1 \qquad (i = 1,2, \ldots),$$

whence (11.2) and (11.3) follow. However, $T = (t_{nm})$ need not satisfy (11.1), even when it is triangular, as shown by example 11.9.

Theorem 11.1 a) was given, for triangular matrices and Hilbert spaces, by V. Ya. Kozlov ([224], theorem 1) and, independently, for triangular matrices and Banach spaces, by B. R. Gelbaum ([124], theorem 16). In the general case, theorem 11.1 a) was observed by M. M. Day in the review [68] of Kozlov's paper and found again also by V. F. Gapoškin ([118], theorem 1); however, M. M. Day [68] has only stated that "the usual proof for Banach bases can be used to prove the theorem..." and V. F. Gapoškin [118] has only stated that one can prove the completeness of $A_1^{(T)}$, by using the isomorphism of all $n$-dimensional Banach spaces. Theorem 11.1 b) was noted by V. F. Gapoškin and M. I. Kadec ([121], remark 4). Problem 11.1 a) was raised in [367], Ch. III, problem 3.18. Although every Banach space $E$ with a $T$-basis has the approximation property*) (see the Notes and remarks to § 10), it is possible that example 10.4 above could be improved to yield a space with a $T$-basis, which has no approximative basis (and hence no basis).

A. Wilansky and K. Zeller [408] observed that the space $c$ (of all convergent sequences) has an $M$-basis which is not a $T$-basis for any consistent matrix $T$. Corollary 11.2 b) answers in the affirmative a question of V. F. Gapoškin [118]. Proposition 11.1 is due to V. F. Gapoškin and M. I. Kadec ([121], p. 3).

Examples 11.1 and 11.3 were noted by V. Ya. Kozlov [224]. As was already mentioned, example 11.2 can be found, implicitly, in Zygmund's monograph [411], Ch. VIII. Problem 11.2 was raised in [367], problem 3.19.

The idea of extending characterizations of bases to characterizations of $T$-bases can be found already in the paper [124] of B. R. Gelbaum,

---

*) Hence, by § 0, there exist separable Banach spaces which have no $T$-basis. This answers a question of V. Ya. Kozlov [224].

where the implication $2° \Rightarrow 1°$ of theorem 11.2 was given ([124], theorem 1′). The set $\mathscr{E}_0^{(T)}$, for any $E$-complete biorthogonal system and any (not necessarily triangular) consistent matrix $T$, was considered in [367], proposition 3.15, where its density in $E$ was observed. The sets $E \setminus \mathscr{E}_2^{(T)}$ and $E \setminus \mathscr{E}_3^{(T)}$, for any triangular consistent matrix $T$, were introduced by J. R. Retherford [334], who proved proposition 11.3 a), b) and, essentially*), the implications $3° \Rightarrow 1°$ and $4° \Rightarrow 1°$ of theorem 11.3.

Lemmas 11.2, 11.3, remark 11.2 and some generalizations of these results to higher order differences can be found in the paper [67] of D. F. Dawson. Theorem 11.4 is due to M. I. Kadec [198]. The simple proof of the implication $1° \Rightarrow 2°$ of this theorem, presented here, is due to P. L. Butzer, R. J. Nessel and W. Trebels ([41], theorem 3.2). Theorem 11.4 was extended to $(C, p)$-bases and $bv^{p+1}$ by the same authors ([42], theorem 7.1 and the remark made after it; see also W. Trebels [392], § 3, theorem 3.3). Various extensions in other directions (e.g., to $T$-bases of subspaces**), to locally convex spaces, etc.) and related results were also given in the above papers and some later papers of the group of mathematicians in Aachen (see e.g. E. Görlich, R. J. Nessel and W. Trebels [134], H. J. Mertens, R. J. Nessel and G. Wilmes [280] and the references therein). For some other results on multipliers and $T$-bases in certain Banach spaces see also the paper [129] of G. Goes.

Let us also mention here the following generalization of Ch. I, § 5, corollary 5.1 to triangular $T$-bases (the proof is similar):

**Proposition 11.13.** *Let $E$ be a Banach space, $\{x_n\}$ an $M$-basis of $E$, with the a.s.f. $\{f_n\} \subset E^*$, and $T = (t_{nm})$ a triangular consistent matrix. The following statements are equivalent (where the numbers $T_{nm}$ are defined by* (11.58)*):*

1°. *$\{x_n\}$ is a $T$-basis of $E$.*

2°. $\sup_{1 \leqslant n < \infty} \left\| \sum_{i=1}^{n} T_{ni} e_i \right\| \leqslant C < \infty$ *(where $\{e_n\}$ and $\|\cdot\|$ are in $M(E, (x_n, f_n))$).*

3°. *There exists a constant $C \geqslant 1$ such that*

$$\sup_{1 \leqslant n < \infty} \left\| \sum_{i=1}^{n} T_{ni} \gamma_i e_i \right\| \leqslant C \|\{\gamma_n\}\| \qquad (\{\gamma_n\} \in M(E, (x_n, f_n))).$$

Theorem 11.5 was proved, for the particular case of Cesàro bases, in [365], theorem 1; in the general case (i.e. for an arbitrary triangular consistent matrix) it has been stated, without proof, in [367], theorem 3.27.

The observation that the known general stability theorems, stating the existence of an isomorphism $u$ such that $u(x_n) = y_n$ $(n = 1,2, \ldots)$, yield also stability theorems for $T$-bases, was made by V. F. Gapoškin

---

*) Namely, with $3° \Rightarrow 1°$ replaced by $\mathscr{E}_0^{(T)} = \mathscr{E}_2^{(T)} \Rightarrow 1°$, since in [334] the set $\mathscr{E}_1^{(T)}$ was not defined.

**) See the Notes and remarks to § 15.

[118]. Some stability theorems for $T$-bases were given by Z. A. Čanturija [43]; one of them ([43], theorem 11), extending part of Ch. I, § 10, theorem 10.2 a), is concerned with "$\omega_T$-linearly independent sequences", which generalize the usual $\omega$-linearly independent sequences.

The first statement of theorem 11.6, and corollary 11.4, were given, without proof, by V. F. Gapoškin ([118], theorems 3 and 5), who also proved corollary 11.3 ([118], theorem 4). In the particular case of a triangular consistent matrix, theorem 11.7 was observed, essentially, by B. R. Gelbaum ([124], theorems 16 and 1′) and in the general case it was proved by V. F. Gapoškin ([118], theorem 2).

For the matrix (11.6) of Cesàro summability, the notion of $T$-boundedly complete $T$-basis was introduced in [365], while in the general case (i.e., for an arbitrary consistent matrix), in [367], definition 3.26. For the particular case of strongly series summable $M$-bases it has been found again by W. B. Johnson [174], who has also noted proposition 11.4 for this particular case.

$T$-unconditional $T$-bases were introduced by V. F. Gapoškin [118], who has given also theorem 11.9 ([118], theorem 12 and [119], theorem 3) and has noted proposition 11.5 for the particular case of the trigonometric system in $E = C_{2\pi}$ ([118], corollary).

Z. A. Čanturija (see [43] and the references therein) has considered some extensions of Besselian and Hilbertian bases, involving $T$-summability. Namely, if $\{e_n\}$ is a fixed $T$-basis of a Banach space $E$ (where $T$ is a consistent matrix), let $\Phi_T$ be*) the linear space of sequences of scalars

$$\Phi_T = \left\{\alpha = \{\alpha_n\} \subset K \,\middle|\, \sum_{i=1}^{\infty} \alpha_i e_i \text{ is } T\text{-summable}\right\}, \tag{11.151}$$

endowed with the norm

$$\|\{\alpha_n\}\| = \sup_{1 \leqslant n < \infty} \|\sigma_n(x)\|, \tag{11.152}$$

where $\sum_{i=1}^{\infty} \alpha_i e_i$ is $T$-summable to $x$. Now, an $M$-basis $\{x_n\}$ of $E$, with the a.s.f. $\{f_n\} \subset E^*$, is called

a) *$\Phi_T$-Besselian*, if $\{f_n(x)\} \in \Phi_T \quad (x \in E)$;

b) *$\Phi_T$-Hilbertian*, if for each $\{\alpha_n\} \in \Phi_T$ there exists a (clearly, unique) element $x \in E$ such that $f_n(x) = \alpha_n \quad (n = 1, 2, \ldots)$.

Some results of Ch. II, § 11, can be extended to $\Phi_T$-Besselian and $\Phi_T$-Hilbertian shrinking $M$-bases in reflexive Banach spaces (Z. A. Čanturija [43]).

The problem of the order of growth of the sequences $\{\|f_n\|\}$, for arbitrary $T$-bases, was raised by V. Ya. Kozlov [224], who also stated, without proof, proposition 11.6 for Hilbert spaces. V. F. Gapoškin [118] has announced, without proof, some more general results. Namely,

---

*) A more appropriate notation would be: $\Phi_{\{e_n\}, T}$ .

a consistent matrix $T$ is said to have the *order of growth* $\varphi(n)$, if for each $T$-summable sequence of scalars $\{\beta_n\} \subset K$ we have*)

$$|\beta_n| = O(\varphi(n)). \qquad (11.153)$$

If $\{x_n\}$ is a normalized $T$-basis of a Banach space $E$, with the a.s.f. $\{f_n\} \subset E^*$ and if $T$ is a consistent matrix having the order of growth $\varphi(n)$, then $\|f_n\|$ grows not faster than $O(\varphi(n))$ ([118], theorem 8). In particular, for the method $T = (C, \alpha)$, where $\alpha \geqslant 0$, it follows that if $\{x_n\}$ is a normalized $(C, \alpha)$-basis of $E$, with the a.s.f. $\{f_n\} \subset E^*$, then $\|f_n\| = O(n^\alpha)$ ([118], theorem 9); obviously, for $\alpha = 1$ this reduces to proposition 11.6. Example 11.5 is due to V. F. Gapoškin ([120], proposition 1 and the remark to it), who has also observed that if in (11.75) we replace $\frac{A}{n}$ by $\frac{A}{n^\alpha}$, where $0 < \alpha < 1$, then the same example yields a $(C, \alpha)$-basis of $E$, with the a.s.f. $\{f_n\} \subset E^*$ satisfying $\|f_n\| \geqslant C' n^\alpha$ $(n = 1, 2, \ldots)$. Example 11.6 was given by V. F. Gapoškin ([118], theorem 11).

The lacunary sequences (the elements of $\Lambda$) are also called "lacunary in the sense of Hadamard". Lemma 11.4 is due to S. B. Stečkin ([384], lemma 1), but in his proof he assumes, in addition, that the sequence $\{n_k\} \in \Lambda_\sigma$ is also monotone. Although his proof can be modified so as to cover also the general case, we have preferred to give here a simpler proof, the idea of which was communicated to us, in a conversation, by K. I. Oskolkov. Lemma 11.5 can be found in the monograph of A. Zygmund ([411], Ch. III, proof of theorem (1.27)). For the particular case of the trigonometric system in $E = C_{2\pi}$, theorem 11.10 a) was proved by S. B. Stečkin [384]; for a Cesàro basis $\{x_n\}$ in a Banach space $E$ with $\sup_{1 \leqslant n < \infty} \|x_n\| \leqslant M < \infty$, it has been given by V. M. Kokilašvili ([222], theorem 1), but in his proof he assumes, in addition, that $\sup_{1 \leqslant n < \infty} \|f_n\| < \infty$ (which, of course, need not hold for Cesàro bases; see example 11.5 and proposition 11.6). In its full generality, theorem 11.10 is due to V. F. Gapoškin ([120], proposition 2 A), B)), who has also given examples 11.7 and 11.8 ([120], proposition 2 C)). For the trigonometric system in $E = C_{2\pi}$, corollary 11.5 can be found in the monograph of N. K. Bari ([13], p. 178—179), but the proof given there cannot be extended to the general case, since it relies, among other things, on the fact that $\lim_{n \to \infty} f_n(x) = 0$ for all $x \in E = C_{2\pi}$ (which, of course, need not hold for Cesàro bases).

The idea of strongly series summable $M$-bases and proposition 11.8 are contained, implicitly, in the paper of O. Frink ([116], § 6), which includes also an example of a non-strongly series summable $M$-basis of $C([-1, 1])$, namely, the system of Legendre polynomials. Strongly

*) We recall that $O(\alpha_n)$, where $\alpha_n > 0$, denotes a quantity such that $\frac{O(\alpha_n)}{\alpha_n}$ is bounded.

series summable $M$-bases were also considered by R. E. Edwards ([94], § 6 and [95], Ch. 6, § 6.8.7) and, independently, by Z. A. Čanturija [44], who called them "$\gamma$-bases with respect to triangular matrices $(\lambda_{nm})$" (one could say, shortly, "triangular $\gamma$-bases") and mentioned, without proof, proposition 11.7. The notion was found again by W. H. Ruckle ([345], § 7), who introduced the term "strongly series summable $M$-basis" and observed proposition 11.8 for the particular case of permutations of bases ([345], § 5). W. B. Johnson [178] used for $(\lambda_{nm})$ the term "summation matrix", but we have replaced it here by "summability matrix", in order to avoid possible confusion with $T$-matrices. Problem 11.3 a) was raised by W. B. Johnson ([178], conjecture 1).

Proposition 11.9 and theorem 11.11 are due to W. B. Johnson ([178], lemma 4 and theorem 1). It is still unknown whether these results remain valid also for real Banach spaces. W. B. Johnson has communicated us in 1970, in a letter, the following observation: the argument given in [178] actually shows that in theorem 11.11 the assumption that $E^*$ has the bounded approximation property can be replaced by the (weaker) assumption that $E$ has a commuting approximative basis $\{u_n\}$ and then one obtains a strongly series summable $M$-basis $\{x_n\}$ of $E$ with coefficient space $[f_n] = \left[\bigcup_{n=1}^{\infty} u_n^*(E^*)\right]$. We shall give now a proof of this result, and we shall show that the converse is also true. To this end, we shall use the following generalization of § 11, proposition 11.9:

**Proposition 11.14.** *Under the assumptions of* § 11, *proposition* 11.9 *and if, in addition,* $\{f_i\}_{i=1}^{n} \subset F$, *where* $F$ *is a total subspace of* $E^*$, *the conclusion of that proposition holds, with the additional property* $\{f_i\}_{i=1}^{n+m} \subset F$.

*Proof.* The proof is similar to that of § 11, proposition 11.9, using, in addition, the fact that every finite-dimensional subspace $G$ of $E$ is*) closed for $\sigma(E, F)$, and hence for each $x \in G$ there exists**) $f \in F$ such that $f(y) = 0$ $(y \in G)$, $f(x) = 1$ (in this way we obtain the additional property $\{f_i\}_{i=1}^{n+m} \subset F$).

Now we are ready to prove

**Theorem 11.13.** *If a complex Banach space* $E$ *has a commuting approximative basis* $\{u_n\}$, *then* $E$ *has a strongly series summable* $M$*-basis* $\{x_n\}$ *with coefficient space* $[f_n] = \left[\bigcup_{n=1}^{\infty} u_n^*(E^*)\right]$. *Conversely, if* $E$ *(complex or real) has a strongly series summable* $M$*-basis* $\{x_n\}$ *with the a.s.f.* $\{f_n\} \subset E^*$, *then* $E$ *has a commuting approximative basis* $\{u_n\}$ *such that* $\left[\bigcup_{n=1}^{\infty} u_n^*(E^*)\right] = [f_n]$.

*) See e.g. [33], Ch. I, § 2.
**) See e.g. [87], p. 422, corollary 12.

*Proof.* Assume that $\{u_n\}$ is a commuting $\lambda$-approximative basis of $E$. Then, by § 9, proposition 9.2, $F = \left[\bigcup_{n=1}^{\infty} u_n^*(E^*)\right]$ is total on $E$. Let $\{y_m\} \subset E$ and $\{g_m\} \subset F$ be dense sequences in $E$ and $F$ respectively, such that $g_1(y_1) = 1$. Put $x_1 = y_1$, $f_1 = g_1$, $k_1 = 1$ and $\lambda_{1i} = \delta_{1i}$ ($i = 1, 2, \ldots$). Assume now that $k_m$, $(x_i, f_i)_{i=1}^{k_m}$ with $\{f_i\}_{i=1}^{k_m} \subset F$ and $\lambda_{ni}$ ($n = 1, \ldots, k_m$; $i = 1,2, \ldots$) have been defined. Then $(x_i, f_i)_{i=1}^{k_m}$ can be extended to a biorthogonal system $(x_i, f_i)_{i=1}^{j_m}$, where $j_m = k_m$ or $k_{m+1}$ or $k_{m+2}$, so that $\{f_i\}_{i=1}^{j_m} \subset F$, $y_{m+1} \in [x_i]_{i=1}^{j_m}$, $g_{m+1} \in [f_i]_{i=1}^{j_m}$; indeed, the proof is similar to that of the corresponding statement in § 11, proof of theorem 11.11, applying, in addition, the two facts used in the above proof of proposition 11.14 (in this way we obtain the additional property $\{f_i\}_{i=1}^{j_m} \subset F$). Now, by lemma 9.4 of the Notes and remarks to § 9 $\left(\text{with } G = [x_i]_{i=1}^{j_m},\ \Gamma = [f_i]_{i=1}^{j_m} \subset F,\ \varepsilon = \frac{1}{m}\right)$, there exists a finite rank operator $w_m$ on $E$ satisfying $w_m(x_i) = x_i$, $w_m^*(f_i) = f_i$ ($i = 1, \ldots, j_m$), $\|w_m\| \leqslant \lambda + \frac{1}{m}$. Hence, using proposition 11.14 above and then continuing as in § 11, proof of theorem 11.11, we obtain a strongly series summable $M$-basis $\{x_n\}$ of $E$ with coefficient space $[f_n] = F = \left[\bigcup_{n=1}^{\infty} u_n^*(E^*)\right]$.

Conversely, assume now that $\{x_n\}$ is a strongly series summable $M$-basis of $E$ (complex or real) with the a.s.f. $\{f_n\} \subset E^*$ and with summability matrix $(\lambda_{ni})$ and let $\tau_n$ be the finite rank operator on $E$ defined by (11.93). Then, by (11.92), $\{\tau_n\}$ is an approximative basis of $E$. Furthermore, similarly to theorem 11.6, we have

$$f = \lim_{n\to\infty} \tau_n^*(f) \qquad \left(f \in \left[\bigcup_{n=1}^{\infty} \tau_n^*(E^*)\right] = [f_n]\right);$$

indeed, this follows from

$$\tau_n^*(f_j) = \sum_{i=1}^{n} \lambda_{ni} f_j(x_i) f_i = \lambda_{nj} f_j \qquad (n \geqslant j),$$

together with (11.96) and $\sup_{1\leqslant n<\infty} \|\tau_n\| < \infty$. Hence*), by proposition 9.9 of the Notes and remarks to § 9, $E$ has a commuting approximative

*) Alternatively, one can observe that

$$\tau_n\tau_m(x) = \tau_m\tau_n(x) = \sum_{i=1}^{\min(n,m)} \lambda_{ni}\lambda_{mi} f_i(x) x_i \qquad (x \in E,\ m \neq n;\ m, n = 1, 2, \ldots),$$

and then apply proposition 9.10, implication $2^\circ \Rightarrow 1^\circ$, and its proof.

basis $\{u_n\}$ such that $\left[\bigcup_{n=1}^{\infty} u_n^*(E^*)\right] = \left[\bigcup_{n=1}^{\infty} \tau_n^*(E^*)\right] = [f_n]$, which completes the proof of theorem 11.13.

From theorem 11.13 it follows, in particular, that for complex Banach spaces problem 11.3 a) is equivalent to § 9, problem 9.5 b). Combining the converse part in theorem 11.13 with the results of § 9 and § 11, one can obtain a number of other consequences, e.g. the following

**Corollary 11.7.** *Every Banach space $E$ with a strongly series summable $M$-basis (hence, in particular, every Banach space $E$ with a triangular $T$-basis) admits an equivalent norm $|||\cdot|||$ such that $(E, |||\cdot|||)$ has a commuting 1-approximative basis.*

Theorem 11.11 shows that if $E$ is a separable complex Banach space and $E^*$ has the bounded approximation property, then $E$ has strongly series summable $M$-bases whose coefficient spaces are "arbitrarily large". However, in contrast with the situation for operational bases (see the Notes and remarks to § 10), it is not true that if $E$ is separable and $E^*$ has the bounded approximation property and if $V$ is a separable subspace of $E^*$, containing the coefficient space of some strongly series summable $M$-basis of $E$, then $V$ itself is the coefficient space of some strongly series summable $M$-basis of $E$; indeed, a counter-example in $E = l^1$ was given by W. B. Johnson ([178], remark 1).

Corollary 11.6 and theorem 11.12 are due to W. B. Johnson ([178], corollary 1 and theorem 2).

Propositions 11.10 and 11.11 were given, with different proofs, by W. H. Ruckle ([345], theorem 7.2 and proposition 7.3); W. H. Ruckle also stated in [345], proposition 7.3, that every Banach space with a strongly series summable $M$-basis has the metric approximation property, but a counter-example (even for bases) is provided e.g. by § 9, remark 9.9 c). Note also that some other characterizations of strongly series summable $M$-bases can be obtained by replacing in proposition 11.13 above the numbers $T_{nm}$ by $\lambda_{nm}$ satisfying (11.96) (and requiring in 2° and 3° the existence of such numbers $\lambda_{nm}$). Example 11.10, which solves a problem raised by W. H. Ruckle ([345], § 7), is due to L. Crone, D. J. Fleming and P. Jessup ([53], example 4.24).

The idea of $\gamma$-bases is contained, implicitly, in the paper of O. Frink ([116], § 6). $\gamma$-bases were considered by R. E. Edwards ([95], Ch. 6, § 6.8.7) and, independently, by Z. A. Čanturija [44]. Let us observe that if $\{x_n\}$ satisfies the conditions of definition 11.6 for some matrix $(\lambda_{nm})$, then necessarily this matrix satisfies (11.96). Indeed, by (11.145) and biorthogonality,

$$\tau_n(x_i) = \sum_{j=1}^{\infty} \lambda_{nj} f_j(x_i) x_j = \lambda_{ni} x_i \qquad (n, i = 1, 2, \ldots),$$

whence, by (11.146),

$$x_i = \lim_{n\to\infty} \tau_n(x_i) = (\lim_{n\to\infty} \lambda_{ni})x_i \qquad (i = 1, 2, \ldots),$$

which, since $x_i \neq 0$ ($i = 1,2, \ldots$), implies (11.96). However, $(\lambda_{nm})$ need not satisfy (11.97), even when it is triangular, as shown by example 11.9 (with $\lambda_{nm}$ defined by (11.147)).

Example 11.11 is contained in the monograph of R. G. Cooke ([50], Ch. 4, § 4.6), with the mention that it is due to P. Vermes. Some analogues of properties of $T$-bases (e.g., stability properties) were proved for $\gamma$-bases by Z. A. Čanturija [44].

R.E. Edwards [94] has considered further generalizations of the notions of strongly series summable $M$-bases and of $\gamma$-bases, by discarding the requirement of biorthogonality of $(x_n, f_n)$ (as was done for quasi-bases, so in this way one obtains also a generalization of quasi-bases; however, the existence of such objects is still equivalent to the existence of quasi-bases). Actually, R. E. Edwards [94] has worked directly in the more general framework in which sums are replaced by integrals on a locally compact space $T$ (see the Notes and remarks to § 15).

**§ 12.** Banach spaces with $\pi_\lambda$-bases (actually, directly Banach spaces with extended $\pi_\lambda$-bases) were considered by J. Lindenstrauss ([235], p. 25), who called them spaces "with the $\lambda$-projection approximation property"; in fact, he used the equivalent condition given in § 18, theorem 18.4 $\beta$) 2°. Such spaces were also found, independently, by D. G. de Figueiredo [110] (see also F. E. Browder and D. G. de Figueiredo [39] and D. G. de Figueiredo [111], Ch. IV, § 1 and the references therein), who introduced them for the study of fixed points of weakly continuous non-linear mappings in Banach spaces and called them spaces "with property $\pi_\lambda$"; the terms "$\pi_\lambda$-space" (E. Michael and A. Pelczynski [282] for $\lambda = 1$) and "$\pi$-space" (W. B. Johnson, H. P. Rosenthal and M. Zippin [187]) are also used. Among other applications of such spaces, we shall see in Vol. III, Ch. IV, that they are a natural framework for the study of functional equations (see e.g. W. V. Petryshin and T. S. Tucker [314] and the references therein). For $\pi_\lambda$-bases the terms "$\pi_\lambda$ decomposition" (W. B. Johnson [173]), "$\pi_\lambda$ system" [377], are also used and for $\pi$-bases W. B. Johnson ([177], [179]) has used the term "Schauder operator basis"; the term "$\pi$ system" is also used [377]. $\Pi$-bases occur, implicitly, without any term for them, in the paper of W. B. Johnson, H. P. Rosenthal and M. Zippin ([187], proposition 1.2, condition (b)), who have also raised problem 12.1 a).

For the case when both $u$ and $v$ are projections, lemma 12.1 a), b) can be found e.g. in [87], p. 514, exercise 23 (i). Corollary 12.1 is contained, essentially, in the paper of W. B. Johnson, H. P. Rosenthal and M. Zippin ([187], proposition 1.2, equivalence (a) $\Leftrightarrow$ (b)). The results of weak duality, mentioned after corollary 12.1, were observed by W. B. Johnson ([173], § I). Problem 12.2 was raised by D. W. Dean

[73]. An example of a $\pi$-basis $\{u_n\}$ of $E = l^p$ $(1 < p < \infty)$, such that $\{u_n^*|_V\}$ is not an approximative basis of $V = \left[\bigcup_{n=1}^{\infty} u_n^*(E^*)\right] = E^*$ $\left(\text{moreover, } \lim_{n\to\infty} \|u_n^*(f)\| = 2^{\frac{1}{q}}\|f\| \text{ for all } f \in E^*, \text{ where } \frac{1}{p} + \frac{1}{q} = 1\right)$, was given by W. B. Johnson [177]; however, we preferred to give here example 12.1, since it works in any Banach space with a normalized basis and since it illustrates the last assertion of remark 12.1.

The term*) "$\varepsilon$-close" was suggested by W. B. Johnson, H. P. Rosenthal and M. Zippin [187], who also proved lemma 12.2 ([187], lemma 2.4 (b)). As mentioned above, the $\lambda$-projection approximation property, in a different form (namely, in the form of condition $\beta$) 2° of § 18, theorem 18.4), was introduced by J. Lindenstrauss ([235], p. 25). Theorem 12.1 $\delta$) and, essentially, theorem 12.1 $\gamma$), equivalences 1° $\Leftrightarrow$ 3° $\Leftrightarrow$ 5°, were given by W. B. Johnson, H. P. Rosenthal and M. Zippin ([187], proposition 1.2); also, remark 12.2 is contained, implicitly, in their paper [187]. Theorem 12.1 $\gamma$), implication 1° $\Rightarrow$ 4°, was proved in the initial version of [86].

Problem 12.3 can be also formulated, equivalently, as follows: Let us call the *$\Pi$-basis constant* (*$\pi$-basis constant*) $\Pi(E)$ (respectively, $\pi(E)$) of a Banach space $E$ the infimum of all numbers $\lambda$ such that $E$ has a $\Pi_\lambda$-basis (respectively, a $\pi_\lambda$-basis). Then, by § 12, theorem 12.1 $\gamma$), equivalence 1° $\Leftrightarrow$ 3°, we have $\Pi(E) = \pi(E)$. Problem 12.3 asks whether for every Banach space $E$ with a $\Pi$-basis this number is attained a) by some $\Pi$-basis of $E$ or b) by some $\pi$-basis of $E$. Clearly, $\Lambda(E) \leqslant \Pi(E)$, where $\Lambda(E)$ is the approximative basis constant of $E$ (see the Notes and remarks to § 9). Let us observe that the isomorph $E$ of $l^2$, with $0 < \Gamma(E) < 1$, constructed by P. Enflo and J. Lindenstrauss [98] (see the Notes and remarks to § 9), has $\Lambda(E) = 1 < \Pi(E) < \infty$, which shows that the metric approximation property and the projection approximation property do not imply the metric projection approximation property.

Example 12.2 was given in [376], theorem 3.

Banach spaces with $\pi_1^\infty$-bases were introduced by E. Michael and A. Pelczynski [282], who called them "$\pi_1^\infty$-spaces". W. B. Johnson ([173], § III) has used for $\pi_1^\infty$-bases the term "$\pi_1^\infty$ decomposition". The characterization of isometries $T: l_m^\infty \to l_n^\infty$, given in lemma 12.3, can be found in [282], lemma 2.3; actually, there an additional condition is also required, but it is superfluous, since it is a consequence of (12.28). Lemmas 12.3—12.5, remark 12.3 and theorem 12.2 are due to E. Michael and A. Pelczynski ([282], lemmas 2.3, 3.2, 2.1, remark 3.3 and theorem 1.1). A slightly weaker version of theorem 12.3 (in which, among other things, $v_{\{x_n\}} = 1$ is replaced by $v_{\{x_n\}} \leqslant 1 + \varepsilon$), was given by

---

*) Note that this notion is essentially a quantitative version of "*PW*-near" (Ch. I, § 10, definition 10.1 a)).

V. I. Gurariĭ ([144], theorem 8 and its proof). Theorem 12.3 was proved by E. Michael and A. Pelczynski ([282], corollary 6.2), who also observed ([282], corollary 6.2) that if $\{x_n\}$ is any monotone basis of a Banach space $E$ with a $\pi_1^\infty$-basis, then the sequence of subspaces $G_n = [x_i]_{i=1}^n$ $(n = 1,2, \ldots)$ satisfies conditions 3° a), b), c) of theorem 12.2. Let us note that the following converse of theorem 12.3 is also valid (so we have actually a characterization of Banach spaces with a $\pi_1^\infty$-basis): *If $E$ has a monotone basis $\{x_n\}$ such that $[x_i]_{i=1}^n \equiv l_n^\infty$ $(n = 1,2, \ldots)$, then $E$ has* a *$\pi_1^\infty$-basis* (namely, the sequence $\{s_n\}$ of partial sum operators*) associated to $\{x_n\}$). Some further characterizations of Banach spaces with a $\pi_1^\infty$-basis follow from the results of V. I. Gurariĭ ([144], § 2), E. Michael and A. Pelczynski ([282], theorem 1.1) and others. We mention here only the following characterization, due to A. J. Lazar and J. Lindenstrauss ([231], theorem 2): *A separable Banach space $E$ has a $\pi_1^\infty$-basis if and only if the conjugate space $E^*$ is linearly isometric to some space $L^1(T, \nu)$.* Characterizations of Banach spaces $E$ with $E^* \equiv L^1(T, \nu)$ were given by A. Grothendieck ([138], [139]), J. Lindenstrauss ([235] and some subsequent papers) and others (some authors call such spaces "Lindenstrauss spaces"); we shall consider some of these results in Vol. III, in connection with $\mathscr{L}_\infty$ spaces. See also the Notes and remarks to § 18.

The distance coefficient $d(A, B)$ (actually, the pseudo-metric $\rho(A,B) = \ln d(A, B)$, where $d(A, B)$ is defined by (12.41)) was introduced by S. Banach and S. Mazur (see [11], the Appendix, Remarks to Ch. XI, § 6). Lemma 12.6 is well known (e.g., it is implicitly assumed in definition 12.3 of the spaces $C_p$), but we could not locate any reference to a direct proof in the literature**). The spaces $C_p$ $(1 \leqslant p \leqslant \infty)$ were introduced by W. B. Johnson ([176], § 3), who also proved theorem 12.4 and raised problem 12.6 ([176], theorem 4 and p. 343, problem).

**§ 13.** Dual $\pi_\lambda$-bases were considered by W. B. Johnson (in [173] he used the term "dual $\pi_\lambda$ decomposition" for them and the term "dual $\pi_\lambda$ space" for spaces having a dual $\pi_\lambda$-basis, while in [177], [179] he used the term "Schauder dual operator basis"; the term "dual $\pi_\lambda$ system" is also used [377]). Dual $\pi_1^\infty$-bases were introduced by W. B. Johnson ([173], § III), who called them "dual $\pi_1^\infty$ decompositions".

Proposition 13.2 and the result of weak duality, mentioned before it, were noted by W. B. Johnson ([173], § I).

---

*) This is even a strict $\pi_1^\infty$-basis (in the sense that it satisfies (12.4)).

**) It is even known that for each $n < \infty$, $\mathscr{F}_n$ is compact for the Blaschke distance $\Delta(A, B) = \Delta(S_A, S_B)$ or, equivalently for the Hausdorff distance $\chi(A, B) = \chi(S_A, S_B)$, where $S_A, S_B$ are the unit balls (see e.g. H. G. Eggleston [96], pp. 59–61, respectively F. A. Valentine [400], p. 36, theorem 3.7), whence $\mathscr{F}_n$ is "compact" also for $d(A, B)$ (indeed, $\Delta$ and $\chi$ are defined by $\Delta(S_A, S_B) = \delta_1 + \delta_2$ and $\chi(S_A, S_B) = \max(\delta_1, \delta_2)$ respectively, where $S_A, S_B$ are considered as closed convex subsets of the same $n$-dimensional space and where $\delta_1 = \inf_{S_A \subset \lambda S_B} \lambda$, $\delta_2 = \inf_{S_B \subset \lambda S_A} \lambda$, whence, clearly, $\rho(A, B) \leqslant \ln \delta_1 + \ln \delta_2 \leqslant \Delta(A, B) \leqslant 2\chi(A, B) \leqslant 2\Delta(A, B)$). The direct proof of the separability of $\mathscr{F}_n$, given here, is different.

The existence of a commuting $\pi$-basis in a Banach space is equivalent to a number of other properties (see theorem 13.1), among which the earliest studied was the existence of finite-dimensional decompositions; for references, see also the Notes and remarks to § 15 and § 14. Commuting $\pi_\lambda$-bases were called, by W. B. Johnson [173], "$\pi_\lambda$-dual $\pi_\lambda$ decompositions". Proposition 13.3 was obtained, with the second proof given here, by W. B. Johnson ([173], theorem 1). This result can be also found, with a sketch of the same proof, in the paper of P. Billard ([28], theorem 1-1), with the mention that the proof was communicated to him by A. Pelczynski. The first part of remark 13.1 was noted by W. B. Johnson ([173], remark 1), with a weaker estimate than (13.27) (which did not converge to 0 as $n\to\infty$); formulae (13.27)—(13.29) were proved in [376], remark 1. The third proof of proposition 13.3, which yields a different commuting $\pi$-basis, and remark 13.3, were given in [376], theorem 1. Let us also mention the following shorter variant of this proof, which was given in the initial version of [376]: By § 15, theorem 15.3, it will be enough to show that the sequence

$$G_n = u_n(E) \cap \operatorname{Ker} u_{n-1} \qquad (n = 1,2, \ldots; u_0 = 0) \qquad (13.114)$$

is a finite-dimensional $(\lambda + \varepsilon)$-decomposition of $E$. Let $y_i \in G_i$ $(i = 1, \ldots, n + m)$. Then, by (13.6), $y_{n+1} \in \operatorname{Ker} u_n, \ldots, y_{n+m} \in \operatorname{Ker} u_{n+m-1} \subset \operatorname{Ker} u_n$, whence, by (13.17),

$$\left\|\sum_{i=1}^{n} y_i - u_n\left(\sum_{i=1}^{n+m} y_i\right)\right\| = \left\|\sum_{i=1}^{n-1} y_i - u_n\left(\sum_{i=1}^{n-1} y_i\right)\right\| <$$

$$< \frac{\varepsilon}{2^{n-1}(\lambda + \varepsilon)}\left\|\sum_{i=1}^{n-1} y_i\right\|.$$

Hence, by (13.3) and induction, we obtain

$$\left\|\sum_{i=1}^{n} y_i\right\| \leqslant (\lambda + \varepsilon)\left\|\sum_{i=1}^{n+m} y_i\right\|,$$

and therefore, by § 15, theorem 15.5, $\{G_n\}$ is a finite-dimensional $(\lambda+\varepsilon)$-decomposition of $E$. Proposition 13.4 is due to W. B. Johnson ([173], corollary 1 and remark 2). Example 13.1 was given in [376], theorem 3. Problem 13.2 was raised by W. B. Johnson ([173], problem 1), who also proved proposition 13.5 under the stronger assumption (13.51) and with the weaker conclusion that $E$ has a commuting $\pi_{\lambda\mu}$-basis ([173], theorem 3). Remark 13.4 was made in [376], remark 4. Proposition 13.6 and remark 13.5 were given in [376], theorem 2. Remarks 13.6 and 13.7 were made in the same paper ([376], remarks 3 and 4). Proposition 13.7 b) can be found in the paper of W. B. Johnson ([173], theorem 2; see also D. J. Fleming and W. Ruckle [113], theorem 5.8), with the mention that the referee has observed another simple proof of it, generalizing the proof of

Ch. II, § 1, proposition 1.3*); for $\Pi_{1+\varepsilon}$ replaced by $\pi_{1+\varepsilon}$ in its assumption, proposition 13.7 a) was also observed by W. B. Johnson ([176], p. 343).

Weak dual $\pi_\lambda$-bases were considered by W. B. Johnson [177], [179]. The qualitative part of proposition 13.8 can be found, with a different proof (using § 14, remark 14.3 b)), in the paper of W. B. Johnson, H. P. Rosenthal and M. Zippin ([187], theorem 1.3); the proof given here is due to W. B. Johnson ([177], Added in proof).

Bases with parentheses were introduced by M. I. Kadec [196]. The equivalent version of problem 13.3, in which bases with parentheses are replaced by finite-dimensional decompositions, has appeared simultaneously with the notion of decomposition (see the corresponding Notes and remarks).

Finite-dimensional decompositions are a special case of decompositions and therefore we refer for comments on them to the Notes and remarks to § 15. Proposition 13.11 is contained, implicitly, e.g. in the paper of W. B. Johnson and H. P. Rosenthal ([186], remark IV.6).

For theorem 13.1 see the preceding Notes and remarks (to the results used in the proof of theorem 13.1). Theorem 13.2 was announced, without proof, by M. I. Kadec ([196], theorem 1 and [197], theorem 3.1); a proof, in the framework of Fréchet spaces, was given by M. I. Kadec and A. Pelczynski ([206], theorem 3). $w^*$-Schauder basic sequences with parentheses and proposition 13.12, in the equivalent version of $w^*$-Schauder finite-dimensional decompositions, are contained, implicitly, in the paper of W. B. Johnson and H. P. Rosenthal ([186], p. 90), who also proved theorem 13.3, remark 13.8 and corollary 13.6 ([186], theorem IV.4). Lemma 13.1 was proved, independently, by M. I. Kadec [195] and V. Klee [218].

Theorem 13.4 can be found, essentially, in [311], footnote (9) (or, in [187], lemma 2.2); see also V. I. Gurariĭ [141], theorem 2 and D. J. Fleming and W. Ruckle [113], theorem 5.4.

Proposition 13.13 and corollary 13.7 were given by W. B. Johnson ([176], Added in proof).

Let us call the *finite-dimensional decomposition constant* $\Delta(E)$ of a Banach space $E$ the infimum of all numbers $\lambda$ such that $E$ has a commuting $\pi_\lambda$-basis (or, equivalently, a finite-dimensional decomposition $\{G_n\}$ with the associated partial sum operators $u_n$ satisfying $\sup_{1\leqslant n<\infty} \|u_n\| \leqslant \lambda$). Then, similarly to § 12, problem 12.3, it is natural to ask whether for every Banach space $E$ with a finite-dimensional decomposition this number is attained. The analogous problem for the basis constant

*) Indeed, if we define $\{G_n\}$ by (13.114), then for any $y_i \in G_i$ $(i = 1, \ldots, n+1)$ we have

$$\left\|\sum_{i=1}^{n} y_i\right\| = \left\|u_n\left(\sum_{i=1}^{n+1} y_i\right)\right\| \leqslant \|u_n\| \left\|\sum_{i=1}^{n+1} y_i\right\| = \left\|\sum_{i=1}^{n+1} y_i\right\|,$$

whence, by § 15, theorem 15.5, $\{G_n\}$ is a monotone finite-dimensional decomposition of $E$.

$\beta(E) = \dfrac{1}{\Gamma(E)}$, of Banach spaces $E$ with a basis, is also open. One can also raise the "easier" problem whether these constants are at least "*asymptotically attained*", in the sense of § 12, theorem 12.1 $\gamma$), implication $1° \Rightarrow 4°$ or $3° \Rightarrow 4°$. By slightly modifying the proof of proposition 13.7 a), we obtain the following result, which implies that the answer is affirmative when $\Delta(E) = 1$:

**Proposition 13.14.** *If* $\Pi(E) = 1$, *then for every sequence* $\varepsilon_n' > 0$ $(n = 1,2, \dots)$ *with* $\lim_{n\to\infty} \varepsilon_n' = 0$ *there exists a commuting* $\pi$*-basis* $\{u_n\}$ *of* $E$ *satisfying*

$$\|u_n\| \leqslant 1 + \varepsilon_n' \qquad (n = 1, 2, \dots). \tag{13.115}$$

*Proof.* Let $\varepsilon_n > 0$ $(n = 1, 2, \dots)$, $\prod_{n=1}^{\infty} (1 + \varepsilon_n) < \infty$. Then, by $\Pi(E) = 1$ and § 12, theorem 12.1 $\gamma$), implication $1° \Rightarrow 4°$, $E$ has a $\pi$-basis $\{v_n\} \subset L(E, E)$ satisfying

$$\|v_n\| \leqslant 1 + \varepsilon_n \qquad (n = 1, 2, \dots). \tag{13.116}$$

Let $\{n_k\}$ be an increasing sequence of positive integers such that

$$\prod_{n=n_k}^{\infty} (1 + \varepsilon_n) \leqslant 1 + \varepsilon_k' \qquad (k = 1, 2, \dots). \tag{13.117}$$

Then, by the proof of proposition 13.5, the sequence

$$u_k(x) = \lim_{j\to\infty} v_{n_k} v_{n_k+1} \cdots v_{j-1} v_j(x) \qquad (x \in E, k = 1,2, \dots)$$

is a commuting $\pi$-basis of $E$. Clearly,

$$\|u_k\| \leqslant \prod_{n=n_k}^{\infty} (1 + \varepsilon_n) \leqslant 1 + \varepsilon_k' \qquad (k = 1, 2, \dots),$$

so $\{u_k\}$ satisfies (13.115), which completes the proof.

Let us note that, by proposition 13.7 a), we have $\Delta(E) = 1$ if (and, obviously, only if) $\Pi(E) = 1$. When $\Delta(E) > 1$, respectively, when $\beta(E) = \dfrac{1}{\Gamma(E)} \geqslant 1$, the above mentioned problems (of whether $\Delta(E)$, $\beta(E)$, are asymptotically attained) are still open. These problems were raised in the initial version of [86], where proposition 13.14 was also proved.

§ **14.** Lemma 14.1 can be found in the paper of W. B. Johnson, H. P. Rosenthal and M. Zippin ([187], lemma 3.1 and corollary 3.2), with the mention that it was essentially proved by W. B. Johnson in [178], lemma 1 (which is nothing else than § 9, lemma 9.3 a)); in [187] a different proof is given, using the "principle of local reflexivity".

Propositions 14.1, 14.2 and remarks 14.1, 14.2 were given by W. B. Johnson, H. P. Rosenthal and M. Zippin ([187], lemmas 4.4, 4.5, 4.2

and 4.3). Let us mention separately the following underlying lemma, which may unify the proofs of some results of § 13 and § 14 (e.g., the third proof of proposition 13.3, and the proofs of propositions 13.6, 13.8, 14.1 and 14.2):

**Lemma 14.5.** *Let $E$ be a Banach space and let $t, p \in L(E, E)$, where $p$ is a projection.*
*a) If $pt = t$, then*

$$q = t + p - tp = I - (I - t)(I - p) \tag{14.62}$$

*is a projection on $E$, satisfying*

$$pq = q, \; q(E) = p(E). \tag{14.63}$$

*b) If $tp = t$, then*

$$q = p + t - pt = I - (I - p)(I - t) \tag{14.64}$$

*is a projection on $E$, satisfying*

$$qp = q, \operatorname{Ker} q = \operatorname{Ker} p, \; q^*(E^*) = p^*(E^*). \tag{14.65}$$

c) *If $pt = p$, then*

$$q = tp \tag{14.66}$$

*is a projection on $E$, satisfying*

$$pq = p, \; \operatorname{Ker} q = \operatorname{Ker} p, \; q^*(E^*) = p^*(E^*). \tag{14.67}$$

d) *If $tp = p$, then*

$$q = pt \tag{14.68}$$

*is a projection on $E$, satisfying*

$$qp = p, \; q(E) = p(E). \tag{14.69}$$

*Proof.* The proofs of a), b) are contained in the proofs of propositions 14.1 and 14.2 (the last two equalities in (14.65) are equivalent). Finally, c) and d) are consequences of a) and b) respectively, applied to $t, p$ and $q$ replaced by $I - t$, $I - p$ and $I - q$ respectively (but, of course, it is also easy to show them directly), which completes the proof.

Theorem 14.1 a), b), d), for $V = E^*$ in a) and with $q_n^*(E^*)$ and $q_n(E)$ only $\varepsilon_n$-close to $v_{m_n}(E^*)$ and $u_{m_n}(E)$ in b) and d) respectively, where $0 <$

$< \varepsilon_n < 1$ $(n = 1,2, \ldots)$, and remark 14.3 b), were given by W. B. Johnson, H. P. Rosenthal and M. Zippin ([187], theorems 1.3 and 4.1). The main part of theorem 14.1 c), with $q_n^*(E^*)$ only $\varepsilon_n$-close to $v_{m_n}(E^*)$ in (14.27), where $0 < \varepsilon_n < 1$ $(n = 1,2, \ldots)$, was proved by D. W. Dean ([73], theorem 2 and corollary 1). The improvements (14.27) and (14.28) of the above mentioned results were given in [377], theorem 1 and corollary 1.

Lemmas 14.2 and 14.3 are well known; for example, their qualitative parts can be found in [140], Ch. I, § 11, corollary 1 and [213], lemma 3 respectively. The main parts of theorem 14.2 a), b), were proved by W. B. Johnson, H. P. Rosenthal and M. Zippin ([187], theorem 1.4 (a)) and D. W. Dean ([73], corollary 2) respectively; the complements (14.39), (14.40) were given in [377], theorem 2. In connection with problem 14.1 b) let us mention that the problem is also of interest in the case of non-locally convex spaces $U$. Thus, for example, N. J. Kalton and J. H. Shapiro have recently proved [210] some results on the (still unsolved) problem of existence of basic sequences in $F$-spaces (see also N. J. Kalton [209]). Lemma 14.4 was given by W. B. Johnson, H. P. Rosenthal and M. Zippin ([187], lemma 2.3), who have omitted the assumption that $\dim G_i < \infty$ $(i = 1,2)$. However, as shown by simple examples (with $G_1 = G_2 = E$, $\operatorname{codim}_E F_i = i$ for $i = 1,2$ and $F_1$ linearly isometric to $F_2$), lemma 14.4 need not hold for infinite dimensional $G_i$; for such $G_i$, the proof given here is no longer valid, since $t$ need not map $G_1$ *onto* $G_2$. Theorem 14.3 and remark 14.4 were given by W. B. Johnson, H. P. Rosenthal and M. Zippin ([187], theorem 1.4 (b), remark 4.10 and theorem 4.9). Corollary 14.1 is due to W. B. Johnson ([176], Added in proof).

Corollary 14.2 was given, with a different proof (applying § 18, theorem 18.7, to $E^*$), by W. B. Johnson ([176], Added in proof). However, his proof for $p = 1$, given in [176], contradicts the result that $C_1^*$ does not have the bounded approximation property. As an answer to this question, W. B. Johnson has observed (verbal communication) that the proof in [176] yields, actually, theorem 14.4, which implies also corollary 14.2 for $1 \leqslant p \leqslant \infty$. Corollary 14.3, remark 14.5 and problem 14.2 are due to W. B. Johnson ([176], Added in proof).

**§ 15.** The notion of decomposition (basis of subspaces) of a Banach space was introduced by M. M. Grinblium [136], who used the term "representation as a direct sum of subspaces". Independently, M. K. Fage [103], [104] has studied this concept in Hilbert space (see also A. S. Markus [258]). According to communications of C. Bessaga (recorded in the paper [353] of B. L. Sanders) and of A. Pelczynski (recorded in [367], p. 718, footnote), the notion was also discussed by S. Mazur, at a seminar in Warsaw, in 1953; however, S. Mazur has not published his results. A further stimulus to the development of the study of decompositions of Banach spaces was given in the 1960's by C. W. McArthur

([272], [273]) and his students at Florida State University, in a number of doctoral dissertations.

Proposition 15.1 is due to B. L. Sanders ([354], theorem 1), who observed that it is also valid for every normed linear space. The sequence of coordinate projections associated to a decomposition was considered by C. W. McArthur [272], who also observed the orthogonality relations (15.6). The distinction between decompositions and Schauder decompositions was made by C. W. McArthur ([272], [273]), who introduced the terms "Schauder decomposition" and "Schauder basis of subspaces" (see also B. L. Sanders [354], § 2 and W. H. Ruckle [343], pp. 543, 549). Furthermore, C. W. McArthur [272] gave the first example of a non-separable Banach space $E$ having a Schauder decomposition (reproduced by B. L. Sanders in [354], example 1), namely, $E=(G \times G \times \ldots)_{c_0}$, where $G$ is a non-separable Banach space). An example of a decomposition of $E=l^\infty$, which is not a Schauder decomposition, was given by B. L. Sanders ([354], example 2). Proposition 15.2, for complete metric linear spaces, can be found, without proof, in the paper of W. H. Ruckle ([343], lemma I.2 and corollary I.4) and a proof can be found in J. T. Marti [263], p. 87. In [273], C. W. McArthur assumed that $\{G_n\}$ is $\omega$-linearly independent and considered, instead of the space $D_1$ defined by (15.8), the space

$$\mathscr{E}_c=\left\{x \in E \,\middle|\, \exists\, \{y_n\} \subset E,\ y_n \in G_n \ (n=1,2, \ldots), \text{ such that } x=\sum_{i=1}^{\infty} y_i\right\}, \tag{15.151}$$

endowed with the norm

$$|||x|||=\sup_{1 \leqslant n<\infty}\left\|\sum_{i=1}^{n} y_i\right\| ; \tag{15.152}$$

in [273], theorem 3.5 (ii) he showed that $\mathscr{E}_c$ is a Banach space and $\{G_n\}$ is a basis of $\mathscr{E}_c$. Proposition 15.3 is contained, implicitly, in W. H. Ruckle [343], theorem I.6. Theorem 15.1 (whence also corollary 15.1), for complete metric linear spaces, essentially with the first proof given here, was shown by C. W. McArthur ([273], theorem 4.3). The second proof of theorem 15.1 can be found, essentially, in J. T. Marti [263], Ch. VII, § 1. A more concise way of writing the spaces $D_1, D_2$ and other similar expressions in § 15 would be to replace $\{y_n\} \subset E, y_n \in G_n$ $(n=1,2, \ldots)$ by: $\{y_n\} \in \prod_{n=1}^{\infty} G_n$.

Proposition 15.4 has appeared in the paper of J. R. Retherford ([335], lemma 3), with the mention that it is due, essentially to B. L. Sanders [354]. Corollary 15.2 b) was observed by J. R. Retherford ([333], corollary 2.4). Lemma 15.1 was proved by A. Sobczyk ([382], p. 946); its proof, presented here, is due to W. A. Veech [401]. Corollary 15.3 (the first statement) was given by B. L. Sanders ([354], theorem 2).

The problem, whether every Banach space $E$ and, in particular, the space $E = l^\infty$, has a Schauder decomposition, was raised by B. L. Sanders ([354], § 1). When this problem was answered in the negative (see theorem 15.2), J. R. Retherford raised problem 15.1*) ([335], problem 1). Some partial results in the direction of theorem 15.2 were obtained by B. L. Sanders, who proved that $E = l^\infty$ has no Schauder decomposition $\{G_n\}$ such that the unit vectors $e_n$ lie in distinct $G_k$'s ([354], theorem 3) and J. R. Retherford, who proved that $E = l^\infty$ has no Schauder decomposition $\{G_n\}$ with each $G_n$ an ideal ([333], theorem 4.1) and the result stated in remark 15.3 b) ([333], theorem 3.1); the latter result was also obtained, independently, by D. W. Dean, who proved theorem 15.2 and made remark 15.3 a) [71]. In connection with remark 15.3 a) let us mention that initially the $\mathscr{P}_\lambda$ spaces had been defined as the Banach spaces $E$ such that for every (Banach) superspace $F$ containing $E$, there exists a projection $u$ of $F$ onto $E$ with $\|u\| \leqslant \lambda$; however, it is well known (see e.g. [158], [383]) that $E$ is a $\mathscr{P}_\lambda$ space if and only if there exists a compact extremally disconnected space $Q$ such that $E$ is linearly isometric to a complemented subspace of $C(Q)$, admitting a projection of norm $\leqslant \lambda$. It is also known [383] that $E$ is complemented in every superspace if and only if there exists a $\lambda \geqslant 1$ such that $E$ is a $\mathscr{P}_\lambda$ space.

The following might be another possible way of introducing the generalized biorthogonal systems: One can say that a pair of sequences of subspaces $(G_n, \Gamma_n)$ $(G_n \subset E, \Gamma_n \subset E^*; n = 1,2, \ldots)$ is a *generalized biorthogonal system* (or a biorthogonal system of subspaces) if we have

$$f_i(y_j) = \delta_{ij} \quad (f_i \in \Gamma_i, \ y_j \in G_j; \ i,j = 1,2, \ldots); \tag{15.153}$$

another modification of definition 15.4 might be to require only $\{v_n\} \subset$ $\subset L\left(\left[\bigcup_{n=1}^{\infty} G_n\right], \left[\bigcup_{n=1}^{\infty} G_n\right]\right)$ instead of $\{v_n\} \subset L(E, E)$. Definition 15.4 is more restrictive and therefore it yields more properties of the generalized biorthogonal systems.

Theorem 15.3 is due to W. H. Ruckle ([343], theorem II.2 and lemma II.3; see also C. W. McArthur [273], theorem 5.5 (ii)). The term "resolution of the identity" means actually "Schauder resolution of the identity", in accordance with the terminology of definition 15.3; however, we have omitted "Schauder", since it is also omitted in the transfinite case (see the Notes and remarks to § 19) and in the theory of operators

---

*) In this connection let us mention (see § 1, remark 1.14) that it is not known whether every Banach space $E$ can be written as the direct sum of two infinite dimensional subspaces; this "weaker" problem had been raised by R. Whitley, long before problem 15.1.

(see the Notes and remarks to § 18). The observation that (15.23) is a Banach space, was made by W. H. Ruckle ([343], lemma I.2).

Theorem 15.4 can be found, in the somewhat different language of "$\pi$-rings", used by S. Yamazaki (see the Notes and remarks to Ch. I, § 5), in the book of J. T. Marti ([263], p. 109, theorem 6). In the form presented here, the notions of multiplier (with respect to a total generalized biorthogonal system) and Cesàro decomposition, as well as theorem 15.4 and the corresponding extension of § 11, theorem 11.4, to Cesàro decompositions, were given by P. L. Butzer, R. J. Nessel and W. Trebels ([41], § 3); for further results in this direction, see also the references mentioned in the Notes and remarks to § 11.

$\omega$-linearly independent sequences of subspaces were considered by C. W. McArthur [273], who called them "$\omega$-independent" and by W. H. Ruckle [343], who called them (in contrast with the terminology*) of A. I. Markuševič for sequences of elements) "strongly linearly independent" and, independently, by A.S.Markus [258] (see also V. N. Viziteĭ [403], who considered minimal sequences of subspaces and observed that every minimal sequence of subspaces is $\omega$-linearly independent). Let us mention that W. H. Ruckle [343] and C. W. McArthur [273] have also considered some other types of linear independence of a sequence of subspaces $\{G_n\}$, such as *weakly linearly independent* (i.e., such that $G_n \neq \{0\}$ for $n = 1,2,\ldots$ and that the relations $y_n \in G_n$ for $n = 1,2,\ldots$ and $\sum_{i=1}^{\infty} f(y_i) = 0$ for all $f \in E^*$ imply $y_n = 0$ for $n = 1,2,\ldots$) and *unordered S-independent* or *unordered strongly linearly independent sequences* (which are the natural extension of unconditionally $\omega$-linearly independent sequences of elements; see Ch. II, § 16, definition 16.1).

The equivalences $1° \Leftrightarrow 3°$ and $1° \Leftrightarrow 2°$ of theorem 15.5 were proved by M. M. Grinblium ([136], theorem 5) and C. W. McArthur ([273], theorem 5.2 (iii)) respectively. The latter has also observed, as a consequence ([273], corollary 5.3), that *if* $\{G_n\}$ *is a Schauder decomposition of* $E$ *and* $F_n \neq \{0\}$ *is a subspace of* $G_n$ $(n = 1,2,\ldots)$, *then* $\{F_n\}$ *is a Schauder decomposition of* $\left[\bigcup_{n=1}^{\infty} F_n\right]$; this yields, in particular, theorem 15.21 a). Corollary 15.4, proposition 15.5 and remark 15.4 are due, essentially, to W. H. Ruckle ([343], corollary I.7 and theorem I.6).

Blockings (i.e., block-Schauder decompositions) with respect to a finite-dimensional decomposition $\{G_n\}$ were considered by W. B. Johnson and M. Zippin [190], [191]. Fully equivalent sequences of subspaces were considered by V. N. Viziteĭ [403], who called them "equivalent".

---

*) See the Notes and remarks to Ch. I, § 6.

Theorem 15.6 is due to C. W. McArthur ([273], corollary 5.6 (ii)). For a sequence of subspaces which is a (Schauder) decomposition of its closed linear span, C. W. McArthur ([273], § 5) has used the term *(Schauder) basic family of subspaces;* one might replace here "family" by "sequence", in analogy with the particular case of basic sequences of elements. Example 15.1 was given by J. R. Retherford ([335], §§ 2–3).

Some parts of proposition 15.7 and a converse to a part of proposition 15.7 c), were given, essentially, by J. J. M. Chadwick ([47], lemmas 3.1, 3.4, 4.2 and corollary 3.2).

For some aspects of the theory of vector sequence spaces, in the framework of locally convex spaces, see e.g. the book of A. Pietsch [318] and the references therein. Vector *BK*-spaces $S(\{G_j\})$ were considered by W. H. Ruckle [344], who has called them "$B$-$\omega(\{G_j\})$ spaces" (where $\omega(\{G_j\}) = s(\{G_j\})$ with our notation) and has also assumed, in the definition, that each $\underbrace{\{0\} \times \ldots \times \{0\}}_{n-1} \times G_n \times \times \{0\} \times \ldots$ belongs to $S(\{G_j\})$ (which, however, is not assumed*) in the scalar case). The $\gamma$-dual, $\beta$-dual, $\gamma$-predual and $\beta$-predual were introduced by W. H. Ruckle [344], who used for the last two ones the terms "sub-$\beta$ dual" and "sub-$\gamma$ dual". Theorem 15.9 is due, essentially, to W. H. Ruckle ([344], p. 124). In the particular case when $\{x_n\}$ is a basis of $E$, the sequence spaces (15.80) and the mappings $A(E, E) \rightarrow a(E)$ defined by $u \rightarrow \{u(x_n)\}$, were considered by W. H. Ruckle ([344], corollary I.7 and theorem I.6). For the case when $\{G_n\}_{n=1}^{m}$ is a *finite* collection of closed subspaces of $E$, some results corresponding to theorem 15.5, corollary 15.4 and remark 15.4 a), were proved by H. A. Kober [221] and M. M. Grinblium ([136], § 3). The spaces of operators (15.81) and conditions under which they are ideals of operators were studied by W. H. Ruckle [349]. One can also define "multipliers" of operators as follows (see W. H. Ruckle [348]): If $u \in L(E, F)$, where $E$, $F$ are Banach spaces and if $S_1 = S_1(\{G_n\})$, $S_2 = S_2(\{F_n\})$ are spaces of sequences in $E$ and $F$ respectively, one can define

$$M(u, (S_1, S_2)) = \{\{\gamma_n\} \subset K \mid \{\gamma_n u(y_n)\} \in S_2 \text{ for all } \{y_n\} \in S_1\}, \quad (15.154)$$

which is a scalar sequence space. However, we shall not make more comments in these directions, since in § 15 the vector sequence spaces were considered mainly in their relation to Schauder decompositions.

Schauder decompositions of general topological linear spaces and weak, bounded weak, weak* and bounded weak* Schauder decompositions for Banach spaces and their duals were considered by J. R. Retherford [332] (see also C. W. McArthur and J. R. Retherford [276], J. R. Retherford [333]). The following "weak basis theorem" for bases

---

*) See Ch. I, § 12, p. 131.

of subspaces, which follows by putting together remark 15.5 and theorem 15.10, was proved by W. H. Ruckle ([343], theorem I. 20): *If* $\{G_n\}$ *is a weak decomposition of a Banach space E, such that each* $G_n$ *is closed, then* $\{G_n\}$ *is a Schauder decomposition of E*; another proof, for complete metric linear spaces, was given by C. W. McArthur [274]. Theorem 15.11 was proved by J. R. Retherford ([332], theorem 3.1), but part of it can be also found in W. H. Ruckle [343], lemma II.4. J. R. Retherford has also extended Ch. I, § 13, theorem 13.1 b) to the case of decompositions ([332], theorems 5.3 and 3.2). Examples 15.2, 15.3 and the remark made after example 15.3, are due to J. R. Retherford ([333], theorem 3.2 and [335], § 4).

Monotone Schauder decompositions were considered by J. P. Russo [350] and others (e.g., A. Pelczynski and P. Wojtaszczyk [313]). Shrinking Schauder decompositions were introduced by B. L. Sanders [353], who also proved the equivalences $1° \Leftrightarrow 2° \Leftrightarrow 3°$ of theorem 15.13; the latter result can be also found in W. H. Ruckle [343], lemma II.6, with the mention that it is a version of theorem VI.1 of the thesis of B. L. Sanders (1962). Boundedly complete Schauder decompositions were introduced in the same thesis of B. L. Sanders (see W. H. Ruckle [343], the remark before theorem II.11). A version of the implication $1° \Rightarrow 3°$ of theorem 15.14 (namely, that if $\{G_n\}$ is a boundedly complete Schauder decomposition of $E$, with the a.s.c.p. $\{v_n\}$, then $E$ is linearly isomorphic to $\left[\bigcup_{n=1}^{\infty} v_n(E^*)\right]^*$), was proved by W. H. Ruckle ([343], theorem II.10). B. L. Sanders ([353], p. 205) raised the problem, whether this result remains valid without the requirement of reflexivity of each $G_n$. Example 15.4 a) and proposition 15.11, which give a negative answer to this question, are due to J. R. Retherford ([335], remark 1), with the mention that, previously, another (more complicated) example, which also gives a negative answer to the question, had been obtained in the thesis of J. B. Richmond (1966); the additional example 15.4 b) can be found in the papers of P. K. Subramanian and S. Rothman ([386], § 7) and J. J. M. Chadwick ([46], example 3.7) respectively. The equivalence $3° \Leftrightarrow 4°$ of theorem 15.15, for any separable total subspace $V$ of $E^*$, was given in [365a], theorem 2.

If $\{e_n\}$ is a symmetric basis of $F$, with symmetric norm $\nu^{(s)}_{\{e_n\}} = 1$ (see Ch. II, § 22, definitions 22.1 and 22.2), then $\Phi(\{\alpha_n\}) = \left\|\sum_{i=1}^{\infty} \alpha_i e_i\right\|$ is a "symmetric gauge function" [356] and conversely; the class of Besselian Schauder decompositions with respect to such a function was studied by V. N. Viziteĭ [403], without using any special term for them. Finite-dimensional $l^p$-decompositions ($1 \leqslant p < \infty$) and $c_0$-decompositions were considered by W. B. Johnson and M. Zippin [190]. Infinite dimensional $l^1$-decompositions were considered by J. J. M. Chadwick [48], who called them "absolute Schauder decompositions", since they generalize the absolutely convergent bases (see Ch. II, § 25) and proved corollary 15.6 for $p = 1$ ([48], theorem 2.2). Theorem 15.17 is due to V. N. Viziteĭ ([403], theorems 1 and 3); the proof of part a), given in [403], shows that one can take

$$\lambda = \frac{1}{4 \sup_{1 \leqslant n < \infty} \left\| \sum_{i=1}^{n} v_i \right\| (1 + \sup_{1 \leqslant n < \infty} \|v_n\|)^2}, \tag{15.155}$$

where $\{v_n\}$ is the a.s.c.p. to $\{G_n\}$. For some other stability theorems for Schauder decompositions, see also A. S. Markus [258] and V. N. Viziteĭ [403].

Following W. B. Johnson and E. Odell [185], a finite-dimensional decomposition*) $\{G_n\}$ of a Banach space $E$ is called *block p-Besselian*, respectively *block p-Hilbertian*, where $1 \leqslant p < \infty$, provided there exists a constant $C > 0$ such that

$$C(\sum \|z_i\|^p)^{\frac{1}{p}} \leqslant \|\sum z_i\|, \tag{15.156}$$

respectively such that

$$\|\sum z_i\| \leqslant C(\sum \|z_i\|^p)^{\frac{1}{p}}, \tag{15.157}$$

for any blocking $\{F_n\}$ of $\{G_n\}$ (see definition 15.10, formula (15.36)) and any $z_i \in F_i$; for the particular case of bases and $p = 2$, these notions were introduced by E. Dubinsky, A. Pelczynski and H. P. Rosenthal ([85], definition 5.2).

The notions of unconditional decomposition and unconditional Schauder decomposition are as old as those of decomposition and Schauder decomposition (see M. M. Grinblium [136] and C. W. McArthur [273]) and several theorems for (Schauder) decompositions were obtained together with their unconditional analogues (and appeared in the same papers). Thus, the equivalence 1° ⇔ 2° of theorem 15.18 was proved by C. W. McArthur ([273], theorem 5.2 (i)) and an unconditional analogue of theorem 15.5, equivalence 1° ⇔ 3°, was proved by M. M. Grinblium ([136], theorem 6); the implication 1° ⇒ 4° of theorem 15.18 was observed by J. Lindenstrauss and A. Pelczynski ([243], § 6). The unconditional analogues of corollary 15.4, proposition 15.5 and remark 15.4 were proved, essentially, by W. H. Ruckle ([343], corollary I.11 and theorem I.10). The observation that the known examples of Banach spaces which cannot be embedded into any Banach space with an unconditional basis have the property that they cannot be embedded into any Banach space with a finite-dimensional decomposition, was made by A. Pelczynski and P. Wojtaszczyk [313]. Theorem 15.19 and the generalization mentioned after it, are due to J. Lindenstrauss and A. Pel-

*) Actually, one can consider here any Schauder decomposition $\{G_n\}$.

czynski ([243], § 6, corollary 8). As was already mentioned above, absolutely convergent Schauder decompositions were considered by J. J. M. Chadwick [48].

$M$-decompositions were considered by G. F. Bachelis and H. P. Rosenthal [10], who called them "complete biorthogonal decompositions"; they have also given proposition 15.13 and remark 15.7 ([10], § 3, proposition and remark). G. F. Bachelis ([9], theorem 3.4, equivalence (1) $\Leftrightarrow$ (4)) proved the following unconditional version of proposition 15.13, which is a generalization of § 7, theorem 7.3: *A sequence* $\{G_n\}$ *of subspaces of a Banach space* $E$, *with* $G_n \neq \{0\}$ $(n = 1, 2, \ldots)$ *and* $\left[\bigcup_{n=1}^{\infty} G_n\right] = E$, *is an unconditional Schauder decomposition of* $E$ *if and only if for every (finite or infinite) sequence of positive integers* $\{n_k\}$ *we have* $E = \left[\bigcup_{k=1}^{\infty} G_{n_k}\right] \oplus [\bigcup_{j \in \mathcal{N} \setminus \{n_k\}} G_j]$.

For finite-dimensional quasi-decompositions, the implications $1° \Rightarrow 2°$ and $1° \Rightarrow 3°$ of theorem 15.20 and their unconditional analogues can be found, in the equivalent language of finite-dimensional expansions of $I_E$, in the paper of A. Pelczynski and P. Wojtaszczyk ([313], lemma 1.2 ($\beta$), ($\gamma$)), who have also observed, in this particular case, that the sequence $\{F_n\}$ defined by (15.15) is an unconditional finite-dimensional decomposition of $D_1^{(u)}$ defined by (15.123) ([313], lemma 1.2 ($\alpha$)). Corollary 15.7 was given, without proof, by A. Pelczynski and P. Wojtaszczyk ([313], theorem 1.1); the remarks made after corollary 15.7 (except the first one) and example 15.5 are due to A. Pelczynski and P. Wojtaszczyk ([313], example 1.1, lemma 1.3 and the remark made after it).

As has been already mentioned above, theorem 15.21 a) is due to C. W. McArthur [273]. Theorem 15.21 b), example 15.6 and theorem 15.22 were given by W. J. Davis [59].

Cofinal bibasic systems were considered by J. J. M. Chadwick [48], who called them "cofinal biorthogonal systems". Propositions 15.14 and 15.15 are due to J. J. M. Chadwick ([48], theorem 2.1 and 2.3).

The idea of generalizing bases, replacing the sequence $\{x_n\}$ by a function $X: t \to X(t)$ from $[0, 1]$ into $E$, with $\|X(t)\| = 1$ $(t \in [0, 1])$, and the series expansions $\sum_{i=1}^{\infty} f_i(x) x_i$ by representations of the elements $x \in E$ as certain integrals $x = \int_0^1 X(t) d\lambda_x(t)$, is due to I. Maddaus [256] (see the Notes and remarks to § 9).

Integral bases and Schauder integral bases $(X, T, \{Q_n\}, \mathfrak{M})$ of inductive limits of Fréchet spaces, were introduced by R. E. Edwards [93], who did not assume condition (iii) (which, as has been mentioned, is posed in order to exclude some trivial cases); however, this condition may be assumed, without restricting the generality (by omitting, if necessary, some of the $Q_n$'s). Independently, another version of integral bases was introduced, for Hilbert spaces, by G. M. Keselman [216] (who called them "continual bases", with the mention that their definition was suggested by B. E. Ljance) and extended to Banach spaces in [368], definition 7.1. Namely, let $E$ be a Banach space, $(T, \nu)$ a positive measure space and $M_\nu$ the linear space of equivalence classes of $\nu$-measurable and $\nu$-almost everywhere finite functions. A linear mapping $v: E^* \to M_\nu$ is called ([216], [368]) a *continual basis* of $E$, if for every $x \in E$ there exists a *unique* element $\xi_x \in M_\nu$ such that

$$f(x) = \int_T \xi_x(t)[v(f)(t)]\, d\nu(t) \qquad (f \in E^*), \tag{15.158}$$

where $\int_T \ldots d\nu(t)$ is defined as $\lim_{n\to\infty} \int_{T_n} \ldots d\nu(t)$, with $\{T_n\}$ a fixed increasing sequence of $\nu$-measurable subsets of $T$, given in advance. In particular, if $\{x_n\}$ is a basis of $E$, with the a.s.c.f. $\{f_n\} \subset E^*$, we can take $T = \mathcal{N} = \{1,2,3, \ldots\}$, $T_n = \{1, \ldots, n\}$ $(n = 1,2, \ldots)$, $\nu(\{1\}) = \nu(\{2\}) = \ldots = 1$ (hence $M_\nu = s$, the linear space of all scalar sequences $\xi = \{\xi(n)\}$) and define $v: E^* \to M_\nu = s$ by

$$v(f) = \{f(x_n)\} \qquad (f \in E^*); \tag{15.159}$$

then, (15.158) means that

$$f(x) = \lim_{n\to\infty} \sum_{i=1}^{n} \xi_x(i)v(f)(i) = \sum_{i=1}^{\infty} \xi_x(i)f(x_i) \qquad (f \in E^*),$$

so it is satisfied with $\xi_x(n) = f_n(x)$ $(n = 1,2, \ldots)$. The uniqueness of $\xi_x$ follows from biorthogonality, since if we have

$$f(x) = \lim_{n\to\infty} \sum_{i=1}^{n} \alpha_i v(f)(i) = \sum_{i=1}^{\infty} \alpha_i f(x_i) \qquad (f \in E^*),$$

then for $f = f_n$ we obtain $f_n(x) = \sum_{i=1}^{\infty} \alpha_i f_n(x_i) = \alpha_n$ $(n = 1,2,\ldots)$. Thus, every basis is a continual basis in the above sense. G. M. Keselman has also defined orthogonal, Besselian and Hilbertian continual bases in Hilbert spaces [216].

Theorems 15.23 and 15.24 for strict inductive limits of Fréchet spaces were given, with different proofs, by R. E. Edwards ([93], theorem 1).

Theorem 15.25 is contained, essentially, in the thesis of D. F. Hale [153], where definitions 15.27 and 15.28 were also given.

For some other types of integral bases see also § 17 and the Notes and remarks to it.

R. E. Edwards [94] considered the extension of strongly series summable $M$-bases and of quasi-bases of $E$, defined by the condition

$$\tau_n(x) = \int_T \lambda_n(t) f_t(x) X(t) d\mu(t) \to x \qquad (x \in E), \qquad (15.160)$$

where $T$ is an infinite locally compact space, $X: t \to X(t)$ a mapping of $T$ into $E$, $t \to f_t$ a mapping of $T$ into $E^*$, $\mu$ a positive Radon measure on $T$ and $\{\lambda_n(\cdot)\}$ a sequence of scalar-valued functions on $T$, called "summability factors" (the integrals being taken in the sense of § 17, hence unconditionally convergent). Taking, in particular, $T = \mathcal{N}$ with the discrete topology, $X(n) = x_n \in E$ and $f_n \in E^*$ such that $\{x_n\}$ is an $M$-basis of $E$, with the a.s.f. $\{f_n\}$, $\mu(\{n\}) = 1$ $(n = 1,2, \ldots)$ and $(\lambda_{nm})$ a triangular matrix, one obtains a strongly series summable $M$-basis $\{x_n\}$ of $E$, while taking $T = \mathcal{N}$ with the discrete topology, $X(n) = x_n \in E$ and $f_n \in E^*$ $(n = 1,2, \ldots)$ arbitrary, $\mu(\{n\}) = 1$ and $\lambda_{nm} = \delta_{nm}$ $(n, m = 1,2, \ldots)$, one obtains a quasi-basis $\{x_n\}$ of $E$. This concept also extends the special class of $\gamma$-bases, in which each series $\tau_n(x) = \sum_{i=1}^{\infty} \lambda_{ni} f_i(x) x_i$ is required to be unconditionally convergent.

§ **16.** Bases for sets were introduced by V. I. Gurariĭ [143] who has not required the condition $\{x_n\} \subset M$ in their definition (see remark 16.1), but has assumed it separately in the results which he obtained in [143]. Acute angled cones and the angular modulus (16.2) of a set were considered by V. I. Gurariĭ [143], who also mentioned, without proof, formula (16.3) and the following additional examples ([143], propositions 3 and 4): a) The set $\mathcal{M} = \{x \in L^p([a, b]) | x \geqslant 0\}$ is an acute angled cone in $E = L^p([a, b])$, with $\theta_{\mathcal{M}}(t) = \sqrt[p]{1 + t^p} - 1$ $(1 \leqslant p < \infty,\ 0 \leqslant t < \infty)$; b) For each $t_0 \in [a, b]$, the set $\mathcal{M} = \{x \in C([a, b]) \mid x(t_0) = \|x\|\}$ is an acute angled cone in $E = C([a, b])$, with $\theta_{\mathcal{M}}(t) = t$ $(0 \leqslant t < \infty)$. Lemmas 16.1, 16.2 and remark 16.2 are due to V. I. Gurariĭ ([143], propositions 1 and 2).

The cone $\mathscr{C}_{\{x_n\}}$ defined by (16.10) is nothing else than the smallest closed convex cone having the origin as extreme point, which contains the sequence $\{x_n\}$ and it is called in [143] the "closure of the conical envelope" of $\{x_n\}$. Acute angled sequences and the angular modulus (16.13) of a sequence were considered in the same paper of V. I. Gurariĭ [143], who has also given, without proof, the following example ([143], proposition 5): For every sequence $\{\varepsilon_n\}$ with $\varepsilon_n = \pm 1$ $(n = 1,2, \ldots)$, the sequence $\{\varepsilon_n r_n\}$, where the $r_n$'s are the Rademacher functions*) on $[a, b]$, is acute angled in $E = L^p([a, b])$ for $1 < p \leqslant \infty$ and not acute angled in $L^1([a, b])$. The equivalence $1° \Leftrightarrow 2°$ of theorem 16.1, and corollary 16.1, are due to V. I. Gurariĭ ([143], theorem 2 and corollary); however, the proof given here for the implication $2° \Rightarrow 1°$ is slightly different, since in [143] it has been claimed that every $x \in \mathscr{C}_{\{x_n\}}$ can be written in the form $x = \lim_{n\to\infty} \sum_{i=1}^{m_n} \lambda_{ni} f_i(x) x_i$, where $\lambda_{ni} \geqslant 0$ (which we know only**) for strong $M$-bases). V. I. Gurariĭ has raised [143] the following problem: Does there exist in every separable Banach space an acute angled and complete sequence $\{x_n\}$ such that $\{x_n\}$ is a basis of $\mathscr{C}_{\{x_n\}}$? For some additional results on acute angled sets and bases of sets, see the same paper [143] of V. I. Gurariĭ.

Theorems 16.2, 16.3 and examples 16.1, 16.2, 16.3 were given in [277] ([277], theorems 1, 2 and remarks 3, 2, 4 respectively).

**§ 17.** The problem of extension of the notion of a basis $\{x_n\}$ of a separable Banach space $E$ to the case when $E$ is not assumed to be separable, replacing the sequence $\{x_n\}$ by a family of elements $\{x_i\}_{i\in I}$ which need not be countable and extending, in some suitable way, the unique series expansion, of each $x \in E$, has preoccupied the specialists soon after the appearance of (usual countable) bases. Several solutions were proposed (some main ones are presented in § 17), but the discussion on the problem is not yet closed and other solutions may still appear. Although historically the first such concept was that of extended unconditional basis (see below), we have followed in § 17 the method of Vol. I, starting with $ER$-sets, which do not involve unconditional convergence and then introducing extended unconditional bases as a special class of $ER$-sets (namely, as the unconditional $ER$-sets).

As mentioned in definition 17.1, $ER$-sets were introduced by P. Enflo and H. P. Rosenthal [101], who called them "Schauder basis sets" and expressed the feeling that this is perhaps the "correct" definition for a basis in the nonseparable situation ([101], p. 326). We have preferred the term "$ER$-set", which was introduced in [379], in order to avoid confusion between "Schauder basis" and "Schauder basis set" when $\{x_i\}_{i\in I}$ is countable. Example 17.1, problem 17.1, proposition 17.1 and corollary 17.1 are due to P. Enflo and H. P. Rosenthal ([101], § 1). Theorems 17.1, 17.2, corollaries 17.2, 17.3, proposition 17.2 and the definitions of coefficient functionals and partial sum operators associated to $ER$-sets, and of shrinking and boundedly complete $ER$-sets, were given in [379]. Problems 17.2 b), c) were raised, essentially, by C. Bessaga (verbal communication).

Extended unconditional bases were introduced by E. R. Lorch [251], who simply called them "bases" and observed that any (countable or

*) See Ch. II, § 14.

**) See the Notes and remarks to § 8.

uncountable) orthonormal set in a Hilbert space $E$ is an extended unconditional basis of E. Also, E. R. Lorch has given ([251], theorem A) the following intrinsic characterization of extended unconditional bases, which is a generalization of § 7, theorem 7.3: *A complete family* $\{x_i\}_{i\in I}$ *in a Banach space E, such that* $x_i \neq 0$ $(i \in I)$, *is an extended unconditional basis of E if and only if for every subset M of I we have* $E=[x_i]_{i\in M}\oplus \oplus[x_i]_{i\in I\setminus M}$. Extended unconditional bases $\{x_i\}_{i\in I}$ were also considered by C. Bessaga and A. Pelczynski [22], who called them "absolute bases of power $|I|$" and proved theorem*) 17.3 ([22], corollaries 2 and 3). More generally, Ch. II, § 15, proposition 15.2 admits the following extension, due to A. Pelczynski [298a]: *Every Banach space E with an extended unconditional basis has property* $(u)$.**) Theorem 17.4 is due to P. Enflo and H. P. Rosenthal ([101], theorem 1.1), who also raised problem 17.3 ([101], § 1). The equivalence $1° \Leftrightarrow 4°$ of theorem 17.5 is due to R. C. James ([159], p. 167), who also observed [159] that if $E$ has an extended unconditional basis, then each separable subspace of $E$ is contained in a complemented separable subspace of $E$ (the proof is a particular case of the proof of proposition 17.1). The implication $1° \Rightarrow 5°$ of theorem 17.5 can be found in the paper of J. Lindenstrauss and A. Pelczynski ([243], p. 299) and some other parts of theorem 17.5 can be found in the paper of J. T. Marti [264], where extended unconditional bases were defined by 2° of theorem 17.5 and called "extended bases". By theorem 17.5, equivalence $1° \Leftrightarrow 4°$, the uniform boundedness of the partial sum operators

$$s_d(x) = \sum_{i\in d} f_i(x)x_i \qquad (x \in E,\ d \in \mathscr{D}) \tag{17.63}$$

associated to an extended biorthogonal system $(x_i, f_i)_{i\in I}$ (where $\mathscr{D}$ is the collection of all finite subsets of $I$) *characterizes extended unconditional bases* among $E$-complete extended biorthogonal systems (this generalizes Ch. II, § 16, theorem 16.1, equivalence $1° \Leftrightarrow 7°$). This fact yields the following characterization, observed by J. Lindenstrauss and M. Zippin ([249], p. 123, remark d): *A family of elements* $\{x_i\}_{i\in I}\subset E$, *with* $[x_i]_{i\in I} = E$, *is an extended unconditional basis of E if and only if for each* $i \in I$ *the formula*

$$p(\{i\})(x_j) = \delta_{ij}x_i \qquad (j \in I) \tag{17.64}$$

*) For Banach spaces $E$ containing a subspace isomorphic to $l^\infty$, the conclusion of theorem 17.3 can be also obtained in a different way. Namely, S. Troyanski [393] has proved that *every Banach space with an extended unconditional basis (and hence every subspace of such a space) admits an equivalent locally uniformly convex norm*, while it is known that $l^\infty$ admits no such equivalent norm (see e.g. J. Lindenstrauss [242], theorem 5.3 or S. Troyanski [395], proposition 4).

**) For the definition of property $(u)$ see Ch. II, p. 442.

*defines a continuous linear projection of E onto the one-dimensional subspace* $\{\alpha x_i | \alpha \in K\}$ *and the Boolean algebra*) of projections generated by* $\{p(\{i\})\}_{i \in I}$ *is uniformly bounded on E.* Indeed, if the condition is satisfied, then $p(\{i\})(x) = f_i(x)x_i$ $(x \in E)$, where $f_i \in E^*, f_i(x_j) = \delta_{ij}$, so the above remark applies. Conversely, if $\{x_i\}_{i \in I}$ is an extended unconditional basis of $E$, then for every subset $A$ of $I$ one can define a natural projection $s_A$ of $E$ onto $[x_i]_{i \in A}$, by

$$s_A(x) = \lim_{d \in \mathscr{D}_A} \sum_{i \in d} f_i(x)x_i \qquad (x \in E), \tag{17.65}$$

where $\mathscr{D}_A$ denotes the collection of all finite subsets of $A$. Then, clearly, for any pair $A, B \subset I$ we have

$$s_A s_B = s_B s_A = s_{A \cap B}, \; I_E - s_A = s_{I \setminus A}, \; s_A + s_B - s_A s_B = s_{A \cup B}, \tag{17.66}$$

and hence the Boolean algebra of projections generated by (17.64) is contained in $\{s_A\}_{A \subset I}$, which is uniformly bounded (by $\sup_{d \in \mathscr{D}} \|s_d\|$). In a subsequent paper [250], J. Lindenstrauss and M. Zippin have defined, more generally, *Banach spaces with sufficiently many Boolean algebras of projections;* we shall consider them in Vol. III, since they are not generalizations of extended unconditional bases, but of Banach spaces with extended unconditional bases. There are also two frequently used related notions which are not generalizations of extended unconditional bases, but of Banach spaces with extended unconditional bases, respectively with extended orthogonal bases, and therefore we shall consider them only in Vol. III, namely, *Banach spaces with local unconditional structure* and, respectively, *Banach lattices* (for the former, see also the Notes and remarks to § 18).

Most of the characterizations of unconditional bases among $M$-bases in terms of properties of the associated cone, presented in Ch. II, § 16, remain valid for extended unconditional bases; in fact, they were given directly in this more general setting (see R. E. Fullerton [117]).

Theorem 17.6 was proved by J. Lindenstrauss and A. Pelczynski ([243], p. 297, corollary 1 and the remark made after it). J. Lindenstrauss and M. Zippin observed ([249], p. 122, remark a) that the converse of theorem 17.6 is also true, i.e. *the only infinite dimensional Banach spaces with an extended unconditional basis, in which all bounded extended unconditional bases are equivalent, are (up to an isomorphism)* $l^2(I)$, $l^1(I)$ *and* $c_0(I)$ *for some set* $I$; in fact [249], this is a consequence of the separable case (see Ch. II, § 24, theorem 24.3). M. G. Arsove and R. E. Edwards [7] (see also J. T. Marti [264]) have given generalizations

---

*) We recall that a Boolean algebra of projections on $E$ is a set $\mathscr{P}$ of commuting projections on $E$, such that for every pair $p_1, p_2 \in \mathscr{P}$ we have $p_1 p_2 (= p_2 p_1) \in \mathscr{P}$ and $I_E - p_1 \in \mathscr{P}$ (whence also $p_1 + p_2 - p_1 p_2 = I_E - (I_E - p_1)(I_E - p_2) \in \mathscr{P}$, $0 = p_1(I_E - p_1) \in \mathscr{P}$ and $I_E = p_1 + (I_E - p_1) - p_1(I_E - p_1) \in \mathscr{P}$). Thus, by (17.66), the Boolean algebra of projections generated by (17.64) is nothing else than the set of all $s_A$ such that either $A$ or $I \setminus A$ is finite (or $\emptyset$). Uniformly bounded Boolean algebras of projections on Banach spaces were considered e.g. by E. R. Lorch [252] (see also [88], [247] and the references therein).

of the theorem on the continuity of the coefficient functionals and of the weak basis theorem and J. T. Marti [265] has given generalizations of the Paley-Wiener and Krein-Milman-Rutman stability theorems, from unconditional bases to extended unconditional bases. The suggestion to use the prefix "extended", for the case of not necessarily countable sets of objects, was made by M. G. Arsove and R. E. Edwards [7].

Normalized extended hyperorthogonal bases were considered, under the name "orthonormal bases", by R. J. Fleming and J. E. Jamison [114] and by N. J. Kalton and G. V. Wood [211]; in the latter paper [211] it was also shown that this concept is equivalent to a concept studied previously by E. Berkson ([15], p. 116). Let us mention that, even in the countable case, some authors use for hyperorthogonal bases the term "unconditionally monotone" bases (see e.g. T. Figiel, W. B. Johnson and L. Tzafriri [109]).

Recently S. Troyanski [396] has introduced extended symmetric bases (he has simply called them "symmetric bases"), as follows: An extended unconditional basis $\{x_i\}_{i\in I}$ of a Banach space $E$ is said to be *symmetric*, if for any two sequences $\{i_n\}$, $\{j_n\} \subset I$ the basic sequences $\{x_{i_n}\}$ and $\{x_{j_n}\}$ are equivalent. This is a natural extension of Ch. II, § 21, definition 21.1 and § 22, theorem 22.1, condition 12°; one could also give an equivalent definition, extending Ch. II, § 22, formula (22.1). S. Troyanski has proved some rather unexpected results for the non-separable case, e.g. the following ([396], corollary 3): *If $E$ is a Banach space containing a subspace isomorphic to $c_0(\Gamma) \times l^1(\Delta)$, where $\Gamma$ and $\Delta$ are infinite sets with $\Gamma \cup \Delta$ uncountable, then $E$ has no extended symmetric basis.* Furthermore ([396], proposition 5), *there exists a Banach space $E_0$ with an extended symmetric basis, such that $E_0$ has no subspace isomorphic to $c_0(\Gamma)$ for uncountable $\Gamma$, but every infinite dimensional subspace of $E_0$ contains a subspace isomorphic to $c_0$.* In the framework of Banach lattices there is another natural generalization of Banach spaces with a symmetric basis, namely, the notion of a *rearrangement invariant function space*, which will be considered in Vol. III.

The following extension of the notion of a basis, to the case of families of elements $\{x_i\}_{i\in I}$, was communicated to us, in a letter, by D.H. Fremlin:

*Definition 17.14.* Let $I$ be a set and $\mathscr{D}$ a family of finite subsets of $I$ such that

(i) $\bigcup_{d\in\mathscr{D}} d = I$;

(ii) for every $d, d' \in \mathscr{D}$, we have $d \cup d' \in \mathscr{D}$.

A family $\{x_i\}_{i\in I}$ of elements in a Banach space $E$ is called an $(I, \mathscr{D})$-*conditional basis* of $E$, if for every $x \in E$ there exists a *unique* family of scalars $\{\alpha_i\}_{i\in I} \subset K$ such that

$$x = \sum_{i\in I}^{\mathscr{D}} \alpha_i x_i, \tag{17.67}$$

in the sense that for every $\varepsilon > 0$ there exists a $d = d(\varepsilon) \in \mathscr{D}$ such that $\|x - \sum_{i\in d'} \alpha_i x_i\| < \varepsilon$ for all $d' \in \mathscr{D}$ with $d' \supset d$.

For example, if $I = \mathcal{N} = \{1,2,3, \ldots\}$ and $\mathcal{D}$ is the family of all finite subsets of $I$ of the form $d_n = \{1, \ldots, n\}$ $(n = 1,2, \ldots)$, then an $(I, \mathcal{D})$-conditional basis is a basis in the usual sense. If $\mathcal{D}$ is the family of all finite subsets of some set $I$, then an $(I, \mathcal{D})$-conditional basis is an extended unconditional basis (see theorem 17.5, equivalence $1° \Leftrightarrow 2°$).

D. H. Fremlin proved that the coefficient functionals associated to an $(I, \mathcal{D})$-conditional basis $\{x_i\}_{i \in I}$ (defined by $f_i(x) = \alpha_i$, where $\alpha_i$ is as in (17.67)), are continuous (clearly, they are also linear) and observed, as a corollary, that if $E$ has an $(I, \mathcal{D})$-conditional basis, then $E$ has the approximation property. Also, D. H. Fremlin raised the problem, whether (ii) of definition 17.14 above can be replaced by the weaker condition that $\mathcal{D}$ is directed upwards.

Monotone transfinite bases were defined by C. Bessaga (in [17], where he called them "monotone bases of type $\vartheta$"), as monotone one-dimensional ordinal resolutions*) of the identity, i.e., as the transfinite sequence $\{u_\lambda\}_{\lambda \leqslant \vartheta}$ of proposition 17.8 (see also the Notes and remarks to § 19). Subsequently, C. Bessaga [18] has replaced the condition of being monotone by the weaker condition of uniform boundedness and has used, for this concept, the term "projection basis of type $\vartheta$"; in the same paper [18], C. Bessaga has also defined "biorthogonal systems $(x_\lambda, f_\lambda)_{\lambda < \vartheta}$ associated to the projection basis $\{u_\lambda\}_{\lambda \leqslant \vartheta}$" by the conditions $x_\lambda \in E$, $f_\lambda \in E^*$, $f_\lambda(x)x_\lambda = (u_{\lambda+1} - u_\lambda)(x)$ $(x \in E, \lambda < \vartheta)$. The definition of the sum of a transfinite series $\sum_{\lambda < \vartheta} y_\lambda$ in a Banach space $E$ and the definition of transfinite bases, as given in § 17, are due to**) L. Dorembus [84], who worked directly with transfinite bases of subspaces (see the Notes and remarks to § 19). The last statement of lemma 17.1 was given, without proof, by L. Dorembus ([84], § 5). For analogues of the space $A_1^{(\vartheta)}$ and of proposition 17.5 and theorem 17.8, in the framework of transfinite bases of subspaces of topological linear spaces, see the Notes and remarks to § 19. Proposition 17.6 is due, essentially, to L. Dorembus ([84], example 5.2), who also observed, without giving details, that $c_0([1, \vartheta])$ has to be modified to be a subspace of $C([1, \vartheta])$. Proposition 17.7 and example 17.2 are due, essentially, to L. Dorembus [84] and so is the weak transfinite basis theorem ([84], § 2); again, we refer to the Notes and remarks to § 19.

The necessity part of proposition 17.8 was given, essentially, by C. Bessaga ([18], proposition 1), with a slightly different proof (namely***), assuming essentially 2° of theorem 17.8 with $C = 1$, C. Bessaga observed that for any $y \in \operatorname{lin} \{x_\lambda\}_{\lambda < \vartheta}$, say $y = \sum_{i=1}^{n} \alpha_i x_{\lambda_i}$, by defining $u_\lambda(y) =$

---

*) Note that in § 19, definition 19.5, they are said to be "of type $\vartheta + 1$".

**) We mention here that L. Dorembus changed his name into L. E. Dor, but we have left here both names as they appear on his papers to which we refer.

***) This argument can be also used to give a different proof of the implication $2° \Rightarrow 1°$ of theorem 17.8.

$= \sum_{\lambda_i < \lambda} \alpha_i x_{\lambda_i}$, one obtains a projection of norm 1 on lin $\{x_\lambda\}_{\lambda<\vartheta}$, so it can be extended uniquely, by continuity, to a projection of norm 1 on $E$, with the required properties). Also, C. Bessaga has shown ([18], proposition 5), that *every Banach space with a transfinite basis* $\{x_\lambda\}_{\lambda<\vartheta}$ *admits an equivalent strictly convex norm in which* $\{x_\lambda\}_{\lambda<\vartheta}$ *is monotone.*

Boundedly complete transfinite bases (one-dimensional ordinal resolutions of the identity) were defined by C. Bessaga ([17], (18]), who has also given, jointly with M. I. Kadec ([19], § 2), the following equivalent definition of unconditional transfinite bases: A transfinite basis $\{x_\lambda\}_{\lambda<\vartheta}$ of a Banach space $E$ is said to be *unconditional*, if there exists a countably additive*) projection-valued vector measure $p(.)$ defined on the $\sigma$-field of all subsets of $[1, \vartheta]$, such that

$$p([1, \lambda)) = s_\lambda \qquad (\lambda \leqslant \vartheta), \tag{17.68}$$

where $\{s_\lambda\}_{\lambda\leqslant\vartheta}$ is the transfinite sequence of partial sum operators associated to $\{x_\lambda\}_{\lambda<\vartheta}$. Let us show now that a transfinite basis $\{x_\lambda\}_{\lambda<\vartheta}$ of $E$ is an extended unconditional basis if and only if it is unconditional in the above sense. If $\{x_\lambda\}_{\lambda<\vartheta}$ is an extended unconditional basis, with the a.s.t.c.f. $\{f_\lambda\}_{\lambda<\vartheta} \subset E^*$, then, putting

$$p(A)(x) = \sum_{\varkappa\in A} f_\varkappa(x)x_\varkappa \qquad (x \in E,\ A \subset [1, \vartheta]), \tag{17.69}$$

we obtain a projection-valued measure $p(.)$ with the required properties. Conversely, assume that there exists a projection-valued measure $p(.)$ with the required properties and let $x \in E$. Then, since $\{x_\lambda\}_{\lambda<\vartheta}$ is a transfinite basis of $E$, by § 17, lemma 17.1 the set

$$\mathscr{S}_x = \{\lambda < \vartheta \mid f_\lambda(x)x_\lambda \neq 0\} = \{\lambda < \vartheta \mid p(\{\lambda\})(x) \neq 0\} \tag{17.70}$$

is countable and hence the series $\sum_{\lambda\in\mathscr{S}_x} f_\lambda(x)x_\lambda = \sum_{\lambda\in\mathscr{S}_x} p(\{\lambda\})(x)$ converges unconditionally to $y = p(\mathscr{S}_x)x$. But then, by biorthogonality, $f_\lambda(y) = f_\lambda(x)$ $(\lambda < \vartheta)$, whence, since $\{f_\lambda\}_{\lambda<\vartheta}$ is total on $E$, $y = x$. Thus, by § 17, theorem 17.5, implication $3^\circ \Rightarrow 1^\circ$, $\{x_\lambda\}_{\lambda<\vartheta}$ is an extended unconditional basis of $E$. Note that in the particular case $\vartheta = \omega$, this characterization of (usual) unconditional bases**) in terms of projection-valued measures was not mentioned in Ch. II. A similar characterization of extended unconditional bases is also valid, with a similar proof, among $ER$-sets $\{x_i\}_{i\in I}$, replacing the $\sigma$-field of all subsets of $[1, \vartheta]$ by the $\sigma$-field of all subsets of $I$ and (17.68) by

$$p(\{i\})(x) = f_i(x)x_i \qquad (x \in E,\ i \in I), \tag{17.71}$$

where $\{f_i\}_{i\in I} \subset E^*$ is the a.f.c.f. to the $ER$-set $\{x_i\}_{i\in I}$ (or by (17.64), which is equivalent, since $\{x_i\}_{i\in I}$ is an $ER$-set).

---

*) In the strong operator topology on $L(E, E)$ (i.e., $p\left(\bigcup_{n=1}^{\infty} A_n\right)(x) = \sum_{n=1}^{\infty} p(A_n)(x)$ for $x \in E$, $A_n \subset [1, \vartheta)$ with $A_i \cap A_j = \varnothing$ whenever $i \neq j$; clearly, the series is unconditionally convergent).

**) Among countable total biorthogonal systems.

The convergence of a transfinite series $\sum_{\lambda<\vartheta} y_\lambda$ for every well ordering of $[1, \vartheta)$, was considered (without using a special term for it) by L. Dorembus ([84], § 5), who also mentioned, without proof, lemma 17.3.

Transfinite basic sequences were considered by C. Bessaga ([17], [18]), who proved theorem 17.10 ([18], proposition 2) and observed the first part of remark 17.6 a) ([17], theorem 1 and [18], theorem 1). The second part of remark 17.6 a) shows that theorem 17.10 covers also the case of weakly compactly generated Banach spaces $E$, but for this case J. Reif ([329], theorem 1) proved, with a different method (using a version of § 19, proposition 19.5 with*) $p_{\lambda+1} \neq p_\lambda$ for all $\omega \leqslant \lambda \leqslant \vartheta$ and choosing**), by transfinite induction, $x_\lambda \in \frac{1}{2}(p_{\lambda+1} - p_\lambda)(\mathscr{K}) \subset \mathscr{K}$ for all $\omega \leqslant \lambda < \vartheta$), that one can even find the transfinite basic sequence $\{x_\lambda\}_{\lambda<\vartheta}$ of theorem 17.10 so that $[x_\lambda]_{\lambda<\vartheta}$ is weakly compactly generated. Also, J. Reif showed ([331], proposition 5) that *every subspace $E$ of a weakly compactly generated Banach space satisfies* (17.58); consequently, by theorem 17.10, $E$ has a transfinite basic sequence of the largest possible cardinality. K. John and V. Zizler ([167], proposition 2) proved that *if $E$ is a non-separable Banach space with an $M$-basis $\{x_i\}_{i\in I}$ such that $r([f_i]_{i\in I}) = 1$, where $\{f_i\}_{i\in I} \subset E^*$ is the a.f.f to $\{x_i\}_{i\in I}$, then $\{x_i\}_{i\in I}$ has a monotone transfinite basic subsequence $\{x_{i_j}\} \subset \{x_i\}_{i\in I}$ with* card $\{x_{i_j}\}$=card $\{x_i\}_{i\in I}$=dens $E$. The proof is based, again, on the existence of a suitable long sequence of projections on $E$ (see the Notes and remarks to § 20). Note that if $\{x_i\}_{i\in I}$ is a norming $M$-basis, then $E$ admits an equivalent norm in which $r([f_i]_{i\in I}) = 1$, so one obtains the same conclusion in the initial norm, except the monotony of $\{x_{i_j}\}$. Let us mention that K. John and V. Zizler ([171], proposition 9) proved the existence, for each non-separable weakly compactly generated space $E$, of a quotient space $E/G$ with dens $E/G$ = dens $E$, such that $E/G$ has a monotone transfinite basis (extending § 1, theorems 1.5 and 1.7). Also, J. Reif [329] and K. John and V. Zizler [171] proved some results on the existence of shrinking and boundedly complete transfinite basic and transfinite $w^*$-Schauder basic sequences in certain non-separable conjugate spaces, which we shall give in Vol. III.

Let us mention that, using the notion of the sum of a transfinite series $\sum_{\lambda<\vartheta} y_\lambda$ in $E$, one can define, in a natural way, $\vartheta$-linearly independent transfinite sequences $\{x_\lambda\}_{\lambda<\vartheta}$ (by the condition that $\sum_{\lambda<\vartheta} \alpha_\lambda x_\lambda = 0$ implies $\alpha_\lambda = 0$ for all $\lambda < \vartheta$), transfinite pseudo-bases, transfinite quasi-bases, etc. We leave these to the reader.

The relations between extended unconditional bases and representations of elements as integrals with respect to scalar-valued measures

---

*) See e.g. § 20, remark 20.6.

**) For a similar construction, see § 20, proof of theorem 20.5 a).

were considered by R. E. Fullerton [117], who made the following observation ([117], theorem 3'): If $\mathscr{K} = \mathscr{K}_{(x_i, f_i)_{i\in I}} = \{x \in E \mid f_i(x) \geqslant 0 \ (i \in I)\}$ is the cone associated to an extended unconditional basis $\{x_i\}_{i\in I}$ of a Banach space $E$, with the a.f.c.f. $\{f_i\}_{i\in I} \subset E^*$, then there exists a one-to-one correspondence between the elements $x$ of $\mathscr{K}$ and certain countably additive atomic non-negative measures $\nu_x$ on $T=\{x_i\}_{i\in I}$, whose support is a countable subset of $T$, the correspondence beeing given by*) $x=\int_T t d\nu_x(t)$. Actually, note that, since $E=\mathscr{K}-\mathscr{K}$ (see Ch. II, § 16, theorem 16.3), the correspondence $x \leftrightarrow \nu_x$ extends to the whole space $E$, by dropping the non-negativity of $\nu_x$. Also, R.E. Fullerton [117] observed that the Choquet theory of integral representations of elements of cones with compact metrizable base cannot be applied for this case (even for countable $I$), since $\mathscr{K}$ has no compact base (see Ch. II, § 10, proposition 10.1 b)).

Strongly unconditional integral bases were considered by R. E. Edwards ([93], § 5), who has not used any special term for them (the term was suggested here because of remark 17.7). Theorem 17.11 was proved by R.E. Edwards ([93], theorem 2).

Let us also mention that J. A. Dyer [90], [91] has introduced two other concepts of integral bases which do not involve countability (which he called "$I$-bases" and "$L$-bases", or "left Cauchy integral bases"), using integrals of Hellinger type. Although some theorems for bases admit natural analogues for these two types of integral bases, J. A. Dyer has observed ([90], [91]) that the two notions are *not* generalizations of bases.

Finally, we mention that R. E. Edwards [94] has defined a generalization of unconditional quasi-bases, by the condition

$$x = \int_T f_t(x) X(t) d\mu(t) \qquad (x \in E), \tag{17.72}$$

where $T$ is an infinite locally compact space, $X: t \to X(t)$ a mapping of $T$ into $E$, $t \to f_t$ a mapping of $T$ into $E^*$ and $\mu$ a positive Radon measure on $T$ and where the integral is taken in the sense of § 17 (hence weakly unconditionally convergent). In particular, for $T=\mathscr{N}$ with the discrete topology, $X(n) = x_n \in E$, $f_n \in E^*$ and $\mu(\{n\}) = 1$ ($n = 1,2, \ldots$), this holds precisely for an unconditional quasi-basis $\{x_n\}$ of $E$. For another generalization of unconditional quasi-bases, involving integral representations, see the Notes and remarks to § 9.

---

*) I. e., $f(x) = \int_T f(t) d\nu_x(t) \ (f \in E)^*$. Indeed, $f(x) = \sum_{i\in I} f(x_i) f_i(x) \ (f \in E^*)$, so $\nu_x(A) = \sum_{x_i \in A} f_i(x) \ (A \subset T)$.

We have seen above that projection-valued measures can be used to characterize extended unconditional bases among $ER$-sets. P. Masani ([266], [267]) has introduced the following generalization of orthogonal bases of Hilbert spaces $E$ as vector-valued measures, with values in a Hilbert space $E$: Let $T$ be a set, $\mathscr{B}$ a $\sigma$-ring over $T$ and $\mu$ a non-negative, $\sigma$-finite, countably additive measure on $\mathscr{B}$ and let $\mathscr{B}_\mu$ be the subring

$$\mathscr{B}_\mu = \{A \in \mathscr{B} \mid \mu(A) < \infty\}. \tag{17.73}$$

A function $\xi\colon \mathscr{B}_\mu \to E$ is called ([266], [267]) an *E-valued countably additive orthogonally scattered measure over* $(T, \mathscr{B}, \mu)$, if $\xi$ is countably additive and if for any pair $A, B \in \mathscr{B}$ with $A \cap B = \emptyset$ we have $\xi(A) \perp \perp \xi(B)$ in $E$ (i.e., $\xi$ is "orthogonally scattered"); $G_\xi = [\xi(A)]_{A \in \mathscr{B}_\mu}$ is called *the subspace* of $\xi$. If $G_\xi = E$, then $\xi$ is called ([266], [267]) a *countably additive orthogonally scattered measure basis* of $E$, or an *orthogonal differential basis* of $E$. In this case, for every $x \in E$ there exists ([266], [267]) a unique $\alpha_x(.) \in L^2(T, \mathscr{B}, \mu)$ such that

$$x = \int_T \alpha_x(t) d\xi(t); \tag{17.74}$$

the integral $\int_T \alpha(t) d\xi(t)$ is defined in [266] first for $\mathscr{B}_\mu$-simple functions and then extended by continuity to all $\alpha(.) \in L^2(T, \mathscr{B}, \mu)$.

In the particular case when $T = \mathscr{N} = \{1,2,3, \ldots\}$, $\mathscr{B} =$ the $\sigma$-ring of all subsets of $\mathscr{N}$, $\mu(\{n\}) = 1$ for $n = 1,2, \ldots$ (so $\mu(A) = \operatorname{card} A$ for all $A \in \mathscr{B}$ and hence $\mathscr{B}_\mu$ is the subring*) of all finite subsets of $\mathscr{N}$), a function $\xi\colon \mathscr{B}_\mu \to E$ is an orthogonal differential basis of $E$ if and only if the sequence

$$x_n = \xi(\{n\}) \qquad (n = 1,2, \ldots) \tag{17.75}$$

is an orthogonal basis of $E = l^2$. Indeed, if $\xi$ is an orthogonal differential basis of $E$, then $x_n \perp x_m$ for all $n \neq m$ and $E = [\xi(A)]_{A \in \mathscr{B}_\mu} = [x_n]$, since

$$\xi(A) = \sum_{n \in A} \xi(\{n\}) = \sum_{n \in A} x_n \qquad (A \in \mathscr{B}_\mu); \tag{17.76}$$

conversely, if $\{x_n\}$ (defined by (17.75)) is an orthogonal basis of $E = l^2$, then $\xi$ satisfies the conditions of the above definition**) of an orthogonal differential basis. In this case, the integral representation (17.74) amounts to the series expansion

$$x = \sum_{n=1}^{\infty} \alpha_x(\{n\}) x_n. \tag{17.77}$$

---

*) $\mathscr{B}_\mu$ is not a $\sigma$-ring.

**) The condition of countably additivity of $\xi$ on $\mathscr{B}_\mu$ holds vacuously, since no countable disjoint union of non-empty members of $\mathscr{B}_\mu$ can belong to $\mathscr{B}_\mu$.

P. Masani [266] has also given the following example of an orthogonal differential basis in the separable Hilbert space $E = L^2(-\infty, \infty)$ (so $\mu$ = the Lebesgue measure):

$$\xi(A) = \chi_A \qquad (A \in \mathscr{B}_\mu), \tag{17.78}$$

the characteristic function of $A$. Although there is no orthogonal basis of $E = L^2(-\infty, \infty)$ associated in a natural way to this $\xi$, P. Masani [266] has pointed out that it has numerous applications (e.g., to the theory of the Fourier-Plancherel transform).

§ **18.** Extended approximative bases of Banach spaces were introduced by R. S. Phillips ([317], § 3, p. 526), who called them "generalized bases". In the same paper, R. S. Phillips [317] also considered the more general case when the operators $u_d$ are compact (but not necessarily of finite rank); the existence of such objects is equivalent to the compact approximation property (see the Notes and remarks to § 0). The equivalences $1° \Leftrightarrow 4° \Leftrightarrow 5°$ of theorem 18.1 are mentioned in the paper of W. B. Johnson, H. P. Rosenthal and M. Zippin ([187], proposition 1.1). The sufficiency part of theorem 18.2 was given in the paper of W. B. Johnson ([180], lemma 3 (b)), with the mention that it is essentially known*); the argument in the proof of the necessity part of theorem 18.2 is also known (see e.g. W. B. Johnson [173], proof of corollary 1 or D. J. Fleming and W. Ruckle [113], proof of theorem 5.1). Problems 18.1 b) and c) (for the approximation property) were raised by A. Pelczynski [306] and J. Lindenstrauss [239] respectively. In the meantime, it has been communicated to us by W. J. Davis that recently A. Szankowski has solved problem 18.1 b) (and hence a) ) in the negative. Proposition 18.1 is due to W. B. Johnson ([178], § II).

Extended weak and weak* approximative bases were considered by D. W. Dean [73], who proved theorem 18.3 ([73], theorem 1 and its proof).

J. Lindenstrauss [239] considered the following condition on a Banach space $E$, which he called "the nonseparable analog of the approximation property": For every separable subspace $G$ of $E$, there exists an operator $v = v_G \in L(E, E)$ with separable range, such that $v(x) = x$ for all $x \in G$ (so $v$ is an extension of the identity operator $I_G$). Let us call this property *the separable range approximation property* (then the usual bounded approximation property might be called "the finite-dimensional range bounded approximation property"). Clearly, every

---

*) In fact, J. Lindenstrauss observed ([236], proposition) that if every separable subspace of a reflexive Banach space $E$ has the metric approximation property, then, by a "simple compactness argument" it follows that $E$ has the metric approximation property.

separable Banach space $E$ has this property (since one can take simply $v = I_E$). However, J. Lindenstrauss [239] observed that *the space* $E = l^\infty$ (which has the metric approximation property) *does not have the separable range approximation property;* indeed, by a theorem of A. Grothendieck [137], every continuous linear operator defined on $l^\infty$ and having a separable range is weakly compact (and hence for any separable nonreflexive subspace $G$ of $E = l^\infty$, there is no $v = v_G$ as above). Also, J. Lindenstrauss observed in [239] that *every weakly compactly generated Banach space E has the separable range approximation property;* indeed, this follows e.g. from the version of § 19, lemma 19.5, given in § 19, remark 19.1 b) (taking $\mathfrak{m} = \aleph_0$).

Let us also mention the following strengthening of the notion of bounded approximation property, introduced by A. Pelczynski and H. P. Rosenthal ([309], § 2): A Banach space $E$ is said to have the *$\lambda$-uniform approximation property* if there exists a function $N(k)$ such that for every $k$-dimensional subspace $G$ of $E$ there exists a finite rank operator $v = v_G \in L(E, E)$, satisfying

$$v(x) = x \qquad (x \in G), \tag{18.51}$$

$$\|v\| \leqslant \lambda, \tag{18.52}$$

$$\dim v(E) \leqslant N(k); \tag{18.53}$$

the space $E$ is said to have the *uniform approximation property* if there exists a $\lambda < \infty$ such that $E$ has the $\lambda$-uniform approximation property. If the finite rank operator $v$ above can be chosen to be a projection, then $E$ is said to have the ($\lambda$-) *uniform projection approximation property.* A. Pelczynski and H. P. Rosenthal [309] showed that the spaces $L^p$ $(1 \leqslant p \leqslant \infty)$ have the uniform projection approximation property; J. Lindenstrauss and L. Tzafriri proved ([248], theorem 4 and corollary) that every reflexive Orlicz space has this property and the $(1 + \varepsilon)$-uniform approximation property for all $\varepsilon > 0$. However, using the existence of a Banach space which fails to have the bounded approximation property, W. B. Johnson observed (see [309]) that there exists a Banach space with a basis (and hence with the bounded approximation property), which does not have the uniform approximation property. Moreover, A. Szankowski showed ([387], proposition 3) that *there exists a uniformly convex Banach space with a symmetric basis, which does not have the uniform approximation property.* J. Lindenstrauss and L. Tzafriri proved ([248], theorem 3) *that a Banach space E has the uniform approximation property if and only if $E^{**}$ has the same property* (hence, in particular, in this case $E^{**}$ has the bounded approximation property and $E$ has an extended commuting approximative basis). Also, they proved ([248], theorem 2) that *a uniformly convex space E has the uniform approximation property if and only if $E^*$ has the same property* and raised the problem whether this result remains valid without the assumption of uniform convexity; concerning this latter problem they observed ([248], the remark after theorem 2) that if $(E, |\cdot|)$ has the 2-uniform approximation property for every equivalent norm $|\cdot|$ on $E$, then $E^*$ has the uniform approximation property (this result is similar to § 9, theorem 9.4).

The comments made in the Notes and remarks to § 12 and § 13, on the notions of (norm, weak, weak*) $\Pi_\lambda$-, $\pi_\lambda$-, dual $\pi_\lambda$-, commuting $\pi_\lambda$-, $\pi_1^\infty$- and dual $\pi_1^\infty$-bases respectively and on Banach spaces having such objects, are also valid for the corresponding "extended" notions. The observation that the spaces $L^p(T, \nu)$ $(1 \leqslant p \leqslant \infty)$ have a $\pi_1$-basis, is due, essentially, to R. S. Phillips [317]; see also A. Grothendieck [139], proof of Ch. I, proposition 41 and D. G. de Figueiredo [111], proposition IV.2 (for $1 \leqslant p < \infty$). Theorem 18.4 $\delta$) and, essentially, theorem 18.4 $\gamma$), equivalences $1° \Leftrightarrow 3° \Leftrightarrow 4°$, can be found in the paper of W. B. Johnson, H. P. Rosenthal and M. Zippin ([187], proposition 2). Theorem 18.5 and corollary 18.2 are due, essentially, to W. B. Johnson ([173], corollary 1 and remark 2). The references given in the Notes and remarks to § 12, on characterizations of Banach spaces with $\pi_1^\infty$-bases, remain valid also for the case of extended $\pi_1^\infty$-bases. Lemmas 18.1 and 18.2 were given by E. Michael and A. Pelczynski ([282], proposition 5.1 and [283], lemma 2.1). Theorem 18.6 was proved by W. B. Johnson ([173], theorem 4). It has been observed by J. Lindenstrauss ([235], p. 21) that the unit vector basis $\{x_n\}$ of $c_0$ and the usual Schauder basis $\{x_n\}$ of $C([0, 1])$ have the property $[x_i]_{i=1}^n \equiv l_n^\infty$ $(n = 1,2, \ldots)$; we have seen in Ch. II, § 1, that these bases are also monotone. Bases of the space $C(Q)$, where $Q$ is a compact metric space, were constructed by F. S. Vaher [398] (as has observed C. Bessaga [16], the construction of [398] works only for uncountable $Q$) and, independently, by C. Bessaga [16] (see also L. A. Gurevič [151], H. W. Ellis and D. G. Kuehner [97], Z. Semadeni [359] and J. Ryll [352]). Corollary 18.3 was announced, mentioning the idea of the proof, by V. I. Gurariĭ ([144], p. 298, footnote) and another proof was given by E. Michael and A. Pelczynski ([282], corollary 1.3); the proof presented here is different and is due to W. B. Johnson ([173], remark 2). We shall return to bases of $C(Q)$ spaces ($Q$ compact metric) in Vol. III.

Theorem 18.7 was proved by W. B. Johnson ([176], theorem 4).

Proposition 18.3 is contained, implicitly, in the paper [73] of D. W. Dean ([73], proof of theorem 3). Theorems 18.8 a), b) are due, essentially, to W. B. Johnson, H. P. Rosenthal and M. Zippin ([187], lemmas 4.4, 4.5 and corollary 4.8); actually, they have not considered extended commuting $\pi$-bases in non-separable spaces. Theorem 18.8 c) appeared in the paper of D. W. Dean ([73], theorem 3 and its proof), with the mention that it was observed by W. B. Johnson, in a conversation. In [377], remark 1, it has been claimed that under the assumptions of proposition 18.3 and theorem 18.8 c) $E$ has also a dual $\pi$-basis (more precisely, a dual $\pi_{\lambda^2+2\lambda+\varepsilon}$-basis for each $\varepsilon > 0$); however, while this can be proved for separable Banach spaces $E$ (see § 13, proposition 13.8 and § 14, theorem 14.1 c)), the non-separable case seems to be still open.

Theorem 18.1, equivalence $1° \Leftrightarrow 5°$ and theorem 18.4 $\gamma$), equivalence $1° \Leftrightarrow 4°$, show that a Banach space $E$ has an extended approximative basis (respectively, an extended $\Pi$-basis) if and only if there exists a constant $\mu$ such that every finite-dimensional subspace $G$ of $E$ is contained in some finite-dimensional subspace*) $F$ of $E$ admitting a continuous linear operator (respectively, projection) $v$ from $E$ onto $F$, of norm $\|v\| \leqslant \mu$. There are a number of other frequently used notions defined in a similar way (by different properties of the superspaces $F$), which we shall consider in Vol. III, since they are not generalizations of bases, but of Banach spaces with bases. Here we only mention that if we require, instead of the above conditions, the existence of a constant $\nu$ such that every finite-dimensional subspace $G$ of $E$ is contained in some finite-dimensional subspace $F$ of $E$, admitting a basis of norm $\leqslant \nu$ (respectively, an unconditional basis of unconditional norm $\leqslant \nu$), we arrive at the notion of**) "Banach spaces with *the finite-dimensional subspace basis property*", introduced by L. R. Pujara [325], [326] (respectively, with a *local unconditional structure*, introduced by***) E. Dubinsky, A. Pelczynski and H. P. Rosenthal [85] and studied by a number of other authors). Furthermore, if we require that every finite-dimensional subspace $G$ of $E$ be contained in some finite-dimensional subspace $F$ of $E$, admitting a basis of norm $\leqslant \nu$ and a projection of $E$ onto $F$, of norm $\leqslant \nu$, then we arrive at the notion of "$B_\nu$ *spaces*", introduced by L. R. Pujara ([325], definition 3.15); clearly, every such space $E$ has an extended $\pi$-basis. L. R. Pujara has shown ([325], example 3.16) that every space with a basis is a $B_\nu$ space and

---

*) Let us observe that any family $\mathscr{F} = \{F\}$ of subspaces of $E$, such that every finite-dimensional subspace $G$ of $E$ is contained in some $F \in \mathscr{F}$, is directed by inclusion (indeed, if $F_1, F_2 \in \mathscr{F}$, then for $G = [F_1 \cup F_2]$ there exists $F \in \mathscr{F}$ with $F \supset G \supset F_1, F_2$) and satisfies $\bigcup_{F \in \mathscr{F}} F = E$. The converse is also true, i.e., if $\mathscr{F} = \{F\}$ is a family of subspaces of $E$, directed by inclusion and satisfying $\bigcup_{F \in \mathscr{F}} F = E$, then every finite dimensional subspace $G$ of $E$ is contained in some $F \in \mathscr{F}$ (indeed, for any basis $\{x_1, \ldots, x_n\}$ of $G$ there exist $F_{d_i} \in \mathscr{F}$ with $F_{d_i} \ni x_i$, $i = 1, \ldots, n$, whence, for any $F \in \mathscr{F}$ with $F \supset F_{d_1}, \ldots, F_{d_n}$ we obtain $F \supset [x_1, \ldots, x_n] = G$). Thus, the condition that every finite-dimensional subspace $G$ is contained in some $F \in \mathscr{F}$ can be replaced by the conditions that $\mathscr{F} = \{F\}$ is directed by inclusion and $\bigcup_{F \in \mathscr{F}} F = E$. Since usually in such definitions every subspace which is sufficiently near to some $F \in \mathscr{F}$ also belongs to $\mathscr{F}$ (by changing $\nu$), the condition $\bigcup_{F \in \mathscr{F}} F = E$ can be replaced by $\overline{\bigcup_{F \in \mathscr{F}} F} = E$.

**) We suggest to use for such spaces, the term "Banach spaces with a *local basis structure*".

***) Let us mention that a different notion of Banach spaces with a local unconditional structure was introduced by Y. Gordon and D. R. Lewis [74].

there exist also non-separable $B_v$ spaces. If we require that, for some constant $\lambda$, every finite-dimensional subspace $G$ of $E$ be contained in some finite-dimensional subspace $F$ of $E$, such that*) $d(F, l_m^p) \leqslant \lambda$, where $m = \dim F$, then we arrive at the notion of "$\mathscr{L}_p$ spaces", mentioned in § 15 (for $p=1, \infty$) and in the Notes and remarks to § 12 (for $p = \infty$).

Extended resolutions of the identity were considered by D. J. Fleming and W. Ruckle [113], who called them "uniformly bounded generalized Schauder decompositions"; however, in general the term "decomposition" is used for families of subspaces and there is no family of subspaces corresponding to an extended resolution of the identity, which would generalize**) the usual decompositions $\{G_n\} = \{(u_n - u_{n-1})(E)\}$ (since $\mathscr{D}$ is only partially ordered). The condition $u_d \neq I_E$ $(d \in \mathscr{D})$ was added by J. J. M. Chadwick [49]. Proposition 18.4 and theorem 18.9 were proved by D. J. Fleming and W. Ruckle ([113], theorem 5.1 and lemmas 3.2 and 4.2). J. J. M. Chadwick observed ([49], theorems 2.2 and 2.1) that, conversely, *if a net* $\{u_d\}_{d \in \mathscr{D}} \subset L(E, E)$ *is such that* $\{u_d^*\}_{d \in \mathscr{D}}$ *is an extended weak* resolution of* $I_{E^*}$, *then* $\{u_d\}_{d \in \mathscr{D}}$ *is an extended resolution of* $I_E$. Also, J. J. M. Chadwick gave an analog of the weak basis theorem for this setting, by observing ([49], theorem 2.1) that $\{u_d\}_{d \in \mathscr{D}}$ *is an extended weak resolution of* $I_E$ *if and only if it is an extended resolution of* $I_E$.

Shrinking and boundedly complete extended resolutions of the identity were considered by D. J. Fleming and W. Ruckle ([113], theorem 3.7, condition (c) and definition 3.5), who replaced (18.49) and (18.50) by the equivalent***) pair

$$u_d(z_{d'}) = z_d \qquad (d \leqslant d'), \tag{18.54}$$

$$u_d(x) = z_d \qquad (d \in \mathscr{D}), \tag{18.55}$$

and proved theorem 18.10 ([114], theorem 3.6); actually, the pair (18.54), (18.55) shows some similarity with the definitions of boundedly complete transfinite bases (§ 17, definition 17.11) and boundedly complete ordinal resolutions of the identity (§ 19, defini-

---

*) For the definition of the distance coefficient $d(A, B)$ see § 12, formula (12.41).

**) Nevertheless, we shall see that the net of subspaces $\{u_d(E)\}_{d \in \mathscr{D}}$ $\left(\text{generalizing the sequence } \sum_{i=1}^{n} \oplus G_i\right)$ can be used as a replacement of such a generalization.

***) See § 15, theorem 15.15, equivalence $1° \Leftrightarrow 2°$ (using there $u_n = \sum_{i=1}^{n} v_i$ and $x_n = \sum_{i=1}^{n} y_i$ for $n = 1, 2, \ldots$).

tion 19.6). J. J. M. Chadwick [47] has defined shrinking extended resolutions $\{u_d\}_{d\in\mathscr{D}}$ of the identity by the condition $\lim_{d\in\mathscr{D}} \|f\|_d = 0$ $(f\in E^*)$, where

$$\|f\|_d = \sup_{\substack{x\in \operatorname{Ker} u_d \\ \|x\|\leqslant 1}} |f(x)| \qquad (f\in E^*), \tag{18.56}$$

and has shown ([47], theorem 2.1) that this condition is equivalent to the condition that $\{u_d^*\}_{d\in\mathscr{D}}$ is an extended resolution of $I_{E^*}$; clearly (by theorem 18.9), this latter condition is equivalent to that of definition 18.6 b). Moreover ([47], theorem 2.2), if $\{u_d\}_{d\in\mathscr{D}}$ is shrinking, then $\{u_d^*\}_{d\in\mathscr{D}}$ is a boundedly complete extended resolution of $I_{E^*}$; if each $u_d(E)$ $(d\in\mathscr{D})$ is reflexive, then, dually, $\{u_d\}_{d\in\mathscr{D}}$ is boundedly complete if and only if $\{u_d^*|_V\}_{d\in\mathscr{D}}$ is a shrinking extended resolution of $I_V$, where $V=\overline{\bigcup_{d\in\mathscr{D}} u_d^*(E^*)}$ ([47], theorem 4.8). Actually, J. J. M. Chadwick [47] has defined boundedly complete extended resolutions $\{u_d\}_{d\in\mathscr{D}}$ of the identity replacing (18.49) by (18.54) above and replacing (18.55) by the condition that the limit

$$x = \lim_{d\in\mathscr{D}} z_d \tag{18.57}$$

exists. He observed [47] that (18.57), $u_d\in L(E, E)$ and (18.54) imply

$$u_d(x) = \lim_{d'\in\mathscr{D}} u_d(z_{d'}) = \lim_{d'\in\mathscr{D}} z_d = z_d \qquad (d\in\mathscr{D}),$$

i.e., (18.55); of course, the converse is also true (indeed, by (18.55) and since $\{u_d\}_{d\in\mathscr{D}}$ is an extended resolution of $I_E$, we obtain $\lim_{d\in\mathscr{D}} z_d = \lim_{d\in\mathscr{D}} u_d(x) = x$), so the two definitions are equivalent. In the same paper, J. J. M. Chadwick observed ([47], lemma 4.1) that if $\{u_d\}_{d\in\mathscr{D}}$ is an extended resolution of $I_E$, then the canonical mapping $u$ of $E$ into $V^*$, where $V=\overline{\bigcup_{d\in\mathscr{D}} u_d^*(E^*)}$, is an isomorphism; hence, as we know from Ch. I, § 12, $r(V)>0$. Furthermore, he has shown ([47], theorem 4.5) that the following converse of theorem 18.10 is also true: *If $\{u_d\}_{d\in\mathscr{D}}$ is an extended resolution of $I_E$, such that $E$ is canonically isomorphic to $V^*$ (where $V=\overline{\bigcup_{d\in\mathscr{D}} u_d^*(E^*)}$), then $\{u_d\}_{d\in\mathscr{D}}$ is boundedly complete and each $u_d(E)$ $(d\in\mathscr{D})$ is reflexive.* D. J. Fleming and W. Ruckle proved ([113], theorem 3.3) that *if $\{u_d\}_{d\in\mathscr{D}}$ is an extended resolution of $I_E$ and if $V=\overline{\bigcup_{d\in\mathscr{D}} u_d^*(E^*)}$, then $V^*$ is isomorphic to the Banach space $\Omega_0$ of nets of functionals*

$$\Omega_0 = \{\{\Phi_d\}_{d\in\mathscr{D}} \subset E^{**} \mid u_d^{**}(\Phi_{d'}) = \Phi_d \ (d\leqslant d'),\ |||\{\Phi_d\}_{d\in\mathscr{D}}||| = \sup_{d\in\mathscr{D}} \|\Phi_d\| < \infty\}, \tag{18.58}$$

*by the mapping* $\Psi \to \{u_d^{**}(\Phi)\}$, *where* $\Psi \in V^*$ *and* $\Phi \in E^{**}$ *is any extension of* $\Psi$ *to* $E^*$. Since $\Phi_d \in u_d^{**}(E^{**})$, it is natural to try $\left(\text{in view of § 15, formula (15.53), applied to } u_n = \sum_{i=1}^{n} v_i\right)$ to replace $\Omega_0$ by a Banach space of nets of elements of $E$. In fact, J. J. M. Chadwick [47] has considered, for an arbitrary extended resolution $\{u_d\}_{d\in\mathscr{D}}$ of $I_E$, the Banach space of nets

$$\Omega = \{\{z_d\}_{d\in\mathscr{D}} \subset E \mid u_d(z_{d'}) = z_d \ (d \leqslant d'),\ |||\{z_d\}_{d\in\mathscr{D}}||| = \sup_{d\in\mathscr{D}} \|z_d\| < \infty\} \tag{18.59}$$

(so here $z_d \in u_d(E)$); this space is "new" even in the particular cases of the one-dimensional decomposition associated to a usual basis $\{x_n\}$ or of a Schauder decomposition $\{G_n\}$, since it reduces respectively to

$$\Omega = \left\{\left\{\sum_{i=1}^{n} \alpha_i x_i\right\} \subset E \ \middle|\ \left|\left|\left|\left\{\sum_{i=1}^{n} \alpha_i x_i\right\}\right|\right|\right| = \sup_{1\leqslant n<\infty} \left\|\sum_{i=1}^{n} \alpha_i x_i\right\| < \infty\right\},$$

$$\Omega = \left\{\left\{\sum_{i=1}^{n} y_i\right\} \subset E \ \middle|\ y_n \in G_n\ (n = 1,2,\ldots),\ \left|\left|\left|\left\{\sum_{i=1}^{n} y_i\right\}\right|\right|\right| = \sup_{1\leqslant n<\infty} \left\|\sum_{i=1}^{n} y_i\right\| < \infty\right\}.$$

Next, J. J. M. Chadwick [47] has defined a linear mapping $t_1$ of $E$ into $\Omega$ by

$$t_1(x) = \{u_d(x)\}_{d\in\mathscr{D}} \qquad (x \in E), \tag{18.60}$$

and has observed that

$$\|t_1(x)\| \leqslant (\sup_{d\in\mathscr{D}} \|u_d\|)\ \|x\| \leqslant \sup_{d\in\mathscr{D}} \|u_d\|\ \|t_1(x)\| \quad (x \in E), \tag{18.61}$$

so $t_1$ *is an isomorphism of* $E$ *into* $\Omega$. Clearly ([47], proposition 5.1), $t_1(E) = \Omega$ *if and only if* $\{u_d\}_{d\in\mathscr{D}}$ *is boundedly complete.*

Furthermore, J. J. M. Chadwick ([47], lemma 5.2) observed that *for each net* $\{z_d\}_{d\in\mathscr{D}} \in \Omega$ *there exists a unique* $\Psi_{\{z_d\}_{d\in\mathscr{D}}} \in V^*$ (*where* $V = \overline{\bigcup_{d\in\mathscr{D}} u_d^*(E^*)}$), *such that**)

$$\Psi_{\{z_d\}_{d\in\mathscr{D}}}(f) = f(z_{d'}) \qquad (f \in u_{d'}^*(E^*),\ d' \in \mathscr{D}), \tag{18.62}$$

*) For the particular case of the one-dimensional decomposition associated to a basis $\{x_n\}$, this is equivalent to the second statement of Ch. I, § 12, proposition 12.2 b) and in the case of a Schauder decomposition $\{G_n\}$, this is equivalent to the particular case $\Psi_n = u(y_n)$ ($y_n \in G_n$, $n = 1, 2, \ldots$) of the second statement of § 15, proposition 15.6 b).

and thus one can define a linear mapping $t_2$ of $\Omega$ into $V^*$ by

$$t_2(\{z_d\}_{d\in\mathscr{D}}) = \Psi_{\{z_d\}_{d\in\mathscr{D}}} \qquad (\{z_d\}_{d\in\mathscr{D}} \in \Omega). \tag{18.63}$$

Using that $r(V) > 0$, it is easy to see ([47], p. 102) that $t_2$ is an isomorphism. J. J. M. Chadwick proved ([47], theorem 5.3) that $t_2(\Omega) = V^*$ *if and only if each* $u_d(E)$ $(d \in \mathscr{D})$ *is reflexive*. Since, clearly, the canonical mapping $u$ of $E$ into $V^*$ is nothing else than

$$u = t_2 t_1, \tag{18.64}$$

these facts throw some new light on theorem 18.10 and its converse mentioned above (and yield another proof for them).

Let us also mention the following existence theorem, due to J. J. M. Chadwick ([49], theorem 3.6): *If $E$ is a non-separable smooth Banach space, then for every subspace $\Gamma$ of $E^*$, with* $\operatorname{dens}\Gamma = \operatorname{dens} E$, *there exists a subspace $F \subset E^*$ with $F \supset \Gamma$, which has an extended resolution of $I_F$*. Hence, in particular, *if $E$ is smooth and* $\operatorname{dens} E = \operatorname{dens} E^*$, *then $E^*$ has an extended resolution of $I_{E^*}$* ([49], corollary 3.7). The proof uses § 19, lemma 19.12 and transfinite induction (similarly to the proof of § 19, proposition 19.7).

The theory of vector sequence spaces can be generalized to this setting, as follows. If $\{G_d\}_{d\in\mathscr{D}}$ is a net of Banach spaces (so $\mathscr{D}$ is a directed set), a *net space* $S = S(\{G_d\}_{d\in\mathscr{D}})$ is a linear subspace of $s(\{G_d\}_{d\in\mathscr{D}}) = \prod_{d\in\mathscr{D}} G_d$ (the collection of all nets $\{y_d\}_{d\in\mathscr{D}}$ with $y_d \in G_d$ for $d \in \mathscr{D}$, endowed with the natural linear operations). The definition of $BK$-spaces can be also extended verbatim from vector sequence spaces to vector net spaces, but in this case $s(\{G_d\}_{d\in\mathscr{D}}) = \prod_{d\in\mathscr{D}} G_d$, endowed with the product topology, is not metrizable when $\mathscr{D}$ is uncountable. Another generalization of vector sequence spaces was proposed by D. J. Fleming and W. Ruckle [113], as follows: Let $\mathfrak{M}$ be a complete metric linear space (not necessarily locally convex), on which there exists a net of continuous linear projections $\{u_d\}_{d\in\mathscr{D}}$ satisfying the following conditions:

a) $z \in \mathfrak{M}$, $u_d(z) = 0$ $(d \in \mathscr{D})$ imply $z = 0$;

b) $u_{d'} u_d = u_d u_{d'} = u_d \qquad (d \leqslant d')$;

c) for each net $\{z_d\}_{d\in\mathscr{D}} \subset \mathfrak{M}$ for which $u_d(z_{d'}) = z_d$ $(d \leqslant d')$, there exists an element $z \in \mathfrak{M}$ satisfying $u_d(z) = z_d$ $(d \in \mathscr{D})$. Then, an *$\mathfrak{M}$-space* is a linear subspace $S$ of $\mathfrak{M}$ and an *$\mathfrak{M}BK$-space* is an $\mathfrak{M}$-space which is a Banach space, such that the inclusion map of $S$ into $\mathfrak{M}$ is continuous (actually, D. J. Fleming and W. Ruckle [113] have used a different terminology and have assumed, in addition, that $S$ contains $\bigcup_{d\in\mathscr{D}} u_d(\mathfrak{M})$ and that each $u_d|_S$ is continuous on the Banach space $S$; however, even the first one of these conditions is not assumed in the case of scalar sequence spaces, as was mentioned in the Notes and remarks to § 15).

C. W. McArthur ([273], definition 2.2) introduced an *extended unconditional Schauder basis of subspaces*, or *extended unconditional (Schauder) decomposition* (he used the term "unordered basis of subspaces", while W. H. Ruckle [343] used the term "unordered Schauder decomposition"), of a Banach space $E$, as a family $\{G_i\}_{i\in I}$ of (closed linear and $\neq \{0\}$) subspaces of $E$ with the property that for each $x \in E$ there exists a unique family of elements $\{y_i\}_{i\in I} \subset E$, with $y_i \in G_i$ $(i \in I)$, such that

$$x = \lim_{d\in\mathscr{D}} \sum_{i\in d} y_i, \tag{18.65}$$

where $\mathscr{D}$ is the collection of all finite subsets of the set $I$, directed by inclusion. Many results on extended unconditional bases and on the usual unconditional Schauder decompositions can be generalized to the setting of extended unconditional Schauder decompositions. For example, C. W. McArthur [273], W. H. Ruckle [343] and G. F. Bachelis [9] have given some intrinsic characterizations of such decompositions, which generalize parts of § 17, theorem 17.8 and § 15, theorem 15.5. Also, G. F. Bachelis has given ([9], theorem 3.4, equivalence $1° \Leftrightarrow 4°$) the following characterization, which contains § 7, theorem 7.3 and its generalizations mentioned in the Notes and remarks to § 17 and § 15: *A family $\{G_i\}_{i\in I}$ of subspaces of a Banach space $E$, with $G_i \neq \{0\}$ $(i \in I)$ and $[\bigcup_{i\in I} G_i] = E$, is an extended unconditional Schauder decomposition of $E$ if and only if for every subset $M$ of $I$ we have* $E = [G_i]_{i\in M} \oplus [G_i]_{i\in I\setminus M}$. Some further generalizations of this result were given by G. F. Bachelis and H. P. Rosenthal ([10], theorem 2 and remark (1)).

One can define the *family of coordinate projections* $\{v_i\}_{i\in I}$ associated to an extended unconditional Schauder decomposition $\{G_i\}_{i\in I}$, by

$$v_i(x) = y_i \qquad (x = \lim_{d\in\mathscr{D}} \sum_{j\in d} y_j \in E,\ i \in I), \tag{18.66}$$

and *the net of partial sum operators* $\{u_d\}_{d\in\mathscr{D}}$ associated to $\{G_i\}_{i\in I}$, by

$$u_d(x) = \sum_{i\in d} y_i \qquad (x = \lim_{d'\in\mathscr{D}} \sum_{j\in d'} y_j \in E,\ d \in \mathscr{D}), \tag{18.67}$$

where $\mathscr{D}$ is the collection of all finite subsets of $I$; thus,

$$u_d = \sum_{i\in d} v_i \qquad (d \in \mathscr{D}). \tag{18.68}$$

Clearly, the net of partial sum operators $\{u_d\}_{d\in\mathscr{D}}$ associated to an extended unconditional Schauder decomposition of $E$ is an extended resolution of $I_E$. It is also obvious that the family of coordinate projections $\{v_i\}_{i\in I}$ associated to an extended unconditional Schauder decomposition $\{G_i\}_{i\in I}$ of a Banach space $E$ satisfies the "orthogonality relations"

$$v_i v_j = \delta_{ij} v_i = \delta_{ij} v_j \qquad (i, j \in I); \tag{18.69}$$

more generally, one can define an extended generalized biorthogonal system*) as a pair of families $(G_i, v_i)_{i\in I}$, where $\{G_i\}_{i\in I}$ is a family of subspaces of $E$ with $G_i \neq \{0\}$ $(i \in I)$ and $\{v_i\}_{i\in I} \subset L(E, E)$ is a family of projections with $v_i(E) = G_i$ $(i \in I)$, satisfying (18.69). Then, the *net of partial sum operators* $\{u_d\}_{d\in\mathscr{D}}$ can be defined for any such system $(G_i, v_i)_{i\in I}$ (by (18.68)), and the uniform boundedness of this net *characterizes extended unconditional Schauder decompositions* of $E$ among $E$-complete extended generalized biorthogonal systems. Similarly to the particular case of extended unconditional bases (see the Notes and remarks to § 17), if $\{G_i\}_{i\in I}$ is an extended unconditional Schauder decomposition of $E$, then for every subset $A$ of $I$ one can define a natural projection $u_A$ of $E$ onto $[\bigcup_{i\in A} G_i]$, by

$$u_A(x) = \lim_{d\in\mathscr{D}_A} \sum_{i\in d} v_i(x) \qquad (x \in E), \tag{18.70}$$

where $\mathscr{D}_A$ denotes the collection of all finite subsets of $A$. The net of projections $\{u_A\}_{A\subset I}$ obtained in this way is uniformly bounded and hence, in particular, *the Boolean algebra of projections generated by* $\{v_i\}_{i\in I}$ (or, what is the same thing, by $\{u_d\}_{d\in\mathscr{D}}$) *is uniformly bounded* (by $\sup_{d\in\mathscr{D}} \|u_d\|$). Furthermore, similarly to the particular case of extended unconditional bases, if $\{G_i\}_{i\in I}$ is an extended unconditional Schauder decomposition of $E$, then, putting

$$p(A)(x) = u_A(x) \qquad (x \in E,\ A \subset I), \tag{18.71}$$

where $u_A$ is defined by (18.70), we obtain a countably additive projection-valued measure $p(.)$ defined on the $\sigma$-field of all subsets of $I$, such that

$$p(\{i\}) = v_i \qquad (i \in I). \tag{18.72}$$

Let us observe that one can also define (*Schauder*) *ER-sets of subspaces* and *$(I, \mathscr{D})$-conditional* (*Schauder*) *bases of subspaces* $\{G_i\}_{i\in I}$

*) See § 15, definition 15.4, for the particular case when $I$ is countable.

(one can also use here the term *decomposition* instead of sets or bases of subspaces), replacing in the definitions of the corresponding notions of bases $\{x_i\}_{i\in I}$ (see § 17 and the Notes and remarks to it), the elements $\alpha_i x_i$ by $y_i$, where $y_i \in G_i$ $(i \in I)$. Clearly, the net of partial sum operators $\{u_d\}_{d\in\mathscr{D}}$ associated to an $(I, \mathscr{D})$-conditional Schauder decomposition $\{G_i\}_{i\in I}$ is an extended resolution of $I_E$. Furthermore, one can discard the requirement of uniqueness of the representation of every $x \in E$ in each of the preceding definitions, arriving thus at the corresponding "pseudo-" and "quasi-" notions (which generalize those of §§ 5, 9, 15, etc.). We leave the other possible generalizations to the reader.

Let us mention that the term "resolution of the identity" is used also in operator theory, with somewhat different meanings. For example, if $E$ is a reflexive Banach space, E. R. Lorch ([252], definition 4) has called a set of projections $\{u_\lambda\}_{\lambda\in(-\infty,\infty)}$ on $E$ a "resolution of the identity" *) if: a) 0 and $I_E$ are in $\{u_\lambda\}_{\lambda\in(-\infty,\infty)}$; b) $u_\lambda u_\mu = u_\mu u_\lambda = u_\mu$ $(\mu \leqslant \lambda)$; c) the projection-valued function $\lambda \to u_\lambda$ is "of bounded variation" in the sense that there exists a constant $C$ such that for any real $\mu_1 \leqslant \lambda_1 \leqslant \leqslant \ldots \leqslant \mu_n \leqslant \lambda_n$ and any complex $\beta_i$ with $|\beta_i| \leqslant 1$ $(i = 1, \ldots, n)$, we have $\left\| \sum_{i=1}^{n} \beta_i (u_{\lambda_i} - u_{\mu_i}) \right\| \leqslant C$. Although $(-\infty, \infty)$ is directed by $\geqslant$, these resolutions of the identity are obviously different from the extended resolutions of the identity considered in § 18, definition 18.5. In the monograph of N. Dunford and J. T. Schwartz ([88], Ch. X, § 1), the "resolutions of the identity" of a Hilbert space $E$ are measures defined on the Borel subsets of the plane, whose values are projections on $E$, with certain properties.

**§ 19.** The difficulties in extending the notions of decompositions and Schauder decompositions from a sequence of subspaces $\{G_n\}$ to a family of subspaces $\{G_i\}_{i\in I}$ (which is not assumed countable) are essentially the same as those for the particular case of bases (i.e., when $\dim G_i = 1$ for all $i \in I$). Therefore it is quite natural that many such extensions were given directly for the case of (Schauder) decompositions and that, in general, even those which were given only for bases, admit further extensions to (Schauder) decompositions. Indeed, we have seen these facts for extended resolutions of the identity and extended unconditional Schauder decompositions, *ER*-sets of subspaces, $(I, \mathscr{D})$-conditional bases of subspaces, etc., in § 18 and the Notes and remarks to § 18 and we shall see them again for ordinal resolutions of the identity and transfinite decompositions, in the present Notes and remarks to § 19.

Monotone ordinal resolutions of the identity and (general) ordinal resolutions of the identity were defined by C. Bessaga ([17], [18]), with the mention ([18], p. 4, footnote to definition 1) that such resolutions of the identity had been applied in the work of J. Lindenstrauss ([236],

*) See also the related notion of "spectral family" in the case of Hilbert spaces, in the monograph of F. Riesz and B.Sz.-Nagy ([338], Ch. VII).

[237]) and D. Amir and J. Lindenstrauss [4]. Let us mention that some authors (e.g., L. Dorembus [84]) use for these notions the terms monotone, respectively general, *Schauder resolutions of the identity* (considering also the non-Schauder case, when the projections need not be continuous).

The notions of transfinite basis of subspaces and transfinite Schauder basis of subspaces were introduced, in the framework of topological linear spaces, by L. Dorembus [84], using his definition of the sum of a transfinite series in such spaces. The set $D_1^{(\vartheta)}$ of proposition 19.1 was considered by L. Dorembus [84], who called it "the summation field" of the transfinite sequence of subspaces $\{G_\lambda\}_{\lambda<\vartheta}$. Theorems 19.1, 19.2 and lemma 19.1 are due to L. Dorembus ([84], corollary 2.2 and facts 1.8, 1.2) and so are proposition 19.3 and theorem 19.3 ([84], lemmas 1.10 and 3.4). For transfinite sequences of subspaces $\{G_\lambda\}_{\lambda<\vartheta}$ satisfying 2° or 3° of theorem 19.3, L. Dorembus has introduced the term "semi-orthogonal", with the mention that the term is new, but the notion itself appears in various forms in the literature (even in the more general framework of topological linear spaces). Proposition 19.4 is due to L. Dorembus ([84], lemma 4.1); this proposition yields simpler proofs of some results of J. J. M. Chadwick [46] on the existence of certain (usual) Schauder decompositions.

The first results on the existence of certain long sequences of projections, for non-separable reflexive Banach spaces, were obtained by J. Lindenstrauss ([236], [237]) and his methods were extended by D. Amir and J. Lindenstrauss [4] to yield existence of such long sequences for non-separable weakly compactly generated spaces; a description of these methods can be also found in the survey paper of J. Lindenstrauss [242]. The proofs of the later results on the existence of ordinal resolutions of the identity, having certain additional properties, in non-separable weakly compactly generated spaces and on the existence of such resolutions for various other classes of non-separable Banach spaces, are also based on (refinements or variations of) the techniques of D. Amir and J. Lindenstrauss [4].

The first statement of lemma 19.2 and its stronger form mentioned in the footnote to this lemma, are due to J. Lindenstrauss [238]. A similar proof of both statements of lemma 19.2 was given by K. John and V. Zizler ([164], lemma 1), but C. Bessaga has observed (verbal communication) that the second statement is a consequence of the first one. Lemmas 19.3, 19.4 and 19.5 were obtained by K. John and V. Zizler ([164], lemmas 2, 3 and proposition 1), based on the above mentioned works of Lindenstrauss ([236], [237]) and Amir-Lindenstrauss [4]; the main additional features of lemmas 19.3—19.5 are that they contain one more norm (in order to ensure the weak* continuity of projections in the case of conjugate spaces) and the inclusions*) $u(G) \subset G$, respec-

*) For $E$ reflexive and for $G$ a weakly compactly generated subspace of a weakly compactly generated space $E$, these inclusions were proved by J. Lindenstrauss ([238], theorem 2 and [242], § 2).

tively $p(G) \subset G$, where $G$ is any given closed linear subspace of $E$. Remark 19.1 b) was made by K. John and V. Zizler ([164], remark 1). Lemma 19.6, with the alternative proof using the sets (19.75), was observed by D. Amir and J. Lindenstrauss ([4], lemma 5). The first statement of proposition 19.5 was obtained by K. John and V. Zizler ([164], proposition 2), the proof being again a variation of the arguments of D. Amir and J. Lindenstrauss ([4], lemma 6); the second part of proposition 19.5 can be found e.g. in the survey paper of J. Lindenstrauss ([242], theorem 2.3). The second statement of remark 19.2 a) was observed by K. John and V. Zizler ([166], lemma 2 ii)). The third statement of remark 19.2 a) can be found, for reflexive spaces, in the paper of J. Lindenstrauss ([236], lemma 3), with the mention that it is well known and it goes back to E. R. Lorch [252]; in the general case (i.e., for weakly compactly generated spaces), it has been proved by D. Amir and J. Lindenstrauss ([4], lemma 7), but the simpler proof presented here is due to K. John and V. Zizler ([166], lemma 2 ii)). Remark 19.2 b) can be found, essentially, in the paper of J. Reif ([330], lemma and proposition). Let us mention that lemma 19.2 (in the more complicated version of John-Zizler mentioned above), lemmas 19.3–19.5, the first part of proposition 19.5 and a variant of the third statement of remark 19.2 a) can be also found in the book of J. Diestel ([77], Ch. V, § 1, lemmas 1–6), with some differences in the proofs. Lemma 19.7 is due to L. Dorembus ([84], lemma 4.4). A weaker version of theorem 19.4, without (19.89), (19.90) and with $\{u_\lambda\}_{\lambda \leqslant \vartheta}$ and (19.91) replaced by $\{u_\lambda\}_{\lambda \leqslant \mu}$, where $\mu$ is a limit ordinal satisfying $\mu \leqslant \vartheta$, and by $\operatorname{dens} u_\lambda(E) < \operatorname{dens} E$ $(\lambda < \mu)$ respectively, was proved by L. Dorembus ([84], theorem 4.5 (i)), who also gave the final part of remark 19.4 a), i.e., that if $\operatorname{dens} E = \aleph_{\tau+1}$, then one can take $\mu = \vartheta$ ([84], theorem 4.5 (ii)). Remark 19.4 b) can be found, implicitly, in the survey paper of J. Lindenstrauss [241] and in several later papers (e.g., K. John and V. Zizler [164], [166], J. Reif [330], etc.), where it is used for constructions (e.g., of $M$-bases) by transfinite induction.

Lemma 19.8 and proposition 19.6 are due to K. John and V. Zizler ([164], propositions 3, 5 and [163], lemma 3).

Lemmas 19.9–19.14 and proposition 19.7 were proved by D. G. Tacon ([389], lemmas 2–7 and theorem 2). Using a different method, K. John and V. Zizler ([168], theorem 1) have shown that if $E$ is a Banach space which admits a continuously Fréchet differentiable function with bounded non-empty support, then the conclusion of proposition 19.7 holds for $E$. Problem 19.1 a) was raised, essentially, by S. M. Gutman [152] (actually, S. M. Gutman [152] asked only about the very smoothness in the initial norm of $E$, to which the answer is clearly negative).

Lemma 19.15 was observed by S. M. Gutman ([152], lemma 4). Proposition 19.8, under the stronger assumption that the norm of $E$ is Fréchet differentiable at every non-zero point, was given by K. John and V. Zizler ([166], lemma 3 and [165], lemma 2); for weakly compactly generated very smooth spaces $E$ it has been given by S. M. Gutman ([152], lemma 5).

Recently, K. John and V. Zizler ([171], proposition 1) have proved the following proposition on subspaces $G$ of weakly compactly generated Banach spaces, admitting an equivalent very smooth norm (actually, in [171] they assumed that $G$ admits an equivalent Fréchet differentiable norm), which they used to give an alternative proof of § 20, theorem 20.6 and corollary 20.6 (with $(F)$ instead of very smooth):

**Proposition 19.9.** *Let $(E, \|\cdot\|)$ be a weakly compactly generated Banach space and let $\vartheta$ be the first ordinal number of cardinality* dens $E$. *Furthermore, let $G$ be a closed linear subspace of $E$, admitting an equivalent very smooth norm $|||\cdot|||$. Then there exists a long sequence of linear projections $\{p_\lambda\}_{\omega\leqslant\lambda\leqslant\vartheta}$ on $E$, satisfying*

$$|||p_\lambda|_G||| = \|p_\lambda\| = 1 \qquad (\omega \leqslant \lambda \leqslant \vartheta), \tag{19.162}$$

$$p_\lambda p_\varkappa = p_\varkappa p_\lambda = p_\varkappa \qquad (\omega \leqslant \varkappa \leqslant \lambda \leqslant \vartheta), \tag{19.163}$$

$$p_\lambda(G) \subset G \qquad (\omega \leqslant \lambda \leqslant \vartheta), \tag{19.164}$$

$$\operatorname{dens} p_\lambda(E) \leqslant \operatorname{card} \lambda \qquad (\omega \leqslant \lambda \leqslant \vartheta), \tag{19.165}$$

$$\operatorname{dens} (p_\lambda|_G)^*(G^*) \leqslant \operatorname{card} \lambda \qquad (\omega \leqslant \lambda \leqslant \vartheta), \tag{19.166}$$

*and such that for each fixed $x \in E$ and $\varphi \in G^*$, the mappings $\lambda \to p_\lambda(x)$ and $\lambda \to (p_\lambda|_G)^*(\varphi)$ from $[\omega, \vartheta]$ into $E$, respectively, into $G^*$, are continuous.*

Thus, proposition 19.9 yields subspaces $p_\omega(G)$ and $(p_{\lambda+1} - p_\lambda)(G)$ $(\omega \leqslant \lambda < \vartheta)$ of $G$, which are suitable for transfinite induction.

Problem 19.2 was raised, essentially, by S. M. Gutman [152]. Let us mention that both D. G. Tacon [389] and S. M. Gutman [152] have used for very smooth spaces the term "spaces with property $(A)$".

Finally, let us mention that, similarly to the Notes and remarks to § 17, one can also define, in a natural way, $\vartheta$-linearly independent transfinite sequences of subspaces $\{G_\lambda\}_{\lambda<\vartheta}$, transfinite pseudo-bases of subspaces, transfinite quasi-bases of subspaces, etc. We leave these to the reader.

§ **20.** The notion of an extended biorthogonal system for a Banach space $E$ and the first part of proposition 20.1 for separable spaces can be found already in the monograph of S. Kaczmarz and H. Steinhaus ([194], Ch. VIII, § 1, where the term "biorthonormal system" was used); the proof for arbitrary spaces, presented here, is completely similar. For the space $L^2([a, b])$ the second part of proposition 20.1 can be also found in the same monograph, with a different proof ([194], Ch. III, § 4).

As has been already mentioned in the Notes and remarks to § 6, maximal biorthogonal systems were considered by J. Dieudonné [81], who also gave proposition 20.2 and theorem 20.1 ([81], propositions 3 and 2). Problem 20.1 was raised by W. J. Davis and W. B. Johnson [63].

Some sufficient conditions for a Banach space $E$ to admit a total extended biorthogonal system were given by J. A. Dyer [92]. Theorem 20.2 (via lemma 20.1 and a variant of proposition 20.3) was proved by A. N. Pličko [320]; the proof presented here is somewhat simpler (in [63] transfinite induction was used instead of Zorn's lemma).

Problem 20.2 a) was raised by W. J. Davis and W. B. Johnson [63], who obtained some partial results concerning problems 20. 2 a), b). Thus, proposition 20.4, theorem 20.3 and corollary 20.1 were given in [63], lemma 2, theorem 2 and remark 3 respectively, with the mentions that the perturbation techniques used in the proof of theorem 20.3 was suggested by the proof of proposition 1 in [372] and that the implication 1° ⇒ 2° of corollary 20.1 is shown by an argument of W. B. Johnson [175]. Furthermore, theorem 20.4 and the remark made after it were given in [63], remark 1, with the mention that the proof of theorem 20.4 is a simple modification of the proof of [374], theorem 1. However, recently A. N. Pličko [321], refining his previous method [320] of the proof of theorem 20.2, has shown, solving thus problem 20.2 b) in the affirmative, that for each $\varepsilon > 0$, every Banach space $E$ has a total extended biorthogonal system $(x_i, f_i)_{i\in I}$ such that $\sup_{i\in I} \|x_i\| \|f_i\| \leqslant 6+\varepsilon$ and has conjectured that this bound can be considerably lowered; subsequently [323], he has shown that this is indeed possible for reflexive spaces (see below, the comments on strictly bounded extended $M$-bases).

J. A. Dyer has shown ([92], theorem 4) that *if $I$ is a set endowed with the discrete topology and $Q$ is a compactification of $I$, then $C(Q)$ admits a norming*) extended biorthogonal system.* Also, J. A. Dyer has shown ([92], theorem 5) that *if $E$ is a $\mathscr{P}_\lambda$-space (for some $\lambda$) and if $I$ is an infinite set, the following statements are equivalent:* 1°. *$E$ has a norming extended biorthogonal system $(x_j, f_j)_{j\in J}$ such that* card $J =$ card $I$; 2°. *$E$ is isomorphic to $l^\infty(I)$.* The problem of the existence of norming extended biorthogonal systems in the general case is still open**).

---

*) This notion is defined similarly to norming $M$-bases (see definition 20.5).

**) See the Appendix.

The results of § 7 on similarity of generalized bases and biorthogonal systems were obtained by M. G. Arsove and R. E. Edwards [7] and W. J. Davis [58], directly in the more general setting of extended total biorthogonal systems and extended biorthogonal systems respectively; for some extensions to more general classes of topological linear spaces see also W. B. Johnson and J. A. Dyer [183].

Extended $M$-bases were considered by A. I. Markuševič [259], who called them "bases in the wide sense"; the term "extended $M$-basis" was proposed by M. G. Arsove and R. E. Edwards [7].

The first example of a Banach space having no extended $M$-basis, namely, $E = l^\infty(I)$ with $I$ uncountable, was given by J. A. Dyer ([89], § 3), using proposition 20.6 ([89], § 3). Proposition 20.5 and corollary 20.2 were proved by W. B. Johnson [175]. Remark 20.3 was made by W. J. Davis and W. B. Johnson ([63], remark 2). Some other examples of Banach spaces having no extended $M$-basis follow from a paper of F. K. Dashiell and J. Lindenstrauss [54]. In fact, they give in [54] an increasing transfinite sequence of pairwise non-isomorphic subspaces of $l^\infty([0, 1])$, which admit no continuous linear mapping into any $c_0(I)$ (and hence, by proposition 20.6, they have no extended $M$-basis); all spaces in this transfinite sequence, except the last one, admit equivalent strictly convex norms ([54], theorems 10 and 9). Problem 20.3 was raised by S. Troyanski at the Colloquium on Geometry of Banach spaces held in Oberwolfach in October 1973 and, independently, by J. Diestel ([76], problem 9). More generally, it seems to be unknown whether every (or, at least, every complemented) subspace $G$ of a Banach space $E$ with an extended $M$-basis must have an extended $M$-basis; a result of this type, involving norming $M$-bases, will be mentioned below (see the comments to problem 20.4 a)). Clearly, an affirmative answer to this problem would yield a negative answer to problem 20.3.

It is also natural to ask whether for every extended $M$-basis $\{y_i\}_{i\in I}$ of a subspace $G$ of a Banach space $E$ with an extended $M$-basis there exists an extended $M$-basis of $E$, containing $\{y_i\}_{i\in I}$ (for separable spaces $E$ this is true by § 8, theorem 8.2). Recently, K. John and V. Zizler ([172], proposition 4) have shown that the answer is affirmative for weakly compactly generated Banach spaces $E$.

Under the more restrictive assumption that $E$ is weakly compactly generated, the conclusions of proposition 20.6 were proved by D. Amir and J. Lindenstrauss ([4], main theorem and theorem 3 (a)), who have not used extended $M$-bases, but worked with the "long sequence" of projections of § 19, proposition 19.5 and transfinite induction*). As we mentioned above, in the general case proposition 20.6 was proved by J. A. Dyer ([89], § 3).

---

*) Also, D. G. Tacon [389] has deduced similar conclusions for the conjugate space $E^*$ of a very smooth space $E$, using § 19, proposition 19.7. For another result of this type, see K. John and V. Zizler [168].

The idea of proving inductively the existence of $M$-bases in spaces which admit decompositions into subspaces of smaller dimension, can be found in the survey paper of J. Lindenstrauss [241], where it has been observed that every weakly compactly generated Banach space $E$ has an extended $M$-basis (see also [373], where it has been proved that every Banach space with a finite-dimensional decomposition admits a strictly bounded $M$-basis). Theorems 20.5 a) and 20.5 b) were given by J. Reif ([330], corollary 1 and [331], proposition 1 (iii)). Unfortunately, in general (in the non-separable case), this method does not yield strictly bounded extended $M$-bases. Recently, A. N. Pličko has shown the existence of strictly bounded extended $M$-bases in all weakly compactly generated spaces, with the bound $2 + \varepsilon$, with a different construction ([322] and [323], theorem and remark 2). The final statement of remark 20.4 a) can be found in the survey paper of J. Lindenstrauss ([242], corollary 1 of theorem 3.3); the proof presented here is due to J. Reif ([330], corollary 2). Corollary 20.4 was proved by D. Amir and J. Lindenstrauss ([4], proposition 2); the simpler proof presented here is due to J. Reif ([330], corollary 3). The first part of corollary 20.5 was given by D. Amir and J. Lindenstrauss ([4], theorem 3 (c), (b)), while the second part is contained, implicitly, in the survey paper of J. Lindenstrauss ([242], § 5).

Problem 20.4 a) was raised, independently, by H. P. Rosenthal ([341], p. 108) and K. John and V. Zizler ([167], p. 688, problem 1). Recently K. John and V. Zizler ([172], proposition 6), have shown that *if a weakly compactly generated Banach space E has a norming extended M-basis, then every subspace G of E also has a norming extended M-basis.* Thus, problem 20.4 a) reduces to the particular case of weakly compactly generated $C(Q)$ spaces (indeed, $E$ is linearly isometric to a subspace of $C(S_{E^*})$, where $S_{E^*}$ is endowed with the $\sigma(E^*, E)$ topology, and it is known*) that if $E$ is weakly compactly generated, then so is $C(S_{E^*})$). Concerning such spaces, K. John and V. Zizler have shown ([172]**), proposition 1) that *every weakly compactly generated $C(Q)$ space has an extended M-basis $\{x_i\}_{i\in I}$ such that* $\operatorname{lin}\{f_i\}_{i\in I}\subset\operatorname{lin}\{\varepsilon_q|q\in Q\}$, where $\{f_i\}_{i\in I}\subset E^*$ is the a.f.f. to $\{x_i\}_{i\in I}$ and $\varepsilon_q(y)=y(q)$ $(y\in C(Q),\ q\in Q)$; furthermore ([172], corollary 1), weakly compactly generated $C(Q)$ spaces admit equivalent norms with nice properties. Also, K. John and V. Zizler have shown [167] that some of the properties of weakly compactly generated Banach spaces and, in particular, of Banach spaces with shrinking extended

---

*) See e.g. [242], theorems 3.3 and 3.2.

**) Actually, proposition 1 of [172] is the following more general result: If $E$ is a weakly compactly generated Banach space and $\mathscr{L}\subset E^*$ is a $\sigma(E^*, E)$-compact subset of $E^*$ with $r([\mathscr{L}])=1$, such that either $\mathscr{L}$ or $\mathscr{L}\setminus\{0\}$ is finitely linearly independent, then $E$ has an extended $M$-basis $\{x_i\}_{i\in I}$ such that $\operatorname{lin}\{f_i\}_{i\in I}\subset\operatorname{lin}\mathscr{L}$, where $\{f_i\}_{i\in I}\subset E^*$ is the a.f.f. to $\{x_i\}_{i\in I}$.

$M$-bases, presented in § 20, can be generalized to Banach spaces with norming extended $M$-bases; we shall give these results of K. John and V. Zizler below, after the comments on the corresponding results on weakly compactly generated Banach spaces and on Banach spaces with shrinking extended $M$-bases.

Under the more restrictive assumption that $\{x_i\}_{i \in I}$ is shrinking, proposition 20.7 was given, without proof (but with the mention that the proof is a simple modification of a proposition of D. Amir and J. Lindenstrauss [4]), by S. Troyanski ([395], proposition 2). In connection with remark 20.6 let us note that J. Reif ([329], lemma 1) has also stated, without proof (but with the mention that it easily follows from the results of D. Amir and J. Lindenstrauss [4]) the existence, on every weakly compactly generated Banach space, of a long sequence of projections $\{p_\lambda\}_{\omega \leqslant \lambda \leqslant \vartheta}$ having the usual properties occurring in § 19 and satisfying, in addition $p_{\lambda+1} \neq p_\lambda$ for all $\omega \leqslant \lambda < \vartheta$.

Shrinking extended $M$-bases were considered by S. Troyanski [395], who also observed proposition 20.8 ([395], lemma 1). In the particular case when $G = E$ and under the stronger assumption that the norm of $E$ is $(F)$, theorem 20.6 was proved by S. Troyanski ([395], proposition 5) with a direct method, which did not use the canonical isomorphisms $\rho_\omega$ and $\rho_{\lambda+1}$. In the general case (when $G \neq E$), still under the assumption that the norm of $E$ is $(F)$, theorem 20.6 and corollary 20.6 were proved by K. John and V. Zizler ([166], lemma 4 and proposition 3). According to H. P. Rosenthal ([341], § 0), in the case when $G \neq E$ and under the assumption that the subspace $G$ has an equivalent $(F)$-norm, corollary 20.6 was obtained by D. Friedland in a manuscript submitted in 1974 (which, apparently, remained unpublished). Finally, we note that corollary 20.6 (and, essentially, theorem 20.6) can be also found, with a different proof, in the book of J. Diestel ([77], Ch. V, § 7), where it is called "the Friedland-John-Zizler theorem"; also, as has been mentioned in the Notes and remarks to § 19, recently K. John and V. Zizler have given new proofs of theorem 20.6 and corollary 20.6 ([171], theorem 2 (i) and corollary 2 (i)). Propositions 20.9 a) and b), formula (20.58), were proved by S. Troyanski ([395], proof of proposition 3 and [395], lemma 2). Lemmas 20.2 and 20.3 a), b) are due to S. Troyanski ([394], lemma and proposition 1 and [395], proof of theorem 1); the proof of lemma 20.2, presented here, is due to P. Morris [290]. Under some additional assumptions (which are not used), lemma 20.3 a) can be also found in the book of M. M. Day ([70], Ch. VII, § 4, lemma 3). Also, lemma 20.2 and, under slightly modified assumptions, lemma 20.3 a), are presented in the seminar notes of R. E. Huff ([157], lemma 1 and theorem 3) and in the book of J. Diestel ([77], Ch. IV, § 2); the latter also contains a mention of lemma 20.3 b). The implication 1° ⇒ 2° of theorem 20.7 is due, for the particular space $E = c_0(I)$, to R. R. Phelps [315] and E. Asplund ([8], theorem 3) and in the general case to S. Troyanski ([395], theorem 1); as shown by the proof of theorem 20.7, the other implications are consequences of previously mentioned results. The implications $3° \Rightarrow 2°$ and $4° \Rightarrow 2°$ were proved by K. John and V. Zizler ([165], theorem 1) and S. M. Gutman ([152], proposition 4) respectively, without going through extended shrinking $M$-bases. The equivalences $1° \Leftrightarrow 2° \Leftrightarrow 3°$ were given in the paper of K. John and V. Zizler ([166], theorem 1), together with some other equivalent conditions, such as: 5°. *E is weakly compactly generated and admits a continuously Fréchet differentiable real-valued function with bounded non-empty support.* Remark 20.7 a) was made by S. Troyanski ([395], proposition 3 and lemma 3). The result mentioned in remark 20.7 b) is nothing else than a well known theorem of S. Troyanski ([394], theorem 1), which has solved in the affirmative a problem raised by J. Lindenstrauss ([242], problem 12); in the same paper, S. Troyanski has also shown ([394], theorem 2), that every abstract $L$-space $E$ is weakly compactly generated and hence isomorphic to a locally uniformly convex space. Corollary 20.7 is due to K. John and V. Zizler ([166], theorem 3 and proposition 2). The two results mentioned after corollary 20.7 were proved by K. John and V. Zizler ([164], corollary of proposition 6 and [163], main result); actually, they deduced the first one from the following more general result ([164], proposition 6): *If both $E$ and $E^*$ are weakly compactly generated and $V \subset E^*$ is a total closed linear subspace of $E^*$, then $E$ has an extended $M$-basis $\{x_i\}_{i \in I}$ with the a.f.f. $\{f_i\}_{i \in I} \subset E^*$ satisfying $[f_i]_{i \in I} = V$* (let us mention that, as has been shown recently by K. John and V. Zizler in [171], proposition 4, this latter result remains also valid if we replace the assumption that $E^*$ is weakly compactly generated by the assumption that $V$ is weakly compactly generated). Furthermore, W. B. Johnson and J. Lindenstrauss ([184], theorem) proved that if *$E$ is a subspace of a weakly compactly generated space and $E^*$ is weakly compactly generated, then $E$ has a shrinking extended $M$-basis (and hence $E$ is weakly compactly generated)*; recently, K. John and V. Zizler have given a different proof of this result ([171], theorem 2 (ii)). Subsequently*⁾, K. John and V. Zizler ([169], theorem and corollary 2) showed that *if both $E$ and $E^*$ are subspaces of weakly compactly generated spaces, then the same conclusion holds and, moreover, for any total closed linear subspace $V$ of $E^*$, the space $E$ has an extended $M$-basis $\{x_i\}_{i \in I}$ with the a.f.f. $\{f_i\}_{i \in I} \subset E^*$ satisfying $[f_i]_{i \in I} = V$*; the proofs are suitable refinements of the techniques of § 19 and § 20. In this connection, let us mention the following problem of J. Lindenstrauss ([242], problem 13; see also [184], problem (2)):

*Problem 20.6.* If $E^*$ is weakly compactly generated, does $E$ admit an equivalent $(F)$-norm?

---

*⁾ Indeed, although their paper [169] appeared earlier than [171], it has been submitted later than [171].

In the above we have seen that the answer is affirmative when $E$ is a subspace of a weakly compactly generated space. W. B. Johnson and J. Lindenstrauss [184] have constructed some examples which are not subspaces of any weakly compactly generated space, for which the answer is still affirmative.*)

Recently K. John and V. Zizler have shown ([172], proposition 5), similarly to the result mentioned above on extended $M$-bases of subspaces of weakly compactly generated spaces (after the comments on problem 20.3), that for every shrinking extended $M$-basis $\{y_i\}_{i \in I}$ of a subspace $G$ of a Banach space $E$ with a shrinking extended $M$-basis, there exists a shrinking extended $M$-basis of $E$, containing $\{y_i\}_{i \in I}$.

Let us mention now, as announced in the above comments on problem 20.4 a), some recent results of K. John and V. Zizler [167] on norming extended $M$-bases. Firstly, they have shown ([167], proposition 1) that *proposition* 20.7 *holds also for any Banach space E with a* 1-*norming***) *extended M-basis* $\{x_i\}_{i \in I}$. In the same paper [167], K. John and V. Zizler have observed ([167], lemma 2) the following generalization of proposition 20.8: *If E is a Banach space with an extended M-basis* $\{x_i\}_{i \in I}$ *satisfying* $\sup_{i \in I} \|x_i\| < \infty$, *then the set* $\{x_i\}_{i \in I} \cup \{0\}$ *is* $\sigma(E, [f_i]_{i \in I})$-*compact* (where $\{f_i\}_{i \in I} \subset E^*$ is the a.f.f. to $\{x_i\}_{i \in I}$). Furthermore, K. John and V. Zizler have shown ([167], proposition 3) that *if the norm of a Banach space E is an* $(F)$-*norm and if E has a* 1-*norming extended M-basis, then E also has a shrinking extended M-basis* (and hence $E$ is weakly compactly generated); this, together with the comment after problem 20.6 above, provides examples of (non-weakly compactly generated) spaces with an $(F)$-norm which have no 1-norming extended $M$-basis. Also, K. John and V. Zizler have proved ([167], theorem) that *if E is a Banach space with a norming extended M-basis* $\{x_i\}_{i \in I}$, *then E admits an equivalent locally uniformly convex norm* $|||\cdot|||$ *which is lower* $\sigma(E, [f_i]_{i \in I})$-*semi-continuous* (where $\{f_i\}_{i \in I} \subset E^*$ is the a.f.f. to $\{x_i\}_{i \in I}$); in the particular case when $E$ is a conjugate space $B^*$ and $[f_i]_{i \in I} = \pi(B)$, the canonical image of $B$ in $B^{**}$, this result yields theorem 20.7, implication $1° \Rightarrow 2°$.

Boundedly complete extended $M$-bases were considered by S. Troyanski [395], who observed corollary 20.8 ([395], corollary 2). K. John and V. Zizler ([171], proposition 6) proved that if a transfinite basis $\{x_\lambda\}_{\lambda < \vartheta}$, regarded as an $M$-basis, is boundedly complete in the sense of definition 20.7, then it is also boundedly complete in the sense of § 17, definition 17.11 b) and that if $r([f_\lambda]_{\lambda < \vartheta}) > 0$ (where $f_\lambda(x_\mu) = \delta_{\lambda\mu}$), then the converse is also true.

---

*) Actually, as has been communicated to us recently by several colleagues, the proof of [184] that the spaces of these examples admit equivalent $(F)$-norms, is not convincing (the norm on $E^*$, given in [184], is not locally uniformly convex), but the authors have corrected the proof (using another norm on $E^*$).

**) I.e., such that $r([f_i]_{i \in I}) = 1$, where $\{f_i\}_{i \in I} \subset E^*$ is the a.f.f. to $\{x_i\}_{i \in I}$.

In the particular case when the norm of $E$ is $(F)$, problem 20.5 was raised by K. John and V. Zizler ([170], problem (a)), who proved ([170], theorem) that *if $E^*$ is locally uniformly convex* (which implies*) that the norm of $E$ is $(F)$), *then the answer is affirmative;* also, K. John and V. Zizler observed [170] that this result cannot be strengthened to obtain an extended $M$-basis of $E^*$, with the a.f.f. contained in $\pi(E)$, the canonical image of $E$ in $E^{**}$ (or, equivalently, to obtain a shrinking extended $M$-basis of $E$), since there exist**) non-weakly compactly generated Banach spaces $E$ such that $E^*$ is locally uniformly convex. S. M. Gutman proved ([152], theorem 1) that if $E$ is a non-separable very smooth Banach space, then the conjugate space $E^*$ admits an equivalent locally uniformly convex norm; however, this latter norm need not be the dual of a norm on $E$. Let us also mention the following two related problems, raised by K. John and V. Zizler ([167], problem 2 and [170], problem (b)):

*Problem 20.7.* Does every Banach space with an extended $M$-basis admit an equivalent locally uniformly convex norm?

*Problem 20.8.* Does every locally uniformly convex Banach space $E$ have an extended $M$-basis?

The results of § 20 and of the above Notes and remarks show that in a number of particular cases the answers to these problems are affirmative (e.g., to problem 20.7 for weakly compactly generated spaces and for spaces with a norming or a boundedly complete extended $M$-basis and to problem 20.8 for locally uniformly convex conjugate spaces, etc.). An affirmative answer to problem 20.7 would strengthen proposition 20.6, which shows the existence of an equivalent strictly convex norm. Also, an affirmative answer to problem 20.7 would imply a negative answer to problem 20.3 (see the comments made after § 20, corollary 20.8 and in the Notes and remarks to § 17, theorem 17.3). On the other hand, in problem 20.8 it is not sufficient to assume that $E$ is strictly convex, since we have seen in the preceding that $E = l^\infty$ admits an equivalent strictly convex norm (but no equivalent locally uniformly convex norm) and has no extended $M$-basis.

The notions of extended biorthogonal system and extended $M$-basis were further generalized by S. Kaplan [212] and J. A. Dyer [89], as follows: Let $E$ be a Banach space. A pair of sets $(A, B)$ $(A \subset E$, $B \subset E^*)$ is called a *biorthogonal system in the wide sense*, or shortly, a *bows*, if it satisfies the following conditions I and II:

I. (a) If $x \in A$ and $f \in B$, then $f(x) = 0$ or 1. (b) For every $x \in A$ there exists $f \in B$ such that $f(x) = 1$ and for every $f \in B$ there exists $x \in A$ such that $f(x) = 1$.

---

*) See § 20, the comments before lemma 20.2.

**) See W. B. Johnson and J. Lindenstrauss [184].

II. The $B$-orthogonal subsets of $A$ are directed by inclusion (a finite subset $\{x_1, \ldots, x_n\} \subset A$ is called *B-orthogonal*, if for each $f \in B$ there exists at most one $i \in \{1, \ldots, n\}$ such that $f(x_i) = 1$).

A bows $(A, B)$ is called a *generalized M-basis* of $E$ if $[A] = E$ and $B$ is total on $E$.

S. Kaplan [212] has shown that the bows defined in this way includes integration theory (by taking $A \subset E = L^\infty$ to be the set of characteristic functions of measurable sets and $B$ the set of functionals defined by these characteristic functions, so the $B$-orthogonal subsets of $A$ are those corresponding to the finite collections of pairwise disjoint measurable sets). J. A. Dyer ([89], theorem 1) has given an extension of proposition 20.6 to generalized $M$-bases and a characterization of Banach spaces isomorphic to $\mathscr{P}_1$ spaces in terms of generalized $M$-bases.

# Bibliography


[1] Aleksandrov, P. S.: Introduction to the general theory of sets and functions. Moscow–Leningrad: OGIZ, Gostehizdat 1948 [Russian].

[2] Altshuler, Z.: Characterizations of $c_0$ and $l_p$ among Banach spaces with symmetric basis. Israel J. Math. **24**, 39–44 (1976).

[3] Altshuler, Z.: A Banach space with a symmetric basis which contains no $l_p$ or $c_0$, and all its symmetric basic sequences are equivalent. Compositio Math. **35**, 189–195 (1977).

[4] Amir, D., Lindenstrauss, J.: The structure of weakly compact sets in Banach spaces. Ann. of Math. **88**, 35–46 (1968).

[5] Andô, T.: On fundamental properties of a Banach space with a cone. Pacific J. Math. **12**, 1163–1169 (1962).

[6] Aronszajn, N., Smith, K. T.: Invariant subspaces of completely continuous operators. Ann. of Math. **60**, 345–350 (1954).

[7] Arsove, M. G., Edwards, R. E.: Generalized bases in topological linear spaces. Studia Math. **19**, 95–113 (1960).

[8] Asplund, E.: Boundedly Krein-compact Banach spaces. Proc. of "Functional Analysis" Week held in Aarhus, 1969, pp. 1–4. Aarhus: Math. Inst. Univ. Aarhus 1969.

[9] Bachelis, G. F.: Homomorphisms of annihilator Banach algebras. Pacific J. Math., **25**, 229–247 (1968).

[10] Bachelis, G. F., Rosenthal, H. P.: On unconditionally converging series and biorthogonal systems in a Banach space. Pacific J. Math. **37**, 1–5 (1971).

[11] Banach, S.: Théorie des opérations linéaires. Warszawa: Monografje Matematyczne 1932.

[12] Banach, S., Mazur, S.: Zur Theorie der linearen Dimension. Studia Math. **4**, 100–112 (1933).

[13] Bari, N. K.: Trigonometrical series. Moscow 1961 [Russian]. English translation, New York 1964.

[14] Bartle, R. G., Graves, L. M.: Mappings between function spaces. Trans. Amer. Math. Soc. **72**, 400–413 (1952).

[15] Berkson, E.: Hermitian projections and orthogonality in Banach spaces. Proc. London Math. Soc. (1) **24**, 101–118 (1972).

[16] Bessaga, C.: Bases in certain spaces of continuous functions. Bull. Acad. Polon. Sci. **5**, 11–14 (1957).

[17] Bessaga, C.: Topological equivalence of unseparable reflexive Banach spaces: Ordinal resolutions of identity and monotone bases. Bull. Acad. Polon. Sci. **15**, 397–399 (1967).

[18] Bessaga, C.: Topological equivalence of non-separable reflexive Banach spaces. Ordinal resolutions of identity and monotone bases. Proc. Sympos. on "Infinite dimensional topology" held in Baton Rouge, Louisiana, 1967, pp. 3–14. Princeton: Ann. of Math. Studies 69, Princeton Univ. Press 1972.

[19] Bessaga, C., Kadec, M. I.: On topological classification of non-separable Banach spaces. Proc. Sympos. on "Infinite dimensional topology" held in Baton Rouge, Louisiana, 1967, pp. 15–24. Princeton: Ann. of Math. Studies 69, Princeton Univ. Press 1972.

[20] Bessaga, C., Pelczynski, A.: Some remarks on conjugate spaces containing subspaces isomorphic to the space $c_0$. Bull. Acad. Polon. Sci. **6**, 249–250 (1958).

[21] Bessaga, C., Pelczynski, A.: On bases and unconditional convergence of series in Banach spaces. Studia Math. **17**, 151–164 (1958).

[22] Bessaga, C., Pelczynski, A.: A generalization of results of R. C. James concerning absolute bases in Banach spaces. Studia Math. **17**, 165–174 (1958).

[23] Bessaga, C., Pelczynski, A.: Properties of bases in spaces of type $B_0$. Prace Mat. **3**, 123–142 (1959) [Polish].

[24] Bessaga, C., Pelczynski, A.: Spaces of continuous functions (IV) (On isomorphical classification of spaces $C(S)$). Studia Math. **19**, 53–62 (1960).

[25] Bessaga, C., Pelczynski, A.: Some remarks on homeomorphism of Banach spaces. Bull. Acad. Polon. Sci. **8**, 757–761 (1960).

[26] Bessaga, C., Pelczynski, A.: Selected topics in infinite-dimensional topology. Warszawa: Monografie Matematyczne 1975.

[27] Bierstedt, K.-D.: Neuere Ergebnisse zum Approximationsproblem von Banach-Grothendieck. Jahrbuch Überblicke Mathematik, pp. 45–72. Mannheim: Bibliographisches Inst. 1976.

[28] Billard, P.: Bases dans $H^1$ et bases de sous-espaces de dimension finie dans $A$. Proc. Conf. on "Linear operators and approximation" held in Oberwolfach, August 1971, pp. 310–324. Basel-Stuttgart: Birkhäuser Verlag 1972.

[29] Bishop, E., Phelps, R. R.: A proof that every Banach space is subreflexive. Bull. Amer. Math. Soc. **67**, 97–98 (1961).

[30] Bočkarev, S. V.: Existence of a basis in the space of functions, analytic in a disc, and some properties of the Franklin system. Matem. Sbornik **95** (137), 3–18 (1974) [Russian].

[31] Bourbaki, N.: Topologie générale. Ch. I–II. Paris: Hermann et Cie 1951.

[32] Bourbaki, N.: Intégration. Ch. I–IV. Paris: Hermann et Cie 1952.

[33] Bourbaki, N.: Espaces vectoriels topologiques. Ch. I–II. Paris: Hermann et Cie 1953.

[34] Brace, J. W.: The topology of almost uniform convergence. Pacific J. Math. **9**, 643–652 (1959).

[35] Brace, J. W.: Approximating compact and weakly compact operators. Proc. Amer. Math. Soc. **12**, 392–393 (1961).

[36] Brass, H.: Grundmengen in normierten Räumen. Proc. Conf. on "Approximation theory" held in Oberwolfach, August 1963, pp. 172–178. Basel: Birkhäuser Verlag 1964.

[37] Braun, Ben-Ami: On the multiplicative completion of certain basic sequences in $L^p$, $1 < p < \infty$. Trans. Amer. Math. Soc. **176**, 499–508 (1973).

[38] Braun, Ben-Ami: On the serial completion of deleted Schauder bases by domain adjustment. Proc. Amer. Math. Soc. **44**, 427–431 (1974).

[39] Browder, F. E., de Figueiredo, D. G.: $J$-monotone nonlinear operators in Banach spaces. Nederl. Akad. Wetensch. Proc. Ser. A 69, 412–420 (1966).

[40] Buhvalov, A. V.: Factorization of compact operators, and an example of a reflexive Banach lattice without the approximation property. Doklady Akad. Nauk SSSR **227**, 528–530 (1976); Errata, ibid. **229** (1976), 528 [Russian].

[41] Butzer, P. L., Nessel, R. J., Trebels, W.: On summation processes of Fourier expansion in Banach spaces. I: Comparison theorems. Tôhoku Math. J. **24**, 127–140 (1972).

[42] Butzer, P. L., Nessel, R. J., Trebels, W.: On summation processes of Fourier expansions in Banach spaces. II: Saturation theorems. Tôhoku Math. J. **24**, 551–569 (1972).

[43] Čanturija, Z. A.: On some properties of biorthogonal systems in a Banach space and their applications in the theory of functions. Candidate thesis, Tbilissi State Univ. 1965 [Russian].

[44] Čanturija, Z. A.: On some properties of biorthogonal systems and $\gamma$-bases in a Banach space. Trudy Tbiliss. Gos. Univ. **117**, 223–247 (1966) [Russian].

[45] Casazza, P. G., Lin, Bor-Luh: A remark on subsymmetric bases in Banach spaces. Rev. Roum. math. pures et appl. **21**, 1025–1027 (1976).

[46] Chadwick, J. J. M.: Schauder decompositions in non-separable Banach spaces. Bull. Austral. Math. Soc. **6**, 133–144 (1972).

[47] Chadwick, J. J. M.: Shrinking and boundedly complete bases of projections. Proc. Royal Irish. Acad., Sect. A **74**, 95–102 (1974).

[48] Chadwick, J. J. M.: Cofinal biorthogonal systems and absolute Schauder decompositions. Proc. Royal Irish Acad. Sect. A **75**, 33–36 (1975).

[49] Chadwick, J. J. M.: Bases of projections in Banach spaces. Quarterly J. Math. **26**, 107–111 (1975).

[50] Cooke, R. G.: Infinite matrices and sequence spaces. London: Macmillan & Co. 1950.

[51] Courage, W. H., Davis, W. J.: A characterization of $M$-bases. Math. Annalen **197**, 1–4 (1972).

[52] Courant, R., Hilbert, D.: Methoden der mathematischen Physik. Berlin: Springer 1931.

[53] Crone, L., Fleming, D. J., Jessup, P.: Fundamental biorthogonal sequences and $K$-norms on $\Phi$. Canad. J. Math. **23**, 1040–1050 (1971).

[53a] Dacunha-Castelle, D.: Contre-exemple à la propriété d'approximation uniforme dans les espaces de Banach (d'après Enflo et Davie). Sémin. Bourbaki Exp. 433, 1–8 (1972/73).

[54] Dashiell, F. K., Lindenstrauss, J.: Some examples concerning strictly convex norms on $C(K)$ spaces. Israel J. Math. **16**, 329–342 (1973).

[55] Daugavet, I. K.: On a property of completely continuous operators in the space $C$. Uspehi Matem. Nauk **18**, 5 (113), 157–158 (1963) [Russian].

[56] Davie, A. M.: The approximation problem for Banach spaces. Bull. London Math. Soc. **5**, 261–266 (1973).

[57] Davie, A. M.: The Banach approximation problem. J. Approx. Theory **13**, 392–394 (1975).

[58] Davis, W. J.: Dual generalized bases in linear topological spaces. Proc. Amer. Math. Soc. **17**, 1057–1063 (1966).

[59] Davis, W. J.: Schauder decompositions in Banach spaces. Bull. Amer. Math. Soc. **74**, 1083–1085 (1968).

[60] Davis, W. J., Dean, D. W., Lin, Bor-Luh: Bibasic sequences and norming basic sequences. Trans. Amer. Math. Soc. **176**, 89–102 (1973).

[61] Davis, W. J., Dean, D. W., Singer, I.: Complemented subspaces and $\Lambda$ systems in Banach spaces. Israel J. Math. **6**, 303–309 (1968).

[62] Davis, W. J., Dean, D. W., Singer. I.: Multipliers and unconditional convergence of biorthogonal expansions. Pacific J. Math. **37**, 35–39 (1971).

[63] Davis, W. J., Johnson, W. B.: On the existence of fundamental and total bounded biorthogonal systems in Banach spaces. Studia Math. **45**, 173–179 (1973).

[64] Davis, W. J., Johnson, W. B.: Basic sequences and norming subspaces in non-quasi-reflexive Banach spaces. Israel J. Math. **14**, 353–367 (1973).

[65] Davis, W. J., Lindenstrauss, J.: On total nonnorming subspaces. Proc.Amer. Math. Soc. **31**, 109–111 (1972).

[66] Davis, W. J., Singer. I.: Boundedly complete $M$-bases and complemented subspaces in Banach spaces. Trans. Amer. Math. Soc. **175**, 187–194 (1973).

[67] Dawson, D. F.: Variation properties of sequences. Mat. Vesnik **6** (21), 437–441 (1969).

[68] Day, M. M.: Review of the paper of V. Ya. Kozlov "On a generalization of the concept of a basis", Math. Reviews **12**, 110 (1951).

[69] Day, M. M.: On the basis problem in normed spaces. Proc. Amer. Math. Soc. **13**, 655–658 (1962).

[70] Day, M. M.: Normed linear spaces. 3d Ed., New York—Heidelberg—Berlin: Springer 1973.

[71] Dean, D. W.: Schauder decompositions in (*m*). Proc. Amer. Math. Soc. **18**, 619—623 (1967).

[72] Dean, D. W.: The equation $L(E, X^{**}) = L(E, X)^{**}$ and the principle of local reflexivity. Proc. Amer. Math. Soc. **40**, 145—148 (1973).

[73] Dean, D. W.: Approximation and weak star approximation in Banach spaces. Proc. Amer. Math. Soc. **79**, 725—728 (1973).

[74] Dean, D. W., Lin, Bor-Luh, Singer, I.: On $k$-shrinking and $k$-boundedly complete bases in Banach spaces. Pacific J. Math. **32**, 323—331 (1970).

[75] Dean, D. W., Singer, I., Sternbach, L.: On shrinking basic sequences in Banach spaces. Studia Math. **40**, 23—33 (1971).

[76] Diestel, J.: Grothendieck spaces and vector measures. Proc. Sympos. on "Vector and operator valued measures and applications" held in Alta, Utah 1972, pp. 97—108. New York: Academic Press 1973.

[77] Diestel, J.: Geometry of Banach spaces—selected topics. Lecture Notes in Math. 485. Berlin—Heidelberg—New York: Springer 1975.

[78] Diestel, J.: The Radon-Nikodym property and spaces of operators. Proc. Confer. on "Measure theory" held in Oberwolfach 1975, pp. 211—227. Lecture Notes in Math. 541. Berlin—Heidelberg—New York: Springer 1976.

[79] Diestel, J., Uhl, J. J., Jr.: The Radon-Nikodym theorem for Banach space valued measures. Rocky Mountain J. Math. **6**, 1—46 (1976).

[80] Dieudonné, J.: Sur la convergence des suites de mesures de Radon. An. Acad. Brasil. Ci. **23**, 21—38, 277—282 (1951).

[81] Dieudonné, J.: On biorthogonal systems. Michigan Math. J. **2**, 7—20 (1953/54).

[82] Dor, L. E.: A note on monotone bases. Israel J. Math. **14**, 285—286 (1973).

[83] Dor, L. E.: On sequences spanning a complex $l_1$ space. Proc. Amer. Math. Soc. **47**, 515—516 (1975).

[84] Dorembus, L.: Transfinite bases of subspaces in Hausdorff linear topological spaces. Math. Ann. **192**, 71—82 (1971).

[85] Dubinsky, E., Pelczynski, A., Rosenthal, H. P.: On Banach spaces $X$ for which $\Pi_2(\mathscr{L}_\infty, X) = B(\mathscr{L}_\infty, X)$. Studia Math. **44**, 617—648 (1972).

[85a] van Dulst, D.: Total bounded extensions of biorthogonal sequences in Banach spaces. Math. Annalen **212**, 45—46 (1974/75).

[86] van Dulst, D., Singer, I.: On Kadec-Klee norms on Banach spaces. Studia Math. **54**, 205—211 (1976).

[87] Dunford, N., Schwartz, J. T.: Linear operators. Part. I: General theory. New York: Intersci. Publ. 1958.

[88] Dunford, N., Schwartz, J. T.: Linear operators. Part. II: Spectral theory. Self adjoint operators in Hilbert space. New York—London: Intersci. Publ. 1963.

[89] Dyer, J. A.: Generalized Markushevich bases. Israel J. Math. **7**, 51—59 (1969).

[90] Dyer, J. A.: Integral bases in linear topological spaces. Illinois J. Math. **14**, 468—477 (1970).

[91] Dyer, J. A.: Left Cauchy integral bases in linear topological spaces. Canad. Math. Bull. **13**, 431—439 (1970).

[92] Dyer, J. A.: Generalized bases in $P$-spaces. Math. Annalen **189**, 319—324 (1970).

[93] Edwards, R. E.: Integral bases in inductive limit spaces. Pacific J. Math. **10**, 797—812 (1960).

[94] Edwards, R. E.: Weak convergence of vector-valued series and integrals. J. Austral. Math. Soc., **3**, 159–166 (1963).

[95] Edwards, R. E.: Functional analysis; theory and applications. New York: Holt, Rinehart and Winston 1965.

[96] Eggleston, H. G.: Convexity. Cambridge: Cambridge Univ. Press 1958.

[97] Ellis, H. W., Kuehner, D. G.: On Schauder bases for spaces of continuous functions. Canad. Math. Bull. **3**, 173–184 (1960).

[98] Enflo, P.: A Banach space with basis constant > 1. Ark. Mat. **11**, 103 – 107 (1973).

[99] Enflo, P.: A counterexample to the approximation problem. Acta Math. **13**, 309–317 (1973).

[100] Enflo, P.: Recent results on general Banach spaces. Proc. Internat. Congress of Mathematicians held in Vancouver, August 1974, Vol. II, pp. 53–55. Canad. Math. Congress 1975.

[101] Enflo, P., Rosenthal, H. P.: Some results concerning $L^p(\mu)$-spaces. J. Functional Anal. **14**, 325–348 (1973).

[102] Erdös, P., Straus, E. G.: On linear independence of sequences in a Banach space. Pacific J. Math. **3**, 689–694 (1953).

[103] Fage, M. K.: Idempotent operators and their rectification. Doklady Akad. Nauk SSSR **73**, 895–897 (1950) [Russian].

[104] Fage, M. K.: The rectification of bases in a Hilbert space. Doklady Akad. Nauk SSSR **74**, 1053–1056 (1950) [Russian].

[105] Figiel, T.: Factorization of compact operators and applications to the approximation problem. Studia Math. **45**, 191–210 (1973).

[106] Figiel, T.: Further counterexamples to the approximation problem. Dittoed notes.

[107] Figiel, T., Johnson, W. B.: The approximation property does not imply the bounded approximation property. Proc. Amer. Math. Soc. **41**, 197–200 (1973).

[108] Figiel, T., Johnson, W. B.: A uniformly convex Banach space which contains no $l_p$. Compositio Math. **29**, 179–190 (1974).

[109] Figiel, T., Johnson, W. B., Tzafriri, L.: On Banach lattices and spaces having local unconditional structure, with applications to Lorentz function spaces. J. Approx. Theory **13**, 395–412 (1975).

[110] de Figueiredo, D. G.: Fixed point theorems for weakly continuous mappings. Math. Res. Center, Tech. Rep. No. 638. Univ. of Wisconsin 1966.

[111] de Figueiredo, D. G.: Topics in nonlinear functional analysis. Lecture Notes. Inst. for Fluid Dynamics and Appl. Math. Univ. of Maryland 1967.

[112] Fihtengolc, G. M.: Course of differential and integral calculus. II. Moscow: Nauka 1966 [Russian].

[113] Fleming, D. J., Ruckle, W.: Continuous decompositions. Rev. Roumaine math. pures et appl. **15**, 1153–1170 (1970).

[114] Fleming, R. J., Jamison, J. E.: Hermitian and adjoint abelian operators on certain Banach spaces. Pacific J. Math. **52**, 67–85 (1974).

[115] Foiaş, C., Singer, I.: Points of diffusion of linear operators and almost diffuse operators in spaces of continuous functions. Math. Zeitschr. **87**, 434–450 (1965).

[116] Frink, O., Jr.: Series expansions in linear spaces. Amer. J. Math. **63**, 87–100 (1941).

[117] Fullerton, R. E.: Geometric structure of absolute basis systems in a linear topological space. Pacific J. Math. **12**, 137–147 (1962).

[118] Gapoškin, V. F.: On summation bases. Naučn. Doklady Vysš. Školy **1**, 24–27 (1958) [Russian].

[119] Gapoškin, V. F.: On unconditionally summable series in Banach spaces. Naučn. Doklady Vysš. Školy **1**, 87–95 (1958) [Russian].

[120] Gapoškin, V. F.: On the order of approximation by partial sums of lacunary series in Banach spaces. Matem. Zametki **8**, 575–582 (1970) [Russian].

[121] Gapoškin, V. F., Kadec, M. I.: Operatorial bases in Banach spaces. Matem. Sbornik **61** (103), 3–12 (1963) [Russian].

[122] Garling, D. J. H.: On symmetric sequence spaces. Proc. London Math. Soc. **16**, 85–106 (1966).

[123] Garling. D. J. H.: On topological sequence spaces. Proc. Cambr. Phil. Soc. **63**, 997–1019 (1967).

[124] Gelbaum, B. R.: Expansions in Banach spaces. Duke Math. J. **17**, 187–196 (1950).

[125] Gelbaum, B. R.: Notes on Banach spaces and bases. An. Acad. Brasil. Ci. **30**, 29–36 (1958).

[126] Giesy, D. P., James, R. C.: Uniformly non-$l^{(1)}$ and $B$-convex Banach spaces. Studia Math. **48**, 61–69 (1973).

[127] Gil de Lamadrid, J.: On finite dimensional approximations of mappings in Banach spaces. Proc. Amer. Math. Soc. **13**, 163–168 (1962).

[128] Gillman, L., Jerison, M.: Rings of continuous functions. Princeton: D. van Nostrand 1960.

[129] Goes, G.: Identische Multiplikatorenklassen und $C_k$-Basen in $C_k$-komplementären Fourierkoeffizientenräumen. Math. Nachrichten **21**, 150–159 (1960).

[130] Gohberg, I. C., Kreĭn, M. G.: The basic propositions on defect numbers, root numbers and indices of linear operators. Uspehi Matem. Nauk **12**, 2(74), 43–118 (1957) [Russian].

[131] Gohberg, I. C., Markus, A. S.: On the stability of bases in Banach and Hilbert spaces. Izv. Akad. Nauk Mold. SSR **5**, 17–35 (1962) [Russian].

[132] Gordon, Y., Lewis, D. R.: Absolutely summing, $L_1$ factorizable operators and their applications. Bull. Amer. Math. Soc. **79**, 1270–1273 (1973).

[133] Gordon, Y., Lewis, D. R.: Absolutely summing operators and local unconditional structures. Acta Math. **133**, 27–48 (1974).

[134] Görlich, E., Nessel, R. J., Trebels, W.: Zur Approximationstheorie für Summationsprozesse von Fourier-Entwicklungen in Banach-Räumen: Vergleichssätze und Ungleichungen vom Bernstein-Typ. Approximation theory, pp. 87–96. Warszawa: PWN 1975.

[135] Grinblium, M. M.: Biorthogonal systems in Banach space. Doklady Akad. Nauk SSSR **47**, 75–78 (1945).

[136] Grinblium, M. M.: On the representation of a space of type $B$ as a direct sum of subspaces. Doklady Akad. Nauk SSSR **70**, 749–752 (1950) [Russian].

[137] Grothendieck, A.: Sur les applications linéaires faiblement compactes d'espaces du type $C(K)$, Canad. J. Math. **5**, 129–173 (1953).

[138] Grothendieck, A.: Une caractérisation vectorielle-métrique des espaces $L^1$. Canad. J. Math. **7**, 552–561 (1955).

[139] Grothendieck, A.: Produits tensoriels topologiques et espaces nucléaires. Memoirs Amer. Math. Soc. **16**, Ch. I, 1–191; Ch. II, 1–140 (1955).

[140] Grothendieck, A.: Espaces vectoriels topologiques. 2[d] Ed., São Paulo 1958.

[141] Gurariĭ, V. I.: On inclinations of subspaces and conditional bases in a Banach space. Doklady Akad. Nauk SSSR **145**, 504–506 (1962) [Russian].

[142] Gurariĭ, V. I.: On bases in spaces of continuous functions. Doklady Akad. Nauk SSSR **148**, 493–495 (1963) [Russian].

[143] Gurariĭ, V. I.: On bases for sets in Banach spaces. Rev. Roumaine math. pures et appl. **10**, 1235–1240 (1965) [Russian].

[144] Gurariĭ, V. I.: Bases in spaces of continuous functions on compacta and some geometric problems. Izv. Akad. Nauk SSSR, Ser. Matem. **30**, 289–306 (1966) [Russian].

[145] Gurariĭ, V. I.: Investigations on the geometry of normed spaces. Doctoral thesis. Harkov State Univ. 1966 [Russian].

[146] Gurariĭ, V. I., Gurariĭ, N. I.: On a sequential property of the space $l_1$. Matem. Issled. Akad. Nauk Mold. SSR **4**, 140–145 (1969) [Russian].

[146a] Gurariĭ, V. I., Gurariĭ, N. I., Liokumovič, V. A.: Basic sequences in the space $l_p$. Funkcional. Anal. 1, pp. 226–231, Uljanovsk. Gos. Ped. Inst., Uljanovsk 1973 [Russian].

[147] Gurariĭ, V. I., Kadec, M. I.: On minimal systems and quasi-complements in a Banach space. Doklady Akad. Nauk SSSR **145**, 256–258 (1962) [Russian].

[148] Gurariĭ, V. I., Macaev, V. I.: Lacunary power sequences in the spaces $C$ and $L_p$. Izv. Akad. Nauk SSSR, Ser. Matem. **30**, 3–14 (1966) [Russian].

[149] Gurariĭ, V. I. Meletidi, M. A.: Stability of completeness of sequences in Banach spaces. Bull. Acad. Pol. Sci. **18**, 533–536 (1970) [Russian].

[150] Gurariĭ, V. I. Meletidi, M. A.: Estimates of the coefficients of polynomials that approximate continuous functions. Funkcional. Anal. i Priložen. **5**, 1, 73–75 (1971) [Russian].

[151] Gurevič, L. A.: On a basis in the space of continuous functions, defined on a closed bounded set of the $n$-dimensional space. Trudy Voronež. Gos. Univ. **27**, 84–87 (1954) [Russian].

[152] Gutman, S. M.: On equivalent norms in certain non-separable $B$-spaces. Teor. Funkciĭ. Funkcional. Anal. i Priložen. **20**, 63–69 (1974) [Russian].

[153] Hale, D. F.: Integral bases in Banach spaces. Dissertation. Ohio State University 1969.

[154] Hall, M., Jr.: Combinatorial theory. Waltham–Toronto–London: Blaisdell Publ. Co. 1967.

[155] Halmos, P. R.: Finite-dimensional vector spaces. Princeton: D. van Nostrand 1958.

[156] Holub, J. R., Retherford, J. R.: A note on unconditional bases. Canad. Math. Bull. **15**, 369–372 (1972).

[157] Huff, R. E.: Troyanski's renorming theorem. Pennsylvania State University Seminar Notes, 12 pp. Winter 1973–74.

[158] Isbell, R., Semadeni, Z.: Projection constants and spaces of continuous functions. Trans. Amer. Math. Soc. **107**, 38–48 (1963).

[159] James, R. C.: Reflexivity and the supremum of linear functionals. Ann. of Math. **66**, 159–169 (1957).

[160] James, R. C.: Separable conjugate spaces. Pacific J. Math. **10**, 563–571 (1960).

[161] James, R. C.: Uniformly non-square Banach spaces. Ann. of Math. **80**, 542–550 (1964).

[162] John, F.: Extremum problems with inequalities as subsidiary conditions. Studies and Essays, Courant Anniversary Volume, pp. 187–204. New York: Interscience Publ. 1948.

[163] John, K., Zizler, V.: A renorming of dual spaces. Israel J. Math. **12**, 331–336 (1972).

[164] John, K., Zizler, V.: Projections in dual weakly compactly generated Banach spaces. Studia Math. **49**, 41–50 (1973).

[165] John, K., Zizler, V.: A note on renorming of dual spaces. Bull. Acad. Polon. Sci. **21**, 47–49 (1973).

[166] John, K., Zizler, V.: Smoothness and its equivalents in the class of weakly compactly generated Banach spaces. J.Functional Anal. **15**, 1–11 (1974).

[167] John, K., Zizler, V.: Some remarks on non-separable Banach spaces with Markuševič basis. Comment. Math. Univ. Carolinae **15**, 679–691 (1974).

[168] John, K., Zizler, V.: Duals of Banach spaces which admit nontrivial smooth functions. Bull. Austral. Math. Soc. **11**, 161–166 (1974).

[169] John, K., Zizler, V.: On the heredity of weak compact generating. Israel J. Math. **20**, 228–236 (1975).

[170] John, K., Zizler, V.: Markuševič bases in some dual spaces. Proc. Amer. Math. Soc. **50**, 293–296 (1975).

[171] John, K., Zizler, V.: Weak compact generating in duality. Studia Math. **55**, 1–20 (1976).

[172] John, K., Zizler, V.: Some notes on Markuševič bases in weakly compactly generated Banach spaces. Compositio Math. **35**, 113–123 (1977).

[173] Johnson, W. B.: Finite-dimensional Schauder decompositions in $\pi_\lambda$ and dual $\pi_\lambda$ spaces. Illinois J. Math. **14**, 642–647 (1970).

[174] Johnson, W. B.: Markuschevich bases and duality theory. Trans. Amer. Math. Soc. **149**, 171–177 (1970).

[175] Johnson, W. B.: No infinite dimensional *P* space admits a Markuschevich basis. Proc. Amer. Math. Soc. **26**, 467–468 (1970).

[176] Johnson, W.B.: Factoring compact operators. Israel J.Math. **9**, 337–345 (1971).

[177] Johnson, W. B.: Finite-dimensional Schauder decompositions in certain Fréchet spaces. Colloq. Math. **23**, 269–272 (1971).

[178] Johnson, W. B.: On the existence of strongly series summable Markuschevich bases in Banach spaces. Trans. Amer. Math. Soc. **157**, 481–486 (1971).

[179] Johnson, W. B.: Operator and dual operator bases in linear topological spaces. Trans. Amer. Math. Soc. **166**, 387–400 (1972).

[180] Johnson, W. B.: A complementably universal conjugate Banach space and its relation to the approximation problem. Israel J. Math. **13**, 301–310 (1972).

[181] Johnson, W. B.: On quasi-complements. Pacific J. Math. **48**, 113–118 (1973).

[182] Johnson, W. B.: Complementably universal separable Banach spaces: An application of counterexamples to the approximation problem. Preprint.

[183] Johnson, W. B., Dyer, J. A.: Isomorphisms generated by fundamental and total sets. Proc. Amer. Math. Soc. **22**, 330–334 (1969).

[184] Johnson, W. B., Lindenstrauss, J.: Some remarks on weakly compactly generated Banach spaces. Israel J. Math. **17**, 219–229 (1974).

[185] Johnson, W. B., Odell, E.: Subspaces of $L_p$ which embed into $l_p$. Compositio Math. **28**, 37–49 (1974).

[186] Johnson, W. B., Rosenthal, H. P.: On $w^*$-basic sequences and their applications to the study of Banach spaces. Studia Math. **43**, 77–92 (1972).

[187] Johnson, W. B., Rosenthal, H. P., Zippin, M.: On bases, finite dimensional decompositions and weaker structures in Banach spaces. Israel J. Math. **9**, 488–506 (1971).

[188] Johnson, W. B., Szankowski, A.: Complementably universal Banach spaces. Studia Math. **58**, 91–97 (1976).

[189] Johnson, W. B., Tzafriri, L.: Some more Banach spaces which do not have local unconditional structure. Houston J. Math. **3**, 55–60 (1977).

[190] Johnson, W. B., Zippin, M.: On subspaces of quotients of $(\Sigma G_n)_{l_p}$ and $(\Sigma G_n)_{c_0}$. Israel J. Math. **13**, 311–316 (1972).

[191] Johnson, W. B., Zippin, M.: Subspaces and quotient spaces of $(\Sigma G_n)_{l_p}$ and $(\Sigma G_n)_{c_0}$. Israel J. Math. **17**, 50–55 (1974).

[192] Josephson, B.: Weak sequential convergence in the dual of a Banach space does not imply norm convergence. Ark. Mat. **13**, 79–89 (1975).

[193] Julia, G.: Exemples de structures des systèmes duaux de l'espace hilbertien. Comptes rendus Acad. Sci. (Paris) **216**, 465–468 (1943).

[194] Kaczmarz, S., Steinhaus, H.: Theorie der Orthogonalreihen. Warszawa–Lwów: Monografje Matematyczne 1935.

[195] Kadec, M. I.: On the connection between weak and strong convergence. Dopovidi Akad. Nauk Ukr. RSR **9**, 942–952 (1969) [Ukrainian].

[196] Kadec, M. I.: On biorthogonal systems and summation bases. Funkcional. Anal. i ego Primen., pp. 106–108, Akad. Nauk Azerb. SSR, Baku 1961 [Russian].

[197] Kadec, M. I.: Some problems of the geometry of Banach spaces. Doctoral thesis, Moscow State Univ. 1963 [Russian].

[198] Kadec, M. I.: Bases and their spaces of coefficients. Dopovidi Akad. Nauk Ukr. RSR **9**, 1139–1141 (1964) [Ukrainian].

[199] Kadec, M. I.: Conditions for the differentiability of the norm in a Banach space. Uspehi Matem. Nauk **20**, 3(123), 183–187 (1965) [Russian].

[200] Kadec, M. I.: Non-linear operatorial bases in a Banach space. Teor. Funkciĭ, Funkcional. Anal. i Priložen. **2**, 128–130 (1966) [Russian].

[201] Kadec, M. I.: Topological equivalence of all separable Banach spaces. Doklady Akad. Nauk SSSR **167**, 23–25 (1966) [Russian].

[202] Kadec, M. I.: A proof of the topological equivalence of all separable infinite dimensional Banach spaces. Funkcional. Anal. i Priložen. **1**, 1, 61–70 (1967) [Russian].

[203] Kadec, M. I.: On complementably universal Banach spaces. Studia Math. **40**, 85–89 (1971).

[204] Kadec, M. I.: The geometry of normed spaces. Itogi Nauki i Tehniki, Ser. Matem. Analiz **13**, pp. 99–127, Moscow 1975 [Russian].

[205] Kadec, M. I., Pelczynski, A.: Bases, lacunary sequences and complemented subspaces in the spaces $L_p$. Studia Math. **21**, 161–176 (1962).

[206] Kadec, M. I., Pelczynski, A.: Basic sequences, biorthogonal systems and norming sets in Banach and Fréchet spaces. Studia Math. **25**, 297–323 (1965) [Russian].

[207] Kadec, M. I., Snobar, M. G.: On certain functionals on the Minkowski compactum. Matem. Zametki **10**, 453–457 (1971) [Russian].

[208] Kahane, J. -P., Salem, R.: Ensembles parfaits et séries trigonométriques. Paris: Hermann 1963.

[209] Kalton, N. J.: Basic sequences in $F$-spaces and their applications. Proc. Edinburgh Math. Soc. **19**, 151–167 (1974).

[210] Kalton, N. J., Shapiro, J. H.: Bases and basic sequences in $F$-spaces. Studia Math. **56**, 47–61 (1976).

[211] Kalton, N. J., Wood, G. V.: Orthonormal systems in Banach spaces and their applications. Proc. Cambr. Phil. Soc. **79**, 493–510 (1976).

[212] Kaplan, S.: Biorthogonality and integration. Proc. Amer. Math. Soc. **7**, 109–114 (1956).

[213] Karlin, S.: Bases in Banach spaces. Duke Math. J. **15**, 971–985 (1948).

[214] Kato, T.: Perturbation theory for linear operators. Berlin–Heidelberg–New York: Springer 1966.

[215] Kelley, J. L.: General topology. Princeton: D. van Nostrand 1955.

[216] Keselman, G. M.: On a generalization of the notion of basis and of theorems of N. K. Bari on bases. Doklady Akad. Nauk SSSR **141**, 300–303 (1961) [Russian].

[217] Klee, V.: On the borelian and projective types of linear subspaces. Math. Scand. **6**, 189–199 (1958).

[218] Klee, V.: Mappings into normed linear spaces. Fund. Math. **49**, 25–34 (1960).

[219] Klee, V.: Shrinkable neighborhoods in Hausdorff linear spaces. Math. Annalen **141**, 281–285 (1960).

[220] Klee, V.: Leray-Schauder theory without local convexity. Math. Annalen **141**, 286–296 (1960).

[221] Kober, H. A.: A theorem on Banach spaces. Compositio Math. **7**, 135–140 (1939).

[222] Kokilašvili, V. M.: On series with gaps. Soobšč. Akad. Nauk Gruz. SSR **51**, 269–274 (1968) [Russian].

[223] Köthe, G.: Topologische lineare Räume. I. Berlin–Heidelberg–New York: Springer 1960.

[224] Kozlov, V. Ya.: On a generalization of the notion of basis. Doklady Akad. Nauk SSSR **73**, 643–646 (1950) [Russian].

[225] Kreĭn, M. G., Krasnoselskiĭ, M. A., Milman, D. P.: On defect numbers of linear operators in a Banach space and on some geometric problems. Sbornik Trud. Inst. Matem. Akad. Nauk Ukr. SSR **11**, 97–112 (1948) [Russian].

[226] Kuratowski, C.: Topologie. Vol. I. 4ème éd. Warszawa: PWN 1958.

[227] Kwapien, S.: On Enflo's example of a Banach space without the approximation property. Sémin. Goulaouic-Schwartz Exp. 8, 1–9 (1972–1973).

[228] Kwapien, S.: Comments to Enflo's construction of Banach space without the approximation property. Sémin. Goulaouic-Schwartz Exp. 9, 1–4 (1972–1973).

[229] Lacey, E.: Separable quotients of Banach spaces. An. Acad. Brasil. Ci. **44**, 185–189 (1972).

[230] Lami Dozo, E.: Opérateurs non-expansifs, *P*-compacts et propriétés géométriques de la norme. Thèse. Univ. Libre de Bruxelles 1970.

[231] Lazar, A. J., Lindenstrauss, J.: On Banach spaces whose duals are $L_1$ spaces. Israel J. Math. **4**, 205–207 (1966).

[232] Lazar, A., Zippin, M.: On finite dimensional subspaces of Banach spaces. Israel J. Math. **3**, 147–156 (1965).

[233] Lewin, S.: Über einige mit der Konvergenz im Mittel verbundenen Eigenschaften von Funktionenfolgen. Math. Zeitschr. **32**, 491–511 (1930).

[234] Lewin, S. S.: Integralgleichungen und Funktionalräume. Matem. Sbornik **39**, 3–72 (1932).

[235] Lindenstrauss, J.: Extension of compact operators. Memoirs Amer. Math. Soc. **48**, 1–112 (1964).

[236] Lindenstrauss, J.: On reflexive spaces having the metric approximation property. Israel J. Math. **3**, 199–204 (1965).

[237] Lindenstrauss, J.: On nonseparable reflexive Banach spaces. Bull. Amer. Math. Soc. **72**, 967–970 (1966).

[238] Lindenstrauss, J.: On a theorem of Murray and Mackey. An. Acad. Brasil. Ci. **39**, 1–6 (1967).

[239] Lindenstrauss, J.: Some aspects of the theory of Banach spaces. Advances in Math. **5**, 159–180 (1970).

[240] Lindenstrauss, J.: On James' paper "Separable conjugate spaces". Israel J. Math. **9**, 279–284 (1971).

[241] Lindenstrauss, J.: Decomposition of Banach spaces. Indiana Univ. Math. J. **20**, 917–919 (1971).

[242] Lindenstrauss, J.: Weakly compact sets—their topological properties and the Banach spaces they generate. Proc. Sympos. on "Infinite dimensional topo-

logy"held in Baton Rouge, Louisiana, 1967, pp. 235–274. Princeton: Ann. of Math. Studies **69**, Princeton Univ. Press 1972.

[243] Lindenstrauss, J., Pelczynski, A.: Absolutely summing operators in $\mathcal{L}_p$ spaces and their applications. Studia Math. **29**, 275–326 (1968).

[244] Lindenstrauss, J., Tzafriri, L.: On Orlicz sequence spaces. Israel J. Math. **10**, 379–390 (1971).

[245] Lindenstrauss, J., Tzafriri, L.: On Orlicz sequence spaces. II. Israel J. Math. **11**, 355–379 (1972).

[246] Lindenstrauss, J., Tzafriri, L.: On Orlicz sequence spaces. III. Israel J. Math. **14**, 368–389 (1973).

[247] Lindenstrauss, J., Tzafriri, L.: Classical Banach spaces. Lecture Notes in Math. 338. Berlin–Heidelberg–New York: Springer 1973.

[248] Lindenstrauss, J., Tzafriri, L.: The uniform approximation property in Orlicz spaces. Israel J. Math. **23**, 142–155 (1976).

[248a] Lindenstrauss, J., Tzafriri, L.: Classical Banach spaces. I: Sequence spaces. Berlin–Heidelberg–New York: Springer 1977.

[249] Lindenstrauss, J., Zippin, M.: Banach spaces with a unique unconditional basis. J. Functional Anal. **3**, 115–125 (1969).

[250] Lindenstrauss, J., Zippin, M.: Banach spaces with sufficiently many Boolean algebras of projections. J. Math. Anal. Appl. **25**, 309–320 (1969).

[251] Lorch, E. R.: Bicontinuous linear transformations in certain vector spaces. Bull. Amer. Math. Soc. **45**, 564–569 (1939).

[252] Lorch, E. R.: On a calculus of operators in reflexive vector spaces. Trans. Amer. Math. Soc. **45**, 217–234 (1939).

[253] Lovaglia, A. R.: Locally uniformly convex Banach spaces. Trans. Amer. Math. Soc. **78**, 225–238 (1955).

[254] Mackey, G. W.: On infinite-dimensional linear spaces. Trans. Amer. Math. Soc. **57**, 155–207 (1945).

[255] Mackey, G. W.: Note on a theorem of Murray. Bull. Amer. Math. Soc. **52**, 322–325 (1946).

[256] Maddaus, I.: On completely continuous linear transformations. Bull. Amer. Math. Soc. **44**, 279–282 (1938).

[257] Maharam, D.: On homogeneous measure algebras. Proc. Nat. Acad. Sci. USA **28**, 108–111 (1942).

[258] Markus, A. S.: On the basis of root vectors of a dissipative operator. Doklady Akad. Nauk SSSR **132**, 524–527 (1960) [Russian].

[259] Markouchevitch, A.: Sur les bases (au sens large) dans les espaces linéaires. Doklady Akad. Nauk SSSR **41**, 227–229 (1943).

[260] Markuševič, A. I.: A generalization of a theorem of D. E. Menšov. Matem. Sbornik **15**(57), 433–436 (1944) [Russian].

[261] Markouchevitch, A. I.: Sur la meilleure approximation. Doklady Akad. Nauk SSSR **44**, 262–264 (1944).

[262] Markuševič, A. I.: On the basis in the space of analytic functions. Matem. Sbornik **17**(59), 211–252 (1945) [Russian].

[263] Marti, J. T.: Introduction to the theory of bases. Berlin–Heidelberg–New York: Springer 1969.

[264] Marti, J. T.: Extended bases for Banach spaces. Illinois J. Math. **15**, 135–143 (1971).

[265] Marti, J. T.: A note on equivalent extended bases. Monatsh. Math. **75**, 250–255 (1971).

[266] Masani, P.: Orthogonally scattered measures. Advances in Math. **2**, 61–117 (1968).

[267] Masani, P.: Quasi-isometric measures and their applications. Bull. Amer. Math. Soc. **76**, 427–528 (1970).

[268] Maurey, B., Pisier, G.: Caractérisation d'une classe d'espaces de Banach par des propriétés de séries aléatoires vectorielles. Comptes rendus Acad. Sci. (Paris) A **277**, 687–690 (1973).

[269] Maurey, B., Pisier, G.: Séries de variables aléatoires indépendantes et propriétés géométriques des espaces de Banach. Studia Math. **58**, 45–90 (1976).

[270] Maurey, B., Rosenthal, H. P.: Normalized weakly null sequences with no unconditional subsequence. Studia Math. **61**, 77–98 (1977).

[271] Mazur, S., Sternbach, L.: Über die Borelschen Typen von linearen Mengen. Studia Math. **4**, 48–53 (1933).

[272] McArthur, C. W.: Infinite direct sums in metric linear spaces. Dittoed notes.

[273] McArthur, C. W.: Generalized Schauder bases. Dittoed notes.

[274] McArthur, C. W.: The weak basis theorem. Colloq. Math. **17**, 71–76 (1967).

[275] McArthur, C. W.: Developments in Schauder basis theory. Bull. Amer. Math. Soc. **78**, 877–908 (1972).

[276] McArthur, C. W., Retherford, J. R.: Uniform and equicontinuous Schauder bases of subspaces. Canad. J. Math. **17**, 207–212 (1965).

[277] McArthur, C. W., Singer, I., Levin, M.: On the cones associated with biorthogonal systems and bases in Banach spaces. Canad. J. Math. **21**, 1206–1217 (1969).

[278] McGivney, R. J., Ruckle, W.: Multiplier algebras of biorthogonal systems. Pacific J. Math. **29**, 375–387 (1969).

[279] Menchoff, D. E.: Sur les sommes partielles des séries de Fourier des fonctions continues. Matem. Sbornik **15**(57), 385–432 (1944).

[280] Mertens, H. J., Nessel, R. J., Wilmes, G.: Über Multiplikatoren zwischen verschiedenen Banach-Räumen im Zusammenhang mit diskreten Orthogonalentwicklungen. Forschungsber. des Landes Nordrhein-Westfalen. Opladen 1976.

[281] Michael, E.: Continuous selections. I. Ann. of Math. **63**, 361–382 (1956).

[282] Michael, E., Pelczynski, A.: Separable Banach spaces which admit $l_n^\infty$ approximations. Israel J. Math. **4**, 189–198 (1966).

[283] Michael, E., Pelczynski, A.: Peaked partition subspaces of $C(X)$. Illinois J. Math. **11**, 555–562 (1967).

[284] Milman, V. D.: On perturbations of sequences of elements of a Banach space. Sibirsk. Mat. Ž. **6**, 398–412 (1965) [Russian].

[285] Milman, V. D.: Some properties of unconditional bases. Doklady Akad. Nauk SSSR **162**, 269–272 (1965) [Russian].

[285a] Milman, V. D.: The spectrum of bounded continuous functions which are given on the unit sphere of a $B$-space. Funkcional. Anal. i Priložen. **3**, 2, 67–79 (1969) [Russian].

[286] Milman, V. D.: James classes of minimal systems and their connection with isometric properties of $B$-spaces. Doklady Akad. Nauk SSSR **192**, 742–745 (1970) [Russian].

[287] Milman, V. D.: The geometric theory of Banach spaces. I. The theory of basic and minimal systems. Uspehi Matem. Nauk **25**, 3(153), 113–174 (1970) [Russian].

[288] Milman, V. D.: The geometric theory of Banach spaces. II. The geometry of the unit sphere. Uspehi Matem. Nauk **26**, 6(162), 73–149 (1971) [Russian].

[289] Milman, D. P., Milman, V. D.: Some properties of non-reflexive Banach spaces. Matem. Sbornik **65**(104), 486–497 (1964) [Russian].

[290] Morris, P.: The Day-Rainwater norm on $c_0(\Gamma)$. Pennsylvania State University Seminar Notes, 12 pp. Winter 1973–74.

[291] Murray, F. J.: On complementary manifolds and projections in spaces $L_p$ and $l_p$. Trans. Amer. Math. Soc. **41**, 138–152 (1937).

[292] Nagumo, M.: Degree of mapping in convex linear topological spaces. Amer. J. Math. **73**, 497–511 (1951).

[293] Nakamura, M., Kakutani, S.: Banach limits and the Čech compactification of a countable discrete set. Proc. Imp. Acad. Japan **19**, 224–229 (1943).

[294] Nissenzweig, A.: $w^*$ sequential convergence. Israel J. Math. **22**, 266–272 (1975).

[295] Olevskiĭ, A. M.: Fourier series of continuous functions with respect to bounded orthonormal systems. Izv. Akad. Nauk SSSR, Ser. Matem. **30**, 387–432 (1966) [Russian].

[296] Oostenbrink, W.: Multiplier functions and the bounded approximation property. Report ZW 7604, pp. 1–21. Univ. of Groningen 1976.

[297] Ovsepian, R. I., Pelczynski, A.: The existence in every separable Banach space of a fundamental total and bounded biorthogonal sequence and related constructions of uniformly bounded orthonormal systems in $L^2$. Studia Math. **54**, 149–159 (1975).

[298] Paley, R. E. A. C.: A remarkable series of orthogonal functions. I. Proc. London Math. Soc. **34**, 241–264 (1932).

[298a] Pelczynski, A.: A connection between unconditional convergence and weakly completeness of Banach spaces. Bull. Acad. Polon. Sci. **6**, 251–253 (1958).

[299] Pelczynski, A.: Projections in certain Banach spaces. Studia Math. **19**, 209–228 (1960).

[300] Pelczynski, A.: A note on the paper of I. Singer "Basic sequences and reflexivity of Banach spaces". Studia Math. **21**, 371–374 (1962).

[301] Pelczynski, A.: A proof of Eberlein-Šmulian theorem by an application of basic sequences. Bull. Acad. Polon. Sci. **12**, 543–548 (1964).

[302] Pelczynski, A.: Some problems on bases in Banach and Fréchet spaces. Israel J. Math. **2**, 132–138 (1964).

[303] Pelczynski, A.: Some open questions in functional analysis (A lecture given to Louisiana State University). Dittoed notes (1966).

[304] Pelczynski, A.: Any separable Banach space with the bounded approximation property is a complemented subspace of a Banach space with a basis. Studia Math. **40**, 239–242 (1971).

[305] Pelczynski, A.: On some problems of Banach. Uspehi Matem. Nauk **28**, 6(174), 67–75 (1973) [Russian].

[306] Pelczynski, A.: On some Banach spaces of operators. A lecture given at the International Conference on Banach spaces held in Wabash College, Crawfordsville (Indiana), 1973.

[307] Pelczynski, A.: All separable Banach spaces admit for every $\varepsilon > 0$ fundamental total and bounded by $1 + \varepsilon$ biorthogonal sequences. Studia Math. **55**, 295–304 (1976).

[308] Pelczynski, A., Figiel, T.: On Enflo's method of construction of Banach spaces without the approximation property. Uspehi Matem. Nauk **28**, 6(174), 95–108 (1973) [Russian].

[309] Pelczynski, A., Rosenthal, H. P.: Localization techniques in $L_p$ spaces. Studia Math. **52**, 263–289 (1975).

[310] Pelczynski, A., Rosenthal, H. P.: in preparation.

[311] Pelczynski, A., Singer, I.: On non-equivalent bases and conditional bases in Banach spaces. Studia Math. **25**, 5–25 (1964).

[312] Pelczynski, A., Sudakov, V. N.: Remark on non-complemented subspaces of the space $m(S)$. Colloq. Math. **9**, 85–88 (1962).

[313] Pelczynski, A., Wojtaszczyk, P.: Banach spaces with finite dimensional expansions of identity and universal bases of finite dimensional subspaces. Studia Math. **40**, 91–108 (1971).

[314] Petryshyn, W. V., Tucker, T. S.: On functional equations involving nonlinear generalized $P$-compact operators. Trans. Amer. Math. Soc. **135**, 343–373 (1969).

[315] Phelps, R. R.: A Banach space characterization of purely atomic measure spaces. Proc. Amer. Math. Soc. **12**, 447–452 (1961).

[316] Phelps, R. R.: Lectures on Choquet's theorem. Princeton: D. van Nostrand 1966.

[317] Phillips, R. S.: On linear transformations. Trans. Amer. Math. Soc. **48**, 516–541 (1940).

[318] Pietsch, A.: Verallgemeinerte vollkommene Folgenräume. Berlin: Akademie-Verlag 1962.

[319] Pličko, A. N.: The existence of a complete $\varepsilon$-orthonormal system in a separable normed space. Dopovidi Akad. Nauk Ukr. SSR, Ser. A 21–22, 93 (1976) [Ukrainian].

[320] Pličko, A. N.: The existence of a total biorthogonal system in a Banach space. Preprint [Russian].

[321] Pličko, A. N.: The existence of a bounded total biorthogonal system in a Banach space. Teor. Funkciĭ, Funkcional. Anal. i Priložen. (to appear) [Russian].

[322] Pličko, A. N.: $M$-bases in separable and reflexive spaces. Ukr. Matem. Ž. **29**, 681–685, 711 (1977) [Russian].

[323] Pličko, A. N.: The existence of a bounded $M$-basis in a WCG-space (to appear) [Russian].

[324] Pták, V.: Biorthogonal systems and reflexivity of Banach spaces. Czechoslovak Math. J. **9** (84), 319–326 (1959).

[325] Pujara, L. R.: $\mathscr{L}_p$ spaces and decompositions in Banach spaces. Thesis. Ohio State University 1971.

[326] Pujara, L. R.: Some local structures in Banach spaces. Math. Japonicae **20**, 49–54 (1975).

[327] Rainwater, J.: Local uniform convexity of Day's norm on $c_0(\Gamma)$. Proc. Amer. Math. Soc. **22**, 335–339 (1969).

[328] Rainwater, J.: Day's norm on $c_0(\Gamma)$. Proc. of "Functional Analysis" Week held in Aarhus, 1969, pp. 46–50. Aarhus: Math. Inst. Univ. Aarhus 1969.

[329] Reif, J.: Subspaces with boundedly complete bases in some nonseparable Banach spaces. Bull. Acad. Polon. Sci. **22**, 1117–1120 (1974).

[330] Reif, J.: A note on Markuševič bases in weakly compactly generated Banach spaces. Comment. Math. Univ. Carolinae **15**, 335–340 (1974).

[331] Reif, J.: Some remarks on subspaces of weakly compactly generated spaces. Comment. Math. Univ. Carolinae **16**, 787–793 (1975).

[332] Retherford, J. R.: $w^*$-bases and $bw^*$-bases in Banach spaces. Studia Math. **25**, 65–71 (1964).

[333] Retherford, J. R.: Some remarks on Schauder bases of subspaces. Rev. Roumaine math. pures et appl. **11**, 787–792 (1966).

[334] Retherford, J. R.: On $T$-bases in Banach spaces. Rev. Roumaine math. pures et appl. **12**, 691–696 (1967).

[335] Retherford, J. R.: Some remarks on Schauder bases of subspaces. II. Rev. Roumaine math. pures et appl. **13**, 521–527 (1968).

[336] Riedrich, T.: Die Räume $L^p(0,1)$ $(0 < p < 1)$ sind zulässig. Wiss. Zeitschr. Techn. Univ. Dresden **12**, 1149–1152 (1963).

[337] Riedrich, T.: Der Raum $S(0,1)$ ist zulässig. Wiss. Zeitschr. Techn. Univ. Dresden **13**, 1–6 (1964).

[338] Riesz, F., Sz.-Nagy, B.: Leçons d'analyse fonctionnelle. 2ème éd. Budapest: Acad. Sci. Hongrie 1953.

[339] Rosenthal, H. P.: On quasi-complemented subspaces of Banach spaces, with an Appendix on compactness of operators from $L^p(\mu)$ to $L^r(\nu)$. J. Functional Anal. **4**, 176–214 (1969).

[340] Rosenthal, H. P.: On subspaces of $L^p$, Ann. of Math. **97**, 344–373 (1973).

[341] Rosenthal, H. P.: The heredity problem for weakly compactly generated Banach spaces. Compositio Math. **28**, 83–111 (1974).

[342] Rosenthal, H. P.: A characterization of Banach spaces containing $l^1$. Proc. Nat. Acad. Sci. USA **71**, 2411–2413 (1974).

[343] Ruckle, W. H.: The infinite sum of closed subspaces of an $F$-space. Duke Math. J. **31**, 543–554 (1964).

[344] Ruckle, W. H.: Decompositions of operator spaces. Rev. Roumaine math. pures et appl. **15**, 119–134 (1970).

[345] Ruckle, W. H.: Representation and series summability of complete biorthogonal sequences. Pacific J. Math. **34**, 511–528 (1970).

[346] Ruckle, W. H.: The tensor product of complete biorthogonal sequences. Duke Math. J. **38**, 681–696 (1971).

[347] Ruckle, W. H.: On the classification of biorthogonal sequences. Canad. J. Math. **26**, 721–733 (1974).

[348] Ruckle, W. H.: The $(q, p)$ multipliers of an operator. Preprint.

[349] Ruckle, W. H.: Vector sequence spaces and ideals of operators. Preprint.

[350] Russo, J. P.: Monotone and $e$-Schauder bases of subspaces. Canad. J. Math. **20**, 233–241 (1968).

[351] Rutowicz, D.: Some parameters associated with finite-dimensional Banach spaces. J. London Math. Soc. **40**, 241–255 (1965).

[352] Ryll, J.: Schauder bases for the space ot continuous functions on an $n$-dimensional cube. Prace Mat. **17**, 201–213 (1973).

[353] Sanders, B. L.: Decomposition and reflexivity in Banach spaces. Proc. Amer. Math. Soc. **16**, 204–208 (1965).

[354] Sanders, B. L.: On the existence of Schauder decompositions in Banach spaces. Proc. Amer. Math. Soc. **16**, 987–990 (1965).

[354a] Saxon, S. A., Wilansky, A.: The equivalence of some Banach space problems. Colloq. Math. **37**, 217–226 (1977).

[355] Schaefer, H. H.: Topological vector spaces. New York: Macmillan 1966.

[356] Schatten, R.: A theory of cross-spaces. Princeton: Ann. of Math. Studies **26**, Princeton Univ. Press 1950.

[357] Schmidt, E.: Entwicklung willkürlichen Funktionen nach Systemen vorgeschriebener. Math. Ann. **63**, 433–476 (1907).

[358] Seever, G. L.: Measures on $F$-spaces. Trans. Amer. Math. Soc. **133**, 267–280 (1968).

[359] Semadeni, Z.: Product Schauder bases and approximation with nodes in spaces of continuous functions. Bull. Acad. Polon. Sci. **11**, 387–391 (1963).

[360] Semadeni, Z.: Banach spaces of continuous functions. I. Warszawa: Monografie Matematyczne 1971.

[361] Singer, I.: Sur l'approximation uniforme des opérateurs linéaires compacts par des opérateurs non linéaires de rang fini. Arch. Math. **11**, 289–293 (1960).

[362] Singer, I.: Sur un problème d'isomorphie de S. Banach. Rendiconti Accad. Naz. Lincei **30**, 343–346 (1961).

[363] Singer, I.: Basic sequences and reflexivity of Banach spaces. Studia Math. **21**, 351–369 (1962).

[364] Singer, I.: Some characterizations of symmetric bases in Banach spaces. Bull. Acad. Polon. Sci. **10**, 185–192 (1962).

[365] Singer, I.: On Cesàro bases in Banach spaces. Rev. math. pures et appl. **7**, 135–142 (1962).

[365a] Singer, I.: On Banach spaces reflexive with respect to a linear subspace of their conjugate space.III. Rev. Roumaine math. pures et appl. **8**, 139–150 (1963).

[366] Singer, I.: Bases in Banach spaces. I. Studii şi cercet. mat. **14**, 539–585 (1963) [Romanian].

[367] Singer, I.: Bases in Banach spaces. II. Studii şi cercet. mat. **15**, 157–208 (1964) [Romanian].

[368] Singer, I.: Bases in Banach spaces. III. Studii şi cercet. mat. **15**, 675–725 (1964) [Romanian].

[369] Singer, I.: Some remarks on domination of sequences. Math. Annalen **184**, 113–132 (1970).

[370] Singer, I.: Bases in Banach spaces. Vol. I. Berlin–Heidelberg– New York: Springer 1970.

[371] Singer, I.: Best approximation in normed linear spaces by elements of linear subspaces. Bucureşti: Editura Acad. R.S. România 1967 [Romanian]. English translation, Bucureşti and Berlin–Heidelberg–New York: Editura Acad. R.S. România and Springer 1970.

[372] Singer, I.: On biorthogonal systems and total sequences of functionals. Math. Annalen **193**, 183–188 (1971).

[373] Singer, I.: On minimal sequences of type $l_+$ and bounded biorthogonal systems in Banach spaces. Czechoslovak Math. J. **23** (98), 11–14 (1973).

[374] Singer, I.: On biorthogonal systems and total sequences of functionals. II. Math. Annalen **201**, 1–8 (1973).

[375] Singer, I.: On pseudo-complemented subspaces of Banach spaces. J. Functional Anal. **13**, 223–232 (1973).

[376] Singer, I.: On separable $\pi_\lambda$ and dual $\pi_\lambda$ spaces. Rev. Roumaine math. pures et appl. **18**, 1109–1121 (1973).

[377] Singer, I.: On some duality theorems for bases, finite dimensional decompositions and $\pi$ systems. Math. Annalen **211**, 337–343 (1974).

[378] Singer, I.: On the extension of basic sequences to bases. Bull. Amer. Math. Soc. **80**, 771–772 (1974).

[379] Singer, I.: Enflo-Rosenthal basis sets in Banach spaces. Serdica Bulg. Math. Publ. **2**, 247–251 (1976).

[380] Singer, I.: On Banach spaces in which every $M$-basis is a generalized summation basis. Proc. Semester on "Approximation theory" held in Warsaw, October–December 1975. Banach Center Publications 4 (to appear).

[381] Sobczyk, A.: Projections in Minkowski and Banach spaces. Duke Math. J. **8**, 78–106 (1941).

[382] Sobczyk, A.: Projections of the space $(m)$ onto its subspace $(c_0)$. Bull. Amer. Math. Soc. **47**, 938–947 (1941).

[383] Sobczyk, A.: Extension properties of Banach spaces. Bull. Amer. Math. Soc. **68**, 217–224 (1962).

[384] Stečkin, S. B.: On absolute convergence of Fourier series. III. Izv. Akad. Nauk SSSR, Ser. Matem. **20**, 385–412 (1956) [Russian].

[385] Sternbach, L.: Bases and quasi-reflexive spaces. Thesis. Ohio State Univ. 1968.

[386] Subramanian, P. K., Rothman, S.: Two-norm spaces and decompositions of Banach spaces. II. Trans. Amer. Math. Soc. **181**, 313–327 (1973).

[387] Szankowski, A.: A Banach lattice without the approximation property. Israel J. Math. **24**, 329–337 (1976).

[388] Taam, C. T.: Compact linear transformations. Proc. Amer. Math. Soc. **11**, 39–42 (1960).

[389] Tacon, D. G.: The conjugate of a smooth Banach space. Bull. Austral. Math. Soc. **2**, 415–425 (1970).

[390] Thorp, E., Whitley, R.: Operator representation theorems. Illinois J. Math. **9**, 595–601 (1965).

[391] Thorp, E., Whitley, R.: Partially bounded sets of infinite width. J. reine u. angew. Math. **248**, 117–122 (1971).

[392] Trebels, W.: Multipliers for $(C, \alpha)$-bounded Fourier expansions in Banach spaces and approximation theory. Lecture Notes in Math. 329. Berlin–Heidelberg–New York: Springer 1973.

[393] Troyanski, S. L.: Equivalent norms in non-separable $B$-spaces with an unconditional basis. Teor. Funkciĭ, Funkcional. Anal. i Priložen. **6**, 59–65 (1968) [Russian].

[394] Troyanski, S. L.: On locally uniformly convex and differentiable norms in certain non-separable Banach spaces. Studia Math. **37**, 173–180 (1971).

[395] Troyanski, S. L.: On equivalent norms and minimal systems in non-separable Banach spaces. Studia Math. **43**, 125–138 (1972) [Russian].

[396] Troyanski, S. L.: On non-separable Banach spaces with a symmetric basis. Studia Math. **53**, 253–263 (1975).

[397] Tsirelson, B. S.: Not in every Banach space can one embed $l_p$ or $c_0$. Funkcional. Anal. i Priložen. **8**, 2, 57–60 (1974) [Russian].

[398] Vaher, F. S.: On a basis in the space of continuous functions defined on a compactum. Doklady Akad. Nauk SSSR **101**, 589–592 (1955) [Russian].

[399] Vaher, F. S.: The local problem of existence of operatorial bases in Banach spaces. Sibirsk. Matem. Ž. **16**, 853–855 (1975) [Russian].

[400] Valentine, F. A.: Convex sets. New York: McGraw Hill 1964.

[401] Veech, W. A.: Short proof of Sobczyk's theorem. Proc. Amer. Math. Soc. **28**, 627–628 (1971).

[402] Vinokurov, V. G.: On biorthogonal systems passing through given subspaces. Doklady Akad. Nauk SSSR **85**, 685–687 (1952) [Russian].

[403] Viziteĭ, V. N.: On the stability of bases of subspaces of a Banach space. Issledovania po Algebre i Matematičeskomu Analizu. Akad. Nauk. Mold. SSR, pp. 32–44. Kišinev: Kartja Moldovenjaska 1965 [Russian].

[404] van der Waerden, B. L.: Algebra I. $8^{th}$ Ed. Berlin–Heidelberg–New York: Springer 1971.

[405] Whitley, R.: Projecting $m$ onto $c_0$. Amer. Math. Monthly **73**, 285–286 (1966).

[406] Wilansky, A.: The basis in Banach space. Duke Math. J. **18**, 795–798 (1951).

[407] Wilansky, A.: Summability: the inset, replaceable matrices, the basis in summability space. Duke Math. J. **19**, 647–660 (1952).

[408] Wilansky, A., Zeller, K.: A biorthogonal system which is not a Toeplitz basis. Bull. Amer. Math. Soc. **69**, 725–726 (1963).

[409] Yosida, K.: Functional analysis. Berlin–Heidelberg–New York: Springer 1966.

[410] Zeller, K.: Theorie der Limitierungsverfahren. Berlin–Göttingen–Heidelberg: Springer 1958.

[411] Zygmund, A.: Trigonometric series. Vol. I. Cambridge: Cambridge Univ. Press 1959.

# Appendix: Complements added in proof

1) *To pages* 38 *and* 726: a) W. B. Johnson's negative solution of § 0, problem 0.1, has appeared in the meantime in the book of J. Lindenstrauss and L. Tzafriri [424], pp. 112—113, example 1.g.7. Recently, W. B. Johnson [421] has constructed another class of examples of Banach spaces $X$ with the following properties: (i) *$X$ has no quotient space containing a subspace isomorphic to $l^2$*; (ii) *every subspace of every quotient space of $X$ has a basis* (solving thus also Ch. I, § 1, problem 1.3) *and the uniform projection approximation property;* (iii) *$X$ is of type* 2 *and of cotype $q$ for all $q > 2$**); (iv) *$X$ has a hyperorthogonal basis.* In the same paper [421], W. B. Johnson has asked whether the space $X$ can be constructed so that, in addition, every subspace of $X$ has an unconditional basis. Also, he has raised [421] the following problem: Does there exist a Banach space $X$ with a symmetric basis, not isomorphic to $l^2$, or a non-reflexive Banach space $E$, such that every subspace of $X$ has the approximation property?

b) For A. Szankowski's affirmative solution of the particular case $X = l^p$, $1 \leqslant p < 2$, of § 0, problem 0.1, and his result on Banach spaces $X$ with $p(X) < 2$ or $q(X) > 2$, mentioned on p. 726, see [436] or [424], pp. 107—111, theorems 1.g.4 and 1.g.5.

c) Another class of Banach sequence spaces $X$, containing a subspace $E$ without the approximation property, has been given by G. W. Johnson and L. V. Petersen [420]. For another example, see A. Pietsch [427], p. 141.

2) *To pages* 47, 296 *and* 718: Proofs of the results of A. Grothendieck [139] on the approximation property, to which we referred on these pages, can be also found in Lindenstrauss-Tzafriri [248a], section 1.e. See also J. Diestel and J. J. Uhl, Jr. [415], Ch. VIII, § 3.

3) *To pages* 84—85: Recently, P. Terenzi has shown [448] that § 1, *problem* 1.6 a) *has an affirmative answer*, with the following short proof: If $\{x_n\} \subset E$ is a basic sequence with codim $[x_n] = m < \infty$, then $\{x_n\}$ can be extended to a basis**) $\{x_n\} \cup \{z_i\}_{i=1}^m$ of $E$. If $\{f_n\} \cup \{h_n\}_{i=1}^m \subset E^*$ is the a.s.c.f., then $(x_n, f_n)$ is a biorthogonal system and, by Ch. I, § 12, theorem 12.1 and Ch. I, § 4, proposition 4.1 a), $\{f_n\}$ is a basic sequence. Assume now that $\{x_n\} \subset E$ is any minimal sequence with codim $[x_n] = \infty$. Then dim $[x_n]^\perp = \infty$, so by § 1, theorem 1.1, $[x_n]^\perp$ contains an infinite basic sequence $\{g_n\}$. Now, by Ch. I, § 10, theorem 10.3, there exists a sequence $\gamma_n > 0$ $(n = 1, 2, \ldots)$ such that every sequence $\{g'_n\} \subset E^*$ satisfying $\|g'_n - g_n\| \leqslant \gamma_n$ $(n = 1, 2, \ldots)$ is a basic sequence. Since

---

*) We have seen on p. 726 that for any Banach space $X$ such that every subspace of $X$ has the approximation property (hence, in particular, for any $X$ with property (ii)), we must have $p(X) = 2$ and $q(X) = 2$. On the other hand, let us also mention that *a Banach space $X$ is of type* 2 *and cotype* 2 *if and only if $X$ is isomorphic to a Hilbert space* (S. Kwapien [423]).

**) If $m = 0$, then $\{z_i\}_{i=1}^m = \emptyset$.

$\{x_n\}$ is minimal, let $\{h_n\} \subset E^*$ be any sequence such that $(x_n, h_n)$ is a biorthogonal system and let

$$f_n = h_n + \frac{2\|h_n\|}{\gamma_n} g_n \qquad (n = 1, 2, \ldots). \tag{1.156}$$

Then, by $\{g_n\} \subset [x_n]^\perp$, $(x_n, f_n)$ is a biorthogonal system. Moreover, we have

$$\left\| \frac{\gamma_n f_n}{2\|h_n\|} - g_n \right\| = \left\| \frac{\gamma_n h_n}{2\|h_n\|} \right\| = \frac{\gamma_n}{2} < \gamma_n \qquad (n = 1, 2, \ldots)$$

and therefore $\left\{\frac{\gamma_n f_n}{2\|h_n\|}\right\}$, whence also $\{f_n\}$, is a basic sequence, which completes the proof. Let us also observe here that one can show, with similar methods, that *the answers* to § 1, *problems* 1.6′ *and* 1.6″ *are affirmative, even for any (not necessarily basic) sequence* $\{X_n\} \subset E/G$ *when* $\dim G = \infty$ *and, respectively, for any sequence* $\{\varphi_n\} \subset G^*$ *when* $\operatorname{codim} G = \infty$. In other words, *if* $\dim G = \infty$, $\{X_n\} \subset E/G$, *then there exist* $x_n \in X_n$ $(n = 1, 2, \ldots)$ *such that* $\{x_n\}$ *is a basic sequence;* hence, *if* $\operatorname{codim} G = \infty$ (so $\dim G^\perp = \infty$) *and* $\{\varphi_n\} \subset G^*(\equiv E^*/G^\perp)$, *then there exists a basic sequence* $\{f_n\} \subset E^*$ *such that* $f_n|_G = \varphi_n$ $(n = 1, 2, \ldots)$. Indeed, if $\dim G = \infty$, it is enough to take an infinite basic sequence $\{y_n\} \subset G$, a sequence $\gamma_n > 0$ $(n = 1, 2, \ldots)$ for it as in Ch. I, § 10, theorem 10.3, and then $z_n \in X_n$, $z_n \neq 0$ $(n = 1, 2, \ldots)$. Then, putting

$$x_n = z_n + \frac{2\|z_n\|}{\gamma_n} y_n \in X_n \qquad (n = 1, 2, \ldots), \tag{1.157}$$

we obtain, as above, that $\{x_n\}$ is a basic sequence, which completes the proof.

4) *To page* 153: Recently, L. Drewnowski, I. Labuda and Z. Lipecki [415a] have proved that every separable topological linear space $E$ has a semi-basis (they use the term: quasi-basis).

5) *To pages* 157—162: a) Actually, the proof of § 6, theorem 6.1 may be regarded as a direct construction (6.19) of a $\{\lambda_i\}$-linearly independent subsequence of a given finitely linearly independent sequence $\{x_n\}$ with $\sup_{1 \leq n < \infty} \|x_n\| < \infty$. The "ad absurdum" start (on p. 158) serves to avoid the repetition of the same argument in cases 1° and 2°.

b) In the proof for case 2° (on p. 161), the estimate $M > \frac{1}{4} |c_{k_1}|$ $(m > m_0)$ is correct only under the additional assumption

$$c_k^{(m)}(a_{l_k, l_k} - b_{l_k}) = 0 \qquad (k = k_1, k_1 + 1), \tag{6.96}$$

since (6.28) gives no information about these terms. In this connection, P. Terenzi has communicated*) us an example showing that in case

*) Subsequently, he included this example in a new version of [445].

2° a subsequence $\{x_{l_k}\}$ selected by (6.28) need not even be $\omega$-linearly independent and has observed that, nevertheless, in case 2° the proof can be corrected by considering, instead of $\{x_{l_k}\}$, the subsequence $\{x_{l_{2k}}\}$ of $\{x_n\}$, where (6.28) still holds (hence now (6.10), (6.22) and (6.21) will be modified accordingly). Indeed, then similar computations hold, for $M$ of p. 161 replaced by

$$M' = \left| \frac{1}{b_{l_{2k_1-1}}} \sum_{k=k_1}^{\infty} c_k^{(m)} a_{l_{2k}, l_{2k_1-1}} - \frac{1}{b_{l_{2k_1+1}}} \sum_{k=k_1+1}^{\infty} c_k^{(m)} a_{l_{2k}, l_{2k_1+1}} \right|, \quad (6.97)$$

in which the terms (6.96) do not occur.

Let us also mention the following simple alternative proof of § 6, theorem 6.1, avoiding case 2°, which resulted from correspondence with P. Terenzi: We may assume, as before, that $E = L^2([0, 1])$ and $\|x_n\| = 1$ $(n = 1, 2, \ldots)$. Then $\{x_n\}$ has a subsequence, which we shall denote again by $\{x_n\}$, converging weakly to an element $x_0 \in E$. If $x_0 = 0$, then, by § 1, corollary 1.1, $\{x_n\}$ has even a basic subsequence. If $x_0 \neq 0$, we may assume that $x_0 \cup \{x_n\}$ is finitely linearly independent (omitting, if neccssary, a suitable element of $\{x_n\}$). Let $\{z_n\}_0^\infty$ be the orthonormal sequence obtained from $x_0 \cup \{x_n\}$ by the Gram-Schmidt process, so

$$x_n = \sum_{k=0}^{n} a_{nk} z_k \qquad (n = 0, 1, 2, \ldots), \quad (6.98)$$

with $a_{nn} \neq 0$ $(n = 0, 1, 2, \ldots)$. Then, by $x_n \xrightarrow{w} x_0$, we obtain

$$b_m = \lim_{n\to\infty} a_{nm} = \lim_{n\to\infty} (x_n, z_m) = (x_0, z_m) = (a_{00} z_0, z_m) = 0 \quad (m = 1, 2, \ldots), \quad (6.99)$$

and hence we can apply the argument of case 1° (p. 160). By that argument, we omit $x_0$ from $x_0 \cup \{x_n\}$ (indeed, in our case $\{n_i\} = \{0\} \cup \cup \mathcal{N}$, $\{n_{i_j}\} = \mathcal{N}$) and we get that $\{x_n\}$ contains a $\{\lambda_i\}$-linearly independent subsequence, which completes the proof of theorem 6.1.

Finally, let us mention that P. Terenzi has given [445] yet another proof of § 6, theorem 6.1, working directly in the initial Banach space $E$.

c) M. I. Kadec has communicated us that V. M. Kadec has constructed in the space $E = s$ a finitely linearly independent sequence $\{x_n\}$ containing no $\omega$-linearly independent subsequence, namely

$$x_n = \{1^n, 2^n, 3^n, \ldots\} \qquad (n = 1, 2, \ldots). \quad (6.100)$$

6) *To page* 232: The argument in the proof of § 8, theorem 8.2 shows only that $\{y_n\}$ can be extended to an $M$-basis $\{y_n\} \cup \{z_n\}$ of $E$, with $\{\omega(z_n)\} \subset \omega(F)$, but it can be continued to show also the existence of an $M$-basis $\{y_n\} \cup \{z'_n\}$ of $E$ with $\{z'_n\} \subset F$, as follows (P. Terenzi [444]): By § 3, theorem 3.2 or Ch. I, § 10, theorem 10.3, let $\{x'_n\} \subset E$ and $\gamma_n > 0$ $(n = 1, 2, \ldots)$ be such that for every $\{y'_n\} \subset E$ with $\|x'_n - y'_n\| \leqslant \gamma_n$ $(n = 1, 2, \ldots)$ we have $[y'_n] = E$. Now, by the argument on p. 232, there exists a sequence $\{z_{n1}\} \subset F$ such that $\{z_{n1} + G\}$ is an $M$-basis of

$E/G$, where $G=[y_n]$. Furthermore, there exists $y_1' \in \text{lin}\{y_n\} + \text{lin}\{z_{n1}\}_1^{p_1}$ (for some $p_1 \in \mathcal{N}$), such that $\|x_1' - y_1'\| \leqslant \gamma_1$ (indeed, since $[z_{n1} + G] = E/G$, there exists $z' \in \text{lin}\{z_{n1}\}_1^{p_1}$ such that $\|x_1' - z' + G\| = \|x_1' + G - (z' + G)\| < \gamma_1$, whence also $g \in G = [y_n]$ with $\|x_1' - z' - g\| < \gamma_1$, whence also $g' \in \text{lin}\{y_n\}$ with $\|x_1' - g' - z'\| \leqslant \gamma_1$). We claim that $\{y_1 + G_2\} \cup \{z_{n1} + G_2\}_1^\infty$ is a finitely linearly independent complete sequence in $E/G_2$, where $G_2=[y_j]_2^\infty$. Indeed, if $\alpha_1 y_1 + \sum_{n=1}^{k} \beta_n z_{n1} + G_2 = G_2$, then $\sum_{n=1}^{k} \beta_n z_{n1} \in [y_j]_1^\infty = G$, whence, since $\{z_{n1} + G\}$ is an $M$-basis of $E/G$, $\beta_n = 0$ $(n = 1, \dots, k)$. Therefore, $\alpha_1 y_1 \in G_2$, whence since $\{y_n\}_1^\infty$ is an $M$-basis of $G$, $\alpha_1 = 0$, so $\{y_1 + G_2\} \cup \{z_{n1} + G_2\}_1^\infty$ is finitely linearly independent. Now, if $x \in E$, $\varepsilon > 0$, then there exists $z' \in \text{lin}\{z_{n1}\}_1^\infty$ such that $\|x - z' + G\| = \|x + G - (z' + G)\| < \varepsilon$, whence also $g \in G = [y_j]_1^\infty$ with $\|x - z' - g\| < \varepsilon$ and hence, writing $g = \alpha_1 y_1 + g_2$, where $g_2 \in G_2$, we obtain $\|x + G_2 - (\alpha_1 y_1 + G_2) - (z' + G_2)\| \leqslant \|x - \alpha_1 y_1 - z' - g_2\| < \varepsilon$, which proves the claim. Consequently, there exists a sequence $\{z_{n2}\}_1^\infty \subset \text{lin}\{z_{n1}\}_{p_1+1}^\infty$, such that $\{y_1 + G_2\} \cup \{z_{n1} + G_2\}_1^{p_1} \cup \{z_{n2} + G_2\}_1^\infty$ is an $M$-basis of $E/G_2$ (indeed, one can extend $\{y_1 + G_2\} \cup \{z_{n1} + G_2\}_1^{p_1}$ to an $M$-basis of $E/G_2$ and then apply Ch. I, § 10, theorem 10.3). Next, as above, there exists $y_2' \in \text{lin}\{y_n\} + \sum_{i=1}^{2} \text{lin}\{z_{ni}\}_1^{p_i}$ (where $p_2 \in \mathcal{N}$), such that $\|x_2' - y_2'\| \leqslant \gamma_2$. Again, as above, $\{y_n + G_3\}_1^2 \cup \bigcup_{i=1}^{2} \{z_{ni} + G_3\}_1^{p_i}$ can be extended to an $M$-basis of $E/G_3$, where $G_3 = [y_j]_3^\infty$, by means of a sequence $\{z_{n3} + G_3\}_1^\infty$ with $\{z_{n3}\}_1^\infty \subset \text{lin}\{z_{n2}\}_{p_2+1}^\infty$. Continuing in this way indefinitely, we obtain two sequences $\{z_n\} = \bigcup_{i=1}^{\infty} \{z_{ni}\}_{n=1}^{p_i} \subset F$ with $\{z_{nm}\} \subset \text{lin}\{z_{n,m-1}\}_{n=p_{m-1}+1}^\infty$ $(m = 2, 3, \dots)$ and $\{y_n'\} \subset E$ with

$$\{y_n'\}_1^{m-1} \subset \text{lin}\{y_n\} + \sum_{i=1}^{m-1} \text{lin}\{z_{ni}\}_1^{p_i} \quad (m = 2, 3, \dots), \tag{8.160}$$

such that $\{y_n + G_m\}_{n=1}^{m-1} \cup \bigcup_{i=1}^{m-1} \{z_{ni} + G_m\}_{n=1}^{p_i} \cup \{z_{nm} + G_m\}_{n=1}^\infty$ is an $M$-basis of $E/G_m$, where $G_m = [y_j]_m^\infty$ $(m = 2, 3, \dots)$ and that

$$\|x_n' - y_n'\| \leqslant \gamma_n \qquad (n = 1, 2, \dots). \tag{8.161}$$

Then, by (8.160), (8.161) and the definition of $\{x_n'\}$, $\{y_n\}$, we have $[\{y_n\} \cup \{z_n\}] = E$. Furthermore, $y_m \notin [\{y_n\}_{n \neq m} \cup \{z_n\}]$ $(m = 1, 2, \dots)$ (indeed, if $y_m \in [\{y_n\}_{n \neq m} \cup \{z_n\}]$, then*) $y_m + G_{m+1} \in [\{y_n + G_{m+1}\}_{n \neq m} \cup$

---

*) We recall that $\bigcup_{i=m+1}^{\infty} \{z_{ni}\}_{n=1}^{p_i} \subset \text{lin}\{z_{n,m+1}\}_{n=1}^\infty$.

$\cup \{z_n + G_{m+1}\}] \subset \left[ \{y_n + G_{m+1}\}_{n=1}^{m-1} \cup \bigcup_{i=1}^{m} \{z_{ni} + G_{m+1}\}_{n=1}^{p_i} \cup \{z_{n,\,m+1} + \right.$ $\left. + G_{m+1}\}_{n=1}^{\infty} \right]$, so $\{y_n + G_{m+1}\}_{n=1}^{m} \cup \bigcup_{i=1}^{m} \{z_{ni} + G_{m+1}\}_{n=1}^{p_i} \cup \{z_{n,\,m+1} +$ $+ G_{m+1}\}_{n=1}^{\infty}$ is not an $M$-basis of $E/G_{m+1}$, a contradiction). Therefore, there exists $\{g_n\} \subset E^*$ such that

$$g_i(y_j) = \delta_{ij},\ g_i(z_j) = 0 \qquad (i, j = 1, 2, \ldots). \tag{8.162}$$

Consequently, since $\overline{G + [z_n]} = E$ and $G \cap [z_n] \subset G \cap F = \{0\}$, there exists, by § 8, theorem 8.3, implication 3° ⇒ 1°, a sequence $\{z_n'\} \subset \subset [z_n] \subset F$ with $[z_n'] = [z_n]$, such that $\{y_n\} \cup \{z_n'\}$ is an $M$-basis of $E$. This completes the proof of theorem 8.2. In the final step one can avoid the use of § 8, theorem 8.3, observing that since $[G \cup \{z_n\}] = E$ and $G \cap \operatorname{lin}\{z_n\} \subset G \cap F = \{0\}$, there exists, by the argument on p. 232, a sequence $\{z_n'\} \subset \operatorname{lin}\{z_n\}$ such that $\{z_n' + G\}$ is an $M$-basis of $E/G$; then, if $\{\psi_n\} \subset (E/G)^*$ is the a.s.f. to $\{z_n' + G\}$ and $\{h_n\} \subset G^\perp$ the image of $\{\psi_n\}$ by the canonical linear isometry $(E/G)^* \equiv G^\perp$, it follows that $\{y_n\} \cup \{z_n'\}$ is an $M$-basis of $E$, with the a.s.f. $\{g_n\} \cup \{h_n\}$.

Finally, let us also mention that A. N. Pličko has communicated us, in a letter, another proof of § 8, theorem 8.2, based on the following lemma: *Under the assumptions of theorem* 8.2, *if* $\{\varphi_n\} \subset [y_n]^*$ *is the a.s.f. to* $\{y_n\}$, *then for each n there exists an extension* $g_n \in E^*$ *of* $\varphi_n$ *such that*

$$\overline{[\{g_n\}_1^\infty + F^\perp]} \cap G^\perp = \{0\}. \tag{8.163}$$

The lemma is proved by an inductive construction (fixing a sequence of $\sigma(E^*, E)$-compact convex sets $\{Q_n\} \subset G^\perp \setminus \{0\}$ such that $\bigcup_{n=1}^{\infty} Q_n = G^\perp \setminus \{0\}$ and applying repeatedly the separation theorem in $E^*$ endowed with $\sigma(E^*, E)$) and it implies § 8, theorem 8.2, as follows: by (8.163), $G$ and $F_0 = (\{g_n\}_1^\infty + F^\perp)_\perp$ are quasi-complementary subspaces of $E$ and, clearly, $g_i(y_j) = \delta_{ij}$, $g_i(F_0) = 0$ $(i, j = 1, 2, \ldots)$, $F_0 \subset F$. Hence, taking any sequence $\{z_n\} \subset F_0$ with $[z_n] = F_0$, by either one of the two arguments after (8.162) we obtain an extension of $\{y_n\}$ to an $M$-basis $\{y_n\} \cup \{z_n'\}$ of $E$, with $\{z_n'\} \subset F_0 \subset F$, which completes the proof.

7) *To pages* 309—310, 541—542 *and* 772: Applying the argument used in the proof of Ch. I, § 15, lemma 15.7, M. Feder [416] has obtained the following results (some of the proofs are also presented in Lindenstrauss-Tzafriri [424], pp. 33—34 and P. Saphar [432]): i) *A reflexive Banach space E, which is a subspace of a Banach space with an unconditional basis, has an unconditional finite-dimensional expansion of* $I_E$ *if (and only if) it has the approximation property.* ii) *A subspace E of a Banach space with a shrinking unconditional basis has a shrinking uncon-*

*ditional finite-dimensional expansion of* $I_E$ *if (and only if) the conjugate space* $E^*$ *has the approximation property.* iii) *Let* $E$ *be a subspace of a Banach space* $X$ *with an unconditional basis* $\{x_n\}$ *and let* $\{f_n\} \subset X^*$ *be the a.s.c.f. to* $\{x_n\}$. a) *If* $\{x_n\}$ *is shrinking and* $(X/E)^*$ *has the approximation property, then* $X/E$ *has a shrinking unconditional finite-dimensional expansion of* $I_{X/E}$. b) *If* $\{x_n\}$ *is boundedly complete and* $E$ *is* $\sigma(X, [f_n])$*-closed, then* $E$ *has an unconditional finite-dimensional expansion of* $I_E$ *if (and only if) it has the approximation property.*

8) *To pages* 599—600 *and* 814: A. N. Pličko has shown ([430], theorem 1) that *the Banach space* $E$ *constructed by W.B. Johnson and J. Lindenstrauss* (see p. 836, the comments to § 20, problem 20.6) *contains no uncountable transfinite basic sequence*. Thus, there exist Banach spaces, even admitting an equivalent ($F$)-norm (but not satisfying (17.58)), for which the conclusion of § 17, theorem 17.10 does not hold.

9) *To pages* 607 *and* 817: For a presentation of the main ideas leading to the proof of A. Szankowski's negative solution of § 18, problem 18.1 a), b), see [437].

10) *To page* 673: Combining § 19, corollary 19.1, with the result of A. N. Pličko [430], mentioned in 8) above, it follows that *the answer to* § 19, *problem* 19.2 *is negative*.

11) *To pages* 675 *and* 685: Recently, S. Shelah has announced the following result [433]: *It follows from* $\diamondsuit_{\aleph_1}$ (which is consistent with the axioms of set theory) *that there exists a non-separable Banach space* $E$ *with the property that for any uncountable family* $\{x_i\}_{i\in I} \subset E$ *there is an* $i_0 \in I$ *such that* $x_{i_0} \in \overline{\text{co}}\, \{x_i\}_{i\in I\setminus\{i_0\}}$, the closed convex hull of $\{x_i\}_{i\in I\setminus\{i_0\}}$ (hence $x_{i_0} \in [x_i]_{i\in I\setminus\{i_0\}}$, so $\{x_i\}_{i\in I}$ is not minimal). Thus, in this space every extended minimal system is countable, giving a negative answer to § 20, problem 20.1, whence also to § 20, problem 20.2 a), c). However, a negative answer to these problems, without assuming $\diamondsuit_{\aleph_1}$, has been given, independently, by A. N. Pličko [429], and, with a different proof, by B. V. Godun and M. I. Kadec [419], who have shown that *if* card $I > 2^{\mathfrak{m}}$, *where* $\mathfrak{m} \geqslant \aleph_0$ ($\mathfrak{m} = \aleph_0$ in [429] and $\mathfrak{m} \geqslant \aleph_0$ in [419]), *then the Banach space* $E = l^\infty_{\mathfrak{m}}(I)$ *has no complete extended minimal system**) $\{x_i\}_{i\in I}$. We recall that $l^\infty_{\mathfrak{m}}(I)$ denotes the subspace of $l^\infty(I)$ consisting of all functions $x \in l^\infty(I)$ such that card $\{i \in I | x(i) \neq 0\} \leqslant \mathfrak{m}$; as was observed in [419], each $l^\infty_{\mathfrak{m}}(I)$ is a Grothendieck space.

12) *To pages* 683, 697, 814, 833 *and* 836—837: A. N. Pličko has shown ([430], theorem 2) that *the Banach space* $E$ *of W. B. Johnson and J. Lindenstrauss* (see 8) above) *admits no norming extended biorthogonal system* (and hence, in particular, no norming extended $M$-basis).

13) *To pages* 685—691, 831 *and* 833: Recently, A. N. Pličko has obtained the following results [428]: i) *If* $E$ *admits an* $E$*-complete (respectively, a total) extended biorthogonal system, then*

---

*) Using this result (and part of § 20, remark 20.3), one can also show that the converse of § 20, remark 20.1 is not valid.

*for each* $\varepsilon > 0$ *it admits also an E-complete (respectively, a total) extended biorthogonal system* $(x_i, f_i)_{i \in I}$ *with* $\sup_{i \in I} \|x_i\| \, \|f_i\| \leqslant 8 + \varepsilon$ *(respectively,* $\leqslant 4 + \varepsilon$*).* Combining this with § 20, theorem 20.2, it follows [428] that *for each* $\varepsilon > 0$, *every Banach space admits a total extended biorthogonal system* $(x_i, f_i)_{i \in I}$ *with* $\sup_{i \in I} \|x_i\| \, \|f_i\| \leqslant 4 + \varepsilon$; this improves the bound $6 + \varepsilon$, mentioned on p. 831. ii) *If E has a norming M-basis* $\{y_i\}_{i \in I}$, *then for each* $\varepsilon > 0$ *it admits also an M-basis* $\{x_i\}_{i \in I}$ *with* $\sup_{i \in I} \|x_i\| \, \|f_i\| \leqslant \frac{2}{r([h_i]_{i \in I})} + \varepsilon$, *where* $\{h_i\}_{i \in I}$, $\{f_i\}_{i \in I} \subset E^*$ *are the a.f.f. to* $\{y_i\}_{i \in I}$ *and* $\{x_i\}_{i \in I}$ *respectively.*

14) *To pages* 687—688, 692 *and* 757: B. V. Godun has observed [417] that the system $\bigcup_{\lambda < \vartheta} (x_n^\lambda, f_n^\lambda)_{n=1}^\infty$ defined by (20.38) — (20.40) is *not* an extended biorthogonal system, since

$$f_{n+2}^\lambda(x_n^\lambda) = (h_{n+1}^\lambda + h_{n+2}^\lambda)(z_n^\lambda - z_{n+1}^\lambda - y_\lambda) = -1. \qquad (20.101)$$

Also, B. V. Godun has shown [417] that, nevertheless, § 20, theorem 20.3 is true and can be proved by extending the techniques used in § 6, proofs of theorems 6.4 and 6.3. Namely, he has proved the following proposition [417], which, together with proposition 20.4, implies immediately theorem 20.3: *Let E be a Banach space, G a subspace of E and* $\bigcup_{i \in I} (Z_n^i, \psi_n^i)_{n=1}^\infty$ $(\{Z_n^i\} \subset E/G, \{\psi_n^i\} \subset (E/G)^*)$ *an E/G-complete extended biorthogonal system with the following properties:* a) $\|Z_n^i\| \, \|\psi_n^i\| \leqslant C < \infty$ $(i \in I, n \in \mathcal{N})$; b) card $I =$ dens $E$; c) *for each* $i \in I$, *the sequence* $\{Z_n^i\}_{n=1}^\infty$ *is not equivalent to the unit vector basis of* $l^1$. *Then, for each* $\varepsilon > 0$, $\bigcup_{i \in I} (Z_n^i, \psi_n^i)_{n=1}^\infty$ *admits a lifting to an E-complete extended biorthogonal system* $\bigcup_{i \in I} (x_n^i, h_n^i)_{n=1}^\infty$, $(x_n^i \in Z_n^i, h_n^i(x) = \psi_n^i(x + G)$ *for all* $i \in I$, $n \in \mathcal{N}$, $x \in E)$ *such that*

$$\|x_n^i\| \, \|h_n^i\| \leqslant 2C + \varepsilon \quad (i \in I, n \in \mathcal{N}). \qquad (20.102)$$

For the proof, assume that $\|Z_n^i\| \leqslant C$, $\|\psi_n^i\| = 1$ $(i \in I, n \in \mathcal{N})$. Let $\varepsilon' = \frac{\varepsilon}{2C + 1}$ and, for each $i \in I$ and $n \in \mathcal{N}$, choose $z_n^i \in Z_n^i$ with $\|z_n^i\| \leqslant C(1 + \varepsilon')$. We claim that *for each* $i \in I$ *there exists a sequence* $\{y_n^i\} \subset G$ *such that* $\{z_n^i - y_n^i\}$ *is not equivalent to the unit vector basis of* $l^1$ *and* $\sup_{1 \leqslant n < \infty} \|z_n^i - y_n^i\| \leqslant 2C(1 + 2\varepsilon')$. Indeed, fix $i \in I$. Then, since $\{Z_n^i\}$ is not equivalent to the unit vector basis of $l^1$, there exists a sequence of scalars $\{\beta_n^i\}$ such that $\sum_{n=1}^\infty \beta_n^i Z_n^i$ converges and $\sum_{n=1}^\infty |\beta_n^i| = \infty$. Thus, there exist a number $d > 0$ and indices $n_k < m_k < n_{k+1}$ $(k = 1, 2, \ldots)$ such that

$$\left\| \sum_{n=n_k}^{m_k} \beta_n^i Z_n^i \right\| \leqslant \frac{1}{2^k} \qquad (k = 1, 2, \ldots), \tag{20.103}$$

$$\sum_{n=n_k}^{m_k} |\beta_n^i| = d_k \geqslant \max\left(d, \frac{1+\varepsilon'}{2^{k+1}C\varepsilon'}\right) \qquad (k = 1, 2, \ldots). \tag{20.104}$$

By (20.103), for each $k \in \mathcal{N}$ choose $g_k \in G$ such that

$$\left\| \sum_{n=n_k}^{m_k} \beta_n^i z_n^i - g_k \right\| \leqslant (1+\varepsilon') \left\| \sum_{n=n_k}^{m_k} \beta_n^i Z_n^i \right\| \leqslant \frac{1+\varepsilon'}{2^k}, \tag{20.105}$$

and let

$$y_n^i = \begin{cases} \left(\dfrac{1}{d_k} \operatorname{sign} \beta_n^i\right) g_k & \text{if} \quad n = n_k, n_k + 1, \ldots, m_k \\ 0 \quad \text{if} \quad n \in \mathcal{N} \setminus \{n_k, n_k + 1, \ldots, m_k\}. \end{cases} \tag{20.106}$$

Then $\{y_n^i\} \subset G$ and $\{z_n^i - y_n^i\}$ is not equivalent to the unit vector basis of $l^1$, since for $\mathcal{N}_0 = \bigcup_{k=1}^{\infty} \{n_k, n_k + 1, \ldots, m_k\}$ we have, by (20.104) and (20.105), $\sum_{n \in \mathcal{N}_0} |\beta_n^i| = \sum_{k=1}^{\infty} d_k = \infty$ and

$$\left\| \sum_{n \in \mathcal{N}_0} \beta_n^i (z_n^i - y_n^i) \right\| \leqslant \sum_{k=1}^{\infty} \left\| \sum_{n=n_k}^{m_k} \beta_n^i z_n^i - \frac{1}{d_k} \sum_{n=n_k}^{m_k} |\beta_n^i| g_k \right\| \leqslant (1+\varepsilon') \sum_{k=1}^{\infty} \frac{1}{2^k}.$$

Also, if $n \notin \{n_k, n_k + 1, \ldots, m_k\}$, then $\|z_n^i - y_n^i\| = \|z_n^i\| \leqslant C(1 + \varepsilon')$. Finally, if $n \in \{n_k, n_k + 1, \ldots, m_k\}$, then, by (20.105) and (20.104),

$$\|z_n^i - y_n^i\| = \left\| z_n^i - \left(\frac{1}{d_k} \operatorname{sign} \beta_n^i\right) g_k \right\| \leqslant C(1+\varepsilon') + \frac{1}{d_k} \|g_k\| \leqslant$$

$$\leqslant C(1+\varepsilon') + \frac{1}{d_k} \left\{ \left\| \sum_{n=n_k}^{m_k} \beta_n^i z_n^i - g_k \right\| + \left\| \sum_{n=n_k}^{m_k} \beta_n^i z_n^i \right\| \right\} \leqslant$$

$$\leqslant C(1+\varepsilon') + \frac{1}{d_k} \left\{ \frac{1+\varepsilon'}{2^k} + C(1+\varepsilon') \sum_{n=n_k}^{m_k} |\beta_n^i| \right\} =$$

$$= 2C(1+\varepsilon') + \frac{1+\varepsilon'}{2^k d_k} \leqslant 2C(1+2\varepsilon'),$$

which proves the claim. Thus, for each $i \in I$ there exists a sequence of scalars $\{\alpha_n^i\}$ such that $\sum_{n=1}^{\infty} \alpha_n^i (z_n^i - y_n^i)$ converges and $\sum_{n=1}^{\infty} |\alpha_n^i| = \infty$. Then, taking a dense family $\{\bar{y}_i\}_{i \in I}$ in $S_G = \{y \in G \mid \|y\| \leqslant 1\}$ and putting

$$x_n^i = z_n^i - y_n^i - (\varepsilon' \operatorname{sign} \alpha_n^i) \bar{y}_i \in Z_n^i \qquad (i \in I, n \in \mathcal{N}), \tag{20.107}$$

the system $\bigcup_{i \in I} (x_n^i, h_n^i)_{n=1}^{\infty}$ has the required properties (indeed, $\|x_n^i\| \leqslant \|z_n^i - y_n^i\| + \varepsilon' \leqslant 2C(1 + 2\varepsilon') + \varepsilon' = 2C + \varepsilon$ and the proof of

$[x_n^i] = E$ is similar to the final part of the proof of § 6, theorem 6.3, using now that $[z_n^i] + G$ is dense in $E$, because of $[Z_n^i] = E/G$). This completes the proof.

From the above it follows that p. 688, corollary 20.3 and p. 692, remark 20.3 (which use theorem 20.3) are also true. Moreover, the argument on p. 757 (which used the techniques of proof of theorem 20.3), becomes also correct if we apply the above techniques of proof to $(Z_n, \psi_n)$, where $\{Z_n\} \subset E/\text{Ker } v$ is the image of the unit vector basis of $c_0$ by the isomorphism $c_0 \cong E/\text{Ker } v$ occurring in that argument.

Following B. V. Godun [417], a strictly bounded extended biorthogonal system $(x_i, f_i)_{i \in I}$ is called an $l^1(\mathfrak{m})$-*system*, for a cardinal $\mathfrak{m} \leqslant \text{card } I$, if every subfamily of $\{x_i\}_{i \in I}$, of cardinality $\mathfrak{m}$, contains a subfamily which is equivalent to the unit vector basis of $l^1(S)$, where $S$ is a set of cardinality $\mathfrak{m}$ (for $\mathfrak{m} = \aleph_0$ the term $l^1$-*system* is used [417]). Such systems have been studied by B. V. Godun [417] and by B. V. Godun and M. I. Kadec [419]. Among other results, B. V. Godun has shown [417] that there are some close connections between complete $l^1(\mathfrak{m})$-systems in $E/G$ and the existence of strictly bounded liftings of $E/G$-complete extended biorthogonal systems.

15) *To page* 717: G. F. Bachelis [413] has shown that if $X$ is a Banach space with the bounded approximation property, containing a subspace $E$ without the approximation property, then $X$ also contains a *subspace* $F$ such that there exists a compact $u \in L(F, E)$ which cannot be approximated uniformly by finite rank operators in $L(F, E)$. If, in addition, $X$ is isomorphic to $X \times X$, then $X$ contains a subspace $F_0$ such that some compact $u \in L(F_0, F_0)$ is not the uniform limit of finite rank operators belonging to $L(F_0, F_0)$. For $X = l^p$, where $2 < p < \infty$, the latter result had been shown, previously, by F. E. Alexander [412], who had also raised the following open problem, with the mention that an affirmative answer seems unlikely [412]: If for a Banach space $E$, every compact $u \in L(E, E)$ is a limit of finite rank operators $u_n \in L(E, E)$, does $E$ have the approximation property?

16) *To pages* 738–751: a) Z. Altshuler has observed [3] that the norms (1.155) of T. Figiel and W. B. Johnson [108] can be given explicitly by the formulas

$$\|x\|_n = 2^n \sup_{1 \leqslant m < \infty} \left\{ \frac{\sum_{i=1}^{m} \hat{\xi}_i}{4^n + m} \right\} \quad (x = \{\xi_n\} \in X,\ n = 1, 2, \ldots), \tag{1.156}$$

where $\{\hat{\xi}_n\}$ is the rearrengement in non-decreasing order of $\{|\xi_n|\}$.

b) The Banach spaces $E$ containing no subspace isomorphic to $c_0$ or $l^p$ $(1 \leqslant p < \infty)$, mentioned on pp. 738–751, including that of Z. Altshuler [3], are reflexive. In this connection, P. G. Casazza, B.-L. Lin and R. H. Lohman [414] have shown that *for each* $n \in \mathcal{N}$ *there exists a Banach space* $E_n$ *with a basis, quasi-reflexive of order* $n$, *which contains no subspace isomorphic to* $c_0$ *or* $l^p$ $(1 \leqslant p < \infty)$ and that *there exist also non-quasi-reflexive Banach spaces E with a basis, containing no isomorph of* $c_0$ *or* $l^p$ $(1 \leqslant p < \infty)$.

17) *To page* 770: For the Radon-Nikodym property, see also the book of J. Diestel and J. J. Uhl, Jr. [415].

18) *To page* 771: Recently, B. V. Godun has shown [418] that the answer to the unconditional analogue of problem c) (whence also b) and a)) is affirmative, namely, *for every Banach space E with an unconditional basis and each* $\varepsilon$ *with* $0 < \varepsilon < 2$, *there exists an equivalent norm* $|||\cdot|||$ *on E such that* $0 < \Gamma^{(u)}_{|||\cdot|||}(E) \leqslant \dfrac{1}{2-\varepsilon}$; here $\Gamma^{(u)}(E) = \sup \gamma^{(u)}_{\{x_n\}}$ (see Ch. I, p. 63 and Ch. II, p. 504). Moreover [418], *a similar result is also valid for unconditional finite-dimensional decompositions.*

19) *To page* 772: The result of J. Lindenstrauss, mentioned in the footnote on p. 772 (namely, that *every Banach space with an unconditional finite-dimensional decomposition is isomorphic to a subspace of a Banach space with an unconditional basis*), is proved in J. Lindenstrauss-L. Tzafriri [248a], pp. 51—52, theorem 1.g.5.

20) *To pages* 776—779: Subspaces $V$ of a conjugate Banach space $E^*$, for which there exists an approximative basis $\{v_n\}$ of $E$, satisfying (10.97) of lemma 10.3, have been also considered, independently (slightly later than in [380]), by V. A. Vinokurov and A. N. Pličko [451], who have called such subspaces *quasi-basic* subspaces of $E^*$ (however, we use the term "quasi-basis" in a different sense in § 9, definition 9.2); they have also raised [451] problem a) of p. 779. A similar notion, with $\{v_n\} \subset L(E, E)$ not necessarily of finite rank, has been introduced by V. A. Vinokurov and L. D. Menihes [450], who have called such subspaces *prebasic subspaces* of $E^*$. L. D. Menihes and A. N. Pličko have shown ([425], corollary 1) that *if E is a separable Banach space with the bounded approximation property, then every total subspace V of* $E^*$, *with* codim $V < \infty$, *is quasi-basic*. Of course, this implies again lemma 10.3 (found also in [425], corollary 2), since, as was observed in the proof of lemma 10.2, for a quasi-reflexive space $E$ every total subspace $V$ of $E^*$ has codim $V < \infty$ (this observation has been made in [434], lemma 4). Moreover, the above result also implies that *the answer to the problem on p.* 779, *concerning non-quasi reflexive spaces E such that every subspace V of* $E^*$ *with* $r(V) > 0$ *has* codim $V < \infty$, is affirmative. For further results on quasi-basic subspaces and for their connections with problems of regularizability of inverses of one-to-one continuous linear operators, see also L. D. Menihes and A. N. Pličko [425], F. S. Vaher and A. N. Pličko [449] (among other results, in [449] it is noted that the example of F. S. Vaher, mentioned on p. 779, has $r(V) > 0$ and $V$ non-quasi-basic) and the book of Yu. I. Petunin and A. N. Pličko [426a].

21) *To page* 814: A. N. Pličko has proved ([428], corollary of lemma 1) that *if a Banach space E admits an E-complete extended biorthogonal system, then E has a quotient space E/G with* dens $E/G =$ dens $E$, *such*

*that $E/G$ has a monotone transfinite basis*. This implies again the similar result of K. John and V. Zizler on weakly compactly generated spaces, mentioned on p. 814. In the opposite direction, B. V. Godun and M. I. Kadec have shown [419], as a corollary of their result mentioned in 11) above, that *if* card $I > 2^{\mathfrak{m}}$, *where* $\mathfrak{m} \geqslant \aleph_0$, *then no quotient space $E/G$ of* $E = l_{\mathfrak{m}}^{\infty}(I)$, *with* dens $E/G$ = dens $E$, *has a complete extended minimal system* (hence, in particular, $E$ has no quotient space $E/G$ with dens $E/G$ = dens $E$ and with a transfinite basis).

22) *To pages* 816—817: Concerning orthogonally scattered measures, see also K. Sundaresan and W. A. Woyczynski [435] and N. J. Kalton, B. Turrett and J. J. Uhl, Jr. [422].

23) *To page* 818: Concerning the uniform projection approximation property, see also C. Samuel [431].

24) A number of results related to those of the present volume, have been obtained in the recent papers of P. Terenzi ([438] — [448]). Let us mention here some sample results: i) In a Banach space, every normalized sequence $\{x_n\}$ containing no basic subsequence has an infinite subsequence $\{y_n\}$ which is either hypercomplete (in $[y_n]$) or can be decomposed as $\{y_n\} = \{z_n\} + \{y_n'\}$, where $\{y_n'\}$ is a basic sequence and $\{z_n\}$ is hypercomplete and such that for each infinite sequence $\{n_k\} \subset \mathcal{N}$ we have $[y_{n_k}] \supset [z_n]$ ([439], corollary I). ii) For hypercomplete sequences, a stability result, similar to § 3, theorem 3.2, is not valid ([447], theorem III); however ([447], remark 6), every hypercomplete sequence has an infinite subsequence which remains hypercomplete for sufficiently near sequences. iii) For every sequence $\{x_n\}$ in a Banach space $E$, with $\dim [x_n] = \infty$, there exist a minimal sequence $\{y_n\} \subset [x_n]$ with $[y_n] = [x_n]$, $y_n = \sum_{i=q_n}^{r_n} \alpha_i^{(n)} x_i$, $\alpha_{q_n}^{(n)} \neq 0$, $q_n < q_{n+1}$ $(n = 1, 2, \ldots)$ and an $M$-basis $\{z_n\}$ of $[x_n]$ with $z_n = \sum_{i=p_n}^{s_n} \beta_i^{(n)} x_i$, $\beta_{p_n}^{(n)} \neq 0$, $p_n \leqslant p_{n+1}$ $(n = 1, 2, \ldots)$, $\lim p_n = \infty$ [442]. iv) There exists [444] a separable Banach space $E$, having two quasi-complementary subspaces $G$ and $F$, such that if $\{y_n\} \subset G$, $\{z_n\} \subset F$ are infinite sequences with either $[y_n] = G$ or $[z_n] = F$, then $\{y_n\} \cup \{z_n\}$ is not uniformly minimal (in the sense of § 4, theorem 4.1, condition 3°). Consequently, § 8, theorem 8.2 and § 8, corollary 8.4 cannot be sharpened so as to hold for strictly bounded $M$-bases [444]. v) In § 8, proposition 8.7, the assumption that $\{x_n\}$ is minimal, cannot be omitted [438]. Furthermore, there exist [438] a Banach space $E$, an $M$-basis $\{x_n\}$ of $E$ and an element $x \in E$ such that for every infinite sequence $\{n_k\} \subset \mathcal{N}$ we have $x \notin [x_{n_k}] + [x_j]_{j \in \mathcal{N} \setminus \{n_k\}}$. vi) If $G$ is a subspace of a Banach space $E$, with $\dim G = \operatorname{codim} G = \infty$, and if $\{y_n\}$ is an $M$-basis of $G$, then exists a subspace $F$ of $E$, quasi-complementary to $G$, such that $\{y_n + F\}$ is hypercomplete in $E/F$ [440]. vii) If $X$ is a Banach space with a basis and $E$ is a subspace of $X$ with $\dim E = \infty$, then there exist a basis $\{z_n\}$ of $X$, an $M$-basis $\{x_n\}$ of $E$ and an infinite sequence $\{n_k\} \subset \mathcal{N}$ such that $x_{n_k} = z_{n_k}$ $(k = 1, 2, \ldots)$, $\{x_j\}_{j \in \mathcal{N} \setminus \{n_k\}} \subset \operatorname{lin} \{z_j\}_{j \in \mathcal{N} \setminus \{n_k\}}$ ([443],

theorem II). This implies again (by embedding any separable space $E$ into $X = C([0, 1]))$ § 8, proposition 8.13 without (8.74) and with (8.75) replaced by $\sup_{1 \leq k < \infty} \|x_{n_k}\| \, \|f_{n_k}\| < \infty$ for some infinite sequence $\{n_k\} \subset \mathcal{N}$ [443]. For other results on the structure of various classes of sequences in Banach spaces, see P. Terenzi [438] — [448].

25) We note also the appearance of the survey paper [426] of A. Pelczynski (in collaboration with C. Bessaga).

26) Finally, we mention that, in the meantime, the following papers of the Bibliography have appeared: [321], **33**, 111–118 (1980); [323], Teor. Funkciĭ, Funkcional. Anal. i Priložen., **32**, 61–69 (1980); [380], 237–240 (1979); also, a summary of [349], 3ème colloque sur l'analyse fonctionnelle tenu à Liège du 14 au 16 septembre 1970, p. 155. Centre Belge de rech. math. Louvain: Vander 1971.

# Bibliography to the Appendix


[412] Alexander, F. E.: Compact and finite rank operators on subspaces of $l_p$. Bull. London Math. Soc. **6**, 341–342 (1974).

[413] Bachelis, G. F.: A factorization theorem for compact operators. Illinois J. Math. **20**, 626–629 (1976).

[414] Casazza, P. G., Lin, Bor-Luh, Lohman, R. H.: On nonreflexive Banach spaces which contain no $c_0$ or $l_p$ (to appear).

[415] Diestel, J., Uhl, J. J., Jr.: Vector measures. Providence: Amer. Math. Soc. 1977.

[415a] Drewnowski, L., Labuda, I., Lipecki, Z.: Existence of quasi-bases for separable topological linear spaces (to appear).

[416] Feder, M.: On subspaces of spaces with an unconditional basis and spaces of operators (to appear).

[417] Godun, B. V.: On complete $l_1$-systems in a Banach space (to appear) [Russian].

[418] Godun, B. V.: A remark on the unconditional basis constant of a Banach space (to appear) [Russian].

[419] Godun, B. V., Kadec, M. I.: On Banach spaces without complete minimal systems (to appear) [Russian].

[420] Johnson, G. W., Petersen, L. V.: A note on Banach spaces without the approximation property. Comment. Math. Univ. Carolinae **18**, 579–589 (1977).

[421] Johnson, W. B.: Banach spaces all of whose subspaces have the approximation property (to appear).

[422] Kalton, N. J., Turrett, B., Uhl, J. J. Jr.: Basically scattered vector measures (preprint).

[423] Kwapien, S.: Isomorphic characterizations of inner product spaces by orthogonal series with vector valued coefficients. Studia Math. **44**, 583–595 (1972).

[424] Lindenstrauss, J., Tzafriri, L.: Classical Banach spaces. II: Function spaces. Berlin–Heidelberg–New York: Springer 1979.

[425] Menihes, L. D., Plička, A. N.: Conditions for linear and finite-dimensional regularizability of linear inverse problems. Doklady Akad. Nauk SSSR **241**, 1027–1030 (1978) [Russian].

[426] Pelczynski, A. (in collaboration with Bessaga, C.): Some aspects of the present theory of Banach spaces. Article in Banach, S.: Oeuvres. Vol. II, pp. 223–302. Warszawa: PWN 1979.

[426a] Petunin, Yu. I., Pličko, A. N.: The theory of characteristics of subspaces and its applications. Kiev: GIIO "Vissa Škola" 1980 [Russian].

[427] Pietsch, A.: Operator ideals. Berlin: VEB Deutscher Verlag Wiss. 1978.

[428] Pličko, A. N.: Construction of bounded fundamental and total biorthogonal systems from unbounded ones. Dopovidi Akad. Nauk Ukr. RSR **A5**, 19–22 (1980) [Russian].

[429] Pličko, A. N.: A Banach space without a complete biorthogonal system (to appear) [Russian].

[430] Pličko, A. N.: Some properties of the space of Johnson-Lindenstrauss (to appear) [Russian].

[431] Samuel, C.: Exemples d'espaces de Banach ayant la propriété de projection uniforme. Sémin. Géom. esp. de Banach Exp. 27, 1–15 (1978).

[432] Saphar, P.: Sur les sous-espaces des espaces de Banach à base inconditionnelle. Sémin. Géom. esp. de Banach Exp. 15, 1–5 (1978).

[433] Shelah, S.: On iterated forcing and on Banach spaces. Notices Amer. Math. Soc. **26**, A-525 (1979).

[434] Singer, I.: Bases and quasi-reflexivity of Banach spaces. Math. Annalen **153**, 199–209 (1964).

[435] Sundaresan, K., Woyczynski, W. A.: $L$-orthogonally scattered measures. Pacific J. Math. **43**, 785–797 (1972).

[436] Szankowski, A.: Subspaces without the approximation property. Israel J. Math. **30**, 123–129 (1978).

[437] Szankowski, A.: The space of all bounded operators on Hilbert space does not have the approximation property. Sémin. Anal. fonctionnelle Exp. 14–15, 1–21 (1979).

[438] Terenzi, P.: Markuschevich bases and quasi complementary subspaces in Banach spaces. Rend. Ist. Lombardo **A 111**, 49–61 (1977).

[439] Terenzi, P.: On the structure, in a Banach space, of the sequences without an infinite basic subsequence. Boll. U.M.I. (5) **15–B**, 32–48 (1978).

[440] Terenzi, P.: Properties of structure and completeness, in a Banach space, of the sequences without an infinite minimal subsequence. Rend. Ist. Lombardo **A 112**, 47–66 (1978).

[441] Terenzi, P.: Biorthogonal systems in Banach spaces. Riv. Mat. Univ. Parma (4) **4**, 165–204 (1978).

[442] Terenzi, P.: Some completeness properties of general sequences in a Banach space. Boll.U.M.I. (5) **15–B**, 743–753 (1978).

[443] Terenzi, P.: A complement to Krein-Milman Rutman theorem, with applications. Rend. Ist. Lombardo **A 113**, 1–14 (1979).

[444] Terenzi, P.: On bounded and total biorthogonal systems spanning given subspaces. Rend. Accad. Naz. Lincei **67**, 1–11 (1979).

[445] Terenzi, P.: Proof of a theorem on $\omega$-linear independence in Banach spaces. Rend. Ist. Lombardo (to appear).

[446] Terenzi, P.: On the structure of the overfilling sequences of a Banach space. Riv. Mat. Univ. Parma (4) **6** (1980) (to appear).

[447] Terenzi, P.: Stability properties of a sequence in a Banach space (to appear).

[448] Terenzi, P.: Convergence in the theory of bases in Banach spaces (to appear).

[449] Vaher, F. S., Pličko, A. N.: The bounded approximation property and linear finite-dimensional regularizability. Ukr. Matem. Ž. 32 (1980) (to appear) [Russian].

[450] Vinokurov, V. A., Menihes, L. D.: Necessary and sufficient conditions for linear regularizability. Doklady Akad. Nauk SSSR **229**, 1292–1294 (1976) [Russian].

[451] Vinokurov, V. A., Pličko, A. N.: On the regularizability of linear inverse problems by linear methods. Doklady Akad. Nauk SSSR **229**, 1037–1040 (1976) [Russian].

# Notation Index

---

# Author Index

# Subject Index

MIX
Papier aus verantwortungsvollen Quellen
Paper from responsible sources
FSC® C105338

If you have any concerns about our products,
you can contact us on
**ProductSafety@springernature.com**

In case Publisher is established outside the EU,
the EU authorized representative is:
**Springer Nature Customer Service Center GmbH**
**Europaplatz 3, 69115 Heidelberg, Germany**

Printed by Libri Plureos GmbH
in Hamburg, Germany